ENGINEERING MECHANICS

AF519488

ENGINEERING MECHANICS

(In S.I. Units)

Second Edition

By

Late Dr. I.S. GUJRAL

B.E. (Hons.), M.E. (Thermal), Ph. D

Fellow Institution of Engineers (India)

Principal

Hitkarini College of Engg. & Technology, Jabalpur

Madhya Pradesh

Formerly, Director,

Shri Ram Institute of Science and Technology,

Jabalpur (M.P.)

(An Imprint of Laxmi Publications Pvt. Ltd.)

An ISO 9001:2008 Company

BENGALURU • CHENNAI • COCHIN • GUWAHATI • HYDERABAD

JALANDHAR • KOLKATA • LUCKNOW • MUMBAI • RANCHI • NEW DELHI

BOSTON (USA) • ACCRA (GHANA) • NAIROBI (KENYA)

ENGINEERING MECHANICS

© by Laxmi Publications (P) Ltd.
All rights reserved including those of translation into other languages. In accordance with the Copyright (Amendment) Act, 2012, no part of this publication may be reproduced, stored in a retrieval system, or transmitted in any form or by any means, electronic, mechanical, photocopying, recording or otherwise. Any such act or scanning, uploading, and or electronic sharing of any part of this book without the permission of the publisher constitutes unlawful piracy and theft of the copyright holder's intellectual property. If you would like to use material from the book (other than for review purposes), prior written permission must be obtained from the publishers.

Printed and bound in India
Typeset at Goswami Associates, Delhi
First Edition : 2008; *Second Edition* : 2011; *Reprint* : 2016
ISBN 978-93-80856-66-7

Limits of Liability/Disclaimer of Warranty: The publisher and the author make no representation or warranties with respect to the accuracy or completeness of the contents of this work and specifically disclaim all warranties. The advice, strategies, and activities contained herein may not be suitable for every situation. In performing activities adult supervision must be sought. Likewise, common sense and care are essential to the conduct of any and all activities, whether described in this book or otherwise. Neither the publisher nor the author shall be liable or assumes any responsibility for any injuries or damages arising here from. The fact that an organization or Website if referred to in this work as a citation and/or a potential source of further information does not mean that the author or the publisher endorses the information the organization or Website may provide or recommendations it may make. Further, readers must be aware that the Internet Websites listed in this work may have changed or disappeared between when this work was written and when it is read.

All trademarks, logos or any other mark such as Vibgyor, USP, Amanda, Golden Bells, Firewall Media, Mercury, Trinity, Laxmi appearing in this work are trademarks and intellectual property owned by or licensed to Laxmi Publications, its subsidiaries or affiliates. Notwithstanding this disclaimer, all other names and marks mentioned in this work are the trade names, trademarks or service marks of their respective owners.

Branches

©	Bengaluru	080-26 75 69 30	
©	Chennai	044-24 34 47 26,	24 35 95 07
©	Cochin	0484-237 70 04,	405 13 03
©	Guwahati	0361-254 36 69,	251 38 81
©	Hyderabad	040-27 55 53 83,	27 55 53 93
©	Jalandhar	0181-222 12 72	
©	Kolkata	033-22 27 43 84	
©	Lucknow	0522-220 99 16	
©	Mumbai	022-24 91 54 15,	24 92 78 69
©	Ranchi	0651-220 44 64	

PUBLISHED IN INDIA BY

UNIVERSITY SCIENCE PRESS
(An Imprint of Laxmi Publications Pvt. Ltd.)

An ISO 9001:2008 Company
113, GOLDEN HOUSE, DARYAGANJ,
NEW DELHI - 110002, INDIA
Telephone : 91-11-4353 2500, 4353 2501
Fax : 91-11-2325 2572, 4353 2528
www.laxmipublications.com info@laxmipublications.com

Contents

Chapters *Pages*

Chapters *Pages*

Chapters | Pages

Chapters *Pages*

Preface to the Second Edition

The author gratefully acknowledge the healthy criticism and encouraging appreciations for the experienced and the worthy teachers and the large number of engineerging students of the country.

To further strengthen the text, the book has been thoroughly revised. The salient features of this revised editions are:

- Several figures have been corrected.
- Typographic errors have been thoroughly edited.
- A new chapter on Virtual Work has been added.

The author hopes that this edition of the book will enhance its utility for students and teachers of Engineering Mechanics of all universities.

Suggestions from the readers for further upgrading the book will be appreciated and thankfully acknowledged.

—Author

Preface to the First Edition

The subject "Engineering Mechanics" is taught in the very first year of undergraduate Engineering curriculam in almost all universities in the world. A student learns to develop his capability to model an actual problem into an engineering problem and obtains its solution by using the fundamental laws of mechanics. Engineering problems, to a great extent, require the drawing of the free body diagrams of the entire body and also of a small portion cut from it. Students are advised to learn drawing of the free-body diagrams. The entire book is divided into two portions—Statics and Dynamics.

Throughout this book, SI systems of units have been used. Analytical methods are emphasized throughout this book but the graphical and semi-graphical methods have not been neglected. Most of the fundamentals needed to learn the basics of Engineering Mechanics have been discussed in First Chapter. Majority of the force systems, a student encounters in his life, have also been discussed in the First Chapter.

Chapter on Trusses has been discussed at length. Here, certain so called "Checks" are presented which a student is advised to use to confirm the correctness of his solution. If the "Check" fails, the student should hurriedly go through his solution to locate the mistake. A very large number of solved examples and the unsolved problems have been given for practice. The graphical method has been given in a very systematic fashion.

Separate chapters have been included on Friction, Centre of Gravity and Moment Inertia of Plane Area and the solid bodies. Standard cases have been given in tabular forms for ready reference.

A separate chapter on Shear Force and Bending Moment Diagrams has been given. A novel concept has been introduced which reduces labour and time of students for solving a problem. The concept of Non-concurrent Coplanar force systems is used to draw the Shear Force Diagrams. Also, the basic approach of changing (or rotating) reference frames is used.

Chapters on Lifting Machines and on Transmission of Power have been given. Thorough preparation of chapters mentioned so far will ensure success in examination. These chapters form part of Statics.

Chapter on Dynamics includes : Kinematics of Particle, Kinematics of Rigid Bodies, Rectilinear and Curvilinear Motion of Particles. Special methods for solving problems on Dynamics have been discussed in chapters on Work-Energy Equations ; Impulse-Moment Method and Collision of Elastic Bodies.

This book has been written to help students in understanding the fundamentals of the subject to develop their ability to deal with any situation.

It is hoped that the material presented in this book will be of value to the students and the teachers as well. It will also help students in preparing for competitive examinations such as GATE, GRE, UPSC, PSC, AMIE and other public sector utilities.

Many standard books have been used in preparation of this book and the author thankfully acknowledges the same. I also thank my colleagues and the students of many colleges affiliated to the Rajiv Gandhi Technical University, Bhopal, Pt. Ravishankar University, Raipur, Nagpur University, Amravati University, Pune University, Bangalore, Punjab University etc., for giving their valuable suggestions which helped in the improvement of the quality of the book.

I also acknowledge the moral support from my wife Smt. Kuldip Kaur and all my children for bearing with me during the long period of writing of this book. They also helped me in reading the proofs.

—Author

1

System of Forces

PART A—FUNDAMENTAL PRINCIPLES AND RELATIONS

1.1. MECHANICS

Mechanics is that branch of science which deals with the state of rest or motion of bodies under the action of forces. No one subject plays a greater role in the engineering analysis and application than mechanics. Modern research and advancements in the fields of stability, strength and design of structures and machines, vibration, robotics, rockets, missiles, aeroplane and spacecraft design, automobiles, automatic control, fluid flow, engine performance, electrical machines and apparatus, transmission towers, superstructures, heavy earth moving machines, locomotives, metro railways, super sonic aircrafts; molecular, atomic and subatomic behaviour, etc., are highly dependent on the basic principles of mechanics. A thorough and clear understanding of this subject is an essential requirement for work in these and many other subjects, not mentioned above.

It is divided into three parts : **Mechanics of Rigid bodies, Mechanics of Deformable Bodies** and **Mechanics of Fluids**. Mechanics of rigid bodies is sub-divided into **Statics** and **Dynamics.** Statics deals with bodies at rest, while Dynamics deals with bodies in motion. In the study of mechanics bodies are assumed to be perfectly rigid. Actual machines and structures are not absolutely rigid and deform under the given loads but these deformations are negligibly small and therefore do not affect their conditions of equilibrium or motion.

The second division of mechanics is the Mechanics of Deformable Bodies, which is further subdivided into Strength of Materials, Theory of Elasticity, Theory of Plasticity. The third division of mechanics is the Mechanics of Fluid, which is further subdivided into Mechanics of Incompressible Fluids and Mechanics of Compressible Fluids.

1.2. FUNDAMENTAL CONCEPTS

Certain concepts and definitions are basic to the study of mechanics and they should be understood before starting the subject. The basics used in mechanics are **space, time, mass, force, particle** and **rigid body.**

(*a*) **Space.** The concept of space is associated with the notion of the position of a body. Space is the geometric region occupied by bodies whose positions are described by linear or angular measurements relative to a co-ordinate system. For three dimensional problems three independent lengths (or co-ordinates) measured with a reference point (known as **origin**) are needed. For two-dimensional problems only two co-ordinates will be required. These co-ordinates are called the **Cartesian co-ordinates** or the **rectangular co-ordinates**. Sometime angular co-ordinates are used. They are known as **cylindrical co-ordinates** for three dimensional problems and **polar co-ordinates** for two-dimensional problems. There are many other co-ordinate systems used for different problems.

(*b*) **Time.** Time is the measure of happening of an event. It is a basic quantity in the study of dynamics.

(*c*) **Mass.** Mass is the measure of the quantity of matter in a body. In Statics mass is also the property of a body by which it experiences force of mutual attraction to other bodies in space. Mass is also the measure of the inertia of a body. And, inertia is the resistance to the change of velocity of a body. In SI units, unit of mass is kilogram (kg).

(*d*) **Force.** Force is the action of one body on another. A force tends to move a body in the direction of its action. It may be exerted either by actual contact or from a distance, as in the case of gravitational forces and magnetic forces. A force is described by :

(*i*) its point of application,

(*ii*) its magnitude, and

(*iii*) its direction.

These quantities are called specifications of a force. A force, having magnitude and direction, is a vector quantity. In SI system of units force is measured in newtons and its symbol is N.

(*e*) **Particle.** A body of negligible dimensions is called a particle, Mathematically, a particle is a body whose dimensions approach zero so that it may be analysed as a point mass. Forces acting on a particle are called concurrent forces.

(*f*) **Rigid Body.** A rigid body may be defined as a definite amount of matter the parts of which are fixed in position relative to one another. Physical bodies, such as we deal with in the design of engineering structures and machine parts, are never absolutely rigid but they deform slightly under the action of load which they are to carry. But, these deformations are very small and do not affect the equilibrium of the rigid body under study in mechanics. The deformations of bodies are studied in subjects like Mechanics of Deformable Bodies, also known as Strength of Materials, Theory of Elasticity, Theory of Plasticity, etc.

1.3. FUNDAMENTAL PRINCIPLES

The study of mechanics rests on six fundamental principles based on experimental evidence. They will be discussed at length in subsequent sections as they are required. These principles are stated below for reference only :

(1) **The Parallelogram Law of Forces.** It states that the resultant of two forces is represented by the diagonal of the parallelogram formed as the sides representing these forces.

The Three Laws of Newton. Sir Isaac Newton (1642-1727) was the first to satisfactorily formulate the fundamental principles of mechanics. Slightly reworded to use the modern terminology, these laws are :

(2) **First Law of Motion.** A particle remains at rest or continues to move in a straight line with a uniform velocity if there is no unbalanced resultant force acting on it.

(3) **Second Law of Motion.** The acceleration is proportional to the resultant force acting on a particle and is in the direction of this force.

Expressed mathematically, it may be stated as :

$$\overline{F} = m\overline{a} \quad ...(1.1)$$

where $\overline{F}$ is the resultant force acting on the particle of mass m and $\overline{a}$ is the resulting acceleration.

(4) **Third Law of Motion.** The forces of action and reaction between the bodies are equal in magnitude, opposite in direction and collinear.

(5) **The Principle of Transmissibility.** This states that the external effect of a force on a rigid body is the same for all points of application of a force along its line of action, *i.e.*, it is independent of its point of application. Its **internal effect** definitely change with the point of application of the force.

Force P, in Fig. 1.3.1(*a*) will result the sliding of the block whether the force pulls the block at A or pushes at B. The local internal effects at A and B will, however, be quite different.

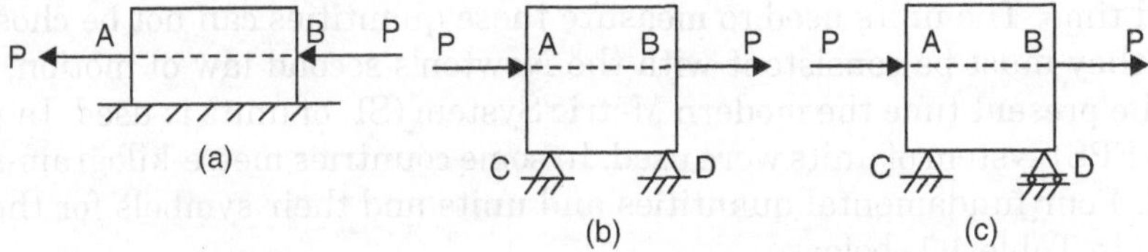

Fig. 1.3.1. Principle of transmissibility is for (a) and (b) but not for (c)

It should be noted that the principle of transmissibility applies to the external effect of a force on the same rigid body. Observe that since the supports at C and D have changed in Figs. (*b*) and (*c*) therefore the principle of transmissibility will not hold good although the force P acts along its line of action through points A and B. The support at D is changed from a hinge at (*b*) to a roller at (*c*), this changes the external condition of equilibrium of the body.

(6) **Newton's Law of Gravitation.** This law states that the two masses M and m situated at a distance r are mutually attracted with equal and opposite forces (Fig. 1.3.2) $\overline{F}$ and F′ of magnitude F given by the formula :

$$F = G \cdot \frac{Mm}{r^2} \qquad ...(1.2)$$

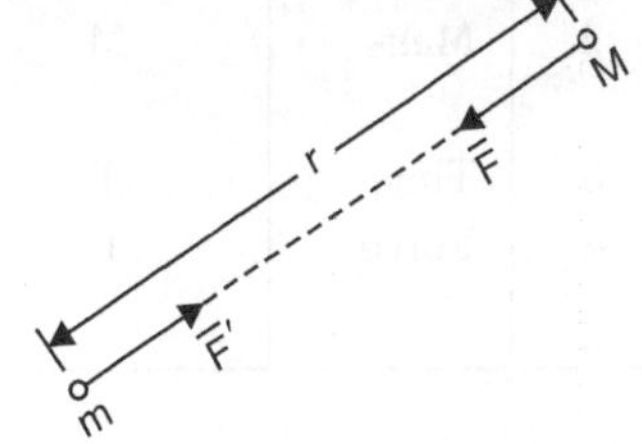

Fig. 1.3.2. Newton's law of gravitation

where G = **Universal constant known** as the constant of gravitation.

By experiments the constant of gravitation is found to be $G = 6.673 \times 10^{-11}$ m^3/(kgs^2).

The mutual forces F obey the law of action and reaction since they are equal and opposite and are directed along the line joining the centres of the particles M and m as shown in Fig. 1.3.2.

The gravitational attraction of the earth on a given body is called the **weight** of the body. This force exits whether the body is at rest or in motion. In SI units the unit of weight is Newton (N). For a body of mass m near the surface of earth, the gravitational attraction on the body may be calculated by using equation (1.1). If the gravitational force or weight has a magnitude W, then since the body falls with an acceleration g, equation (1.2) gives

$$W = mg \qquad ...(1.3)$$

The weight is in newtons (N) and mass in kilograms (kg) and g in m/s^2. The standard value of g = 9.81 m/s^2 is generally used.

In equation (1.2), let M_E = mass of earth in kg and r = radius of earth in metres (m), we get

$$F = \frac{G \times (M_E)}{r^2} \times m = gm,$$

where, $$g = \frac{G(M_E)}{r^2} = 9.81 \times \left[\frac{m^3}{kg \cdot s^2}\right] \cdot \left[\frac{kg}{m^2}\right] = 9.81 \text{ m/s}^2.$$

1.4. UNITS

As discussed is Sec 1.2, Mechanics deals with four fundamental quantities—length, mass force and time. The units used to measure these quantities can not be chosen independently because they must be consistent with the Newton's second law of motion, expressed by eqn. (1.1). In the present time the modern Metric System (SI) of units is used. In past, the foot-pound-second (FPS) system of units were used. In some countries metre-kilogram-second (MKS) units are used. Four fundamental quantities and units and their symbols for these three systems are given in Table 1.1, below.

Table 1.1. Different Systems of Units

S.No.	*Quantity*	*Dimensional symbol*	*SI units*		*MKS units*		*FPS units*	
			Basic units	*Symbol*	*Basic units*	*Symbol*	*Basic units*	*Symbol*
1.	Length	L	metre	m	metre	m	foot	ft
2.	Mass	M	kilogram	kg	kilogram mass	kg_m	slugs	slugs
3.	Time	T	second	s	second	sec	second	sec
4.	Force	F	newton	N	kilogram force	kg_f	Pound	lb

PART B—CONCURRENT COPLANAR FORCE SYSTEMS

1.5. SYSTEMS OF FORCES

In the study of mechanics various systems of forces are encountered. When all forces meet at one point, they are known as **concurrent forces**. When all concurrent forces are lying in one plane they belong to coplanar concurrent force system. This force system is illustrated in Fig. 1.5.2 (*b*), in which all forces are situated in the XY plane and meet at a point A.

If the concurrent forces are situated in different planes in space, they belong to **concurrent spatial forces**. In Fig. 1.5.2(*e*) forces F_1 and F_3 are situated in XZ plane ; forces F_1 and F_4 are situated in YZ plane ; forces F_2 and F_5 do not lie in any plane. All these forces meet at the corner C of the cube.

If all forces acting at a point lying in any plane are acting along the same straight line they form **coplanar collinear force system**, as shown in Fig. 1.5.2(*a*).

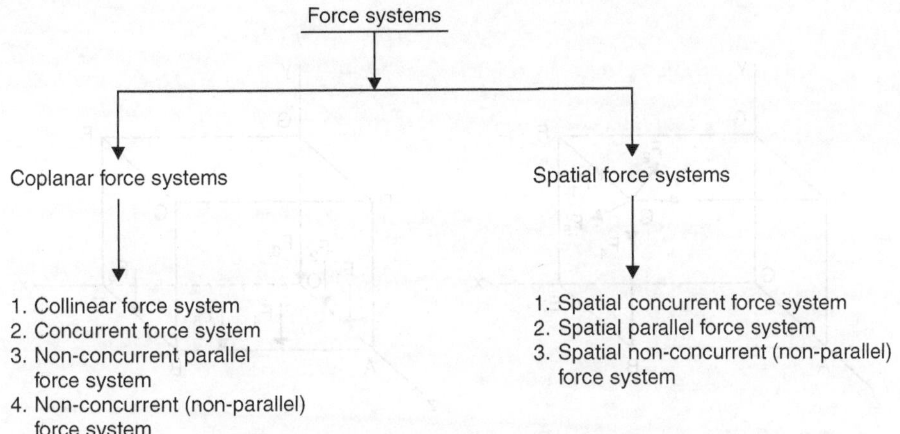

Fig. 1.5.1. Various force systems

Various force systems listed in Fig. 1.5.1 are illustrated in Fig. 1.5.2 below.

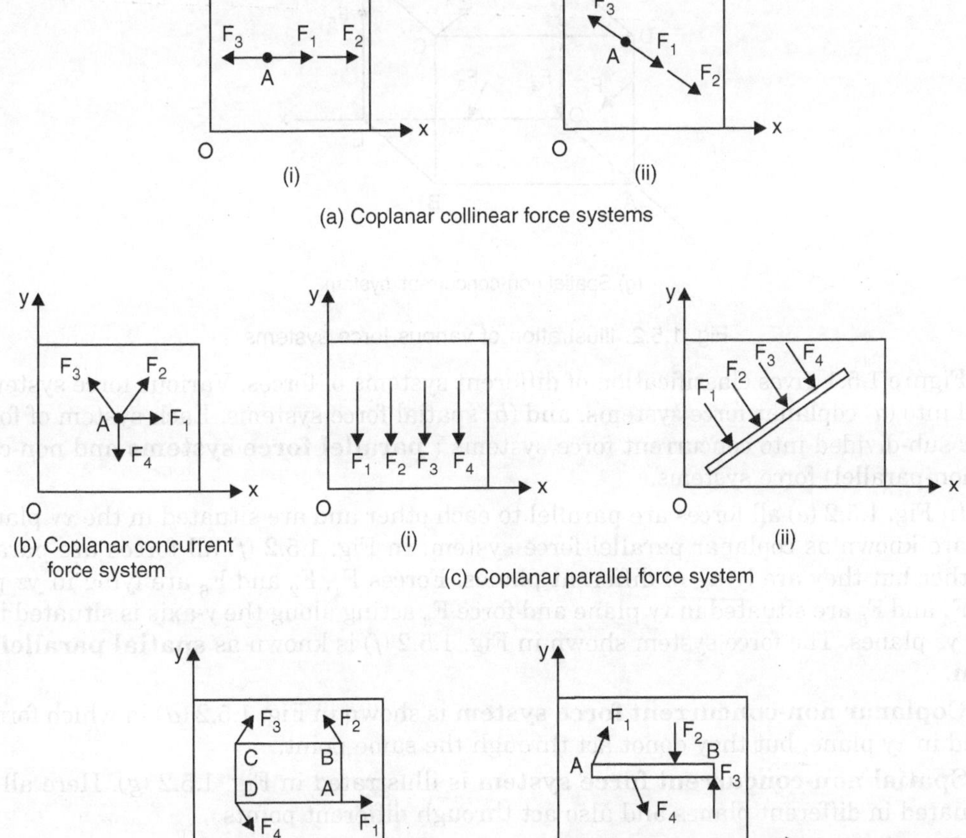

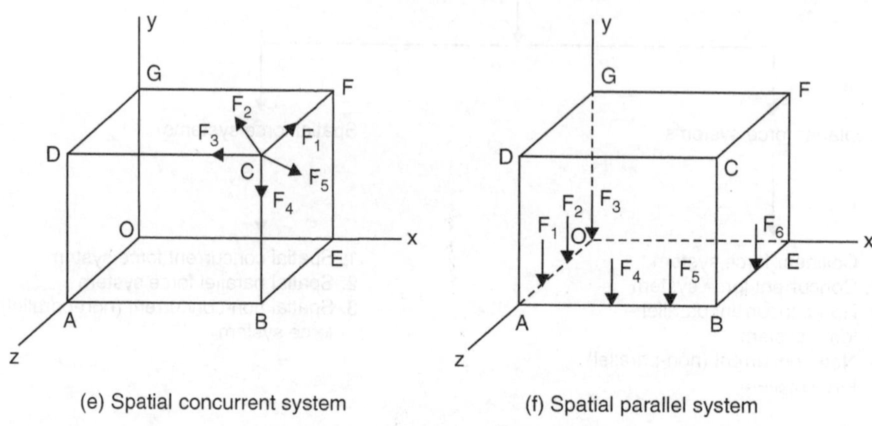

(e) Spatial concurrent system (f) Spatial parallel system

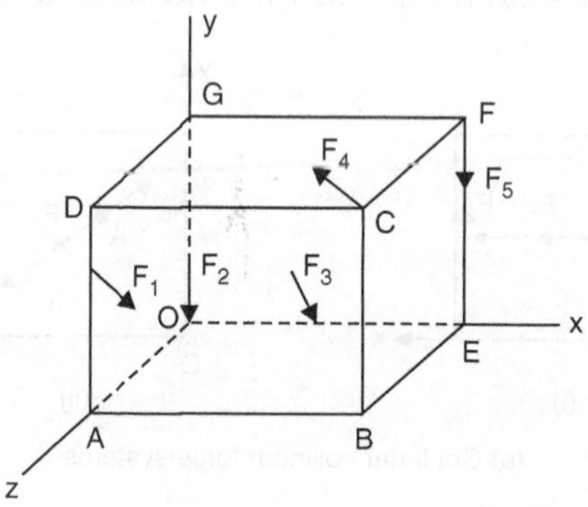

(g) Spatial non-concurrent system

Fig. 1.5.2. Illustration of various force systems

Figure 1.5.1 gives classification of different systems of forces. Various force systems are divided into (*a*) coplanar force systems, and (*b*) spatial force systems. Each system of forces is further sub-divided into concurrent force systems ; **parallel force systems** and non-concurrent (non-parallel) force systems.

In Fig. 1.5.2 (*c*) all forces are parallel to each other and are situated in the *xy* plane and hence are known as coplanar parallel force system. In Fig. 1.5.2 (*f*) all forces are parallel to each other but they are lying in different planes. Forces F_1, F_2 and F_6 are lying in *yz* plane ; forces F_4 and F_5 are situated in *xy* plane and force F_3 acting along the *y*-axis is situated in both *xy* and *yz* planes. The force system shown in Fig. 1.5.2 (*f*) is known as **spatial parallel force system.**

Coplanar non-concurrent force system is shown in Fig. 1.5.2 (*d*) in which forces are situated in *xy* plane, but they donot act through the same point.

Spatial non-concurrent force system is illustrated in Fig. 1.5.2 (*g*). Here all forces are situated in different planes and also act through different points.

The above force systems, their applications, equilibrium etc. will be discussed in subsequent sections of this book.

1.6. LAW OF PARALLELOGRAM OF FORCES

This law is used to determine the resultant of two coplanar forces acting on a point.

It states that, **if two forces acting on a point be represented in magnitude and direction by the two adjacent sides of a parallelogram, then their resultant is represented in magnitude and direction by the diagonal of the parallelogram passing through that point.**

This law is based on the experimental results. It cannot be proved analytically but can be verified experimentally.

The law of parallelogram of forces is represented in Fig. 1.6 (*a*), where the resultant (R) of force F_1 and F_2 inclined at an angle α and acting at O is given by the diagonal of the parallelogram passing through O. The resultant R is inclined at an angle θ with the force F_1.

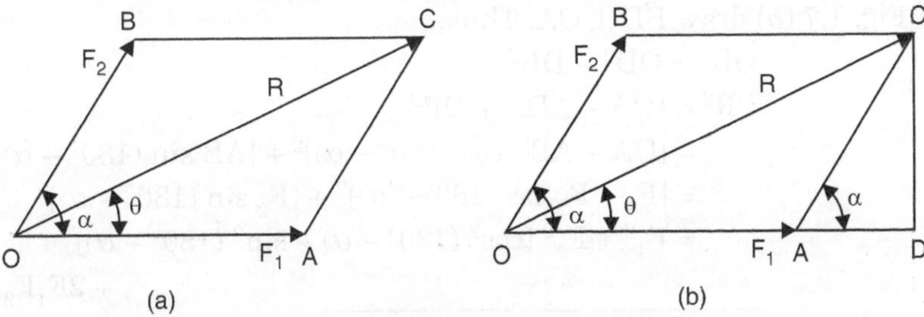

Fig. 1.6. Parallelogram law of forces

Draw the parallelogram OACB as shown.

Draw CD ⊥ OA as shown in Fig. 1.6(*b*). Then from ΔCOD, we have

$$OC^2 = OD^2 + DC^2 \quad \text{or} \quad R^2 = (OA + AD)^2 + DC^2$$

In the Fig. 1.6, F_1 is represented by OA and F_2 by OB. Since AC is parallel to OB, therefore AC also represents F_2.

$$\therefore \quad R^2 = (OA + AC\cos\alpha)^2 + (AC\sin\alpha)^2$$
$$= (F_1 + F_2\cos\alpha)^2 + F_2^2\sin^2\alpha$$
$$= F_1^2 + F_2^2\cos^2\alpha + 2F_1F_2\cos\alpha + F_2^2\sin^2\alpha$$
$$= F_1^2 + F_2^2(\cos^2\alpha + \sin^2\alpha) + 2F_1F_2\cos\alpha$$
$$= F_1^2 + F_2^2 + 2F_1F_2\cos\alpha$$

or
$$R = \sqrt{F_1^2 + F_2^2 + 2F_1F_2\cos\alpha} \quad \text{...(1.4)}$$

And inclination θ of the resultant R is given by :

$$\tan\theta = \frac{CD}{OD} = \frac{AC\sin\alpha}{OA + AC\cos\alpha} = \frac{F_2\sin\alpha}{F_1 + F_2\cos\alpha}$$

or
$$\theta = \tan^{-1}\left(\frac{F_2\sin\alpha}{F_1 + F_2\cos\alpha}\right) \quad \text{...(1.5)}$$

1.7. TRIANGLE LAW OF FORCES

This law states that, "If two forces acting at a point are represented in magnitude and direction by the two sides of a triangle, taken in order, then the third side represents their resultant in magnitude and direction".

In Fig. 1.7.1, the side OA of the ΔOAB represents force F_1 and the side AB represents the force F_2. Then, the side OB will represent the resultant R in direction and magnitude. Here the two forces F_1 and F_2 are represented by the sides OA and AB of the ΔOAB.

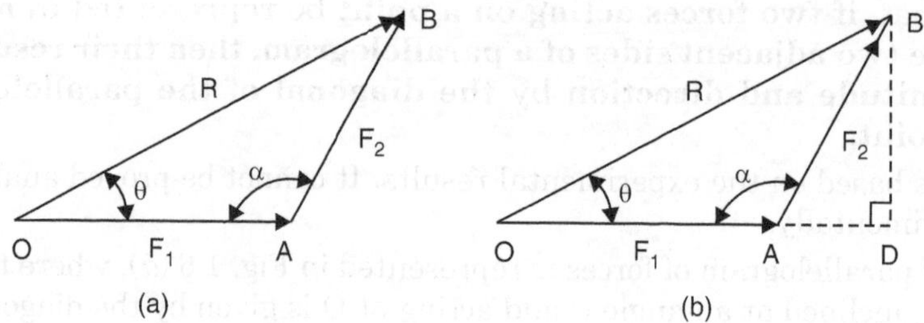

Fig. 1.7.1. Triangle law of forces

Thus, in Fig. 1.7 (*b*) draw BD ⊥ OA. Thus,

$$OB^2 = OD^2 + DB^2$$

or

$$R^2 = (OA + AD)^2 + DB^2$$
$$= [OA + AB \cdot \cos(180° - \alpha)]^2 + [AB \sin(180° - \alpha)]^2$$
$$= [F_1 + F_2 \cos(180° - \alpha)]^2 + [F_2 \sin(180° - \alpha)]^2$$
$$= F_1^2 + F_2^2 [\cos^2(180° - \alpha) + \sin^2(180° - \alpha)] + 2F_1F_2 \cos(180° - \alpha)$$

$$\therefore \quad \boxed{R^2 = F_1^2 + F_2^2 - 2F_1F_2 \cos \alpha} \quad ...(1.6)$$

Equation (1.6) is known as the **cosine line.**

The angle θ which the resultant R makes with force F_1 can be obtained by the sine rule. Thus from triangles OAB and ADB, we get

$$BD = OB \sin\theta = AB \sin(180° - \alpha)$$
$$= R \sin\theta = F_2 \sin(180° - \alpha)$$

$$\therefore \quad \boxed{\frac{R}{\sin(180° - \alpha)} = \frac{F_2}{\sin\theta}} \quad ...(1.7)$$

or

$$\sin\theta = \frac{F_2}{R} \sin(180° - \alpha) \quad ...(1.8)$$

Converse of the triangle law of forces :

If three concurrent forces acting at a point be represented in magnitude and direction by the three sides of the triangle taken in order, they will be in equilibrium.

Thus, in general, if the three sides of a triangle are A, B and C as shown in Fig. 1.7.2. Then, the following formulae can be applied :

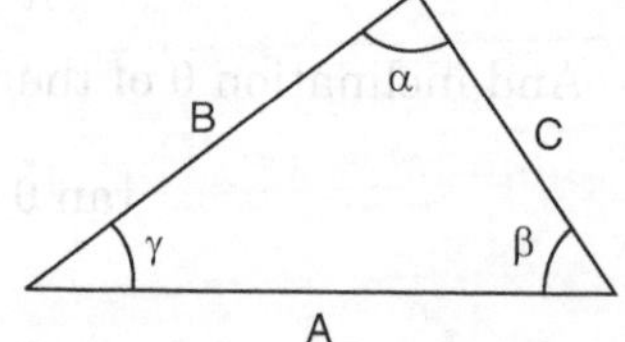

Fig. 1.7.2. The triangle law

$$\frac{A}{\sin\alpha} = \frac{B}{\sin\beta} = \frac{C}{\sin\gamma} \quad ...(1.9)$$

$$A^2 = B^2 + C^2 - 2B \cdot C \cdot \cos\alpha \quad ...(1.10(a))$$

$$B^2 = A^2 + C^2 - 2A \cdot C \cdot \cos\beta \quad ...(1.10(b))$$
$$C^2 = A^2 + B^2 - 2A \cdot B \cdot \cos\gamma \quad ...(1.10(c))$$

1.8. POLYGON LAW OF FORCES

It states that if a number of concurrent forces acting on a particle are represented in magnitude and direction by the sides of a polygon, taken in order, then their resultant is represented in magnitude and direction by the closing side of the polygon, taken in opposite order.

In Fig. 1.8 (*a*) a system of concurrent coplanar forces acting on a particle are shown.

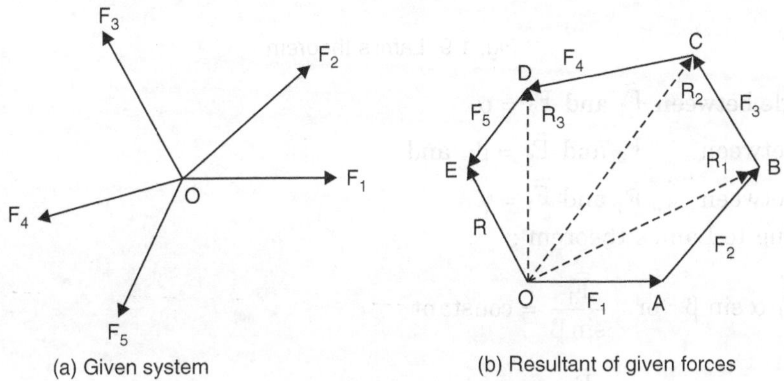

Fig. 1.8. Polygon law of forces

In Fig. 1.8 (*b*) :

$\overline{OA}$ represents force F_1 and $\overline{AB}$ represents force F_2. Using the triangle law of forces, $\overline{OB}$ represent the resultant of F_1 and F_2, say R_1.

$\overline{BC}$ represents force F_3. Then, in ΔOBC, the side $\overline{OC}$ represents the resultant of F_3 and R_1, say R_2.

$\overline{CD}$ represents F_4, and then in ΔOCD, using the triangle law of forces side $\overline{OD}$ represents the resultant of F_4 and R_2, say R_3.

$\overline{DE}$ represents F_5. Then in the ΔODE, using the triangle law of forces, the side $\overline{OE}$ represents the resultant of F_5 and R_3, say R.

Instead of constructing sets of triangles, as in Fig. 1.8 (*b*), the above resultant R can be obtained by constructing the force polygon OABCDE with its sides representing given forces in magnitude and direction. Then the last side $\overline{OE}$ will give the resultant of the given forces.

Important. It is important to note that the resultant is always directed from the starting point O towards the last point.

1.9. LAMI'S THEOREM

It states that, **"If a body is in equilibrium under the action of three forces, then each force will be proportional to the sine of the angle between the other two forces".**

Suppose the three forces $\overline{F}_1$, $\overline{F}_2$ and $\overline{F}_3$ are acting on a point O and they are in equilibrium as shown in Fig. 1.9 (*a*).

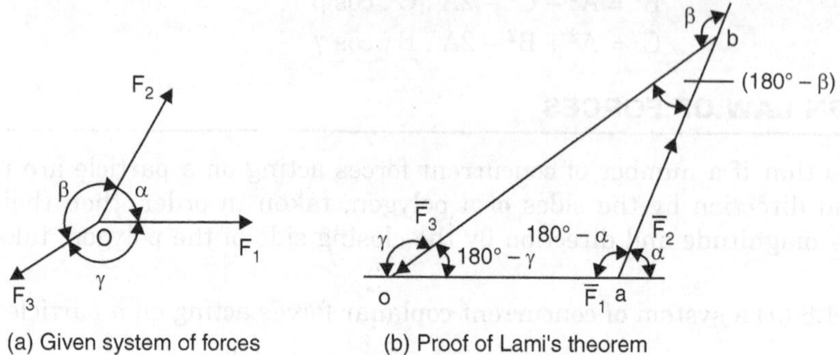

(a) Given system of forces (b) Proof of Lami's theorem

Fig. 1.9. Lami's theorem

Let angle between $\overline{F}_1$ and $\overline{F}_2 = \alpha$,

angle between $\overline{F}_2$ and $\overline{F}_3 = \beta$ and

angle between $\overline{F}_3$ and $\overline{F}_1 = \gamma$.

According to Lami's theorem :

$$\overline{F}_1 \propto \sin\beta \quad \text{or} \quad \frac{\overline{F}_1}{\sin\beta} = \text{constant} \qquad ...(1)$$

$$\overline{F}_2 \propto \sin\gamma \quad \text{or} \quad \frac{\overline{F}_2}{\sin\gamma} = \text{constant} \qquad ...(2)$$

and, $$\overline{F}_3 \propto \sin\alpha \quad \text{or} \quad \frac{\overline{F}_3}{\sin\alpha} = \text{constant} \qquad ...(3)$$

or, combining the above equations, we get

$$\frac{\overline{F}_1}{\sin\beta} = \frac{\overline{F}_2}{\sin\gamma} = \frac{\overline{F}_3}{\sin\alpha} \qquad ...(1.11)$$

Proof of Lami's Theorem. Draw the Δoab as shown in Fig 1.9 (*b*) representing the system of forces shown in Fig. 1.9 (*a*). The external angle between the forces $\overline{F}_1$ and $\overline{F}_2$ is α, that between $\overline{F}_2$ and $\overline{F}_3$ is β and that between $\overline{F}_3$ and $\overline{F}_1$ is γ. Then applying sine law, we have

$$\frac{\overline{F}_1}{\sin(180° - \beta)} = \frac{\overline{F}_2}{\sin(180° - \gamma)} = \frac{\overline{F}_3}{\sin(180° - \alpha)}$$

or simply, $$\frac{F_1}{\sin\beta} = \frac{F_2}{\sin\gamma} = \frac{F_3}{\sin\alpha}. \quad \textbf{(Proved).}$$

1.10. RESOLUTION AND COMPOSITION OF FORCES

This is an analytical method of finding the resultant of a given system of many coplanar forces acting at a point. In this method we first resolve all forces in two mutually perpendicular directions, say horizontal and vertical directions ; or in the *x*- and *y*-directions. Then we find the algebraic sum of these components and find the required resultant, as explained below.

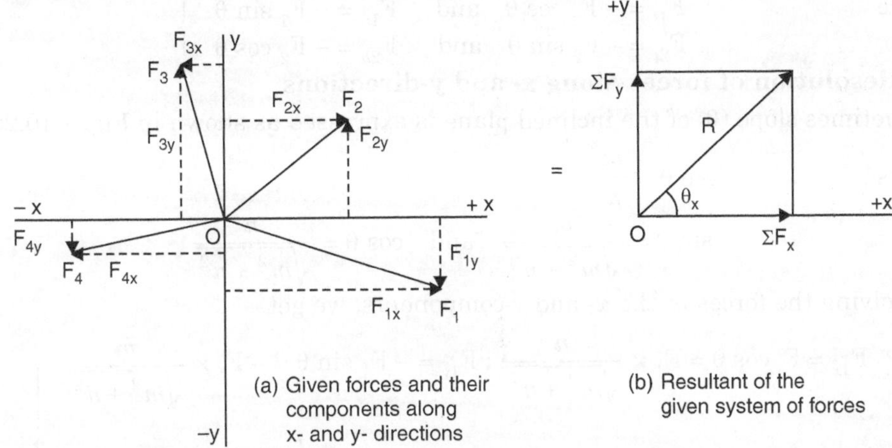

Fig. 1.10.1. Resolution of forces for finding resultant

(*a*) By Resolution of forces along x- and y-directions

Let F_1, F_2, F_3 and F_4 shown in Fig. 1.10.1 (*a*) be acting at a point O. It is required to find the resultant R of these forces.

Step 1. (*a*) Find the *x*-components of all forces. Let these be F_{1x}, F_{2x}, F_{3x} and F_{4x}. Let the algebraic sum of these components be ΣF_x.

(*b*) Find the *y*-components of all forces. Let these be F_{1y}, F_{2y}, F_{3y} and F_{4y}. Let the algebraic sum of these components be ΣF_y.

(**Note.** In the example shown in Fig. 1.10.1 (*a*) F_{1y}, F_{4y}, F_{3x} and F_{4x} are negative).

Step 2. The resultant of the given forces is given by

$$R = \sqrt{\Sigma F_x^{\,2} + \Sigma F_y^{\,2}} \quad \text{...(1.12)}$$

and its inclination with *x*-axis given by

$$\tan \theta_x = \frac{\Sigma F_y}{\Sigma F_x} \quad \text{or} \quad \theta_x = \tan^{-1}\left(\frac{\Sigma F_y}{\Sigma F_x}\right) \quad \text{...(1.13)}$$

Also, note that $\Sigma F_x = R \cos \theta_x = R_x$ ...(1.14)

and $\Sigma F_y = R \sin \theta_x = R_y$ ...(1.15)

Thus, ΣF_x and ΣF_y are the *x*- and *y*-components R*x* and R*y* respectively of the resultant.

(*b*) Resolving forces along the n- and t-directions

Let the given forces of Fig. 1.10.2 (*a*) be resolved in the normal direction and tangential direction (*i.e.*, along *n*- and *t*-directions)

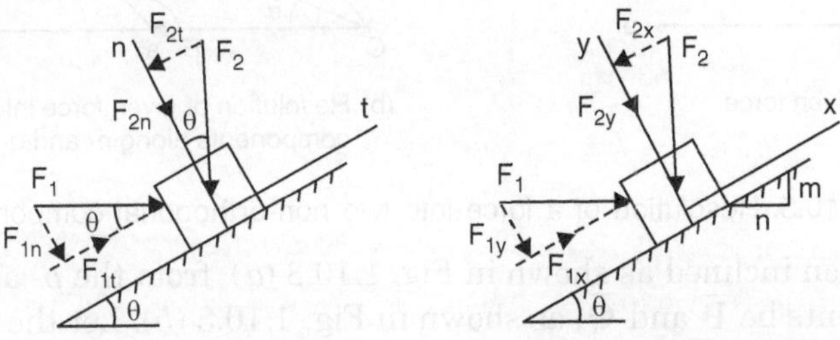

Fig. 1.10.2. Resolution of forces

Here $F_{1t} = + F_1 \cos\theta$ and $F_{1n} = - F_1 \sin\theta$

Also, $F_{2t} = - F_2 \sin\theta$ and $F_{2n} = - F_2 \cos\theta$ ...(1.16)

(*c*) **Resolution of forces along x- and y-directions**

Sometimes slope (θ) of the inclined plane is expressed as shown in Fig. 1.10.2(*b*).

Thus $\tan\theta = \dfrac{m}{n}$;

$$\therefore \quad \sin\theta = \frac{m}{\sqrt{m^2+n^2}} \quad \text{and} \quad \cos\theta = \frac{n}{\sqrt{m^2+n^2}} \qquad ...(1.17)$$

Resolving the forces in the *x*- and *y*-components, we get

$$\left.\begin{aligned} F_{1x} &= F_1 \cos\theta = F_1 \times \frac{n}{\sqrt{m^2+n^2}} ; \; F_{1y} = - F_1 \sin\theta = - F_1 \times \frac{m}{\sqrt{m^2+n^2}} \\ F_{2x} &= - F_2 \sin\theta = - F_2 \times \frac{m}{\sqrt{m^2+n^2}} ; \; F_{2y} = - F_2 \cos\theta = - F_2 \times \frac{n}{\sqrt{m^2+n^2}} \end{aligned}\right\} \qquad ...(1.18)$$

For example, if m = 3 units and n = 4 units,

Then $\tan\theta = \dfrac{3}{4}$; $\therefore$ $\sin\theta = \dfrac{3}{\sqrt{3^2+4^2}} = \dfrac{3}{5}$ and $\cos\theta = \dfrac{4}{5}$

$\therefore$ $F_{1x} = F_1 \cos\theta = \dfrac{4}{5} F_1$; $F_{1y} = -F_1 \sin\theta = -\dfrac{3}{5} F_1$

$F_{2x} = - F_2 \sin\theta = -\dfrac{3}{5} F_2$; $F_{2y} = - F_2 \cos\theta = -\dfrac{4}{5} F_2$

(*d*) **Resolving forces along two non-orthogonal components**

So far forces have been resolved in two mutually perpendicular (or orthogonal) directions. But sometimes, it is required to resolve a given force into two given directions which are not perpendicular to each other. These directions are known as non-orthogonal directions.

Thus, in Fig. 1.10.3, a given force F is required to be resolved into two directions *p* and *q* as shown.

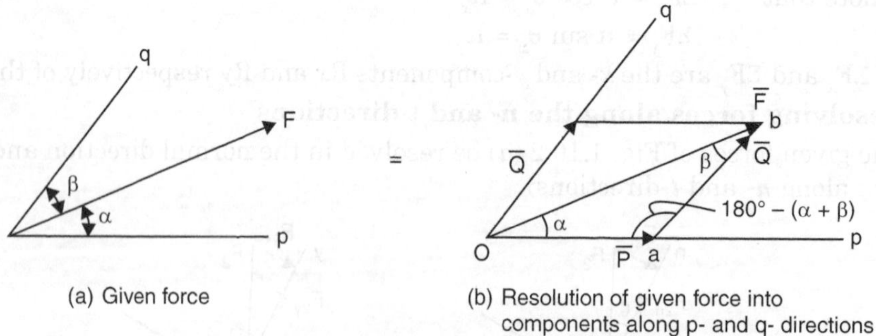

(a) Given force

(b) Resolution of given force into components along p- and q- directions

Fig. 1.10.3. Resolution of a force into two non-orthogonal components

A force F is given inclined as shown in Fig. 1.10.3 (*a*) from the *p*- and *q*- directions. Let the required components be $\overline{P}$ and $\overline{Q}$, as shown in Fig. 1.10.3 (*b*). Let the Δ*oab* show the force F and its components $\overline{P}$ and $\overline{Q}$. Then, by using the triangle law of forces, we get

Thus, from the Δoab,

$$\left.\begin{aligned} \frac{\overline{P}}{\sin\beta} = \frac{\overline{Q}}{\sin\alpha} &= \frac{\overline{F}}{\sin(180^\circ - \alpha - \beta)}, \\ \text{Thus,} \quad \overline{P} &= \overline{F} \times \frac{\sin\beta}{\sin(180^\circ - \alpha - \beta)}, \\ \text{and} \quad \overline{Q} &= \overline{F} \times \frac{\sin\alpha}{\sin(180^\circ - \alpha - \beta)} \end{aligned}\right\} \quad ...(1.19)$$

Conclusion. The following definitions for composition and resolution of forces should be remembered.

(1) **Composition of forces.** The reduction of a given system of forces to the simplest system that will be equivalent to the given system of forces is called the problem of **composition of forces.**

(2) **Resolution of a force.** The replacement of a single force by several components which will be equivalent in action to the given force is called the problem of **resolution of a force.**

Problem 1.1. *Component of a force F along x-axis is 819.15 N. The force F makes an angle of 35° to the x-axis. Find the components along P and Q of the force F. The components P and Q are inclined at 20° and 65°, respectively, with the x-axis.*

Sol. The force F and its components P and Q are shown in Fig. P.1.1 (*a*).

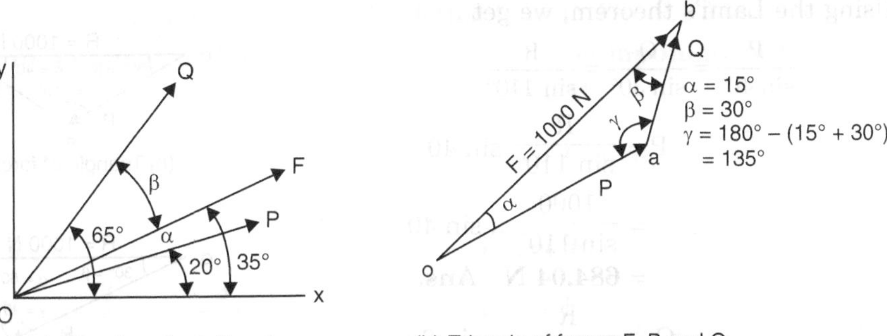

(a) The given force F and its components P and Q

(b) Triangle of forces F, P and Q

Fig. P.1.1

From Fig. P.1.1 (*a*), we find :

α = angle between force F and its component P = 35° – 20° = 15°

β = angle between force F and its component Q = 65° – 35° = 30°

The Δoab in Fig. P.1.1 (*b*) represents forces P, Q and F.

Given : $F_x \equiv 819.15$ N

$\therefore \quad F\cos 35^\circ = F_x$

$$\therefore \quad F = \frac{F_x}{\cos 35^\circ} = \frac{819.15}{\cos 35^\circ} = 1000 \text{ N}.$$

Applying the Lami's theorem on Δoab, we write :

$$\frac{oa}{\sin 30^\circ} = \frac{ab}{\sin 15^\circ} = \frac{ob}{\sin 180^\circ - (30 + 15)^\circ},$$

Problem 1.4. *Find the x and y components of force system shown in Fig. P.1.4(a). Also find resultant of the given forces.*

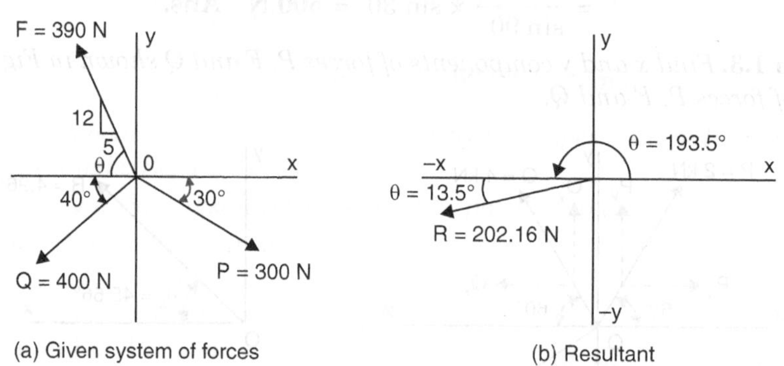

(a) Given system of forces (b) Resultant

Fig. P. 1.4

Sol. x-components : $P_x = + P \cos 30° = 300 \times \cos 30° =$ **259.8 N Ans.**

(along $+x$ direction)

$$F_x = - F \times \frac{5}{\sqrt{12^2 + 5^2}} = - F \times \frac{5}{13}$$

$$= - 390 \times \frac{5}{13} = \mathbf{- 150\ N} \quad \textbf{Ans.} \quad \text{(along } -x \text{ direction)}$$

$$Q_x = - Q \cos 40° = - 400 \cos 40°$$

$$= \mathbf{- 306.4\ N} \quad \textbf{Ans.} \quad \text{(along } - x \text{ direction)}$$

∴ Total of x components $= R_x = 259.8 - 150 - 306.4 =$ **– 196.6 N**

y-components : $P_y = - P \sin 30° = - 300 \sin 30° =$ **– 150 N** (along – y-axis) **Ans.**

$$F_y = + F \times \frac{12}{\sqrt{12^2 + 5^2}}$$

$$= + 390 \times \frac{12}{13} = \mathbf{360\ N} \quad \text{(along + } y\text{-axis)} \quad \textbf{Ans.}$$

$$Q_y = - Q \sin 40° = - 400 \times \sin 40°$$

$$= - \mathbf{257.16\ N} \text{ (along } - y\text{-axis)} \quad \textbf{Ans.}$$

∴ Total of y components

$$= R_y = - 150 + 360 - 257.16 = \mathbf{- 47.16\ N} \quad \text{(along } - y\text{-axis)}$$

Hence, Resultant $R = \sqrt{R_x^2 + R_y^2} = \sqrt{(- 196.6)^2 + (- 47.16)^2} =$ **202.16 N Ans.**

And, $\tan \theta_x = \dfrac{- 47.16}{- 196.6} = 0.2399.$

∴ $\theta_x = 13.5°$ passing through third quadrant as shown in Fig. P.1.4(b), or $\theta_x = 193.5°$ counterclockwise from x-axis.

Problem 1.5. *Find x and y components of the given force system shown in Fig. P.1.5. Also find their resultant.*

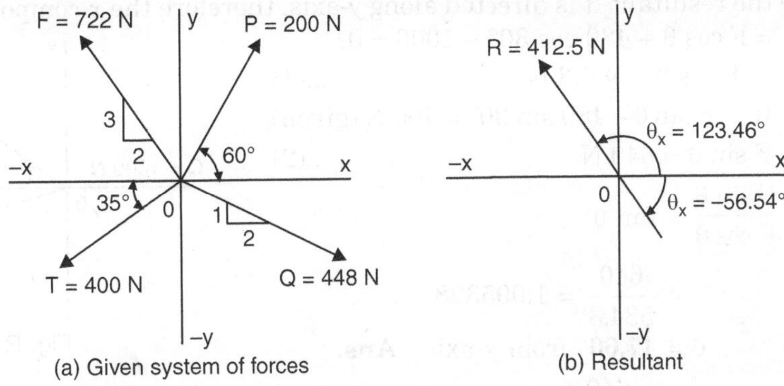

Fig. P. 1.5

Sol. See Fig. P.1.5 (*a*).

x-components of forces

Force P :

$P_x = + 200 \cos 60° = +$ **100 N Ans.**

Force Q :

$$Q_x = + 448 \times \frac{2}{\sqrt{1^2 + 2^2}} = \frac{448 \times 2}{\sqrt{5}} = \mathbf{+ 400.7\ N}$$ **Ans.**

Force F :

$$F_x = - 722 \times \frac{2}{\sqrt{2^2 + 3^2}} = \frac{722 \times 2}{\sqrt{13}} = \mathbf{- 400.5\ N}$$ **Ans.**

Force T :

$T_x = - 400 \cos 35° =$ **– 327.66 N Ans.**

Sum of *x*-components

$= R_x = 100 + 400.7 - 400.5 - 327.66$ N

$=$ **– 227.46 N**

y-components of forces

Force P :

$P_y = + 200 \sin 60° =$ **+ 173.2 N Ans.**

Force Q :

$$Q_y = - 448 \times \frac{1}{\sqrt{5}} = \mathbf{- 200.35\ N}$$ **Ans.**

Force F :

$$F_y = + 722 \times \frac{3}{\sqrt{13}} = \mathbf{+ 600.74\ N}$$ **Ans.**

Force T :

$T_y = - 400 \sin 35° =$ **– 229.43 N Ans.**

Sum of *y*-components

$= R_y = 173.2 - 200.35 + 600.74 - 229.43$

$=$ **+ 344.16 N**

$$\text{Resultant} = R = \sqrt{R_x{}^2 + R_y{}^2}$$

$$= \sqrt{(-227.46)^2 + (344.16)^2}\ \text{N}$$

$=$ **412.5 N Ans.**

Slope of the resultant from *x*-axis is

$$\theta_x = \tan^{-1}\left(\frac{344.16}{-227.46} = -1.5131\right) = -56.54° \text{ from } + x\text{-axis}$$

or $(180 - 56.54)° =$ **123.46°**.

The resultant passes through second quadrant, as shown in Fig. P.1.5 (*b*)

Problem 1.6. *Find the force F and angle θ when the resultant of the force system shown in Fig. P.1.6 is 400 N directed along y-axis.*

Sol. Since the resultant R is directed along y-axis, therefore the x-component R_x is equal to zero. Thus, $R_x = F \cos \theta + 480 \cos 30° - 1000 = 0$

$\therefore$ $F \cos \theta = 584.3 \text{ N}$ ...(1)

And, $R_y = F \sin \theta - 480 \sin 30° = 400 \text{ N (given)}$

$\therefore$ $F \sin \theta = 640 \text{ N}$ ...(2)

$$\therefore \quad \frac{F \sin \theta}{F \cos \theta} = \tan \theta$$

$$= \frac{640}{584.3} = 1.095328$$

$\therefore$ $\theta = \mathbf{47.60°}$ from x-axis **Ans.**

Fig. P.1.6

And, from (2), $F = \dfrac{640}{\sin 47.60} = \mathbf{866.67\ N}$ **Ans.**

Problem 1.7. *Find x and y components and resultant of forces shown in Fig. P.1.7 (a).*

Sol. x-components of forces :

$$P_x = +\,800 \times \frac{3}{\sqrt{2^2 + 3^2}} = + \frac{800 \times 3}{\sqrt{13}} = \mathbf{+\,665.64\ N} \quad \textbf{Ans.}$$

$$Q_x = + \frac{300 \times 4}{\sqrt{4^2 \times 3^2}} = 300 \times \frac{4}{5} = \mathbf{+\,240\ N} \quad \textbf{Ans.}$$

$$T_x = -\,600 \times \frac{1}{\sqrt{1^2 + 2^2}} = - \frac{600 \times 1}{\sqrt{5}} = \mathbf{-\,268.33\ N} \quad \textbf{Ans.}$$

$$S_x = -\,1000 \times \frac{12}{\sqrt{5^2 + 12^2}} = - \frac{1000 \times 12}{13} = -\,923.08 \text{ N}$$

$\therefore$ x-component of resultant

$$= R_x = 665.64 + 240 - 268.33 - 923.08 = \mathbf{-\,285.77\ N}$$

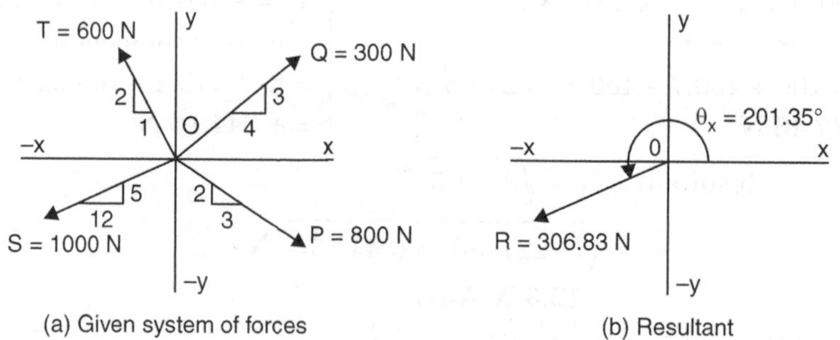

(a) Given system of forces (b) Resultant

Fig. P.1.7

y-component of forces :

$$P_y = -\,800 \times \frac{2}{\sqrt{13}} = -\,443.76 \text{ N}$$

$$O_y = +\,300 \times \frac{3}{5} = +\,180 \text{ N}$$

$$T_y = +600 \times \frac{2}{\sqrt{5}} = +536.66 \text{ N}$$

$$S_y = -1000 \times \frac{5}{13} = -384.62 \text{ N}$$

∴ Y-component of resultant

$$= R_y = -443.76 + 180 + 536.66 - 384.62 \text{ N}$$
$$= -111.72 \text{ N}$$

Hence, resultant $= R = \sqrt{R_x^2 + R_y^2} = \sqrt{(-285.77)^2 + (-111.72)^2}$ N

$$= \sqrt{94145.8513} = \mathbf{306.83\ N} \quad \textbf{Ans.}$$

Since R_x and R_y both are negative, therefore the resultant R will pass through third quadrant as shown in Fig. P.1.7(*b*). Its inclination with + *x*-axis is :

$$\theta_x = \tan^{-1}\left(\frac{R_y}{R_x} = \frac{-111.72}{-285.77} = 0.39094\right) = \mathbf{21.35°} \quad \text{or} \quad 201.35°$$

The correct angle θ_x = **201.35° Ans.**

Problem 1.8. *Find x and y components of the resultant of the force system shown in Fig. P.1.8.*

Sol. The *x*-component of the resultant

$$= R_x = P \times \frac{2}{\sqrt{5}} + T \times \frac{5}{13} - Q \times \frac{2}{\sqrt{5}} - F \times \cos 60°$$

$$= 8 \times \frac{2}{\sqrt{5}} + 12 \times \frac{5}{13} - 4 \times \frac{2}{\sqrt{5}} - 6 \cos 60°$$

= **5.193 kN** along + *x* direction.

Similarly, *y*-component of resultant

$$= R_y = \frac{P \times 1}{\sqrt{5}} - T \times \frac{12}{13} + Q \times \frac{1}{\sqrt{5}} - F \sin 60°$$

$$= 8 \times \frac{1}{\sqrt{5}} - 12 \times \frac{12}{13} + 4 \times \frac{1}{\sqrt{5}} - 6 \sin 60°$$

$$= \mathbf{-10.91\ kN}$$

along – *y* direction. Components R_x and R_y of the resultant are shown in Fig. P.1.8.

Resultant $R = \sqrt{R_x^2 + R_y^2}$

$$= \sqrt{(5.193)^2 + (-10.91)^2} \text{ kN}$$

$$= \sqrt{145.995} = \mathbf{12.083\ kN} \quad \textbf{Ans.}$$

Inclination of resultant

$$= \theta_x = \tan^{-1}\left(\frac{R_y}{R_x} = \frac{-10.91}{5.193} = -2.1009\right)$$

= **– 64.55° Ans.**

or $\theta_x = 64.55°$

clockwise from *x*-axis, as shown in Fig. P.1.8.

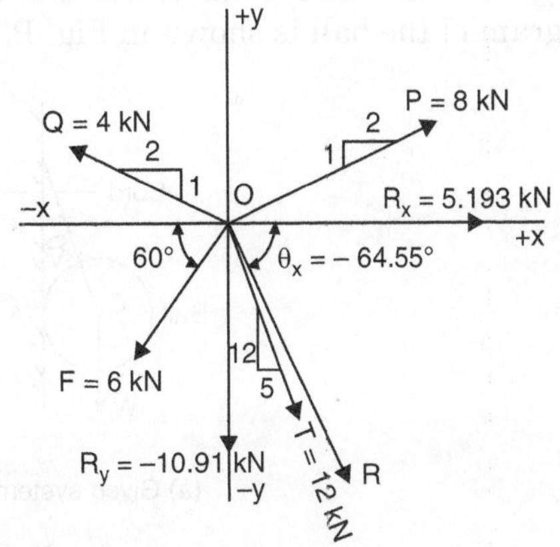

Fig. P.1.8

Problem 1.9. *Fig. P.1.9 (a) shows forces acting at a point O. Determine the magnitude and direction of the resultant.*

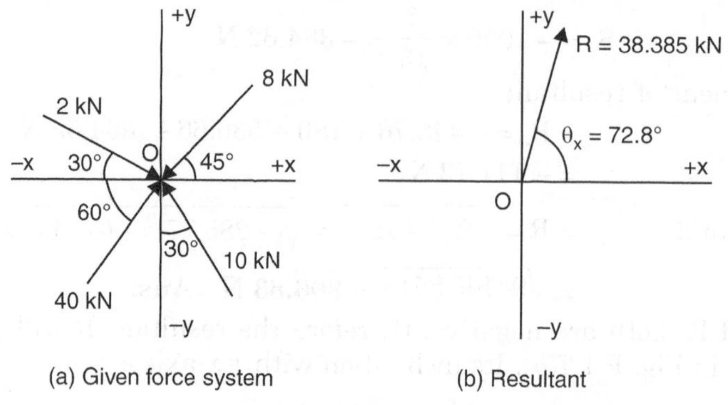

Fig. P.1.9

Sol. x-component of the resultant

$$= R_x = -8\cos 45° - 10\sin 30° + 2\cos 30° + 40\cos 60° \text{ kN}$$

$$= \mathbf{11.34\ kN} \quad \text{along } +x \text{ direction}$$

y-component of the resultant

$$= R_y = -8\sin 45° + 10\cos 30° - 2\sin 30° + 40\sin 60°$$

$$= \mathbf{36.644\ kN} \quad \text{along } +y \text{ direction.}$$

Resultant, $$R = \sqrt{R_x{}^2 + R_y{}^2} = \sqrt{(11.34)^2 + (36.644)^2} \text{ kN}$$

$$= \sqrt{1471.378} \text{ kN} = \mathbf{38.385\ kN\ \ Ans.}$$

Inclination, $$\theta_x = \tan^{-1}\left(\frac{R_y}{R_x} = \frac{36.644}{11.34} = 3.2314\right) = \mathbf{72.80°}\quad \textbf{Ans.}$$

counterclockwise from x-axis.

The resultant is shown in Fig. P.1.9 (*b*).

Problem 1.10. *A ball of weight 50 N is tied to a smooth wall by a cord making an angle of 20° to the wall. Find the tension in the cord and the reaction of wall to the ball.*

Sol. The system of ball and cord are shown in Fig. P.1.10 (*a*), while the free-body diagram of the ball is shown in Fig. P.1.10 (*b*). See section 1.14 for free-body diagrams.

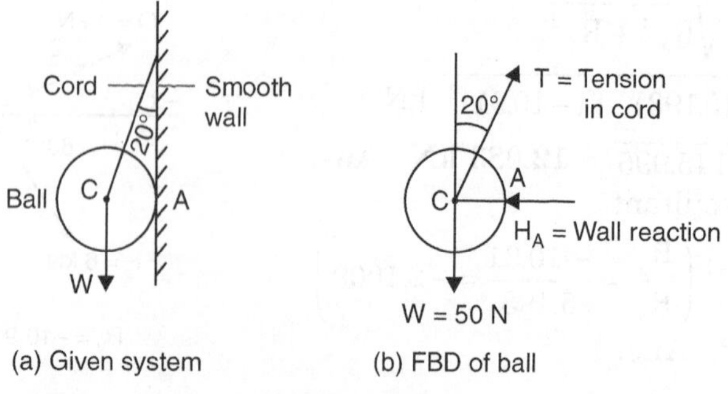

Fig. P.1.10

Method I. By Lami's theorem :

We can write :

$$\frac{H_A}{\sin 20°} = \frac{T}{\sin 90°} = \frac{W}{\sin (90 - 20)°}$$

$$\therefore \qquad H_A = \frac{W}{\sin 70°} \times \sin 20° = \frac{50}{\sin 70°} \times \sin 20° = \mathbf{18.2\ N} \quad \textbf{Ans.}$$

And, $$T = \frac{W}{\sin 70°} \times \sin 90° = \frac{50}{\sin 70°} \times \sin 90° = \mathbf{53.21\ N} \quad \textbf{Ans.}$$

This method is preferred to the method II given below :

Method II. By equations of equilibrium :

By $\Sigma F_v = 0$, we get

$$T \cos 20° - W = 0$$

$$\therefore \qquad T = \frac{W}{\cos 20°} = \frac{50}{\cos 20°} = \mathbf{53.21\ N} \quad \textbf{Ans.}$$

By $\Sigma F_h = 0$, we get

$$H_A - T \sin 20 = 0$$

$$\therefore \qquad H_A = T \sin 20° = 53.21 \times \sin 20° = \mathbf{18.2\ N} \quad \textbf{Ans.}$$

Problem 1.11. *Find the horizontal force P required to keep a 1000 N block in position on a 20° smooth inclined plane. Also find reaction R of the plane.*

Sol. Fig. P.1.11(*a*) shows the block lying on the smooth plane. Fig. 1.11 (*b*) shows the free-body diagram of the block. The block is in equilibrium under the action of (*a*) the weight, W = 1000 N, of the block, (*b*) the reaction R of the plane, and (*c*) the applied for P. R is normal to the plane.

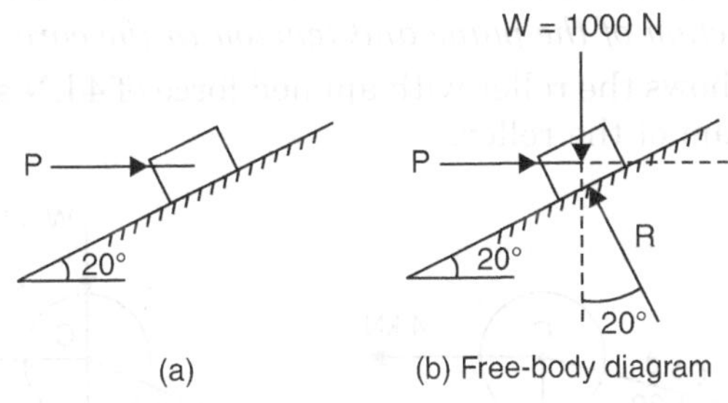

Fig. P.1.11

By $\Sigma F_v = 0$, we get

$$R \times \cos 20° - 1000 = 0$$

$$\therefore \qquad R = \frac{1000}{\cos 20°} = \mathbf{1064.18\ N} \quad \textbf{Ans.}$$

By $\Sigma F_h = 0$, we get

$$P - R \sin 20° = 0$$

$$\therefore \qquad P = R \sin 20° = \mathbf{363.97\ N} \quad \textbf{Ans.}$$

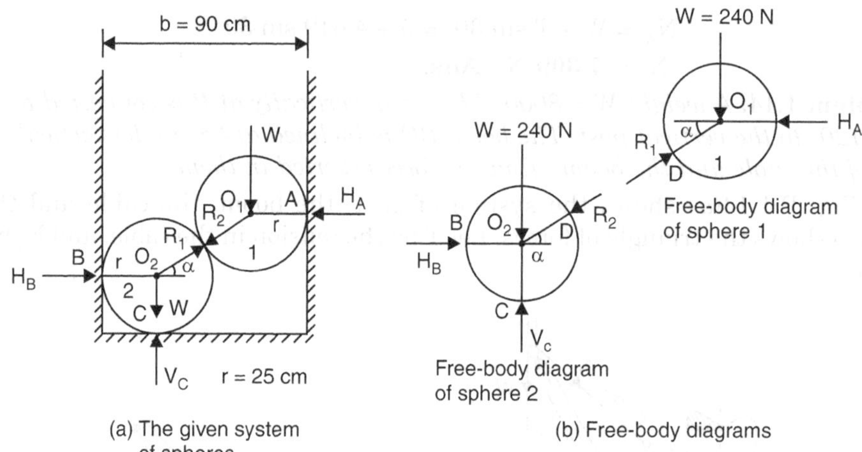

Fig. P.1.15

Fig. P.1.15 (*b*) shows the free-body diagrams of the two spheres. The contact pressures R_1 and R_2 at point C are equal to each other.

The inclination α of R_1 and R_2 is given by :

$$b = 2r + 2r \cos \alpha$$

$$\therefore \quad \cos \alpha = \frac{b}{2r} - 1 \qquad \text{...(1)}$$

or $$\cos \alpha = \frac{90}{2 \times 25} - 1 = 0.8 \quad \therefore \quad \alpha = 36.87^\circ$$

$$\therefore \quad \sin \alpha = \sin 36.87^\circ = 0.6.$$

For sphere 1 :

$$\frac{H_A}{\sin (90^\circ - \alpha)} = \frac{R_1}{\sin 90^\circ} = \frac{W}{\sin \alpha}$$

$$\therefore \quad H_A = \frac{W}{\sin \alpha} \times \sin (90^\circ - \alpha) = \frac{240}{0.6} \times 0.8 = \mathbf{320\ N} \quad \textbf{Ans.}$$

$$R_1 = \frac{W}{\sin \alpha} \times \sin 90^\circ = \frac{240}{0.6} \times 1 = \mathbf{400\ N} \quad \textbf{Ans.}$$

For sphere 2 :

By $\Sigma F_v = 0$, we get

$$V_c - W - R_2 \sin \alpha = 0 \quad \text{where} \quad R_2 = R_1 = 400\ N$$

$$\therefore \quad V_c = W + R_2 \sin \alpha = 240 + 400 \times 0.6$$

or $$V_c = \mathbf{480\ N} \quad \textbf{Ans.}$$

By $\Sigma F_h = 0$, we get

$$H_B - R_2 \cos \alpha = 0$$

$$\therefore \quad H_B = R_2 \cos \alpha = 400 \times 0.8$$

or $$H_B = \mathbf{320\ N} \quad \textbf{Ans.}$$

Problem 1.16. *A uniform wheel weighing 1000 N and of 40 cm diameter rests against a rectangular block 10 cm thick as shown in Fig. P.1.16. The wheel is pulled by a horizontal force P applied at the end of a cord wound tightly around the circumference of the wheel.*

(a) Determine the magnitude of P required to start the wheel move over the rectangular block.

(b) Also find the magnitude and inclination of the least pull required to start the wheel move over the block.

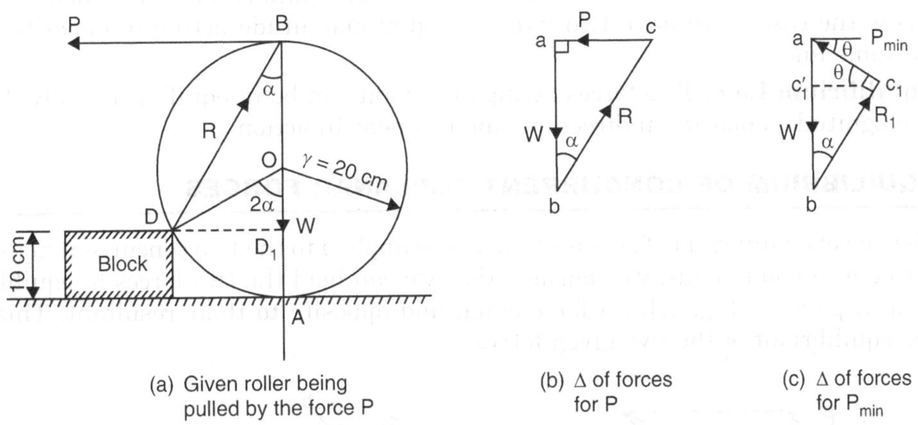

Fig. P.1.16

Sol. (*a*) Fig. P.1.16 (*a*) shows the roller of weight W resting against the block, the pull P is applied horizontally at the circumference through the point B. The weight W also passes through point B, and therefore the reaction R of the block must also pass through B.

The $\angle DOD_1 = 2\alpha$ given by

$$\cos(2\alpha) = \frac{OD_1}{OD} = \frac{AO - AD_1}{OD} = \frac{20 - 10}{20} = 0.5$$

$\therefore$ $2\alpha = 60°$, here $\alpha = \angle DBO = 30°$.

Fig. P.1.16 (*b*) shows the triangle *abc* of forces for finding the pull P. When the roller rolls over the block, the roller is lifted off the point of contact A. The roller is acted upon by forces shown in Fig. P.1.16(*b*). From which we find that the pull

$$P = W \tan\alpha = 1000 \tan 30° = \mathbf{577.35\ N} \quad \textbf{Ans.}$$

And, the reaction R of the block is

$$R = \frac{W}{\cos\alpha} = \frac{1000}{\cos 30°} = \mathbf{1154.7\ N} \quad \textbf{Ans.}$$

(*b*) For the pull to be the minimum, it should be perpendicular to the resultant R_1 of the block, as shown in Fig. P.1.16(*c*).

Thus, we get $P_{min} = W \sin\alpha = 1000 \sin 30° = \mathbf{500\ N}$ **Ans.**

And, its inclination θ with the horizontal is given by

$$\theta = \angle ac_1c_1' = 90° - \angle bc_1c_1' = 90° - (90° - \alpha) = \alpha$$

or $\theta = 30°$ with the horizontal. **Ans.**

1.11. LAW OF ACTION AND REACTION

Any pressure or force on a support causes an equal and opposite pressure or force from the support so that action and reaction are two equal and opposite forces. This principle of

1.15. EQUATIONS OF EQUILIBRIUM

Equations of equilibrium used in the analysis of problems of Statics and the unknown reaction components (or the reactive forces) in a problem are quite inter-related. The following discussion will be helpful to the reader.

(*a*) **Concurrent coplanar force system.** The following equations of motion, which are based on discussion on the drawing of the free-body diagrams, can be easily written :

$$\Sigma\overline{F} = 0 \quad \text{or} \quad \overline{F} = \bar{i}\,\Sigma F_x + \bar{j}\,\Sigma F_y = 0 \qquad \text{...(1.20)}$$

or
$$\Sigma F_x = 0 \qquad \text{...(1.21 (a))}$$

and
$$\Sigma F_y = 0 \qquad \text{...(1.21 (b))}$$

Equations 1.21 (*a*) and (*b*) indicate that in case of concurrent coplanar force system we can find the magnitudes of the two unknown reactive forces. Thus unknown forces H_A and T in Fig. 1.14.2(*b*) can be determined by using eqns. (1.21).

(*b*) **Concurrent spatial (or noncoplanar) force system.** For the condition of equilibrium of this system of forces the essential condition again is :

$$\Sigma\overline{F} = 0, \quad \text{or} \quad \overline{F} = \bar{i}\,\Sigma F_x + \bar{j}\,\Sigma F_y + \bar{k}\,\Sigma F_z = 0 \qquad \text{...(1.22)}$$

Collecting terms in the *x*, *y*, and *z*-directions the analytical equations as follows :

$$\Sigma F_x = 0 \qquad \text{...(1.23 (a))}$$

$$\Sigma F_y = 0 \qquad \text{...(1.23 (b))}$$

$$\Sigma F_z = 0 \qquad \text{...(1.23 (c))}$$

Thus, using eqns. (1.22) or (1.23) we can find three unknown reactive forces or reaction components involved in the problems in Statics.

(*c*) **Non-concurrent coplanar force system.** In order to have the given body in equilibrium the following equations (to be discussed in subsequent Sections) must be satisfied :

$$\Sigma\overline{F} = 0, \quad \text{or} \quad \overline{F} = \bar{i}\,\Sigma F_x + \bar{j}\,\Sigma F_y = 0 \qquad \text{...(1.24)}$$

and
$$\Sigma M_z = 0 \qquad \text{...(1.25)}$$

or
$$\Sigma F_x = 0 \qquad \text{...(1.26 (a))}$$

$$\Sigma F_z = 0 \qquad \text{...(1.26 (b))}$$

and
$$\Sigma M_z = 0 \qquad \text{...(1.26 (c))}$$

In this case, the forces are coplanar and are confined in the *x-y* plane. Eqn. 1.26 (*c*) indicates that the sum of the moments about any point in the *x-y* or about any point in the plane of forces should be equal to zero.

A special case arises when the given or applied forces are vertical. In such case, the equations of equilibirum (1.26) reduce to :

$$\Sigma F_y = 0 \quad \text{or} \quad \Sigma F_v = 0 \qquad \text{...(1.27 (a))}$$

and
$$\Sigma M_z = 0 \qquad \text{...(1.27 (b))}$$

In this case, the *x*-components or the horizontal components of forces do not exist.

Thus, only two unknown can be determined by using eqns.1.27 (*a*) and 1.27 (*b*).

(*d*) **Non-concurrent spatial force system.** For the equilibrium of non-concurrent spatial force system, we should have the following equations of equilibrium :

$$\Sigma\overline{F} = 0, \quad \text{or} \quad \overline{F} = \bar{i}\,\Sigma F_x + \bar{j}\,\Sigma F_y + \bar{k}\,\Sigma F_z = 0 \qquad \text{...(1.28)}$$

and
$$\Sigma\overline{M} = 0\,; \quad \text{or} \quad \overline{M} = \bar{i}\,\Sigma M_x + \bar{j}\,\Sigma M_y + \bar{k}\,\Sigma M_z = 0 \qquad \text{...(1.29)}$$

Problem 1.16. *A uniform wheel weighing 1000 N and of 40 cm diameter rests against a rectangular block 10 cm thick as shown in Fig. P.1.16. The wheel is pulled by a horizontal force P applied at the end of a cord wound tightly around the circumference of the wheel.*

(a) Determine the magnitude of P required to start the wheel move over the rectangular block.

(b) Also find the magnitude and inclination of the least pull required to start the wheel move over the block.

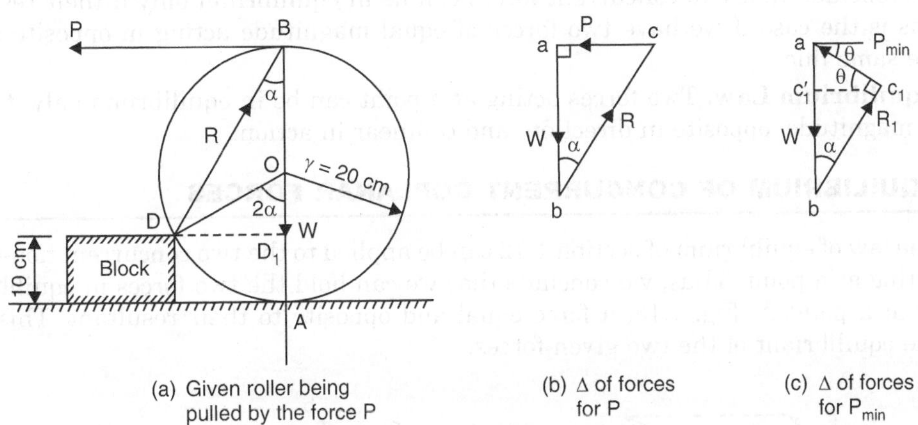

(a) Given roller being pulled by the force P

(b) Δ of forces for P

(c) Δ of forces for P_{min}

Fig. P.1.16

Sol. (*a*) Fig. P.1.16 (*a*) shows the roller of weight W resting against the block, the pull P is applied horizontally at the circumference through the point B. The weight W also passes through point B, and therefore the reaction R of the block must also pass through B.

The $\angle DOD_1 = 2\alpha$ given by

$$\cos(2\alpha) = \frac{OD_1}{OD} = \frac{AO - AD_1}{OD} = \frac{20 - 10}{20} = 0.5$$

$\therefore$ $2\alpha = 60°$, here $\alpha = \angle DBO = 30°$.

Fig. P.1.16 (*b*) shows the triangle *abc* of forces for finding the pull P. When the roller rolls over the block, the roller is lifted off the point of contact A. The roller is acted upon by forces shown in Fig. P.1.16(*b*). From which we find that the pull

$$P = W \tan\alpha = 1000 \tan 30° = \mathbf{577.35\ N} \quad \textbf{Ans.}$$

And, the reaction R of the block is

$$R = \frac{W}{\cos\alpha} = \frac{1000}{\cos 30°} = \mathbf{1154.7\ N} \quad \textbf{Ans.}$$

(*b*) For the pull to be the minimum, it should be perpendicular to the resultant R_1 of the block, as shown in Fig. P.1.16(*c*).

Thus, we get $P_{min} = W \sin\alpha = 1000 \sin 30° = \mathbf{500\ N}$ **Ans.**

And, its inclination θ with the horizontal is given by

$$\theta = \angle ac_1c_1' = 90° - \angle bc_1c_1' = 90° - (90° - \alpha) = \alpha$$

or $\theta = 30°$ with the horizontal. **Ans.**

1.11. LAW OF ACTION AND REACTION

Any pressure or force on a support causes an equal and opposite pressure or force from the support so that action and reaction are two equal and opposite forces. This principle of statics is nothing but is reworded Newton's third law of motion. This helps in the analysis of various problems of statics subjected to any system of forces—collinear, coplanar or spatial.

1.12. EQUILIBRIUM OF COLLINEAR FORCES

From the principle of parallelogram of forces, it follows that the two forces applied at a point can always be replaced by their resultant which is equivalent to the applied forces. Thus, we conclude that two concurrent forces can be in equilibrium only if their resultant is zero. This is the case if we have two forces of equal magnitude acting in opposite direction along the same line.

Equilibrium Law. Two forces acting at a point can be in equilibrium only if they are equal in magnitude, opposite in direction, and collinear in action.

1.13. EQUILIBRIUM OF CONCURRENT COPLANAR FORCES

The law of equilibrium of section 1.12 can be applied to the two concurrent non-collinear forces acting at a point. Thus, we conclude that we can hold the two forces in equilibrium by applying at a point A. Fig. 1.13, a force equal and opposite to their resultant. This force is called the equilibriant of the two given forces.

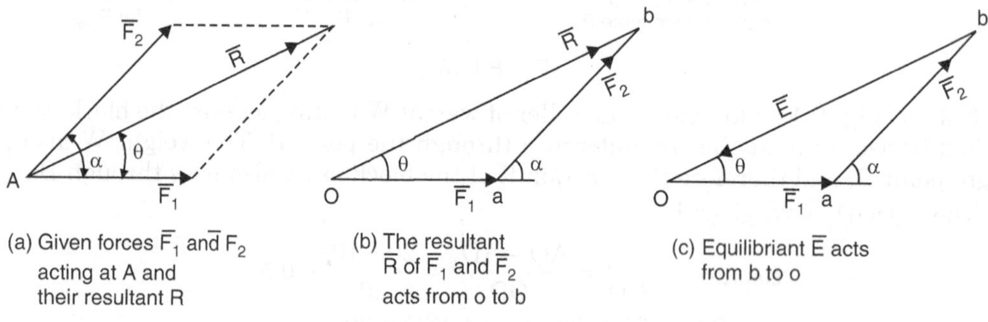

Fig. 1.13. The equilibriant

In Fig. 1.13(*a*), two forces $\overline{F}_1$ and $\overline{F}_2$ acting at A have their resultant $\overline{R}$, given as the diagonal of the parallelogram drawn on sides $\overline{F}_1$ and $\overline{F}_2$. In Fig. 1.13 (*b*) vector $\overline{oa}$ represents force $\overline{F}_1$ and the vector $\overline{ab}$ represents the force $\overline{F}_2$. Their resultant $\overline{R}$ is given by vector $\overline{ob}$. Thus, the two forces $\overline{F}_1$ and $\overline{F}_2$ acting at point A can be replaced by their resultant $\overline{R}$. Thus, under the action of the two forces $\overline{F}_1$ and $\overline{F}_2$ or their resultant $\overline{R}$, the particle at A has tendency to move in the direction of $\overline{ob}$ or in the direction of $\overline{R}$. Hence, in order to present particle to move, a force $\overline{E}$ equal and opposite to $\overline{R}$ should be applied at particle A. This force is known as the **equilibrium** and is directed from *b* to *o* or its direction is opposite to the direction of the resultant $\overline{R}$, as shown in Fig. 1.13 (*c*).

1.14. FREE-BODY DIAGRAMS

To study the equilibrium of a **constrained body**, we shall imagine that we remove the supports and replace them by the **reactions** which they exert on the body. Consider the ball resting on a horizontal surface, shown in Fig. 1.14.1 (*a*). Horizontal support below the ball does not allow the ball to move downward. Thus, the downward motion of the ball is constrained. The horizontal support, through point of contact A, offers a vertical reaction V_a perpendicular to the support surface. This reaction V_a is equal and opposite to the weight W of the ball. Thus, when the support below the ball is removed we show the support reaction V_a and the weight W of the ball, as shown in Fig. 1.14.1 (*b*). Fig. 1.14.1 (*b*) is called the *free-body diagram* of the ball.

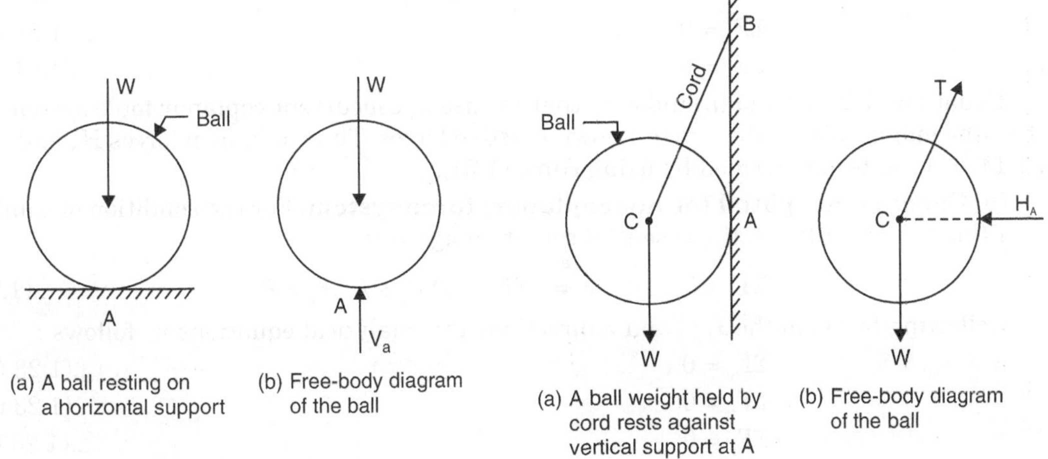

(a) A ball resting on a horizontal support
(b) Free-body diagram of the ball

Fig. 1.14.1

(a) A ball weight held by cord rests against vertical support at A
(b) Free-body diagram of the ball

Fig. 1.14.2

In the case of the ball in Fig. 1.14.2 (*a*), we again remove the supports and isolate it as a free-body as shown in Fig. 1.14.2 (*b*). Then, apart from the weight W of the ball acting at its centre C, we have two reactive forces to apply, one replacing the wall at A and another replacing the cord. Since the cord is attached to the ball at C and since a cord can pull only along its length, we have the reactive force T applied at C and parallel to BC. Its magnitude is unknown. If the surface of the wall is smooth the reaction of wall at A will be normal to wall. Here the reaction at A is horizontal and hence it is represented by H_A – its magnitude remains unknown. Here, we shall not discuss the method of finding the magnitudes of T and H_A. But it is necessary to so proportion them that their resultant is equal and opposite to the vertical forces W.

It is important to note that, we shall always obtain two kinds of forces acting on the body : the given forces, called **active forces**, such as the weight W, and **reactive forces** replacing the supports and its connections to other adjoining bodies such as H_A and the tension T in the cord BC in Fig. 1.14.2. To have equilibrium of a body, it is necessary that the active forces and the reactive forces together form a system of forces which keep the body in equilibrium. Thus, by the free-body diagram we define the system of forces with which we investigate the conditions of equilibrium of any constrained body.

Thus, the essential problem we encounter in Statics may be as follows :

We have a body, either completely or partially constrained, which remains at rest under the action of applied forces. We isolate the body from its supports and from its connections with other bodies and show all active and reactive forces acting on it by the vectors. Then, we consider which conditions this system of forces must satisfy for its equilibrium, *i.e.*, in order that there is no resultant of the forces. The necessary equations of equilibrium and the free-body diagrams generally encountered are discussed in the subsequent Sections.

1.15. EQUATIONS OF EQUILIBRIUM

Equations of equilibrium used in the analysis of problems of Statics and the unknown reaction components (or the reactive forces) in a problem are quite inter-related. The following discussion will be helpful to the reader.

(*a*) **Concurrent coplanar force system.** The following equations of motion, which are based on discussion on the drawing of the free-body diagrams, can be easily written :

$$\Sigma\overline{F} = 0 \quad \text{or} \quad \overline{F} = \bar{i}\,\Sigma F_x + \bar{j}\,\Sigma F_y = 0 \qquad ...(1.20)$$

or $$\Sigma F_x = 0 \qquad ...(1.21\ (a))$$

and $$\Sigma F_y = 0 \qquad ...(1.21\ (b))$$

Equations 1.21 (*a*) and (*b*) indicate that in case of concurrent coplanar force system we can find the magnitudes of the two unknown reactive forces. Thus unknown forces H_A and T in Fig. 1.14.2(*b*) can be determined by using eqns. (1.21).

(*b*) **Concurrent spatial (or noncoplanar) force system.** For the condition of equilibrium of this system of forces the essential condition again is :

$$\Sigma\overline{F} = 0, \quad \text{or} \quad \overline{F} = \bar{i}\,\Sigma F_x + \bar{j}\,\Sigma F_y + \bar{k}\,\Sigma F_z = 0 \qquad ...(1.22)$$

Collecting terms in the *x*, *y*, and *z*-directions the analytical equations as follows :

$$\Sigma F_x = 0 \qquad ...(1.23\ (a))$$

$$\Sigma F_y = 0 \qquad ...(1.23\ (b))$$

$$\Sigma F_z = 0 \qquad ...(1.23\ (c))$$

Thus, using eqns. (1.22) or (1.23) we can find three unknown reactive forces or reaction components involved in the problems in Statics.

(*c*) **Non-concurrent coplanar force system.** In order to have the given body in equilibrium the following equations (to be discussed in subsequent Sections) must be satisfied :

$$\Sigma\overline{F} = 0, \quad \text{or} \quad \overline{F} = \bar{i}\,\Sigma F_x + \bar{j}\,\Sigma F_y = 0 \qquad ...(1.24)$$

and $$\Sigma M_z = 0 \qquad ...(1.25)$$

or $$\Sigma F_x = 0 \qquad ...(1.26\ (a))$$

$$\Sigma F_z = 0 \qquad ...(1.26\ (b))$$

and $$\Sigma M_z = 0 \qquad ...(1.26\ (c))$$

In this case, the forces are coplanar and are confined in the *x*-*y* plane. Eqn. 1.26 (*c*) indicates that the sum of the moments about any point in the *x*-*y* or about any point in the plane of forces should be equal to zero.

A special case arises when the given or applied forces are vertical. In such case, the equations of equilibirum (1.26) reduce to :

$$\Sigma F_y = 0 \quad \text{or} \quad \Sigma F_v = 0 \qquad ...(1.27\ (a))$$

and $$\Sigma M_z = 0 \qquad ...(1.27\ (b))$$

In this case, the *x*-components or the horizontal components of forces do not exist.

Thus, only two unknown can be determined by using eqns.1.27 (*a*) and 1.27 (*b*).

(*d*) **Non-concurrent spatial force system.** For the equilibrium of non-concurrent spatial force system, we should have the following equations of equilibrium :

$$\Sigma\overline{F} = 0, \quad \text{or} \quad \overline{F} = \bar{i}\,\Sigma F_x + \bar{j}\,\Sigma F_y + \bar{k}\,\Sigma F_z = 0 \qquad ...(1.28)$$

and $$\Sigma\overline{M} = 0\ ; \quad \text{or} \quad \overline{M} = \bar{i}\,\Sigma M_x + \bar{j}\,\Sigma M_y + \bar{k}\,\Sigma M_z = 0 \qquad ...(1.29)$$

or, analytical equations of equilibrium are :

$$\Sigma F_x = 0 \qquad ...(1.30\ (a))$$

$$\Sigma F_y = 0 \qquad ...(1.30\ (b))$$

$$\Sigma F_z = 0 \qquad ...(1.30\ (c))$$

And, also

$$\Sigma M_x = 0 \qquad ...(1.31\ (a))$$

$$\Sigma M_y = 0 \qquad ...(1.31\ (b))$$

$$\Sigma M_z = 0 \qquad ...(1.31\ (c))$$

Thus, there are six unknown quantities that can be determined by using eqns. (1.28) and (1.29) or their equivalent (1.30) and (1.31).

1.16. STATICALLY DETERMINATE AND INDETERMINATE PROBLEMS

In Statics we are required to find the unknown forces and reaction components by using the equations of equilibrium discussed in section (1.15).

Let, R be the number of unknown reaction components or the unknown forces that are encountered in a problem.

Also, E = number of equations of equilibrium available for determining the values of R. If the number of unknown reaction components or forces (R) is equal to the number of equations of equilibrium (E), the problem is known as **statically stable and determinate problem.**

If R is less than E, then the problem is known as **determinate and statically unstable.** And, if R is greater than E, then the problem is known as **statically stable and indeterminate.**

Further, let (R-E) be defined as the **degree of indeterminacy**, that is

$$\boxed{D = R - E} \qquad ...(1.32)$$

Thus,

(*a*) D = 0, or R = E for statically determinate and stable problems.

(*b*) D < 0, *i.e.,* R < E for statically determinate but unstable problems.

(*c*) D > 0, *i.e.,* R > E for statically stable but indeterminate problems.

(*i*) Cases of D > 0, or R > E are beyond the scope of the subject of mechanics and will not be discussed.

(*ii*) Cases of D = 0, or R = E will be discussed in Statics in this book or in any book on Mechanics.

(*iii*) Cases of D < 0, or R < E are the problems which are discussed in Dynamics. This as an example, is case of motion of an automobile, an over-head crane, a hand cart, etc. in which the rigid body moves in the direction of the unbalanced horizontal force. This force, known as an *effort or tractive effort*, pulls a rigid body in its own direction. This is generally studied in Dynamics of Machines.

Table 1.2 gives some cases of supports and free-body diagrams. It also presents the values of the degrees of indeterminacy. This information will help the reader in drawing the free-body diagrams, and also is writing down the equations for determining the unknown reactive forces/reaction components/other unknown forces by using the proper equations of equilibrium as discussed in Section (1.15).

Table 1.2. Some Free-Body Diagrams

S.No.	*Support / Connection / type of body isolated*	*Free-body diagram of isolated body*	*Unknowns to be found*	*No. of unknowns (R)*	*Eq. of equilibrium used (E)*	*D = R – E*
1.	Ball resting on a smooth floor		V_A	1	$\Sigma F_v = 0$ or $\Sigma F_y = 0$	0
2.	Suspended ball resting against a smooth wall		H_A and T	2	$\Sigma F_h = 0$ $\Sigma F_v = 0$ or $\Sigma F_x = 0$ $\Sigma F_y = 0$	0
3.	Suspended ball resting on a smooth floor and also against smooth wall		T, V_A, H_B	3	$\Sigma F_x = 0$ $\Sigma F_y = 0$	3 – 2 = 1
4.	A weight suspended by string		T_1 and T_2	2	$\Sigma F_x = 0$ $\Sigma F_y = 0$	0
5.	Roller, Rocker, Frictionless surface		V	1	$\Sigma F_v = 0$	0
6.	Roller on inclined surface		N	1	$\Sigma F_y = 0$	0
7.	Hinge; Hinge or smooth pin		H_A and V_A	2	$\Sigma F_v = 0$ $\Sigma F_h = 0$	0

8.	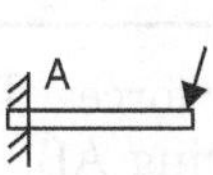 Fixed support	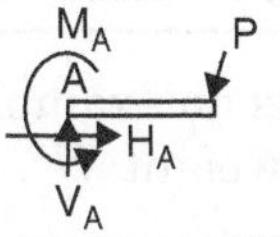	V_A , H_A M_A	3	$\Sigma F_v = 0$ $\Sigma F_h = 0$ $\Sigma M = 0$	0
9.	Ball and socket joint 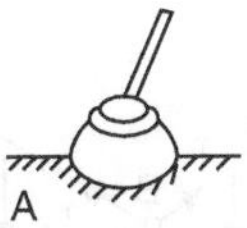	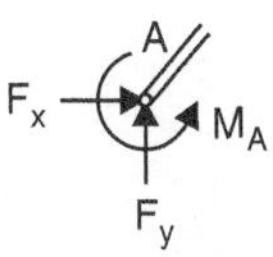	F_x, F_y , and M_A	3	$\Sigma F_x = 0$ $\Sigma F_y = 0$ $\Sigma M = 0$	0
10.	Ledder resting on smooth floor and against smooth wall 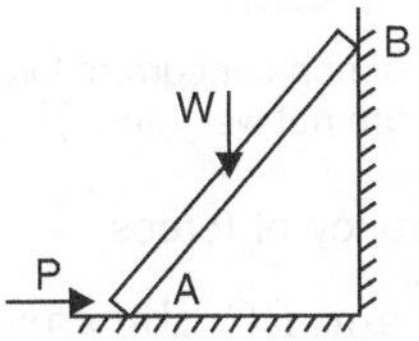	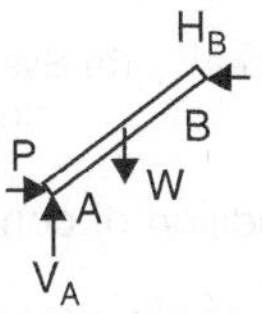	P, V_A and H_B	3	$\Sigma F_v = 0$ $\Sigma F_h = 0$ $\Sigma M = 0$	0
11.	Ladder resting on rough floor and against rough wall 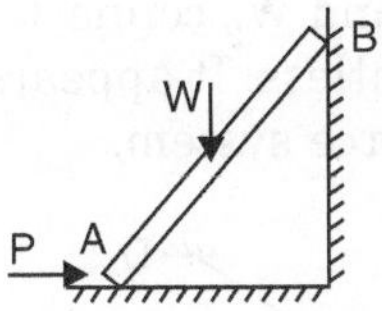	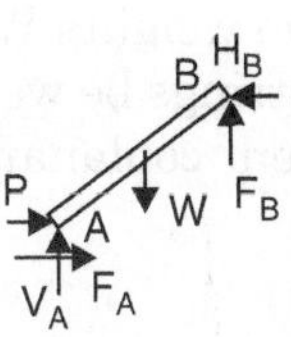	P; (V_A and F_A); and (H_B and F_B) Here $F_A = \mu_A . V_A$ and $F_B = \mu_B . H_B$	3	$\Sigma F_v = 0$ $\Sigma F_h = 0$ $\Sigma M = 0$	0
12.	Simply supported beam, hinged at A and roller at B. 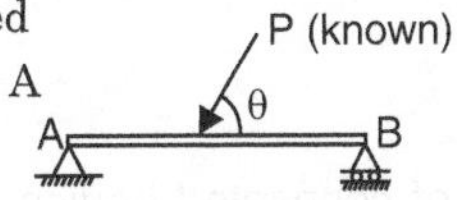	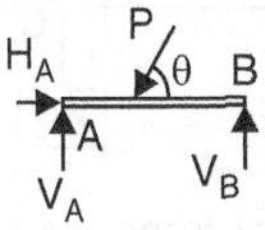	V_A , H_A and V_B	3	$\Sigma F_v = 0$ $\Sigma F_h = 0$ $\Sigma M = 0$	0
13.	Simply supported beam hinged at A and B 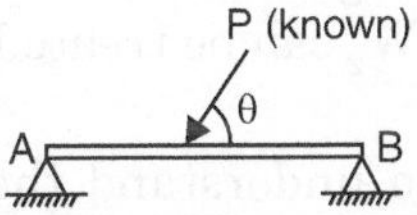	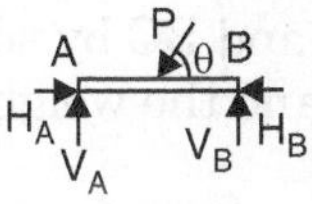	V_A, V_B, H_A , H_B	4 (R = 4)	$\Sigma F_v = 0$ $\Sigma F_h = 0$ (E = 3)	4 – 3 = 1
14.	Propped cantilever fixed at A and roller at B. 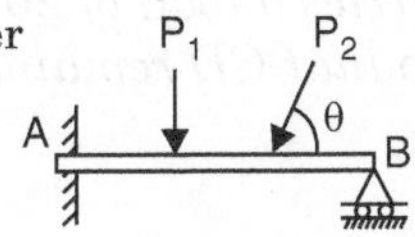	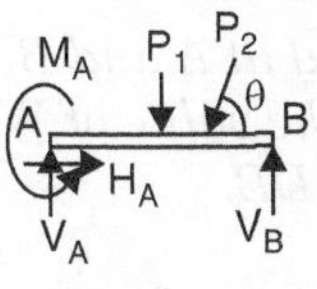	V_A, H_A M_A and V_B	4 R = 4	$\Sigma F_v = 0$ $\Sigma F_h = 0$ $\Sigma M = 0$ (E = 3)	4 – 3 = 1
15.	Propped cantilever with vertical loading 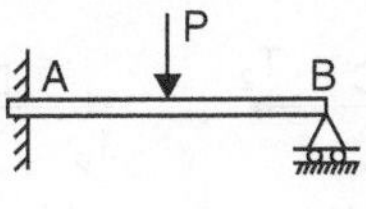	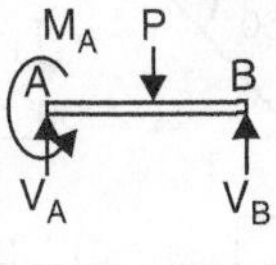	V_A, V_B and M_A	3 R = 3	$\Sigma F_v = 0$ $\Sigma M = 0$ (E = 2)	3 – 2 = 1
16.	Fixed beam 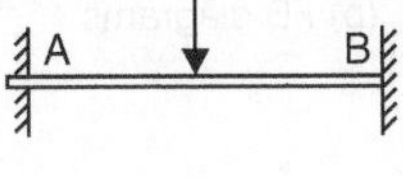	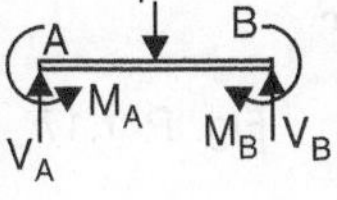	V_A , V_B M_A and M_A	4 R = 4	$\Sigma F_v = 0$ $\Sigma M = 0$ (E = 2)	4 – 2 = 2

1.17. ANALYSIS OF CONNECTED BODIES

The system shown in Fig.1.17(*a*) is an example of concurrent coplanar forces. The three forces meeting at point B are : (*i*) the weight W , (*ii*) the tension T_1 in string AB, and (*iii*) tension T_2 in string BC.

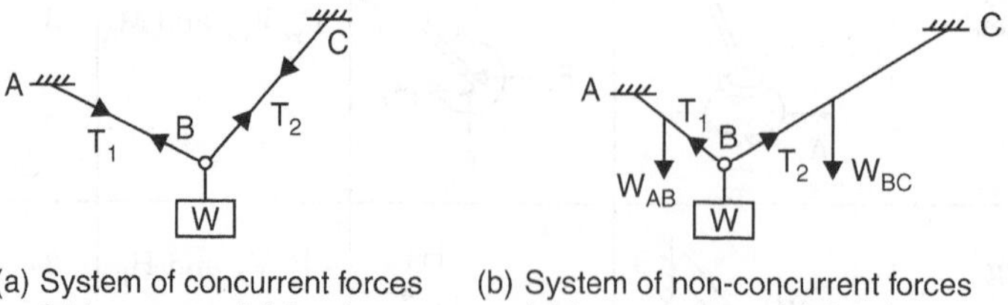

(a) System of concurrent forces (strings are weightless) (b) System of non-concurrent forces (strings are not weightless)

Fig. 1.17.1. Condition of concurrency of forces

But, when we consider the weight of strings AB and BC, the various forces do not meet at one point and it becomes a non-concurrent coplanar force system, as shown in Fig. 1.17.1 (*b*).

As a second example, consider two weights W_1 and W_2 connected by strings at points B and C, as shown in Fig. 1.17.2. Let the strings be weightless. It appears to be a non-concurrent coplanar force system, but it is concurrent coplanar force system.

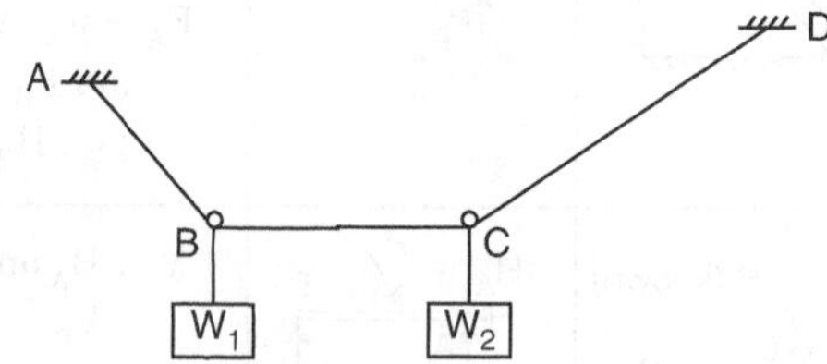

Fig. 1.17.2. Concurrent force system of connected bodies

We first find force in strings AB and BC by solving for forces meeting at point B. Then, the two tensions in strings BC and CD and the weight W_2 can be treated another set of concurrent coplanar force system.

The following examples will help the reader to understand the procedure of solving problems of this nature.

Problem 1.17. *A string supported at A and B carries a load of 20 kN at C and a load W at D as shown in Fig. P.1.17 (a). Find the value of W so that CD remains horizontal. Also find tension in string segments AC, CD and DB.*

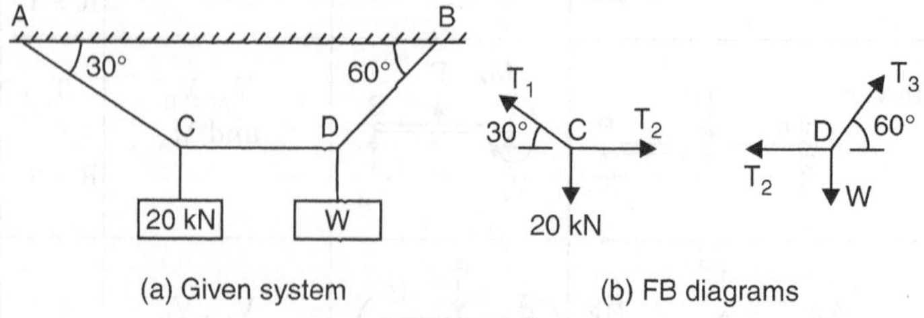

(a) Given system (b) FB diagrams

Fig. P. 1.17

Sol. Fig. P.1.17 (*b*) gives the free-body diagrams for joints C and D.

For joint C. By $\Sigma F_v = 0$, we get

$$T_1 \sin 30° - 20 = 0$$

$$\therefore \qquad T_1 = \frac{20}{\sin 30°} = \mathbf{40\ kN} \quad \textbf{Ans.}$$

By $\Sigma F_h = 0$, we get

$$T_1 \cos 30° - T_2 = 0,$$

$$\therefore \qquad T_2 = T_1 \cos 30° = \frac{40\sqrt{3}}{2} = \mathbf{34.64\ kN\ Ans.}$$

For joint D. By $\Sigma F_h = 0$, we get

$$T_3 \cos 60° - T_2 = 0, \quad \therefore \quad T_3 = \frac{T_2}{\cos 60°} = \frac{34.64}{0.5} = \mathbf{69.28\ kN} \quad \textbf{Ans.}$$

By $\Sigma F_v = 0$, we get $\quad T_3 \sin 60° - W = 0$

$$\therefore \qquad W = T_3 \sin 60° = 69.28 \times \sin 60° = \mathbf{60\ kN} \quad \textbf{Ans.}$$

Problem 1.18. *In a four-bar mechanism shown in Fig. P.1.18(a), bars are hinged at A, B, C and D. Find the force P that will prevent movement of bars. Also find forces in bars AB, BC and CD.*

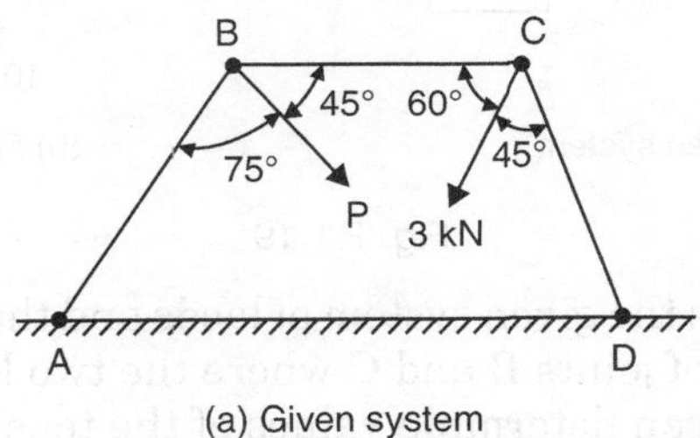

(a) Given system

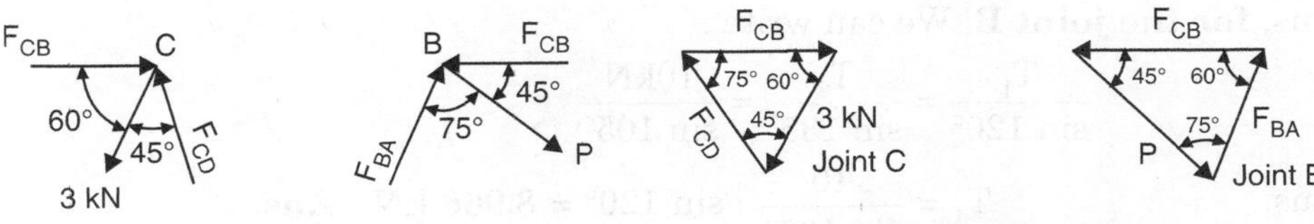

(b) FBDs for joints C and B

(c) Triangles of forces

Fig. P.1.18

Sol. Fig. P.1.18 (*b*) shows the free-body diagrams for joints C and B , while Fig. 1.18 (*c*) shows the corresponding triangles of forces (not drawn to scale).

For joint C. By using the triangle of forces, we can write :

$$\frac{F_{CB}}{\sin 45°} = \frac{F_{CD}}{\sin 60°} = \frac{3\text{kN}}{\sin 75°}$$

$\therefore$ We get $\qquad F_{CB} = \dfrac{3}{\sin 75°} . \sin 45° = 2.196$ kN (compression) $\quad$ **Ans.**

And $\qquad F_{CD} = \dfrac{3}{\sin 75°} . \sin 60° = 2.69$ kN (compression) $\quad$ **Ans.**

For joint B : By using the triangle of forces, we can write :

$$\frac{P}{\sin 60°} = \frac{F_{BA}}{\sin 45°} = \frac{F_{CB}}{\sin 75°}$$

$\therefore$
$$P = \frac{F_{CB}}{\sin 75°} . \sin 60° = \frac{2.196}{\sin 75°} . \sin 60°$$
$$= 1.97 \text{ kN (compression)} \quad \textbf{Ans.}$$

and
$$F_{BA} = \frac{F_{CB}}{\sin 75°} . \sin 45° = \frac{2.196}{\sin 75°} . \sin 45°$$
$$= 1.608 \text{ kN (compression)} \quad \textbf{Ans.}$$

Problem 1.19. *Two vertical loads are supported by the system of flexible and weightless cables shown in Fig. P.1.19 (a). Find forces in all cable segments.*

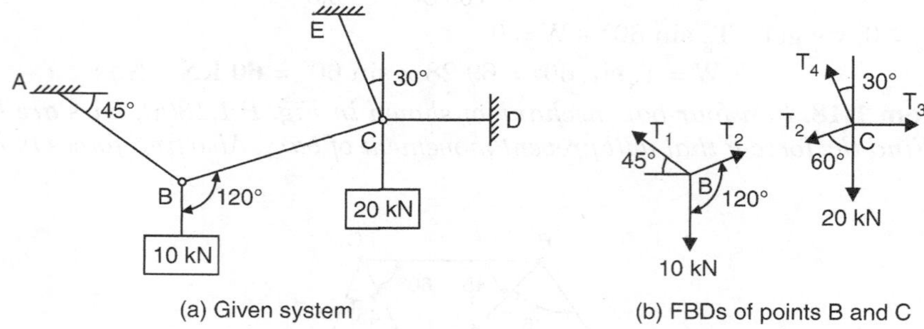

(a) Given system (b) FBDs of points B and C

Fig. P.1.19

Sol. Fig. P.1.19 (*a*) shows the given system of laods and the cable segments. Fig. P.1.19(*b*) shows the free-body diagrams of joints B and C where the two loads are suspended. Using the Lami's theorem for joint B we can determine values of the tensions in cable segments AB and BC.

Thus, **for the joint B.** We can write :

$$\frac{T_1}{\sin 120°} = \frac{T_2}{\sin 135°} = \frac{10\text{kN}}{\sin 105°}$$

Thus,
$$T_1 = \frac{10}{\sin 105°} . \sin 120° = \textbf{8.966 kN} \quad \textbf{Ans.}$$

and
$$T_2 = \frac{10}{\sin 105°} . \sin 135° = \textbf{7.3205 kN} \quad \textbf{Ans.}$$

For joint C. By $\Sigma F_v = 0$, we get

$$T_4 \cos 30° - T_2 \cos 60° - 20 = 0$$

$\therefore$
$$T_4 = \frac{1}{\cos 30°} [T_2 \cos 60° + 20] = \frac{1}{\cos 30°} [7.3205 \times \cos 60° + 20]$$

or
$$T_4 = \textbf{27.3205 kN} \quad \textbf{Ans.}$$

By $\Sigma F_h = 0$, we get $T_3 - T_2 \sin 60° - T_4 \times \sin 30° = 0$

or,
$$T_3 = T_2 \sin 60° + T_4 \sin 30° = 7.3205 \sin 60° + 27.3205 \sin 30°$$
$$= \textbf{20 kN} \quad \textbf{Ans.}$$

Problem 1.20. *Two cylinders, A of weight 6 kN and B of weight 4 kN rest on smooth inclined surfaces shown in Fig. P.1.20 (a). They are connected by a bar of negligible weight* *hinged to each cylinder at its centre by smooth pins. Find the force F parallel to the 60° incline required, as shown, to hold the system in the given position. Also find the reactions of inclined planes to the cylinder A and B.*

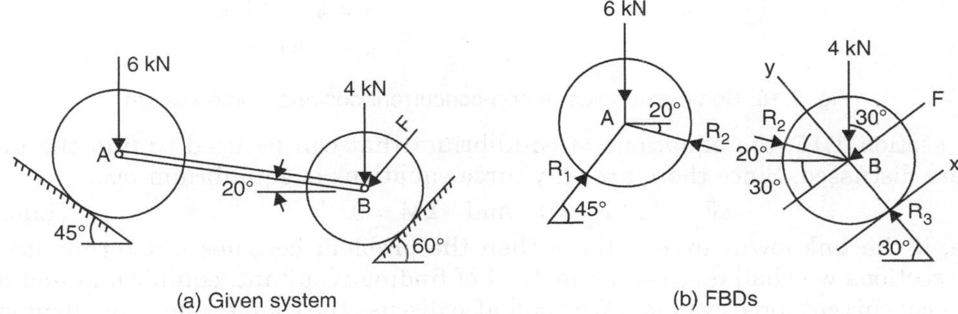

Fig. P.1.20

Sol. Fig. P.1.20 (*b*) shows the free-body diagrams of cylinder A and B. Applying Lami's theorem to cylinder A, we get

$$\frac{R_1}{\sin 110°} = \frac{R_2}{\sin 135°} = \frac{6 \text{ kN}}{\sin 115°}$$

$$\therefore \quad R_1 = \frac{6}{\sin 115°} \cdot \sin 110° = \mathbf{6.2210 \text{ kN}} \quad \textbf{Ans.}$$

and

$$R_2 = \frac{6}{\sin 115°} \cdot \sin 135° = \mathbf{4.6812 \text{ kN}} \quad \textbf{Ans.}$$

Cylinder B. Let us take x-axis parallel to and y-axis perpendicular to the 30° inclined plane as shown in Fig. 1.20(*b*).

By $\Sigma F_y = 0$, we get

$$R_3 - 4 \cos 60° - R_2 \cos 40° = 0$$

$$\therefore \quad R_3 = 4 \cos 60° + 4.6812 \cos 40° = \mathbf{5.586 \text{ kN}} \quad \textbf{Ans.}$$

And, by $\Sigma F_x = 0$, we get

$$F + 4 \sin 60° - R_2 \sin 40° = 0$$

$$\therefore \quad F = R_2 \sin 40° - 4 \sin 60° = 4.6812 \sin 40° - 4 \sin 60°$$

$$= \mathbf{2.5909 \text{ kN}} \quad \textbf{Ans.}$$

PART C—NON-CONCURRENT COPLANAR FORCE SYSTEM

1.18. INTRODUCTION

As discussed in Section 1.5 the non-concurrent coplanar forces are confined in one plane and they do not meet at a single point. Some examples of the non-concurrent coplanar force system are shown in Fig. 1.18.

In Fig. 1.18 (*a*), the loads P_1 and P_2 are inclined and therefore reaction at A will also be inclined. All active forces (P_1 and P_2) and reactive forces (R_A and R_B) act through different points. In Fig. 1.18 (*b*) all forces are parallel to each other. In Fig. 1.18 (*c*) all forces act through different points and do not meet at any single point.

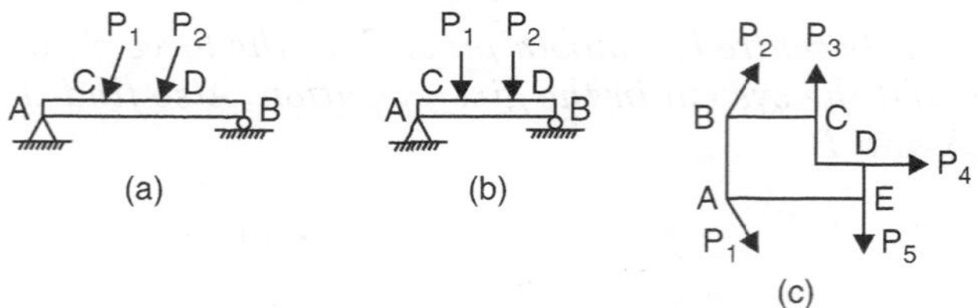

Fig. 1.18. Some examples of non-concurrent coplanar force system

In section 1.15, the equations of equilibrium that can be used to find the unknown forces were discussed. Since there are only three equations of equilibrium *viz.*,

$$\Sigma F_x = 0, \Sigma F_y = 0 \quad \text{and} \quad \Sigma M = 0. \qquad \text{[Eqn. (1.27)]}$$

therefore, if the unknowns exceed three then the problem becomes indeterminate. In the following sections we shall discuss the method of finding resultant, equilibrium and moment of the non-concurrent force systems. We shall also discuss the couple, and equivalent systems.

1.19. MOMENT OF FORCE

The **moment of a force** about a point is defined as the product of the magnitude of the force and the perpendicular distance between the point and the line of action of the force. In Fig. 1.19.1 the perpendicular distance *d* is known as the **moment arm.** The point O about which the moment of the force F is taken is known as the **moment centre.**

The tendency of a force to make a (rigid) body rotate about an axis is measured by the moment of the force about that axis. The **moment axis** is perpendicular to the plane containing the force F and the moment arm *d* and passes through the **moment centre**. Here, *y*-axis through O is the moment axis.

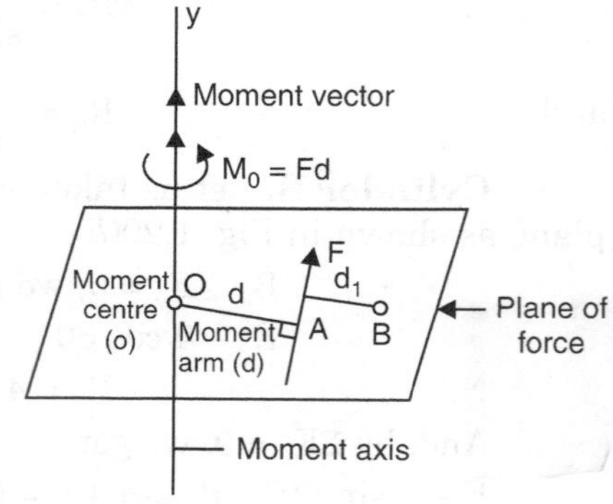

Fig. 1.19.1. Moment of a force

Thus, the moment of F about O, is

$$M_o \text{ or } M_o^F = Fd \qquad ...(1.33)$$

In SI system of units, where distance is expressed in metres (*m*), and force in newtons (N), the moment will be expressed in newton metres (N-m).

The moment of force has magnitude as well as the sense. The moment of a force will be either clockwise or counterclockwise. Thus, the moment of the force F about O , in Fig. 1.19.1, is counter-clockwise, while its moment about point B (*i.e.*, M_B^F) is clockwise. Generally, counterclockwise moment is considered positive, and the clockwise moment is considered negative. Moments can therefore be added algebraically. Figs. 1.19.2 and 1.19.3 illustrate the **sign conventions of the positive and negative moments.**

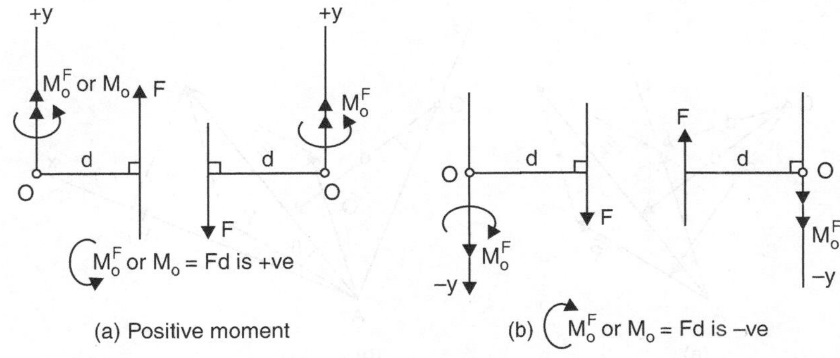

Fig. 1.19.2. Sign conventions for positive and negative moments. OY is the moment axis

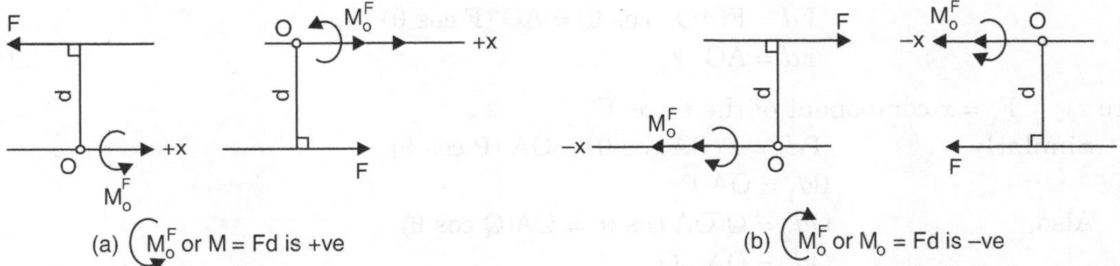

Fig. 1.19.3. Sign conventions for +ve moment and –ve moment. OX is the moment axis

The moment vector generally is represented by double arrows, as shown in Fig. 1.19.1, 1.19.2 and 1-19.3. In Fig. 1.19.2, the *y*-axis through the moment centre O is the moment axis for the moment of the force F acting the Y-axis. In Fig. 1.19.3, the force F is acting along the X-axis and the OX is the moment axis. Sign conventions for both positive moments and the negative moments are illustrated in Figs. 1.19.2 and 1.19.3.

1.20. VARIGNON'S THEOREM

An important theorem of statics is due to the French mathematician Varignon (1654-1722). It states that **the moment of a force about any point is equal to the sum of moments of its components about that point.**

Proof. Consider a force $\overline{F}$ acting at a point A. Let its components be $\overline{P}$ and $\overline{Q}$ in any two directions, as shown in Fig. 1.20.1 (*a*). The moment of the force $\overline{F}$ about any arbitrary point O is F*d*, where *d* is the perpendicular distance from O to the line of action of $\overline{F}$.

Similarly, the moments of $\overline{P}$ and $\overline{Q}$ about the point O are Pd_1 and Qd_2, respectively, where d_1 and d_2 are the perpendicular distances from O to the lines of action of $\overline{P}$ and $\overline{Q}$. Thus, we have to prove that

$$Fd = Pd_1 + Qd_2 \qquad ...(1.34)$$

We shall take X and Y axis through point A. Let AO be Y-axis then AX perpendicular to Y-axis is the X-axis. Component of the force $\overline{F}$ along the *x*-axis is F_x, as shown in Fig. 1.20.1 (*b*). The angle between F_x and $\overline{F}$ is θ, and θ is also angle between Y-axis the perpendicular from O to the line of action of the force $\overline{F}$. We can write

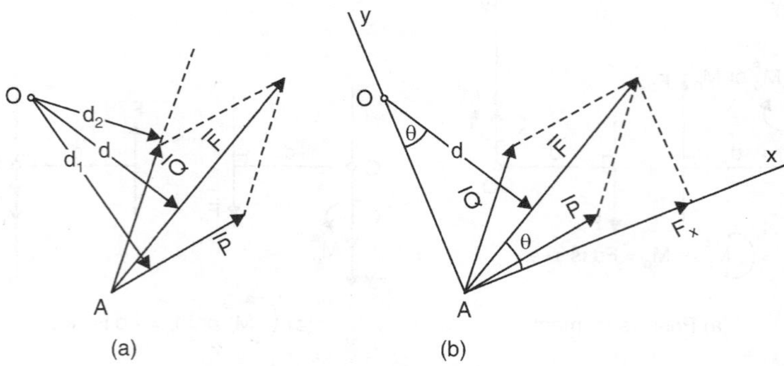

Fig. 1.20.1. Proof of Varignon's theorem

$$Fd = F(AO\ .\ \cos\theta) = AO\ (F \cos\theta)$$

or $$Fd = AO.\ F_x \quad ...(1.35)$$

where F_x = x-component of the force $\overline{F}$.

Similarly, $$Pd_1 = P(OA \cos\theta) = OA\ (P \cos\theta)$$

or $$Pd_1 = OA\ P_x \quad ...(i)$$

Also, $$Qd_2 = Q(OA \cos\theta) = OA(Q \cos\theta)$$

or $$Qd_2 = OA\ .\ Q_x \quad ...(ii)$$

Combining equations (*i*) and (*ii*), we get

$$Pd_1 + Qd_2 = OA(P_x + Q_x) \quad ...(1.36)$$

But the sum $P_x + Q_x$ of the x-components of the two forces $\overline{P}$ and $\overline{Q}$ is equal to the x-component of the F_x of the resultant force $\overline{F}$.

Thus, eqn. (1 . 36) gives

$$Pd_1 + Qd_2 = OA(F_x) \quad ...(1.37)$$

Thus, equating the left hand sides of eqns. (1.35) and (1.37), we get

$$Fd = Pd_1 + Qd_2 \quad ...(1.38)$$

which is the same as eqn. (1.35).

If the system of forces consists of more than two forces, the above result can be extended as below.

Let $\overline{F}_1, \overline{F}_2, \overline{F}_3$ and $\overline{F}_4$ be four concurrent forces acting at A, and let $\overline{R}$ be their resultant.

Let d_1, d_2, d_3 and d_4 be the perpendicular distances of the lines of action of forces $\overline{F}_1, \overline{F}_2, \overline{F}_3$ and $\overline{F}_4$ from the moment centre O. Also, let the perpendicular distance of $\overline{R}$ from O be a.

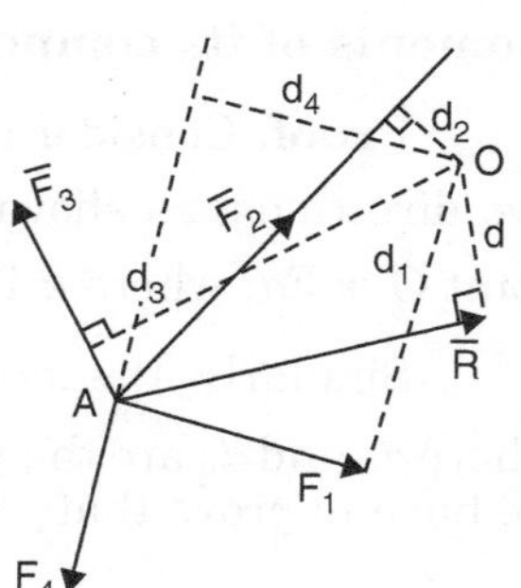

Fig. 1.20.2. Application of Varignon's theorem to more than two concurrent forces

If $\overline{R}_1$ is the resultant of $\overline{F}_1$ and $\overline{F}_2$ and its distance from O is a_1, then by applying Varignon's theorem we can write :

$$R_1a_1 = F_1d_1 + F_2d_2 \quad ...(iii)$$

If $\overline{R}_2$ be the resultant of $\overline{R}_1$ and $\overline{F}_3$ (and hence of $\overline{F}_1$ and $\overline{F}_2$) and if its distance from O is a_2, then applying Varignon's theorem we can write :

$$R_2 \cdot a_2 = F_3 \cdot d_3 + (R_1 \cdot a_1)$$

Or, using eqn. (*iii*), we get

$$R_2 \cdot a_2 = F_3 \cdot d_3 + (F_1 \cdot d_1 + F_2 \cdot d_2) \quad ...(iv)$$

Similarly, by applying the Varignon's theorem to the forces $\overline{F}_4$ and $\overline{R}_2$ and the result $\overline{R}$ of all forces, we get

$$Ra = F_4 \cdot d_4 + (R_2 \cdot a_2)$$
$$= F_4 \cdot d_4 + (F_3 \cdot d_3 + F_1 d_1 + F_2 d_2)$$

Or, $$R \cdot a = F_1 d_1 + F_2 d_2 + F_3 d_3 + F_4 d_4 \quad ...(1.39)$$

Eqn. (1.39) states that **the moment of the resultant of a number of forces about a moment centre is equal to the sum of moments of its components about that moment centre.**

1.21. COUPLE

Two forces having the same magnitude parallel lines of action, oppositely directed and separated by a defined distance are said to form a **couple**. The sum of the forces forming a couple in any direction is zero. This means that there is no translatory effect of a couple.

Two such forces $\overline{F}$ and $\overline{F}'$, of the same magnitude F, are shown acting in Fig. 1.21 (*a*). The sum of the *x*-components and of the *y*-components of $\overline{F}$ and $\overline{F}'$ are zero. The sum of their moments about a point (or an axis through a point) A, however, is not zero. Thus, the two forces will not move the body away, they will tend to make it turn or rotate.

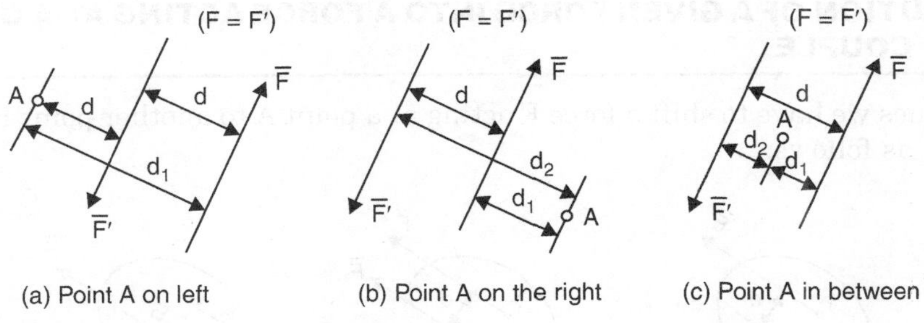

(a) Point A on left (b) Point A on the right (c) Point A in between

Fig. 1.21. A couple

Let the perpendicular distances of $\overline{F}$ and $\overline{F}'$ from the point A be d_1 and d_2. Noting that the counterclock-wise moment is positive, we write the sum M of the two forces about A :

$$+M = Fd_1 - Fd_2 = F(d_1 - d_2)$$

Since, $d = (d_1 - d_2)$, therefore we get

$$M = Fd \quad ...(1.40)$$

The sum M is called the **moment of the couple.** It is important to note **that M does not depend upon the location of the point A** ; M will have the same magnitude and the same sense regardless of the location of the point A.

This is illustrated in Fig. 1.21 (*b* and *c*).

In Fig. 1.21 (*b*) : $M = Fd_2 - Fd_1 = F(d_2 - d_1) = Fd$.

In Fig. 1.21 (*c*) : $M = Fd_1 + Fd_2 = F(d_1 + d_2) = Fd$.

Properties of a couple

Since the moment centre A and the position of the forces $\overline{F}$ and $\overline{F}'$ do not affect the couple, therefore they may be chosen arbitrarily. Thus

(*a*) The couple is a free vector, it may be placed anywhere parallel to itself.

(*b*) The sense of rotation of the couple is found by inspection.

(*c*) The couple does not give a translatory motion to the body. It causes only the effect of rotation.

(*d*) The couple is proportional to the magnitude of the two equal and opposite parallel forces and also on the perpendicular distance between the forces.

The effect of the couple does not change if :

(1) The couple is rotated through any angle in the plane of forces.

(2) The couple is shifted to any other position in the plane of forces.

(3) The couple is shifted to a parallel plane.

(4) The couple is replaced by another pair of forces in its plane whose product F*d* and sense of rotation is unchanged.

Important :

(1) Various couples of different sense and magnitude can be added algebraically to get their resultant.

(2) The two couples having the same moment (same magnitude and same sense) are equivalent.

1.22. RESOLUTION OF A GIVEN FORCE INTO A FORCE ACTING AT A GIVEN POINT AND A COUPLE

Sometimes we have to shift a force F acting at a point A to another point B. The effect of this is studied as follows.

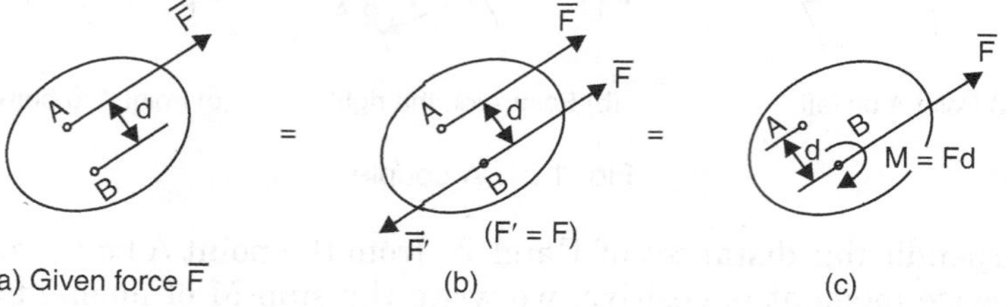

Fig. 1.22. Resolution of a given force into a force acting at a given point and a couple

In Fig. 1.22 (*a*), a force $\overline{F}$ is given acting at a given point A. It is required to shift this force $\overline{F}$ to another point B. The distance between A and B is *d*.

Apply two equal and opposite forces $\overline{F}$ and $\overline{F}'$ of the same magnitude F to the point B, as shown in Fig. 1.22 (*b*). The two forces $\overline{F}$ and $\overline{F}'$ form couple M = F*d*, clockwise in sense.

Thus, the given system of force $\overline{F}$ acting at point A can be replaced by a force $\overline{F}$ acting at point B and a clockwise couple M = F*d*, as shown in Fig. 1.22 (*c*).

This is known as the *resolution of a force into a force and a couple.*

1.23. WRENCH

The resolution we have discussed in Section 1.22 may be performed in reverse. That is, a force $\overline{F}$ and a couple $\overline{M}$ acting at a point B may be combined into a single force $\overline{F}$. This is done by moving the force $\overline{F}$ until its moment about B becomes equal to the moment M of the couple to be eliminated. The magnitude and direction of F are unchanged, but its new line of action will be at a distance $d = \dfrac{M}{F}$ from the point B. This transformation of a force $\overline{F}$ and a collinear couple M into a single force $\overline{F}$ is known as a *Wrench*.

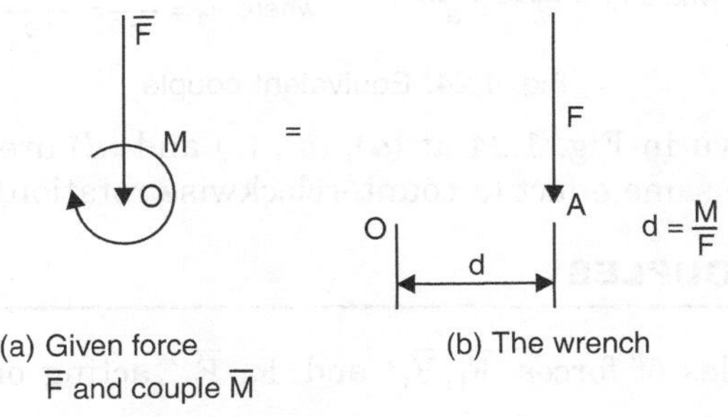

Fig. 1.23. The wrench

An example of such a transformation is given in a Fig. 1.23. In Fig. 1.23(*a*), a force $\overline{F}$ and couple M are acting at a point O. This system is replaced by a force $\overline{F}$ acting a point A which is at a distance $d = \dfrac{M}{F}$ from O, as shown in Fig. 1.23 (*b*).

1.24. EQUIVALENT COUPLES

Consider a thin rectangular plate ABCD whose sides AB = CD = *a* and sides AD = BC = *b* while the diagonals AC = BD = $c = \sqrt{a^2 + b^2}$. In Fig. 1.24 (*a*) couple due to forces F_1 gives moment $M_1 = + F_1 \cdot b$ (*i.e.* counterclockwise). The couple due to forces F_2 in Fig. 1.24 (*b*) is equivalent to that of F_1.

The moment due $F_2 = M_2 = F_2 \times a$. The moment $M_2 = M_1$ if $F_2 = \dfrac{M_1}{a} = \dfrac{F_1 b}{a}$.

It is important to note that forces F_2 act at corners B and D, each perpendicular to the sides AB and CD. Further, the forces F_3 applied at corners A and C perpendicular to the diagonal, as shown in Fig. 1.24 (*c*), create the moment M_3 which is equal to M_1. Thus magnitude of the forces $F_3 = \dfrac{M_1}{c} = \dfrac{F_1 \cdot b}{c}$, where $c = \sqrt{a^2 + b^2}$. Similarly, the couple due to forces F_4 acting perpendicular to the diagonal BD and through corners B and D also gives the moment which is equal to the moment M_1 if the forces $F_4 = \frac{M_1}{c} = \frac{F_1 . b}{c} = \frac{F_1 . b}{\sqrt{a^2 + b^2}}$.

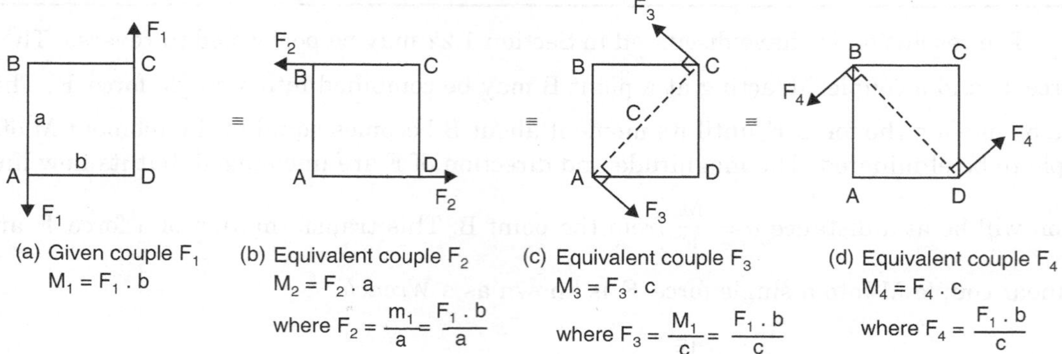

Fig. 1.24. Equivalent couple

Thus, couples shown in Fig. 1.24 at (*a*), (*b*), (*c*) and (*d*) are equivalent to each other. Each couple produces the same effect (a counterclockwise rotation) on the thin plate ABCD.

1.25. ADDITION OF COUPLES

Consider two couples of forces $\overline{F}_1, \overline{F}_1'$ and $\overline{F}_2, \overline{F}_2'$ acting on the same rigid body, as shown in Fig. 1.25.1 (*a*). The forces $\overline{F}_2$ and $\overline{F}_2'$ may be replaced by a couple of forces $\overline{F}_3$ and $\overline{F}_3'$ which have respectively, the same line of action as $\overline{F}_1$ and $\overline{F}_1'$, as shown in Fig. 1.25.1 (*b*). The magnitude F_3 of the new forces $\overline{F}_3$ and $\overline{F}_3'$ must be such that

$$M_3 = M_2 \quad \text{or} \quad F_3 . d_1 = F_2 . d_2 \quad \text{or} \quad F_3 = \frac{F_2 . d_2}{d_1} \qquad ...(1.41)$$

Adding forces along the same line of action, we get a single couple consisting of the forces $\overline{F}_1 + \overline{F}_3$ and $\overline{F}_1' + \overline{F}_3'$, as shown in Fig. 1.25.1 (*c*).

The moment of this couple is

$$M = (F_1 + F_3)\, d_1 = F_1 d_1 + F_3 . d_1 = F_1 d_1 + F_2 d_2 \qquad ...(1.42)$$

The moment M is therefore found to be the algebraic sum of the moments $F_1 d_1$ and $F_2 d_2$ of the two original couples.

It should be noted that, if the two couples are of opposite senses the moments will have opposite signs and we get the difference of the two couples, as shown in Fig. 1.25.2 (*c*).

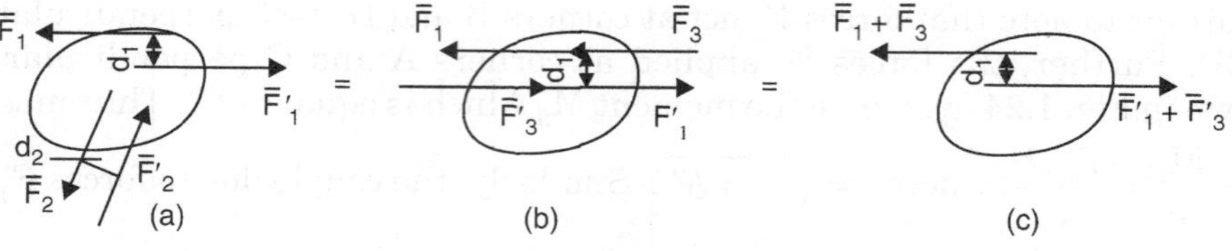

Fig. 1.25.1 Addition of couples

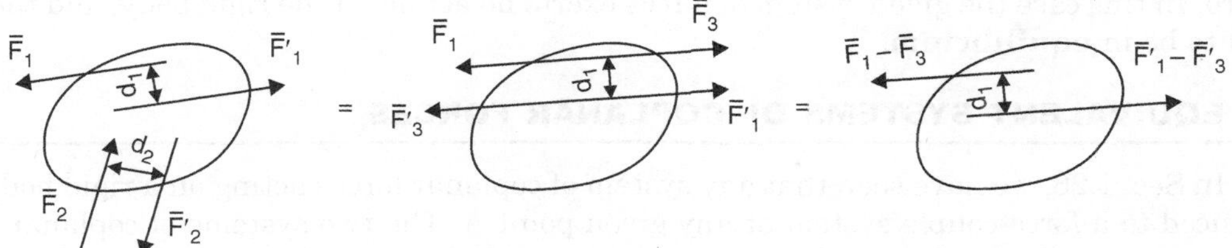

Fig. 1.25.2. Subtraction of couples

Thus we conclude : Given two coplanar couples may be replaced by a single couple of moment equal to the algebraic sum of moments of the given couples.

The **Principle of algebraic addition of two couples** may be applied to many couples lying in the same plane.

1.26. RESULTANT OF A SYSTEM OF COPLANAR FORCES

Consider a system of coplanar forces $\bar{F}_1, \bar{F}_2, \bar{F}_3$, etc., acting on a rigid body as shown in Fig. 1.26 (*a*). As seen in Sec. 1.22, force $\bar{F}_1$ may be moved to a given point A alongwith a moment M_1 of the force $\bar{F}_1$ about A. Repeating this procedure with forces $\bar{F}_2$, $\bar{F}_3$, etc., we obtain the system shown in Fig. 1.26 (*b*). Now since all forces act at a single point A, they can be replaced by their resultant $\bar{R}$ as shown in Fig. 1.26 (*c*). On the other hand, the couples $\bar{M}_1, \bar{M}_2, \bar{M}_3$, etc., may be replaced by a single couple $\bar{M}$ of moment M equal to the algebraic sum of moments. Thus, any system of coplanar forces may thus be reduced to an equivalent force-couple system acting at a given point A.

When the resultant force R is not equal to zero, the system obtained may be reduced to a single force $\bar{R}$, called the resultant of the given system, as shown in Fig. 1.26 (*d*). This is done, as shown in Sec. 1.22, by moving $\bar{R}$ until its moment about A becomes equal to $\bar{M}$, thus eliminating the couple $\bar{M}$. The distance from A to the line of action of $\bar{R}$ is d = M/R.

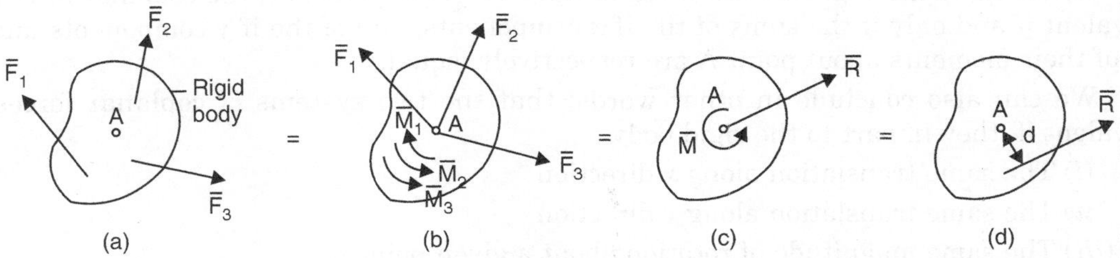

Fig. 1.26. Resultant of a system of forces

When $\bar{R}$ is equal to zero, the force couple system reduces to the couple $\bar{M}$. Thus, the given system of forces is reduced to a single couple $\bar{M}$, called the resultant couple or the pure couple of the system.

Thus, we conclude that any given system of coplanar forces may be reduced to a **single force $\bar{R}$** or a **single couple $\bar{M}$**, as the case may be. The only exception is when both $\bar{R}$ and $\bar{M}$

are zero. In this case the given system of forces exerts no action on the rigid body, and the rigid is said to be in **equilibrium**.

1.27. EQUIVALENT SYSTEMS OF COPLANAR FORCES

In Sec. 1.26, we have seen that any system of coplanar forces acting on a rigid body may be reduced to a force-couple system at any given point A. The two systems of coplanar forces are equivalent to each other, if they may be reduced to the same force-couple system at any given point A.

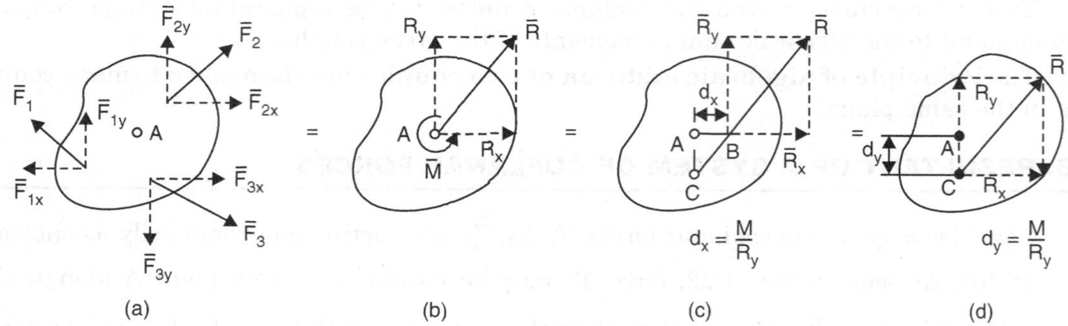

Fig. 1.27. Equivalent system of coplanar forces

The given forces $\overline{F}_1$, $\overline{F}_2$ and $\overline{F}_3$ are resolved into their x components ($\overline{F}_{1x}$, $\overline{F}_{2x}$ and $\overline{F}_{3x}$) and into their y components ($\overline{F}_{1y}$, $\overline{F}_{2y}$ and $\overline{F}_{3y}$) as shown in Fig. 1.27(a). The algebraic sums ΣF_x and ΣF_y are respectively equal to the x and y components $\overline{R}_x$ any $\overline{R}_y$ as shown in Fig. 1.27(b). The given forces create a resultant couple $\overline{M}$ as well. Let the resultant $\overline{R}$ pass through points B and C as shown in Fig. 1.27 (c) and (d). The distance of point B from the given point A is given by $d_x = \dfrac{M}{R_y}$ and the distance of point C from the given point A is given $d_y = \dfrac{M}{R_x}$.

Referring to the Fig. 1.27 (b), we find that : The two systems of the coplanar forces are equivalent if and only if the sums of the if x components, and of the if y components and the sum of their moments about point A are respectively, equal.

We can also conclude in other words, that the two systems of coplanar forces are equivalent if, they impart to the rigid body :

(i) The same translation along x direction

(ii) The same translation along y direction

(iii) The same magnitude of rotation about a given point A

(iv) The same sense of rotation of the rigid body.

1.28. THE X AND Y INTERCEPTS OF THE RESULTANT

As discussed in Sec 1.26, the resultant $\overline{R}$ produces the same moment about a given point, as does the given system of coplanar forces. Sometime, it is required to find the points of intersection of $\overline{R}$ with the x-axis and with the y-axis.

Let the resultant $\overline{R}$ be at a perpendicular distance d from the given point O, as shown in Fig. 1.28. Also, let θ_x be its inclination with x-axis. The resultant $\overline{R}$ intersects x-axis at point A and the y-axis at point B.

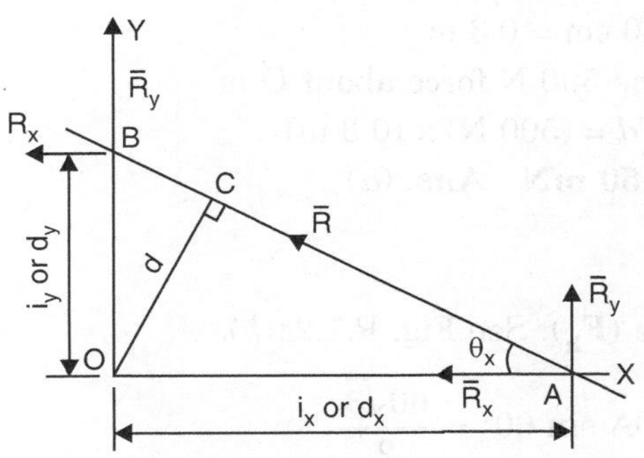

Fig. 1.28. The x and y intercepts of resultant

Since OC = d is perpendicular to $\overline{R}$, therefore, we get

$$\text{The } x \text{ intercept } i_x \text{ or } d_x = \frac{d}{\sin \theta_x} \quad ...(1.41(a))$$

$$\text{and, the } y \text{ intercept } i_y \text{ or } d_y = \frac{d}{\cos \theta_x} \quad ...(1.41(b))$$

Another method of finding these intercepts is as follows. Let R_x and R_y be the x and y components of the resultant $\overline{R}$, as shown in Fig. 1.28. Taking the moment of $\overline{R}$ about O, as the sum of moments of R_x and R_y, we get

Since $M_o = Rd$,

$\therefore \quad M_o = (R_x) \times 0 + (R_y)(d_x)$

$$\therefore \quad \left. d_x = \frac{M_o}{R_y} \text{ or } i_x = \frac{M_o}{R_y} \right\} \quad ...(1.42(a))$$

Also, $M_o = (R_x)\, d_y + (R_y) \times 0.$

$$\therefore \quad \left. d_y = \frac{M_o}{R_x} \text{ or } i_y = \frac{M_o}{R_x} \right\} \quad ...(1.42(b))$$

Problem 1.21. *A 500 N vertical force is applied to a 60 cm long bar OA hinged at O and inclined at 60° to the horizontal.*

Determine :

(a) The moment of the 500 N force about O,

(b) The magnitude of the horizontal force F applied at A which gives the same moment about O,

(c) The smallest force applied at A which gives the same moment about O,

(d) At what distance from O, a vertical 1500 N force should be applied which gives the same moment about O.

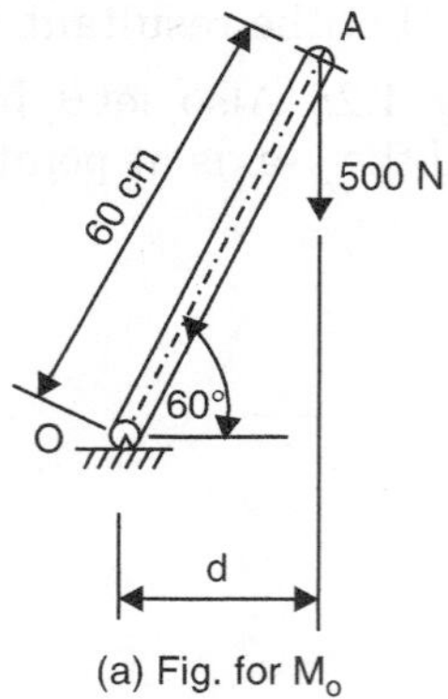

(a) Fig. for M_o

Sol. Fig. P.1.21 (*a*) shows the bar OA and the applied force.

(*a*) **Moment about O.** The perpendicular distance *d* from the point O to the line of action of the 500 N vertical force is

$$d = \text{OA} \cos 60° = 60 \times \cos 60°$$
$$= 30 \text{ cm} = 0.3 \text{ m}$$

∴ The moment of the 500 N force about O is

$$M_o = Fd = (500 \text{ N}) \times (0.3 \text{ m})$$
$$= \mathbf{150 \text{ mN}} \quad \textbf{Ans. } (a)$$

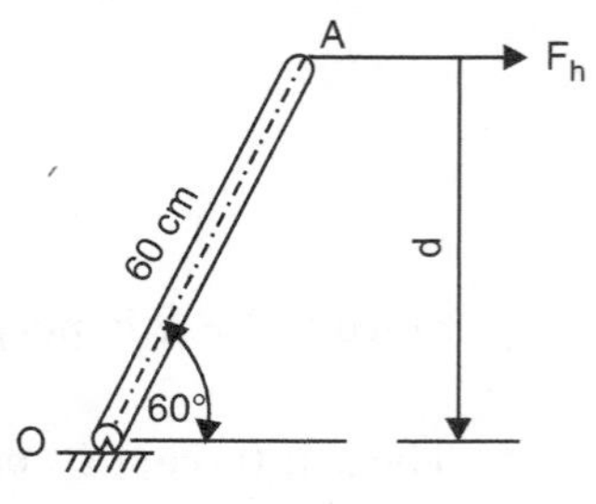

(b) Fig. for F_h

(*b*) **Horizontal force ($\mathbf{F_h}$).** See Fig. P.1.21(*b*).

Here $$d = \text{OA} \sin 60° = \frac{60\sqrt{3}}{2}$$
$$= 30\sqrt{3} \text{ cm} = 0.3\sqrt{3} \text{ m}$$

Since the moment about O must be 150 Nm, we write

$$M_o = F_h d$$
$$F_h = \frac{M_o}{d} = \frac{150}{0.3\sqrt{3}}$$
$$= \mathbf{288.675 \text{ N}} \quad \textbf{Ans. } (b)$$

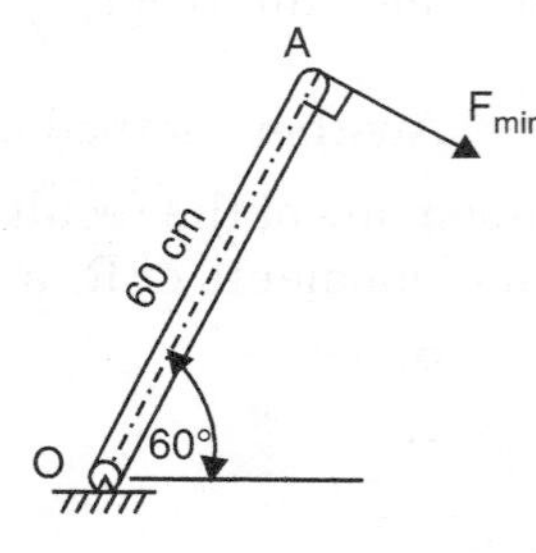

(c) Fig. for F_{min}

(*c*) **The smallest force at A.** Since $M_o = Fd$, the smallest force F_{min} occurs when *d* is the maximum. The maximum *d* = OA = 0.6 m

Hence, $$F_{min} = \frac{M_o}{\text{OA}}$$
$$= \frac{150}{0.6} = \mathbf{250 \text{ N}} \quad \textbf{Ans. } (c)$$

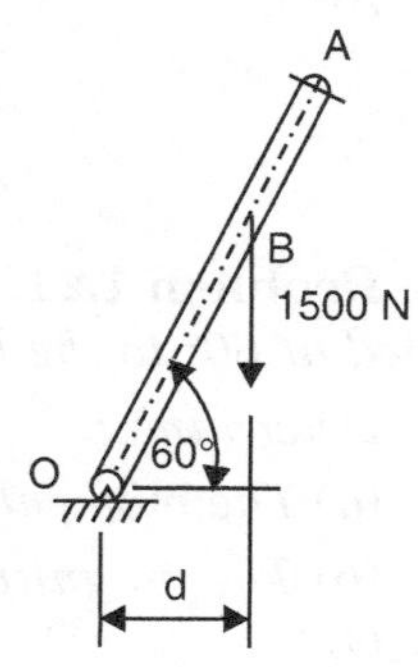

(d) Fig. for case (d)

(*d*) **1500 N Vertical force.** Here $M_o = Fd$ gives

$$150 \text{ Nm} = (1500)\, d$$

∴ $$d = \frac{150}{1500} = 0.1 \text{ m}$$

But OB cos 60° = 0.1

∴ $$\text{OB} = \frac{0.1}{\cos 60°} = 0.2 \text{ m}$$
$$= \mathbf{20 \text{ cm}} \quad \textbf{Ans.}$$

Note. None of the forces in parts (*b*), (*c*) and (*d*) is equivalent to the given 500 N force, because they do not have the same F_x and F_y components. However, each force tends to rotate the bar about O in the same sense and with same moment. Each force pulls the bar in a different way.

Fig. P. 1.21

Problem 1.22. *A thin plate is subjected to coplanar system of forces shown in Fig. P.1.22 (a). In the graph or the grid each square has a side of 0.1 m. Find the resultant force and its x and y intercepts.*

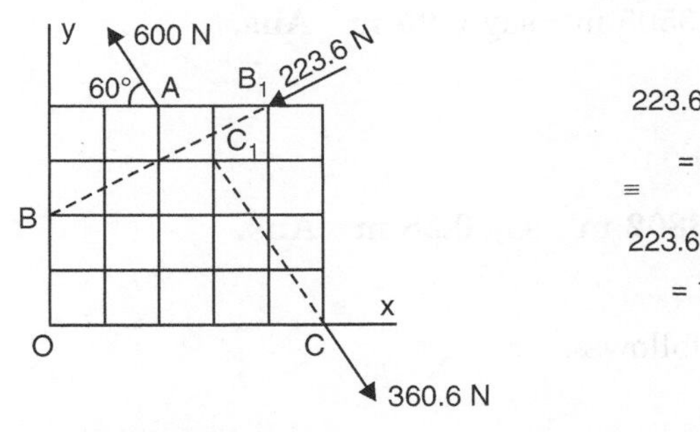

(a) Given system of forces

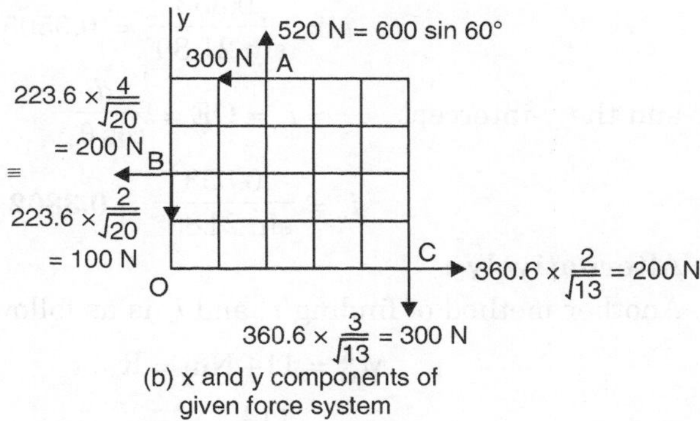

(b) x and y components of given force system

Fig. P. 1.22

Sol. We first resolve all forces in their *x* and *y* components, and draw the Fig. P.1.22 (*b*) showing these components. The effects of resultant are as given below :

$[\overset{+}{\rightarrow} R_x = \Sigma F_x]$ $\quad\therefore\quad R_x = -300 - 200 + 200 = -300 \text{ N} \leftarrow$

$[+\uparrow R_y = \Sigma F_y]$ $\quad\therefore\quad R_y = 520 - 100 - 300 = +120 \text{ N} \uparrow$

$[+\circlearrowleft M_o^R = \Sigma M_o]$ $\quad M_o^R = -300 \times 0.4 - 520 \times 0.2 - 200 \times 0.2 + 300 \times 0.5 = -114 \text{ Nm}$

Combining these effects, the resultant has a magnitude of

$$R = \sqrt{R_x^2 + R_y^2} = \sqrt{(-300)^2 + (120)^2} = 323.11 \text{ N}$$

up to the left at an angle $\theta_x = \tan^{-1}\left(\frac{120}{-300}\right) = -21.80°$

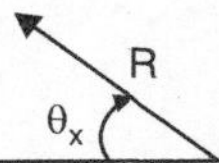

with X axis and passing through II and IV quadrants. The arm of moment

$$d = \frac{M_o^R}{R} = \frac{114}{323.11} \text{ m} = 0.353 \text{ m from O.}$$

From Fig. P. 1.22 (*c*), we get

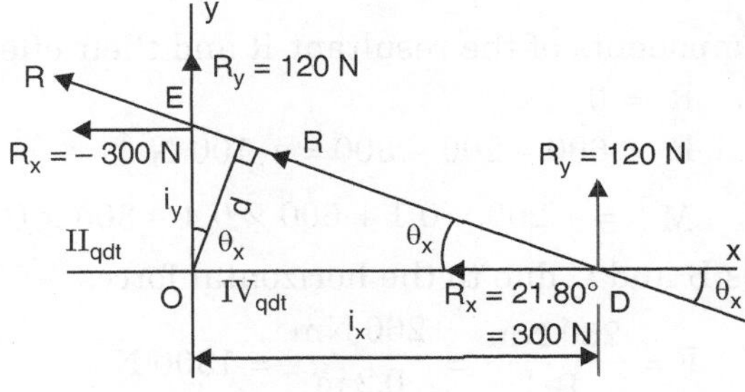

Fig. P. 1.22 (c) x and y intercepts of resultant

The x and y intercepts of the resultant.

The x-intercept $= i_x = \text{OD} = \dfrac{d}{\sin \theta_x}$

or $i_x = \dfrac{0.353}{\sin 21.80^\circ} = 0.9505$ m say **0.95 m Ans.**

and the y-intercept $= i_y = \text{OE} = \dfrac{d}{\cos \theta_x}$

or $i_y = \dfrac{0.353}{\sin 21.80^\circ} = $ **0.3802 m** say **0.38 m Ans.**

Alternatively :

Another method of finding i_x and i_y is as follows :

$$M_O^R = 114 \text{ Nm} = R_y \,.\, i_x$$

$\therefore$ $i_x = \dfrac{114}{120} = $ **0.95 m** as shown **Ans.**

and $M_O^R = 114 \text{ Nm} = R_x \,.\, i_y$

$\therefore$ $i_y = \dfrac{114}{R_x} = \dfrac{114}{300} = $ **0.38 m Ans.**

Problem 1.23. *Replace the system of forces acting on a thin frame as shown in Fig. P.1.23 (a), by a resultant force R through a point A and a couple acting through B and C.*

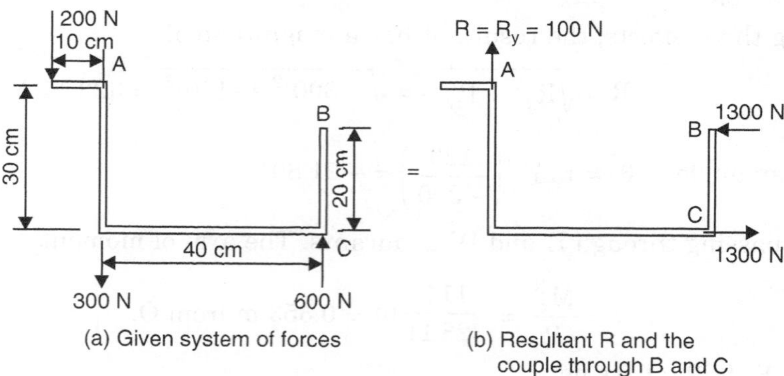

(a) Given system of forces

(b) Resultant R and the couple through B and C

Fig. P. 1.23

Sol. The x and y components of the resultant R and their effects are given below :

$[+ \rightarrow R_x = \Sigma F_x]$ $\therefore$ $R_x = 0$

$[+ \uparrow R_y = \Sigma F_y]$ $\therefore$ $R_y = 600 - 200 - 300 = +\,100 \text{ N} \uparrow$

$[+ \circlearrowleft M_A^R = \Sigma M_A]$ $\therefore$ $M_A^R = +\,200 \times 0.1 + 600 \times 0.4 - 300 \times 0 = 260 \text{ Nm}$

The couple at points B and C due to the horizontal forces

$$F = \frac{260 \text{ Nm}}{BC} = \frac{260 \text{ Nm}}{0.2 \text{ m}} = 1300 \text{ N}$$

giving a 260 Nm CCW couple.

1300 N towards left at B and 1300 N towards right at C, as shown in Fig. P.1.23 (*b*).

The resultant $R = R_y = 100 \text{ N} \uparrow$ **Ans.**

Problem 1.24. *Find moment about the point A of the inclined forces shown in Fig. P.1.24 (a). Also find the position of the resultant R which will create this moment about A.*

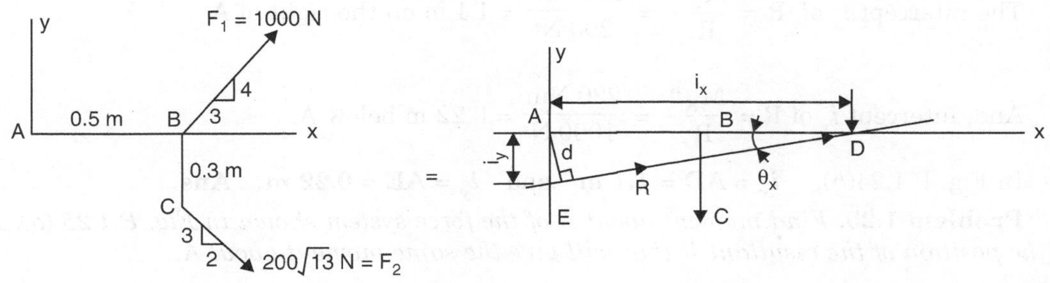

Fig. P. 1.24

Sol. The x and y components of the forces F_1 and F_2 acting at points B and C are as under.

For force F_1 = 1000 N at B

$$F_{1x} = +\,1000 \times \frac{3}{5} = 600 \text{ N} \rightarrow$$

$$F_{1y} = +\,1000 \times \frac{4}{5} = 800 \text{ N} \uparrow$$

For force $F_2 = 200\sqrt{13}$ N,

$$F_{2x} = 200\sqrt{13} \times \frac{2}{\sqrt{13}} = 400 \text{ N} \rightarrow$$

$$F_{2y} = -\,200\sqrt{13} \times \frac{3}{\sqrt{13}} = -\,600 \text{ N} \downarrow$$

$$\therefore \quad M_A^R = +\,F_{1y} \times 0.5 + F_{2x} \times 0.3 - F_{2y} \times 0.5 \text{ Nm}$$

$$= 800 \times 0.5 + 400 \times 0.3 - 600 \times 0.5 \text{ Nm}$$

or $$M_A^R = +\,\mathbf{220 \text{ Nm CCW}} \quad \textbf{Ans.}$$

$$R_x = \Sigma F_x = +\,600 + 400 = 1000 \text{ N} \rightarrow$$

$$R_y = \Sigma F_y = +\,800 - 600 = 200 \text{ N} \uparrow$$

$\therefore$ Resultant $$R = \sqrt{{R_x}^2 + {R_y}^2} = \sqrt{(1000)^2 + (200)^2} = \mathbf{1019.8 \text{ N}} \quad \textbf{Ans.}$$

Its inclination θ_x with x-axis is given by

$$\tan\theta_x = \frac{R_y}{R_x} = \frac{200}{1000} = 0.2$$

$$\therefore \quad \theta_x = \mathbf{11.31°} \quad \textbf{Ans.}$$

The moment arm d from A = $\dfrac{{M_A}^R}{R} = \dfrac{220 \text{ Nm}}{1019.8 \text{ N}} = \mathbf{0.216 \text{ m}}$

These results are illustrated in Fig. 1.24 (*b*)

$$\text{The intercept } i_x \text{ of R} = \frac{M_A{}^R}{R_y} = \frac{220 \text{ Nm}}{200 \text{ N}} = 1.1 \text{ m on the right of A}$$

$$\text{And, intercept } i_y \text{ of R} = \frac{M_A{}^R}{R_x} = \frac{220 \text{ Nm}}{1000 \text{ N}} = 0.22 \text{ m below A.}$$

In Fig. P.1.24(*b*), $\boldsymbol{i_x}$ **= AD = 1.1 m** and $\boldsymbol{i_y}$ **= AE = 0.22 m. Ans.**

Problem 1.25. *Find moment about A of the force system shown in Fig. P.1.25 (a). Also find the position of the resultant R that will give the same moment about A.*

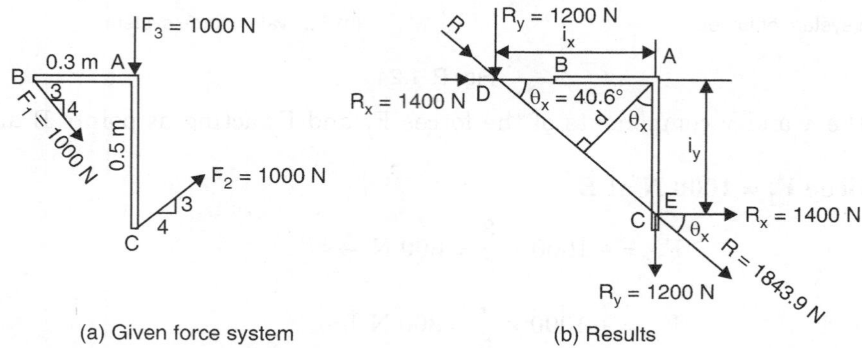

(a) Given force system (b) Results

Fig. P. 1.25

Sol. The x- and y-components of forces F_1, F_2 and F_3 are :

Force F_1 : $F_{1x} = +\,1000 \times \frac{3}{5} = 600 \text{ N} \rightarrow$

$F_{1y} = -\,1000 \times \frac{4}{5} = -\,800 \text{ N} \downarrow$

Force F_2 : $F_{2x} = +\,1000 \times \frac{4}{5} = 800 \text{ N} \rightarrow$

$F_{2y} = +\,1000 \times \frac{3}{5} = +\,600 \text{ N} \uparrow$

Force F_3 : $F_{3x} = 0; \quad F_{3y} = -\,1000 \text{ N}$

$\therefore \quad R_x = \Sigma F_x = 600 + 800 + 0 = 1400 \text{ N} \rightarrow$

and $\quad R_y = \Sigma F_y = -\,800 + 600 - 1000 = -\,1200 \text{ N} \downarrow$

$$\therefore \quad R = \sqrt{R_x{}^2 + R_y{}^2} = \sqrt{(1400)^2 + (-\,1200)^2} = 1843.9 \text{ N}$$

And, its inclination θ_x with x-axis is given by

$$\tan \theta_x = \frac{R_y}{R_x} = \frac{-\,1200}{1400} = -\,0.8571$$

$\therefore \quad \theta_x = \mathbf{-\,40.60°}$ **Ans.**

Moment about A = $M_A^R = F_{1y} \times 0.3 + F_{2x} \times 0.5$

$= 800 \times 0.3 + 800 \times 0.5 =$ **640 Nm CCW Ans.**

Its arm from A = $d = \dfrac{M_A^R}{R} = \dfrac{640 \text{ Nm}}{1843.9 \text{ N}} =$ **0.347 m Ans.**

x intercept of R = $i_x = \dfrac{M_A^R}{R_y} = \dfrac{640}{1200} =$ **0.533 m** left of A. **Ans.**

and y intercept of R = $i_y = \dfrac{M_A^R}{R_x} = \dfrac{640}{1400} =$ **0.4571 m below A. Ans.**

Results are illustrated in Fig. P.1.25 (b).

Problem 1.26. *Replace the force and the couple system at O by a single force applied at a point (a) on diameter AB, and (b) on diameter CD. Find the position of the force in each case. See Fig. P.1.26 (a).*

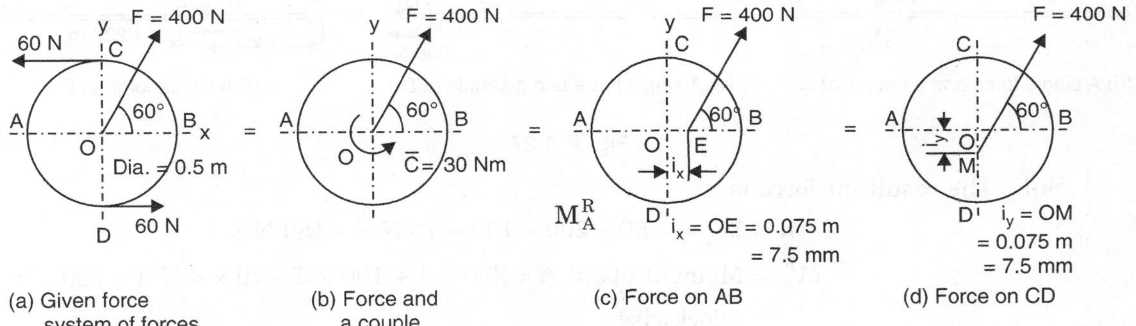

Fig. P. 1.26

Sol. Fig. P.1.26 (a) shows the force of 400 N at O and the couple due to force of 60 N at C and D.

Couple at O = C = 60 N × 0.5 m = 30 Nm CCW.

Fig.P.1.26 (b) shows equivalent system with force of 400 N at O and the couple C = 30 Nm.

(a) Fig.P.1.26 (c) gives the force of 400 N applied at point E giving the effect of Fig. P.1.26 (b). The x-intercept i_x is given by

$$i_x = \frac{\overline{C}}{F_y} = \frac{30 \text{ Nm}}{400 \text{ N} \sin 60^\circ} = 0.0866 \text{ m on the right of O.}$$ **Ans.** (a)

(b) Fig.P.1.26 (d) gives the force F applied at point M giving the effect of Fig.P.126 (b). The y-intercept i_y is given by

$$i_y = \frac{\overline{C}}{F_x} = \frac{30 \text{ Nm}}{400 \text{ N} \cos 60^\circ} = 0.15 \text{ m below point O.}$$ **Ans.** (b).

Problem 1.27. *A system of parallel non-concurrent forces is acting on a rigid bar AB as shown in Fig. P.1.27 (a). Reduce this system of forces to :*

(a) a single force R and a couple at A.

(b) a single force R and a couple at B.

(c) a single force R and a couple at C.

(d) a single force R and a couple at D.

(e) a single force R and its position.

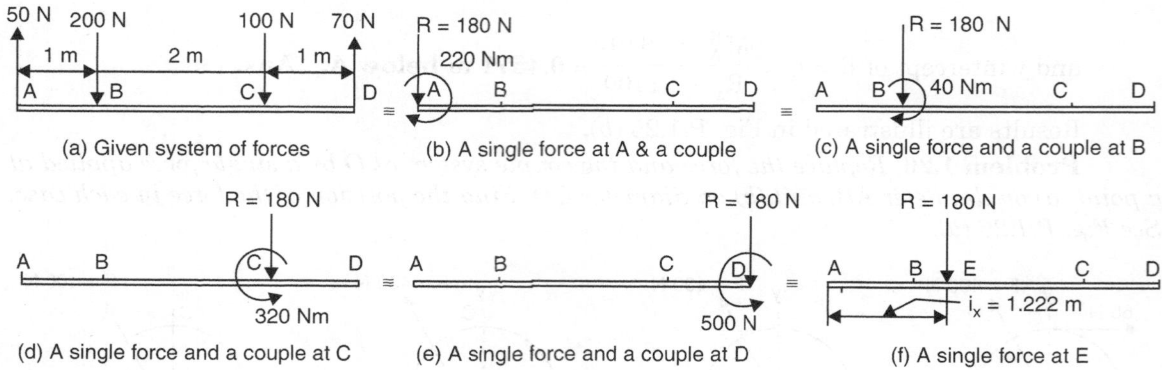

Fig. P. 1.27

Sol. The resultant force is

$$R = \Sigma F_y = +50 - 200 - 100 + 70 \text{ N} = -180 \text{ N} \downarrow$$

$$M_A^R = \text{Moment about A} = 200 \times 1 + 100 \times 3 - 70 \times 4 \text{ Nm} = 220 \text{ Nm clockwise}$$

(*a*) If the given system of forces is to be replaced by a single force

R = 180 N ↓ and a couple of 220 Nm C.W., then the equivalent loading is illustrated in Fig. P.1.27 (*b*)

We first find the moment of forces at B, C, and D.

Thus, $$M_B^R = 50 \times 1 + 100 \times 2 - 70 \times 3 = 40 \text{ Nm C.W.}$$

$$M_C^R = 50 \times 3 - 200 \times 2 - 70 \times 1 = -320 \text{ Nm CCW}$$

$$M_D^R = 50 \times 4 - 200 \times 3 - 100 \times 1 = -500 \text{ Nm CCW}$$

(*b*) Fig.P.1.27 (*c*) shows a single force R = 180 N ↓ and a couple of 40 Nm C.W. at B. Its effect is similar to that of the given force system.

(*c*) Fig.P.1.27 (*d*) shows a single force R = 180 N ↓ and a couple of 320 Nm CCW at C.

(*d*) Fig.P.1.27 (*e*) shows a single force R = 180 N ↓ and a couple 500 Nm CCW at D.

(*e*) The single force R = 180 N ↓ giving the couple M_A^R = 220 Nm should be applied at a point E such that $i_x = AE = \dfrac{M_A^R}{R} = \dfrac{220 \text{ Nm}}{180 \text{ N}} = \mathbf{1.222 \text{ m}}$. **Ans.**

This case is illustrated in Fig. P.1.27 (*f*).

Problem 1.28. *Find the resultant of forces shown in Fig. P.1.28 (a). Find its position and direction.*

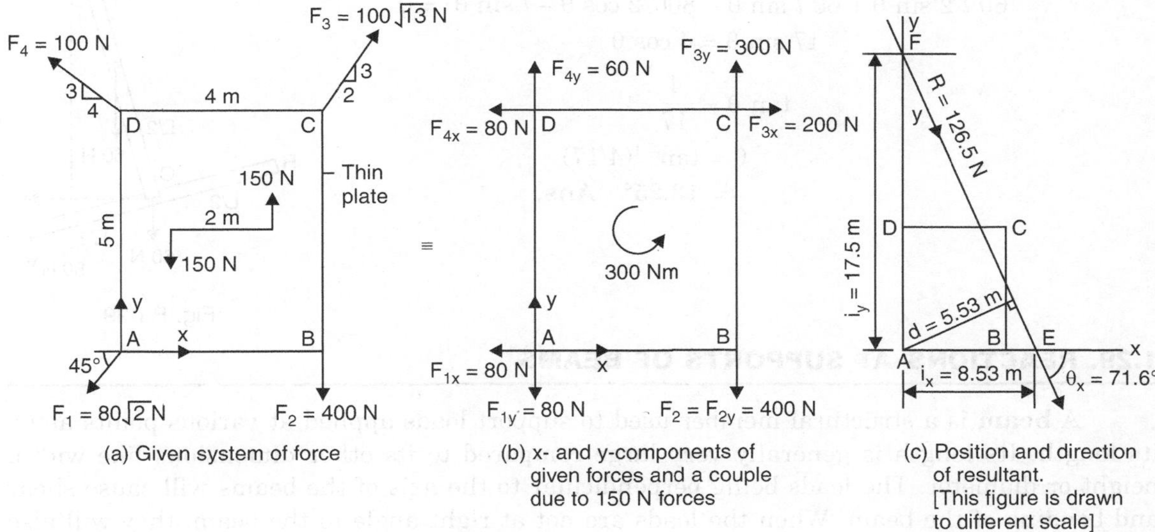

Fig. P. 1.28

Sol. Components of given forces are shown in Fig. P.1.28 (*b*).

$$R_x = \Sigma F_x = 200 - 80 - 80 = +40 \text{ N} \rightarrow$$

$$R_y = \Sigma F_y = 300 + 60 - 80 - 400 = -120 \text{ N} \downarrow$$

$$M_A^R = 400 \times 4 + 200 \times 5 - 300 \times 4 - 80 \times 5 - 150 \times 2 = 700 \text{ Nm C.W.}$$

Resultant $R = \sqrt{R_x{}^2 + R_y{}^2} = \sqrt{(40)^2 + (-120)^2} = \mathbf{126.5\ N}$

$$\theta_x = \tan^{-1}\frac{R_y}{R_x} = \tan^{-1}\frac{-120}{40} = \tan^{-1}(-3) = -71.6° \text{ from } x\text{-axis.}$$

Arm of moment d from A $= \dfrac{M_A^R}{R} = \dfrac{700 \text{ Nm}}{126.5 \text{ N}} = \mathbf{5.53\ m}$ **Ans.**

Intercepts along x and y are shown in Fig.P.1.28 (*c*).

$$i_x = \frac{M_A^R}{R_y} = \frac{700}{120} = \mathbf{5.83\ m} \text{ to the right of A = AE} \quad \textbf{Ans.}$$

$$i_y = \frac{M_A^R}{R_x} = \frac{700}{40} = \mathbf{17.5\ m} \text{ above A = AF} \quad \textbf{Ans.}$$

Problem 1.29. *Two identical prismatic bars PQ and RS each weighing 60 N are welded together to form a Tee and are suspended in a vertical plane as shown in Fig. P.1.29. Calculate the angle that the bar PQ will make with the vertical, when a vertical load of 80 N is applied at S.*

[RGPV-2000 Summer]

Sol. Let the length of each bar be 'l'.

Taking moments about P, we have

$$60\, l/2 \sin \theta + 60\, l \sin \theta - 80(l/2 \cos \theta - l \sin \theta) = 0$$

$$\therefore \qquad 17 \sin \theta = 4 \cos \theta$$

$$\therefore \qquad \tan \theta = \frac{4}{17}$$

$$\theta = \tan^{-1}(4/17)$$

$$= \mathbf{13.25°} \quad \textbf{Ans.}$$

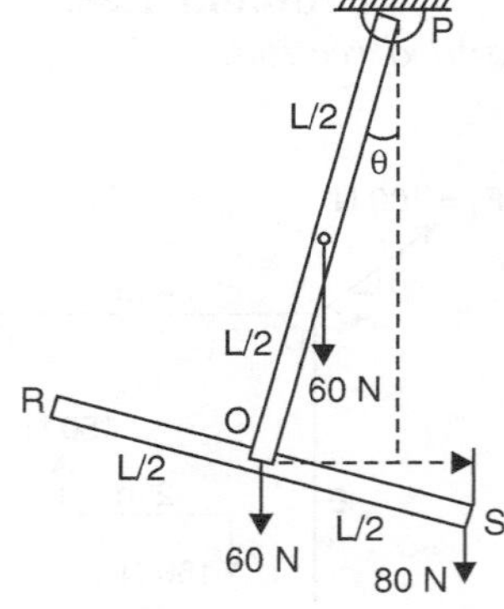

Fig. P.1.29

1.29. REACTIONS AT SUPPORTS OF BEAMS

A **beam** is a structural member used to support loads applied at various points along its length. Its length is generally very large compared to its other dimensions like width, height or diameter. The loads being perpendicular to the axis of the beams will cause shear and bending of the beam. When the loads are not at right angle to the beam, they will also produce axial forces in the beam. Axial forces may usually be neglected in the design of beams, since the ability of the beam to resist shear and bending is more critical than its ability to resist axial forces produced due to the externally applied forces on the beam.

Beams are generally long, straight and prismatic (*i.e.* of the same cross-sectional area throughout the length of the beam).

Types of Supports. Beams are supported on roller, hinged or fixed supports as shown in Fig. 1.29.

(1) **Simple Support.** If one end of the beam rests an a fixed support, the support is known as simple support. The reaction of the simple support is always perpendicular to the surface of support. The beam is free to slide and rotate at the simple support. See Fig. 1.29 (*a*).

(2) **Roller Support.** Here one end of the beam is supported on a roller. The only reaction of the roller support is normal to the surface on which the roller rolls without friction. See Fig. 1.29 (*b*) in which four possible situations are illustrated. Examples of roller supports are wheels of a motorcycle, or a handcart, or an over-head crane, or of a car, etc.

(3) **Hinged Support.** At the hinged support [see Fig. 1.29 (*c*)] the beam does not move either along or normal to its axis. The beam, however, may rotate at the hinged support. The total support reaction is R and its horizontal and vertical components are H and V, respectively. Since the beam is free to rotate at the hinged support, no resisting moment will exist. The hinged support behaves like the hinges provide to doors and windows.

(4) **Fixed Support.** At the fixed support, the beam is not free to rotate or slide along the length of the beam or in the direction normal to the beam. Therefore, there are three reaction components, *viz.*, vertical reaction component (V), horizontal reaction component (H) and the moment (M), as shown in Fig. 1.29 (*d*). Fixed support is also known as built-in support.

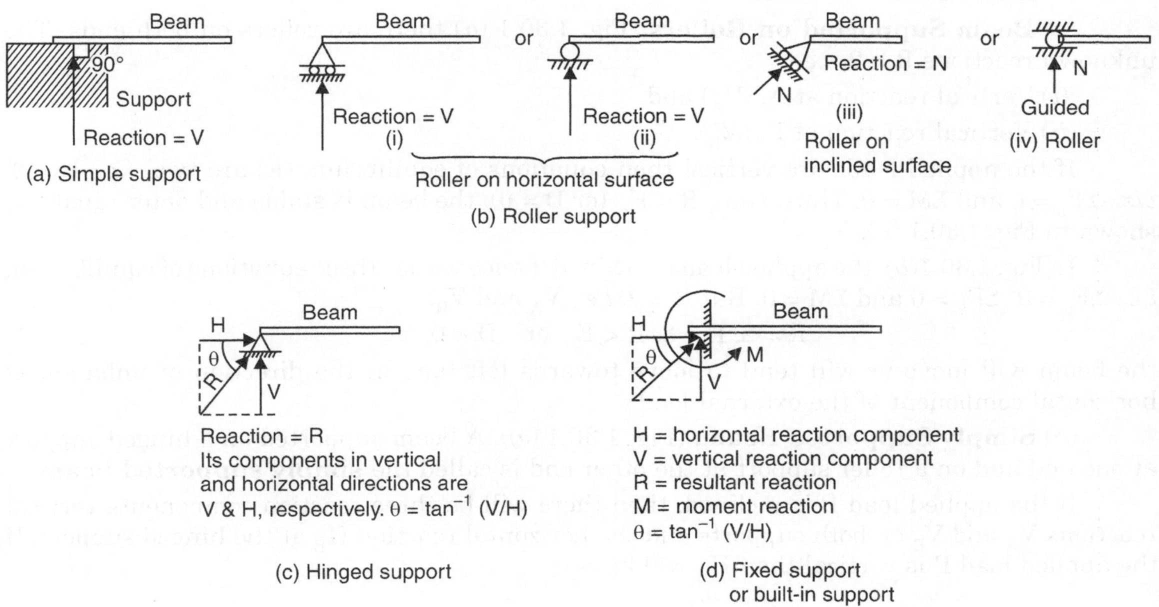

Fig. 1.29. Types of supports

1.30. TYPES OF BEAMS

Beams are classified on the manner they are supported. Several types of beams frequently used are shown in Fig. 1.30. The distance L between the two supports is called **span**. Since there are only three equations of equilibrium for the non-concurrent coplanar force system, *viz.*,

$$\Sigma F_x = 0, \Sigma F_y = 0 \quad \text{and} \quad \Sigma M_z = 0 \qquad ...(1.43\ (a))$$

or

$$\Sigma F_h = 0, \Sigma F_v = 0 \quad \text{and} \quad \Sigma M = 0 \qquad ...(1.43\ (b))$$

hence using these equations maximum three unknown reactions can be determined. Beams with maximum three unknown reactions are called **determinate beams**. Beams in which unknown reactions are more than three are known as **indeterminate beams**. The methods of Statics are not sufficient to determine the reactions of supports of indeterminate beams.

Let R = the number of unknown support reactions, and E = the number of equations of equilibrium available/used.

D = Degree of indeterminacy or redundancy

i.e., $D = R - E$ ...(1.44)

For determinate beams $D = 0$, *i.e.,* $R = E$ ...(1.45)

For indeterminate beams, $D > 0$, *i.e.,* $R > E$ ...(1.46)

Both, determinate and indeterminate beams are stable. When $D < 0$, *i.e.,* $R < E$, the beams are externally unstable but the support reactions can be determined by equations of equilibrium.

Beams supported on rollers are examples of unstable determinate beams, for which $R < E$. Examples of such beams are motor bikes, hand carts, bicycles, automobiles, overhead cranes, etc. All these things move in the direction of pull or effort acting in the direction parallel to the surface of rollers.

(1) **Determinate Beams**

(*a*) **Beam Supported on Rollers.** Fig. 1.30.1 (*a*) there are rollers on both ends. The unknown reactions R = 2, *i.e.*

(*i*) Vertical reaction at A, (V_A) and

(*ii*) Vertical reaction at B, (V_B).

If the applied loads are vertical then equations of equilibrium (E) are two, *i.e.*, E = 2, *viz.*, $\Sigma F_v = 0$ and $\Sigma M = 0$. Thus, since R = E, (or D = 0), the beam is stable and determinate as shown in Fig. 1.30.1 (*a*).

In Fig. 1.30.1(*b*), the applied load is inclined, hence we use three equations of equilibrium, *i.e.*, $\Sigma F_v = 0$, $\Sigma F_h = 0$ and $\Sigma M = 0$. But R = 2, *i.e.*, V_A and V_B,

$$\because \quad R = 2, E = 3, R < E \quad \text{or} \quad D < 0,$$

the beam will move or will tend to move towards left (*i.e.*, in the direction of unbalanced horizontal component of the external load).

(*b*) **Simply Supported Beam.** (Fig. 1.30.1 (*c*)). A beam supported on a hinged support at one end and on a roller support at the other end is called the **simply supported beam.**

If the applied load P is inclined, then there will be three reaction components-vertical reactions V_A and V_B on both supports and the horizontal reaction H_B at the hinged support. If the applied load P is vertical then H_B will be zero.

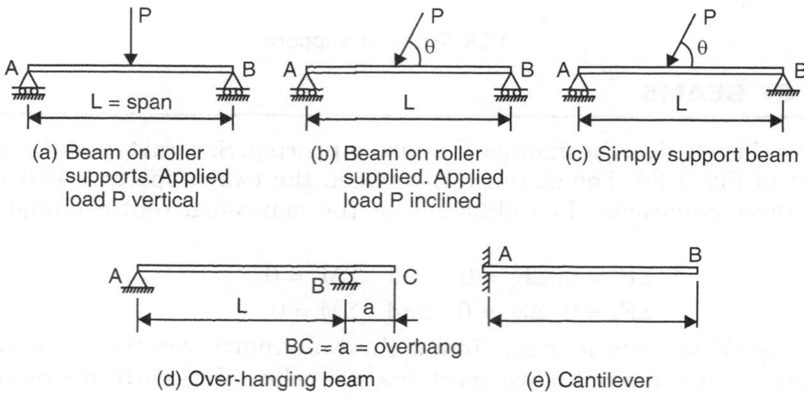

Fig. 1.30.1 Determinate beams

It is important to note that the roller supported end on the beam is free to move on the surface of the roller. When the loads are applied, the beam sags and the roller support moves inwards, while it moves outwards when the loads are gradually removed. Thus, the roller support introduces elasticity to the beam.

(*c*) **Over Hanging Beams.** It is a beam longer than its span. Thus, some portion of the beam is either on the left of its left support or on the right of its right support or both. In Fig. 1.30.1 (*d*) there is some portion of the beam on the right of the right support. The portion on the outside of the support is called an **overhang**. For vertical external loads there are only two reactions, *viz.*, V_A and V_B.

(*d*) **Cantilever.** When one end of the beam is free (*i.e.*, unsupported) and the other end is fixed or built-in then beam is known as a **cantilever**. As shown in Fig. 1.30.1 (*e*), end A is fixed while B is free. For the inclined external loads there will be three reactions as the fixed end, *viz.*, H_A, V_A and M_A. When the external loads are vertical the horizontal reaction H_A becomes zero.

(2) **Indeterminate Beams**

As mentioned above, when the number of unknown reactions in a beam exceed the number of equations of equilibrium (*i.e.*, R > E), the beam becomes indeterminate. In such a beam the difference R – E is more than zero. Fig. 1.30.2 shows examples of indeterminate beams.

(*a*) **Propped Cantilever.** When the free end of the cantilever is supported on a roller or a hinged supported, we get a propped cantilever. When the applied loads are vertical, as shown Fig. 1.30.2 (*a*), there are two reactions at the fixed end (*viz.*, V_A and M_A) and one reaction V_B at the propped end B. Thus, there are a total of three reactions (R = 3), while there are only two equations of equilibrium (E = 2), *viz.*, $\Sigma F_v = 0$ and $\Sigma M_B = 0$. Thus the difference R – E = 3 – 2 = 1. Hence, the degree of indeterminacy (D = R – E) is one.

When the applied load is inclined, the unknown reactions become 4

(V_A, V_B, M_A and H_A) while E = 3,

(*i.e.*, $\Sigma F_v = 0$, $\Sigma F_h = 0$ and $\Sigma M = 0$), and again R – E = 4 – 3 = 1. See Fig. 1.30.2(*b*).

If the propped end B is supported on a hinged support, then for the inclined loads R = 5 (V_A, V_B, H_A, H_B and M_A), while E = 3, and R – E = 2. See Fig. 1.30.2 (*c*).

(*b*) **Fixed Beams.** In a fixed beam both ends are built-in walls. For vertical loads there are four unknown reactions (R = 4 *i.e.*, V_A, M_A, V_B and M_B), while E = 2 (*i.e.*, $\Sigma F_v = 0$ and $\Sigma M = 0$), thus R – E = 4 – 2 = 2, as shown in Fig. 1.30.2 (*d*).

When the applied load is inclined, R = 6, E = 3 and D = 6 – 3 = 3, as shown in Fig. 1.30.2 (*e*).

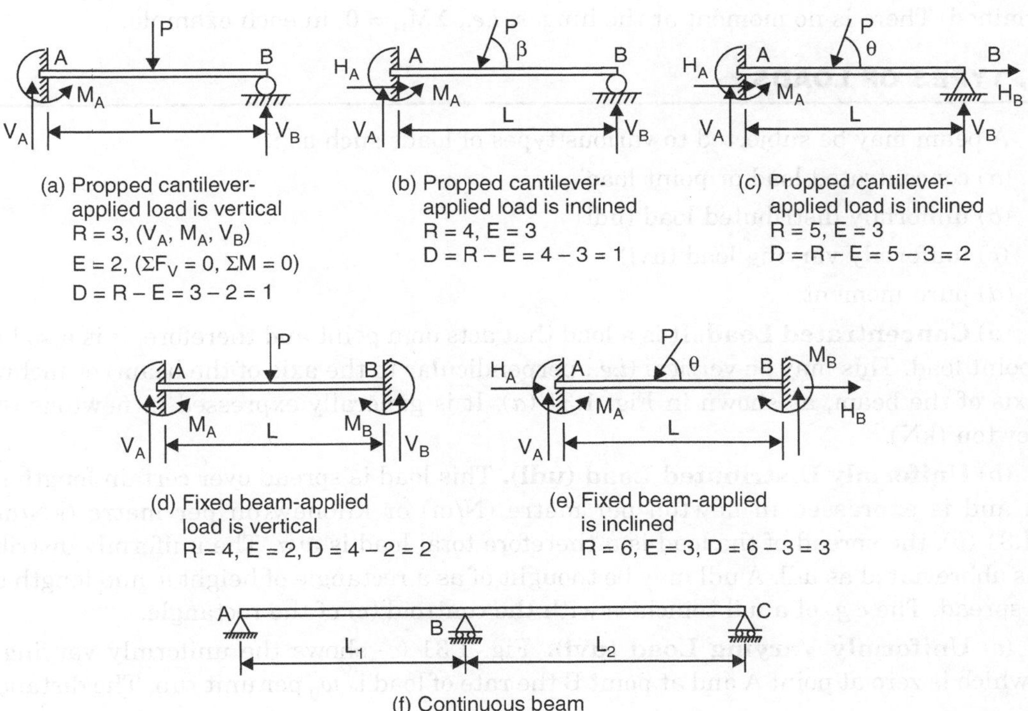

Fig. 1.30.2. Indeterminate beams

(*c*) **Continuous Beams.** When a beam is supported on more than two supports it is known as a continuous beam. See Fig. 1.30.2 (*f*).

Note. The determination of reactions of the supports of indeterminate beams is beyond the scope of any book on Engineering Mechanics.

(3) **Combined Beams.** Sometime two or more determinate beams are connected by hinges to form a single continuous structure. In Fig. 1.30.3, three examples of combined beams are shown. In Fig. 1.30.3 (*a*) there are two beams of lengths L_1 and L_2 connected by a hinge H.

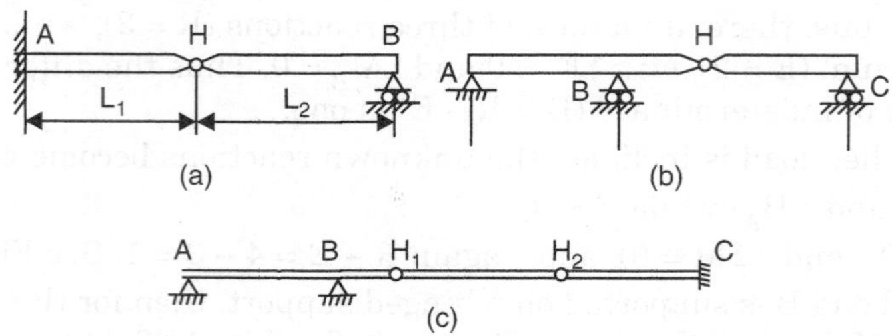

Fig. 1.30.3. Combined beams

In Fig. 1.30.3 (*b*), an overhanging beam ABC is connected to another beam by a hinge H. In Fig. 1.30.3 (*c*), there are two hinges H_1 and H_2. In each case the reactions at the supports cannot be determined by considering the free-body diagram of the entire beam. However, by considering the free-body diagram of each beam separately, reactions of all supports can be determined. There is no moment at the hinges, *i.e.*, $\Sigma M_H = 0$, in each example.

1.31. TYPES OF LOADS

A beam may be subjected to various types of loads such as :

(*a*) concentrated load or point load

(*b*) uniformly distributed load (udl)

(*c*) uniformly varying load (uvl)

(*d*) pure moment.

(*a*) **Concentrated Load.** It is a load that acts on a point and therefore, it is also known as a point load. This may be vertical (*i.e.*, perpendicular to the axis of the beam) or inclined to the axis of the beam, as shown in Fig. 1.31 (*a*). It is generally expressed in newtons (N), or kilonewton (kN).

(*b*) **Uniformly Distributed Load (udl).** This load is spread over certain length of the beam and is expressed in newton per metre (N/m) or kilonewton per metre (kN/m). In Fig. 1.31 (*b*), the spread of the load is *a* therefore total load is *w.a.* The uniformly distributed load is abbreviated as udl. A udl may be thought of as a rectangle of height *w* and length equal to its spread. The *c.g.* of a udl coincides with the centroid (*c*) of the rectangle.

(*c*) **Uniformly Varying Load (uvl).** Fig. 1.31 (*c*) shows the uniformly varying load (uvl) which is zero at point A and at point B the rate of load is w_0 per unit run. The distance AB is called the spread of the load. The total load $W = \Delta ABB_1 = \frac{1}{2} BB_1 \times AB = \frac{1}{2} w_0 l$. The *cg.* of the load coincides with the centroid of the ΔABB_1. Here, *c.g.* of the load is C and its distance from

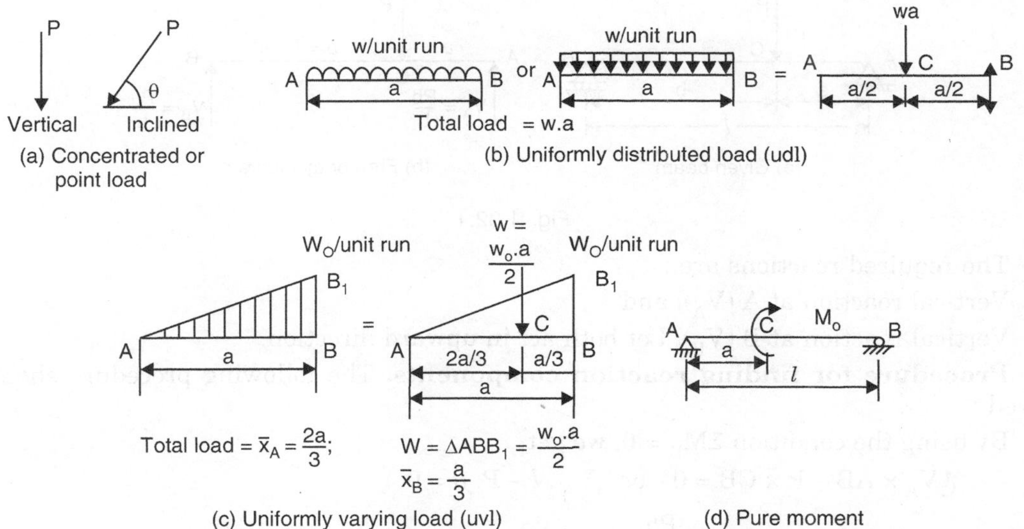

Fig. 1.31. Types of loads

$$B = \bar{x}_B = \frac{AB}{3} = \frac{a}{3} \text{ and that from A it is } \bar{x}_A = \frac{2}{3} \times AB = \frac{2a}{3}.$$

(*d*) **Pure Moment.** The pure moment M_o acting at any point C is shown in Fig. 1.31 (*d*). The units of pure moment are Nm or kNm.

A beam may be subjected to any combination of the loads discussed above.

1.32. METHODS OF FINDING REACTION AT SUPPORTS OF BEAMS

In sec. 1.15 we discussed the equations of equilibrium for non-concurrent coplanar force systems. These equations are

$$\Sigma F_v = 0 \quad \text{or} \quad \Sigma F_y = 0 \qquad \text{...(1.47)}$$

$$\Sigma F_h = 0 \quad \text{or} \quad \Sigma F_x = 0 \qquad \text{...(1.48)}$$

$$\Sigma M = 0 \quad \text{or} \quad \Sigma M_z = 0 \qquad \text{...(1.49)}$$

Thus, we see that there are only three equations of equilibrium available for solving problems of non-concurrent coplanar force systems. Hence, only three unknown reactions can be determined by using above equations. There are two methods for finding support reactions of beams, namely :

(*a*) **Analytical Method,**

(*b*) **Graphical Method.**

A few examples will help to understand the analytical method.

(*a*) **ANALYTICAL METHOD.**

Case I. A simply supported beam with a vertical point load. Consider a beam with a hinged support at A and a roller support at B. Only one load P acts on the beam at point C, as shown in Fig. 1.32.1.

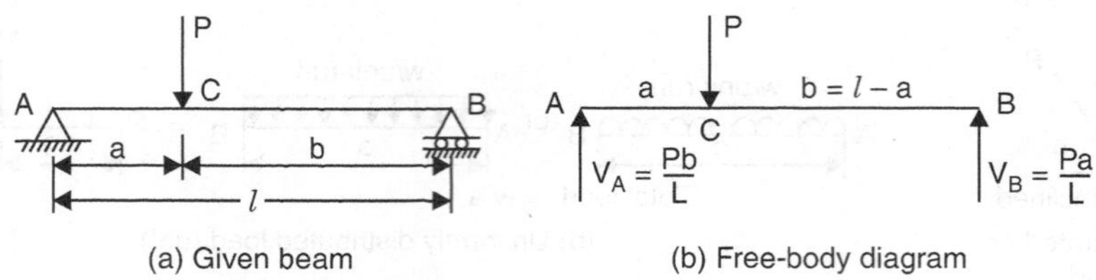

(a) Given beam (b) Free-body diagram

Fig. 1.32.1

The required reactions are :

Vertical reaction at A (V_A), and

Vertical reaction at B (V_B). Let both act in upward direction.

Procedure for finding reaction components. The following procedure should be followed :

By using the condition $\Sigma M_B = 0$, we get

$$\curvearrowleft V_A \times AB - P \times CB = 0 \quad \text{or} \quad V_A \,.\, l - P \,.\, b = 0$$

$$\therefore \qquad V_A = \frac{Pb}{l} \uparrow \qquad ...(1.50)$$

The +ve sign of the right hand side of the above equation shows that V_A is acting in the upward direction.

Now, by using the condition $\Sigma F_v = 0$, we get

$$(V_A + V_B)\uparrow - P = 0$$

$$\therefore \qquad V_B = P - V_A = P - \frac{Pb}{l} = \frac{P}{l}(l - b)$$

But $\qquad (l - b) = a,$

$$\therefore \qquad V_B = \frac{P \,.\, a}{l} \qquad ...(1.51)$$

These reaction components are shown in the free-body diagram of the beam in Fig. 1.32.1 (*b*)

Note. If $a = b = l/2$, then we get $V_A = V_B = P/2$. ...(1.52)

Case II. A simply supported with an inclined load. Fig. 1.32.2 shows a simply supported beam with a hinged support at A and a roller at B. A single inclined load P acts on the beam.

Given beam

Fig. 1.32.2

Here there are three unknown reaction components, viz., vertical reaction at A (V_A), vertical reaction at B (V_B), and horizontal reaction at A (H_A). Let V_A and V_B act in upward direction. Thus, three equations of equilibrium, *viz.*, $\Sigma F_v = 0$, $\Sigma F_h = 0$ and $\Sigma M = 0$, will be used for finding the unknown reaction components.

The given load has two components : $P_v = P \sin \theta$ and $P_h = P \cos \theta$.

Procedure. By using the condition $\Sigma M_B = 0$, we get

$$\curvearrowleft V_A \times AB - P_v \,.\, CB = 0 \quad \text{or} \quad V_A \,.\, l - (P \sin \theta) \,.\, b = 0$$

$$\therefore \qquad V_A = \frac{Pb}{l} \,.\, \sin \theta \uparrow \qquad ...(1.53)$$

By using the condition $\Sigma F_v = 0$, we get

$$(V_A + V_B)\uparrow - P_v = 0$$

$$\therefore \quad V_B = P_v - V_A = P \sin\theta - \frac{Pb}{l}\sin\theta = \left(\frac{P\sin\theta}{l}\right)(l - b)$$

But $l - b = a$, $\therefore$ we get

$$V_B = \frac{Pa}{l}\sin\theta\uparrow \qquad ...(1.54)$$

Lastly, by using the condition $\Sigma F_h = 0$, we get

$$H_A \rightarrow - P_h = 0$$

$$\therefore \quad H_A = P_h = P\cos\theta. \rightarrow \qquad ...(1.55)$$

Case III. A simply supported beam carrying a udl over its span. A simply supported beam AB carrying a uniformly distributed load w/unit length is shown in Fig. 1.32.3. The support reactions are V_A and V_B, both directed upward.

Total load = W = rate of loading × length (or spread) of load.

i.e., $$W = w \times l$$

The *c.g.* of the load wl is at a distance $l/2$ from A and from B.

Fig. 1.32.3

Procedure. By using the condition $\Sigma M_B = 0$, we get

$$\circlearrowleft V_A \times AB - (w \times AB) \times \frac{AB}{2} = 0$$

or $$V_A \cdot l - wl \cdot l/2 = 0$$

$$\therefore \quad V_A = \frac{wl}{2} \qquad ...(1.56)$$

By $\Sigma F_v = 0$, we get

$$(V_A + V_B)\uparrow - wl = 0$$

$$V_B = wl - V_A = wl - \frac{wl}{2} \quad \text{or} \quad V_B = \frac{wl}{2} \qquad ...(1.57)$$

Case IV. A simply supported beam partially loaded by a udl. Fig. 1.32.4 shows a simply supported beam AB of span L. It carries a udl (w) over a portion AC of length l_1.

The total load = wl_1.

Its *c.g.* is at a distance $\bar{x}_A = \dfrac{l_1}{2}$ from A.

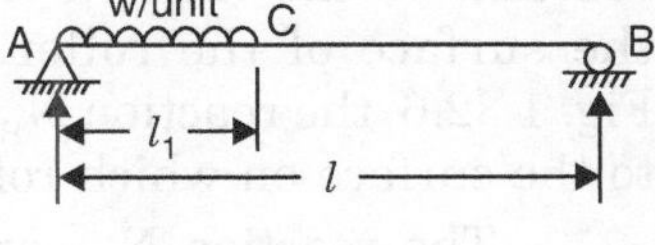

Fig. 1.32.4

Procedure. By using the condition $\Sigma M_A = 0$, we get

$$\circlearrowleft V_B \times AB - (w \times AC) \times \bar{x}_A = 0$$

or $$V_B \cdot l - (wl_1) \cdot \left(\frac{l_1}{2}\right) = 0$$

$$\therefore \quad V_B = \frac{wl_1^2}{2l} \qquad ...(1.58)$$

And by using the condition $\Sigma F_v = 0$, we get

$$(V_A + V_B)\uparrow - w \times l_1 = 0$$

$$\therefore \quad V_A = wl_1 - V_B = wl_1 - \frac{wl_1^2}{2l} \quad \text{or} \quad V_A = \frac{\omega \, . \, l_1}{2l}(2l - l_1) \quad \text{...(1.59)}$$

Case V. A simply supported beam partially loaded by a udl. A simply supported beam of span l carries a uniformly distributed load (w) over a portion CD of length c, shown in Fig. 1.32.5.

Total load = W = $w \times$ CD = $w.\ c$.

Its distance of *c.g.* from A = $\bar{x}_A$

$$\bar{x}_A = AC + \frac{CD}{2} = (a + c/2)$$

Fig. 1.32.5

The distance *c.g.* of the load from B

$$= \bar{x}_B = DB + \frac{CD}{2} = \left(b + \frac{c}{2}\right).$$

Procedure. By using the condition $\Sigma M_B = 0$, we get

$$\circlearrowleft V_A \times AB - (w \times CD) \times (\bar{x}_B) = 0 \quad \text{or} \quad V_A \, . \, l - (wc)\left(b + \frac{c}{2}\right) = 0$$

$$\therefore \quad V_A = \frac{wc}{l}\left(b + \frac{c}{2}\right) \quad \text{or} \quad V_A = \frac{wc}{2l}(2b + c) \quad \text{...(1.60)}$$

By using the condition $\Sigma F_v = 0$, we get

$$(V_A + V_B)\uparrow - (wc) = 0.$$

$$\therefore \quad V_B = wc - V_A = wc - \frac{wc}{2l}(2b + c) = \frac{wc}{2l}(2l - 2b - c)$$

But $\quad 2l - 2b = 2(l - b) = 2(a + b + c - b) = 2(a + c).$

$$\therefore \quad V_B = \frac{wc}{2l}(2a + 2c - c) \quad \text{or,} \quad V_B = \frac{wc}{2l}(2a + c) \quad \text{...(1.61)}$$

Case VI. A simply supported beam with roller resting on a smooth incline and carrying a vertical load. When a simply supported beam has the roller support resting on a smooth incline, the reaction on the roller is in the direction perpendicular to the surface of the roller. Thus, in the beam shown in Fig. 1.32.6, the reaction N_B on the roller is along the normal to the surface on which roller rests.

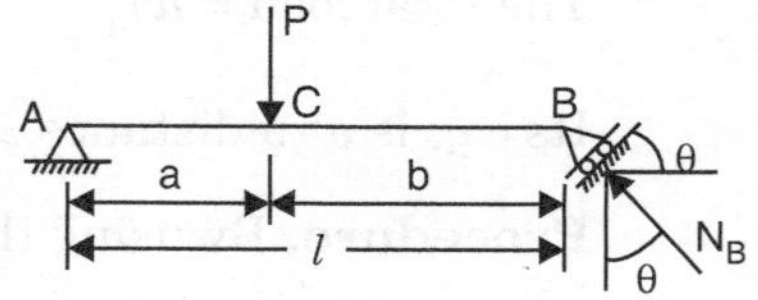

Fig. 1.32.6

The reaction N_B can be resolved in horizontal (H_B) and vertical (V_B) components. Thus

$$H_B = N_B \, . \, \sin\theta \text{ and } V_B = N_B \, . \, \cos\theta \quad \text{...(1)}$$

Procedure. By using the condition $\Sigma M_A = 0$, we get

$$V_B \, . \, AB - P\ AC = 0 \quad \text{or} \quad V_B\, l = P \, . \, a$$

$$\therefore \quad V_B = \frac{P \, . \, a}{l}$$

But, $\because$ $V_B = N_B \cos\theta$ from eqns. (1).

$\therefore$ We get $N_B . \cos\theta = \dfrac{Pa}{l}$

$$\therefore \quad N_B = \frac{Pa}{l \cos\theta} \qquad \text{...(1.62)}$$

$$\therefore \quad H_B = N_B . \sin\theta = \left(\frac{Pa}{l\cos\theta}\right)\sin\theta$$

or $$H_B = \frac{P.a}{l}\tan\theta \leftarrow \qquad \text{...(1.63)}$$

By using the condition $\Sigma F_v = 0$, we get

$$(V_A + V_B)\uparrow - P = 0$$

$$\therefore \quad V_A = P - V_B = P - \frac{Pa}{l} = \frac{P}{l}(l-a)$$

or $$V_A = \frac{P}{l}.b \qquad \text{...(1.64)}$$

Lastly, by using the condition $\Sigma F_h = 0$, we get

$$H_A + H_B = 0$$

$$\therefore \quad H_A + \frac{Pa}{l}.\tan\theta = 0 \quad \text{or} \quad H_A = -\frac{Pa}{l}.\tan\theta$$

The negative sign for H_A indicates that H_A is directed towards right ($\rightarrow$).

Thus, $$H_A = \frac{Pa}{l}.\tan\theta \rightarrow \qquad \text{...(1.65)}$$

Case VII. A simply supported beam carrying uvl on its entire span. A simply supported beam carrying uniformly varying load is shown in Fig. 1.32.7. The rate of loading at A is zero while that at B it is w_0/unit length. Total load = W = $\Delta ABB_1 = \frac{1}{2} BB_1 . AB = \frac{1}{2} w_0 l$.

C.G. of the load coincides with the centroid of the triangle ABB_1. Hence the *c.g.* of the load is $x_A = \frac{2}{3} \times AB = \frac{2}{3} l$ from A and $x_B = \frac{l}{3}$ from B.

Fig. 1.32.7

By using the condition $\Sigma M_B = 0$, we get

$$\circlearrowleft V_A . AB - \left(\frac{1}{2} \times w_0 l\right)(l/3) = 0$$

$$\therefore \quad V_A \times l - \frac{w_0 l^2}{6} = 0$$

$$\therefore \quad V_A = \frac{w_0 l}{6} = \frac{W}{3} \qquad \text{...(1.66)}$$

By $\Sigma F_v = 0$, we get

$$(V_A + V_B)\uparrow - \frac{w_0\, l}{2} = 0$$

$$\therefore \qquad V_B = \frac{w_0 \,.\, l}{2} - V_A = \frac{w_0 \,.\, l}{2} - \frac{w_0 \,.\, l}{6} \quad \text{or} \quad V_B = \frac{w_0 \,.\, l}{3} = 2V_A \qquad ...(1.67)$$

Case VIII. A simply supported beam partially loaded by a uvl. Fig. 1.32.8 show a simply supported beam carrying uniformly varying load on a small portion of its span.

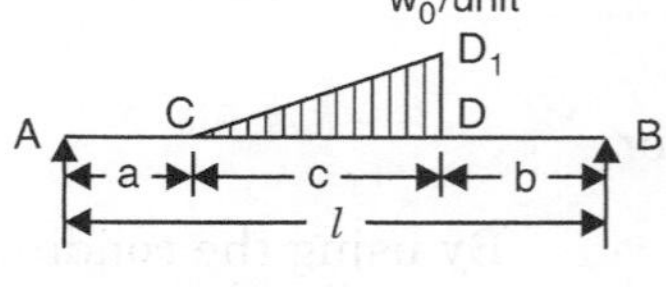

Fig. 1.32.8

$$\text{Total load} \quad = W = \Delta CDD_1 = \frac{1}{2} \times DD_1 \times CD = \frac{w_0 \,.\, c}{2}$$

c.g. of the load from A is at a distance $\bar{x}_A = a + \frac{2c}{3}$ and that from point B = $\bar{x}_B = b + \frac{c}{3}$.

$\therefore$ By using the condition $\Sigma M_B = 0$, we get

$$\curvearrowleft V_A \times l - W \,.\, \bar{x}_B = 0 \quad \text{or} \quad V_A \times l - \frac{w_0 \,.\, c}{2}\left(b + \frac{c}{3}\right) = 0$$

$$\therefore \qquad V_A = \frac{w_0 \,.\, c}{2l}(b + c/3) \qquad ...(1.68)$$

By using the condition $\Sigma F_v = 0$, we get

$$(V_A + V_B)\uparrow - W = 0$$

$$\therefore \qquad V_B = W - V_A = \frac{w_0 \,.\, c}{2} - \frac{w_0 \,.\, c}{2l}(b + c/3)$$

$$= \frac{w_0 \,.\, c}{2l}\left(l - b - \frac{c}{3}\right) = \frac{w_0 \,.\, c}{2l}\left(a + b + c - b - \frac{c}{3}\right)$$

$$V_B = \frac{w_0 \,.\, c}{2l}\left(a + \frac{2c}{3}\right) \qquad ...(1.69)$$

Case IX. A simply supported beam carrying a pure moment. A simply supported beam of span l with hinged support at A and a roller at B and carrying a pure moment M_0 at point C, as shown in Fig. 1.32.9.

By using the condition $\Sigma M_B = 0$, we get

$$\curvearrowleft V_A \times l + M_0 = 0$$

$$\therefore \qquad V_A = -\frac{M_0}{l}\downarrow \qquad ...(1.70)$$

Fig. 1.32.9

The reaction V_A is directed vertically downwards.

By using the condition $\Sigma F_v = 0$, we get

$$V_A\downarrow + V_B = 0$$

$$\therefore \qquad V_B = -V_A = -\left(\frac{-M_0}{l}\right) \quad \text{or} \quad V_B = \frac{M_0}{l}\uparrow \qquad ...(1.71)$$

V_B is directed vertically upwards.

Problem 1.30. *Find reactions at supports A and B of the beam shown in Fig. P.1.30 (a).*

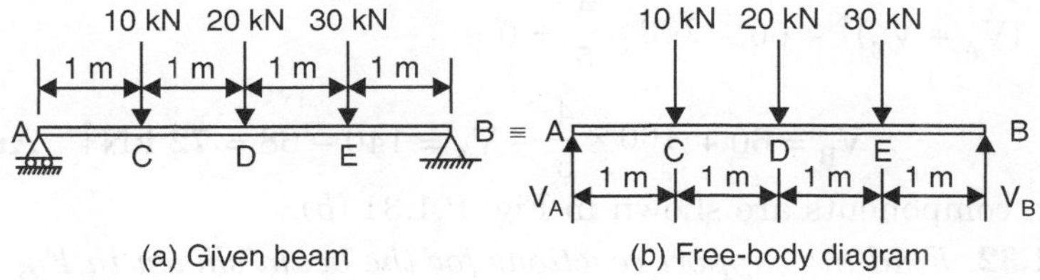

(a) Given beam (b) Free-body diagram

Fig. P. 1.30

Sol. The free-body diagram of the given beam is given in Fig. P.1.30 (*b*). Let support reactions at A and B be V_A and V_B, respectively.

By using the, condition, $\Sigma M_B = 0$, we get

$$\circlearrowleft + V_A \times AB - 10 \times CB - 20 \times DB - 30 \times EB = 0$$

or $$4V_A - 10 \times 3 - 20 \times 2 - 30 \times 1 = 0$$

or $$V_A = \frac{1}{4}[30 + 40 + 30] = \mathbf{25\ kN}\uparrow \quad \textbf{Ans.}$$

By $\Sigma F_v = 0$, we get

$$(V_A + V_B)\uparrow - (10 + 20 + 30) = 0$$

$\therefore$ $$V_B = 60 - V_A = 60 - 25 = \mathbf{35\ kN}\uparrow \quad \textbf{Ans.}$$

Problem 1.31. *Find reactions of supports for the beam shown in Fig. P.1.31 (a).*

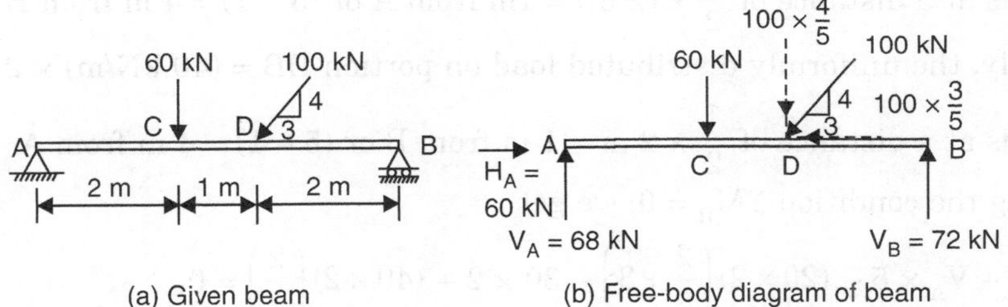

(a) Given beam (b) Free-body diagram of beam

Fig. P. 1.31

Sol. The inclined load of 100 kN may be resolved into its vertical and horizontal components as shown in Fig. P. 1.31 (*b*)

The vertical component of 100 kN lood = $100 \times \frac{4}{5}$ kN↑

and the horizontal component of 100 kN lood = $100 \times \frac{3}{5}$ kN←

By using $\Sigma F_h = 0$, we get

$$\overset{+}{\rightarrow} H_A - 100 \times \frac{3}{5} = 0, \quad \therefore \quad H_A = 100 \times \frac{3}{5} = \mathbf{60\ kN} \rightarrow \quad \textbf{Ans.}$$

By $\Sigma M_B = 0$, we get

$$\circlearrowleft + V_A \times 5 - 60 \times 3 - \left(100 \times \frac{4}{5}\right) \times 2 = 0 \quad \text{or,} \quad 5V_A - 180 - 160 = 0$$

$\therefore$ $$V_A = \frac{1}{5} \times (180 - 160) = \mathbf{68\ kN}\uparrow \quad \textbf{Ans.}$$

And, finally by using $\Sigma F_v = 0$, we get

$$(V_A + V_B)\uparrow - 60 - 100 \times \frac{4}{5} = 0$$

$$\therefore \quad V_B = 60 + 100 \times \frac{4}{5} - V_A = 140 - 68 = \mathbf{72\ kN}\uparrow \quad \textbf{Ans.}$$

All reaction components are shown in Fig. P.1.31 (*b*).

Problem 1.32. *Find the support reactions for the beam shown in Fig. P.1.32 (a).*

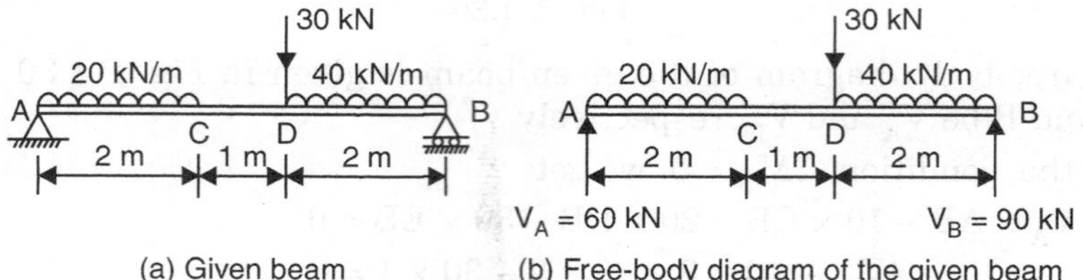

Fig. P.1.32

Sol. Since there is no inclined load acting on the beam, therefore, the reactions at supports A and B will be vertical. Only two equations, *viz.*, $\Sigma M_B = 0$ and $\Sigma F_v = 0$, will be used to find V_A and V_B.

The uniformly distributed load on portion AB = (20 kN/m) × 2 m = 40 kN.

Its *c.g.* is at a distance of $\frac{1}{2} \times (2\text{ m}) = 1\text{m}$ from A or (5 – 1) = 4 m from B.

Similarly, the uniformly distributed load on portion DB = (40 kN/m) × 2 m = 80 kN.

Its *c.g.* is at a distance of $\frac{1}{2} \times 2\text{ m} = 1\text{ m}$ from B or (5 – 1) = 4 m from A.

By using the condition $\Sigma M_B = 0$, we get

$$\circlearrowleft + V_A \times 5 - (20 \times 2)\left(\frac{2}{2} + 3\right) - 30 \times 2 - (40 \times 2)\left(\frac{2}{2}\right) = 0,$$

$$\text{or,} \quad 5V_A - 40 \times 4 - 60 - 80 \times 1 = 0$$

$$\text{or} \quad V_A = \frac{1}{5}(40 \times 4 + 60 + 80) = \frac{1}{5}(300) = \mathbf{60\ kN}\uparrow \quad \textbf{Ans.}$$

By using the condition $\Sigma F_v = 0$, we get

$$(V_A + V_B)\uparrow - 20 \times 2 - 30 - 40 \times 2 = 0$$

$$\therefore \quad V_B = 20 \times 2 + 30 + 40 \times 2 - V_A$$

$$= 40 + 30 + 80 - 60\text{ kN}$$

$$\text{or,} \quad V_B = \mathbf{90\ kN}\uparrow \quad \textbf{Ans.}$$

Fig. P.1.32(*b*) shows the free-body diagram of the given beam.

Problem 1.33. *Determine the support reactions for the beam shown in Fig. P.1.33.*

Sol. There are two uniformly distributed loads (udls) *viz.*, (*i*) a 20 kN/m load over a length of 5 m, and (*ii*) a 10 kN/m load over portion CB = 2 m long.

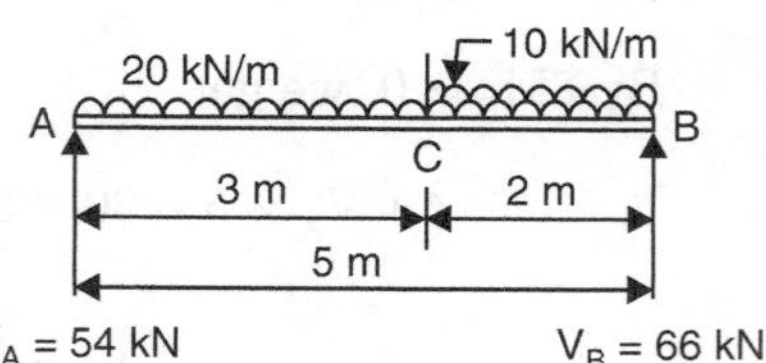

Fig. P.1.33

By using the condition $\Sigma M_B = 0$, we get

$$\circlearrowleft + V_A \times 5 - (20 \text{ kN/m} \times 5 \text{ m}) \times \left(\frac{5}{2}\right) - (10 \text{ kN/m} \times 2 \text{ m}) \times \frac{2}{2} = 0$$

or, $$5V_A - 100 \times 2.5 - 20 \times 1 = 0$$

or, $$V_A = \frac{1}{5}(250 + 20) = \mathbf{54\ kN\uparrow} \quad \textbf{Ans.}$$

By using condition $\Sigma F_v = 0$, we get

$$(V_A + V_B)\uparrow - (20 \times 5) - (10 \times 2) = 0$$

$\therefore$ $$V_B = 20 \times 5 + 10 \times 2 - V_A = 100 + 20 - 54$$

or $$V_B = \mathbf{66\ kN\uparrow} \quad \textbf{Ans.}$$

The reactions V_A and V_B are shown in Fig. P.1.33.

Problem 1.34. *Find support reactions of beam shown in Fig. P.1.34.*

Sol. Since the given loads are vertical, therefore the reactions at the supports will be vertical. These are V_A and V_B.

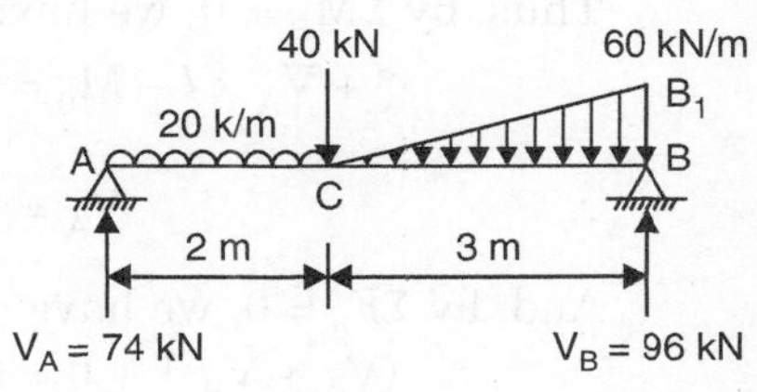

Fig. P. 1.34

The uniformly distributed load (udl) of 20 kN/m acts on the portion AC = 2m long. Therefore, the total load on portion AC = 20 kN/m × 2m = 40 kN.

Its *c.g.* is at a distance of $\frac{1}{2} \times AC = \frac{1}{2} \times 2 = 1$ m from A

or $$(5 - 1) = 4 \text{ m from B.}$$

The total load due to uniformly varying load (uvl) of rate 60 kN/m

$$= \text{Area of the } \Delta CBB_1$$

$$= \frac{1}{2} BB_1 \times CB = \frac{1}{2} \times (60 \text{ kN/m}) \times 3 \text{ m} = 90 \text{ kN.}$$

Its *c.g.* is at a distance of $\frac{1}{3} \times (CB) = \frac{1}{3} \times 3 \text{ m} = 1$ m from B.

Hence, by using the condition $\Sigma M_B = 0$, we get

$$\circlearrowleft + V_A \times 5 - (20 \times 2)\left(\frac{2}{2} + 3\right) - 40 \times 3 - \left(\frac{1}{2} \times 60 \times 3\right) \times \left(\frac{1}{3} \times 3\right) = 0$$

or $$5V_A - 40 \times 4 - 120 - 90 \times 1 = 0$$

or $$V_A = \frac{1}{5}[160 + 120 + 90] = \frac{1}{5}[370] \text{ kN}$$

or $$V_A = \mathbf{74\ kN\uparrow} \quad \textbf{Ans.}$$

By using the condition $\Sigma F_v = 0$, we get

$$(V_A + V_B)\uparrow - 20 \times 2 - 40 - \frac{1}{2} \times 60 \times 3 = 0$$

or $$V_B = 20 \times 2 + 40 + \frac{1}{2} \times 60 \times 3 - V_A$$

$$= 40 + 40 + 90 - 74$$

or $$V_B = \mathbf{96\ kN}\uparrow \quad \textbf{Ans.}$$

Problem 1.35. *Find support reactions for the beam shown in Fig. P.1.35 (a).*

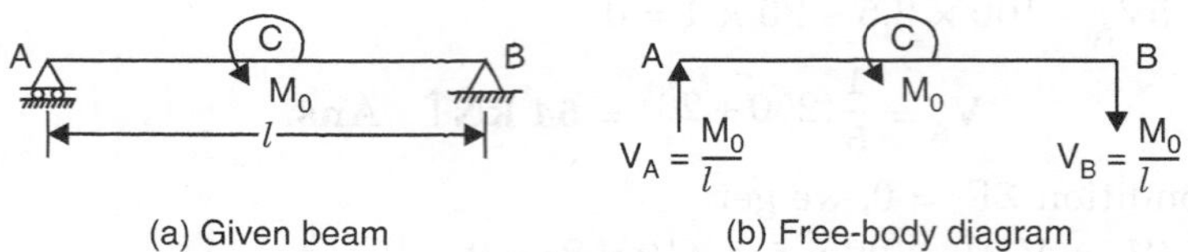

(a) Given beam (b) Free-body diagram

Fig. P.1.35

Sol. Here there are no horizontal or vertical forces applied on the beam. The applied bending couple M_0 is counter balanced by the reactions of the supports. Since the applied couple is counterclockwise, the couple due to support reactions must be clockwise in sense.

Thus, by $\Sigma M_B = 0$, we have

$$\circlearrowleft + V_A \times l - M_0 = 0$$

$$\therefore \qquad V_A = \frac{M_0}{l}\uparrow \quad \textbf{Ans.}$$

And, by $\Sigma F_v = 0$, we have

$$(V_A + V_B)\uparrow + 0 = 0$$

$$\therefore \qquad V_B = -V_A = -\frac{M_0}{l}\downarrow \quad \textbf{Ans.}$$

Fig. P.1.35 (*b*) shows the free-body diagram of the beam.

Problem 1.36. *Find support reactions of the beam shown in Fig. P.1.36 (a).*

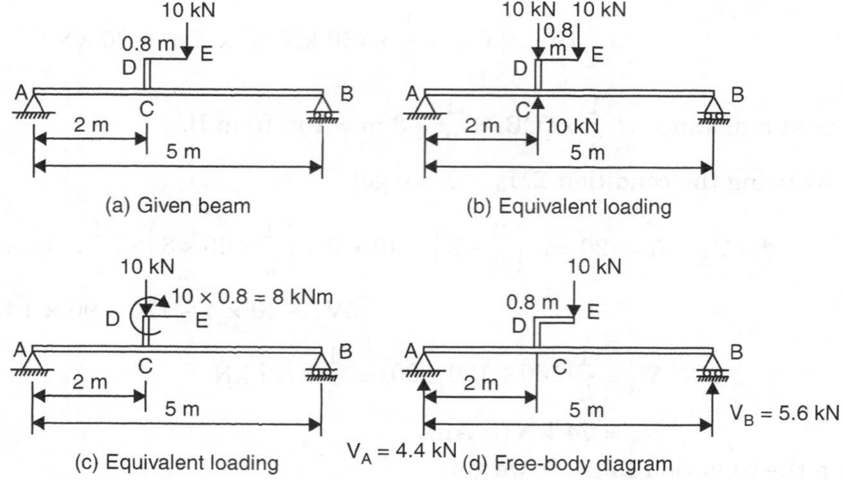

(a) Given beam (b) Equivalent loading

(c) Equivalent loading (d) Free-body diagram

Fig. P.1.36

Sol. The given loading system on the extended portion DE of the beam can be replaced by a vertical load of 10 kN at D plus a clockwise couple equal to 10 kN × 0.8 m = 8 kNm applied at D as shown in Figs. (*b*) and (*c*). Now the equations of equilibrium (*i*) $\Sigma M = 0$, and (*ii*) $\Sigma F_v = 0$, can be applied to the beam shown in Fig. P.1.36 (*c*). Free body diagram of the given beain is shown in Fig. P. 1.36 (*d*).

Thus, by using $\Sigma M_B = 0$, we get

$$\circlearrowleft + V_A \times 5 - 10 \times 3 + 8 = 0$$

$$\therefore \quad V_A = \frac{1}{5}(30 - 8) = \frac{22}{5} \text{ kN} \quad \text{or} \quad V_A = \mathbf{4.4\ kN}\uparrow \quad \textbf{Ans.}$$

And, by using $\Sigma F_v = 0$, we get

$$(V_A + V_B)\uparrow - 10 = 0$$

$$\therefore \quad V_B = 10 - V_A = 10 - 4.4 \text{ kN}$$

$$\therefore \quad V_B = \mathbf{5.6\ kN}\uparrow \quad \textbf{Ans.}$$

Problem 1.37. *Determine the support reactions of the beam shown in Fig. P.1.37(a).*

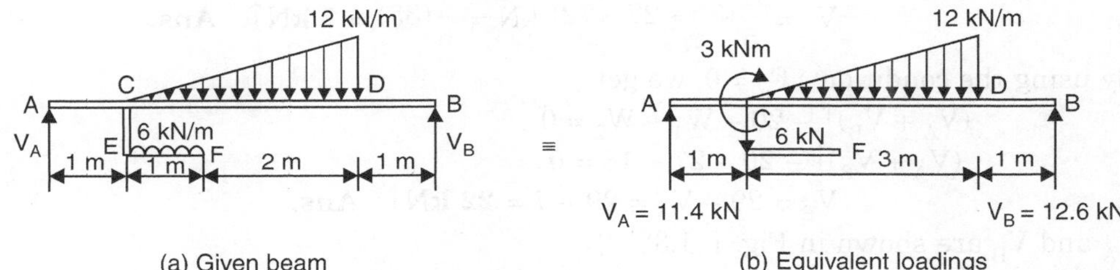

(a) Given beam (b) Equivalent loadings

Fig. P.1.37

Sol. The load on the portion EF may be replaced by a force of 6 × 1 kN at C and a couple $= \frac{1}{2} \times 6 \times 1^2 = 3$ kNm clockwise in sense, as shown in Fig. P.1.37 (*b*).

By using the condition $\Sigma M_B = 0$, we get

$$\circlearrowleft + V_A \times 5 + 3 - 6 \times 4 - \frac{1}{2} \times (12 \times 3) \times \left(\frac{3}{3} + 1\right) = 0$$

or $$5V_A + 3 - 24 - 18 \times 2 = 0$$

or $$V_A = \frac{1}{5}(24 + 36 - 3) = \frac{57}{5} = \mathbf{11.4\ kN}\uparrow \quad \textbf{Ans.}$$

By using the condition $\Sigma F_v = 0$, we get

$$(V_A + V_B)\uparrow - 6 - \frac{1}{2} \times 12 \times 3 = 0$$

or $$(V_A + V_B)\uparrow - 6 - 18 = 0$$

$$\therefore \quad V_B = 24 - V_A = 24 - 11.4 \text{ kN} \quad \text{or} \quad V_B = \mathbf{12.6\ kN}\uparrow \quad \textbf{Ans.}$$

V_A and V_B are shown in Fig. P.1.37 (*b*).

Problem 1.38. *Find support reactions of the beam shown in the Fig. P.1.38.*

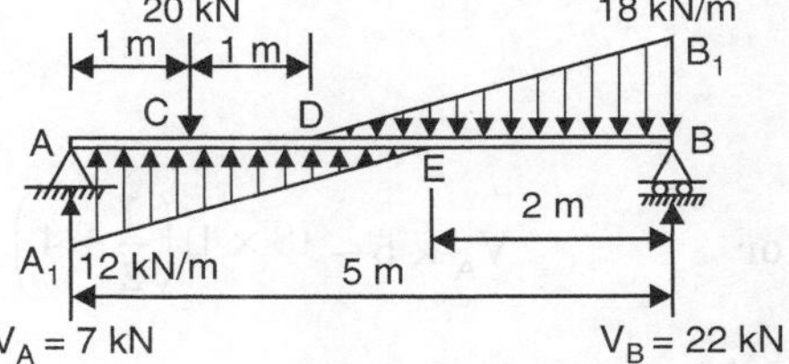

Fig. P.1.38

Sol. The downward acting uniformly varying load given by the

$$\Delta DBB_1 = W_1 = \frac{1}{2} \times BB_1 \times DB$$

$$= \frac{1}{2} \times (18 \text{ kN/m}) \times (3 \text{ m}) = 27 \text{ kN}$$

Its *c.g.* is at a distance of $\frac{1}{3} \times 3$ m = 1 m from B.

The upward acting uniformly varying load given by the

$$\Delta EAA_1 = \frac{1}{2} \times AA_1 \times AE = \frac{1}{2} \times 12 \times 3 = 18 \text{ kN} = W_2$$

Its *c.g.* is at a distance of $\frac{1}{3} \times 3$ = 1m from A or (5 – 1) = 4m from B.

Thus, by using the condition $\Sigma M_B = 0$, we get

$$\circlearrowleft + V_A \times 5 - 20 \times 4 - W_1 \times 1 + W_2 \times 4 = 0$$

or, $$5V_A - 80 - 27 \times 1 + 18 \times 4 = 0$$

$\therefore$ $$V_A = \frac{1}{5}(80 + 27 - 72) \text{ kN} = \frac{1}{5}(35) = \mathbf{7 \text{ kN}\uparrow} \quad \textbf{Ans.}$$

By using the condition $\Sigma F_v = 0$, we get

$$(V_A + V_B)\uparrow - 20 - W_1 + W_2 = 0$$

or, $$(V_A + V_B)\uparrow - 20 - 27 + 18 = 0$$

$\therefore$ $$V_B = 29 - V_A = 29 - 7 = \mathbf{22 \text{ kN}\uparrow} \quad \textbf{Ans.}$$

V_A and V_B are shown in Fig. P.1.38.

Problem 1.39. *Determine support reactions for the beam shown in Fig. P.1.39(a).*

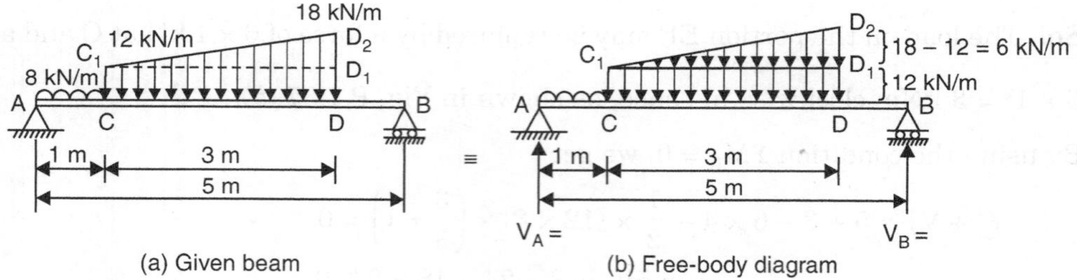

Fig. P.1.39

Sol. The trapezoidal load CC_1D_2D can be subdivided into (*a*) a udl given by the rectangle CC_1D_1D of height equal to 12 kN/m, and (*b*) a uniformly varying load (uvl) given by the $\Delta C_1D_1D_2$ of height $D_1D_2 = DD_2 - DD_1 = 18 - 12 = 6$ kN/m, as shown in Fig. P.1.39 (*b*).

Thus, by using the conditions $\Sigma M_B = 0$, we get

$$\circlearrowleft + V_A \times AB - (8 \text{ kN/m} \times AC)\left(\frac{AC}{2} + CB\right) - (CC_1 \times CD) \times \left(\frac{CD}{2} + DB\right)$$
$$- \left(\frac{1}{2} \times D_1D_2 \times C_1D_1\right) \times \left(\frac{C_1D_1}{3} + DB\right) = 0$$

or $$V_A \times 5 - (8 \times 1)\left(\frac{1}{2} + 4\right) - (12 \times 3)\left(\frac{3}{2} + 1\right) - \left(\frac{1}{2} \times 6 \times 3\right) \times \left(\frac{3}{3} + 1\right) = 0$$

or, $$5V_A - 8 \times 4.5 - 36 \times 2.5 - 9 \times 2 = 0$$

or, $5V_A - 36 - 90 - 18 = 0$

$$\therefore \quad V_A = \frac{1}{5}(36 + 39 + 18) = \mathbf{28.8\ kN}\uparrow \quad \textbf{Ans.}$$

By using the condition $\Sigma F_v = 0$, we get

$$(V_A + V_B)\uparrow - \left(8 \times 1 + 12 \times 3 + \frac{1}{2} \times 6 \times 3\right) = 0 \quad \text{or,} \quad 28.8 + V_B - 53 = 0$$

or $$V_B = 58 - 28.8 = \mathbf{24.2\ kN}\uparrow \quad \textbf{Ans.}$$

Problem 1.40. *Determine the intensity of the uniformly distributed load w if reaction at support C of the beam is zero. Also find the reaction of support B. See Fig. P.1.40.* [Univ. of Pune, May 1998]

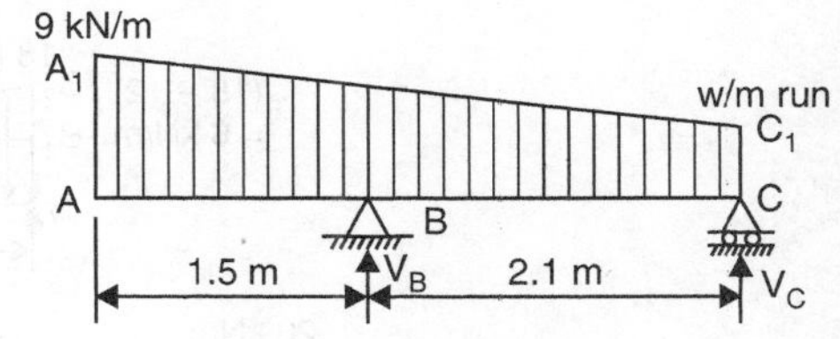

Fig. P.1.40

Sol. The trapezoidal load can be resolved into two triangular loads, *i.e.*, ΔAA_1C and ΔCC_1A.

By using the condition $\Sigma M_C = 0$, we get

$$V_B \times 2.1 - (\Delta AA_1C) \cdot \bar{x}_{C1} - (\Delta ACC_1) \cdot \bar{x}_{C2} = 0$$

or, $$2.1V_B - \frac{9 \times 3.6}{2} \times 2 \times \frac{3.6}{3} - \frac{w \times 3.6}{2} \times \frac{1}{3} \times 3.6 = 0$$

or, $$2.1V_B - \frac{3.6 \times 3.6}{6}(18 + w) = 0$$

or, $$V_B = \frac{3.6 \times 3.6}{2.1 \times 6}(18 + w) \quad \text{or,} \quad V_B = \frac{7.2}{7}(18 + w) \qquad ...(1)$$

By using the condition $\Sigma F_v = 0$, we get

$$(V_B + V_C)\uparrow - \text{area } AA_1C_1C = 0$$

where $V_C = 0$, given,

$$\therefore \quad \text{We get} \quad V_B + 0 - \frac{9 + w}{2} \times 3.6 = 0$$

$$\therefore \quad V_B = 1.8(9 + w) \qquad ...(2)$$

Equating the RHS of eqns. (1) and (2), we get

$$\frac{7.2}{7} \times (18 + w) = 1.8(9 + w) \quad \text{or,} \quad \frac{4}{7} \times (18 + w) = (9 + w)$$

$$\therefore \quad 3w = 72 - 63 = 9$$

$$\therefore \quad w = \frac{9}{3} = \mathbf{3\ kN/m} \quad \textbf{Ans.}$$

From eqn. (2), we have

$$V_B = 1.8(9 + w)$$

$$\therefore \quad V_B = 1.8(9 + 3) = 1.8 \times 12 \text{ kN} \quad \text{or,} \quad V_B = \mathbf{21.6\ kN}\uparrow \quad \textbf{Ans.}$$

Problem 1.41. *Find reactions at A, B, C, D and E for the beam shown in Fig. P.1.41 (a). There are hinges at B and C.*

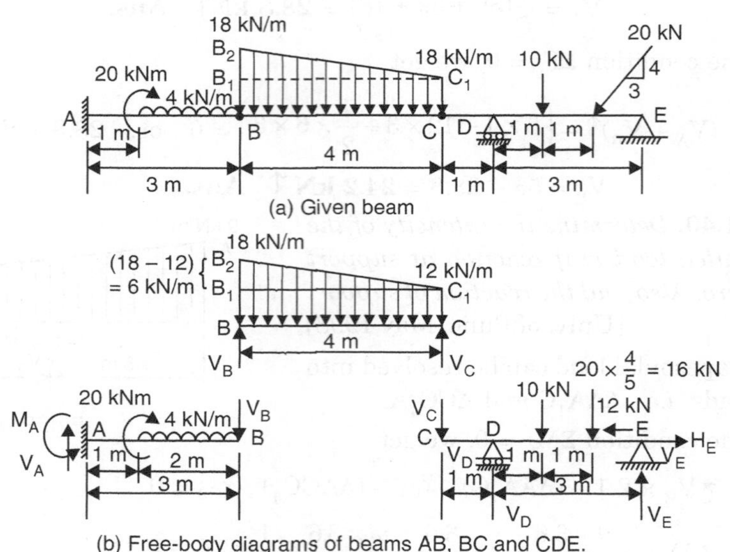

(a) Given beam

(b) Free-body diagrams of beams AB, BC and CDE.

Fig. P.1.41

Sol. The given beam of Fig. P.1.41(*a*) can be resolved into (*i*) a cantilever AB, (*ii*) a simply supported beam BC, and (*iii*) an over hanging beam CDE. The free-body diagrams of the three beams are shown in Fig. P.1.41(*b*). Let vertical reactions at the hinges B and C be VB and V_C, respectively. The beam BC is supported at B and C, we first find the reactions V_B and V_C at hinges. The reaction V_B acts as external load on the cantilever AB. Also, the reaction V_C acts as external load on the overhanging beam CE, shown in Fig. P. 1.41(*b*).

For simply supported beam BC :

By $\Sigma M_C = 0$, we get

$$\circlearrowleft + V_B \times 4 - (12 \times 4) \times \frac{4}{2} - \left(\frac{1}{2} \times 6 \times 4\right) \times \frac{2}{3} \times 4 = 0, \quad \text{or,} \quad 4V_B - 96 - 32 = 0$$

$$\therefore \qquad V_B = \frac{128}{4} = \mathbf{32\ kN}\uparrow \quad \textbf{Ans.}$$

By $\Sigma F_v = 0$, we get

$$(V_B + V_C)\uparrow - 12 \times 4 - \frac{1}{2} \times 6 \times 4 = 0$$

or

$$(V_B + V_C) - 48 - 12 = 0$$

$$\therefore \qquad V_C = 60 - V_B = 60 - 32 = \mathbf{28\ kN}\uparrow \quad \textbf{Ans.}$$

For cantilever AB : By $\Sigma F_V = 0$, we get

$$V_A - 4 \times 2 - V_B = 0$$

$$\therefore \qquad V_A = 8 + V_B = 8 + 32 = \mathbf{40\ kN}\uparrow \quad \textbf{Ans.}$$

$$M_B = -(4 \times 2) \times \left(\frac{2}{2} + 1\right) - V_B \times 3 - 20 \text{ kNm}$$

$$= -16 - 32 \times 3 - 20 \text{ kNm}$$

or, $$M_B = -16 - 96 - 20 = \mathbf{132 \text{ kNm}} \quad \textbf{Ans.}$$

For over-hanging beam CDE :

By $\Sigma F_h = 0$, we get

$$H_E \rightarrow -20 \times \frac{3}{5} = 0 \quad \text{or} \quad H_E = \frac{20 \times 3}{5} = \mathbf{12 \text{ kN}} \rightarrow \quad \textbf{Ans.}$$

By $\Sigma M_E = 0$, we get

$$+V_D \times 3 - V_C \times 4 - 10 \times 2 - 16 \times 1 = 0$$

or $$3V_D - 28 \times 4 - 20 - 16 = 0$$

or, $$3V_D - 112 - 20 - 16 = 0$$

$\therefore$ $$V_D = \frac{148}{3} = \mathbf{49.33 \text{ kN}}\uparrow \quad \textbf{Ans.}$$

By $\Sigma F_V = 0$, we get

$$(V_D + V_E)\uparrow - V_C - 10 - 16 = 0$$

or, $$49.33 + V_E - 28 - 10 - 16 = 0$$

$\therefore$ $$V_E = 54 - 49.33 = \mathbf{4.67 \text{ kN}}\uparrow \quad \textbf{Ans.}$$

Problem 1.42. *A bar AB of negligible weight is subjected to a vertical force of 600 N and a horizontal force of 300 N applied as shown in Fig. P.1.42. Assuming smooth inclined surfaces, find angle θ at which equilibrium exists.* [Univ. of Pune, Dec. 1999]

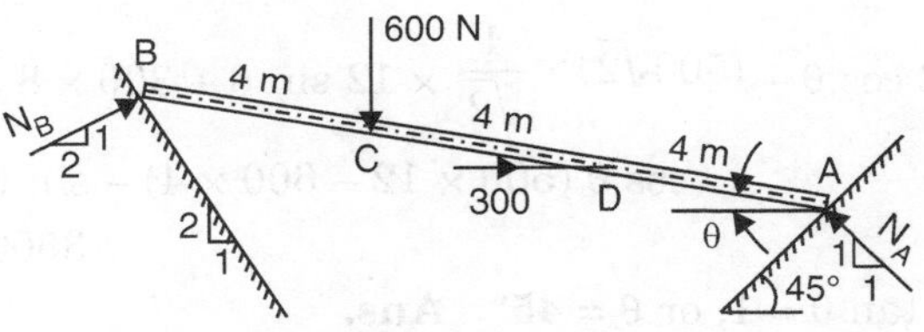

Fig. P.1.42

Sol. For the 45° inclined plane at A, let the normal reaction be N_A.

$\therefore$ x and y components of N_A are

$$N_{Ax} = N_A \times \frac{1}{\sqrt{2}} \text{ and } N_{Ay} = N_A \times \frac{1}{\sqrt{2}}$$

Similarly, since slope of inclined plane at B is as shown (1 unit horizontal : 2 units vertical),

$\therefore$ The slope of the normal reaction N_B is as shown in Figure (2 units horizontal : 1 unit vertical).

$\therefore$ $$N_{Bx} = \frac{N_B \times 2}{\sqrt{1^2 + 2^2}} = \frac{2}{\sqrt{5}} N_B \text{ and } N_{By} = N_B \times \frac{1}{\sqrt{5}}$$

By using the condition $\Sigma M_B = 0$, we get

$$\circlearrowleft + N_{Ay} \times AB \, . \cos\theta - N_{Ax} \, . \, AB \sin\theta + 300 \times BO \sin\theta - 600 \times BC \, . \cos\theta = 0$$

or, $$\frac{N_A}{\sqrt{2}} \times 12 \cos\theta - \frac{N_A}{\sqrt{2}} \times 12 \sin\theta + 300 \times 8 \times \sin\theta - 600 \times 4 \cos\theta = 0 \quad ...(1)$$

where θ = inclination of bar AB from horizontal.

By $\Sigma F_x = 0$, we get

$$N_{Bx \rightarrow} + 300 - N_{Ax}\frac{1}{\sqrt{2}} = 0 \quad \text{or,} \quad N_B \cdot \frac{2}{\sqrt{5}} + 300 - \frac{N_A}{\sqrt{2}} = 0 \qquad ...(2)$$

And, by $\Sigma F_y = 0$, we get

$$N_{By} + N_{Ay} - 600 = 0$$

or,
$$N_B \times \frac{1}{\sqrt{5}} + N_A \times \frac{1}{\sqrt{2}} - 600 = 0 \qquad ...(3)$$

Adding eqns. (2) and (3), we get

$$\frac{N_B}{\sqrt{5}}(2+1) + 300 - 600 = 0$$

$$\therefore \quad N_B = \frac{300\sqrt{5}}{3} = \mathbf{100\sqrt{5}\ N = 223.61\ N} \quad \textbf{Ans.}$$

From eqn. (3), $N_A = \left[600 - \frac{N_B}{\sqrt{5}}\right]\sqrt{2} = \left(600 - \frac{100\sqrt{5}}{\sqrt{5}}\right) \times \sqrt{2} = \mathbf{500\sqrt{2}\ N}$

$$= \mathbf{707.11\ N} \quad \textbf{Ans.}$$

And, from eqn. (1), we get

$$(500\sqrt{2}) \times \frac{1}{\sqrt{2}} \times 12\cos\theta - (500\sqrt{2}) \times \frac{1}{\sqrt{2}} \times 12\sin\theta + 300 \times 8\sin\theta - 600 \times 4\cos\theta = 0$$

or
$$\cos\theta\,(500 \times 12 - 600 \times 4) - \sin\theta\,(500 \times 12 - 300 \times 8) = 0$$

or
$$3600(\cos\theta) - 3600\sin\theta = 0$$

$$\therefore \quad \tan\theta = 1, \text{ or } \boldsymbol{\theta = 45°} \quad \textbf{Ans.}$$

Problem 1.43. *A hollow right cylinder of radius 0.9 m is open at both ends and rests on a smooth horizontal surface as shown in Fig. P.1.43 (a). Two spheres of weights 2 kN and 3 kN and radii 0.5 m and 0.7 m, respectively, are kept inside the cylinder. Find reactions at all contact surfaces. Also find the minimum weight W of the cylinder for which it will not tip over.*

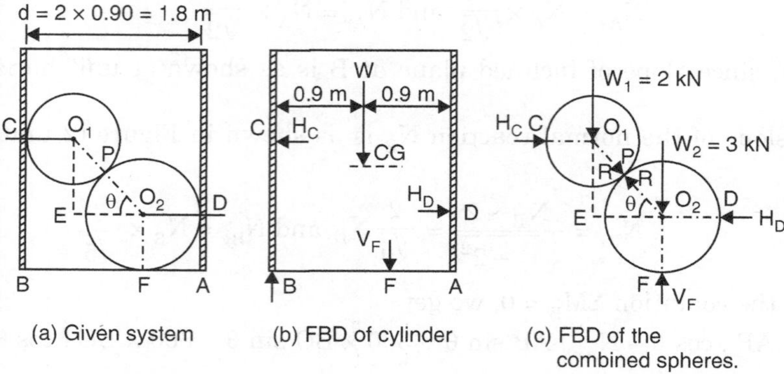

(a) Given system (b) FBD of cylinder (c) FBD of the combined spheres.

Fig. P.1.43

Sol. Fig. P.1.43 (*a*) shows the given system of cylinder and the two spheres. The points of contact between cylinder and the two spheres are C and D. Sphere 2 is in contact with the horizontal smooth surface at point F and the two spheres touch each other at point P. The reactions are represented in Figs. P.1.43 (*b*) and (*c*).

Drop a vertical line O_1E through O_1 which meets the horizontal line from O_2 at point E. In ΔO_1O_2E :

$$O_1O_2 = r_1 + r_2 = 0.5 + 0.7 = 1.2 \text{ m}$$

$$O_2E = \text{Diameter of cylinder} - O_1C - O_2D$$

$$= 2 \times 0.9 - 0.5 - 0.7 = 1.8 - 1.2 = 0.6$$

∴ Angle $EO_2O_1 = \theta$ can be obtained as follows :

$$\cos\theta = \frac{O_2E}{O_1O_2} = \frac{0.6}{1.2} = 0.5, \theta = 60°$$

∴ $$\sin\theta = \sin 60° = \frac{\sqrt{3}}{2}$$

From the equilibrium of the two spheres (combined) we can write

By $\Sigma M_{O_2} = 0$,

$$H_C \times O_1E - W_1 \times O_2E = 0$$

or, $$H_C \times O_1O_2 \,.\, \sin\theta - (2 \text{ kN}) \times O_1O_2 \cos\theta = 0$$

or $$H_C \times 1.2 \times \frac{\sqrt{3}}{2} - 2 \times 1.2 \times 0.5 = 0$$

∴ $$H_C = \frac{(2 \times 0.5) \times 2}{\sqrt{3}} = \frac{2}{\sqrt{3}} = \mathbf{1.1547 \text{ kN}} \quad \textbf{Ans.}$$

By $\Sigma F_h = 0$, we get

$$H_C - H_D = 0$$

∴ $$H_D = \mathbf{1.1547 \text{ kN}} \quad \textbf{Ans.}$$

By $\Sigma F_V = 0$, we get

$$V_F - W_1 - W_2 = 0$$

∴ $$V_F = W_1 + W_2 = 2 + 3 = \mathbf{5 \text{ kN}} \quad \textbf{Ans.}$$

From the equilibrium of sphere 1, we write

$$\frac{H_C}{\sin 30°} = \frac{R}{\sin 90°} = \frac{W_1}{\sin 60°}$$

∴ $$R = \frac{W_1 \sin 90°}{\sin 60°} = \frac{2 \times 1}{\sin 60°} = \mathbf{2.3094 \text{ kN}} \quad \textbf{Ans.}$$

Let us now consider the equilibrium of the cylinder shown in its FBD of Fig. (*b*). When the cylinder is about to tip over B, there is no reaction at A. There will be only reaction at B. Thus, by $\Sigma M_B = 0$, we get

$$H_C \times CB - W \times 0.9 - H_D \times DA = 0$$

or, $$\frac{2}{\sqrt{3}} \times (O_1E + FO_2) - W \times 0.9 - \frac{2}{\sqrt{3}} \times O_2F = 0$$

or, $$\frac{2}{\sqrt{3}} \times (O_1O_2 \sin\theta + r_2) - W \times 0.9 - \frac{2}{\sqrt{3}} \times r_2 = 0$$

or $$\frac{2}{\sqrt{3}}(1.2 \sin 60° + 0.7) - W \times 0.9 - \frac{2}{\sqrt{3}} \times 0.7 = 0$$

or, $$\frac{2}{\sqrt{3}} \times 1.2 \times \frac{\sqrt{3}}{2} + 0.7 \times \frac{2}{\sqrt{3}} - 0.9\,W - \frac{2}{\sqrt{3}} \times 0.7 = 0$$

$$1.2 - 0.9\,W = 0$$

or $$W = \frac{1.2}{0.9} = \mathbf{1.333\ kN}\quad \textbf{Ans.}$$

Problem 1.44. *A bar AB of weight W and length l is supported by a small roller at C and pressed against a smooth vertical wall at A, as shown in Fig. P.1.44. If an additional load W_1 is applied at the end B of the bar, find the position of equilibrium of the bar as given by the angle θ.*

Sol. Neglecting friction at A and C, the reactions R_a and R_c will be in the directions shown, R_a is horizontal and R_c is perpendicular to AB.

The point of intersection of their lines of action is O.

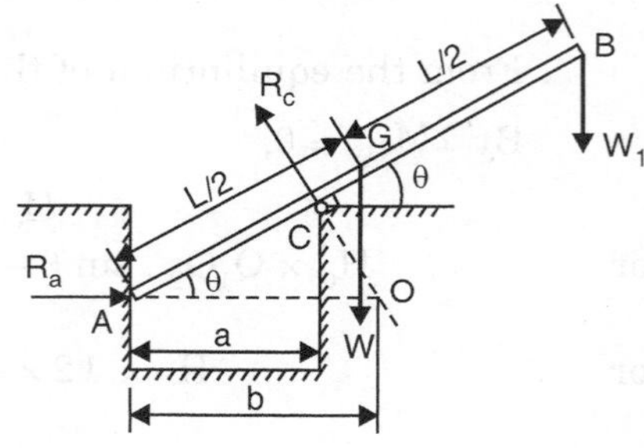

Fig. P.1.44

From Fig. P.1.44, distance $AC = \dfrac{a}{\cos\theta}$

and the distance $$b = AO = \frac{AC}{\cos\theta} = \frac{a}{\cos^2\theta}$$

The arm of moment (a_1) of the weight W of the bar about

$$\text{Point O} = a_1 = AO - AG\cos\theta = \left(\frac{a}{\cos^2\theta} - \frac{l}{2}\cos\theta\right)$$

Similarly, the distance between the line of weight

$$W_1 \text{ and point O} = a_2 = AB\cos\theta - b = l\cos\theta - \frac{a}{\cos^2\theta}$$

Hence by taking moments about point O, we get

$$W \times a_1 - W_1 \times a_2 = 0$$

or $$W\left(\frac{a}{\cos^2\theta} - \frac{l}{2}\cos\theta\right) - W_1\left(l\cos\theta - \frac{a}{\cos^2\theta}\right) = 0$$

or $$\frac{W}{\cos^2\theta}\left(a - \frac{l}{2}\cos^3\theta\right) = \frac{W_1}{\cos^2\theta}\left(l\cos^3\theta - a\right)$$

or $$W\left(a - \frac{l}{2}\cos^3\theta\right) = W_1(l\cos^3\theta - a)$$

or $$\cos^3\theta\left(l\,W_1+\frac{l}{2}W\right)=W\,.\,a+W_1\,.\,a \quad \text{or,} \quad \frac{l\cos^3\theta}{2}(2W_1+W)=a(W+W_1)$$

or, $$\cos^3\theta=\frac{2a\,(W+W_1)}{l\,(2W_1+W)}$$

$\therefore$ $$\cos\theta=\sqrt[3]{\frac{2a\,(W+W_1)}{l\,(2W_1+W)}} \quad \textbf{Ans.} \qquad ...(i)$$

If the additional weight W_1 is not applied, then $W_1 = 0$, hence equation (*i*) reduces to

$$\cos\theta=\sqrt[3]{\frac{2a}{l}} \quad \textbf{Ans.} \qquad ...(ii)$$

It is assumed, that $a < l/2$.

Problem 1.45. *Three spheres A, B and C having weights 20 N, 40 N and 15 N and radii 50 mm, 80 mm and 30 mm respectively are piled in a cylindrical channel of 200 mm diameter, as shown in Fig. P.1.45 (a). Find reactions to the walls and base of the channel.*

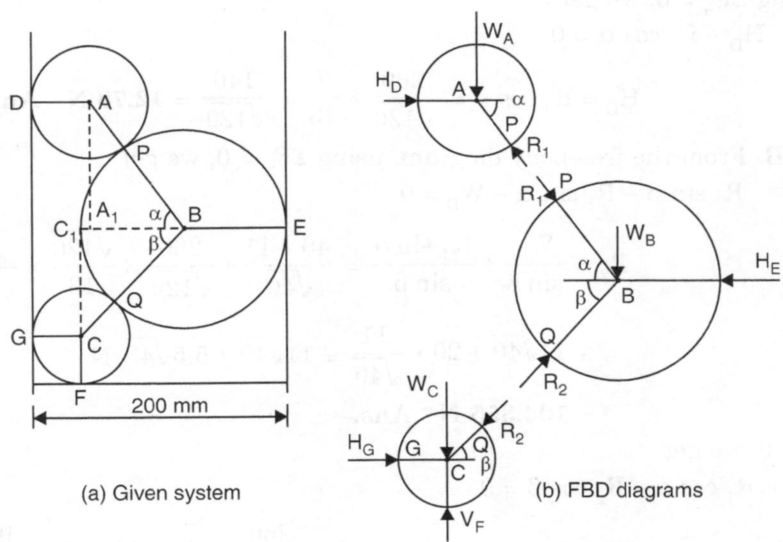

Fig. P.1.45

Sol. In Fig. P.1.45(*a*) :

$$W_A = 20\text{ N},\ W_B = 40\text{ N},\ W_C = 15\text{ N},$$
$$r_A = DA = AP = 50\text{ mm},\ r_B = BE = BP = BQ = 80\text{ mm},$$
$$r_C = CQ = CG = CF = 30\text{ mm}.$$

Draw vertical line AA_1 intersecting EB at A_1. Also draw vertical CC_1 intersecting EB at C_1.

Then, $BA_1 = 200 - DA - EB = 200 - 50 - 80 = 70$ mm,

and $BC_1 = 200 - EB - CG = 200 - 80 - 30 = 90$ mm.

$AB = AP + BP = r_A + r_B = 50 + 80 = 130$ mm, and

$BC = BQ + CQ = 80 + 30 = 110$ mm.

Let $\angle ABA_1 = \alpha$ and $\angle CBC_1 = \beta$.

Then, $\cos\alpha = \dfrac{BA_1}{BA} = \dfrac{70}{130} = \dfrac{7}{13}$.

$\therefore$ $\sin\alpha = \dfrac{\sqrt{13^2 - 7^2}}{13} = \dfrac{\sqrt{120}}{13}$

And, $\cos\beta = \dfrac{BC_1}{BC} = \dfrac{90}{110} = \dfrac{9}{11}$

$\therefore$ $\sin\beta = \dfrac{\sqrt{11^2 - 9^2}}{11} = \dfrac{\sqrt{40}}{11}$.

Sphere A. From the free-body diagram, using $\Sigma F_v = 0$, we get

$$R_1 \sin\alpha - W_A = 0$$

$\therefore$ $R_1 = \dfrac{W_A}{\sin\alpha} = \dfrac{20}{\sqrt{120}} \times 13 = \dfrac{260}{\sqrt{120}} = \mathbf{23.7345\ N}$ **Ans.**

And, using $\Sigma F_h = 0$, we get

$$H_D - R_1 \cos\alpha = 0$$

$\therefore$ $H_D = R_1 \cos\alpha = \dfrac{260}{\sqrt{120}} \times \dfrac{7}{13} = \dfrac{140}{\sqrt{120}} = \mathbf{12.78\ N}$ **Ans.**

Sphere B. From the free-body diagram, using $\Sigma F_v = 0$, we get

$$R_2 \sin\beta - R_1 \sin\alpha - W_B = 0$$

$\therefore$ $R_2 = \dfrac{W_B}{\sin\beta} + \dfrac{R_1 \sin\alpha}{\sin\beta} = \dfrac{40 \times 11}{\sqrt{40}} + \dfrac{260}{\sqrt{120}} \times \dfrac{\sqrt{120}}{13} \times \dfrac{11}{\sqrt{40}}$ N

$= 11\sqrt{40} + 20 \times \dfrac{11}{\sqrt{40}} = 11\sqrt{40} + 5.5\sqrt{40}$ N

$= \mathbf{104.355\ N}$ **Ans.**

By $\Sigma F_h = 0$, we get

$$H_E - R_1 \cos\alpha - R_2 \cos\beta = 0$$

$\therefore$ $H_E = R_1 \cos\alpha + R_2 \cos\beta = \dfrac{260}{\sqrt{120}} \times \dfrac{7}{13} + 104.355 \times \dfrac{9}{11}$ N

$= \mathbf{98.161\ N}$ **Ans.**

Sphere C. From free-body diagram, using $\Sigma F_h = 0$, we get

$$H_G - R_2 \cos\beta = 0$$

$\therefore$ $H_G = R_2 \cos\beta = 104.355 \times \dfrac{9}{11} = \mathbf{80.314\ N}$ **Ans.**

By $\Sigma F_v = 0$, we get

$$V_F = W_C - R_2 \sin\beta = 0$$

$\therefore$ $V_F = W_C - R_2 \sin\beta$

$= 15 + 104.355 \times \dfrac{\sqrt{40}}{11}$ N $= \mathbf{75\ N}$ **Ans.**

Thus, reaction at the walls of the cylindrical channel are :

$$H_D = 12.78 \text{ N} ; H_E = 98.161 \text{ N} ; H_G = \mathbf{80.315 \text{ N}} \quad \textbf{Ans.}$$

Problem 1.46. *A uniform rod AB of negligible weight is hinged at end A and is held by an inextensible string at B, as shown in Fig. P.1.46. Find the angle θ for the equilibrium of the rod. The string passes over a smooth pulley C and carries a weight Q.Take P = 2Q, and AC = AB = l.*

Sol. Let $\angle A = \theta$,

$\therefore \quad \angle B = \angle C = \alpha = (90° - \theta/2)$

For the equilibrium of rod, by

$\Sigma M_A = 0$, we get

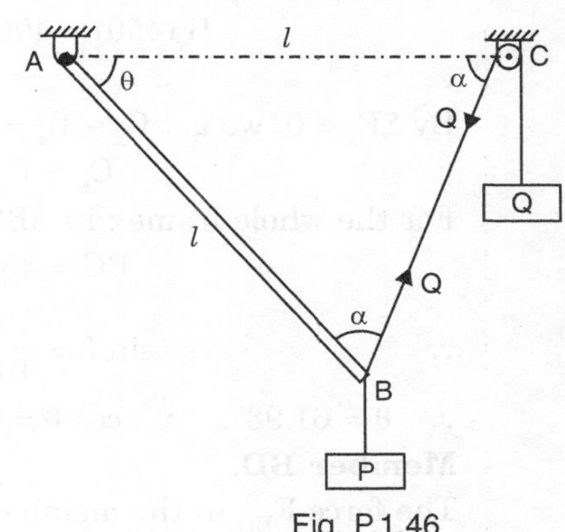

Fig. P.1.46

$$Q \times l \sin \alpha - Pl \cos \theta = 0$$

or, $$Q \times l \sin (90° - \theta/2) - Pl \cos \theta = 0$$

or $$Q\, l \cos \theta/2 - (2Q) \times l \cos \theta = 0$$

$\therefore$ $$2 \cos \theta - \cos \theta/2 = 0$$

or, $$2(2 \cos^2 \theta/2 - 1) - \cos \theta/2 = 0$$

or, $$4 \cos^2 \theta/2 - \cos \theta/2 - 2 = 0$$

or, $$\cos^2 \theta/2 - \frac{1}{4} \cos \theta/2 - \frac{1}{2} = 0$$

$\therefore$ $$\cos \theta/2 = \frac{1}{2}\left[\frac{1}{4} \pm \sqrt{(1/4)^2 + 4 \times 1 \times \frac{1}{2}}\right] = \frac{1}{2}\left[\frac{1}{4} \pm 1.4361\right]$$

$$= 0.84305; -0.59305$$

$\therefore$ $$\theta/2 = 32.5364°, -126.3737°$$

$\therefore$ $$\theta = \mathbf{65.0728° ; 252.7475°}$$

Taking $\theta < 90°$, we get

$$\theta = \mathbf{65.0728°} \quad \textbf{Ans.}$$

Problem 1.47. *Determine the force in the member BD of the frame shown in Fig. P.1.47 (a). Also find the components of reaction at C.* [Pune Univ. Dec. 1998]

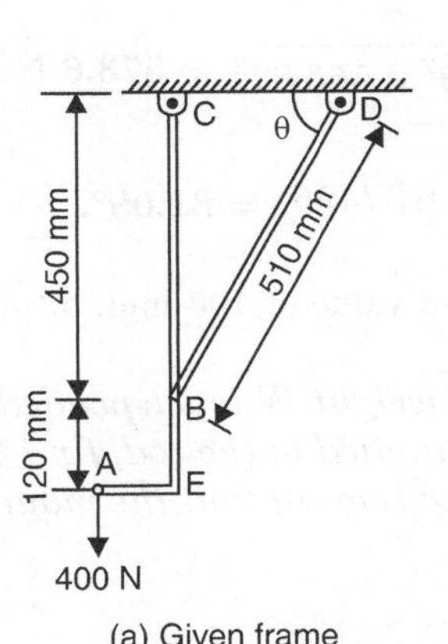

(a) Given frame

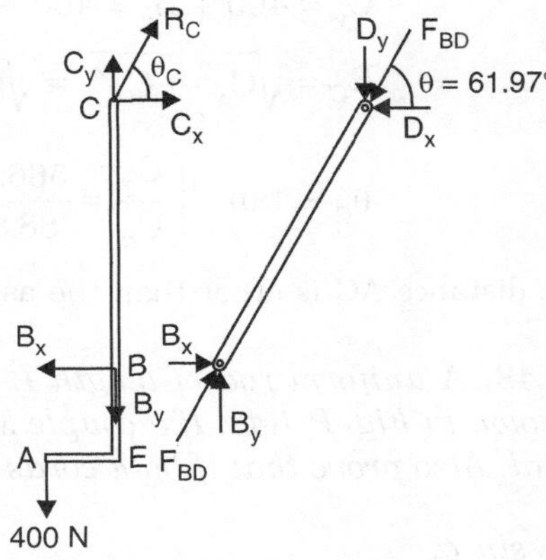

(b) Free-body diagrams

Fig. P.1.47

Sol. The dimension AE is not given in the problem. Let us assume AE = 100 mm. Fig. P.1.47(*b*) shows the free-body diagrams of members ABC and BD. Let the reaction components at supports C and D be C_x, C_y and D_x, D_y, respectively. The reaction components at B are B_x and B_y.

Member ABC.

By $\Sigma M_C = 0$, we get

$$B_x(450) - 400(100) = 0$$

$$\therefore \quad B_x = \textbf{88.89 N} \quad \textbf{Ans.}$$

By $\Sigma F_x = 0$, we get $C_x - B_x = 0$

$$\therefore \quad C_x = B_x = \textbf{88.89 N} \quad \textbf{Ans.}$$

For the whole frame : in ΔBCD,

$$BC = 450 \text{ mm}, BD = 510 \text{ mm},$$

$$\therefore \quad \sin\theta = \frac{BC}{BD} = \frac{450}{510} = 0.882353$$

$\therefore \quad \theta = 61.93°, \quad \therefore \quad \cos\theta = 0.47059$ and $\tan\theta = 1.875.$

Member BD.

The force F_{BD} in the member can be found as follows :

$$\because \quad F_{BD} \cdot \cos\theta = B_x, \quad \therefore \quad F_{BD} = \frac{B_x}{\cos\theta} = \frac{88.89}{\cos 61.93°} = \textbf{188.89 N} \quad \textbf{Ans.}$$

And, $\therefore$ $\quad B_y = F_{BD} \sin\theta = 188.89 \sin 61.93° = \textbf{166.67 N} \quad \textbf{Ans.}$

By $\Sigma F_y = 0$, $D_y - B_y = 0$, $\therefore$ $D_y = \textbf{166.67 N}$ **Ans.**

By $\Sigma F_x = 0$, we get

$$D_x - B_x = 0$$

$$\therefore \quad D_x = B_x = \textbf{88.89 N} \quad \textbf{Ans.}$$

In member ABC.

$$B_y = \textbf{166.67 N.} \quad \textbf{Ans.}$$

By $\Sigma F_y = 0$, $C_y - 400 - B_y = 0$

$$\therefore \quad C_y = 400 + B_y = 400 + 166.67 = \textbf{566.67 N} \quad \textbf{Ans.}$$

$$\therefore \quad R_C = \sqrt{{C_x}^2 + {C_y}^2} = \sqrt{88.89^2 + 566.67^2} = \textbf{573.6 N}$$

$$\text{And,} \quad \theta_C = \tan^{-1}\left(\frac{C_y}{C_x} = \frac{566.67}{88.89} = 6.37495\right) = \textbf{81.09°.}$$

[**Note.** If the distance AC is other than the assumed value of 100 mm, all above answers will change].

Problem 1.48. *A uniform rod of length L and weight W is suspended by two cords of equal length, as shown in Fig. P.1.48. If a couple M is applied to the rod, find the angle θ of the rod to the horizontal. Also prove that if both cords are to remain taut the magnitude of M must be less than $\frac{1}{2}$ WL sin α.*

Sol. Refer Fig. P.1.48. The rod AB is in equilibrium under the action of forces : Weight (W), tension T_1, T_2 and the couple M.

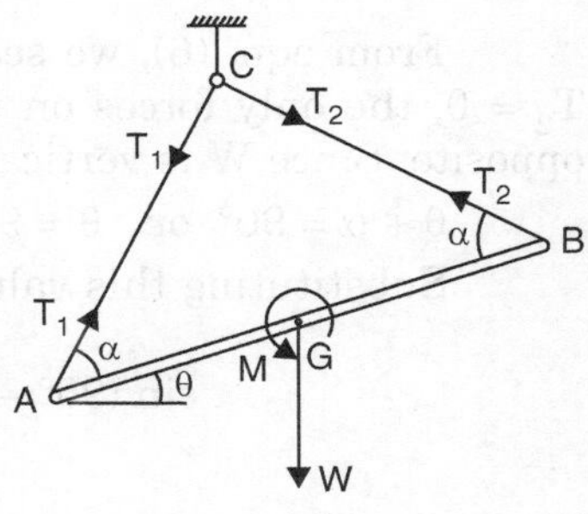

Fig. P. 1.48

Resolving forces horizontally, and by using the condition $\Sigma F_h = 0$, we get

$$T_1 \cos(\alpha + \theta) - T_2 \cos(\alpha - \theta) = 0 \quad ...(1)$$

By $\Sigma F_v = 0$, we get

$$T_1 \sin(\alpha + \theta) + T_2 \sin(\alpha - \theta) - W = 0 \quad ...(2)$$

From (1), $T_2 = T_1 \dfrac{\cos(\alpha+\theta)}{\cos(\alpha-\theta)}$ ...(3)

∴ Eqn. (2) gives

$$T_1 \sin(\alpha + \theta) + T_1 \frac{\cos(\alpha+\theta)}{\cos(\alpha-\theta)} \times \sin(\alpha - \theta) = W$$

$$\therefore \quad T_1[\sin(\alpha + \theta) \,.\, \cos(\alpha - \theta) + \cos(\alpha + \theta) \,.\, \sin(\alpha - \theta)] = W \cos(\alpha - \theta)$$

or, $$T_1[\sin(\alpha + \theta + \alpha - \theta)] = W \cos(\alpha - \theta)$$

or, $$T_1 \times \sin 2\alpha = W \cos(\alpha - \theta)$$

or, $$T_1 = \frac{W \,.\, \cos(\alpha-\theta)}{\sin 2\alpha} \quad ...(4)$$

And, from eqn. (3), we get

$$T_2 = \frac{W \,.\, \cos(\alpha-\theta)}{\sin 2\alpha} \times \frac{\cos(\alpha+\theta)}{\cos(\alpha-\theta)}$$

or, $$T_2 = \frac{W \,.\, \cos(\alpha+\theta)}{\sin 2\alpha} \quad ...(5)$$

By taking moments about G, we get

$$T_1 \,.\, \frac{L}{2} \sin\alpha - T_2 \,.\, \frac{L}{2} \sin\alpha - M = 0$$

or, $$(T_1 - T_2)\frac{L}{2} \sin\alpha = M \quad \text{or,} \quad T_1 - T_2 = \frac{2M}{L \sin\alpha} \quad ...(6)$$

Substituting the values of T_1 and T_2 eqn. (6) becomes,

$$\frac{W}{\sin 2\alpha}[\cos(\alpha-\theta) - \cos(\alpha+\theta)] = \frac{2M}{L\sin\alpha}$$

or, $$\frac{W}{\sin 2\alpha} \times (2 \sin\alpha \sin\theta) = \frac{2M}{L\sin\alpha}$$

$$\therefore \quad \sin\theta = \frac{2M}{L\sin\alpha} \times \frac{\sin 2\alpha}{W} \times \frac{1}{2\sin\alpha}$$

$$= \frac{2M}{LW} \times \frac{2\sin\alpha\cos\alpha}{2\sin\alpha\sin\alpha} = \frac{2M}{LW} \times \frac{\cos\alpha}{\sin\alpha}$$

or, $$\sin\theta = \frac{2M\cot\alpha}{WL} \quad ...(7)$$

From eqn. (6), we see that as M increases, $(T_1 - T_2)$ increases or T_2 decreases. When $T_2 = 0$, the only forces on the rod are T_1 and W. For equilibrium, their directions must be opposite. Since W is vertical, T_1 must also be vertical. Hence

$$\theta + \alpha = 90°, \text{ or } \quad \theta = 90° - \alpha \qquad ...(8)$$

Substituting this value of θ into eqn. (7), we find

$$\sin(90° - \alpha) = \frac{2M}{WL} \cot \alpha$$

or,
$$\cos \alpha = \frac{2M}{WL} \times \left(\frac{\cos \alpha}{\sin \alpha}\right)$$

∴
$$\sin \alpha = \frac{2M}{WL} \qquad ...(9)$$

or,
$$M = \frac{1}{2} WL \sin \alpha \qquad ...(10)$$

If $M < \frac{1}{2} WL \sin \alpha$, both strings will remain taut.

Problem 1.49. *Two rollers of weights P and Q are connected by a flexible string AB. The rollers rest on two smooth planes DE and EF at right angles, as shown in Fig. P.1.49 (a). Find the tension in the string and the angle θ for the system to be in equilibrium.*

Find tension and θ when P = 100 N, Q = 200 N and α = 30°.

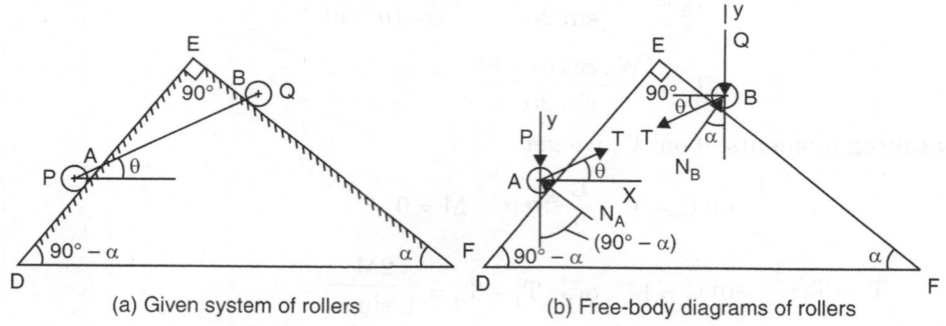

Fig. P.1.49

Sol. Free-body diagrams of the two rollers are shown in Fig. P.1.49 (*b*). Let us assume *x* and *y* axes to be horizontal and vertical. Let T be the tension in the string. Also let the reactions to the rollers be N_A and N_B each acting normal to the respective plane, as shown.

Roller P.

By $\Sigma F_x = 0$, $T \cos \theta - N_A \cdot \sin(90° - \alpha) = 0$ or $T \cos \theta = N_A \cdot \cos \alpha$ ...(1)

By $\Sigma F_y = 0$, $T \sin \theta - P + N_A \cdot \cos(90° - \alpha) = 0$ or $T \sin \theta = P - N_A \cdot \sin \alpha$ ...(2)

Roller Q.

By $\Sigma F_x = 0$, $T \cos \theta - N_B \cdot \sin \alpha = 0$ or $T \cos \theta = N_B \cdot \sin \alpha$ ...(3)

By $\Sigma F_y = 0$, $-T \sin \theta + N_B \cdot \cos \alpha - Q = 0$ or, $T \sin \theta = N_B \cdot \cos \alpha - Q$ ...(4)

From eqn. (1) $N_A = \dfrac{T \cos \theta}{\cos \alpha}$ ...(5)

$\therefore$ From eqn. (2),

$$P = T \sin \theta + N_A \,.\, \sin \alpha$$

$$= T \sin \theta + \frac{T \cos \theta}{\cos \alpha} \,.\, \sin \alpha$$

or $$P = T(\sin \theta + \cos \theta \,.\, \tan \alpha) \quad ...(6)$$

From eqn. (3), $N_B = \dfrac{T \cos \theta}{\sin \alpha}$ ...(7)

$\therefore$ From eqn. (4), $Q = N_B \,.\, \cos \alpha - T \,.\, \sin \theta = \dfrac{T \cos \theta}{\sin \alpha} \,.\, \cos \alpha - T \sin \theta$

or $$Q = T(\cos \theta \,.\, \cot \alpha - \sin \theta) \quad ...(8)$$

By dividing eqn. (8) by eqn. (6), we get

$$\frac{Q}{P} = \frac{T\,(\cos \theta \,.\, \cot \alpha - \sin \theta)}{T\,(\sin \theta + \cos \theta \,.\, \tan \alpha)}$$

$\therefore$ $Q \sin \theta + Q \cos \theta \,.\, \tan \alpha = P \cos \theta \cot \alpha - P \sin \theta$

or, $\sin \theta\,(Q + P) = \cos \theta\,(P \cot \alpha - Q \tan \alpha)$

$\therefore$ $$\tan \theta = \frac{P \cot \alpha - Q \,.\, \tan \alpha}{P + Q} \quad ...(9)$$

Substituting $P = 100$ N ; $Q = 200$ N and $\alpha = 30°$,

We get $$\tan \theta = \frac{100 \cot 30° - 200 \tan 30°}{100 + 200} = \frac{57.735}{300} = 0.19245$$

$\therefore$ $\theta = 10.893°$ **Ans.**

And from eqn. (6)

$$T = \frac{P}{\sin \theta + \cos \theta \tan \alpha} = \frac{100}{\sin 10.893° + \cos 10.893° \,.\, \tan 30°} \text{ N}$$

or $T = $ **132.29 N Ans.**

PART D—GRAPHICAL METHODS

1.33. INTRODUCTION

The **graphical methods** at one time were very popular and were very widely used. But, they are not preferred now a days. In most cases, the care, labour and the time needed to obtain results graphically are much greater than are required for the same results by the use of analytical methods. Moreover, the graphical methods are less accurate than the analytical methods.

In spite of these defects and objections the graphical methods are still used for the following reasons :

(1) Graphical method can be used to check the analytical calculations.

(2) These checks are fast and sufficiently accurate.

(3) These checks may even be done by using the free-hand diagrams drawn to an approximate scale.

(4) The analytical calculations are generally long, complex and cumbersome.

(5) The degree of accuracy of graphical methods generally lies within the permissible limits for engineering purposes.

1.34. GRAPHICAL METHOD

The graphical method consists of the following simple steps :

(1) Drawing of **space diagram,**

(2) Use of **Bow's notation,**

(3) Drawing of **force polygon**, or **vector diagram**,

(4) Drawing of **string polygon,**

(5) Determination of reactions, for beams and trusses,

(6) Drawing of **Maxwell diagram** for trusses.

The above steps are discussed below.

1.35. SPACE DIAGRAM

The given system of forces is drawn to some suitable scale. Space of the part or structure or beam or truss is drawn accurately by using drawing instruments. Fig. 1.35 shows some examples of the space diagrams.

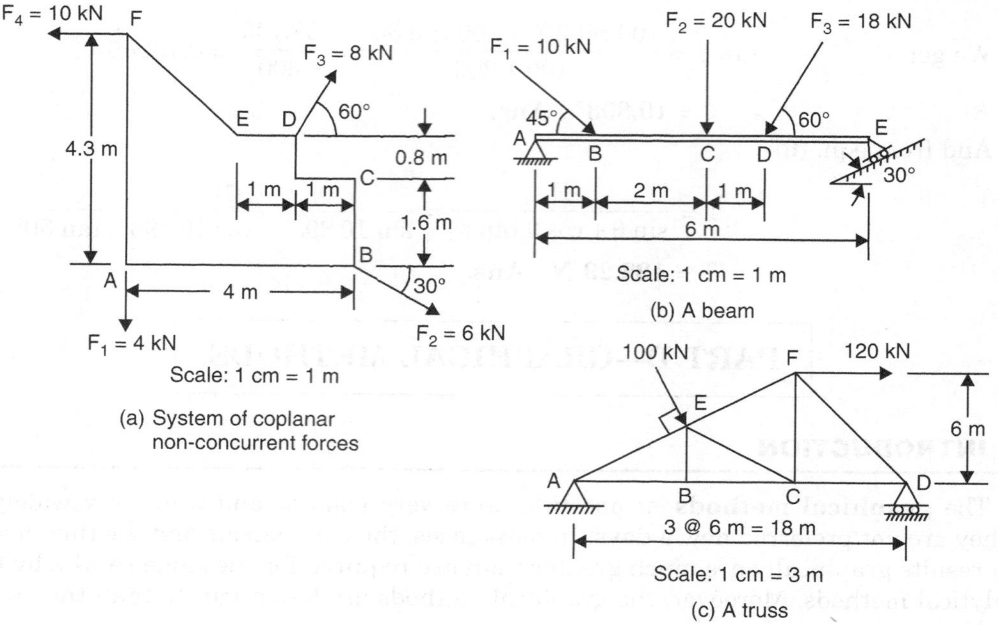

Fig. 1.35. Space diagrams

All cases shown in Fig. 1.35, represent coplanar non-concurrent force systems. In Fig. 1.35(*a*) a scale of 1 cm = 1 m is used to draw the machine part ABCDEF. All forces are drawn in proper directions by using the protractor. **Freehand and unproportionate diagrams are not used.** We can replace this given system of force by a single force known as resultant with the help of graphical method.

Fig. 1.35(*b*) is an example of a beam. The beam is drawn to a scale of 1 cm = 1 m and all forces are drawn in the given directions with the help of protractor. We can find the resultant R of the given force system and the support reactions both with the help of graphical method.

Fig. 1.35(*c*) is an example of truss. The truss is drawn to the size by using a scale 1 cm = 3 m. And, all forces are accurately drawn in proper direction. Here the graphical method can be used (*a*) to find reactions of supports, and (*b*) to find forces (or stresses) in various members of the truss. This is done in the Chapter 2.

1.36. BOW'S NOTATION

Different loads and reactions are named by two capital letters, placed on either side of the load in the space diagram. This procedure is known as **Bow's notation**. Fig. 1.36 illustrates a few cases of Bow's notation.

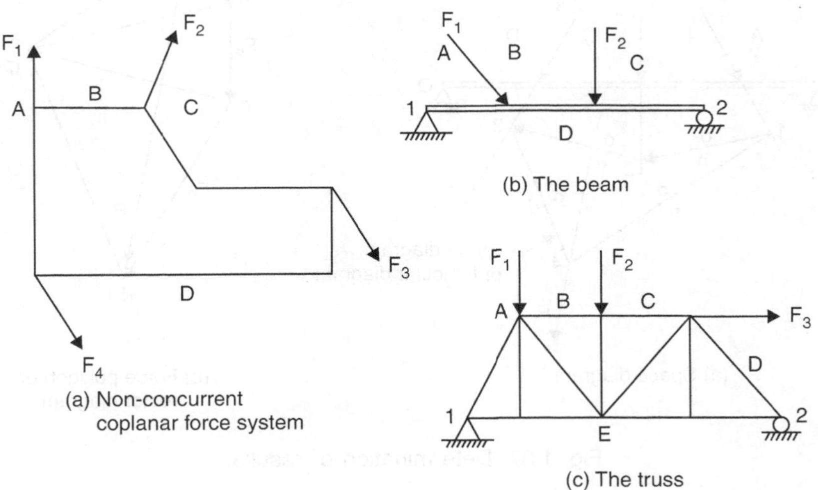

Fig. 1.36. Bow's notation

In each case of Fig. 1.36(*a*), (*b*) and (*c*) the space on left of force F_1 is named as A while that on its right the space is named as B. Similarly, space on the right of force F_2 is C and the space on the other side of F_3 is D.

Force F_1 is named as AB, and force F_2 as BC. In Fig. 1.36(*b*), the reaction R_2 at roller 2 will be named as CD and the reaction R_1 at the hinged support 1 will be given by DA. Method of finding R_1 and R_2 will be discussed in section 1.38. And the resultant of the given system of forces will be discussed in section 1.37. In Fig. 1.36 (*c*) the reaction R_2 on roller 2 will be given by DE while the reaction of hinged support 1 will be given by EA.

1.37. TO DETERMINE RESULTANT OF THE GIVEN FORCE SYSTEM

Step 1. Draw the given force system to a suitable scale. This diagram is called the **space diagram**. See Fig. 1.37(*a*).

Step 2. Using Bow's notation, name space on the left of force F_1 as A and that on the right of F_1 as B. The space on the right of F_2 is C. Similarly, the space on the right of force F_3 is D. Thus, the force F_1 is given by AB, F_2 is given by BC and the force F_3 is given by CD.

Step 3. Using a suitable scale, draw vector *ab* to represent force F_1. Also, vector *bc* represents force F_2 and vector *cd* represents force F_3. These vectors are parallel to the respective forces. Lastly, draw vector *ad* to represent the resultant R of the given forces F_1, F_2 and F_3, as shown in Fig. 1.37(*b*). The polygon *abcd* is known as **force polygon** or **vector diagram**.

Step 4. Take a suitable point *o*, known as the **pole** and draw **rays** *oa*, *ob*, *oc* and *od*.

Step 5. Take a point 1 on the line of action of force F_1 and draw a **string** parallel to ray *ob* from it. This string intersects the line of force F_2 at point 2. From 2 draw a string parallel to ray *oc*, which intersects the line of action of force F_3 at point 3.

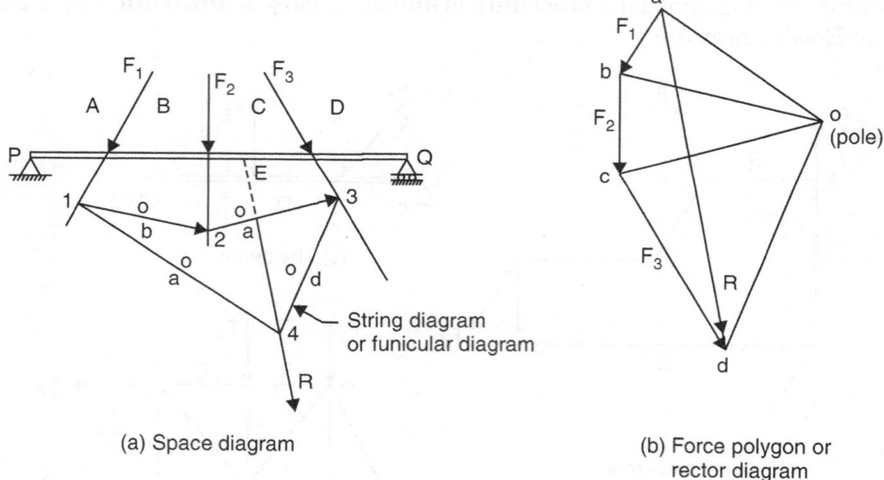

(a) Space diagram

(b) Force polygon or rector diagram

Fig. 1.37. Determination of resultant

Step 6. Through point 1 draw a string parallel to ray *oa* and through point 3 draw a string parallel to ray *od*. These lines intersect at point 4. The polygon 12341 is called string polygon or funicular polygon.

Step 7. Through point 4 draw a line parallel to the resultant R.

Magnitude and direction of the resultant R is given the vector *ad* and its position is given by point E where the vector R drawn through point 4 intersects the given beam.

1.38. TO DETERMINE REACTIONS OF THE BEAM

Step 1. Draw space diagram with suitable scale, as shown in Fig. 1.38(*a*). It consists of beam with hinged support P and a roller support Q. Three forces are acting on the beam.

Step 2. Name spaces A, B, C and D to name forces F_1, F_2 and F_3. The reaction R_q at roller Q will be named as DE and reaction R_p at the hinge P will be EA.

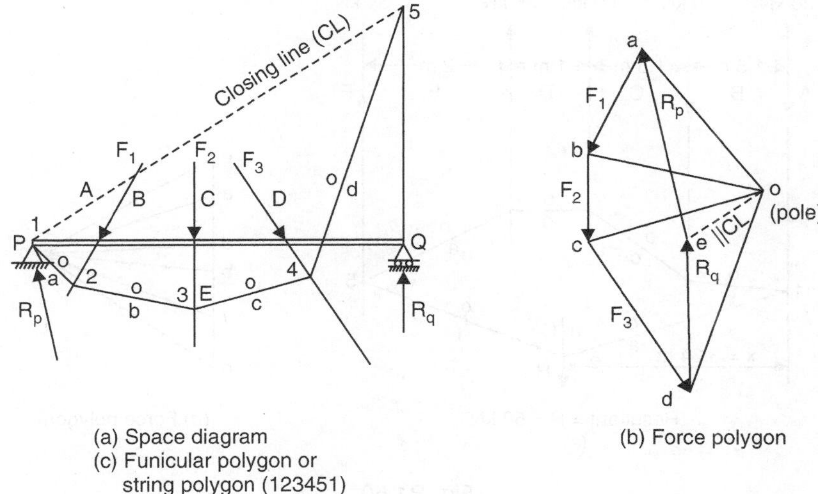

(a) Space diagram
(c) Funicular polygon or string polygon (123451)

(b) Force polygon

Fig. 1.38. Determination of support reactions

Step 3. Draw force polygon as shown in Fig. 1.38(*b*). Vector *ab* represents force F_1, vector *bc* represents force F_2 and vector *cd* represents force F_3. A suitable scale is assumed to show these forces.

Step 4. Choose a suitable point *o* as pole and draw rays *oa*, *ob*, *oc* and *od*.

Step 5. Take point 1 on the hinge and draw line 12 parallel to ray *oa*, intersecting the line of force F_1 at point 2.

Step 6. From point 2 draw a line 23 parallel to the ray *ob* which intersects line of action of force F_2 at point 3.

Step 7. From point 3 draw a line 34 parallel to the ray *oc* which intersects the line of action of force F_3 at point 4.

Step 8. From point 4 draw a line 45 parallel to the ray *od* which intersects the line of reaction R_q at point 5. Here, since roller at P moves on the horizontal surface, therefore the line of action of reaction R_p is vertical, as shown in Fig. 1.38(*a*).

Step 9. Join points 5 and 1. The line 51 is called the **closing line** (CL). From the pole *o*, draw the dotted ray parallel to the closing line. From point *d* draw a line parallel to the reaction R_q. The two lines intersect at point *e*. Joint *e* with *a*.

Step 10. Reaction R_q at Q is given by vector *de* and the reaction R_p at P is given by vector *ea*.

Note. Polygon 123451 is known as **funicular polygon** or **string polygon.**

Problem 1.50. *Determine graphically the resultant of the force system shown in Fig. P.1.50(a).*

Sol. Using Bow's notation, name various spaces as shown in Fig. P.1.50(*a*). The force polygon in Fig. P.1.50(*b*) is obtained by drawing to scale a tip-to-tail addition of the force vectors comprising of the given force system. Since given forces are parallel and vertical, the force polygon is a straight line. The sense of a force is indicated by reading the letters in alphabetical order. Thus, the 30 kN downward force whose line AB is read alphabetically downward from *a* to *b* in the force polygon. The 35 kN upward force DE is read alphabetically upward from *d* to *e* in the force polygon. The resultant R is represented by the vector *af* drawn from the tail *a* of the first vector to the tip *f* of the last vector. From the force polygon *af* scales 50 kN, and its direction is downward from *a* to *f*.

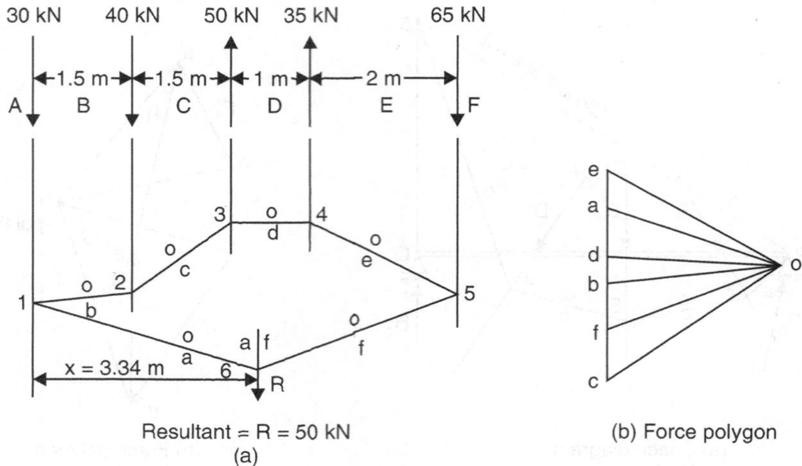

Fig. P.1.50

Choose a suitable pole *o* and draw rays *oa*, *ob*, *co*, etc., as shown in Fig. P.1.50 (*b*). Take point 1 on force AB of 30 kN. From point 1 draw a string 12 parallel to the ray *ob* which intersects force BC at point 2. From point 2 draw a string 23 parallel to the ray *oc* which intersects the force CD at point 3. From point 3 draw a string 34 parallel to the ray *od* intersecting the force DE at point 4. From point 4 draw a string 45 parallel to the ray *oe* intersecting the force EF at point 5.

From point 1 draw a string parallel to ray *oa*, and from point 5 draw a string parallel to ray *of*. The two strings intersect at point 6. Point 6 is point on the line of action of resultant R. In Fig. P.1.50 (*b*), resultant R is shown passing through point 6. From Fig. P.1.50(*a*) the distance *x* from the line of action of AB to the line of action of R is found, by scaling, to be 3.34 m. **Ans.**

Problem 1.51. *Determine graphically the resultant of the coplanar non-concurrent force system shown in Fig. P.1.51 (a).*

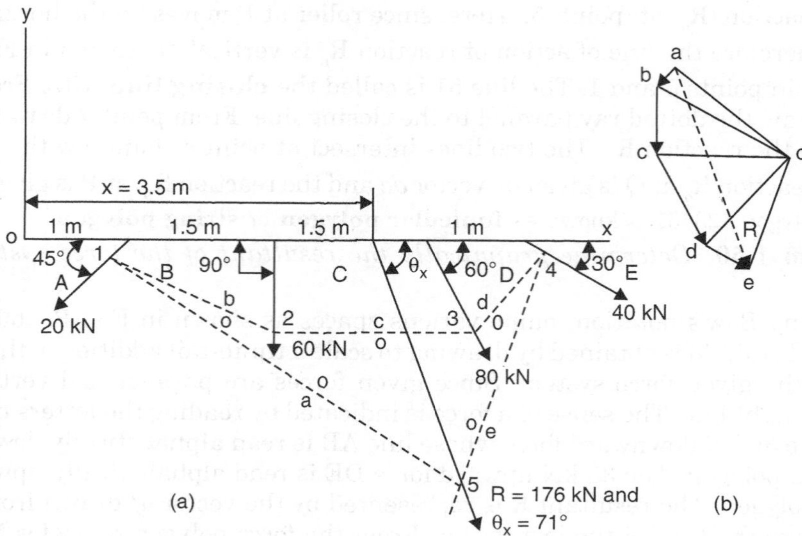

Fig. P. 1.51

Sol. Draw the space diagram to scale as shown in Fig. P.1.51 (*a*) and name forces like AB, BC, etc. Draw force polygon to some suitable scale as shown in Fig. P.1.51 (*a*). The resultant R is vector *ab* which scales 176 kN.

Choose a suitable pole *o* and draw rays *ao*, *bo*, etc. Starting from any point 1 on the line of action of force AB in Fig. P.1.51 (*a*), draw strings *ao* and *bo*. String *bo* intersects the line of force BC at point 2. Through point 2 draw string *co* parallel to the ray *co*, intersecting the line of force CD at point 3. Similarly, the string *do* through point 3 intersects the line of force DE at point 4.

Through point 4, draw string *oe* parallel to ray *oe* intersecting the string *oa* at point 5. The resultant R. passes through point 5. Draw resultant R parallel to *ae* in the force polygon, through point 5. The line of action of R meets the *ox* line at a distance 3.5 m as shown in Fig. P.1.51 (*a*). The resultant R is inclined at $\theta_x = 71°$, as shown.

Problem 1.52. *Determine graphically the reactions on the beam shown in Fig. P.1.52 (a).*

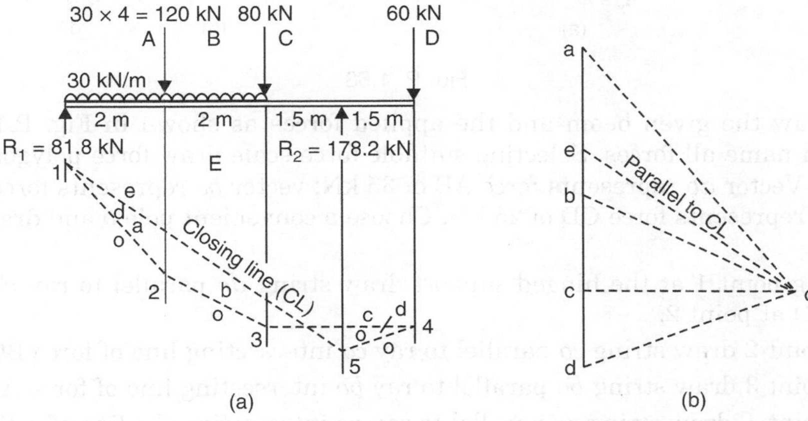

Fig. P.1.52

Sol. We replace the uniformly distributed load by its resultant of 30 × 4 = 120 kN having its line of action AB passing through the mid point of the uniformly distributed load. We draw the force polygon as shown in Fig. P.1.52 (*b*). Choosing a suitable pole *o* we draw rays *oa*, *ob*, etc. The reaction R_2 is DE and R_1 is EA. We have to locate point *e* in the force polygon.

Selecting point 1 on the line of action of the reaction R_1, draw string *ao* parallel to ray *oa*, intersecting line of action of force AB at point 2. Through point 2 draw string *bo* parallel to ray *ob* intersecting line of action of force BC at point 3. Through point 3 draw string *co* parallel to ray *oc* intersecting the line of action of force CD at point 4. Through point 4 draw string *do* parallel to ray *od* intersecting the line of action of reaction R_2 at point 5. Join point 1 and point 5. The line 15 is called the closing line of the string polygon or the funicular polygon.

Through the pole *o* draw a line parallel to the closing line intersecting the force line of R_2 drawn through point *d* at point *e*.

Then, reaction R_2 is given by *de*, acting vertically upward. And, the reaction R_1 is equal to *ea*, acting vertically upward.

By scaling from the force polygon, we find :

R_2 = **178.2 kN** and R_1 = **81.8 kN Ans.**

Problem 1.53. *Determine graphically the support reactions for the beam shown in Fig. P.1.53 (a).*

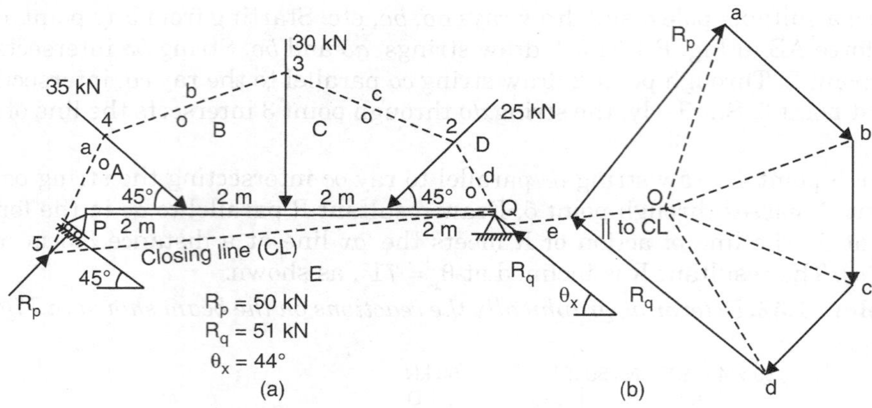

Fig. P. 1.53

Sol. Draw the given beam and the applied forces as shown in Fig. P.1.53(*a*). Using Bow's notation name all forces. Selecting suitable force scale draw force polygon as shown in Fig. P.1.53(*b*). Vector *ab* represents force AB of 35 kN; vector *bc* represents force BC of 30 kN and vector *cd* represents force CD of 25 kN. Choose a convenient pole *o* and draw rays *oa*, *ob*, *oc* and *od*.

Choosing point 1 at the hinged support draw string *do* parallel to ray *do* intersecting line of force CD at point 2.

From point 2 draw string *co* parallel to ray *co* intersecting line of force BC at point 3.

From point 3 draw string *bo* parallel to ray *bo* intersecting line of force AB at point 4.

From point 4 draw string *ao* parallel to ray *ao* intersecting the line of action of reaction R_p of the roller support P at point 5. Reaction R_p is perpendicular to the surface of roller.

Draw closing line 51 of the string or funicular polygon. From the pole *o* draw a ray parallel to the closing line, and from point *a* draw a line parallel to the reaction R_p of the roller support. The two lines intersect at point *e*. Draw vectors *de* and *ea*. Vector *de* represents reaction R_q of the hinged support in magnitude and direction. Vector *ea* represents reaction R_p of the roller support P. Scaling from the force polygon, we get R_p = **50 kN,** perpendicular to the surface of roller R_q = **51 kN,** θ_x = **44°. Ans.**

1.39. GRAPHICAL CONDITIONS OF EQUILIBRIUM

In the light of discussion is Sec. 1.33 to 1.38, the graphical conditions of equilibrium can be drawn. Two cases are important. The analytical equations of equilibrium and the graphical conditions for (*a*) concurrent coplanar force system, and (*b*) Non-concurrent coplanar force system are given below :

Case I. Coplanar concurrent force system.

(*a*) Analytical Equations of Equilibrium

$$\left.\begin{aligned} \Sigma F_x = 0 \quad &\text{or} \quad \Sigma F_h = 0 \\ \Sigma F_y = 0 \quad &\text{or} \quad \Sigma F_v = 0 \end{aligned}\right\} \qquad ...(1.72)$$

(*b*) Graphical Condition of Equilibrium.

(*i*) Force polygon should closed ...(1.73)

Case II. Coplanar non-concurrent force system.

(*a*) Analytical Equations of Equilibrium

$$\left.\begin{array}{l} \Sigma F_x = 0 \quad \text{or} \quad \Sigma F_h = 0 \\ \Sigma F_y = 0 \quad \text{or} \quad \Sigma F_v = 0 \\ \Sigma M_z = 0 \end{array}\right\} \qquad ...(1.74)$$

(*b*) Graphical Conditions of Equilibrium.

(*i*) Force polygon should be closed. ...(1.75)

(*ii*) Funicular diagram should be closed. ...(1.76)

EXERCISE I

A. Theoretical Questions

1. Define "mechanics" and subdivide it into various parts.
2. What is the scope of "mechanics" ?
3. Define space, time, mass and force.
4. Define particle, rigid body and the specifications of a force.
5. What are fundamental principles of mechanics ?
6. State the parallelogram law of forces.
7. State Newton's laws of motion.
8. State and discuss the principle of transmissibility.
9. State and discuss Newton's law of gravitation.
10. What are the units of the four fundamental quantities in systems of SI units, MKS units and FBS units ?
11. Discuss various coplanar force systems.
12. Illustrate (*a*) coplanar collinear force system, (*b*) coplanar concurrent force system, (*c*) coplanar parallel force system, and (*d*) coplanar non-concurrent force system.
13. Illustrate (*a*) concurrent spatial force system, (*b*) Spatial parallel force system, and (*c*) spatial non-concurrent force system.
14. State parallelogram law of forces. Derive the expression for finding resultant of two forces acting at a point.
15. State triangle law of forces.
16. State converse of the triangle law of forces.
17. State and prove polygon law of forces.
18. State and prove Lami's theorem.
19. Describe methods of resolution of forces.
20. Define equilibriant.
21. What do you mean by "free-body diagram" ? Explain with the help of at least three examples.
22. What do you mean by equations of equilibrium ? Write a set of equations of equilibrium for (*a*) a concurrent coplanar force system, and (*b*) a spatial concurrent force system.
23. Write down the set of equations of equilibrium for (*a*) a coplanar non-concurrent force system, and (*b*) a spatial non-concurrent force system.
24. What are statically determinate and indeterminate problems ?
25. What do you understand by moment of a force ? Illustrate with the help of suitable sketches the terms : force, moment centre, moment arm and moment axis.
26. State Varignon's theorem of moments.

27. State and prove Varignon's theorem of moments.

28. What do you understand by a couple ? Prove that the moment of couple does not depend upon the location of the point about which moments of the couple are taken.

29. State properties of a couple.

30. Prove that a given force can be resolved into a force and a couple acting at a given point.

31. What do you mean by the term "wrench" ?

32. What do you understand by the term "equivalent couple" ? Explain with the help of suitable example and neat sketches.

33. Explain the method of reducing the system of coplanar non-concurrent forces into their resultant force R and a resultant couple M.

34. Explain the equivalent system of coplanar forces.

35. Explain the x- and y intercepts of the resultant of non-concurrent coplanar force system.

36. Discuss various types of supports used in beam. Explain with the help of suitable sketches.

37. Discuss various types of beams, with the help of suitable sketches.

38. What do you mean by determinate and indeterminate beams ?

39. What different types of loads are known to you ? Explain with the help of suitable sketches.

B. Numerical Problems

1.1. Two forces of magnitude 15 N and 12 N are acting at a point. If the angle between the forces is 60°, find the resultant force and its inclination with 15 N force. [**Ans.** 27 N ; 26.33°]

1.2. The resultant of two forces, when they act at an angle of 60° is 28 N. If the same forces act at right angles, their resultant is $2\sqrt{136}$ find the magnitude of forces. [**Ans.** 20, 12 or 12, 20]

1.3. A point O is subjected to three forces, shown in Fig. E.1.3. Find the angle α for which the resultant of the three forces is (*a*) vertical, (*b*) horizontal, also find the resultant.

[**Ans.** (*a*) α = – 18.4349° ; R = – 116.23 N ; (*b*) α = 32.33° or 69.21°; R = 244.94 N or 315.96 N]

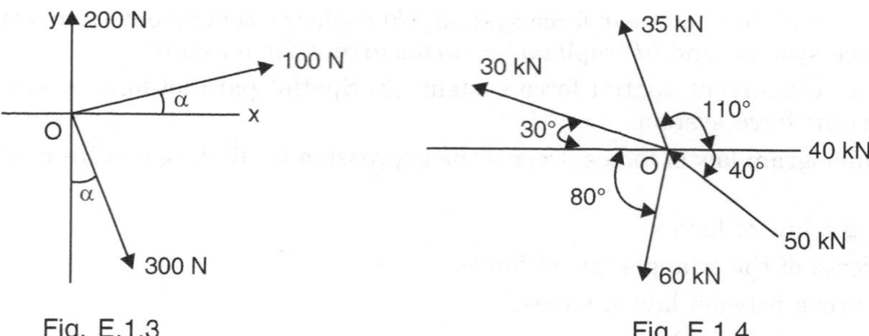

Fig. E.1.3 Fig. E.1.4

1.4. Determine the equilibriant of the system of forces shown in Fig. E.1.4.

[**Ans.** Equilibriant = 86.133 kN ; θ_x = – 18.46°]

1.5. A 200 N force is to be resolved into components along lines $a - a'$ and $b - b'$ as shown in Fig. E.1.5,

(*i*) Determine the angle α knowing that the component along $a - a'$ is to be 150 N.

(*ii*) What is the corresponding value of the component along $b - b'$. [*Pune Univ. Dec. 2003*]

[**Ans.** (*i*) α = 70.02° (*ii*) P = 63.5 N]

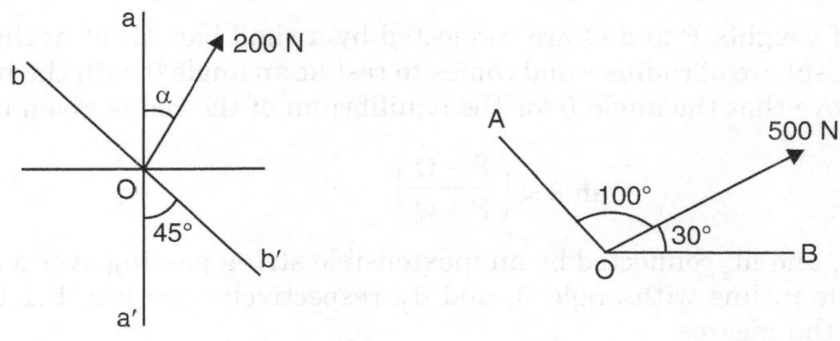

Fig. E. 1.5 Fig. E. 1.6

1.6. Resolve the force of 500 N into its components along the directions OA and OB shown in Fig. E.1.6.

[Pune Univ., May, 2003]

[**Ans.** Along OA force = 642.79 N ; along OB force = 326.35 N]

1.7. A rod of negligible weight is hinged at end O and carries a load of 60 N at end A as shown in Fig. E.1.7. It is held in position by a string at A. (*a*) Find the force in the rod (*b*) Also find the range of values of the tension in the string when the rod is in the vertical position. Assume pulley B to be smooth. [**Ans.** (*a*) 50 N ; (*b*) 78.1 N]

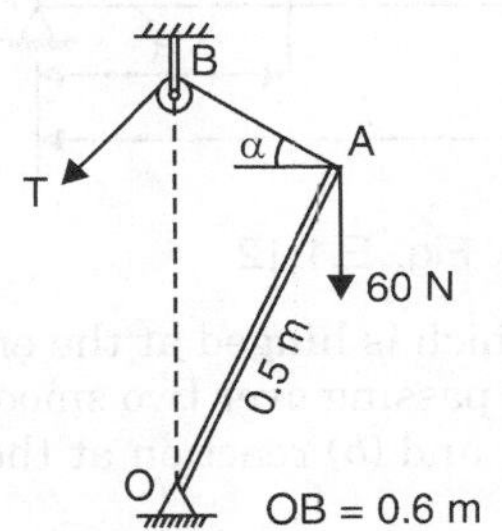

Fig. E. 1.7

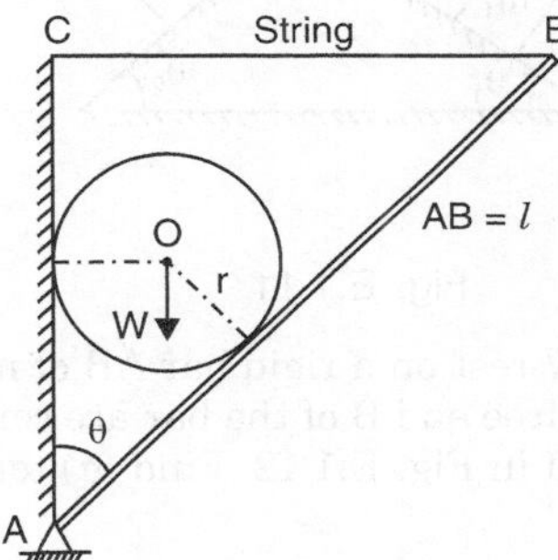

Fig. E. 1.8

1.8. A cylinder of weight W and radius r is supported against a smooth vertical wall and rests on a bar AB of length l and of negligible weight, as shown in Fig. E.1.8. The bar is hinged at A and is supported by a horizontal string BC, as shown. Find the angle θ that the bar will make with the wall so that the tension in the string is minimum. Also find the tension in the string.

$$\left[\textbf{Ans. } \theta = 60°; \text{Tension} = \frac{4\,Wr}{l}\right]$$

1.9. A uniform rod AB of weight W in vertical plane is resting on two frictionless surfaces OA and OB as shown in Fig. E.1.9. Find the angle θ for which the rod is in equilibrium. Also find reactions at the two ends of bar.

$$\left[\textbf{Ans. } \theta = 60°; N_A = \frac{\sqrt{3}W}{2} \text{ and } N_B = \frac{W}{2}\right]$$

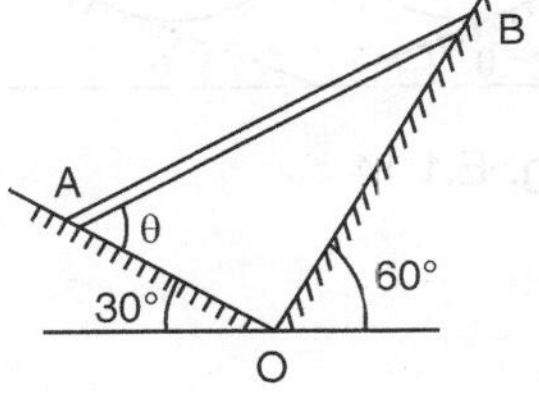

Fig. E.1.9

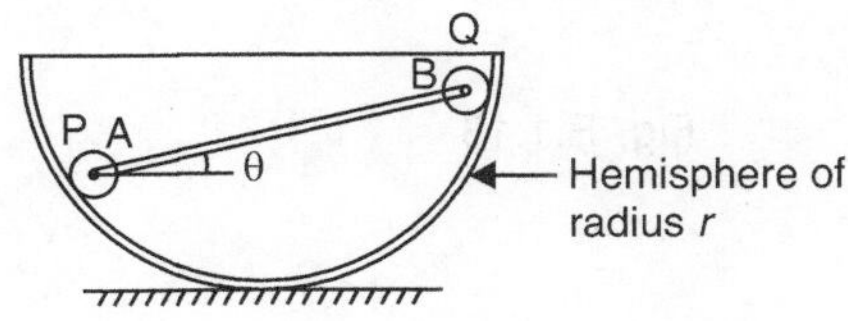

Fig. E.1.10

1.10. Two spheres of weights P and Q are connected by a rigid bar AB of negligible weight. The bar slides in a hemisphere of radius r and comes to rest at an angle θ with the horizontal as shown in Fig. E.1.10. Prove that the angle θ for the equilibrium of the rod is given by :

$$\tan \theta = \left(\frac{P - Q}{P + Q}\right).$$

1.11. Two masses m_1 and m_2 connected by an inextensible string passing over a smooth pulley rest on a smooth double incline with angle θ_1 and θ_2, respectively. See Fig. E.1.11. Prove that for the equilibrium of the masses

$$\frac{m_1}{m_2} = \frac{\sin \theta_2}{\sin \theta_1}.$$

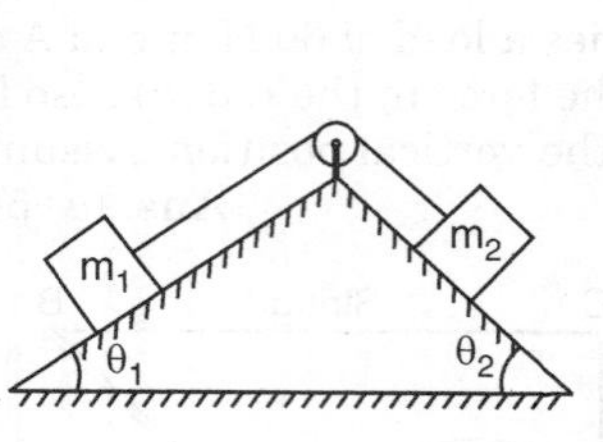

Fig. E.1.11

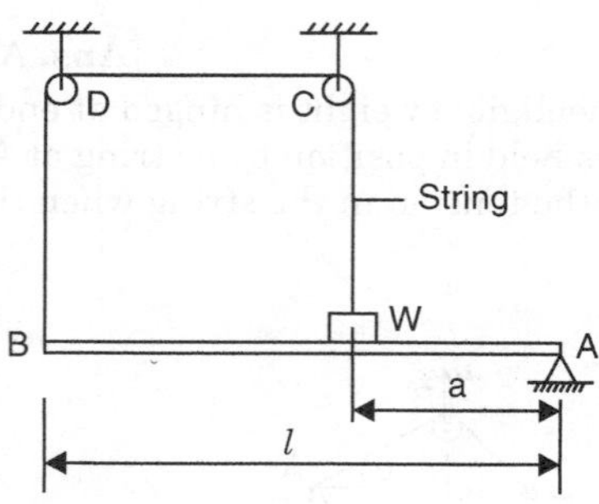

Fig. E.1.12

1.12. A weight W rest on a rigid bar AB of negligible weight which is hinged at the end A. The weight W and the free end B of the bar are connected by a string passing over two smooth pulleys C and D as shown in Fig. E.1.12. Find (*a*) tension in the string, and (*b*) reaction at the hinge.

$$\left[\textbf{Ans. } (a) \text{ Tension} = \frac{W \cdot a}{l + a}; (b)\ R_A = V_A = \frac{W(l - a)}{l + a}\right]$$

1.13. Two cylindrical rollers weighing 10 kN and 6 kN lying on two smooth inclines 60° and 30° respectively are connected by an inextensible string as shown in Fig. E.1.13. Find the angle θ for the equilibrium of the masses. Also find tension in the string. [**Ans.** θ = 40.893° ; T = 9.165 N]

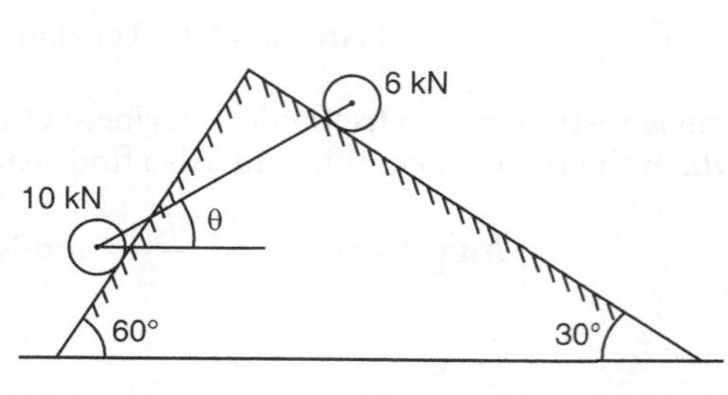

Fig. E.1.13

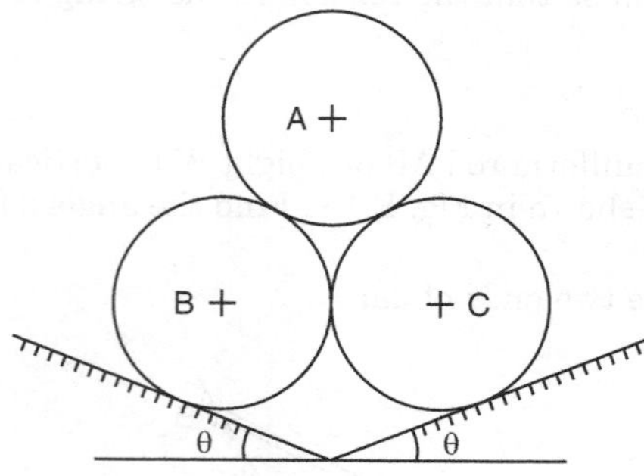

Fig. E.1.14

1.14. Three identical spheres, each of radius r and each weighing W, are stacked on smooth surfaces each inclined at an angle θ with the horizontal, as shown in Fig. E.1.14. Determine the least value of angle θ to prevent stack from falling. *[Pune Univ., Dec. 1999]*

[**Ans.** θ = 10.91°]

1.15. Two round solid identical cylinders, each of radius 100 mm and weight 50 N, tied together by a thread of length 200 mm, rest on a horizontal surface. They hold a third homogeneous cylinder of radius 125 mm and weight 75 N, placed as shown in Fig. E.15. Calculate the tension in the thread and the reactions at all contact surfaces. *[Pune Univ., May 2003]*

[**Ans.** T = 18.61 N ; R_p = 41.86 N; R_D = 87.5 N]

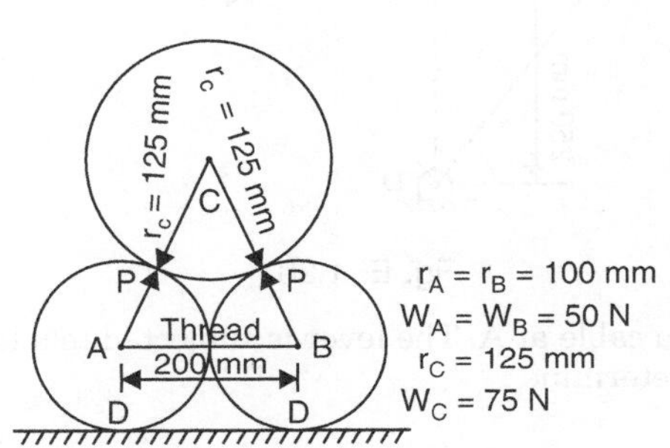

Fig. E. 1.15

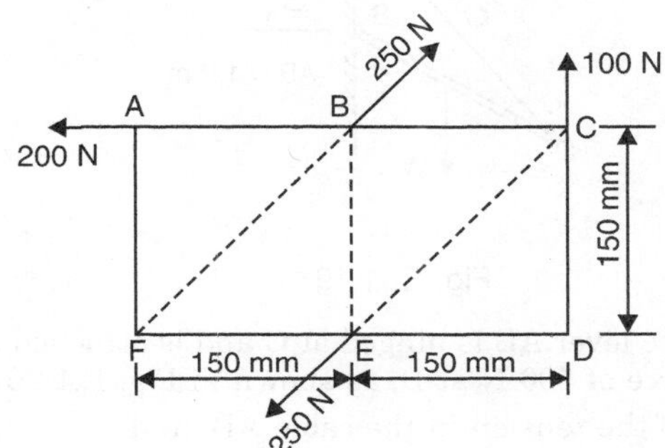

Fig. E. 1.16

1.16. A 150 × 300 mm plate is subjected to four loads as shown in Fig. E.1.16. Find the resultant of the four forces and the two points at which the line of action of the resultant intersects edges of the plate. *[Pune Univ., May 1998]*

[**Ans.** Resultant, R = 223.61 N ; θ_x = – 26.56° ; i_x = 265 mm left of C along CA and i_y = 133 mm below C along CD]

1.17 A homogeneous wire ABCD is bent as shown in Fig. E.1.17 and is attached to a hinge at C. Determine the length L for which portion AB is horizontal. *[Pune Univ., May 1998]*

[**Ans.** L = 248.75 mm]

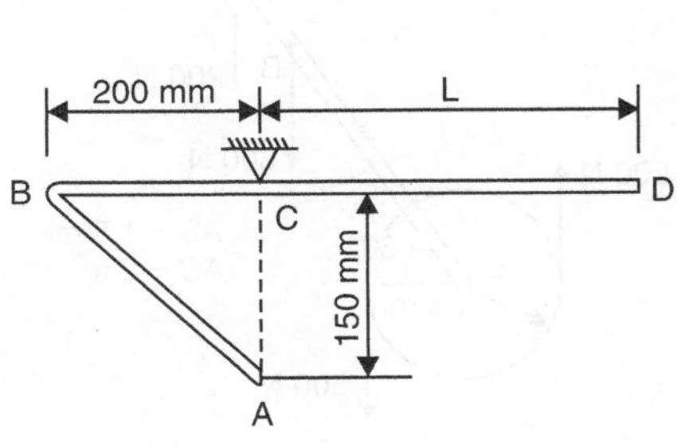

Fig. E.1.17

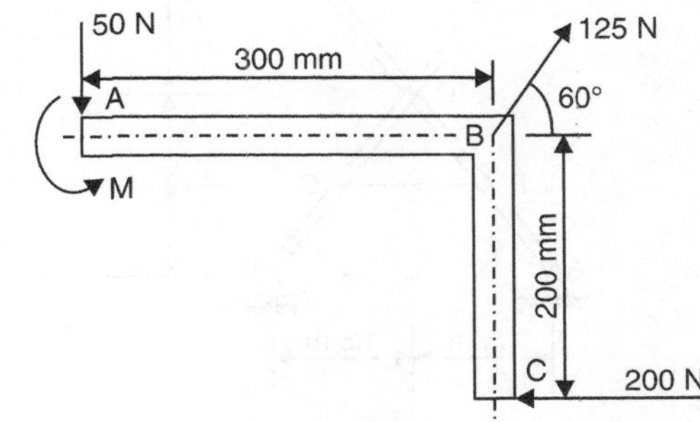

Fig. E.1.18

1.18. The three forces as shown in Fig. E.1.18 and a couple M are applied to an angle bracket. Determine the moment of the couple M if the line of action of the resultant of this force system is to pass through (*i*) B, (*ii*) C. *[Pune Univ., Dec. 1998]*

[**Ans.** (*i*) M = 25000 Nmm = 25 Nm; (*ii*) M = 2500 Nmm = 2.5 Nm]

1.19. A 1.2 m long rod, of uniform cross-section, is held in equilibrium as shown in Fig. E.1.19 with one end against a frictionless wall and the other end attached to the cord. Determine the length of the cord. *[Pune Univ., Dec. 1998]*

[**Ans.** AC = 1.588 m]

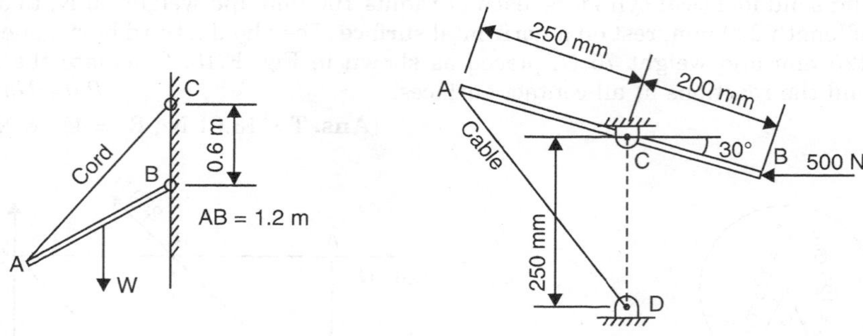

Fig. E. 1.19 Fig. E. 1.20

1.20. The lever AB is hinged at C and is attached to a cable at A. The lever is subjected to a horizontal force of 500 N at B, as shown in Fig. E.1.20. Determine :

(*a*) the tension in the cable AD, and

(*b*) the reaction at C. *[Pune Univ., May 1999]*

[**Ans.** (*a*) Tension = 400 N ; (*b*) C_v = 346.41 N ; C_x = 300 N ; R_c = 458.26 N ; θ_{cx} = 49.11°].

1.21. ACE and BCD are two rigid rods connected by a string FG and a pin at C. Find reactions at A and B, the force on the pin C and the tension in string FG for the frame shown in Fig. E.1.21.

[**Ans.** B_y = 23.333 kN↑ ; A_y = 13.333 kN↓ ; A_x = 10 kN← ; C_y = 23.333 kN↓ ; C_x = 5 kN→; Tension = 15 kN]

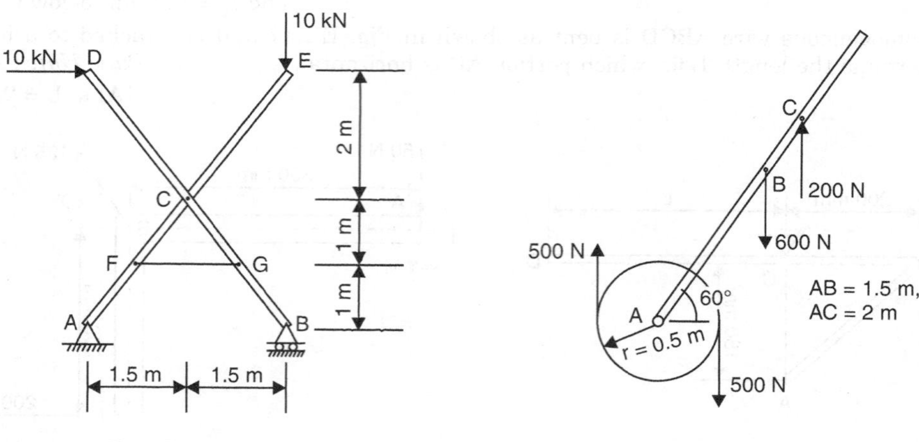

Fig. E. 1.21 Fig. E. 1.22

1.22. Replace couple-force system shown in Fig. E.1.22 by an equivalent force applied to the lever. Determine the distance from the centre of the shaft to the point of application of this equivalent force. [**Ans.** Resultant R = 400 N↓ ; x = 3.75 m]

1.23. Find resultant of forces shown in Fig. E.1.23. Hence, replace this force system by a three forces acting along the sides of an equilateral triangle ABC of sides 50 cm.

[**Ans.** F_{AB} = 7.735 N from A to B ; F_{CB} = 10.735 N from C to B ; F_{AC} = 50 N from A to C ; Resultant R_x = 0 ; R_y = 100 N↑]

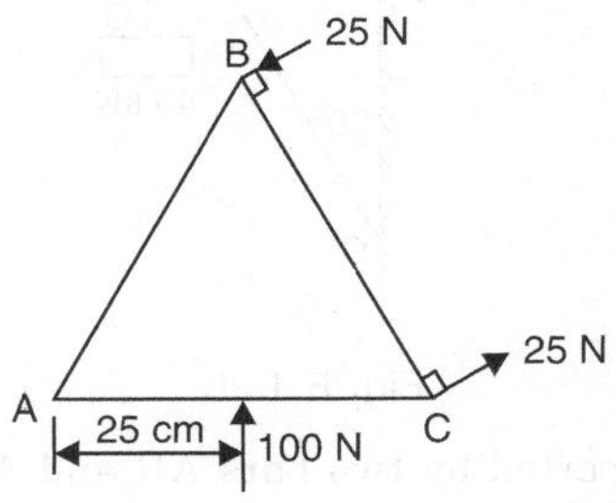

Fig. E. 1.23

Fig. E. 1.24

1.24. A rectangular plate ABCD is subjected to forces as shown in Fig. E.1.24. Find resultant force acting on the plate. Also find the magnitude and direction of the forces which should be applied (*a*) at points B and C, and (*b*) at A and B to keep the plate in the original state.

[**Ans.** Resultant = 0 ; (*a*) force at B = F = 171.43 N← and that at C = 171.43 N→ ; (*b*) force at A = F′ = 120 N↑, and that at B = 120 N↓]

1.25. The three forces shown in Fig. E.1.25 create a vertical resultant acting through point B. If P = $100\sqrt{13}$ N, find values of forces F and T. Each small square is of 1 cm sides.

[**Ans.** F = $300\sqrt{5}$ N = 670.82 N ; T = $800\sqrt{10}$ N = 2529.82 N, Resultant R_x = 0 ; R_y = 2400 N↑]

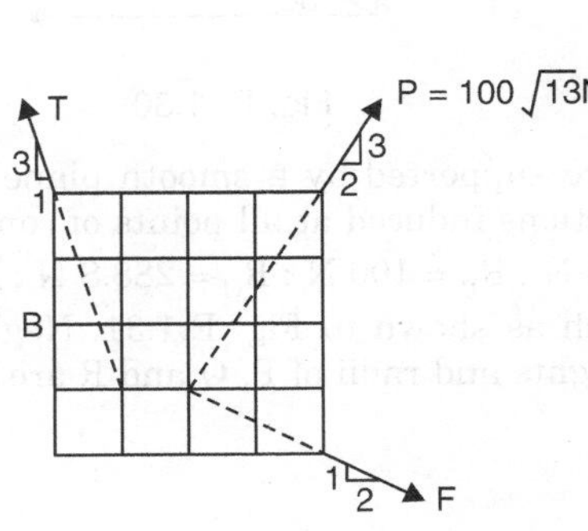

Fig. E.1.25

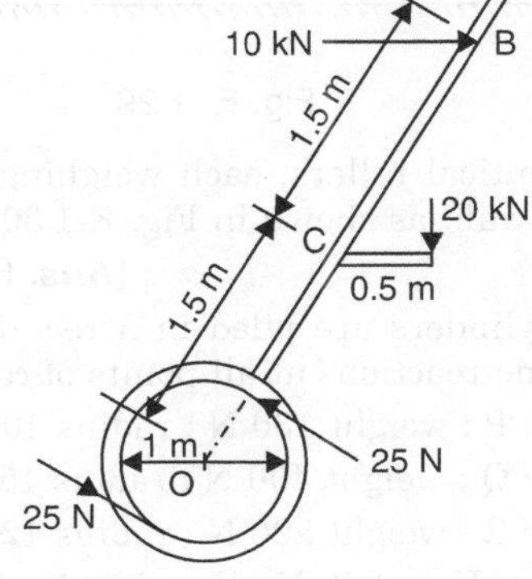

Fig. E.1.26

1.26. A machine part is subjected to forces as shown in Fig. E.1.26. Find the resultant of forces in direction and magnitude. Also locate the point where the resultant will intersect the centreline of the bar.

[**Ans.** Resultant = $10\sqrt{5}$ N = 22.361 N ; θ_x = – 63.435° ; 1.392 m from 0 or 0.108 m below point C]

1.27. Two cylinders, A of weight 2000 N and B of weight 1000 N, rest on smooth inclines as shown in Fig. E.1.27. They are connected by a bar of negligible weight hinged to each cylinder at its geometric centre by smooth pins. Find the force P to be applied such that it will hold the system in the given position. [**Ans.** 535.9 N]

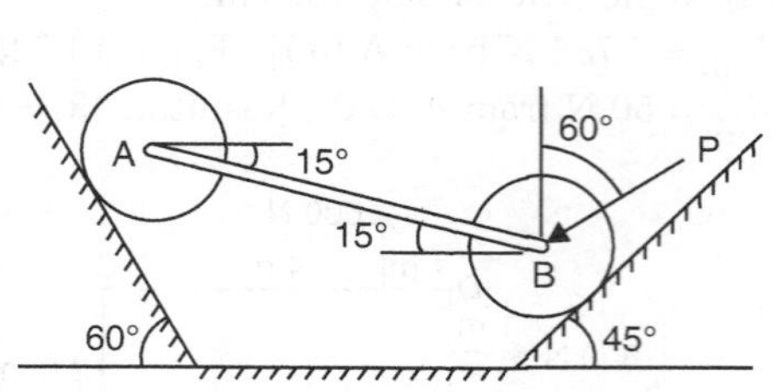

Fig. E.1.27

Fig. E.1.28

1.28. A frictionless pulley, as shown in Fig. E.1.28, is supported by two bars AB and AC which are hinged at B and C to a vertical wall. A flexible cable DG passing over a small pulley supports a load of 40 kN at G. The angles between various members are shown in the figure. Determine the force in the bars AB and AC. [**Ans.** $F_{AB} = 0$, $F_{AC} = 69.282$ N]

1.29. Three bars, hinged at A and D and pinned at B and C as shown in Fig. E.1.29, for a four linked mechanism. Determine the value of force P that will prevent movement of bars.

[**Ans.** P = 6094.4 N]

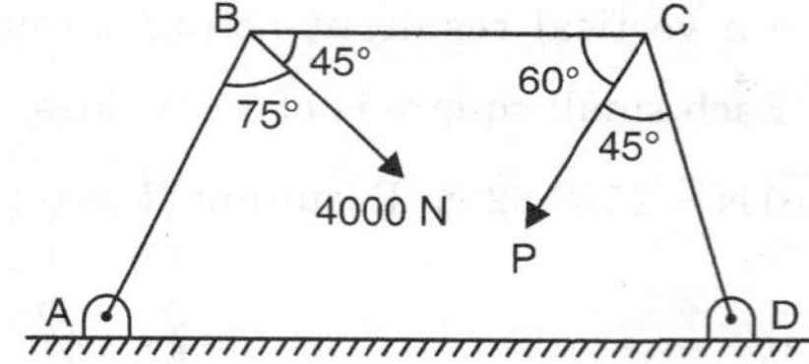

Fig. E. 1.29

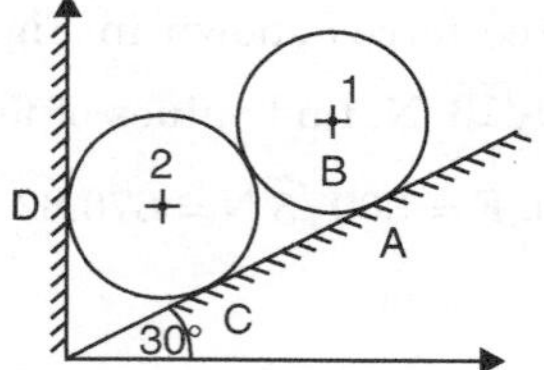

Fig. E. 1.30

1.30. Two identical rollers, each weighing 200 N, are supported by a smooth plane and a smooth vertical wall, as shown in Fig. E-1.30. Find reactions induced at all points of contact.

[**Ans.** $R_A = 173.34$ N ; $R_B = 100$ N ; $R_C = 288.8$ N ; $R_D = 231.04$ N]

1.31. Three cylinders are piled in a rectangular ditch as shown in Fig. E.1.31. Neglecting friction determine reactions at all points of contact. Weights and radii of P, Q and R are :

Cylinder P : weight 130 N ; radius 100 mm

Cylinder Q : weight 400 N ; radius 150 mm

Cylinder R : weight 200 N ; radius 125 mm

[**Ans.** $H_A = 800$ N ; $V_B = 750$ N ; $H_C = 965$ N ; $H_D = 259.27$ N ; $R_1 = 1000$ N ; $R_2 = 407.43$ N]

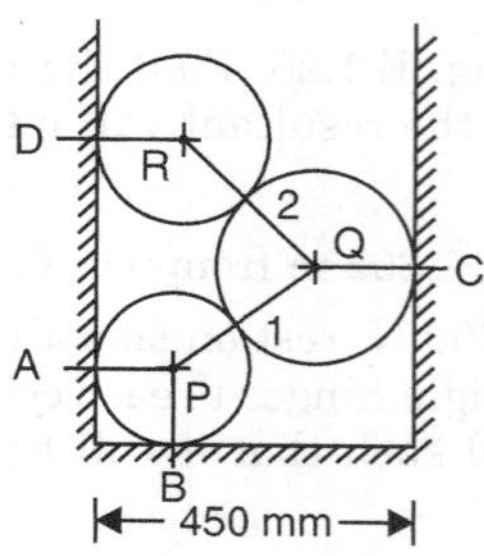

Fig. E. 1.31

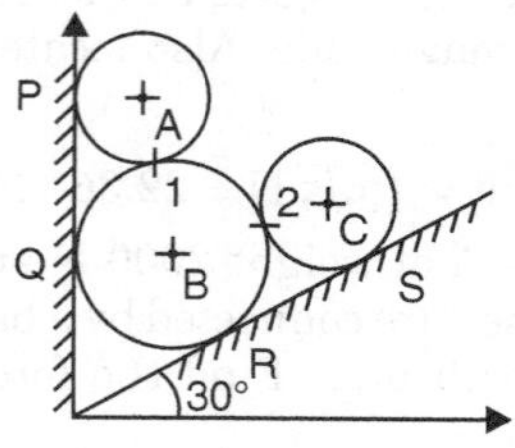

Fig. E. 1.32

1.32. Three uniform, homogeneous and smooth spheres A, B and C weighing 100 N, 200 N and 100 N and having radii 400 m, 600 mm and 400 mm respectively, are placed in a trench as shown in Fig. E.1.32. Determine reactions at all points of contact.

[**Ans.** H_P = 20.41 N ; H_Q = 105.12 N ; N = 133.84 **N**
N_S = 96.81 N ; R_1 = 102.06 N ; R_2 = 51.03 N]

1.33. Two identical cylinders, each of weight 1000 N, are supported by an inclined plane and a vertical wall as shown in Fig. E.1.33. Assuming smooth surfaces find the reactions at the points of contact A, B and C. [**Ans.** R_A = 866 N ; R_B = 1442.14 N and R_C = 1154.8 N]

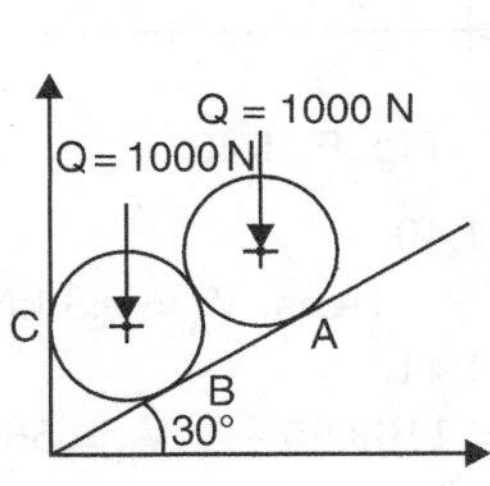

Fig. E. 1.33

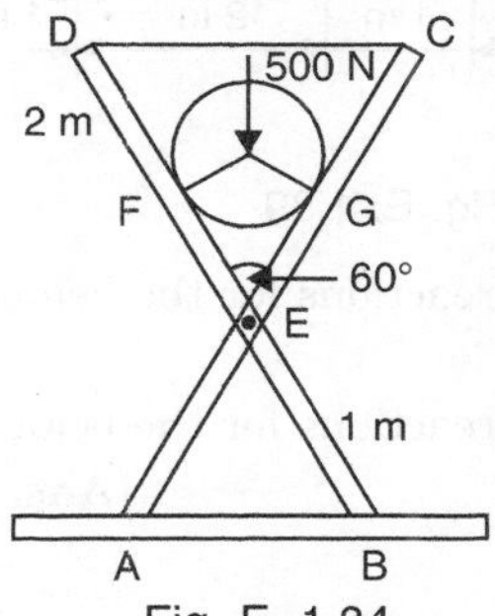

Fig. E. 1.34

1.34. A cylinder of diameter 1 m weighing 500 N is supported by two cross bars each 3 m long and hinged together at point E, 1 m from the lower ends of the bars as shown in Fig. E.1.34. A horizontal string connects the upper ends of the bars and thus preventing the opening out of the cross bars. If the angle between the bars is 60° and if the floor is smooth find tension in the string and the reaction components at the hinge.

[**Ans.** $V_A = V_B$ = 250 N ; $N_F = N_G$ = 600 N ; T= 481.98 N ; H_E = 181.98 N ; V_E = 50 N]

1.35. Two equal loads of 3000 N are supported by a flexible string ABCD at points B and C. Find tension in the portions AB, BC and CD of the string. Refer Fig. E.1.35.

[**Ans.** T_1 = 5196 N ; T_2 = .T_3 = 3000 N]

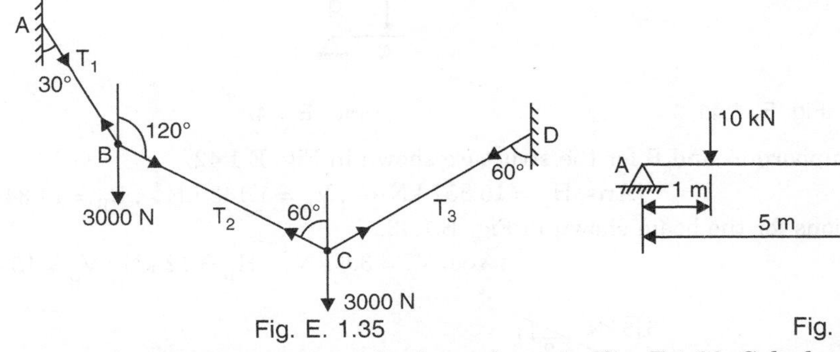

Fig. E. 1.35

Fig. E. 1.36

1.36. An overhanging beam carries loads as shown in Fig. E.1.36. Calculate support reactions.

[**Ans.** V_A = 10.8 kN ↑ ; V_B = 19.2 kN↑]

1.37. An overhanging beam carrying load as shown in Fig. E.1.37. Find reactions at the two supports.

[**Ans.** V_A = 9 kN ↑ ; V_B = 16 kN ↑]

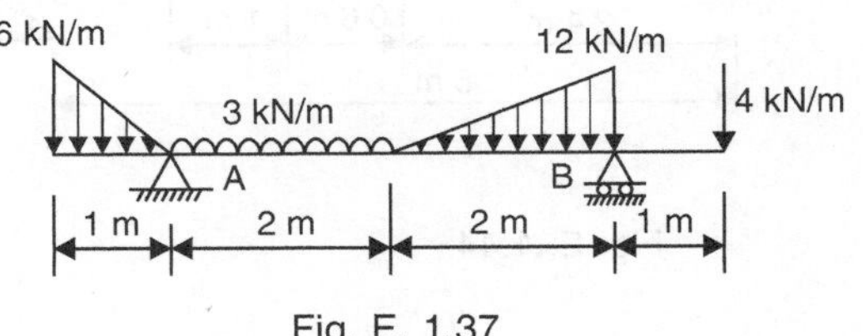

Fig. E. 1.37

Fig. E. 1.38

1.38. A beam is shown in Fig. E.1.38. Determine the reactions at both supports.

[**Ans.** $V_A = 2$ kN ↑ ; $V_B = 31$ kN ↑]

1.39. Find support reactions of the beam shown in Fig. E.1.39. [**Ans.** $V_A = 1.6$ kN ↑ ; $V_B = 8.4$ kN ↑]

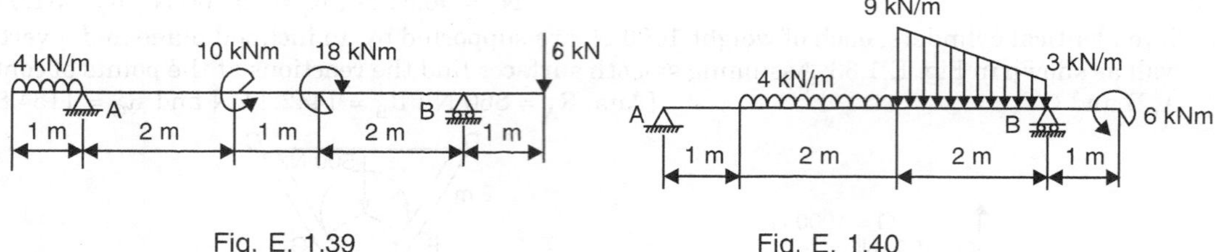

Fig. E. 1.39 Fig. E. 1.40

1.40 Find support reactions for the beam shown in Fig. E.1.40.

[**Ans.** $V_A = 8.8$ kN ↑ ; $V_B = 11.2$ kN ↑]

1.41. Find support reactions for the beam shown in Fig. E.1.41.

[**Ans.** $V_A = 640$ N ; $H_A = 1169.95$ → ; $V_B = 560$ N ; $H_B = 969.95$ N]

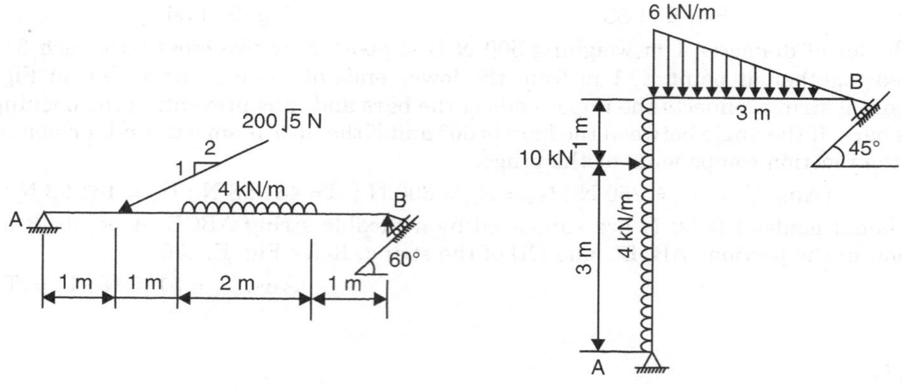

Fig. E. 1.41 Fig. E.1.42

1.42. Find reactions at supports A and B for the structure shown in Fig. E.1.42.

[**Ans.** $H_A = 15.857$ kN ← ; $V_A = 1.143$ kN↓ ; $R_B = 14.344$ kN]

1.43. Find support reactions for the beam shown in Fig. E.1.43.

[**Ans.** $V_A = 3.6$ kN↑ ; $H_B = 12$ kN ; $V_B = 15.4$ kN]

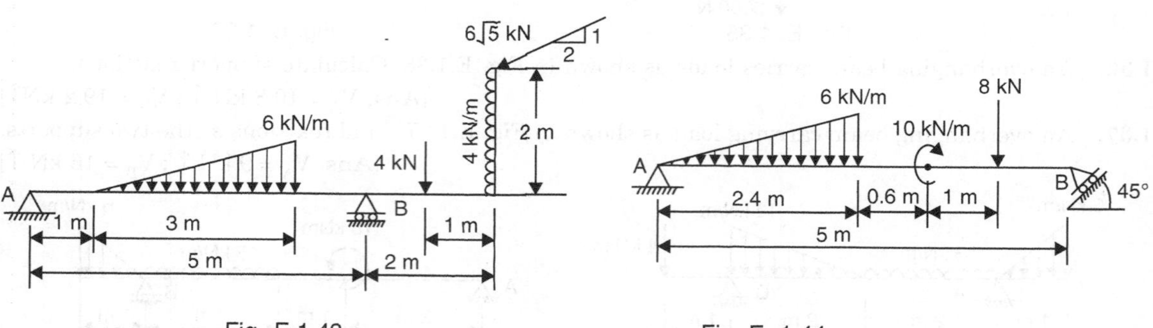

Fig. E.1.43 Fig. E. 1.44

1.44. Find support reactions for the beam shown in Fig. E.1.44.

[**Ans.** V_A = 4.496 kN ; H_A = 10.704 kN→ ; R_A = 11.61 kN ; θ_x = 22.74° ; V_B = 10.704 kN↑; H_B = 10.704 kN← ; R_B = 15.138 kN]

1.45. Find support reactions for the beam shown in Fig. E.1.45.

[**Ans.** V_A = 45.76 kN ; H_A = 58 kN ; V_B = 30.24 kN]

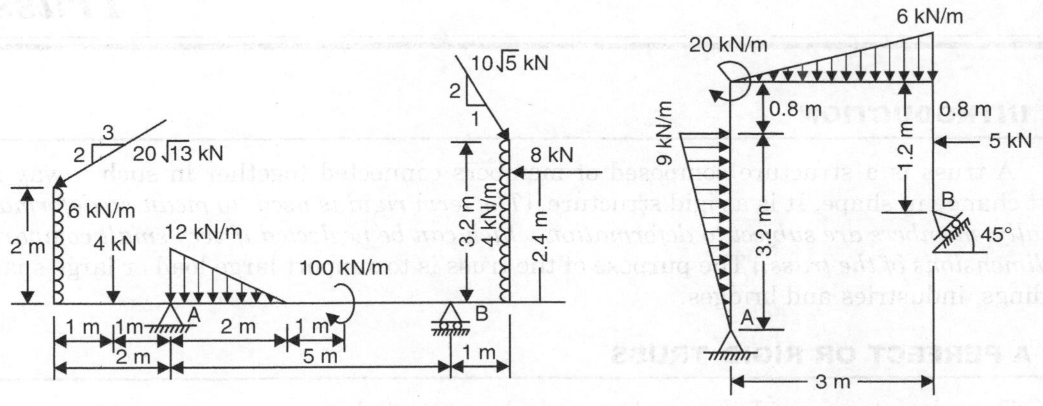

Fig. E. 1.45 Fig. E. 1.46

1.46. Find support reactions for the portal frame shown in Fig. E.1.46.

[**Ans.** V_A = 1.544 kN↓ ; H_A = 4.744 kN→ ; V_B = 10.544 kN↑ ; H_B = 10.544 kN←].

1.47. A long rod BC of length l and weight W is held by two cables AB and CD as shown in Fig. E.1.47. If cable AB is horizontal find inclination θ of the cable CD and the tension in each cable.

[**Ans.** θ = 49.11° ; T_1 = 0.866 W = Tension in AB ; T_2 = 1.323 W = tension in CD]

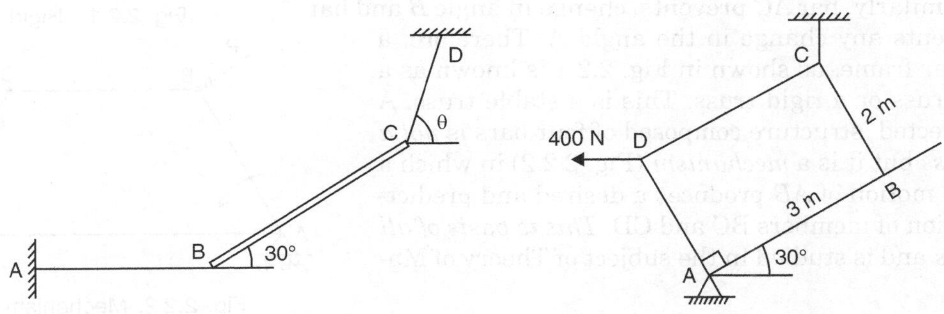

Fig. E.1.47 Fig. E.1.48

1.48. A 2 kN uniform rectangular plate ABCD is held vertical by a hinge at A and vertical cable at C, as shown in Fig. E.1.48. Find reaction at the hinge A and tension in cable at C.

[**Ans.** V_A = 1433.686 N ; H_A = 400 N ; T = 566.314 N]

2

Trusses

2.1. INTRODUCTION

A truss is a structure composed of members connected together in such a way as to resist change in shape. It is a rigid structure. [*The term rigid is used to mean no deformation. Actually, members are subject to deformation which can be neglected if very small compared to the dimensions of the truss.*] The purpose of the truss is to support large load or large spans in buildings, industries and bridges.

2.2. A PERFECT OR RIGID TRUSS

To make a truss rigid, its members must be connected together in such a way to prevent any movement between them. The simplest pin-connected structure consists of three bars forming the triangular truss shown in Fig. 2.2.1. Since all bars are rigid, their deformations (*i.e.*, changes in their lengths) under the external load *P* are very small and negligible, as discussed earlier.

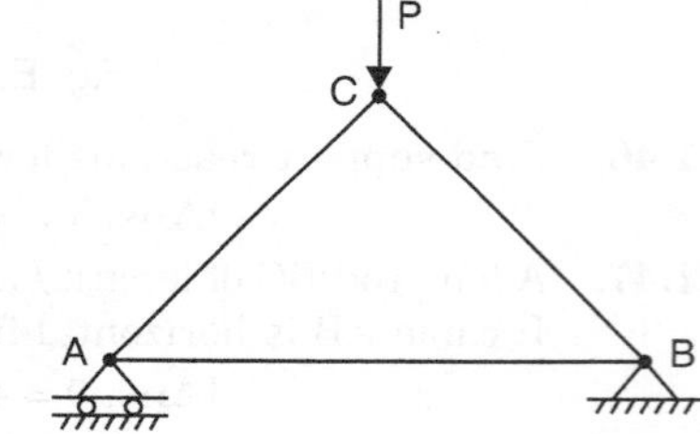

Fig. 2.2.1. Rigid truss

Bar *AB* does not allow any change in the opposite angle *C* ; similarly, bar *AC* prevents change in angle *B* and bar *BC* prevents any change in the angle *A*. Therefore, a triangular frame, as shown in Fig. 2.2.1 is known as a perfect truss or a rigid truss. This is a stable truss. A pin-connected structure composed of four bars is *not a rigid truss* but it is a *mechanism* (Fig. 2.2.2) in which a specified motion of *AB* produces a desired and predictable motion of members BC and CD. *This is basis of all machines* and is studied in the subject of Theory of Machines.

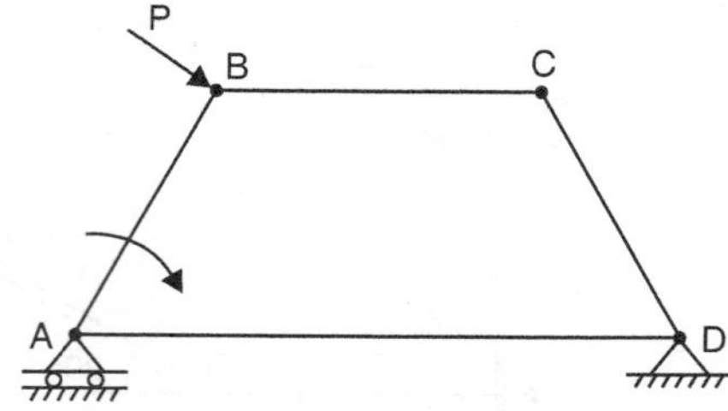

Fig. 2.2.2. Mechanism

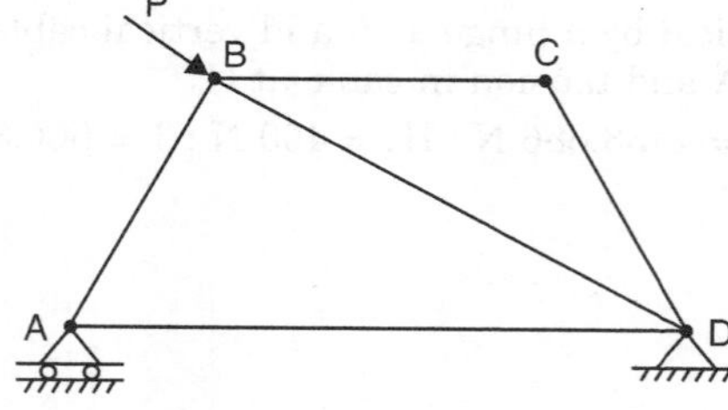

Fig. 2.2.3. Just-rigid truss

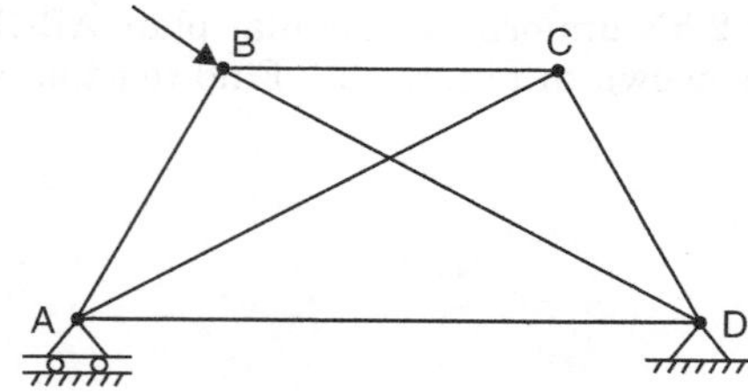

Fig. 2.2.4. Over-rigid truss
[Either AC or BD is redundant member]

Here, all angles A, B, C and D change and therefore it is collapsible mechanism used for transmission of power and motion from the driving member AB to the coupler BC and the driven member CD. This is also known as a four-bar mechanism.

In Fig. 2.2.3, adding a bar BD to the 4-bar mechanism of Fig. 2.2.2 restores rigidity, while by adding two bars AC and BD as shown in Fig. 2.2.4 we get an over-rigid and statically indeterminate truss. Either member AD or BD is called a redundant (not necessary) member. By removing one of the bars from this over-rigid truss we get the trusses shown in (*a*) and (*b*) of Fig. 2.2.5, whereas in part (*c*) two bars are removed but stability is restored by adding a hinge at A in place of a roller support.

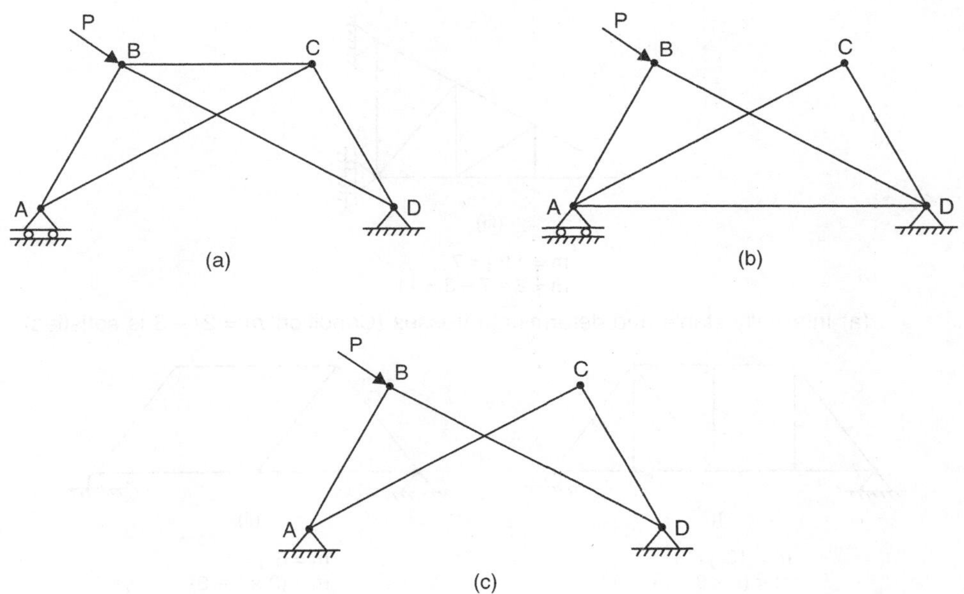

Fig. 2.2.5. Just rigid trusses derived from Fig. 2.2.4

2.3. RELATION BETWEEN NUMBER OF MEMBERS AND NUMBER OF JOINTS

From the manner in which simple trusses are constructed, we can deduce that the relation between m members and j joints is

$$m = 2j - 3 \qquad ...(2.1)$$

This relation is developed by noting that to the three joints of the original base triangle ABC of Fig. 2.2.1, two additional members BC and CD are needed for the joint C of Fig. 2.2.3.

A truss satisfying equation 2.1 is internally stable and determinate.

If $$m < (2j - 3) \qquad ...(2.2)$$

we get a 4-bar mechanism similar to the one shown in Fig. 2.2.2.

We get internally unstable structure.

If $$m > (2j - 3) \qquad ...(2.3)$$

we get internally stable and indeterminate truss, as shown in Fig. 2.2.4.

Some more examples giving relationship between m and j are shown in Fig. 2.3.

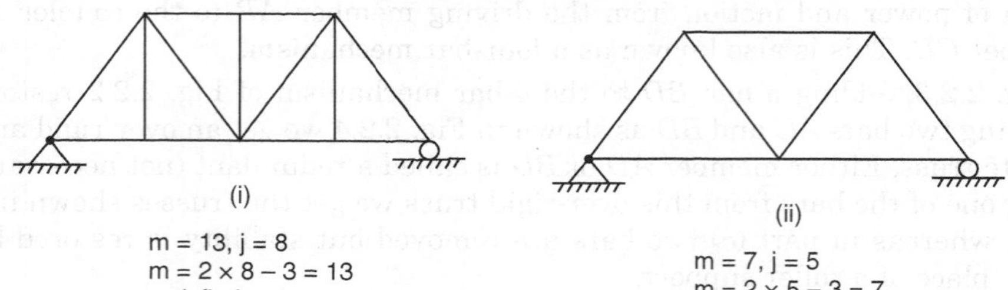

(i)

m = 13, j = 8
m = 2 × 8 – 3 = 13
satisfied

(ii)

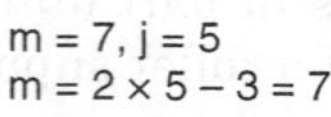

m = 7, j = 5
m = 2 × 5 – 3 = 7

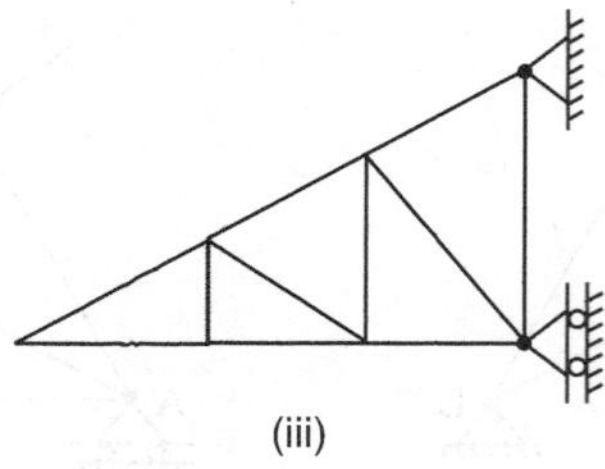

(iii)

m = 11, j = 7
m = 2 × 7 – 3 = 11

(a) Internally stable and determinate trusses (Condition $m = 2j - 3$ is satisfied)

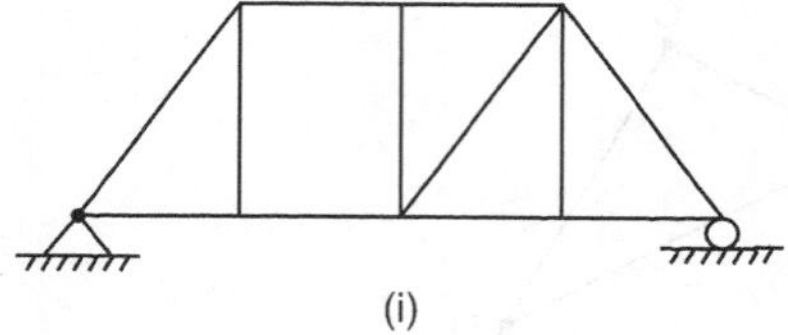

(i)

m = 12, j = 8
m < (2 × 8 – 3)

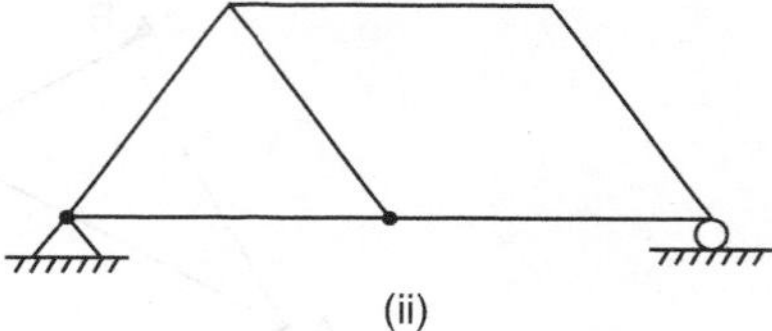

(ii)

m = 6, j = 5
m < (2 × 5 – 3)

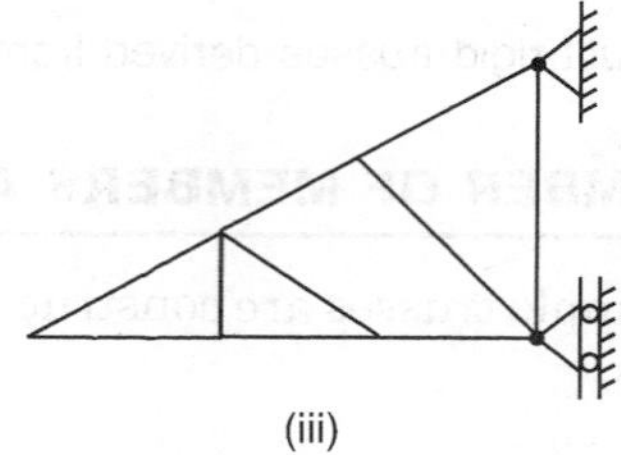

(iii)

m = 10, j = 7
m < (2 × 7 – 3)

(b) Internally unstable structure [$m < (2j - 3)$]

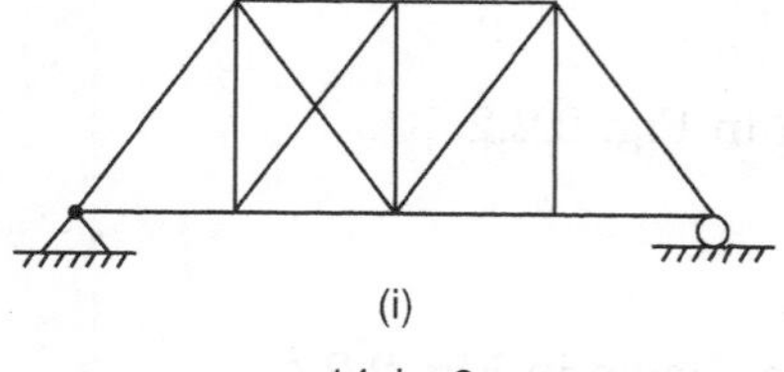

(i)

m = 14, j = 8
m > (2 × 8 – 3)

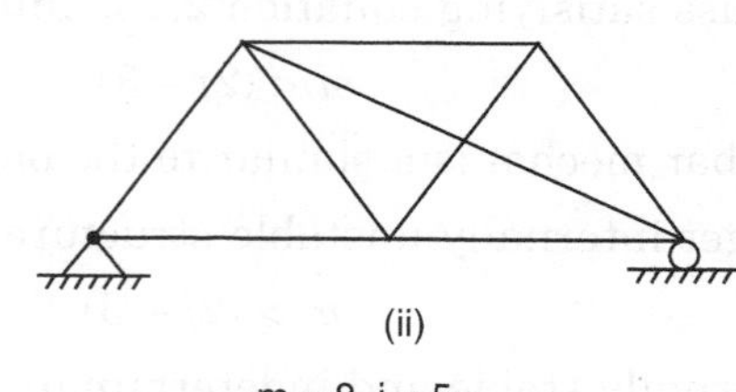

(ii)

m = 8, j = 5
m > (2 × 5 – 3)

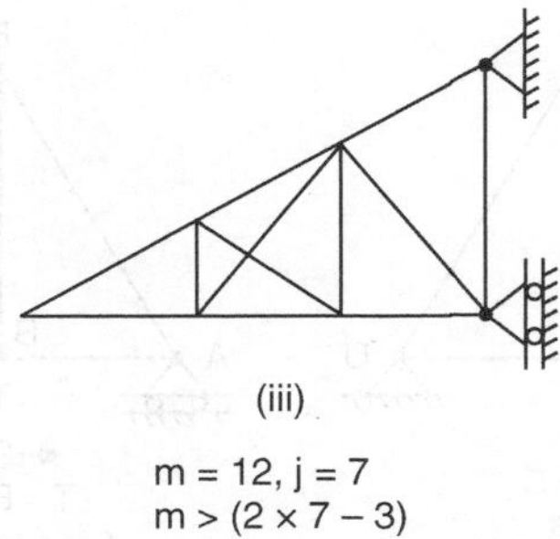

(iii)

m = 12, j = 7
m > (2 × 7 – 3)

(c) Internally stable and indeterminate trusses [$m > (2j - 3)$].

Fig. 2.3. Relationship between m and j. (a) $m = (2j - 3)$, (b) $m < (2j - 3)$, (c) $m > (2j - 3)$

Note. (1) Students are advised to first check the relation $m \gtreqless (2j - 3)$. And, attempt only those problems for which the given truss has $m = 2j - 3$. That is, solve only determinate trusses.

(2) Students are further advised to immediately ask for any correction in the problem on trusses in which either $m < (2j - 3)$ or $m > (2j - 3)$.

2.4. ASSUMPTIONS

The following assumptions are made to simplify the analysis of the truss.

(1) Members of the truss are pin-connected to each other.

(2) Loads act only at joints.

(3) Members of the truss are not subjected to bending moments.

(4) Members of the truss are rigid.

(5) Self weights of the members are negligible.

(6) All members are of uniform cross-section.

However, in actual practice members are connected to each other by bolting, riveting or by welding.

In most of the trusses loads act at joints. Even if a load is not acting at a joint, it can be replaced by its reactions at the joints and a bending moment on the member.

In most of the trusses the self weight is generally small compared to the loads they carry. Hence, self weight of members can be neglected.

While fabricating the trusses, the engineer-in-charge has to ensure that members are straight and centroids of the cross-section of the members lie along their axis so that the member is in equilibrium under the action of only axial forces.

2.5. TENSILE AND COMPRESSIVE MEMBERS

Members which are stretched are said to be in tension while those that are contracted are said to in compression. Generally, member in tension is called a **tie** and a member in compression is called a **strut.**

In Fig. 2.5, if member BC were in tension, isolating it from the truss would give the free-body diagram as shown in (b), and its effect on the joints of the truss would appear as forces T which pull the joints B and C. Similarly, member EF assumed to be in compression, would be represented by forces C which push the joints E and F. Thus, from these diagrams we deduce the following rule :

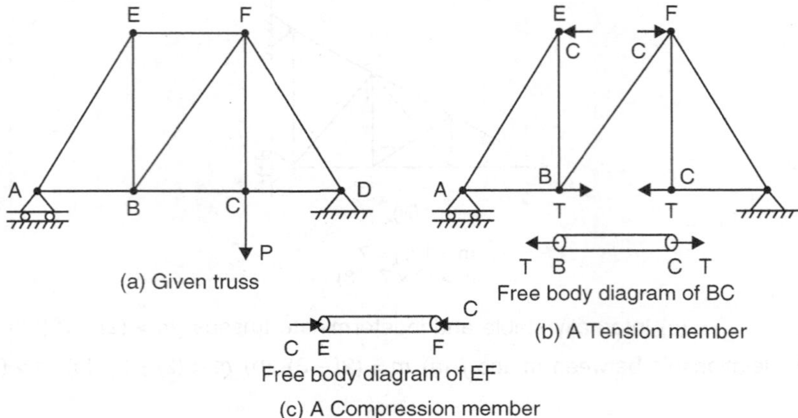

Fig. 2.5. Representation of tension and compression members of the truss.

A member in tension causes forces which **pull** away its end joints whereas a member in compression causes forces which **push** toward its end joints.

2.6. TRUSS NOMENCLATURE

Fig. 2.6 shows a typical truss. Joints of the truss are numbered from left to the right. The prefix L is used for lower joints while the prefix u is used for the upper joints.

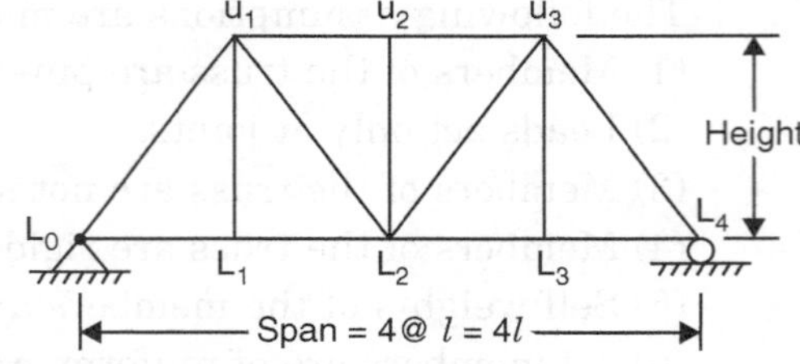

Fig. 2.6. Nomenclature of a truss

Various members of the truss may be grouped as :

(*a*) **Lower chord members,** such as L_0L_1, L_1L_2, L_2L_3 and L_3L_4.

(*b*) **Upper chord members,** such u_1u_2, u_2u_3.

(*c*) **End members,** such as L_0u_1 and L_4u_3.

(*d*) **Verticals :** Vertical members such as L_1u_1, L_2u_2 and L_3u_3.

(*e*) **Diagonals :** Such L_2u_1 and L_2u_3

(*f*) **Panel length :** The length of members L_0L_1, L_1L_2 is called panel length.

(*g*) **Span :** If all lower chord members are of the same length l, then the span is written as n @ $l = nl$, where n is the number of panels.

In the Fig. 2.6 the span is written as 4 @ $l = 4l$.

(*h*) **Height :** The distance of highest joint above the lower joints is called the height of the truss. It is also called the **rise**.

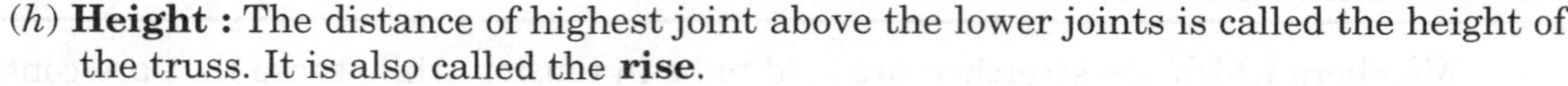

2.7. NATURE OF FORCES IN DIFFERENT MEMBERS OF A TRUSS

For simply supported trusses, the lower members are in tension ; upper chord members are in compression ; end members are in compression. These are shown in Fig. 2.7.1 (*a*) and (*b*).

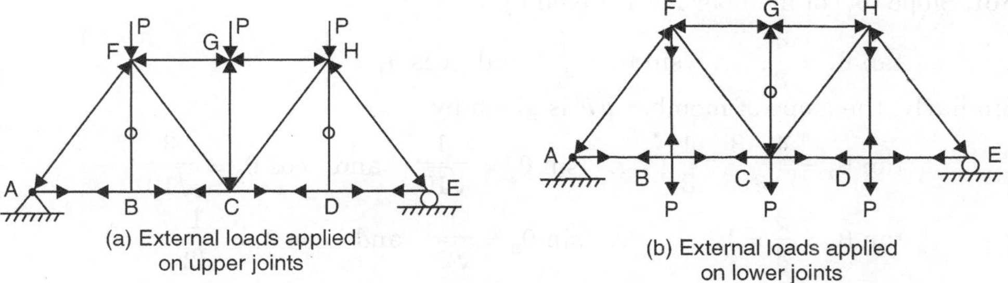

Fig. 2.7.1. Nature of forces in members of simply supported trusses

Thus, the nature of forces or stresses in lower and upper chord members can be found by observation. End members of the simply supported trusses are under compression.

(*a*) When external loads are applied on upper joints then vertical member *GC* will be under compression while verticals *BF* and *DH* carry no force. See Fig. 2.7.1 (*a*).

(*b*) When external loads are applied on the lower joints, as shown in Fig. 2.7.1(*b*), the verticals *BF* and *DH* are in tension while the diagonal *CG* will carry zero force.

(*c*) For the cantilever truss shown in Fig. 2.7.2, the upper chord members *AB, BC,* and *CD* are under tension and members *AE, EF* and *FG* are in compression. Since there is no external load applied on the joints *B* and *C*, therefore force in vertical members *BE* and *CF* is zero.

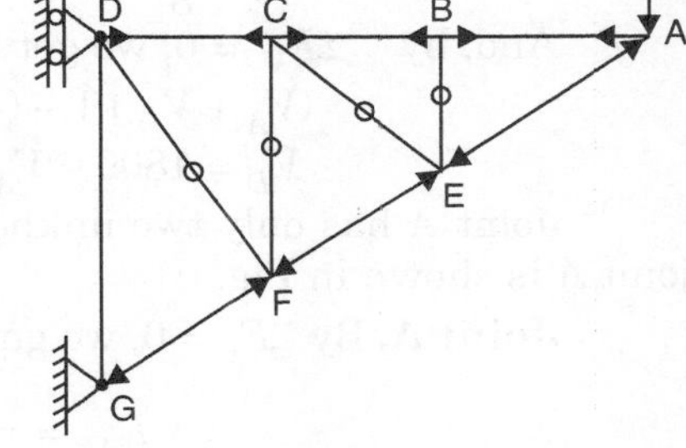

Fig. 2.7.2. Nature of forces in a cantilever truss

Similarly, by resolving forces acting on joint *E* in the direction perpendicular to the inclined members *EA* and *EF*, we find that force in member *EC* is zero. For equilibrium of joint *C*, since $F_{CE} = 0$, $\therefore$ by $\Sigma F_y = 0$, we get $F_{CF} = 0$. Also, by resolving forces on the joint *F*, along the normal to the members *FE* and *FG*, we find that the force F_{FD} in member *FD* is zero.

2.8. METHODS OF ANALYSING TRUSSES

There are three methods of finding forces in various members of the truss. These include :

(1) Method of Joints, (2) Method of Sections

(3) Graphical Method.

Method of joints is explained with the help of examples which follow here under.

Problem 2.1. *Find forces in all members of the truss shown in Fig. P.2.1(a).*

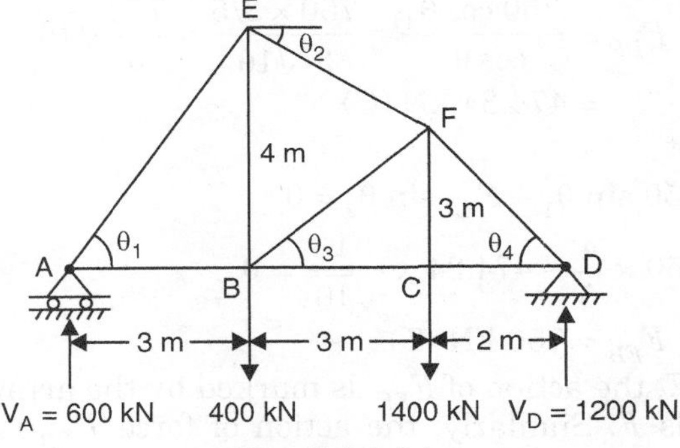

Fig. P. 2.1 (a).

Sol. Slope (θ_1) of member AE is given by

$\because \quad \tan\theta_1 = \frac{4}{3}, \quad \therefore \quad \sin\theta_1 = \frac{4}{5} \quad \text{and} \quad \cos\theta_1 = \frac{3}{5}$

Similarly, the slope of member EF is given by

$\because \quad \tan\theta_2 = \frac{4-3}{3} = \frac{1}{3}\,; \quad \therefore \quad \sin\theta_2 = \frac{1}{\sqrt{10}} \quad \text{and} \quad \cos\theta_2 = \frac{3}{\sqrt{10}},$

also, $\because \quad \tan\theta_3 = \frac{3}{3} = 1 \quad \therefore \quad \sin\theta_3 = \frac{1}{\sqrt{2}} \quad \text{and} \quad \cos\theta_3 = \frac{1}{\sqrt{2}}$

and $\quad \tan\theta_4 = \frac{3}{2}, \quad \therefore \quad \sin\theta_4 = \frac{3}{\sqrt{13}} \quad \text{and} \quad \cos\theta_4 = \frac{2}{\sqrt{13}}$

By $\Sigma M_D = 0$, we get, $\quad V_A \times 8 - 400 \times 5 - 1400 \times 2 = 0$

$\therefore \quad V_A = \frac{1}{8}(2000 + 2800) = 600 \text{ kN} \uparrow$

And, by $\quad \Sigma F_V = 0$, we get

$(V_A + V_D)\uparrow - (400 + 1400) = 0$

$V_D = 1800 - V_A = 1800 - 600 = 1200 \text{ kN} \uparrow$

Joint A has only two unknown members AB and AE. The FBD (Free-body-diagram) of joint A is shown in Fig. (*i*)

Joint A. By $\Sigma F_V = 0$, we get $F_{AE} \sin\theta_1 \downarrow - 600 = 0$ [See Fig. (*i*)]

$\therefore \quad F_{AE} = \frac{600}{\sin\theta_1} = \frac{600 \times 5}{4} = 750 \text{ kN (C)}$

By $\Sigma F_h = 0$, we get

$F_{AB} - F_{AE}\cos\theta_1 = 0,$

$\therefore \quad F_{AB} = F_{AE}\cos\theta_1$

$= 750 \times \frac{3}{5} = 450 \text{ kN } (T)$

Fig. (i) FBD of joint A

After solving joint A, the action of members AB and AE indicating respectively tension (T) and compression (C), on their end pins may now be shown by arrows in Fig. P. 2.1 (*b*). These arrows should be inserted on the original truss, but to indicate the marked and unmarked members only clearly, the truss has been redrawn (but of reduced size) as in Fig. P.2.1 (*b*). The arrow for member AB at A pulls the joint, and the arrow at B also pulls the joint B. Similarly, the arrow for member AE marked at E pushes the joint E.

Joint E. By $\Sigma F_h = 0$, we get [See Fig. (*ii*)]

$F_{EF}\cos\theta_2 - 750\cos\theta_1 = 0$

$\therefore \quad F_{EF} = \frac{750\cos\theta_1}{\cos\theta_2} = \frac{750 \times 3/5}{3/\sqrt{10}} = \frac{750}{5}\sqrt{10}$

$= 474.34 \text{ kN (C)}$

By $\quad \Sigma F_V = 0$, we get

$F_{EB} - 750\sin\theta_1 - F_{EF}\sin\theta_2 = 0$

$\therefore \quad F_{EB} - 750 \times \frac{4}{5} - 474.34 \times \frac{1}{\sqrt{10}} = 0$

$\therefore \quad F_{EB} = 750 \text{ kN } (T)$

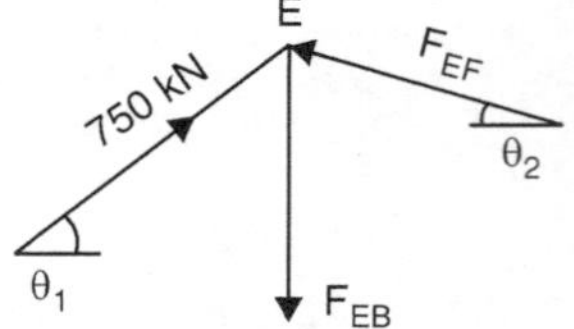

Fig. (ii) FBD of joint E

After solving joint E, the action of F_{EF} is marked by the arrow acting upwards towards E and downwards towards F. Similarly, the action of force F_{EB} is marked by arrow acting downwards at E and by an arrow acting upwards at B, see Fig. P. 2.1 (*c*). All these arrows can be marked on the original truss of Fig. P. 2.1 (*a*).

Joint B. Joint B is in equilibrium under the action of forces 450 kN (←), applied force 400 kN ↓, 750 kN ↑, F_{BF} (*C*) and F_{BC} (*T*). [See Fig. (*iii*)]

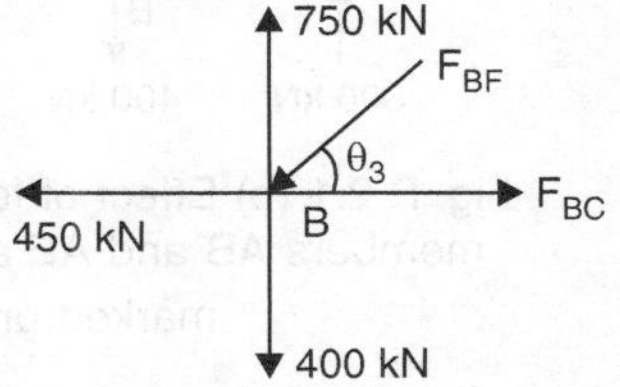

Fig. (iii) FBD of joint B

By $\Sigma F_V = 0$, we get

$$F_{BF} \sin \theta_3 - 750 + 400 = 0$$

$$\therefore \quad F_{BF} = \frac{1}{\sin \theta_3} \times 350$$

$$= \frac{350 \times \sqrt{2}}{1} = 495 \text{ kN } (C)$$

By $\Sigma F_h = 0$; we get

$$F_{BC} - F_{BF} \cos \theta_3 - 450 = 0$$

$$\therefore \quad F_{BC} - 495 \times \frac{1}{\sqrt{2}} - 450 = 0$$

$$\therefore \quad F_{BC} = 800 \text{ kN } (T)$$

The action of forces F_{BC} and F_{BF} has been marked in Fig. P. 2.1 (*d*).

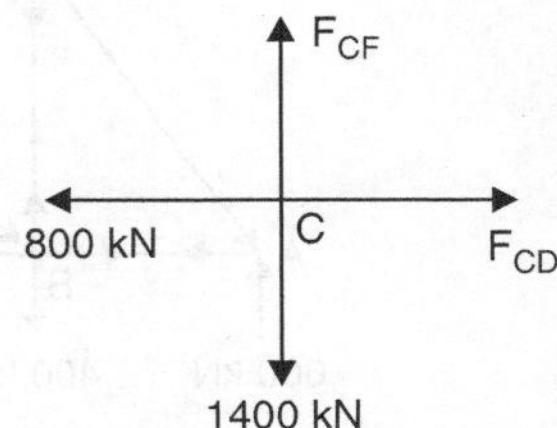

Fig. (iv) FBD of joint C

Joint C. By $\Sigma F_V = 0$, we get [See (Fig. *iv*)]

$$F_{CF} - 1400 = 0$$

$$\therefore \quad F_{CF} = 1400 \text{ kN}(T),$$

By $\Sigma F_h = 0$, we get

$$F_{CD} - 800 = 0$$

$$\therefore \quad F_{CD} = 800 \text{ kN } (T)$$

After solving joint C, action of forces in members CF and CD are marked in Fig. P. 2.1 (*e*)

Joint D. By $\Sigma F_V = 0$, we get [See Fig. (*v*)].

F_{DF}

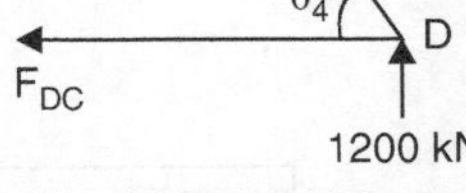

Fig. (v) FBD of joint D

$$F_{DF} \sin \theta_4 - 1200 = 0.$$

$$F_{DF} = \frac{1200}{\sin \theta_4} = \frac{1200\sqrt{3}}{3}$$

$$= \mathbf{1442.2 \text{ kN}} \ (C)$$

Forces in all members of the truss have been found. To check the correctness of the solution, let us assume that the force in member DC ($=F_{DC}$) is unknown.

By the condition $\Sigma F_h = 0$, at joint D the force F_{DC} is determined. The force in member CD has already been found to be equal to $F_{CD} = 800$ kN(*T*). The two values of force in member CD determined by solving joints C and D should be the same. ***If so, the entire, solution is correct.***

Here by $\Sigma F_h = 0$, we get

$$F_{DC} - F_{DF} \cos \theta_4 = 0$$

$$\therefore \quad F_{DC} = F_{DF} \cos \theta_4 = 1442.2 \times \frac{2}{\sqrt{13}}$$

$$= 800 \text{ kN } (T)$$ **Check of entire solution.**

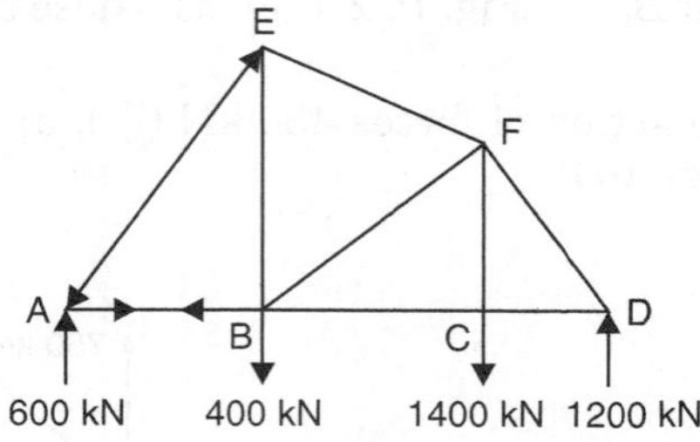

Fig. P. 2.1 (b) Effect of forces F_{AB} and F_{AE} in members AB and AE after solving joint A marked on truss

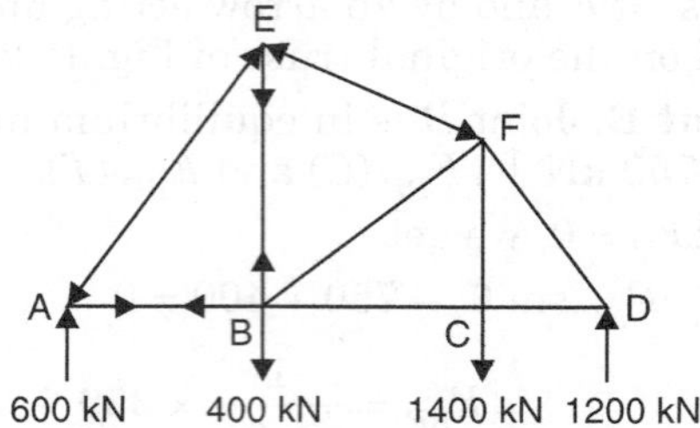

Fig. P. 2.1 (c) Effect of forces F_{EB} and F_{EF} is members EB and EF after solving joint E have been marked on the truss

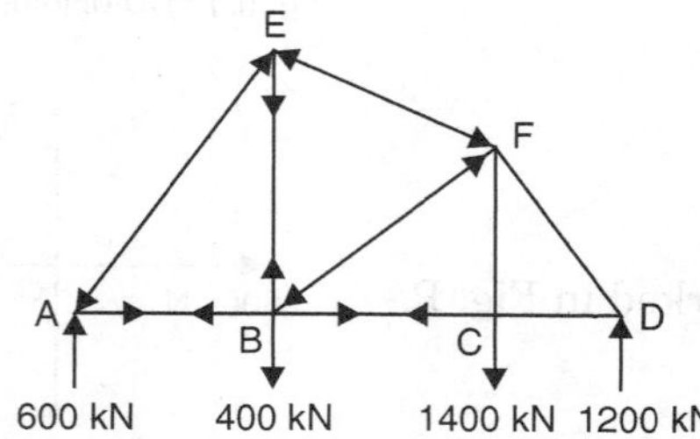

Fig. P. 2.1 (d) Truss marked after finding forces F_{BC} and F_{BF} in members BC and BF

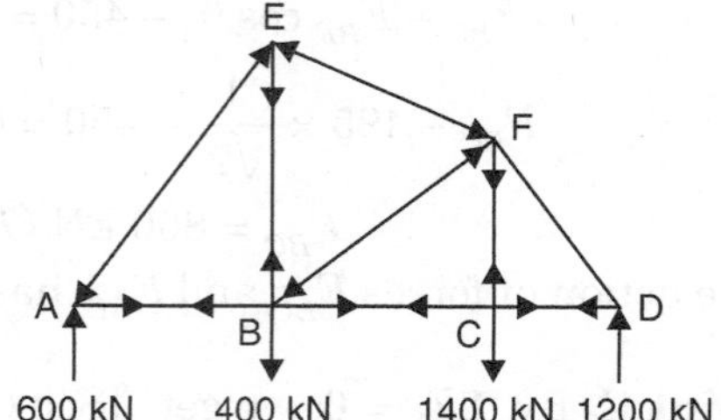

Fig. P. 2.1 (e) Effect of forces F_{CD} and F_{DF} in members CD and CF marked on truss after solving joint E

Note. While solving a problem on trusses, Figs. P. 2.1 (*b*, *c*, *d*, *e*) should not be drawn and arrow heads alongwith the numerical values of the forces in various members are directly marked and entered in the original truss shown in Fig. P. 2.1 (*a*). After entering all forces the original truss will look like the one shown in Fig. P. 2.1 (*f*).

Table of Results

+ for Tension ; – for compression

Member	*Force*	*Types of members*
AB *BC* *CD*	+ 450 kN + 800 kN + 800 kN	Lower chord Members
AE *DF*	– 750 kN – 1442.2 kN	End members
EF	– 474.34 kN	upper chord member
BE *CF*	+ 750 kN + 1400 kN	verticals
BF	– 495 kN	Diagonal member

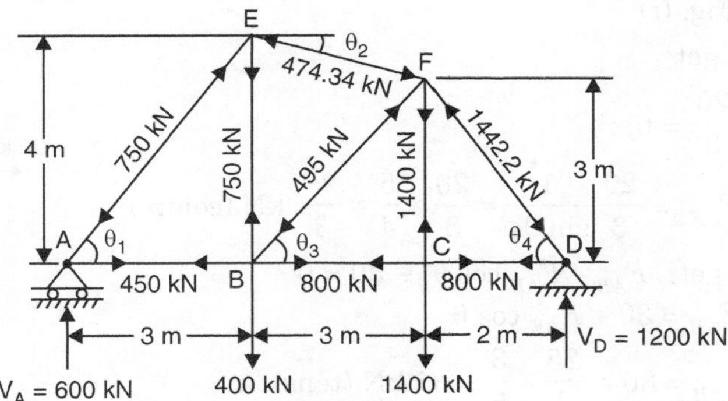

Fig. P. 2.1. (f) Given truss with forces introduced in various members.

Problem 2.2. *Find forces in all members of the truss shown in Fig. P. 2.2.*

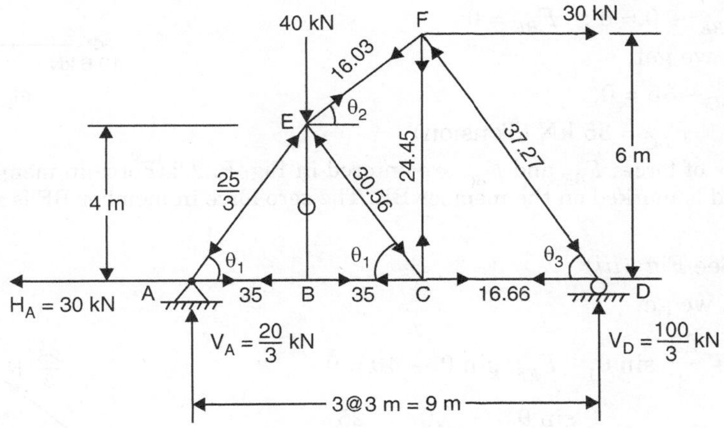

Fig. P. 2.2

Sol. $\because \tan\theta_1 = \dfrac{BE}{AB} = \dfrac{4}{3} \quad \therefore \quad \sin\theta_1 = \dfrac{4}{5}$ and $\cos\theta_1 = \dfrac{3}{5}$

$$\because \tan\theta_2 = \frac{CF - BE}{BC} = \frac{6-4}{3} = \frac{2}{3} \quad \therefore \quad \sin\theta_2 = \frac{2}{\sqrt{13}} \text{ and } \cos\theta_2 = \frac{3}{\sqrt{13}}$$

and, $\because \tan\theta_3 = \dfrac{CF}{CD} = \dfrac{6}{3} = \dfrac{2}{1} \quad \therefore \quad \sin\theta_3 = \dfrac{2}{\sqrt{5}}$ and $\cos\theta_3 = \dfrac{1}{\sqrt{5}}$

By $\Sigma F_h = 0$, we get

$$H_A - 30 = 0, ; H_A = 30 \text{ kN} \leftarrow$$

By $\Sigma M_A = 0$, we get, $V_D \times 9 - 30 \times 6 - 40 \times 3 = 0$

$$\therefore \quad V_D = \frac{1}{9}[180 + 120] = \frac{300}{9} = \frac{100}{3} \text{ kN} \uparrow$$

By $\Sigma F_V = 0$, we get, $(V_A + V_D) \uparrow - 40 = 0$

$$\therefore \quad V_A = 40 - V_D = 40 - \frac{100}{3} = \frac{20}{3} \text{ kN} \uparrow.$$

Joint A : See Fig. (*i*)

By $\Sigma F_V = 0$, we get

$$F_{AE} \sin \theta_1 - \frac{20}{3} = 0, :$$

$$F_{AE} = \frac{20}{3} \frac{1}{\sin \theta_1} = \frac{20}{3} \times \frac{5}{4} = \frac{25}{3} \text{ kN (comp.)}$$

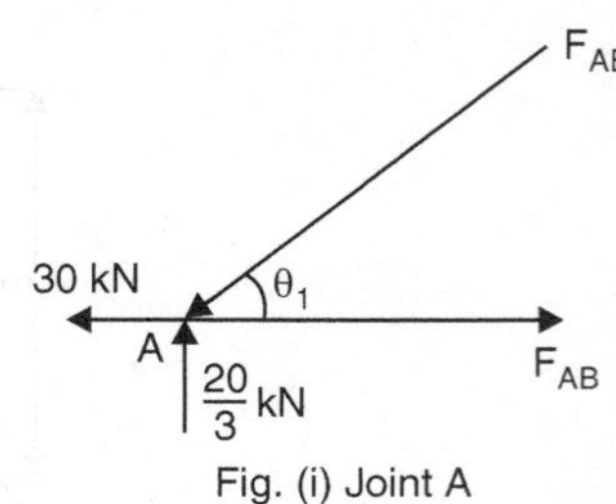

Fig. (i) Joint A

By $\Sigma F_h = 0$, we get $F_{AB} - F_{AE} \cos \theta_1 - 30 = 0$

$$\therefore \quad F_{AB} = 30 + F_{AE} \cos \theta_1$$

or, $$F_{AB} = 30 + \frac{25}{3} \times \frac{3}{5} = 35 \text{ kN (tension)}$$

Note. Values of F_{AE} and F_{AB} are entered in Fig. P. 2.2

Joint B : (See Fig. (*ii*))

By $\Sigma F_V = 0$, we get

$$F_{BE} + 0 = 0 ; \quad F_{BE} = 0$$

By $\Sigma F_h = 0$, we get,

$$F_{BC} - 35 = 0,$$

$$\therefore \quad F_{BC} = 35 \text{ kN (Tension)}$$

F_{BE}
19.6 kN
B
F_{BC}

Fig. (ii) Joint B

Note. Values of forces F_{BE} and F_{BC} are entered in Fig. P. 2.2. Force in member BE is zero and hence no arrow-head is marked on the member BE. The zero force in member BF is marked as ϕ on the member.

Joint E : See Fig. (*iii*)

By $\Sigma F_V = 0$, we get

$$F_{EC} \sin \theta_1 + \frac{25}{3} \sin \theta_1 - F_{EF} \sin \theta_2 - 40 = 0$$

$$\therefore \quad F_{EC} - F_{EF} \frac{\sin \theta_2}{\sin \theta_1} = \frac{40}{\sin \theta_1} - \frac{25}{3}$$

or $$F_{EC} - F_{EF} \times \frac{2}{\sqrt{13}} \times \frac{5}{4} = \frac{40 \times 5}{4} - \frac{25}{3}$$

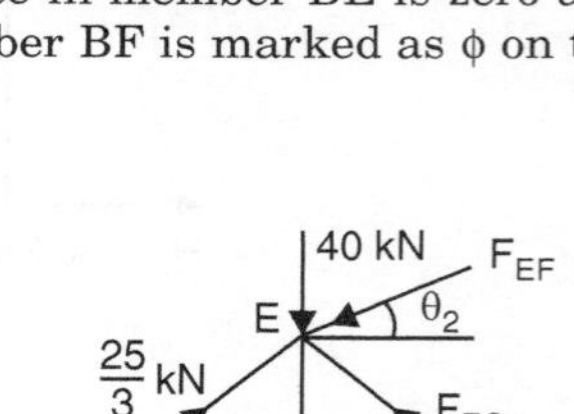

Fig. (iii) Joint E

$$F_{EC} - \frac{2.5}{\sqrt{13}} F_{EF} = \frac{125}{3}$$

or $$F_{EC} - 0.693 F_{EF} = 41.67 \quad \text{...(1)}$$

By $\Sigma F_h = 0$, we get

$$F_{EC} \cos \theta_1 + F_{EF} \cos \theta_2 - \frac{25}{3} \times \cos \theta_1 = 0$$

$$\therefore \quad F_{EC} + F_{EF} \frac{\cos \theta_2}{\cos \theta_1} = \frac{25}{3}$$

$$\therefore \quad F_{EC} + F_{EF} \times \frac{3}{\sqrt{13}} \times \frac{5}{3} = \frac{25}{3}$$

$$\therefore \quad F_{EC} + 1.387 F_{EF} = 8.33 \quad \text{...(2)}$$

By (1) – (2), we get

$$F_{EF}\ (-\,0.693 - 1.387) = 41.67 - 8.33$$

or

$$-\,2.080\ F_{EF} = 33.34$$

$$F_{EF} = -\,16.03 \text{ kN},$$

hence member *EF* is in tension, and not in compression.

i.e., $F_{EF} = 16.03$ kN (Tension)

And, from (1) $F_{EC} = 41.67 + 0.693 \times F_{EF}$

$$= 41.67 + 0.693 \times (-\,16.03) \text{ kN}$$

$\therefore$ F_{EC} = **30.56 kN** (Compression).

Note. Forces F_{EC} and F_{EF} have been entered in the truss shown in Fig. P. 2.2.

Joint C : See Fig. (*iv*)

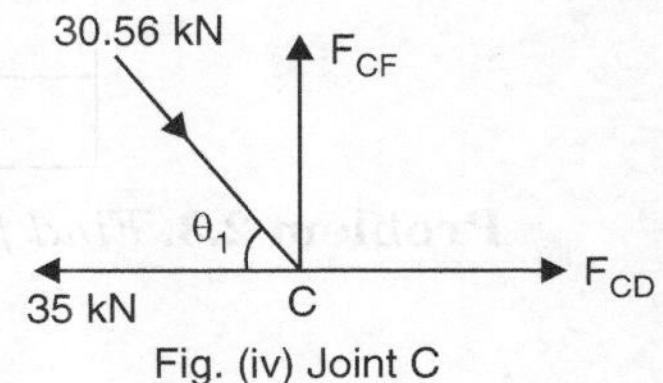

Fig. (iv) Joint C

By $\Sigma F_V = 0$, we get

$$F_{CF} - 30.56 \sin \theta_1 = 0$$

$$F_{CF} = 30.56 \times \frac{4}{5}$$

= **24.45 kN** (Tension)

By $\Sigma F_h = 0$, we get

$$F_{CD} + 30.56 \cos \theta_1 - 35 = 0$$

$$\therefore \quad F_{CD} + 30.56 \times \frac{3}{5} - 35 = 0$$

$\therefore$ F_{CD} = **16.66 kN** (Tension)

Joint D : See Fig. (*v*)

Accuracy and correctness of all of the above calculations can be checked by assuming the force (F_{DC}) in the member *CD* to be unknown. If this force is the same as determined earlier by solving joint *C*, the correctness of the entire solution will be ascertained. If the two values of F_{CD} or F_{DC} determined independently by solving joints *C* and *D* differ, then the earlier solution should be checked again.

Fig. (v) Joint D

By $\Sigma F_V = 0$, we get

$$F_{DF}\ \sin \theta_3 - \frac{100}{3} = 0,$$

$\therefore$ $F_{DF} = \dfrac{100}{3 \sin \theta_3} = \dfrac{100}{3} \times \dfrac{\sqrt{5}}{2}$ = **37.27 kN** (*Compression*)

By $\Sigma F_h = 0$, we get

$$F_{DC} - F_{DF}\ \cos \theta_3 = 0,$$

$\therefore$ $F_{DC} = F_{DF} \cos \theta_3 = \left(\dfrac{100\sqrt{5}}{3 \times 2}\right) \times \dfrac{1}{\sqrt{5}}$ = **16.67 kN** (Tension). **Check**

Note. Since the two values F_{CD} and F_{DC} determined by solving joints *C* and *D* are the same, the entire solution is correct.

Table of Results

\+ For tension ; – for compression

Member	*Force*
AB	+ 35 kN
BC	+ 35 kN
CD	+ 16.66 kN
CE	– 25/3 = – 8.33 kN
DF	– 37.27 kN
EF	+ 16.03 kN
BE	0
CF	+ 24.45 kN
CE	– 30.56 kN

Problem 2.3. *Find forces in all members of the truss shown in Fig. P.2.3 (a).*

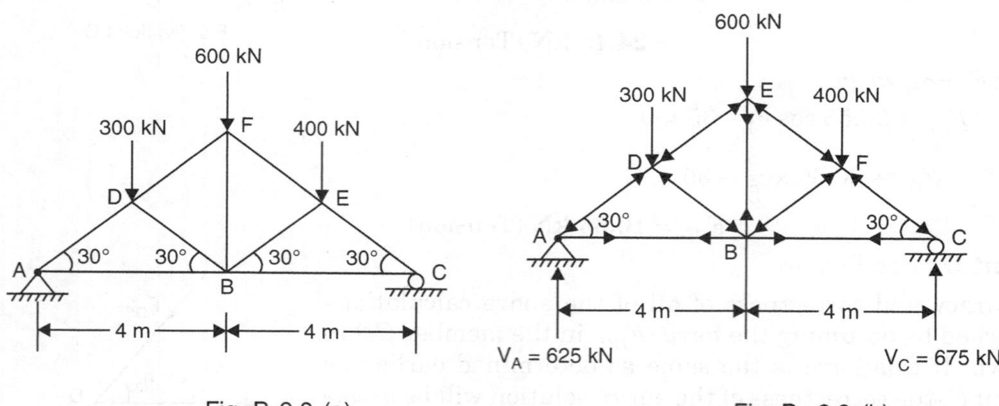

Fig. P. 2.3 (a) Fig. P. 2.3 (b)

Sol. Here there are two unknown reaction components *viz.*, V_A and V_C which can be determined by using two conditions of equilibrium *i.e.*, $\Sigma M = 0$ and $\Sigma F_V = 0$. V_A and V_C both may be assumed acting upward.

To find V_A, use $\Sigma M_C = 0$ $\therefore$ we get,

$$V_A \times 8 - 300 \times 6 - 600 \times 4 - 400 \times 2 = 0$$

$$V_A = \frac{1}{8}[1800 + 2400 + 800] = 625 \text{ kN} \uparrow.$$

By using the condition $\Sigma F_V = 0$, we get

$$(V_A + V_C) \uparrow - (300 + 600 + 400) = 0$$

$$\therefore \quad V_C = 1300 - 625 = 675 \text{ kN} \uparrow.$$

By using method of joints we can find forces in all members.

Joint A :

The free body diagram of joint *A* is shown in Fig. (*i*)

By $\Sigma F_V = 0$, we get

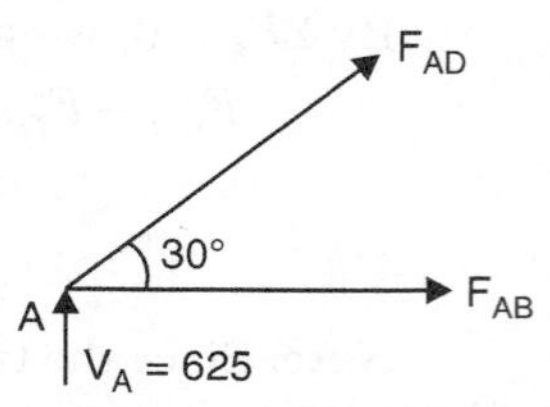

Fig. (i) Free body diagram (FBD) of joint A

$\uparrow F_{AD} \sin 30° + 625 = 0$

$$\therefore \quad F_{AD} = \frac{-625}{\sin 30°} = -1250 \text{ kN}$$

Since the sign of F_{AD} is negative therefore the nature *of* F_{AD} as shown in Fig. (*i*) is opposite.

$$\therefore \quad F_{AD} = 1250 \text{ kN (compressive)}$$

By $\Sigma F_H = 0$, we get

$\rightarrow \quad F_{AB} + F_{AD} \cos 30° = 0$

$$F_{AB} = -F_{AD} \cos 30° = -(-1250) \times \frac{\sqrt{3}}{2} = 625\sqrt{3} \text{ kN (tensile)}$$

direction of the forces are transferred in Fig. P.2.3 (*c*)

We move to the joint *D*, because there are two unknown forces concurrent at joint *D*.

Joint D : See Fig. (*ii*)

By $\Sigma F_V = 0$, we get

$\uparrow F_{DB} \sin 30° + 1250 \sin 30° - F_{DE} \sin 30° - 300 = 0$

$$\therefore \quad F_{DB} - F_{DE} = \frac{300}{\sin 30°} - 1250$$

$$\therefore \quad F_{DB} - F_{DE} = 600 - 1250 = -650 \quad \text{...(1)}$$

By $\Sigma F_H = 0$, we get

$\leftarrow F_{DB} \cos 30° + F_{DE} \cos 30° - 1250 \cos 30° = 0$

$$\therefore \quad F_{DB} + F_{DE} = 1250 \quad \text{...(2)}$$

Adding (1) and (2) $\quad 2F_{DB} = 600$

$$\therefore \quad F_{DB} = 300 \text{ kN (compressive)}$$

and from (2), $\quad F_{DE} = 1250 - 300 = 950$ kN (compressive)

Fig. (ii) FBD of joint D

Forces F_{DB} and F_{DE} are transferred in Fig. P.2.3 (*c*).

Joint E : See Fig. (*iii*)

By $\Sigma F_H = 0$, we get

$\leftarrow F_{EF} \cos 30° - 950 \cos 30° = 0$

$$\therefore \quad F_{EF} = 950 \text{ kN (compressive)}$$

By $\Sigma F_V = 0$, we get

$\uparrow F_{EB} + 950 \sin 30° + F_{EF} \sin 30° - 600 = 0$

$$\therefore \quad F_{EB} = 600 - 950 \sin 30° - 950 \sin 30°$$

$$= 600 - 950 = -350$$

$$\therefore \quad F_{EB} = 350 \text{ (tensile)}$$

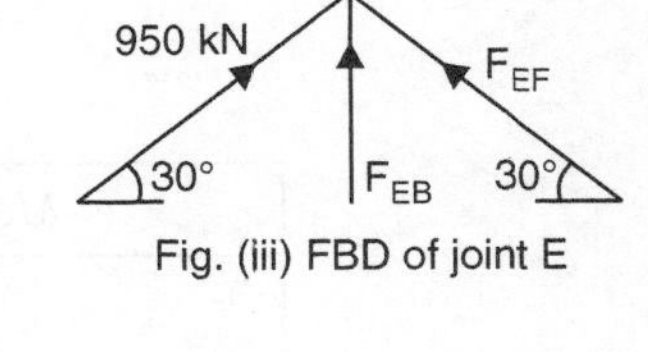

Fig. (iii) FBD of joint E

Correct values and direction of forces in members *EB and EF* are entered on the truss shown in Fig. P. 2.3 (*c*).

Joint B : see Fig. (*iv*)

By $\Sigma F_V = 0$, we get

$\downarrow F_{BF} \sin 30° + 300 \sin 30° - 350 = 0$

$$F_{BF} = \frac{350}{\sin 30} - 300$$

$$= 400 \text{ kN (compressive)}$$

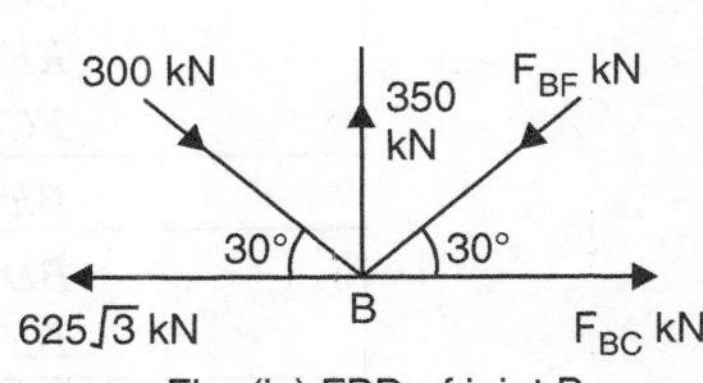

Fig. (iv) FBD of joint B

By $\Sigma F_H = 0$, we get

$$\rightarrow F_{BC} - F_{BF} \cos 30° + 300 \cos 30° - 625\sqrt{3} = 0$$

$$F_{BC} - 400 \cos 30° + 300 \cos 30° - 625\sqrt{3} = 0$$

$$F_{BC} - 100 \cos 30° - 625\sqrt{3} = 0$$

$$F_{BC} = 675\sqrt{3} \text{ kN (tensile)}$$

These forces are also transferred in Fig. 2.17(*c*).

Joint C : See Fig. (*v*)

Assume that the force in the member *CB* to be unknown, solve joint *C* for the unknown forces F_{CB} and F_{CE},

By $\Sigma F_V = 0$, we get

$$\downarrow F_{CF} \sin 30 - 675 = 0$$

$$\therefore \quad F_{CF} = \frac{675}{\sin 30} = 1350 \text{ kN (compressive)}$$

By $\Sigma F_H = 0$, we get

$$\leftarrow F_{CB} - F_{CF} \cos 30° = 0 \quad \therefore \quad F_{CB} - 1350 \times \frac{\sqrt{3}}{2} = 0$$

$$\therefore \quad F_{CB} = 675\sqrt{3} \text{ (Tensile) } \textbf{check.}$$

Fig. (*v*) FBD of joint C

Check. Force in the member *BC* has been found by independently solving joints *B* and *C*. Since solution of both joints give the same force $F_{BC} = 675\sqrt{3}$ kN (tensile), therefore the entire solution is correct. This check should be applied by the students in the examination hall to ensure the correctness of the solution. If the two values of force in member *BC* differ, then check to find the mistake incurred in the solution of various joints.

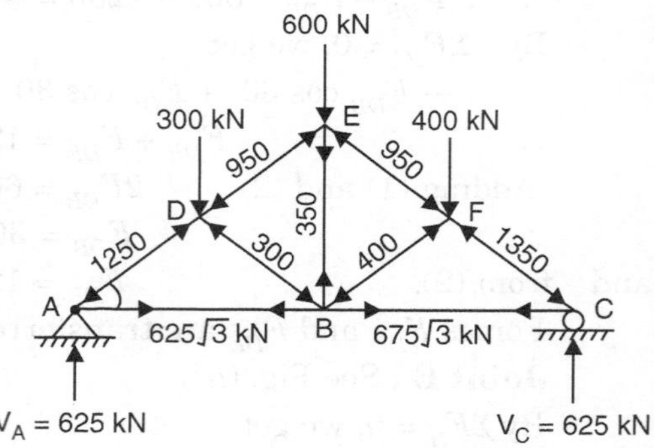

Fig. P. 2.3 (c) Forces induced in members of truss.

Table of Results

\+ for Tension ; – for compression

Member	*Force*
AB	$+ 625\sqrt{3} = + 1082.53$ kN
BC	$+ 675\sqrt{3} = + 1169.13$ kN
AD	– 1250 kN
DE	– 950 kN
EF	– 950 kN
FC	– 1350 kN
BE	+ 350 kN
BD	– 300 kN
BF	– 400 kN

Problem 2.4. *Find force in each member of the truss shown in Fig. P. 2.4(a).*

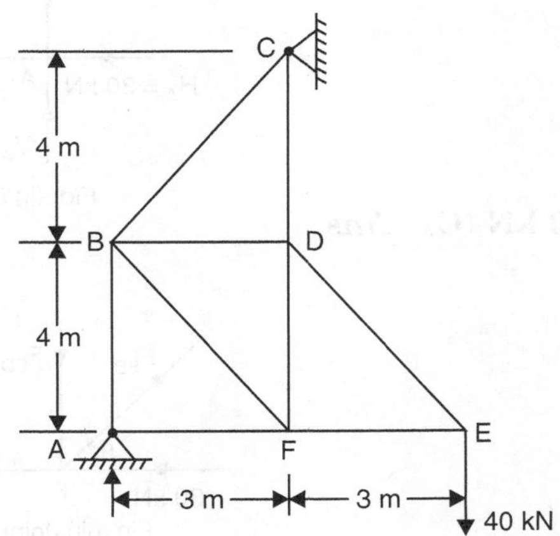

Fig. P. 2.4 (a) Given truss

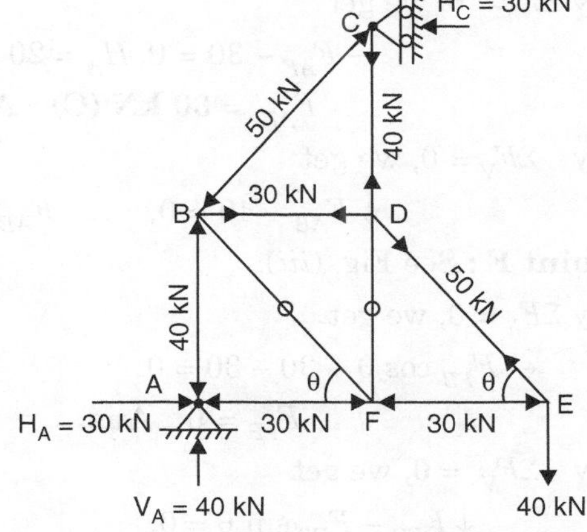

Fig. P 2.4 (b) Forces in members of the truss

Sol. Here, $\tan\theta = 4/3$,

$\therefore$ $\sin\theta = 4/5$; $\cos\theta = 3/5$.

By $\Sigma M_A = 0$, we get

$$H_C \times 8 - 40 \times 6 = 0$$

$$\therefore \quad H_C = \frac{40 \times 6}{8} = 30 \text{ kN} \leftarrow$$

By $\Sigma F_h = 0$, we get

$$\leftarrow H_A + H_C = 0,$$

$$\therefore \quad H_A = -H_C = -30 \text{ kN} \quad \text{or} \quad H_A = 30 \text{ kN} \rightarrow$$

By $\Sigma F_V = 0$, we get

$$\uparrow V_A - 40 = 0, \quad \therefore \quad V_A = 40 \text{ kN} \uparrow.$$

All reaction components, *viz.*, V_A, H_A and H_C, are shown in Fig. P.2.4(*b*)

Joint E : See Fig. (*i*)

By $\Sigma F_V = 0$, we get

$$F_{ED} \sin\theta - 40 = 0$$

$$\therefore \quad F_{ED} = \frac{40}{\sin\theta} = \frac{40 \times 5}{4}$$

$$= \mathbf{50\ kN\ (T)} \quad \textbf{Ans.}$$

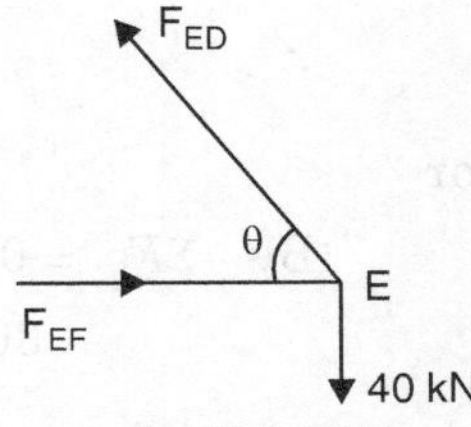

Fig. (i) Joint E

By $\Sigma F_h = 0$, we get

$$F_{EF} - F_{ED} \cos\theta = 0$$

$$\therefore \quad F_{EF} = F_{ED} \cos\theta = 50 \times \frac{3}{5}$$

$$= \mathbf{30\ kN\ (C)} \quad \textbf{Ans.}$$

Joint A : See Fig. (*ii*).

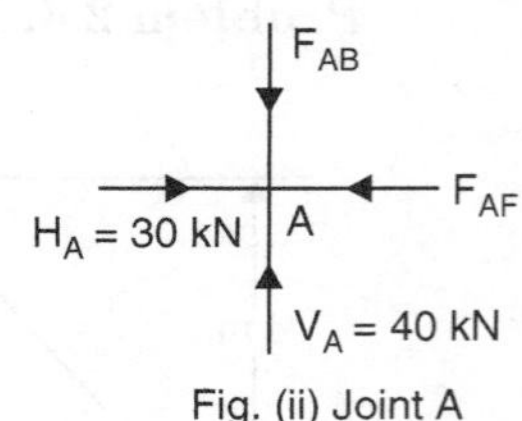

Fig. (ii) Joint A

By $\Sigma F_h = 0$, we get

$$\leftarrow F_{AF} - 30 = 0, H_A = 20 \text{ kN}$$

$\therefore$ F_{AF} = **30 kN (C) Ans.**

By $\Sigma F_V = 0$, we get

$$\downarrow F_{AB} - 40 = 0, \quad \therefore \quad F_{AB} = \textbf{40 kN (C)} \quad \textbf{Ans.}$$

Joint F : See Fig. (*iii*).

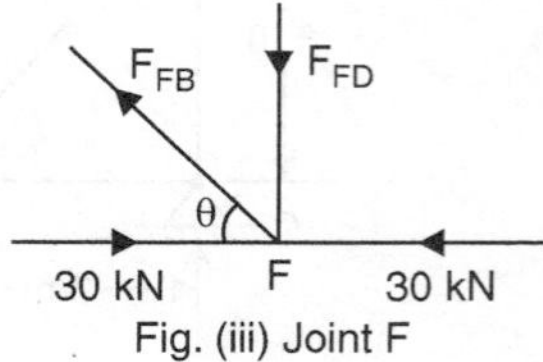

Fig. (iii) Joint F

By $\Sigma F_h = 0$, we get

$$\leftarrow F_{FB} \cos\theta + 30 - 30 = 0.$$

$\therefore$ $F_{FB} = 0$ **Ans.**

By $\Sigma F_V = 0$, we get

$$\downarrow F_{FD} - F_{FB} \sin\theta = 0,$$

$\therefore$ $F_{FD} = F_{FB} \sin\theta = 0$ **Ans.**

Joint B : See Fig. (*iv*).

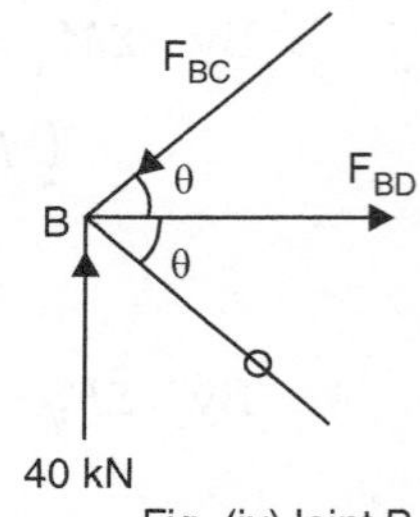

Fig. (iv)Joint B

By $\Sigma F_V = 0$, we get

$$\downarrow F_{BC} \sin\theta - 40 = 0$$

$$\therefore \quad F_{BC} = \frac{40}{\sin\theta} = \frac{40 \times 5}{4}$$

$$= \textbf{50 kN (C)} \quad \textbf{Ans.}$$

By $\Sigma F_h = 0$, we get

$$\rightarrow F_{BD} - F_{BC} \cos\theta = 0,$$

$$\therefore \quad F_{BD} = F_{BC} \cos\theta$$

$$= 50 \times \frac{3}{5} = \textbf{30 kN (T)} \quad \textbf{Ans.}$$

Joint D : See Fig. (*v*).

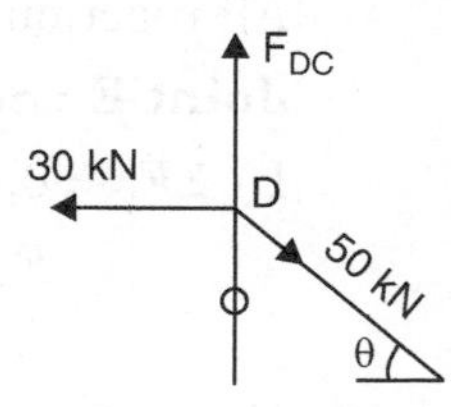

Fig. (v)Joint D

By $\Sigma F_V = 0$, we get $\uparrow F_{DC} - 50 \sin\theta = 0$

$$\therefore \quad F_{DC} = 50 \sin\theta = 50 \times \frac{4}{5} \text{ kN}$$

or F_{DC} = **40 kN (T) Ans.**

By $\Sigma F_h = 0$, we get

$$30 - 50 \cos\theta = 0$$

$$\therefore \quad 30 - 50 \times \frac{3}{5} = 0$$

$\therefore$ $0 = 0$ **check Ans.**

This *check* confirms that forces in various members of the truss are 100% correct. All results are shown in Fig. P.2.4 (*b*).

Problem 2.5. *Find force in each member of the truss shown in Fig. P.2.5 (a).*

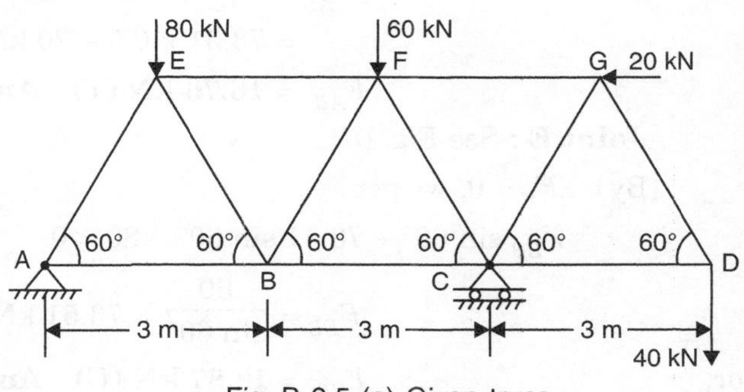

Fig. P. 2.5 (a) Given truss

Sol. By $\Sigma F_h = 0$, we get

$$H_A - 20 = 0$$

$$\therefore \quad H_A = 20 \text{ kN} \rightarrow$$

By $\Sigma M_A = 0$, we get

$$V_C \times 6 + 20 \times (3 \sin 60°) - 80 \times 1.5 - 60 \times 4.5 - 40 \times 9 = 0$$

$$\therefore \quad V_C = \frac{1}{6}\left[40 \times 9 + 80 \times 1.5 + 60 \times 4.5 - 20 \times \frac{3 \times \sqrt{3}}{2}\right] \text{ kN}$$

$$= \frac{1}{6}\,[360 + 120 + 270 - 30\sqrt{3}\,] \text{ kN}$$

$$\therefore \quad V_C = 116.34 \text{ kN} \uparrow$$

By $\Sigma F_V = 0$, we get $(V_A + V_C)\uparrow - (80 + 60 + 40) = 0.$

$$\therefore \quad V_A = 180 - V_C = 180 - 116.34 = 63.66 \text{ kN} \uparrow$$

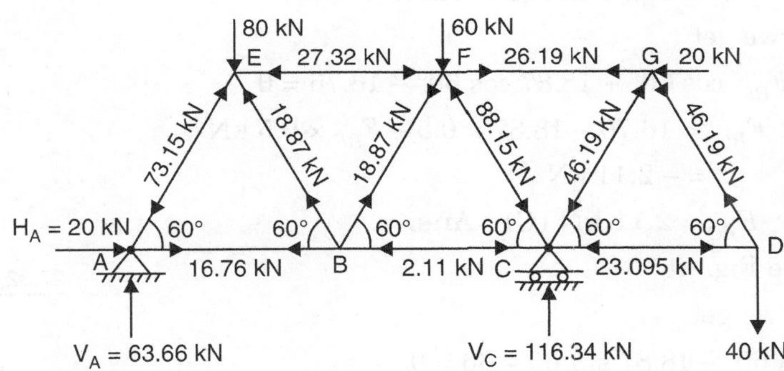

Fig. P. 2.5 (b) The results

Joint A : See Fig. (*i*)

By $\Sigma F_V = 0$, we get

$$F_{AE} \sin 60° - 63.34 = 0$$

$$\therefore \quad F_{AE} = \frac{63.34}{\sin 60°} = \mathbf{73.51 \text{ kN (C)}} \quad \textbf{Ans.}$$

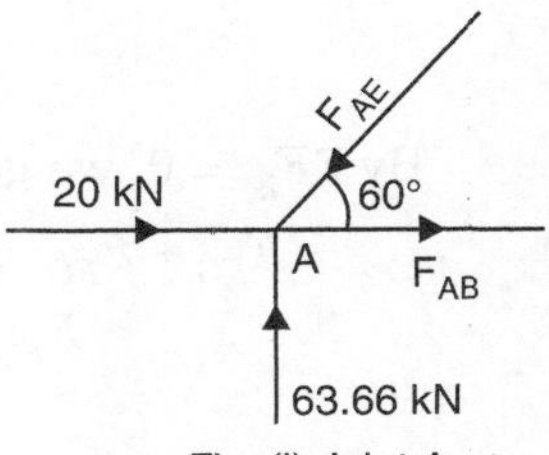

Fig. (i) Joint A

By $\Sigma F_V = 0$, we get

$$F_{AB} + 20 - F_{AE} \cos 60° = 0$$

$\therefore \qquad F_{AB} = F_{AE} \cos 60° - 20$

$= 73.51 \times 0.5 - 20$ kN

F_{AB} = **16.76 kN (T) Ans.**

Joint E : See Fig. *(ii)*

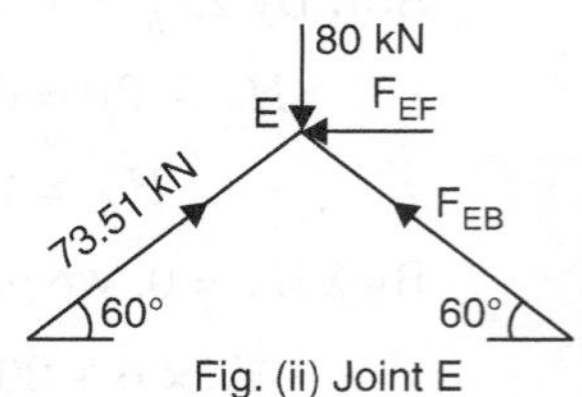

Fig. (ii) Joint E

By $\Sigma F_V = 0$, we get

$$F_{EB} \sin 60° + 73.51 \sin 60° - 80 = 0$$

$$\therefore \qquad F_{EB} = \frac{80}{\sin 60°} - 73.51 \text{ kN.}$$

or, F_{EB} = **18.87 kN (C) Ans.**

By $\Sigma F_h = 0$, we get

$$F_{EF} + F_{EB} \cos 60° - 73.51 \cos 60° = 0.$$

$$\therefore \qquad F_{EF} = 73.51 \cos 60° - F_{EB} \cos 60°$$

$$= (73.51 - 18.87) \times \cos 60°$$

F_{EF} = **27.32 kN (Compression)**

Joint B : See Fig. *(iii)*

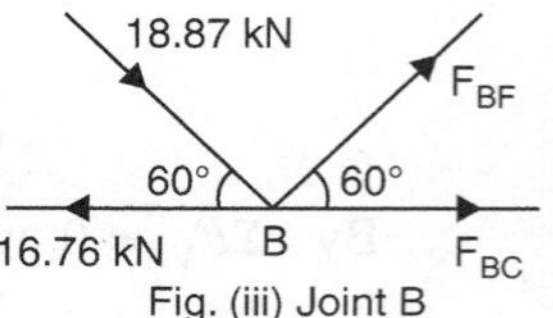

Fig. (iii) Joint B

By $\Sigma F_V = 0$, we get

$$F_{BF} \sin 60° - 18.87 \sin 60° = 0$$

$\therefore \qquad F_{BF}$ = **18.87 kN (T) Ans.**

By $\Sigma F_h = 0$, we get

$$F_{BC} + F_{BF} \cos 60° + 18.87 \cos 60° - 16.76 = 0$$

$$\therefore \qquad F_{BC} = 16.76 - 18.87 \times 0.5 - F_{BF} \times 0.5 \text{ kN}$$

$$= -2.11 \text{ kN}$$

$\therefore \qquad F_{BC}$ = **2.11 kN (C) Ans.**

Joint F : See Fig. *(iv)*

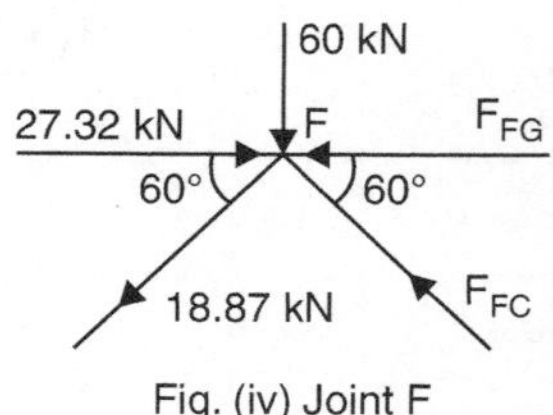

Fig. (iv) Joint F

By $\Sigma F_V = 0$, we get

$$F_{FC} \sin 60° - 18.87 \sin 60 - 60 = 0$$

$$\therefore \qquad F_{FC} = \frac{60}{\sin 60°} + 18.87$$

= **88.15 kN (C) Ans.**

By $\Sigma F_h = 0$, we get

$$F_{FG} + F_{FC} \cos 60° + 18.87 \cos 60° - 27.32 = 0$$

$$F_{FG} = 27.32 - F_{FC} \cos 60° - 18.87 \cos 60°$$

$$= -26.19 \text{ kN}$$

$\therefore \qquad F_{FG}$ = **26.19 kN (T) Ans.**

Joint G : See Fig. *(v)*

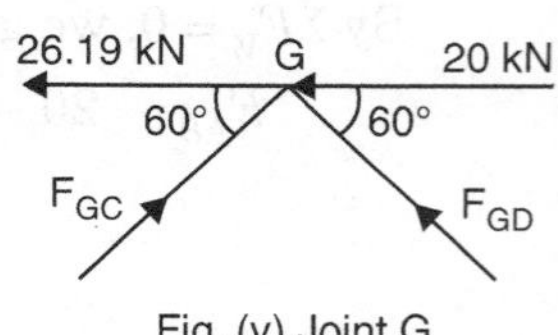

Fig. (v) Joint G

By $\Sigma F_V = 0$, we get

$$F_{GC} \sin 60° + F_{GD} \sin 60° = 0$$

$\therefore \qquad F_{GC} + F_{GD} = 0 \qquad \text{...(1)}$

By $\Sigma F_h = 0$, we get

$$F_{GC} \cos 60° - F_{GD} \cos 60° - 26.19 - 20 = 0$$

$$\therefore \qquad F_{GC} - F_{GD} = \frac{46.19}{\cos 60°}$$

or, $\qquad F_{GC} - F_{GD} = 92.38 \qquad \text{...(2)}$

Adding (1) and (2), we get

$$2F_{GC} = 92.38 \quad \therefore \quad F_{GC} = \mathbf{46.19\ kN\ (C)} \quad \textbf{Ans.}$$

from (1), $\qquad F_{GD} = - F_{GC} = - 46.19$ kN.

$$\therefore \qquad F_{GD} = \mathbf{46.19\ kN\ (T)} \quad \textbf{Ans.}$$

Joint D : See Fig. (*vi*)

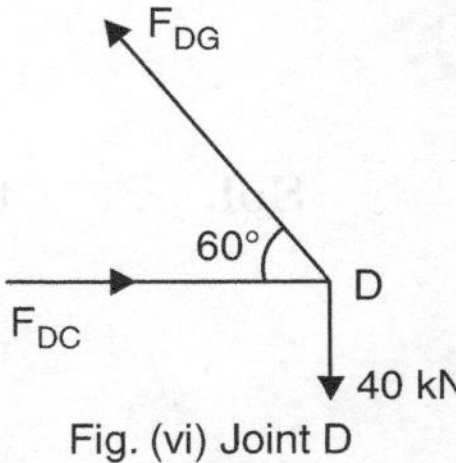

Fig. (vi) Joint D

To check the correctness of solution, the force F_{DG} in the member DG is taken as unknown. If the solution of joint D gives the same value of F_{DG} as was obtained by solving joint G, then the entire solution till this stage will be correct. If not, then the entire solution needs a check-up.

Therefore, by $\Sigma F_v = 0$, we get

$$F_{DG} \sin 60° - 40 = 0$$

$$\therefore \qquad F_{DG} = \frac{40}{\sin 60°} = \mathbf{46.19\ kN\ (T)} \quad \textbf{Ans.} \text{ (check).}$$

Till this stage the entire solution of the truss is correct.

Now, by $\Sigma F_h = 0$, we get

$$F_{DC} - F_{DG} \cos 60° = 0$$

$$\therefore \qquad F_{DC} = F_{DG} \cos 60° = 46.19 \times 0.5 \text{ kN}$$

$$\therefore \qquad F_{DC} = \mathbf{23.095\ kN\ (C)} \quad \textbf{Ans.}$$

The results are indicated in Fig. P.2.5(*b*).

All reaction components have been shown in Fig. P.2.5 (*b*).

Problem 2.6. *Find forces in all members of the cantilever truss shown in Fig. P.2.6(a)* .

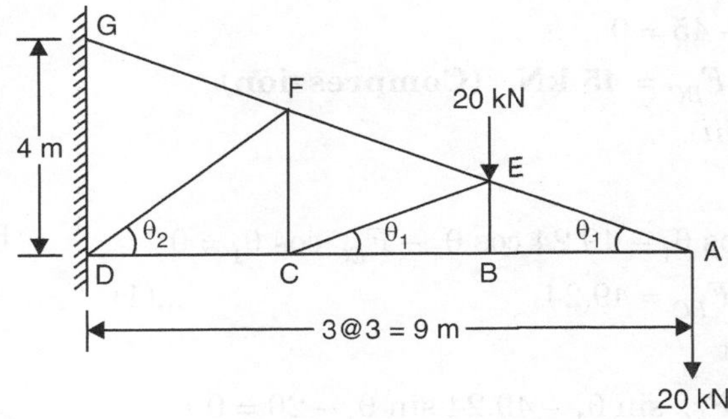

Fig. P. 2.6 (a) Given Truss

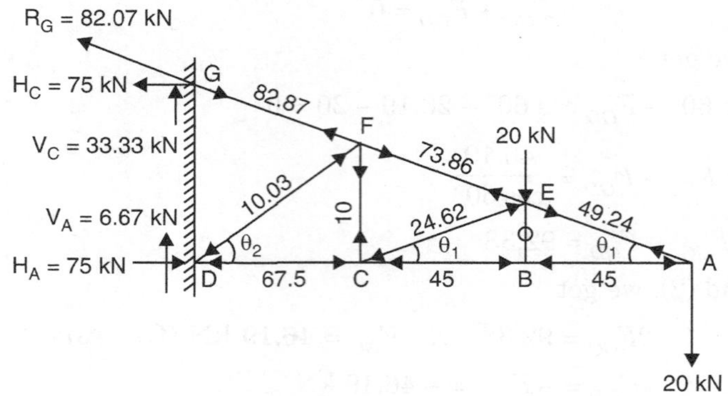

Fig. P. 2.6 (b) Forces in members

Sol. $\because \quad \tan\theta_1 = \dfrac{4}{9}; \quad \therefore \quad \sin\theta_1 = \dfrac{4}{\sqrt{97}}$ and $\cos\theta_1 = \dfrac{9}{\sqrt{97}}$

$\because \quad \tan\theta_2 = \dfrac{FC}{DC} = \dfrac{DG \times 2/3}{3} = \dfrac{4 \times 2/3}{3} = \dfrac{8}{9}$

$\therefore \quad \sin\theta_2 = 8/\sqrt{81+64} = 8/\sqrt{145}$ and $\cos\theta_2 = 9/\sqrt{145}$

Joint A : See Fig. (*i*).

By $\Sigma F_V = 0$, we get $\uparrow F_{AE} \sin\theta_1 - 20 = 0$,

$$F_{AE} = \frac{20}{\sin\theta_1} = \frac{20}{4/\sqrt{97}}$$

$\therefore \quad F_{AE} =$ **49.24 kN** (Tension)

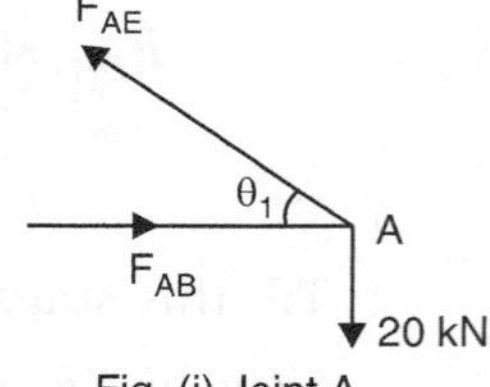

Fig. (i) Joint A

By $\Sigma F_h = 0$, we get

$$\rightarrow F_{AB} - F_{AE} \cos\theta_1 = 0$$

$\therefore \quad F_{AB} = 49.24 \times \dfrac{9}{\sqrt{97}} = 44.996 = 45$ kN

$\therefore \quad F_{AB} =$ **45 kN (Compression)**

Joint B : See Fig. (*ii*).

By $\Sigma F_V = 0$, we get

$$F_{BE} - 0 = 0, \quad \therefore \quad F_{BE} = 0$$

By $\Sigma F_h = 0$, we get

$$F_{BC} - 45 = 0$$

$\therefore \quad F_{BC} =$ **45 kN (Compression)**

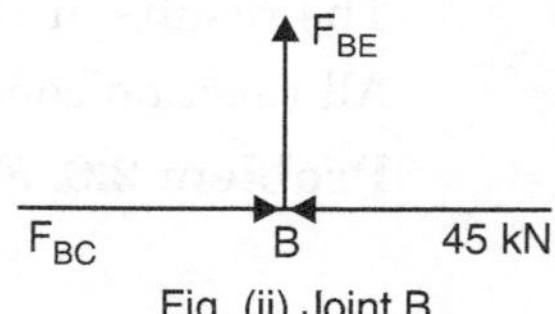

Fig. (ii) Joint B

Joint E : See Fig. (*iii*).

By $\Sigma F_h = 0$, we get

$$\leftarrow F_{EF} \cos\theta_1 - 49.24 \cos\theta_1 - F_{EC} \cos\theta_1 = 0$$

$\therefore \quad F_{EF} - F_{EC} = 49.24 \quad \quad ...(1)$

By $\Sigma F_V = 0$, we get

$$\uparrow F_{EF} \sin\theta_1 + F_{EC} \sin\theta_1 - 49.24 \sin\theta_1 - 20 = 0$$

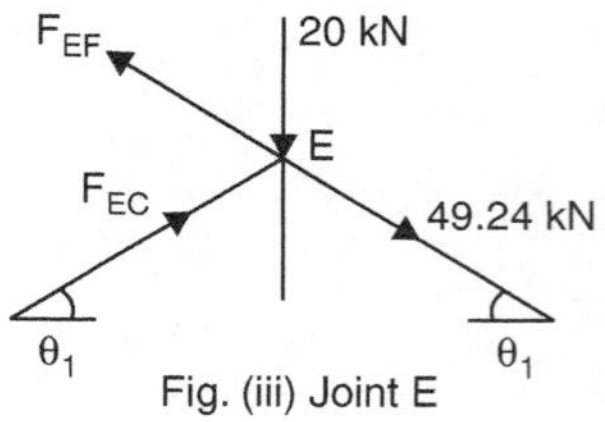

Fig. (iii) Joint E

$$F_{EF} + F_{EC} - 49.24 - \frac{20}{\sin \theta_1} = 0$$

$$\therefore \quad F_{EF} + F_{EC} = 49.24 + \frac{20}{4} \times \sqrt{97}$$

$$\therefore \quad F_{EF} + F_{EC} = 98.48 \qquad \ldots(2)$$

Adding (1) and (2), we get

$$2F_{EF} = 98.48 + 49.24$$

$$\therefore \quad F_{EF} = \mathbf{73.86\ kN} \quad \text{(Tension)}$$

From (2), we get $F_{EC} = 98.48 - (73.86)$

or $F_{EC} = 24.62$ kN (Compression)

Joint C : See Fig. (*iv*).

By $\Sigma F_V = 0$, we get

$$\uparrow F_{CF} - 24.62 \sin \theta_1 = 0$$

$$F_{CF} = 24.62 \times 4/\sqrt{97}$$

$$= \mathbf{10\ kN} \quad \text{(Tension)}$$

Fig. (iv) Joint C

By $\Sigma F_h = 0$, we get

$$\rightarrow F_{CD} - 45 - 24.62 \cos \theta_1 = 0$$

$$\therefore \quad F_{CD} = 45 + 24.62 \cos \theta_1$$

$$= \mathbf{67.5\ kN} \text{ (Compression)}$$

Joint F : See Fig. (*v*).

By $\Sigma F_h = 0$, we get

$$\leftarrow F_{FG} \cos \theta_1 - 73.86 \cos \theta_1 - F_{FD} \cos \theta_2 = 0$$

$$\therefore \quad F_{FG} = 73.86 + \frac{F_{FD} \cos \theta_2}{\cos \theta_1} = 73.86\ T\ \frac{9}{\sqrt{145}} \times \frac{\sqrt{97}}{9} F_{FD}$$

or $$F_{FG} - \sqrt{\frac{97}{145}}\, F_{FD} = 73.86 \text{ kN} \qquad \ldots(3)$$

By $\Sigma F_V = 0$

$$F_{FD} \sin \theta_2 + F_{FG} \sin \theta_1 - 73.86 \sin \theta_1 - 10 = 0$$

$$F_{FG} + F_{FD} \frac{\sin \theta_2}{\sin \theta_1} = 73.86 + \frac{10}{\sin \theta_1}$$

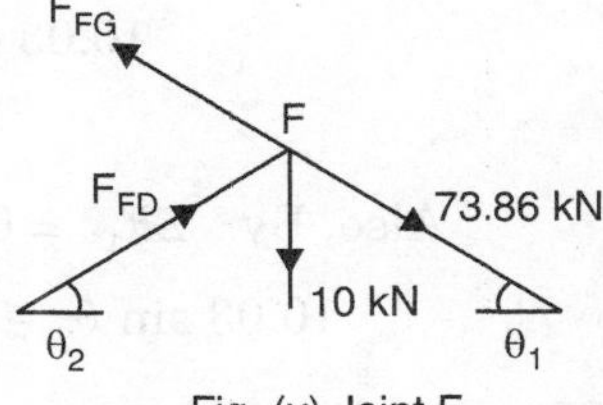

Fig. (v) Joint F

or $$F_{FG} + \frac{8}{\sqrt{145}} \times \frac{\sqrt{97}}{4} F_{FD} = 73.86 + \frac{10 \times \sqrt{97}}{4}$$

or $$F_{FG} + \frac{2\sqrt{97}}{\sqrt{145}} F_{FD} = 98.48$$

Dividing by 2, we get

$$0.5\ F_{FG} + \sqrt{\frac{97}{145}}\ F_{FD} = 49.24 \qquad ...(4)$$

Adding (3) and (4), we get

$$1.5\ F_{FG} = 73.86 + 49.24 = 123.10$$

$\therefore$ $F_{FG} =$ **82.07 kN** (Tension)

$\therefore$ $F_{FD} = \sqrt{\frac{145}{97}}\ (49.24 - 0.5\ F_{FG})$ from (4)

$\therefore$ $F_{FD} =$ **10.03** kN (Compression)

Support Reactions : Since there is one member (*FG*) at support *G*.

$\therefore$ Reaction at $G = F_{GF} \leftarrow$

$\therefore$ $R_G = 82.07$ kN inclined at an angle of θ_1 to the horizontal,

$\therefore$ $H_G = R_G \cos\theta_1 = 82.07 \times \frac{9}{\sqrt{97}} = 75\ \text{kN} \leftarrow$

And, $V_G = R_G \cdot \sin\theta_1 = 82.07 \times \frac{4}{\sqrt{97}} = 33.33\ \text{kN} \uparrow$

For the entire truss,

By $\Sigma F_h = 0$, we get

$$H_G - H_D = 0,$$

$\therefore$ $H_D = 75\ \text{kN} \rightarrow$

By $\Sigma F_V = 0$, we get

$$V_D + V_G - 20 - 20 = 0$$

$\therefore$ $V_D = 40 - V_G = 40 - 33.33 = 6.67\ \text{kN} \uparrow$

Joint D : See Fig. (*vi*).

By $\Sigma F_H = 0$, we get

$$10.03 \cos\theta_2 - 75 + 67.5 = 0$$

$\therefore$ $0 = 0$ **Check**

Also, By $\Sigma F_V = 0$, we get

$$10.03 \sin\theta_2 - 6.67 = 0$$

or $0 = 0$ **Check**

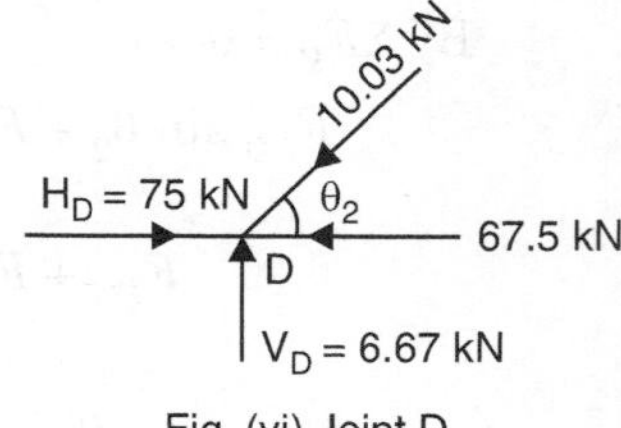

Fig. (vi) Joint D

Note. Forces in various members of the truss are shown in Fig. P.2.6 (*b*).

Problem 2.7. *Find forces in all members of the truss shown in Fig. P.2.7 (a).*

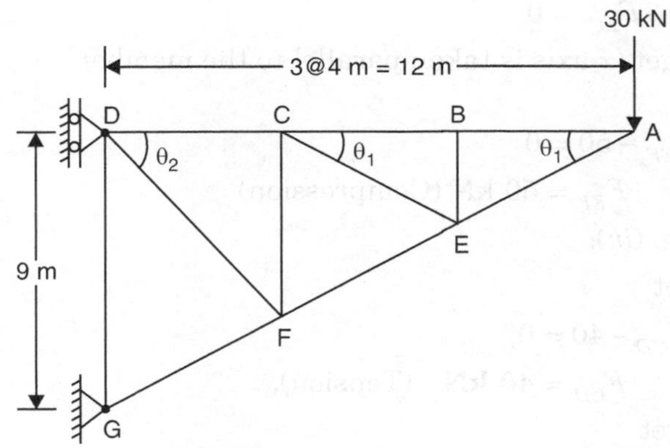

Fig. P. 2.7 (a) Given truss.

Sol. $\because \quad \tan \theta_1 = \frac{9}{12} = \frac{3}{4}$

$\therefore \quad \sin \theta_1 = \frac{3}{5}$ and $\cos \theta_1 = \frac{4}{5}$

Also, $\because \quad \tan \theta_2 = \frac{CF}{CD} = \frac{6}{4} = \frac{3}{2},$

$\therefore \quad \sin \theta_2 = \frac{3}{\sqrt{13}}$ and $\cos \theta_2 = \frac{2}{\sqrt{13}}.$

Joint A : See Fig. (*i*)

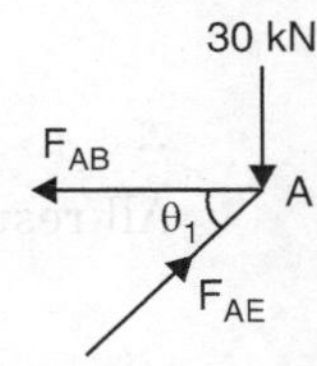

Fig. (i) Joint A

By $\Sigma F_V = 0$, we get

$$\uparrow F_{AE} \sin \theta_1 - 30 = 0,$$

$$\therefore \quad F_{AE} = \frac{30}{\sin \theta_1} = 30 \times \frac{5}{3}$$

$$= \mathbf{50\ kN}\ \text{(Compression)}$$

By $\Sigma F_h = 0$, we get

$$\leftarrow F_{AB} - F_{AE} \cos \theta_1 = 0$$

$$\therefore \quad F_{AB} = F_{AE} \cos \theta_1 = 50 \times \frac{4}{5}$$

$$= \mathbf{40\ kN}\ \text{(Tension)}.$$

Joint B : See Fig. (*ii*).

F_BC B 40 kN

F_BE

Fig. (ii) Joint B

By $\Sigma F_h = 0$, we get

$$F_{BC} - 40 = 0,$$

$$\therefore \quad F_{BC} = 40\ \text{kN (Tension)}$$

By $\Sigma F_V = 0$, we get

$$F_{BE} + 0 = 0 \quad \therefore \quad F_{BE} = 0.$$

Joint E : See Fig. (*iii*).

Let us equate the forces acting perpendicular to the member *EA* to zero, *i.e.*, by $\Sigma F_y = 0$, we get

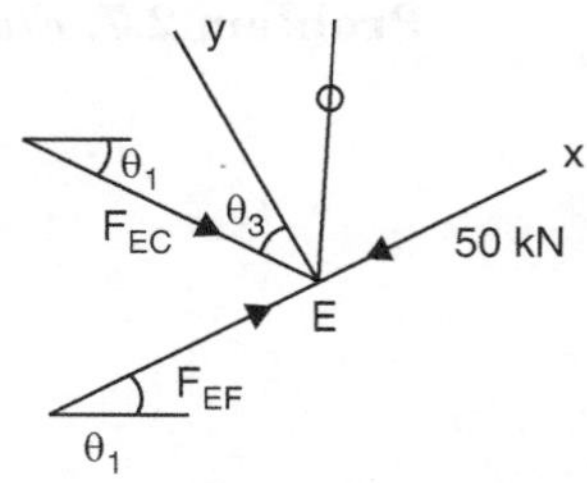

Fig. (iii) Joint E

$$F_{EC} \cos \theta_3 + 0 = 0$$

$\therefore \qquad F_{EC} = 0$

By $\Sigma F_x = 0$, we get (x-axis is taken parallel to the member EA)

$$F_{EF} - 50 = 0$$

$\therefore \qquad F_{EF} = 50$ kN (Compression)

Joint C : See Fig. (*iv*).

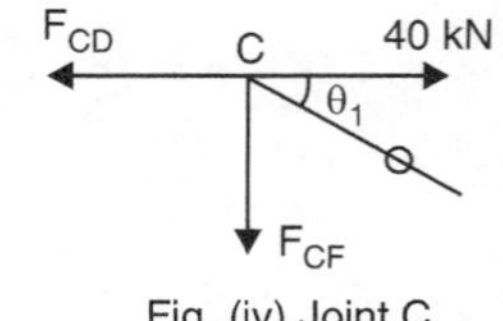

Fig. (iv) Joint C

By $\Sigma F_h = 0$, we get

$$F_{CD} - 40 = 0,$$

$\therefore \qquad F_{CD} =$ **40 kN** (Tension),

By $\Sigma F_V = 0$, we get

$$F_{CF} + 0 = 0.$$

$\therefore \qquad F_{CF} = 0$

Joint F : See Fig. (*v*).

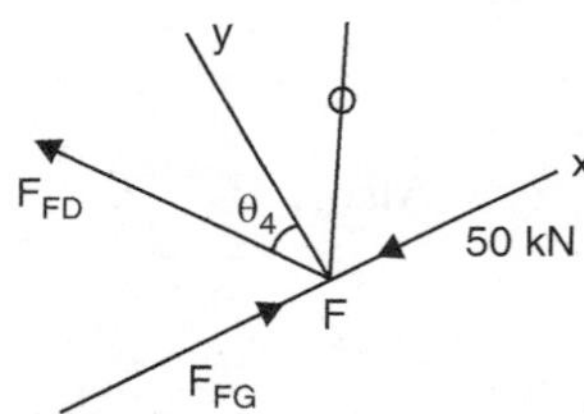

Fig. (v) Joint F

Let us take x-axis along member FE and y-axis perpendicular to member FE.

$\therefore$ By $\Sigma F_y = 0$, we get

$$F_{FD} \cos \theta_4 + 0 = 0,$$

$\therefore \qquad F_{FD} = 0.$

and by $\Sigma Fx = 0$, we get

$$F_{FG} - 50 = 0,$$

$\therefore \qquad F_{FG} =$ **50 kN** (Compression)

All results and reactions are shown in Fig. P. 2.7 (*b*).

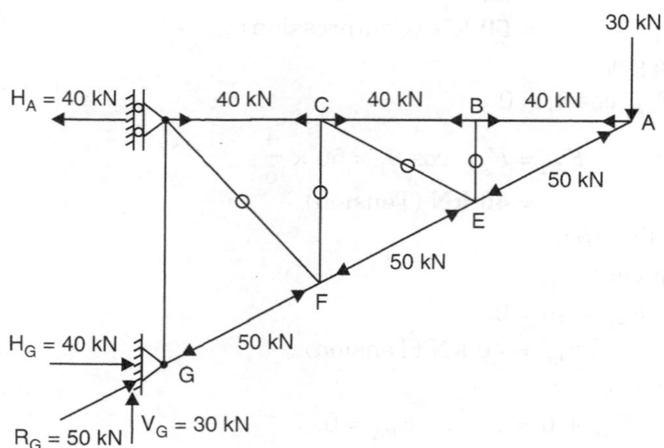

Fig. P. 2.7 (b) Reactions and forces in all members of the truss

Problem 2.8. *Find forces in all members of the truss shown in Fig. P. 2.8 (a).*

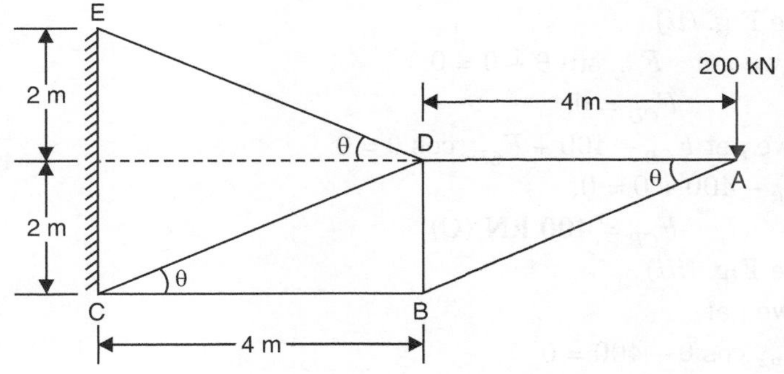

Fig. P. 2.8 (a) Given truss

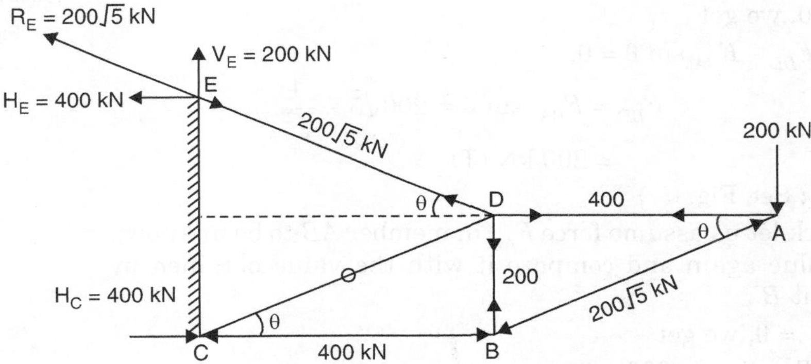

Fig. 2.8 (b) Results : forces in members of the truss, and support reactions

Sol. $\because \quad \tan\theta = \frac{2}{4} = \frac{1}{2}$

$\therefore \quad \sin\theta = \frac{1}{\sqrt{5}}$ and $\cos\theta = \frac{2}{\sqrt{5}}$

By $\Sigma M_E = 0$, we get $H_C \times 4 - 200 \times 8 = 0,$

$\therefore \quad H_C = 400 \text{ kN} \rightarrow$

$\therefore$ By $\Sigma F_h = 0$, we get $H_E - H_C = 0$

$\therefore \quad H_E = H_C = 400 \text{ kN} \leftarrow$

Joint E : See Fig. (*i*)

Let vertical reaction at E be V_E

$\therefore$ by $\Sigma F_h = 0$ at joint E, we get

$F_{ED} \cos\theta - 400 = 0$

$$\therefore \quad F_{ED} = \frac{400}{\cos\theta} = \frac{400\sqrt{5}}{2} = \mathbf{200\sqrt{5}\ kN\ (T)}$$

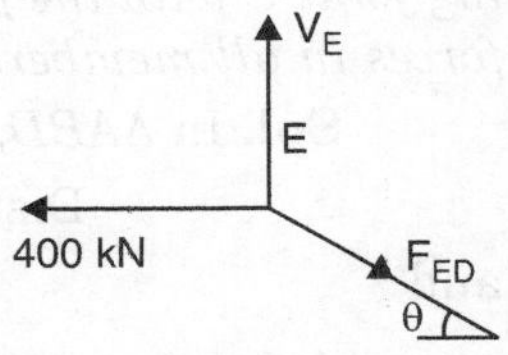

Fig. (i) Joint E

By $\Sigma F_V = 0$, we get $V_E - F_{ED} \sin\theta = 0$

$$\therefore \quad V_E = F_{ED} \sin\theta = (200\sqrt{5}) \times \frac{1}{\sqrt{500}} = 200 \text{ kN} \uparrow$$

$\therefore$ For the entire truss, By $\Sigma F_V = 0$, we get $(V_C + V_E) \uparrow - 200 = 0$

$\therefore$ $V_C = 200 - V_E = 200 - 200 = 0$

Joint C : See Fig. (*ii*)

By $\Sigma F_V = 0$, we get $F_{CD} \sin\theta + 0 = 0.$

$\therefore$ $F_{CD} = 0.$

By $\Sigma F_h = 0$, we get $F_{CB} - 400 + F_{CD} \cos\theta = 0$

$\therefore$ $F_{CB} - 400 + 0 = 0,$

$\therefore$ $F_{CB} = \mathbf{400\ kN\ (C)}$

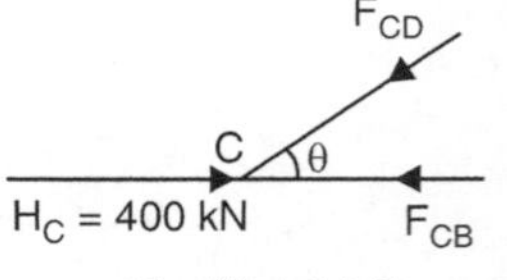

Fig. (ii) Joint C

Joint B : See Fig. (*iii*)

By $\Sigma F_h = 0$, we get

$\leftarrow F_{BA} \cos\theta - 400 = 0$

$\therefore$ $F_{BA} = \dfrac{400}{\cos\theta} = \dfrac{400\sqrt{5}}{2} = \mathbf{200\sqrt{5}\ kN\ (C)}$

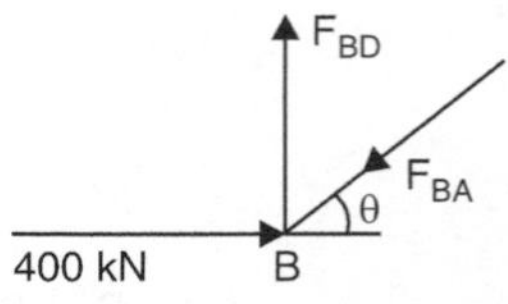

Fig. (iii) Joint B

By $F_V = 0$, we get

$\uparrow F_{BD} - F_{BA} \sin\theta = 0,$

$\therefore$ $F_{BD} = F_{BA} \sin\theta = 200\sqrt{5} \times \dfrac{1}{\sqrt{5}}$

$= 200$ kN (T)

Joint A : See Fig. (*iv*)

As a check let us assume force F_{AB} in member *AB* to be unknown and find its value again and compare it with the value obtained by considering joint *B*.

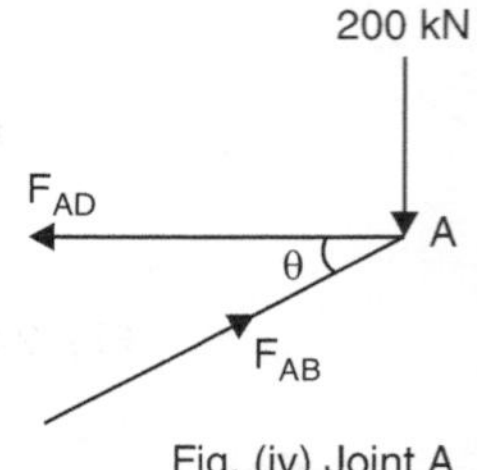

Fig. (iv) Joint A

By $\Sigma F_V = 0$, we get

$\uparrow F_{AB} \sin\theta - 200 = 0$

$\therefore$ $F_{AB} = \dfrac{400}{\cos\theta} = \dfrac{400\sqrt{5}}{2} = \mathbf{200\sqrt{5}\ kN(C)\ Check}$

By $\Sigma F_h = 0$, we get

$F_{AD} - F_{AB} \cos\theta = 0$

$\therefore$ $F_{AD} = F_{AB} \cos\theta = 200\sqrt{5} \times \dfrac{2}{\sqrt{5}}$

$= \mathbf{400\ kN\ (T)}$

Forces in all members are shown in Fig. P.2.8(*b*). **Ans.**

Problem 2.9. *In the truss shown in Fig. P.2.9(a), A is pin-jointed and the string connecting joint C with the fixed wall keeps the member AD horizontal. Find tension in the string and forces in all members of the truss.*

Sol. In ΔABD, $\angle ABD = 90°$, $\therefore$ $DB = AD \sin 30°$ or $DB = 2.4 \sin 30° = 1.2$ m.

$\therefore$ Distance $BB_1 = DB \sin \angle BDB_1 = 1.2 \sin 30° = 0.6$ m.

and $DB_1 = BD \cos 30° = 1.2 \cos 30° = 1.039$ m.

$\therefore$ $CB_1 = 1.8 - 1.039 = 0.761$ m.

$\therefore$ $\tan\theta = \dfrac{BB_1}{CB_1} = \dfrac{0.6}{0.761} = 0.7884$

$\therefore$ $\theta = \mathbf{38.25°}$

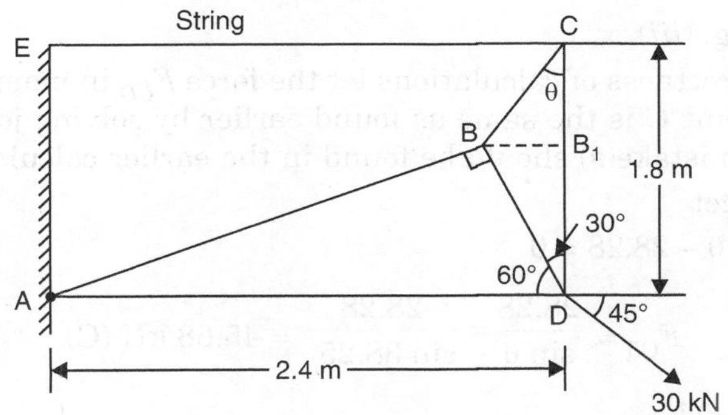

Fig. P. 2.9. (a) Given truss

Let T be the pull in the string.

$\therefore$ By $\Sigma M_A = 0$, we get

$$T \times CD - 30 \sin 45° \times AD = 0$$

$$T \times 1.8 - 30 \sin 45° \times 2.4 = 0$$

$$\therefore \quad T = \frac{30}{\sqrt{2}} \times \frac{2.4}{1.8} = \mathbf{28.28\ kN} \quad \textbf{Ans.}$$

By $\Sigma F_h = 0$, we get, $\quad H_A - T + 30 \cos 45° = 0$

$\therefore \quad H_A - 28.28 + 21.21 = 0,$

$\therefore \quad HA = 7.07 \text{ kN} \rightarrow$

By $\Sigma F_V = 0$, we get

$$V_A - 30 \sin 45° = 0$$

$\therefore \quad V_A = 30 \sin 45° = 21.21 \text{ kN} \uparrow$

Joint A : See Fig. (*i*).

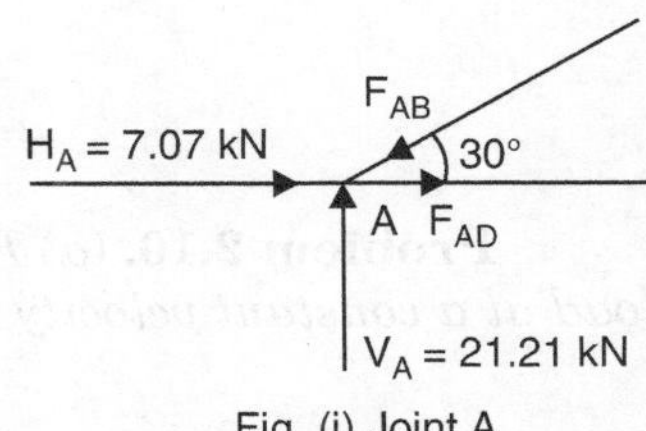

Fig. (i) Joint A

By $\Sigma F_V = 0$, we get

$\downarrow F_{AB} \sin 30° - 21.21 = 0$

$\therefore \quad F_{AB} = \frac{21.21}{\sin 30°} = \mathbf{42.42\ kN}$ (C)

By $\Sigma F_h = 0$, we get

$\rightarrow F_{AD} + 7.07 - F_{AB} \cos 30° = 0$

$\therefore \quad F_{AD} + 7.07 - 42.42 \cos 30° = 0$

$\therefore \quad F_{AD} = \mathbf{29.67\ kN}$ (T)

Joint D : See Fig. (*ii*).

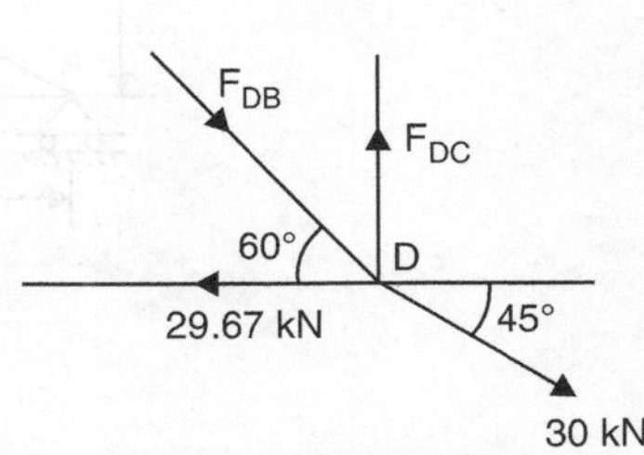

Fig. (ii) Joint D

By $\Sigma F_h = 0$, we get

$\rightarrow F_{DB} \cos 60° - 29.67 + 30 \cos 45° = 0$

$$\therefore \quad F_{DB} = \frac{29.67}{\cos 60°} - \frac{30 \cos 45°}{\cos 60°}$$

$= \mathbf{16.91\ kN}$ (C)

By $\Sigma F_V = 0$, we get

$\uparrow F_{DC} - F_{DB} \sin 60° - 30 \sin 45° = 0$

$\therefore \quad F_{DC} - 16.91 \sin 60° - 30 \sin 45° = 0$

$\therefore \quad F_{DC} = \mathbf{35.86\ kN}$ (T).

Joint C : See Fig. (*iii*).

To cheek the correctness of calculations let the force F_{CD} in member *CD* be unknown. If F_{CD} found by solving joint *C* is the same as found earlier by solving joint *D*, then results are correct, otherwise the mistake(s) should be found in the earlier calculations.

By, $\Sigma F_h = 0$, we get

$$\rightarrow F_{CB} \sin\theta - 28.28 = 0$$

$$\therefore \quad F_{CB} = \frac{28.28}{\sin\theta} = \frac{28.28}{\sin 38.25^\circ} = 45.68 \text{ kN (C)}$$

Fig. (iii) Joint C

By $\Sigma F_V = 0$, we get

$$\downarrow F_{CD} - F_{CB} \cos\theta = 0$$

$$\therefore \quad F_{CD} = F_{CB} \cos 38.25^\circ$$

$$= 45.68 \cos 38.25^\circ = 35.87 \text{ kN (T), as before}$$

Check. All calculations are correct.

Forces in all members of the truss are indicated in Fig. P.2.9 (*b*).

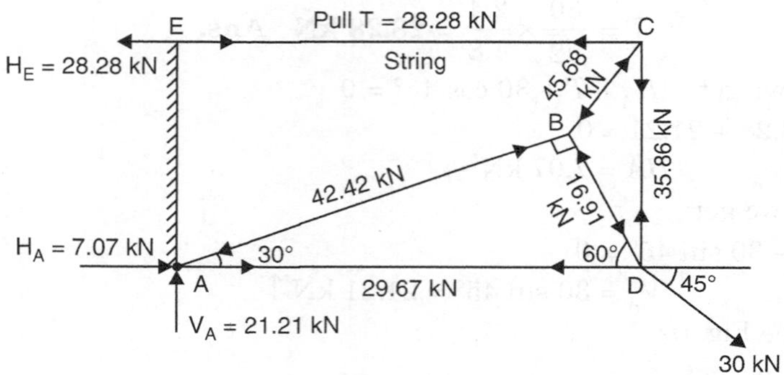

Fig. P. 2.9. (b) Results

Problem 2.10. (*a*) *Find force in all members of the truss induced by lifting the 120 kN load at a constant velocity of 2 m/s.*

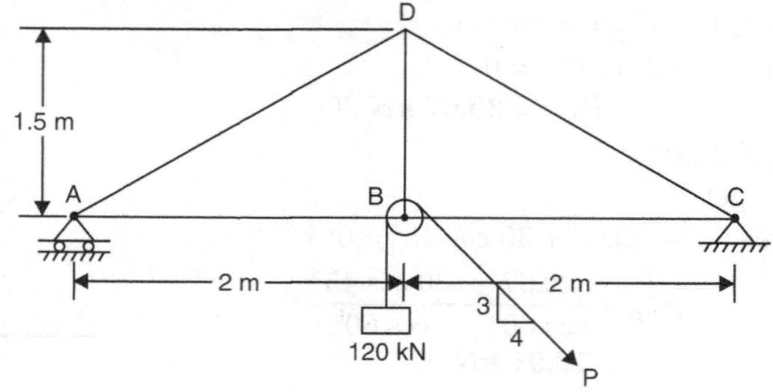

Fig. P. 2.10 (a) Given truss

(*b*) *Also find forces in each member when roller support is at A and the hinge support at C.*

Sol. (*a*) If the pulley at B is smooth, then the pull P will also be equal to 120 kN. And, the total vertical force acting at the joint

$$B = 120 + P \times \frac{3}{5} = 120 + 120 \times \frac{3}{5} = 192 \text{ kN} \downarrow$$

And, the horizontal force at $B = P \times \frac{4}{5} = 96 \text{ kN} \rightarrow$ as shown in Fig. P. 2.10 (*b*)

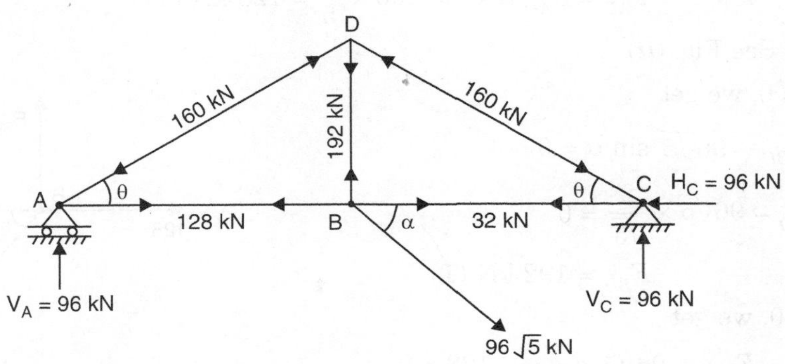

Fig. P. 2.10 (b) Forces

$\therefore$ Resultant force at $B = \sqrt{96^2 + 192^2} = 96\sqrt{1+4} = 96\sqrt{5}$ kN

and $$\tan\alpha = \frac{192}{96} = 2 \quad \therefore \quad \sin\alpha = \frac{2}{\sqrt{5}} \quad \text{and} \quad \cos\alpha = \frac{1}{\sqrt{5}}$$

Also, $$\tan\theta = \frac{1.5}{2} = \frac{3}{4}\ ; \ \therefore \quad \sin\theta = \frac{3}{5} \quad \text{and} \quad \cos\theta = \frac{4}{5}$$

By $\Sigma M_C = 0$, we get

$$V_A \times AC - (96\sqrt{5}) \sin\alpha \times BC = 0$$

$$\therefore \quad V_A \times 4 - 96\sqrt{5} \times \frac{2}{\sqrt{5}} \times 2 = 0$$

$$\therefore \quad V_A = 96 \text{ kN} \uparrow$$

By $\Sigma F_V = 0$, we get $(V_A + V_C) \uparrow - 96\sqrt{5}\ \sin\theta = 0$

$$\therefore \quad V_C = 96\sqrt{5} \times \frac{2}{\sqrt{5}} - V_A = 192 - 96 = 96 \text{ kN} \uparrow$$

By $\Sigma F_h = 0$, we get

$$\leftarrow H_C - 96\sqrt{5}\ \cos\alpha = 0$$

$$\therefore \quad H_C = 96\sqrt{5} \times \frac{1}{\sqrt{5}} = 96 \text{ kN} \leftarrow$$

Joint A : See Fig. (*i*).

By $\Sigma F_V = 0$, we get

$$\downarrow F_{AD} \sin\theta - 96 = 0$$

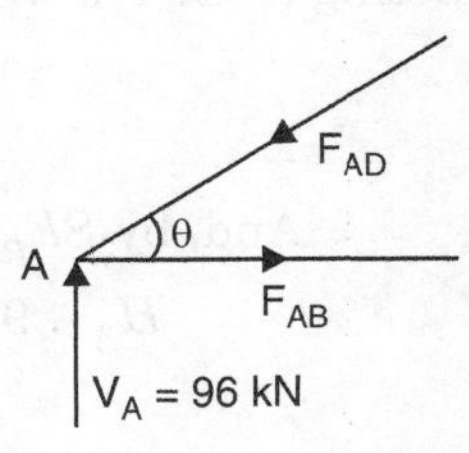

Fig. (i) Joint A

$$\therefore \qquad F_{AD} = \frac{96}{\sin\theta} = \frac{96 \times 5}{3} \text{ kN}$$

$$\therefore \qquad F_{AD} = \mathbf{160\ kN}\ (C)$$

and by $\Sigma F_R = 0$, we get

$$\rightarrow F_{AB} - F_{AD} \cos\theta = 0$$

$$\therefore \qquad F_{AB} = F_{AD} \cos\theta = 160 \times \frac{4}{5} = 128 \text{ kN (T)}$$

Joint B : See Fig. (*ii*).

By $\Sigma F_V = 0$, we get

$$\uparrow F_{BD} - 96\sqrt{5}\,\sin\alpha = 0$$

$$\therefore \qquad F_{BD} - 96\sqrt{5} \times \frac{2}{\sqrt{5}} = 0$$

$$\therefore \qquad F_{BD} = 192 \text{ kN (T)}$$

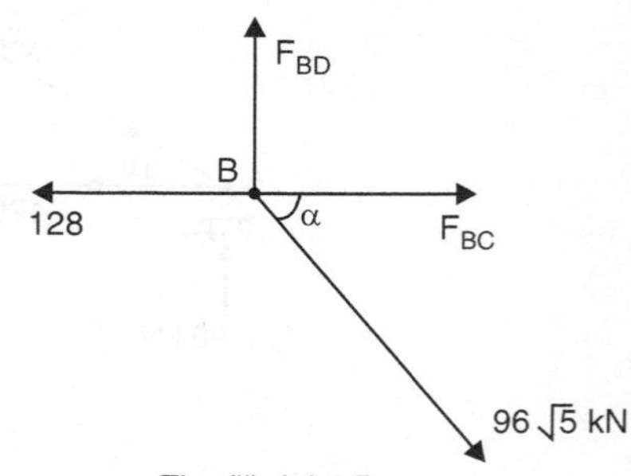

Fig. (ii) Joint B

By $\Sigma F_h = 0$, we get

$$F_{BC} + 96\sqrt{5}\,\cos\alpha - 128 = 0,$$

$$\therefore \qquad \rightarrow F_{BC} + 96\sqrt{5} \times \frac{1}{\sqrt{5}} - 128 = 0$$

$$\therefore \qquad F_{BC} = 128 - 96 = 32 \text{ kN (T)}$$

Joint C : See Fig. (*iii*).

Let F_{CB} be unknown.

$\therefore$ by $\Sigma F_V = 0$, we get

$$\downarrow F_{CD} \sin\theta - 96 = 0$$

$$\therefore \qquad F_{CD} = \frac{96}{\sin\theta} = \frac{96 \times 5}{3} = \mathbf{160\ kN}\ (C)$$

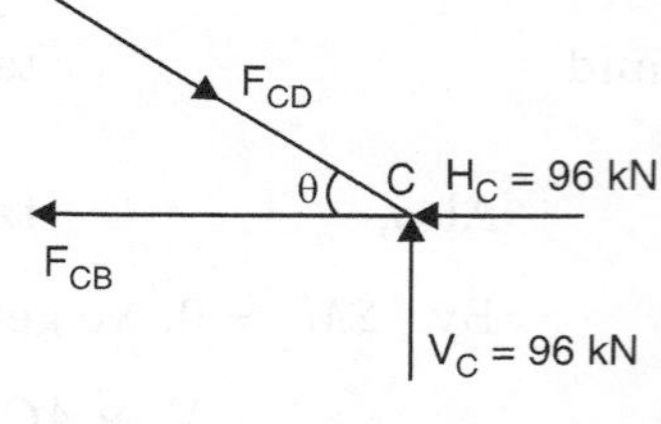

Fig. (iii) Joint C

And, by $\Sigma F_h = 0$, we get

$$\leftarrow F_{CB} + 96 - F_{CD} \cos\theta = 0,$$

$$\therefore \qquad F_{CB} + 96 - 160 \times \frac{4}{5} = 0$$

$$\therefore \qquad F_{CB} = 128 - 96 = \mathbf{32\ kN}\ \text{(T), as before (Check)}$$

Results are shown in Fig. P.2.10 (*b*).

(*b*) When support *C* is on rollers, there will be only vertical reaction component V_C acting at *C*. The vertical load at *B* will be equally shared by supports *A* and *C*. Thus,

$$V_A + V_C = 192 \text{ kN} = 2V_A = 2V_C$$

$$\therefore \qquad V_A = V_C = 96 \text{ kN} \uparrow$$

And, by $SF_R = 0$, we have

$$H_A - 96\sqrt{5} \times \cos\alpha = 0$$

$$\therefore \qquad H_A = 96\sqrt{5} \times \frac{1}{\sqrt{5}} = 96 \text{ kN} \leftarrow$$

Joint C : See Fig. (*iv*).

By $\Sigma F_V = 0$, we get $F_{CD} \sin\theta - 96 = 0$

$$\therefore \qquad F_{CD} = \frac{96}{\sin\theta} = \frac{96}{3} \times 5 = \mathbf{160\ kN}\ \text{(C)}.$$

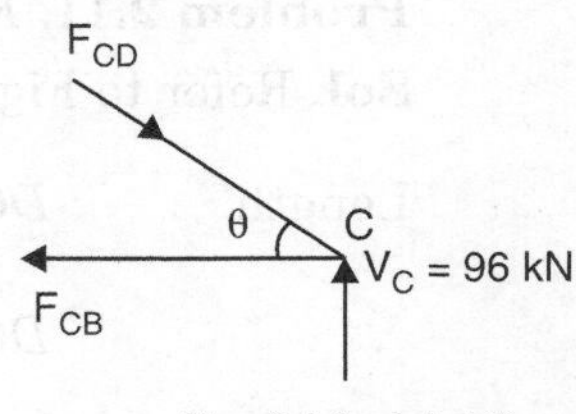

Fig. (iv) for joint C

By $\Sigma F_h = 0$, we get

$$F_{CB} - F_{CD} \cos\theta = 0$$

$$\therefore \qquad F_{CB} = F_{CD} \cos\theta = 160 \times \frac{4}{5} = \mathbf{128\ kN}\ \text{(T)}$$

Jont B : See Fig. (*v*).

By $\Sigma F_V = 0$, we get

$$F_{BD} - 96\sqrt{5}\ \sin\alpha = 0$$

$$\therefore \qquad F_{BD} = 96\sqrt{5}\ \sin\alpha = 96\sqrt{5} \times \frac{2}{\sqrt{5}} = \mathbf{192\ kN}\ \text{(T)} \quad \textbf{Ans.}$$

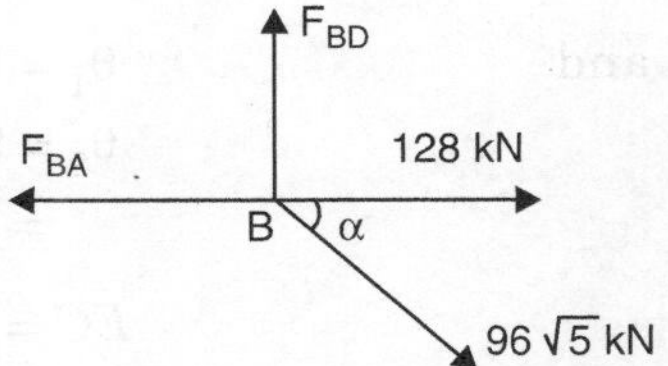

Fig. (v) for joint B

By $\Sigma F_h = 0$, we get

$$F_{BA} - 128 - 96\sqrt{5}\ \cos\alpha = 0$$

$$\therefore \qquad F_{BA} - 128 - 96\sqrt{5} \times \frac{1}{\sqrt{5}} = 0$$

$$\therefore \qquad F_{BA} = 128 + 96 = \mathbf{224\ kN}\ \text{(T)}$$

Joint A : See Fig. (*vi*).

By $\Sigma F_v = 0$, we get

$$F_{AD} \sin\theta - 96 = 0$$

$$\therefore \qquad F_{AD} = \frac{96}{\sin\theta} = \frac{96}{3} \times 5 = \mathbf{160\ kN}$$

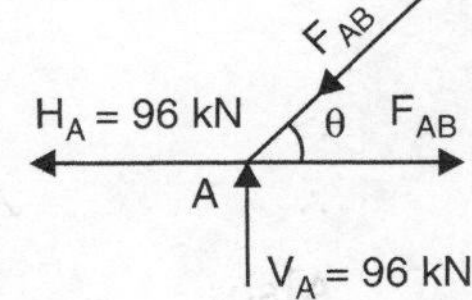

Fig. (vi) for joint A

By $\Sigma F_h = 0$, we get

$$F_{AB} - 96 - F_{AD} \cos\theta = 0$$

$$\therefore \qquad F_{AD} = 96 + F_{AD} \times \frac{4}{5} = 96 + \frac{160 \times 4}{5}$$

$$= 96 + 128 = \mathbf{224\ kN}\ \text{(T)}$$

Since value of F_{AD} found by solving two joints A and B is the same, therefore all results are correct. Forces in various members of the truss for case (*b*) are shown in Fig. P.2.10(*c*).

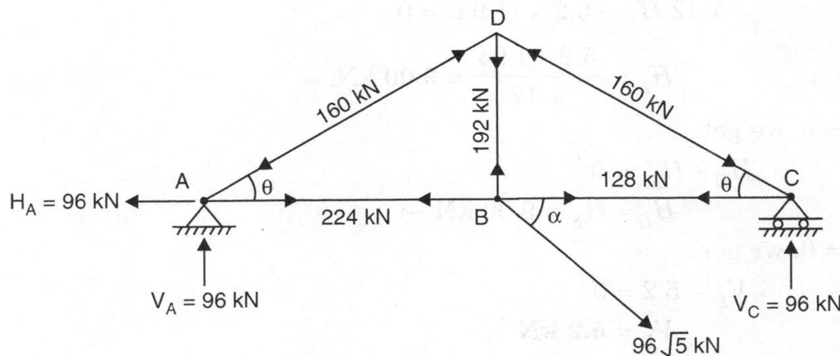

Fig. P. 2.10(c). Results for case (b)

Problem 2.11. *Find force in each member of the truss shown in Fig. P. 2.11(a).*

Sol. Refer to Fig. P. 2.11(*a*)

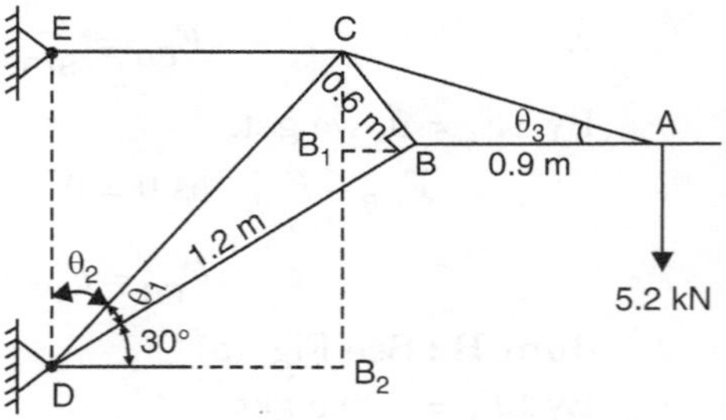

Length, $DC = \sqrt{BD^2 + BC^2}$

$\therefore \quad DC = \sqrt{1.2^2 + 0.6^2} = 0.6\sqrt{5}$ m

$\because \quad \tan\theta_1 = \dfrac{BC}{DB} = \dfrac{0.6}{1.2} = \dfrac{1}{2}$;

$\therefore \quad \sin\theta_1 = \dfrac{1}{\sqrt{5}}$ and $\cos\theta_2 = \dfrac{2}{\sqrt{5}}$

and $\theta_1 = \tan^{-1} 0.5 = 26.57°$

$\therefore \quad \theta_2 = 90° - (\theta_1 + 30°)$

$= 90° - 56.57° = \mathbf{33.43°}$

Fig. P. 2.11 (a) Given truss

$EC = DC \sin\theta_2 = 0.6\sqrt{5} \times \sin 33.43° = 0.739$ m.

$\because \quad \angle DBB_1 = 30°, \quad \therefore \quad \angle CBB_1 = 90° - 30° = 60°.$

$CB_1 = BC \sin 60° = 0.6 \sin 60° = 0.3\sqrt{3}$ m $= 0.52$ m

$DB_2 = DC \cos\angle CDB_2 = 0.6\sqrt{5} \cos 56.57° = 0.74$ m

$BB_1 = CB \cos 60° = 0.6 \cos 60° = 0.3$ m.

$\therefore \quad AB_1 = 0.9 + 0.3 = 1.2$ m.

$\therefore \quad \tan\angle CAB = \tan\theta_3 = CB_1/AB_1 = \dfrac{0.52}{1.2} = 0.4333$

$\therefore \quad \theta_3 = \mathbf{23.43°}$

$\therefore \quad AB_1 + DB_2 = 1.2 + 0.74 = \mathbf{1.94\ m}$

Also, $CB_2 = DE = DC \cos\theta_2 = (0.6\sqrt{5}) \cos 33.43° = \mathbf{1.12\ m}$

Since, there is one member connected to support E, therefore the support reaction at E will be along the member EC, which is horizontal.

$\therefore$ by $\Sigma M_D = 0$, we get

$$H_E \times ED - 5.2 \times (AB_1 + B_2D) = 0$$

or $$1.12\, H_E - 5.2 \times (1.94) = 0$$

$\therefore \quad H_E = \dfrac{5.2 \times 1.94}{1.12} = \mathbf{9.00}$ kN. $\leftarrow$

By $\Sigma F_h = 0$, we get

$$H_D - H_E = 0$$

$\therefore \quad H_D = H_E = 9.00$ kN $\rightarrow$

By $\Sigma F_V = 0$, we get

$$V_D - 5.2 = 0$$

$\therefore \quad V_D = \mathbf{5.2\ kN} \uparrow.$

Joint E :

By $\Sigma F_h = 0$, we get

$$F_{EC} - H_E = 0$$

$$\therefore \quad F_{EC} = H_E = \mathbf{9.00\ kN}\ \text{(T)}$$

Joint A : See Fig. (*i*)

By $\Sigma F_V = 0$, we get

$$\uparrow F_{AC} \sin \theta_3 - 5.2 = 0$$

$$\therefore \quad F_{AC} = \frac{5.2}{\sin \theta_3} = \frac{5.2}{\sin 23.43^\circ}$$

$$= \mathbf{13.08\ kN}\ \text{(T)}$$

Fig. (i) Joint A

By $\Sigma F_h = 0$, we get

$$\rightarrow F_{AB} - F_{AC} \cdot \cos \theta_3 = 0$$

$$\therefore \quad F_{AB} = F_{AC} \cdot \cos \theta_3 = 13.08 \cos 23.43^\circ$$

$$= \mathbf{12.0\ kN}\ \text{(C)}$$

Joint B : See Fig. (*ii*)

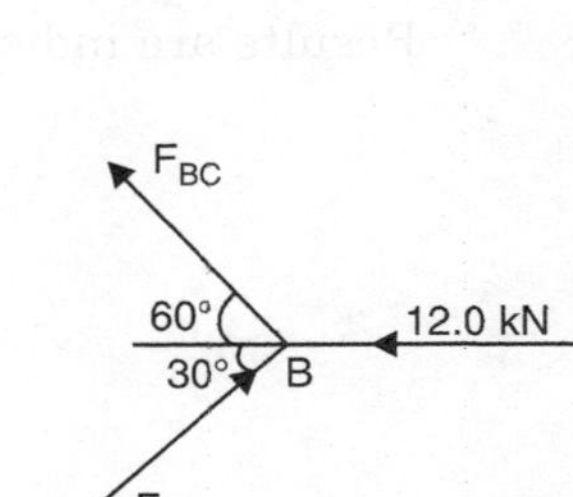

Fig. (ii) Joint B

By $\Sigma F_V = 0$, we get

$$\uparrow F_{BC} \sin 60^\circ + F_{BD} \sin 30^\circ = 0$$

$$F_{BD} = - F_{BC} \cdot \frac{\sin 60^\circ}{\sin 30^\circ}$$

or

$$F_{BD} = - 2F_{BC} \times \sin 60^\circ$$

$$= - 2F_{BC} \times \frac{\sqrt{3}}{2} = - F_{BC} \sqrt{3} \quad \text{...(1)}$$

By $\Sigma F_h = 0$

$$F_{BC} \cos 60^\circ - F_{BD} \cos 30^\circ + 12 = 0$$

$$\therefore \quad F_{BC} - \frac{F_{BD} \cos 30^\circ}{\cos 60^\circ} + \frac{12}{\cos 60^\circ} = 0$$

or

$$F_{BC} = \frac{F_{BD}\ \sqrt{3}/2}{\frac{1}{2}} - 12 \times 2$$

$$F_{BC} = \sqrt{3} F_{BD} - 24 \quad \text{...(2)}$$

By (1) into (2), we get

$$F_{BC} = \sqrt{3} \times (- \sqrt{3}\ F_{BC}) - 24$$

$$= - 3\ F_{BC} - 24$$

$$\therefore \quad 4F_{BC} = - 24$$

$$\therefore \quad F_{BC} = - 6 \text{ kN} \quad \text{or} \quad F_{BC} = \mathbf{6\ kN\ (C)}$$

and

$$F_{BD} = - \sqrt{3} F_{BC} = - \sqrt{3} \times (- 6) = 6\sqrt{3} = \mathbf{10.39\ kN}\ \text{(C)}.$$

Joint D : See Fig. (*iii*).

By $\Sigma F_V = 0$, we get

$$F_{DC} \cos \theta_2 - 10.39 \sin 30° + 5.2 = 0$$

$$\therefore \qquad F_{DC} = \frac{10.39 \times \sin 30°}{\cos \theta_2} - \frac{5.2}{\cos \theta_2}$$

$$= \frac{10.39 \times \frac{1}{2}}{\cos 33.43°} - \frac{5.2}{\cos 33.43°} = 0$$

$$\therefore \qquad \mathbf{F_{DC} = 0}$$

Fig. (iii) Joint D

By $\Sigma F_h = 0$, we get

$$H_D + F_{DC} \sin \theta_2 - 10.39 \cos 30° = 0$$

$$\therefore \qquad H_D = 10.39 \cos 30° - (0) \sin 33.43°$$

$$\therefore \qquad H_D = 9.0 \text{ kN} \rightarrow. \textbf{ (Check).}$$

Results are indicated in Fig. P.2.11 (*b*)

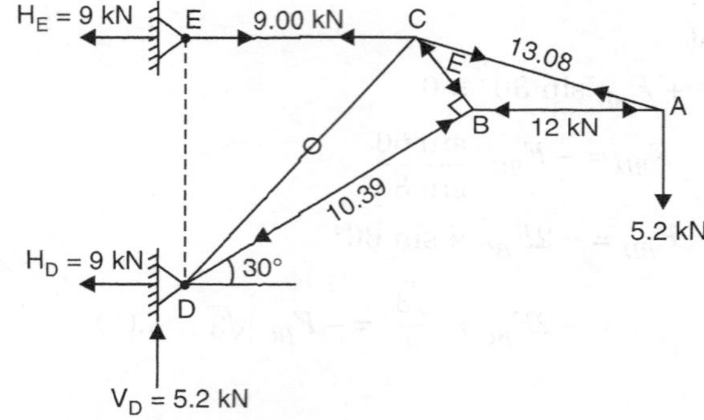

Fig. P. 2.11 (b) Forces in members of the truss

Problem 2.12. *Find force in all members of the truss shown in Fig. P.2.12 (a).*

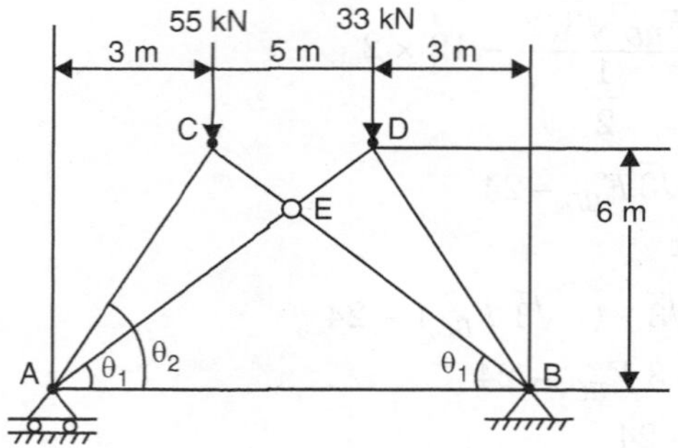

Fig. P. 2.12 (a) Given truss

Sol. $\tan\theta_1 = \frac{6}{8} = \frac{3}{4}$,

$\therefore$ $\sin\theta_1 = 3/5 = 0.6$ and $\cos\theta_1 = 4/5 = 0.8$

And, $\tan\theta_2 = \frac{6}{3} = \frac{2}{1}$, $\therefore$ $\sin\theta_2 = 2/\sqrt{5}$ and $\cos\theta_2 = 1/\sqrt{5}$.

By $\Sigma M_B = 0$, we get

$$V_A \times 11 - 55 \times 8 - 33 \times 3 = 0$$

$$\therefore \quad V_A = \frac{1}{11}(55 \times 8 + 33 \times 3) = 49 \text{ kN} \uparrow$$

By $\Sigma F_V = 0$, we get

$$(V_A + V_B) \uparrow - (55 + 33) = 0,$$

$$\therefore \quad V_B = 88 - V_A = 88 - 49 = 39 \text{ kN} \uparrow$$

Joint C : See Fig. (*i*).

By $\Sigma F_h = 0$, we get

$$\rightarrow F_{CA}\cos\theta_2 - F_{CE}\cos\theta_1 = 0$$

$$\therefore \quad F_{CA} \times \frac{1}{\sqrt{5}} = F_{CE} \times \frac{4}{5}$$

$$\therefore \quad F_{CA} = \frac{4}{5} \times \sqrt{5}\, F_{CE} = \frac{4}{\sqrt{5}} F_{CE} \quad \text{...(1)}$$

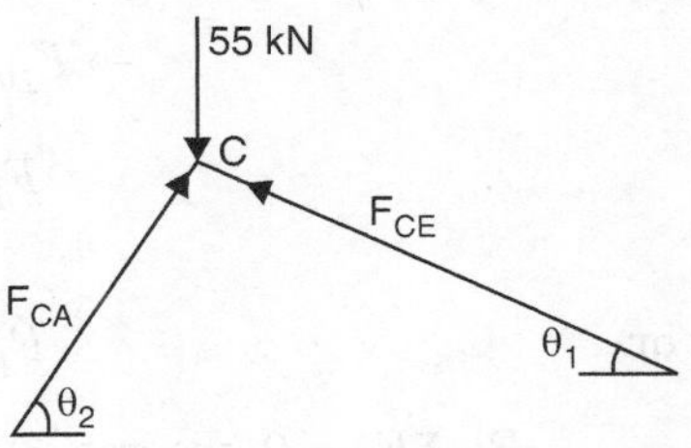

Fig. (i) Joint C

By $\Sigma F_V = 0$, we get

$$\uparrow F_{CA}\sin\theta_2 + F_{CE}\sin\theta_1 - 55 = 0,$$

$$\therefore \quad F_{CA} \times \frac{2}{\sqrt{5}} + F_{CE} \times \frac{3}{5} = 55 \quad \text{...(2)}$$

By substituting the value of F_{CA} from eqn. (1), into eqn. (2), we get

$$\left(\frac{4}{\sqrt{5}} F_{CE}\right) \times \frac{2}{\sqrt{5}} + \frac{3}{5} F_{CE} = 55$$

or

$$\frac{F_{CE}}{5}(8 + 3) = 55$$

$$\therefore \quad F_{CE} = \frac{55 \times 5}{11} = 25 \text{ kN (C)}$$

$\therefore$ From (1), we get

$$F_{CA} = \frac{4}{\sqrt{5}} \times (25) = \mathbf{20\sqrt{5}\ kN\ (C)}$$

Joint A : See Fig. (*ii*).

By $\Sigma F_V = 0$, we get

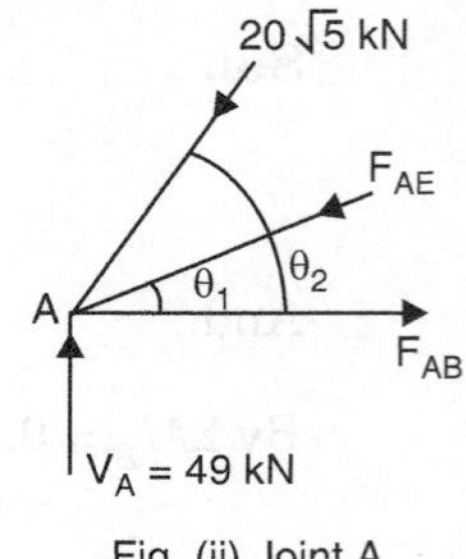

Fig. (ii) Joint A

$$\downarrow F_{AE} \sin\theta_1 + 20\sqrt{5}\ \sin\theta_2 - 49 = 0$$

$\therefore$
$$F_{AE} = \frac{49}{\sin\theta_1} - \frac{20\sqrt{5}\sin\theta_2}{\sin\theta_1}$$

$$= \frac{49}{0.6} - \frac{20\sqrt{5}\times 2/\sqrt{5}}{0.6}$$

or
$$F_{AE} = \frac{1}{0.6}(49-40) = \frac{9}{0.6} = \mathbf{15\ kN\ (C).}$$

By $\Sigma F_H = 0$, we get

$$\rightarrow F_{AB} - 20\sqrt{5}\ \cos\theta_2 - F_{AE}\ \cos\theta_1 = 0$$

$$F_{AB} - 20\sqrt{5}\times\frac{1}{\sqrt{5}} - 15\times\frac{4}{5} = 0$$

$\therefore$
$$F_{AB} - 20 - 12 = 0,$$

$\therefore$
$$F_{AB} = \mathbf{32\ kN}\ \text{(T)}$$

Joint D : See Fig. (*iii*).

By $\Sigma F_h = 0$, we get

$$\rightarrow F_{DE}\cos\theta_1 - F_{DB}\cos\theta_2 = 0$$

$\therefore$
$$F_{DE} = F_{DB}\frac{\cos\theta_2}{\cos\theta_1} = F_{DB}\times\frac{1}{\sqrt{5}}\times\frac{5}{4}$$

or
$$F_{DE} = \frac{\sqrt{5}}{4}F_{DB} \qquad ...(3)$$

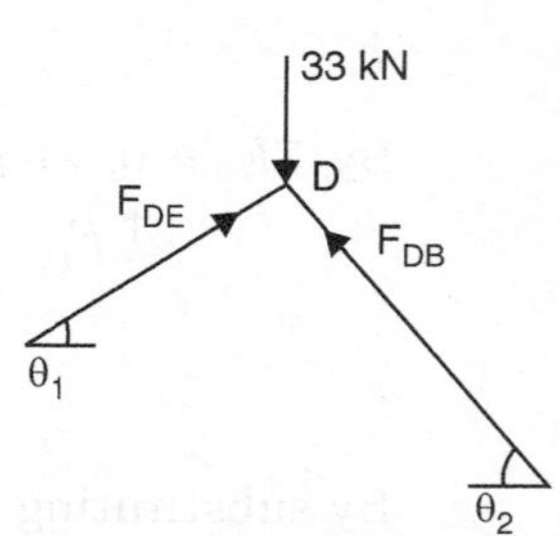

Fig. (iii) Joint D

By $\Sigma F_V = 0$, we get

$$\uparrow F_{DE}\sin\theta_1 + F_{DB}\sin\theta_2 - 33 = 0$$

$\therefore$
$$F_{DE}\times\frac{3}{5} + F_{DB}\times\frac{2}{\sqrt{5}} = 33$$

or,
$$F_{DE} + \frac{2}{\sqrt{5}}\times\frac{5}{3}F_{DB} = \frac{33\times 5}{3} \qquad ...(4)$$

Adding (3) and (4), we get

$$F_{DB}\left(\frac{\sqrt{5}}{4} + \frac{2\times 5}{3\sqrt{5}}\right) = \frac{33\times 5}{3}$$

or,
$$F_{DB}\left(\frac{\sqrt{5}}{4} + \frac{2}{3}\sqrt{5}\right) = 55$$

$\therefore$
$$\frac{\sqrt{5}}{12}F_{DB}(3+8) = 55$$

$\therefore$
$$F_{DB} = \frac{55\times 12}{\sqrt{5}\times 11} = \mathbf{12\sqrt{5}\ \ kN\ (C)}$$

$\therefore$ From (3), $F_{DE} = \frac{\sqrt{5}}{4} F_{DE} = \frac{\sqrt{5}}{4} \times 12\sqrt{5}$

$= \mathbf{15\ kN}$ (C)

Joint B : See Fig. (*iv*).

By $\Sigma F_V = 0$, we get

$$\downarrow F_{BE} \sin \theta_1 + 12\sqrt{5} \sin \theta_2 - 39 = 0$$

$$\therefore \quad F_{BE} = \frac{1}{\sin \theta_1}\left(39 - 12\sqrt{5} \times \frac{2}{\sqrt{5}}\right) = \frac{15}{0.6}$$

or $F_{BE} = \mathbf{25\ kN\ (C)}$

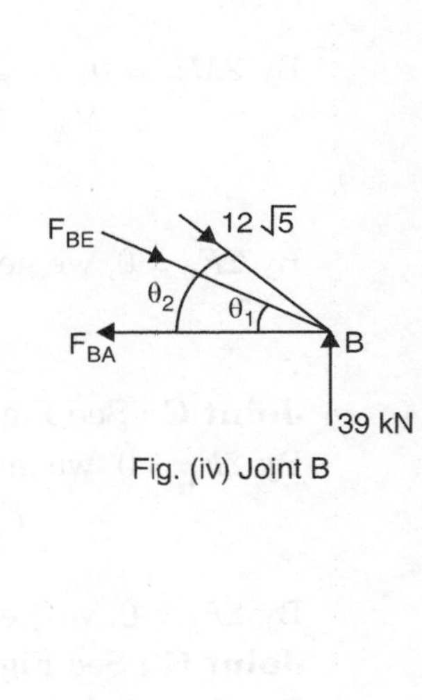

Fig. (iv) Joint B

By $\Sigma F_h = 0$, we get

$$\leftarrow F_{BA} - 12\sqrt{5} \cos \theta_2 - F_{BE} \cos \theta_1 = 0$$

$$\therefore \quad F_{BA} - 12\sqrt{5} \times \frac{1}{\sqrt{5}} - 25 \times \frac{4}{5} = 0$$

$$\therefore \quad F_{BA} - 12 - 20 = 0$$

$\therefore$ $F_{BA} = \mathbf{32\ kN\ (T)}$ **(Check)**

Results, are indicated in Fig. P. 2.12 (*b*).

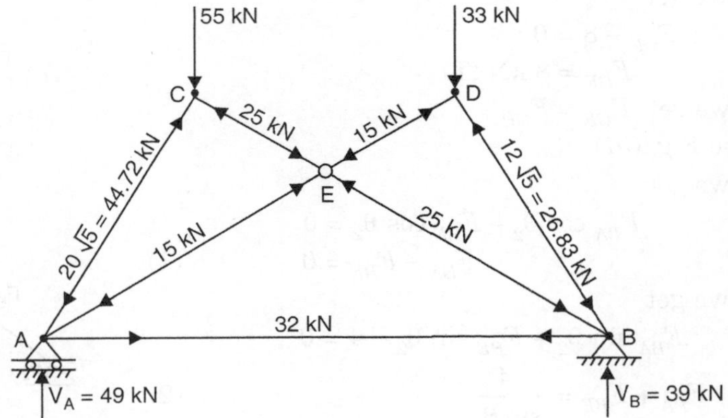

Fig. P. 2.12 (b) Forces in all members of truss

Problem 2.13. *Find forces in all members of the truss shown in Fig. P.2.13 (a).*

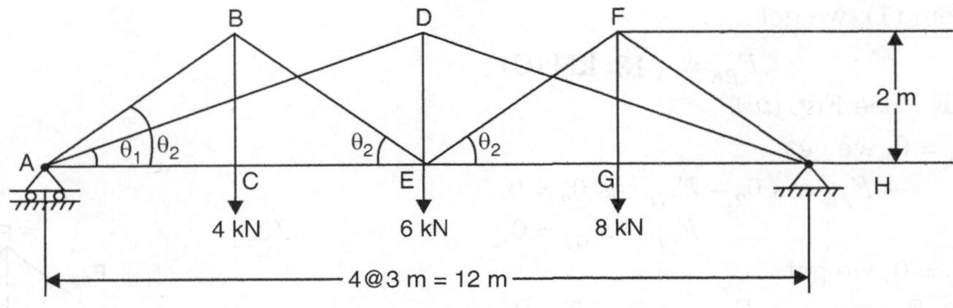

Fig. P.2.13 (a)

Sol. $\because \quad \tan\theta_1 = \frac{2}{6} = \frac{1}{3}$; $\therefore \quad \sin\theta_1 = \frac{1}{\sqrt{10}}$; $\cos\theta_1 = \frac{3}{\sqrt{10}}$

Also, $\because \quad \tan\theta_2 = \frac{2}{3}$; $\therefore \quad \sin\theta_2 = \frac{2}{\sqrt{13}}$; $\cos\theta_2 = \frac{3}{\sqrt{13}}$.

By $\Sigma M_H = 0$, we get

$$V_A \times 12 - 4 \times 9 - 6 \times 6 - 8 \times 3 = 0.$$

$$\therefore \quad V_A = \frac{1}{12}(36 + 36 + 24) = 8 \text{ kN} \uparrow$$

By $\Sigma F_V = 0$, we get

$$(V_A + V_H) \uparrow - (4 + 6 + 8) = 0$$

$$\therefore \quad V_H = 18 - V_A = 18 - 8 = 10 \text{ kN} \uparrow$$

Joint C : See Fig. (*i*).

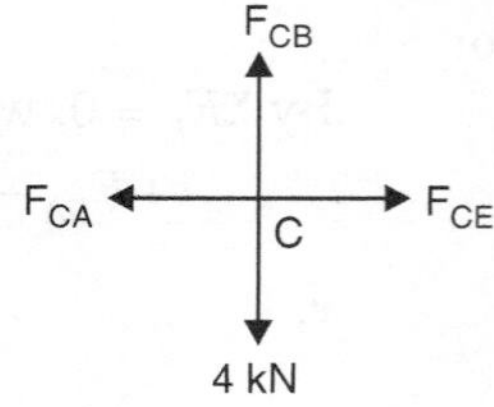

Fig. (i) Joint C

By $\Sigma F_V = 0$, we get

$$F_{CB} - 4 = 0,$$

$$\therefore \quad F_{CB} = 4 \text{ kN (T)}.$$

By $\Sigma F_h = 0$, we get $F_{CA} = F_{CE}$

Joint G : See Fig. (*ii*).

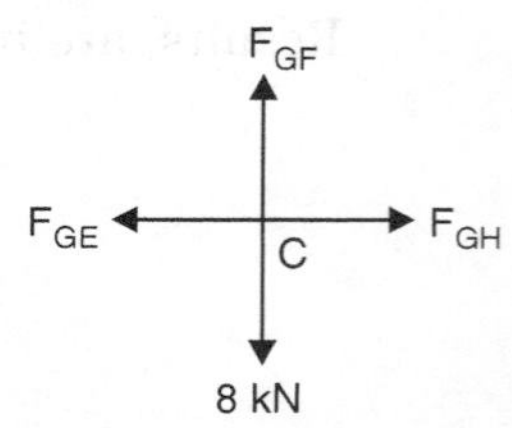

Fig. (ii) Joint G

By $\Sigma F_V = 0$, we get

$$F_{GF} - 8 = 0 ;$$

$$\therefore \quad F_{GF} = 8 \text{ kN (T)}$$

By $\Sigma F_h = 0$, we get $F_{GE} = F_{GH}$.

Joint B : See Fig. (*iii*).

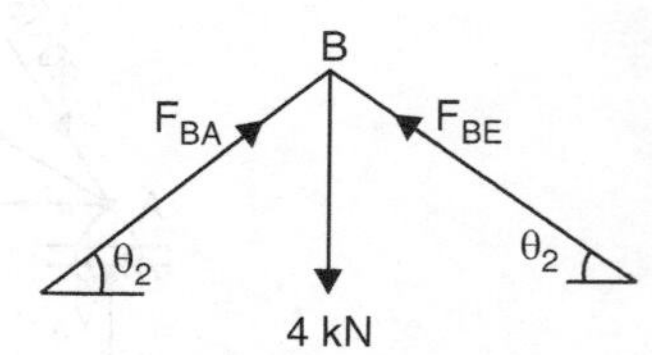

Fig. (iii) Joint B

By $\Sigma F_h = 0$, we get

$$F_{BA}\cos\theta_2 - F_{BE}\cos\theta_2 = 0$$

$$\therefore \quad F_{BA} - F_{BE} = 0 \qquad \text{...(1)}$$

By $\Sigma F_V = 0$, we get

$$F_{BA}\sin\theta_2 + F_{BE}\sin\theta_2 - 4 = 0$$

$$\therefore \quad F_{BA} + F_{BE} = \frac{4}{\sin\theta_2} \qquad \text{...(2)}$$

Adding (1) and (2), we get

$$2F_{BA} = \frac{4}{\sin\theta_2}; \quad \therefore \quad F_{BA} = \frac{2\sqrt{13}}{2} = \sqrt{13} \text{ kN (C)}$$

And from (1), we get

$$F_{BE} = \mathbf{\sqrt{13}\ kN} \text{ (C)}$$

Joint F : See Fig. (*iv*).

By $\Sigma F_h = 0$, we get

$$\rightarrow F_{FE}\cos\theta_2 - F_{FH}\cos\theta_2 = 0$$

$$\therefore \quad F_{FE} - F_{FH} = 0 \qquad \text{...(3)}$$

By $\Sigma F_V = 0$, we get

$$\uparrow F_{FE}\sin\theta_2 + F_{FH}\sin\theta_2 - 8 = 0$$

$$\therefore \quad F_{FE} + F_{FH} = \frac{8}{\sin\theta_2} \qquad \text{...(4)}$$

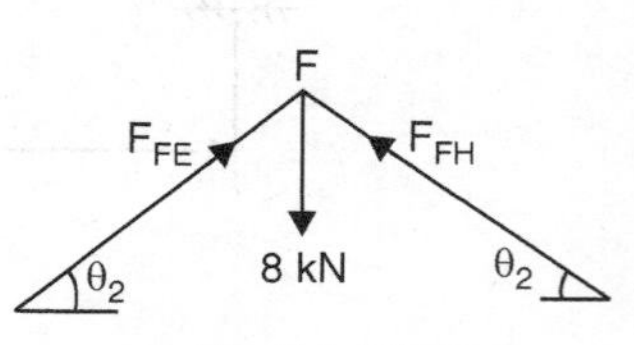

Fig. (iv) Joint F

Adding (3) and (4), we get

$$2F_{FE} = \frac{8}{\sin \theta_2},$$

$$\therefore \quad F_{FE} = \frac{4\sqrt{13}}{2} = \mathbf{2\sqrt{13}\ kN}\ (C)$$

Also, from (3) ; $F_{FH} = \mathbf{2\sqrt{13}\ kN}$ (C)

Joint A : See Fig. (*v*).

By $\Sigma F_V = 0$, we get

$$\downarrow F_{AD} \sin \theta_1 + \sqrt{13} \sin \theta_2 - 8 = 0$$

$$\therefore \quad F_{AD} = \frac{8}{\sin \theta_1} - \frac{\sqrt{13} \sin \theta_2}{\sin \theta}$$

$$= \frac{\sqrt{10}}{1}\left(8 - \sqrt{13} \times \frac{2}{\sqrt{13}}\right)$$

$$= \mathbf{6\sqrt{10}\ kN}\ (C)$$

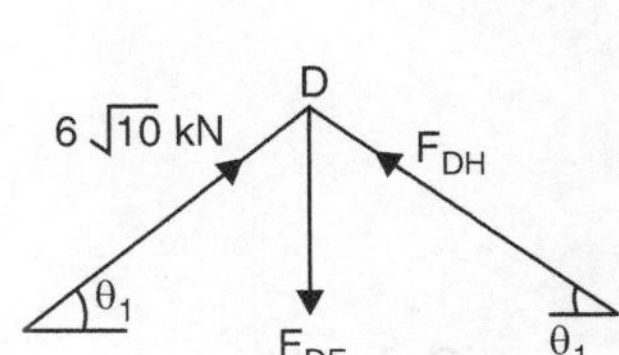

Fig. (v) Joint A

By $\Sigma F_h = 0$, we get

$$\rightarrow F_{AC} - \sqrt{13} \cos \theta_2 - F_{AD} \cos \theta_1 = 0$$

$$\therefore \quad F_{AC} = \sqrt{13} \cos \theta_2 + F_{AD} \cos \theta_1$$

$$= \sqrt{13} \times \frac{3}{\sqrt{13}} + 6\sqrt{10} \times \frac{3}{\sqrt{10}}$$

$$\therefore \quad F_{AC} = \mathbf{21}\ \text{kN (T)}$$

Joint D : See Fig. (*vi*).

By $\Sigma F_h = 0$, we get

$$\leftarrow F_{DH} \cos \theta_1 - 6\sqrt{10} \cos \theta_1 = 0$$

$$\therefore \quad F_{DH} = 6\sqrt{10}\ \text{kN (C)}$$

By $\Sigma F_V = 0$, we get

$$\downarrow F_{DE} - 6\sqrt{10} \sin \theta_1 - F_{DH} \sin \theta_1 = 0$$

$$\therefore \quad F_{DE} - 6\sqrt{10} \times \frac{1}{\sqrt{10}} - 6\sqrt{10} \times \frac{1}{\sqrt{10}} = 0$$

$$\therefore \quad F_{DE} = \mathbf{12\ kN}\ (T)$$

Fig. (vi) Joint D

Joint H : See Fig. (*vii*).

By $\Sigma F_V = 0$, we get

$$F_{HD} \sin \theta_1 + 2\sqrt{13} \sin \theta_2 - 10 = 0$$

$$\therefore \quad F_{HD} \times \frac{1}{\sqrt{10}} + 2\sqrt{13} \times \frac{2}{\sqrt{13}} - 10 = 0$$

$$\therefore \quad F_{HD} = \sqrt{10}\ (10 - 4) = \mathbf{6\sqrt{10}\ kN}\ (C)$$

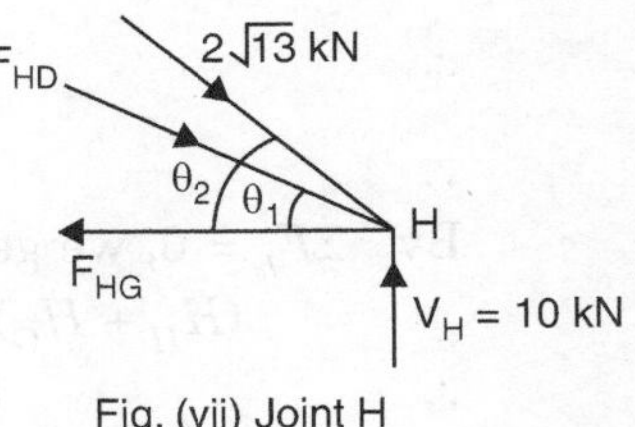

Fig. (vii) Joint H

This value is the same as found earlier by solving joint D. (**Check**)

By $\Sigma F_h = 0$, we get

$$F_{HG} - 2\sqrt{13} \cos \theta_2 - F_{HD} \cos \theta_1 = 0$$

$$\therefore \quad F_{HG} - 2\sqrt{13} \times \frac{3}{\sqrt{13}} - 6\sqrt{10} \times \frac{3}{\sqrt{10}} = 0$$

$\therefore \qquad F_{HG} = 6 + 18 = \mathbf{24\ kN}$ (T)

Also, $\qquad F_{GE} = \mathbf{24\ kN}$ (T)

Results are indicated in Fig. P. 2.13 (*b*).

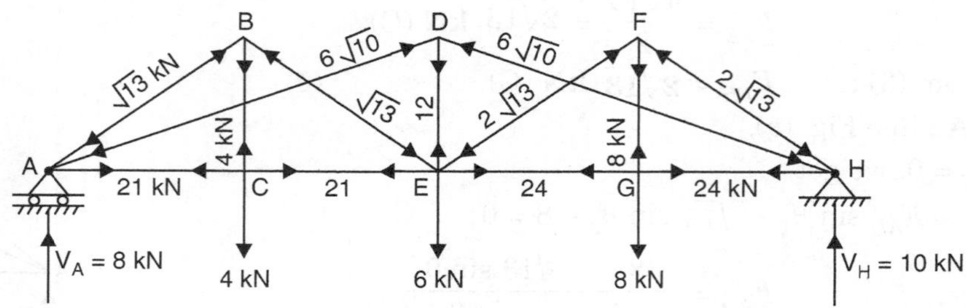

Fig. P. 2.13 (b) Results

Problem 2.14. *Find force in each member of the truss shown in Fig. P. 2.14 (a).*

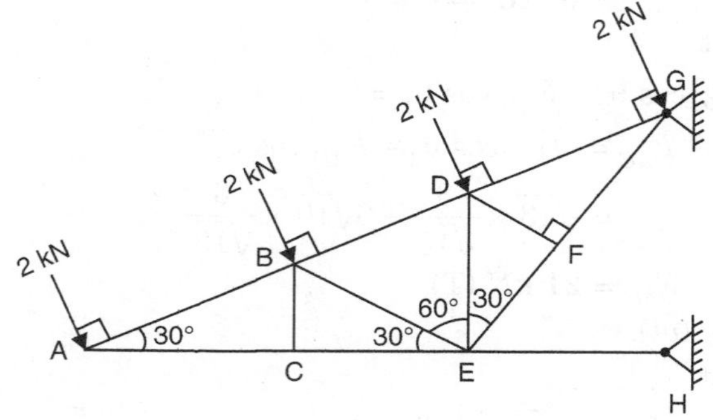

Fig. P. 2.14 (a) Given truss

Sol. Let $\qquad AB = a = BD = DG$

By $\Sigma M_G = 0$, we get

$$H_H \times GH - 2 \times AG - 2 \times BG - 2 \times DG = 0$$

$$H_H \times GH - 2 \times (3a) - 2 \times (2a) - 2 \times a = 0$$

$$\therefore \qquad H_H = \frac{1}{GH} \times 2 \times 6a = \frac{1}{AG \sin 30^\circ} \times 2 \times 6a = \frac{1}{3a \times 0.5} \times 12a$$

$$\therefore \qquad H_H = 8 \text{ kN} \leftarrow$$

By $\Sigma F_h = 0$, we get

$$(H_H + H_G) \leftarrow -\ 8 \times \sin 30^\circ = 0$$

$$\therefore \qquad 8 + H_G - 8 \times 0.5 = 0$$

$$\therefore \qquad H_G = 4 - 8 = -\ 4 \text{ kN}$$

$$\therefore \qquad H_G = 4 \text{ kN} \rightarrow$$

By $\Sigma F_V = 0$, we get

$$V_G - 8 \times \cos 30^\circ = 0$$

$$\therefore \qquad V_G = \frac{8 \times \sqrt{3}}{2} = 4\sqrt{3} \text{ kN} \uparrow$$

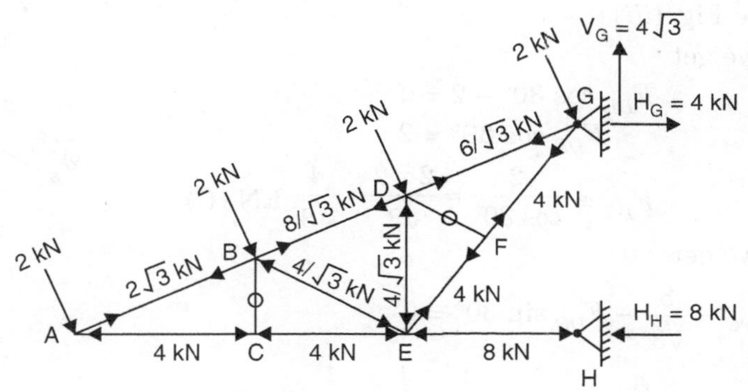

Fig. P. 2.14 (b) Results

By $\Sigma F_V = 0$, at joint C, we get

$$F_{CB} + 0 = 0, \quad \therefore \quad F_{CB} = 0.$$

By $\Sigma F_N = 0$ at F, we get

$$F_{FD} + 0 = 0\ ; \quad \therefore \quad F_{FD} = 0$$

Joint A : See Fig. (*i*).

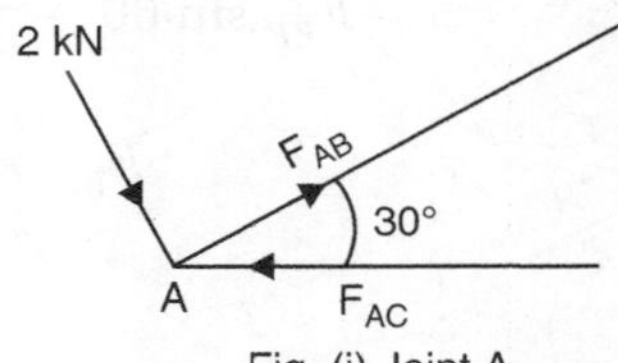

Fig. (i) Joint A

By $\Sigma F_V = 0$, we get

$$F_{AB} \sin 30° - 2 \cos 30° = 0$$

$$\therefore \quad F_{AB} = \frac{2 \times \sqrt{3}/2}{1/2} = \mathbf{2\sqrt{3}\ kN}\ \text{(T)}$$

and by $\Sigma F_h = 0$, we get

$$F_{AC} - F_{AB} \cos 30° - 2 \sin 30° = 0,$$

$$\therefore \quad F_{AC} = 2 \times \frac{1}{2} + F_{AB} \times \frac{\sqrt{3}}{2} = 1 + 2\sqrt{3} \times \frac{\sqrt{3}}{2} = 4$$

$$F_{AC} = 4 \text{ kN (C)}$$

Joint B : See Fig. (*ii*).

By $\Sigma F_y = 0$, we get

$$-2 + F_{BE} \sin 60° = 0$$

$$\therefore \quad F_{BE} = \frac{2}{\sin 60°} = \frac{2}{\sqrt{3}} \times 2$$

$$= \frac{4}{\sqrt{3}} \text{ kN (C)} = \mathbf{2.31\ kN}\ \text{(C)}$$

Fig. (ii) Joint B

By $\Sigma F_x = 0$, we get

$$F_{BD} - 2\sqrt{3} - F_{BE} \cos 60° = 0$$

$$F_{BD} - 2\sqrt{3} - \frac{4}{\sqrt{3}} \times \frac{1}{2} = 0$$

$$\therefore \quad F_{BD} = 2\sqrt{3} + \frac{2}{\sqrt{3}} = \frac{8}{\sqrt{3}} \text{ kN (T)}$$

$$= \mathbf{4.62\ kN}\ \text{(T)}$$

Joint D : See Fig. (*iii*)

By $\Sigma F_y = 0$, we get

$$F_{DE} \cos 30° - 2 = 0$$

$$\therefore \quad F_{DE} \cos 30° = 2$$

$$\therefore \quad F_{DE} = \frac{2}{\cos 30°} = \frac{2 \times 2}{\sqrt{3}} = \frac{4}{\sqrt{3}} \textbf{ kN (C)}$$

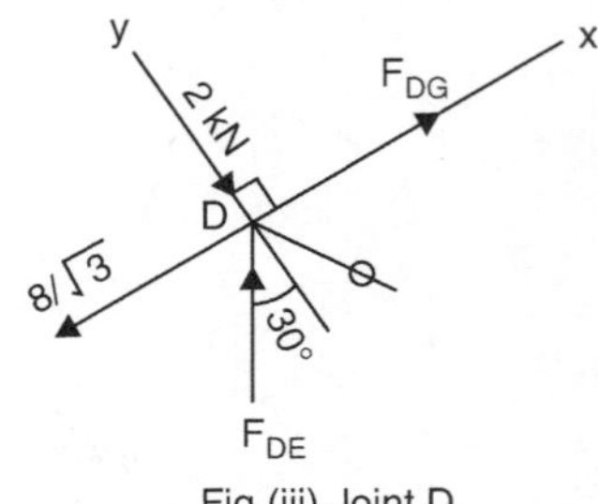

Fig.(iii) Joint D

By $\Sigma F_x = 0$, we get

$$F_{DG} - \frac{8}{\sqrt{3}} + F_{DE} \sin 30° = 0$$

$$\therefore \quad F_{DG} = \frac{8}{\sqrt{3}} - \frac{4}{\sqrt{3}} \times \frac{1}{2} = \frac{1}{\sqrt{3}}(8 - 2)$$

$$= \frac{6}{\sqrt{3}} \text{ kN (T)} = \textbf{3.464 kN} \text{ (T)}$$

Joint E : See Fig. (*iv*)

By $\Sigma F_V = 0$, we get

$$F_{EF} \sin 60° - \frac{4}{\sqrt{3}} - \frac{4}{\sqrt{3}} \sin 30° = 0$$

$$\therefore \quad F_{EF} \times \frac{\sqrt{3}}{2} - \frac{4}{\sqrt{3}} - \frac{2}{\sqrt{3}} = 0$$

$$\therefore \quad F_{EF} = \frac{2}{\sqrt{3}}\left(\frac{4}{\sqrt{3}} + \frac{2}{\sqrt{3}}\right)$$

$$= \frac{2}{\sqrt{3}} \times \frac{6}{\sqrt{3}} = \textbf{4 kN (T)}$$

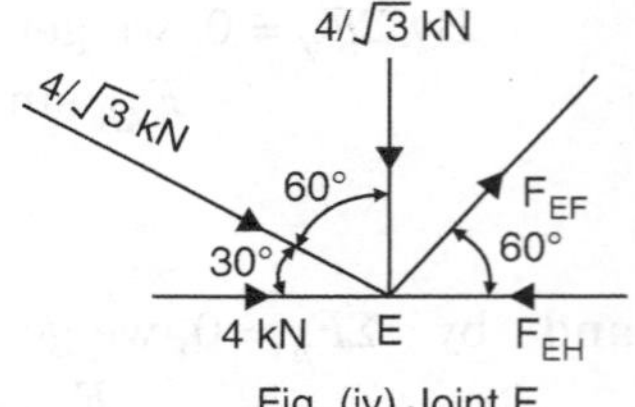

Fig. (iv) Joint E

By $\Sigma F_h = 0$, we get

$$F_{EH} - F_{EF} \cos 60° - \frac{4}{\sqrt{3}} \cos 30° - 4 = 0$$

$$\therefore \quad F_{EH} - 4 \times \frac{1}{2} - \frac{4}{\sqrt{3}} \times \frac{\sqrt{3}}{2} - 4 = 0$$

$$\therefore \quad F_{EH} - 2 - 2 - 4 = 0$$

$$\therefore \quad F_{EH} = \textbf{8 kN (C). Check}$$

Joint F : See Fig. (*v*).

By $\Sigma F_x = 0$, we get

$$F_{FG} - 4 = 0,$$

$$\therefore \quad F_{FG} = \textbf{4 kN} \text{ (T)}$$

Forces in all members of the truss are shown in Fig. P.2.14 (*b*).

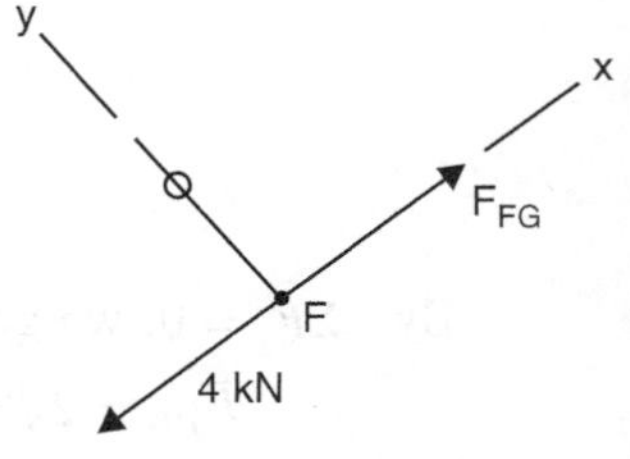

Fig. (v) Joint F

2.9. METHOD OF SECTIONS

We have studied that the method of joints deals with analysing trusses by applying the principles of equilibrium to the concurrent force systems acting at all joints of the truss. The principle of equilibrium for non-concurrent force systems can also be applied for the analysis of trusses ; the procedure is popularity known as **"Method of Sections".** Its use allows us to find *directly* the force in almost any member instead of proceeding to that member by the joint-to-joint analysis, as discussed earlier.

In the method of sections, the truss is cut in two portions by passing an **imaginary section or cutting plane**. Then the free-body diagram of either portion, thus isolated, is drawn showing all external loads, support reactions and unknown internal forces induced in the members cut by the cutting plane. Since there are three equations of equilibrium (*viz.* $\Sigma F_x = 0$, $\Sigma F_y = 0$, $\Sigma M_z = 0$ or $\Sigma F_h = 0$, $\Sigma F_V = 0$ and $\Sigma M = 0$), therefore, only three unknowns can be determined by using these equations of equilibrium. Hence, the selected section or the cutting plane should cut not more than three members of the truss. Forces in the members cut by the section plane are exerted by one part on the other. The complete procedure of using the method of sections in the solution of trusses is given in the following problems.

Problem 2.15. *Using the method of sections find force in members BC, CF and FG of the truss shown in Fig. P. 2.15 (a). Also find force in members CD, DH and HE.*

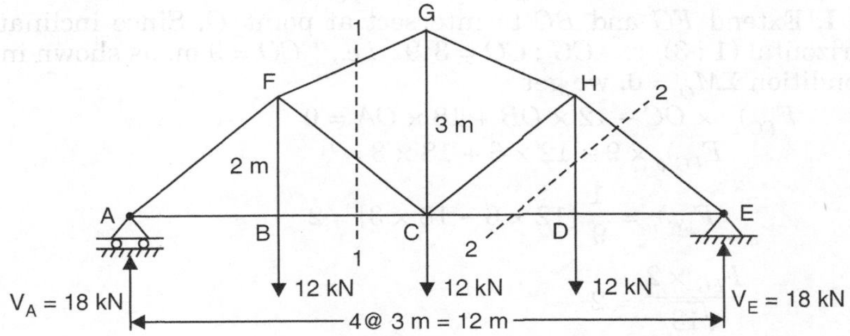

Fig. P. 2.15 (a) Given truss

Sol. The required forces in members *BC, CF* and *FG* can be determined by passing the cutting plane 1-1 cutting these members. (See Fig. P.2.15 (*a*)).

Consider the equilibrium of the portion of the truss which is on the left of the cutting plane 1–1. See Fig. P.2.15 (*b*).

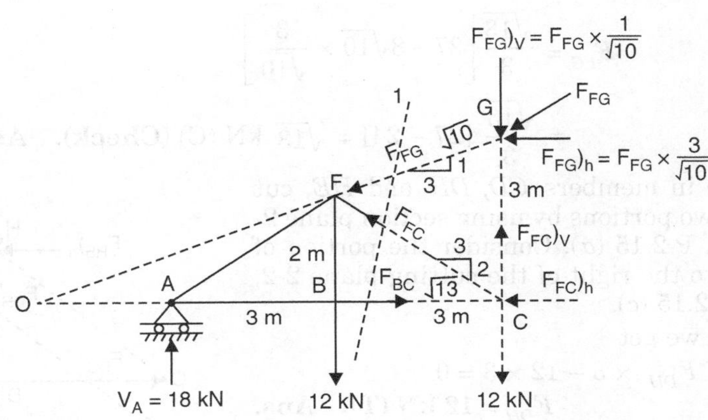

Fig. P.2.15 (b)

To find the value of F_{BC}, we take moments of all forces about point *F*. This eliminates the unknown forces F_{FC} and F_{FG}.

Thus, by $\Sigma M_F = 0$, we get $F_{BC} \times 2 - 18 \times 3 = 0$,

$$\therefore \quad F_{BC} = \frac{18 \times 3}{2} = \mathbf{27\ kN}\ (T). \quad \textbf{Ans.}$$

To find value of F_{FG}, use the condition :

$$\Sigma M_C = 0.$$

$\therefore$ we get

$$F_{FG})_h \times 3 - 18 \times 6 + 12 \times 3 = 0$$

$$\therefore \quad F_{FG})_h = \frac{1}{3}(18 \times 6 - 12 \times 3) = 24 \text{ kN}$$

or,
$$F_{FG} \times \frac{3}{\sqrt{10}} = 24$$

$$\therefore \quad F_{FG} = 8\sqrt{10} \text{ kN (Compression)} = \mathbf{25.30\ kN\ (C)}. \quad \textbf{Ans.}$$

There are two methods of finding the force F_{FC} in the member FC.

Method I. Extend FG and BC to intersect at point O. Since inclination of FG is 1 vertical to 3 horizontal (1 : 3), $\therefore$ $CG : CO = 3:9$ *i.e.,* $CO = 9$ m, as shown in Fig. P. 2.15(*b*). By using the condition $\Sigma M_O = 0$, we get

$$F_{FC})_v \times OC - 12 \times OB + 18 \times OA = 0$$

$$\therefore \quad F_{FC})_v \times 9 - 12 \times 6 + 18 \times 3 = 0$$

$$\therefore \quad F_{FC})_v = \frac{1}{9}[12 \times 6 - 18 \times 3] = 2$$

or,
$$\frac{F_{FC} \times 2}{\sqrt{13}} = 2$$

$$\therefore \quad F_{FC} = \sqrt{13} \text{ kN} = \mathbf{3.61\ kN\ (C)} \quad \textbf{Ans.}$$

Method II. This method is preferred. By equating to zero the sum of horizontal components of forces acting on the portion of the truss which is on the left of section 1-1, we can find $F_{FC})_h$. Thus by $\Sigma F_h = 0$, we get $F_{FC})_h + F_{FG})_h - F_{BC} = 0$

or,
$$F_{FC} \times \frac{3}{\sqrt{13}} + F_{FG} \times \frac{3}{\sqrt{10}} - 27 = 0$$

or,
$$F_{FC} = \frac{\sqrt{13}}{3}\left[27 - 8\sqrt{10} \times \frac{3}{\sqrt{10}}\right]$$

$$= \frac{\sqrt{13}}{3}[27 - 24] = \sqrt{13} \text{ kN (C) } \textbf{(Check)}. \quad \textbf{Ans.}$$

To find force in members CD, DH and HE, cut the given truss in two portions by using section plane 2-2, as shown in Fig. P.2.15 (*a*). Consider the portion of the truss which is to the right of the cutting plane 2-2, as shown in Fig. P.2.15 (*c*).

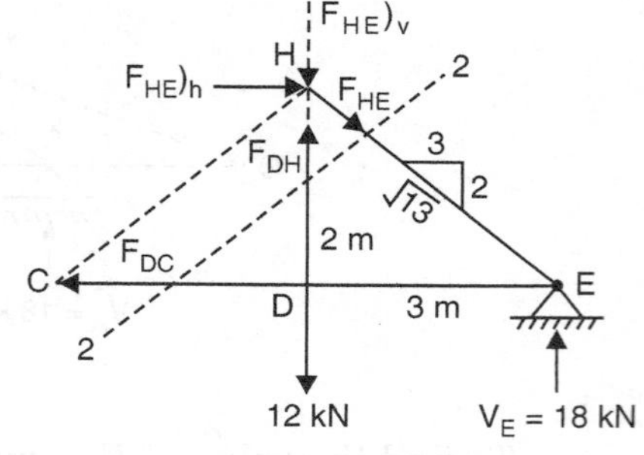

Fig. P.2.15 (c)

By $\Sigma M_E = 0$, we get

$$\uparrow F_{DH} \times 3 - 12 \times 3 = 0$$

$$\therefore \quad F_{DH} = \mathbf{12\ kN\ (T)}. \quad \textbf{Ans.}$$

By $\Sigma M_H = 0$, we get

$$F_{DC} \times 2 - 18 \times 3 = 0$$

$$\therefore \quad F_{DC} = \mathbf{27\ kN}\ \text{(Tension)}. \quad \textbf{Ans.}$$

By $\Sigma M_D = 0$, we get

$$F_{HE})_h \times 2 - 18 \times 3 = 0$$

$$\therefore \quad F_{HE})_h = 27$$

or,

$$F_{HE} \times \frac{3}{\sqrt{13}} = 27$$

$$\therefore \quad F_{HE} = \mathbf{9\sqrt{13}\ kN}\ (C) = \mathbf{32.45\ kN(C).\quad Ans.}$$

Alternatively, by $\Sigma F_V) = 0$, we get

$$F_{HE})_V - F_{DH} + 12 - 18 = 0$$

$$\therefore \quad F_{HE})_V = 18 - 12 + 12 = 18$$

$$\therefore \quad \frac{2}{\sqrt{13}} \times F_{HE} = 18$$

$$\therefore \quad F_{HE} = \mathbf{9\sqrt{13}\ (C)\quad (Check)}$$

Results :

Member	*Force*
BC	27 kN (T)
CF	3.61 kN (C)
FG	25.3 kN (C)
CD	27 kN (T)
HE	32.45 kN (C)
DH	12 kN (T)

Problem 2.16. *Find force in members BC, CE, EF, CF and CD of the truss shown in Fig. P.2.16 (a).*

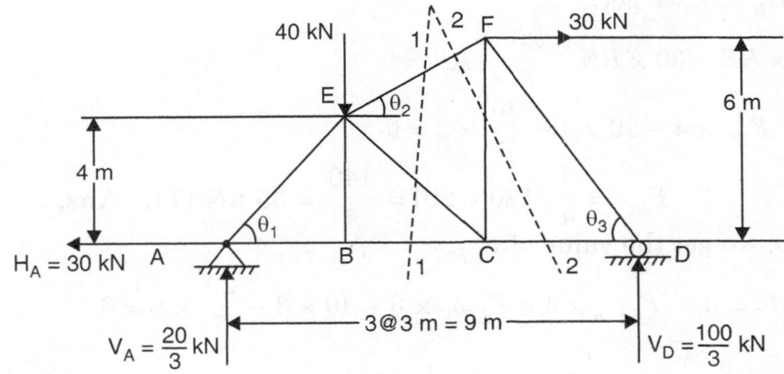

Fig. P.2.16 (a)

Sol. We have $\tan\theta_1 = \frac{4}{3}$, $\quad\therefore\ \sin\theta_1 = \frac{4}{5}$ and $\cos\theta_1 = \frac{3}{5}$.

$$\tan\theta_2 = \frac{6-4}{3} = \frac{2}{3}; \quad \therefore\ \sin\theta_2 = \frac{2}{\sqrt{13}} \text{ and } \cos\theta_2 = \frac{3}{\sqrt{13}}$$

and,

$$\tan\theta_3 = \frac{6}{3} = \frac{2}{1}; \quad \therefore\ \sin\theta_3 = \frac{2}{\sqrt{5}} \text{ and } \cos\theta_3 = \frac{1}{\sqrt{5}}.$$

By $\Sigma M_D = 0$, we get

$$V_A \times 9 - 40 \times 6 + 30 \times 6 = 0$$

$$\therefore \qquad V_A = \frac{1}{9}[240 - 180] = \frac{60}{9} = \frac{20}{3} \text{ kN} \uparrow$$

By $\Sigma F_V = 0$, we get $(V_A + V_D) \uparrow - 40 = 0$

$$\therefore \qquad V_D = 40 - V_A = 40 - \frac{20}{3} = \frac{100}{3} \text{ kN} \uparrow$$

By $\Sigma F_h = 0$, we get

$$H_A - 30 = 0, \quad \therefore \quad H_A = 30 \text{ kN} \leftarrow$$

Consider portion of the truss which is on the left of section 1-1, Fig. P.2.16 (*b*).

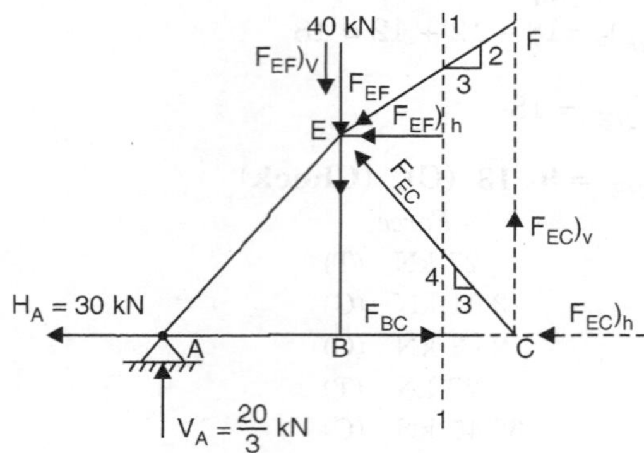

Fig. P. 2.16 (b)

For finding F_{BC}, use $\Sigma M_E = 0$.

$\therefore$ by $\Sigma M_E = 0$, we get

$$F_{BC} \times BE - 30 \times BE - \frac{20}{3} \times AB = 0$$

$$\therefore \qquad F_{BC} \times 4 - 30 \times 4 - \frac{20}{3} \times 3 = 0$$

$$\therefore \qquad F_{BC} = \frac{1}{4}[120 + 20] = \frac{140}{4} = \mathbf{35 \text{ kN (T)}}. \quad \textbf{Ans.}$$

By $\Sigma M_C = 0$, we get the value of F_{EF}.

$$\therefore \quad \text{by } \Sigma M_C = 0, \quad F_{EF})_h \times 4 + F_{EF})_V \times 3 + 40 \times 3 - \frac{20}{3} \times 6 = 0$$

or,
$$\left(F_{EF} \times \frac{3}{\sqrt{13}}\right) \times 4 + \left(F_{EF} \times \frac{2}{\sqrt{13}}\right) \times 3 + 120 - 40 = 0$$

$$\therefore \qquad \frac{F_{EF}}{\sqrt{13}}(12 + 6) = -80$$

$$\therefore \qquad F_{EF} = \frac{-80\sqrt{13}}{18} = -16.03 \text{ kN}$$

or
$$F_{EF} = \mathbf{16.03 \text{ kN (T)}}. \quad \textbf{Ans.}$$

By $\Sigma F_V = 0$ for the portion of the truss which is on the left of section 1-1,

we get,
$$\uparrow F_{EC})_V + \frac{20}{3} - 40 - F_{EF})_V = 0.$$

$$\therefore \quad F_{EC} \times \frac{4}{5} + \frac{20}{3} - 40 - \left(-\frac{80\sqrt{13}}{18}\right) \times \frac{2}{\sqrt{13}} = 0$$

$$\therefore \quad F_{EC} = \frac{5}{4}\left(40 - \frac{20}{3} - \frac{80\sqrt{13}}{18\times\sqrt{13}}\times 2\right) = \frac{5}{4}\left[40 - \frac{20}{3} - \frac{80}{9}\right]$$

$$= \frac{5}{4}\times\frac{360-60-80}{9} = \frac{5}{4}\times\frac{220}{9}\text{ kN}$$

or F_{EC} = **30.56 kN** (Compressive). **Ans.**

To find force in members *CD* and *CF*, cut the truss by section plane 2-2 and consider the portion which is on the right of section 2-2, as shown in Fig. P.2.16 (*c*).

By $\Sigma M_F = 0$, we get

$$F_{CD}\times 6 - \frac{100}{3}\times 3 = 0.$$

$$\therefore \quad F_{CD} = \frac{100}{3}\times\frac{3}{6} = 0$$

$$\therefore \quad F_{CD} = \frac{50}{3} = \mathbf{16.67\ kN}\ (T).\quad \textbf{Ans.}$$

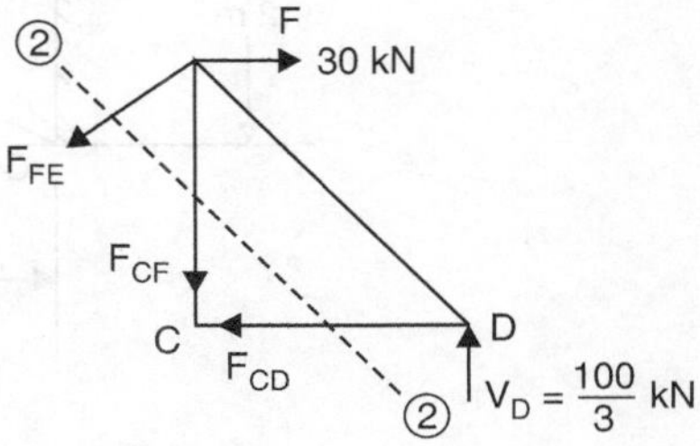

Fig. P.2.16 (c)

By $\Sigma M_C = 0$, we get

$$F_{FE})_h \times 6 - 30\times 6 + \frac{100}{3}\times 3 = 0$$

$$\therefore \quad F_{EF})_h = \frac{1}{6}[180-100] = \frac{80}{6}\text{ kN}$$

or, $$F_{FE}\times\frac{3}{\sqrt{13}} = \frac{80}{6}$$

$$\therefore \quad F_{FE} = \frac{80\sqrt{13}}{3\times 6} = \mathbf{16.03\ kN}$$

or, F_{FE} = **16.03 kN** (T). **Ans.** (as before)

By $\Sigma F_V = 0$, we get

$$\downarrow F_{CF} + F_{FE})_V - \frac{100}{3} = 0$$

$$\therefore \quad F_{CF} + \frac{80\sqrt{13}}{18}\times\frac{2}{\sqrt{13}} - \frac{100}{3} = 0$$

$$\therefore \quad F_{CF} + \frac{80}{9} - \frac{100}{3} = 0$$

$$\therefore \quad F_{CF} = \frac{-80+300}{9} = +\frac{220}{9} = \mathbf{+\ 24.45\ kN}\ \text{(Tension)}$$

$$\therefore \quad F_{CF} = \mathbf{24.45\ kN}\ (T).\quad \textbf{Ans.}$$

Note. All of the above results are the same as obtained in the solution of Problem 2.2 which was solved by Method of Joints.

Results :

Member	*Force*
BC	35 kN (T)
CE	30.56 kN (C)
EF	16.03 kN (T)
CF	24.45 kN (T)
CD	16.67 kN (T)

Problem 2.17. *Find force in members BC, CD, DE and BD of the truss shown in Fig. P.2.17 (a).*

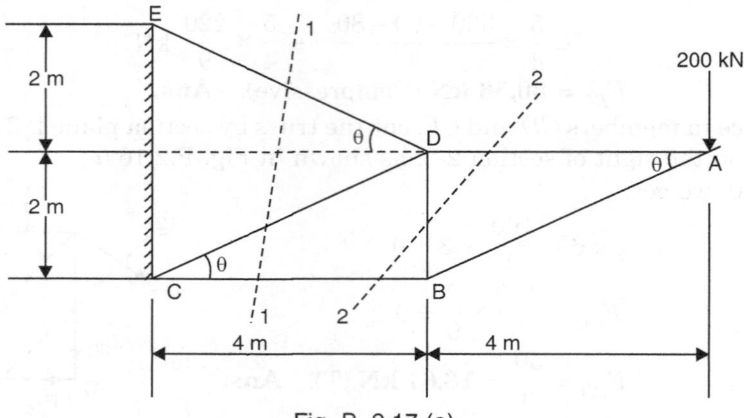

Fig. P. 2.17 (a)

Sol. $\tan\theta = \frac{2}{4} = \frac{1}{2}$; $\therefore$ $\sin\theta = \frac{1}{\sqrt{5}}$ and $\cos\theta = \frac{2}{\sqrt{5}}$.

Consider the portion of the truss which is on the right of the section 1-1 as shown in Fig. P.2.17 (*b*)

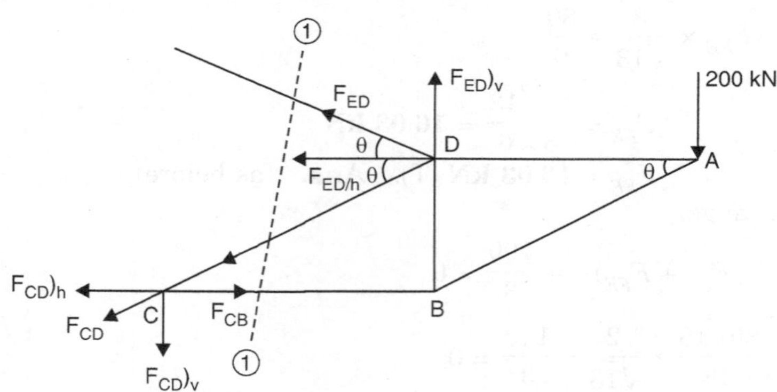

Fig. P. 2.17 (b)

By $\Sigma M_C = 0$, we get (this condition is used to find F_{ED})

$$F_{ED})_h \times 2 + F_{ED})_V \times 4 - 200 \times 8 = 0$$

$$\therefore \quad F_{ED}\cos\theta \times 2 + F_{ED}\sin\theta \times 4 = 200 \times 8$$

$$\therefore \quad F_{ED} \times \frac{2}{\sqrt{5}} \times 2 + F_{ED} \times \frac{1}{\sqrt{5}} \times 4 = 200 \times 8$$

$$\therefore \quad F_{ED}\left(\frac{8}{\sqrt{5}}\right) = 200 \times 8$$

$$\therefore \quad F_{ED} = \mathbf{200\sqrt{5}\ kN} \text{ (Tension).} \quad \textbf{Ans.}$$

By $\Sigma M_D = 0$, we get (this condition is used to find F_{CB}).

$$F_{CB} \times 2 - 200 \times 4 = 0$$

$$\therefore \quad F_{CB} = \mathbf{400\ kN} \text{ (C).} \quad \textbf{Ans.}$$

2.10. GRAPHICAL METHOD

In Chapter 1, the basic procedure for finding the reactions of a beam and a truss was discussed in details. That included discussion and application of use of Bow's notation, drawing of space diagram, force diagram and string/funicular polygon. The procedure for finding the resultant of the given system of forces was also discussed.

In this Chapter, we shall discuss the procedure of finding forces (or stresses) in member of the truss. We shall discuss the Maxwell's diagram here. The full graphical method as used for finding forces in members of a given truss consists of :

(*a*) Drawing of **space diagram.** (to some suitable linear scale)

(*b*) Use of **Bow's notation** for naming the given loads and the members of the truss.

(*c*) Selection of a suitable **pole** and drawing of force diagram/polygon. (to some suitable force scale).

(*d*) Construction of **string/funicular polygon.**

(*e*) Determination of support reactions in direction and magnitude.

(*f*) Construction of **Maxwell diagram.**

(*g*) Determination of magnitude and **nature** (tension or compression) **of forces** in members of the truss.

As mentioned above steps from (*a*) to (*e*) have been discussed in details in Chapter 1. Only one or two problems will be solved here as a review of the underlined procedure.

2.11. SEMI-GRAPHICAL METHOD

Sometimes the reactions are determined by using the equations of equilibrium (*viz.*, $\Sigma F_v = 0$, $\Sigma F_h = 0$ and $\Sigma M = 0$) as discussed in analytical method, and only Maxwell's diagram is drawn for finding forces/stresses in members of the truss. This procedure is known as **semi-graphical method**. Sometimes this is referred to as graphical method.

Note 1. It is important to note that space diagram is compulsorily drawn to some suitable linear scale, for example 1 cm = 1 m ; 1 cm = 2 m ; 1 cm = 3 m, etc., and the inclination of forces, members of the truss and the surface of the roller support should be drawn by using the protractor.

Note 2. The force diagram and the Maxwell's diagram should be drawn by selecting suitable force-scale, for example 1 cm = 50 kN, 1 cm = 10 kN, 1 cm = 50 N etc., and the inclination of the given or applied forces should be drawn with the help of protractor.

Problem 2.18. *Graphically determine support reactions of the beam shown in Fig. P.2.18 (a).*

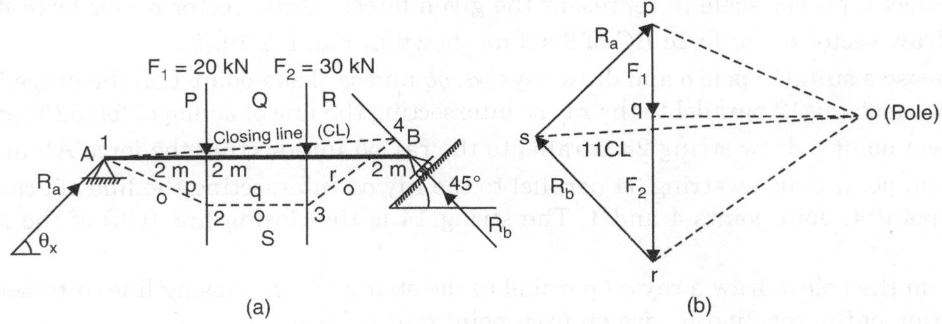

Fig. P.2.18

Sol. Draw beam to the scale. Name spaces on the left and right of the 20 kN force as P and Q, as shown in Fig. P.2.18 (*a*). The space on the right of 30 kN force is named as R. Reaction R_b at roller support B is named as RS and the reaction R_a at A is named as SP.

Using a suitable scale to represent forces, represent the force PQ by the vector pq and the force QR by the vector qr, as shown in Fig. P.2.18 (*b*). Take some convenient point o as pole and draw the rays op, oq and or.

Starting from point 1 on the hinge draw the string 12 parallel to the ray po intersecting line of action of 20 kN force at point 2. From point 2 draw string 23 parallel to the ray qo intersecting the force QR at point 3.

From point 3 draw the string 34 parallel to the ray ro intersecting the line of action of reaction R_b at point 4. Join points 4 and 1. This line is known as the closing line of the funicular polygon.

From the point o draw a ray parallel to the closing line 41 of the funicular polygon intersecting a line parallel to the action line of reaction R_b at point s. Join point s with p. The vector rs gives magnitude and direction of reaction R_b. Similarly, the vector sp gives direction and magnitude of reaction R_a.

Scaling from the force polygon we get

R_b = 36 kN, $\perp^{ar}$ to the surface of roller. **Ans.**

Also, R_a = 33 kN, inclined at angle θ_x = 42°. **Ans.**

Problem 2.19. *Determine graphically the reactions of supports of the beam shown in Fig. P.2.19 (a).*

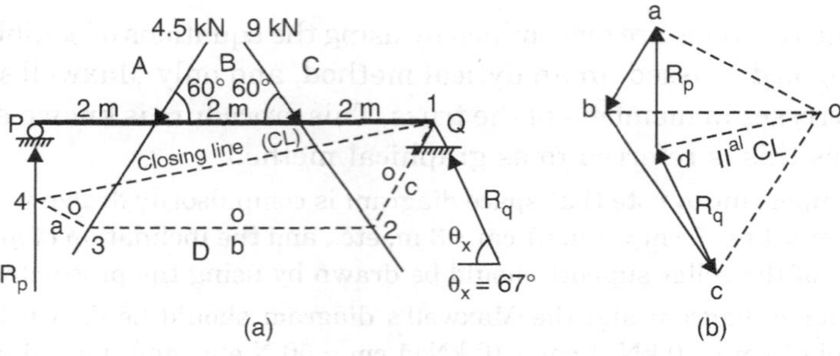

Fig. P.2.19

Sol. Draw space diagram accurately by using suitable linear scale and protractor, as shown in Fig. P.2.19 (*a*). Using Bow's notation name various spaces such as A, B, C and D as shown. Select a proper scale to represent the given forces. Draw vector ab for force AB of 4.5 kN and draw vector bc for force BC of 9 kN as shown in Fig. P.2.19 (*b*).

Choose a suitable pole o and draw rays oa, ob and oc. Take point 1 on the hinged support Q and draw a string 12 parallel to the ray oc intersecting the line of action of force BC at point 2.

From point 2 draw string 23 parallel to the ray ob intersecting the force AB at point 3.

From point 3 draw string 34 parallel to the ray oa intersecting the line of action of the roller at point 4. Join points 4 and 1. The string 14 is the closing line (CL) of the funicular polygon.

From the pole o draw a ray od parallel to the string 14 (the closing line) intersecting the line of action of the reaction R_p drawn from point a at point d.

Draw vector *cd* representing the reaction R_q of the hinge support. Vector *da* represents the reaction R_p of the roller support *P*.

Ans. Magnitude of R_P = 3.6 kN, vertical, and R_q = 6.6 kN inclined at an angle θ_x = 67° to the horizontal. (These answers are scaled off from Fig. P.2.19 (*b*)).

Problem 2.20. *Determine graphically the support reactions for the beam shown in Fig. P.2.20 (a).*

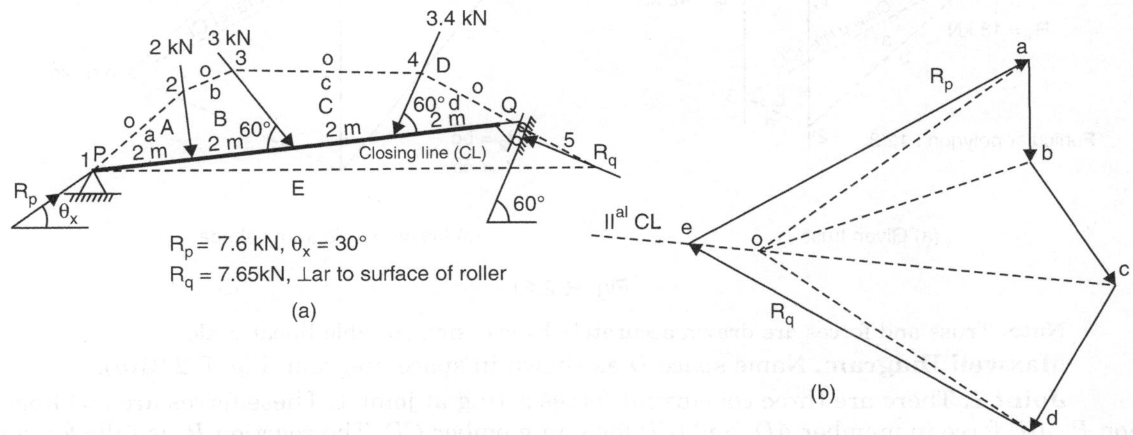

Fig. P. 2.20

Sol. Draw space diagram by choosing a suitable linear scale and by drawing forces with accurate inclinations, as shown in Fig. P. 2.20 (*a*). Using Bow's notations name forces such as *AB*, *BC* and *CD* for 2 kN force, 3 kN force and 3.4 kN force, respectively.

Using a suitable scale for representing forces draw the force diagram shown in Fig. P.2.20 (*b*). Select a convenient pole *o* and draw rays *oa*, *ob*, *oc* and *od*. Taking point 1 on the hinged support draw string 12 parallel to ray *oa* intersecting the line of force *AB* at point 2.

Through point 2 draw string 23 parallel to the ray *ob* intersecting line of force *BC* at point 3.

Through point 3 draw string 34 parallel to ray *oc* intersecting line of force *CD* at point 4.

Through point 4 draw string 45 parallel to the ray *od* intersecting line of reaction R_q (perpendicular to the surface of roller) at point 5.

Join point 5 with point 1. Line 51 is called the closing line (CL) of the funicular polygon 12345.

Through the pole *o* draw a ray *oe* parallel to the closing line intersecting the line parallel to the direction of the reaction R_q drawn through point *d* at the point *e*. Vector *de* represents R_q. Join point *e* with point *a*. Vector *ea* represents reaction R_p at support *P*.

Magnitude and inclination of reactions R_p and R_q are scaled from the Fig. P.2.20 (*b*), and are given as under :

Ans. R_P = 7.6 kN, θ_x = 30°

R_q = 7.65 kN, perpendicular to the surface of roller as shown in Fig. P.2.20 (*a*).

Problem 2.21. *Graphically find force (stress) in members of the truss shown in Fig. P.2.21(a).*

Sol. Find graphically the reactions R_1 and R_2 of the supports 1 and 2, respectively. Scaling from the force polygon R_2 represented by vector *bc* is equal to 42 kN, and R_1 represented by vector *ca* is equal to 18 kN, both acting vertically upward. The procedure for finding R_1 and is self-explanatory.

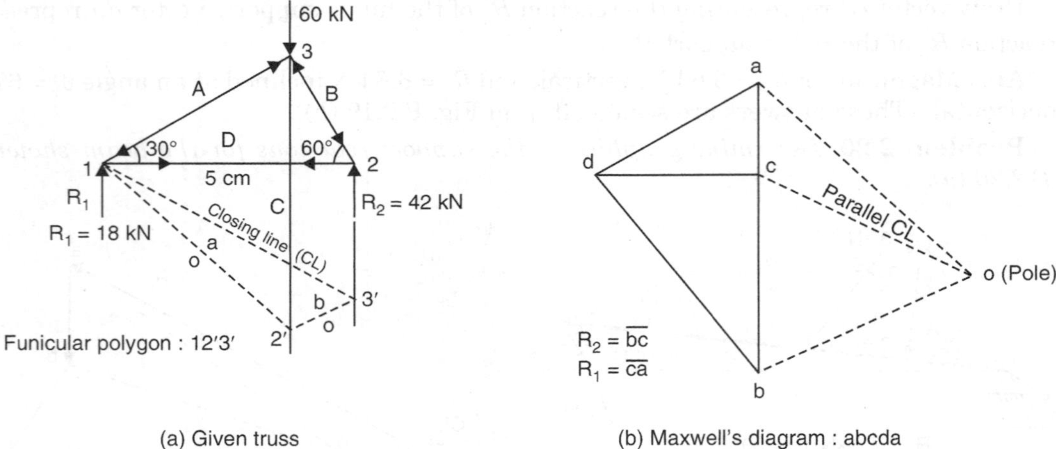

(a) Given truss (b) Maxwell's diagram : abcda

Fig. P. 2.21

Note. Truss and forces are drawn accurately by choosing suitable linear scale.

Maxwell Diagram. Name space *D* as shown in space diagram, Fig. P.2.21(*a*).

Joint 1. There are three concurrent forces acting at joint 1. These forces are : (*i*) Reaction R_1, (*ii*) force in member *AD*, and (*iii*) force in member *CD*. The reaction R_1 is fully known in direction and magnitude. To find the two unknown forces proceed as follows :

Through point *a* of the force diagram draw a line parallel to member *AD*. Through point *d* draw a line parallel to member *CD*. These two lines intersect at the required point *d*.

Magnitude and nature of forces in members AD and CD. The three forces acting at joint 1 are given by the Δacd. Starting from point *a* we go to *d* in direction *a* to *d* ; from *d* to point *c* in direction *dc* and finally from *c* to *a* in direction *ca*, thus returning to the starting point *a*. Thus, force in member *AD* is acting in direction *ad*, force in member *DC* acts in direction *dc*. Draw arrows at joint 1, as shown in Fig. P.2.21 (*a*). Thus, force in member $AD = F_{ad} = 37$ kN compression (*i.e.*, towards joint 1).

And, force in member $DC = F_{dc} = 32.5$ kN tension (*i.e.*, away from joint 1), as shown in Fig. P.2.21 (*b*).

Draw arrow heads in member *AD* in opposite direction at joint 3 and also in member *CD* in opposite direction at joint 2, *ab* shown in Fig. P.2.21 (*a*).

Joint 2. Forces acting at joint 2 are : reaction R_2, force F_{cd} in member *CD* and force F_{bd} in member *BD*. F_{cd} and R_2 are known in direction and magnitude. To find force F_{bd} in member *BD* proceed as follows :

Forces acting at joint 2 form a closed triangle Δbcd. Start from point *b* in the Maxwell's diagram and go to point *c* moving along the direction of R_2, from *c* to *d* towards left (*i.e.*, along the leftward arrow in member *CD* and finally return to starting point *b*. The force in member *DB* is given by the line *db* both in magnitude and direction. Put arrow along *bd* on member *BD* at joint 3 in Fig. P.2.21 (*a*). Draw arrow in opposite direction in member *BD* at joint 3.

From Fig. P.2.21 (*b*), we find that $F_{bd} = 53$ kN compression.

Results. Following results are found graphically :

R_1 = 18 kN ↑ ; R_2 = 42 kN ↑.

F_{ad} = 37 kN (compression) ; F_{cd} = 32.5 kN (tension) ;

F_{bd} = 53 kN (compression).

Note. If reactions R_1 and R_2 are determined analytically, we directly locate c in the force polygon. After locating point c, we can construct the Maxwell's diagram $adcb$ and find the forces F_{bd}, F_{dc} and F_{bd}.

Problem 2.22. *Graphically find forces in all members of truss shown in Fig. P.2.22 (a).*

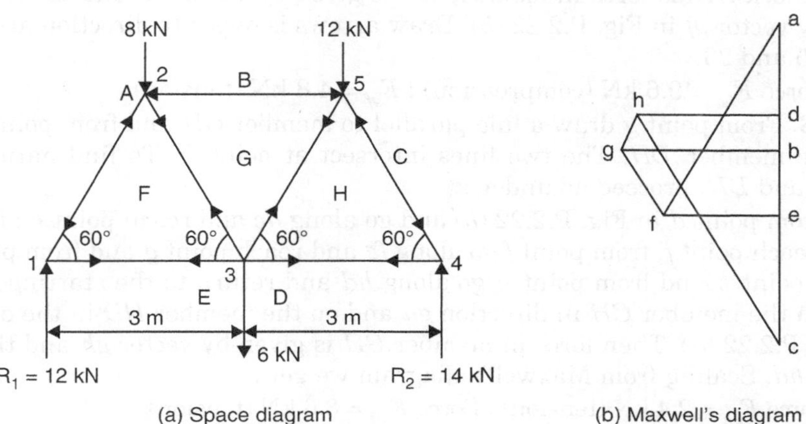

(a) Space diagram (b) Maxwell's diagram

Fig. P. 2.22

Sol. Draw the given truss to a suitable scale (*i.e.*, 1 cm = 1 m) and show all given forces acting on it as shown in Fig. P.2.22 (*a*). Let us determine reactions R_1 and R_2 analytically.

By using the condition $\Sigma M_4 = 0$, we get

$$R_1 \times 6 - 8 \times \left(\frac{3}{2} + 3\right) - 12\left(\frac{3}{2}\right) - 6 \times 3 = 0$$

or $$6R_1 - 8 \times 4.5 - 12 \times 1.5 - 18 = 0$$

$\therefore$ $$R_1 = \frac{1}{6}(36 + 18 + 18) = \mathbf{12\ kN} \uparrow \quad \textbf{Ans.}$$

And, using the $\Sigma F_v = 0$, we get

$$(R_1 + R_4) \uparrow - (8 + 12 + 6) = 0$$

$\therefore$ $$12 + R_4 - 26 = 0$$

$\therefore$ $$\mathbf{R_4 = 14\ kN} \uparrow \quad \textbf{Ans.}$$

Draw the force polygon with some suitable scale ab represents force AB, bc represents force BC, cd represents R_4, de represents force DE and ea represents reaction R_1.

Using Bow's notation name spaces F, G and H within the truss.

Joint 1. From point a draw a line parallel to the member AF and from point e draw a line parallel to the member EF. The two lines intersect at point f.

To find forces in magnitude and nature in all members meeting at the joint 1 proceed from point a, go to f, then go from f to e and return to point a in the same order and draw arrows on member AF in direction af ; on member FE in direction fe ; as shown in Fig.P.2.21(*a*).

Thus, by scaling from Maxwell's diagram, we get :

Ans. F_{af} = 13.6 kN (compression) ; F_{ef} = 7.2 kN (tension).

Draw arrows in opposite direction at the other ends of members AF and EF, and go to joint 2.

Joint 2. Start from point a go to point b and from point b draw a line parallel to member BG. From point f draw a line parallel to member FG. The two lines intersect at point g.

To find nature of forces in members *BG* and *GF* proceed as follows :

Starting from point *b*, go along *bg* and reach point *g*, from point *g*, go along *af* and reach point *f*, from point *f*, go along *fa* and return to point *a*. Put arrows on member *BG* in direction *b* to *g* at joint 2 ; on member *GF* in direction *g* to *f* and in member *FA* in direction *f* to *a* (already marked as a check). Thus force in member *BG* is given by vector *bg* and the force in member *GF* is given by vector *gf* in Fig. P.2.22 (*b*). Draw arrows is opposite direction at the other ends of members 25 and 23.

Ans. Force F_{bg} = 9.6 kN (compression) ; F_{gf} = 4.8 kN (tension).

Joint 3. From point *g* draw a line parallel to member *GH* and from point *d* from a line parallel to the member *DH*. The two lines intersect at point *h*. To find nature of forces in members *GH* and *DH*, proceed as under :

Start from point *d* in Fig. P.2.22 (*b*) and go along *de* and reach point *e* ; from point *e* go along *ef* and reach point *f*, from point *f* go along *fg* and reach point *g* and from point *g* go along *gh* and reach point *h* and from point *h* go along *hd* and return to the starting point *d*. Thus, draw arrow on the member *GH* in direction *gh* and on the member *HD* in the direction *hd*, as shown in Fig. P.2.22 (*a*). Then force in member *GH* is given by vector *gh* and that in member *HD* by vector *hd*. Scaling from Maxwell's diagram we get :

Ans. Force F_{gh} = 2.4 kN (tension) ; Force F_{hd} = 8.6 kN (tension).

Draw arrows in opposite direction at the other ends of members 34 and 34.

Joint 4. From point *h* draw a line parallel to member *CH*. Then force in member *CH* is given by vector *hc*, mark arrow on the member *HC* in the direction *hc*. Scaling from Maxwell's diagram we get :

Ans. Force F_{hc} or F_{ch} = 17.2 kN (compression)

Problem 2.23. *Find graphically the forces in all members of the truss shown in Fig. P.2.23 (a).*

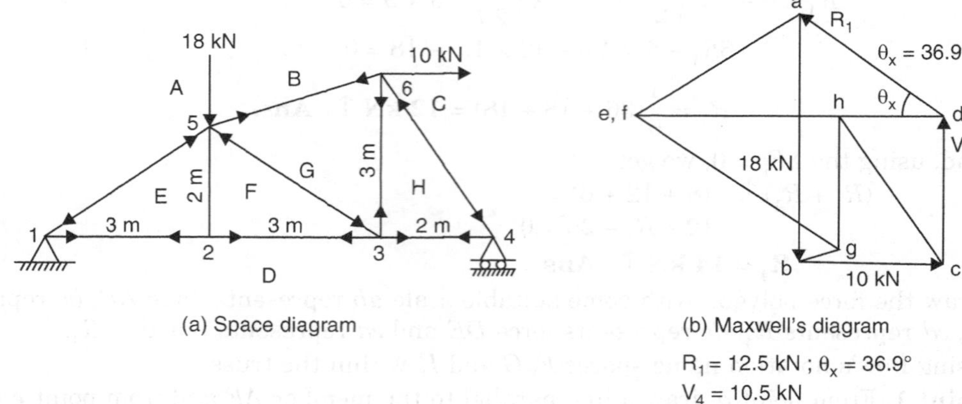

(a) Space diagram

(b) Maxwell's diagram

R_1 = 12.5 kN ; θ_x = 36.9°

V_4 = 10.5 kN

Fig. P.2.23

Sol. Select a suitable linear scale (*e.g.* 1 cm = 1 m) and draw truss as shown in Fig. P.2.23 (*a*). Mark various forces acting on the truss.

Find reactions H_1, V_1 and V_4 by analytical method.

By $\Sigma F_h = 0$, we get $H_1 - 10 = 0$

$\therefore \quad H_1 = 10 \text{ kN} \leftarrow$

By $\Sigma M_4 = 0$, we get

$V_1 \times 8 - 18 \times 5 + 10 \times 3 = 0$

$$\therefore \quad V_1 = \frac{1}{8}(90 - 30 = 60) = 7.5 \text{ kN} \uparrow$$

Hence, $R_1 = \sqrt{H_1^2 + V_1^2} = \sqrt{10^2 + 7.5^2} = 12.5 \text{ kN} \uparrow$

$$\theta_x = \tan^{-1} (V_1/H_1 = \frac{7.5}{10} = 0.75) = 36.9°$$

Finally, by $\Sigma F_v = 0$, we get

$$(V_1 + V_4) \uparrow - 18 = 0$$

$\therefore$ $V_4 = 18 - V_1 = 18 - 7.5 = 10.5 \text{ kN} \uparrow$

Selecting a suitable force scale (*e.g.,* 1 cm = 4 kN) draw vector *ab* to represent vertical force *AB* ; draw vector *bc* to represent horizontal force *BC*. From point *c* draw vector *cd* to represent vertical reaction V_4 at the roller support 4. Then vector *da* represents the reaction R_1 at the hinged support both magnitude and direction (θ_x = 36.9°). See Fig. P.2.23 (*b*).

Maxwell's Diagram. Name spaces *E*, *F*, *G* and *H* in the truss as shown in Fig. P.2.23 (*a*).

Joint 1. From point *a* in Fig. P.2.23 (*b*) draw line parallel to the member *AE*, and from point *d* draw a line parallel to member *DE*. These lines intersect at point *e*.

Joint 2. From point *e* draw a line parallel to member *EF* and from point *d* draw a line parallel to member *DF*. The two lines intersect at point *f* which coincides with point *e*.

Joint 5. From point *f* draw a line parallel to the member *FG*, and from point *b* draw a line parallel to member *BG*. The two lines intersect at point *g*.

Joint 6. From point *g* draw a line parallel to the member *GH*, and from point *C* draw a line parallel to the member *CH*. The two lines intersect at point *b*.

This completes the Maxwell's diagram *aefghdc*.

Magnitude and nature of forces in members of the truss :

Joint 1. Force in member *AE* is given by vector *ae*, the force in member *DE* is given by vector *de*. These are shown by arrows drawn on members *AE* and *DE*. Member *AE* is in compression and member *DE* is in tension.

Joint 2. Force in member *DF* is given by vector *fd* and it is tensile. Force in member *EF* is given by vector *ef*. Since vector *ef* is of zero length, therefore force in member *EF* is zero. It is marked on the member itself.

Joint 5. Force in member *BG* is given by vector *bg*. This member is in tension. Force in member *GF* is given by vector *gf*. Member *GF* is in compression.

Joint 3. Force in member *GH* is given by vector *gh*. Thus, member *GH* is in tension. Force in member *HD* is given by vector *hd*. Thus, member *HD* is in tension.

Joint 4. Force in member *HC* is given by vector *hc*. Thus, member *HC* is in compression.

Results. Magnitude and nature of forces in various members of the truss are tabulated below :

Table of Results

Member	*Force*	*Nature*
12 or *DE*	22 kN	Tension
23 or *DF*	22 kN	Tension
34 or *DH*	7.4 kN	Tension
15 or *AE*	13.6 kN	Compression
56 or *BG*	3 kN	Tension
64 or *CH*	15 kN	Compression
25 or *EF*	0	—
35 or *FG*	17.4 kN	Compression
36 or *GH*	7 kN	Tension

Reactions : $R_1 = 12.5$ kN, $\theta_x = 36.9°$ as shown

$V_4 = 10.5$ kN vertical.

Problem 2.24. *Graphically determine forces in all members of the truss shown in Fig. P.2.24(a).*

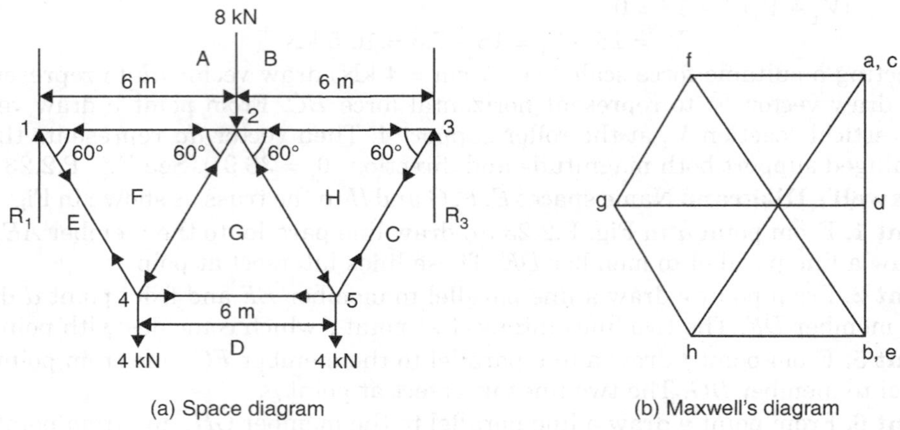

(a) Space diagram (b) Maxwell's diagram

Fig. P.2.24

Sol. Draw space diagram of the truss by using suitable linear scale (for example 1 cm = 2 m) as shown in Fig. P.2.24 (*a*).

Let us find reactions R_1 and R_3 by the analytical method, thus by using the condition $\Sigma M_3 = 0$, we get

$$R_1 \times 12 - 8 \times 6 - 4 \times 9 - 4 \times 3 = 0$$

$$\therefore \qquad R_1 = \frac{1}{12}(48 + 36 + 12) = 8 \text{ kN} \uparrow$$

By $\Sigma F_v = 0$, We get

$$\uparrow (R_1 + R_3) - (8 + 4 + 4) = 0$$

$$\therefore \qquad R_3 = 16 - R_1 = 16 - 8 = 8 \text{ kN} \uparrow.$$

Name various spaces like *F*, *H* and *G* inside the truss.

Force Polygon. Using a suitable force-scale draw vector *ab* to represent force *AB* ; draw vector *bc* to represent reaction R_3 ; draw vector *cd* to represent force *CD* ; draw vector *de* to represent force *DE* and lastly vector *ea* represents reaction R_1, as shown in Fig. P.2.24 (*b*).

Maxwell's Diagram. From point *a* draw a line parallel to the member *AF* and from point *e* draw a line *ef* parallel to the member *EF*. The two lines intersect at point *f*.

From point *f* draw a line parallel to member *FG* and from point *d* draw a line parallel to the member *DG*. The two lines intersect at point *g*.

From point *b* draw a line parallel to the member *BH* and from point *g* draw a line parallel to the member *GH*. The two lines intersect at point *h*.

Draw vector *ch*. This completes the construction of Maxwell's diagram.

Magnitude and nature of forces in members of the truss :

Joint 1. Force in member *AF* is compressive and is given by vector *af*. Force in member *FE* is tensile and is given by vector *fe*.

Joint 4. Force in member *FG* is compressive and is given vector *fg*. Force in member *GD* is tensile and is given by vector *gd*.

Joint 5. Force in member *GH* is compressive is given by vector *gh*. Force in member *HC* is tensile and is given by vector *hc*.

Joint 3. Force in member *HB* is compressive and is given by vector *hb*.

Results are tabulated below :

Table of Results

Member	*Force*	*Nature*
AF, BF	5.2 kN	Compressive
EF, CH	9.6 kN	Tensile
FG, GH	5.6. kN	Compressive
GD	7.5 kN	Tensile

Problem 2.25. *Find force in each member of truss shown in Fig. P.2.25 (a).*

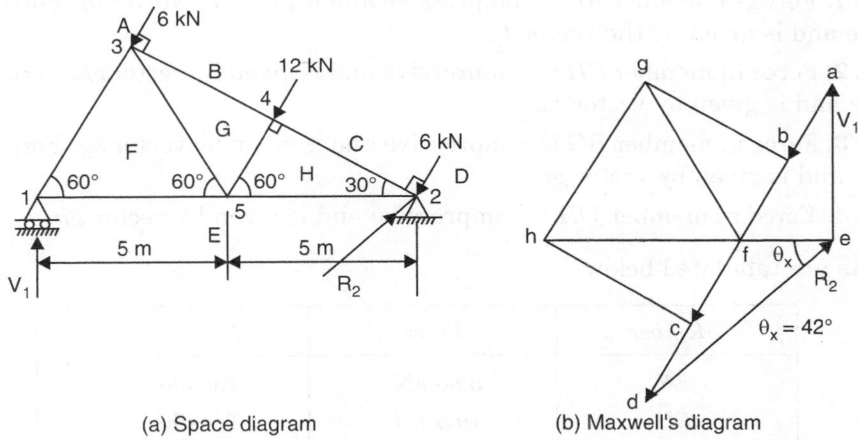

(a) Space diagram (b) Maxwell's diagram

Fig. P.2.25

Sol. Draw space diagram with the help of some suitable linear scale, say 1 cm = 1 m, or 4 cm = 5 m, etc.

Let us find reactions analytically.

Since, Δ123 is a right triangle, and side 12 = 10 m.

$$\therefore \quad \text{distance } (3-2) = 10 \cos 30° = \frac{10\sqrt{3}}{2} = 5\sqrt{3} \text{ m}$$

Also, $\quad \text{distance } (3-4) = \text{distance } (4-2) = \frac{1}{2} \times 5\sqrt{3} = \frac{5\sqrt{3}}{2}$ m.

By using $\Sigma M_2 = 0$, we get $\quad V_1 \times 10 - 6 \times 5\sqrt{3} - 12 \times \frac{5}{2}\sqrt{3} = 0$

$$\therefore \quad V_1 = \frac{1}{10} \times (12 \times 5\sqrt{3}) = 6\sqrt{3} \text{ kN, vertical} = \textbf{10.39 kN. Ans.}$$

Force Polygon. Choose some suitable scale for representing forces. Vector *ab* represents force *AB* ; vector *bc* represents force *BC* and *cd* represents force *CD*. From point *a* draw a line *ae* parallel to the reaction V_1 at roller 1, measuring 10.39 kN. Join points *d* and *e*. Then vector *de* represents reaction R_2 of support 2, as shown in Fig. P.2.25 (*b*).

Maxwell's Diagram. Name spaces *F*, *G* and *H* within the truss as shown in Fig. P.2.25 (*a*).

Joint 2. From point *c* draw a line parallel to the member *CH* and from point *e* draw a line parallel to the member *EH*. The two lines intersect at point *h*.

Joint 1. From point *a* draw a line parallel to member *AF* and from point *e* draw a line parallel to member *EF*. The two lines intersect at point *f*.

Joint 3. From point *b* draw a line parallel to the member *BG* and from point *f* draw a line parallel to member *FG*. The two lines intersect at point *g*.

This completes the construction of Maxwell's diagram.

Magnitude and nature of forces in members of the truss :

Joint 1. Force in member *AF* is compressive and is given by vector *af*. Force in member *EF* is tensile and is given by the vector *fe*.

Joint 2. Force in member *CH* is compressive and is given by vector *ch*. Force in member *HE* is tensile and is given by vector *he*.

Joint 3. Force in member *BG* is compressive and is given by vector *bg*. Force in member *GF* is tensile and is given by vector *gf*.

Joint 4. Force in member *GH* is compressive and is given by vector *gh*.

Results are tabulated below

Member	*Force*	*Nature*
EF	5.85 kN	Tensile
EH	19.5 kN	Tensile
AF	12 kN	Compressive
BG	8.85 kN	Compressive
CH	8.85 kN	Compressive
GF	12 kN	Tensile
GH	12 kN	Compressive

$$V_1 = 10.39 \text{ kN, vertical}$$

R_2 is given by vector *de*.

$$R_2 = 15.9 \text{ kN}$$

$$\theta_x = 42°. \quad \textbf{Ans.}$$

Problem 2.26. *Determine graphically forces in all members of the centilever truss shown in Fig. P.2.26 (a).*

Sol. In case of a cantilever truss, as discussed earlier, it is not necessary to find support reactions for finding forces in its members.

Draw space diagram as shown in Fig. P.2.26 (*a*). Using Bow's notations, name all spaces, as shown.

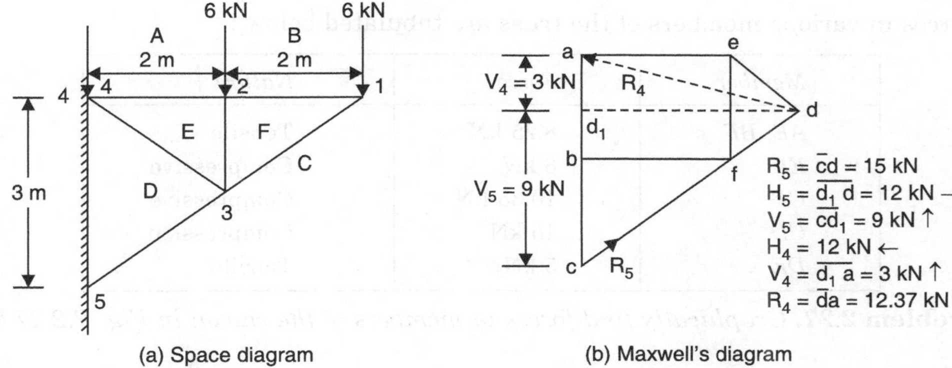

Fig. P. 2.26

Using some suitable force scale draw vector *ab* to represent force *AB* and vector *bc* to represent force *BC*.

Maxwell's Diagram :

Joint 1. From point *b* draw a line parallel to member *BF* and from point *c* draw a line parallel to member *cf*. The two lines intersect at point *f*.

Joint 2. From point *a* draw a line parallel to member *AE* and from point *f* draw a line parallel to the member *FE*. The two lines intersect at point *e*.

Joint 3. From point *e* draw a line parallel to member *ED* and from point *c* draw a line parallel to the member *CD*. The two lines intersect of point *d*. This completes construction of Maxwell's diagram.

Magnitude and nature of forces in members of the truss :

Joint 1. Force in member *CF* is compressive and is given by vector *cf*. Force in member *FB* is tensile and is given by vector *fb*.

Joint 2. Force in member *AE* is tensile and is given by vector *ae*. Force in member *FE* is compressive and is given by vector *fe*.

Joint 3. Force in member *CD* is compressive and is given by vector *cd*. Force in member *DE* is tensile and given by vector *de*.

Reactions :

Reaction at support 5 = R_5 = force in member *CD* and acts in direction of the vector *cd*.

$$\therefore \quad R_5 = 15 \text{ kN} \nearrow$$

Horizontal component of reaction R_5

$$= H_5 = 12 \text{ kN} \rightarrow.$$

H_5 is given by vector $\overline{d_1 d}$

Vertical component of R_5

$$= V_5 = cd_1 = 9 \text{ kN} \uparrow$$

Reaction at joint 4 = R_4, it is given by vector $\overline{da}$ = 12.37 kN.

$$H_4 = \text{Horizontal component of } R_4 = 12 \text{ kN} \leftarrow$$

Vertical component of R_4

$$= V_4 = \overline{d_1 a} = 3 \text{ kN} \uparrow$$

R_4, H_4 and R_5 are shown by dotted lines in Fig. P.2.26 (*b*).

Forces in various members of the truss are tabulated below :

Member	*Force*	*Nature*
AE, BF	8.25 kN	Tensile
EF	6 kN	Compressive
CF	10.35 kN	Compressive
CD	15 kN	Compression
DE	5 kN	Tensile

Problem 2.27. *Graphically find forces in members of the shown in Fig. P.2.27 (a).*

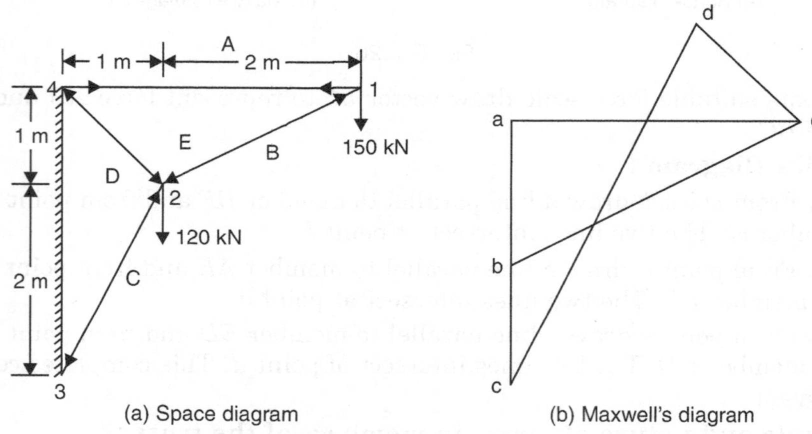

(a) Space diagram (b) Maxwell's diagram

Fig. P.2.27

Sol. Draw truss to some suitable linear scale and using Bow's notations name various spaces shown in Fig. P. 2.27(*a*). Taking a convenient force-scale draw vector *ab* to represent force *AB* and vector *bc* to represent for *BC*, as shown in Fig. P.2.27 (*b*).

Maxwell's Diagram :

Joint 1. From point *a* draw a line parallel to the member *AE* and from point *b* draw a line parallel to the member *BE*. These lines intersect at point *e*.

Joint 2. From point *e* draw a line parallel to the member *ED* and from point *c* draw a line parallel to the member *CD*. The two lines intersect at point *d*.

This completes the Maxwell's diagram, as shown in Fig. P.2.27 (*b*).

Magnitude and nature of forces in members the truss :

Joint 1. Force in member *AE* is tensile and is given by vector *ea*. Force in member *BE* is compressive and is given by vector *be*.

Joint 2. Force in member *CD* is compressive and is given by vector *cd*. Force in member *DE* is compressive and is given by vector *de*. Results are tabulated below :

Member	*Force*	*Nature*
AE	287.5 kN	Tensile
BE	325 kN	Compressive
CD	415 kN	Compressive
DE	150 kN	Compressive

Problem 2.28. *Graphically determine forces in all members of the truss shown in Fig. P.2.28.*

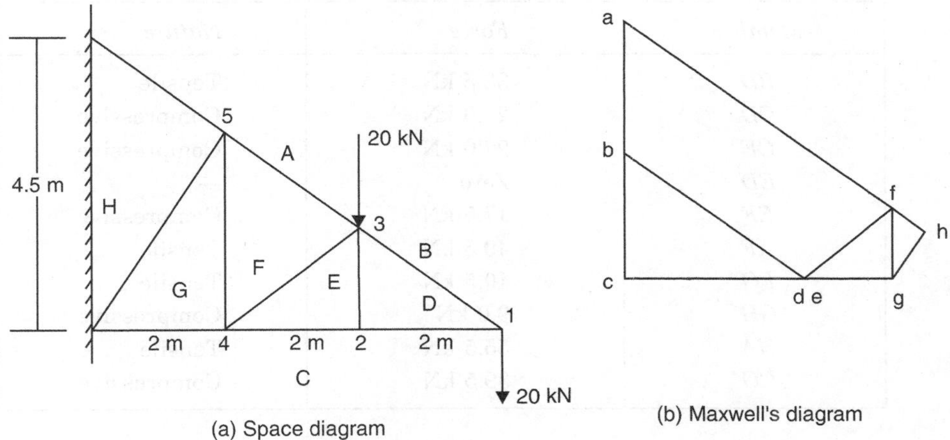

(a) Space diagram

(b) Maxwell's diagram

Fig. P.2.28

Sol. Using some convenient linear scale draw the given truss as shown in Fig. P.2.28 (*a*). Using Bow's notations name all spaces as shown.

Take some convenient force-scale to draw force diagram. Select a point *a* and draw a vector *ab* to represent force *AB*. From point *b* draw vector *bc* to represent force *BC*.

Maxwell's Diagram :

Joint 1. From point *b* draw a line parallel to the member *BD* and from point *c* draw a line parallel to the member *CD*. The two lines intersect at point *d*.

Joint 2. From point *d* draw a line parallel to member *DE* and from point *c* draw a line parallel to member *CE*.The two lines intersect at point *e*. Here point *e* coincides with point *d*.

Joint 3. From point *e* draw a line parallel to the member *EF* and from point *a* draw a line parallel to member *AF*. The two lines intersect at point *f*.

Joint 4. From point *f* draw a line parallel to the member *FG* and from point *c* draw a line parallel to the member *CG*. The two lines intersect at point *g*.

Joint 5. From point *g* draw a line parallel to the member *GH* and from point *a* draw a line parallel to the member *AH*. The two lines intersect at point *h*.

This completes the construction of the Maxwell's diagram, as shown in Fig. P.2.28 (*b*).

Magnitude and nature of forces in members of the truss :

Joint 1. Force in member *BD* is tensile and is given by vector *bd*. Force in member *CD* is compressive and is given by vector *cd*.

Joint 2. Force in member *ED* is zero.

Force in member *CE* is compressive and is given by vector *ce*.

Joint 3. Force in member *AF* is tensile and is given by vector *af*.

Force in member *EF* is compressive and is given by vector *ef*.

Joint 4. Force in member *CG* is compressive and is given by vector *cg*.

Force in member *GF* in tensile and is given by vector *gf*.

Joint 5. Force in member *AH* is tensile and is given by vector *ha*.

Force in member *GH* is compressive and is given by vector *gh*.

Results are tabulated below :

Table of Results

Member	*Force*	*Nature*
BD	33.5 kN	Tensile
CD	27.0 kN	Compressive
CE	27.0 kN	Compressive
ED	Zero	—
EF	17.5 kN	Compressive
AF	49.5 kN	Tensile
FG	10.5 kN	Tensile
GH	9.0 kN	Compressive
HA	55.5 kN	Tensile
CG	39.5 kN	Compressive

EXERCISE II

A. Theoretical Questions

1. Describe, what do you mean by a truss ?

2. What is the condition of a perfect truss ? Explain with suitable examples.

3. What is indeterminate truss ? Explain with examples.

4. What is the difference between a truss and a mechanism ?

5. Draw a simple truss and show the following in the diagram :
span, rise, lower chord member, upper chord members, end members, verticals and diagonals.

6. Explain the types of supports.

7. What methods are used for the analysis of trusses ? Give a list.

8. What equations are used in the method of joints ? Explain.

9. What is the concept of method of sections ?

10. How many equations are used in the method of sections.

11. What is space diagram as used in graphical method ? Explain with two examples.

12. What is the Bow's method ?

13. What is string diagram or funicular polygon.

14. What are the equations of equilibrium for coplanar concurrent force system ? What graphical condition is equivalent to the above equations ?

15. What are the equations of equilibrium for the coplanar non-concurrent force system ? What are the corresponding graphical conditions ?

16. Explain with suitable example the method of finding resultant of coplanar non-concurrent forces.

17. What do you understand by Maxwell's diagram ?

18. How would you find nature of the forces in members of a truss ? Explain with a simple example.

19. What do you mean by a null member ? Explain with three examples.

B. Numerical Problems

2.1. Find forces in all members of the truss shown in Fig. E.2.1. Use method of joints.

[**Ans.** $F_{AD} = F_{BD}$ = 20.5 kN (T) ; $F_{AC} = F_{BC}$ = 28.0 kN (C) ; F_{CD} = 10 kN (T)]

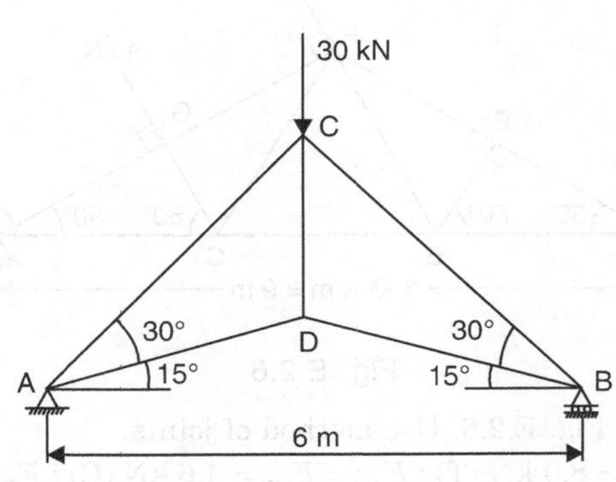

Fig. E.2.1

Fig. E.2.2

2.2. Find force in each member of the truss shown in Fig. E.2.2. Use method of joints.

[**Ans.** $F_{AB} = F_{CD}$ = 23 kN (T) ; $F_{AE} = F_{ED}$ = 10 kN (C) ; $F_{BE} = F_{CE}$ = 11.2 kN (C), F_{BC} = 15.2 kN (T)]

2.3. Find force in each member of the truss shown in Fig. E.2.3. Use method of joints.

[**Ans.** F_{AB} = 5.55 kN (C) ; F_{BC} = 9 kN (C) ; F_{CD} = 7.125 kN (T) ; F_{ED} = 7.125 kN (T) ; F_{BD} = 0 ; F_{BE} = 6.9 kN (T)]

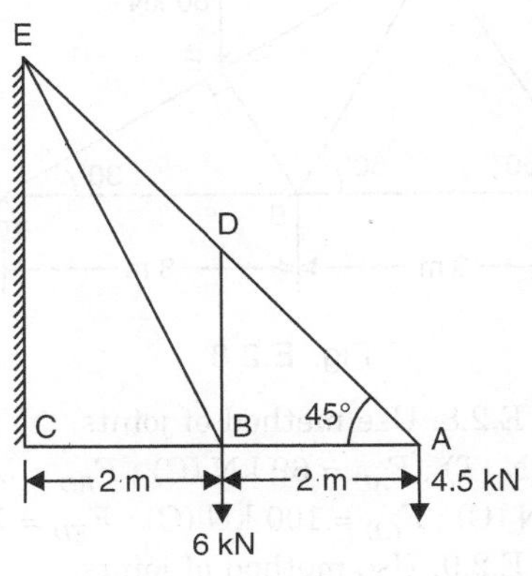

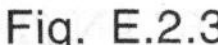

Fig. E.2.3

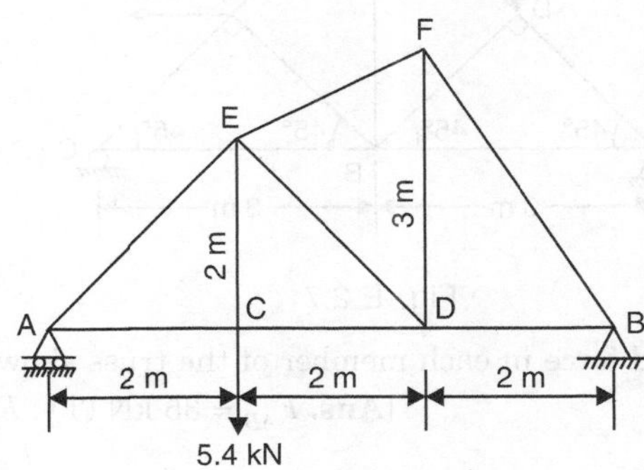

Fig. E.2.4

2.4. Find force in the members of the truss shown in Fig. E.2.4. Use method of joints.

[**Ans.** F_{AB} = 5.1 kN (C) ; F_{AC} = 3.6 kN (T) = F_{AC} ; F_{EF} = 13.5 kN (C), F_{BF} = 2.16 kN (C) ; F_{BD} = 1.2 kN (T)]

2.5. Find force in each member of the truss shown in Fig. E.2.5. Use method of joints.

[**Ans.** F_{AG} = 9.67 kN (T) ; $F_{GF} = F_{FE}$ = 4.186 kN (T) ; F_{AB} = 11.186 kN (C) ; F_{BC} = 9.67 kN (C) ; $F_{CD} = F_{DE}$ = 4.834 kN (C) ; F_{BG} = 2.67 kN (C) ; F_{GC} = 8.37 kN (T) ; $F_{FD} = F_{FC}$ = 0]

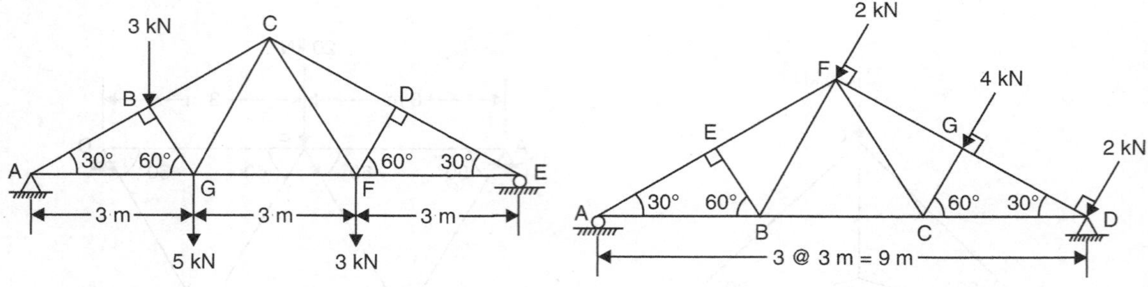

Fig. E.2.5

Fig. E.2.6

2.6. Find force in each member of the truss shown in Fig. E.2.6. Use method of joints.

[**Ans.** F_{AB} = 4.0 kN (T) = F_{BC} ; F_{CD} = 8.0 kN (T) ; $F_{AE} = F_{EF}$ = 4.6 kN (C) ; F_{EB} = 0 ; F_{BF} = 0 ; F_{CG} = 4 kN(C) ; $F_{DG} = F_{FG}$ = 5.78 kN (C), F_{CF} = 4 kN (T)]

2.7. Find force in each member of the truss shown in Fig. E.2.7. Use method of joints.

[**Ans.** F_{AB} = 4.5 kN (T) ; F_{BC} = 1.5 kN (T) ; F_{AD} = 3.54 kN (C) ; F_{BD} = 2.54 kN (C) ; F_{BF} = 1.42 kN (T) ; F_{BE} = 1.0 kN (T) ; F_{DE} = 0.72 kN (C) ; F_{EF} = 0.70 kN (C) ; F_{CF} = 1.12 kN (C)]

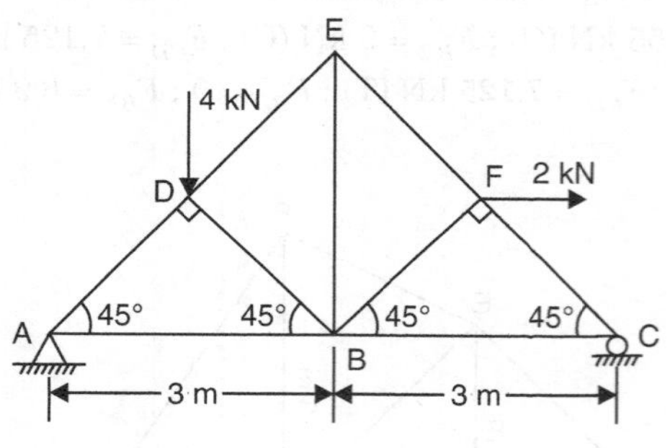

Fig. E.2.7

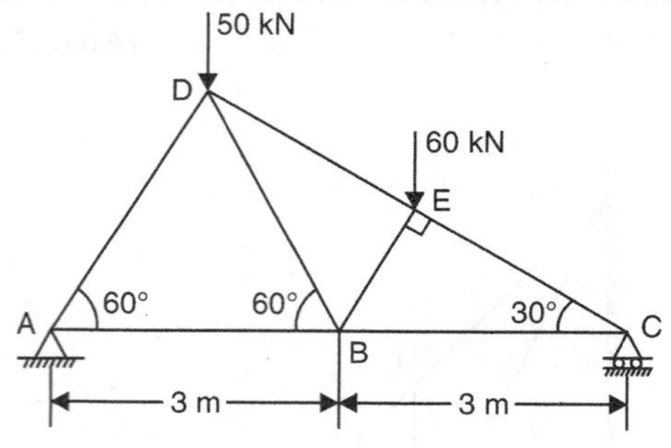

Fig. E.2.8

2.8. Find force in each member of the truss shown in Fig. E.2.8. Use method of joints.

[**Ans.** F_{AB} = 35 kN (T) ; F_{BC} = 87 kN (T) ; F_{AD} = 69 kN (C) ; F_{BD} = 52 kN (T) ; F_{BE} = 52 kN (C) ; F_{CE} = 100 kN (C) ; F_{ED} = 70 kN (C)]

2.9. Find force in each member of the truss shown in Fig. E.2.9. Use method of joints.

[**Ans.** $F_{AB} = F_{BC}$ = 60 kN (T) ; F_{CD} = 26.67 kN (T) ; F_{AE} = 28.28 kN (C) ; F_{EF} = 14.9 kN (T) ; F_{DF} = 48.07 kN (C) ; F_{BE} = 0 ; F_{CE} = 47.14 kN (C) ; F_{CF} = 33.33 kN (T)]

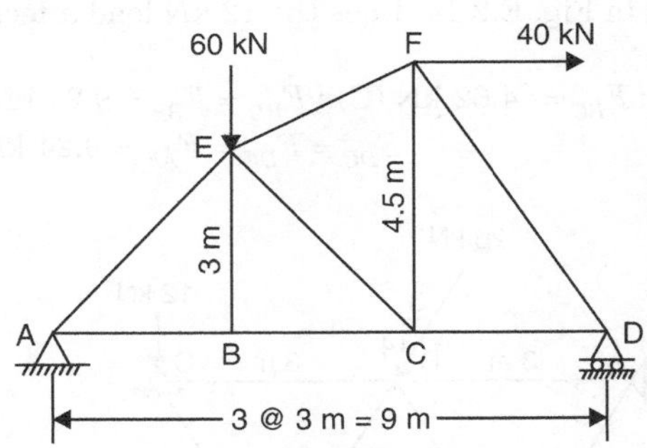

Fig. E.2.9

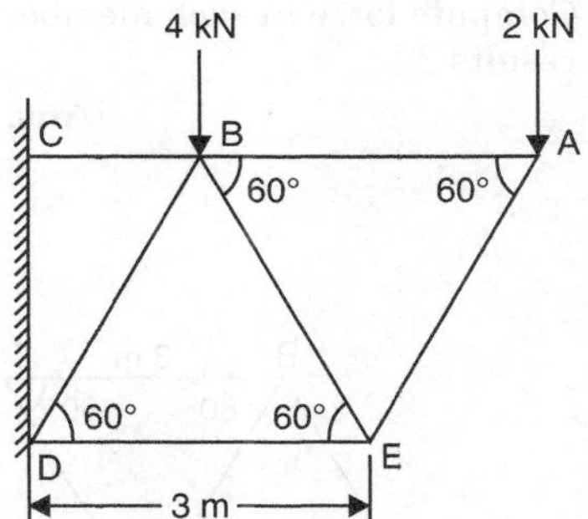

Fig. E.2.10

2.10. Find force in each member of the truss shown in Fig. E.2.10. Use method of joints.

[**Ans.** F_{AB} = 1.15 kN (T) ; F_{BC} = 5.7 kN (T) ; F_{AE} = 2 kN (C), F_{EB} = 2.3 kN (T) ; F_{ED} = 1.30 kN (C) ; F_{BD} = 7.0 kN (C)]

2.11. Find force in each member of the truss shown in Fig. E.2.11. Use method of joints.

[**Ans.** $F_{AB} = F_{BC}$ = 24.04 kN (T) ; F_{AD} = 20 kN (C) ; F_{CD} = 33.32 kN (C), F_{BD} = 13.32 kN (T)]

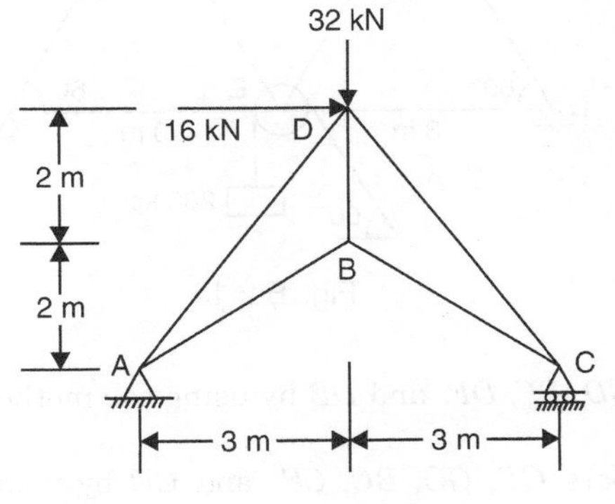

Fig. E.2.11

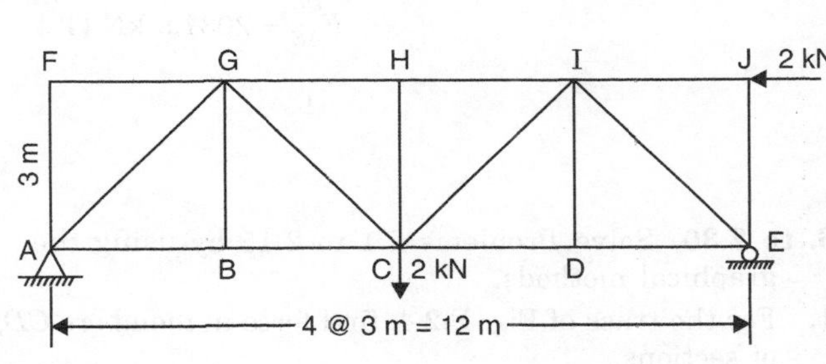

Fig. E.2.12

2.12. Find the force in each member of the truss shown in Fig. E.2.12. Use method of joints.

[**Ans.** $F_{AB} = F_{BC} = F_{CD} = F_{DE}$ = 0.5 kN (T) ; $F_{AF} = F_{EJ}$ = 0 ; $F_{BG} = F_{DI} = F_{DH} = F_{FG}$ = 0 ; $F_{GH} = F_{HI}$ = 3.0 kN (C) ; F_{CG} = 2 kN (T), F_{CI} = 0.7 kN (T) ; F_{AG} = 2 kN (C) ; F_{EI} = 0.70 kN (C) ; F_{IJ} = 2 kN (C)]

2.13. Determine force in each member of the truss shown in Fig. E.2.13. Use method of joints. Also find tension in the tie *DG*.

[**Ans.** F_{AB} = 69.28 kN (T), F_{AC} = 34.64 kN (C) ; F_{BC} = 69.28 kN (C) ; F_{BD} = 69.28 kN (T), F_{CD} = 115.48 kN (T). F_{CE} = 127.02 kN (C) ; F_{DE} = 23.10 kN (C) ; Tension in the tie *DG* = 160 kN]

2.14. Compute force in each member of the truss shown in Fig. E.2.14. Does the 12 kN load affect your results ?

[**Ans.** F_{AB} = 7.38 kN (T) ; F_{BC} = 4.62 kN (C) ; $F_{BD} = F_{BE}$ = 9.24 kN (C) ; $F_{DC} = F_{DE} = F_{AE}$ = 9.24 kN (T)]

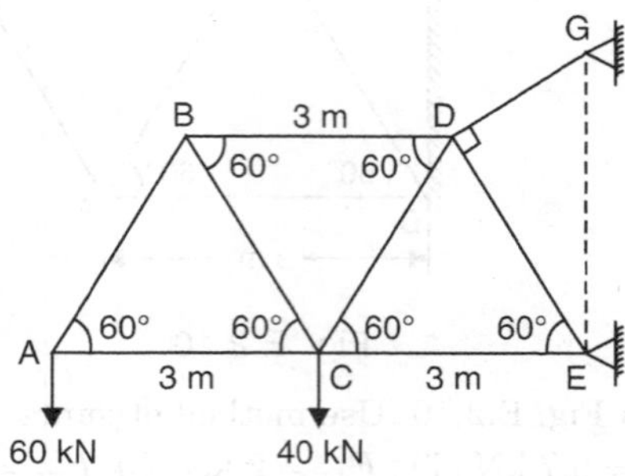

Fig. E.2.13

Fig. E.2.14

2.15. Find force in each member of the truss shown in Fig. E.2.15. The pulley mounted at joint E is smooth.

[**Ans.** $F_{AB} = F_{DC} = F_{BC}$ = 2113.8 kN (C) ;
$F_{CE} = F_{BE}$ = 2113.8 kN (T) ;
F_{DE} = 1056.9 kN (T) ;
F_{AE} = 2037.9 kN (T)]

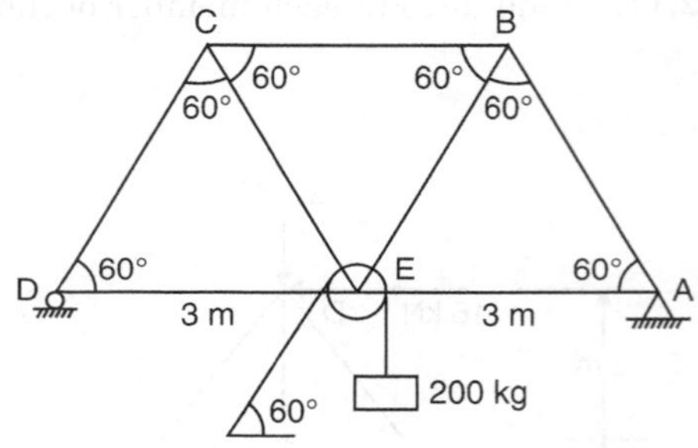

Fig. E.2.15

2.16. to 2.30. Solve Problems 2.1 to 2.15 by using the graphical methods.

2.31. For the truss of Fig. E.2.4, find force in members *CD, ED, EF, DF,* and *DB* by using the method of sections.

2.32. For the truss shown in Fig. E.2.5, find force in members *GF, GC, BC, CF* and *CD* by using method of sections.

2.33. For the truss shown in Fig. E.2.7 find by using the method of section the force in members *BD, BE, DE, BF* and *EF*.

2.34. For the truss shown in Fig. E.2.13 find force in members *AC*, *BC*, *BD* and *DF* by using method of sections.

2.35. By using method of sections find force in members *AE, BE* and *BC* for the truss shown in Fig. E.2.15.

2.36. By using method of sections find force in members *BC, BD, BF, DF* and *AF* of truss of solved problem 2.4 (Fig. P.2.4).

2.37. For the truss shown in Fig. P.2.8 (Solved Problem 2.8). Find in members *BD, BC, CD* and *DE* by using method of sections.

2.38. For the truss of solved problem 2.11 (Fig. P.2.11) find force in members *BD, BC, CD* and *CE* by using method of sections.

2.39. Find the force in members *AC*, *AE*, *AB*, *CE*, *DE* and *BE* of the truss shown in Fig. P.2.12. Use method of sections.

2.40. Find force in members *AD*, *CE, DE, DH* and *BE* of the truss shown in Fig. P.2.13. Use method of sections.

3

Centroids and Moments of Inertia

PART A—CENTROIDS AND CENTRE OF GRAVITY

3.1. INTRODUCTION

We already know that attraction of earth could be represented by a single force W, directed towards the centre of the earth. This force W, called the weight of the body, is to be applied at the **centre of gravity** of the body. Since the body of weight W consists of several particles and the gravity of earth exerts a force on each of these particles forming the body. Thus, the action of earth on a rigid body can be represented by a large number of small and parallel forces distributed over the entire body. And, all these forces can be replaced by a single force W, passing through a point called the centre of gravity.

3.2. CENTRE OF GRAVITY OF A FLAT PLATE

The analytical method of locating the centre of gravity is simply an application of the principle of moments, or the Varignon's theorem, which states that the moment of the resultant force is equal to the sum of moments of its parts. For example, consider a thin uniform plate of irregular shape as shown in Fig. 3.2. The plate is divided into small elements of weights $w_1, w_2, w_3, \ldots w_n$ acting through their own centres of gravity. The total weight is W, acting through the centre of gravity of the entire plate. These gravity forces represent a parallel force system. Let the co-ordinates of the particles be (x_1, y_1), (x_2, y_2), (x_i, y_i), ... etc. of weights W_1, W_2, W_i, ... , respectively. And, the co-ordinates of centre of gravity of the entire plate be $(\bar{x}, \bar{y})$ for the weight W.

The magnitude of W is equal to the sum of the weights of various particles. Therefore, by $\Sigma F_z = 0$, we get

$$W = W_1 + W_2 + \ldots\ldots + W_n$$

or $$W = \Sigma W_i \qquad \ldots(3.1)$$

By taking moments of the weights about y-axis, we get

i.e., By $\Sigma M_y = 0$, we get

$$W.\bar{x} = W_1.x_1 + W_2.x_2 + \ldots\ldots + W_n \cdot x_n$$

or $$W\,\bar{x} = \Sigma W_i.x_i \qquad \ldots(3.2)$$

While by taking moments about x-axis.

i.e., by $\Sigma M_x = 0$, we get

$$W\,\bar{y} = W_1 \cdot y_1 + W_2 \cdot y_2 + \ldots\ldots + W_n \cdot y_n$$

or $$W\;\bar{y} = \Sigma W_i.y_i \qquad \ldots(3.3)$$

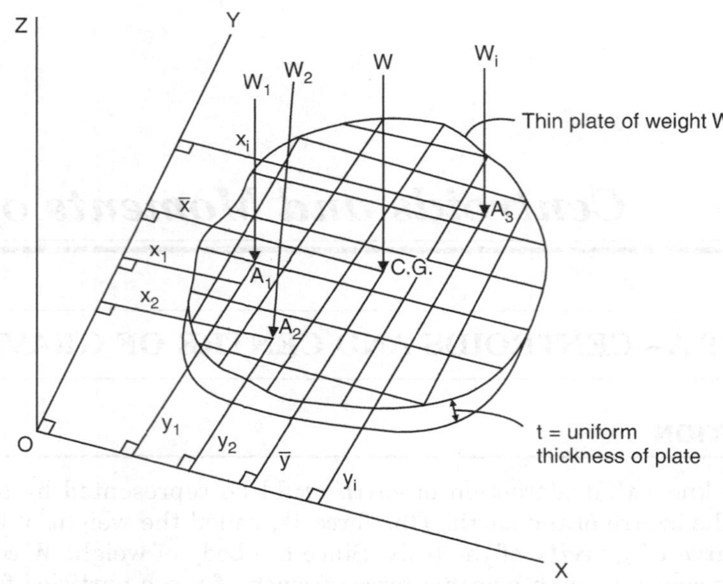

Fig. 3.2. Co-ordinates of the centre of gravity $(\bar{x}, \bar{y})$ and of other particles (x_1, y_1), etc.

If we increase the number of elements into which the plate is divided and simultaneously decrease each element to size tending to zero, the above expressions become

$$\begin{aligned} W &= \int dW \\ \bar{x}\,W &= \int x dW \\ \bar{y}\,W &= \int y dW \end{aligned} \qquad \text{...(3.4)}$$

In eqn. (3.4) it is assumed that the boundary of plate is some mathematical function.

3.3. CENTROIDS OF AREAS AND LINES

In the case of a homogeneous plate of uniform thickness t, the weight ΔW of an element may be expressed as

$$\Delta W = \rho g t . \Delta A \qquad \text{...(3.5)}$$

where ρ = gravity (*i.e.*, mass per unit volume) of plate material (kg/m^3)

$g = 9.81$ m/s^2

t = thickness of plate, m.

A_i = area of the ith element, m^2.

Thus, equation (3.1) yields

$$\rho g t A = \rho g t\ (A_1 + A_2 + ... + A_n)$$

or
$$A = \Sigma A_i \qquad \text{...(3.6)}$$

And equations (3.2) and (3.3) yield,

By $\Sigma M_y = 0$, $\quad \bar{x}A = A_1.x_1 + A_2.x_2 + ... + A_n \,.\, x_n$

or
$$\bar{x}A = \Sigma A_i.x_i \qquad \text{...(3.7)}$$

or By $\Sigma M_x = 0$, $\quad \bar{y}A = A_1.y_1 + A_2.y_2 + ... + A_n.y_n$

or
$$\bar{y}A = \Sigma A_i.y_i \qquad \text{...(3.8)}$$

When the boundary of the plate is some mathematical function equations (3.6), (3.7)

(3.8) become :

$$A = \int dA \qquad \bar{x}A = \int x dA \qquad \bar{y}A = \int y dA \qquad \text{...(3.9)}$$

From equations (3.7) and (3.8), we get

$$\bar{x} = \frac{\Sigma A_i \,.\, x_i}{A} \qquad \bar{y} = \frac{\Sigma A_i \cdot y_i}{A} \qquad \text{...(3.10)}$$

Equations (3.7) and (3.8) define the co-ordinates $\bar{x}$ and $\bar{y}$ of the centre of gravity of a homogeneous plate of uniform thickness. The point of co-ordinates $\bar{x}$ and $\bar{y}$ is also known as the **centroid of** the area A of the plate as shown in Fig. 3.3.1

$A = \Sigma A_i$

By ΣM_y : $\bar{x}A = \Sigma x_i A_i$

By ΣM_x : $\bar{y}A = \Sigma y_i A_i$

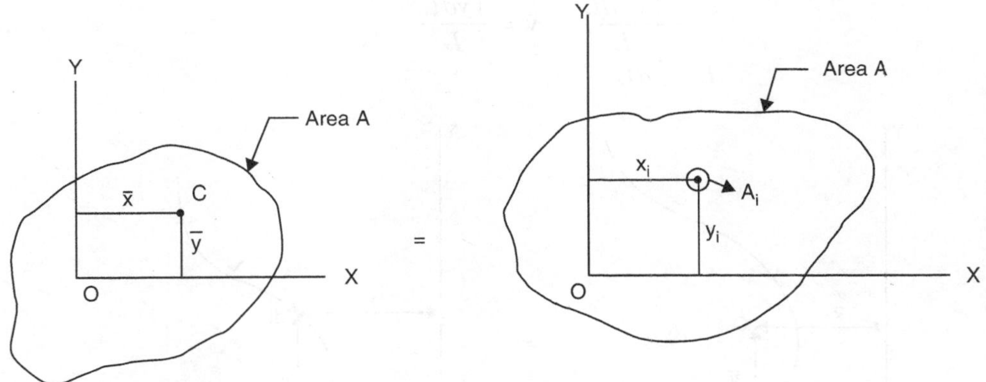

Fig. 3.3.1 Centroid of an area.

The integral $\int$ **x dA is known as the first moment of the area A with respect to y-axis,** and sometimes it is represented by Q_y. Similarily, $\int$ **ydA defines the first moment of the area A with respect to x-axis,** and is sometimes denoted by Q_x. Thus, we get

and

$$Q_y = \bar{x}A = \int x dA \qquad Q_x = \bar{y}A = \int y dA \qquad \text{...(3.11)}$$

Hence, we get

$$\bar{x} = \frac{\int x\, dA}{A} \qquad \bar{y} = \frac{\int y\, dA}{A} \qquad \text{...(3.12)}$$

It is important to note, that if the centroid of an area is located on a co-ordinate axis, the first moment of the area with respect to that axis is zero. Thus,

if $\bar{x} = 0$, then $\int x dA = Q_y = 0$ and also, if

$\bar{y} = 0$, then $\int y dA = Q_x = 0$

Eqns. (3.10) are used when the given area A can be divided into simple areas of known properties. And, equations (3.12) are used when the boundary of the given area is given by some mathematical functions of x and y, such as $y = kx^2$ for parabola ; $\frac{x^2}{a^2} + \frac{y^2}{b^2} = 1$ for ellipse, or $y = kx^n$ for general spandrel, etc.

In case of a homogeneous wire of uniform cross-section, weight ΔW of element may be expressed as

$$\Delta W = \rho g a \Delta L \quad \text{...(3.13)}$$

where a = cross-sectional area, m^2

ΔL = length of the element, m.

The centre of gravity of the wire then coincides with the centroid C of the line L describing the shape of the wire (Fig. 3.3.2). The co-ordinates $\bar{x}$ and $\bar{y}$ of the centroid C of the line L can be obtained from equations :

$$\bar{x} L = \int x.dL, \quad \bar{y} L = \int y dL \quad \text{...(3.14)}$$

or

$$\bar{x} = \frac{\int x \, . \, dL}{L}, \quad \bar{y} = \frac{\int y dL}{L} \quad \text{...(3.15)}$$

where $L = \int dL$.

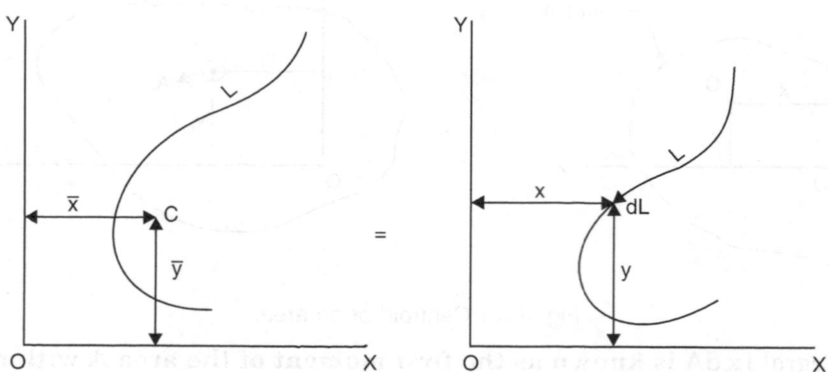

Fig. 3.3.2 Centroid of the uniform wire or line.

Let the centre of ith area A_i have co-ordinates (x_i, y_i)

Then by ΣM_y : $\bar{x} L = \Sigma x_i L_i$; or $\bar{x} L = \int x dL$

and by ΣM_x : $\bar{y} L = \Sigma y_i L_i$; or $\bar{y} L = \int y dL$

3.4. CENTROIDS OF SYMMETRICAL FIGURES

A line (L) or an area (A) may be symmetrical about one axis or two axes. The distance of the centroid from the axis of symmetry (AS) is zero.

A circle, a rectangle, a square, an ellipse (Fig. 3.4.1) have two **axis of symmetry.** The centroids for these areas lie at the intersection of the axes of symmetry. The co-ordinate axes

X_0, and Y_0 passing through the centroid are called **centroidal axes**. Intersection of the two orthogonal axes of symmetry (C-X_0 and C-Y_0) gives the point of symmetry C of the area or a line.

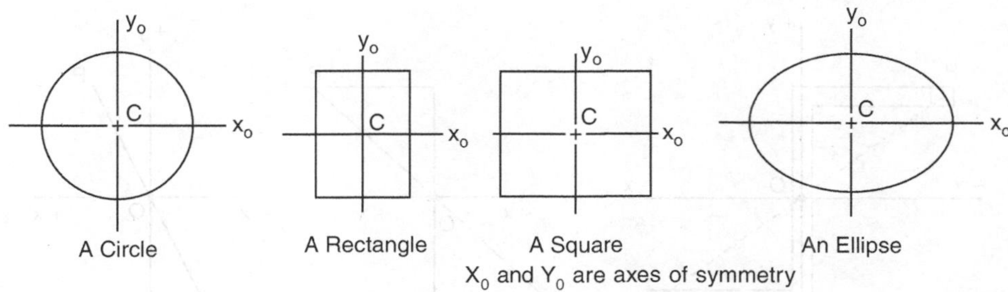

Fig. 3.4.1 Areas having two axes of symmetry. Here $\bar{x} = 0$; $\bar{y} = 0$ w.r.t. X_0 and Y_0 axes.

Areas and lines having X_0-axis of symmetry are shown in Fig. 3.4.2.

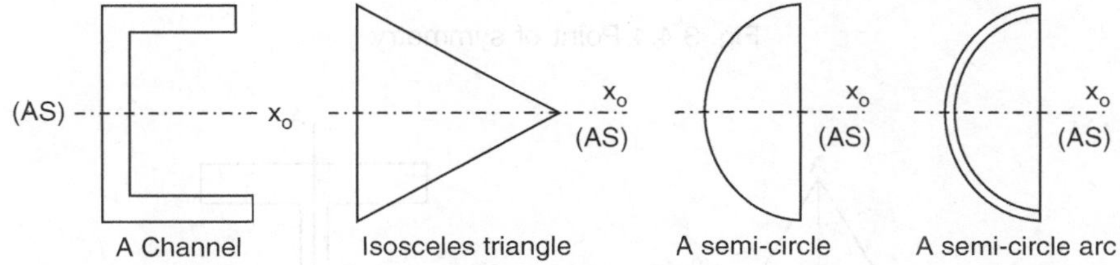

Fig. 3.4.2 Areas having X_0 as the axis of symmetry ($\bar{y}$ – 0) w.r.t. X_0-axis.

Other axes of symmetry of areas are shown in Fig. 3.4.3, Fig. 3.4.4 and Fig. 3.4.5.

An area A is said to be symmetrical about an axis AS if to every point P of the area corresponds a point P' of the same area such that the line PP' is perpendicular to A-S and is divided into two equal parts by A-S (Fig. 3.4.3). A line is said to be symmetrical about A-S if it satisfies similar conditions as stated above.

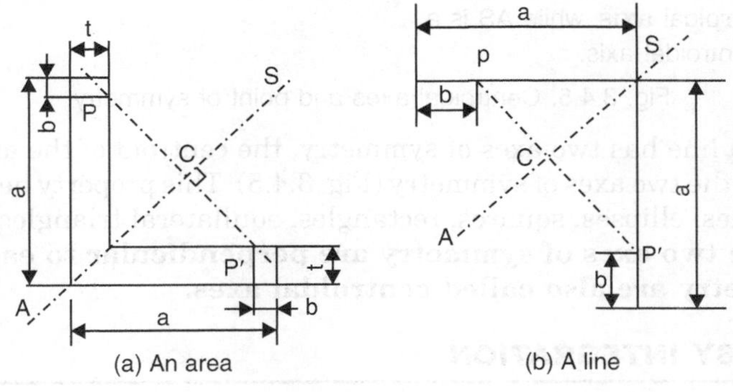

PP′ ⊥ar AS

AS = Axis of symmetry

C = Centroid

Fig. 3.4.3 Axis of symmetry (AS) perpendicular to PP′.

The symmetry of the figure may be about a point. An area (or a line) is said to be symmetrical about a point O if every area dA (or length dL on a line) of co-ordinates x- and y

corresponds to an area dA (or length dL) with co-ordinates $-x$ and $-y$. Fig. 3.4.4 shows cases of symmetrical about a point O.

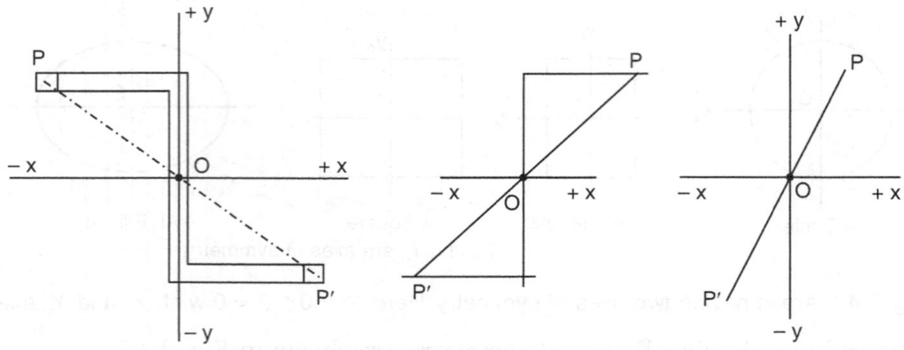

0 is the point of symmetry of the figures

Fig. 3.4.4 Point of symmetry.

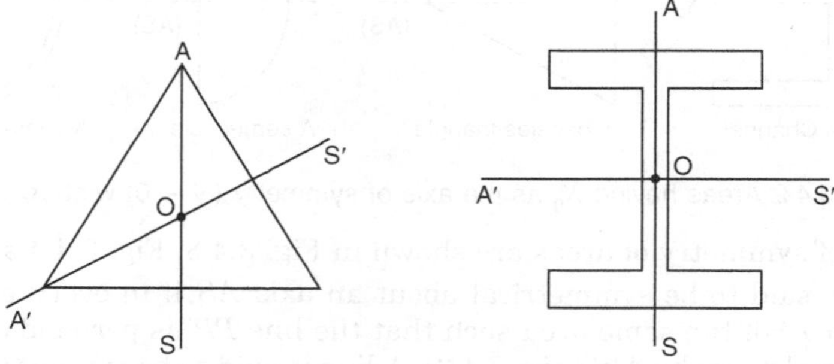

(a) AS not perpendicular to A′S′. O is not point of symmetry but A′S′ is not centroidal axis while AS is a centroidal axis.

(b) AS perpendicular A′S′. O is point of symmetry and A′S′ and AS are centroidal axes.

Fig. 3.4.5. Centroidal axes and point of symmetry.

If an area or a line has two axes of symmetry, the centroid of the area or line is located at the intersection of the two axes of symmetry (Fig. 3.4.5). This property helps to locate centroid of areas such as circles, ellipses, squares, rectangles, equilateral triangles, or other symmetrical areas. **When the two axes of symmetry are perpendicular to each other, then the two axes of symmetry are also called centroidal axes.**

3.5. CENTROIDS BY INTEGRATION

Here we shall locate centroids of some standard areas and line segments shown in Table 3.1 by the method of integration.

3.5.1. Centroid of a circular arc

The circular arc of Fig. 3.5.1 is symmetrical about x-axis.

$\therefore\ \bar{y} = 0$. The arc subtends an angle of 2α at the centre O.

The length of the differential element

$$= dL = r d\theta.$$

Total length, $L = \int dl = \int_{-\alpha}^{+\alpha} r d\theta = 2\int_{0}^{\alpha} r d\theta$

$$L = 2r\alpha \quad ...(i)$$

The distance x of the centre of the element from y-axis $= x = r \cos \theta$.

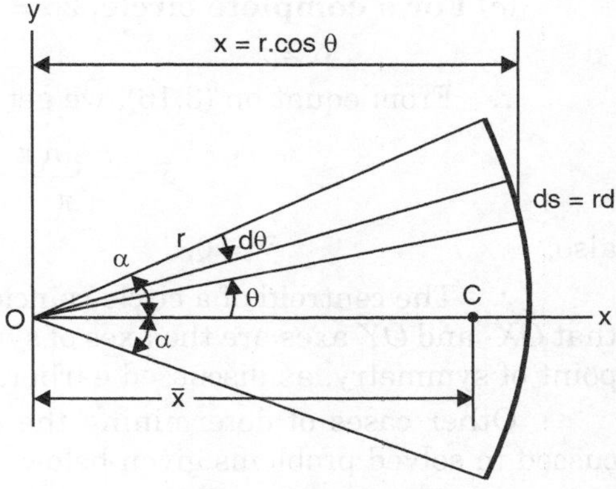

Fig. 3.5.1 Arc of circle.

$$\therefore \quad \int x dL = \int_{-\alpha}^{\alpha} (r \cos \theta) \times r d\theta = \bar{x} L$$

$$= 2\int_{0}^{\alpha} r^2 \cos \theta . d\theta = 2r^2 \sin \theta \Big|_{0}^{\infty}$$

or $\int x dL = 2r^2 \sin \alpha = \bar{x} L$

$$\therefore \quad \bar{x} = \frac{\int x dL}{L} = \frac{2r^2 \sin \alpha}{2r\alpha}$$

$$\therefore \quad \boxed{\bar{x} = \frac{r \sin \alpha}{\alpha}} \quad ...(3.16)$$

(*a*) **For a semi-circle,** $2\alpha = 180°$, $\therefore$ $\alpha = 90° = \pi/2$ radians

$\therefore$ From equation 3.16, we get

$$\bar{x} = \frac{r \sin \pi/2}{\pi/2} = \frac{2r}{\pi}$$

$$\therefore \quad \boxed{\bar{x} = \frac{2r}{\pi}} \quad ...(3.17)$$

(*b*) **For a quarter-circle,** $2\alpha = 90°$ or $\alpha = 45° = \pi/4$,

$$\therefore \quad \bar{x} = \frac{r \sin \pi/4}{\pi/4} = \frac{4r}{\pi} \times \frac{1}{\sqrt{2}}$$

or $\boxed{\bar{x} = \frac{4r}{\sqrt{2}\,\pi}} \quad ...(3.18)$

When x-axis is the axis of symmetry, as in Fig. 3.5.1. When the quarter circle is as shown in Fig. 3.5.2, then, $OC = \frac{r \sin \pi/4}{\pi/4}$

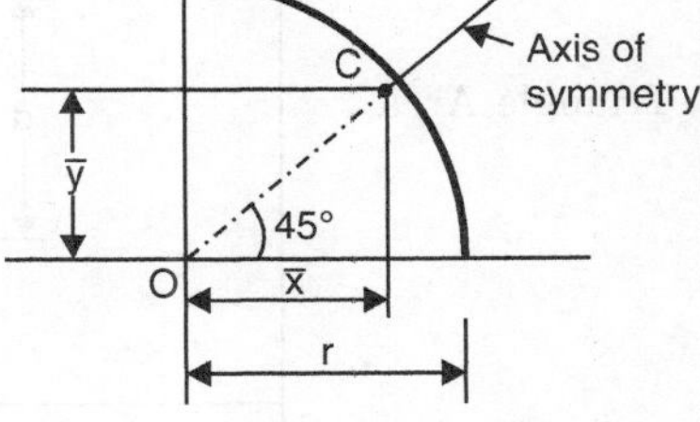

Fig. 3.5.2. A quarter circle.

$$\therefore \quad \bar{x} = \bar{y} = OC \times \sin \pi/4 = \left(\frac{r \sin \pi/4}{\pi/4}\right) \times \sin \pi/4 = \frac{r}{\pi/4} \times \left(\frac{1}{\sqrt{2}}\right)^2$$

or $\boxed{\bar{x} = \bar{y} = \frac{2r}{\pi}} \quad ...(3.19)$

(*c*) **For a complete circle,** $2\alpha = 360° = 2\pi$,

$\therefore \quad \alpha = \pi$,

$\therefore$ From equation (3.16), we get

$$\bar{x} = \frac{r \sin \pi}{\pi} = 0$$

also, $\quad \bar{y} = 0$,

$\therefore$ The centroid of a circle coincides with its centre *O*. It is also important to note here that *OX*- and *OY*-axes are the axes of symmetry perpendicular to each other, therefore *O* is the point of symmetry, as discussed earlier.

Other cases of determining the centroid by using the method of integration are discussed in solved problems given below.

Table 3.1. Standard cases of Areas and Lines

Name	*Shape*	*A*	$\bar{x}$	$\bar{y}$
Rectangular Area	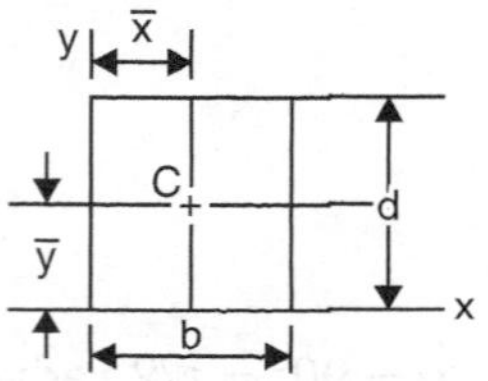	*bd*	*b*/2	*d*/2
Parallelogram	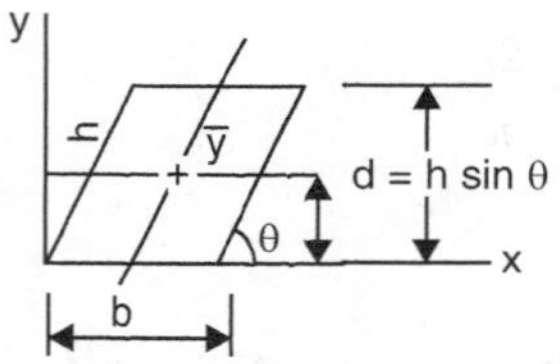	*bd*	—	*d*/2
Triangle Area	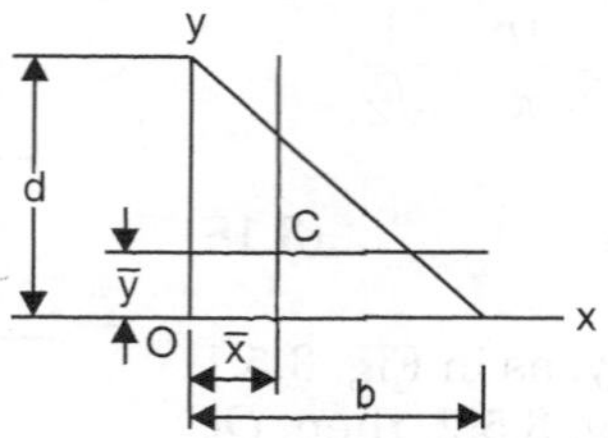	$\frac{bd}{2}$	$\frac{b}{3}$	$\frac{d}{3}$
	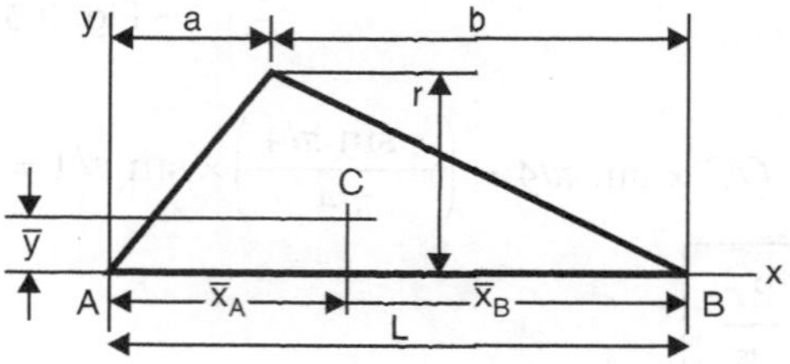	$\frac{ld}{2}$	$\bar{x}_A = \frac{l+a}{3}$ $\bar{x}_B = \frac{l+b}{3}$	$\frac{d}{3}$

Name	*Shape*	A	$\bar{x}$	$\bar{y}$
Circular Areas	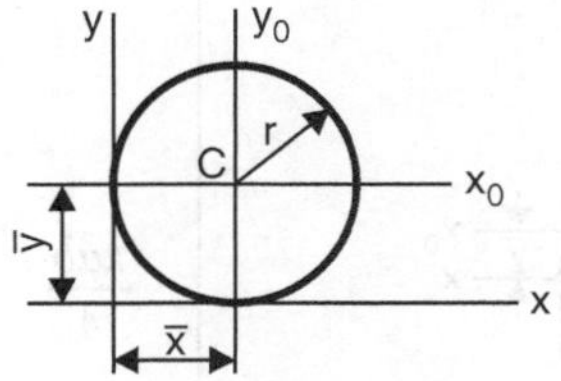 Circle	πr^2	r	r
	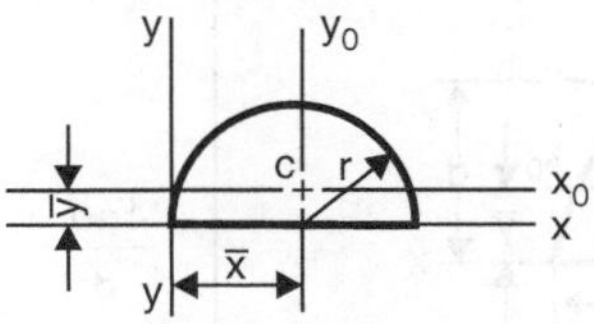 Semi-circle	$\frac{\pi r^2}{2}$	r	$\frac{4r}{3\pi}$
	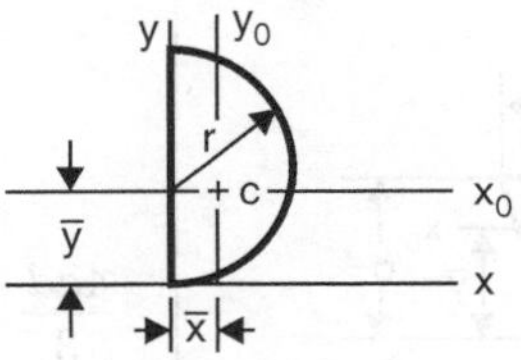 Semi-circle	$\frac{\pi r^2}{2}$	$\frac{4r}{3\pi}$	r
	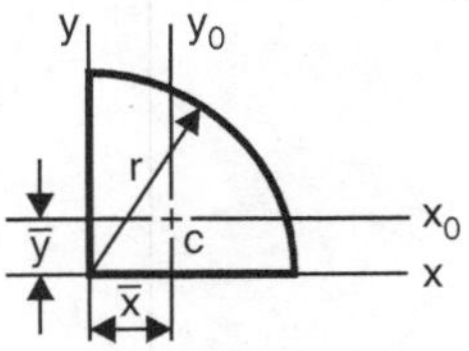	$\frac{\pi r^2}{4}$	$\frac{4r}{3\pi}$	$\frac{4r}{3\pi}$
Elliptical Areas	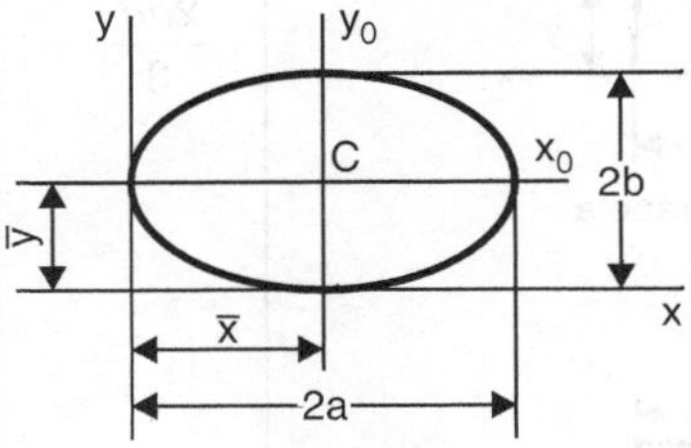 Ellipse	πab	a	b
	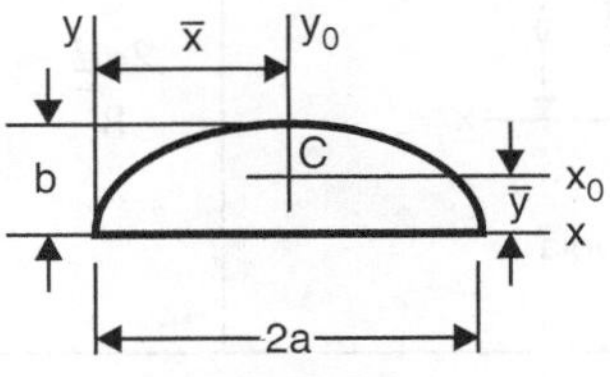 Semi-ellipse	$\frac{\pi ab}{2}$	a	$\frac{4b}{3\pi}$

Name	Shape	A	$\bar{x}$	$\bar{y}$
	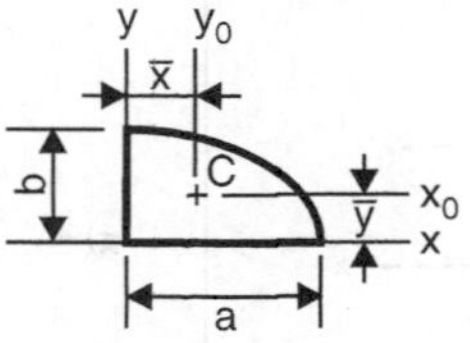 Quarter-ellipse	$\frac{\pi ab}{4}$	$\frac{4a}{3\pi}$	$\frac{4b}{3\pi}$
Parabolic Areas	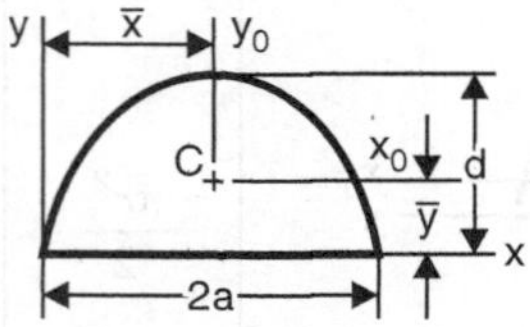 Parabolic Area	$\frac{4ad}{3}$	a	$\frac{2d}{5}$
	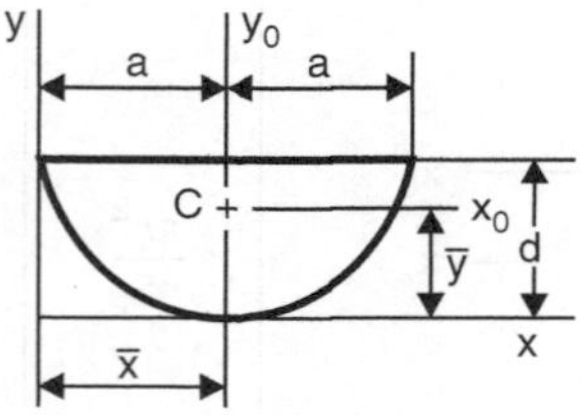 Parabolic Area	$\frac{4ad}{3}$	a	$\frac{3d}{5}$
	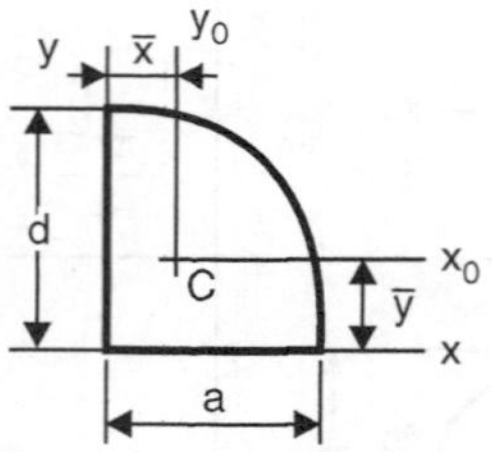 Quarter Parabola	$\frac{2ad}{3}$	$\frac{3a}{8}$	$\frac{2d}{5}$
	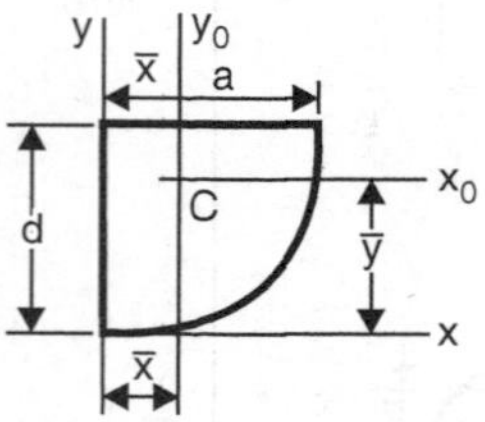 Quarter Parabola	$\frac{2ad}{3}$	$\frac{3a}{8}$	$\frac{3d}{5}$

Name	*Shape*	*A*	$\bar{x}$	$\bar{y}$
General Spandrel	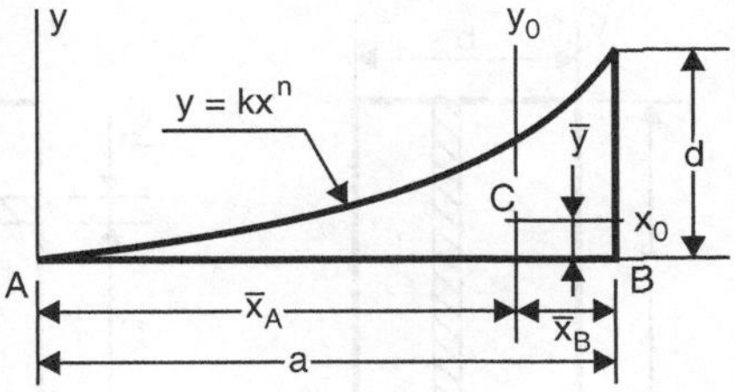 y, y_0, $y = kx^n$, $\bar{y}$, d, C, x_0, A, B, $\bar{x}_A$, $\bar{x}_B$, a	$\dfrac{ad}{n+1}$	$\bar{x}_A = \dfrac{n+1}{n+2}.a$ $\bar{x}_B = \dfrac{a}{n+2}$	$\dfrac{n+1}{4n+2}.d$
Parabolic Spandrel	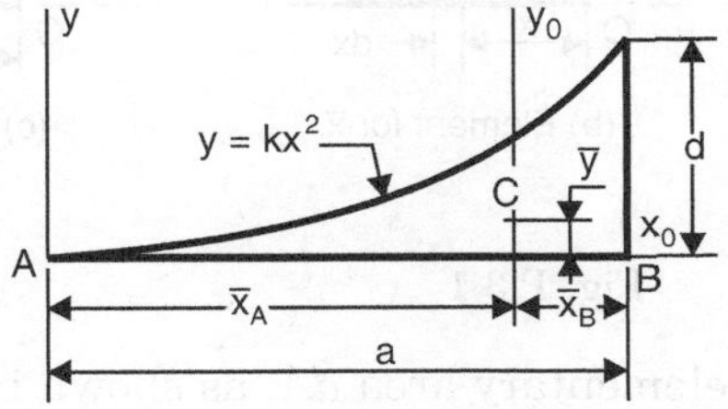 y, y_0, $y = kx^2$, $\bar{y}$, d, C, x_0, A, B, $\bar{x}_A$, $\bar{x}_B$, a	$\dfrac{ad}{3}$	$\bar{x}_A = \dfrac{3a}{4}$ $\bar{x}_B = \dfrac{a}{4}$	$\dfrac{3d}{10}$
Circular Sector	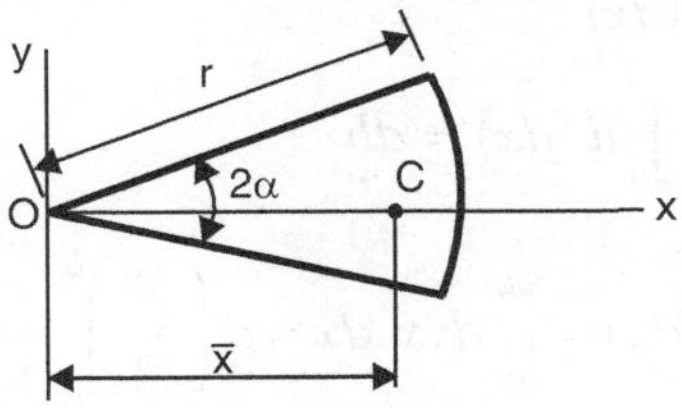 y, r, 2α, C, O, x, $\bar{x}$	αr^2	$\dfrac{2r \sin \alpha}{3\alpha}$	0
Circular Arcs	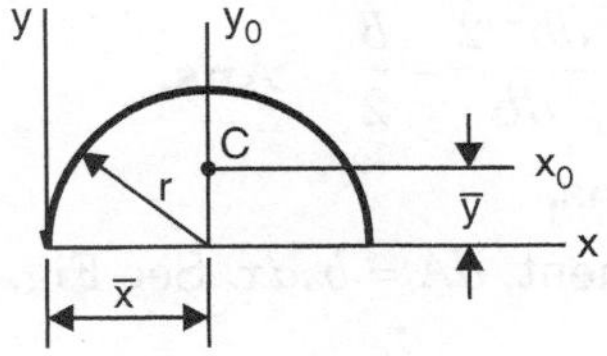 y, y_0, C, r, x_0, $\bar{y}$, x, $\bar{x}$ Semi-circular Arc	$L = \pi r$	r	$\dfrac{2r}{\pi}$
	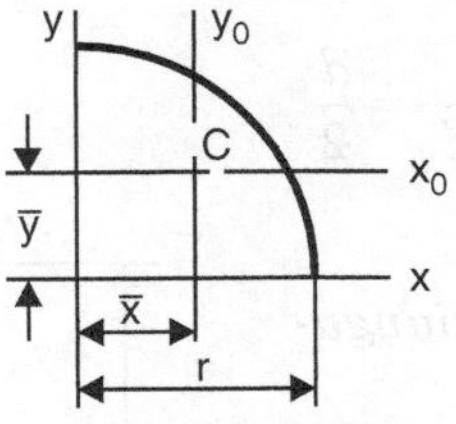 y, y_0, C, x_0, $\bar{y}$, x, $\bar{x}$, r Quarter-circular Arc	$L = \dfrac{\pi r}{2}$	$\dfrac{2r}{\pi}$	$\dfrac{2r}{\pi}$
	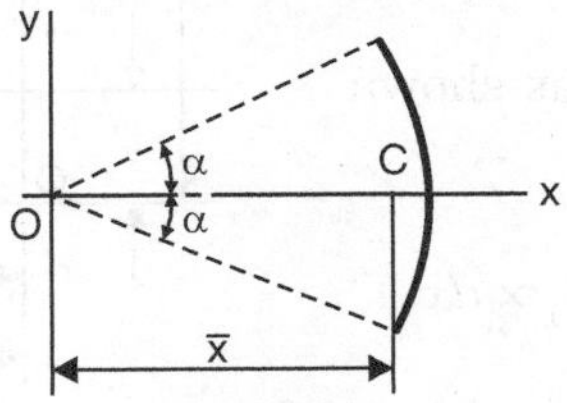 y, α, α, C, O, x, $\bar{x}$ Arc of Circle	$L = 2\alpha r$	$\dfrac{r \sin \alpha}{\alpha}$	0

Problem 3.1. *Locate centroid of the rectangular area shown in Fig. P.3.1 (a).*

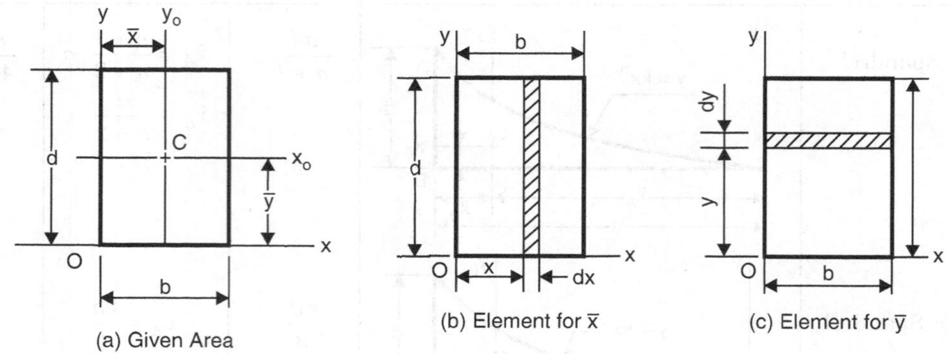

Fig. P.3.1

Sol. For $\bar{\mathbf{x}}$. Consider a vertical elementary area *dA*, as shown in Fig. P.3.1(*b*).

∴ Area of the element $= dA = d(dx)$

∴ $$A = \int dA = \int_0^b d\,(dx) = db$$

$$\int x dA = \int_0^b x\,d.(dx) = \int_0^b d.x.dx = d\left(\frac{x^2}{2}\right)_0^b = \frac{db^2}{2}$$

∴ $$\bar{x} = \frac{\int x\,dA}{\int dA} = \frac{db^2/2}{db} = \frac{b}{2}.$$ **Ans.**

∴ $\boxed{\bar{x} = b/2.}$ **Ans.**

For $\bar{\mathbf{y}}$. Consider a horizontal element, $dA = b.dx$. See Fig. P. 3.1(*c*).

$$\int y.dA = \int_0^d y \times b.dx = bd^2$$

∴ $$\bar{y} = \frac{\int y\,dA}{\int dA} = \frac{\frac{bd^2}{2}}{bd} = \frac{d}{2}$$

∴ $\boxed{\bar{y} = d/2.}$ **Ans.**

Problem 3.2. *Locate centroid of the triangular area shown in Fig. P.3.2 (a).*

Sol. Let the co-ordinates of the centroid *C* be $\bar{x}$ and $\bar{y}$, as shown in Fig. P.3.2(*a*).

For $\bar{x}$ consider a vertical element as shown in Fig. P.3.2 (*b*).

Area of the element

$$= dA = A_1C_1 \times dx$$

Since $\Delta BA_1C_1 \equiv \Delta BAC$

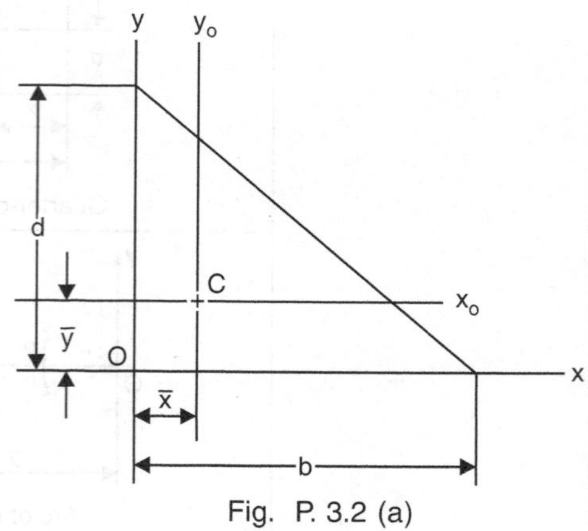

Fig. P. 3.2 (a)

$$\therefore \quad \frac{A_1C_1}{BA_1} = \frac{AC}{BA}$$

$$\therefore \quad A_1C_1 = \frac{AC}{BA} \times BA_1$$

$$\therefore \quad A_1C_1 = \frac{d}{b} \times (b - x)$$

$$\therefore \quad dA = \frac{d}{b}(b - x).dx$$

$$\therefore \quad A = \int dA = \frac{d}{b}\int_0^b (b - x).dx = \frac{d}{b}\left(bx - \frac{x^2}{2}\right)_0^b = \frac{d}{b} \times \left(b^2 - \frac{b^2}{2}\right)$$

$$= \frac{db^2}{2b}$$

or

$$\boxed{A = \frac{bd}{2}}$$

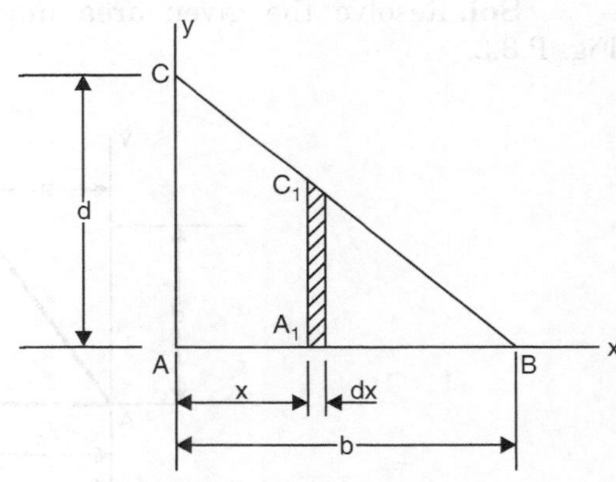

Fig. P. 3.2 (b) Element for $\bar{x}$

And,
$$\bar{x} = \frac{\int x\,dA}{A} = \frac{1}{A}\int_0^b x.\frac{d}{b}(b - x)\,dx$$

$$= \frac{2}{bd}\int_0^b \frac{d}{b}(bx - x^2).dx$$

$$= \frac{2}{b^2}\left[\frac{bx^2}{2} - \frac{x^3}{3}\right]_0^b$$

$$= \frac{2}{b^2}.b^3\left[\frac{3-2}{6}\right]$$

$$\therefore \quad \boxed{\bar{x} = \frac{b}{3}.} \quad \textbf{Ans.}$$

For $\bar{y}$ consider a horizontal element as shown in Fig.P. 3.2 (*c*).

Area of the element = dA

$$= A_1.B_1 \times dy$$

$$\because \quad \Delta CA_1B_1 \equiv \Delta CAB$$

$$\therefore \quad \frac{A_1B_1}{CA_1} = \frac{AB}{CA}$$

$$\therefore \quad A_1B_1 = \frac{AB}{CA} \times CA_1 = \frac{b}{d}(d - y)$$

$$\therefore \quad dA = \frac{b}{d}(d - y).dy$$

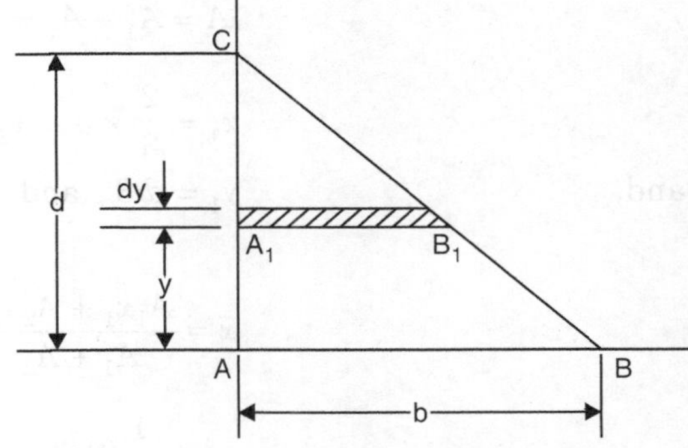

Fig. P. 3.2 (c) Element for $\bar{y}$

$$\therefore \qquad \int y.dA = \int_0^d y \times \frac{b}{d}(d-y)\,dy$$

$$= \frac{b}{d}\int_0^d (yd - y^2)\,dy = \frac{b}{d}\left[\frac{d \times y^2}{2} - \frac{y^3}{3}\right]_0^d = \frac{b}{d} \times \frac{d^3}{6} = \frac{bd^2}{6}$$

$$\therefore \qquad \bar{y} = \frac{\int y\,dA}{A} = \frac{bd^2/6}{bd/2} = \frac{d}{3}$$

$$\therefore \qquad \boxed{\bar{y} = \frac{d}{3}.} \quad \textbf{Ans.}$$

Problem 3.3. *Locate centroid of the triangular area shown in Fig. P.3.3.*

Sol. Resolve the given area into two triangles *viz.*, ΔACC_1 and ΔCC_1B, as shown in Fig. P.3.3.

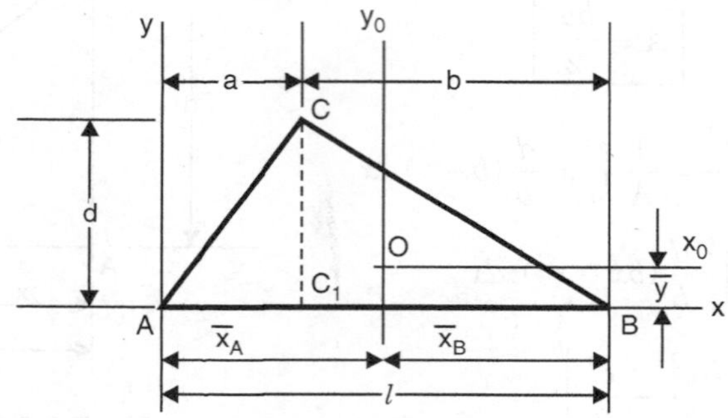

Fig. P. 3.3

$$\therefore \qquad A = A_1 + A_2 = \frac{ad}{2} + \frac{bd}{2} = \frac{d}{2}(a+b) = \frac{dl}{2}$$

$$\therefore \qquad x_1 = \frac{2}{3} \times a, \quad x_2 = a + b/3 \text{ from } y\text{-axis}$$

and, $\qquad y_1 = d/3$ and $y_2 = d/3$ from x-axis.

$$\therefore \qquad \bar{x} = \frac{A_1x_1 + A_2x_2}{A_1 + A_2} = \frac{(ad/2).\frac{2a}{3} + \left(\frac{bd}{2}\right)(a + b/3)}{ld/2} = \frac{\frac{2a^2}{3} + \frac{b}{3}(3a+b)}{l}$$

$$= \frac{1}{3l}(2a^2 + 3ab + b^2) = \frac{1}{3l}[2a(a+b) + b(a+b)]$$

$$= \frac{1}{3l} \times (a+b)(2a+b) = \frac{1}{3l} \times l(2a+b) = \frac{1}{3}(a+b+a) = \frac{l+a}{3}$$

$$\therefore \qquad \boxed{\bar{x} = \frac{l+a}{3} = \bar{x}_A \text{ from point } A.} \quad \textbf{Ans.}$$

And from point B,

$$\therefore \quad \bar{x}_B = l - \bar{x}_A = l - \frac{l+a}{3} = \frac{3l - l - a}{3} = \frac{2l - a}{3} = \frac{l + (l-a)}{3} = \frac{l+b}{3}$$

$$\therefore \quad \boxed{\bar{x}_B = \frac{l+b}{3}.} \quad \textbf{Ans.}$$

And,
$$\bar{y} = \frac{A_1 y_1 + A_2 y_2}{A_1 + A_2} = \frac{(ad/2).d/3 + \left(\frac{bd}{2}\right).d/3}{\frac{l \times d}{2}} = \frac{ad + bd}{3l} = \frac{d}{3l}(a+b)$$

$$= \frac{d(l)}{3l}$$

$$\therefore \quad \boxed{\bar{y} = \frac{d}{3}, \text{ from } x\text{-axis.}} \quad \textbf{Ans.}$$

Problem 3.4. *Locate centroid of the quadrant of a circle of radius R.*

Sol. There are many methods of locating the centroid of the quarter circular area.

Method I. See Fig. P.3.4 (*a*). Consider an element as shown. The area of the element may be treated as a triangle.

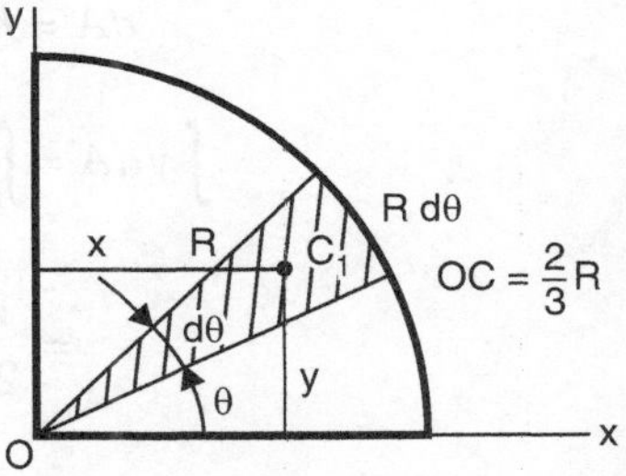

$$\therefore \quad \text{Area} \quad dA = \frac{1}{2} R.ds$$

its distance OC_1 from O is $\frac{2}{3}R$.

Fig. P. 3.4 (a)

$$\therefore \quad dA = \frac{1}{2} R.ds = \frac{1}{2} R \times (Rd\theta) = \frac{R^2 d\theta}{2}$$

$$\therefore \quad A = \int_0^{\pi/2} \frac{R^2 d\theta}{2} = \frac{R^2 d\theta}{2}\Bigg|_0^{\pi/2} = \frac{R^2 \pi}{4} \quad \text{or} \quad \boxed{A = \frac{\pi R^2}{4}}$$

Centroid of the element is given by $x = \left(\frac{2}{3}R\right)\cos\theta$; $y = \left(\frac{2}{3}R\right)\sin\theta$

$$\therefore \quad \int x.dA = \int_0^{\pi/2} \left(\frac{2}{3} R \cos\theta\right) \times \frac{R^2 d\theta}{2} = \frac{R^3}{3}\int_0^{\pi/2} \cos\theta.d\theta = \frac{R^3}{3}\sin\theta\Bigg|_0^{\pi/2}$$

$$\therefore \quad \int x.dA = \frac{R^3}{3}$$

$$\therefore \quad \bar{x} = \frac{\int x\, dA}{A} = \frac{R^3/3}{\frac{\pi R^2}{4}}$$

$$\therefore \quad \boxed{\bar{x} = \frac{4R}{3\pi}.} \quad \textbf{Ans.}$$

Also, $$\int y\,dA = \int_0^{\pi/2}\left(\frac{2}{3}R\sin\theta\right)\times\frac{R^2 d\theta}{2} = \frac{R^3}{3}\int_0^{\pi/2}\sin\theta.d\theta$$

$$= \frac{R^3}{3}\Big(-\cos\theta\Big)_0^{\pi/2} = -\frac{R^3}{3}(0-1)$$

$$\therefore \quad \int ydA = \frac{R^3}{3}$$

$$\bar{y} = \frac{\int ydA}{A} = \frac{R^3/3}{\pi R^2/4} = \frac{4R}{3\pi}$$

$$\therefore \quad \boxed{\bar{y} = \frac{4R}{3\pi}.} \quad \textbf{Ans.}$$

Method II. Vertical Element Method

For finding $\bar{x}$ and $\bar{y}$ take vertical element as shown in Fig. P.3.4 (*b*)

$$dA = hdx = \sqrt{R^2 - x^2}\,.dx$$

$$\therefore \quad \int y\,dA = \int_0^R \left(\frac{h}{2}\right)\sqrt{R^2 - x^2}\,.dx = \frac{1}{2}\int_0^R (\sqrt{R^2 - x^2})\sqrt{(R^2 - x^2)}\,dx$$

$$= \frac{1}{2}\int_0^R (R^2 - x^2)\,dx = \frac{1}{2}\left[R^3 - \frac{R^3}{3}\right]$$

$$= \frac{1}{2}\,\frac{2R^3}{3} = \frac{R^3}{3}$$

$$\therefore \quad \bar{y} = \frac{\int y\,dA}{A} = \frac{R^3/3}{\pi R^2/4}$$

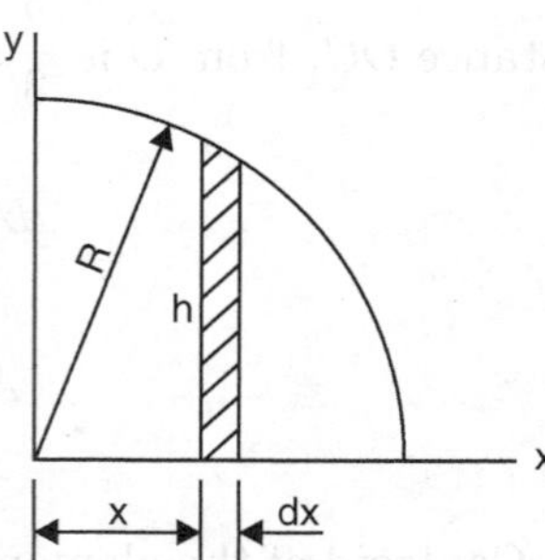

Fig. P. 3.4 (b) The vertical element.

$$\therefore \quad \boxed{\bar{y} = \frac{4R}{3\pi}}$$

We have seen that

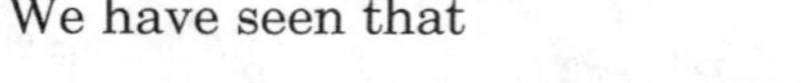

$$\because \quad dA = h.dx = \sqrt{R^2 - x^2}\,.dx$$

$$\therefore \quad \int x\,dA = \int dA = \int x\sqrt{R^2 - x^2}\,dx$$

$$= \frac{1}{3}\left[-(R^2 - x^2)^{3/2}\right]_0^R = +\frac{1}{3}[\mathrm{R}^3]$$

$$\therefore \quad \bar{x} = \frac{\int x\,dA}{A} = \frac{R^3/3}{\frac{\pi R^2}{4}}$$

$$\therefore \quad \boxed{\bar{x} = \frac{4R}{3\pi}}$$

Method III. The Horizontal Element Method : Take horizontal element shown in Fig. P.3.4 (*c*).

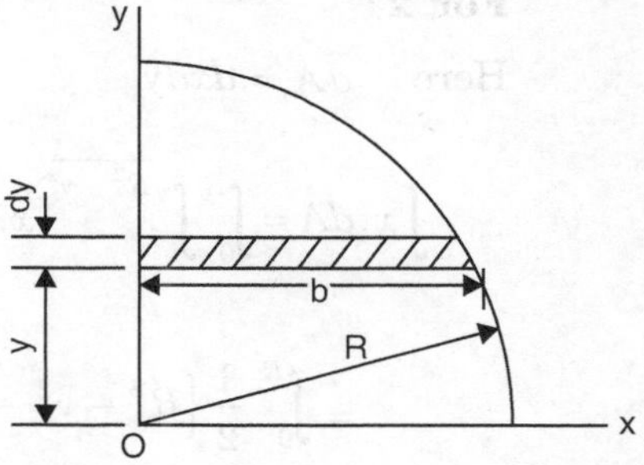

For $\bar{\mathbf{x}}$:

Here, $dA = b.dy = \sqrt{R^2 - y^2}\, dy$

$$\therefore \quad \int x\, dA = \int \frac{x}{2} \sqrt{R^2 - y^2}\, dy$$

$$= \frac{1}{2}\int_0^R (\sqrt{R^2 - y^2}).(\sqrt{R^2 - y^2}).dy$$

$$= \frac{1}{2}\int_0^R (R^2 - y^2)\,dy$$

$$= \frac{1}{2}\left[R^2 y - \frac{y^3}{3}\right]_0^R = \frac{R^3}{3}$$

$$\because \quad \bar{x}A = \int x\, dA$$

$$\therefore \quad \bar{x} = \frac{\int x\, dA}{A} = \frac{\dfrac{R^3}{3}}{\pi R^2/4}$$

$$\therefore \quad \boxed{\bar{x} = \frac{4R}{3\pi}.} \quad \textbf{Ans.}$$

Fig. P. 3.4 (c) The horizontal element.

For $\bar{\mathbf{y}}$: $\quad dA = bdy = \sqrt{R^2 - y^2}\,.dy$

$$\therefore \quad \int y dA = \int_0^R y.\sqrt{R^2 - y^2}\,.dy = \left[-\frac{(R^2 - y^2)^{3/2}}{3}\right]_0^R = \frac{R^3}{3}$$

$$\because \quad \bar{y}A = \int y dA$$

$$\therefore \quad \bar{y} = \frac{\int y dA}{A} = \frac{R^3/3}{\pi R^2/4}$$

$$\therefore \quad \boxed{\bar{y} = \frac{4R}{3\pi}}$$

Method IV. See Fig.P.3.4 (*d*).

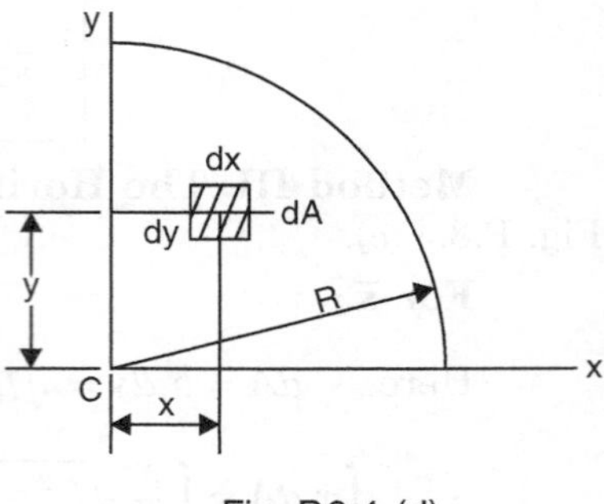

Fig. P.3.4 (d)

$$= \int_0^R \sqrt{R^2 - y^2}\,.dy$$

For $\bar{\mathbf{x}}$:

Here $\quad \bar{d}A = dxdy$

$$\therefore \quad \int x.dA = \int_0^R \int_0^{\sqrt{R^2 - y^2}} x.dx.dy = \int_0^R \left[\frac{x^2}{2}\right]_0^{\sqrt{R^2-y^2}} dy$$

$$= \int_0^R \frac{1}{2}\left\{R^2 - y^2 - 0\right\} dy = \frac{1}{2}\left[R^2.y - \frac{y^3}{3}\right]_0^R$$

$$= \frac{1}{2}\left[R^3 - \frac{R^3}{3}\right] = \frac{R^3}{2\times 3}(3-1)$$

$$\therefore \quad \int x.dA = \frac{R^3}{3}$$

$$\because \quad \bar{x}A = \int x.dA$$

$$\therefore \quad \bar{x} = \frac{(R^3/3)}{(\pi R^2/4)} = \frac{4R}{3\pi}$$

$$\therefore \quad \boxed{\bar{x} = \frac{4R}{3\pi}}$$

For $\bar{\mathbf{y}}$:

$$\int y.dA = \int_0^R \int_0^{\sqrt{R^2 - x^2}} y\,dy.dx = \int_0^R \left[\frac{y^2}{2}\right]_0^{\sqrt{R^2 - x^2}} .dx$$

$$= \frac{1}{2}\int_0^R (R^2 - x^2)\,dx = \frac{1}{2}\left[R^2x - \frac{x^3}{3}\right]_0^R = \frac{R^3}{3}$$

But $\because \quad \bar{y}A = \int y.dA$

$$\therefore \quad \bar{y} \times \frac{\pi R^2}{4} = \frac{R^3}{3}$$

$$\therefore \quad \boxed{\bar{y} = \frac{4R}{3\pi}}$$

Method V. Consider an element at radius *r* and of radial thickness *dr*, as shown in Fig. P.3.4 (*e*).

Here $y = r\sin\theta$ and $x = r\cos\theta$ and $dA = dr.(rd\theta)$

$$\therefore \quad A = \int dA = \int_0^R \int_0^{\pi/2} r\,dr\,d\theta$$

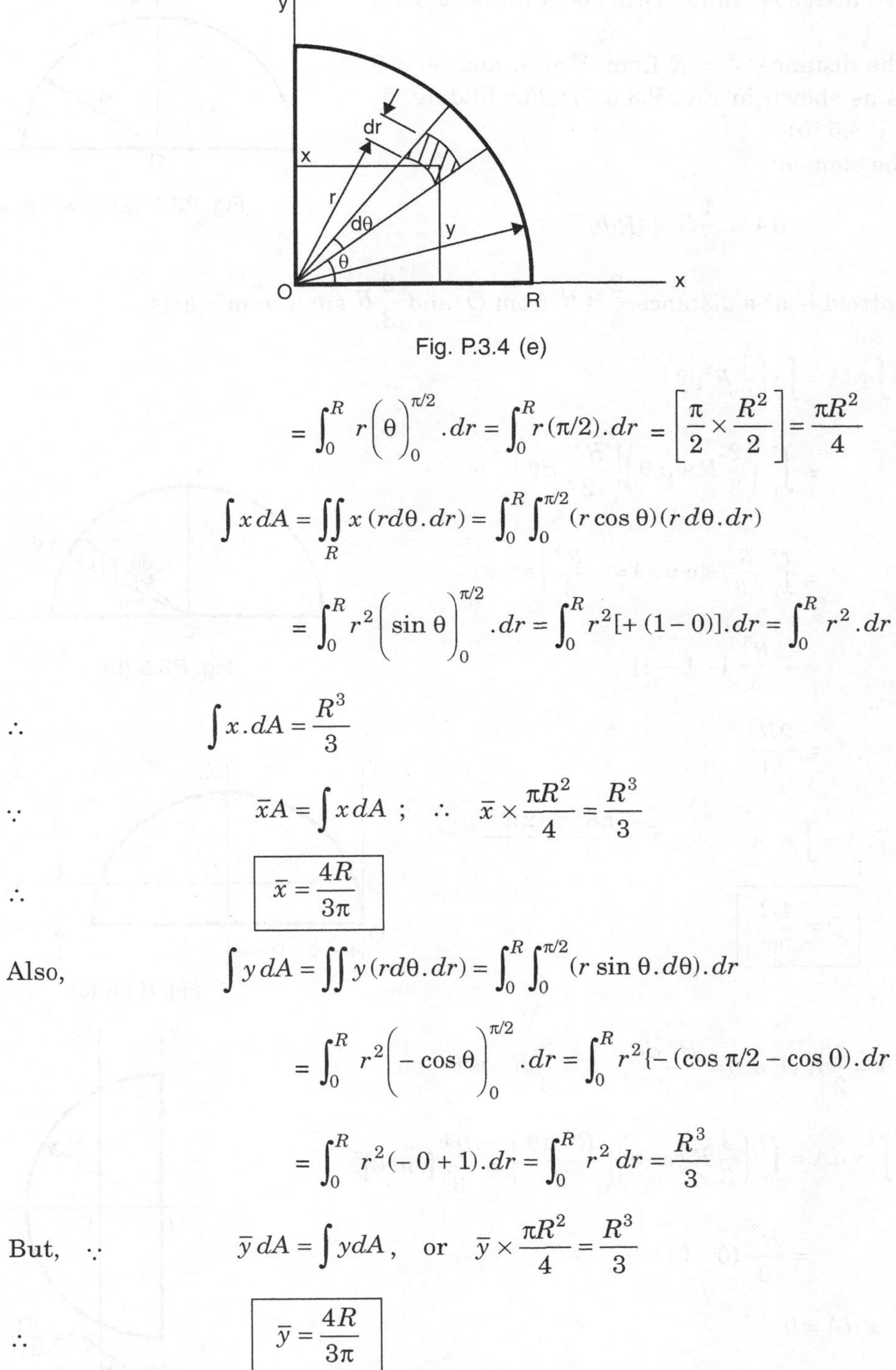

Fig. P.3.4 (e)

$$= \int_0^R r\left(\theta\right)_0^{\pi/2}.dr = \int_0^R r(\pi/2).dr = \left[\frac{\pi}{2}\times\frac{R^2}{2}\right] = \frac{\pi R^2}{4}$$

$$\int x\,dA = \iint_R x\,(rd\theta.dr) = \int_0^R \int_0^{\pi/2} (r\cos\theta)(r\,d\theta.dr)$$

$$= \int_0^R r^2\left(\sin\theta\right)_0^{\pi/2}.dr = \int_0^R r^2[+(1-0)].dr = \int_0^R r^2.dr$$

$\therefore$ $$\int x.dA = \frac{R^3}{3}$$

$\because$ $$\bar{x}A = \int x\,dA\ ; \quad \therefore \quad \bar{x}\times\frac{\pi R^2}{4} = \frac{R^3}{3}$$

$\therefore$ $$\boxed{\bar{x} = \frac{4R}{3\pi}}$$

Also, $$\int y\,dA = \iint y\,(rd\theta.dr) = \int_0^R \int_0^{\pi/2} (r\sin\theta.d\theta).dr$$

$$= \int_0^R r^2\left(-\cos\theta\right)_0^{\pi/2}.dr = \int_0^R r^2\{-(\cos\pi/2 - \cos 0).dr$$

$$= \int_0^R r^2(-0+1).dr = \int_0^R r^2\,dr = \frac{R^3}{3}$$

But, $\because$ $$\bar{y}\,dA = \int y dA, \quad \text{or} \quad \bar{y}\times\frac{\pi R^2}{4} = \frac{R^3}{3}$$

$\therefore$ $$\boxed{\bar{y} = \frac{4R}{3\pi}}$$

Problem 3.5. *Locate centroid of a semi-circular area of radius shown in Fig. P.3.5 (a).*

Sol. Any one of methods discussed in the solution of problem 3.4 can be used for finding the value of $\bar{y}$.

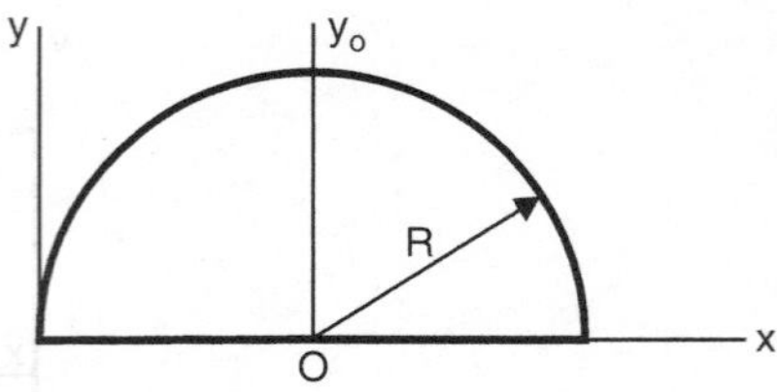

Fig. P.3.5 (a) Given area.

The given area is symmetrical about the centroidal axis y_0.

$\therefore$ The distance $\bar{x} = R$ from Y-axis, and $\bar{x} = 0$ from Y_0-axis as shown in Fig. P.3.5 (*a*). For finding $\bar{y}$, refer to Fig. P.3.5 (*b*).

For the element,

$$dA = \frac{1}{2}R \times (Rd\theta)$$

Its centroid is at a distance $\frac{2}{3} \times R$ from O, and $\frac{2}{3}R$ sin θ from x-axis.

$$\therefore \quad \int y dA = \int y\left(\frac{1}{2}R^2 d\theta\right)$$

$$= \int_0^{\pi}\left(\frac{2}{3}R\sin\theta\right)\left(\frac{R^2}{2}.d\theta\right)$$

$$= \int_0^{\pi}\frac{R^3}{3}\sin\theta\, d\theta = -\frac{R^3}{3}\left(\cos\theta\right)_0^{\pi}$$

$$= -\frac{R^3}{3}[-1-1]$$

$$= \frac{2R^3}{3}$$

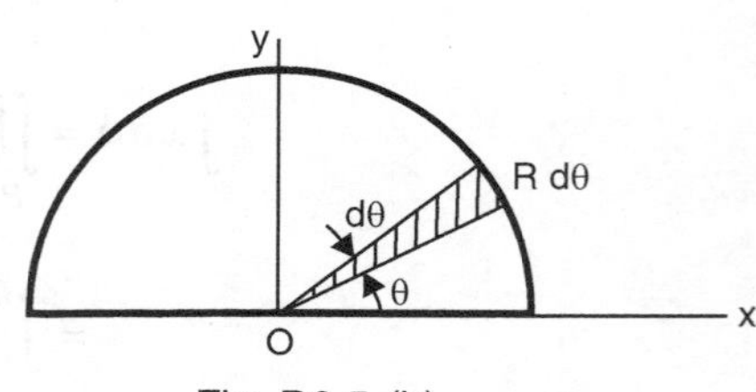

Fig. P.3.5 (b)

$$\because \quad \bar{y}.A = \int y dA, \quad \therefore \quad \bar{y} \times \frac{\pi R^2}{2} = \frac{2R^3}{3}$$

$$\therefore \quad \boxed{\bar{y} = \frac{4R}{3\pi}}$$

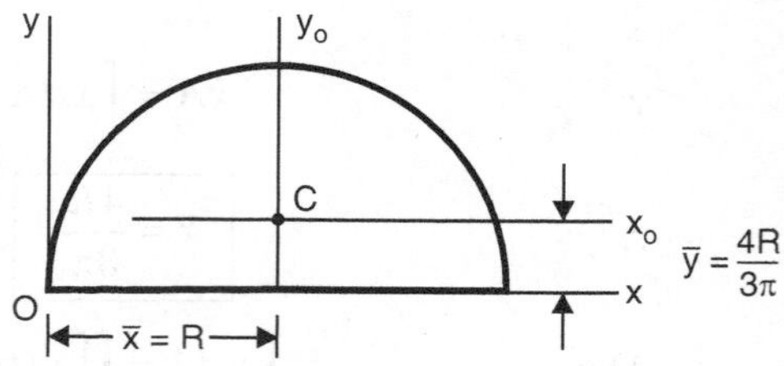

Fig. P.3.5 (c)

For $\bar{x}$:

$$dA = \frac{1}{2}R\,(R\,d\theta) = \frac{R^2\,d\theta}{2}, \quad x = \frac{2}{3}R\cos\theta$$

$$\therefore \quad \int x\,dA = \int_0^{\pi}\left(\frac{2}{3}R\cos\theta\right)\left(\frac{R^2\,d\theta}{2}\right) = \frac{R^3}{3}[\sin\theta]_0^{\pi}$$

$$= \frac{R^3}{3}(0-0) = 0$$

$$\therefore \quad \int x\,dA = 0$$

$$\therefore \quad \bar{x} = \frac{\int x\,dA}{A} = \frac{0}{A} = 0 \text{ from } y\text{-axis}$$

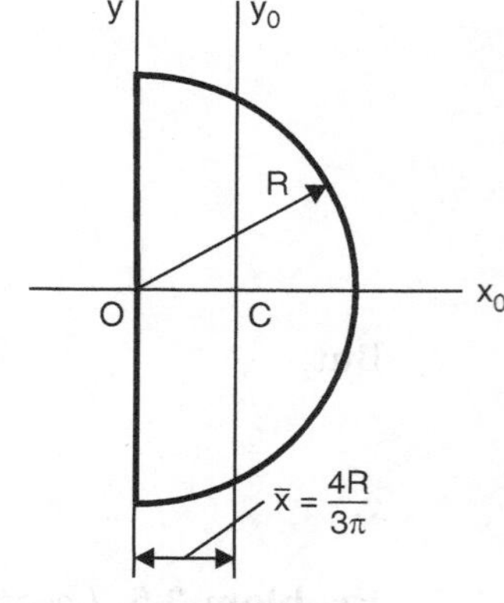

Fig. P.3.5 (d)

$\therefore \qquad \bar{x}.\dfrac{\pi R^2}{2} = 0$

$\therefore \qquad \bar{x} = 0$ from Y_0-axis, as in Fig. P.3.5 (*c*).

$\therefore \qquad \bar{x} = R$ from *y*-axis as in Fig. P.3.5 (*c*).

Hence for Fig. P.3.5 (*a*), the centroid is shown in Fig. P.3.5 (*c*).

For the position of the semi-circle shown in Fig. P.3.5 (*d*) centroid is given as under.

$$\boxed{\bar{x} = \frac{4R}{3\pi} \; ; \quad \bar{y} = 0.}$$

Problem 3.6. *Locate centroid of the parabola $y = kx^2$ shown in Fig. P.3.6 (a).*

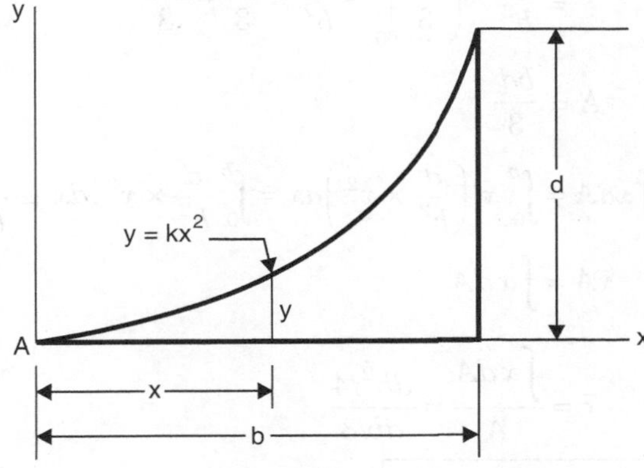

Fig. P.3.6 (a) Given Parabolic area.

Sol. Consider an element at a distance *x*, as shown in Fig. P.3.6 (*b*).

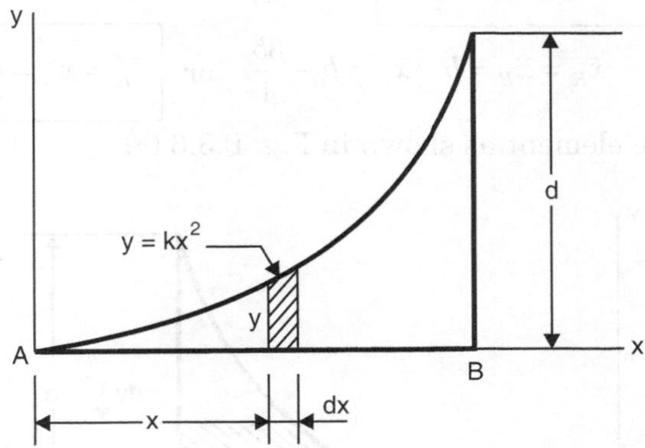

Fig. P.3.6 (b) Parabolic area element for $\bar{x}$.

First, we shall find the constant *k* in $y = k.x^2$ by using the following condition.

When, $\quad x = b, y = d$

$\therefore$ Equation $y = kx^2$, ...(1)

becomes $d = k \,.\, b^2$,

$\therefore$ $k = d/b^2$

$\therefore$ Equation (1) becomes, $y = \dfrac{d}{b^2} . x^2$ ...(2)

$$\therefore \quad dA = y.dx = \left(\frac{d}{b^2} . x^2\right) . dx \qquad ...(3)$$

$$\therefore \quad A = \int dA = \int_0^b \frac{d}{b^2} \times x^2 . dx = \frac{d}{b^2} \int_0^b x^2 . dx$$

$$= \frac{d}{b^2} \times \left(\frac{x^3}{3}\right)_0^b = \frac{d}{b^2} \times \frac{b^3}{3} = \frac{bd}{3}$$

$$\therefore \quad A = \frac{bd}{3} \qquad ...(4)$$

$$\int x dA = \int_0^b x.\left(\frac{d}{b^2} \times x^2\right) dx = \int_0^b \frac{d}{b^2} \times x^3 . dx = \frac{d}{b^2} \times \frac{b^4}{4} = \frac{db^2}{4} \qquad ...(5)$$

$$\because \quad \bar{x} A = \int x \, dA$$

$$\therefore \quad \bar{x} = \frac{\int x \, dA}{A} = \frac{db^2/4}{db/3}$$

$$\therefore \quad \boxed{\bar{x} = \frac{3b}{4}} \qquad ...(6)$$

or

$$\boxed{\bar{x}_A = \bar{x}_L = \frac{3b}{4}} \qquad ...(6a)$$

and

$$\bar{x}_R = \bar{x}_B = b - \bar{x}_L = b - \frac{3b}{4} \quad \text{or} \quad \boxed{\bar{x}_R = \bar{x}_B = \frac{b}{4}} \qquad ...(7)$$

For $\bar{y}$, consider the element as shown in Fig. P.3.6 (*c*)

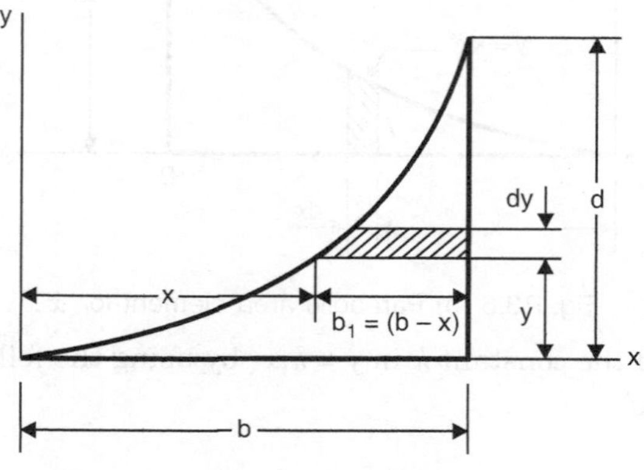

Fig. P.3.6 (c) Element for $\bar{x}$ and $\bar{y}$.

Here $dA = b_1 \times dy$

or $dA = (b - x).dy$

$$\therefore \quad A = \int dA = \int (b - x).\left(\frac{d}{b^2}.\, 2xdx\right)$$

$$\therefore \quad A = \frac{bd}{3}$$

$$\because \quad y = \frac{d}{b^2}\, x^2, \quad \therefore \quad dy = \frac{d}{b^2}.\, 2x\, dx, \text{ and } \therefore \quad x = \sqrt{\frac{b^2 y}{d}}$$

$$\therefore \quad ydA = y(b - x)dy = \left(by - y \times \sqrt{\frac{b^2 y}{d}}\right) dy$$

or
$$\int ydA = \int_0^d \left(by - \frac{b}{\sqrt{d}}.\, y^{3/2}\right) dy = \left[\frac{b \times y^2}{2} - \frac{b}{\sqrt{d}}\,\frac{2y^{5/2}}{5}\right]_0^d$$

$$= \frac{bd^2}{2} - \frac{b}{\sqrt{d}} \times \frac{2}{5} \times d^{5/2} = \frac{bd^2}{2} - \frac{2bd^2}{5} = \frac{bd^2}{10}$$

$$\therefore \quad \int ydA = \frac{bd^2}{10} \qquad \text{...(8)}$$

$$\therefore \quad \bar{y}A = \int ydA$$

$$\therefore \quad \bar{y} \times \frac{bd}{3} = \frac{bd^2}{10}$$

$$\therefore \quad \boxed{\bar{y} = \frac{3d}{10}.} \quad \textbf{Ans.} \qquad \text{...(9)}$$

Problem 3.7. *Locate centroid of the area under the curve given by $y = ax^2 + bx + c$.*

Sol. Area bounded by the parabola

$$y = ax^2 + bx + c \qquad \text{...(1)}$$

is shown in Fig. P.3.7 (*a*).

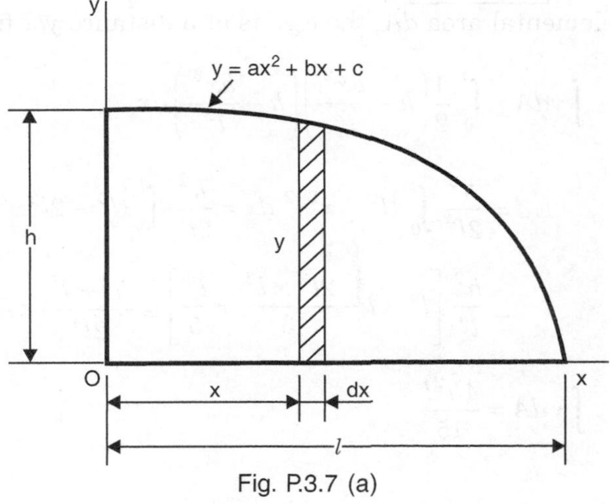

Fig. P.3.7 (a)

Let the base be l and the height be h. First find the constants a, b and c by using the following conditions.

(*a*) When $x = 0$, $y = h$, $\therefore$ we get

$$h = a(0)^2 + b(0) + c$$

$$\therefore \quad c = h \qquad \text{...(a)}$$

(*b*) When $x = l$, $y = 0$,

$$\therefore \quad 0 = al^2 + bl + c$$

$$\therefore \quad al^2 + bl + h = 0 \qquad \text{...(b)}$$

(*c*) When $x = 0$, $\frac{dy}{dx} = 0$, $\therefore \left.\frac{dy}{dx}\right]_{x=0} = 2ax + b + 0\Big]_{x=0} = 0$

$$\therefore \quad 2a \times (0) + b = 0$$

$$\therefore \quad b = 0 \qquad \text{...(c)}$$

$\therefore$ Eqn. (*b*) gives, $al^2 + 0 + h = 0$

$$\therefore \quad a = -h/l^2 \qquad \text{...(d)}$$

$\therefore$ Eqn. (1) becomes $y = \frac{-h}{l^2} \cdot x^2 + h$

or

$$y = h - \frac{hx^2}{l^2} = h\left(1 - \frac{x^2}{l^2}\right) \qquad \text{...(2)}$$

$\therefore$ Elemental area, $dA = y.dx$

$$\therefore \quad A = \int dA = \int y.dx = \int_0^l \left(h - \frac{hx^2}{l^2}\right) dx$$

$$= h\left[x - \frac{x^3}{3l^2}\right]_0^l = h\left[l - \frac{l^3}{3l^2}\right]$$

$$\therefore \quad \boxed{A = \frac{2hl}{3}} \qquad \text{...(3)}$$

For $\bar{y}$: For the elemental area dA, the *e.g.*, is at a distance $y/2$ from x-axis.

$$\int ydA = \int_0^l \frac{1}{2}\left(h - \frac{hx^2}{l^2}\right)\left(h - \frac{hx^2}{l^2}\right) dx$$

$$= \frac{h^2}{2l^4} \int_0^l (l^2 - x^2)^2 \, dx = \frac{h^2}{2l^4} \int_0^l (l^4 - 2l^2x^2 + x^4) \, dx$$

$$= \frac{h^2}{l^4}\left[l^4 \times l - \frac{2l^2 \times l^3}{3} + \frac{l^5}{5}\right] = \frac{h^2 \times l^5}{2l^4} \times \frac{15 - 10 + 3}{15} = \frac{8h^2 l}{2 \times 15}$$

or

$$\int y\,dA = \frac{4h^2 l}{15} \qquad \text{...(4)}$$

Since $\bar{y}A = \int y\,dA$

$\therefore$ $\bar{y} \times \dfrac{2hl}{3} = \dfrac{4}{15}h^2 l$

$\therefore$ $\boxed{\bar{y} = \dfrac{2h}{5}}$ ...(5)

Also, **for $\bar{\mathbf{x}}_{\mathbf{L}}$:** $\int x.dA = \int_0^l x\left(h - \dfrac{hx^2}{l^2}\right).dx = \dfrac{h}{l^2}\int_0^l (l^2 x - x^3).dx$

$$= \frac{h}{l^2}\left[\frac{l^2 \times l^2}{2} - \frac{l^4}{4}\right] = \frac{hl^4}{4l^2} \times (2-1)$$

$\therefore$ $\int x dA = \dfrac{hl^2}{4}$ ...(6)

$\because$ $\bar{x}_L A = \int x.dA,$ [$\bar{x}_L$ = distance $\bar{x}_A$ measured from point A or from left end]

$\therefore$ $\bar{x}_L \times \dfrac{2}{3}hl = \dfrac{hl^2}{4}$

$\therefore$ $\boxed{\bar{x}_L = \dfrac{3l}{8}}$ ...(7)

or $\boxed{\bar{x}_R = l - \bar{x}_L = \dfrac{5l}{8}}$ ...(8)

$\therefore$ Results are

$$\boxed{\begin{aligned} A &= \frac{2}{3} \times hl \\ \bar{x}_L &= \frac{3l}{8} \\ \bar{x}_R &= \frac{5l}{8} \\ \bar{y} &= \frac{2}{5}h. \end{aligned}}$$

Problem 3.8. *Locate the centroid of a half parabolic area $y = kx^2$, bounded by the co-ordinates (a, b) as shown in Fig. P.3.8 (a).*

Sol. The equation of the given parabola is $y = kx^2$ ...(1)

First find constant k, by using the condition

$$y = b \text{ at } x = a$$

Therefore, we get $b = k.a^2$

$\therefore$ $k = \dfrac{b}{a^2}$

$\therefore$ Eqn. (1) becomes, $y = \dfrac{b}{a^2}.x^2$ ...(2)

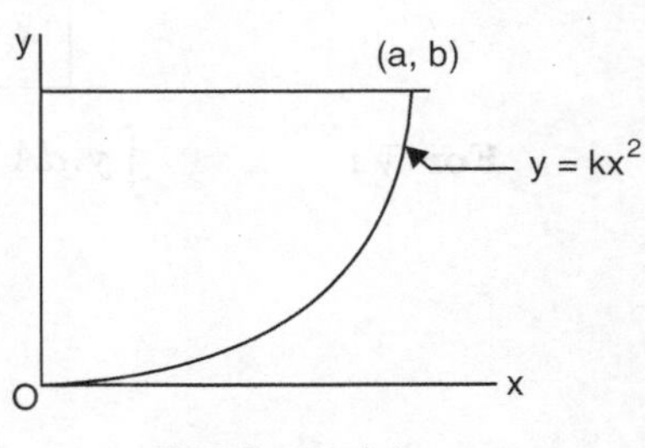

Fig. P. 3.8 (a)

Consider a horizontal element as shown in Fig. P.3.8 (*b*).

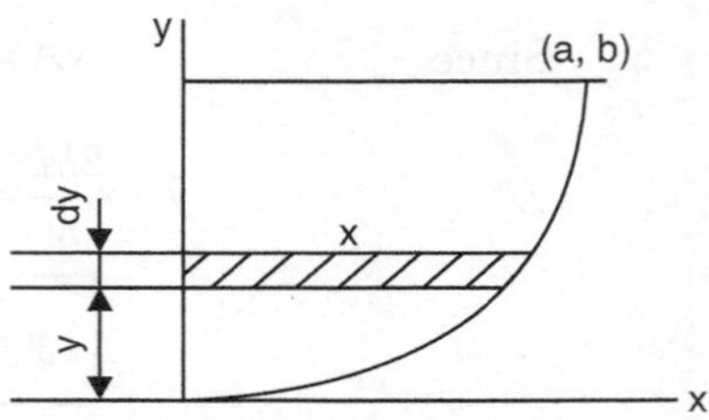

Fig. P. 3.8 (b) Horizontal element

For $\overline{\mathbf{x}}$:

Area of element = $dA = x.dy$

From eqn. (2), $y = \dfrac{b}{a^2}.x^2$, we get

$$dy = \frac{2b}{a^2}\,x.dx$$

$$\therefore \quad dA = x.dy = (x) \times \left\{\frac{2b}{a^2}\,.\,x\,.\,dx\right\}$$

or
$$dA = \frac{2b}{a^2}\,x^2\,dx$$

$$\therefore \quad A = \int dA = \int_0^a \frac{2b}{a^2}\,.\,x^2\,dx \qquad ...(3)$$

$$= \frac{2b}{a^2}.\frac{a^3}{3}$$

or
$$A = \frac{2}{3} \times ab \qquad ...(4)$$

And
$$\int x.dA = \int\left(\frac{x}{2}\right) \times dA \int_0^a \left(\frac{x}{2}\right)\left(\frac{2b}{a^2}.x^2\,dx\right) \qquad \text{from (3)}$$

$$= \frac{b}{a^2}\int_0^a x^3.dx = \frac{b}{a^2} \times \frac{a^4}{4}$$

$$\therefore \quad \int x\,dA = \frac{a^2 b}{4} \qquad ...(5)$$

$$\because \quad \bar{x}A = \int x\,dA$$

$$\therefore \quad \bar{x} \times \frac{2}{3}\,ab = \frac{a^2 b}{4}$$

$$\therefore \quad \boxed{\bar{x} = \frac{3a}{8}}$$

For $\overline{\mathbf{y}}$:
$$\int y.dA = \int y \times \left(\frac{2b}{a^2}\,x^2\,dx\right)$$

$$= \int_0^a \left(\frac{b}{a^2}\,x^2\right)\left(\frac{2b}{a^2}\,x^2\,dx\right) \qquad \text{from (2) and (3)}$$

$$= \frac{2b^2}{a^4}\left[\frac{x^5}{5}\right]_0^a = \frac{2b^2}{a^4}\,\frac{a^5}{5}$$

or $$\int y.dA = \frac{2}{5}.ab^2 \qquad ...(6)$$

$\because$ $$\bar{y}A = \int y.dA$$

$\therefore$ $$\bar{y} \times \frac{2}{3}ab = \frac{2}{5} \times ab^2$$

$\therefore$ $$\boxed{\bar{y} = \frac{3b}{5}.}$$

Problem 3.9. *Determine the position of the centroid of area formed by the intersection of a parabola $ay = x^2$ and a straight line $y = x$. See Fig. P.3.9.*

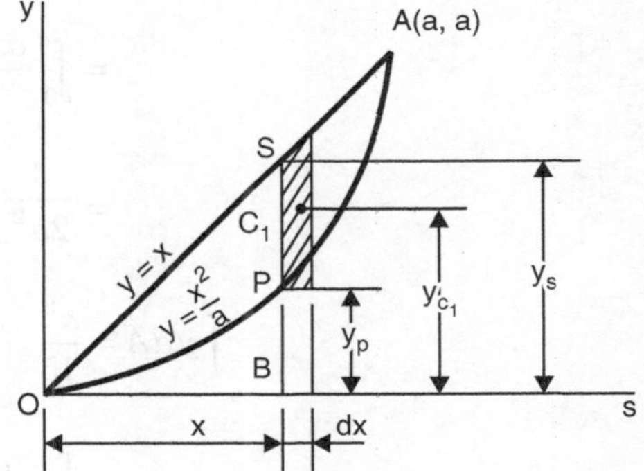

Fig. P. 3.9

Sol. The given equations of parabola and the straight line are :

For parabola, $y = \dfrac{x^2}{a}$ ...(*i*)

and for straight line, $y = x$ ...(*ii*)

Solving (*i*) and (*ii*), we find that

Points of intersection A has co-ordinates (a, a) and other point of intersection is 0.

The element shown has area $dA = PS \times dx$

$\therefore$ $$dA = (BS - BP)\,dx$$

or $$dA = (y_s - y_P).dx \qquad ...(iii)$$

Now, $$BS = y_s = x$$

and $$BP = y_P = \frac{x^2}{a}$$

$\therefore$ $$dA = \left(x - \frac{x^2}{a}\right).dx$$

$\therefore$ $$A = \int dA = \int_0^a \left(x - \frac{x^2}{a}\right).dx = \left[\frac{x^2}{2} - \frac{x^3}{3a}\right]_0^a = \frac{3a^2 - 2a^2}{6}$$

or $$A = \frac{a^2}{6} \qquad ...(iv)$$

And, $$\int x\,dA = \int_0^a x\left(x - \frac{x^2}{a}\right).dx$$

$$= \int_0^a \left(x^2 - \frac{x^3}{a} \right) . dx = \frac{1}{a} \left[\frac{ax^3}{3} - \frac{x^4}{4} \right]_0^a = \frac{1}{12\,a} (4a^4 - 3a^4)$$

$\therefore$ $$\int x dA = \frac{a^3}{12} \qquad ...(v)$$

The distance $$yc_1 = \frac{y_p + y_s}{2} = \frac{x + \dfrac{x^2}{a}}{2} = \frac{ax + x^2}{2a}$$ (See Fig. P.3.9)

And, $$\int y dA = \int y_{c_1}\, dA = \int \left(\frac{ax + x^2}{2a} \right) \left(x - \frac{x^2}{a} \right) . dx$$

$$= \int_0^a \frac{ax + x^2}{2a} \times \frac{(ax - x^2)}{a} \,.\, dx = \frac{1}{2a^2} \int_0^a (a^2 x^2 - x^4)\, dx$$

$$= \frac{1}{2a^2} \left[\frac{a^5}{3} - \frac{a^5}{5} \right] = \frac{2a^5}{2 \times 15a^2} = \frac{a^3}{15}$$

$\therefore$ $$\int y\, dA = \frac{a^3}{15} \qquad ...(vi)$$

$\therefore$ $$\bar{x} = \frac{\int x\, dA}{A} = \frac{\dfrac{a^3}{12}}{\dfrac{a^2}{6}} \quad \text{from } (v)$$

$\therefore$ $$\boxed{\bar{x} = \frac{a}{2}.}$$ **Ans.**

and from (*vi*) $$\bar{y} = \frac{\int y dA}{A} = \frac{\dfrac{a^3}{15}}{\dfrac{a^2}{6}} = \frac{6}{15} a$$

or $$\bar{y} = \frac{2a}{5}$$

or $$\boxed{\bar{y} = \frac{2a}{5}.}$$ **Ans.**

$\therefore$ Centroid is, $C\left(\dfrac{a}{2}, \dfrac{2a}{5} \right)$. **Ans.**

Problem 3.10. *Find the position of centroid of the angle section shown in Fig. P.3.10.*

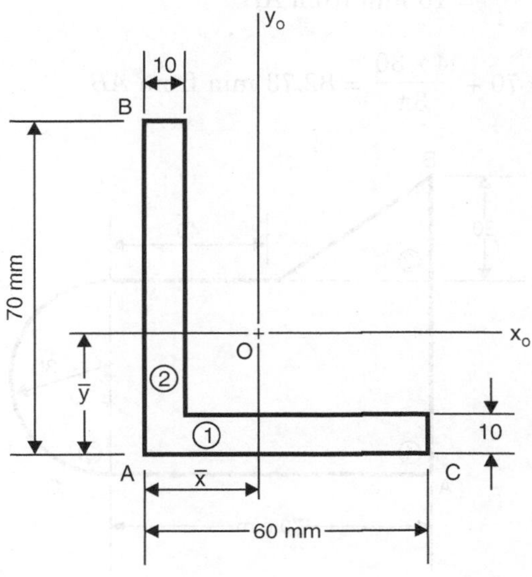

Fig. P.3.10

Sol. Area $A_1 = 60 \times 10 = 600 \text{ mm}^2$

$A_2 = (70 - 10) \times 10 = 600 \text{ mm}^2$

$x_1 = \dfrac{60}{2} = 30$ mm from AB

$x_2 = \dfrac{10}{2} = 5$ mm from AB

$y_1 = \dfrac{10}{2} = 5$ mm from AC

$y_2 = 10 + \dfrac{(70 - 10)}{2} = 40$ mm from AC

$\therefore$ $A = A_1 + A_2 = 600 + 600 = 1200 \text{ mm}^2$

$\therefore$ $\bar{x} = \dfrac{A_1x_1 + A_2x_2}{A_1 + A_2} = \dfrac{600 \times 30 + 600 \times 5}{600 + 600}$ = **17.5 mm** from AB. **Ans.**

and, $\bar{y} = \dfrac{A_1y_1 + A_2y_2}{A_1 + A_2} = \dfrac{600 \times 5 + 600 \times 40}{600 + 600}$ = **22.5 mm** from AC. **Ans.**

Problem 3.11. *Locate centroid of the area shown in Fig. P.3.11.*

Sol. $A_1 = 70 \times 60 = 4200 \text{ mm}^2$

$A_2 = \dfrac{30 \times 30}{2} = 450 \text{ mm}^2$

$A_3 = \dfrac{\pi \times 30^2}{2} = 141.39 \text{ mm}^2$

$x_1 = 35$ mm from AB

$$x_2 = \frac{30}{3} = 10 \text{ mm from } AB$$

$$x_3 = 70 + \frac{4 \times 30}{3\pi} = 82.73 \text{ mm from } AB$$

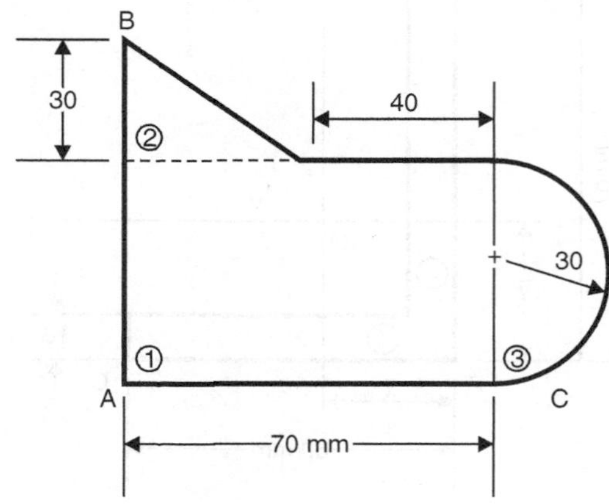

Fig. P.3.11

$$y_1 = \frac{60}{2} \text{ mm} = 30 \text{ mm from } AC$$

$$y_2 = 60 + \frac{30}{3} = 70 \text{ mm from } AC$$

$$y_3 = 30 \text{ mm from } AC$$

$\therefore$ $$A = A_1 + A_2 + A_3 = 4200 + 450 + 141.39 = 4791.39 \text{ mm}^2$$

$$\bar{x} = \frac{A_1x_1 + A_2x_2 + A_3x_3}{A_1 + A_2 + A_3} = \frac{4200 \times 35 + 450 \times 10 + 141.39 \times 82.73}{4791.39}$$

$$= \frac{163197.1947}{4791.39} = \mathbf{34.06 \text{ mm}} \text{ from } AB. \quad \textbf{Ans.}$$

$$\bar{y} = \frac{A_1y_1 + A_2y_2 + A_3y_3}{A_1 + A_2 + A_3} = \frac{4200 \times 30 + 450 \times 70 + 141.39 \times 30}{4791.39}$$

$$= \frac{161741.7}{4791.39} = \mathbf{33.76 \text{ mm}} \text{ from } AC. \quad \textbf{Ans.}$$

Problem 3.12. *Locate centroid of the area shown in Fig. P.3.12.*

Sol. $$A_1 = \frac{1}{2} \times 24 \times 90 = 1080 \text{ mm}^2$$

$$A_2 = 30 \times 90 = 2700 \text{ mm}^2$$

$$A_3 = \frac{1}{2} \times (110 - 54) \times 90 = 2520 \text{ mm}^2$$

$$A_4 = -\frac{\pi \times 40^2}{4} = -400\,\pi = -1256.8 \text{ mm}^2$$

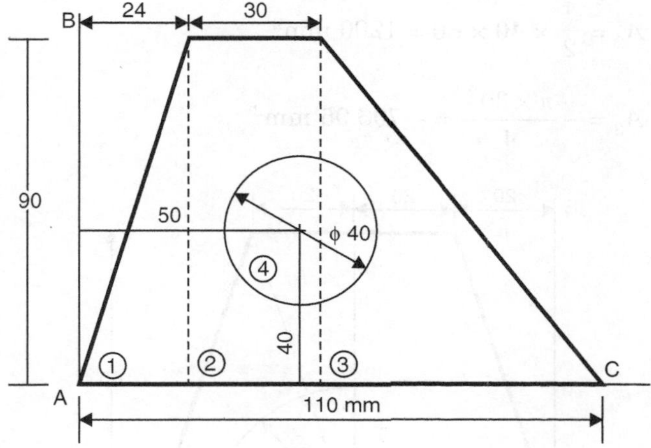

Fig. P.3.12

Note. A_4 is taken negative because of removal of area by drilling a hole of 40 mm diameter.

From the reference line *AB*,

$$x_1 = \frac{2}{3} \times 24 = 16 \text{ mm} \; ; \quad x_2 = 24 + 30 \times \frac{1}{2} = 39 \text{ mm}$$

$$x_3 = 24 + 30 + \frac{1}{3}(110 - 54) = 72.67 \text{ mm}$$

$$x_4 = 50 \text{ mm}$$

From the reference line *AC*,

$$y_1 = \frac{1}{3} \times 90 = 30 \text{ mm} \; ; \quad y_2 = 90 \times \frac{1}{2} = 45 \text{ mm}$$

$$y_3 = \frac{1}{3} \times 90 = 30 \text{ mm} \; ; \quad y_4 = 40 \text{ mm}$$

$$A = A_1 + A_2 + A_3 + A_4 = 1080 + 2700 + 2520 + (-1256.8)$$
$$= 5043.2 \text{ mm}^2$$

$$\therefore \quad \bar{x} = \frac{A_1x_1 + A_2x_2 + A_3x_3 + A_4x_4}{A_1 + A_2 + A_3 + A_4} = \frac{1}{5043.2} \times [1080 \times 16 + 2700 \times 39 + 2520 \times 72.67 + (-1256.8) \times 50] \text{ mm}$$

$$= \frac{242868.4}{5043.2} = \mathbf{48.16} \text{ mm from AB.} \quad \textbf{Ans.}$$

and

$$\bar{y} = \frac{A_1y_1 + A_2y_2 + A_3y_3 + A_4y_4}{A_1 + A_2 + A_3 + A_4} = \frac{1080 \times 30 + 2700 \times 45 + 2520 \times 30 + (-1256.8) \times 40}{5043.2}$$

$$= \frac{179228}{5043.2} = \mathbf{35.54 \text{ mm}} \text{ from } AC. \quad \textbf{Ans.}$$

Problem 3.13. *Locate centroid of the area shown in Fig. P.3.13.*

Sol. The given area can be divided into one parallelogram (40 × 60) ; one triangle (40 × 60) and one quarter circle of radius 30 mm.

$$A_1 = 40 \times 60 = 2400 \text{ mm}^2$$

$$A_2 = \frac{1}{2} \times 40 \times 60 = 1200 \text{ mm}^2$$

$$A_3 = -\frac{\pi \times 30^2}{4} = -706.95 \text{ mm}^2$$

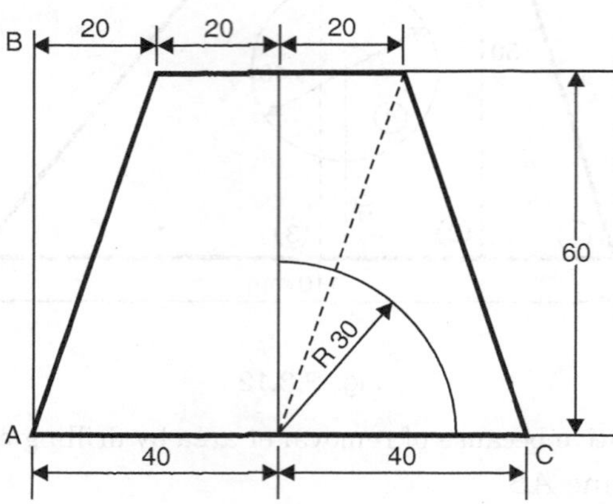

Fig. P.3.13

$$x_1 = \frac{1}{2}(20 + 40) = 30 \text{ mm from } AB,$$

$$x_2 = 40 + \frac{40}{2} = 60 \text{ mm from } AB,$$

$$x_3 = 40 + \frac{4 \times 30}{3\pi} = 52.73 \text{ mm from } AB$$

$$y_1 = 30 \text{ mm from } AC,$$

$$y_2 = \frac{60}{3} = 20 \text{ mm from } AC$$

$$y_3 = \frac{4 \times 30}{3\pi} = 12.73 \text{ mm from } AC$$

$$A = A_1 + A_2 + A_3 = 2400 + 1200 + (-706.95) = 2893.05$$

$$\therefore \quad \bar{x} = \frac{A_1x_1 + A_2x_2 + A_3x_3}{A_1 + A_2 + A_3} = \frac{2400 \times 30 + 1200 \times 60 + (-706.95) \text{ mm}^2 \times 52.73}{2893.05}$$

$$= \frac{106722.5265}{2893.05} = \mathbf{36.89} \text{ mm from } AB. \quad \textbf{Ans.}$$

and, $$\bar{y} = \frac{A_1y_1 + A_2y_2 + A_3y_3}{A_1 + A_2 + A_3} = \frac{2400 \times 30 + 1200 \times 20 + (-706.95) \times 12.73}{2893.05}$$

$$= \frac{87000.5265}{2893.05} = \mathbf{30.07 \text{ mm}} \text{ from } AC. \quad \textbf{Ans.}$$

Problem 3.14. *A square hole is cut out of a circular plate of diameter D as shown in the Fig. P.3.14. Show that the centroid of the remainder area is at a distance of* $\dfrac{4\pi - 3}{8\pi - 4} \times D$ *from the extreme left end of the horizontal diameter of the plate.*

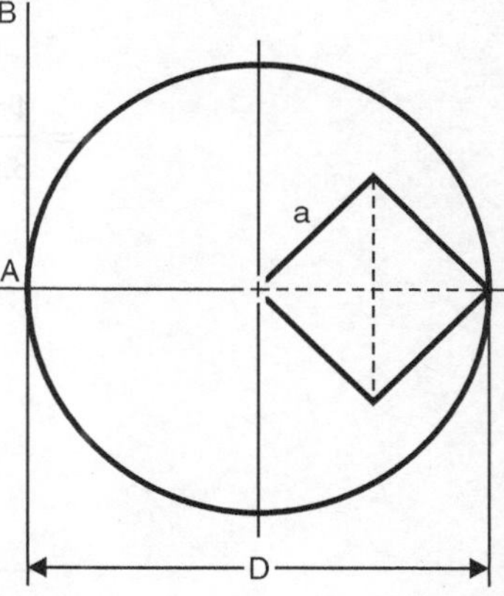

Fig. P.3.14

Sol. For the circular plate

$$A_1 = \frac{\pi D^2}{4} \; ; x_1 = D/2 \text{ from } AB.$$

For the square hole :

Sides of the hole a is given by

$$a\sqrt{2} = D/2 \; ; \quad \therefore \quad a = \frac{D}{2} \times \frac{1}{\sqrt{2}}$$

$$\therefore \qquad A_2 = -a^2 = -\left(\frac{D}{2\sqrt{2}}\right)^2 = -\left(\frac{D}{8}\right)^2$$

$$x_2 = \frac{D}{2} + \frac{D}{2} \times \frac{1}{2} = 3D/4$$

$$A = A_1 + A_2 = \frac{\pi D^2}{4} - \frac{D^2}{8} = \frac{D^2}{8}(2\pi - 1)$$

$$\therefore \qquad \bar{x} = \frac{A_1 x_1 + A_2 x_2}{A_1 + A_2} = \frac{\pi/4 \times D^2 \times D/2 + \left(-\dfrac{D^2}{8}\right)(3D/4)}{\dfrac{D^2}{8}(2\pi - 1)}$$

$$\therefore \qquad \bar{x} = \frac{D^3}{32} \times \frac{4\pi - 3}{\dfrac{D^2}{8}(2\pi - 1)} = \frac{4\pi - 3}{8\pi - 4} D \quad \text{from } AB \left(= \frac{4\pi - 3}{4\pi - 2} \right) R. \quad \textbf{Ans.}$$

Problem 3.15. *A circular hole of radius R/2 is cut off a circular plate of radius R. Find the distance between the centres of the plate and the hole, if the centre of the mass of the remainder lies on the extreme left point on the circumference of the hole.*

Sol. See Fig. P.3.15. Let the distance between the centre O_1 of the plate and the centre O_2 of the hole be C.

Then $\quad A_1$ = area of plate = πR^2

$$A_2 = \text{area of hole} = -\pi(R/2)^2 = -\frac{\pi R^2}{4}$$

$x_1 = AO_1 = R$ from reference point A, and

$x_2 = AO_2 = R + C$ from point A.

Then, $$\bar{x} = OA = AO_2 - R/2 = (R + C) - R/2 = \frac{R}{2} + C$$

$$\therefore \qquad A = A_1 + A_2 = \pi R^2 - \frac{\pi R^2}{4} = \frac{3\pi R^2}{4}$$

$$\therefore \quad \bar{x} = \frac{A_1x_1 + A_2x_2}{A_1 + A_2} = \frac{\pi R^2 \times R + \left(-\dfrac{\pi R^2}{4}\right) \times (R + C)}{\dfrac{3\pi R^2}{4}}$$

$$= \frac{4\pi R^2}{3\pi R^2} \times \left(R - \frac{R+C}{4}\right) = \frac{4}{3} \times \frac{4R - R - C}{4} = \frac{3R - C}{3}$$

Fig. P.3.15

$$\therefore \quad \bar{x} = \frac{3R - C}{3}. \quad ...(1)$$

But $$\bar{x} = \frac{R}{2} + C \quad ...(2)$$

Equating, we get $\dfrac{R}{2} + C = \dfrac{3R - C}{3}$, $\therefore \quad C = \dfrac{3}{8} R$. **Ans.**

Problem 3.16. *Locate C.G. of the plane area ABCDEE$_1$F$_1$F shown in Fig. P.3.16, such that CE = EF = FD = a/3.*

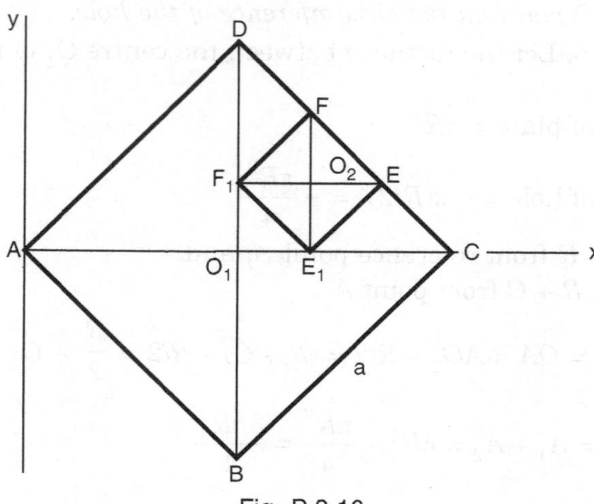

Fig. P.3.16

Sol. Area of bigger square = $A_1 = a^2$; $x_1 = a/\sqrt{2}$ from y-axis, and $y_1 = 0$ from x-axis.

Area of the smaller square = $A_2 = -(a/3)^2 = -\dfrac{a^2}{9} = \dfrac{A_1}{9}$.

$$x_2 = AO_1 + O_1E_1 = a/\sqrt{2} + a/3\sqrt{2} = \frac{a}{\sqrt{2}}\left(\frac{4}{3}\right) = \frac{2\sqrt{2}\,a}{3} \text{ from } y\text{-axis}$$

$$y_2 = E_1O_2 = \frac{a/3}{\sqrt{2}} = \frac{a}{3\sqrt{2}} \text{ from } x\text{-axis.}$$

$$\therefore \qquad A = A_1 + A_2 = a^2 - \frac{a^2}{9} = \frac{8a^2}{9}$$

$$\bar{x} = \frac{A_1x_1 + A_2x_2}{A_1 + A_2} = \frac{a^2 . a/\sqrt{2} - (a^2/9) \times \dfrac{2\sqrt{2}a}{3}}{8a^2/9} = \frac{a}{27\sqrt{2}} \frac{[27 - 2 \times 2]}{\dfrac{8}{9}}$$

$$= \frac{23a \times 9}{27 \times \sqrt{2} \times 8} = \frac{23}{24\sqrt{2}} a \text{ from } y\text{-axis.} \quad \textbf{Ans.}$$

or

$$\bar{y} = \frac{A_1y_1 + A_2y_2}{A_1 + A_2} = \frac{a^2 \times (0) - \dfrac{a^2}{9} \times \dfrac{a}{3\sqrt{2}}}{\dfrac{8a^2}{9}} = \frac{-a}{24\sqrt{2}} \text{ from } x\text{-axis.} \quad \textbf{Ans.}$$

Problem 3.17. *Locate centroid of area shown in Fig. P.3.17.*

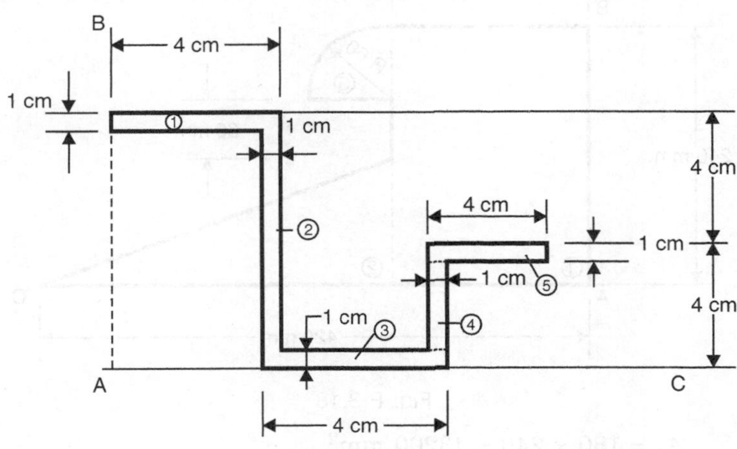

Fig. P.3.17

Sol. $A_1 = 4 \times 1 = 4 \text{ cm}^2$; $A_2 = (8 - 2) \times 1 = 6 \text{ cm}^2$; $A_3 = 4 \times 1 \text{ cm}^2 = 4 \text{ cm}^2$

$A_4 = (4 - 2) \times 1 = 2 \text{ cm}^2$; $A_5 = 4 \times 1 = 4 \text{ cm}^2$

$\therefore \quad A = A_1 + A_2 + A_3 + A_4 + A_5 = 4 + 6 + 4 + 2 + 4 = 20 \text{ cm}^2$

From line AB : $\quad x_1 = 2$ cm ; $x_2 = 4 - 0.5 = 3.5$ cm ; $x_3 = (4 - 1) + \dfrac{4}{2} = 5$ cm.

$$x_4 = (4 + 4 - 1) - 0.5 = 6.5 \text{ cm}, \; x_5 = (4 + 4 - 1 - 1) + \frac{4}{2} = 8 \text{ cm}$$

From line AC : $\quad y_1 = 8 - 0.5 = 7.5$ cm ; $\; y_2 = 1 + \dfrac{6}{2} = 4$ cm ; $\; y_3 = 0.5$ cm,

$$y_4 = 1 + \frac{2}{2} = 2 \text{ cm} ; \quad y_5 = 1 + 2 + 0.5 = 3.5 \text{ cm}$$

$$\therefore \quad A = A_1 + A_2 + A_3 + A_4 + A_5 = 4 + 6 + 4 + 2 + 4 = 20 \text{ cm}^2$$

$$\therefore \quad \bar{x} = \frac{A_1x_1 + A_2x_2 + A_3x_3 + A_4x_4 + A_5x_5}{A_1 + A_2 + A_3 + A_4 + A_5}$$

$$= \frac{1}{20}[4 \times 2 + 6 \times 3.5 + 4 \times 5 + 2 \times 6.5 + 4 \times 8] \text{ cm}$$

$$= \frac{86}{20} = \mathbf{4.3 \text{ cm}} \text{ from } AB. \quad \textbf{Ans.}$$

and $$\bar{y} = \frac{A_1y_1 + A_2y_2 + A_3y_3 + A_4y_4 + A_5y_5}{A_1 + A_2 + A_3 + A_4 + A_5}$$

$$= \frac{1}{20}[4 \times 7.5 + 6 \times 4 + 4 \times 0.5 + 2 \times 2 + 4 \times 3.5]$$

$$= \frac{74}{20} = \mathbf{3.7 \text{ cm}} \text{ from } AC. \quad \textbf{Ans.}$$

Problem 3.18. *Locate centroid of the area shown in Fig. P.3.18.*

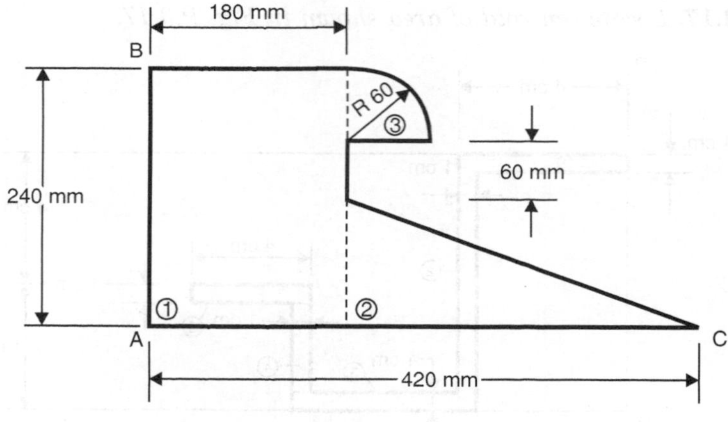

Fig. P.3.18

Sol. $$A_1 = 180 \times 240 = 43200 \text{ mm}^2$$

$$A_2 = \frac{1}{2} \times (420 - 180) \times (240 - 60 - 60) = 120 \times 120$$

$$= 14{,}400 \text{ mm}^2$$

$$A_3 = \frac{\pi \times 60^2}{4} = 900\,\pi \text{ mm}^2 = 2827.43 \text{ mm}^2$$

From AB : $x_1 = 90$ mm ; $x_2 = 180 + (420 - 180) \times \frac{1}{3} = 260$ mm

$$x_3 = 180 + \frac{4 \times 60}{3\pi} = 180 + 25.465 \text{ mm} = 205.465 \text{ mm.}$$

$$A = A_1 + A_2 + A_3 = 43200 + 14400 + 2827.43 = 60427.43 \text{ mm}^2$$

$$\bar{x} = \frac{A_1x_1 + A_2x_2 + A_3x_3}{A_1 + A_2 + A_3} = \frac{43200 \times 90 + 14400 \times 260 + 2827.43 \times 205.465}{60427.43}$$

$$= \frac{8212937.905}{60427.43} = \mathbf{135.914 \text{ mm}} \text{ from } AB. \textbf{ Ans.}$$

Form line AC : $y_1 = \frac{240}{2} = 120$ mm ; $y_2 = \frac{240 - 120}{3} = 40$ mm ;

$$y_3 = (240 - 120) + 60 + \frac{4 \times 60}{3\pi} = 180 + 25.465 = 205.465 \text{ mm.}$$

$$\therefore \quad \bar{y} = \frac{A_1y_1 + A_2y_2 + A_3y_3}{A_1 + A_2 + A_3} = \frac{43200 \times 120 + 14400 \times 40 + (2827.43 \times 205.465)}{60427.43}$$

$$= \frac{6340937.905}{60427.43} = \mathbf{104.935 \text{ mm}} \text{ from } AC. \textbf{ Ans.}$$

Problem 3.19. *Locate centroid of the area shown in Fig. P.3.19.*

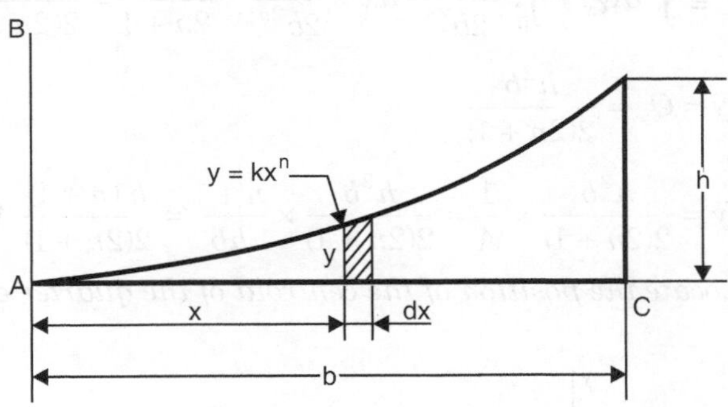

Fig. P.3.19

Sol. Given : $y = kx^n$ ...(1)

At C, $x = b$ and $y = h$,

$\therefore$ We get $h = kb^n$; $\therefore$ $k = h/b^n$ ...(2)

$\therefore$ equation (1), yields, $y = \frac{h}{b^n} x^n$ ...(3)

Area of element $= dA = ydx = \frac{h}{b^n} x^n dx$

$\therefore$ Total area $= A = \int dA = \int_0^b \frac{h}{b^n} x^n dx = \frac{h}{b^n} \times \frac{b^{n+1}}{n+1}$

$$\therefore \quad \boxed{A = \frac{hb}{n+1}} \quad \text{...(4)}$$

Moment of the element about $AB = xdA$

$$= dQ_y = x.\frac{h}{b^n}.x^n dx = \frac{h}{b^n}.x^{n+1}dx$$

$$\therefore \quad \int dQ_y = Q_y = A\bar{x}$$

$$\therefore \quad A\bar{x} = \int_0^b \frac{h}{b^n}.x^{n+1}\,dx = \frac{hb^{n+2}}{b^n \times (n+2)} = \frac{hb^2}{n+2}$$

$$\therefore \quad \bar{x} = \frac{hb^2/(n+2)}{A} = \frac{hb^2/(n+2)}{hb/(n+1)}$$

$$\therefore \quad \bar{x} = \frac{b\,(n+1)}{(n+2)} \text{ from } AB. \quad \textbf{Ans.}$$

and

$$\bar{x}_c = b - \bar{x} = b - \frac{b(n+1)}{n+2} = \frac{b}{n+2} \quad \textbf{Ans.}$$

Moment of the element about $AC = \frac{y}{2}.dA$

$$= dQ_x = \frac{y}{2} \times \frac{h}{b^n}x^n dx = \frac{1}{2}\left(\frac{hx^n}{b^n}\right).\frac{h}{b^n}.x^n.dx$$

$$\therefore \quad Q_x = \int dQ_x = \int_0^b \frac{h^2x^{2n}}{2b^{2n}}dx = \frac{h^2}{2b^{2n}} \times \frac{b^{2n+1}}{2n+1} = \frac{h^2b}{2(2n+1)}$$

or

$$A\bar{y} = Q_x = \frac{h^2b}{2(2n+1)}$$

$$\therefore \quad \bar{y} = \frac{h^2b}{2(2n+1)} \times \frac{1}{A} = \frac{h^2b}{2(2n+1)} \times \frac{n+1}{hb} = \frac{h\,(n+1)}{2(2n+1)} \text{ from } AC. \quad \textbf{Ans.}$$

Problem 3.20. *Locate the position of the centroid of the quarter elliptical lamina shown in Fig. P.3.20.*

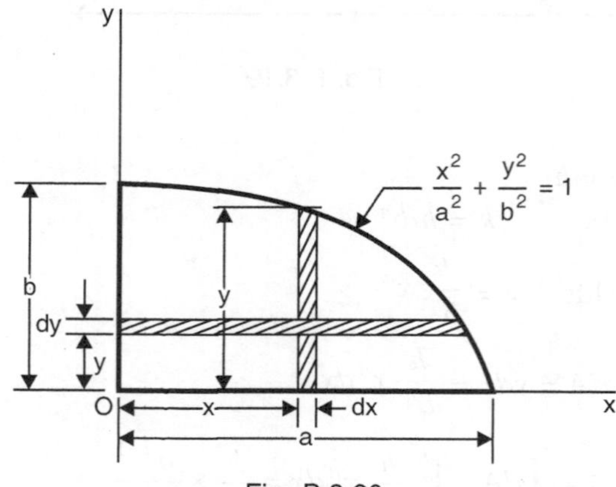

Fig. P.3.20

Sol. For $\bar{x}$, consider a vertical element of area $= dA = ydx$

from the equation of $\frac{x^2}{a^2} + \frac{y^2}{b^2} = 1$

$$\therefore \quad \frac{y^2}{b^2} = 1 - \frac{x^2}{a^2} = \frac{a^2 - x^2}{a^2}$$

$$\therefore \quad y = \frac{b}{a}\sqrt{a^2 - x^2}$$

$$\therefore \quad dA = ydx = \frac{b}{a}\sqrt{a^2 - x^2}\, dx$$

$\therefore$ Area,
$$A = \int dA = \int_0^a \frac{b}{a}\sqrt{a^2 - x^2}\, dx = \frac{b}{a}\left[\frac{x}{2}\sqrt{a^2 - x^2} + \frac{a^2}{2}\sin^{-1}\frac{x}{a}\right]_0^a$$

or
$$A = \frac{b}{a}\left[0 + \frac{a^2}{2}\frac{\pi}{2}\right] = \frac{ab\pi}{4}$$

$$\int xdA = \int_0^a xydx = \int_0^a x\frac{b}{a}\sqrt{a^2 - x^2}\,.dx$$

$$= \frac{b}{a}\int_0^a x\sqrt{a^2 - x^2}\, dx = \frac{b}{a}\left[\int_0^a\left(-\frac{1}{2}\right)\sqrt{a^2 - x^2} \times (-2x)\, dx\right]$$

$$= \frac{-b}{2a} \times \frac{2}{3}\left[\left\{a^2 - x^2\right\}^{3/2}\right]_0^a = \frac{b \times a^3}{3a} = \frac{a^2 b}{3}$$

$$\therefore \quad \bar{x} = \frac{\int x\,dA}{A} = \frac{\frac{a^2 b}{3}}{\frac{ab\pi}{4}} = \frac{4a}{3\pi}$$

or
$$\boxed{\bar{x} = \frac{4a}{3\pi}}$$

For $\bar{y}$ consider a horizontal element of area $dA = x\, dy$

From the equation

$$\frac{x^2}{a^2} + \frac{y^2}{b^2} = 1, \quad x = a\sqrt{1 - \frac{y^2}{b^2}}$$

or
$$x = \frac{a}{b}\sqrt{b^2 - y^2}$$

$$\therefore \quad dA = xdy = \frac{a}{b}\sqrt{b^2 - y^2}\, dy$$

and
$$\int ydA = \int_0^b y.\frac{a}{b}\sqrt{b^2 - y^2}\, dy = \frac{a}{b}.\left(-\frac{1}{2}\right)\int_0^b \sqrt{b^2 - y^2} \times (-2y)\, dy$$

$$= +\frac{a}{b}\int_0^b (b^2 - y^2)^{1/2}\, dy = +\frac{a}{2} \times \frac{2}{3b}\left[\left(b^2 - y^2\right)_0^{3/2}\right]_0^b$$

$$= \frac{a}{3b}[(b^2)^{3/2}] = \frac{ab^3}{3b} = \frac{ab^2}{3} = \bar{y}A$$

$$\therefore \qquad \bar{y} = \frac{\int y\,dA}{A} = \frac{ab^2/3}{\frac{\pi ab}{4}} = \frac{4b}{3\pi}$$

or
$$\boxed{\bar{y} = \frac{4b}{3\pi}}$$

Problem 3.21. *Locate the centroid of the wire of uniform cross-section shown in Fig. P.3.21 with respect to X and Y-axes.*

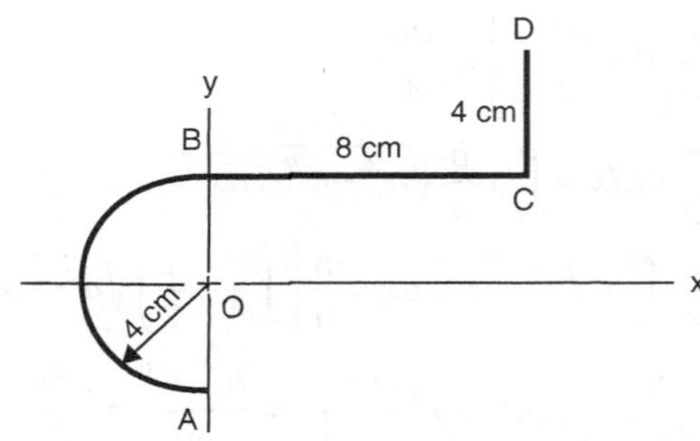

Fig. P.3.21

Sol. The wire is considered equivalent to three line segments : a semi-circle of 4 cm radius ; a straight horizontal line *BC*, 8 cm long and a straight vertical line *CD*, 4 cm long, as shown in Fig. P.3.21.

$$L = L_1 + L_2 + L_3 = \pi \times 4 + 8 + 4 = 4\pi + 12 \text{ cm} = 24.57 \text{ cm}$$

with respect to y-axis :

$$x_1 = -\frac{2 \times 4}{\pi} = -\frac{8}{\pi} \text{ cm on the left of } y\text{-axis.}$$

x_2 = 8/2 cm = 4 cm the right of y-axis.

x_3 = 8 cm on the right of y-axis

with respect to x-axis :

y_1 = 0 *i.e.*, on the x-axis

y_2 = 4 cm above x-axis

$$y_3 = 4 + \frac{4}{2} = 6 \text{ cm above } x\text{-axis}$$

$$\therefore \qquad \bar{x} = \frac{L_1x_1 + L_2x_2 + L_3x_3}{L_1 + L_2 + L_3} = \frac{(4\pi) \times (-8/\pi) + 8 \times 4 + 4 \times 8}{24.57} \text{ cm} = \frac{32}{24.57} \text{ cm}$$

$\therefore$ $\bar{x}$ = **1.30 cm** on the right of y-axis. **Ans.**

and
$$\bar{y} = \frac{L_1y_1 + L_2y_2 + L_3y_3}{L_1 + L_2 + L_3} = \frac{(4\pi \times 0) + (8 \times 4) + (4 \times 6)}{24.57} \text{ cm} = \frac{56}{24.57} \text{ cm}$$

$\therefore$ $\bar{y}$ = **2.28 cm. above x-axis. Ans.**

Problem 3.22. *Find the position of the centroid of the homogeneous wire bent in the form shown in Fig. P.3.22 with reference to the X- and Y-axes shown.*

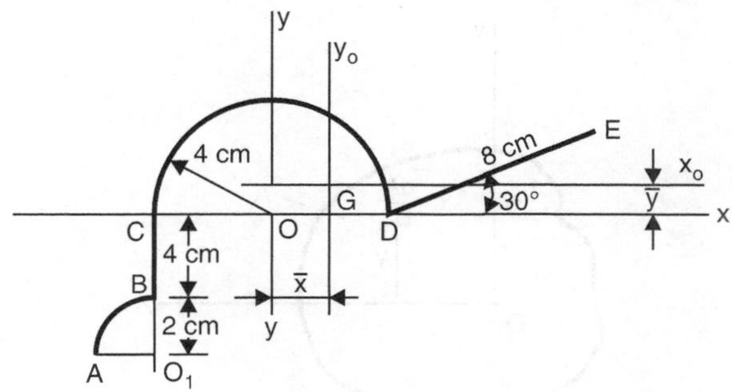

Fig. P.3.22

Sol. There are four wire segments, *viz.*, quadrant AB of radius 2 cm ; straight portion BC ; semi-circle CD and straight line DE, as shown in Fig. P.3.22.

$$L_1 = \text{arc } AB = \frac{2\pi}{2} = \pi \text{ cm}$$

$$L_2 = 4 \text{ cm} ; \quad L_3 = 4\pi \text{ cm} ; \quad L_4 = 8 \text{ cm}.$$

With respect to Y-axis :

$$x_1 = -\left(4 + \frac{2 \times 2}{\pi}\right) = -5.27 \text{ cm} ; \quad x_2 = -4 \text{ cm},$$

$$x_3 = 0 ; \quad x_4 = 4 + \frac{8}{2} \cos 30° = 7.46 \text{ cm}$$

With respect to X-axis :

$$y_1 = -\left[CB + \left(BO_1 - \frac{2 \times BO_1}{\pi}\right)\right] = -[4 + (4 - 4/\pi)] = -6.73 \text{ cm}$$

$$y_2 = -2 \text{ cm} ; \quad y_3 = +\frac{2 \times 4}{\pi} = +2.55 \text{ cm} ;$$

$$y_4 = (8/2) \sin 30° = 2 \text{ cm}.$$

$$\therefore \quad \bar{x} = \frac{l_1x_1 + l_2x_2 + l_3x_3 + l_4x_4}{l_1 + l_2 + l_3 + l_4} = \frac{\pi \times (-5.27) + 4(-4) + 4\pi \times (0) + 8 \times (7.46)}{\pi + 4 + 4\pi + 8}$$

$$= \frac{27.12}{27.71} = \mathbf{0.98} \text{ cm. on the right of } \mathbf{y\text{-axis}}$$

$$\bar{y} = \frac{l_1y_1 + l_2y_2 + l_3y_3 + l_4y_4}{l_1 + l_2 + l_3 + l_4} = \frac{\pi \times (-6.73) + 4 \times (-2) + (4\pi)\left(\frac{2 \times 4}{\pi}\right) + 8 \times 2}{27.71}$$

$$= \frac{18.857}{27.71} = +\mathbf{0.68 \text{ cm}} \text{ above X-axis.} \quad \textbf{Ans.}$$

Point G (0.98, 0.68) represents the centroid of the given wire with respect to X and Y-axes. **Ans.**

3.6. MOMENT OF INERTIA

Moment of inertia of a plane area is defined as follows :

Consider an area A and an elemental area dA at a distance x from y-axis and y from x-axis, as shown in Fig. 3.6.

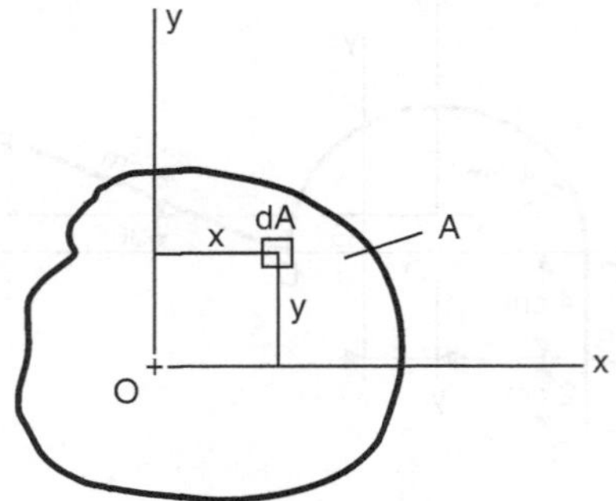

Fig. 3.6. Definitions of moment of inertia.

Then the second moment of the dA about x-axis is defined as dI_x.

Thus, $dI_x = y^2.dA$

$$\therefore \quad \int dI_x = I_x = \int_A y^2 dA \quad \text{or} \quad \boxed{I_x = \int_A y^2 dA} \qquad ...(3.20)$$

Similarly, $dI_y = x^2 dA$

$$\therefore \quad \boxed{\int dI_y = I_y = \int_A x^2 dA} \qquad ...(3.21)$$

Sometimes, moment of inertia of area is called the *second moment of area.*

3.7. UNITS AND SIGN OF MOMENT OF INERTIA

It is easy to find that units of moment of inertia of area is m^4 or mm^4 in SI units ; cm^4 in M.K.S. units ; in^4 in FBS units.

The sign of I is always positive because x^2 or y^2 is always positive. But, when some part is removed or cut from the given area, then the, area removed or cut is negative.

3.8. PARALLEL AXES THEOREM

Consider an area A whose centroid is C. Then X_0 and Y_0 are centroidal axes. Also, let X-axis be parallel to X_0 axis and at a distance h from it.

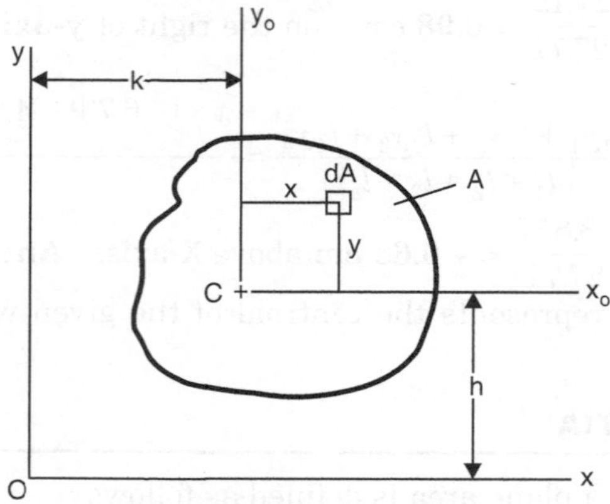

Fig. 3.8. Parallel axes theorems.

Similarly, Y-axis is parallel to Y_0-axis and at a distance k from it.

Consider an elementary area dA at x from Y_0-axis and y from X_0-axis, as shown in Fig. 3.8.

Then by definition (eqn. 3.19), we can write

$$\int dI_x = I_x = \int_A (y+h)^2 .dA$$

or $$I_x = \int_A (y^2 + 2hy + h^2).dA$$

or $$I_x = \int_A y^2 dA + 2h\int_A ydA + \int_A h^2 dA \qquad ...(3.22)$$

$\because$ $$\int_A y^2 dA = I_{x_0} \text{ or } \bar{I}_x$$

also $$\int_A ydA = A\bar{y} \text{ and } \int_A dA = A,$$

$\therefore$ We have $I_x = I_{x_0} + 2h(A\bar{y}) + Ah^2$

Since y is measured from the centroidal axis X_0, $\therefore$ $\bar{y} = 0$. Hence $A\bar{y} = 0$

$\therefore$ Eqn. (3.22) yields

$$I_x = I_{x_0} + 2hA(0) + Ah^2$$

$\therefore$ $$\left.\begin{aligned} I_x &= I_{x_0} + Ah^2 \\ I_x &= \bar{I}_x + Ah^2 \end{aligned}\right\} \qquad ...(3.23)$$

or

Eqn. (3.23) is called the parallel-axis theorem. It states that *"for any area the moment of inertia I_x with respect to X-axis in the plane of the area is equal to the moment of inertia with respect to a parallel centroidal X_0-axis plus the product of the area (A) and the square of the distance h between the X_0- and X-axes".*

Similarly, the moment of inertia with respect to Y-axis is :

$$\int dI_y = I_y = \int_A (x+k)^2 dA = \int_A (x^2 + 2kx + k^2)\, dA$$

or $$I_y = \int_A x^2 dA + 2k\int_A x\, dA + k^2 \int_A dA \qquad ...(3.24)$$

Again, we know that $\int_A x^2 dA = I_{y_0}$ or $\bar{I}_y$ from eqn. (3.20) ; $\int_A x\,dA = A\bar{x}$, and since Y_0 is the centroidal axis, $\therefore$ $\bar{x} = 0$; also $\int_A dA = A$.

$\therefore$ Eqn. (3.24) yields

$$I_y = I_{y_0} + 2k(\bar{x}.A) + Ak^2 = I_{y_0} + 2k(0).A + Ak^2$$

$\therefore$ $$I_y = I_{y_0} + Ak^2 \text{ or } I_y = \bar{I}_y + Ak^2 \qquad ...(3.25)$$

Eqn. (3.25) is also called the parallel-axis equation in terms of Y_0- and Y-axes.

Eqn. (3.25) states that :

The moment of inertia I_y of any area A is equal to the moment of inertia I_{y_0} with respect to the centroidal axis Y_0 plus the product of the area A and square of the distance between the two parallel axes Y- and Y_0.

3.9. PERPENDICULAR AXES THEOREM

Consider an area A whose centroid is C. X_0 and Y_0 are the centroidal axes. The third centroidal axis is Z_0 which is perpendicular to the area A appears as a point Z_0 and is co-incident with C. Let an elemental area dA have co-ordinates x and y, as shown in Fig. 3.9.

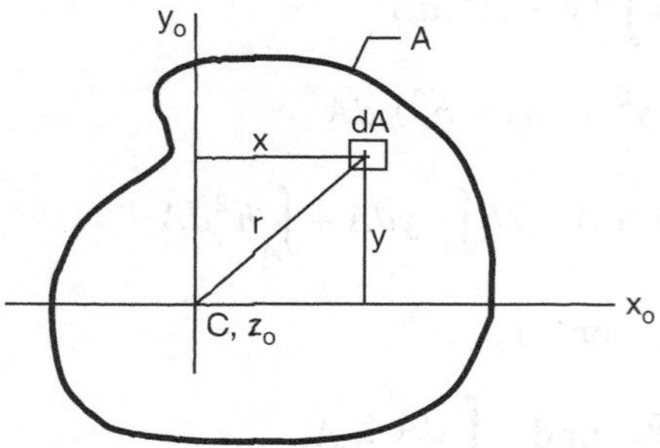

Fig. 3.9. Perpendicular axes theorem.

We know that by definition.

$$I_{y_0} = \int_A x^2\, dA \quad ...(3.26)$$

$$I_{x_0} = \int_A y^2\, dA \quad ...(3.27)$$

$\therefore$ By adding eqns. (3.26) and (3.27), we get

$$I_{y_0} + I_{x_0} = \int_A x^2\, dA + \int_A y^2\, dA = \int_A (x^2 + y^2)\, dA \quad ...(3.28)$$

Let the element dA be at the shortest distance r from the centroid or Z_0-axis, therefore, eqn. (3.28) becomes

$$I_{y_0} + I_{x_0} = \int_A r^2\, dA \qquad \text{(where } r^2 = x^2 + y^2\text{)} \quad ...(3.29)$$

$\int_A r^2\, dA$ is known as the **polar moment of inertia** and is represented by I_{z_0} or I_P or $\bar{J}$. This term is very important in the study of torsion of rotating shaft used for the transmission of power.

Therefore, eqn. (3.28) becomes

$$\left.\begin{aligned} \bar{J} &= I_{x_0} + I_{y_0} \\ \text{or} \quad I_{z_0} &= I_{x_0} + I_{y_0} \\ \text{or} \quad \bar{I}_z &= \bar{I}_x + \bar{I}_y \end{aligned}\right\} \quad ...(3.30)$$

Eqns. (3.30) are called *perpendicular axes theorem* and *it states that for an area the sum of two moments of inertia with respect to centroidal axes X_0 and Y_0 is equal to the moment of inertia about the third centroidal axis Z_0.*

3.10. PARALLEL-AXIS AND PERPENDICULAR-AXIS THEOREM COMBINED

Consider an area A as shown in Fig. 3.10. Its centroidal axes X_0 and Y_0, are as shown. The third axis Z_0 appears as a point and is coincident with the centroid C.

The X and Y axes are respectively parallel to X_0 and Y_0 and meet at the origin O. And, Z which is perpendicular to X- and Y-axis, is coincident with O.

Let the distance between X and X_0 be h ; that between Y and Y_0 be k ; and the distance between C and O or Z_0 and Z be λ.

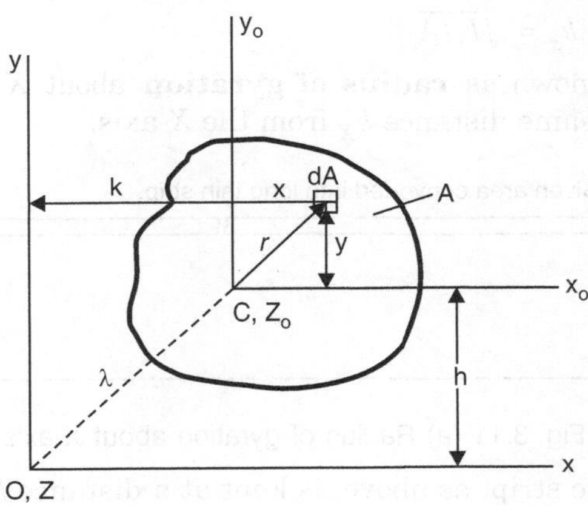

Fig. 3.10. Parallel and perpendicular axes theorems combined.

The parallel axes theorem between X_0- and X-axes states

$$I_x = \bar{I}_x + Ah^2 \quad ...(3.31)$$

The parallel axes theorem between Y_0- and Y-axes states.

$$I_y = \bar{I}_y + Ak^2 \quad ...(3.32)$$

Adding (3.31) and (3.32), we get

$$I_x + I_y = (\bar{I}_x + \bar{I}_y) + A(h^2 + k^2) \quad ...(3.33)$$

From eqn. (3.30), we have

$\because$ $\bar{I}_x + \bar{I}_y = \bar{I}_z$ ...(3.34)

And, $h^2 + k^2 = \lambda^2$

Hence $I_x + I_y = \bar{I}_z + A\lambda^2$ ...(3.35)

But left hand side $I_x + I_y = I_z$ from eqn. (3.30).

$\therefore$ Eqn. (3.34) can be written as

$$I_z = \bar{I}_z + A\lambda^2 \quad ...(3.36)$$

or $$(I_x + I_y) = (\bar{I}_x + \bar{I}_y) + A\lambda^2 \text{ or } \boxed{I_z = I_{z_0} + A\lambda^2} \quad ...(3.37)$$

Eqns. (3.36) and (3.37) are the combinations of perpendicular axes the parallel axes theorems.

3.11. RADIUS OF GYRATION

Let the area A be converted into a very long strip of very small width. If this strip is kept at a distance k_x from X-axis, then the moment of inertia of this strip with respect to x-axis will be given by

$$I_x = k_x^2 A \quad ...(3.38)$$

But $$I_x = \int y^2 \, dA \quad ...(3.39)$$

Thus, we get $$k_x = \sqrt{I_x/A} \quad ...(3.40)$$

The distance k_x is known as **radius of gyration** about X-axis. See Fig. 3.11 (*a*). All points of this area are the same distance k_X from the X-axis.

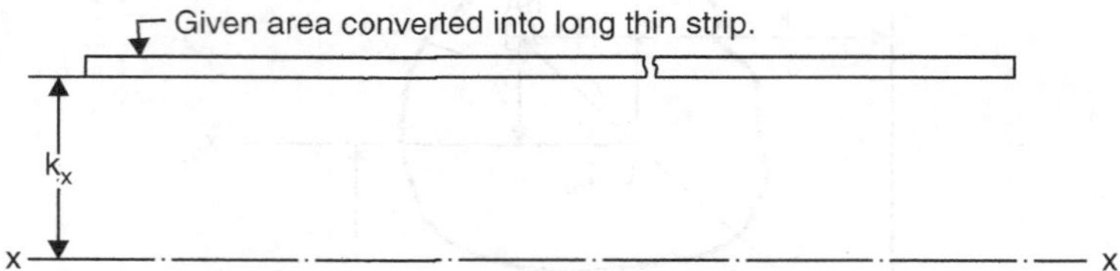

Fig. 3.11 (a) Radius of gyration about X-axis

Similarly, if the same strip, as above, is kept at a distance k_y from the Y-axis, then we can write

$$I_y = k_y^2 .A \quad ...(3.41)$$

But $$I_y = \int x^2 dA \quad ...(3.42)$$

$\therefore$ we get, $$k_y = \sqrt{I_y/A} \quad ...(3.43)$$

For k_y see Fig. 3.11 (*b*).

The parallel axes theorem given by eqns. (3.23) and (3.25) can be rewritten for radius of gyrations.

We can write

$$\bar{I}_x = A\bar{k}_x^2, \text{ and } I_x = Ak_x^2, \text{ then eqn. (3.23)}$$

$$I_x = \bar{I}_x + Ah^2 \quad \text{from (3.23)}$$

Can be written as $Ak_X^2 = A\bar{k}_x^2 + Ah^2$

or $$\boldsymbol{k_x^2 = \bar{k}_x^2 + h^2 \text{ or } k_x^2 = k_{x_0}^2 + h^2} \quad ...(3.44)$$

Similarly, $$I_y = \bar{I}_y + Ak^2$$

can be rewritten as follows

$$Ak_y^2 = A\bar{k}_y^2 + Ak^2$$

or, $$\boldsymbol{k_y^2 = \bar{k}_y^2 + k^2 = k_{y_0}^2 + k^2} \quad ...(3.45)$$

where k = distance between Y-axis and Y_0-axis. In a similar fashion perpendicular axes eqns. (3.30) and (3.36) can be rewritten as follows :

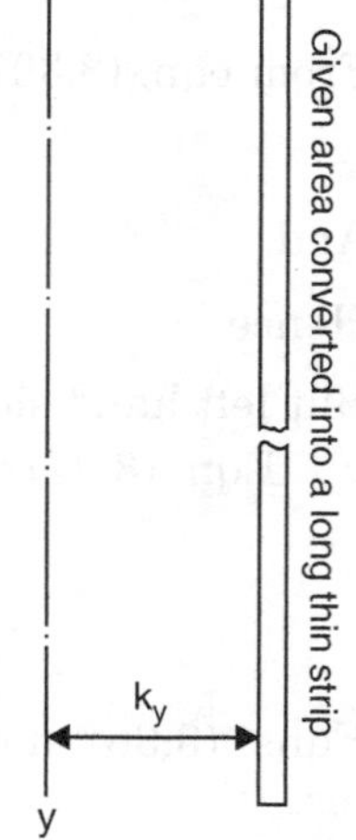

Fig. 3.11 (b) Radius of gyration about Y-axis.

Eqn. (3.30),

$$\left.\begin{aligned} \bar{I}_z &= \bar{I}_x + \bar{I}_y \\ I_z &= I_{x_0} + I_{y_0} \end{aligned}\right\}$$

Can be rewritten as follows :

$$A\bar{k}_z^{\ 2} = A\bar{k}_x^{\ 2} + A\bar{k}_y^{\ 2}$$

Cancelling A, we get

or $$\bar{k}_z^{\ 2} = \bar{k}_x^{\ 2} + \bar{k}_y^{\ 2} \quad \text{or} \quad k_{z_0}^{\ 2} = k_{x_0}^{\ 2} + k_{y_0}^{\ 2} \qquad \text{...(3.46)}$$

Also, equation $$I_z = \bar{I}_z + A\lambda^2 \qquad \text{...(3.47)}$$

yields $$Ak_z^{\ 2} = A\bar{I}_z^{\ 2} + A\lambda^2 \quad \text{or} \quad k_z^{\ 2} = \bar{k}_z^{\ 2} + \lambda^2 \qquad \text{...(3.48)}$$

3.12. MOMENTS OF INERTIA BY INTEGRATION

Expressions for moments of inertia I_{x_0}, I_{y_0} (or $\bar{I}_x, \bar{I}_y$) for some standard areas such as rectangle, triangle, circle, etc. can be developed by method of integration.

(*a*) **A Rectangular Area**

For $\bar{I}_x$ or I_{x_0}, consider a horizontal element as shown in Fig. 3.12.1 (*a*). Here $dA = b.dy$,

$$\therefore \qquad d\bar{I}_X = y^2.dA = y^2(bdy)$$

Integrating, we get

$$\int d\bar{I}_X = \bar{I}_X = \int_{-d/2}^{d/2} y^2(b.dy)$$

or $$\bar{I}_X = b\int_0^{d/2} 2 \times y^2 dy = 2b\left[\frac{y^3}{3}\right]_0^{d/2} = \frac{2}{3}b \times \left(\frac{d}{2}\right)^3$$

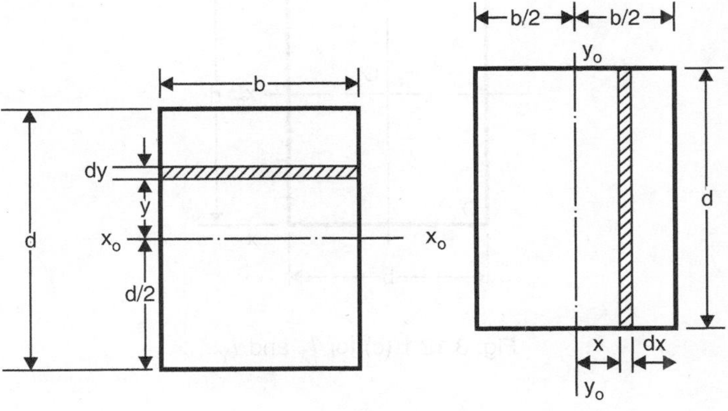

(a) Element for I_{x_0}. (b) Element for I_{y_0}.

Fig. 3.12.1

$$\therefore \quad \boxed{\bar{I}_x = \frac{bd^3}{12}} \quad \text{or} \quad \boxed{I_{x_0} = \frac{bd^3}{12}} \quad \text{...(3.49)}$$

For $\bar{I}_y$, consider a vertical element as shown in Fig. 3.38 (*b*).

Here $dA = (d)\, dx,$

and by definition $d\bar{I}_y = x^2\, dA$

or $d\bar{I}_y = x^2 \times (d.dx)$

Integrating, we get $\bar{I}_y = \int d\bar{I}_y = \int_{-b/2}^{b/2} x^2 .d.dx$

or $\bar{I}_y = 2d \int_0^{b/2} x^2 .dx$

or $\bar{I}_y = 2d.\left[\frac{x^3}{3}\right]_0^{b/2} = \frac{2d}{3}.(b/2)^3$

$$\therefore \quad \boxed{\bar{I}_y = \frac{db^3}{12}} \quad \text{or} \quad \boxed{I_{y_0} = \frac{db^3}{12}} \quad \text{...(3.50)}$$

For the indicated *X*- and *Y*-axes, we can find the values of I_x and I_y by using the parallel-axis theorems as discussed in Art. 3.8.

Thus, for the X_0- and *X*-parallel axes, we have,

$I_x = I_{x_0} + Ah^2,$ from (3.23)

where h = distance between X_0- and *X*-axes = $d/2$ and $A = bd$. Therefore, we get

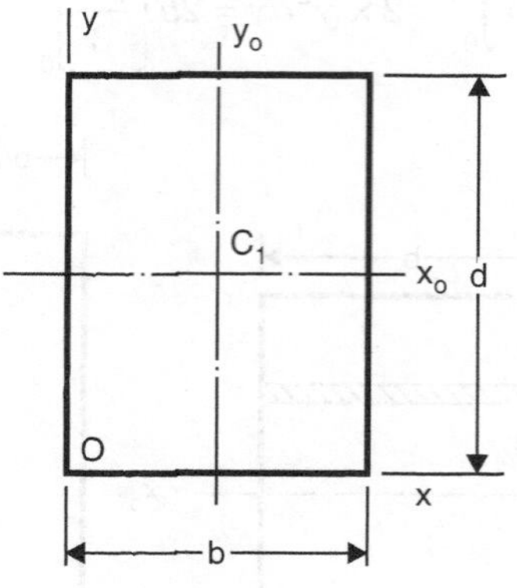

Fig. 3.12.1 (c) for I_X and I_Y.

$$I_X = \frac{bd^3}{12} + (bd)\left(\frac{d}{2}\right)^2 \quad ...(3.51)$$

or $$I_X = bd^3\left(\frac{1}{12}+\frac{1}{4}\right) = \frac{4bd^3}{12}$$

or $$\boxed{I_X = \frac{bd^3}{3}} \quad ...(3.52)$$

Similarly in $$I_Y = \bar{I}_Y + Ak^2 \quad \text{from (3.25)}$$

$$\bar{I}_Y = \frac{db^3}{12}, \text{ and } k = \bar{x} = b/2, \text{ and } A = bd$$

$\therefore$ we get $$I_y = \frac{db^3}{12} + (bd)(b/2)^2 = db^3\left(\frac{1}{12}+\frac{1}{4}\right)$$

$\therefore$ $$\boxed{I_y = \frac{db^3}{3}} \quad ...(3.53)$$

The radius of gyration k_{x_0}, k_{y_0} and k_x, k_y can be found by the definitions (3.37) and (3.40) of radius of gyration.

Thus $\because$ $$\bar{I}_x = \bar{k}_x^{\,2} A \quad \text{from (3.38)}$$

$\therefore$ $$\bar{k}_x^{\,2} = \frac{\bar{I}_x}{A} = \frac{\left(\dfrac{bd^3}{12}\right)}{bd}$$

or $$\bar{k}_x = \sqrt{\frac{d^2}{12}} = \frac{d}{2\sqrt{3}}$$

or $$\boxed{\bar{k}_x = k_{x_0} = \frac{d}{2\sqrt{3}}} \quad ...(3.54)$$

Also, $\because$ $$\bar{I}_y = \bar{k}_y^{\,2} \times A \quad \text{from (3.41)}$$

$\therefore$ we get $$\bar{k}_y^{\,2} = \frac{\bar{I}_y}{A} = \frac{db^3/12}{bd} = \frac{b^2}{12}$$

$\therefore$ $$\boxed{\bar{k}_y = \frac{b}{2\sqrt{3}} = k_{y_0}} \quad ...(3.55)$$

Also, eqn. (3.52) gives

$$k_x^{\,2} = \frac{I_x}{A} = \frac{\left(\dfrac{bd^3}{3}\right)}{bd}$$

$$\therefore \qquad k_x = \sqrt{\frac{d^2}{3}}$$

or
$$\boxed{k_x = \frac{d}{\sqrt{3}}} \qquad ...(3.56)$$

Finally, eqn. (3.53) gives

$$k_y{}^2 = I_y/A = \frac{db^3/3}{bd}$$

$$\therefore \qquad \boxed{k_y = \frac{b}{\sqrt{3}}} \qquad ...(3.57)$$

Important : The parallelogram and the rectangular area (shown dotted in Fig. 3.12.1 (*d*)) have the same area and same value of I_{X_0} and I_X.

Thus, we have $\qquad A = bd,$

$$I_{x_0} = \frac{bd^3}{12} \quad \text{and} \quad I_x = \frac{bd^3}{3}$$ for the parallelogram of Fig. 3.12.1 (*d*).

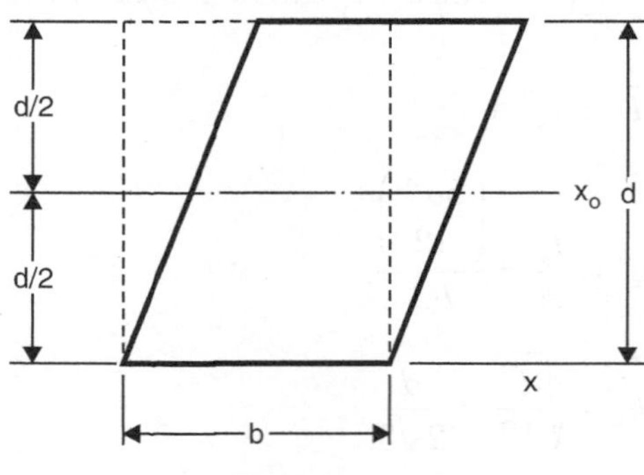

Fig. 3.12.1 (d)

(*b*) **A Circle:**

Method I. Consider a circle of radius *r*. See Fig. 3.12.2 (*a*). Expressions for I_{X_0}, I_{Y_0} and I_{Z_0} or J_0 can be derived by using different methods. The simplest method is by using the perpendicular axes theorem. Consider an annular elemental area of dA of radius ρ and radial thickness $d\rho$. Therefore, the area, $dA = 2\pi\rho.d\rho$

Perpendicular axes theorem,

$$I_{z_0} = I_{x_0} + I_{y_0}, \text{ can be written as}$$

or,
$$I_{z_0} = 2I_{x_0} = 2I_{y_0} \qquad ...(a)$$

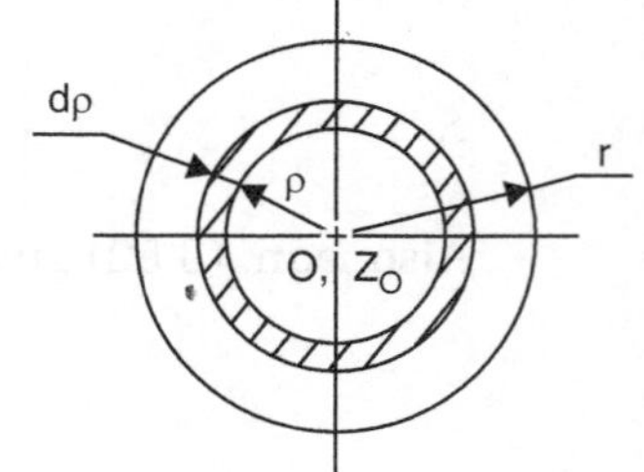

Fig. 3.12.2 (a)

because here $I_{X_0} = I_{Y_0}$ due to symmetry about X_0 and Y_0 axes.

Writing (*a*), as follows,

$$\int dI_{z_0} = \int \rho^2 . dA$$

or $$I_{z_0} = \int (\rho^2)\, dA \quad \text{or} \quad I_{z_0} = \int_0^r \rho^2 \times (2\pi\rho\, d\rho) \qquad ...(3.58)$$

$\therefore$ $$I_{z_0} = 2\pi \frac{r^4}{4}$$

$\therefore$ $$\boxed{I_{z_0} = \frac{\pi r^4}{2}} \qquad ...(3.59)$$

But, $$I_{z_0} = 2I_{x_0} = 2I_{y_0}$$

$\therefore$ $$\boxed{I_{x_0} = I_{y_0} = \frac{I_{z_0}}{2} = \frac{\pi r^4}{4}} \qquad ...(3.60)$$

or $$\boxed{\bar{I}_x \quad \text{or} \quad \bar{I}_y = \frac{\pi r^4}{4}}$$

For hollow circle of inner radius r_1 and outer radius r_2, eqn. (3.58) gives

$$I_{z_0} = \int_{r_1}^{r_2} \rho^2\, (2\pi\rho\, d\rho) = 2\pi \left[\frac{\rho^4}{4}\right]_r^{r_2}$$

or $$I_{z_0} = \frac{\pi}{2}\, (r_2{}^4 - r_1{}^4) \qquad ...(3.61)$$

and eqn. (3.60) gives,

$$I_{z_0} \text{ or } \bar{I}_z = \frac{\pi}{2} (r_2{}^4 - r_1{}^4) \qquad ...(3.62)$$

Radius of gyration, $\bar{k}_z$ can be obtained from eqns. (3.59) and (3.61).

$\because$ $$I_{z_0} = Ak_{z_0}{}^2 \quad \text{or} \quad \bar{I}_z = A\bar{k}_z{}^2\ ;$$

$\therefore$ $$\bar{k}_z{}^2 = \frac{\bar{I}_z}{A} = \frac{\pi r^4/2}{\pi r^2}$$

$\therefore$ For a circle $$\boxed{\bar{k}_z = k_{z_0} = \frac{r}{\sqrt{2}}} \qquad ...(3.63)$$

And, for a hollow circle,

$$Ak_{z_0}^2 = I_{z_0} \quad \text{or} \quad k_{z_0} = \sqrt{\frac{I_{z_0}}{A}} = \sqrt{\frac{\left(\frac{\pi}{2}\right)(r_2^4 - r_1^4)}{\pi\, (r_2^2 - r_1^2)}}$$

$\therefore$ $$k_{z_0} \text{ or } \bar{k}_z = \sqrt{\frac{(r_2^4 - r_1^4)}{2\, (r_2^2 - r_1^2)}} \quad \text{or} \quad \boxed{k_{z_0} \text{ or } \bar{k}_z = \sqrt{\frac{(r_2{}^2 + r_1{}^2)}{2}}} \qquad ...(3.64)$$

Above equations can also be expressed in terms of diameter d for a circle,
Since, $r = d/2$, $\therefore$ we get

$$I_{z_0} = \bar{I}_z = \frac{\pi r^4}{2} = \frac{\pi}{2} \times (d/2)^4$$

$\therefore$ $$\boxed{I_{z_0} = \bar{I}_z = \frac{\pi d^4}{32}}$$ from (3.59)

$$\boxed{I_{x_0} = \bar{I}_{y_0} = \frac{\pi}{4}(r^4) = \frac{\pi d^4}{64}}$$ from (3.60)

$$= \bar{I}_x \text{ or } \bar{I}_y$$

For a hollow circle, we get :

$$I_{z_0} = \bar{I}_z = \frac{\pi}{2}[(d_2/2)^4 - (d_1/2)^4]$$

$\therefore$ $$I_{z_0} = \bar{I}_z = \frac{\pi}{32}(d_2^4 - d_1^4)$$ from (3.61)

And $$I_{x_0} = \bar{I}_{y_0} = \frac{\pi}{64}(d_2^4 - d_1^4)$$ from (3.62)

Method II. See Fig. 3.12.2 (b)

Consider an element at a distance y from x_0-axis. The thickness of the element is dy. Area of the element is $dA = 2x \cdot dy = 2(r \cdot \cos\theta) \cdot dy$.

$$y = r\sin\theta, \quad \therefore \quad dy = r\cos\theta \cdot d\theta$$

$\therefore$ $$dA = 2(r\cos\theta) \cdot r\cos\theta \cdot d\theta = 2r^2\cos^2\theta \cdot d\theta.$$

$\therefore$ $$dI_{x0} = y^2 \cdot dA = (r\sin\theta)^2 \cdot 2r^2\cos^2\theta \cdot d\theta$$

$\therefore$ $$Ix_0 = \int dIx_0 = \int_{-\pi/2}^{\pi/2} 2r^4 \cdot \sin^2\theta \cdot \cos^2\theta \cdot d\theta$$

$$= 4 \times r^4 \int_0^{\pi/2} \sin^2\theta \cdot \cos^2\theta \cdot d\theta$$

$$= 4r \frac{\Gamma\frac{2+1}{2} \cdot \frac{2+1}{2}}{2\Gamma\frac{2+2+2}{2}} r^4 = 4\frac{\frac{1}{2} \cdot \sqrt{\pi} \cdot \frac{1}{2} \cdot \sqrt{\pi}}{2 \cdot 2 \cdot 1\sqrt{1}} \times r^4$$

$\therefore$ $$Ix_0 = \frac{\pi \cdot r^4}{4}$$

Also, $$Iy_0 = Ix_0 = \frac{\pi \cdot r^4}{4}$$

Method III. See Fig. 3.12.2(*c*).

Consider an element at a radius ρ, of thickness $d\rho$ subtending an angle $d\theta$ at the centre of the cicle. The area of the element is

$$dA = (\rho \,.\, d\theta) \,.\, dr$$

$$\therefore \quad A = \int dA = \int_0^{\pi/2}\int_0^{r} 4\,\rho \,.\, d\theta \,.\, d\rho = 4\int_0^{\pi/2} \frac{r^2}{2} \,.\, d\theta$$

$$= 2r^2\,\theta\Big]_0^{\pi/2}$$

$$\therefore \quad A = \pi r^2$$

$$\therefore \quad dIx_0 = y^2 \,.\, dA = (\rho \,.\, \sin\theta)^2\,\rho \,.\, d\theta \,.\, d\rho$$

$$\therefore \quad Ix_0 = \int dIx_0 = \iint \rho^2\, d\rho \,.\, \sin^2\theta \,.\, d\theta$$

$$= 4\int_0^{\pi/2}\int_0^{r} \sin^2\theta \,.\, \rho^3 \,.\, d\theta \,.\, d\rho$$

$$= \frac{4r^4}{4}\int_0^{\pi/2} \sin^2\theta \,.\, d\theta = r^4\int_0^{\pi/2}\left(\frac{1-\cos 2\theta}{2}\right)d\theta$$

$$= \frac{r^4}{2}\left[\theta - \frac{\sin 2\theta}{2}\right]_0^{\pi/2}$$

$$\therefore \quad Ix_0 = \frac{\pi \,.\, r^4}{2}$$

Note. The easiest is method *I* for finding Ix_0 and Iy_0.

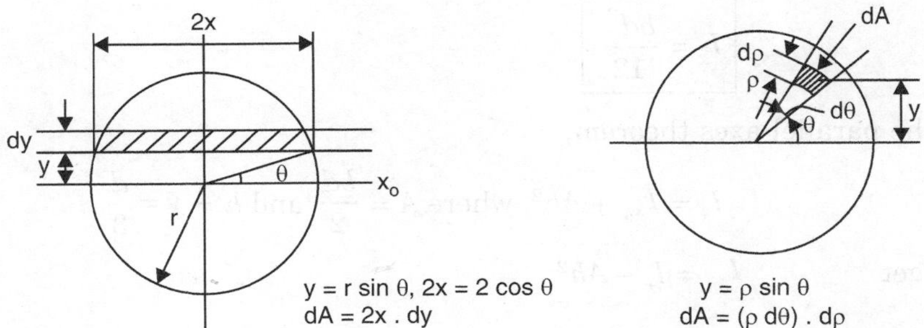

Fig. 3.12.2 (b) Element for I_{x_0} Fig. 3.12.2 (c) Element for I_{x_0}

(*c*) Right Triangular Area

Here we first find I_x and then by using the parallel axes theorem we find I_{x_0}. Consider an element as shown in Fig. 3.12.3 (*a*).

Area of element $= dA = A_1B_1 \times dy = b_1 \times dy.$

$$\because \quad \Delta CA_1B_1 \equiv \Delta CAB\;;$$

$$\therefore \quad \frac{A_1B_1}{CA_1} = \frac{AB}{CA}$$

$\therefore$ $$A_1B_1 = \frac{AB}{CA} \times CA_1 = \frac{b}{d}(d - y).$$

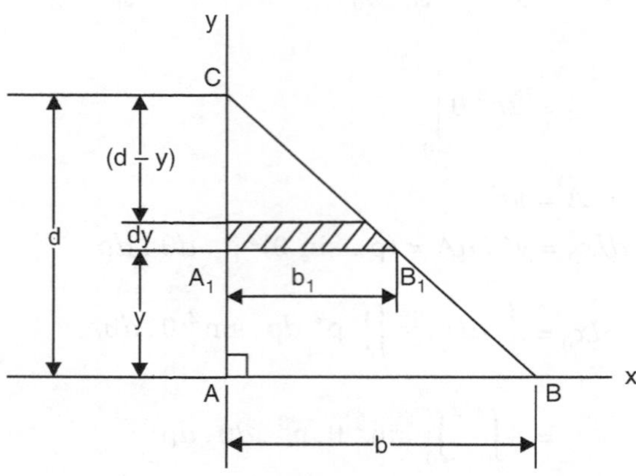

Fig. 3.12.3 (a) For I_x.

$\therefore$ $$dA = \frac{b}{d}(d - y).dy$$

$\therefore$ $$dI_x = y^2 dA = y^2 \times \frac{b}{d}(d - y).dy$$

$\therefore$ $$I_x = \int dI_x = \frac{b}{d}\int_0^d (dy^2 - y^3)\, dy = \frac{b}{d}\left[d \times \frac{d^3}{3} - \frac{d^4}{4}\right].$$

$\therefore$ $$\boxed{I_x = \frac{bd^3}{12}} \qquad ...(3.65)$$

Using the parallel axes theorem,

$$I_x = I_{x_0} + Ah^2, \text{ where } A = \frac{bd}{2} \text{ and } h = \bar{y} = \frac{d}{3}$$

$\therefore$ we get $$I_{x_0} = I_x - Ah^2$$

$$= \frac{bd^3}{12} - \left(\frac{bd}{2}\right)(d/3)^2 = bd^3\left(\frac{1}{12} - \frac{1}{18}\right)$$

$\therefore$ $$\boxed{I_{x_0} = \frac{bd^3}{36}} \qquad ...(3.66)$$

Similarly, for finding I_y consider the vertical element as shown in Fig. 3.12.3 (*b*).

Here, $dA = A_1C_1.dx$ or $dA = d_1 \times dx$

$\because$ $\Delta BA_1C_1 \equiv \Delta BAC,$

$\therefore$ $$\frac{A_1C_1}{BA_1} = \frac{AC}{BA}$$

$$\therefore \quad A_1C_1 = \frac{AC}{BA} \times BA_1 \quad \text{or} \quad d_1 = \frac{d}{b} \times (b - x)$$

$$\therefore \quad dA = \frac{d}{b}(b - x).dx.$$

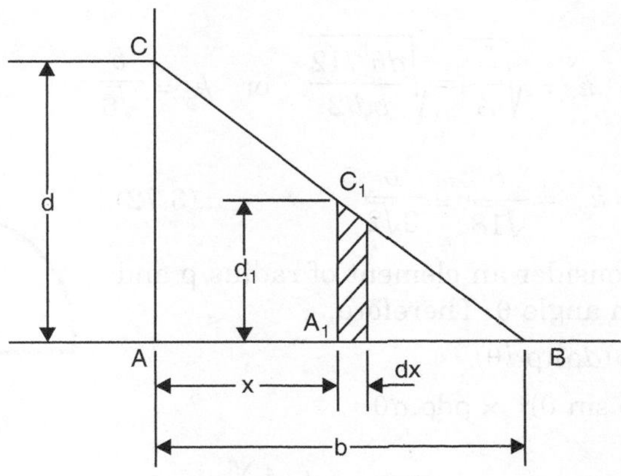

Fig. 3.12.3 (b) Element for I_y.

And, $dI_y = x^2.dA$

$$\therefore \quad I_y = \int x^2 dA = \int_0^b x^2 \times \frac{d}{b}(b - x).dx = \frac{d}{b}\left[\frac{bx^3}{3} - \frac{x^4}{4}\right]_0^b$$

or

$$\boxed{I_y = \frac{db^3}{12}} \quad \text{...(3.67)}$$

And, by using the parallel-axis theorem between Y- and Y_0-axes, we get

$$I_y = I_{y_0} + Ak^2 \quad \text{where} \quad k = \bar{x} = \frac{b}{3}$$

$$\therefore \quad I_{y_0} = I_y - Ak^2 = \frac{db^3}{12} - \left(\frac{bd}{2}\right)\left(\frac{b}{3}\right)^2$$

$$\therefore \quad \boxed{I_{y_0} = \frac{db^3}{36}} \quad \text{...(3.68)}$$

From the above relations we get the expressions for radius of gyration $\bar{k}_x$, k_x, $\bar{k}_y$ and k_y, as follows :

From $I_x = Ak_x^2$, we get

$$\boxed{k_x = \sqrt{\frac{I_x}{A}} = \sqrt{\frac{bd^3/12}{bd/2}} = \frac{d}{\sqrt{6}} \quad \text{or} \quad k_x = \frac{d}{\sqrt{6}}} \quad \text{...(3.69)}$$

And, $$\bar{k}_x = \sqrt{\frac{I_x}{A}} = \sqrt{\frac{bd^3/36}{bd/2}}$$

$$\therefore \quad \bar{k}_x = k_{x_0} = \frac{d}{\sqrt{18}} = \frac{\sqrt{d}}{3\sqrt{2}} \qquad \text{...(3.70)}$$

Similarly, $$k_y = \sqrt{\frac{I_y}{A}} = \sqrt{\frac{db^3/12}{bd/2}} \quad \text{or} \quad k_y = \frac{b}{\sqrt{6}} \qquad \text{...(3.71)}$$

And, $$\bar{k}_y = k_{y_0} = \frac{b}{\sqrt{18}} = \frac{b}{3\sqrt{2}} \qquad \text{...(3.72)}$$

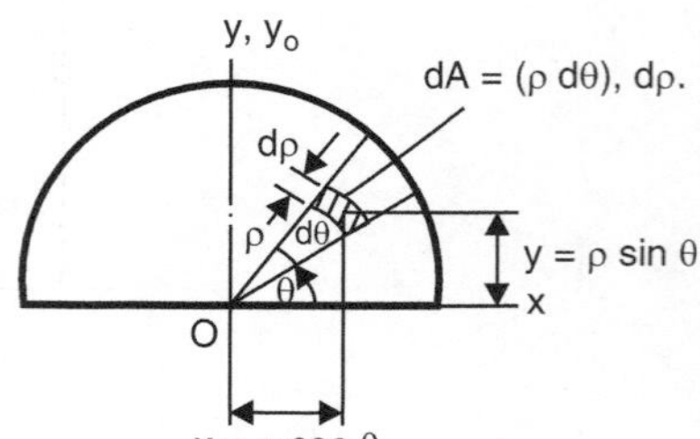

Fig. 3.12.4 (a) Element for I_x, I_y and I_{y_0}

(*d*) **Semi-circle.** Consider an element of radius ρ and thickness $d\rho$ located at an angle θ. Therefore,

$$dA = (d\rho).ds = (d\rho)(\rho d\theta).$$

$$\therefore \quad dI_x = y^2\, dA = (\rho \sin\theta)^2 \times \rho d\rho . d\theta$$

$$\therefore \quad I_x = \int_0^r \int_0^\pi \rho^3 \sin^2\theta . d\rho . d\theta = \int_0^\pi \sin^2\theta \left(\frac{\rho^4}{4}\right)_0^r . d\theta$$

$$= \frac{r^4}{4}\int_0^\pi \left(\frac{1-\cos 2\theta}{2}\right) d\theta = \frac{r^4}{8}\left[\theta - \frac{\sin 2\theta}{2}\right]_0^\pi = \frac{r^4}{8}[\pi - 0]$$

$$\therefore \quad \boxed{I_x = \frac{\pi r^4}{8}} \qquad \text{...(3.73)}$$

Here $$I_y = I_{y_0} = \int x^2 . dA = \int_0^r \int_0^\pi (\rho\cos\theta)^2 . \rho d\rho . d\theta$$

$$= \int_0^\pi \cos^2\theta \left(\frac{\rho^4}{4}\right)_0^r . d\theta = \frac{r^4}{4}\int_0^\pi \left(\frac{1+\cos 2\theta}{2}\right) d\theta$$

$$= \frac{r^4}{4\times 2}\left[\theta + \frac{\sin 2\theta}{2}\right]_0^\pi = \frac{r^4}{8}[\pi + 0].$$

$$\therefore \quad \boxed{I_y = I_{y_0} = \frac{\pi r^4}{8}} \qquad \text{...(3.74)}$$

Using parallel-axis theorem the value of I_{x_0} can be found. See Fig. 3.12.4(*b*).

Fig. 3.12.4 (b) For I_{x_0}.

Using $\quad I_x = I_{x_0} + Ah^2$

and writing $\quad A = \dfrac{\pi r^2}{2}$,

and $\quad h = \bar{y} = \dfrac{4r}{3\pi}$, we get

$$\therefore \quad I_{x_0} = I_x - Ah^2 = \frac{\pi r^4}{8} - \left(\frac{\pi r^2}{2}\right) \times \left(\frac{4r}{3\pi}\right)^2$$

$$\therefore \quad \boxed{I_{x_0} = 0.11 r^4} \qquad \text{...(3.75)}$$

(*e*) **Quarter circle.** See Fig. 3.12.5. For the element

$$dA = (\rho d\theta)\, d\rho$$

$$y = \rho \sin \theta \quad \text{and} \quad x = \rho \cos \theta$$

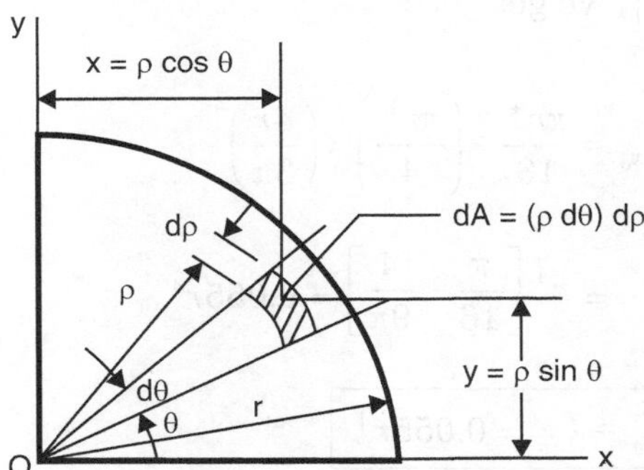

Fig. 3.12.5. Element for I_x and I_y.

$$\therefore \quad I_x = \int dI_x = \int y^2 . dA = \iint (\rho \sin \theta)^2 \, \rho d\rho d\theta$$

$$I_x = \int_0^r \int_0^{\pi/2} \rho^3 \sin^2 \theta \, d\rho \, d\theta$$

$$= \int_0^{\pi/2} \int_0^r \sin^2 \theta \, \rho^3 \, d\rho . d\theta$$

$$= \int_0^{\pi/2} \sin^2 \theta \left[\frac{\rho^4}{4}\right]_0^r d\theta = \frac{r^4}{4} \int_0^{\pi/2} \left(\frac{1 - \cos 2\theta}{2}\right) d\theta$$

$$= \frac{r^4}{4}\left[\frac{1}{2}\left(\theta - \frac{\sin 2\theta}{2}\right)_0^{\pi/2}\right]$$

$$= \frac{r^4}{4 \times 2}\left[\pi/2 - \frac{1}{2}(0)\right]$$

$$\therefore \quad \boxed{I_x = \frac{\pi r^4}{16}} \qquad ...(3.76)$$

It can also be proved that $\boxed{I_y = \frac{\pi r^4}{16}}$ ...(3.77)

Using parallel axes theorem,

$$I_x = I_{x_0} + Ah^2, \text{ we get}$$

$$I_{x_0} = I_x - Ah^2$$

where $A = \frac{\pi r^2}{4}$; $h = \left(\frac{4r}{3\pi}\right)$, we get

$$I_{x_0} = \frac{\pi r^4}{16} - \left(\frac{\pi r^2}{4}\right) \times \left(\frac{4r}{3\pi}\right)^2$$

$$= r^4\left[\frac{\pi}{16} - \frac{4}{9\pi}\right] = 0.055r^4$$

$$\therefore \quad \boxed{I_{x_0} = I_{y_0} = 0.055r^4} \qquad ...(3.78)$$

Also, $$k_x = k_y = \sqrt{\frac{\pi r^4/16}{\pi r^2/4}} = \frac{r}{2}$$

Hence $$\boxed{k_x = k_y = \frac{r}{2}} \qquad ....(3.79)$$

And, $$\bar{k}_x = \bar{k}_y = \sqrt{\frac{0.055r^4}{\pi r^2/4}} = (0.469)r$$

$$\boxed{\bar{k}_x = \bar{k}_y = 0.469r} \qquad ...(3.80)$$

Note. Moments of inertia and radii of gyration of various standard areas are given in Table 3.2.

Table 3.2. Standard Cases of Moment of Inertia of Areas

S.No.	*Name of Area*	*Figure*	*Area and centroid*	*Moments of inertia, radius of gyration, P_{xy}, etc.*
1.	Rectangle	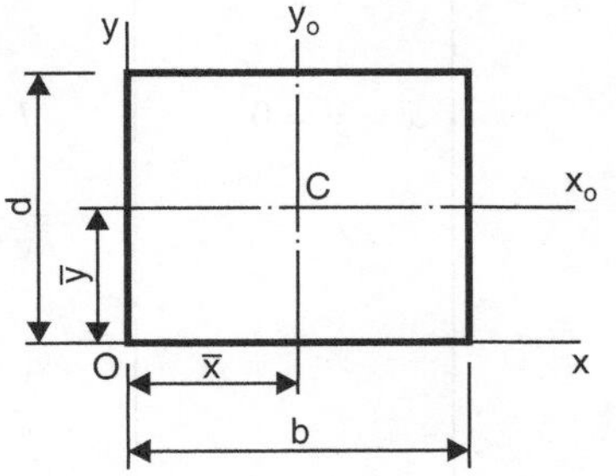	$A = bd$ $\bar{x} = b/2$ $\bar{y} = d/2$	$I_x = \frac{bd^3}{3}$; $I_y = \frac{db^3}{3}$ $\bar{I}_x = I_{x_0} = \frac{bd^3}{12}$; $\bar{I}_y = I_{y_0} = \frac{db^3}{12}$ $I_z = J = \frac{bd}{3}(d^2 + b^2)$ $I_{z_0} = J_0 = \frac{bd}{12}(d^2 + b^2)$ $\bar{k}_x = k_{x_0} = \frac{d}{\sqrt{12}} = \frac{d}{2\sqrt{3}}$ $\bar{k}_y = k_{y_0} = \frac{b}{\sqrt{12}} = \frac{b}{2\sqrt{3}}$ $\bar{P}_{xy} = \theta$; $P_{xy} = \frac{b^2d^2}{4}$
2.	Triangle	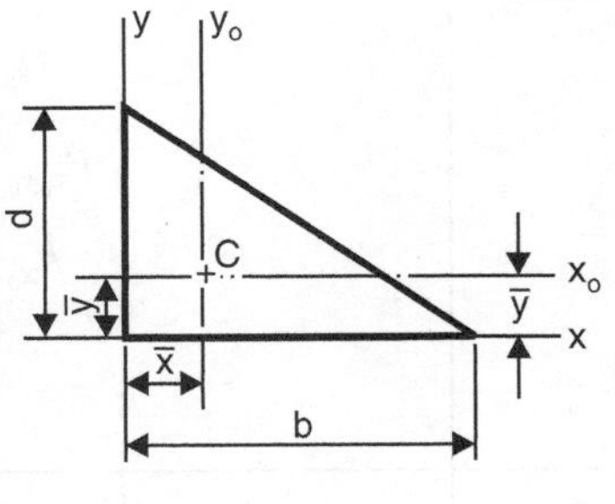	$A = \frac{bd}{2}$ $\bar{x} = \frac{b}{3}$ $\bar{y} = \frac{d}{3}$	$I_x = \frac{bd^3}{12}$; $I_y = \frac{db^3}{12}$ $I_{x_0} = \frac{bd^3}{36} = \bar{I}_x$ $I_{y_0} = \frac{db^3}{36} = \bar{I}_y$ $k_{x_0} = \bar{k}_x = \frac{d}{\sqrt{18}} = \frac{d}{3\sqrt{2}}$ $k_{y_0} = \bar{k}_y = \frac{b}{\sqrt{18}} = \frac{b}{3\sqrt{2}}$ $k_x = \frac{d}{\sqrt{6}}$; $k_y = \frac{b}{\sqrt{6}}$ $P_{xy} = \frac{b^2d^2}{24}$ $P_{x_0y_0} = \overline{P_{xy}} = -\frac{b^2d^2}{72}$

S.No.	Name of Area	Figure	Area and centroid	Moments of inertia, radius of gyration, P_{xy}, etc.
3.	Circle	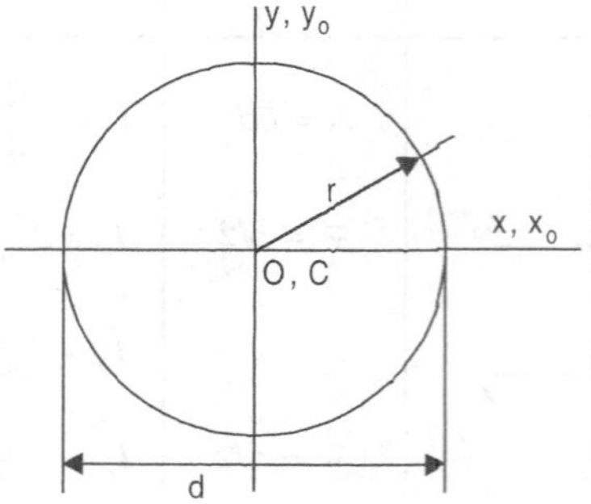 y, y_o; r; x, x_o; O, C; d	$A = \pi r^2 = \frac{\pi d^2}{4}$ $\bar{x} = \bar{y} = 0$	$I_x = I_y = I_{x_0} = I_{y_0}$ $= \frac{\pi r^4}{4} = \frac{\pi d^4}{64}$ $I = \bar{I_z} = I_{z_0} = \frac{\pi r^4}{2} = \frac{\pi d^4}{32}$ $\bar{k_x} = k_{x_0} = \frac{r}{2}$ $\bar{k_y} = k_{y_0} = \frac{r}{2}$ $\bar{k_z} = k_{z_0} = \frac{r}{\sqrt{2}}$ $P_{x_0 y_0} = \bar{P}_{xy} = 0$
4.	Hollow Circle	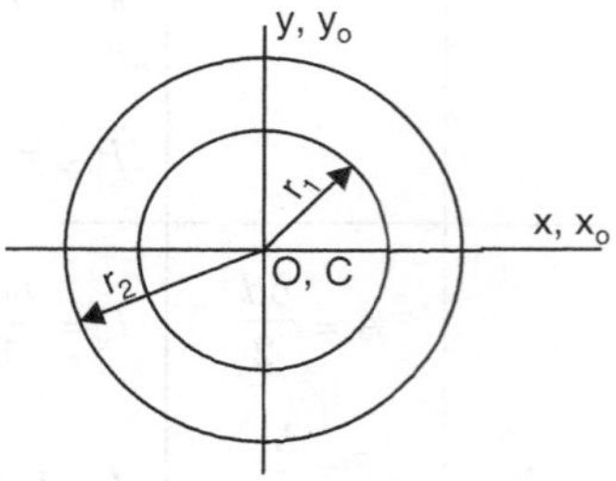 y, y_o; r_1; r_2; x, x_o; O, C	$A = \pi(r_2^2 - r_1^2)$ $= \frac{\pi}{4}(d_2^2 - d_1^2)$ $\bar{x} = \bar{y} = 0$	$\bar{I}_x = I_{x_0} = \bar{I}_y = I_{y_0}$ $= \frac{\pi}{4}(r_2^4 - r_1^4)$ $= \frac{\pi}{64}(d_2^4 - d_1^4)$ $I_{z_0} = \bar{I}_z = \frac{\pi}{2}(r_2^4 - r_1^4)$ $P_{x_0 y_0} = \overline{P_{xy}} = 0$ $k_{x_0} = \bar{k}_x = k_{y_0} = \bar{k}_y$ $= \frac{\sqrt{\pi(r_2^2 - r_1^2)}}{2}$
5.	Semi-circle	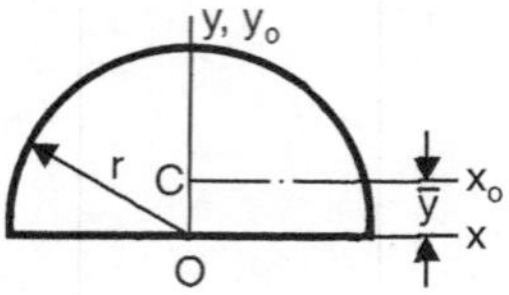 y, y_o; r; C; x_o; $\bar{y}$; x; O	$A = \frac{\pi r^2}{2}$ $\bar{x} = 0$ $\bar{y} = \frac{4r}{3\pi}$	$I_x = I_{y_0} = \bar{I}_y = \frac{\pi r^4}{8}$ $I_{x_0} = \bar{I}_x = \frac{\pi r^4}{8} - \frac{8r^4}{9\pi}$ $= 0.11r^4$ $\bar{P}_{xy} = \overline{P_{x_0 y_0}} = 0$

S.No.	*Name of Area*	*Figure*	*Area and centroid*	*Moments of inertia, radius of gyration, P_{xy}, etc.*
		y, y_o, $\bar{x}$, O, C, r, x, x_o	$A = \frac{\pi r^2}{2}$ $\bar{x} = \frac{4r}{3\pi}$ $\bar{y} = 0$	$I_{x_0} = I_x = I_y = \frac{\pi r^4}{8}$ $I_{y_0} = \bar{I}_y = 0.11r^4$ $P_{x_0 y_0} = \overline{P_{xy}} = 0$
6.	Quarter Circle	y, y_o, C, r, x_o, $\bar{y}$, O, $\bar{x}$, x	$A = \frac{\pi r^2}{4}$ $\bar{x} = \bar{y} = \frac{4r}{3\pi}$	$I_{x_0} = I_{y_0} = \bar{I}_x = \bar{I}_y$ $= r^4\left(\frac{\pi}{16} - \frac{4}{9\pi}\right) = 0.055\ r^4$ $I_x = I_y = \frac{\pi r^4}{16}$ $I_z = J = \frac{\pi r^4}{8}$ $I_{z_0} = \bar{J} = r^4\left(\frac{\pi}{8} - \frac{8}{9\pi}\right)$ $= 0.11r^4$ $k_x = k_y = \frac{r}{2}$ $k_{x_0} = k_{y_0} = \left(\frac{2 \times 0.055\, r^4}{\pi r^2 / 4}\right)^{1/2}$ $= 0.374\ r$ $k_{x_0} = \bar{k}_x = k_{y_0} = \bar{k}_y$ $P_{xy} = \frac{r^4}{8}$; $\overline{P_{xy}} = P_{x_0 y_0}$ $= -\left(\frac{1}{8} - \frac{4}{9\pi}\right) r^4$ $= -\ 0.0165r^4$
7.	Circular Sector	y, y_o, r, α, α, O, C, x, x_o	$A = \alpha r^2$ $\bar{x} = \frac{2r \sin \alpha}{3\alpha}$ $\bar{y} = 0$	$I_x = \frac{r^4}{4}\left(\alpha - \frac{1}{2}\sin 2\alpha\right)$ $I_y = \frac{r^4}{4}\left(\alpha + \frac{1}{2}\sin 2\alpha\right)$ $I_z = \frac{1}{2} r^4 . \alpha$

S.No.	*Name of Area*	*Figure*	*Area and centroid*	*Moments of inertia, radius of gyration, P_{xy}, etc.*
8.	Elliptical Quadrant	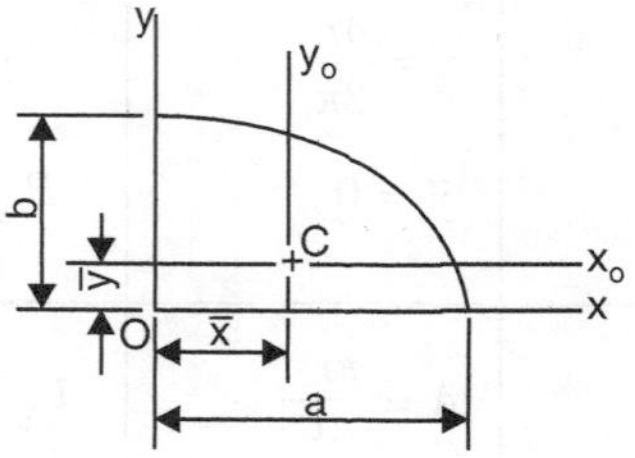	$A = \frac{\pi ab}{4}$ $\bar{x} = \frac{4a}{3\pi}$ $\bar{y} = \frac{4b}{3\pi}$	$I_x = \frac{\pi ab^3}{16}$; $I_y = \frac{\pi ba^3}{16}$ $\bar{I}_x = \left(\frac{\pi}{16} - \frac{4}{9\pi}\right) ab^3 = 0.055\ ab^3$ $\bar{I}_y = \left(\frac{\pi}{16} - \frac{4}{9\pi}\right) a^3b = 0.055\ a^3b$ $I_z = \frac{\pi\, ab}{16}(a^2 + b^2)$ $\bar{I}_z = ab\left[\frac{\pi}{16} - \frac{4}{9\pi}\right](b^2 + a^2)$ $= ab(a^2 + b^2)\left(\frac{\pi}{16} - \frac{4}{9\pi}\right)$ $= 0.055 ab(a^2 + b^2)$ $k_x = \frac{a}{2}$; $k_Y = \frac{b}{2}$
9.	Sub-parabolic Area	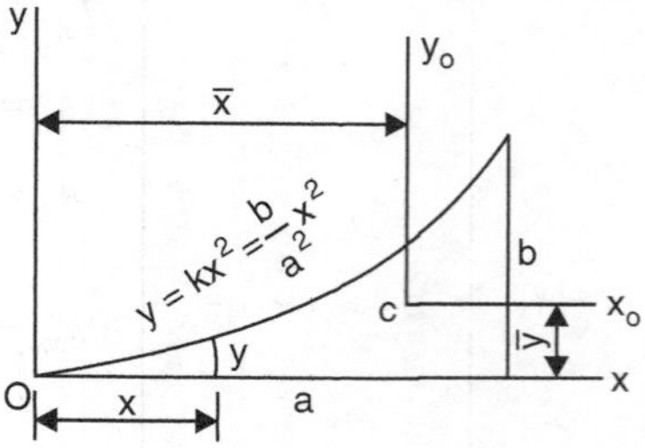	$A = \frac{ab}{3}$ $\bar{x} = \frac{3a}{4}$ $\bar{y} = \frac{3}{10} b$	$I_x = \frac{ab^3}{21}$; $I_y = \frac{a^3b}{5}$ $I_z = ab\left(\frac{b^2}{21} + \frac{a^2}{5}\right)$ $\bar{I}_x = 0.0176ab^3$ $\bar{I}_y = 0.0125\ a^3b$
10.	Parabolic Area	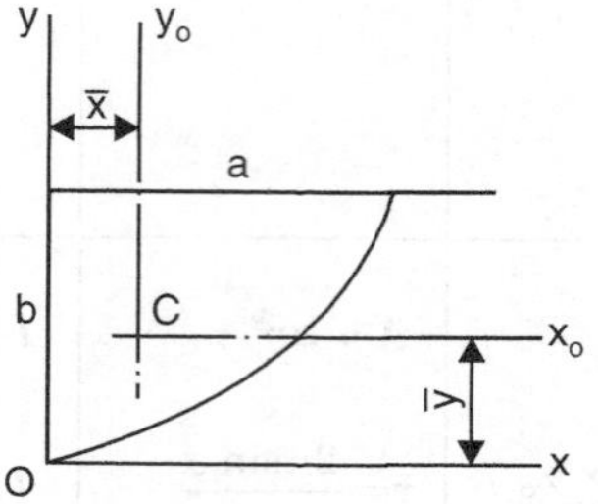	$A = \frac{2}{3} ab$ $\bar{x} = \frac{3}{8} a$ $\bar{y} = \frac{3}{5} b$	$I_x = \frac{2ab^3}{17}$; $I_y = \frac{2a^3b}{15}$ $I_z = 2ab\left(\frac{a^2}{15} + \frac{b^2}{17}\right)$ $\bar{I}_x = 2ab^3\left(\frac{1}{7} - \frac{3}{25}\right) = 0.457ab^3$ $\bar{I}_y = 2a^3b\left(\frac{1}{15} - \frac{3}{64}\right) = 0.0396a^3b$

Problem 3.23. *Find $\bar{I}_x$, $\bar{I}_y$, I_x, I_y for the trapezoid section shown in Fig. P.3.23.*

Sol. The given area can be divided into a rectangle and a triangle, minus the circular hole.

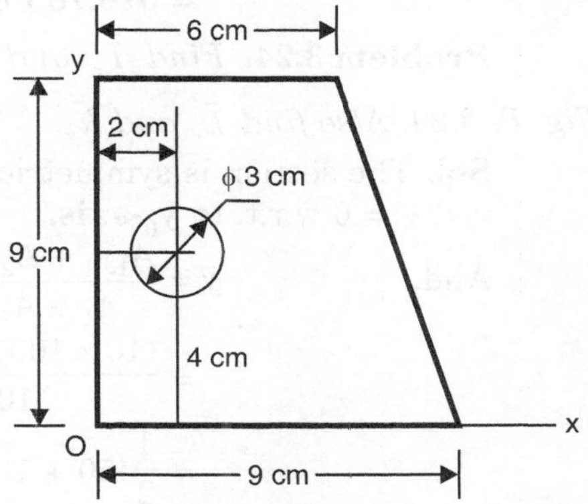

6 cm
y
2 cm
ϕ 3 cm
9 cm
4 cm
O
x
9 cm

Fig. P.3.23

For the rectangle

Area, $A_1 = 6 \times 9 = 54 \text{ cm}^2$

$x_1 = 3$ cm from y-axis

$y_1 = 4.5$ cm from x-axis

For the triangle

Area, $A_2 = \dfrac{(9-6)\times 9}{2} = 13.5 \text{ cm}^2$

$x_2 = 6 + \dfrac{9-6}{3} = 7$ cm from y-axis

$y_2 = 3$ cm from x-axis.

∴ **For the circle**

Area, $A_3 = -\dfrac{\pi \times 3^2}{4} = -2.25\pi \text{ cm}^2$

$x_3 = 2$ cm from y-axis

and $y_3 = 4$ cm from x-axis

∴ $$\bar{x} = \frac{A_1x_1 + A_2x_2 + A_3x_3}{A_1 + A_2 + A_3} = \frac{54 \times 3 + 13.5 \times 7 + (-2.25\pi) \times 2}{54 + 13.5 + (-2.25\pi)}$$

$$= \frac{242.363}{60.43} = \textbf{4.01 cm} \text{ from } y\text{-axis}$$

and $$\bar{y} = \frac{A_1y_1 + A_2y_2 + A_3y_3}{A_1 + A_2 + A_3} = \frac{54 \times 4.5 + 13.5 \times 3 + (-2.25\pi) \times 4}{60.43} = \frac{255.23}{60.43} \text{ cm}$$

$= 4.22$ cm from x-axis

$$I_x = I_{x_1} + I_{x_2} + I_{x_3} = \frac{b_1d_1^3}{3} + \frac{b_2d_2^3}{12} - \left[\frac{\pi \times 3^4}{64} + \left(\frac{\pi}{4} \times 3^2\right) \times 4^2\right]$$

$$= \frac{6 \times 9^3}{3} + \frac{(9-6) \times 9^3}{12} - 117.07 = 1640.25 - 117.07 = \mathbf{1523.18}$$

or $I_x = \mathbf{1523.18\ cm^4}$. **Ans.**

∴ $I_{x_0} = I_x - A\bar{y}^2 = 1523.18 - (60.43) \times (4.22)^2 \text{ cm}^4$

$I_{x_0} = \mathbf{1050.81\ cm^4}$. **Ans.**

$$I_y = I_{y_1} + I_{y_2} + I_3 = \frac{d_1b_1^3}{3} + \left(\frac{d_2b_2^3}{36} + A_2 \times x_2^2\right) - \left[\frac{\pi \times r_3^4 + A_3x_3^2}{4}\right]$$

$$= \left(\frac{9 \times 6^3}{3}\right) + \left[\frac{9 \times 3^3}{36} + 13.5 \times 7^2\right] - \left[\frac{\pi \times 3^4}{64} + (2.25\pi) \times 2^2\right] \text{cm}^4$$

$= 648 + 668.25 - 32.25 \text{ cm}^4$

∴ $I_y = \mathbf{1348.50\ cm^4}$. **Ans.**

and $$I_{y_0} = I_y - A\bar{x}^2 = 1348.50 - 60.43 \times (4.01)^2 \text{ cm}^4$$
$$= \mathbf{376.78 \text{ cm}^4}. \quad \textbf{Ans.}$$

Problem 3.24. *Find $\bar{I}_x$ and $\bar{I}_y$ for the section shown in Fig. P. 3.24. Also find $\bar{k}_x$ and $\bar{k}_y$.*

Fig. P.3.24

Sol. The section is symmetrical about Y-axis

$\therefore$ $\bar{x} = 0$ w.r.t. to $\mathbf{y_0}$**-axis.**

And,
$$\bar{y} = \frac{A_1 y_1 + A_2 y_2}{A_1 + A_2}$$
$$= \frac{(10 \times 100) \times 50 + (100 \times 10) \times (105)}{10 \times 100 + 100 \times 10} \text{ mm from AB}$$
$$= \frac{1}{2}(50 + 105) = 77.5 \text{ mm from } AB.$$

$\therefore$
$$I_{AB} = I_{AB})_1 + I_{AB})_2$$
$$= \frac{1}{3} \times (10) \times (100)^3 + \left[\frac{100 \times 10^3}{12} + (100 \times 10) \times (105)^2\right] \text{mm}^4$$
$$= 1000\left[\frac{10000}{3} + \frac{100}{12} + (105)^2\right] \text{mm}^4 = 14.37 \times 10^6 \text{ mm}^4$$

$\because$
$$I_{AB} = \bar{I}_x + A\bar{y}^2 \;;$$

$\therefore$
$$\bar{I}_x = I_{AB} - A\bar{y}^2 = 14.37 \times 10^6 - 2000 \times (77.5)^2$$
$$= 2357500 \text{ mm}^4. \quad \textbf{Ans.}$$
$$\bar{I}_y = \frac{1}{12} \times 100 \times 10^3 + \frac{10 \times (100)^3}{12} = \mathbf{841666.7 \text{ mm}^4}$$

$\because$
$$\bar{I}_x = A\bar{k}_x^2 \;; \quad \therefore \quad \bar{k}_x = \sqrt{\frac{\bar{I}_x}{A}} = \sqrt{\frac{2357500}{2000}} = \mathbf{34.33 \text{ mm.}} \quad \textbf{Ans.}$$

And
$$\bar{k}_y = \sqrt{\frac{\bar{I}_y}{A}} = \sqrt{\frac{841666.7}{2000}} = \mathbf{20.51 \text{ mm.}} \quad \textbf{Ans.}$$

Problem 3.25. *Find moment of inertia of the rectangle ABCD shown in Fig. P.3.25 (a) about the X-axis and also about X_0-axis.*

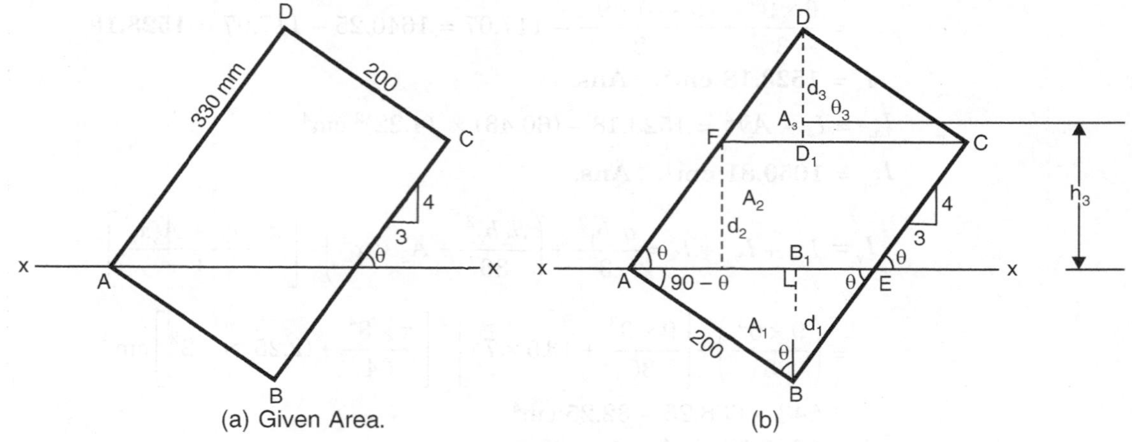

(a) Given Area. (b)

Fig. P.3.25

Sol. The given area can be divided into three areas as shown in Fig. P. 3.25 (*b*)

$$\sin\theta = \frac{4}{5},\quad \cos\theta = 3/5 \text{ and } \tan\theta = 4/3$$

$$\because\quad \frac{AB}{BE} = \tan\theta\,;\quad \therefore\ BE = \frac{AB}{\tan\theta} = \frac{200}{4}\times 3 = 150\,\text{mm} = DF$$

$$\therefore\quad EC = 330 - 150 = 180 \text{ mm} = AF$$

$$AE = \frac{BE}{\cos\theta} = \frac{150}{3}\times 5 = 250 \text{ mm} = CF$$

Perpendicular, $d_1 = d_3 = 200\cos\theta = 200 \times \frac{3}{5} = 120$ mm

$\therefore$ Area $\quad A_1 = A_3 = \frac{1}{2}\times 200 \times 150 = 15{,}000 \text{ mm}^2$

$$d_2 = AF\sin\theta = 180\sin\theta = 180 \times \frac{4}{5} = 144 \text{ mm}$$

$$h_3 = d_2 + \frac{d_3}{3} = 144 + \frac{120}{3} = 184 \text{ mm}$$

$$I_x = (I_x)_1 + (I_x)_2 + (I_x)_3$$

$$= \left[\frac{AE\times BB_1^3}{12}\right] + \left[\frac{AE\times d_2^3}{3}\right] + \left[\frac{FC\times DD_1^3}{36} + A_3\times h_3^2\right]$$

$$= \left[\frac{250\times 120^3}{12}\right] + \left[\frac{250\times 144^3}{3}\right] + \left[\frac{250\times 120^3}{36} + \left(\frac{250\times 120}{2}\right)\times 184^2\right]$$

$= \mathbf{804672000\ mm^4} = 804.672 \times 10^6$ mm. **Ans.**

$$I_x = I_{x_0} + Ah^2, \text{ where } h = d_2/2 = \frac{144}{2} = 72 \text{ mm}$$

$\therefore\quad I_{x_0}$ **= 804.672 × 10⁶ – (330 × 200) × (72)² = 462.528 × 10⁶ mm⁴. Ans.**

Problem 3.26. *Find moment of inertia of the area shown in Fig. P. 3.26 about the horizontal centroidal axis.*

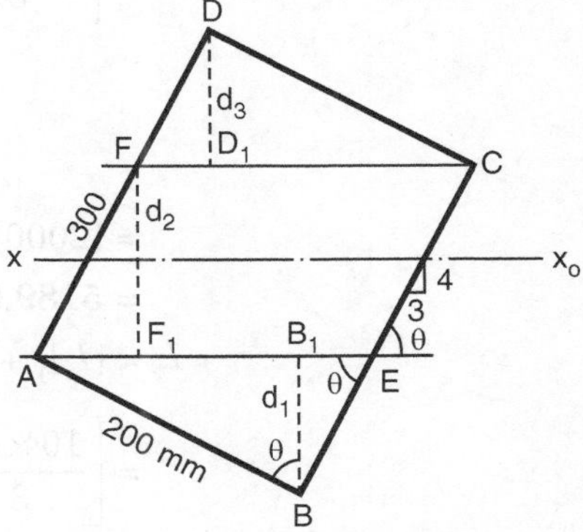

Fig. P.3.26

Sol. Here $\sin\theta = 4/5$, $\cos\theta = 3/5$ $\therefore$ $\tan\theta = 4/3$,

$$\therefore\quad BE = \frac{AB}{\tan\theta} = \frac{200\times 3}{4} = 150 \text{ mm} = DF$$

$$EC = 300 - 150 = 150 \text{ mm}$$

$$AE = \frac{BE}{\cos\theta} = \frac{150\times 5}{3} = 250 \text{ mm} = CF$$

$$AF = CE = 150 \text{ mm}$$

$$BB_1 = AB\cos\theta = 200\times 0.6 = 120 \text{ mm} = DD_1$$

or $\quad d_1 = d_3 = 120$ mm

$$FF_1 = AF\sin\theta = 150\times 0.8 = 120 \text{ mm} = d_2$$

$$A_1 = A_2 = \frac{BE\times AB}{2} = \frac{150\times 200}{2} = 15{,}000 \text{ mm}^2$$

$$\bar{I}_x = \bar{I}_{x_1} + \bar{I}_{x_3} + \bar{I}_{x_2} = 2\bar{I}_{x_1} + \bar{I}_{x_2} \qquad \Big|\because\ \bar{I}_{x_1} = \bar{I}_{x_3}$$

$$= 2\left[\frac{AE \times BB_1^3}{36} + A_1 \times \left(\frac{BB_1}{3} + \frac{FF_1}{2}\right)^2\right] + \left[\frac{AE \times (FF_1)^3}{12}\right]$$

$$= 2\left[\frac{250 \times 120^3}{36} + 15{,}000 \times \left(\frac{120}{3} + \frac{120}{2}\right)^2\right] + \left[\frac{250 \times (120)^3}{12}\right] \text{mm}^4$$

$= \mathbf{360 \times 10^6\ mm^4}$. **Ans.**

Problem 3.27. *Find I_x and I_y for the area shown in Fig. P.3.27.*

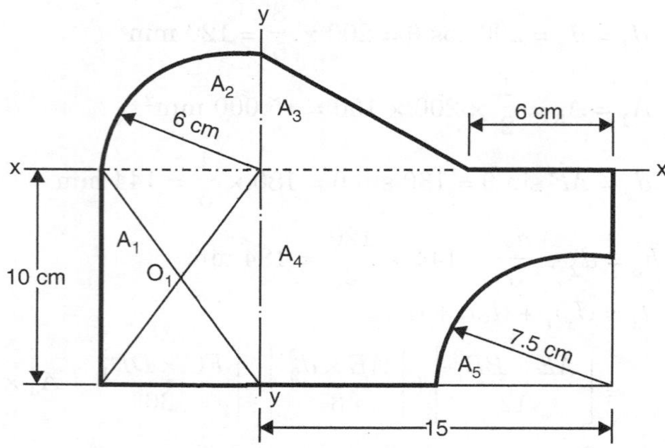

Fig. P.3.27

Sol. $I_x = I_{x_1} + Ix_2 + Ix_3 + Ix_4 - Ix_5$

$= [\bar{I}_x]_1 + [\bar{I}_x]_2 + [\bar{I}_x]_3 + [\bar{I}_x]_4 - [\bar{I}_x + Ah^2]_5$

$$= \left[\frac{6 \times 10^3}{3}\right] + \left[\frac{\pi \times 6^4}{16}\right] + \left[\frac{(15-6) \times 6^3}{12}\right] + \left[\frac{15 \times 10^3}{3}\right]$$

$$- \left[0.055 \times 7.5^4 + \frac{\pi \times (7.5)^2}{4} \times \left(10 - \frac{4 \times 7.5}{3\pi}\right)^2\right]$$

$= [2000] + [254.5] + [162] + [5000] - [2227.01]\ \text{cm}^4$

$= \mathbf{5189.5\ cm^4}$. **Ans.**

$I_y = [I_y]_1 + [I_y]_2 + [I_y]_3 + [I_y]_4 - [I_y]_5$

$$= \left[\frac{10 \times 6^3}{3}\right] + \left[\frac{\pi \times 6^4}{16}\right] + \left[\frac{6 \times (15-6)^3}{12}\right] + \left[\frac{15 \times 10^3}{3}\right]$$

$$- \left[0.055 \times 7.5^4 + \frac{\pi \times (7.5)^2}{4} \times \left(15 - \frac{4 \times 7.5}{3\pi}\right)^2\right] \text{cm}^4$$

$= [720] + [254.5][+ [364.5] + [15000] - [6343.1]$

$= \mathbf{9995.9\ cm^4}$. **Ans.**

Problem 3.28. *Find* I_{x_1} and I_{x_2} *for the area shown in Fig. P.3.28.*

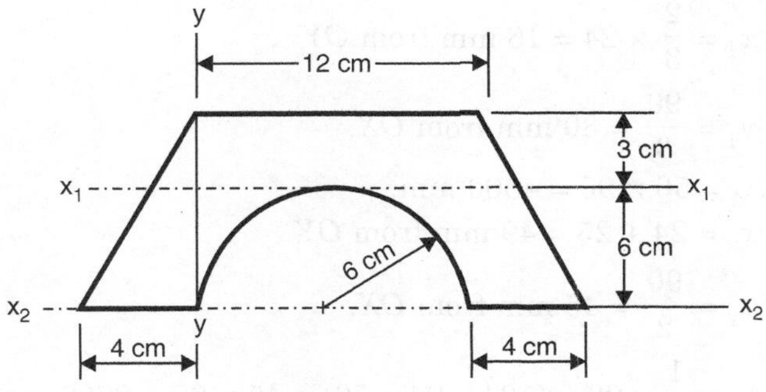

Fig. P.3.28

Sol. The given area can be split into 4 areas *viz.*, one rectangle (12 × 9 cm), two triangles (size 4 cm × 9 cm) ; one semi circle (6 cm radius).

Using the parallel-axis theorem, we get

$$I_{x_1} = \left[\frac{12\times 9^3}{12} + (12\times 9)(6-4.5)^2\right] + 2\left[\frac{4\times 9^3}{36} + \left(\frac{4\times 9}{2}\right)\times\left(6-\frac{9}{3}\right)^2\right]$$

$$-\left[0.11\times 6^4 + \frac{\pi\times 6^2}{2}\times\left(6-\frac{4\times 6}{3\pi}\right)^2\right]$$

$$= 729 + 243 + 243 \times 2 - (817) = \mathbf{641\ cm^4}. \quad \textbf{Ans.}$$

$$I_{x_2} = \left[\frac{12\times 9^3}{3}\right] + 2\left[\frac{4\times 9^3}{12}\right] - \left[\frac{\pi\times 6^4}{8}\right]\text{cm}^4$$

$$= 2916 + 486 - 508.94 = \mathbf{2893.06 \quad cm^4}.$$

Problem 3.29. *Locate centroid of the area shown in Fig. P.3.29 and find* I_{x_0} *and* I_{y_0}. *The area about the two centroidal axes.*

Sol. The four areas are shown in Fig. P.3.29.

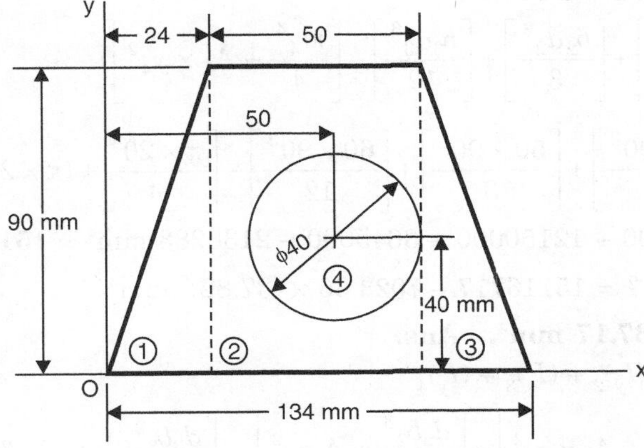

Fig. P.3.29

$$\text{Area } A_1 = \frac{24 \times 90}{2} = 1080 \text{ mm}^2$$

$$x_1 = \frac{2}{3} \times 24 = 16 \text{ mm from } OY$$

$$y_1 = \frac{90}{3} = 30 \text{ mm from } OX.$$

$$A_2 = 50 \times 90 = 4500 \text{ mm}^2$$

$$x_2 = 24 + 25 = 49 \text{ mm from } OY$$

$$y_2 = \frac{90}{2} = 45 \text{ mm from } OX.$$

$$A_3 = \frac{1}{2} \times 90 \times (134 - 24 - 50) = 45 \times 60 = 2700 \text{ mm}^2$$

$$x_3 = 24 + 50 + \frac{60}{3} = 94 \text{ mm from } OY$$

and $$y_3 = \frac{90}{3} = 30 \text{ mm from } OX :$$

$\therefore$ $$A_4 = -\pi \times 20^2 = -1256.64 \text{ mm}^2$$

$$x_4 = 50 \text{ mm from } OY$$

$$y_4 = 40 \text{ mm from } OX$$

$\therefore$ $$A = A_1 + A_2 + A_3 + A_4 = 1080 + 4500 + 2700 + (-1256.64) = 7023.36 \text{ mm}^2$$

$\therefore$ $$\bar{x} = \frac{A_1x_1 + A_2x_2 + A_3x_3 + A_4x_4}{A_1 + A_2 + A_3 + A_4} = \frac{1080 \times 16 + 4500 \times 49 + 2700 \times 94 - 1256.64 \times 50}{7023.36}$$

$$= \frac{428748}{7023.36} \text{ mm} = \mathbf{61.05 \text{ mm}} \text{ from } OY$$

$$\bar{y} = \frac{A_1y_1 + A_2y_2 + A_3y_3 + A_4y_4}{A_1 + A_2 + A_3 + A_4} = \frac{1080 \times 30 + 4500 \times 45 + 2700 \times 30 - 1256.64 \times 40}{7023.36}$$

$$= \frac{265634.4}{7023.36} = \mathbf{37.82 \text{ mm}} \text{ from } OX. \quad \textbf{Ans.}$$

$$I_x = I_x]_1 + I_x]_2 + I_x]_3 + I_x]_4$$

$$I_x = \left[\frac{b_1d_1^3}{12}\right] + \left[\frac{b_2d_2^3}{3}\right] + \left[\frac{b_3d_3^3}{12}\right] - \left[\frac{\pi r_4^4}{4} + A_4 \times y_4^2\right]$$

$$= \left[\frac{24 \times 90^3}{12}\right] + \left[\frac{50 \times 90^3}{3}\right] + \left[\frac{60 \times 90^3}{12}\right] - \left[\frac{\pi \times 20^4}{4} + (\pi \times 20^2) \times 40^2\right]$$

$$= 1458000 + 12150000 + 3645000 - 2136283 \text{ mm}^4 = \mathbf{15116717 \text{ mm}^4}$$

$\therefore$ $$I_{x_0} = I_x - A\bar{y}^2 = 15116717 - 7023.36 \times (37.82)^2 \text{ mm}^4$$

$$= \mathbf{5070837.17 \text{ mm}^4}. \quad \textbf{Ans.}$$

$$I_y = (I_y)_1 + (I_y)_2 + (I_y)_3 + (I_y)_4$$

$$= \left[\frac{d_1b_1^3}{36} + A_1 \times x_1^2\right] + \left[\frac{d_2b_2^3}{12} + A_2x_2^2\right] + \left[\frac{d_3b_3^3}{36} + A_3 \cdot x_3^2\right] - \left[\frac{\pi \times r_2^4}{4} + A_4x_4^2\right]$$

$$= \left[\frac{90 \times (24)^3}{36} + 1080 \times (16)^2\right] + \left[\frac{90 \times (50)^3}{12} + 4500 \times (49)^2\right]$$

$$+ \left[\frac{90 \times (60)^3}{36} + 2700 \times (94)^2\right] - \left[\frac{\pi \times (20)^4}{4} + 1256.64 \times (50)^2\right]$$

$$= 311040 + 11742000 + 24397200 - 3267263.7$$

$$= \mathbf{33182976.3\ mm^4}$$

and $\quad I_{y_0} = I_x - A\bar{x}^2 = 38528976.3 - 7023.36 \times (61.05)^2$

$$= 33182976.3 - 26176782.6 = \mathbf{7006193.7\ mm^4}. \quad \textbf{Ans.}$$

Problem 3.30. *Find moment of inertia of the area shown in Fig. P.3.30 : about line AB parallel to the centroidal axis X_0. Also find I_{x_0}.*

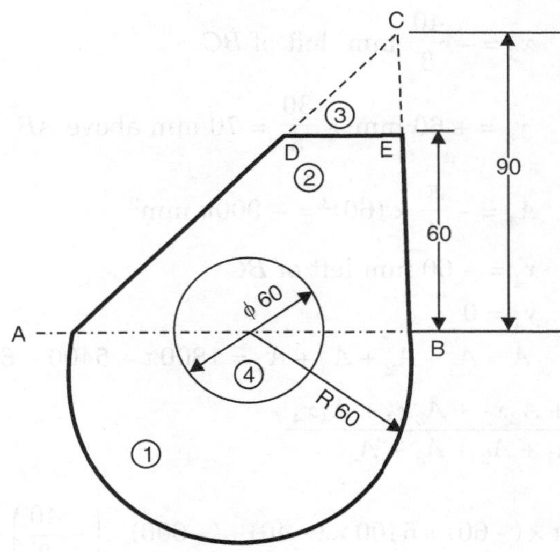

Fig. P.3.30

Sol. There are four areas, *viz.,*

A_1 = semi-circle of radius 60 mm.

A_2 = ΔABC.

A_3 = ΔDEC

A_4 = circle of diameter 60 mm.

$$A_1 = \frac{\pi \times 60^2}{2} = 1800\,\pi \text{ mm}^2$$

x_1 = – 60 mm left of *BC*.

$$y_1 = \frac{-4 \times 60}{3\pi} = -\frac{80}{\pi} \text{ mm below } AB.$$

$$A_2 = \frac{1}{2} \times 120 \times 90 = 5400 \text{ mm}^2$$

$$x_2 = -\frac{120}{3} = -40 \text{ mm left of } BC.$$

$$y_2 = +\frac{90}{3} = +30 \text{ mm above } AB.$$

$\because$ $\Delta CDE \equiv \Delta CAB$

$\therefore$ $$\frac{DE}{90-60} = \frac{AB}{90}$$

$\therefore$ $$DE = \frac{AB}{90} \times 30 = \frac{120}{90} \times 30 = 40 \text{ mm}$$

$\therefore$ $$A_3 = -\frac{40 \times 30}{2} = -600 \text{ mm}^2$$

$$x_3 = -\frac{40}{3} \text{ mm left of } BC$$

$$y_3 = +60 \text{ mm} + \frac{30}{3} = 70 \text{ mm above } AB.$$

$$A_4 = -\frac{\pi}{4} \times (60)^2 = -900\pi \text{ mm}^2$$

$$x_4 = -60 \text{ mm left of } BC$$

$$y_4 = 0$$

$\therefore$ $$A = A_1 + A_2 + A_3 + A_4 = 1800\pi + 5400 - 600 - 900\pi = 7627.43 \text{ mm}^2$$

$\therefore$ $$\bar{x}_4 = \frac{A_1x_1 + A_2x_2 + A_3x_3 + A_4x_4}{A_1 + A_2 + A_3 + A_4}$$

$$= \frac{1800\pi \times (-60) + 5400 \times (-40) + (-600) \times \left(-\frac{40}{3}\right) + (-900\pi) \times (-60)}{7627.43}$$

$$= \frac{377646}{7627.43} \text{ mm} = \mathbf{-49.51 \text{ mm}} \text{ on the left } BC.$$

or $$\bar{y} = \frac{A_1y + A_2y_2 + A_3\, y_3 + A_4y_4}{A_1 + A_2 + A_3 + A_4}$$

$$= \frac{1800\pi \times \left(-\frac{80}{\pi}\right) + 5400 \times (30) - 600 \times (70) + (-900\pi) \times 0}{7627.43} = \frac{-24000}{7627.43} \text{ mm}$$

$$= \mathbf{-3.15 \text{ mm}} \text{ below } AB$$

$$I_{AB} = I_{AB} \text{ for semi-circle} + I_{AB} \text{ for } \Delta ABC - I_{AB} \text{ for } \Delta CDE - I_{AB} \text{ for circle}$$

$$= \left[\frac{\pi \times 60^4}{8}\right] + \left[\frac{120 \times 90^3}{12}\right] - \left[\frac{40 \times 30^3}{36} + \frac{40 \times 30}{2} \times (70)^2\right] - \left[\frac{\pi \times 30^4}{4}\right] \text{ mm}^4$$

$= \mathbf{8773207.6 \text{ mm}^4}$. **Ans.**

$$\therefore \quad I_{x_0} = I_{AB} - A\bar{y}^2 = 8773207.6 - 7627.43 \times (-3.15)^2 \text{ mm}^4$$

$$= \mathbf{8697524.426\ mm^4.} \quad \textbf{Ans.}$$

Problem 3.31. *Find the moment of inertia about the Y-axis and X-axis for the area shown in Fig. P.3.31. Also find* I_{x_0}.

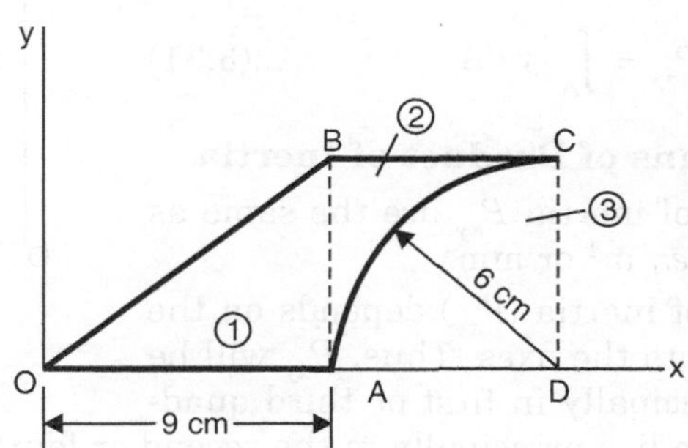

Fig. P.3.31

Sol. $I_y = I_y$ for $\Delta OAB + I_y$ for square $ABCD - I_y$ for quarter circle

$$= \left[\frac{6 \times 9^3}{36} + \frac{9 \times 6}{2} \times \left(\frac{2}{3} \times 9\right)^2\right] + \left[\frac{6 \times 6^3}{12} + (6 \times 6) \times \left(9 + \frac{6}{2}\right)^2\right]$$

$$- \left[0.055 \times 6^4 + \frac{\pi \times 6^2}{4} \times \left(15 - \frac{4 \times 6}{3\pi}\right)^2\right]$$

$$= \mathbf{1023.5 + 5292 - 423.4 = 5962.1\ mm^4.} \quad \textbf{Ans.}$$

$I_x = I_x$ for $\Delta OAB + I_x$ for square $ABCD - I_x$ for quarter circle

$$= \frac{9 \times 6^3}{12} + \frac{6^4}{3} - \left[\frac{\pi \times 6^4}{16}\right] \text{ mm}^4$$

$$= 162 + 432 - 254.57 = \mathbf{339.43\ mm^4.\ Ans.}$$

$$A_1 = \frac{9 \times 6}{2} = 27 \text{ mm}^2\ ;\ y_1 = \frac{6}{3} = 2 \text{ mm from } Ox$$

$$A_2 = 6 \times 6 = 36 \text{ mm}^2\ ;\ y_2 = 3 \text{ cm from } Ox$$

$$A_3 = \frac{-\pi \times 6^2}{4} \text{ mm}^2\ ;\ y_3 = \frac{4 \times 6}{3\pi} = \frac{8}{\pi} \text{ mm from } Ox$$

$$\therefore \quad \bar{y} = \frac{A_1 y_1 + A_2 y_2 + A_3 y_3}{A_1 + A_2 + A_3} = \frac{27 \times 2 + 36 \times 3 - 9\pi \times \frac{8}{\pi}}{27 + 36 - 9\pi} \text{ mm}$$

$$= \frac{90}{34.73} = \mathbf{2.59} \text{ mm from } Ox.$$

$$\therefore \quad I_{x_0} = I_x - A\bar{y}^2 = 339.53 - 34.73 \times (2.59)^2 \text{ mm}^4$$

$$= \mathbf{106.56\ mm^4.} \quad \textbf{Ans.}$$

3.13. PRODUCT OF INERTIA

The product of inertia for an area is the sum of the product of each differential area and its moment arms about two perpendicular reference axes. Thus, for the area shown in Fig. 3.13, the product of inertia with respect to *X*- and *Y*- axes is given by

$$P_{xy} = \int_A xy\, dA \qquad ...(3.81)$$

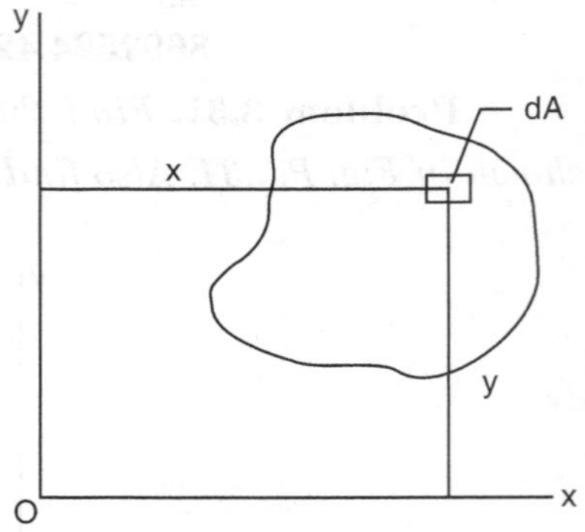

Fig. 3.13. The product of Inertia, P_{xy}.

3.13.1 Units and Signs of Product of Inertia

The *units* of product of inertia P_{xy} are the same as that of moment of inertia, *i.e.*, m^4 or mm^4

The *sign* of product of inertia (P_{xy}) depends on the location of the area relative to the axes. Thus, P_{xy} will be positive if the area lies principally in first or third quadrant and negative if the area lies principally in the second or fourth quadrant with respect to *X* and *Y* axes. Fig. 3.13.1 shows the effect of location of area on P_{xy}.

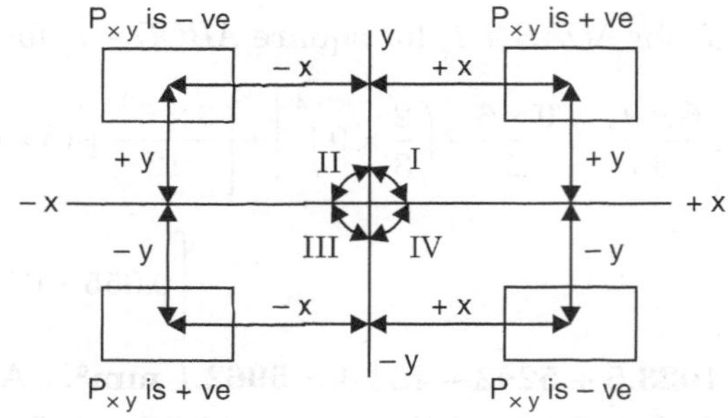

Fig. 3.13.1 Effect of location of area on P_x.

3.14. PRODUCT OF INERTIA IS ZERO IF THE AREA HAS ONE OR TWO AXES OF SYMMETRY

If the given area has one centroidal axis as the axis of symmetry, then product of inertia is zero with respect to this centroidal axis.

In Fig. 3.14.1 the centroidal axes X_0 and Y_0 are axes of symmetry, then the product of inertia of area *dA* at $A = xy\, dA$ and the product of inertia of area *dA* at $B = (-x)\, y\, dA$.

Adding, we get their sum equal to zero. Thus, the sum of the products *xy dA* for each such pair of symmetrically situated elements at *A* and *B* will be zero. Thus, it follows that the value of $\int xy\, dA$ for the entire area will be zero if there is either one or both reference axes are axes of symmetry. Therefore $\overline{P}_{xy}$ for the areas shown in Fig. 3.14.2 is zero.

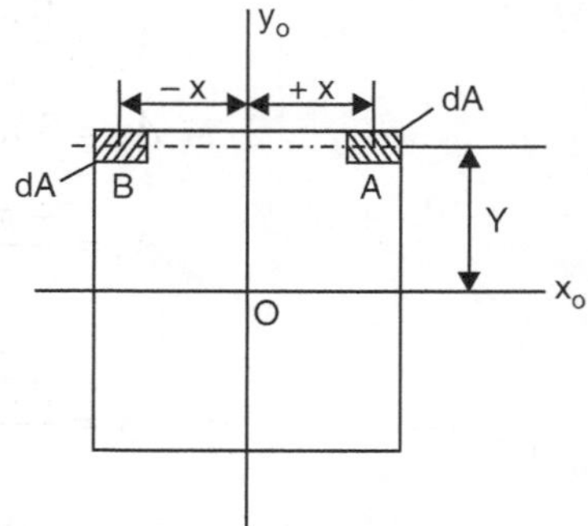

Fig. 3.14.1 $\overline{P}_{xy}$ is zero.

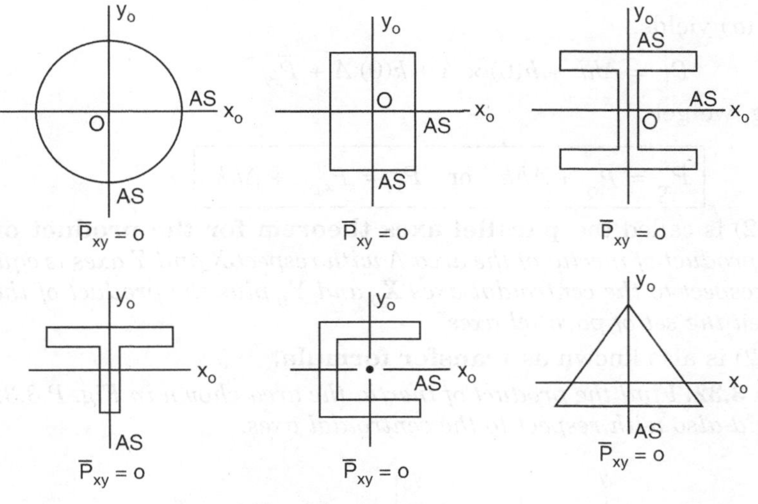

Fig. 3.14.2. Examples of areas having one or two axes of symmetry for which P_{xy} is equal to zero.

3.15. PARALLEL AXES THEOREM FOR PRODUCT OF INERTIA

Consider an area, such as shown in fig 3.15, whose area is A, with X_0 and Y_0 as the centroidal axes. Thus relative to X_0 and Y_0, product of inertia = $\overline{P}_{xy}$.

Let there be a set of X- and Y-axes located as shown in Fig. 3.15.

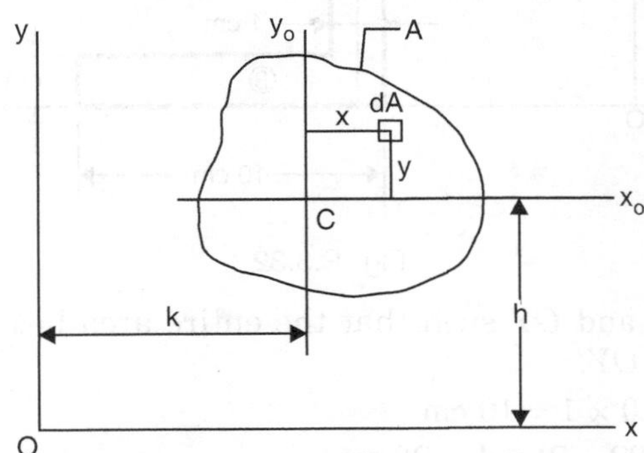

Fig. 3.15. The set of parallel axes.

From the definition, product of inertia with respect to X- and Y-axes is

$$P_{xy} = \int_A (h + y)(k + x)\, dA = \int_A [hk + hx + ky + xy]\, dA$$

$$= \int_A hk\, dA + h\int_A x\, dA + k\int_A y\, dA + \int_A xy\, dA$$

or $$P_{xy} = Ahk + h(\bar{x}A) + k(\bar{y}A) + \overline{P}_{xy} \qquad ...(a)$$

$\because$ X_0 is the centroidal axis, $\bar{y} = 0$ with respect to X_0-axis

Also, $\because$ Y_0 is the centroidal axis, $\bar{x} = 0$, with respect to Y_0-axis.

∴ Eqn. (*a*) yields.

$$P_{xy} = Ahk + h(0) \times A + k(0)\,A + \overline{P}_{xy}$$

or rearranging, we get

$$\boxed{P_{xy} = \overline{P}_{xy} + Ahk \quad \text{or} \quad P_{xy} = P_{x_0 y_0} + Ahk} \qquad ...(3.82)$$

Eqn. (3.82) is called the **parallel axes theorem for the product of inertia,** and it states that "*the product of inertia of the area A with respect X and Y axes is equal to the product of inertia with respect to the centroidal axes X_0 and Y_0 plus the product of the area A and the distances between the set of parallel axes*".

Eqn. (3.82) is also known as **transfer formula.**

Problem 3.32. *Find the product of inertia the area shown in Fig. P.3.32, with respect to the XOY axes and also with respect to the centroidal axes.*

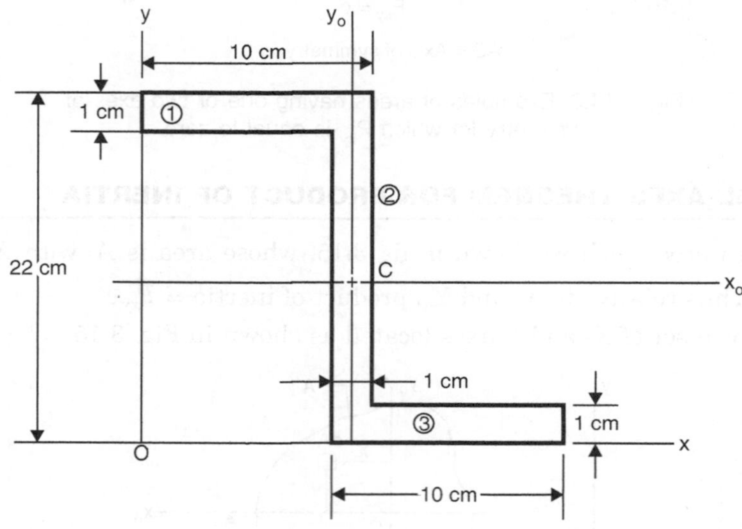

Fig. P.3.32

Sol. Draw lines *OX* and *OY* such that the entire area lies in the first quadrant with respect to the axes *OX* and *OY*.

$A_1 = 10 \times 1 = 10 \text{ cm}^2$

$A_2 = (22 - 2) \times 1 = 20 \text{ cm}^2$

$A_3 = 10 \times 1 = 10 \text{ cm}^2$

$x_1 = 5$ cm from *OY*

$x_2 = 10 - 0.5 = 9.5$ cm from *OY*,

$x_3 = 9 + 5 = 14$ cm from *OY*,

$y_1 = 22 - .05 = 21.5$ cm from *OX*,

$y_2 = 1 + 10 = 11$ cm from *OX*

and $y_3 = 0.5$ cm from *OX*

$$\therefore \quad \overline{x} = \frac{A_1x_1 + A_2x_2 + A_3x_3}{A_1 + A_2 + A_3} = \frac{10 \times 5 + 20 \times 9.5 + 10 \times 14}{10 + 20 + 10} = \frac{380}{40} = 9.5 \text{ cm from } OY$$

$$\bar{y} = \frac{A_1y_1 + A_2y_2 + A_3y_3}{A_1 + A_2 + A_3} = \frac{10 \times 21.5 + 20 \times 11 + 10 \times 0.5}{40} = \frac{440}{40} = 11 \text{ cm from } OX.$$

Using the transfer formula for product of inertia, we have

$$P_{xy} = (\bar{P}_{xy} + Ahk)_1 + (\bar{P}_{xy} + Ahk)_2 + (\bar{P}_{xy} + Ahk)_3$$
$$= [0 + A_1\, x_1y_1] + [0 + A_2x_2y_2] + [0 + A_3\, x_3y_3]$$
$$= [10 \times 5 \times 21.5] + [20 \times 9.5 \times 11] + [10 \times 14 \times 0.5] \text{ cm}^4$$

$\therefore$ P_{xy} = **3210 cm^4. Ans.**

But $P_{xy} = \bar{P}_{xy} + Ahk$ for the entire area. Where $h = \bar{y} = 11$ cm and $k = \bar{x} = 9.5$ cm.

$\therefore$ $\bar{P}_{xy} = P_{xy} - Ahk = 3210 - (40)(11)(9.5)$ $\therefore$ P_{xy} = **– 970 cm^4 Ans.**

Problem 3.33. *Derive expression for the product of inertia of the right triangle shown in Fig. P.3.33 (a) with respect to X- and Y-axes.*

Sol. Take an element as shown in Fig. P.3.33 (*a*).

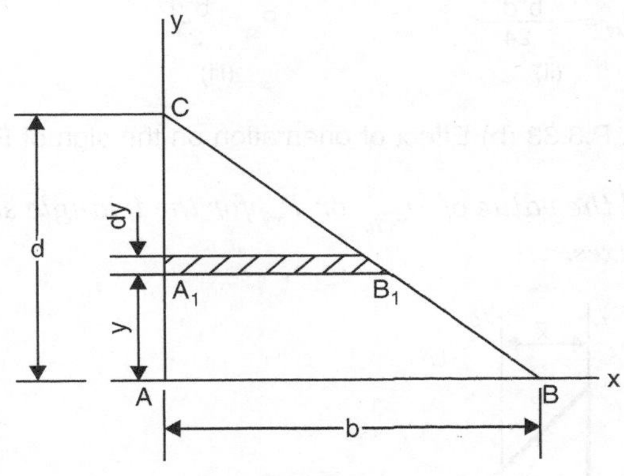

Fig. P. 3.33 (a)

$\therefore$ $dA = A_1B_1 \times dy$

$\because$ $\Delta CA_1B_1 \equiv \Delta CAB,$

$\therefore$ $\dfrac{A_1B_1}{CA_1} = \dfrac{AB}{CA}$

$\therefore$ $A_1B_1 = \dfrac{AB}{CA} \times CA_1 = \dfrac{b}{d}(d - y)$

$\therefore$ $dA = \dfrac{b}{d}(d - y) \,.\, dy$

$\therefore$ $dP_{xy} = \left[\dfrac{b}{d}(d - y)\, dy\right] \times \left(\dfrac{A_1B_1}{2}\right) \times y$

$$= \frac{b}{d}(d - y) \,.\, dy \times \frac{b}{2d}(d - y) \,.\, y = \frac{b^2}{2d^2}\, y(d - y)^2 \,.\, dy$$

$\therefore$ $P_{xy} = \displaystyle\int d\,P_{xy} = \int_0^d \frac{b^2}{2d^2} \,.\, y\,(d - y)^2\, dy$

$$= \frac{b^2}{2d^2} \int_0^d .\,(d^2y - 2dy^2 + y^3)\, dy = \frac{b^2}{2d^2}\left[\frac{d^2 \times d^2}{2} - \frac{2d \times d^3}{3} + \frac{d^4}{4}\right]$$

$$= \frac{b^2}{2d^2} \times d^4 \left(\frac{6 - 8 + 3}{12} \right) \quad \therefore \quad \boxed{P_{xy} = \frac{b^2 d^2}{24}} \text{. } \textbf{Ans.}$$

Expression for P_{xy} for different positions (rather orientations) of this triangle are given in Fig. P.3.33 (*b*)

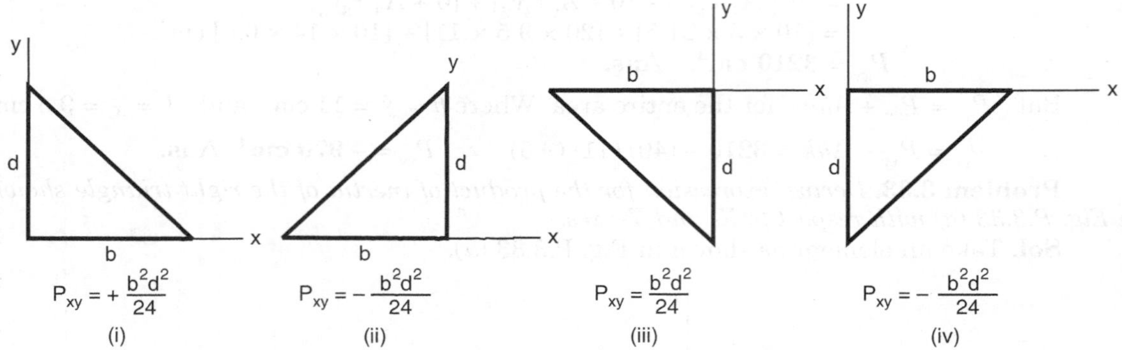

Fig. P.3.33 (b) Effect of orientation on the sign of P_{xy}.

Problem 3.34. *Find the value of* $P_{x_0 y_0}$ *or* $\overline{P}_{xy}$ *for the triangle shown in Fig. P.3.34 (a).* X_0 *and* Y_0 *are the centroidal axes.*

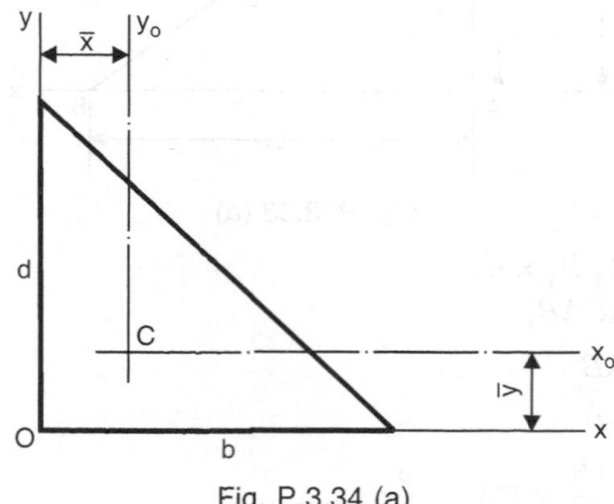

Fig. P.3.34 (a)

Sol. We know that P_{xy} with respect to the X and Y-axes $= \frac{b^2 d^2}{24}$.

Using the parallel axes theorem, we can write

$$P_{xy} = P_{x_0 y_0} + A\overline{x}\overline{y}$$

$$\therefore \quad P_{x_0 y_0} = P_{xy} - A \times \left(\frac{b}{3}\right) \times \left(\frac{d}{3}\right) = \frac{b^2 d^2}{24} - \frac{bd}{2} \times \frac{bd}{9}$$

$$= \frac{b^2 d^2}{24} - \frac{b^2 d^2}{18} = \frac{b^2 d^2}{72}(3 - 4)$$

$$\therefore \quad \boxed{P_{x_0 y_0} \text{ or } \overline{P}_{xy} = -\frac{b^2 d^2}{72}}$$

Values of $\bar{P}_{xy}$ or $P_{x_0y_0}$ for different positions of the triangle of Fig. P.3.34 (*a*) are given in Figs. P.3.34 (*b*) (*i*) to (*iv*).

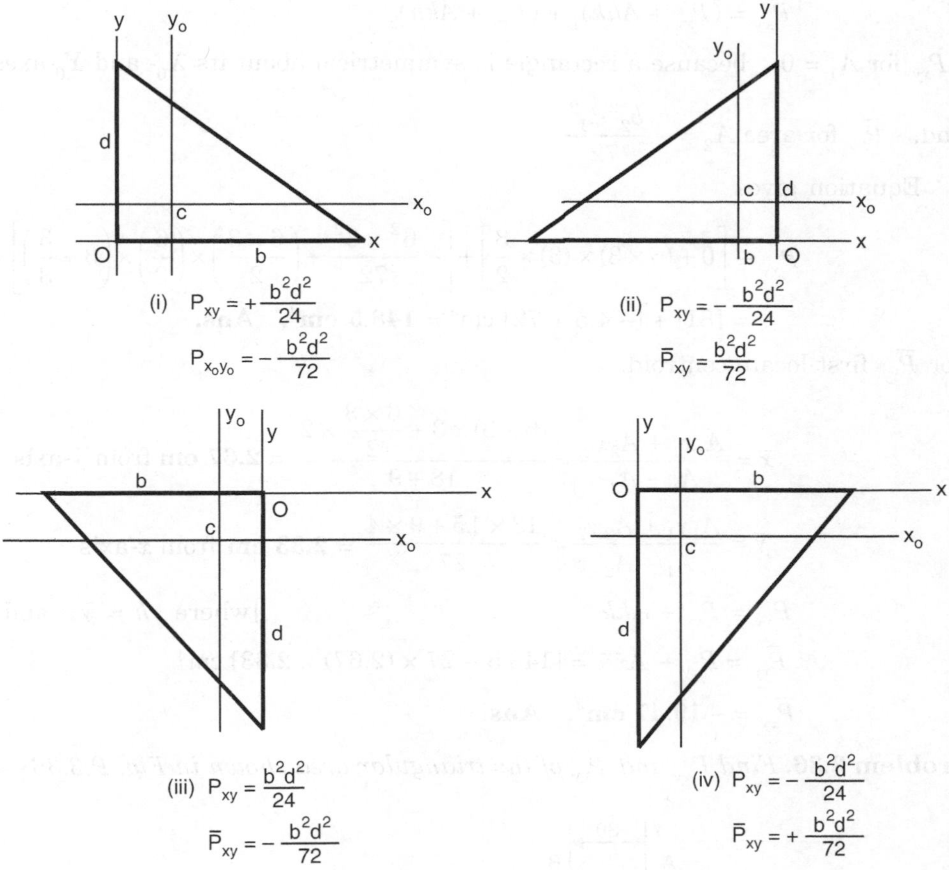

(i) $P_{xy} = +\dfrac{b^2d^2}{24}$

$P_{x_oy_o} = -\dfrac{b^2d^2}{72}$

(ii) $P_{xy} = -\dfrac{b^2d^2}{24}$

$\bar{P}_{xy} = \dfrac{b^2d^2}{72}$

(iii) $P_{xy} = \dfrac{b^2d^2}{24}$

$\bar{P}_{xy} = -\dfrac{b^2d^2}{72}$

(iv) $P_{xy} = -\dfrac{b^2d^2}{24}$

$\bar{P}_{xy} = +\dfrac{b^2d^2}{72}$

Fig. P.3.34 (b) Effect of rotation of triangle on the values of P_{xy} and $\bar{P}_{xy}$.

Problem 3.35. *Find the product of inertia with respect to the X and Y-axes for the area shown in Fig. P.3.35. Also find* $\bar{P}_{xy}$.

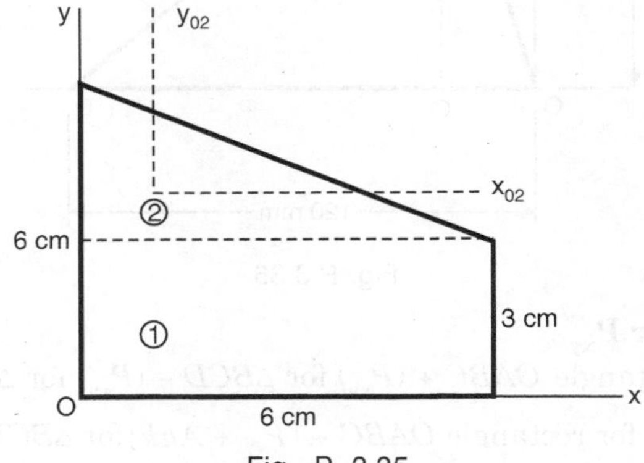

Fig. P. 3.35

Sol. The given area can be divided into a rectangular (area A_1) and a triangle (area A_2) : Using the parallel-axis theorem for the product of inertia, we get

$$P_{xy} = (\overline{P}_{xy} + Ahk)_1 + (\overline{P}_{xy} + Akh)_2 \quad ...(1)$$

where $\overline{P}_{xy}$ for $A_1 = 0$, because a rectangle is symmetrical about its X_0- and Y_0-axes

And, $\overline{P}_{xy}$ for area $A_2 = -\dfrac{b_2^{\,2} d_2^{\,2}}{72}$

$\therefore$ Equation gives

$$P_{xy} = \left[0 + (6 \times 3) \times (3) \times \frac{3}{2}\right] + \left[-\frac{6^2 \times 3^2}{72} + \left(\frac{6 \times 3}{2}\right) \times \left(\frac{6}{3}\right) \times \left(3 + \frac{3}{3}\right)\right] \text{cm}^4$$

$$= [81] + [-4.5 + 72] \text{ cm}^4 = \mathbf{148.5\ cm^4}. \quad \textbf{Ans.}$$

For $\overline{P}_{xy}$ first locate centroid.

$$\bar{x} = \frac{A_1 x_1 + A_2 x_2}{A_1 + A_2} = \frac{(6 \times 3) \times 3 + \frac{6 \times 3}{2} \times 2}{18 + 9} = \mathbf{2.67\ cm} \text{ from } y\text{-axis}$$

$$\bar{y} = \frac{A_1 y_1 + A_2 y_2}{A_1 + A_2} = \frac{18 \times 1.5 + 9 \times 4}{27} = \mathbf{2.33\ cm} \text{ from } x\text{-axis}$$

$\because \quad P_{xy} = \overline{P}_{xy} + Ahk \quad$ [where $h = \bar{y}$ and $k = \bar{x}$]

$\therefore \quad \overline{P}_{xy} = P_{xy} - A\bar{x}\bar{y} = [148.5 - 27 \times (2.67) \times 2.33] \text{ cm}^4$

$\therefore \quad \overline{P}_{xy} = \mathbf{-19.47\ cm^4}$. **Ans.**

Problem 3.36. *Find P_{xy} and $\overline{P}_{xy}$ of the triangular area shown in Fig. P.3.36.*

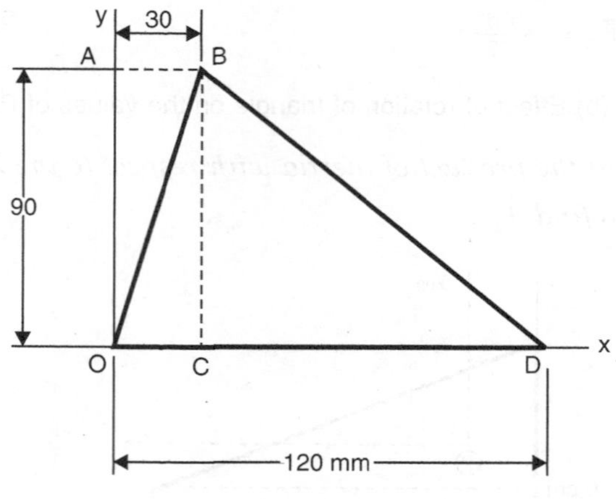

Fig. P.3.36

Sol. Method I. For P_{xy}

$P_{xy} = (P_{xy})$ for rectangle $OABC$ + (P_{xy}) for ΔBCD – (P_{xy}) for ΔOAB

$= [\overline{P}_{xy} + Ahk]$ for rectangle $OABC$ + $[\overline{P}_{xy} + Ahk]$ for ΔBCD – $[\overline{P}_{xy} + Ahk]$ for ΔOAB

$$= [0 + (90 \times 30)(15 \times 45)] + \left[\frac{-(120-30)^2 . (90)^2}{72}\right.$$

$$\left. + \frac{(120-30)\times(90)}{2} \times \left(30 + \frac{90}{3}\right)\left(\frac{90}{3}\right)\right]$$

$$- \left[+\frac{30^2 \times 90^2}{72} + \frac{30 \times 90}{2} \times \left(\frac{2}{3} \times 90\right)\left(\frac{30}{3}\right)\right] \text{mm}^4$$

$= 1822500 + (-911250) + 7290000 - 101250 - 810000$ mm^4

= **7290000 mm^4. Ans.**

Method II. For P_{xy}.

$P_{xy} = (P_{xy})$ for ΔOBC + (P_{xy}) for ΔBCD

$= [\overline{P}_{xy} + Ahk]$ for ΔOBC + $[\overline{P}_{xy} + Ahk]$ for ΔBCD

$$= \left[+\frac{30^2 \times 90^2}{72} + \left(\frac{30 \times 90}{2}\right)\left(\frac{2}{3} \times 30\right) \times (30)\right]$$

$$+ \left[-\frac{90^2 \times 90^2}{72} + \frac{90 \times 90}{2} \times \left(30 + \frac{1}{3} \times 90\right)\left(\frac{90}{3}\right)\right] \text{mm}^4$$

$= (101250 + 810000] + [-911250 + 7290000] = 7290000$

For $\overline{P}_{xy}$ first find $\bar{x}$ and $\bar{y}$ with respect to OX- and OY-axes

$$\bar{x} = \frac{l+a}{3} = \frac{120+30}{3} = 50 \text{ mm from } Y\text{-axis}$$

$$\bar{y} = \frac{90}{3} = 30 \text{ mm from } X\text{-axis}$$

$\because$ $P_{xy} = \overline{P}_{xy} + Ahk$, $\therefore$ We have $\overline{P}_{xy} = P_{xy} - Ahk$, where $h = \bar{y}$ and $k = \bar{x}$

$\therefore$ $\overline{P}_{xy} = 7290000 - \frac{120 \times 90}{2} \times 50 \times 30$ = **$-$ 810000 mm^4. Ans.**

Problem 3.37. *Find product of inertia of the area shown in Fig. P.3.37 with respect to x- and y-axes. Also find product of inertia with respect to the centroidal axes x_0 and y_0.*

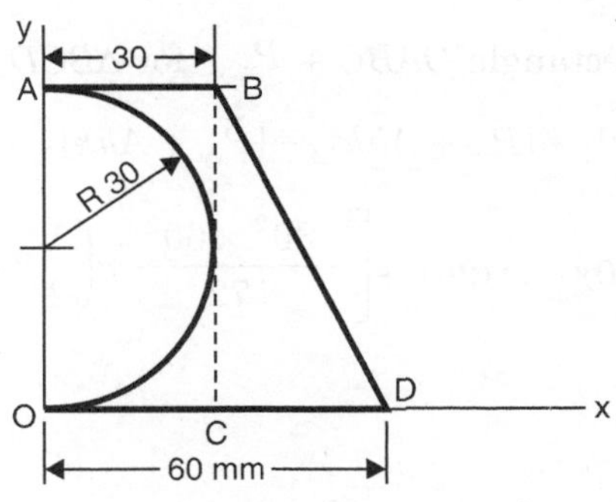

Fig. P.3.37

Sol. $P_{xy} = P_{xy}$ for rectangle $OABC$ + P_{xy} for ΔBCD – P_{xy} for semi-circle.

$= [\bar{P}_{xy} + Ahk]$ for rectangle + $[\bar{P}_{xy} + Ahk]$ for ΔBCD

$- [\bar{P}_{xy} + Ahk]$ for semi-circle

$$= [0 + (30 \times 60)(15 \times 30)] + \left[-\frac{30^2 \times 60^2}{72} + \left(\frac{30 \times 60}{2}\right)\left(30 + \frac{30}{3}\right)\left(\frac{60}{3}\right)\right]$$

$$-\left[0 + \left(\frac{\pi + 30^2}{2}\right)\left(\frac{4 \times 30}{3\pi}\right)(30)\right] \text{mm}^4 = [810{,}000] + [675{,}000] - [540{,}000] \text{ mm}^4$$

or $\quad P_{xy} = \mathbf{945{,}000\ mm^4}$. **Ans.**

$$\bar{x} = \frac{A_1x_1 + A_2x_2 + A_3x_3}{A_1 + A_2 + A_3}$$

$$= \frac{(30 \times 60) \times 15 + \left(\frac{1}{2} \times 30 \times 60\right) \times \left(30 + \frac{30}{3}\right) + \left(-\frac{\pi \times 30^2}{2} \times \frac{4 \times 30}{3\pi}\right)}{30 \times 60 + \frac{1}{2} \times 30 \times 60 + \left(-\frac{\pi}{2} \times 30^2\right)}$$

$$= \frac{45000}{1286.28} = 35 \text{ mm from } OA.$$

$$\bar{y} = \frac{A_1y_1 + A_2y_2 + A_3y_3}{A_1 + A_2 + A_3} = \frac{(30 \times 60) \times 30 + \left(\frac{1}{2} + 30 \times 60\right) \times \frac{60}{3} + \left(-\frac{\pi \times 30^2}{2}\right) \times 30}{1286.28}$$

$$= \frac{29588.5}{1286.28} = 23 \text{ mm from } OB.$$

$h_1 = \bar{y} - y_1 = 23 - 30 = -7$ mm ; $h_2 = 23 - 20 = 3$ mm , $h_3 = 23 - 30 = -7$ mm

$k_1 = \bar{x} - x_1 = 35 - 15 = 20$ mm ; $k_2 = 35 - 40 = -5$ mm ;

$$k_3 = 35 - \frac{4 \times 30}{3\pi} = 35 - 12.73 = 22.27 \text{ mm}$$

$\therefore \quad P_{x_0y_0} = P_{x_0y_0}$ for rectangle $OABC$ + $P_{x_0y_0}$ for ΔBCD – $P_{x_0y_0}$ for semi-circle.

$$= [\bar{P}_{xy} + Ahk]_1 + [\bar{P}_{xy} + Ahk]_2 - [\bar{P}_{xy} + Ahk]_3$$

$$= [0 + 30 \times 60)(-7)(20)] + \left[-\frac{30^2 \times 60^2}{72} + \left(\frac{1}{2} \times 30 \times 60\right)(3)(-5)\right]$$

$$-\left[0 + \frac{\pi \times 30^2}{2} \times (-7)(22.27)\right]$$

$= [-252{,}000] + [-45{,}000 - 13{,}500] - [-220{,}304.3]$ mm^4

$= \mathbf{-90{,}195.7\ mm^4}$. **Ans**

Problem 3.38. *Find P_{xy} and $\overline{P}_{xy}$ for the quarter circular area shown in Fig. P.3.38.*

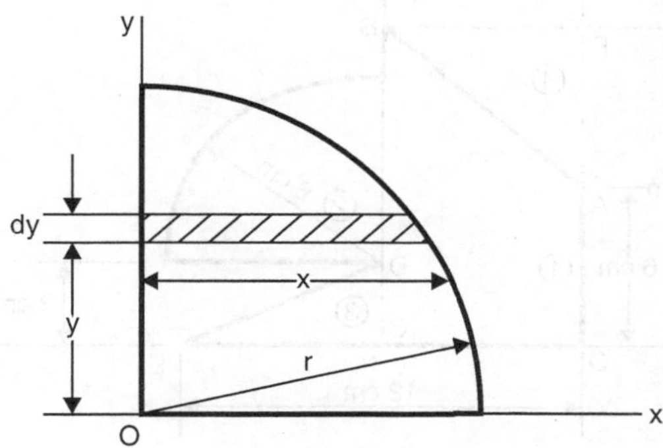

Fig. P.3.38

Sol. Consider the element shown in the figure. Area of the element = $dA = x\,dy$.

And, $$dP_{xy} = \frac{x}{2}\,y\,.\,dA = \frac{xy}{2}(x.dy) \quad \text{where} \quad x = \sqrt{r^2 - y^2}$$

$$\therefore \quad dP_{xy} = \frac{x^2}{2}\,ydy = \left(\frac{r^2 - y^2}{2}\right) y\,dy$$

$$\therefore \quad P_{xy} = \int_0^r \frac{1}{2}(r^2y - y^3)\,dy = \frac{1}{2}\left[\frac{r^2 \times r^2}{2} - \frac{r^4}{4}\right] = \frac{r^4}{2} \times \frac{2-1}{4}$$

$$\therefore \quad \boxed{\mathbf{P_{xy} = \frac{r^4}{8}.}} \quad \textbf{Ans.}$$

Using the parallel axis theorem,

$$P_{xy} = \overline{P}_{xy} + Ahk$$

or $$\overline{P}_{xy} = P_{xy} - Ahk \qquad \text{where } h = \overline{y} = \frac{4r}{3\pi}$$

$$\therefore \quad \overline{P}_{xy} = \frac{r^4}{8} - \frac{\pi r^2}{4} \times \frac{4r}{3\pi} \times \frac{4r}{3\pi} \qquad \text{and } k = \overline{x} = \frac{4r}{3\pi}$$

$$= r^4\left(\frac{1}{8} - \frac{4}{9\pi}\right)$$

or $$\boxed{\mathbf{\overline{P}_{xy} = -\,0.0165\,r^4.}} \quad \textbf{Ans.}$$

Problem 3.39. *Find product of inertia P_{xy} of the area shown in Fig. P.3.39, with respect to X and Y axes. Also find $\overline{P}_{xy}$ or $P_{x_0y_0}$ with respect to the centroidal axes X_0 and Y_0.*

Sol. Given area can be divided into :

A_1 = rectangle *OFBC* ;
A_2 = quarter-circle of radius 6 cm.
A_3 = triangle *CDE*.
A_4 = triangle *AFB* (–ve)
$A_1 = 6 \times 12 = 72\ \text{cm}^2$

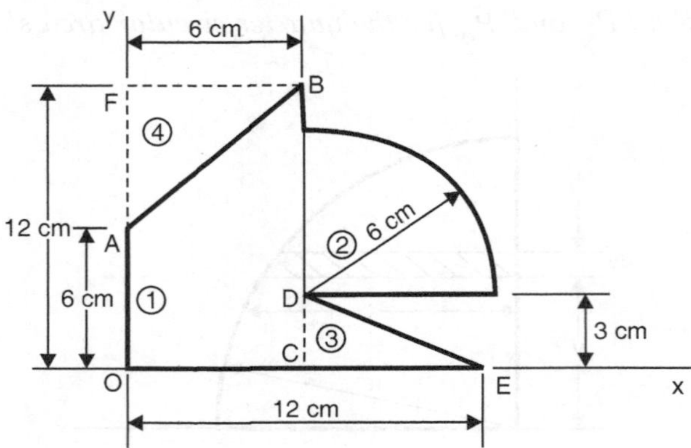

Fig. P. 3.39

$$A_2 = \frac{\pi \times 6^2}{4} = 9\pi \text{ cm}^2$$

$$A_3 = \frac{3 \times 6}{2} = 9 \text{ cm}^2$$

And, $$A_4 = -\frac{1}{2} \times 6 \times 6 = -18 \text{ cm}^2$$

From OY : $x_1 = 3$ cm ; $x_2 = 6 + \frac{4 \times 6}{3\pi} = 6 + \frac{8}{\pi} = 8.55$ cm.

$$x_3 = 6 + \frac{6}{3} = 9 \text{ cm.}$$

$$x_4 = \frac{6}{3} = 2 \text{ cm}$$

From OX : $y_1 = 6$ cm ; $y_2 = 3 + \frac{4 \times 6}{3\pi} = 5.55$ cm

$$y_3 = \frac{3}{3} = 1 \text{ cm} ; y_4 = 6 + \frac{2}{3} \times 6 = 10 \text{ cm}$$

$$\therefore \quad \bar{x} = \frac{A_1 x + A_2 x_2 + A_3 x_3 + A_4 x_4}{A_1 + A_2 + A_3 + A_4}$$

$$= \frac{72 \times 3 + 9\pi \times 8.55 + 9 \times 9 - 18 \times 2}{72 + 9\pi + 18 - 18} = \frac{502.746}{100.27}$$

= **5.014 cm** from OY. **Ans.**

$$\bar{y} = \frac{A_1 y_1 + A_2 y_2 + A_3 y_3 + A_4 y_4}{A_1 + A_2 + A_3 + A_4}$$

$$= \frac{72 \times 6 + 9\pi \times 5.55 + 9 \times 1 - 18 \times 10}{100.27} = \frac{417.92}{100.27} \text{ cm}$$

= **4.17 cm** from OX. **Ans.**

$P_{xy} = (P_{xy})$ for rectangle + (P_{xy}) for quarter circle + (P_{xy}) for ΔCDE – (P_{xy}) for triangle AFB.

$\therefore \quad P_{xy} = [\bar{P}_{xy} + A_1 x_1 y_1]$ for rectangle $OFBC$ + $[\bar{P}_{xy} + A_2 x_2 y_2]$ for semi-circular + $[\bar{P}_{xy} + A_3 x_3 y_3]$ for ΔCDE – $[\bar{P}_{xy} + A_4 x_4 y_4]$ for AFB

$$= [0 + (6 \times 12)(3 \times 6)] + [-0.0165 \times 6^4 + (9\pi)(8.55) \times 5.55]$$

$$+ \left[\frac{-3^2 \times 6^2}{72} + 9 \times 9 \times 1\right] - \left[\frac{6^2 \times 6^2}{72} + 18 \times 2 \times 10\right] \text{cm}^4 = \mathbf{2314.8\ cm^4.\ Ans.}$$

And from $\quad P_{xy} = \bar{P}_{xy} + A\overline{xy}$, we get

$$\bar{P}_{xy} = P_{xy} - Axy = 2314.8 - 100.27 \times (5.014)(4.17) \text{ cm}^4 = \mathbf{218.32\ cm^4.\ \ Ans.}$$

Problem 3.40. *Find product of inertia of the area shown in Fig. P.3.40 with respect to OX and OY axes.*

Sol. The given area may be divided into three quarter circles (A_1), and a semi-circle A_2, as shown in Fig. P.3.40.

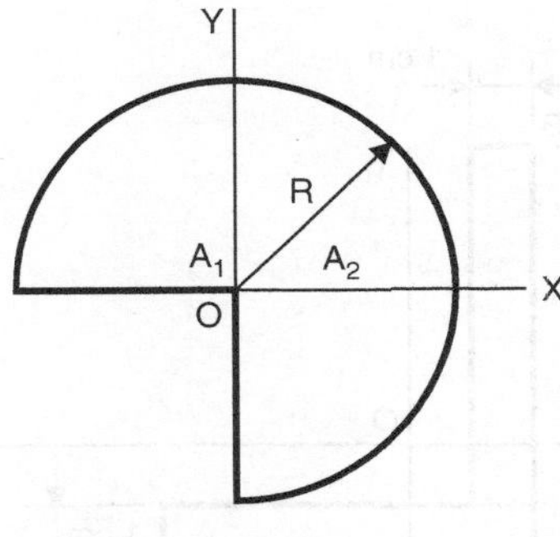

Fig. P. 3.40

$\therefore \qquad P_{xy} = [\bar{P}_{xy} + Ahk]$ for quarter circle A_1 + $[\bar{P}_{xy} + Ahk]$ for the semi-circle A_2

$$= \left[-\frac{R^4}{8}\right] + [0]$$

$\therefore \qquad$ $= \mathbf{-\ 0.125\ R^4.\quad Ans.}$

Problem 3.41. *Find P_{xy} for the area shown in Fig. P.3.41.*

Sol. The given area can be divided into four simple areas :

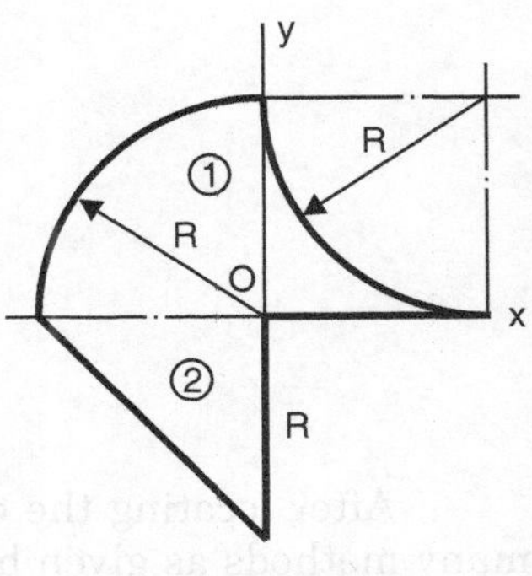

Fig. P.3.41

$$A_1 = \text{the quarter circle (shown)} = \frac{\pi R^2}{4}$$

$$A_2 = \text{the triangle} = \frac{R^2}{2}$$

$$A_3 = \text{the square} = R^2$$

and, $\quad A_4 = \text{the quarter circle} = -\frac{\pi R^2}{4}$

$$\therefore \quad A = A_1 + A_2 + A_3 + A_4 = \frac{\pi R^2}{4} + \frac{R^2}{2} + R^2 - \frac{\pi R^2}{4} = \frac{3R^2}{2}$$

$$P_{xy} = \left[\overline{P}_{x_1y_1} + A_1x_1 \cdot y_1\right] + \left[\overline{P}_{x_2y_2} + A_2x_2y_2\right] + \left[\overline{P}_{x_3y_3} + A_3 \cdot x_3y_3\right] - \left[\overline{P}_{x_4y_4} + A_4x_4y_4\right]$$

$$P_{xy} = \left[-\frac{R^4}{8}\right] + \left[\frac{R^4}{24}\right] + \left[0 + (R^2)\left(\frac{R}{2}\right)\left(\frac{R}{2}\right)\right] - \left[-0.0165\,R^4 + \frac{\pi R^2}{4} \times \left(R - \frac{4R}{3\pi}\right)\left(R - \frac{4R}{3\pi}\right)\right]$$

$$= R^4\left[-\frac{1}{8} + \frac{1}{24} + \frac{1}{4} + 0.0165 + 0.26020\right]$$

$= \mathbf{-\ 0.077\ R^4}$. **Ans.**

Problem 3.42. *Find $\bar{I}_x$, $\bar{I}_y$, I_{AC} and I_{AB} for the angle section shown in Fig. P.3.42. (a). Also find $\bar{k}_x$ and $\bar{k}_y$.*

Sol. The given area can be divided into two rectangles (6 × 1) cm^2 and (7 – 1) × 1 cm^2.

$A_1 = 6 \times 1 = 6$ cm^2

$x_1 = 3$ cm from AB

$y_1 = 0.5$ cm from AC

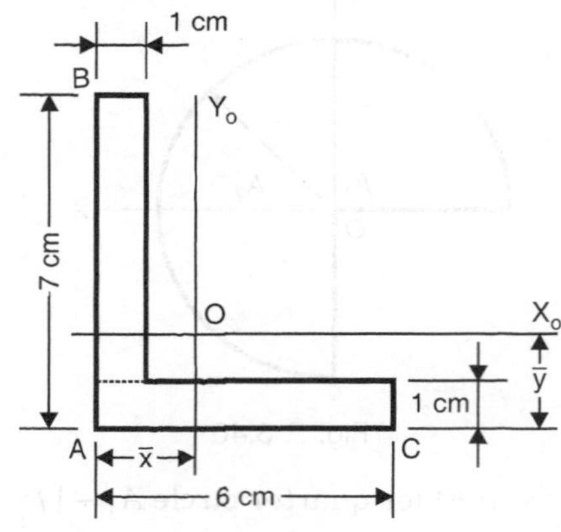

Fig. P.3.42 (a)

$A_2 = 1 \times (7 - 1) = 6$ cm^2

$x_2 = 0.5$ cm from AB

$y_2 = 1 + \dfrac{7-1}{2} = 4$ cm from AC

$$\bar{x} = \frac{A_1x_1 + A_2x_2}{A_1 + A_2} = \frac{6 \times 3 + 6 \times 0.5}{6 + 6} = \mathbf{1.75\ cm} \text{ from } AB$$

$$\bar{y} = \frac{A_1y_1 + A_2y_2}{A_1 + A_2} = \frac{6 \times 0.5 + 6 \times 4}{6 + 6} = \mathbf{2.25\ cm} \text{ from } AC.$$

After locating the centroid C (1.75, 2.25 cm) the rest of the problem can be solved in many methods as given below.

Method I

$$I_{AC} = \frac{b_1{d_1}^3}{3} + \left[\frac{b_2{d_2}^3}{12} + (A_2 \times {y_2}^2)\right]$$

$$= \frac{6 \times 1^3}{3} + \left[\frac{1 \times 6^3}{12} + 6 \times 4^2 \right] \text{cm}^4 = \mathbf{116\ cm^4.\ \ Ans.}$$

Now, $I_{AC} = \bar{I}_x + A\bar{y}^2$

$\therefore$ $\bar{I}_x = I_{AC} - A\bar{y}^2 = 116 - 12 \times (2.25)^2 \text{ cm}^4$

$\therefore$ $\bar{I}_x = \mathbf{55.25\ cm^4.}$ **Ans.**

$$I_{AB} = \frac{d_1 b_1^{\,3}}{3} + \frac{d_2 b_2^{\,3}}{3} = \frac{1 \times 6^3}{3} + \frac{6 \times 1^3}{3} \text{ cm}^4 \quad \therefore \quad I_{AB} = \mathbf{74\ cm^4.\ \ Ans.}$$

$\because$ $I_{AB} = \bar{I}_y + A\bar{x}^2$

$\therefore$ $\bar{I}_y = I_{AB} - A\bar{x}^2 = 74 - 12 \times 1.75^2 \text{ cm}^4$

$\therefore$ $\bar{I}_y = \mathbf{37.25\ cm^4.}$ **Ans.**

Since $\bar{I}_x = A\bar{k}_x^{\,2}$

$$\therefore \quad \bar{k}_x = \sqrt{\frac{\bar{I}_x}{A}} = \sqrt{\left(\frac{55.25}{12}\right)} \text{ cm}$$

$\therefore$ $\bar{k}_x = \mathbf{2.146\ cm.}$ **Ans.**

And
$$\bar{k}_y = \sqrt{\frac{\bar{I}_y}{A}} = \sqrt{\frac{37.25}{12}} \text{ cm}$$

$\therefore$ $\bar{k}_y = \mathbf{1.762\ cm.}$ **Ans.**

Method II

We have, $\bar{I}_x = (\bar{I}_x + Ah^2)_1 + (\bar{I}_x + Ah^2)_2$ ...(1)

and $\bar{I}_y = (\bar{I}_y + Ak^2)_1 + (\bar{I}_x + Ak^2)_2$ ...(2)

where $h_1 = \bar{y} - y_1 = 2.25 - 0.5 = +\ 1.75$ cm

$h_2 = \bar{y} - y_2 = 2.25 - 4 = -\ 1.75$ cm

$k_1 = \bar{x} - x_1 = 1.75 - 3 = -\ 1.25$ cm

and $k_2 = \bar{x} - x_2 = 1.75 - 0.5 = +\ 1.25$ cm

$\therefore$ From (1)

$$\bar{I}_x = \left[\frac{6 \times 1^3}{12} + (6 \times 1) \times (1.75)^2 \right] + \left[\frac{1 \times 6^3}{12} + (1 \times 6) \times (-\ 1.75)^2 \right] \text{cm}^4 = \mathbf{55.25\ cm^4.\ \ Ans.}$$

$$\bar{I}_y = \left[\frac{1 \times 6^3}{12} + (6 \times 1) \times (-\ 1.25)^2 \right] + \left[\frac{6 \times 1^3}{12} + (6 \times 1) \times (1.25)^2 \right] \text{cm}^4 = \mathbf{37.25\ cm^4.\ \ Ans.}$$

$\therefore$ $I_{AC} = \bar{I}_x + A\bar{y}^2 = 55.24 + (6 + 6) \times (2.25)^2 \text{ cm}^4 = \mathbf{116\ cm^4.}$ **Ans.**

and $I_{AB} = \bar{I}_y + A\bar{x}^2 = 37.25 + (6 + 6) \times (1.75)^2 \text{ cm}^4 = \mathbf{74\ cm^4.}$ **Ans.**

Method III. See Fig. P.3.42 (*b*)

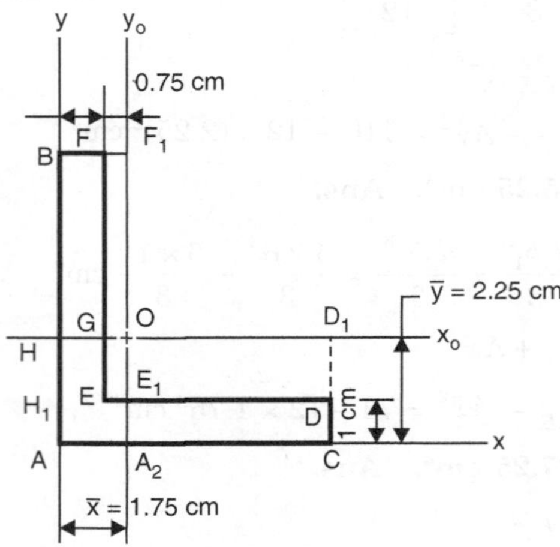

Fig. P. 3.42 (b)

$$\bar{I}_x = [\bar{I}_x \text{ of Rectangular } ACD_1H - I_x \text{ of Rectangle } DED_1G] + [\bar{I}_x \text{ of Rectangle } BFGH]$$

$$= \frac{AC \times CD_1^3}{3} - \frac{(DE) \times (DD_1)^3}{3} + \frac{(BF) \times (EG)^3}{3}$$

$$= \frac{6 \times (2.25)^3}{3} - \frac{(6-1) \times (2.25-1)^3}{3} + \frac{1 \times (7-2.25)^3}{3} \text{ cm}^4$$

$$= \frac{1}{3}[68.34375 - 9.765625 + 107.171875] \text{ cm}^4$$

$$= \frac{1}{3} \times 165.75 = \mathbf{55.25 \text{ cm}^4}. \quad \textbf{Ans.}$$

$$\bar{I}_y = \bar{I}_y \text{ of Rectangle } A_2CDE_1 + \bar{I}_y \text{ of } A_2ABF_1 - \bar{I}_y \text{ of Rectangle } FF_1E_1E$$

$$= \frac{1}{3} \times (6 - 1.75)^3 + \frac{1}{3}(7 \times 1.75^3) - \frac{1}{3}(7-1) \times (1.75-1)^3 \text{ cm}^4$$

$$= \frac{1}{3}[76.765625 + 37.515625 - 2.53125] = 111.75 \times \frac{1}{3} \text{ cm}^4$$

$$= \mathbf{37.25 \text{ cm}^4}. \quad \textbf{Ans.}$$

3.16. MOMENT OF INERTIA WITH RESPECT TO INCLINED AXES

The product of inertia is useful when we need to find the moment of inertia of an area about inclined axes. As in Fig. 3.16, we need to find I_x' and I_y' in terms of I_x, I_y, P_{xy} and the angle between *X*-axis and *X′*-axis.

Here co-ordinates of *dA* are (x, y) with respect to *X*-axis *Y*-axis ; and also (x', y') with respect to *X′* and *Y′* axis. We find that

$$x' = ON = ON_1 + N_1N = ON_2 \cos\theta + N_2N_3$$
$$= x\cos\theta + PN_2 \sin\theta = x\cos\theta + y\sin\theta$$

or $$x' = y\sin\theta + x\cos\theta$$

Also, $y' = PN = PN_3 - NN_3 = PN_3 - N_1N_2 = y\cos\theta - x\sin\theta$

$\therefore$ we get

$$\left.\begin{aligned} x' &= y\sin\theta + x\cos\theta \\ \text{and} \quad y' &= y\cos\theta - x\sin\theta \end{aligned}\right\} \qquad ...(3.83)$$

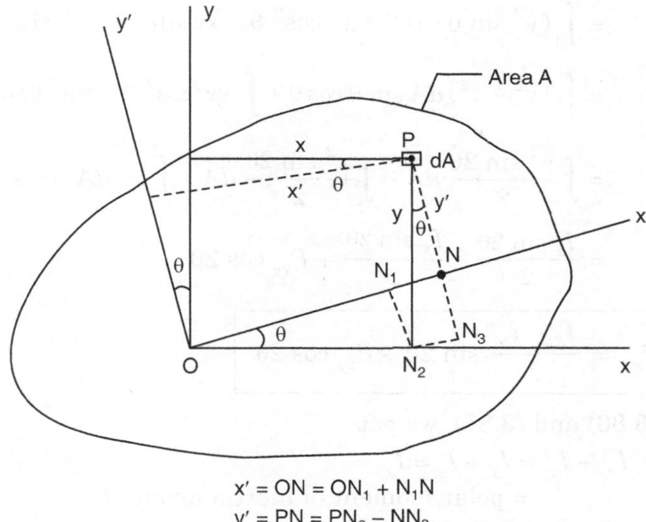

x' = ON = ON$_1$ + N$_1$N
y' = PN = PN$_3$ – NN$_3$

Fig. 3.16. Axes of rotation

$$\therefore \quad I_x' = \int_A y'^2 \, dA = \int_A (y \cos\theta - x \sin\theta)^2 \, dA$$

$$= \int_A (y^2 \cos^2\theta + x^2 \sin^2\theta - 2yx \cos\theta \sin\theta) \, dA$$

$$= \int_A (y^2 \, dA) \cos^2\theta + \int_A x^2 \sin^2\theta \,.\, dA - \int_A xy \sin 2\theta \, dA$$

$$\therefore \quad I_x' = \bar{I}_x \cos^2\theta + \bar{I}_y \sin^2\theta - P_{xy} \sin 2\theta. \qquad ...(3.84)$$

And,
$$I_y' = \int_A (x')^2 \, dA = \int_A (y \sin\theta + x \cos\theta)^2 \, dA$$

$$= \int (y^2 \sin^2\theta + x^2 \cos^2\theta + 2xy \sin\theta \cos\theta) \, dA$$

$$\therefore \quad I_y' = I_x \sin^2\theta + I_y \cos^2\theta + P_{xy} \sin 2\theta \qquad ...(3.85)$$

$$\because \quad \cos 2\theta = \cos^2\theta - \sin^2\theta$$

or
$$\cos 2\,\theta = 1 - 2\sin^2\theta = 2\cos^2\theta - 1$$

$$\therefore \quad \cos^2\theta = \frac{1 + \cos 2\theta}{2} \quad \text{and} \quad \sin^2\theta = \frac{1 - \cos 2\theta}{2}$$

∴ From eqns. (3.84) and (3.85), we get

$$I_x' = \frac{I_x + I_y}{2} + \frac{I_x - I_y}{2} \cos 2\theta - P_{xy} \sin 2\theta \qquad ...(3.86)$$

$$I_y' = \frac{I_x + I_y}{2} - \frac{I_x - I_y}{2} \cos 2\theta + P_{xy} \sin 2\theta \qquad ...(3.87)$$

And
$$P_{x'y'}' = \int x'y' \, dA = \int (y \cos\theta - x \sin\theta)(y \sin\theta + x \cos\theta) \, dA$$

$$= \int_A (y^2 \sin\theta\cos\theta + xy\cos^2\theta - xy\sin^2\theta - x^2\sin\theta\cos\theta)\, dA$$

$$= \int (y^2 - x^2)\, dA \sin\theta\cos\theta + \int xy\,(\cos^2\theta - \sin^2\theta)\, dA$$

$$= \int \frac{y^2 \sin 2\theta}{2} dA - \int \frac{x^2 \sin 2\theta}{2} dA + \int xy\, dA\,.(\cos 2\theta)$$

$$= \frac{I_x \sin 2\theta}{2} - \frac{I_y \sin 2\theta}{2} + P_{xy}\cos 2\theta$$

or

$$\boxed{P_{x'y'} = \frac{I_x - I_y}{2}\sin 2\theta + P_{xy}\cos 2\theta} \quad ...(3.88)$$

Adding eqns. (3.86) and (3.87), we get

$$I_x' + I_y' = I_x + I_y = I_z \quad ...(3.89)$$

= polar moment of inertia about O.

When value of I_x, I_y and P_{xy} are known, eqns. (3.86), (3.87) and (3.88) help to find the values of I_x', I_y' and $P_{x'y'}$ with respect to inclined axes X' and Y', and the angle θ without integration.

3.16.1. The Maximum and Minimum Moments of Inertia. From eqns. (3.86), (3.87), it can be seen that the values of I_x' and I_y' vary with the angle θ for the given values of I_x, I_y and P_{xy}. Certainly there exists some value of θ for which the value of I_x' will be maximum or minimum. The value of θ for I_{max} and I_{min} can be found by differentiating eqn. (3.86) with respect to 2θ and setting the derivative equal to zero. Thus, we get by

$$\frac{d}{d(2\theta)}(I_x') = 0 - \frac{I_x - I_y}{2}\sin 2\theta - P_{xy}\cos 2\theta = 0.$$

$$\therefore \quad \tan 2\theta_{1,2} = -\frac{P_{xy}}{\left(\dfrac{I_x - I_y}{2}\right)} = m_1 \text{ (say)} \quad ...(3.90)$$

$$\therefore \quad \sin 2\theta_{1,2} = \mp\frac{P_{xy}}{\sqrt{\left(\dfrac{I_x - I_y}{2}\right)^2 + P_{xy}^2}} = \mp\frac{P_{xy}}{R_I} \quad ...(3.91)$$

and

$$\cos 2\theta_{1,2} = \pm\frac{(I_x - I_y)/2}{\sqrt{\left(\dfrac{I_x - I_y}{2}\right)^2 + P_{xy}^2}} = \pm\frac{(I_x - I_y)/2}{R_I} \quad ...(3.92)$$

where,

$$R_I = \sqrt{\left(\frac{I_x - I_y}{2}\right)^2 + P_{xy}^2} \quad ...(3.93)$$

By substituting values of cos 2θ, and sin 2θ, in eqns. (3.86) and (3.87),

we get

$$I_{max,\,min} = \frac{I_x + I_y}{2} + \frac{I_x - I_y}{2} \times \left(\pm\frac{I_x - I_y}{2RL}\right) - P_{xy}\cdot\left(\pm\frac{P_{xy}}{R_I}\right)$$

$$= \frac{I_x + I_y}{2} \pm \frac{1}{R_I}\left\{\left(\frac{I_x - I_y}{2}\right)^2 + P_{xy}^2\right\} = \frac{I_x + I_y}{2} \pm \frac{R_I^2}{R_I}$$

or $$I_{\max} = I_u = \left(\frac{I_x + I_y}{2} + R_I\right) = \frac{I_x + I_y}{2} + \sqrt{\left(\frac{I_x - I_y}{2}\right)^2 + P_{xy}^2} \quad ...(3.94)$$

and $$I_{\min.} = I_v = \left(\frac{I_x + I_y}{2} - R_I\right) = \frac{I_x + I_y}{2} - \sqrt{\left(\frac{I_x - I_y}{2}\right)^2 + P_{xy}^2} \quad ...(3.95)$$

And, $$P_{uv} = \left(\frac{I_x - I_y}{2}\right) \times \left\{\mp \frac{P_{xy}}{R_I}\right\} + P_{xy}\left\{\mp \frac{(I_x - I_y)/2}{R_I}\right\}.$$

$\therefore$ $$P_{uv} = 0 \quad ...(3.96)$$

The u and v axes are inclined at angles θ_1 and ($\theta_2 = \theta_1 + 90°$) to the x-axis, and define the axes about which the moment of inertia of the given area is maximum ($= I_u$) and minimum ($= I_v$), respectively. The u-axis and v-axis are known as the *Principal moment of inertia axes*. And, I_u and I_v are called the *Principal moments of inertia* of the area.

Eqn. (3.96) shows that the product of inertia (P_{uv}) with respect to the principal moments of inertia axes is always equal to zero.

3.16.2. Maximum and Minimum Product of Inertia. To find maximum and minimum product of inertia $P_{x'y'})_{\max}$ and $P_{x'y'})_{\min}$ differentiate eqn. (3.88) with respect to 2θ and set the derivative equal to zero. Thus, we get

$$\frac{d}{d(2\theta)}(P_{x'y'}) = \frac{I_x - I_y}{2} \times \cos 2\theta - P_{xy} \sin 2\theta = 0$$

$\therefore$ $$\tan 2\theta = \frac{(I_x - I_y)/2}{P_{xy}} = m_2 \text{ (say)} \quad ...(3.97)$$

$\therefore$ $$\sin 2\theta = \pm \frac{(I_x - I_y)/2}{\sqrt{\left(\frac{I_x - I_y}{2}\right)^2 + P_{xy}^2}} = \pm \frac{(I_x - I_y)/2}{R_I} \quad ...(3.98)$$

and $$\cos 2\theta = \pm \frac{P_{xy}}{\sqrt{\left(\frac{I_x - I_y}{2}\right)^2 + P_{xy}^2}} = \pm \frac{P_{xy}}{R_I} \quad ...(3.99)$$

Thus, by substituting values of sin 2θ and cos 2θ in eqn. (3.88), we get

$$P_{x'y'})_{\substack{\max \\ \min}} = \frac{(I_x - I_y)}{2} \times \left\{\pm \frac{(I_x - I_y)/2}{R_I}\right\} + P_{xy} \times \left\{\frac{\pm P_{xy}}{R_I}\right\}$$

$$= \pm \frac{1}{R_I}\left\{\frac{(I_x - I_y)^2}{2} + P_{xy}^2\right\}$$

or $$P_{x'y'})_{\substack{\max \\ \min}} = \pm \sqrt{\frac{(I_x - I_y)^2}{2} + P_{xy}^2} = \pm R_I \quad ...(3.100)$$

Multiplying eqns. (3.90) and (3.97), it is clear that

$$m_1 \times m_2 = -1 \quad \text{...(3.101)}$$

This states the angle 2θ between the principal moment of inertia axes and the axes of maximum and minimum product of inertia axes is equa to 90°. Therefore the true angle between these axes is 45°.

3.17. MOHR'S CIRCLE FOR MOMENT OF INERTIA

Eqns. (3.86) and (3.88) can be rewritten as under :

$$\left(I_x' - \frac{I_x + I_y}{2}\right) = \frac{I_x - I_y}{2} \cos 2\theta - P_{xy} \sin 2\theta \quad \text{...(3.102)}$$

and
$$(P_{xy}'' - 0) = \frac{I_x - I_y}{2} \sin 2\theta + P_{xy} \cos 2\theta \quad \text{...(3.103)}$$

Squaring and adding we get

$$\left(I_x' - \frac{I_x + I_y}{2}\right)^2 + (P_{xy}'' - 0)^2 = \left(\frac{I_x - I_y}{2}\right)^2 (\cos^2 2\theta + \sin^2 2\theta) + P_{xy}^2 (\sin^2 2\theta + \cos^2 2\theta)$$

$$- 2\left(\frac{I_x - I_y}{2}\right) \cos 2\theta \,.\, P_{xy} \,.\, \sin 2\theta + 2\left(\frac{I_x - I_y}{2}\right) \sin 2\theta \,.\, P_{xy} \cos 2\theta$$

∴ we get
$$\left(I_x' - \frac{I_x + I_y}{2}\right)^2 + (P_{xy}'' - 0)^2 = \left(\frac{I_x - I_y}{2}\right)^2 + P_{xy} \quad \text{....(3.104)}$$

Eqn. (3.104) is similar to the equation of a circle $(x - a)^2 + (y - b)^2 = x^2$ ...(3.105)
whose radius is r and whose centre C is at $x = a$ and $y = b$ (see Fig. 3.17 (*a*)). Comparing eqn. (3.104) with eqn. (3.105).

We find that radius of the circle
$$= R_I = \sqrt{\left(\frac{I_x - I_y}{2}\right)^2 + P_{xy}^2} \quad \text{...(3.106)}$$

and the centre C of this circle has co-ordinates

$\left\{\left(\frac{I_x + I_y}{2}\right), 0\right\}$ with respect to *I*-axis and *P*-axis of the circle of Fig. 3.17 (*b*).

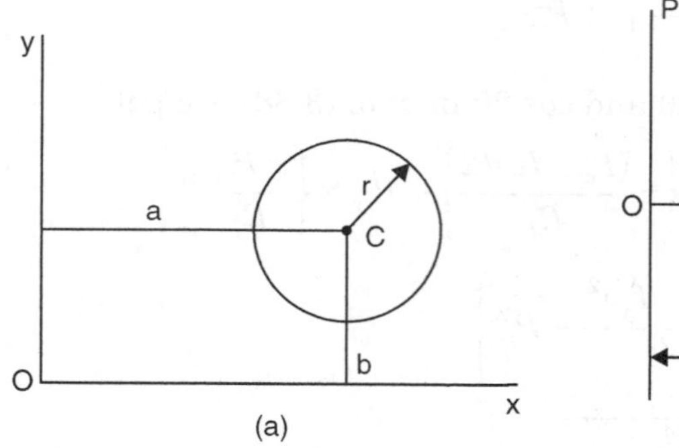

(a)

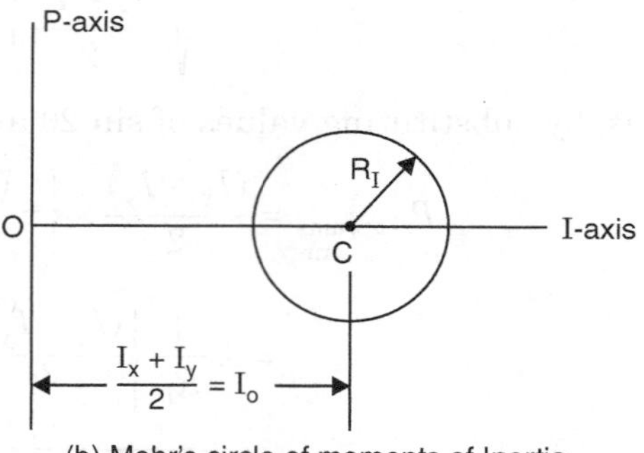

(b) Mohr's circle of moments of Inertia

Fig. 3.17

This circle is known as **Mohr's circle for moments of inertia** and is shown in Fig. 3.17 (*b*).

3.17.1. Construction of Mohr's Circle of Inertia

Given. I_x, I_y and P_{xy}, it is required to construct the Mohr's circle of moments of inertia.

Step (*i*) Draw two axes : *I*-axis and *P*-axis as shown in Fig. 3.17.1.

(*ii*) Let $I_x > I_y$ and let P_{xy} be positive. Then to a suitable scale, I_x is represented by *OM* and I_y is represented by *ON*.

(*iii*) Take $MD = P_{xy}$ and $NE = -P_{xy}$.

(*iv*) Join points *D* and *E*. The line *DE* intersects the *I*-axis at point *C*. Then

$$OC = \frac{OM + ON}{2} = \frac{I_x + I_y}{2} = I_0 \qquad ...(3.107)$$

(*v*) With *C* as centre and radius = *CD* or *CE* draw a circle which intersects the *I*-axis at points *A* and *B*.

(*vi*) Then *CD* represents *X*-axis of given area and *CE* represents *Y*-axis of the given area.

Since Eqns. (3.102) and (3.103) above are in terms of angle 2θ, therefore the *X*-axis and *Y*-axis which are at right and to each other appear at 2 × 90° = 180° . Therefore all angles in the Mohr's circle of Fig. 3.17.1 appear to be doubled.

(*vii*) Thus, we get $\quad CM = CN = \dfrac{MN}{2} = \dfrac{I_x - I_y}{2}$,

and
$$CD = CE = \sqrt{CM^2 + DM^2} = \sqrt{\left(\frac{I_x - I_y}{2}\right)^2 + P_{xy}^2} = R_I \qquad ...(3.108)$$

(*viii*) From the Mohr's circle of inertia, we find :

$$OA = I_{\max} = I_u = OC + CA = \frac{I_x + I_y}{2} + \sqrt{\left(\frac{I_x - I_y}{2}\right)^2 + P_{xy}^2} = I_0 + R_I \qquad ...(3.109)$$

$$OB = I_{\min} = I_v = OC - CB = \frac{I_x + I_y}{2} - \sqrt{\left(\frac{I_x - I_y}{2}\right)^2 + P_{xy}^2} = I_0 - R_I \qquad ...(3.110)$$

The angle $2\theta_1 = \angle DCA$ clockwise is given by

$$\tan 2\theta_1 = \frac{DM}{CM} = \frac{-P_{xy}}{\left(\dfrac{I_x - I_y}{2}\right)} \qquad ...(3.111)$$

Angle 2θ, is negative because it is clockwise from *x*-axis.

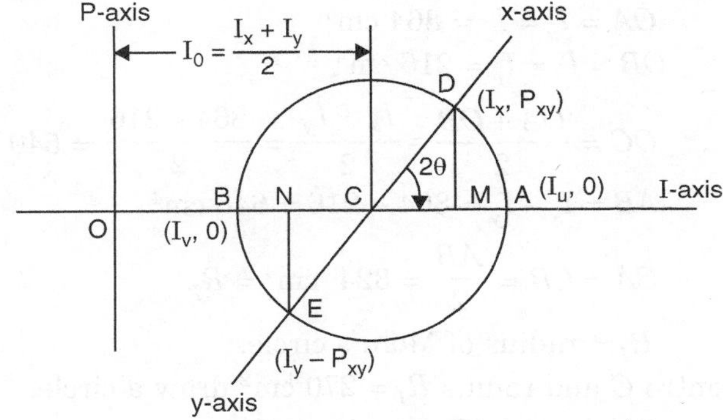

Fig. 3.17.1. Mohr's circle of Moment of Inertia.

CA represents *u*-axis and *CB* represents *v*-axis.

Solved examples will help the reader to understand the procedure used in solving problems.

Problem 3.43. *For the rectangular area shown in Fig. P.3.43 (a) find graphically the values of* I_{x_1}, I_{y_1} *and* $P_{x_1 y_1}$ *with respect to the* x_1 *and* y_1 *axes inclined at an angle of* $\theta' = 30°$ *counterclockwise to the x-and y-axis respectively.*

Also, find I_{x_2}, I_{y_2} *and* $P_{x_2 y_2}$ *with respect to the* x_2, *and* y_2 *axes inclined at an angle of* $\theta = 15°$ *clockwise to the x and y-axes respectively.* (*See Fig. P.3.43 (a)*).

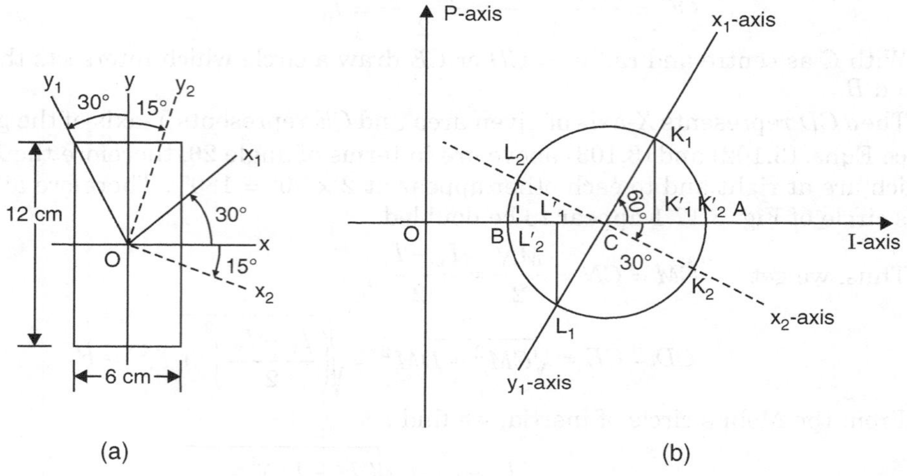

Fig. P. 3.43 Mohr's circle of Inertia.

Sol. For the rectangular area,

$$I_x = \frac{1}{12} \times 6 \times 12^3 = 864 \text{ cm}^4$$

$$I_y = \frac{1}{12} \times 12 \times 6^3 = 216 \text{ cm}^4.$$

Since rectangle has two axes of symmetry, therefore $P_{xy} = 0$.

Therefore, $\quad I_x = I_u = 864 \text{ cm}^4 = I_{max}$

and $\quad I_y = I_v = I_{min} = 216 \text{ cm}^4$.

To draw the Mohr's circle of inertia, shown in Fig. P. 3.43 (*b*) take

$$OA = I_x = I_u = 864 \text{ cm}^4$$

$$OB = I_y = I_v = 216 \text{ cm}^4$$

and

$$OC = \frac{OA + OB}{2} = \frac{I_x + I_y}{2} = \frac{864 + 216}{2} = 540 \text{ cm}^4$$

$$\therefore \quad AB = I_x - I_y = 864 - 216 = 648 \text{ cm}^4$$

$$\therefore \quad CA = CB = \frac{AB}{2} = 324 \text{ cm}^4 = R_I.$$

R_I = radius of Mohr's circle.

Therefore with centre *C* and radius $R_I = 270 \text{ cm}^2$ draw a circle.

Here *CA* represents *X*-axis and *CB* represents *Y*-axis.

To find I_{x_1}, I_{y_1} and $P_{x_1y_1}$, draw a line CK_1 inclined at angle of 2 × 30° = 60° ccw from CA. Draw diameter K_1L_1 as shown. Then CK_1 represents X_1-axis and CL, represents Y_1-axis. Thus, from the Mohr's circle of inertia, we get.

$$I_{x_1} = OK_1' = OC + R_I \cdot \cos 60° = 540 + 304 \times 0.5 = 540 + 162 = 702 \text{ cm}^4. \quad \textbf{Ans}$$

$$I_{y_1} = OL_1' = OC - R_I \cos 60° = 540 - 324 \times 0.5 = 540 - 162 = 378 \text{ cm}^4. \quad \textbf{Ans.}$$

and $$P_{x_1y_1} = K_1K_1' = I_R \sin 60° = 324 \times \frac{\sqrt{3}}{2} = 281 \text{ cm}^4. \quad \textbf{Ans.}$$

To find values of I_{x_2}, I_{y_2}, and $P_{x_2y_2}$ draw a line CK_2 at an angle of 2 × 15 or 30° clockwise from the x-axis (here from CA). Draw the diameter K_2L_2. Draw normal K_2K_2' from K_2 to the I-axis and also draw normal L_2L_2' from L_2 to the I-axis. Therefore, we get

$$I_{x_2} = OK_2' = OC + CK_2' = OC + R_I \cos 30°$$

$$= 540 + 324 \cos 30° = \textbf{820.6 cm}^4. \quad \textbf{Ans.}$$

$$I_{y_2} = OL_2' = OC - CL_2' = OC - R_I \cos 30°$$

$$= 540 - 324 \cos 30° = \textbf{259.4 cm}^4. \quad \textbf{Ans.}$$

And, the product of inertia

$$P_{x_2y_2} = K_2K_2' = CK_2 \sin(-2 \times 15) = 324 \sin(-30°)$$

$$\therefore \quad P_{x_2y_2} = \textbf{-162 cm}^4. \quad \textbf{Ans.}$$

Above results have been graphically shown in Fig. P.3.43 (*b*).

Problem 3.44. *For a certain area the value of I_x = 100 cm^4 . I_y = 60 cm^4 and the product of inertia P_{xy} = + 15 cm^4 with respect to X and Y axes (a) graphically find the maximum and minimum values of moments of inertia of the given area, and the position of the principal moment of inertia axes.*

(b) Also find the position of the axes along which the product of inertia has maximum and minimum values.

(c) Also find I_{x_1}, I_{y_1} and $P_{x_1y_1}$ with respect to the x and y axes which are inclined at 48.5° clockwise to the x and y-axes respectively.

Sol. Draw I-axis and P-axis with origin O as shown in Fig. P. 3.44(*a*). To certain scale mark

$$OM = I_x = 100 \text{ cm}^4$$

$$ON = I_y = 60 \text{ cm}^4$$

Draw $MD = P_{xy} = 15$ cm^4 and $NE = -P_{xy} = -15$ cm^4, as shown. Join D and E. The line DE intersects the I-axis at point C which is the centre of the Mohr's circle. The radius of the Mohr's circle of inertia is equal to CD or CE. Draw this circle, which intersects the I-axis at points A and B. Thus, we get $MN = OM - ON = 100 - 60 = 40$ cm^4.

$$\frac{MN}{2} = CM = CN = 20 \text{ cm}^4$$

radius of Mohr's circle $= R_I = \sqrt{CN^2 + MD^2} = \sqrt{20^2 + 15^2} = 25$ cm^4.

Also we find $$OC = \frac{OM + ON}{2} = \frac{100 + 60}{2} = 80 \text{ cm}^4$$

(*a*) Radius *CD* represents *X*-axis and radius *CE* represents *y*-axis.

Clockwise $\angle DCA = 2\theta_1 = 37°$.

While clockwise $\angle DCB = 2\theta_1 + 180° = 217°$

$\therefore$ The radius *CA* represents *u*-axis while the radius *CB* represents *v*-axis.

Therefore from the Fig. P. 3.44(*a*), we measure and find.

$$I_u = OA = OC + CA = 80 + 25 = 105 \text{ cm}^4.$$

$$I_v = OB = OC - CB = 80 - 25 = 55 \text{ cm}.$$

Results for the given area are shown in Fig. P.3.44 (*b*).

(*b*) Draw the vertical diameter *GH* of the Mohr's circle. Then the radius $CG = P_{xy})_{max} = R_I = +\ 25 \text{ cm}^4$, and radius $CH = -\ P_{xy})_{max} = P_{xy})_{min} = -\ 25 \text{ cm}^4$.

The $\angle ACG = 90°$ counter clockwise while $\angle ACH = 90°$ clockwise

$\therefore$ The axes *x′* and *y′* are given by radii *CG* and *CH* respectively which are inclined at + 45° and – 45° to the *u*-axis.

Results for the given area are shown in Fig. P.3.44 (*c*).

(*c*) Draw a line, CK_1 at an angle of 2 × 48.5 or 97° clockwise from *CD* or the *X*-axis. Draw the diameter *KL*. Draw perpendicular KK_1 and LL_1 to the *I*-axis. Then we get

$$\angle ACK = \angle LCB = 2(48.5) - 37° = 60°,\ OK_1 = I_{X_1} = OC + CK_1 \cos \angle ACK_1$$

$$= 80 + 25 \cos (60°) = 80 + 12.5 = 92.5 \text{ cm}^4. \quad \textbf{Ans.}$$

$$OL_1 = I_{y_1} = OC - CL \cos BCL = 80 - 25 \cos 60°$$

$$= 80 - 12.5 = 67.5° \text{ cm}^4. \quad \textbf{Ans.}$$

Than $P_{x_1y_1} = K_1K = -\ CK \sin 60 = -\ 25 \sin 60° = \mathbf{-\ 21.65\ cm^4}$. **Ans.**

These results are shown in Fig. P.3.44 (*d*).

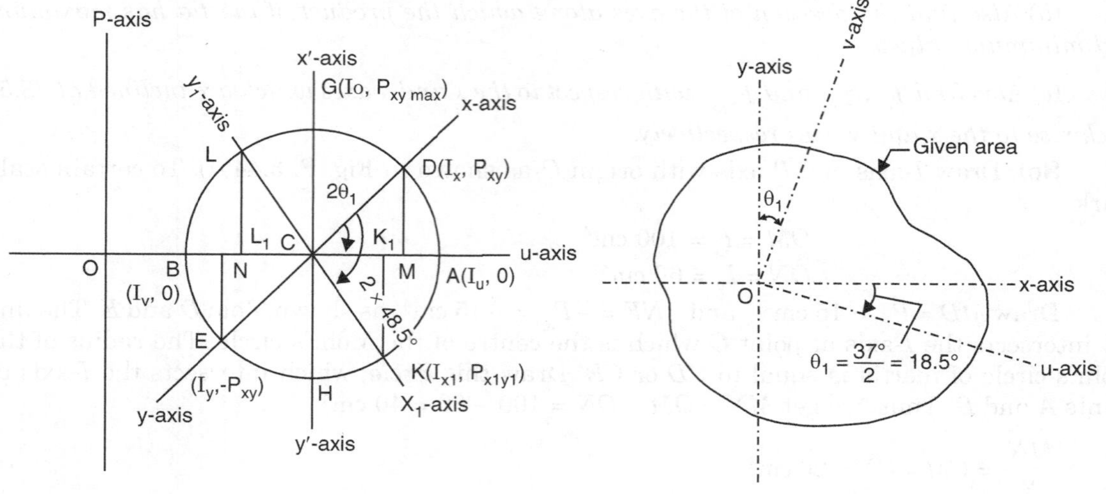

(a) Mohr's circle of inertia

(b) Location of u- and v-axes of maximum and minimum moments of inertia reaction to x and y-axis. $P_{uv} = 0$ on respect to the principal axis

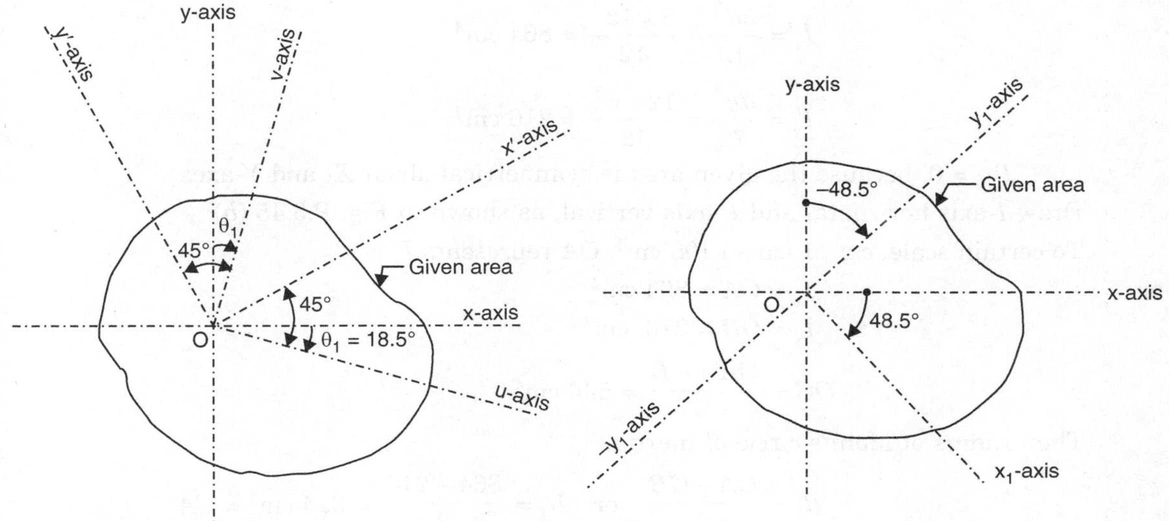

(c) Location of x′ and y′ axes with respect to which the product of inertia $P_{x'y'}$ is maximum and minimum position of the principal axes of inertia u-and v-axis

(d) Location of x_1 and y_1 axes with respect to x and y axes inclined at 48.5° clockwise with respect to x-axis

Fig. P.3.44

Problem 3.45. *For the rectangle shown in Fig. P.3.45 (a) find moments of inertia* I_x', I_y' *and* $P_{x'y'}'$ *with respect to x′ and y′ axes which are inclined at 30° counterclockwise to x- and y-axes, respectively.*

Sol. For the rectangular area shown in Fig. P.3.45 (*a*), we have

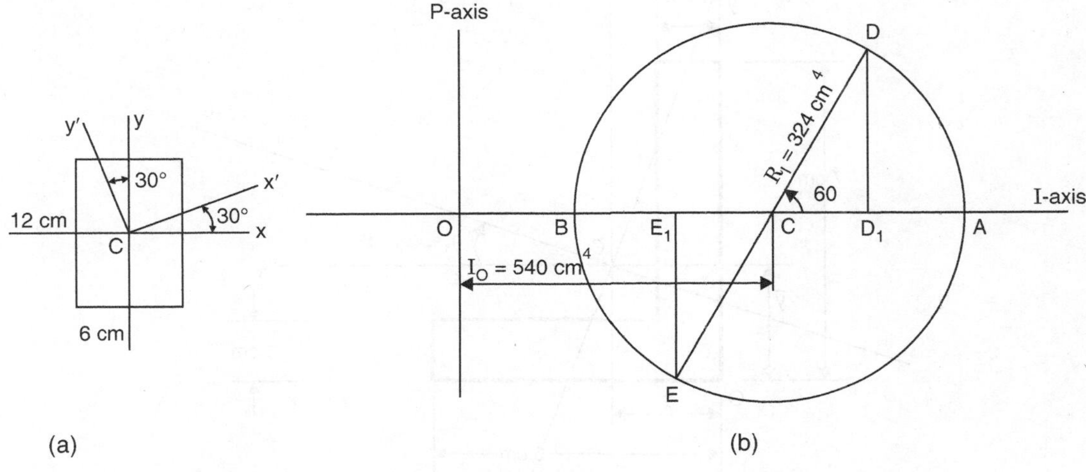

Fig. P.3.45

$$I_x = \frac{bd^3}{12} = \frac{6 \times 12^3}{12} = 864 \text{ cm}^4$$

$$I_y = \frac{db^3}{12} = \frac{12 \times 6^3}{12} = 216 \text{ cm}^4$$

$P_{xy} = 0$, because the given area is symmetrical about X- and Y-axes.

Draw I-axis horizontal and P-axis vertical, as shown in Fig. P.3.45 (*b*).

To certain scale, *e.g.*, 1 cm → 108 cm^4, OA represents I_x

or $$I_x = OA = 864 \text{ cm}^4$$

$$I_y = OB = 216 \text{ cm}^4$$

∴ $$OC = \frac{OA + OB}{2} = 540 \text{ cm}^4.$$

Then radius of Mohr's circle of inertia,

$$R_I = \frac{OA - OB}{2} \quad \text{or} \quad R_I = \frac{864 - 216}{2} = 324 \text{ cm}^4 = CA$$

with C as centre and radius equal to R_I draw a circle. This circle is known as Mohr's circle of moments of inertia. Here CA represents X-axis and CB represents Y-axis.

Draw a line CD inclined at (2 × 30)° or 60° counter-clockwise. Draw a diameter DE, as shown. Project D and E on I-axis as points D_1 and E_1.

Then, $P_x' = OD_1 = OC + CD \cos 2 \times 30° = 540 + 162 =$ **702 cm^4. Ans**

$P_y' = OE_1 = OC - CE \cos 2 \times 30° = 540 - 324 \cos 60° =$ **378 cm^4. Ans**

And $P_{x'y'} = DD_1 = OD \sin 60° = 324 \sin 2 \times 30° =$ **281 cm^4. Ans.**

Problem 3.46. *Find the value of product of inertia $\overline{P}_{xy}$ with respect to the centroidal axes X_0 and Y_0. Also find maximum and minimum values of moment of inertia and their position with respect to X_0-axis. See Fig. P.3.46.*

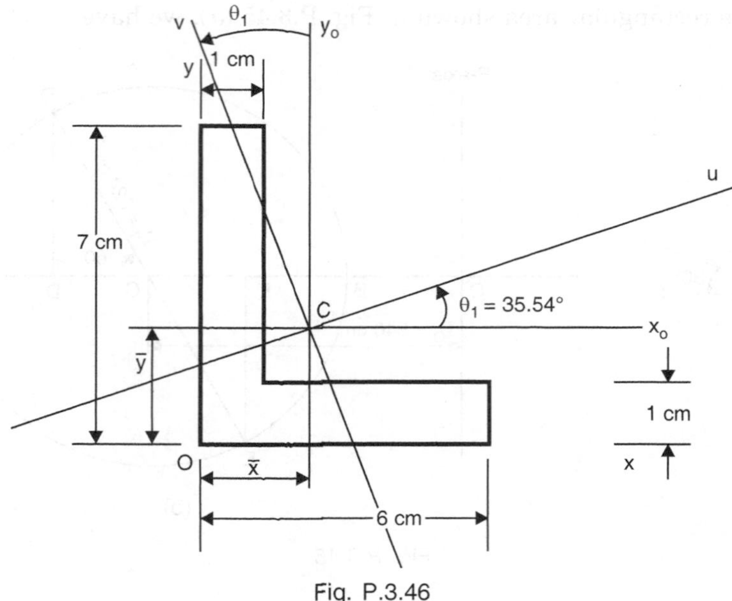

Fig. P.3.46

Sol. Area $A_1 = 6 \times 1 = 6 \text{ cm}^2$

$A_2 = (7 - 1) \times 1 = 6 \text{ cm}^2$

$$\therefore \quad \bar{x} = \frac{A_1x_1 + A_2x_2}{A_1 + A_2} = \frac{6 \times 3 + 6 \times 0.5}{6+6} = 1.75 \text{ cm for } Y\text{-axis}$$

and

$$\bar{y} = \frac{A_1y_1 + A_2y_2}{A_1 + A_2} = \frac{6 \times 0.5 + 6 \times (1+3)}{6+6} = 2.25 \text{ cm from } X\text{-axis.}$$

$\therefore$ Using $I_x = (\bar{I}_x + Ah^2)_1 + (\bar{I}_x + Ah^2)_2$

$$I_x = \left[\frac{6 \times 1^3}{3}\right] + \left[\frac{1 \times 6^3}{12} + 6 \times 4^2\right] = 116 \text{ cm}^4$$

Using $I_x = \bar{I}_x + Ah^2$, we get where $h = \bar{y}$ | $h = \bar{y}$

$$\bar{I}_x = I_x - Ah^2 = 116 - 12 \times (2.25)^2 \text{ cm}^4$$

or $\bar{I}_x = \mathbf{55.25\ cm^4.}$ **Ans.**

$$I_y = (\bar{I}_y + Ak^2)_1 + (\bar{I}_y + Ak^2)_2$$

$$= \frac{6 \times 1^3}{3} + \frac{1 \times 6^3}{3} \quad \text{or} \quad I_y = \mathbf{74\ cm^4.}$$ **Ans.**

But $I_y = \bar{I}_y + Ak^2$ where $k = \bar{x}$

$\therefore$ $\bar{I}_y = I_y - Ak^2 = 74 - 12 \times (1.75)^2 = \mathbf{37.25\ cm^5.}$ **Ans.**

$$P_{xy} = (\bar{P}_{xy} + Ahk)_1 + (\bar{P}_{xy} + Ahk)_2 = [0 + A_1\, x_1y_1] + [0 + A_2\, x_2y_2]$$

$$= [0 + 6 \times 3 \times 0.5] + [0 + 6 \times 0.5 \times 4] \text{ cm}^4 = \mathbf{21\ cm^4}$$

But $P_{xy} = \bar{P}_{xy} + Ahk$ where $h = \bar{y}$ and $k = \bar{x}$

$$\therefore \quad \bar{P}_{xy} = P_{xy} - Ahk = 21 - 12 \times 2.25 \times 1.75 \text{ cm}^4$$

or $\bar{P}_{xy} = \mathbf{-\ 26.25\ cm^4.}$ **Ans.**

$$I_0 = \frac{\bar{I}_x + \bar{I}_y}{2} = \frac{55.25 + 37.25}{2} = \mathbf{46.25\ cm^4}$$

$$\frac{\bar{I}_x - \bar{I}_y}{2} = \frac{1}{2}(55.25 - 37.25) = 9 \text{ cm}^4.$$

Radius of Mohr's circle of inertia $= R_I = \sqrt{\left(\frac{\bar{I}_x - \bar{I}_y}{2}\right)^2 + \bar{P}_{xy}^{\ 2}}$

or $R_I = \sqrt{9^2 + (26.25)^2} = \mathbf{27.75\ cm^4}$

$\therefore$ $\bar{I}_{max} = \bar{I}_u = \bar{I}_0 + \bar{R}_I = 46.25 + 27.75 = \mathbf{74.0\ cm^4.}$ **Ans.**

$\bar{I}_{min} = \bar{I}_v = \bar{I}_0 - \bar{R}_I = \mathbf{18.5\ cm^4.}$ **Ans.**

And, from $\tan 2\theta_1 = -\dfrac{P_{xy}}{\left(\dfrac{I_x - I_y}{2}\right)}$

or $\tan 2\theta_1 = -\dfrac{-26.25}{9} = +2.917$

$\therefore$ $2\theta_1 = 71.08°$

$\therefore$ $\theta_1 = \mathbf{35.54°}$ ccw from the x_0-axis. **Ans.**

EXERCISES IV

A. Theoretical Questions

1. Define centre of gravity and centroid.
2. Explain the term "axis of symmetry" as applied to the plane areas.
3. Find the position of centroid of a quarter circle of radius r.
4. Find the position of centroid of a quarter elliptical area having semi major and semi minor axes equal to a and b respectively.
5. Prove that the position of the centroid of a semi parabola of half base b and height h is given by $\bar{x} = \dfrac{3}{8}b$ and $\bar{y} = \dfrac{3h}{5}$. The reference axes X and Y pass through straight sides of the area.
6. A parabolic spandrel $y = kx^2$ has base b and height h. Prove that
$$A = \frac{bh}{3},\ \bar{x} = \frac{3b}{4} \text{ and } \bar{y} = \frac{3h}{10}$$
7. Define the term "second moment of area" or "moment of inertia of an area".
8. State and prove the parallel-axis theorem.
9. State and prove the perpendicular axes theorems.
10. What do you understand by the term "radius of gyration" of a plane area ?
11. Find expression for the moment of inertia of a rectangular area, base b and height d, about the base (x-axis) and a vertical side (y-axis). Also find the values of moment of inertia about the centroidal axes.
12. Find the moment of inertia about the centroidal axes of a right triangle, base b and height d.
13. Find expressions for the radius of gyration for a rectangular area, base b and height d, with respect to its centroidal axes.
14. Derive expressions for moment of inertia of a circle of radius r.
15. Derive expressions for radius of gyration of a circle with respect to its centroidal axes.
16. Derive expressions for I_{x_0} and I_{y_0} (about the centroidal axes) for a semi-circle of radius r lying with its diameter horizontal.
17. Derive expressions for moments of inertia about the centroidal axes of a quarter circle of radius r, with its straight sides coincident with the X and Y axes.
18. Define "Product of Inertia" of a plane area.
19. State and prove parallel axes theorem for product of inertia.
20. State and prove perpendicular axes theorem for the product of inertia of the plane area.
21. Prove that the product of inertia of an area having one or two centroidal axes as the axes of symmetry is equal to zero.
22. Derive expressions for product of inertia with respect to the centroidal axes for a right triangle of base b and height d. Discuss effect of position of the triangle when it is placed in four different quadrants.

23. Derive expression for product of inertia of a quarter circle of radius r placed in first quadrant with x-and y-axes containing the straight sides of the quarter circle. Discuss the effect of orientation/position of area in different quadrants.
24. Derive expression for product of inertia with respect to the centroidal axes of a quarter circle of radius r placed in the first quadrant with x-and y-axes containing the straight sides of the quarter circle.
25. What do you mean by the principal moment of inertia of an area and the principal moment of inertia axes?
26. Derive expressions for $\bar{I}_x{}'$, $\bar{I}_y{}'$ and $\bar{P}_{x'y'}$ for an area for which the angle between x-and x' and that between y-and y' is θ.
27. Explain the construction of Mohr's circle of inertia for an area.

B. Numerical Problems

3.1. Locate the position of centroid of the area shown in Fig. E.3.1.

[**Ans.** $\bar{x}$ = 13.875 cm ; $\bar{y}$ = 8.5625 cm]

3.2. Locate the centroid of the area shown in Fig. E.3.2. [**Ans.** $\bar{x} = \frac{11}{12}a$; $\bar{y} = a$]

3.3. Find the position of centroid of the area shown in Fig. E.3.3. [**Ans.** $\bar{x}$ = 30 mm ; $\bar{y}$ = 88 mm]

3.4. Locate centroid of the area shown in Fig. E.3.4. [**Ans.** $\bar{x}$ = 5.593 cm ; $\bar{y}$ = 7.297 cm]

3.5. Locate centroid of the area shown in Fig. E.3.5. [**Ans.** $\bar{x}$ = 4.964 cm ; $\bar{y}$ = 6.839 cm]

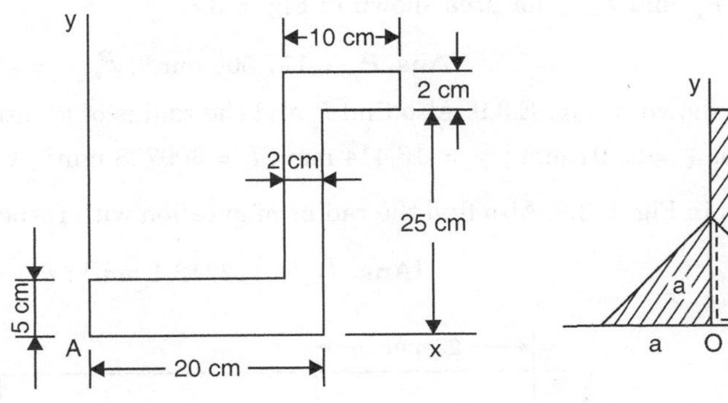

Fig. E.3.1

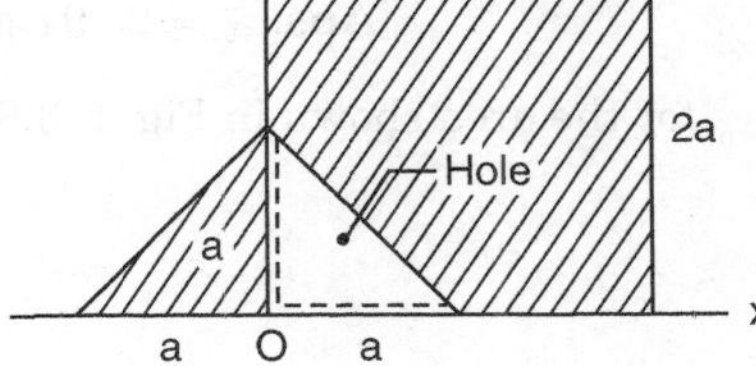

Fig. E.3.2

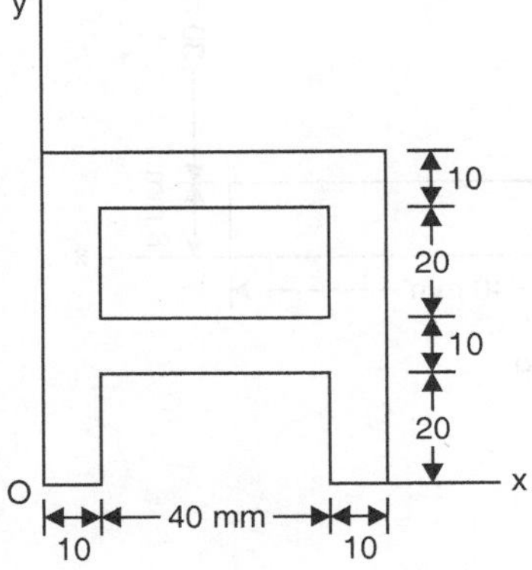

Fig. E.3.3

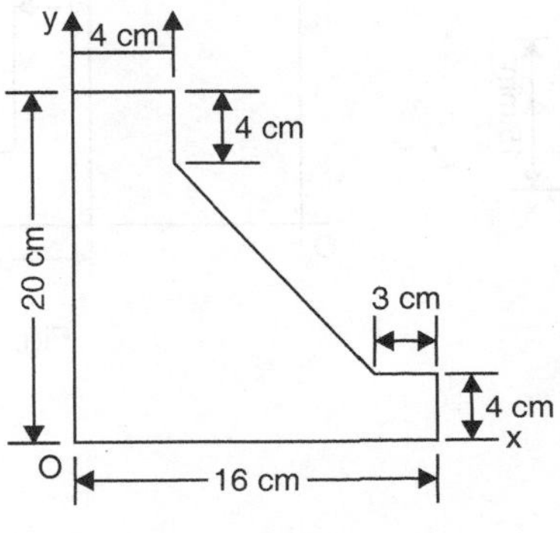

Fig. E.3.4

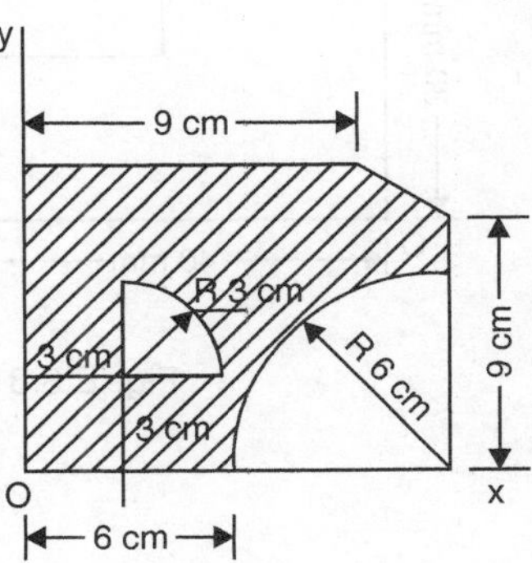

Fig. E.3.5

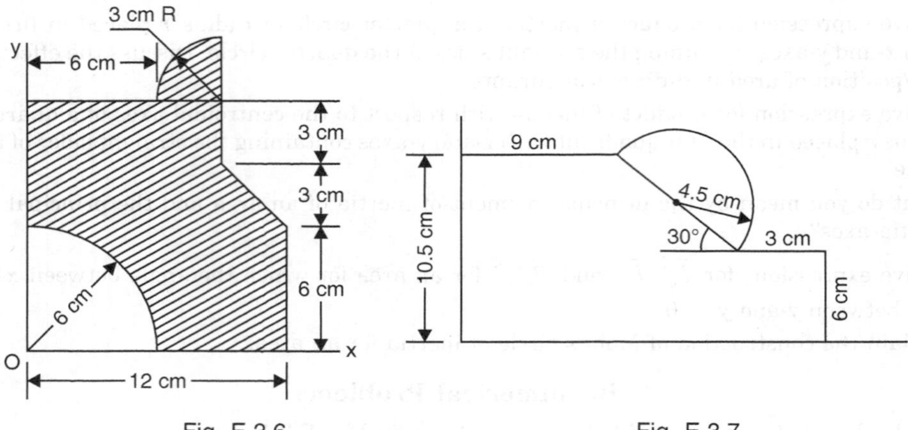

Fig. E.3.6 Fig. E.3.7

3.6. Locate centroid of the area shown in Fig. E.3.6. [**Ans.** $\bar{x}$ = 6.725 cm ; $\bar{y}$ = 7.23 cm]

3.7. Locate centroid of the area shown in Fig. E.3.7. [**Ans.** $\bar{x}$ = 9.6 cm ; $\bar{y}$ = 5.45 cm]

3.8. Locate centroid of the area shown in Fig. E.3.8. Also find I_x and I_{x_0}.

[**Ans.** $\bar{x} = \frac{50}{3}$ = 16.67 mm ; $\bar{y}$ = 12.5 mm ; I_x = 202500 mm^4, I_{x_0} = 61875 mm^4]

3.9. Find I_y and I_{y_0} for the area shown in Fig. E.3.8. [**Ans.** I_y = 360,000 mm^4 ; I_{y_0} = 110,000 mm^4]

3.10. Find the product of inertia P_{xy} and $P_{x_0y_0}$ for area shown in Fig. E.3.8.

[**Ans.** $\boldsymbol{P_{xy}}$ = 157,500 mm^4 ; $P_{x_0y_0}$ = – 30,000 mm^4]

3.11. Locate centroid of the area shown in Fig. E.3.9. Also find I_x and the radius of gyration k_x.

[**Ans.** $\bar{x}$ = 25.91 mm ; $\bar{y}$ = 16.414 mm ; I_x = 309728 mm^4 ; k_x = 21.1 mm]

3.12. Find I_{x_0} for the area shown in Fig. E.3.9. Also find the radius of gyration with respect to x_0 axis.

[**Ans.** I_{x_0} = 122212.1 mm^4 ; k_{x_0} = 13.25 mm]

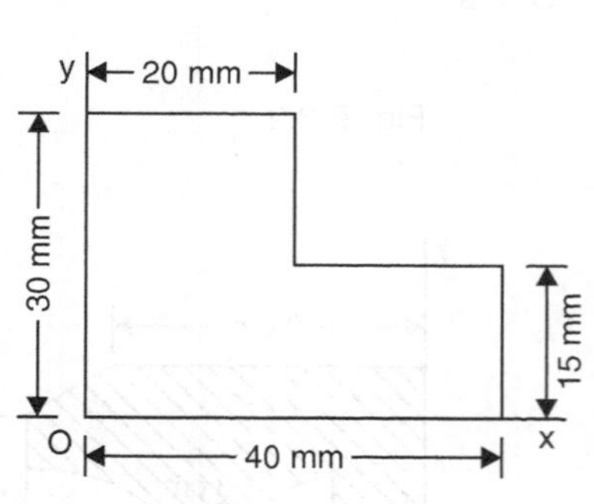

Fig. E.3.8

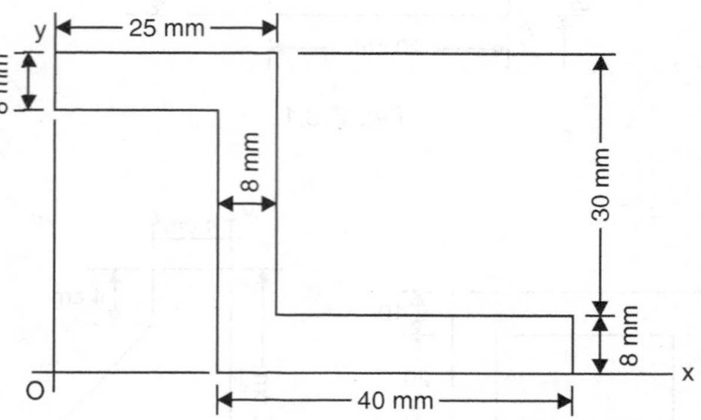

Fig. E.3.9

3.13. Find the product of inertia P_{xy} with respect to the x and y axes shown in the area of Fig. E.3.9. Also find $P_{x_0y_0}$ with respect to the centroidal axes.

[**Ans.** P_{xy} = 202584 mm^4; $P_{x_0y_0}$ = – 93415.57 mm^4]

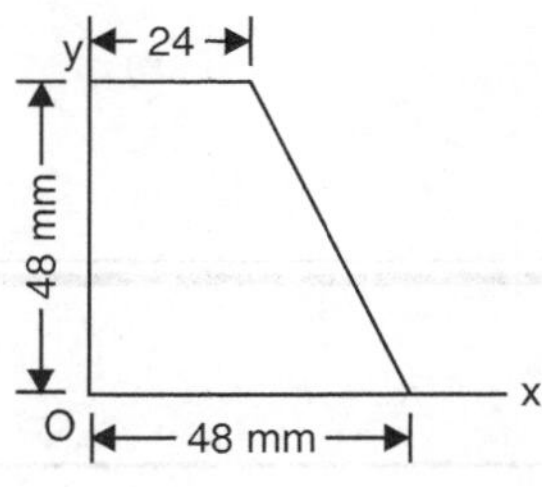

Fig. E.3.14

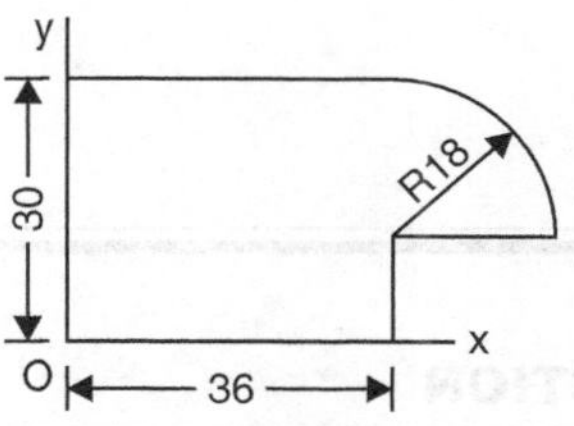

Fig. E.3.17

3.14. Find I_{y_0} and I_y for the area shown in Fig. E.3.9. Also find the radius of gyration k_y and k_{y_0}.

[**Ans.** I_y = 600968 mm^4 ; I_{y_0} = 133723.64 mm^4 ; k_y = 29.385 mm ; k_{y_0} = 13.861 mm^2]

3.15. For the area shown in Fig. E.3.14 locate the centroid. Also find I_x and I_{x_0} for this area. Also find the radius of gyration from X-axis and X_0 axis.

[**Ans.** $\bar{x}$ = 18.67 mm ; $\bar{y}$ = 21.33 mm ; I_x = 1105920 mm^4 ;
I_{x_0} = 319488 mm^4 ; k_x = 25.3 mm ; k_{x_0} = 13.6 mm]

3.16. Find I_y and I_{y_0} for the area shown in Fig. E.3.14. Also find the radius of gyration k_y and k_{y_0}.

[**Ans.** I_y = 829440 mm^4 ; I_{y_0} = 602327.1 mm^4 ; k_y = 21.91 mm ; k_{y_0} = 18.67 mm]

3.17. Find product of inertia P_{xy} and $P_{x_0y_0}$ for the area shown in Fig. E.3.14.

[**Ans.** P_{xy} = 608256 mm^4 ; $P_{x_0y_0}$ = 79872 mm^4]

3.18. Locate the centroid of the area shown in Fig. E.3.17. Also find I_x, I_{x_0}, k_x and k_{x_0} for this area.

[**Ans.** $\bar{x}$ = 19.43 mm ; $\bar{y}$ = 15.26 mm ; I_x = 354312.72 mm^4,
I_{x_0} = 88000.7 mm^4 ; k_x = 17.60 mm ; k_{x_0} = 8.77 mm]

3.19. Find I_y and I_{y_0} for the area shown in Fig. E.3.17. Also find k_y and k_{y_0}.

[**Ans.** I_y = 593489.53 mm^4 ; I_{y_0} = 161744.5 mm^4 ; k_y = 22.78 mm ; k_{y_0} = 11.89 mm]

3.20. Find P_{xy} and $P_{x_0y_0}$ for the area shown in Fig. E.3.17.

[**Ans.** P_{xy} = 344393.6 mm^4 ; $P_{x_0y_0}$ = 5309.03 mm^4]

3.21. Find the centroidal principal axes of inertia for the area shown in Fig. E.3.8. Also find the maximum and minimum moment of inertia, I_u and I_v respectively.

[**Ans.** θ_1 = – 25.63° from x_0-axis ; I_u = 124395.32 mm^4, I_v = 47479.68 mm^4]

3.22. Find the principal axes of inertia passing through the centroid of the area shown in Fig. E.3.9. Also find maximum and minimum values of moment of inertia, I_u and I_v, respectively.

[**Ans.** θ_1 = – 43.24° from x_0-axis ; I_u = 22156.59 mm^4 ; I_v = 34375.15 mm^4]

4

Friction

4.1. INTRODUCTION

Friction is generally defined as the resistance exerted by one body upon the second body in contact which moves or tends to move relative to the first body. Since friction is resistant to motion, therefore, it is a retarding force always acting opposite to the motion or the tendency to move. Obviously, friction exists primarily because of the roughness of the two surfaces in contact. Friction reduces as the contact surfaces become smoother and smoother.

In some applications, friction is an asset while in other situations it is liability. When friction causes loss of power and increases wear, it is undesirable. On the other hand, friction is essential for various holding and fastening devices and also for friction drives and brakes. Friction is also required in providing traction for automobiles. Walking is not possible without friction between shoes and surface of floor or road. Without friction at joints in our bodies, we cannot even stand up. Thus, without friction life in Universe is not possible. Thus there is no development without friction.

4.2. TYPES OF FRICTION

Different types of frictional resistances encountered in engineering mechanics are :

(*a*) Dry friction, (*b*) Fluid friction, and (*c*) Internal friction.

We briefly discuss each type of friction in this article.

(*a*) **Dry Friction.** When two unlubricated surfaces are in contact under a condition of sliding or tendency to slide, dry friction is called into play. A force of friction tangent to the surfaces in contact is developed which always acts in direction opposite to the relative motion between the two surfaces. This type of friction is also called **Coulomb friction.** The principles of dry or Coulomb friction were developed from the experiments of Coulomb in 1781 and as a result of studies of Morin from 1831 to 1834. Sometime it is also called the **solid friction.** Upto the state of impending motion (*i.e.*, motion is just about to start) the frictional resistance developed is known as **static friction** and when the motion just starts the frictional resistance setup is known as **kinetic or dynamic friction.** We shall discuss both frictional resistances in subsequent Articles in this Chapter.

(*b*) **Fluid Friction.** When the two solid bodies in contact are separated by a thin layer of a fluid (gas or oil) the friction is called **fluid friction.** It is also known as **fluid film lubrication.** When the intervening fluid film is very thin it is called **boundary lubrication** and when an abundance of lubricating fluid is available it is termed as **hydrodynamic lubrication.** The magnitude of frictional resistance in this case depends upon the magnitude of the relative velocity. This frictional force also depends on the viscosity of the intervening lubricating fluid present between the two bodies.

(*c*) **Internal Friction.** Internal friction plays an important roll when a material is subjected to cyclic loading. Materials which are highly elastic their recovery from deformation due to external loading occurs with very little loss of energy caused by internal friction. On the other hand, materials with low elasticity and which are under large plastic deformations during loading, the effect of internal friction is considerable and the loss of energy is substantial.

4.3. LAWS OF FRICTION

(*a*) **Laws of Dry Friction :**

Laws of limiting friction for dry surfaces may be stated as follows :

1. When two bodies are in contact, the direction of force of friction on one of them is opposite to the direction in which this body has the tendency to move.
2. The frictional force is independent of the area of contact of the surfaces.
3. When one body is just on the point of sliding over the other, the maximum force of friction is being exerted. This is called the **limiting friction.**
4. The force of friction is dependent upon the types of materials of the two bodies in contact. The limiting frictional force is proportional to the normal force that acts between the bodies in contact. The ratio of the frictional force to the normal reaction is constant and is known as **co-efficient of friction.** (Values of the co-efficient of friction between some typical materials in contact are given in Table 4.1.)
5. Limiting static friction is somewhat greater than the kinetic friction.
6. The kinetic friction is independent of the relative velocity of the bodies in contact.

(*b*) **Laws of Friction for Lubricated Surfaces :**

Laws of friction for lubricated surfaces may be stated as follows :

1. The frictional force is nearly independent of the normal force between the surfaces in contact.
2. The force of friction is nearly independent of the nature of the frictional surfaces.
3. The force of friction depends primarily upon the viscosity of the lubricant present between the two bodies.
4. The force of friction decreases as the temperature increases.
5. The force of friction decreases as the relative velocity between the bodies increases.

Table 4.1 Co-efficient of Friction for some Materials

S. No.	*Materials*	*Co-efficient of friction* (μ)	
		Dry (static) μ_s	*Sliding (kinetic)* μ_k
1.	Hard steel on hard steel	0.78	0.42
2.	Mild steel on mild steel	0.74	0.57
3.	Hard steel on graphite	0.21	—
4.	Hard steel on babbitt	0.71	0.33
5.	Copper on mild steel	0.53	0.36
6.	Brass on mild steel	0.51	0.44
7.	Glass on nickel	0.78	0.56
8.	Copper on glass	0.68	0.53
9.	Glass on glass	0.94	0.40

4.4. THEORY OF DRY FRICTION

Let a block of weight W rest on a rough surface and assume a horizontal force P (called **effort**) to be applied to the block as shown in Fig. 4.4.1. When P is zero, the block is in equilibrium and the frictional resistance F is also zero. As the magnitude of P increases, the frictional resistance F also increases.

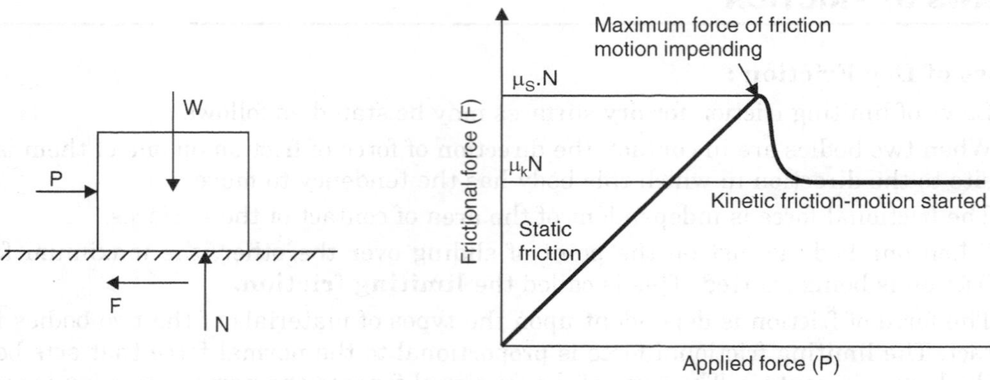

Fig. 4.4.1. Frictional resistance F. Fig. 4.4.2. Frictional force as a function of applied force P.

At some instance, the block is on the verge of moving (*i.e.*, motion impending), the frictional force F attains its maximum value. As P increases further the motion starts, and the frictional resistance F drops rapidly to a kinetic value which remains almost constant thereafter.

The surfaces of the two bodies in contact are rough and the irregularities of the two bodies get interlocked. The maximum frictional resistance is equal to the force required to break the contact between the bodies. After the maximum resistance between the two bodies is reached the motion starts and the frictional resistance drops suddenly, as described above.

The magnitude of the frictional force depends on the normal force N between the contacting surfaces. As a result, the maximum frictional resistance is proportional to the normal pressure, we get the relation

$$F \propto N \quad ...(4.1)$$

or,

$$F_{\max} = \mu_s N \quad ...(4.2)$$

where the constant of proportionality is called the **coefficient of friction**. The maximum frictional resistance (F_{max}) is developed when the motion is impending. When the maximum value of frictional resistance is developed the coefficient of friction is maximum and is equal to μ_s, known as the coefficient of static friction, therefore equation (4.2) becomes

$$F_{\max} = \mu_s \,.\, N \quad \text{or} \quad F_s = \mu_s \,.\, N \quad ...(4.3)$$

In case of the moving bodies, the resistance is defined as the **kinetic friction force** and is given by

$$F_k = \mu_k N \quad ...(4.4)$$

4.5. ANGLE OF FRICTION

Fig. 4.4.1 is redrawn as Fig. 4.5 to show that frictional force F and the normal N are the two components of the resultant reaction R developed between the two bodies. The magnitude of the angle ϕ between the R and N depends on the value of frictional resistance F. If F is zero, this angle ϕ will be zero. As F increases, so does the angle ϕ. The particular value of the angle ϕ when maximum frictional resistance is acting is defined as the **angle of friction.** When motion is impending ϕ is maximum.

From Fig. 4.5, we find that

$$\tan \phi = \frac{F}{N} \quad ...(4.5)$$

Comparing equation (4.2) and (4.5), we find that

$$\tan \phi = \mu_s \quad ...(4.6)$$

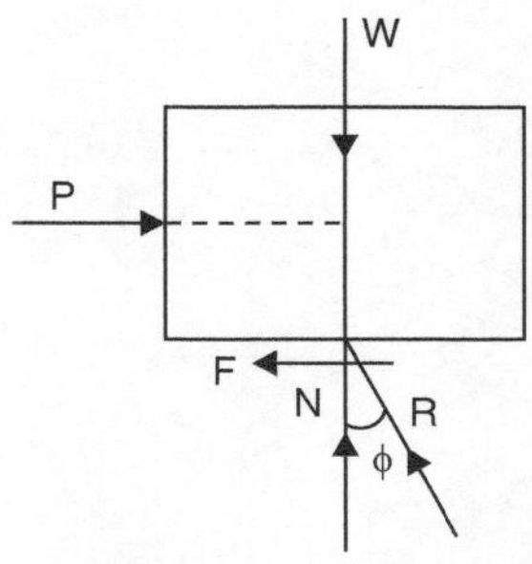

Fig. 4.5. Angle of friction ϕ.

4.6. ANGLE OF REPOSE

A simple method of finding the angle of friction is to place a block of weight W upon an tiltable inclined plane whose inclination θ can be increased, gradually from zero to a maximum value θ at which the block is just on the verge of sliding down the plane. This inclination θ is then equal to the angle of friction. See Fig. 4.6. This angle θ is called the **angle of repose.**

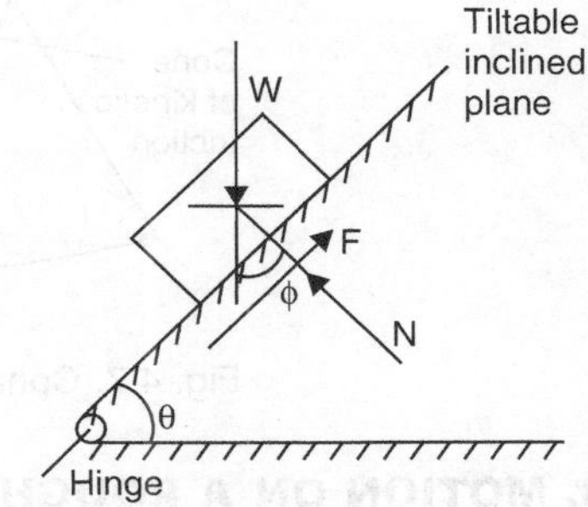

Fig. 4.6. Angle of friction ϕ is equal to the slope θ of the inclined plane.

4.7. CONE OF FRICTION

When we consider coplanar forces, N, R, W etc., in order that motion does not occur, the static reaction R must lie within the angle ABC in Fig. 4.7, whereas for cases of non coplanar forces, it must be contained within the cone generated by revolving line AB about the normal BN. The cone, thus formed, is called the **cone of friction** and is shown in Fig. 4.7.

The Fig. 4.7 shows static and kinetic cones of friction. The generators of static cone are inclined at an angle ϕ_s to the normal N whereas the generators of kinetic cone are inclined at an angle ϕ_k to the normal N. Values of ϕ_s and ϕ_k are given by :

$$\phi_s = \tan^{-1} \mu_s = \tan^{-1} \frac{F_s}{N} \quad ...(4.7)$$

or

$$\phi_k = \tan^{-1} \mu_k = \tan^{-1} \frac{F_k}{N} \quad ...(4.8)$$

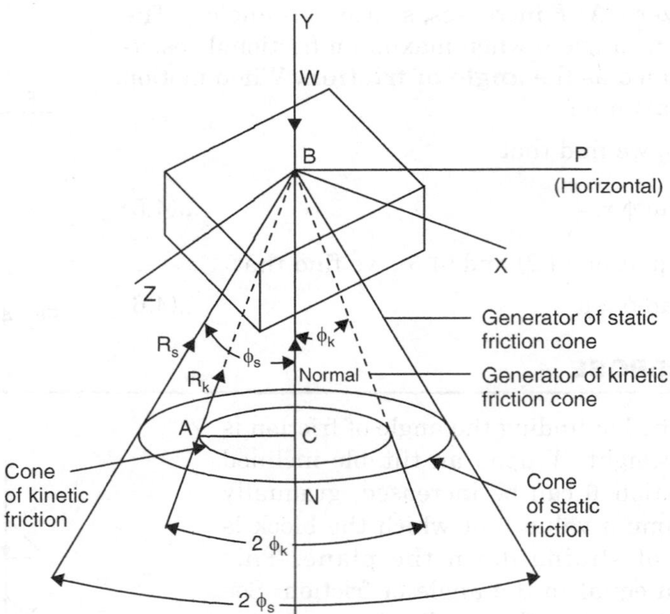

Fig. 4.7. Cone of static friction and the cone of kinetic friction.

4.8. MOTION ON A ROUGH INCLINED PLANE

We have seen that if the inclination of the plane with the horizontal is less than the angle of repose, the body will be in equilibrium with the aid of an external force (P).

Let us analyse various cases of motion on the inclined plane. These include :

(*i*) **External force *P* parallel to the plane :**

(*a*) Motion up the plane, and

(*b*) Motion down the plane.

(*ii*) **External force *P* horizontal :**

(*a*) Motion up the plane, and

(*b*) Motion down the plane.

(*iii*) **External force *P* inclined to the plane (but not horizontal) :**

(*a*) Motion up the plane, and

(*b*) Motion down the plane.

Let W = mg be the weight of the body

m = mass of the body

α = inclination of the plane to the horizontal

θ = inclination of the force (or effort P)

ϕ = the limiting angle of friction (= $\tan^{-1} \mu_s$) for the contact surfaces.

P = the external force (known as the effort) for the impending motion

N = normal reaction

R = resultant reaction

$F = \mu N$ = maximum force of friction

P_0 = idle effort when plane is smooth.

Case I. Force parallel to the inclined plane

(a) **Motion down the plane**

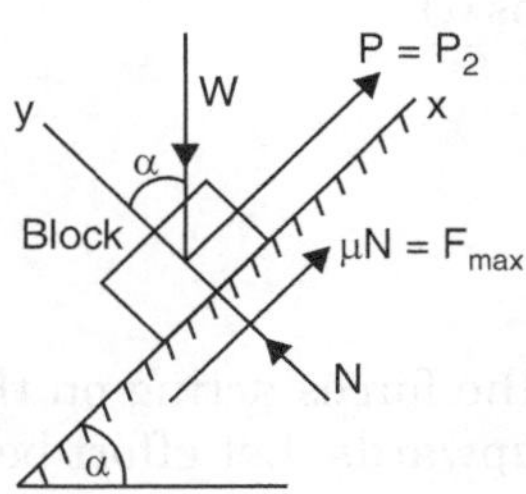

(a) The block just on verge of sliding down.

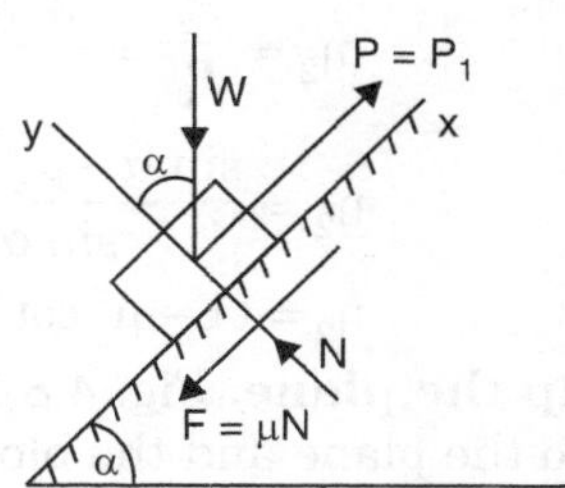

(b) Motion up the plane.

Fig. 4.8.1. Effort P parallel to plane.

When the block is just on verge of sliding down, the frictional force F is directed upward, as shown in Fig. 4.8.1(*a*). As the block is on the verge of sliding down the maximum force of friction F will be equal to $\mu_s N$, *i.e.*, $F_{max} = \mu_s N$.

P_2 is the magnitude of the minimum effort P required. By resolving forces parallel and perpendicular to the plane, we get

By $\quad \Sigma F_y = 0 \quad N = W \cos \alpha$ ...(*i*)

[Note that the direction X is parallel to plane and Y is perpendicular to the plane].

And, by $\quad \Sigma F_x = 0, P_2 - W \sin \alpha + F_{max} = 0$ ...(*ii*)

or, $\quad P_2 + \mu_s N = W \sin \alpha$ $\quad [F_{max} = \mu_s N]$

Substituting (*i*) into (*ii*), we get

$\therefore \quad P_2 + \mu_s(W \cos \alpha) = W \sin \alpha$

$\therefore \quad P_2 = W(\sin \alpha - \mu_s \cos \alpha)$ ...(4.9)

Eqn. (4.9) gives the effort P_2 for the motion down the plane.

Eqn. (4.9) can be modified by writing

$$\tan \phi_s = \mu_s \quad \text{or} \quad \mu_s = \frac{\sin \phi_s}{\cos \phi_s}$$

$$\therefore \quad P_2 = W\left(\sin \alpha - \frac{\sin \phi_s \cos \alpha}{\cos \phi}\right)$$

or $$P_2 = \frac{W \sin (\alpha - \phi_s)}{\cos \phi_s} \quad ...(4.10)$$

When the plane is smooth, the idle effort P_0 can be determined by writing $\phi_s = 0$ or $\mu_s = 0$. Thus, from eqn. (4.9) and also from eqn. (4.10), we get

$$P_0 = W \sin \alpha \quad ...(4.11)$$

It is clear that P_0 is greater than P_2. Therefore the ratio $\frac{P_2}{P_0} < 1$ for downward motion.

The ratio $\frac{P_2}{P_0}$ is called the **efficiency of the plane** for motion down the plane.

i.e., $$\eta_2 = \frac{P_2}{P_0} \quad ...(4.12)$$

or $$\eta_2 = \frac{P_2}{P_0} = \frac{W(\sin\alpha - \mu_s \cos\alpha)}{W \sin\alpha}$$

or $$\eta_2 = \frac{\sin\alpha - \mu_s \cos\alpha}{\sin\alpha}$$

or $$\eta_2 = (1 - \mu_s \cot\alpha) \quad ...(4.13)$$

(*b*) **Motion up the plane.** Fig. 4.8.1(*b*) shows the forces acting on the block when the effort P is parallel to the plane and the block moving upwards. Let effort be P_1, in this case.

By $\Sigma F_y = 0$, we get $N = W \cos\alpha$ ...(*iii*)

By $\Sigma F_x = 0$, $P_1 - W\sin\alpha - \mu_s \,.\, N = 0$ ...(*iv*)

Eliminating N, we get

$$P_1 = W\sin\alpha + \mu_s \times (W\cos\alpha)$$

or, $$P_1 = W(\sin\alpha + \mu_s \cos\alpha) \quad ...(4.14)$$

or, $$P_1 = W(\sin\alpha + \tan\phi_s \cos\alpha) = W\left(\sin\alpha + \frac{\sin\phi_s \cos\alpha}{\cos\phi_s}\right)$$

or $$P_1 = \frac{W\sin(\alpha + \phi_s)}{\cos\phi_s} \quad ...(4.15)$$

Again, when the plane is smooth, the effort is idle and is given by

$$P_0 = W\sin\alpha$$

And, the efficiency of the plane is

$$\eta_1 = \frac{P_0}{P_1} = \frac{W\sin\alpha}{W(\sin\alpha + \mu_s\cos\alpha)} \quad ...(4.16)$$

We note that

$P_0 < P_1$ for upward motion of the block, and

$P_0 > P_2$ for downward motion of the block.

An alternative analysis can be carried out with the aid of Lame's equations. We note that the block is in equilibrium under the action of three forces, *viz.* W, P and the resultant reaction R of the inclined plane. Also, R is inclined at an angle ϕ_s to the normal to the plane.

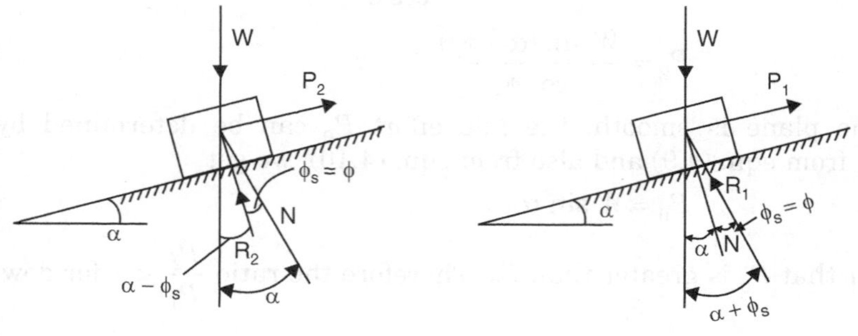

(a) Motion down the plane (b) Motion up the plane

Fig. 4.8.2 Application of triangle law of forces for effort P parallel to the plane.

The resultant of force of friction $\mu_s N$ and the normal reaction N is inclined at an angle ϕ_s to the normal N to the block.

(*a*) Motion Down the Plane :

When the motion is down the plane the resultant R_2 of N and $\mu_s N$ makes angle $(\alpha - \phi_s)$ with the vertical. This is in equilibrium due to three forces : the weight W, the resultant R_2 and the effort P_2. The Δabc_2 is the triangle of forces for W, R_2 and P_2. Side ab represents W, side bc_2 represents R_2 and the side c_2a represents P_2.

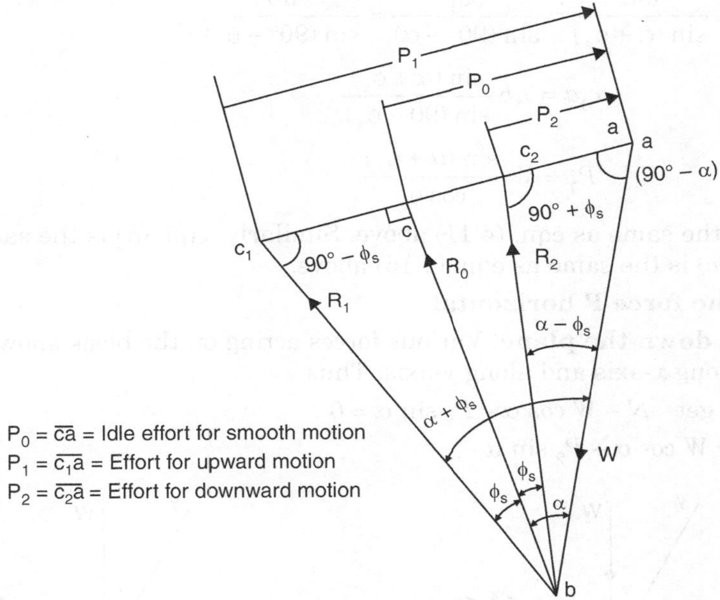

Fig. 4.8.2 (c) The triangles of forces.

(a) Δ abc$_2$ for motion of block down the rough plane, (b) Δabc$_1$ for motion of block up the rough plane and (c) Δabc for smooth plane.

When the plane is smooth, the angle $\phi_s = 0$, then the vector bc is perpendicular to the ac From b draw vectors bc_1 and bc_2 inclined to bc at angle ϕ_s but on its either side. Then c_2a is equal to effort P_2 required for effort moving down the plane. The vector c_1a is equal to effort P_1 for the upward motion of the block and ca is equal to the idle effort (P_0).

$\because \quad \angle acb = 90°,$

$\therefore \quad \angle c_1c_2b = 90° - \phi_s = \angle c_2c_1b$

Therefore, **for the smooth plane,** the idle effort is given by the solution of Δabc. Lame's equations are

$$\frac{ca}{\sin \alpha} = \frac{cb}{\sin (90° - \alpha)} = \frac{ab}{\sin 90°}$$

$$\therefore \quad ca = \frac{ab}{\sin 90°} \cdot \sin \alpha$$

or $$P_0 = W \sin \alpha \qquad ...(a)$$

For downward motion from Δabc_2, we get

$$\frac{c_2a}{\sin (\alpha - \phi_s)} = \frac{c_2b}{\sin (90° - \alpha)} = \frac{ab}{\sin (90° + \phi_s)}$$

$$\therefore \qquad c_2a = \frac{ab \sin(\alpha - \phi_s)}{\sin(90° + \phi_s)} = \frac{ab \sin(\alpha - \phi_s)}{\cos \phi_s}$$

or,
$$P_2 = W \frac{\sin(\alpha - \phi_s)}{\cos \phi_s} \qquad ...(b)$$

(*b*) Motion up the plane :

For upward motion, from Δabc_1, we get

$$\frac{c_1a}{\sin(\alpha + \phi_s)} = \frac{bc_1}{\sin(90° - \alpha)} = \frac{ab}{\sin(90° - \phi_s)}$$

$$\therefore \qquad c_1a = ab \frac{\sin(\alpha + \phi_s)}{\sin(90 - \phi_s)}$$

or,
$$P_1 = W \frac{\sin(\alpha + \phi_s)}{\cos \phi_s} \qquad ...(c)$$

Eqn. (*a*) is the same as eqn. (4.11) above. Similarly, eqn. (*b*) is the same as eqn. (4.10), above. And, eqn. (*c*) is the same as eqn. (4.15) above.

Case II. The force P horizontal

(*a*) **Motion down the plane.** Various forces acting on the block shown in Fig. 4.8.3 (*a*) can be resolved along *x*-axis and along *y*-axis. Thus

by $\Sigma Fy = 0$, we get $N - W\cos\alpha - P_2 \sin\alpha = 0$

$$\therefore \qquad N = W\cos\alpha + P_2\sin\alpha \qquad ...(v)$$

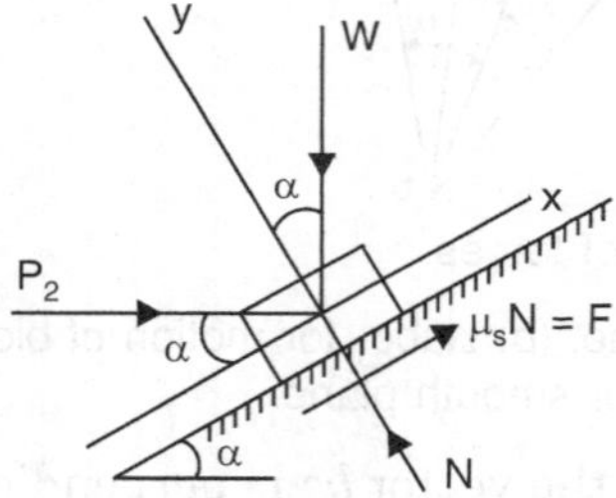

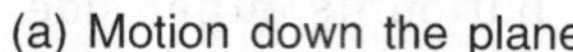
(a) Motion down the plane

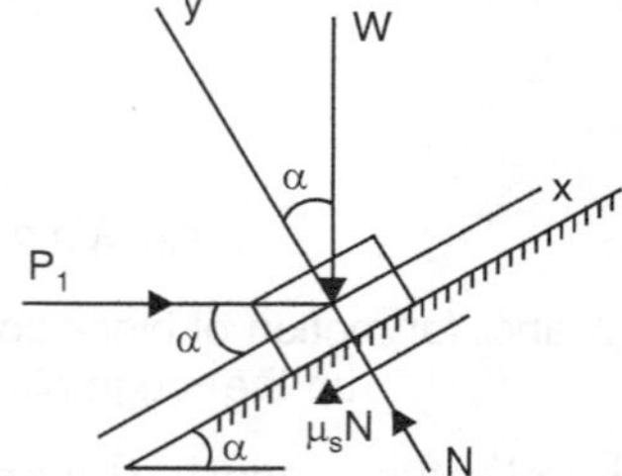

(b) Motion up the plane

Fig. 4.8.3. Force P horizontal analysis by equations of equilibrium.

And, by $\Sigma Fx = 0$, we get $P_2\cos\alpha - W\sin\alpha + \mu_s.N = 0$ ...(*vi*)

$\therefore$ Eliminating *N* from (*v*) and (*vi*), we get

$$P_2\cos\alpha - W\sin\alpha + \mu_s(W\cos\alpha + P_2\sin\alpha) = 0$$

$$\therefore \quad P_2(\cos\alpha + \mu_s\sin\alpha) = W(\sin\alpha - \mu_s\cos\alpha)$$

$$\therefore \qquad P_2 = \frac{W(\sin\alpha - \mu_s.\cos\alpha)}{(\cos\alpha + \mu_s.\sin\alpha)} \qquad ...(4.17)$$

$$\because \qquad \mu_s = \tan\phi_s = \frac{\sin\phi_s}{\cos\phi_s}$$

$\therefore$ we get,
$$P_2 = \frac{W\left(\sin\alpha - \dfrac{\sin\phi_s.\cos\alpha}{\cos\phi_s}\right)}{\cos\alpha + \dfrac{\sin\phi_s}{\cos\phi_s}.\sin\phi_s}$$

or, $$P_2 = W \frac{\sin(\alpha - \phi_s)}{\cos(\alpha - \phi_s)}$$

$\therefore$ $$P_2 = W \tan(\alpha - \phi_s) \quad \text{...(4.18)}$$

when, plane is smooth, $\phi_s = 0$, and the idle effort, $P_0 = W \tan \alpha$ ...(4.19)

(*b*) **Motion up the plane.** Various forces acting on the block for the motion up the plane are shown in Fig. 4.8.3 (*b*)

By $\Sigma F_y = 0$, we get

$$N - W\cos\alpha - P_1 \sin\alpha = 0 \quad \text{or} \quad N = W\cos\alpha + P_1 \sin\alpha \quad \text{...}(vii)$$

By $\Sigma F_x = 0$, we get

$$P_1 \cos\alpha - W \sin\alpha - \mu_s N = 0 \quad \text{...}(viii)$$

By eliminating N from (*vii*) and (*viii*), we get

$$P_1 \cos\alpha - W\sin\alpha - \mu_s(W\cos\alpha + P_1 \sin\alpha) = 0$$

$\therefore$ $$P_1(\cos\alpha - \mu_s \sin\alpha) = W(\sin\alpha + \mu_s \cos\alpha)$$

$$P_1 = W \frac{\sin\alpha + \mu_s \,.\cos\alpha}{\cos\alpha - \mu_s \,.\sin\alpha} \quad \text{...(4.20)}$$

$\because$ $\mu_s = \tan\phi_s = \dfrac{\sin\phi_s}{\cos\phi_s}$, eqn. (4.20) becomes

$$P_1 = W \frac{\sin\alpha\cos\phi_s + \sin\phi_s\cos\alpha}{\cos\alpha\cos\phi_s - \sin\phi_s\sin\alpha}$$

or, $$P_1 = W \frac{\sin(\alpha + \phi_s)}{\cos(\alpha + \phi_s)}$$

$\therefore$ $$P_1 = W\tan(\alpha + \phi_s) \quad \text{...(4.21)}$$

Efficiency of the plane

(*a*) For downward motion, $\eta_2 = \dfrac{P_2}{P_0} = \dfrac{\tan(\alpha - \phi_s)}{\tan\alpha}$ ...(4.22)

(*b*) For upward motion, $\eta_1 = \dfrac{P_0}{P_1} = \dfrac{\tan\alpha}{\tan(\alpha + \phi_s)}$ ...(4.23)

Method II. Application of Lame's equations. When the plane is smooth the forces acting on the block are W, P_0 and R_0. These forces are given by Δabc. For motion of the block down the plane the triangle of forces is Δabc_2. See Fig. 4.8.4 (*c*).

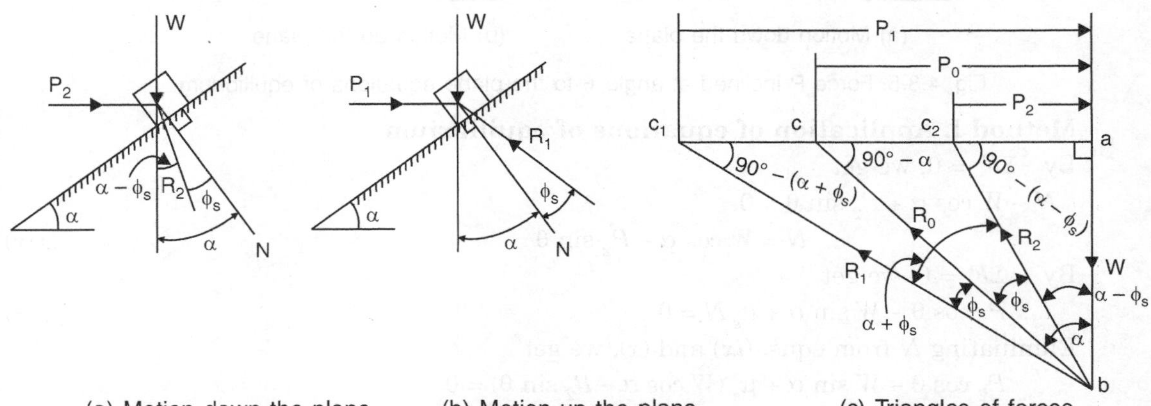

(a) Motion down the plane (b) Motion up the plane (c) Triangles of forces

Fig. 4.8.4

In Δabc_2, side ab represents the weight W, side c_2a represents the effort P_2 and the side bc_2 represents the resultant reaction R_2 of the plane to the block.

$\therefore$ We can write
$$\frac{c_2a}{\sin \angle c_2ba} = \frac{bc_2}{\sin \angle c_2ab} = \frac{ab}{\sin \angle bc_2a}$$

or
$$\frac{P_2}{\sin(\alpha - \phi_s)} = \frac{R_2}{\sin 90^\circ} = \frac{W}{\sin 90^\circ - (\alpha - \phi_s)}$$

or,
$$\frac{P_2}{\sin(\alpha - \phi_s)} = \frac{W}{\sin 90^\circ - (\alpha - \phi_s)}$$

$$\therefore \quad P_2 = \frac{W \sin(\alpha - \phi_s)}{\cos(\alpha - \phi_s)} = W \tan(\alpha - \phi_s) \quad ...(d)$$

For smooth plane $ca = \dfrac{ab \sin \alpha}{\sin(90 - \alpha)}$

or,
$$P_0 = W \tan \alpha \quad ...(e)$$

And for upward motion, from Δabc, we can write
$$\frac{c_1a}{\sin(\alpha + \phi_s)} = \frac{bc_1}{\sin 90^\circ} = \frac{ab}{\sin 90 - (\alpha + \phi_s)}$$

or,
$$c_1a = ab\frac{\sin(\alpha + \phi_s)}{\cos(\alpha + \phi_s)} = ab \tan(\alpha + \phi_s)$$

$$\therefore \quad P_1 = W \tan(\alpha + \phi_s) \quad ...(f)$$

Eqns. (d), (e) and (f) are respectively, the same as eqns. (4.18), (4.19) and (4.21), above.

Case III. Force inclined to the plane at angle θ

(a) **Motion down the plane.** Various forces acting on the plane are shown in Fig. 4.8.5(a).

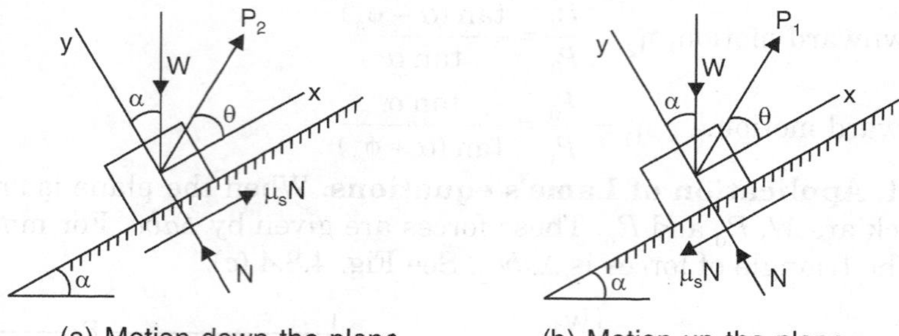

(a) Motion down the plane (b) Motion up the plane

Fig. 4.8.5. Force P inclined at angle θ to the plane equations of equilibrium.

Method I. Application of equations of equilibrium

By $\Sigma F_y = 0$, we get

$N - W \cos \alpha + P_2 \sin \theta = 0$

$$\therefore \quad N = W \cos \alpha - P_2 \sin \theta \quad ...(ix)$$

By $\Sigma F_x = 0$, we get

$$P_2 \cos \theta - W \sin \alpha + \mu_s N = 0 \quad ...(x)$$

Eliminating N from eqns. (ix) and (x), we get

$$P_2 \cos \theta - W \sin \alpha + \mu_s (W \cos \alpha - P_2 \sin \theta) = 0$$

$$\therefore \quad P_2(\cos \theta - \mu_s \sin \theta) = W(\sin \alpha - \mu_s \cos \alpha)$$

$$\therefore \quad P_2 = W \cdot \frac{(\sin\alpha - \mu_s \cos\alpha)}{(\cos\theta - \mu_s \sin\theta)} \quad ...(4.24)$$

Writing $\mu_s = \tan\phi_s = \dfrac{\sin\phi_s}{\cos\phi_s}$, we get

$$P_2 = W \times \frac{\sin\alpha\cos\phi_s - \sin\phi_s\cos\alpha}{\cos\theta\cos\phi_s - \sin\phi_s\sin\theta}$$

or,

$$P_2 = W \frac{\sin(\alpha - \phi_s)}{\cos(\theta + \phi_s)} \quad ...(4.25)$$

When plane is smooth, $\phi_s = 0$ or $\mu_s = 0$,

$\therefore$ From eqns. (4.24) or (4.25), we get the idle effort

$$P_0 = W \frac{\sin\alpha}{\cos\theta} \quad ...(4.26)$$

(*b*) **Motion up the plane.** By resolving forces acting on the block in Fig. 4.8.5 (*b*), perpendicular and parallel to the plane, we get

by $\Sigma F_y = 0, \quad N - W\cos\alpha + P_1 \sin\theta$

or, $N = W\cos\alpha - P_1\sin\theta$ ...(*xi*)

By $\Sigma F_x = 0, P_1 \cos\theta - W\sin\alpha - \mu_s N = 0$ ...(*xii*)

By eliminating N from eqns. (*xi*) and (*xii*), we get

$$P_1 \cos\theta - W\sin\alpha - \mu_s(W\cos\alpha - P_1\sin\theta) = 0$$

$$\therefore \quad P_1(\cos\theta + \mu_s \sin\theta) = W(\sin\alpha + \mu_s\cos\alpha)$$

$$\therefore \quad P_1 = W \times \frac{\sin\alpha + \mu_s\cos\alpha}{\cos\theta + \mu_s \,.\, \sin\theta} \quad ...(4.27)$$

$\because \quad \mu_s = \tan\phi_s = \sin\phi_s/\cos\phi_s$, Eqn. (4.27) yields,

$$P_1 = W \frac{\sin\alpha\cos\phi_s + \sin\phi_s\cos\alpha}{\cos\theta\cos\phi_s + \sin\phi_s\sin\theta}$$

or,

$$P_1 = W \frac{\sin(\alpha + \phi_s)}{\cos(\theta - \phi_s)} \quad ...(4.28)$$

Method II. Application of Lame's equations

Various forces and the corresponding triangles of forces for this case are shown in Fig. 4.8.6.

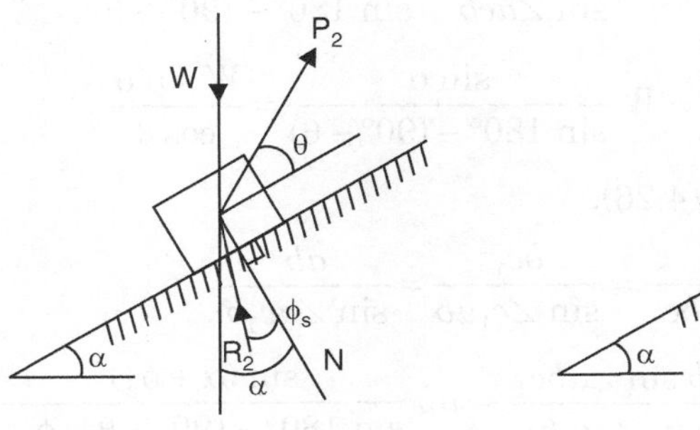

(a) Motion down the plane (b) Motion up the plane

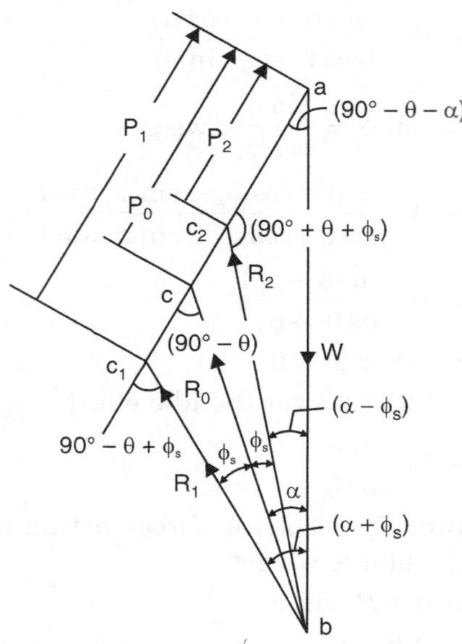

(c) Force triangles for case III

Fig. 4.8.6

Values of efforts P_2, P_0 and P_1 can be obtained from the Fig. 4.8.5(*c*).

From Δabc_2, we get $\dfrac{c_2a}{\sin \angle abc_2} = \dfrac{bc_2}{\sin \angle c_2ab} = \dfrac{ab}{\sin \angle ac_2b}$

$$\therefore \quad c_2a = \frac{ab \sin \angle abc_2}{\sin \angle ac_2b} = \frac{W \sin (\alpha - \phi_s)}{\sin (90° + \theta + \phi_s)} = P_2$$

or, effort $$P_2 = W \frac{\sin (\alpha - \phi_s)}{\cos (\theta + \phi_s)} \qquad ...(g)$$

This is the same as eqn. (4.25).

From Δabc, we get $\dfrac{ca}{\sin \angle abc} = \dfrac{bc}{\sin \angle cab} = \dfrac{ab}{\sin \angle acb}$

or, $$ca = ab \frac{\sin \angle abc}{\sin \angle acb} = \frac{ab \sin \alpha}{\sin 180° - (90° - \theta)} = P_0$$

$\therefore$ Idle effort, $$P_0 = W \frac{\sin \alpha}{\sin 180° - (90° - \theta)} = \frac{W \sin \alpha}{\cos \theta} \qquad ...(h)$$

Eqn. (*h*) is the same as eqn. (4.26).

From Δabc_1, we get $\dfrac{c_1a}{\sin \angle abc_1} = \dfrac{bc_1}{\sin \angle c_1ab} = \dfrac{ab}{\sin \angle ac_1b}$

$$\therefore \quad c_1a = \frac{ab \sin \angle abc_1}{\sin \angle ac_1b} = ab \frac{\sin (\alpha + \phi_s)}{\sin 180° - (90° - \theta - \phi_s)} = P_1$$

or, $$P_1 = W \frac{\sin (\alpha + \phi_s)}{\sin [180 - (90° - \theta + \phi_s)]} = W \frac{\sin (\alpha + \phi_s)}{\cos (\theta - \phi_s)}$$

or, effort,
$$P_1 = W \frac{\sin(\alpha + \phi_s)}{\cos(\theta - \phi_s)} \qquad ...(i)$$

Eqn. (*i*) is the same as eqn. (4.28).

Efficiency for the downward motion :

$$\eta' = \frac{P_2}{P_0} = \frac{\sin(\alpha - \phi_s)}{\cos(\theta + \phi_s)} \cdot \frac{\cos\theta}{\sin\alpha} \qquad ...(4.29)$$

And, efficiency for motion up the plane,

$$\eta = \frac{P_0}{P_1} = \frac{\sin\alpha}{\cos\theta} \times \frac{\cos(\theta - \phi_s)}{\sin(\alpha + \phi_s)} \qquad ...(4.30)$$

Problem 4.1. *Determine the maximum angle θ which the adjustable inclined plane may have with the horizontal before the block of mass m begins to slip. The coefficient of static friction is μ_s.*

Sol. Fig. P. 4.1 (*a*) shows the adjustable plane and the block. The forces acting on the block are :

Weight ($W = mg$) of the block,

Normal reaction N of the plane and the force of friction $F = \mu N$.

(a) Given inclined plane and the block

Since the block is in equilibrium, we get

by $\Sigma F_x = 0$, $mg \sin\theta - F = 0$

∴ $F = mg \sin\theta$ ...(1)

and by $\Sigma F_y = 0$, $N - mg\cos\theta = 0$,

∴ $N = mg\cos\theta$ ...(2)

By dividing (2) by (1), we get

$$\frac{F}{N} = \frac{mg\sin\theta}{mg\cos\theta}$$

or $\frac{F}{N} = \tan\theta$ ...(3)

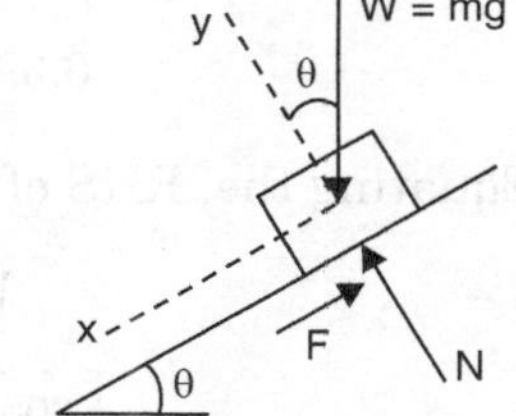

(b) Free body diagram of the block

Fig. P. 4.1

Since the maximum angle θ_{max} occurs when = $F = F_{max} = \mu_s N$, for impending motion,

∴ We have from (3)

$$\frac{\mu_s N}{N} = \tan\theta_{max}$$

∴ $\mu_s = \tan\theta_{max}$

∴ $\theta_{max} = \tan^{-1}\mu_s$. **(Ans.)**

Note. The maximum angle θ_{max} is called the **angle of repose.**

Problem 4.2. *Two blocks A and B of weights 1 kN and 2 kN, respectively are in equilibrium position as shown in Fig. P. 4.2 (a).*

If coefficient of friction between the two blocks as well as the block B and the floor is 0.3, find the force P required to move the block B. Also find the force in the string.

[UPSC Engg. Services, 1985]

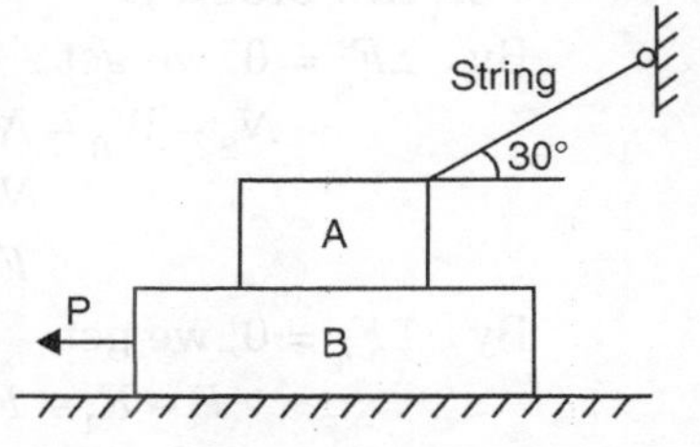

Fig. P. 4.2 (a) Given blocks.

Sol. Method I. Free body diagrams for blocks A and B are shown in Figs. P. 4.2 (b) and (c). Let T be the force in the string.

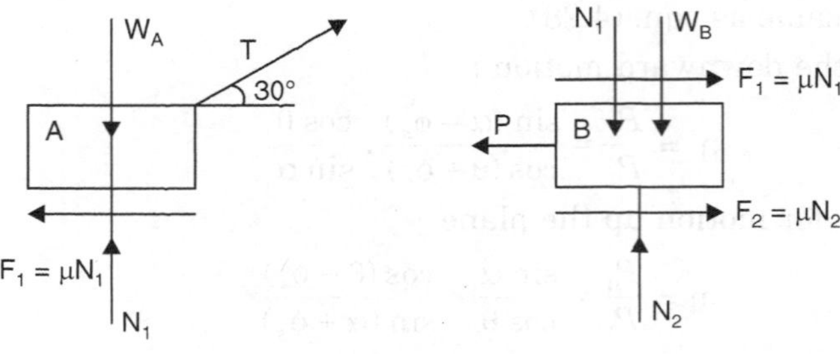

(b) FBD for block A (c) FBD for block B

Fig. P. 4.2

For the block A

By $\Sigma F_v = 0$, we get $T \sin 30° + N_1 - W_A = 0$

$\therefore \quad N_1 = W_A - T \sin 30°$ ...(1)

By $\Sigma F_h = 0$, we get $F_1 - T \cos (30°) = 0$

$\therefore \quad F_1 = T \cos 30°$ ...(2)

But $F_1 = \mu N_1 = 0.3 N_1$

$\therefore$ Eqn. (2) gives

$$0.3 N_1 = T \cos 30° \quad \text{or} \quad N_1 = \frac{T \cos 30°}{0.3} \qquad ...(3)$$

Equating the, RHS of eqns. (1) and (3), we get

$$W_A - T \sin 30° = \frac{T \cos 30°}{0.3}$$

$$\therefore \quad T\left(\frac{\cos 30°}{0.3} + \sin 30°\right) = W_A$$

$$\therefore \quad T\left(\frac{\sqrt{3}}{2 \times 0.3} + 0.5\right) = 1$$

$\therefore$ Tension in string,

$$T = \mathbf{0.2953\ kN.} \quad \textbf{Ans.}$$

Also, $$N_1 = \frac{T \cos 30°}{0.3} = \mathbf{0.8525\ kN}$$

and $$F_1 = 0.3 N_1 = 0.2557 \text{ kN}$$

For the block B

By $\Sigma F_v = 0$, we get

$$N_2 - W_B - N_1 = 0$$

$$N_2 = W_B + N_1 = 2 + 0.8525 = \mathbf{2.8525\ kN}$$

$$\therefore \quad F_2 = \mu N_2$$

By $\Sigma F_h = 0$, we get

$$P - F_1 - F_2 = 0$$

$\therefore \qquad P = F_1 + F_2 = 0.2557 + 2.8525 \times 0.3$

Effort, $\qquad P =$ **1.11145 kN. Ans.**

Method II. Forces acting on the block A, as shown in Fig P. 4.2 (b) can alternatively be shown in Fig. P. 4.2(d)

R_1 = resultant of N_1 and F_1

and $\qquad \phi = \tan^{-1} \mu = \tan^{-1} 0.3 = 16.7°$

The Δabc of forces acting on block A are shown in Fig. P. 4.2(d). Side ab represents W_A, side bc represents R_1 and side ca represents the tension T in string.

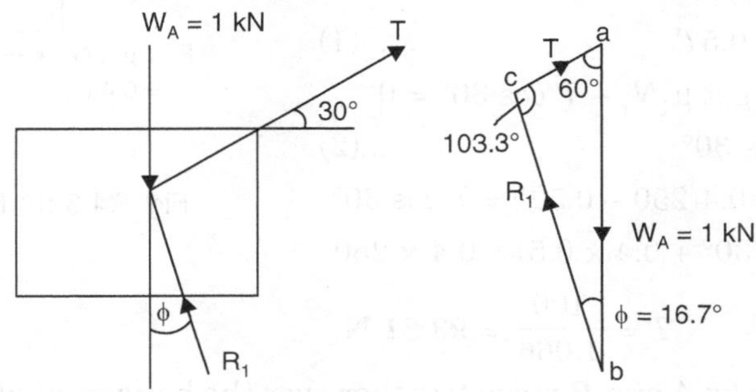

Fig. P. 4.2 (d)

We have $\qquad \angle b = \phi = 16.7°\ ; \quad \angle a = 90° - 30° = 60°$

$\therefore \qquad \angle c = 180 - (16.7 + 60°) = 180 - 76.7° = 103.3°$

$\therefore$ we get $\qquad \dfrac{T}{\sin 16.7°} = \dfrac{R_1}{\sin 60°} = \dfrac{1\,\text{kN}}{\sin 103.3°}$

$\therefore \qquad T = \dfrac{1\,\text{kN}}{\sin 103.3°} \times \sin 16.7°$

$=$ **0.2953 kN. Ans.**

and, $\qquad R_1 = \dfrac{1\,\text{kN}}{\sin 103.3°} \times \sin 60°$

$=$ **0.890 kN**

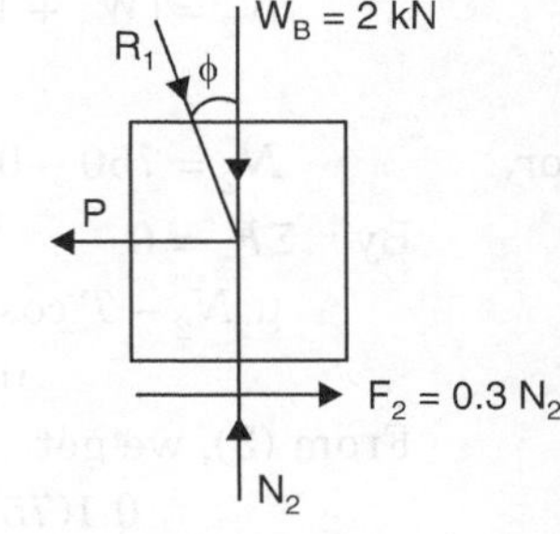

Fig. P. 4.2 (e)

Forces acting on the block B are shown in Fig. P. 4.2(e)

By $\quad \Sigma F_v = 0, \quad$ we get $\quad N_2 - W_B - R_1 \cos \phi = 0$

$\therefore \qquad N_2 = 2 + 0.89 \cos 16.7° =$ **2.8525 kN**

By $\quad \Sigma F_h = 0,$ we get $\quad P - 0.3N_2 - R_1 \sin \phi = 0$

$\therefore \qquad P = 0.3 \times 2.8525 + 0.89 \sin 16.7°$

$=$ **1.1115 kN. Ans.**

Problem 4.3. *In Fig. P. 4.3 (a), the block A weighs 250 N, B weighs 500 N. The co-efficients of friction are 0.4 between A and B, 0.1 between B and the plane and 0.3 between the fixed drum C and the cable. Determine the least weight of the block D for motion of the block D to impend.*

[Nagpur University]

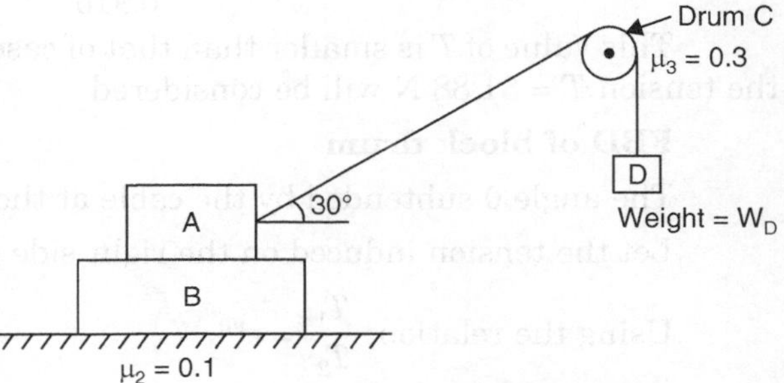

Fig. P.4.3 (a) Given system.

Sol. Weight, $W_A = 250$ N

Weight, $W_B = 500$ N

Let W_D be the weight of block D.

Case I. Let block A move over block B. See Fig. P. 4.3 (*b*)

Let $F_1 = \mu_1 N_1 = 0.4\, N_1$

By $\Sigma F_y = 0$, we get $N_1 - W_A + T \sin 30° = 0$

$\therefore \quad N_1 = W_A - 0.5T$

or $\quad N_1 = 250 - 0.5T \quad$ ...(1)

By $\Sigma F_x = 0$, we get $\mu_1 N_1 - T \cos 30° = 0$

$\therefore \quad 0.4N_1 = T \cos 30° \quad$ ...(2)

From (1), we get $0.4(250 - 0.5T) = T \cos 30°$

$\therefore \quad T(\cos 30° + 0.4 \times 0.5) = 0.4 \times 250$

or, $\quad T = \dfrac{100}{1.066} = \mathbf{93.81\ N}$

y | W_A = 250 N

T

A 30° x

$F_1 = \mu_1 . N_1$
= 0.4 N_1

N_1

Fig. P.4.3 (b) FBD of block A.

Case II. Let blocks A and B move together over the horizontal plane. See Fig. P. 4.3 (*c*) for *FBD* of blocks A and B taken together. Let tension in the cable be T when both blocks A and B move together.

Here, force of friction $F_2 = \mu_2 \times N_2 = 0.1\, N_2$

By $\Sigma F_y = 0$, we get

$N_2 - (W_A + W_B) + T \sin 30° = 0$

$\therefore \quad N_2 = (W_A + W_B) - 0.5T$

$\mid W_A + W_B = 250 + 500 = 750$ N

or, $\quad N_2 = 750 - 0.5T \quad$ ...(3)

By $\Sigma F_x = 0$

$\mu_2 N_2 - T \cos 30° = 0$

$0.1 N_2 = T \cos 30° \quad$ ...(4)

From (3), we get

$0.1(750 - 0.5T) = T \cos 30°$

$\therefore \quad T(\cos 30° + 0.1 \times 0.5) = 0.1 \times 750$

$\therefore \quad T = \dfrac{750 \times 0.1}{0.916} = \mathbf{81.88\ N}$

y | $W_A + W_B$ = 750 N

T

A + B 30° x

$\mu_2 . N_2$

N_2

Fig. P. 4.3 (c) FBD of blocks.

This value of T is smaller than that of case I. Hence for finding value of least weight W_D, the tension T = 81.88 N will be considered.

FBD of block drum

The angle θ subtended by the cable at the centre of drum = θ = 90° + 30° = 120°

Let the tension induced on the right side of the cable = T' = weight of the block D = W_D

Using the relation, $\dfrac{T_1}{T_2} = e^{\mu\theta}$ ...(5)

(For details see chapter 5).

Let slack side tension = $T_2 = T = 81.88$ N,

and T_1 = tight side tension required to cause motion of block D to impend

$= W_D$

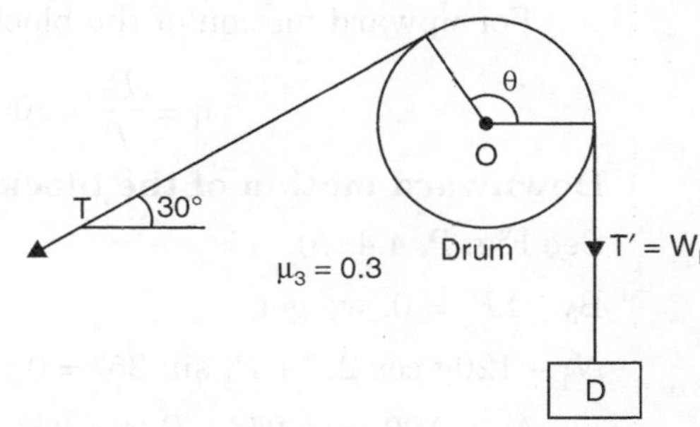

Fig. P. 4.3 (d) FBD of drum.

Then, from eqn. (5), we get

$$\frac{W_D}{81.88} = e^{\left(0.3 \times 120° \times \frac{\pi}{180°}\right)}$$

$$= e^{0.62832} = \mathbf{1.8745}$$

$$\therefore \quad W_D = 1.8745 \times 81.88$$

$$= \mathbf{153.484\ N}$$

$\therefore$ Weight of block D = **153.484 N. Ans.**

Problem 4.4. *Find the effort P required for the 120 kg block resting on a 25° inclined plane to just move upward. What is the efficiency of the plane? What is the pull for the downward motion of the block to impend? Also find efficiency of the plane for the motion down it. Take μ = 0.36. In each case the applied force is inclined at 35° to the plane. What is range of effort P required ?*

Sol. For the upward motion to impend, see Fig. P. 4.4(a).

By $\Sigma F_y = 0$, we get $N - 120g \cos 25° + P \sin 35° = 0$

$\therefore \quad N = 120g \cos 25° - P \sin 35°$ ...(1)

By $\Sigma F_x = 0$, we get $0.36N + 120g \sin 25° - P \cos 35° = 0$

$\therefore \quad 0.36N = P \cos 35° - 120g \sin 25°$ ...(2)

By substituting (1) into (2), we get

$0.36\,(120g \cos 25° - P \sin 35°)$

$= P \cos 35° - 120g \sin 25°$

$\therefore \quad P(\cos 35° + 0.36 \sin 35°)$

$= 120\,g\,(0.36 \cos 25° + \sin 25°)$

or, $\quad 1.0256\,P = 120\,g \times 0.749$

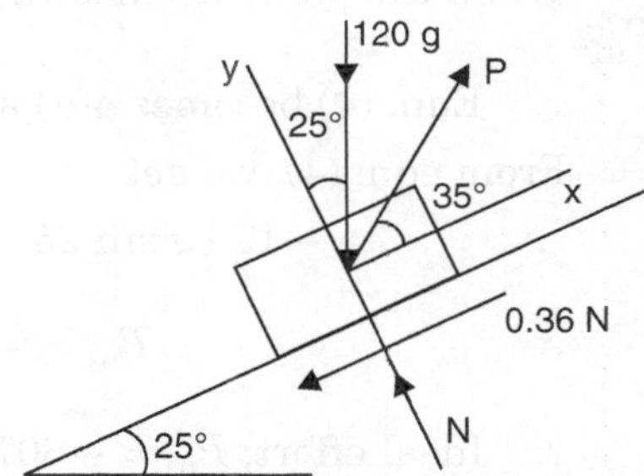

Fig. P. 4.4 (a) FBD of Block moving up the plane.

$$\therefore \quad P = \frac{(120 \times 9.81) \times (0.749)}{1.026}\ N$$

$\therefore$ Effort = P = **859.25 N. Ans.**

When plane is smooth, the value of the effort is P_0. Then the term $0.36N = F$ in eqn. (2) will be zero.

Thus, eqns. (1) and (2) become (By $\Sigma F_y = 0$)

$N = 120g \cos 25° - P_0 \sin 35°$ ...(1a)

(By $\Sigma F_x = 0$), $\quad 120g \sin 25° - P_0 \cos 35° = 0$ ...(2a)

From (2a), $$P_0 = \frac{120g \sin 25°}{\cos 35°}$$

or $\quad P_0$ = **607.34 N** = ideal effort. **Ans.**

∴ For upward motion of the block, efficiency of the plane is

$$\eta = \frac{P_0}{P} \times 100 = \frac{607.34}{859.25} \times 100 = \mathbf{70.68\%.} \quad \textbf{Ans.}$$

Downward motion of the block :

See Fig. P. 4.4 (*b*)

By $\Sigma F_y = 0$, we get

$N_1 - 120g \cos 25^{\circ} + P_1 \sin 35^{\circ} = 0$

∴ $N_1 = 120g \cos 25^{\circ} - P_1 \sin 35^{\circ}$ ...(3)

By $\Sigma F_x = 0$, we get

$0.36N_1 - 120g \sin 25^{\circ} - P_1 \cos 35^{\circ} = 0$ ...(4)

∴ By (3) in (4), we get

$0.36[120g \cos 25^{\circ} - P_1 \sin 35^{\circ}] - 120g \sin 25^{\circ} - P_1 \cos 35^{\circ} = 0$

∴ $P_1(0.36 \sin 35^{\circ} + \cos 35^{\circ}) = 120g(0.36 \cos 25^{\circ} - \sin 25^{\circ})$

$$\therefore \quad P_1 = \frac{-120g \times 9.81 \times (0.09635)}{1.02564}$$

∴ Effort $P_1 = \mathbf{-110.59\ N.}$

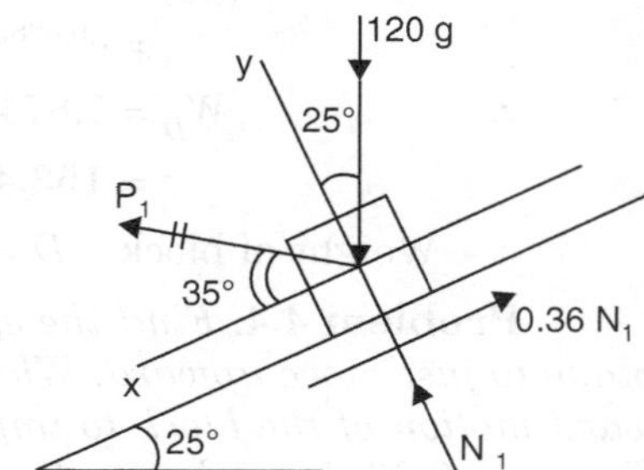

Fig. P. 4.4 (b) FBD of the block for downward motion

The –ve sign of P_1 indicates that its direction is **upwards.**

∴ Effort = P_1 = **110.59 N upwards. Ans.**

When the plane is smooth, the frictional force (0.36 N) in eqn. (4) is zero, and the effort is P_{01}.

∴ Eqn. (4) becomes 4(*a*) as given below

From eqn. (4), we get

$$-120\, g \sin 25^{\circ} - P_{01} \cos 35^{\circ} = 0$$

$$\therefore \quad P_{01} = -\frac{120 \times 9.81 \sin 25^{\circ}}{\cos 35^{\circ}} = -607.34 \text{ N}$$

∴ Ideal effort, $P_{01} = +607.34N$ directed up the plane. **Ans.**

Efficiency of the plane,

$$\eta' = \frac{P_1}{P_{01}} \times 100 = \frac{110.59}{607.34} \times 100 = 18.21\%. \quad \textbf{Ans.}$$

Note. When the plane is smooth the ideal pull P_{01} is the force which is required to hold the block from slipping it down the plane.

Range of the Effort. The maximum effort required = P = 859.25 N for upward motion of block to impend.

The minimum effort required = P_1 = 110.59 N for the downward motion of the block to impend. Therefore, the range of effort

= **110.59 to 859.25 N Ans.**

Problem 4.5. *Determine the effort required to tip the box of Fig. P. 4.5 (a) about A. Also find the minimum coefficient of friction if the box tips before it slides.*

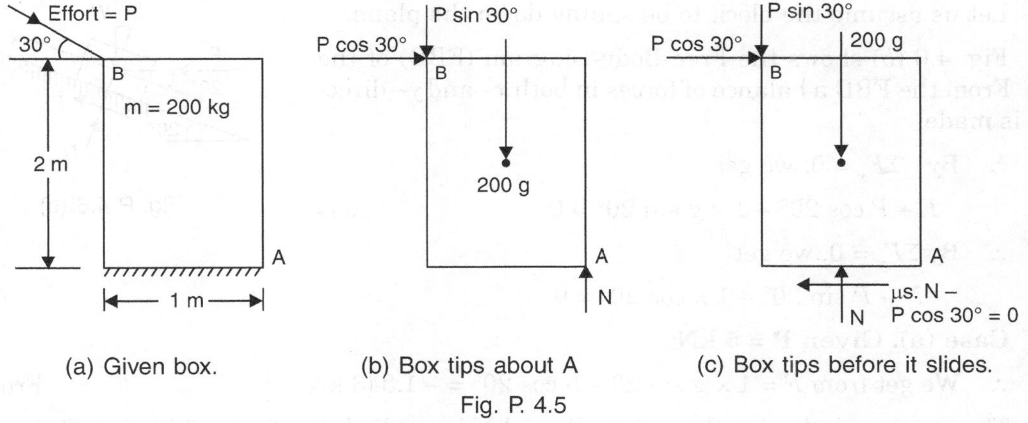

(a) Given box. (b) Box tips about A (c) Box tips before it slides.

Fig. P. 4.5

Sol. When the box tips (*i.e.*, overturns) about *A*, the contact between box and the support exists only at *A* and only normal force *N* acts through *A*. There is no frictional force. Let *P* be the magnitude of the effort required to tip box about *A*. The value of *P* can be found by taking moments about *A*. Thus, by

$$\Sigma M_A = 0, \text{ we have}$$

$$(P \cos 30°) \times 2 - (P \sin 30°) \times 1 - (200g) \times 0.5 = 0$$

$$\therefore \qquad P = \frac{100 \times 9.81}{1.232} = \mathbf{796.2\ N.} \quad \textbf{Ans.}$$

Let μ_s be the minimum value of coefficient of friction between the contacting surfaces when the box tips before it slides.

By $\Sigma F_v = 0$, we get $N - P \sin 30° - 200g = 0$

$$\therefore \qquad N = 0.5P + 200 \times 9.81 = 0.5 \times 796.2 + 1962$$

$$= 2360.1 \quad \text{Newtons}$$

And, by $\Sigma F_h = 0$, we get $\mu_s N - P \cos 30° = 0$

$$\therefore \qquad \mu_s N = P \cos 30° = 689.62 \text{ Newtons}$$

$$\therefore \qquad \mu_s = \frac{689.62}{N} = \frac{689.62}{2360.15} = \mathbf{0.292.} \quad \textbf{Ans.}$$

Thus, if the given coefficient of friction between the contact surfaces is less than 0.292, the tipping will not take place.

Problem 4.6. *Find the magnitude and direction of the force of friction acting on the 1000 g block lying on a 20° inclined plane, if, (a) P = 5 kN and, (b) P = 1 kN. Take μ_s = 0.2 and μ_k = 0.17. Forces are applied with block initially at rest, and P is horizontal.*

W = 1000 g
P
20°

Fig. P. 4.6 (a)

Sol. Fig. P. 4.6 (*a*) shows the given 1000 kg block lying on a 20° inclined plane.

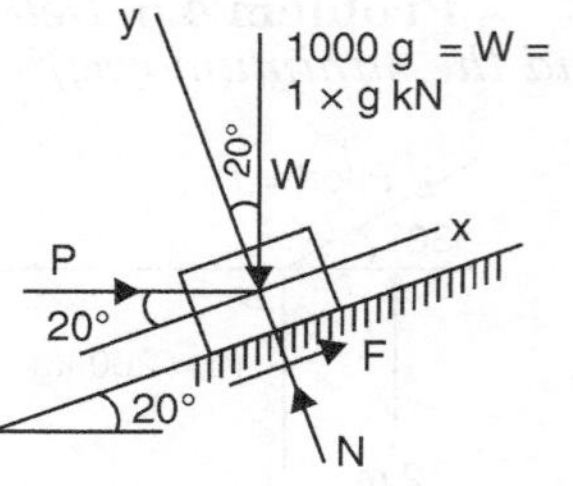

Fig. P. 4.6 (b)

Let us assume the block to be sliding down the plane.

Fig. 4.6 (*b*) shows the Free-Body-Diagram (FBD) of the block. From the FBD a balance of forces in both *x*– and *y*– directions is made.

∴ By $\Sigma F_x = 0$, we get

$$F + P\cos 20° - 1 \times g \sin 20° = 0 \quad ...(1)$$

∴ By $\Sigma F_y = 0$, we get

$$N - P\sin 20° - 1 \times \cos 20° = 0 \quad ...(2)$$

Case (a). Given P = 5 kN

∴ We get from $F = 1 \times g \sin 20 - 5\cos 20° = -1.343$ kN From (1)

The –ve sign indicates that when $P = 5$ kN is applied, the force of friction F should be directed down the plane for the equilibrium of the block.

∴ From eqn. (2), the normal reaction N is given by

$$N = P\sin 20° + 1 \times g\cos 20°$$

$$= 5\sin 20° + 9.81\cos 20° = \mathbf{10.93\ kN}$$

∴ The maximum force of static friction

$$F_{max} = F_s = \mu_s . N = 0.2 \times 10.93 = \mathbf{2.186\ kN.}$$

Since F_{max} is greater than 1.343 kN,

∴ we find that the block is in equilibrium and the force of friction **F = 1.343 kN is acting down the plane. Ans.**

Case (b). When P = 1 kN, from equation (1), we get

$$F = 9.81\sin 20° - P\cos 20°$$

$$= 9.81 \times \sin 20° - 1 \times \cos 20°$$

$$F = 2.42 \text{ kN}$$

and $$N = P\sin 20° + 9.81\cos 20° \quad \text{From eqn. (2)}$$

or $$N = 9.56 \text{ kN}$$

∴ The maximum force of friction, $F_{max} = \mu_s N$

∴ $$F_{max} = 0.2 \times 9.56 = 1.912 \text{ kN}$$

Since $F = 2.42$ kN is $> F_{max}$ ∴ $F = 2.42$ kN cannot be supported. Therefore, block is not in equilibrium, and therefore correct value of force of friction is equal to $\mu_k N$

or $$F = 0.17 \times 9.56 = \mathbf{1.6252\ kN} \text{ up the plane.} \quad \textbf{Ans.}$$

Problem 4.7. *Two blocks A and B weighing 180 N and 270 N are supported on a 30° rough plane. Block A is tied up by a horizontal string and a downward pull P, parallel to the plane, is applied to the block B. If μ = 0.4 find the tension in string and the pull P.*

Sol. The system of blocks and the FBDs of blocks is shown in Fig. P. 4.7.

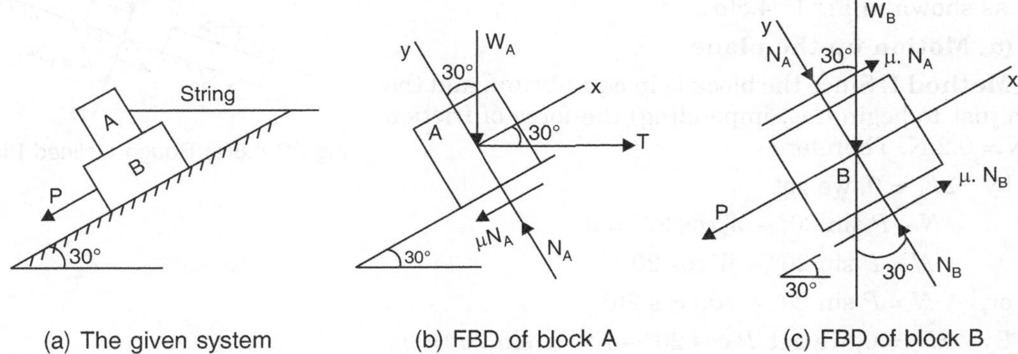

(a) The given system (b) FBD of block A (c) FBD of block B

Fig. P. 4.7

For block A, let T be the tension in string

By $\Sigma F_y = 0$, we get $N_A - W_A \cos 30° - T \sin 30° = 0$

$\therefore \quad N_A = W_A \cos 30° + T \sin 30°$ ...(1)

By $\Sigma F_x = 0$, we get

$$T \cos 30° - \mu N_A - W_A \sin 30° = 0 \quad ...(2)$$

By substituting value of N_A into eqn. (2), we get

$$T \cos 30° - \mu(W_A \cos 30° + T \sin 30°) - W_A \sin 30° = 0$$

$\therefore \; T(\cos 30° - \mu \sin 30°) = W_A(\sin 30° + \mu \cos 30°)$

$$\therefore \quad T = \frac{W_A(\sin 30° + \mu \cos 30°)}{\cos 30° - \mu \sin 30°} = \frac{180(\sin 30° + 0.4 \cos 30°)}{(\cos 30° - 0.4 \sin 30°)} = \frac{152.354}{0.666}$$

$\therefore \quad T$ = **228.75 N. Ans.**

And from (1), $N_A = 180 \cos 30° + 228.75 \sin 30°$

= **270.26 N**

For the block B

By $\Sigma F_y = 0$, we get $N_B - N_A - W_B \cos 30° = 0$

$\therefore \quad N_B = N_A + W_B \cos 30° = 270.26 + 270 \cos 30°$

$\therefore \quad N_B = 504.09$ N

By $\Sigma F_x = 0$, we get

$$P - \mu N_B - \mu N_A + W_B \sin 30° = 0$$

$\therefore \quad P = \mu(N_A + N_B) - W_B \sin 30° = 0.4(270.26 + 504.09) - 270 \times 0.5$ N

$\therefore$ Pull P = **174.74 N. Ans.**

Problem 4.8. *A 400 N block is at rest on a 30° inclined plane*

(*a*) *Determine the horizontal force P that will cause motion to impend up the inclined plane.*

(*b*) *Also find the efficiency of the plane. Take μ = 0.20 between the block and the plane.*

Sol. The *FBD* of the block is shown in Fig. P. 4.8 (*a*). As the motion is impending up the plane therefore the maximum friction force $F = \mu N$ will be directed down the plane. Let us take x-axis in the direction of impending motion (*i.e.,* upward) and the y-axis perpendicular to the plane, as shown in Fig.P. 4.8(*a*).

Fig. P. 4.8(a) Rough Inclined Plane.

(*a*) **Motion up the plane**

Method I. Since the block is in equilibrium and the motion just to begin (*i.e.,* impending) the force of friction $F = \mu N = 0.2$ N. Therefore

by $\Sigma F_y = 0$, we get

$$N - P \sin 20° - W \cos 20° = 0$$

$$\therefore \quad N = P \sin 20° + W \cos 20°$$

$$\text{or,} \quad N = P \sin 20° + 400 \cos 20° \quad \text{...(1)}$$

By $\Sigma F_x = 0$, we get $P \cos 20° - F - W \sin 20° = 0$

$$\therefore \quad F = P \cos 20° - W \sin 20°$$

$$\text{or} \quad \mu N = P \cos 20° - 400 \sin 20° \quad \text{...(2)}$$

From (1), we get $\mu[P \sin 20° + 400 \cos 20°] = P \cos 20° - 400 \sin 20°$

$$\text{or,} \quad P[\mu \sin 20° - \cos 20°] = - \mu \times 400 \cos 20° - 400 \sin 20°$$

$$\therefore \quad P[0.2 \times \sin 20° - \cos 20°] = -[0.2 \times 400 \cos 20° + 400 \sin 20°]$$

$$\therefore \quad P = \frac{-211.9835}{-0.8713} = \mathbf{243.3\ N.} \quad \textbf{Ans.}$$

Method II. The resultant reaction R of the plane to the block is inclined at $\phi = \tan^{-1} 0.2 = 11.3°$ to the normal N, directed as shown in Fig. P. 4.8 (*b*).

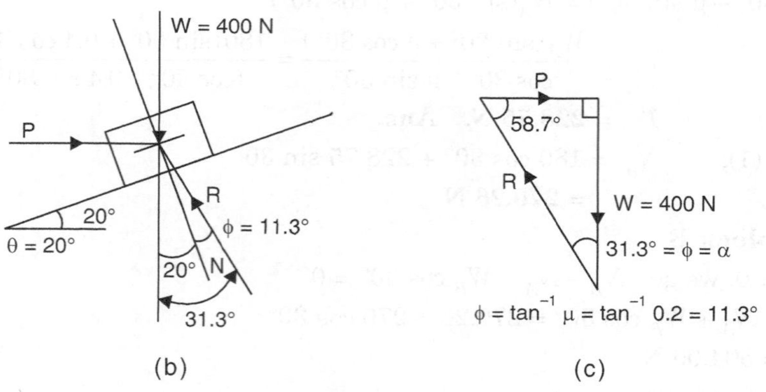

Fig. P. 4.8

The three forces W, P and R keep the block in equilibrium. The resultant R is inclined at $\theta + \phi = (20 + 11.3)° = 31.3°$ to the vertical along which W is directed. The triangle of forces is shown in Fig. P. 4.8 (*c*). Thus, we find that

$$P = W \tan(\alpha + \phi) = 400 \tan 31.3°$$

$$\therefore \quad P = \mathbf{243.3\ N.} \quad \textbf{Ans.}$$

(*b*) **Efficiency of the inclined plane**

Method I. When the inclined plane is smooth $\mu = 0$ and therefore the friction force $F = 0$. Effort when $\mu = 0$, is called idle effort P_0. Fig. P. 4.8 (*d*) shows the *FBD* for the block resting on smooth plane.

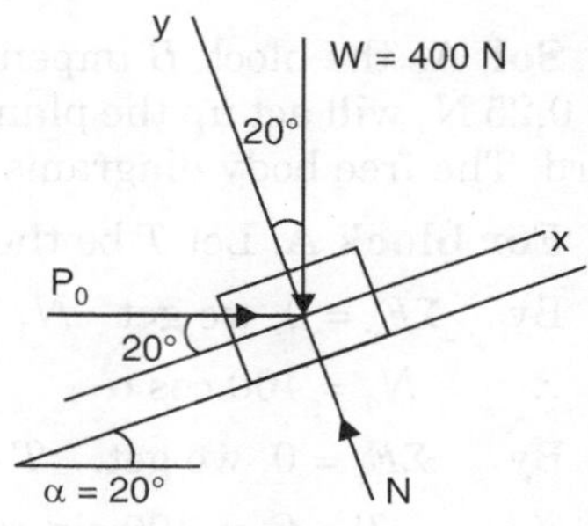

Fig. P. 4.8 (d) Smooth Inclined Plane.

And, $\Sigma F_x = 0$, we get

$$P_0 \cos 20° - W \sin 20° = 0$$

$$\therefore \quad P_0 = W \tan 20° = 400 \tan 20° \quad ...(3)$$

Idle effort,

$$P_0 = 400 \tan 20° = \mathbf{145.6\ N} \quad \textbf{Ans.}$$

Efficiency of plane,

$$\eta = \frac{\text{Idle effort}}{\text{Actual effort}} = \frac{P_0}{P} \quad ...(4)$$

$$\therefore \quad \eta = \frac{145.6}{243.3} \times 100 = \mathbf{45.84\%}. \quad \textbf{Ans.}$$

Method II. We know that, effort

$$P = W \tan (\alpha + \phi)$$

when $\phi = 0$, idle effort $P_0 = W \tan \alpha$

$$\therefore \quad P_0 = 400 \tan 20° = 145.6 \text{ N}$$

$$\therefore \quad \eta = \frac{P_0}{P} = \frac{400 \tan \alpha}{400 \tan (\alpha + \phi)} = \frac{\tan 20°}{\tan 31.3} = \mathbf{45.84\%}. \quad \textbf{Ans.}$$

Problem 4.9. *Find the inclination* α *of the inclined plane shown in Fig. P. 4.9 (a) for which the block B impends sliding down the plane* $\mu = 0.25$ *for all surfaces in contact. Also find tension in string.*

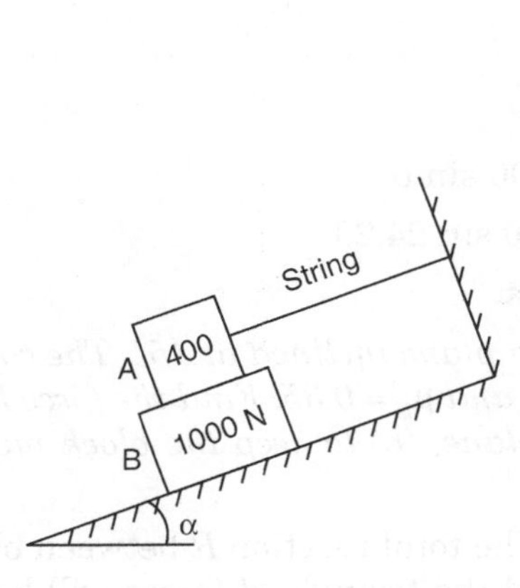

(a) Given System

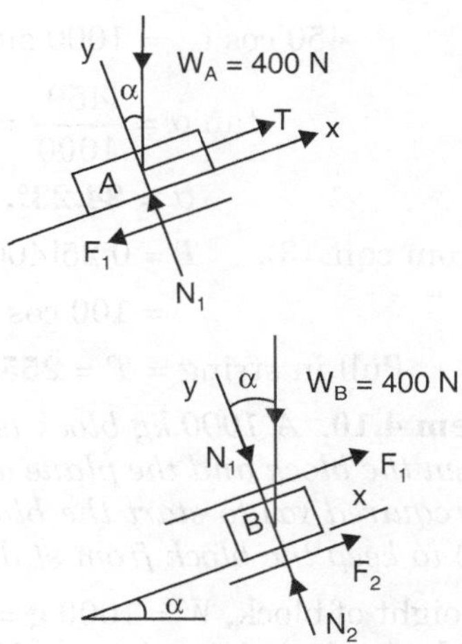

(b) FBDs of Blocks A and B

Fig. P. 4.9

Sol. As the block B impends to slide down the plane, therefore the frictional force $F_2 = \mu N_2 = 0.25\ N_2$ will act up the plane. The force of friction F_1 between the blocks A and B also acts upward. The free body diagrams of blocks A and B are shown in Fig. P.4.9(b).

For block A. Let T be the pull in the string.

By $\Sigma F_y = 0$, we get $N_1 - W_A \cos \alpha = 0$ | $W_A = 400$ N

$\therefore$ $N_1 = 400 \cos \alpha$ ...(1)

By $\Sigma F_x = 0$, we get, $T - F_1 - W_A \sin \alpha = 0$

or, $T = F_1 + 400 \sin \alpha$ ...(2)

But $F_1 = \mu N_1 = 0.25\ N_1$ $\therefore$ we get

$T = 0.25 N_1 + 400 \sin \alpha$

or $T = 0.25\ (400 \cos \alpha) + 400 \sin \alpha$ ...(3)

For the block B

By $\Sigma F_y = 0$, we get $N_2 - N_1 - W_B \cos \alpha = 0$

$\therefore$ $N_2 = W_B \cos \alpha + N_1 = 1000 \cos \alpha + 400 \cos \alpha$

or, $N_2 = 1400 \cos \alpha$ ...(4)

By $\Sigma F_x = 0$, we get

$F_2 + F_1 - W_B \sin \alpha = 0$

$\therefore$ $F_2 = W_B \sin \alpha - F_1$

or, $0.25 N_2 = 1000 \sin \alpha - 0.25 \times N_1$

or, $0.25 N_2 = 1000 \sin \alpha - 0.25 \times (400 \cos \alpha)$ $\therefore N_1 = 400$ N and $N_2 = 1400$ N

or, $0.25 \times 1400 \cos \alpha = 1000 \sin \alpha - 100 \cos \alpha$

$\therefore$ $350 \cos \alpha + 100 \cos \alpha = 1000 \sin \alpha$

$\therefore$ $450 \cos \alpha = 1000 \sin \alpha = 450 \cos \alpha$

$\therefore$ $\tan \alpha = \dfrac{450}{1000} = 0.45$

$\therefore$ $\alpha =$ **24.23°. Ans.**

And, from eqn. (3), $T = 0.25[400 \cos \alpha] + 400 \sin \alpha$

$= 100 \cos 24.23° + 400 \sin 24.23$

$\therefore$ Pull in string $= T =$ **255.35 N. Ans.**

Problem 4.10. *A 1000 kg block is pulled up a plane inclined at 15°. The co-efficients of friction between the block and the plane are $\mu_s = 0.40$ and $\mu_k = 0.35$. Find the force P inclined at 25° to plane required (a) to start the block up the plane, (b) to keep the block moving up the plane, and (c) to keep the block from sliding down.*

Sol. Weight of block, $W = 1000$ g $= 9.81$ kN ; The total reaction R between block and the plane will be determined in each case. The *FBD* and the triangle of forces will be drawn and the required effort P will be determined.

(a) When the block starts moving up, the *FBD* and the triangle of forces are represented as in Fig. P. 4.10 (a).

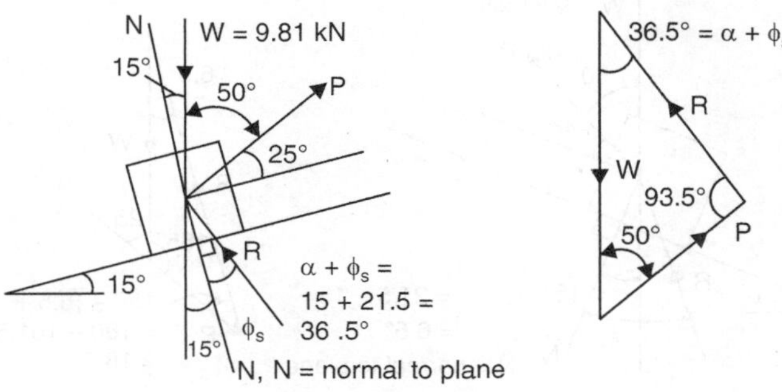

Fig. P. 4.10 (a)

Here $\phi_s = \tan^{-1} \mu_s = \tan^{-1} 0.40 = 21.5$

By triangle law of forces

$$\frac{P}{\sin 36.5°} = \frac{R}{\sin 50°} = \frac{W}{\sin [180 - (36.5 + 50)]°}$$

$$\therefore \quad P = \frac{W \sin 36.5°}{\sin 93.5°} = \frac{9.81 \sin 36.5°}{\sin 93.5°} = \mathbf{5.85\ kN.} \quad \textbf{Ans.}$$

(*b*) **When the block keeps moving up the plane,** the kinetic co-efficient of friction ($\mu_k = 0.35$) will be applicable. Thus, we get

$$\phi_k = \tan^{-1} \mu_k = \tan^{-1} 0.35 = 19.29°$$

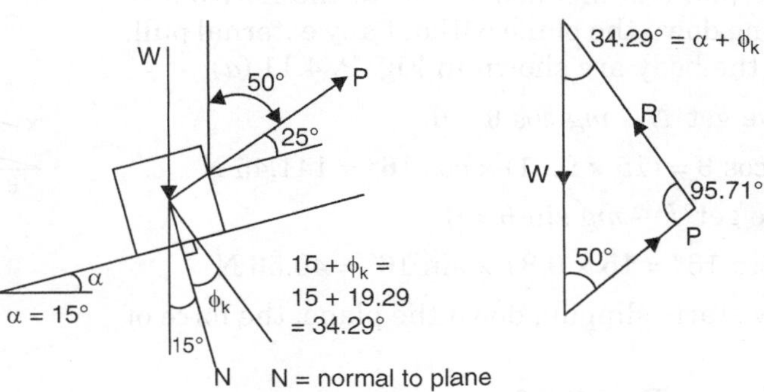

Fig. P. 4.10 (b)

The *FBD* and the triangle of forces are shown in Fig. P. 4.10 (*b*).

From the triangle of forces, we get

$$\frac{P}{\sin 34.29°} = \frac{R}{\sin 50°} = \frac{W}{\sin 95.71°}$$

$$\therefore \quad P = \frac{W \sin 34.29}{\sin 95.71°} = \frac{9.81 \sin 34.29°}{\sin 95.71°}$$

$$\therefore \quad P = \mathbf{5.554\ kN.} \quad \textbf{Ans.}$$

(*c*) **The force P to keep the block from sliding down.** When the block has a tendency to slide down the plane the resultant *R* is inclined at $\phi_s - \alpha = 21.5° - 15° = 6.5°$ from the vertical as shown in Fig. P. 4.10 (*c*). From the triangle of forces, it is clear that the effort *P* is acting downwards.

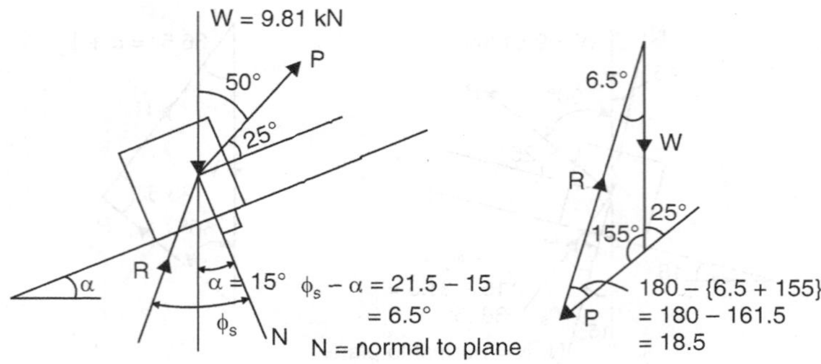

Fig. P. 4.10 (c)

From the triangle of force, we have

$$\frac{P}{\sin 6.5°} = \frac{R}{\sin 155°} = \frac{W}{\sin \{180 - (6.5° + 155°)\}}$$

$$\therefore \qquad P = \frac{W}{\sin 18.5°} \times \sin 6.5° = \frac{9.81}{\sin 18.5°} \times \sin 6.5° = \mathbf{3.5\ kN.} \quad \textbf{Ans.}$$

Since P is directed downward therefore the block will not slide down the plane.

Problem 4.11. *Find the horizontal force required to pull a body of mass 15 kg along a horizontal plane. If the plane on which the body is placed, is gradually tilted up till the body starts sliding down the plane when plane makes an angle of 16° with the horizontal.*

Sol. When the plane is inclined at 16° to the horizontal, the body starts slipping down the plane without any external pull. The forces acting on the body are shown in Fig. P. 4.11 (*a*).

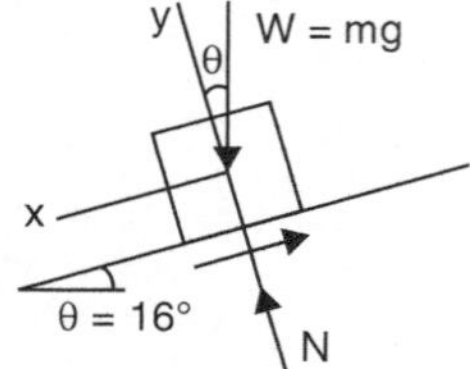

(a)

By $\Sigma F_y = 0$, we get $N - mg \cos \theta = 0$

$\therefore \quad N = mg \cos \theta = (15 \times 9.81) \times \cos 16° = 141.45$ N

By $\Sigma F_x = 0$, we get $F - mg \sin \theta = 0$

$\therefore \quad F = mg \sin 16° = 15 \times 9.81 \times \sin 16° = 40.56$ N

When the body starts slipping down the plane, the force of friction $F = \mu N$

$$\therefore \qquad \mu = \frac{F}{N} = \frac{40.56}{141.45} = \mathbf{0.2867}$$

When the above body is placed on the horizontal plane, the effort (P) required to pull the body is equal to the force of friction.

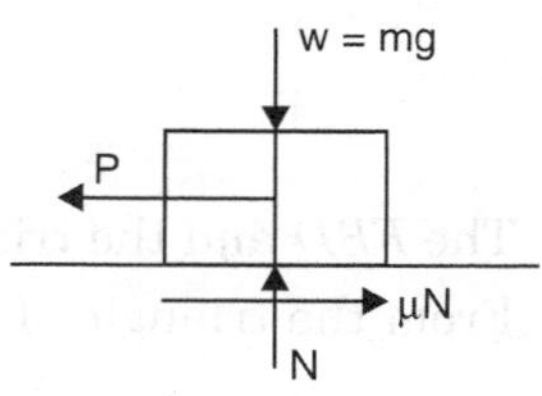

(b)

Fig. P. 4.11

Here the normal reaction is

$$N = W = mg = 15 \times 9.81 \text{ N}$$

$$= \mathbf{147.15\ N}$$

and, the effort $\qquad P = \mu N = 0.2867 \times 147.15 = \mathbf{42.19\ N.}$ **Ans.**

Problem 4.12. *A body of weight 1.5 kN is resting on a rough inclined plane making an angle of 25° with the horizontal. The coefficient of friction between the contact surfaces is 0.3. What is the range of the effort required parallel to the plane ?*

Sol. When the motion of the block is impending up the plane,

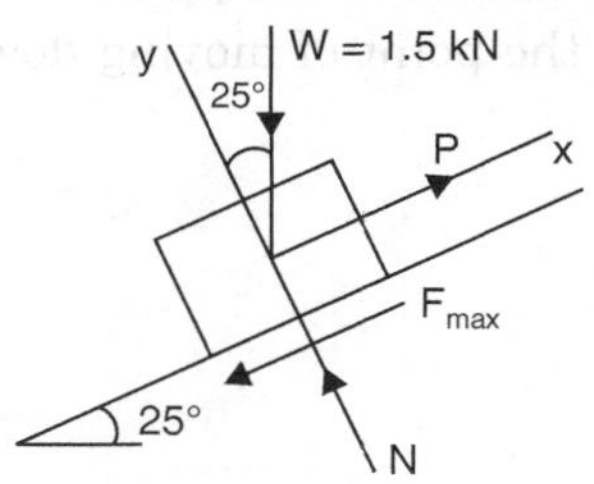

(a) Motion impending up the plane

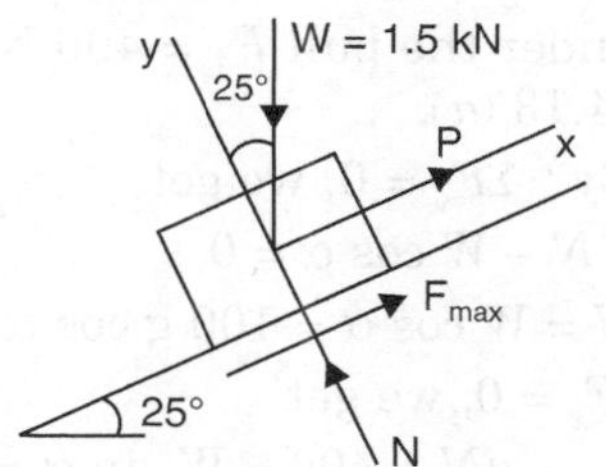

(b) Motion impending down the plane.

Fig. P. 4.12

The force of friction = $F_{max} = \mu_s N$, as shown in Fig. P. 4.12 (*a*)

By $\Sigma F_y = 0$, we get $N - W \cos 25° = 0$

$\therefore \quad N = W \cos 25° = 1.5 \times \cos 25 = 1.3595$ kN

By $\Sigma F_x = 0$, we have $P - F_{max} - W \sin 25° = 0$

$\therefore \quad P = F_{max} + W \sin 25° = 0.3 \times (1.3595) + 1.5 \sin 25°$

$\therefore \quad P = 1.042$ kN

When the block is impending to move down the plane, the *FBD* is shown in Fig. P. 4.12 (*b*).

Since $\mu = 0.3 \tan \phi_s$, $\therefore \quad \phi_s = \tan^{-1} 0.3 = 16.7°$. Here since the inclination of the plane ($\alpha = 25°$), therefore an upward acting force *P* is required to avoid the slipping of the block down the plane, as shown in Fig. P. 4.12 (*b*).

By $\Sigma F_y = 0$, we get $N - W \cos 25° = 0$,

$\therefore \quad N = 1.5 \times \cos 25° = 1.3595$ kN

By $\Sigma F_x = 0$, we get $P + F_{max} - W \sin 25° = 0$

$\therefore \quad P = W \sin 25° - F_{max} = 1.5 \sin 25° - 0.3 \times 1.3595$ kN

$\therefore \quad P = 0.226$ kN

$\therefore$ **Range of required effort is 0.226 kN to 1.042 kN. Ans.**

Problem 4.13. *A 100 kg block just starts moving down an inclined plane when an upward acting pull of 400 N parallel to the plane is applied and it is just at the point of moving up when the pull is increased to 600 N. Find the inclination of the plane and the co-efficient of friction between the block and the plane.*

Sol. *FBD*s of the two cases are shown in Figs. P. 4.13 (*a*) and (*b*). Let α be the inclination of the plane and μ be the co-efficient of friction.

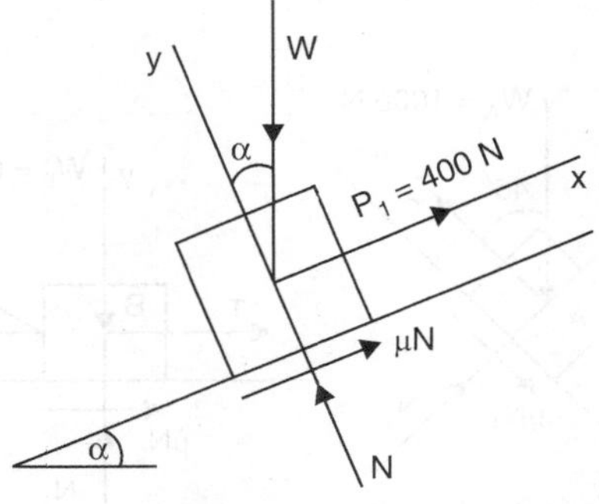

(a) Block just starts moving down the plane

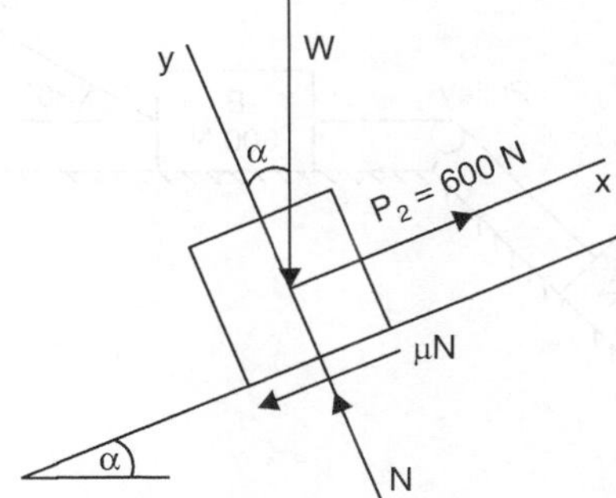

(b) Block just starts moving up the plane

Fig. P. 4.13

Let x-axis be parallel to plane and y-axis be perpendicular to the plane.

(*a*) Under the pull $P_1 = 400$ N, the block is just on the point of moving down the plane See Fig. P. 4.13 (*a*).

$\therefore$ By $\Sigma F_y = 0$, we get

$N - W \cos \alpha = 0,$

$\therefore \quad N = W \cos \alpha = 100 \text{ g} \cos \alpha \quad \quad ...(1)$

By $\Sigma F_x = 0$, we get

$\therefore \quad \mu N + 400 - W \sin \alpha = 0$

$\therefore \quad \mu N = 100 \text{ g} \sin \alpha - 400 \quad \quad ....(2) \mid W = 100 \text{ g}$

(*b*) Under the pull of $P_2 = 600$ N the block is just on the point of moving up the plane See Fig. P. 4.13. (*b*).

$\therefore$ By $\Sigma F_y = 0$, we get $N - W \cos \alpha = 0$

$\therefore \quad N = W \cos \alpha \quad \text{or} \quad N = 100 \text{ g} \cos \alpha \quad \quad$ from (1)

By $\quad \Sigma F_x = 0$, we get $\mu N + W \sin \alpha - 600 = 0$

$\therefore \quad \mu N = 600 - 100 \text{ g} \sin \alpha \quad \quad ...(3) \mid W = 100 \text{ g}$

Equating the right hand sides of eqns. (2) and (3), we get

$100 \text{ g} \sin \alpha - 400 = 600 - 100 \text{ g} \sin \alpha$

$\therefore \quad 200 \text{ g} \sin \alpha = 1000$

$$\therefore \quad \sin \alpha = \frac{1000}{200 \times 9.81} = 0.5097$$

$\therefore \quad \alpha = \sin^{-1} 0.5097 = \mathbf{27°}.$ **Ans.**

And, from eqns. (1) and (2), we get

$\mu(N) = 100 \text{ g} \sin \alpha - 400$

or $\quad \mu(100 \text{ g} \cos \alpha) = 100 \text{ g} \sin \alpha - 400$

$$\therefore \quad \mu = \frac{100 \text{ g} \sin \alpha}{100 \text{ g} \cos \alpha} - \frac{400}{100 \text{ g} \cos \alpha} = \tan \alpha - \frac{4}{9.81 \cos \alpha}$$

$\therefore \quad \mu = \mathbf{0.052}.$ **Ans.**

Problem 4.14. *For what value of the pull P in the system shown in Fig. P.4.14 (a) the motion will impend ? Assume the pulley to be smooth and* $\mu = 0.30$ *between other surfaces in contact.*

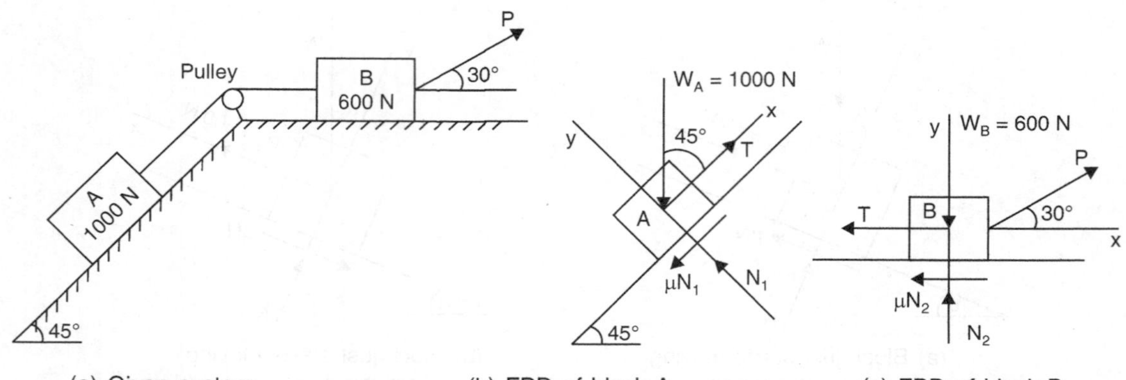

Fig. P. 4.14

Sol. From the *FBD* of block *A*, we can write the equations of equilibrium [Fig.P.4.14 (*b*)]

By $\Sigma F_y = 0$, we have

$$N_1 - W_A \cos 45° = 0$$

$\therefore \quad N_1 = W_A \cos 45° = 1000 \cos 45°$

By $\Sigma F_x = 0$, we have $T - \mu N_1 - W_A \sin 45° = 0$

$\therefore \quad T = \mu N_1 + W_A \sin 45° = 0.3 \times 1000 \cos 45° + 1000 \sin 45°$

$\therefore \quad T = \mathbf{919.24\ N}$

FBD of block *B* is shown Fig. P. 4.14 (*c*).

By $\Sigma F_y = 0$, we get

$$N_2 + P \sin 30° - W_B = 0$$

$\therefore \quad N_2 = W_B - P \sin 30° = 600 - 0.5P$

By $\Sigma F_x = 0$, we get

$$P \cos 30° - T - \mu N_2 = 0$$

$\therefore \quad P \cos 30° = T + \mu N_2 = 919.24 + 0.3\ (600 - 0.5P)$

$\therefore \quad P(\cos 30° + 0.3 \times 0.5) = 919.24 + 180$

$\therefore \quad P = \dfrac{1099.24}{1.0160} = \mathbf{1081.93\ N.}$ **Ans.**

Problem 4.15. *Two blocks A and B, connected by a link AB of negligible weight, are supported on rough surfaces, shown in Fig. P. 4.15 (a). The co-efficients of static friction for blocks A and B are 0.4 and 0.35, respectively. If block B weighs 7.50 kN, what is the least weight of the block A for which the system is in equilibrium ?*

Sol. The given system and the *FBD*s of the two blocks are shown in Fig. P. 4.15.

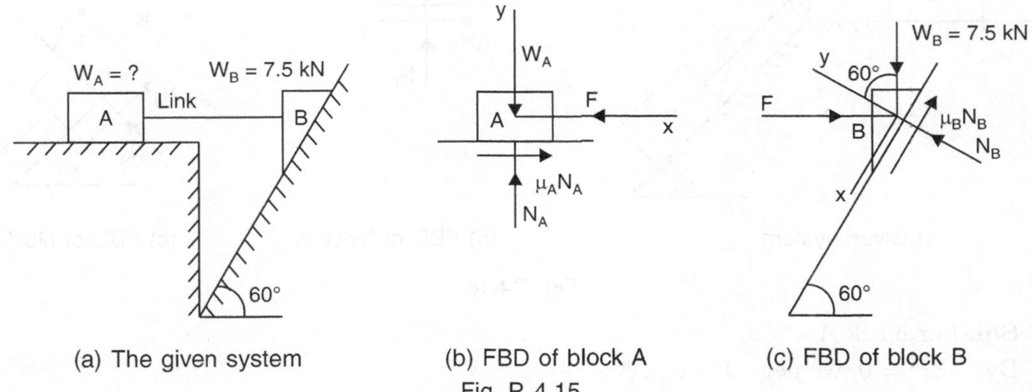

(a) The given system (b) FBD of block A (c) FBD of block B

Fig. P. 4.15

From the equilibrium of **block B,** we can write

by $\quad \Sigma F_y = 0, N_B - W_B \cos 60° - F \cos 30° = 0$

$\therefore \quad N_B = W_B \cos 60° + F \cos 30° \quad$ ...(1)

And by $\quad \Sigma F_x = 0$, we get

$$W_B \sin 60° - \mu_B N_B - F \sin 30° = 0$$

$\therefore \quad F \sin 30° = W_B \sin 60° - \mu_B N_B \quad$ ...(2)

By substituting value of N_B from eqn. (1) into eqn. (2), we

$$F \sin 30° = W_B \sin 60° - \mu_B (W_B \cos 60° + F \cos 30°)$$

or, $$F(\sin 30° + \mu_B \cos 30°) = W_B(\sin 60° - \mu_B \cos 60°)$$

or, $$F(\sin 30° + 0.35 \cos 30°) = 7.50 \times (\sin 60° - 0.35 \times \cos 60°)$$

$$\therefore \quad F = \frac{5.1827}{0.8031} = \frac{7.50 \times 0.691}{0.8031} = \mathbf{6.453\ kN} \text{ (compression)}$$

For block A, By $\Sigma F_y = 0$, we get

$$N_A = W_A \quad ...(3)$$

and by $\Sigma F_x = 0$, we get

$$\mu_A . N_A - F = 0, \quad \therefore \quad N_A = \frac{F}{\mu_A} \quad ...(4)$$

$$\therefore \quad N_A = \frac{6.453}{0.4} \text{ kN} = 16.1325 \text{ kN}$$

$$\therefore \quad W_A = N_A = \mathbf{16.1325\ kN.} \quad \textbf{Ans.}$$

Problem 4.16. *Two blocks of the same weight are connected by a string passing over a smooth pulley. One block rests on a horizontal plane while the other on an inclined plane making an angle α with the horizontal. If the coefficient of friction is 0.25 for all contacting surfaces, (a) find the value of angle α for the motion to impend. (b) Find α when $W_A = 0.5\ W_B$. (c) Also find α when $\mu_A = 0.25$ and $\mu_B = 0.3$ and $W_A = 0.5\ W_B$.*

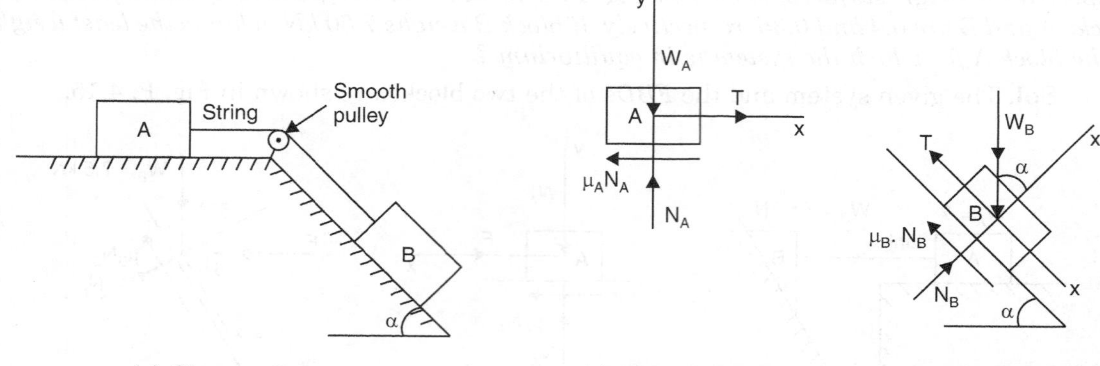

(a) Given system (b) FBD of block A (c) FBD of block B

Fig. P.4.16

Sol. For block *A*

By $\Sigma F_x = 0$, we get $T = \mu_A . N_A$ ...(1)

By $\Sigma F_y = 0$, $N_A = W_A$ ...(2)

$\therefore$ $T = \mu_A W_A$ ...(3)

For block *B*, by $\Sigma F_x = 0$, we get

$$T + \mu_B N_B - W_B \sin \alpha = 0$$

$\therefore$ $W_B \sin \alpha = T + \mu_B N_B$ ...(4)

By $\Sigma F_y = 0$, we get $N_B - W_B \cos \alpha = 0$

$\therefore$ $N_B = W_B \cos \alpha$ ...(5)

By substituting eqns. (5) and (3) in (4), we get

$$W_B \sin\alpha = \mu_A W_A + \mu_B \times (W_B \cos\alpha)$$

$$W_B (\sin\alpha - \mu_B \cos\alpha) = \mu_A W_A \quad \text{....(6)}$$

(*a*) **when $W_A = W_B = W$, and $\mu_A = \mu_B = 0.25 = \mu$,** eqn. (6) gives

$$W(\sin\alpha - \mu\cos\alpha) = \mu W$$

$$\therefore \quad \sin\alpha - \mu\cos\alpha = \mu$$

$$\therefore \quad \sin\alpha = \mu(1 + \cos\alpha)$$

$$\therefore \quad \mu = \frac{\sin\alpha}{1+\cos\alpha} \quad \text{...(7)}$$

Writing $\sin\alpha = 2\sin(\alpha/2)\cos(\alpha/2)$

and $1 + \cos\alpha = 2\cos^2\left(\frac{\alpha}{2}\right)$, eqn. (7) becomes

$$\mu = \frac{2\sin\alpha/2\cos\alpha/2}{2\cos^2\alpha/2}$$

or $$\mu = \frac{\sin\alpha/2}{\cos\alpha/2} = \tan\alpha/2 \quad \text{...(8)}$$

when $\mu = 0.25$, we get $0.25 = \tan\alpha/2$

$$\therefore \quad \alpha/2 = \tan^{-1} 0.25 = 14.036°$$

$$\therefore \quad \alpha = \mathbf{28.072°}. \quad \textbf{Ans.}$$

(*b*) When $W_A = 0.5\, W_B$, eqn. (6) gives

$$W_B(\sin\alpha - \mu_B\cos\alpha) = \mu_A (W_B \times 0.5)$$

$$\therefore \quad \sin\alpha - \mu_B\cos\alpha = 0.5\,\mu_A \quad \text{...(9)}$$

when $\mu_A = \mu_B = 0.25$, we get from (9)

$$\sin\alpha - 0.25\cos\alpha = 0.5 \times 0.25$$

$$\sin\alpha = 0.25(0.5 + \cos\alpha)$$

$$\therefore \quad 0.25 = \frac{\sin\alpha}{0.5+\cos\alpha} \quad \text{or,} \quad \sin\alpha - 0.25\cos\alpha - 0.125 = 0 \quad \text{...(10)}$$

We solve eqn. (10) by trial and error

Rewriting eqn. (10) we get

$$0.125 - \sin\alpha + 0.25\cos\alpha = 0 \quad \text{...(11)}$$

When $\alpha = 30°$, the L.H.S. of eqn. (11) = – 0.1585

For $\alpha = 15°$, L.H.S. = + 0.1077

For $\alpha = 25°$ L.H.S. = – 0.071

For $\alpha = 20°$ L.H.S. = 0.018

Plot values of L.H.S. for

$\alpha = 20°$ and $\alpha = 25°$ as shown in Fig.P.4.16(*d*)

Here $AA' = +\,0.0180$

or $BB' = 0.071$

$\because \quad \Delta AA'C \equiv \Delta BB'C, \quad \therefore$ we get

$$\frac{AC}{AA'} = \frac{BC}{BB'}$$

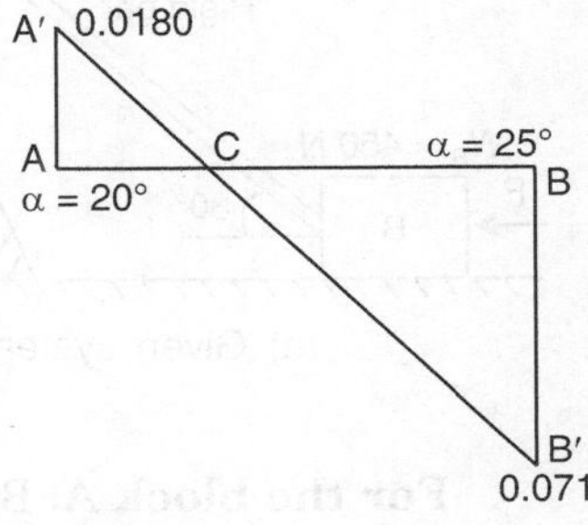

Fig. P. 4.16 (d)

or $$\frac{AC}{AA'} = \frac{AB - AC}{BB'}, \quad \therefore \quad AC.\, BB' = AA'\,.\,AB — AA'\,.\,AC$$

$\therefore \quad AC(BB' + AA') = AB \times AA'$

$$\therefore \quad AC = AB \times \frac{AA'}{BB' + AA'} = (25 - 20) \times \frac{0.018}{0.071 + 0.018}$$

$$\therefore \quad AC = 5 \times \frac{0.018}{0.089} = 1.01°$$

$\therefore \quad \alpha = 20° + 1.01 = $ **21.01°. Ans.**

(*c*) When $W_A = 0.5\ W_B$, $\mu_A = 0.25$ and $\mu_B = 0.3$, we get from eqn. (6), $W_B\ (\sin \alpha - \mu_B \cos \alpha) = \mu_A\ W_A$

$W_B\ (\sin \alpha - 0.3 \cos \alpha) = 0.25 \times 0.5\ W_B$

$\therefore \quad \sin \alpha - 0.3 \cos \alpha = 0.125$

$\therefore \quad 0.125 - \sin \alpha + 0.3 \cos \alpha = 0 \qquad ...(12)$

For $\alpha = 25°$, L.H.S. of eqn. (12) $= - 0.0257$

For $\alpha = 20°$, L.H.S. $= + 0.0649$

Fig. P. 4.16 (e)

Plot these values as shown in Fig. P. 4.16(*e*).

$\because \quad \Delta AA'C \equiv \Delta\, BB'C$, we get

$$\frac{AC}{AA'} = \frac{BC}{BB'}$$

or $$\frac{AC}{AA'} = \frac{AB - AC}{BB'}, \quad \therefore \quad AC(BB' + AA') = AB \times AA'$$

$$\therefore \quad AC = AB \times \frac{AA'}{BB' + AA'} = (25 - 20) \times \frac{0.0649}{0.0649 + 0.0257} = \frac{5 \times 0.0649}{0.0906}$$

$\therefore \quad AC = 3.58°$;

$\therefore \quad \alpha = 20 + 3.58 = $ **23.58°. Ans.**

Problem 4.17. *A block A weighing 160 N rests on a rough plane inclined at 60° to the horizontal. Block A is connected to block B weighing 450N resting on a rough horizontal plane by a rigid bar of negligible weight. Find the horizontal force P required for the motion of the blocks to just begin. Take μ = 0.25.*

Sol. See Fig. P.4.17. Let F be the force in the bar.

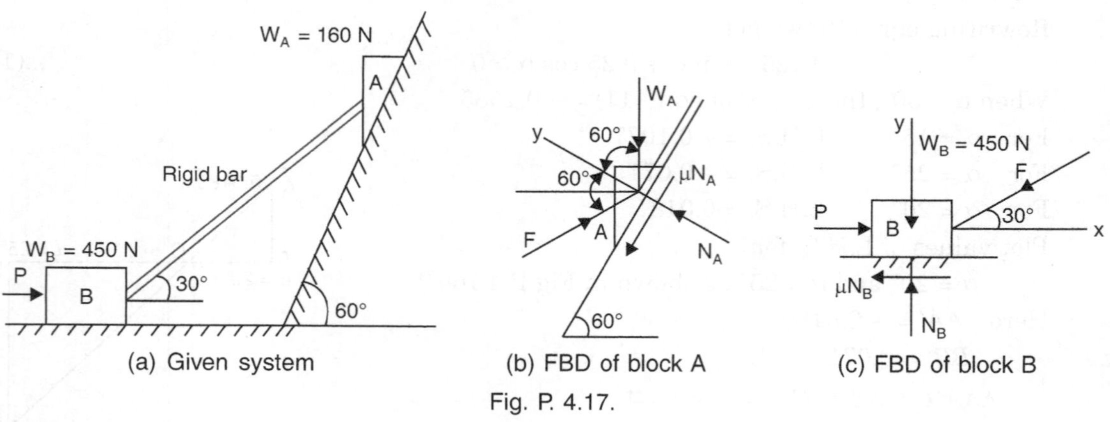

Fig. P. 4.17.

For the block A. By $\Sigma F_y = 0$, we get

$N_A - W_A \cos 60° - F \cos 60° = 0$

$\therefore \quad N_A = W_A \cos 60° + F \cos 60° \qquad ...(1)$

By $\Sigma F_x = 0$, we get

$$-\mu N_A - W_A \sin 60° + F \sin 60° = 0$$

$$\therefore \quad \mu N_A = F \sin 60° - W_A \sin 60° \quad ...(2)$$

By substituting eqn. (1) into eqn. (2), we get

$$\mu [W_A \cos 60° + F \cos 60°] = F \sin 60° - W_A \sin 60°$$

$$F[\mu \cos 60° - \sin 60°] = - W_A [\sin 60° + \mu \cos 60°]$$

$$\therefore \quad F = \frac{W_A (\sin 60° + \mu \cos 60°)}{\sin 60° - \mu \cos 60°}$$

$$= \frac{160 (\sin 60° + 0.25 \cos 60°)}{(\sin 60° - 0.25 \cos 60°)} = \frac{160 \times 0.991}{0.741} = \mathbf{213.98\ N.}$$

For block B. By $\Sigma F_y = 0$, we get $N_B - W_B - F \sin 30° = 0$

$\therefore \quad N_B = W_B + F \sin 30° = 450 + 213.98 \sin 30° = 556.99$ N

By $\Sigma F_x = 0$, we get $P - \mu N_B - F \cos 30° = 0$

$\therefore \quad P = \mu N_B + F \cos 30° = 0.25 \times 556.99 + 213.98 \times \cos 30°$

$\therefore \quad$ **P = 324.56 N. Ans.**

Problem 4.18. *Two blocks A and B of mass m_A and m_B respectively are connected by a string as shown in Fig. P. 4.18 (a). Find the direction and magnitude of the least force of pull P applied to block B so as to cause it slide towards right. Co-efficient of friction μ is the same for all surfaces in contact. The string connecting the blocks is inclined at α to the horizontal.*

Sol. The *FBD* of the system of blocks is shown in Fig. P.4.18 (*b*).

Since the entire system is in equilibrium,

$\therefore \quad$ By $\Sigma F_v = 0$, we get

$$P \sin \theta + (N_A + N_B) - (m_A + m_B)g = 0 \quad ...(1)$$

By $\Sigma F_h = 0$, we get

$$\mu N_A + \mu N_B - P \cos \theta = 0$$

$$\therefore \quad (N_A + N_B) = \frac{P \cos \theta}{\mu} \quad ...(2)$$

Substituting eqn. (2) in eqn. (1), we get

$$P \sin \theta + \frac{P \cos \theta}{\mu} - (m_A + m_B)g = 0$$

$$\therefore \quad P = \frac{\mu (m_A + m_B) g}{\mu \sin \theta + \cos \theta}$$

or, $$P = \frac{\mu \, (m_A + m_B) g \cos \phi}{(\sin \phi \sin \theta + \cos \theta \cos \phi)}$$

or $$P = \frac{(m_A + m_B) g \cos \phi \times \dfrac{\sin \phi}{\cos \phi}}{\cos (\theta - \phi)}$$

or $$P = \frac{(m_A + m_B) \, g \sin \phi}{\cos (\theta - \phi)} \quad ...(3)$$

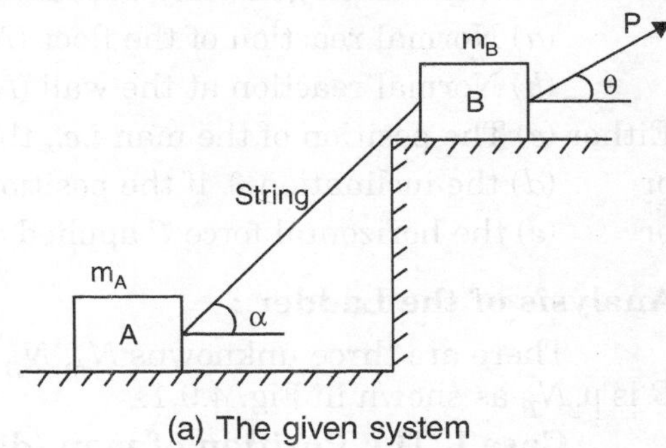

(a) The given system

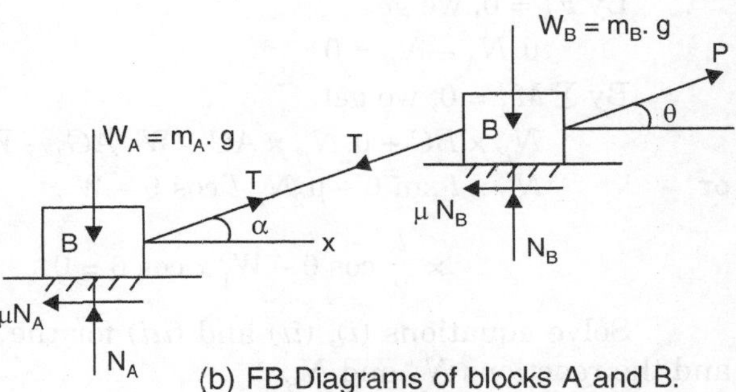

(b) FB Diagrams of blocks A and B.

Fig. P. 4.18

For P to be P_{min}, cos $(\theta - \phi)$ should have the maximum value.

$\therefore \quad \cos(\theta - \phi) = 1$

$\therefore \quad \theta - \phi = 0$

$\therefore \quad \boldsymbol{\theta = \phi}$. or $\theta = \tan^{-1}\mu$. **Ans.**

$\therefore$ Eqn. (3) gives,

$$P_{min} = \frac{(m_A + m_B)g \times \sin\phi}{1}$$

or $$P_{min} = (m_A + m_B)\,g \sin\phi = \boldsymbol{(W_A + W_B)\sin\phi}. \quad \textbf{Ans.} \qquad ...(4)$$

4.9. EQUILIBRIUM OF LADDER

Consider a ladder AB of weight W and length L lying on a rough horizontal floor AC and resting against a rough vertical wall BC as shown in Fig. 4.9.1. Let the co-efficient of friction between ladder and floor be μ_1 and that between ladder and wall be μ_2. Let a man of weight W_1 stand on the ladder at point D at a distance $AD = x$ from the bottom. Let the inclination of the ladder with the floor be θ.

This is a non-concurrent coplanar force system. Hence, only three equations of static equilibrium are available, *e.g.*,

$$\Sigma F_v = 0,$$

$$\Sigma F_h = 0 \quad \text{and} \quad \Sigma M = 0$$

Hence, only three unknown quantities can be determined using the above equations. The unknowns which generally encounter are :

(*a*) Normal reaction of the floor (N_A)

(*b*) Normal reaction at the wall (N_B) , and any one of the following :

Either (*c*) The position of the man *i.e.*, the distance x if the angle θ is given,

or (*d*) the inclination θ, if the position of man (x) is known,

or (*e*) the horizontal force P applied at A to avoid slipping of ladder if given are θ, x, etc.

Analysis of the Ladder :

There are three unknowns N_A, N_B and x. The force of friction at A is $\mu_1 N_A$, and that at B is $\mu_2 N_B$ as shown in Fig. 4.9.1.

Case I. The Position of man (distance *x*) is not known :

By $\Sigma F_v = 0$, we get

$$N_A + \mu_2 N_B - W_1 - W = 0 \qquad ...(i)$$

By $F_h = 0$, we get

$$\mu_1 N_A - N_B = 0 \qquad ...(ii)$$

By $\Sigma M_A = 0$, we get

$$N_B \times BC + \mu_2 N_B \times AC - W \,.\, AG_1 - W_1 \,.\, AD_1 = 0$$

or $$N_B \times l \sin\theta + \mu_2 N_B\, l \cos\theta - W \times \frac{l}{2}\cos\theta - W_1\, x \cos\theta = 0 \qquad ...(iii)$$

Solve equations (*i*), (*ii*) and (*iii*) for the distance x, and the reactions N_A and N_B.

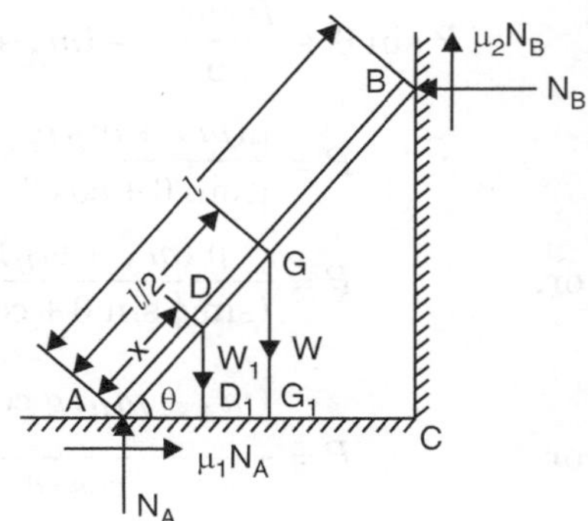

Fig. 4.9.1 Equilibrium of a ladder.

Case II. The inclination θ of the ladder not given :

Solve equations (*i*), (*ii*) and (*iii*) to find θ for the equilibrium of the ladder.

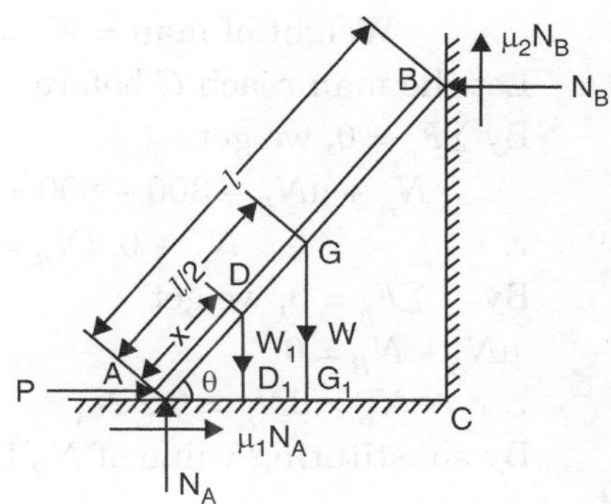

Fig. 4.9.2 Equilibrium of a ladder by application of force P at A.

Case III. Given θ, *x*

To find : The force *P* applied at *A* for the equilibrium of the ladder,

See Fig. 4.9.2

By $\Sigma F_v = 0$, we get

$$N_A + \mu_2 N_B - W - W_1 = 0 \quad ...(i) \text{ (as before)}$$

By $F_h = 0$, we get

$$P + \mu_1 N_A - N_B = 0 \quad ...(iv)$$

By $\Sigma M_A = 0$, we get

$$N_B \times l \sin\theta + \mu_2 N_B\, l \cos\theta - W\, l/2 \cos\theta - W_1\, x \cos\theta = 0 \quad ...(iii) \text{ (as before)}$$

Thus, to find the force *P*, N_A and N_B solve equations (*i*), (*iv*) and (*iii*) simultaneously.

Case IV. Tension in the Rope Tied to the Ladder (as shown in Fig. 4.9.3) for its equilibrium :

Here a rope is tied to the rope at a point *E* at a distance *AE* = *z* to prevent the ladder against slipping.

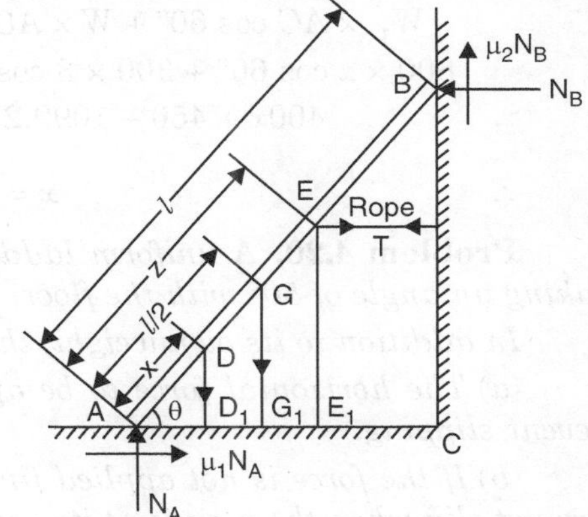

Fig. 4.9.3 Tension in the rope.

Let the tension in the rope be *T*.

Given : The inclination θ; the position of man (*x*) and the position of the rope (the distance *z* = *AE*).

The equations of equilibrium are :

By $\Sigma F_v = 0$, we get

$$N_A + \mu_2 N_B - W - W_1 = 0 \quad ...(i) \text{ (as before)}$$

By $\Sigma F_h = 0$, we get

$$T + \mu_1 N_A - N_B = 0 \quad ...(v)$$

By $\Sigma M_A = 0$, we get

$$N_B \,.\, l \sin\theta + \mu_2 N_B\ l\cos\theta - W\, l/2 \,.\, \cos\theta - W_1 \,.\, x \cos\theta_1 - T \,.\, z \sin\theta = 0 \quad ...(vi)$$

To find the tension *T* in the rope, solve equations (*i*), (*v*) and (*vi*) simultaneously. The other unknown quantities to be found are N_A and N_B. Solved examples given below explain the procedure involved for determination of the required unknowns.

Problem 4.19. *A 6 m long uniform ladder weighing 300 N is placed against a vertical wall. Its inclination with the ground is 60°. At what distance from the lower end of the ladder a man weighing 800 N should climb before slipping of the ladder starts. The co-efficient of friction between contact surfaces is 0.2.*

Sol. Length of ladder = *AB* = 6 m, *G* = *CG* of ladder

∴ *AG* = *GB* = 3 m.

Weight of ladder = 300 N = W

Weight of man = W_1 = 800 N

Let the man reach C before slip occurs.

By $\Sigma F_v = 0$, we get

$$N_A + \mu N_B - 300 - 800 = 0$$

$$\therefore \quad N_A + 0.2 N_B = 1100 \quad \text{...(1)}$$

By $\Sigma F_h = 0$, we get

$$\mu N_A - N_B = 0$$

$$\therefore \quad N_B = \mu N_A = 0.2\, N_A \quad \text{...(2)}$$

By substituting value of N_B from eqn. (2) into (1), we get

$$N_A + 0.2 \times (0.2\, N_A) = 1100$$

$$\therefore \quad N_A = \frac{1100}{1.04} = \mathbf{1057.69\ N.}$$

and $\quad N_B = 0.2\, N_A = 0.2 \times 1057.69 = \mathbf{211.54\ N.}$

Fig. P. 4.19

By $\Sigma M_A = 0$, we get

$$W_1 \times AC_1 + W\ AG_1 - N_B \times BO - \mu N_B \times AO = 0$$

or $\quad W_1 \times AC \cos 60° + W \times AG \cos 60° - N_B \times AC \sin 60° - \mu\, N_B\, AB \cos 60° = 0$

or, $\quad 800 \times x \cos 60° + 300 \times 3 \cos 60° - 211.54 \times 6 \sin 60° - 0.2 \times 211.54 \times 6 \cos 60° = 0$

$$\therefore \quad 400x + 450 - 1099.2 - 126.9 = 0$$

$$\therefore \quad x = \frac{776.1}{400} = \mathbf{1.94\ m.} \quad \textbf{Ans.}$$

Problem 4.20. *A uniform ladder 4 m long weighs 200 N. It is placed against a wall making an angle of 50° with the floor.*

In addition to its own weight, the ladder supports a man 800 N at its top. Calculate

(a) The horizontal force to be applied, to the ladder at its end resting on the floor to prevent slipping.

(b) If the force is not applied find the inclination of the ladder with the floor so that it does not slip when the man is at its top.

The coefficient of friction between the ladder and floor is 0.4 and that between the ladder and the wall is 0.3.

Sol. For part (*a*) the *FBD* of the ladder is shown in Fig. P. 4.20 (*a*).

Let $\quad W_l$ = weight of ladder = 200 N,

W_m = weight of man, 800 N,

AB = 4 m = length of ladder

$AG = BG$ = 2 m,

Man is at B and the G is the $C.G.$ of the ladder.

Let P = force applied at A to prevent slipping of ladder.

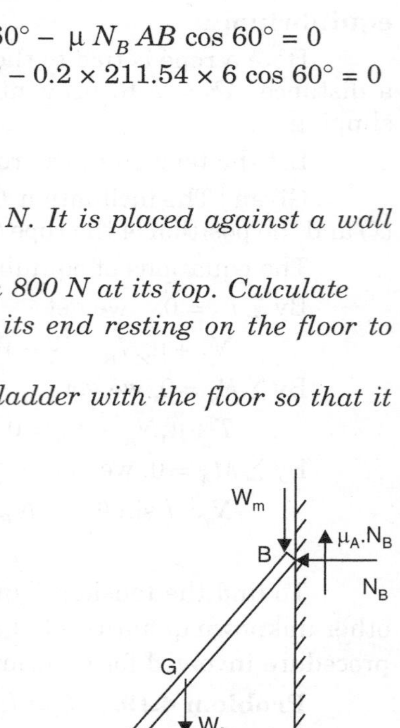

Fig. P. 4.20 (a) FBD of ladder for part (a) of the problem 4.20.

(*a*) The ladder is in equilibrium.

$\therefore \quad$ by $\quad \Sigma F_v = 0$, we get

$$N_A + \mu_B N_B - W_l - W_m = 0$$

$\therefore$ $$N_A + 0.3N_B - 800 - 200 = 0$$

or $$N_A + 0.3N_B = 1000 \quad \text{...(1)}$$

By $\Sigma F_h = 0$, we get

$$\mu_A.N_A + P - N_B = 0$$

$\therefore$ $$P = N_B - \mu_A.N_A \quad \text{or} \quad P = N_B - 0.4\,N_A \quad \text{...(2)}$$

by $\Sigma M_A = 0$, we get

$$N_B \times OB + \mu_B N_B \times OA - W_m \times OA - W_l \times AG_1 = 0$$

or $$N_B \times 4 \sin 50° + 0.3\,N_B \times 4 \cos 50° - 800 \times 4 \cos 50° - 200 \times 2 \cos 50° = 0$$

$\therefore$ $$3.836 \times N_B = 2314.035$$

$\therefore$ $$N_B = 603.24 \text{ N}$$

$\therefore$ $$N_A = 1000 - 0.3\,N_B \quad \text{from (1)}$$

or $$N_A = \mathbf{819.03\ N.}$$

$\therefore$ From (2), $$P = 603.24 - 0.4 \times 819.03$$

or $$P = \mathbf{275.628\ N.} \quad \textbf{Ans.}$$

(*b*) When force P is not applied let the ladder have inclination θ with the floor, as shown in Fig. P. 4.20(*b*).

Since the ladder is in equilibrium, therefore by $\Sigma F_v = 0$, we have

$$N_A + \mu_B.N_B - W_l - W_m = 0$$

or $$N_A + 0.3N_B - 200 - 800 = 0$$

$\therefore$ $$N_A + 0.3\,N_B = 1000 \quad \text{...(3)}$$

By $\Sigma F_h = 0$, we have

$$\mu_A.N_A - N_B = 0$$

$\therefore$ $$N_B = \mu_A.N_A$$

or $$N_B = 0.4\,N_A \quad \text{...(4)}$$

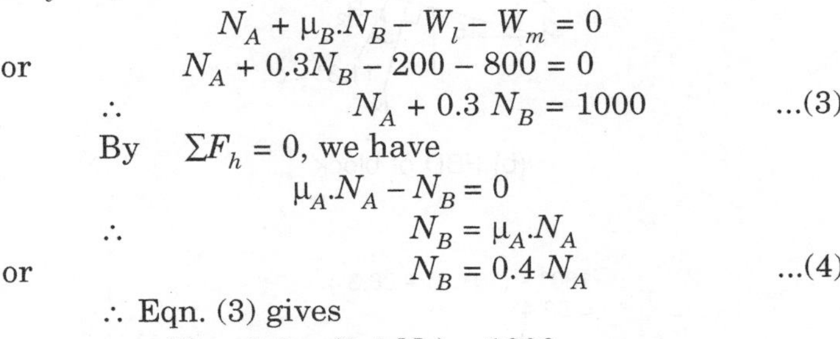

$\therefore$ Eqn. (3) gives

$$N_A + 0.3 \times (0.4\,N_A) = 1000$$

$\therefore$ $$N_A = \frac{1000}{1.12} = \mathbf{892.86\ N.} \quad \textbf{Ans.}$$

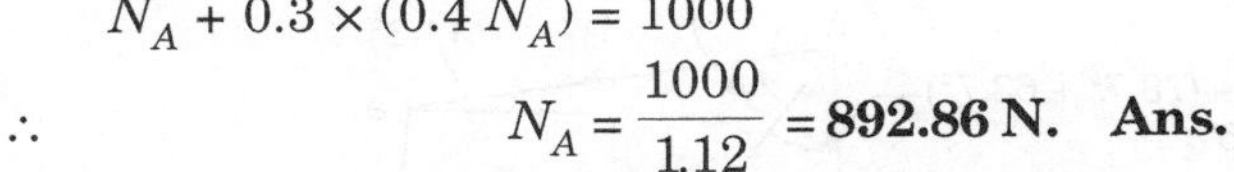

Fig. P. 4.20 (b) FBD for part (b) of problem 4.20.

and from (4), $$N_B = 0.4 \times 892.86 = \mathbf{357.14\ N.} \quad \textbf{Ans.}$$

By $\Sigma M_A = 0$, we have

$$N_B \times OB + 0.3\,N_B \times OA - W_m \times OA - W_l \times AG_1 = 0$$

$\therefore$ $$357.14 \times (4 \times \sin\theta) + 0.3 \times 357.14 \times (4 \times \cos\theta) - 800 \times 4 \times \cos\theta - 200 \times 2 \times \cos\theta = 0$$

$\therefore$ $$1428.56 \sin\theta - 3171.43 \cos\theta = 0$$

$\therefore$ $$\tan\theta = \frac{3171.43}{1428.56} = 2.22$$

$\therefore$ $$\theta = \tan^{-1} 2.22 = \mathbf{65.75°.} \quad \textbf{Ans.}$$

4.10. THE WEDGE

Wedge is a small piece of material with its two adjacent surfaces inclined at some small angle α. It is used (*a*) to life heavy blocks or heavy machines through small distances, (*b*) to slightly slide one end of the beam relative to another end, (*c*) to increase the pressure between two materials which are to be glued to each other, etc. The weight of the wedge is generally small compared to the weight lifted. Hence, almost in all problems the weight of wedge is neglected. Thus, instead of treating the normal reaction and the frictional force independently, it is more convenient to treat their resultant which is inclined at an angle ϕ to the normal reaction. This reduces the con-current coplanar force system of more than three forces into the one of three forces. Thus, the block being lifted, the wedge can be analysed by using the Lami's equations.

The following solved examples give the complete procedure used for the analysis of wedges.

Problem 4.21. *A block of weight 5000 N is to be raised by a 15° wedge. Assuming coefficient of friction between all contact surfaces to be 0.2, determine what minimum horizontal force P should be applied to the block. See Fig. P.4.21 (a).*

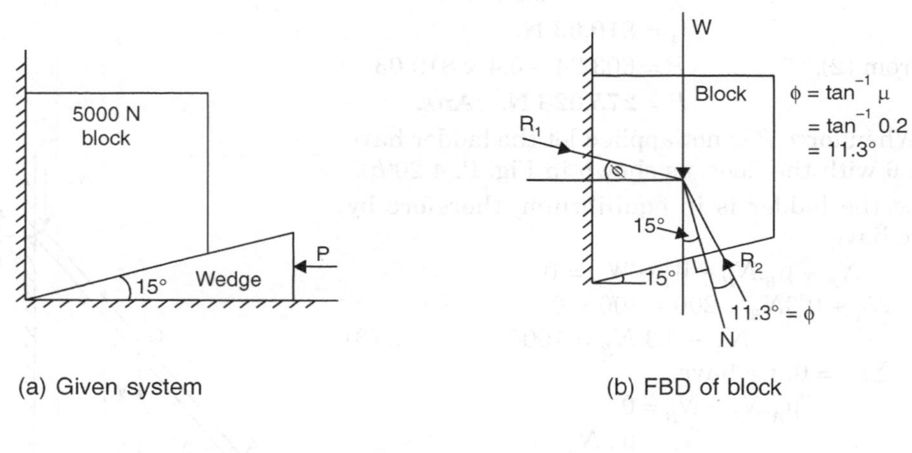

(a) Given system (b) FBD of block

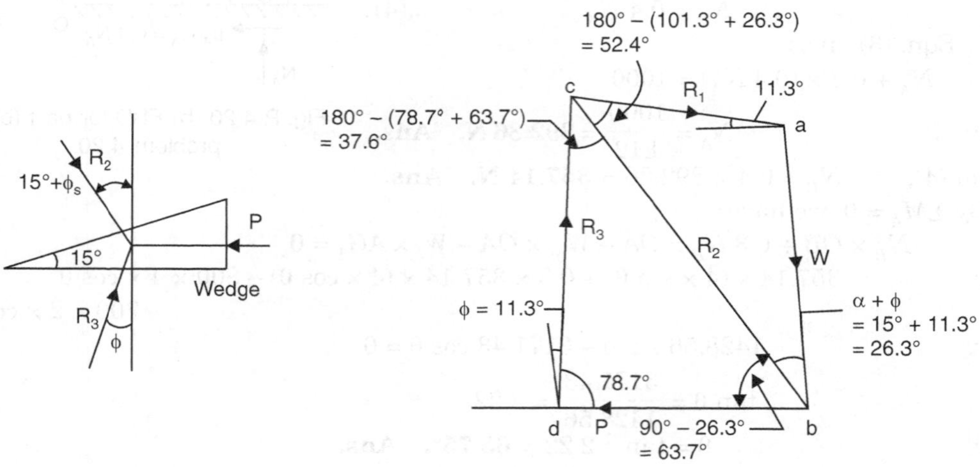

(c) FBD of wedge (d) Triangles of forces for forces shown in FBDs of Figs. (a) and (b)

Fig. P. 4.21

Sol. Fig. P.4.21 (*a*) shows the block resting on the wedge. *FBDs* are shown in Fig. P. 4.21(*b*) and (*c*). Three forces W, R_1, and R_2 acting on the block are represented by the Δabc. Similarly, the three forces P, R_2 and R_3 acting on the wedge are represented by the sides of the Δbcd.

For the block. The resultant reaction R_1 of the vertical wall to the block is given by R_1 which is inclined at $\phi = 11.3°$ to the normal (which is horizontal).

Similarly the resultant reaction R_2 of the wedge to block is inclined at $(\alpha + \phi) = 15° + 11.3 = 26.3$ to the vertical as shown in Fig. P. 4.21(*c*). Side *ab* is vertical.

$$\angle abc = \alpha + \phi = 15 + 11.3° = 26.3°,\ \angle cab = 90° + \phi = 90° + 11.3° = 101.3°,$$

$$\angle acb = 180° - (\angle abc + \angle cab) = 180° - (26.3° + 101.3°) = 52.4°.$$

$\therefore$ From Δabc, we have

$$\frac{cb}{\sin \angle cab} = \frac{ca}{\sin \angle cba} = \frac{ab}{\sin \angle acb}$$

or
$$\frac{R_2}{\sin 101.3°} = \frac{R_1}{\sin 26.3°} = \frac{W}{\sin 52.4°}$$

$$\therefore \quad R_2 = \frac{W}{\sin 52.4°} \times \sin 101.3° = \frac{5000}{\sin 52.4°} \times \sin 101.3° = \mathbf{6188.49\ N.}$$

For the wedge. As mentioned earlier, the sides of the Δcbd represent the three forces acting on the wedge. Side *bd* represents force P; side *bc* represents R_2 and side *dc* represents R_3.

$$\angle cbd = 90° - 26.3° = 63.7°,$$

$$\angle bdc = 90° - \phi = 90° - 11.3° = 78.7°$$

and
$$\angle bcd = 180° - (\angle cbd + \angle bdc) = 180° - (63.7° + 78.7°) = 37.6°.$$

$\therefore$ We can write

$$\frac{dc}{\sin \angle cbd} = \frac{bd}{\sin \angle bcd} = \frac{cb}{\sin \angle bdc}$$

$$\frac{R_3}{\sin 63.7°} = \frac{P}{\sin 37.6°} = \frac{R_2}{\sin 78.7°}$$

$$\therefore \quad P = \frac{R_2}{\sin 78.7°} \times \sin 37.6° = \frac{6188.49}{\sin 78.7°} \times \sin 37.6°$$

$\therefore$ Effort, P = **3850.52 N. Ans.**

Problem 4.22. *A block of weight 5000 N is to be raised by a 15° wedge weighing 500 N as shown in Fig. P.4.22 (a). What minimum horizontal force P should be applied to raise the block ? Co-efficient of friction for all contacting surfaces is 0.2.*

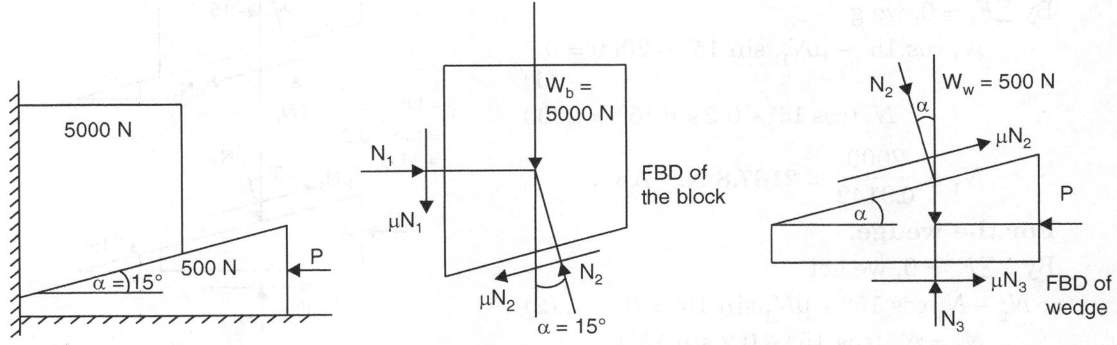

(a) The given system of block and the wedge

(b) FBDs of the block and the wedge.

Fig. P. 4.22

Solution. For the block.

By $\Sigma F_h = 0$, we get,

$$N_1 - \mu N_2 \cos 15° - N_2 \sin 15° = 0 \qquad ...(1)$$

By $\Sigma F_v = 0$, we get

$$N_2 \cos 15° - \mu N_2 \sin 15° - \mu N_1 - 5000 = 0 \qquad ...(2)$$

By multiplying (1) by μ and adding to (2), we get

$$(N_2 \cos 15° - \mu\, N_2 \sin 15 - \mu N_1 - 5000) + (\mu N_1 - \mu^2 N_2 \cos 15° - \mu N_2 \sin 15°) = 0$$

or $$N_2 (\cos 15° - \mu \sin 15° - \mu^2 \cos 15° - \mu \sin 15°) - 5000 = 0$$

or $$N_2 (0.8238) = 5000$$

or $$N_2 = \frac{5000}{0.8238} = 6069.43 \text{ N}.$$

For the wedge.

By $\Sigma F_v = 0$, we get

$$N_3 - 500 - N_2 \cos 15° + \mu N_2 \sin 15° = 0 \qquad ...(3)$$

or $$N_3 - 500 - 6069\,.\,43 \cos 15° + 0.2 \times 6069.43 \sin 15° = 0$$

$\therefore$ $$N_3 - 6048.44 = 0$$

$\therefore$ $$N_3 = 6048.44 \text{ N}$$

By $\Sigma F_h = 0$, we get $P - \mu N_3 - \mu N_2 \cos 15° - N_2 \sin 15° = 0 \qquad ...(4)$

$\therefore$ $$P = \mu N_3 + \mu N_2 \cos 15° + N_2 \sin 15°$$

$$= 0.2 \times 6048.44 + 0.2 \times 6069.43 \cos 15° + 6069.43 \sin 15°$$

$\therefore$ P = **3953.1 N. Ans.**

Problem 4.23. *Two wedges A and B are used to raise a 4000 N block C by applying horizontal force P as shown in Fig. P. 4.23 (a). Find the effort P when μ = 0.20 for each surface in contact.*

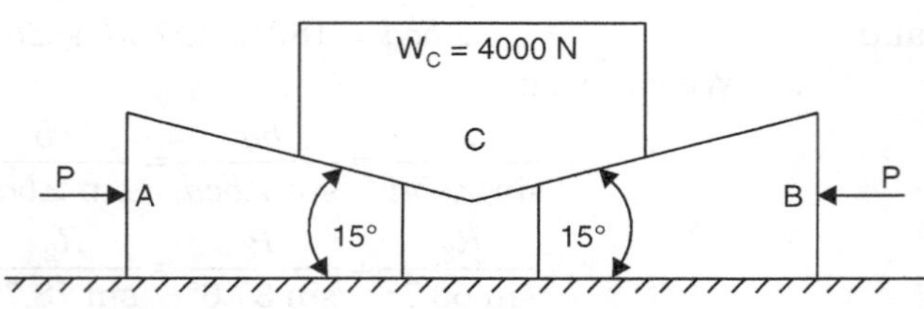

(a) The given system of wedges and the block C.

Sol. Since both wedges are similar therefore it is better to consider half of the weight of block *C* (= 1/2 W_C = 2000 N) to be raised by each wedge. Let us consider the *FBDs* of half of the block and one wedge, as shown in Fig. P. 4.23 (*b*).

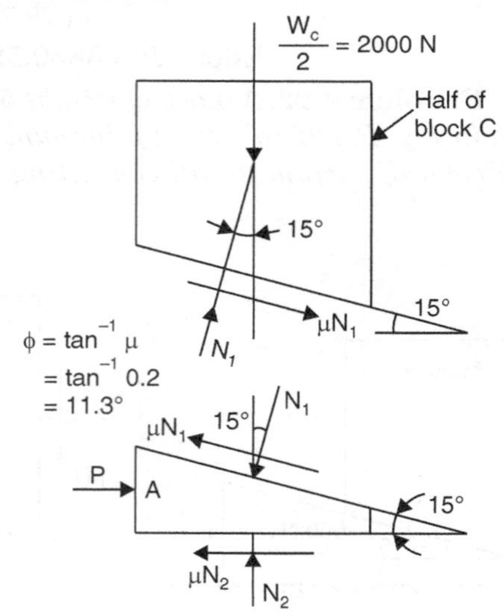

(b) FBDs of wedge A and half of the block C.

Fig. P. 4.23

For block C.

By $\Sigma F_v = 0$, we get

$$N_1 \cos 15° - \mu N_1 .\, \sin 15° - 2000 = 0 \qquad ...(1)$$

$\therefore$ $$N_1 (\cos 15° - 0.2 \sin 15°) = 2000$$

$\therefore$ $N_1 = \dfrac{2000}{0.9142}$ = **2187.8 N. Ans.**

For the wedge.

By $\Sigma F_v = 0$, we get

$$N_2 - N_1 \cos 15° + \mu N_1 \sin 15° = 0 \qquad ...(2)$$

$\therefore$ $$N_2 = N_1 (\cos 15° - 0.2 \sin 15°)$$

or $$N_2 = 2187.8 \times (0.9142)$$

$\therefore \quad N_2 = \mathbf{2000\ N.}$

And, by $\Sigma F_h = 0$, we get

$$P - \mu N_2 - N_1 \sin 15^\circ - \mu N_1 \cos 15^\circ = 0 \quad ...(3)$$

$$\therefore \quad P - 0.2 \times 2000 - 2187.8 \times \sin 15^\circ - 0.2 \times 2187.8 \cos 15^\circ = 0$$

$$\therefore \quad P = \mathbf{1.388.9\ N.} \quad \textbf{Ans.}$$

Problem 4.24. *Determine the minimum value of the force P which is applied to the wedge to move 2 kN block. μ = 0.30 at all contacting surfaces. See Fig. P.4.24 (a).*

Sol. The *FBDs* of the block and the wedge are shown in Fig. P.4.24 (*b*) and (*c*).

For the block.

By $\Sigma F_h = 0$, we get $\quad N_2 - \mu N_1 = 0 \quad ...(1)$

By $\Sigma F_v = 0$, we get $\quad N_1 - \mu N_2 - 2000 = 0 \quad ...(2)$

By (2) × μ + (1), we get

$$(N_2 - \mu N_1) + (\mu N_1 - \mu^2 N_2 - 2000\ \mu) = 0$$

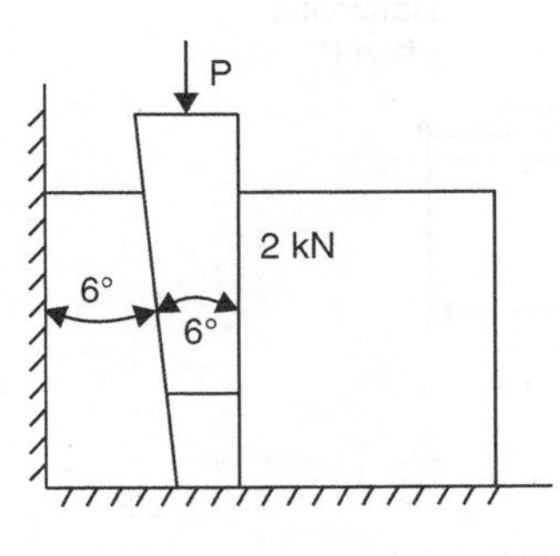

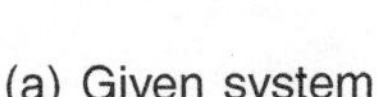

(a) Given system

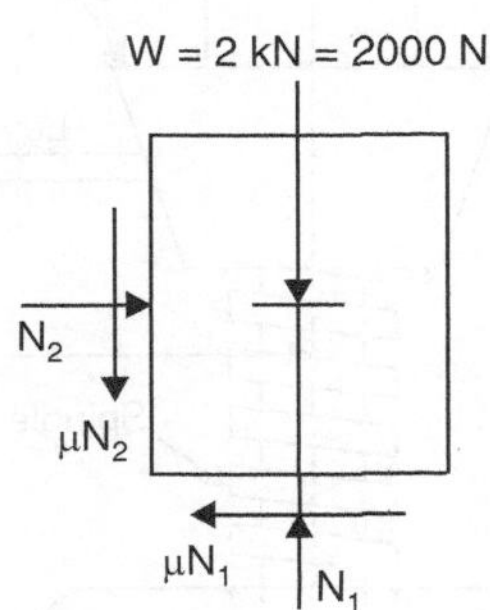

(b) FBD of the block

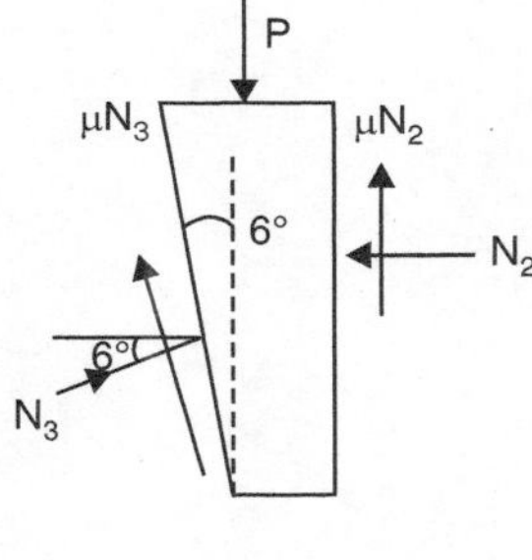

(c) FBD of wedge

Fig. P. 4.24

$$N_2(1 - \mu^2) = 2000\mu$$

or
$$N_2 = \frac{2000\ \mu}{1-\mu^2} = \frac{2000 \times 0.3}{1-(0.3)^2} = \frac{600}{1-0.09} = \frac{600}{0.91}$$

$$N_2 = \mathbf{659.34\ N.}$$

For the wedge.

By $\Sigma F_h = 0, N_3 \cos 6^\circ - N_2 + \mu N_3 \sin 6^\circ = 0$

$\therefore \quad N_3 (\cos 6^\circ + \mu \sin 6^\circ) = N_2$

$$\therefore \quad N_3 = \frac{N_2}{\cos 6^\circ + \mu \sin 6^\circ} = \frac{659.34}{\cos 6^\circ + 0.3 \sin 6^\circ} = \frac{659.34}{1.026} = \mathbf{642.71\ N.}$$

By $\quad \Sigma F_v = 0$, we get $P - \mu N_2 - N_3 \sin 6^\circ - \mu N_3 \cos 6^\circ = 0$

$\therefore \quad P - 0.3 \times 659.34 - 642.71 \sin 6^\circ - 0.3 \times 642.71 \cos 6^\circ = 0$

$\therefore \quad P - 456.74 = 0,$

$$\therefore \quad P = \mathbf{456.74\ N.} \quad \textbf{Ans.}$$

4.11. SCREW JACK

Screw jack is a device used for raising or lifting heavy loads by applying a small effort at a long lever or handle. It works on the principle of inclined plane with horizontal effort *P* for raising or lowering heavy loads. The screw or the spindle may have either square threads or *V*-threads. Square threads are known as power screws and are used to lift heavy loads. *V*-threads are useful for comparatively lighter loads.

(*a*) **Square Threads.** A screw jack consists of a spindle or a screw with square threads which fits into a nut in the form of **frame** of the screw jack. A load *W* is raised or lowered by the application of a horizontal effort applied at the end of a lever of length *l*. [(Fig. 4.11.1 (*a*)].

When spindle turns through one turn the load is raised through an amount either (*i*) equal to pitch of the threads in case of a single start threads, or (*ii*) equal to lead (l) = $n \times$ pitch in case of *n* start threads. Let p = pitch of the thread. If we develop one turn of the threads, [Fig. 4.11.1 (*b*)] we get a plane which is inclined at an angle α to the horizontal, given by :

$$\tan \alpha = \frac{p}{\pi d} \qquad \text{...(4.31)}$$

where, α = helix angle.

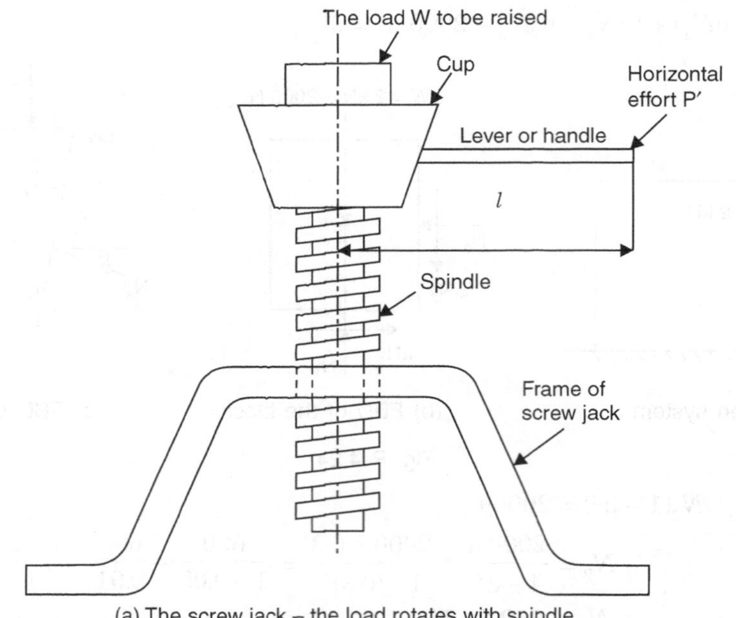

(a) The screw jack – the load rotates with spindle.

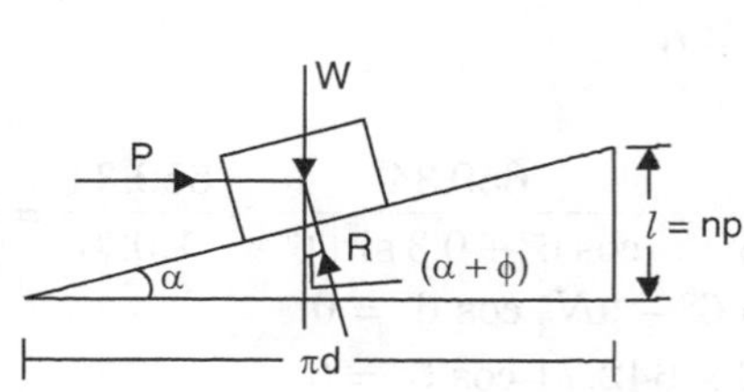

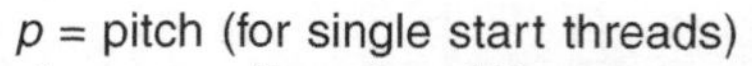

p = pitch (for single start threads)
d = mean diameter of the screw on spindle
n = number of starts of threads, *l* = lead

(b) Development of one turn of spindle with weight W to be lifted by effort P.

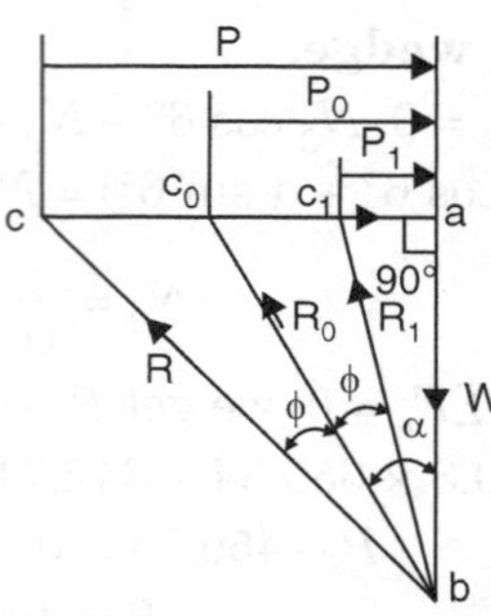

R = resultant reaction at screw threads.

(c) Triangle of forces.

Fig. 4.11.1

Case I. The load W rotates with the spindle

The load to be lifted is placed on the inclined plane which is inclined at an angle equal to the helix angle, as shown in Fig. 4.11 .1 (*a*). The forces acting on the load are : the weight (*W*), the effort (*P*) and the reaction *R* of the inclined plane. These faces form Δabc and is shown in Fig. 4.11. 1 (*c*). *P* is the effort applied on the surface of the spindle or screw threads of mean diameter *d*. Therefore, the effort *P* can be determined as under :

(*i*) **For Raising the load P:** The effort *P* is represented by the side *ca* of the Δabc.

$$\therefore \quad P = W \tan \angle abc = W \tan (\alpha + \phi) \quad ...(4.32)$$

in which

$$\phi = \tan^{-1} \mu \quad ...(4.33)$$

We find α and ϕ with the help of eqns. (4.31) and (4.33) and then find the value of effort *P* to be applied on the mean surface of threads.

Let the actual effort *P ′* is applied at the end of the lever of length *L*. By taking moments of *P* and *P ′* about the axis of the spindle, we get

$$P' \times l = P \times \frac{d}{2}$$

$$\therefore \quad P' = \frac{Pd}{2l} \quad ...(4.34)$$

When screw threads are perfectly smooth, $\phi = 0$, and the effort at the mean surface of spindle is

$$P_0 = W \tan \alpha \quad ...(4.35)$$

The efficiency of screw jack is

$$\eta = \frac{P_0}{P} = \frac{\tan \alpha}{\tan (\alpha + \phi)} \quad ...(4.36)$$

(*ii*) **For lowering the load W,** the effort to be applied on the mean surface of the screw threads is P_1, given by side c_1a of the Δabc_1.

$$\therefore \quad P_1 = W \tan \angle abc_1 = W \tan (\alpha - \phi) \quad \text{[See Fig. 4.11.1 (c)]...(4.37)}$$

And, the effort at the end of the lever is given by

$$P'' = \frac{Pd}{2l} \quad ...(4.38)$$

And, efficiency of screw jack for lowering the load is

$$\eta_1 = \frac{P_1}{P_0} = \frac{\tan (\alpha - \phi)}{\tan \alpha} \quad ...(4.39)$$

Case II. Load does not rotate with the spindle. When the load does not rotate with the spindle, there is a pivot bearing, Fig. 4.11.2 between the spindle and the cup on which load *W* is placed.

The bearing surface at the pivot has a mean radius

$$r_m = \frac{r_1 + r_2}{2} \quad ...(4.40)$$

When the wear rate is uniform, the frictional torque

$$T_f = \mu W r_m = \frac{\mu W (r_1 + r_2)}{2} \quad ...(4.41)$$

For raising the load, torque required at the mean surface of spindle

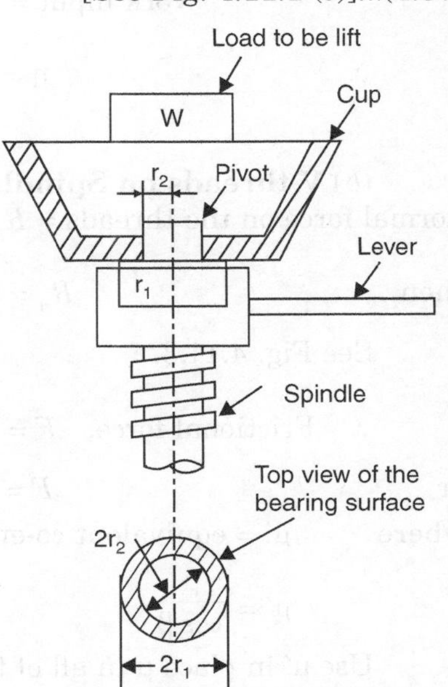

Fig. 4.11.2 Screw Jack arrangement when the load does not rotate with spindle.

$$T' = \frac{P \times d}{2} = \frac{Wd}{2} \tan(\alpha + \phi) \quad ...(4.42)$$

$\therefore$ Total torque required = $T = T' + T_f$

or Total torque = $T_t = \frac{Wd}{2} \tan(\alpha + \phi) + \frac{\mu W (r_1 + r_2)}{2}$ ...(4.43)

For lowering the load,

Total torque = $T_{t_2} = T_1 + T_f$

or $T_{t_2} = \frac{Wd}{2} \tan(\alpha - \phi) + \frac{\mu W (r_1 + r_2)}{2}$ ...(4.44)

Efficiency of screw jack when load does not rotate with the spindle.
For one turn of spindle,

Work out put = $W \times p$ ...(4.45)

Work input = $T_{t_1} \times 2\pi$ *for raising the load* ...(4.46)

$$\therefore \quad \text{Efficiency} = \eta = \frac{\text{work output}}{\text{work input}}$$

$$\therefore \quad \eta = \frac{W \times p}{2\pi T_{t_1}} = \frac{W \times p}{2\pi\left[\frac{Wd}{2}\tan(\alpha + \phi) + \frac{\pi W (r_1 + r_2)}{2}\right]}$$

or, $\eta = \frac{Wp}{2\pi\left[\frac{Wd}{2}\tan(\alpha + \phi) + \frac{\mu W}{2}(r_1 + r_2)\right]}$ ...(4.47)

For lowering the load

Work input = $T_{t_2} \times 2\pi$ for one turn of spindle ...(4.48)

$\therefore \quad \eta = \frac{Wp}{2\pi\,[T_{t_2}]} = \frac{Wp}{2\pi\left[\frac{Wd}{2}\tan(\alpha - \phi) + \mu W \frac{(r_1 + r_2)}{2}\right]}$ ...(4.49)

(*b*) **V-threads on Spindle.** Let the screw angle = 2β. The normal force on the threads = R_n

then, $R_n = \frac{W}{\cos\beta}$ ...(4.50)

See Fig. 4.11.3.

$\therefore$ Frictional force, $F = \mu R_n = \frac{\mu W}{\cos\beta}$

or $F = \mu' W$ ...(4.51)

where μ' = equivalent co-efficient of friction

$\mu' = \frac{\mu}{\cos\beta}$ ...(4.52)

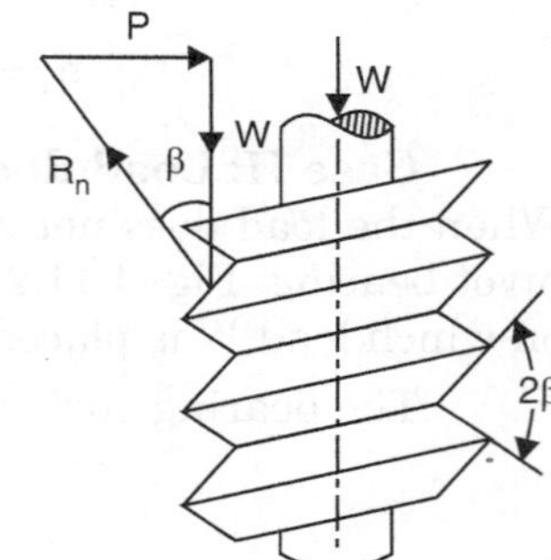

Fig. 4.11.3. V-threads.

Use μ' in place μ in all of the equations for screw jack with square threads.

4.12. (a) CONDITIONS OF MAXIMUM EFFICIENCY FOR SQUARE THREADS FOR RAISING LOAD

We have seen in section 4.11 that the efficiency of the screw jack with square threads for raising the load is given by eqn. (4.36), rewritten as below.

$$\eta = \frac{\tan \alpha}{\tan (\alpha + \phi)} \qquad \text{...(4.53)}$$

For the maximum efficiency $\frac{d\eta}{d\alpha} = 0$,

or, $$\frac{d}{d\alpha}\left[\frac{\tan \alpha}{\tan (\alpha + \phi)}\right] = 0$$

or, $$\frac{\sec^2 \alpha \tan (\alpha + \phi) - \sec^2 (\alpha + \phi) \tan \alpha}{\tan^2 (\alpha + \phi)} = 0$$

or, $$\sec^2 \alpha \tan (\alpha + \phi) - \sec^2 (\alpha + \phi) \tan \alpha = 0$$

$\therefore$ $$\frac{1}{\cos^2 \alpha} \cdot \frac{\sin (\alpha + \phi)}{\cos (\alpha + \phi)} = \frac{1}{\cos^2 (\alpha + \phi)} \cdot \frac{\sin \alpha}{\cos \alpha}$$

$\therefore$ $$\frac{\sin (\alpha + \phi)}{\cos \alpha} = \frac{\sin \alpha}{\cos (\alpha + \phi)}$$

or $$\sin (\alpha + \phi) \cos (\alpha + \phi) = \sin \alpha \cos \alpha$$

$\therefore$ $$\sin 2(\alpha + \phi) = \sin 2\alpha.$$

$\therefore$ $$2(\alpha + \phi) = \text{either } (2\alpha) \text{ or } (\pi - 2\alpha)$$

If $2\alpha + 2\phi = 2\alpha$, then $2\phi = 0$, or $\phi = 0$,

Since the plane is not smooth $\therefore \phi \neq 0$,

$\therefore$ We have $2(\alpha + \phi) = \pi - 2\alpha$

$\therefore$ $$4\alpha = \pi - 2\phi$$

or $$\alpha = \frac{\pi}{4} - \frac{\phi}{2} \qquad \text{...(4.54)}$$

Eqn. (4.54) states the condition for η_{max}.

By substituting value of α into eqn. (4.53), we get

$$\eta_{max} = \frac{\tan\left(\frac{\pi}{4} - \frac{\phi}{2}\right)}{\tan\left(\frac{\pi}{4} - \frac{\phi}{2} + \phi\right)}$$

or, $$\eta_{max} = \frac{\tan\left(\frac{\pi}{4} - \frac{\phi}{2}\right)}{\tan\left(\frac{\pi}{4} + \frac{\phi}{2}\right)} \qquad \text{...}(a)$$

or, $$\eta_{max} = \frac{[(\tan \pi/4 - \tan \phi/2)]/(1 + \tan \pi/4 \,.\, \tan \phi/2)}{[(\tan \pi/4 + \tan \phi/2)]/(1 - \tan \pi/4 \,.\, \tan \phi/2)}$$

$$= \frac{(1-\tan\phi/2)/(1+\tan\phi/2)}{(1+\tan\phi/2)/(1-\tan\phi/2)} = \frac{(1-\tan\phi/2)^2}{(1+\tan\phi/2)^2}$$

$$= \frac{1+\tan^2\phi/2 - 2\tan\phi/2}{1+\tan^2\phi/2 + 2\tan\phi/2} = \frac{\sec^2\phi/2 - 2\tan\phi/2}{\sec^2\phi/2 + 2\tan\phi/2}$$

$$= \frac{\dfrac{1}{\cos^2\phi/2} - \dfrac{2\sin\phi/2}{\cos\phi/2}}{\dfrac{1}{\cos^2\phi/2} + \dfrac{2\sin\phi/2}{\cos\phi/2}} = \frac{1-2\sin\phi/2\,.\cos\phi/2}{1+2\sin\phi/2\,.\cos\phi/2}$$

or $$\eta_{max} = \frac{1-\sin\phi}{1+\sin\phi} \qquad ...(4.55)$$

4.12(b) CONDITION FOR η_{MAX} FOR V-THREADS FOR RAISING LOAD

For *V*-threads we have seen that equivalent or modified co-efficient of friction μ' is given by eqn. (4.52), and is rewritten below as eqn. (4.56) :

$$\mu' = \frac{\mu}{\cos\beta} \qquad ...(4.56)$$

∴ We have $$\phi' = \tan^{-1}\mu' \qquad ...(4.57)$$

Following the above procedure, we get efficiency of the screw jack with *V*-threads on the spindle will be obtained by substituting ϕ' in place of ϕ is eqn. (4.53).

Hence, $$\eta = \frac{\tan\alpha}{\tan(\alpha+\phi')} \qquad ...(4.58)$$

For η_{max}, $\dfrac{d\eta}{d\alpha} = 0$,

∴ $$\frac{d}{d\alpha}\left[\frac{\tan\alpha}{\tan(\alpha+\phi')}\right] = 0$$

and, ∴ As above we shall get $$\alpha = \frac{\pi}{4} - \frac{\phi'}{2} \qquad ...(4.59)$$

From eqn. (4.54)

and, from eqn. (4.55), we get

$$\eta_{max} = \frac{1-\sin\phi'}{1+\sin\phi'} \qquad ...(4.60)$$

Problem 4.25. *The pitch of a screw jack is 12 mm, the mean diameter of the square thread on the spindle is 60 mm, and the lever is 900 mm long. If μ_s = 0.08, find the necessary effort at the end of the lever (a) for lifting a load of 30 kN, (b) for lowering a load of 30 kN, (c) efficiency of the screw jack for raising the load, and (d) efficiency of the screw jack for lowering the load.*

Sol. Given : Pitch, $p = 12$ mm

mean diameter, $d = 60$ mm

$\mu_s = 0.08$

length of lever, $l = 900$ mm

Weight, $W = 30$ kN.

$\because \quad \tan\alpha = \frac{p}{\pi d}; \alpha = \tan^{-1}\left(\frac{p}{\pi d}\right) = \tan^{-1}\left(\frac{12}{60\pi}\right) = \tan^{-1} 0.06366$

$\therefore \quad \alpha = 3.64°; \phi = \tan^{-1}\mu_s = \tan^{-1} 0.08 = 4.57°.$

$\therefore \quad \phi + \alpha = 4.57 + 3.64 = 8.21°; (\phi - \alpha) = (4.57 - 3.64) =$ **0.93°.**

(*a*) Effort on the surface of spindle = $P = W \tan(\alpha + \phi)$

$\therefore \quad P = (30 \times 1000) \times \tan 8.21° =$ **4328.4 N** for raising the load.

Effort at the end of lever $= P' = \frac{P \times d}{2l} = \frac{4328.4 \times 60}{2 \times 900} =$ **144.28 N. Ans.**

(*b*) Effort on the surface of spindle for lowering the load $= P_1 = W \tan(\phi - \alpha)$

$\therefore \quad P_1 = (30{,}000) \tan 0.93° =$ **486.99 N.**

$\therefore$ Effort at the end of lever to lower the load $= P_1' = \frac{P_1 \times d}{2l} = \frac{486.99 \times 60}{2 \times 900}$ N

or $P_1' =$ **16.23 N. Ans.**

(*c*) Ideal effort $= P_0 = W \tan\alpha = 30{,}000 \times \tan 3.64° =$ **1908.47 N**

$\therefore$ Efficiency for raising the load $= \eta = \frac{P_0}{P} \times 100 = \frac{1908.47}{4328.4} \times 100 =$ **44.09%. Ans.**

(*d*) Efficiency for lowering the load $= \eta' = \frac{P_1}{P_0} \times 100 = \frac{486.99}{1908.47} \times 100 =$ **25.52%. Ans.**

Problem 4.26. *A load of 25 kN is lifted through 80 mm by a screw jack with square threads on its spindle. Mean diameter of threads is 72 mm and the maximum diameter is 78 mm; μ_s = 0.08°. Find the work done in lifting the load (a) when the load rotates with the spindle, and (b) when the load does not rotate with the spindle. The mean effective radius of the loose head is 50 mm and the co-efficient of friction between the loose head and the spindle is 0.2.*

Sol. Given : $W = 25$ kN = 25,000 N

y = distance through the load is lifted = 80 mm

mean diameter, $d = 72$ mm

maximum diameter, $d_1 = 78$ mm

$\mu_s = 0.08.$

The relationship between maximum diameter and the mean diameter is shown in Fig. P. 4.26.

Let d_2 = minimum diameter and p = pitch of threads.

For the square threads

$$d_1 - d_2 = 2 \times \frac{p}{2}$$

$$\therefore \quad p = d_1 - d_2 \quad \text{...(1)}$$

Also, $$d_2 = d - 2 \times \left(\frac{p}{4}\right) = d - \frac{p}{2} \quad \text{...(2)}$$

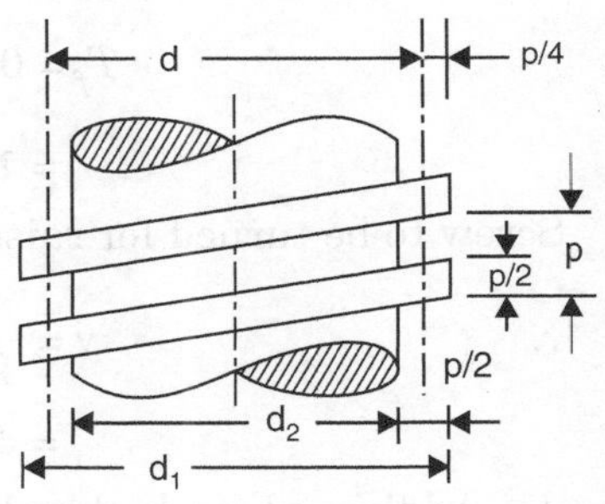

Fig. P. 4.26. Relation between d_1, d_2, d and p.

and $$d_1 = d + 2 \times \left(\frac{p}{4}\right) = d + \frac{p}{2} \qquad ...(3)$$

Here, we have been given d_1 and d,

∴ By using relation (3), we get

$$\frac{p}{2} = d_1 - d$$

∴ Pitch, $p = 2 \times (d_1 - d) = 2 \times (78 - 72) = 12$ mm.

∴ Helix angle, $\alpha = \tan^{-1} \frac{p}{\pi d} = \tan^{-1}\left(\frac{12}{72\pi}\right) = \tan^{-1}\left(\frac{1}{6\pi}\right) = \tan^{-1} 0.05305$

∴ $\alpha = 3.037°$

Friction angle, $\phi = \tan^{-1} \mu_s = \tan^{-1} 0.08 = 4.574°$

∴ $\alpha + \phi = 3.037° + 4.574° = \mathbf{7.611°}$.

(*a*) The load does not rotate with spindle.

Effort = $P = W \tan(\alpha + \phi) = 25000 \tan 7.611° = 3340.6$ N

Ideal effort = $P_0 = W \tan \alpha = 25000 \tan 3.037°$

$= \mathbf{1326.38\ kN}$

∴ Efficiency, $\eta = \frac{P_0}{P_1} = \frac{1326.38}{3340.6} = \mathbf{0.3970 = 39.7\%}$

But, $$\eta = \frac{\text{Useful work done by load}}{\text{Work supplied by effort}}$$

∴ Work supplied by the effort = $\frac{\text{Work done by load}}{\eta}$

$$= \frac{25000 \times \left(\frac{80}{1000}\right)}{0.3970} \text{ Nm} = \mathbf{5037.78\ Nm.}\quad \textbf{Ans.}$$

(*b*) **When the load does not rotate with the spindle**, the additional frictional couple resisting motion at the bearing surface at the cup = $\mu W r_m = T_f$

or $$T_f = 0.2 \times 25000 \times \frac{\left(\frac{50}{2}\right)}{100} \text{ Nm} \qquad (\because \text{ Mean dia.} = 50 \text{ mm.})$$

$= \mathbf{125\ Nm}$

Screw to be turned for raising the load through 80 mm = θ

∴ $$y = \frac{p}{2\pi}\theta; \quad \therefore \quad \theta = y \times \frac{2\pi}{p} = 80 \times \frac{2\pi}{12}$$

$= 41.888$ radians.

∴ Additional work absorbed in friction at bearing below cup

$= W_f = T_f \times \theta = 125 \times 41.888 = \mathbf{5236\ Nm}$

∴ Total work supplied by the effort to lift the load without its rotation with spindle

$$= (5037.8 + 5236) \text{ Nm}$$

$$= \mathbf{10273.8 \text{ Nm.}} \quad \textbf{Ans.}$$

Problem 4.27. *A screw jack with square threads has a spindle of minimum diameter 56 mm and pitch 12 mm. The coefficient of friction at the screw threads in 0.05. A load of 600 N being lifted by the screw jack revolves with the spindle.*

(a) Find the force required at the end of 350 mm long lever for raising the load,

(b) State whether the screw jack is self-locking. If it is self-locking find the force to be applied at the end of the lever for holding the load against lowering down.

Sol. Given : Load = 600 N;

$$d_2 = \text{minimum diameter} = 56 \text{ mm};$$

$$\text{pitch, } p = 12 \text{ mm};$$

$$\mu = 0.05,$$

$$\text{lever length, } l = 350 \text{ mm}.$$

By using the formula,

$$d_2 = d - \frac{p}{2}, \text{ we find the mean diameter, } d$$

$$\therefore \quad d = d_2 + \frac{p}{2} = 56 + \frac{12}{2} = 62 \text{ mm}.$$

$$\therefore \quad \alpha = \tan^{-1}\left(\frac{p}{\pi d}\right) = \tan^{-1}\left(\frac{12}{\pi \times 62}\right) = \tan^{-1}(0.06161)$$

$$\therefore \quad \alpha = \mathbf{3.526°}.$$

Also $\quad \phi = \tan^{-1} \mu = \tan^{-1} 0.05 = 2.862°$

$$\therefore \quad \phi + \alpha = 3.526 + 2.862 = 6.388°$$

or $\quad \phi \sim \alpha = 3.526 \sim 2.862 = 0.664°$

(*a*) Effort required at the surface of spindle, for raising the load,

$$p = W \tan(\phi + \alpha) = 600 \tan 6.388°$$

$$= \mathbf{67.174 \text{ N.}} \quad \textbf{Ans.}$$

Effort to be applied at the end of the lever

$$= P' = \frac{P \times d/2}{l} = \frac{67.174 \times 62/2}{350} = \mathbf{5.95 \text{ N.}} \quad \textbf{Ans.}$$

(*b*) Efficiency of the screw jack,

$$\eta = \frac{\tan \alpha}{\tan(\phi + \alpha)} \times 100\%$$

or $\quad \eta = \dfrac{\tan 3.526}{\tan 6.388} \times 100 = \mathbf{55.04\%.} \quad \textbf{Ans.}$

Since the efficiency is more than 50% the screw jack is **not self-locking,** and a holding torque or holding force is required to keep the load at the raised position.

The effort required for lowering the load,

$$P_1 = W \tan(\alpha \sim \phi) = 600 \tan(0.664°)$$

$$= \mathbf{5.954 \text{ N.}} \quad \textbf{Ans.}$$

∴ Effort required at the end of the lever,

$$P_1' = \frac{P_1 \times d/2}{l} = \frac{5.954 \times 52/2}{350} = \mathbf{0.5274\ N},$$

∴ Holding torque = $P_1' \times l$ = 0.5274 × 350 = **184.574 Nmm. Ans.**

Problem 4.28. *A C-clamp is used to hold two wooden pieces together. The clamp has a double or two-start square threads of mean diameter 12 mm with a pitch of 2 mm. If a maximum torque of 60 Nm is applied in tightening the clamp, determine (a) the compressive force exerted on the pieces of wood, (b) efficiency of the clamp during tightening, (c) the torque required for loosening the clamp, and (d) the efficiency of the clamp during loosening. Take μ = 0.18.*

Sol. The mean diameter of screw, d = 12 mm ;

pitch = p = 2 mm ;

∴ lead, l = pitch × number of starts of threads = 2 × 2 = 4 mm

∴ The helix angle, $\alpha = \tan^{-1}\frac{l}{\pi d} = \tan^{-1}\left(\frac{4}{\pi \times 12}\right) \tan^{-1}\left(\frac{1}{3\pi}\right) = \tan^{-1}(0.1061)$

∴ α = **6.0566°.**

$\phi = \tan^{-1}\mu = \tan^{-1} 0.18 = 10.2040°$

∴ $\phi + \alpha$ = 10.2040 + 5.0566 = 16.2606°

and $\phi \sim \alpha$ = 10.2040 ~ 5.0566 = 5.1474°

(*a*) The tightening torque, $T = P \times r$ applied at the surface of the threads

where P = tangential force required on the surface of threads

$$\therefore \quad P = \frac{T}{r} = \frac{60 \times 1000}{\left(\frac{12}{2}\right)} = \frac{60 \times 1000}{6} = 10{,}000 \text{ N}$$

∴ The axial force (W) exerted on the wooden pieces can be obtained from,

$$P = W \tan(\alpha + \phi)$$

$$\therefore \quad W = \frac{P}{\tan(\alpha + \phi)} = \frac{10{,}000}{\tan 16.2606°} = \mathbf{34284.8\ N.}$$

= **34.2848 kN. Ans.**

(*b*) Efficiency during tightening,

$$\eta = \frac{\tan\alpha}{\tan(\alpha + \phi)} \times 100 = \frac{\tan 6.0566°}{\tan 16.2606°} \times 100$$

or η = **36.38%. Ans.**

(*c*) Effort required for loosening the clamp,

$P_1 = W \tan(\phi \sim \alpha)$ = 34284.8 tan 5.1474°

= **3088.43 N. Ans.**

(*d*) Efficiency of the clamp during loosening,

$$\eta' = \frac{\tan(\phi - \alpha)}{\tan\alpha} \times 100 = \frac{\tan 5.1474}{\tan 6.0566} \times 100$$

= **84.9%. Ans.**

Problem 4.29. *A 160 mm steam stop valve for a boiler pressure of 2 MPa has a 60 mm maximum diameter of spindle with square threads of 6 mm pitch cut on it. Find the torque required to turn the spindle and the efficiency of the screw. Take $\mu = 0.15$.*

Sol. Given : Pressure $p = 2$ MPa $= 2$ N/mm^2 ;

$d = 60$ mm = mean diameter of spindle

pitch = 6 mm ;

$\mu = 0.15$

Diameter of value $= d_v = 160$ mm

Axial force on the spindle due to steam $= W = p \times \frac{\pi}{4} d^2$

or $$W = (2 \times 10^6) \times \frac{\pi}{4} \times (0.160)^2 = 0.0402 \times 10^6 \text{ MN} = 40.2 \times 10^3 \text{ kN.}$$

For spindle, pitch = 6 mm = 0.006 m.

Maximum diameter, $d_1 = 60$ mm,

$\therefore$ minimum diameter, $d_2 = d_1 - 2 \times \frac{p}{2} = 60 - 6 = 54$ mm

$$\therefore \quad \text{mean diameter} = \frac{1}{2}(d_1 + d_2) = \frac{1}{2}(60 + 54) = 57 \text{ mm.}$$

$$\therefore \quad \tan \alpha = \frac{p}{\pi d} = \frac{6}{57 \times \pi} = 0.033506$$

$$\therefore \quad \alpha = \tan^{-1}\left(\frac{p}{\pi d}\right) = \tan^{-1}(0.033506) = \mathbf{1.919°}$$

and $$\phi = \tan^{-1} \mu = \tan^{-1} 0.15 = 8.531°$$

$$\therefore \quad \alpha + \phi = 1.919 + 8.531 = 10.450°$$

Tangential force P to be applied to the surface of the spindle

$$P = W \tan(\alpha + \phi) = (40.2 \times 10^3) \times \tan 10.450° \text{ N}$$

$$= \mathbf{7414.35 \text{ N}}$$

$$\therefore \quad \text{Torque required} = T = P \times r = 7414.35 \times \frac{57}{2} \text{ N-mm}$$

or $$T = 211308.95 \text{ N-mm}$$

or $$T = \mathbf{211.31 \text{ Nm.}} \quad \textbf{Ans.}$$

Ideal effort, $P_0 = W \tan \alpha$

and efficiency, $$\eta = \frac{P_0}{P} \times 100 = \frac{W \tan \alpha}{W \times \tan(\alpha + \phi)} \times 100$$

$$\therefore \quad \eta = \frac{\tan \alpha}{\alpha + \phi} \times 100 = \frac{\tan 1.919}{\tan 10.450} \times 100 = \mathbf{18.17\%.} \quad \textbf{Ans.}$$

Problem 4.30. *A turn buckle is used to tighten a stay wire. The turn buckle has right and left handed single start square threads of 4 mm pitch and the mean diameter 20 mm. The coefficient of friction between the screw threads and the nut is 0.2.*

If the initial tension in the stay-wire is 2.4 kN, find the work done during further tension of the stay wire by 1.5 turns of the turn buckle which increases the tension in the stay wire by 3.6 kN. How much of this work has been wasted in overcoming friction ?

Sol. Total stretch of stay wire in 1.5 turns = $1.5 \times p = \delta$

or $$\delta = 1.5 \times 4 = 6 \text{ mm}.$$

Initial tension in stay wire = 2.4 kN,

Final tension in stay wire = 3.6 kN.

Mean tension in stay wire $= W = \dfrac{2.4 + 3.6}{2} = 3$ kN

$$\alpha = \tan^{-1}\left(\frac{p}{\pi d}\right) = \tan^{-1}\frac{4}{20\pi} = \tan^{-1} 0.063662 = 3.64°$$

$$\phi = \tan^{-1}\mu = \tan^{-1}\ 0.2 = 11.31°$$

$\therefore$ $$\alpha + \phi = (3.64° + 11.31°) = 14.95°$$

$\therefore$ Effort to be applied to the turn buckle,

or $$P = 2W \tan(\alpha + \phi) = 2 \times 3 \times \tan 14.95° = \mathbf{1.60208\ kN}$$

$$= \mathbf{1602.08\ N.}$$

[**Note.** Since there are left-hand threads on one side and right hand threads on the other side of the turn buckle therefore the effort required to over come friction at threads is doubled].

Torque on the turn buckle $= P \times \left(\dfrac{d}{2}\right)$

or $$T = 1602.08 \times \frac{20}{2} = 16020.8 \text{ N-mm} = 16.0208 \text{ Nm}$$

$\therefore$ Work done on the turn buckle to tighten the stay wire = $T \times \theta$

or $$\text{Work} = 16.0208 \times (1.5 \times 2\pi) = \mathbf{150.99\ Nm.} \quad \textbf{Ans.}$$

Work required to stretch the stay wire = $W \times \delta = W(np)$, where n = no. of turns

$$= 3 \times (1.5 \times 4) = 3 \times 6 \text{ mm} = 18 \text{ kN-mm} = \mathbf{18\ Nm}$$

$\therefore$ Work wasted in friction = 150.99 – 18 = **132.99 Nm. Ans.**

Problem 4.31. *The screw jack has spindle of whitworth threads of 20 mm mean diameter and 5 mm pitch and the angle of V is 55°. The axial load to be raised is 2000 N. The co-efficient of friction between the screw and the nut is 0.1, the mean radius of bearing surface is 30 mm and the co-efficient of friction at the bearing surfaces is 0.16. Find the force required at the end of a 400 mm long lever for raising the load when the load does not rotate with the spindle.*

Sol. Given : Mean diameter of threads, d = 20 mm,

Pitch, p = 5 mm ; V-angle = $2\beta = 55°$,

Axial load = W = 2000 N,

Mean radius of bearing surface = R_{eff} = 30 mm,

Co-efficient of friction at the bearing = μ_1 = 0.16,

Length of lever = l = 400 mm.

The co-efficient of friction between nut and the spindle = μ = 0.1.

$\therefore$ The equivalent co-efficient of friction $= \mu' = \dfrac{\mu}{\cos\beta}$

where $\beta = \dfrac{55}{2} = 27.5°$

$\therefore$ $\mu' = \dfrac{0.1}{\cos 27.5°} = 0.11274 = \tan\phi'$

$\therefore$ $\phi' = \tan^{-1}\mu = \tan^{-1} 0.11274 = 6.432°$

$\tan\alpha = \dfrac{p}{\pi d} = \dfrac{5}{\pi \times 20} = \dfrac{1}{4\pi}$

$\therefore$ $\alpha = \tan^{-1}\dfrac{1}{4\pi} = 4.550°.$

$\therefore$ $\alpha + \phi' = 4.550° + 6.432° = 10.982°$

Force required at the surface of spindle, $P = W\tan(\alpha + \phi)$

$\therefore$ $P = 2000 \tan 10.982° = 388.11$ N

and torque, $T_1 = P \times \dfrac{d}{2} = 388.11 \times \dfrac{20}{2} = 3881.1$ N-mm

Additional torque to overcome friction at the bearing surface,

$T_f = \mu_1 . W\, R_{eff.} = 0.16 \times 2000 \times 30$ N-mm

or, $T_f = 9600$ N-mm.

$\therefore$ Total torque, $T = T_1 + T_f = 3881.1 + 9600 = 13481.1$ N-mm

But $T = P' \times l$

$\therefore$ The effort required at the end of the lever, $P' = \dfrac{T}{l}$

and $P' = \dfrac{13481.1}{400} = \mathbf{33.703\ N.}$ **Ans.**

4.13. JOURNAL BEARINGS, AXLE FRICTION

Journal bearings are used to provide lateral (*i.e.*, perpendicular to the axis) support to rotating shafts and axles. Thrust bearings, described in the next section, are used to give axial support to the shafts. When journal bearing is fully lubricated, the frictional resistance depends on the speed of rotation of shaft, the radial clearance between shaft and the bearing and the viscosity of the lubricant. Such problems are studied in fluid mechanics. In Engineering Mechanics, the study is restricted to the unlubricated and partially lubricated journal bearings, in which the axle and the bearing are in direct contact along a single straight line. Fig. 4.13 (*a*).

As the shaft rotates in the clockwise direction it rolls up the inner surface of the bearing until slippage occurs. Here it remains in a more or less fixed position during rotation. The torque M required to maintain rotation and the radial load W on the shaft will cause a reaction R at the contact point B. For vertical equilibrium R must equal W but will not be collinear with it, Fig. 4.13 (*b*). Thus, R will be tangent to a small circle of radius r_f called the **friction circle.** The angle between R and its normal component N is the friction angle ϕ. Equating the sum of the moments about B to zero gives

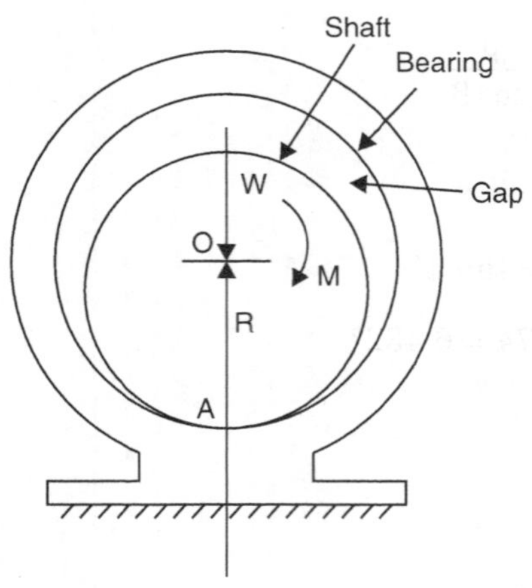

(a) Shaft at rest-contact with bearing at A

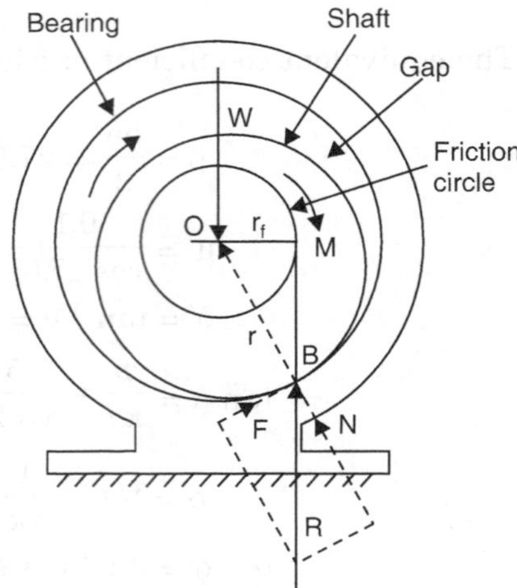

(b) Shaft rotating clockwise contact at B

Fig. 4.13. Axle friction

$$M = W.\, r_f = Lr \sin \phi \qquad ...(4.61)$$

For the small coefficient of friction the angle ϕ is small, and the $\sin \phi = \tan \phi$ may be interchanged with only small error. Since $\mu = \tan \phi$, a good approximate torque M is given by

$$M = \mu W r \qquad ...(4.62)$$

This relation gives the amount of torque or moment applied to the shaft which is necessary to overcome friction for a dry or partially lubricated journal bearing.

4.14. THRUST BEARINGS, DISK FRICTION

Thrust bearings are used to provide axial support to the rotating shafts and axles. They are of three types :

(1) End bearings (Flat type) [Fig. 4.14 (*a*)]

(2) Collar bearings, [Fig. 4.14 (*b*)], and

(3) Conical bearings [Fig. 4.14 (*c*)].

In case of end bearings, friction takes place over full circular area, or over the ring-shaped area when the shafts are hollow.

In case of the collar bearings, friction forces are developed between the two ring-shaped areas which are in contact.

The magnitude of frictional torque in case of the conical type end bearing is very large compared to the flat type end bearing of type (1) or the collar bearings of type (2).

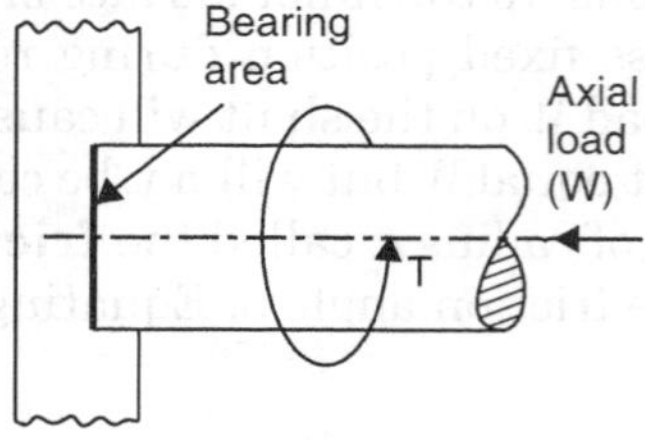

(a) End bearing

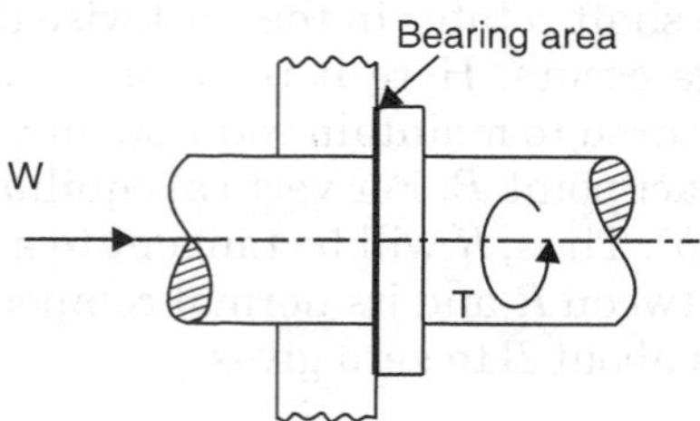

(b) Collar bearing

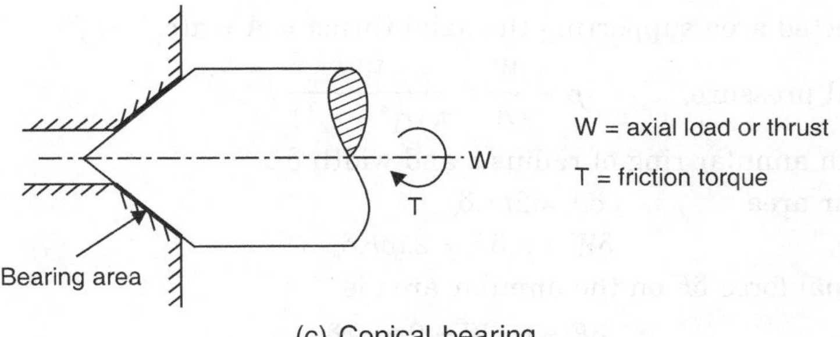

(c) Conical bearing

Fig. 4.14. Different types of thrust bearings.

The frictional force developed between circular areas is also known as **disc friction.** Disc friction also occurs in other mechanisms, such as **disc clutches.**

If the shaft is vertical, the end bearings may be called as foot step bearings.

There are two mechanisms or two assumptions which are used. They are :

(*a*) Uniform pressure distribution, *i.e.*, $p = c$, and

(*b*) Uniform rate of wear, *i.e.*, $pr = c$.

For bearings and for new clutches the assumption of uniform pressure distribution gives safe and conservative results. On the other hand, the assumption of uniform wear rate is used for worn-out clutches.

4.15. PIVOTS AND COLLARS

Force acting along the axis of a shaft is known as **thrust** or the **axial thrust**. This produces lateral motion along the axis, which is prevented by providing one or more bearing surfaces, at right angles to the axis of rotation.

Sometimes the axis of shaft is vertical and then the axial thrust is equal to the weight of machine parts mounted on the shaft.

Examples for axial thrust are : shafts of steam and gas turbines, ship propellers, water turbines and centrifugal pumps.

The following cases have been analysed. The equations establed below are applicable to the analysis of both clutches and bearings.

1. End bearing-flat pivot [Fig. 4.15.1]

A **hollow shaft** of external radius r_1 and internal radius r_2 is supported by an end bearing shown in Fig. 4.15.1 (*a*). **Here the shaft is horizontal. If the shaft is vertical, the bearing is called foot-step bearing,** as mentioned earlier. Let the axial force on the shaft be W, and the frictional torque be T. Let p be the pressure between contact surfaces.

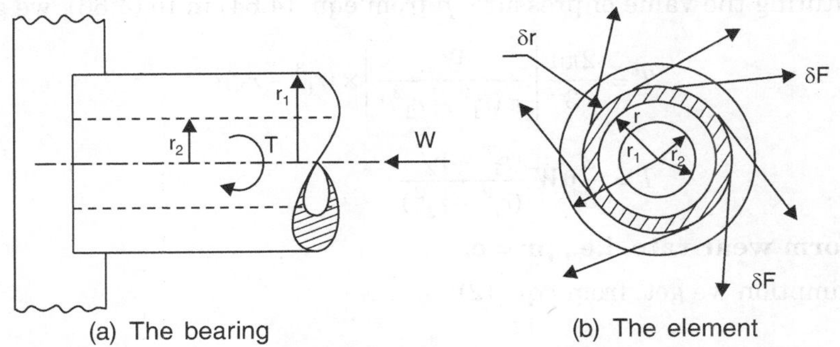

(a) The bearing (b) The element

Fig. 4.15.1 Flat end bearing

Total projected area supporting the axial thrust = $A = \pi(r_1^2 - r_2^2)$

$\therefore$ Normal pressure, $$p = \frac{W}{A} = \frac{W}{\pi(r_1^2 - r_2^2)} \quad ...(1)$$

Consider an annular ring of radius r and width δ_r.

$\therefore$ annular area $\delta A = 2\pi r.\delta_r$

Axial force, $\delta W = p.\delta A = 2\pi pr.\delta_r$ ...(2)

The frictional force δF on the annular area is

$$\delta F = \mu.\delta W = 2\pi\mu pr\delta r \quad ...(3)$$

And, the frictional torque on the element,

$$\delta T = \delta F \times r = 2\pi\mu pr^2\delta r \quad ...(4)$$

Eqns. (2), (3) and (4) can be integrated to find the axial force W, frictional force F and the total friction torque T applied on the entire bearing area. Two assumptions, as described above, can be applied on eqns. (2), (3) and (4).

(*a*) **For uniform distribution of pressure, i.e., p = c.**

By integrating eqns. (2), (3) and (4), we get W, F and T.

From eqn. (2), $$W = \int dW = \int_{r_2}^{r_1} 2\pi pr dr$$

or, $$W = 2\pi p\left[\frac{r^2}{2}\right]_{r_2}^{r_1} \quad \text{or} \quad W = \pi p(r_1^2 - r_2^2) \quad ...(4.63)$$

$\therefore$ $$p = \frac{W}{\pi(r_1^2 - r_2^2)} \quad ...(4.64)$$

which is the same as eqn. (1)

$$F = \int dF = \int_{r_2}^{r_1} 2\pi\mu pr.dr \quad \text{from (3)}$$

$\therefore$ $$F = 2\pi\mu p\frac{(r_1^2 - r_2^2)}{2} \quad \text{or} \quad F = \pi\mu p(r_1^2 - r_2^2) = \mu W \quad ...(4.65)$$

And from eqn. (4), we get

$$T = \int dr = \int_{r_2}^{r_1} 2\pi\mu pr^2 dr \quad \text{or} \quad T = 2\pi\mu p\frac{(r_1^3 - r_2^3)}{3} \quad ...(4.66)$$

or, by substituting the value of pressure p from eqn. (4.64) in to (4.66), we get

$$T = \frac{2\pi\mu}{3}\left[\frac{W}{\pi(r_1^2 - r_2^2)}\right] \times (r_1^3 - r_2^3)$$

or, $$T = \frac{2}{3}\mu W\frac{(r_1^3 - r_2^3)}{(r_1^2 - r_2^2)} \quad ...(4.67)$$

(*b*) **Uniform wear rate, i.e., pr = c.**

Using this assumption we get, from eqn. (2)

$$W = \int dW = \int_{r_2}^{r_1} 2\pi pr\, dr = 2\pi \int_{r_2}^{r_1} c\, dr$$

$\therefore$ $$W = 2\pi c\,(r_1 - r_2) \quad ...(4.68)$$

$\therefore$ $$c = \frac{W}{2\pi(r_1 - r_2)} \quad ...(4.69)$$

From eqn. (3), we get

$$F = \int dF = \int_{r_2}^{r_1} 2\pi\mu prdr = 2\pi \int_{r_2}^{r_1} \mu c\, dr$$

$\therefore$ $$F = 2\pi c\,(r_1 - r_2)\mu = \mu W \quad ...(4.70)$$

And from eqn. (4), we get

$$T = \int dT = \int_{r_2}^{r_1} 2\pi\mu pr^2\, dr = 2\pi\mu \int_{r_2}^{r_1} cr\, dr$$

$\therefore$ $$T = 2\pi\mu c\,\frac{r_1^{\,2} - r_2^{\,2}}{2} \quad \text{or} \quad T = \pi\mu c\,(r_1^{\,2} - r_2^{\,2}) \quad ...(4.71)$$

By substituting the value of the constant c from eqn. 4.69 into eqn. 4.71, we get

$$T = \pi\mu \left(\frac{W}{2\pi\,(r_1 - r_2)}\right)(r_1^{\,2} - r_2^{\,2})$$

or $$T = \mu W \left(\frac{r_1 + r_2}{2}\right) \quad ...(4.72)$$

When the shaft is solid,

$$r_2 = 0, \text{ and } r_2 = r_1 = \sqrt{2}, \text{ thus, we get}$$

From eqn. (4.67), for $p = c$, we get

$$T = \frac{2}{3}\mu W\,(r) \quad ...(4.73\ )$$

and from eqn. (4.72), for $pr = c$, we get

$$T = \frac{\mu Wr}{2} \quad ...(4.74)$$

2. Collar bearings. The above equations, which are established for hollow shaft are valid for a collar bearing. See Fig. 4.15.2

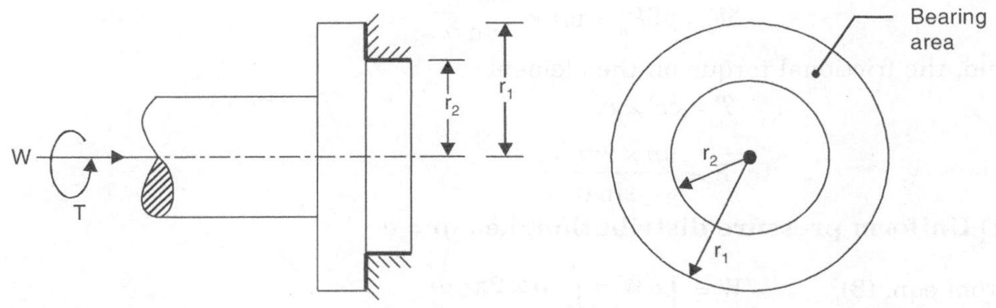

Fig. 4.15.2 Collar bearing.

3. Conical pivot. Let cone angle be 2α.

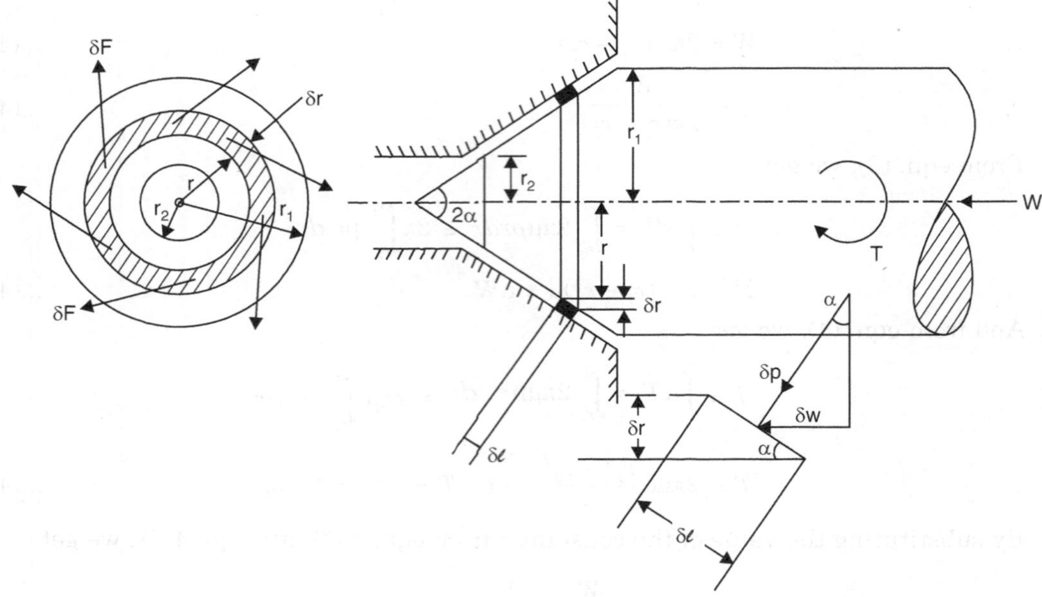

Fig. 4.15.3 Conical end bearing.

Consider elemental ring of radius r, and radial thickness δr.

Area of the element perpendicular to the bearing surface $= \delta A_n = 2\pi r.\delta l$ ...(5)

but length of element, $\delta l = \dfrac{\delta r}{\sin \alpha}$

$\therefore \qquad \delta A_n = \dfrac{2\pi r \delta r}{\sin \alpha}$ ...(6)

If p is the normal pressure on the bearing surface, then the normal force on the element $= \delta P_n = p \times \delta A_n$

or, $\qquad \delta P_n = p \times \dfrac{2\pi r \delta r}{\sin \alpha}$ ...(7)

We find that, axial force on the element is

$$\delta W = \delta P_n.\ \sin a = p \times 2\pi r.\delta r \qquad ...(8)$$

Frictional force on the element is

$$\delta F = \mu \delta P_n = \mu p \times \frac{2\pi r.\delta r}{\sin \alpha} \qquad ...(9)$$

And, the frictional torque on the element

$$T = \delta F \times r$$

or, $\qquad \delta = \dfrac{\mu p \times 2\pi r^2 \delta r}{\sin \alpha}$ ...(10)

(*a*) **Uniform pressure distribution, i.e., p = c**

From eqn. (8), $\qquad W = \int dW = \int_{r_2}^{r_1} p \times 2\pi r\, dr$

$$\therefore \quad W = \frac{2\pi p (r_1^2 - r_2^2)}{2} \quad \text{or} \quad W = \pi p(r_1^2 - r_2^2) \quad ...(4.75)$$

$$\therefore \quad p = \frac{W}{\pi(r_1^2 - r_2^2)} = \frac{W}{\text{Projected area}} \quad ...(4.76)$$

From eqn. (9)

$$F = \int dF = \int_{r_2}^{r_1} \frac{\mu p \times 2\pi r dr}{\sin \alpha} = \frac{\mu p \pi (r_1^2 - r_2^2)}{\sin \alpha} \quad ...(4.77)$$

or

$$F = \frac{\mu W}{\sin \alpha} \quad ...(4.78)$$

and from eqn, (10),

$$T = \int dT = \int_{r_2}^{r_1} \frac{\mu p \times 2\pi r^2 dr}{\sin \alpha} = \frac{2}{3} \frac{\mu p \pi (r_1^3 - r_2^3)}{\sin \alpha} \quad ...(4.79)$$

By substituting value of p from (4.76) into eqn. (4.79), we get

$$T = \frac{2}{3} \mu \left[\frac{W}{\pi(r_1^2 - r_2^2)} \right] \times \frac{\pi(r_1^3 - r_2^3)}{\sin \alpha}$$

or,

$$T = \frac{2}{3} \frac{\mu W}{\sin \alpha} \left(\frac{r_1^3 - r_2^3}{r_1^2 - r_2^2} \right) \quad ...(4.80)$$

(*b*) **Uniform wear rate, i.e., pr = c**

From eqn. (8),

$$W = \int dW = \int_{r_2}^{r_1} 2\pi pr \, dr$$

or,

$$W = 2\pi c \int_{r_2}^{r_1} dr = 2\pi c (r_1 - r_2) \quad ...(4.81)$$

$$\therefore \quad c = \frac{W}{2\pi (r_1 - r_2)} \quad ...(4.82)$$

From eqn. (9),

$$F = \int dF = \int_{r_2}^{r_1} \frac{2\pi \mu pr \, dr}{\sin \alpha}$$

or

$$F = \frac{2\pi c \mu}{\sin \alpha} (r_1 - r_2) \quad ...(4.83)$$

or

$$F = 2\pi \left[\frac{W}{2\pi (r_1 - r_2)} \right] . \frac{\mu (r_1 - r_2)}{\sin \alpha}$$

i.e.,

$$F = \frac{\mu W}{\sin \alpha} \quad ...(4.84)$$

And, from eqn. (10)

$$T = \int dT = \int_{r_2}^{r_1} \frac{2\pi \mu pr^2 \, dr}{\sin \alpha} = \int_{r_2}^{r_1} \frac{2\pi \mu cr \, dr}{\sin \alpha}$$

$$\therefore \quad T = \frac{2\pi \mu c}{\sin \alpha} \times \frac{r_1^2 - r_2^2}{2}$$

or

$$T = \frac{\pi \mu c (r_1^2 - r_2^2)}{\sin \alpha} \quad ...(4.85)$$

By substituting value of c from eqn. (4.82) in eqn. (4.85), we get

$$T = \frac{\pi\mu}{\sin\alpha}\left[\frac{W}{2\pi(r_1 - r_2)}\right](r_1^2 - r_1^2)$$

or $$T = \frac{1}{2}\frac{\mu W}{2\sin\alpha}(r_1 + r_2) \quad ...(4.86)$$

4.16. MEAN EFFECTIVE RADIUS OF BEARING SURFACE

Expressions for frictional torque for all of the above cases can be rewritten in the following form :

$$T = \mu W r_m \quad ...(4.87)$$

where r_m = mean effective radius of the friction surface of the bearing.

Case I. For solid flat shaft thrust bearings

(*a*) For $p = c$

From eqn. (4.73)

$$r_m = \frac{2}{3}r \quad ...[4.88(a)]$$

(*b*) For $pr = c$

From eqn. (4.74)

$$r_m = \frac{1}{2}r \quad ...[4.88(b)]$$

Case II. For collar bearings

(*a*) For $p = c$

From eqn. (4.67),

$$r_m = \frac{2}{3}\left(\frac{r_1^3 - r_2^3}{r_1^2 - r_2^2}\right) \quad ...[4.89(a)]$$

(*b*) For $pr = c$

From eqn. (4.72),

we get $$r_m = \frac{r_1 + r_2}{2} \quad ...[4.89(b)]$$

Case III. For conical bearings

(*a*) For $p = c$

From eqn. (4.80)

$$r_m = \frac{2}{3\sin\alpha} \times \left(\frac{r_1^3 - r_2^3}{r_1^2 - r_2^2}\right) \quad ...[4.90(a)]$$

(*b*) For $pr = c$

From eqn. (4.86)

$$r_m = \frac{1}{2\sin\alpha}(r_1 + r_2) \quad ...[4.90(b)]$$

Problem 4.32. *A flywheel of weight 800 N is mounted at the middle of a horizontal shaft which is supported by the two journal bearings. Diameter of the shaft is 60 mm. If a 4 Nm couple is required to keep the shaft to rotate at uniform speed of 600 rpm, find the co-efficient of kinetic friction* μ_k *and the radius of the friction circle. Also find the power lost in friction.*

Sol. Total load on each bearing $= W = \frac{1}{2} \times 800 = 400$ N

Shaft radius, $r = \frac{60}{2} = 30$ mm; torque, $T = 4$ Nm $= 4000$ Nmm

Torque, $T = W \times r \times \mu_k$

$\therefore$ $4000 = 400 \times 30 \times \mu_k$

$\therefore$ $\mu_k = \frac{4000}{400 \times 30} = \mathbf{0.333} = \mathbf{\frac{1}{3}}$. **Ans.**

Radius of the friction circle,

$$r_f = \mu_k r = \frac{1}{3} \times 30 = \mathbf{10\ mm}. \quad \textbf{Ans.}$$

Speed of shaft, $N = 600$ *rpm*

Angular velocity, $\omega = \frac{2\pi N}{60} = \frac{2\pi \times 600}{60} = 20\pi$ rad/s

$\therefore$ Power lost in friction, $P = T\omega = 4 \times 20\pi$ watts

$= 80\pi$ watts

$= \mathbf{251.327\ watts}$. **Ans.**

Problem 4.33. *A bell-crank lever fits over a 80 mm diameter shaft which is fixed and cannot rotate. A horizontal force F is required to maintain equilibrium of the crank under the action of a vertical force P = 120 N [Fig. P. 4.33 (a)] Determine the maximum and minimum values that the force F may have without causing the crank to rotate in either direction. The coefficient of static friction between the shaft and the bearing is 0.21.*

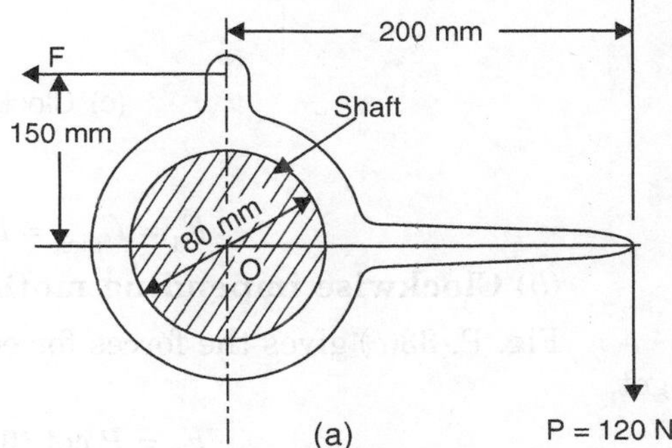

Sol. $\phi = \tan^{-1} \mu = \tan^{-1} 0.21 = 11.86°$

Radius of friction circle, $r_f = r \sin \phi$

$= 40 \sin 11.86°$

$= \mathbf{8.22\ mm}$

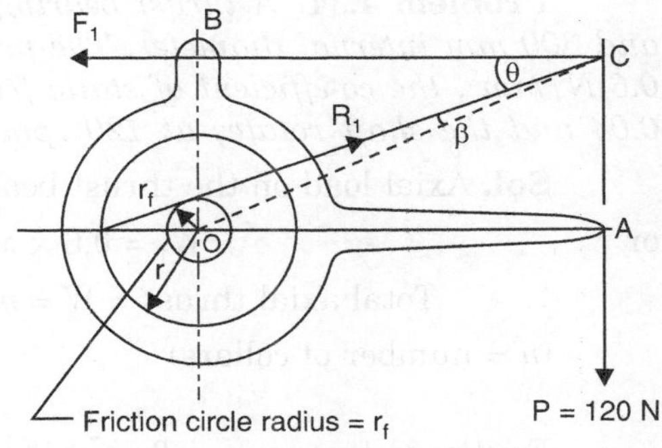

Impending motion of rotation occurs when the reaction R of the fixed shaft of the bell crank makes an angle $\phi = \tan^{-1} \mu$ with the normal to the bearing surface and is, therefore tangent to the friction circle of radius $r_f = r \sin \phi$. For equilibrium the three forces P, F and R meet at a point C, as shown in Fig. P.4.33(*b*).

From Fig. P. 4.33 (*b*), angle $\beta = \sin^{-1} \frac{r_f}{OC}$

and $OC = \sqrt{OA^2 + OB^2} = \sqrt{(200)^2 + (150)^2}$

$= 250$ mm

$\therefore \quad \beta = \sin^{-1} \dfrac{r_f}{OC} = \sin^{-1}\left(\dfrac{8.22}{250}\right)$

$= \sin^{-1} 0.03288$

$= 1.88°$

Angle $\theta = \tan^{-1}\left(\dfrac{150}{200}\right) = 36.87°$

$\therefore \quad \theta - \beta = 36.87° - 1.88° = 34.99°$

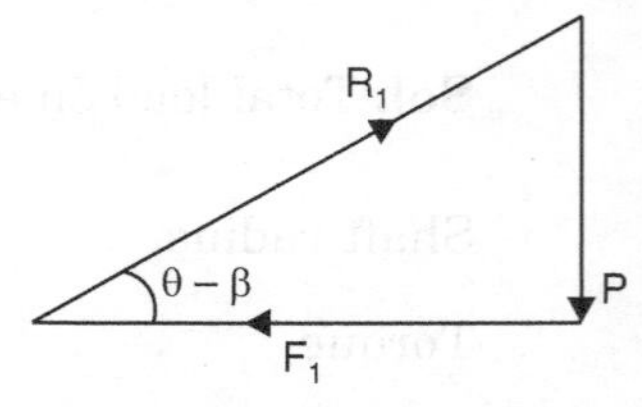

(b) Counter clockwise motion is impending.

Fig. P. 4.33.

(*a*) **Impending motion in counter clockwise direction**

The triangle of forces for equilibrium are shown in Fig. P.4.33(*b*). Thus, we get

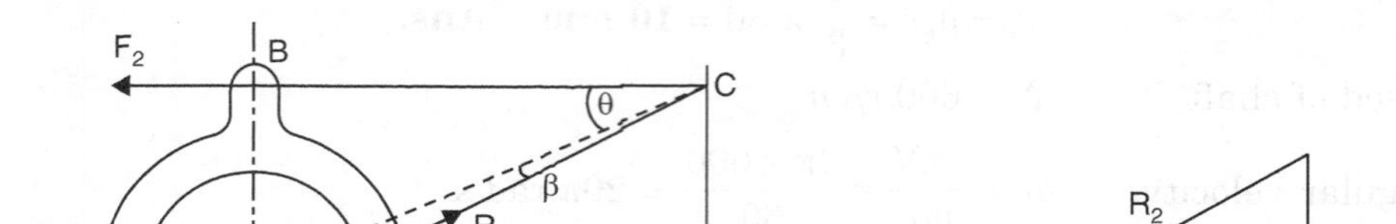

(c) Clockwise impending motion.

Fig. P. 4.33

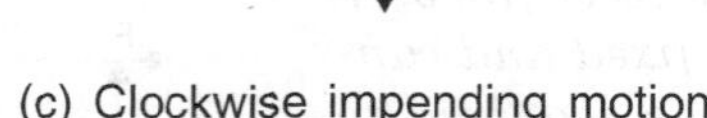

$$F_1 = F_{max} = P \cot(\theta - \beta) = 120 \cot 34.99° = \mathbf{171.44\ N}$$

(*b*) **Clockwise impending motion**

Fig. P. 33(*c*) gives the forces for equilibrium for clockwise impending motion. Thus, we get

$$F_2 = P \cot(\theta + \beta) = 120 \times \cot(36.87 + 1.88)$$

$$\therefore \quad F_2 = F_{min} = \mathbf{149.52\ N.} \quad \textbf{Ans.}$$

Problem 4.34. *A thrust bearing takes the thrust on 4 collars each 400 mm external and 300 mm internal diameter. The pressure on the bearing surfaces is constant and equals 0.6 N/mm², the co-efficient of static friction is 0.06 and the co-efficient of kinetic friction is 0.04 and the shaft rotates at 120 rpm. Calculate the power absorbed by friction.*

Sol. Axial load on the thrust bearing, $W_1 = p\pi(r_1^2 - r_2^2)$ per collar

or $$W_1 = 0.6 \times \pi(200^2 - 150^2) = 32986.7 \text{ N}$$

$\therefore$ Total axial thrust $= W = nW_1 = 4 \times 32986.7 = 131946.8$ N

(n = number of collars)

Frictional torque, $$T = \frac{2}{3}\mu W\left(\frac{r_1^3 - r_2^3}{r_1^2 - r_2^2}\right)$$

$$= \frac{2}{3} \times 0.04 \times 131946.8\left(\frac{200^3 - 150^3}{200^2 - 150^2}\right) \text{ Nmm}$$

$$= 2789732.343 \text{ N-mm} = \mathbf{2.79\ kNm}$$

$$\text{Angular speed of shaft} = \omega = \frac{2\pi N}{60} \text{ rad/s}$$

or
$$\omega = \frac{2\pi \times 120}{60} = 4\pi \text{ rad/s}$$

$\therefore$ Power absorbed in friction,

$$P = T\omega = 2.79 \times 4\pi \text{ kW}$$

$$= \mathbf{35.06\ kW.} \quad \textbf{Ans.}$$

Problem 4.35. *Calculate power lost in overcoming friction and the number of collars required for a thrust bearing of radii 160 mm and 240 mm. The co-efficient of friction is 0.06. The axial load on the shaft is 40 kN. Maximum intensity of pressure is 3 bar (1 bar = 10^5 N/m^2). The speed of shaft is 450 rpm.*

Sol. Friction torque, $T = \frac{2}{3}\mu W \left(\frac{r_1^3 - r_2^3}{r_1^2 - r_2^2}\right)$

where $W = 40 \text{ kN} = 40,000 \text{ N}$; $r_1 = 240$ mm ; $r_2 = 160$ mm and $\mu = 0.06$.

$$\therefore \quad T = \frac{2}{3} \times 0.06 \times (40,000) \times \frac{240^3 - 160^3}{240^2 - 160^2} \text{ Nmm}$$

$$= 486.4 \times 10^3 \text{ N-mm} = \mathbf{486.4\ Nm}$$

$$\omega = \frac{2\pi N}{60} = \frac{2\pi}{60} \times 450 = 15\pi \text{ rad/s}$$

$\therefore$ Power lost in friction,

$$P = T\omega = 486.4 \times 15\pi \text{ watts}$$

$$= 22921.06 \text{ watts} = \mathbf{22.92106\ kW.} \quad \textbf{Ans.}$$

$$\textbf{Number of collars} = \frac{\text{Total load}}{\text{Load per collar}}$$

$$\text{Load per collar} = p \times \pi\,(r_1^2 - r_2^2)$$

$$= (3 \times 10^5) \times \pi \left[\left(\frac{240}{1000}\right)^2 - \left(\frac{160}{1000}\right)^2\right]$$

$$= \frac{30159.3 \times 10^5}{(10)^6} \text{ N} \simeq 30159.3 \text{ N}$$

$$\therefore \quad \text{Number of collars} = \frac{40 \times 1000}{30159.3} = 1.3263 = \textbf{2 collars.} \quad \textbf{Ans.}$$

Problem 4.36. *A load of 30 kN is supported by a conical pivot. The angle of cone is 120° and the intensity of pressure is not to exceed 4 bar. The external radius is 3 times the internal radius. Find radii of the bearing surface. Find power lost in friction if μ = 0.05 and speed of shaft is 300 rpm.*

Sol. We know that axial load,

$$W = \int_{r_2}^{r_1} p \times 2\pi r.\, dr$$

or $W = p \times \text{projected area}$

$= p \times \pi\,(r_1^2 - r_2^2)$

$\therefore$ $30 \times 10^3 = 4 \times 10^5 \times \pi[(3r_2)^2 - r_2^2]$

$\therefore$ $= 4 \times 10^5 \times \pi r_2^2\,[9 - 1]$

$\therefore$ $r_2^2 = \dfrac{30 \times 10^3}{4 \times 10^5 \times \pi \times 8} = 2.984155 \times 10^{-3}\ \text{m}^2$

$\therefore$ $r_2 = 0.05463\ \text{m} =$ **54.63 mm. Ans.**

and $r_1 = 3r_2 =$ **163.89 mm. Ans.**

Frictional torque, $T = \dfrac{2}{3}\dfrac{\mu W}{\sin\alpha}\left(\dfrac{r_1^3 - r_2^3}{r_1^2 - r_2^2}\right)$ | $2\alpha = 120°$ $\therefore$ $\alpha = 60°$

$= \dfrac{2}{3} \times \dfrac{0.05 \times (30 \times 1000)}{\sin 60°} \times \left[\dfrac{(163.89)^3 - (54.63)^3}{(163.89)^2 - (54.63)^2}\right]$

$= 205014.1938$ N-mm = **205.01 Nm**

$\omega = \dfrac{2\pi N}{60} = \dfrac{2\pi \times 300}{60} = 10\pi$ rad/s

$\therefore$ Power, $P = T\omega = 205.01 \times 10\pi =$ **6440.6 Watts**

= **6.4406 kW. Ans.**

Problem 4.37. *A car engine has rated output of 16 kW. Maximum torque developed is 160 Nm. A single plate clutch with both sides effective is used. Axial pressure is not to exceed 0.8 bar. External diameter of the clutch plates is 1.4 times its internal diameter. Find dimensions of the clutch plate and the axial spring force. Take μ = 0.3.*

Sol. For clutch assumption of uniform wear rate ($pr = c$) is used.

The maximum pressure will be at the least radius.

$\because$ $p_1 r_1 = p_2 r_2$

$\therefore$ $p_2 = p_{max} = 0.8\ \text{bar} = 0.8 \times 10^5$ N

$\therefore$ $c = p_2 r_2 = 0.8 \times 10^5 \times r_2$

Axial force, $W = 2\pi c(r_1 - r_2) = 2\pi \times 0.8 \times 10^5\, r_2 \times (1.4r_2 - r_2)$

or $W = 2.01062\, r_2^2 \times 10^5$ ...(1)

Frictional torque with both sides effective,

$T = \left[\dfrac{1}{2}\mu W(r_1 + r_2)\right] \times 2 = \dfrac{1}{2} \times 0.3 \times W \times (1.4r_2 + r_2)$

or $T = 0.15 \times 2.4r_2 \times 2$ ...(2)

or $T = 0.15 \times (2.01062\, r_2^2) \times 10^5 \times 2.4r_2 \times 2$

or $T = 0.7238 \times r_2^3 \times 10^5 \times 2$ ...(3)

But $T = 160$ Nm (given)

$\therefore$ From eqn. (3) ;

$160 = 0.7238 \times r_2^3 \times 10^5 \times 2$

$$\therefore \qquad r_2 = \frac{160}{0.7238 \times 10^5 \times 2} = 221.0555 \times 10^{-5}$$

$$\therefore \qquad r_2 = 0.130267 \text{ m} = \mathbf{130.267 \text{ mm}}$$

and $$r_1 = 130.267 \times 1.4 = \mathbf{182.374 \text{ mm}}$$

$\therefore$ From eqn. (1), axial load,

$$W = 2.01062\, r_2^2 \times 10^5$$
$$= 2.01062 \times (0.130267)^2 \times 10^5 \text{ N} = \mathbf{3411.82 \text{ N.}} \quad \textbf{Ans.}$$

EXERCISES IV

A. Theoretical Questions

1. Explain the following :
Friction, Force of Friction, Coefficient of friction, Limiting Friction, Angle of Friction, Angle of Repose.

2. What are different types of friction ? Under what conditions will each type occur ?

3. State and explain laws of dry friction.

4. Establish equations for finding the effort and efficiency of the plane when a block of weight W lying on an inclined plane (inclination with horizontal being α) is acted upon by a horizontal force. Consider (*a*) upward motion of the block and (*b*) downward motion of the block.

5. Analyse the motion of block of weight W lying on an inclined plane and acted upon by an effort parallel to the plane. Consider (*a*) upward motion and (*b*) downward motion of the block.

6. A block of weight W is resting on an inclined plane. An effort inclined at an angle θ to the plane is applied to the block. Find effort and the efficiency of the plane considering (*a*) upward motion and (*b*) downward motion of the block.

7. Describe a screw jack.

8. Derive expressions for finding effort for raising a weight using a screw jack when the weight rotates with the spindle. Also establish equation for efficiency of the screw jack.

9. Derive expressions for effort for raising a weight with the help of the screw jack when the weight does not rotate with the spindle. Also establish equation for finding efficiency of the screw jack.

10. What is helix angle ? Explain in context with screw jack.

11. What do you understand by a wedge ? How would you find the effort required to raise a heavy block using a wedge ?

12. How would you analyse a ladder ? Explain in the light of different situations.

13. What is a friction circle ?

14. What changes are necessary to be made in different equations derived for square threads in connection with friction if V-threads are used ? Hence explain why square threads are used for transmission of power.

15. Derive the condition for maximum efficiency of V-threads.

16. Neglecting collar friction, prove that the maximum efficiency of screw jack using square threaded screw is given by

$$\eta_{max} = \frac{1 - \sin \phi}{1 + \sin \phi}$$

17. Neglecting collar friction, prove that for maximum efficiency of square threaded screw jack, the helix angle α is given by

$$\alpha = \frac{\pi}{4} - \frac{\phi}{2}.$$

18. Derive an expression for frictional torque of a collar thrust bearing assuming

(*a*) uniform pressure, (*b*) uniform wear.

19. Derive an expression for frictional torque of a conical pivot bearing assuming

(*a*) uniform pressure, (*b*) uniform wear. How will the expressions be modified, if the conical pivot is truncated ?

B. Numerical Problems

4.1. A man wishing to slide a stone block of weight 1000 N over a horizontal concrete floor, ties a rope to the block and pulls it in a direction upwards at an angle of 20° to the horizontal. Calculate the minimum pull necessary to slide the block if the co-efficient of friction m = 0.6. Calculate also the pull required if the inclination of the pull with the horizontal is equal to the angle of friction and prove that this is the least force required to slide the block. [*AMIE Winter 1990*]

[**Ans.** 524N ; 514.44 N]

4.2. Find the least force required to pull a body of weight 2000 N, placed on a rough inclined plane having inclination of 20° to the horizontal. The force applied to the body is inclined at an angle $\theta = 40°$ to the plane and the body is (*a*) on the point of moving up the plane, and (*b*) on the point of moving down the plane. Take $\mu = 0.25$. [**Ans.** (*a*) 1155.03 N, (*b*) 538.61 N]

4.3. A block of weight 1000 N is pulled up an inclined plane inclined at 20° to the horizontal at a steady speed. If the co-efficient of friction between the block and the plane is 0.28, find the force parallel to the plane. If the distance travelled by the block is 20 m along the plane, find the workdone on the block. Also find the efficiency of the plane. What force will be required to pull the block down the plane .[**Ans.** 605.1 N ; 56.52 % ; 342.02 N ; Workdone = 1210.2 Nm or joules]

4.4. An effort of 300 N is required just to move a certain body up an inclined plane of angle 18°, the force acting parallel to the plane. If the angle of inclination is increased to 25°, the effort required parallel to the plane becomes 300 N. Find the weight of the body and the co-efficient of friction between the body and the plane. [**Ans.** $\mu = 0.294$; $W = 509.66$ N]

4.5. Calculate the effort required to pull a block of weight 1000 N up the plane

(*i*) When the effort is applied horizontally,

(*ii*) When the effort applied is parallel to the plane, and

(*iii*) When the effort applied is inclined to the plane through an angle of 30°.

Also, find efficiency of the plane in each case. Take co-efficient of friction, $\mu = 0.25$ and inclination of the plane, $\alpha = 18°$.

[**Ans.** (*i*) $P = 625.84$ N ; $\eta = 51.92\%$; (*ii*) $P = 546.85$ N ; $\eta = 56.51\%$;
(*iii*) $P = 557.7$ N ; $\eta = 44.85\%$]

4.6. Two bodies of weights 600 N and 1200 N placed on an inclined plane are connected by a cord which is parallel to the plane. The co-efficient of friction for the 600 N body is 0.15 while for the 1200 N body it is 0.40. The lighter body is below the heavier one. Find the inclination of the plane and tension in the cord when the two bodies are about to start moving down the plane.

[**Ans.** $\alpha = 17.57°$; tension = 95.32 N]

4.7. A block *A* weighing 300 N rests on a rough inclined plane inclined at 45° to the horizontal. Block *A* is connected to block *B* weighing 900 N, resting on the rough horizontal plane, by a weightless rigid bar inclined at 30° to the horizontal. What horizontal force be applied to the block *B* to start the block *A* to move up the plane. Take $\mu = 0.268$ for all surfaces. [**Ans.** 541 N]

4.8. Referring to the Fig. E .4.8, determine the least value of the force P to cause motion to impend rightwards. Assume the co-efficient of friction under the blocks to be 0.2 and pulley to be frictionless. [*AMIE Winter 1988*]

[**Ans.** P = 168.165]

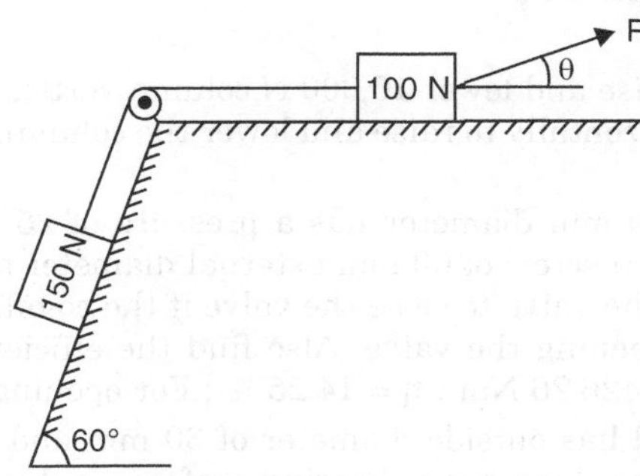

Fig. E.4.8

4.9. A uniform ladder of length 5.4 m and weighing 25 N is placed against a smooth vertical wall with its lower end 2 m from the wall. Show that the ladder will remain in equilibrium in this position. What frictional force acts between the lower end of the ladder and the floor. Take $\mu = 0.3$ at the floor end of the ladder. [**Ans.** Frictional force = 7.5 N]

4.10. A uniform ladder of weight 34 N and length 3 m rests on a horizontal floor and leans against a smooth vertical wall. The ladder makes an angle of 60° to the floor. When a man weighing 600 N stands on the ladder at a distance 1 m from the lower end, the ladder is at point of sliding. Find the co-efficient of friction between the ladder and the floor. [**Ans.** $\mu = 0.198$]

4.11. A uniform ladder of weight 60 N and length 4 m rests on a horizontal floor and leans against a rough vertical wall . The ladder makes an angle of 30° to the horizontal. When a man weighing 600 N reaches 1 m from the lower end, horizontal force P applied at the lower end of the ladder prevents it from sliding. Take $\mu = 0.3$ at the lower end and $\mu = 0.15$ at the upper end of ladder. Find the force P. [**Ans.** 60.61 N]

4.12. A uniform ladder 6 m long and weighing 100 N rests on rough floor and leans against a rough vertical wall. A man weighing 600 N when reaches 2 m from the floor end, the ladder starts slipping.

(*a*) Find the inclination of the ladder with the floor. The co-efficient of friction between the ladder and the floor and the wall is 0.3.

(*b*) Also find inclination of the ladder if the wall is smooth.

[**Ans.** (*a*) $\theta = 44.93°$; (*b*) $\theta = 57.8°$]

4.13. A uniform ladder 6 m long and weighing 100 N rests on a rough floor and leans against smooth wall. It is inclined at 45° to the floor when the ladder starts sliding when a 600 N man reach 2 m from the ground. Find the co-efficient of friction between the ladder and the floor.

[**Ans.** $\mu = 0.476$]

4.14. A uniform ladder 6 m long and weighing 100 N rests on a rough floor and makes an angle of 45° to the floor when its other end rests against a smooth wall. A horizontal rope connects the wall and the ladder at a distance of 2 m from the lower end. When a man weighing 600 N reaches a distance 5 m from the floor, the rope tension T is induced in it. Find this tension in the rope.

[**Ans.** T = 510 N]

4.15. A block over lying a 10° wedge on a horizontal floor and leaning against a vertical wall and weighing 1500 N is to be raised by applying a horizontal force to the wedge. Assuming co-efficient of friction between all the surfaces in contact to be 0.3, determine the horizontal force to be applied to raise the block. [*AMIE Winter 1982*]

[**Ans.** 1699.39 N]

4.16. The movable jaw of a bench vice is at the upper end of a hinged arm 0.5 m long, the centre line of the screw being 35 cm above the hinge. The screw is of 2.5 cm outside diameter and has 1.6 square threads per cm. The mean radius of the thrust collar is 3 cm. Find the tangential force to be applied to the screw at a radius of 28 cm to produce a force of 6000 N at the jaw. Find the mechanical efficiency of the vice. [*Annamalai University*]

[**Ans.** 95 N, 32%]

4.17. A 7.5° wedge is used to raise and lower 10,000 N column vertically between the guides. Find the force which is applied horizontally to raise and lower the column. The co-efficient of friction at all moving surfaces is 0.18. [**Ans.** 5295 N ; 2290 N]

4.18. A steam stop valve of 180 mm diameter has a pressure of 25 bar acting on it. It is closed by means of a square threaded screw of 60 mm external diameter and 6 mm pitch. Find the torque exerted on the handle of the valve to close the valve if the co-efficient of friction is 0.2. Also find the torque required for opening the valve. Also find the efficiency of the screw for closing the valve. [**Ans.** Torque = 426.26 Nm ; η = 14.25 % ; For opening the valve torque = 299.83 Nm]

4.19. A bolt with square thread has outside diameter of 30 mm and pitch 5 mm. It is tightened by screwing a nut whose mean diameter of bearing surfaces is 6 cm. The co-efficient of friction for nut and bolt is 0.18 and for nut and bearing is 0.2. The nut is tightened by a 40 cm long spanner. The load on the bolt is 15 kN. Find the force required at the end of the spanner. What is the efficiency of the bolt. Also find force at the end of the spanner and the efficiency for loosening the bolt. [**Ans.** 348.92 N ; 8.56% ; 287.34 N ; 10.39%]

4.20. A ladder 5 m and 250 N weight is placed against a vertical wall in a position where its inclination to the vertical is 30°. A man weighing 800 N climbs the ladder. At what position x will he induce slipping ? The co-efficient of friction for both contact surfaces of the ladder, *i.e.,* with the wall and the floor is 0.2 . [*AMIE Summer 1983*]

[**Ans.** x = 1.647 m from floor]

4.21. The following particulars refer to a screw jack :

Load on the screw = 30 kN,

Mean diameter of the screw = 2.8 cm,

pitch of the screw = 1 cm,

length of the lever where the effort is applied = 1 m.

Mean radius of the contact surface between the swivel and head = 4 cm,

co-efficient of friction between the screw threads and bearing = 0.15.

Co-efficient of friction between the swivel head and bearing surface = 0.1

Find the force on the lever end while raising the load and lowering the load.

Also determine the efficiency of the screw jack in both the cases. Also determine all above values ignoring the friction at swivel head.

[**Ans.** 232.695 N; 20.52% for raising the load ; 134.96 N, 35.38 % ; 112.7 N; 42.37% ; 14.96 N ; 31.31%]

4.22. The spindle of a square thread has a root diameter of 27.5 mm and pitch 5 mm. A load of 15 kN is being raised by applying a horizontal force at the end of 50 cm long rod. The load does not rotate with the spindle. A bearing surface of mean radius 3 cm is provided at the swivel head. Find (*a*) the effort required for lifting the load, (*b*) the effort required for lowering the load, (*c*) efficiency of the screw jack for lifting and for lowering the load. Co-efficient of friction at the spindle is 0.15 and that at the bearing is 0.20.

[**Ans.** (*a*) 272.126 N ; (*b*) 223.25 N ; (*c*) 8.78 % and 10.7 %. respectively for raising and lowering the load]

4.23. A shaft 6 cm diameter, carrying an axial load of 30 kN is supported by a flat foot-step bearing. Find the power lost in friction when the shaft runs at 3000 rpm (*a*) for uniform distribution of pressure on the bearing surface, and (*b*) for uniform wear rate of bearing surface. Take co-efficient of friction as 0.05. [**Ans.** (*a*) 9.425 kW ; (*b*) 7.069 kW]

4.24. Calculate power lost in friction for the thrust bearing whose contact surfaces are 12 cm inner diameter and 18 cm outer diameter. The intensity of pressure is not to exceed 4 bar. Speed of shaft is 600 rpm. Take $\mu = 0.08$. [**Ans.** 1705.468 Watts]

4.25. Power lost in friction of collar thrust bearing rotating at 180 rpm carrying an axial thrust of 60 kN having outer diameter of 24 cm and inner radius of 15 cm is 4 kW. Find the co-efficient of friction at the bearing. [**Ans.** $\mu = 0.01782$]

4.26. A shaft has a number of collars integral with it. Outer diameter of collars is 18 cm and the shaft diameter 8 cm. If uniform pressure intensity is 12 bar and the co-efficient is 0.06. Find

(*a*) Power absorbed in overcoming friction when the shaft runs at 1800 rpm and carries a load of 10 kN.

(*b*) The number of collars required. [**Ans.** 7.714 kW ; 17]

4.27. 120 kW is transmitted at 3600 rpm by a multiple disc friction clutch. The plates run in oil and have the friction surfaces of steel and phosphor bronze alternatively, $\mu = 0.06$, and the axial intensity of pressure is not to exceed 2.4 bar. External diameter is 1.25 times the internal diameter-the external diameter being 25 cm. Determine the number of plates needed to transmit the required power. Assume condition of uniform wear. [**Ans.** 13 plates]

4.28. Following data relates to a cone clutch :

Maximum radius of contact surface = 18 cm

Minimum radius of contact surface = 15 cm

semi-cone angle = 20°

Co-efficient of friction = 0.3

Allowable normal pressure = 1.5 bar

Speed of shaft = 2400 rpm

Determine (*a*) the axial thrust acting on the clutch, and (*b*) Power transmitted by the clutch.

[**Ans.** (*a*) 2827.43 N ; (*b*) 102.85 kW]

5

Transmission of Power

PART A—BELT AND ROPE DRIVES

5.1. INTRODUCTION

The power from one shaft to another can be transmitted by means of belts, ropes and chains running over pulleys or by gears mounted on the two shafts.

Belt drive. A belt drive consists of driving and driven pulleys and the belt which is mounted on the pulleys with a certain amount of tension and transmits power (rotary motion) by friction. As shown in Fig. (5.1), pulley A is keyed to the rotating shaft and is known as *driving pulley*. The pulley B is keyed to the shaft, to which we wish to transmit the power and is known as *follower pulley*. When the driving pulley A rotates, it carries the belt, which in turn rotates driven (follower) pulley. The pulleys carry the belt due to friction between the surfaces of the pulleys and the belt.

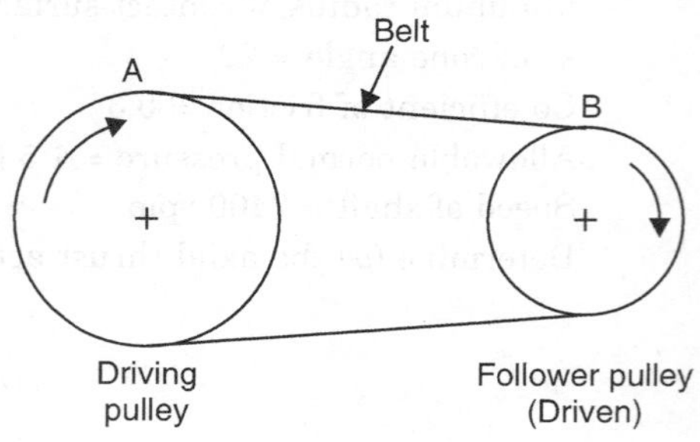

Fig. 5.1 Belt Drive.

5.2. TYPES OF BELT DRIVES

(*i*) (*a*) Open belt drive (*b*) Crossed belt drive

Open belt drive. In this type of belt drive, the direction of rotation of driving and driven pulleys is the same. See Fig. 5.2 (*a*).

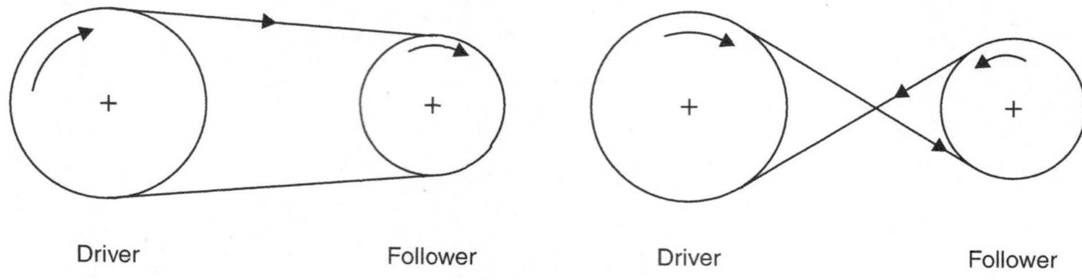

(a) Open belt drive (b) Crossed drive

Fig. 5.2 (a) Open Drive, (b) Crossed Drive.

Crossed belt drive. In this type of belt drive the direction of rotation of driving and driven pulleys are different.

(*ii*) (*a*) Simple belt drive (*b*) Compound belt drive

Simple belt drive. When power is transmitted from one shaft to another through single driving and driven pulleys then it is known as simple belt drive : See Fig. 5.2 (*c*)

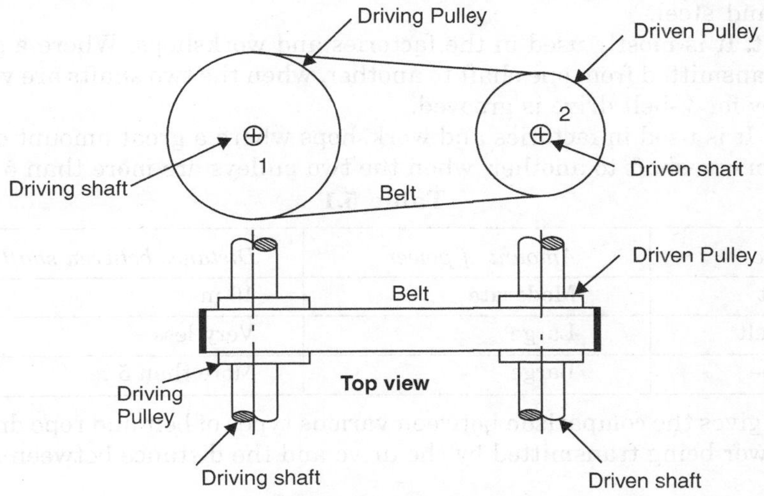

Fig. 5.2 (c) Simple Drive.

Compound belt drive. When power is transmitted from one shaft to another through a number of driving and driven pulleys then it is termed as compound belt drive. There is one pulley each the driving and the driven pulleys and two pulleys on the intermediate shafts. See Fig. 5.2 (*d*).

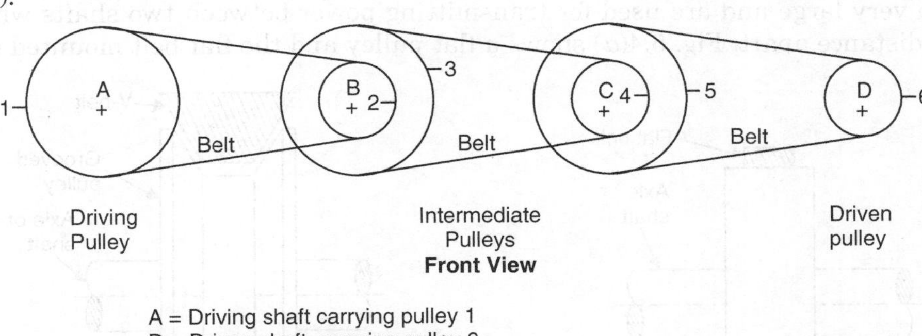

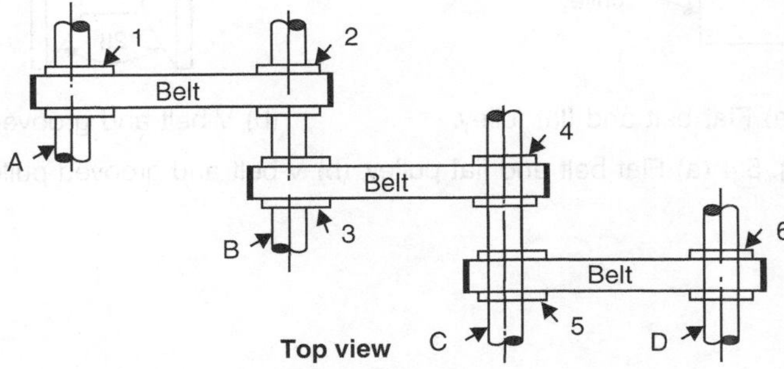

Fig. 5.2 (d) Compound Drive.

5.3. TYPES OF BELTS

(*a*) **Flat Belt.** It is mostly used in the factories and workshops, where a moderate amount of power is to be transmitted from one shaft to another when two shafts are not more than 10 m apart. The surface of pulley is also flat. They are made of leather, rubber, textile, balata and steel.

(*b*) **V-Belt.** It is mostly used in the factories and workshops. Where a great amount of power is to be transmitted from one shaft to another, when the two shafts are very near to each other. The pulley for *V*-belt drive is grooved.

(*c*) **Rope.** It is used in factories and workshops where a great amount of power is to be transmitted from one shaft to another, when the two pulleys are more than 5 m apart.

Table 5.1

Types	*Amount of power*	*Distance between shafts*
Flat	Moderate	10 m
V-belt	Large	Very less
Rope	Large	More than 5 m

Table 5.1 gives the comparison between various types of belt and rope drives in the light of amount of power being transmitted by the drive and the distance between shafts.

5.4. TYPES OF PULLEYS

The surface of pulleys should be such that it can carry the belt or the rope mounted on it. There are three types of pulleys, as given bellow (Fig. 5.4) :

(*a*) **Flat Pulleys.** The flat pulleys are used to carry flat belt in a belt drive. The flat pulleys are very large and are used for transmitting power between two shafts which are at very large distance apart. Fig. 5.4(*a*) shows a flat pulley and the flat belt mounted on it.

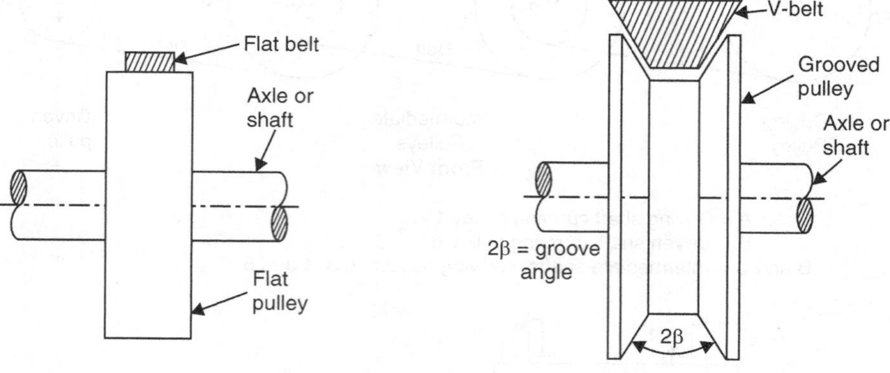

(a) Flat belt and flat fulley. (b) V-belt and grooved pulley.

Fig. 5.4 (a) Flat belt and flat pulley, (b) V-belt and grooved pulley.

(*b*) **Grooved Pulley for V-Belt.** *V*-shaped groove is provided in the pulley to accommodate a *V*-belt. Fig. 5.4(*b*) shows a grooved pulley used for the *V*-belt. The chance of slip between belt and the pulley is reduced substantially in this pulley. The groove angle 2β is generally between 35° to 45°.

(*c*) **Grooved Pulley for Rope.** The grooved pulley for the rope drive is shown in Fig. 5.4(*c*). The groove angle is generally between 30° to 45°.

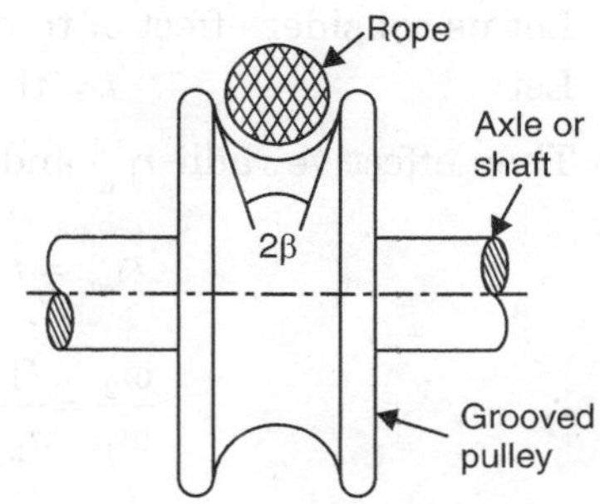

Fig. 5.4. (c) Groove pulley and the rope.

5.5. VELOCITY RATIO

Case I. Simple drive without slip. Fig. 5.5 (*a*)

Velocity ratio is the ratio between the velocities of the driven and driving pulleys.

Let d_1 (= $2r_1$) = diameter of driver.

d_2 (= $2r_2$) = diameter of follower.

N_1 = speed of the driver. (Revolutions/min or rpm)

N_2 = speed of the follower. (rpm)

ω_1 = angular velocity of the driver = $\dfrac{2\pi N_1}{60}$ radians/s

ω_2 = angular velocity of the follower = $\dfrac{2\pi N_2}{60}$ radians/s

If thickness of the belt is negligible, and there is no slip between the pulley and the belt ; the peripheral speed of the pulleys must be the same.

Velocity of the driver $V_1 = \omega_1 r_1$

Since there is no slip

Velocity of the belt $V_b = V_1$ and the velocity of the follower $V_2 = V_b = \omega_2 r_2$.

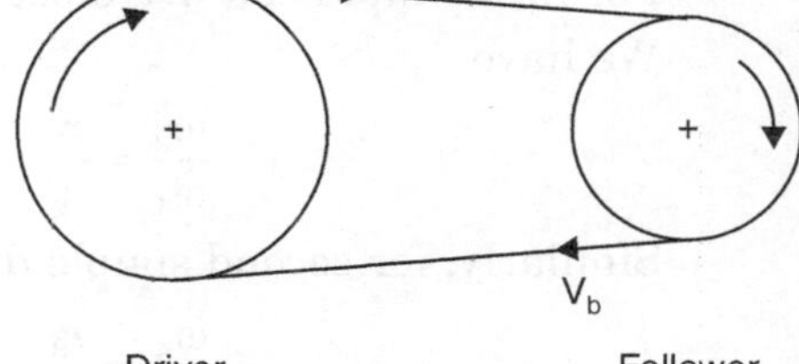

Fig. 5.5. (a) Simple drive.

or, $$\omega_2 r_2 = \omega_1 r_1$$

$$\therefore \quad \text{Velocity ratio} = VR = \frac{\omega_2}{\omega_1} = \frac{r_1}{r_2}$$

or $$VR = \frac{2\pi N_2/60}{2\pi N_1/60} = \frac{r_1}{r_2} \quad \text{or} \quad \text{VR} = \frac{N_2}{N_1} = \frac{r_1}{r_2}$$

$$\boxed{VR = \frac{N_2}{N_1} = \frac{d_1}{d_2}} \qquad ...(5.1)$$

We can say N*d* = constant.

Let us consider effect of thickness of the belt with no slip condition.

Let t = thickness of the belt

Then effective radii $r_{1_{eff}}$ and $r_{2_{eff}}$ will be

$$\therefore \quad r_{1_{eff}} = r_1 + \frac{t}{2} \quad \text{and} \quad r_{2_{eff}} = r_2 + t/2$$

$$\therefore \quad \frac{\omega_2}{\omega_1} = \frac{r_{1_{eff}}}{r_{2_{eff}}} = \frac{r_1 + t/2}{r_2 + t/2}$$

$$\frac{\omega_2}{\omega_1} = \frac{d_1 + t}{d_2 + t}$$

or

$$\boxed{VR = \frac{N_2}{N_1} = \frac{d_1 + t}{d_2 + t}} \quad \text{...(5.2)}$$

Case II. Compound drive without slip. Let d_1, d_2, d_3, ... be the diameters of pulleys 1, 2, 3, ... respectively, and N_1, N_2, N_3 , ... the speeds in revolutions per minute of the pulleys 1, 2, 3, ... See Fig. 5.5 (*b*).

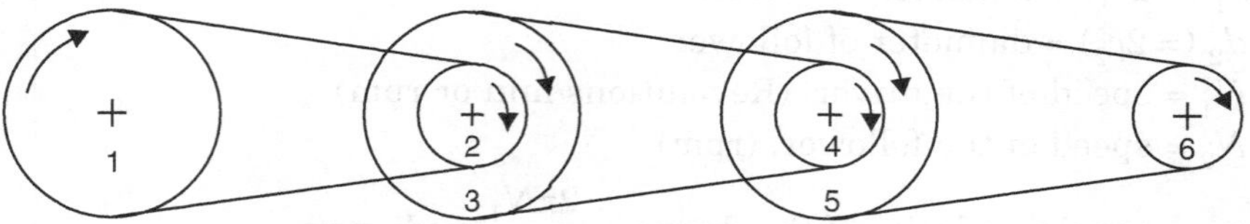

Fig. 5.5 (b) Compound drive.

Similarly ω_1, ω_2, ω_3 ,... are the angular velocities of pulleys 1, 2, 3, ... respectively. Here, there are three simple drives.

For first simple belt drive between pulley 1 and 2,

We have

$$\frac{\omega_2}{\omega_1} = \frac{r_1}{r_2} \quad \text{...}(i)$$

Similarly, for second simple drive between pulleys 3 and 4,

$$\frac{\omega_4}{\omega_3} = \frac{r_3}{r_4} \quad \text{...}(ii)$$

and for third simple drive between pulleys 5 and 6,

$$\frac{\omega_6}{\omega_5} = \frac{r_5}{r_6} \quad \text{...}(iii)$$

Multiplying eqns. (*i*), (*ii*) and (*iii*),

we get

$$\frac{\omega_2}{\omega_1} \times \frac{\omega_4}{\omega_3} \times \frac{\omega_6}{\omega_5} = \frac{r_1}{r_2} \times \frac{r_3}{r_4} \times \frac{r_5}{r_6} \quad \text{...}(iv)$$

Since pulleys 2 and 3 are mounted on the same shaft

$$\therefore \quad \omega_2 = \omega_3$$

and also since pulleys 4 and 5 are mounted on the same shaft

$$\therefore \quad \omega_4 = \omega_5$$

Hence eqn. (*iv*) gives
$$\frac{\omega_6}{\omega_1} = \frac{r_1 r_3 r_5}{r_2 r_4 r_6}$$

or
$$\boxed{VR = \frac{N_6}{N_1} = \frac{d_1 d_3 d_5}{d_2 d_4 d_6}} \quad ...(5.3)$$

Case III. Simple drive with slip. So far we have assumed that there is a firm frictional grip between the pulley and the belt. But sometimes, the frictional grip becomes insufficient due to the presence of thin layer of air between them. This may cause some forward motion of the driver without carrying the belt with it. This may also cause some forward motion of the belt without carrying the follower pulley with it. This is called *slip* of the belt and is generally expressed as a percentage. See Fig. 5.5 (*c*)

Velocity of pulley 1
$$= V_1 = \omega_1 r_1$$
because of the slip between the pulley 1 and the belt, the belt velocity is reduced by
$$V_1 \times \frac{S_1}{100}$$
(S_1 = % slip between belt and pulley 1)

∴ Velocity of belt leaving the pulley 1
$$= V_b = V_1 - V_1 \times \frac{S_1}{100}$$

or
$$V_b = V_1\left(1 - \frac{S_1}{100}\right) \quad ...(v)$$

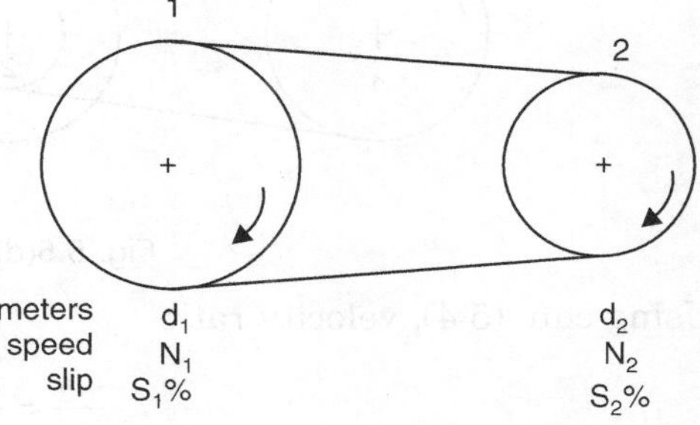

Fig. 5.5.(c) Simple drive with slip.

Similarly, velocity of pulley 2
$$= V_2 = \omega_2 r_2 = V_b\left(1 - \frac{S_2}{100}\right) \quad ...(vi)$$
(S_2 – % slip between belt and pulley 2)

∴
$$\omega_2 r_2 = V_1\left(1 - \frac{S_1}{100}\right)\left(1 - \frac{S_2}{100}\right)$$
(on putting value of V_b from eqn.(*v*))

∴ Velocity ratio
$$\frac{\omega_2}{\omega_1} = \frac{r_1}{r_2}\left(1 - \frac{S_1}{100}\right)\left(1 - \frac{S_2}{100}\right) \quad ...(5.4)$$
$$= \frac{r_1}{r_2}\left(1 - \frac{S_1 + S_2}{100} + \frac{S_1 S_2}{100 \times 100}\right)$$

Neglecting $\frac{S_1 \cdot S_2}{100 \times 100}$, we get
$$VR = \frac{N_2}{N_1} \simeq \frac{r_1}{r_2}\left(1 - \frac{S_1 + S_2}{100}\right) \quad ...[5.4(a)]$$
$$VR = \frac{\omega_2}{\omega_1} = \frac{N_2}{N_1} = \frac{r_1}{r_2}\left(1 - \frac{S_1 + S_2}{100}\right)$$

or
$$VR = \frac{N_2}{N_1} = \frac{r_1}{r_2}\left(1 - \frac{S}{100}\right)$$
($S = S_1 + S_2$ = total percentage slip in the drive)

$$\boxed{\frac{N_2}{N_1} = \frac{d_1}{d_2}\left(1 - \frac{S}{100}\right)} \quad \ldots(5.5)$$

Case IV. Compound drive with slip : See Fig. 5.5 (*d*)

Let d_1 (= $2r_1$) = dia. of pulley 1

d_2 (= $2r_2$) = dia. of pulley 2

d_3 (= $2r_3$) = dia. of pulley 3

d_4 (= $2r_4$) = dia. of pulley 4, etc, etc.

and ω_1, ω_2, ω_3, ω_4, ω_5, ω_6 are angular velocities of the pulleys 1, 2, 3, 4, 5 and 6 respectively.

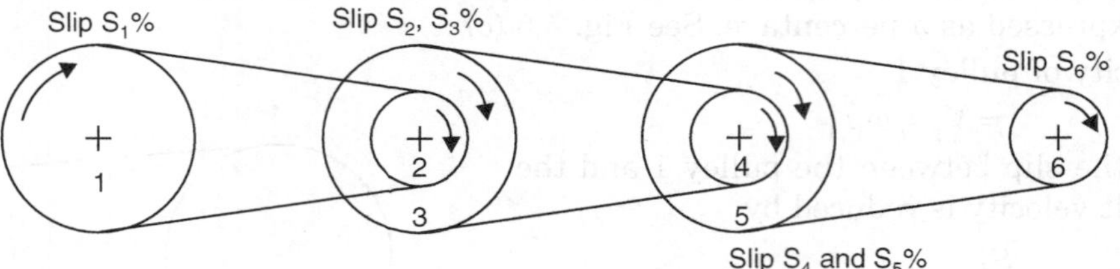

Fig. 5.5(d) Compound drive with slip.

Using eqn. (5.4), velocity ratio

$$\frac{\omega_2}{\omega_1} = \frac{r_1}{r_2}\left(1 - \frac{S_1}{100}\right)\left(1 - \frac{S_2}{100}\right) \quad \ldots(vii)$$

Similarly

$$\frac{\omega_4}{\omega_3} = \frac{r_3}{r_4}\left(1 - \frac{S_3}{100}\right)\left(1 - \frac{S_4}{100}\right) \quad \ldots(viii)$$

and

$$\frac{\omega_6}{\omega_5} = \frac{r_5}{r_6}\left(1 - \frac{S_5}{100}\right)\left(1 - \frac{S_6}{100}\right) \quad \ldots(ix)$$

Multiplying eqn. (*vii*), (*viii*) and (*ix*)

$$\frac{\omega_2}{\omega_1} \times \frac{\omega_4}{\omega_3} \times \frac{\omega_6}{\omega_5} = \frac{r_1}{r_2} \times \frac{r_3}{r_4} \times \frac{r_5}{r_6}\left(1 - \frac{S_1}{100}\right)\left(1 - \frac{S_2}{100}\right)\left(1 - \frac{S_3}{100}\right)\cdots\left(1 - \frac{S_6}{100}\right)$$

$$\frac{\omega_2}{\omega_1} \times \frac{\omega_4}{\omega_3} \times \frac{\omega_6}{\omega_5} = \frac{r_1\, r_3\, r_5}{r_2\, r_4\, r_6}\left(1 - \frac{S_1 + S_2}{100} + \frac{S_1 S_2}{100 \times 100}\right)\left(1 - \frac{S_3}{100}\right)$$

Since, pulleys 2 and 3 ; and pulleys 4 and 5 keyed to the same shaft

$\therefore$ $\omega_2 = \omega_3$; and also $\omega_4 = \omega_5$ and neglecting small terms

We get

$$VR = \frac{\omega_6}{\omega_1} = \frac{r_1\, r_3\, r_5}{r_2\, r_4\, r_6}\left(1 - \frac{S_1 + S_2 + \ldots\ldots + S_6}{100}\right) \quad \ldots(5.6)$$

Also,

$$VR = \frac{\omega_6}{\omega_1} = \frac{r_1\, r_3\, r_5}{r_2\, r_4\, r_6}\left(1 - \frac{S}{100}\right) \quad \ldots(5.7)$$

where $S = S_1 + S_2 + \ldots\ldots + S_6$ = total percentage slip in the compound drive

Also, $$\frac{\omega_6}{\omega_1} = \frac{N_6}{N_1} = \frac{d_1\, d_3\, d_5}{d_2\, d_4\, d_6}\left(1 - \frac{S}{100}\right)$$

$$\therefore \qquad \boxed{VR = \frac{N_6}{N_1} = \frac{d_1\, d_3\, d_5}{d_2\, d_4\, d_6}\left(1 - \frac{S}{100}\right)} \qquad \text{...(5.8)}$$

5.6. LENGTH OF BELT

Case I. Open belt drive

Let O_1 and O_2 = centres of two pulleys,

r_1 and r_2 = radii of the pulley 1 and pulley 2 respectively.

l = distance between centres O_1 and O_2

L_o = total length of the belt of open drive ;

Draw O_2A_1, parallel to AB, from the geometry of the Fig. 5.6 (a).

Points A, B, C and D are points of tangency.

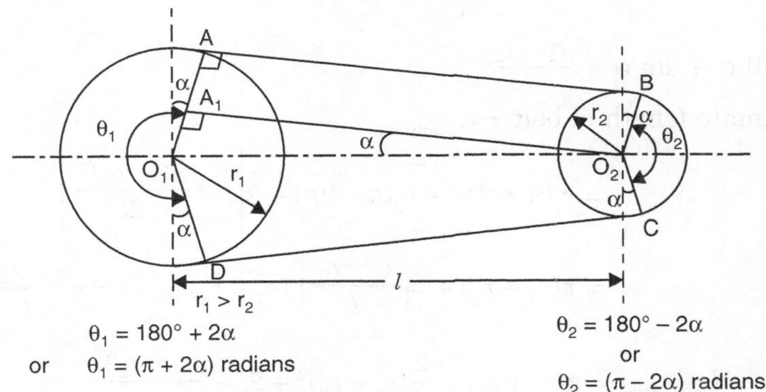

Fig. 5.6 (a) Length of belt for open drive.

Draw a line O_2A_1 parallel to AB which is $\perp^{ar}$ to O_1A.

$\therefore \qquad \angle A_1O_2O_1 = \alpha$

Angle subtended by arc AD on the centre O_1

$$= \theta_1 = (180 + 2\alpha)^\circ = \pi + 2\alpha \text{ radians} \qquad \text{...(5.9)}$$

Similarly, angle subtended by the arc BC on the centre O_2

$$= \theta_2 = (180 - 2\alpha)^\circ = \pi - 2\alpha \text{ radians} \qquad \text{...(5.10)}$$

Since $O_1A_1 \perp^{ar}$ to O_2A_1, $\therefore$ $\sin\alpha = \dfrac{O_1A_1}{O_1O_2} = \dfrac{O_1A - AA_1}{O_1O_2} = \dfrac{O_1A - O_2B}{O_1O_2}$

or $$\sin\alpha = \frac{r_1 - r_2}{l} \qquad \text{...(5.11)}$$

$\therefore$ Length of belt for open drive, L_0 = Arc AD + Arc BC + Free length AB + Free length CD ...(5.12)

or $$L_o = r_1\,\theta_1 + r_2\theta_2 + 2AB \qquad \text{...(5.13)}$$

or $$L_o = r_1(\pi + 2\alpha) + r_2(\pi - 2\alpha) + 2l \cos \alpha \qquad ...(5.14)$$

Eqns. (5.13) and (5.14) give the exact length of belt for the open drive.

Approximate length of belt for open drive

$\because$ $$\sin \alpha = \frac{r_1 - r_2}{l}$$

we get $$\cos \alpha = \sqrt{1 - \sin^2 \alpha} = \sqrt{1 - \left(\frac{r_1 - r_2}{l}\right)^2}$$

Expanding by using Binomial theorem

we get $$\cos \alpha = 1 - \frac{1}{2}\left(\frac{r_1 - r_2}{l}\right)^2 + \frac{1}{4}\left(\frac{r_1 - r_2}{l}\right)^4 ......$$

neglecting small terms, we get

$$\cos \alpha \simeq 1 - \frac{1}{2}\left(\frac{r_1 - r_2}{l}\right)^2$$

$\therefore$ $$l \cos \alpha = l - \frac{(r_1 - r_2)^2}{2l}$$

and since α is small $\alpha \simeq \sin \alpha \simeq \dfrac{r_1 - r_2}{l}$

$\therefore$ Approximate length of belt = $L_{o\ approx.}$

$$= r_1(\pi + 2\alpha) + r_2(\pi - 2\pi) + 2\left[l - \frac{(r_1 - r_2)^2}{2l}\right]$$

$$= \pi(r_1 + r_2) + 2\left(\frac{r_1 - r_2}{l}\right)(r_1 - r_2) + \left(2l - \frac{(r_1 - r_2)^2}{l}\right)$$

$\therefore$ $$L_{o\ approx} = \pi(r_1 + r_2) + \frac{2}{l}(r_1 - r_2)^2 + 2l - \frac{(r_1 - r_2)^2}{l}$$

$$\boxed{L_{o\ approx} = \pi(r_1 + r_2) + 2l + \frac{(r_1 - r_2)^2}{l}} \qquad ...(5.15)$$

Case II. Cross belt

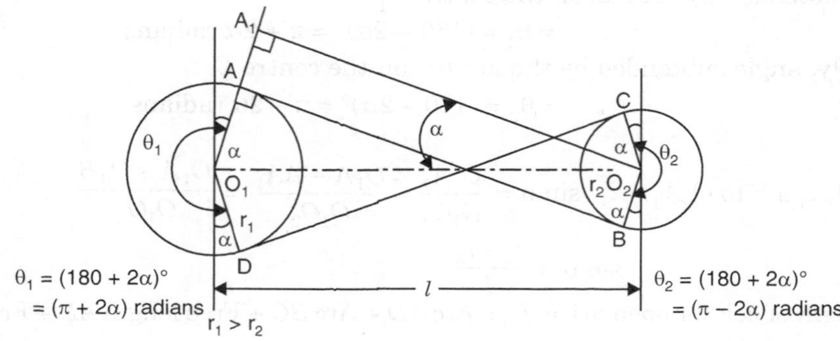

Fig. 5.6(b) Length of belt for crossed drive.

Draw O_2A_1 parallel to AB, which is perpendicular to O_1A.

Since O_2A_1 is perpendicular to O_1A

$\therefore$ $\angle O_1O_2A_1 = \alpha$

$$\therefore \quad \theta_1 = \theta_2 = (180 + 2\alpha)^\circ \qquad \text{...(5.16)}$$

or
$$\theta_1 = \theta_2 = (\pi + 2\alpha) \text{ radians} \qquad \text{...(5.17)}$$

where
$$\sin\alpha = \frac{O_1A_1}{O_1O_2} = \frac{O_1A + AA_1}{O_1O_2} = \frac{O_1A + O_2B}{O_1O_2} = \frac{r_1 + r_2}{l}$$

or
$$\sin\alpha = \left(\frac{r_1 + r_2}{l}\right) \qquad \text{...(5.18)}$$

$\therefore$ Length of belt for cross drive

$$= L_c = \text{Arc } AD + \text{Arc } BC + \text{free length } AB + \text{free length } CD \qquad \text{...(5.19)}$$

$$= r_1\theta_1 + r_2\theta_2 + 2AB \qquad | \because \; AB = CD$$

where
$$\theta_1 = \theta_2 = \theta = \pi + 2\alpha \text{ for cross drive}$$

$$\therefore \quad L_c = r_1\theta + r_2\theta + 2l\cos\alpha$$

$$\boxed{L_c = (r_1 + r_2)\theta + 2l\cos\alpha} \qquad \text{...(5.20)}$$

$$= (r_1 + r_2)(\pi + 2\alpha) + 2l\cos\alpha$$

or
$$L_c = \pi(r_1 + r_2) + 2\alpha(r_1 + r_2) + 2l\cos\alpha \qquad \text{...(5.21)}$$

Eqns. (5.19), (5.20) and (5.21) give exact length of belt for crossed drive.

Approximate length of belt for cross drive

Since
$$\sin\alpha = \frac{r_1 + r_2}{l}$$

$$\therefore \quad \cos\alpha = \sqrt{1 - \sin^2\alpha}$$

$$\simeq 1 - \frac{1}{2}\sin^2\alpha + \frac{1}{4}\sin^4\alpha \text{ ......}$$

$$\simeq 1 - \frac{1}{2}\sin^2\alpha$$

$$\simeq 1 - \frac{1}{2}\left(\frac{r_1 + r_2}{l}\right)^2$$

And, $\therefore$
$$2l\cos\alpha = 2l - \frac{(r_1 + r_2)^2}{l}$$

also since $\sin\alpha$ is very small $\therefore$ $\alpha \simeq \sin\alpha \simeq \dfrac{r_1 + r_2}{l}$

$$\therefore \quad L_{c\text{ approx}} = \pi(r_1 + r_2) + 2\left(\frac{r_1 + r_2}{l}\right)(r_1 + r_2) + 2l - \frac{(r_1 + r_2)^2}{l}$$

$$= \pi(r_1 + r_2) + 2\frac{(r_1 + r_2)^2}{l} + 2l - \frac{(r_1 + r_2)^2}{l}$$

$$\boxed{L_{c\text{ approx}} = \pi(r_1 + r_2) + \frac{(r_1 + r_2)^2}{l} + 2l} \qquad \text{...(5.22)}$$

Note. 1. In the open belt drive the difference between the exact length L_o and the approximate length $L_{o\ approx}$ lies between 2% to 5% of the exact length.

2. In case of cross drive α is not small and therefore the difference between the exact length L_c and the approximate length $L_{c\ approx}$ will be substantial between 2% to 6%.

3. Exact length is always greater than the approximate length for both open drive and cross drives.

5.7. DIRECTION OF ROTATION OF DRIVEN PULLEY

(1) **Open drive.** Fig. 5.7 (*a*) shows an open belt drive with pulley 1 as driver and pulley 2 as driven.

When driving pulley rotates in the clockwise direction the endless belt connecting the two pulleys moves from *A* towards *B* to reach pulley 2. The belt leaving pulley 2 returns back to the driving pulley 1 at point *D*. Since, the belt leaving pulley 2 moves from *C* to *D*, therefore the tension in the portion *CD* is greater than that in the portion *AB*. Therefore portion *CD* is known as **tight side** and portion *AB* is known as **slack side** as shown in Fig. 5.7 (*a*). The two pulleys rotate in the **same** direction.

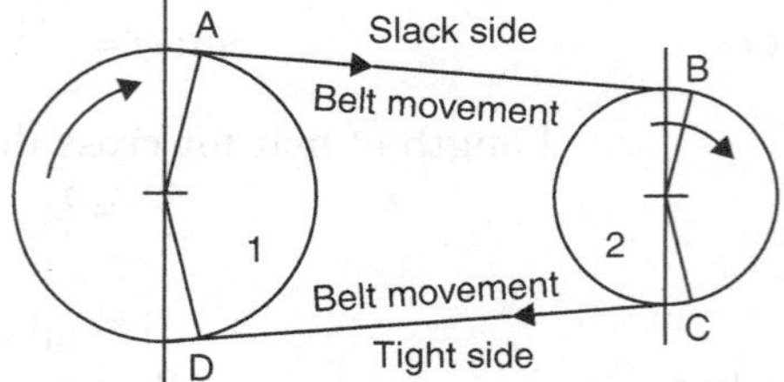

(a) Direction of rotation of driven pulley for open drive.

(2) **Crossed drive.** Fig. 5.7 (*b*) In this case also the driving pulley rotates in the clockwise direction. Therefore, the belt leaving pulley 1 and reaching pulley 2 moves from *A* to *B*. Similarly, the belt leaving pulley 2 and reaching pulley 1 is subjected to larger tension. Therefore, the greater tension acts in the belt in the direction from *C* to *D*. This will rotate the driven pulley 2 in the counter clockwise (*CCW*) direction. Thus, in the crossed drive, the two pulleys rotate in *opposite* direction.

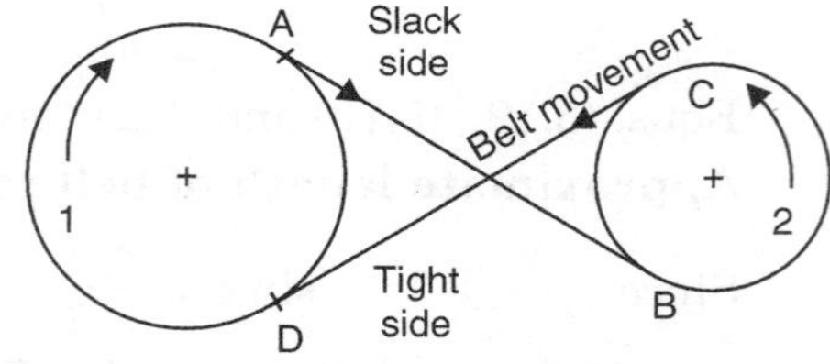

(b) Direction of rotation of driven pulley for crossed drive.

Fig. 5.7. Direction of rotation of driven pulley.

5.8. POWER TRANSMITTED BY A BELT DRIVE

As shown in the Fig. 5.8. pulley *A* is a driving pulley and pulley *B* is the follower. The driving pulley pulls the belt from one side and delivers the same to the other. The tension in the former side (*i.e.,* tight side) will be more than that in the latter side (*i.e.,* slack side).

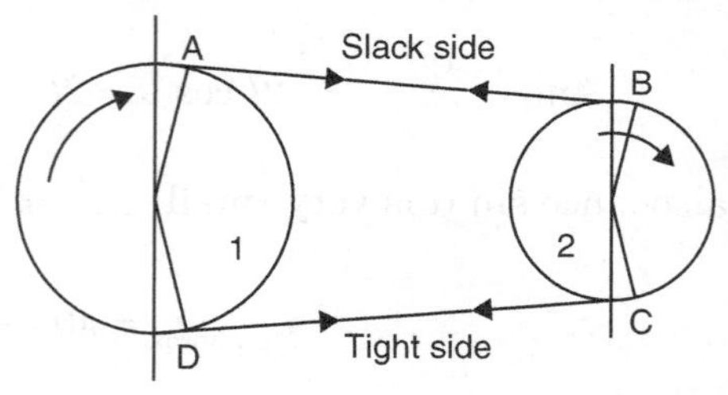

Fig. 5.8. Power transmitted by the belt drive.

Let T_1 = tension in the tight side,

T_2 = tension in the slack side,

V_b = velocity of the belt, m/s :

Effective driving force at the circumference of the follower = $T_{eff} = (T_1 - T_2)$ ...(5.23)

$\therefore$ Work done = force × distance

Work done = $(T_1 - T_2) \times V_b\, N$ m/s ...(5.24)

Power = $(T_1 - T_2) \times V_b$ J/s or Watts

$$\boxed{\text{Power} = (T_1 - T_2)V_b} \quad \text{Watts.} \qquad \ldots(5.25)$$

5.9. RATIO OF BELT TENSIONS FOR FLAT BELT

Let T_1 = tension on the tight side

T_2 = tension on the slack side

R = radial reaction of the pulley on the belt

μ = co-efficient of friction between the belt and the pulley rim.

θ = angle of lap of the belt round the pulley in radians = $\angle AOB$ (Clockwise)

Consider an element *mn* which subtends an angle $\delta\phi$ at the centre, as shown in Fig. 5.9 (*a*).

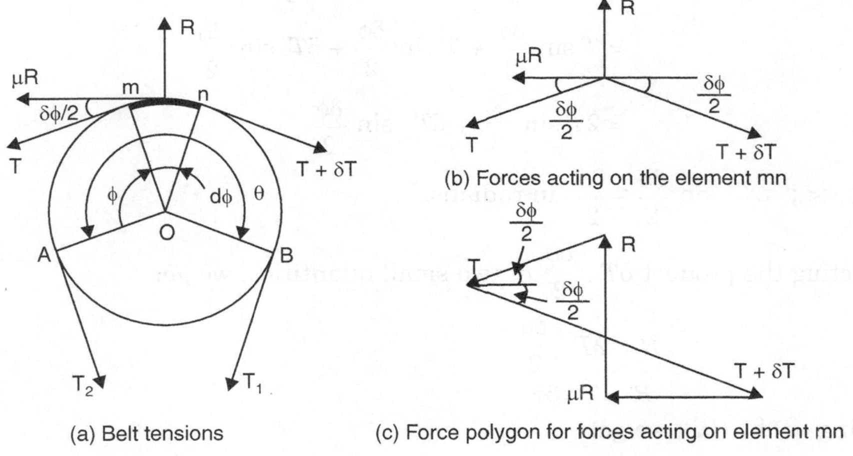

(a) Belt tensions

(b) Forces acting on the element mn

(c) Force polygon for forces acting on element mn

Fig. 5.9. Belt tension ratio for flat belt.

The forces acting on the element *mn* of the belt are shown in Fig. 5.9 (*a*). The small element *mn* subtends an angle $\delta\phi$ at the centre of the pulley and the forces acting on it are :

(1) Tension T on the Slack side, at point *m*

(2) Tension $T + \delta T$ on the tight side at point *n*,

(3) The normal reaction R and (4) tangential friction force μR between the belt element and the rim of pulley. When length of *mn* or $\delta\phi$ is very small these forces become con-current as shown in Fig. 5.9 (*b*). The force polygon for the above forces is shown in Fig. 5.9 (*c*).

When the motion is impending the frictional force becomes maximum and it is equal to μR.

Resolving the forces tangentially, and using the condition $\Sigma F_t = 0$, we have

$$(T + \delta T) \cos \frac{\delta\phi}{2} - T \cos \frac{\delta\phi}{2} - \mu R = 0$$

or,
$$T \cos \frac{\delta\phi}{2} + \delta T \cos \frac{\delta\phi}{2} - T \cos \frac{\delta\phi}{2} - \mu R = 0$$

$\therefore$
$$\mu R = \delta T \,.\, \cos \frac{\delta\phi}{2} \qquad \ldots(i)$$

Since $\frac{\delta\phi}{2}$ is very small $\cos \frac{\delta\phi}{2} \simeq 1$

$\therefore$
$$\mu R = \delta T \qquad \ldots(ii)$$

Resolving the forces radially and using the condition $\Sigma F_r = 0$, we have,

$$R - T \sin \frac{\delta\phi}{2} - (T + \delta T) \sin \frac{\delta\phi}{2} = 0$$

or
$$R = T \sin \frac{\delta\phi}{2} + (T + \delta T) \sin \frac{\delta\phi}{2}$$

$$= T \sin \frac{\delta\phi}{2} + T \sin \frac{\delta\phi}{2} + \delta T. \sin \frac{\delta\phi}{2}$$

$$= 2T \sin \frac{\delta\phi}{2} + \delta T \,.\, \sin \frac{\delta\phi}{2}$$

$\therefore$ $\delta\phi <<$; $\therefore$ $\sin \frac{\delta\phi}{2} \simeq \frac{\delta\phi}{2}$ in radians.

Neglecting the product $\delta T \,.\, \frac{\delta\phi}{2}$ of two small quantities, we get

$$R = 2T \,.\, \frac{\delta\phi}{2}$$

$$R = T \,.\, \delta\phi \qquad \ldots(iii)$$

Dividing (*ii*) by (*iii*), we get

$$\frac{\mu R}{R} = \frac{\delta T}{T \,.\, \delta\phi}$$

or
$$\mu \delta\phi = \frac{\delta T}{T}$$

Integrating between the limits 0 to ϕ, and from T_2 to T_1,

We get
$$\int_0^{\theta} \mu \, d\phi = \int_{T_2}^{T_1} \frac{dT}{T}$$

$$\mu\theta = \log_e \frac{T_1}{T_2}$$

$$e^{\mu\theta} = \frac{T_1}{T_2} \qquad \ldots(iv)$$

or $$\boxed{\frac{T_1}{T_2} = e^{\mu\theta}}$$ ...(5.26)

(1) In the eqn. (5.26), θ is the angle of contact at the smaller pulley for the open drive.

Or, for open belt drive, the angle of contact $\theta = (180 - 2\alpha)° \times \frac{\pi}{180°}$ radians

where $\alpha = \sin^{-1}\left(\frac{r_1 - r_2}{l}\right)$ ...[5.27(*a*)]

and l is distance between centres of the two pulleys in the drive.

(2) In a cross belt drive, the angle of contact $\theta = (180 + 2\alpha)° \times \frac{\pi}{180°}$ radians

where $\alpha = \sin^{-1}\left(\frac{r_1 + r_2}{l}\right)$ ...[5.27(*b*)]

θ is the same for both pulleys.

5.10. TENSION RATIO FOR V-BELT OR ROPE

A *V*-grooved pulley is shown in the above Fig. 5.10 (*a*).

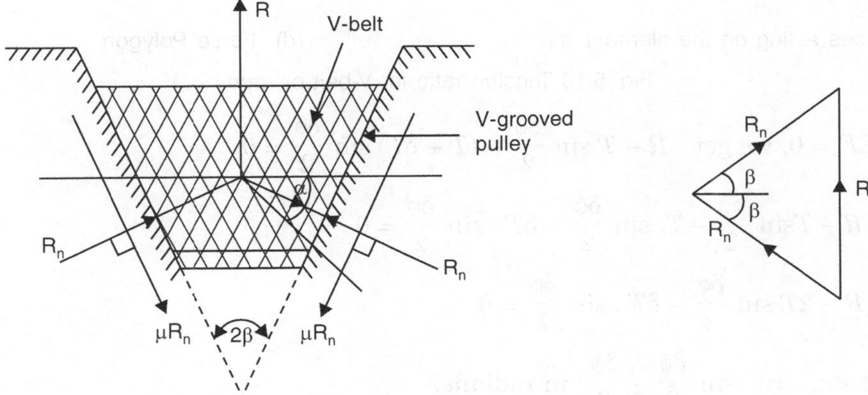

(a) Forces acting on grooved pulley

(b) Relation between radial force R and the force R_n normal to the groove

Fig. 5.10

R_n = normal reaction offered by the sides of the *V*-grooved pulley to the belt or rope.

2β = groove angle

μ = co-efficient of friction between belt and pulley.

R = radial reaction of pulley to belt or rope.

From Fig. 5.10(*b*), resolving forces vertically [*i.e.*, by $\Sigma F_v = 0$], we get

$$R = R_n \sin\beta + R_n \sin\beta = 2\, R_n \sin\beta$$

$$R_n = \frac{R}{2 \sin \beta} \quad \text{...(i)}$$

Frictional force, $F = \mu R_n + \mu R_n = 2\mu \times R_n$

$$\therefore \quad F = \frac{\mu R}{\sin \beta} \quad \text{...(ii)}$$

Forces acting on the element *mn* are shown in Fig. 5.10 (*c*) and the corresponding force polygon is shown in Fig. 5.10 (*d*).

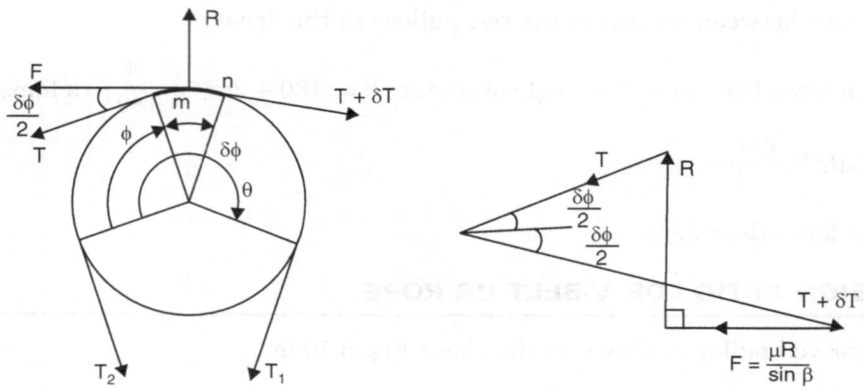

(c) Forces acting on the element *mn* (d) Force Polygon

Fig. 5.10 Tension ratio for V-belt or rope.

By $\Sigma F_r = 0$, we get $R - T \sin \frac{\delta\phi}{2} - (T + \delta T) \sin \frac{\delta\phi}{2} = 0$

or, $R - T\sin \frac{\delta\phi}{2} - T. \sin \frac{\delta\phi}{2} - \delta T \,.\, \sin \frac{\delta\phi}{2} = 0$

or, $R - 2T \sin \frac{\delta\phi}{2} - \delta T \,.\, \sin \frac{\delta\phi}{2} = 0$ ...(*iii*)

$\because \; \delta\phi <<, \; \therefore \; \sin \frac{\delta\phi}{2} \simeq \frac{\delta\phi}{2}$ in radians,

$\therefore$ (*iii*) becomes, $R - 2T \,.\, \frac{\delta\phi}{2} - \delta T \,.\, \frac{\delta\phi}{2} = 0$

Neglecting $\frac{\delta T \,.\, \delta\phi}{2}$, we get

$$R = T\delta\phi \quad \text{...(iv)}$$

By $\Sigma F_t = 0$, we get

$$(T + \delta T) \cos \frac{\delta\phi}{2} - T \cos \frac{\delta\phi}{2} - F = 0$$

or, $$T \cos \frac{\delta\phi}{2} + \delta T \,.\, \cos \frac{\delta\phi}{2} - T .\cos \frac{\delta\phi}{2} = F$$

$\because$ $\delta\phi <<,$ $\therefore$ $\cos \frac{\delta\phi}{2} \simeq 1,$

$\therefore$ we get $$\delta T = F \qquad \ldots(v)$$

or, $$\delta T = \frac{\mu R}{\sin \beta} \quad \text{or} \quad \frac{\mu R}{\sin \beta} = \delta T \qquad \ldots(vi)$$

By $(vi) \div (iv)$, we get

$$\frac{\mu R}{(\sin \beta) \,.\, R} = \frac{\delta T}{T \,.\, \delta\phi}$$

or $$\frac{\mu \delta\phi}{\sin \beta} = \frac{\delta T}{T} \qquad \ldots(vii)$$

Integrating (vii), we get

$$\int_0^{\theta} \frac{\mu d\phi}{\sin \beta} = \int_{T_2}^{T_1} \frac{dT}{T}$$

$$\therefore \quad \frac{\mu\theta}{\sin \beta} = \log_e \frac{T_1}{T_2} \qquad \ldots(viii)$$

$$\therefore \quad \frac{T_1}{T_2} = e^{\frac{\mu\theta}{\sin \beta}} = e^{\mu'\theta} \qquad \ldots(5.28)$$

where $$\mu' = \frac{\mu}{\sin \beta} \qquad \ldots(5.29)$$

5.11. CENTRIFUGAL TENSION

In sections 5.9 and 5.10 only the effect of friction between belt and pulley was considered.

Due to the mass and the speed of the belt, the belt is subjected to centrifugal force, which tends to lift the belt from the pulley rim, thereby reducing the normal reaction and hence the frictional resistance. The centrifugal force produces additional tension in the belt which is known as **centrifugal tension**. At low speeds this tension is negligible, but at speeds more than, 750 m/min, or 15 m/s it becomes a very important factor.

Consider a small portion mn of the belt as shown in Fig. 5.11(a).

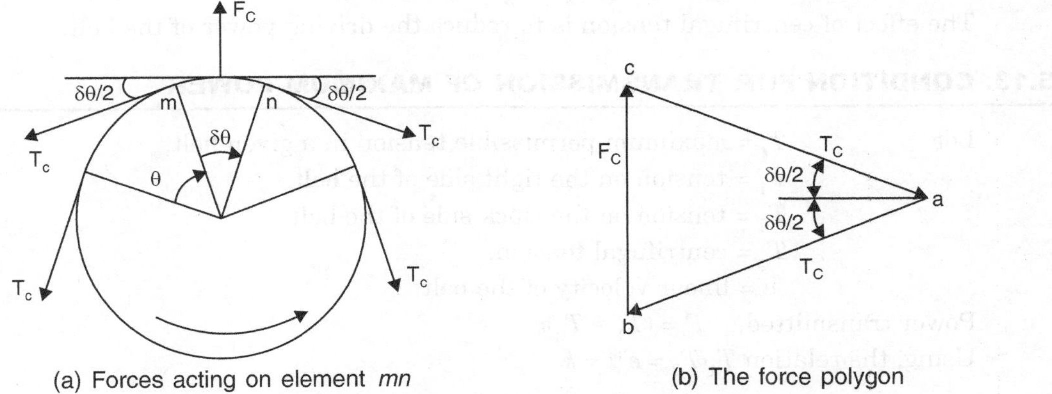

(a) Forces acting on element mn (b) The force polygon

Fig. 5.11. Centrifugal tension.

Let m = mass of the per metre length of belt, kg/m

v = linear velocity of the belt, m/s

r = radius of the pulley over which the belt runs, m

T_c = centrifugal tension acting tangentially at all fibres in the belt.

$\delta\theta$ = angle subtended by the belt element mn at the centre of the pulley,

F_c = centrifugal force, N

Length of the belt $mn = \delta_s = r \,.\, \delta\theta$ ...(i)

and, mass of element, $\delta_m = m.(r\,\delta\theta)$ ...(ii)

Centrifugal force on the belt element

$$F_c = \delta_m \times \frac{v^2}{r}$$

or

$$F_c = (m\,.\,r.\,\delta\theta)\,.\,\frac{v^2}{r}$$

or,

$$F_c = m\,.\,v^2.\,\delta\theta \qquad ...(iii)$$

Fig. 5.11 (*b*) shows the forces acting on the small element mn of belt due to centrifugal force at radius r.

Resolving the forces (F_c and T_c) vertically, and equating the same to zero, we get

$$2T_c \sin\frac{\delta\theta}{2} = F_c$$

Since $\delta\theta$ is very small, $\sin\frac{\delta\theta}{2} = \frac{\delta\theta}{2}$,

$$2T_c\frac{\delta\theta}{2} = F_c \qquad ...(iv)$$

$$T_c\,.\,\delta\theta = m\,.\,v^2\,.\,\delta\theta \quad \text{from } (iii) \text{ and } (iv)$$

$$\therefore \qquad T_c = m.v^2 \qquad ...(5.30)$$

From the eqn. (5.30), we find that the centrifugal tension is independent of the angle of lap, and the tensions in the slack and tight side. It only depends on the velocity of the belt.

5.12. TOTAL TENSIONS IN BELT DRIVE

If T_c is the centrifugal tension in belt,

total tension on the tight side, $T_t = T_1 + T_c$ ...(5.31)

total tension on the slack side, $T_s = T_2 + T_c$ ...(5.32)

The effect of centrifugal tension is to reduce the driving power of the belt.

5.13. CONDITION FOR TRANSMISSION OF MAXIMUM POWER

Let T_t = maximum permissible tension in a given belt,

T_1 = tension on the tight side of the belt.

T_2 = tension on the slack side of the belt.

T_c = centrifugal tension.

v = linear velocity of the belt.

Power transmitted, $P = (T_1 - T_2)v$ ...(5.33)

Using, the relation $T_1/T_2 = e^{\mu\theta} = k$ ...(i)

we get, $$P = T_1\left(1-\frac{T_2}{T_1}\right)v = T_1\left(1-\frac{1}{e^{\mu\theta}}\right)v$$

or $$P = T_1\left(1-\frac{1}{k}\right)v = T_1(k-1)\frac{v}{k} \quad \text{...(ii)}$$

But, $$T_t = T_1 + T_c$$

$\therefore$ $$T_1 = T_t - T_c$$

$$P = (T_t - T_c)\left(1-\frac{1}{e^{\mu\theta}}\right)v_b \quad \text{or,} \quad P = (T_t - T_c)(k-1)\frac{v}{k}$$

or $$P = (T_t.\, v - mv^3)\frac{(k-1)}{k} \quad \text{...(5.34)}$$

For finding the maximum power (P_{max}), use the condition $\frac{dP}{dv} = 0$.

$\therefore$ $$\frac{dP}{dv} = \frac{d}{dv}\left[(T_t.v - mv^3)\frac{(k-1)}{k}\right] = 0$$

or, $$(T_t - 3mv^2)\frac{(k-1)}{k} = 0 \quad \text{...(iii)}$$

$\because$ $k \neq 1$ and $k \neq \infty$,

$\therefore$ $T_t - 3mv^2 = 0$

$\therefore$ Maximum tension, $T_t = 3mv^2 = 3T_c$ ...(5.35)

And, the corresponding belt speed,

$$v = \sqrt{\frac{T_t}{3m}} \quad \text{...(5.36)}$$

Hence, the power transmitted shall be maximum

when, (*a*) the centrifugal tension is $\frac{1}{3}$rd of the belt strength, *i.e.*, when $T_c = T_t/3$ and (*b*) the belt runs at velocity

$$v_1 = \sqrt{\frac{T_t}{3m}} \quad \text{...(5.37)}$$

Eqn. 5.34 represents a cubic parabola shown in Fig. 5.13 (*a*).

Maximum power (P_{max}) is transmitted by the drive at the belt velocity

$$v_1 = \sqrt{\frac{T_t}{3m}}$$

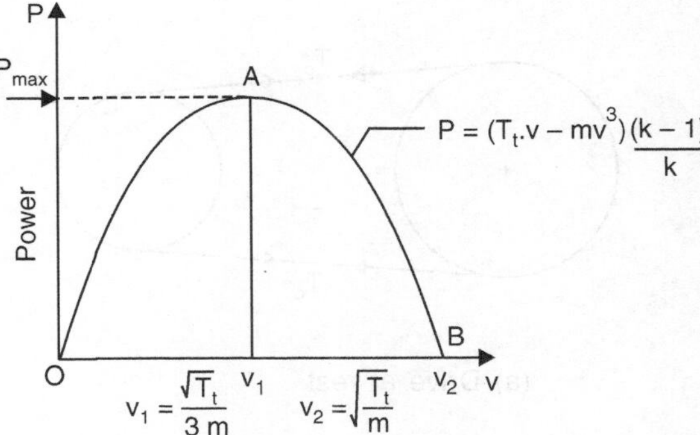

Fig. 5.13 (a). Variation of power transmitted by the belt drive as a function of the belt velocity.

Power increases with belt velocity till it becomes P_{max} at v_1 (as given by point *A*). As the belt velocity increases further the power goes on decreasing and it becomes zero at point *B*. Thus, from Eqn. 5.34, for zero power,

$$P = (T_t.\, v - mv^3)\frac{(k-1)}{k} = 0$$

$\because$ $k \neq 1$ and $k \neq \infty$,

$\therefore \quad T_t \cdot v - mv^3 = 0$

or $\quad T_t = mv^2 \quad$ for zero power ....(5.38)

i.e., for $\quad P = 0$

$\therefore$ At point B, if belt velocity is v_2, we have

$$mv_2^2 = T_t$$

or
$$v_2 = \sqrt{\frac{T_t}{m}} \quad ...(5.39)$$

On comparing eqns. (5.37) and (5.39), it is evident that

$$\frac{v_2}{v_1} = \frac{\sqrt{T_t/m}}{\sqrt{T_t/3m}} = \sqrt{3} \quad ...(5.40)$$

At point B belt velocity $= v_2 = \sqrt{T_t/m}$, and the no power is transmitted by the drive. This is because at the belt velocity v_2 there is no contact between belt and the pulley, as shown in Fig. 5.13(*b*). The driven pulley 2 does not rotate at belt velocity $v_2 = \sqrt{T_t / m}$.

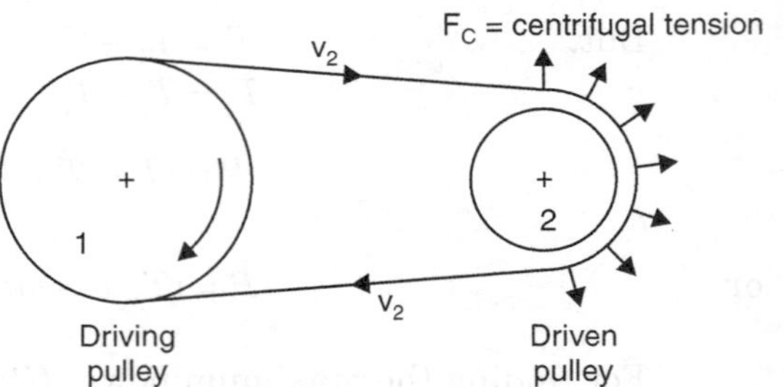

Fig. 5.13(b). At velocity $v_2 = \sqrt{\frac{T_t}{m}}$, there is no contact between belt and the driven pulley.

5.14. INITIAL TENSION

The motion of the belt over the driver and the follower is governed by a firm grip due to friction between the belt and the pulleys. In order to increase this grip the belt is tightened up. Even when the pulleys are stationary, the belt is subjected to some tension, called initial tension (T_0).

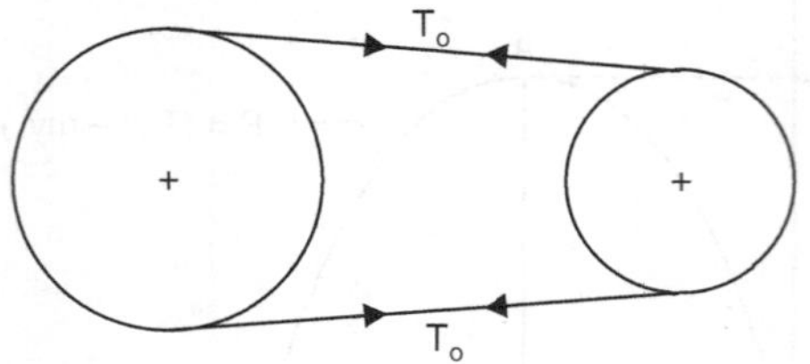

(a) Drive at rest

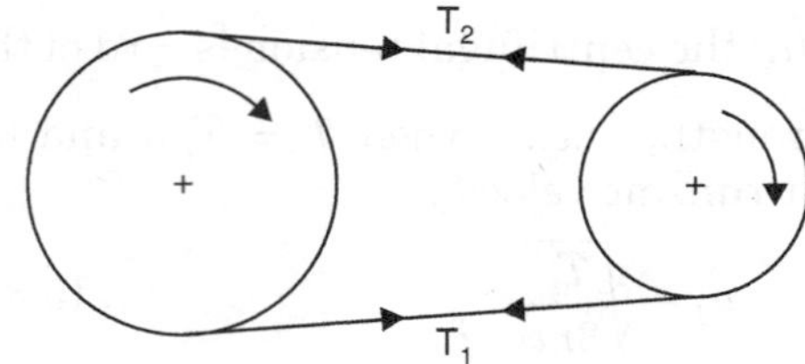

(b) Drive transmitting power

Fig. 5.14. Initial tension.

Let, initial tension in the belt $= T_0$

Tension in the tight side of the belt $= T_1$

Tension in the slack side of belt $= T_2$

When the power is supplied to one of the pulleys and transmitted to the other, the tensions in the two free lengths of the belt will be changed.

The tight side of the belt stretches until the pull is increased from T_0 to T_1 while the slack side shortens until the pull is decreased from T_0 to T_2.

Since the length of the belt remains unchanged, the mean tension must also remain unchanged.

i.e., since $\delta l_{\text{tight side}} = \delta l_{\text{slack side}}$ or $\dfrac{(T_1 - T_0)\, l}{AE} = \dfrac{(T_0 - T_2)\, l}{AE}$

or $(T_1 - T_0) = (T_0 - T_2)$

or $2T_0 = T_1 + T_2$

$$\therefore \quad T_0 = \frac{T_1 + T_2}{2} \qquad \text{...(5.41)}$$

If T_c is considered, then eqn. (5.41) becomes,

$$T_o = \frac{T_1 + T_2 + 2T_c}{2} \qquad \text{...(5.42)}$$

5.15. ANGLE OF LAP OF CONTACT

When two pulleys are connected by an open belt drive and they are of unequal sizes, the angle θ must be taken as minimum angle of lap of contact *i.e.*, angle of contact on the smaller pulley.

Obviously slip will occur on the smaller pulley where frictional force is smaller due to smaller angle of contact.

Thus for open base drive : $\theta = (180 - 2\alpha)° = (\pi - 2\alpha)$ radians ...(5.43)

where $\sin \alpha = \dfrac{r_1 - r_2}{l}$ ...(5.44)

For crossed belt drive,

$$\theta = (\pi + 2\alpha)\text{radians} \qquad \text{...(5.45)}$$

where $\sin \alpha = \dfrac{r_1 + r_2}{l}$ ...(5.46)

Problem 5.1. *An engine shaft running at 250 rpm is required to drive a machine shaft by means of a belt. The pulley on the engine shaft is of 2 m diameter and that of the machine shaft is of 1 m. If the belt thickness is 4.5 mm, find the speed of the belt when :*

(i) there is no slip, and

(ii) there is a slip of 2 percent.

Sol. Given : speed of the shaft, $N_1 = 250$ rpm.

Diameter of the pulley on the engine shaft, $d_1 = 2$ m,

Diameter of the pulley on the machine shaft, $d_2 = 1$ m,

Thickness of the belt, $t = 4.5$ mm $= 0.0045$ m.

Speed of the machine shaft, $N_2 = ?$

Case I. There is no slip :

We know that, $\dfrac{N_2}{N_1} = \dfrac{d_1 + t}{d_2 + t}$

On putting values of N_1, N_2, d_1, d_2 and t in the above equation, we get

$$\frac{N_2}{250} = \frac{2 + 0.0045}{1 + 0.0045} = \frac{2.0045}{1.0045}$$

$$\therefore \qquad N_2 = \frac{2.0045}{1.0045} \times 250 = 498.880 \text{ rpm.}$$

Case II. There is a slip of 2%.

We know that $\dfrac{N_2}{N_1} = \dfrac{d_1 + t}{d_2 + t}\left(\dfrac{100 - S}{100}\right)$

On putting values of N_1, d_1, d_2, t and S in the above equation, we get

$$\frac{N_2}{250} = \frac{2 + 0.0045}{1 + 0.0045}\left(\frac{100 - 2}{100}\right) = \frac{2.0045}{1.0045}\left(\frac{98}{100}\right)$$

$$\therefore \qquad N_2 = \frac{2.0045}{1.0045} \times \frac{98}{100} \times 250 = 488.90 \text{ rpm.}$$

Problem 5.2. *Two parallel shafts 10 metres apart are provided with 800 mm and 250 mm diameter pulleys and are connected by means of a cross belt.*

The direction of rotation of the follower pulley is to be reversed by changing over to an open belt drive. How much length of the belt has to be reduced ?

Sol. Given : Radius of the larger pulley,

$$r_1 = \frac{800}{2} = 400 \text{ mm} = 0.40 \text{ mm}$$

Radius of the smaller pulley,

$$r_2 = \frac{250}{2} = 125 \text{ mm} = 0.125 \text{ m}$$

Distance between the centres of the two pulleys, $l = 10$ m,

We know that, length of cross belt,

$$L_c = \pi(r_1 + r_2) + \frac{(r_1 + r_2)^2}{l} + 2l$$

On putting values of r_1, r_2, and l in the above equation,

we get $$L_c = \pi(0.40 + 0.125) + \frac{(0.40 + 0.125)^2}{10} + 2 \times 10$$

$$= 1.649 + 0.027 + 20 = 21.676 \text{ m.}$$

Length of open belt,

$$L_o = \pi(r_1 + r_2) + \frac{(r_1 - r_2)^2}{l} + 2l$$

On putting values of r_1, r_2 and l in the above equation, we get

$$L_o = \pi(0.40 + 0.125) + \frac{(0.40 - 0.125)^2}{10} + 2 \times 10$$

$$= 1.649 + 7.562 \times 10^{-3} + 20 = 21.656 \text{ m}$$

$\therefore$ The length of the belt to be reduced,

$$= L_c - L_o$$

$$= 21.676 - 21.656 = 0.02 \text{ m} = 20 \text{ mm.}$$

Problem 5.3. *Two pulleys of diameters 250 mm and 700 mm mounted on two parallel shafts 2 m apart, are connected by leather belt 200 mm wide. If the maximum safe tension of the belt is 15 N/mm width, determine the power transmitted in case of (i) open belt drive, (ii) cross belt-drive.*

Assume the speed of the belt as 500 m/min. and the co-efficient of friction between the belt and the pulley rim, 0.30.

Sol. Radius of the larger pulley,

$$r_1 = \frac{700}{2} = 350 \text{ mm}, = 0.350 \text{ m}.$$

Radius of the smaller pulley,

$$r_2 = \frac{250}{2} = 125 \text{ mm} = 0.125 \text{ m}.$$

Distance between the centres of the two shafts, l = 2m.

Maximum safe tension $= f = 15$ N/mm width of belt

Width of the leather belt $= b = 200$ mm,

$\therefore$ Maximum tension, $T_1 = fb = 15 \times 200 = 3000$ N

Speed of the belt, $v_b = 500 \text{ m/min} = \frac{500}{60}$ m/s

Co-efficient of friction, $\mu = 0.30$

Maximum power transmitted, P_{max} = ?

Case I. For open belt drive

we know that, $$\frac{T_1}{T_2} = e^{\mu\theta} \quad \ldots(1)$$

for open belt, $$\theta = \pi - 2\alpha = 180° - 2\alpha$$

and $$\sin \alpha = \frac{r_1 - r_2}{l}$$

$$\therefore \quad \sin \alpha = \frac{0.350 - 0.125}{2} = 0.1125 \text{ m}$$

$$\therefore \quad \alpha = \sin^{-1} (0.1125) = 6.459°$$

$$\therefore \quad \theta = 180 - 2 \times 6.459 = 167.081°$$

Substituting values of T_1, μ and θ in equation 1.

$$\frac{3000}{T_2} = e^{0.30 \times 167.081 \times \frac{\pi}{180}} = e^{0.874}$$

$$\frac{3000}{e^{0.874}} = T_2$$

$$\therefore \quad T_2 = \frac{3000}{2.396} = 1252.086 \text{ N}$$

$\therefore$ Power transmitted, $P = (T_1 - T_2)\, v_b$

$$P = (3000 - 1252.086) \times \frac{500}{60} \text{ Nm/s}$$

$$= 1747.913 \times 8.333 \text{ J/s}$$

$$= 14565.941 \text{ J/s}$$

$$= 14.565 \text{ kW}$$

Case II. For cross belt.

$$\theta = 180° + 2\alpha$$

$$\sin\alpha = \frac{r_1 + r_2}{l}$$

$$\therefore \quad \sin\alpha = \frac{0.35 + 0.125}{2}$$

$$\sin\alpha = 0.2375$$

$$\therefore \quad \alpha = \sin^{-1}(0.2375) = 13.739°$$

$$\therefore \quad \theta = 180 + 2 \times 13.739 = 207.47°$$

$$\frac{T_1}{T_2} = e^{\mu\theta}$$

Substituting the values of T_1, μ, θ in the above equation,

$$\frac{3000}{T_2} = e^{0.30 \times 207.47 \times \frac{\pi}{180}} = e^{1.086}$$

$$\therefore \quad T_2 = \frac{3000}{e^{1.086}} = \frac{3000}{2.963}$$

$$= 1012.487 \text{ N}$$

$\therefore$ Power transmitted, $P = (T_1 - T_2)V_b$

$$P = (3000 - 1012.487) \times \frac{500}{60} \text{ Nm/s}$$

$$= 1987.512 \times 8.33 \text{ J/s}$$

$$= 16562.6 \text{ J/s}$$

$$= 16.56 \text{ kW.}$$

Problem 5.4. *A flat belt is required to transmit 35 kW from a pulley of 1.5 m effective diameter running at 200 rpm. The angle of contact is spread over* $\frac{11}{24}$ *of the circumference and the co-efficient of friction between belt and pulley surface is 0.3. Taking centrifugal tension into account, determine the width of the belt. Take belt thickness as 10 mm, density 1000 kg/m*^{3} *and permissible stress as 2.5 MN/m*^{2}.

Sol. Given data-power required to be transmitted, $P = 35$ kW,

Diameter of the pulley, $d = 1.5$ m.

Speed of the pulley, $N = 200$ rpm.

Angle of contact, $\theta = \frac{11}{24} \times 2\pi = 2.878 \text{ rad} = 165°.$

Co-efficient of friction, $\mu = 0.3$

Thickness of the belt, $t = 10 \text{ mm} = 1 \times 10^{-2}$ m.

Density of the belt material, $\rho = 1000$ kg/m^3.

Permissible stress, $f = 2.5$ MN/m^2

Width of the belt = b = ?

Let T_1 = tension on the tight side of the belt, and

T_2 = tension on the slack side of the belt

$$v_b = \frac{\pi d N}{60} = \frac{\pi \times 1.5 \times 200}{60} = 15.7 \text{ m/s}$$

Power transmitted $P = 35 \times 10^3 = (T_1 - T_2)\, V_b$

$35 \times 10^3 = (T_1 - T_2) \times 15.7$

$(T_1 - T_2) = 2229.29 \text{ N} = 2.229 \text{ kN}$ ...(1)

We know that $\dfrac{T_1}{T_2} = e^{\mu\theta}$

$\dfrac{T_1}{T_2} = e^{0.3 \times 2.878} = e^{0.8634} = 2.371$

or $T_1 = 2.371\, T_2$

Substituting value of T_1 in eqn. (1)

$(2.371\, T_2 - T_2) = 2229.29$

$1.371\, T_2 = 2229.29$

$T_2 = 1626.032 \text{ N}$

$\therefore$ $T_1 = 2.371 \times 1626.032 = 3855.321 \text{ N}$

Maximum tension in the belt

$T = f \times b \times t = 2.5 \times 10^6 \times b \times 1 \times 10^{-2} = 25{,}000\, b$

m = area × length × density

$= b \times 0.01 \times 1 \times 1000 = 10\, b$ kg/m.

T_c = centrifugal tension $= mv_b^2$

$T_c = 10\, b \times (15.7)^2 = 2464.9\, b$

Tension on the tight side of the belt T_1

$3855.32 = 25000\, b - 2464.9\, b$

$3855.32 = 22535.1b$

$\therefore$ $b = 0.171$ m. = **171 mm. Ans.**

Problem 5.5. *A leather belt 200 m × 10 mm has a maximum permissible tension as 2 MN/m². If the ratio of tensions is 1.8, determine at what velocity should it be run so as to transmit maximum power. Also determine the maximum value of the power. Take the density of belt material = 1100 kg/m³.*

Sol. Width of the belt, $b = 200 \text{ mm} = 0.2 \text{ m}$

Thickness of the belt, $t = 10 \text{ mm} = 0.01 \text{ m}$,

Maximum permissible tension $= f = 2 \text{ MN/m}^2 = 2 \times 10^6 \text{ N/m}^2$

$\dfrac{T_1}{T_2} = 1.8,$

Density of belt material, $\rho = 1100 \text{ kg/m}^3$.

Velocity of belt, $V_b = ?$

Maximum permissible tension of the belt,

$T = f \times b \times t = 2 \times 10^6 \times 0.2 \times 0.01 = 4000 \text{ N}$

For maximum transmission of power

$T = 3T_c$

$\therefore$ $T_c = \dfrac{T}{3} = \dfrac{4000}{3} \text{N}$

mass of 1 m length of belt, $m = \rho bt \times 1 = 1100 \times 0.2 \times 0.01 \times 1$ kg/m

$= 0.2 \times 0.01 \times 1 \times 1100 = 2.2$ kg/m

$\therefore \quad T_c = mv_b^2 = 2.2 \times v_b^2, \quad i.e., \quad \frac{4000}{3} = 2.2 \times v_b^2$

$\therefore \quad v_b^2 = 606.06, \quad \text{or} \quad v_b = 24.61 \text{ m/s}$

$\because \quad T = T_1 + T_c, \quad \therefore \quad T_1 = T - T_c$

$$= T - \frac{T}{3} = \frac{2T}{3}$$

$\therefore \quad T_1 = \frac{2 \times 4000}{3} = \frac{8000}{3} \text{ N}$

Maximum power transmitted,

$$P = (T_1 - T_2)\, v_b = T_1 \left(1 - \frac{T_2}{T_1}\right) v_b$$

$$= T_1 \left(1 - \frac{1}{1.8}\right) 24.61 = \frac{8000}{3}\left(1 - \frac{1}{1.8}\right) \times 24.6$$

$$= 29155 \text{ Nm/s} = 29.15 \text{ kW.}$$

Problem 5.6. *A leather belt connects a 1.2 m diameter pulley on a shaft running at 25 rad/sec with another pulley running at 50 rad/sec, the angle of lap on latter pulley being 150°. The maximum permissible load is 1200 N and the co-efficient of friction between the belt and pulley is 0.25. If the initial tension in the belt may have any value between 800 N and 960 N, what is the maximum power which the belt can transmit ?*

[Roorkee Univ., 1995, 1996 ; P.U. 1999]

Sol. Given data :

$d_1 = 1.2 \text{ m}, r_1 = 0.6 \text{ m},$

$\omega_1 = 25 \text{ rad/s}$

$\omega_2 = 50 \text{ rad/sec.},$

$\theta = 150° \text{ (smaller pulley)}$

$T = 1200 \text{ N},$

$\mu = 0.25,$

$T_0 = 800 \text{ N to } 960 \text{ N}$

$v = \omega_1 r_1 = 25 \times 0.6 = 15 \text{ m/sec.}$

$T_0 = \frac{T_1 + T_2}{2}$, where T_0 = initial tension in the belt

T_1 = tension in tight side of the belt

T_2 = tension in slack side of the belt

$$\frac{T_1 + T_2}{2} = 800 \quad \text{...(1)}$$

We know that $\quad \frac{T_1}{T_2} = e^{\mu\theta} = e^{0.25 \times \frac{150}{180}\pi} = 1.924$

$$\therefore \quad T_1 = 1.924\, T_2 \quad \text{...(2)}$$

Substituting value of T_1 from eqn. (2) to eqn. (1)

$1.924T_2 + T_2 = 2 \times 800$

$\therefore \quad T_2 = 547.20 \text{ N}$

$\therefore \quad T_1 = 1052.80 \text{ N}$

Power transmitted $\quad P = (T_1 - T_2)v = (1052.80 - 547.20) \times 15 = 7584 \text{ W}$

When initial tension, $T_0 = 960$ N,

$$T_0 = \frac{T_1 + T_2}{2}$$

$$\therefore \quad \frac{T_1 + T_2}{2} = 960 \quad \text{or} \quad T_1 + T_2 = 1920 \text{ N} \qquad ...(3)$$

$$\frac{T_1}{T_2} = e^{\mu\theta} = 1.924 \quad \text{from (2)}$$

From eqns. (2) and (3), we get

$$1.924\, T_2 + T_2 = 1920$$

$$\therefore \quad T_2 = 656.63 \text{ N}$$

and $\quad T_1 = 1263.37 \text{ N} > 1200 \text{ N}$

as T_1 is greater than maximum permissible load 1200 N therefore

Let us take $\quad T_1 = 1200$ N

$$\therefore \quad T_2 = \frac{T_1}{1.924} = \frac{1200}{1.924} = 623.7 \text{ N}$$

$$\therefore \quad T_0 = \frac{T_1 + T_2}{2} = \frac{1200 + 623.7}{2} = 911.85 \text{ N}$$

Power transmitted $\quad P = (T_1 - T_2)\, v = (1200 - 623.7) \times 15 = 8644.5$ W

This is the maximum power transmitted for which initial tension is 911.85 N.

Problem 5.7. *A shaft running at 100 rpm. is to drive another shaft at 240 rpm. and transmit 11 kW. The distance between the shafts is 2.5 m and the smaller pulley is of 60 cm diameter. The flat belt employed is 114 mm wide and 12 mm thick and the coefficient of friction between the belt and the pulley is 0.25.*

Calculate the stress in the belt if it is an open belt drive. [U.P.S.C. 1976]

Sol. Given data :

$N_1 = 100$ r.p.m. $\quad N_2 = 240$ r.p.m.

$P = 11$ kW, $\quad l = 2.5$ m, $\quad d_2 = 60$ cm. (smaller pulley)

$\therefore \quad r_2 = 30$ cm, $\quad b = 114$ mm, $t = 12$ mm

$\mu = 0.25$

Stress = ?

Using

$$\frac{N_1}{N_2} = \frac{d_2}{d_1}$$

$$\frac{100}{240} = \frac{60}{d_1}$$

$$\therefore \quad d_1 = 144 \text{ cm}, \quad \therefore \quad r_1 = 72 \text{ cm.}$$

for open belt drive $\quad \sin\alpha = \frac{r_1 - r_2}{l} = \frac{72 - 30}{2.5 \times 100} = 0.168$ rad.

angle of contact $\quad \theta = 3.14 - 2 \times 0.168 = 2.8056$ rad.

$$\frac{T_1}{T_2} = e^{\mu\theta}$$

$$\frac{T_1}{T_2} = e^{0.25 \times 2.8056} = 2.0165$$

Velocity of the belt, $v = \frac{\pi d_1 N_1}{60} = \frac{\pi \times 144 \times 100}{60}$ cm/s

$v = 7.54$ m/s

$P = (T_1 - T_2)\, v$

$11 \times 10^3 = (T_1 - T_2) \times 7.54$

$\therefore \quad T_1 - T_2 = 1458.88$ N ...(1)

$\because \quad \frac{T_1}{T_2} = 2.0165$ ...(2)

From eqns. (1) and (2)

$2.0165\, T_2 - T_2 = 1458.88$

$\therefore \quad T_2 = 1435.2$ N

$\therefore \quad T_1 = 2894.08$ N.

Maximum tension is $T_1 = \sigma\, bt$

[Since density of belt is not given therefore $T_c = mv^2$ is considered.]

where σ = stress in the belt

$2894.08 = \sigma \times 0.114 \times 12 \times 10^{-3}$

$\therefore \quad \sigma = 2.115 \times 10^6$ N/m^2 = 2.115 MPa

Problem 5.8. *The following data is given for a rope drive which transmits 60 H.P.*
Dia. of pulley = 40 cm, speed = 200 rpm
Angle of groove = 45°, angle of lap on smaller pulley = 160°
Coefficient of friction = 0.30, No. of ropes = 20
Find initial tension in each rope.

Sol. Given data : $D = 40$ cm, $N = 200$ r.p.m.,

Groove angle $2\beta = 45°$

Power transmitted = 60 H.P.

$\theta = 160°$,

$\mu = 0.30$,

Number of ropes = 20

H.P. transmitted per rope $= \frac{60}{20} = 3$

$v = \frac{\pi DN}{60} = \frac{\pi \times 0.40 \times 200}{60} = 4.18$ m/s

H.P./rope $= \frac{(T_1 - T_2)v}{75} = \frac{60}{20} = 3$

$\therefore \quad T_1 - T_2 = \frac{3 \times 75}{4.18} = 53.82$ N ...(1)

$\frac{T_1}{T_2} = e^{\mu\theta/\sin\beta} = e^{0.30 \times \frac{160 \times \pi}{180 \sin 22.5°}} = e^{2.18917} = 8.928$

$\therefore \quad T_1 = 8.928\, T_2$ ...(2)

$8.928\, T_2 - T_2 = 53.82$

$\therefore \quad T_2 = 6.79$ N

$\therefore \quad T_1 = 8.928 \times 6.79 = 60.62$ N

Initial tension in each rope

$$T_0 = \frac{T_1 + T_2}{2} = \frac{60.62 + 6.79}{2} = 33.705 \text{ N.}$$

Problem 5.9. *2.5 kW of power is transmitted by an open belt drive. The linear velocity of the belt in 2.5 m/s. The angle of lap on the smaller pulley is 165°. The co-efficient of friction is 0.3.*

Determine the effect on power transmission in the following cases

(a) Initial tension in the belt is increased by 10%

(b) Initial tension in the belt is decreased by 10%

(c) Angle of lap is increased by 10% by the use of an idler pulley, for the same speed and the tension on the tight side.

(d) Coefficient of friction is increased by 10% by suitable dressing to the friction surface of the belt.

Sol. $P = 2.5$ kW, $v = 2.5$ m/s, $\theta = 165°$, $\mu = 0.3$

$$P = (T_1 - T_2)\, v$$

$$2500 = (T_1 - T_2) \times 2.5$$

$$T_1 - T_2 = 1000 \text{ N} \quad \ldots(1)$$

$$\frac{T_1}{T_2} = e^{\mu\theta} = e^{0.30 \times \frac{165\pi}{180}} = 2.37$$

$$T_1 = 2.37\, T_2 \quad \ldots(2)$$

From eqns. (1) and (2)

$$2.37\, T_2 - T_2 = 1000 \quad \text{or} \quad T_2 = 729.92 \text{ N}$$

$$\therefore \quad T_1 = 2.37 \times 729.92 = 1729.91 \text{ N}$$

Initial tension $\quad T_0 = \dfrac{T_1 + T_2}{2} = \dfrac{1729.91 + 729.92}{2} = 1229.91$ N

(*a*) **Initial tension is increased by 10%**

$\therefore$ New initial tension $= T_0 = 1229.91 \times 1.10 = 1352.90$ N

or
$$\frac{T_1 + T_2}{2} = 1352.90$$

$$T_1 + T_2 = 2705.8 \text{ N}$$

as μ and θ remain unchanged, $e^{\mu\theta}$ or $\dfrac{T_1}{T_2}$ is same.

$$2.37\, T_2 + T_2 = 2705.8$$

$$T_2 = 802.9 \text{ N}$$

$$\therefore \quad T_1 = 2.37 \times 802.9 = 1902.87 \text{ N}$$

$$\therefore \quad P = (T_1 - T_2)\, v = (1902.87 - 802.9) \times 2.5 = 2749.93 = 2.74 \text{ kW}$$

$\therefore$ Increase in power $= \dfrac{2.74 - 2.5}{2.5} = 9.6\%.$

(*b*) **When initial tension is decreased by 10%**

$\therefore$ New initial tension $= T_0 = 1229.91(1 - 0.10) = 1106.91$ N

$$\frac{T_1 + T_2}{2} = 1106.91$$

$$T_1 + T_2 = 2213.82 \quad \text{as} \quad T_1 = 2.37\, T_2$$

$$2.37\ T_2 + T_2 = 2213.82$$

$$3.37\ T_2 = 2213.82$$

$$T_2 = 656.91 \text{ N},\ T_1 = 1556.90 \text{ N}$$

$$P = (1556.90 - 656.91) \times 2.5 = 2249.97 = 2.24 \text{ kW}.$$

Decrease in power $= \dfrac{2.5 - 2.24}{2.5} = 10.4\%.$

(*c*) **θ is increased 10%**

$$\frac{T_1}{T_2} = e^{\mu\theta}$$

T_1 is the same as before whereas θ increase by 10%

$$\frac{1729.9}{T_2} = e^{0.3 \times \frac{165 \times 1.10 \times \pi}{180}} = 2.58$$

$\therefore$ $T_2 = 670.50$ N

$$P = (1729.9 - 670.5) \times 2.5 = 2648.49 = 2.64 \text{ kW}$$

Increase in power $= \dfrac{2.64 - 2.5}{2.5} = 5.93\%$

(*d*) **μ is increased by 10%**

$$\frac{T_1}{T_2} = e^{\mu\theta} = e^{0.3 \times 1.10 \times \frac{165 \times \pi}{180}} = 2.58$$

or $T_1 = 2.58\ T_2$

Since T_0 remains the same,

$\therefore$ $T_1 + T_2 = 2T_0$

$\therefore$ $T_1 + T_2 = 2 \times 1229.9 = 2458.8$ N

$$T_2 = 687.09 \text{ N}$$

and $T_1 = 687.09 \times 2.58 = 1772.70$ N

Hence, $P = (1772.70 - 687.09) \times 2.5 = 2714.03 = 2.71$ kW

Increase in power $= \dfrac{2.71 - 2.5}{2.5} = 8.4\%.$

Problem 5.10. *A pulley 20 cm in diameter mounted on induction motor shaft running at 950 rpm drives a workshop main shaft at 400 rpm by belt drive. The belt is 5 mm thick and 15 cm wide. Determine the size of the pulley on the main shaft if there is slip of 4% between the belt and each pulley.*

Sol. d_1 = effective dia. of pulley = $d + t$ = 20 + 0.5 = 20.5 cm.

Linear speed of the belt running on the driving pulley

$$v_b = \frac{\pi(d_1 + t)N_1}{60} \times \left(1 - \frac{s}{100}\right)$$

$$= \frac{\pi \times 20.5 \times 950}{60} \times \left(\frac{100 - 4}{100}\right)$$

$$= 978.92 \text{ cm/s} = 9.79 \text{ m/s}$$

Linear speed of the belt on driven pulley will be the same. But due to slip over the driven pulley, the linear speed of the driven pulley = $9.79 \times \left(\frac{100-4}{100}\right)$

$$\therefore \quad 9.79\left(\frac{96}{100}\right) = \frac{\pi(d_2+t)N_2}{60}$$

$$9.79 \times 0.96 = \frac{\pi(d_2+t)\times 400}{60}$$

$$d_2 + t = 0.448$$

$$d_2 = 0.448 - 0.005 = 0.443 \text{ m}$$

$$= 44.3 \text{ cm.}$$

5.16. EFFECT OF INITIAL TENSION ON MAXIMUM POWER

Let $\quad T_1/T_2 = e^{\mu\theta} = k \quad \ldots(i)$

When the effect of centrifugal tension T_c is not considered, the initial tension in belt is given by

$$T_0 = \frac{T_1+T_2}{2} \quad \ldots(ii)$$

When the effect of centrifugal tension $T_c = mv^2$ is considered, the belt tensions in the tight and slack sides of the belt are

and $$\left.\begin{aligned} T_t &= T_1 + T_c \\ T_s &= T_2 + T_c \end{aligned}\right\} \quad \ldots(iii)$$

and, therefore initial tension will be given by

$$T_0 = \frac{T_t+T_s}{2} = \frac{T_1+T_2+2T_c}{2} \quad \ldots(iv)$$

These equations have been derived in Art. 4.14.

Eqn (iv) yields,

$$T_1 + T_2 = 2(T_0 - T_c)$$

or $$T_2(T_1/T_2 + 1) = 2(T_0 - T_c)$$

or $$T_2(k+1) = 2(T_0 - T_c)$$

$$\therefore \quad T_2 = \frac{2(T_0 - T_c)}{k+1} \quad \ldots(5.47)$$

and $$T_1 = T_2 k$$

or $$T_1 = \frac{2k(T_0 - T_c)}{k+1} \quad \ldots(5.48)$$

Power transmitted is given by

$$P = (T_1 - T_2)v = T_2(T_1/T_2 - 1)v = T_2(k-1)v$$

or $$P = 2(T_0 - T_c)\frac{(k-1)v}{k+1} \quad \text{where } T_c = mv^2$$

$$\therefore \quad P = \frac{2(k-1)}{k+1}\left(vT_0 - mv^3\right) \quad \ldots(5.49)$$

For maximum power $\frac{dP}{dv} = 0 = \frac{2(k-1)}{k+1}(T_0 - 3\,mv^2)$ ...(*iv*)

$\because \quad k \neq 1, \quad \text{and } k \neq \infty$

$\therefore \quad T_0 - 3\,mv^2 = 0 \quad \text{for maximum power}$

or $$T_0 = 3T_c \qquad ...(5.50)$$

Hence, the belt velocity for maximum power is

$$v = \sqrt{\frac{T_0}{3m}} = v' \text{ (say)} \qquad ...(5.51)$$

$\therefore$ Maximum power, $P_{max} = \frac{2(k-1)}{(k+1)}(3T_c - T_c)\,v'$

or $$P_{max} = 4T_c\,v'\,\frac{(k-1)}{k+1} \qquad ...(5.52)$$

Also, $$P_{max} = \frac{2(k-1)}{k+1}(T_0 - T_c)\,v'$$

$$= \frac{2(k-1)}{k+1} \times (2T_c)\sqrt{\frac{T_0}{3m}}$$

$$= \frac{4(k-1)}{(k+1)} \times m(v')^2\sqrt{\frac{T_0}{3m}}$$

$$= \frac{4(k-1)}{(k+1)} \times m \times \frac{T_0}{3m} \times \sqrt{\frac{T_0}{3m}}$$

$\therefore$ $$P_{max} = \frac{4(k-1)}{(k+1)} \times \left(\frac{T_0}{3m}\right)^{3/2}.\,m \qquad ...(5.53)$$

When belt is at rest, $T_c = 0$, then maximum tension in belt is

$$T_t = T_1 + 0 = T_1$$

and $$T_1 = \frac{2k}{k+1}T_0 \quad \text{and} \quad T_2 = \frac{2}{k+1}T_0 \qquad ...(5.54)$$

While the tension T_1, when the belt is running with velocity is given as

$$T_1 = \frac{2k}{k+1}(T_0 - T_c) \quad \text{and } T_2 = \frac{2}{k+1}(T_0 - T_c) \qquad ...(5.55)$$

and for maximum power condition $T_0 = 3T_c$,

$$T_1 = \frac{2k}{k+1} \times (3T_c - T_c) = \frac{4kT_c}{(k+1)} \quad \text{and } T_2 = \frac{4T_c}{(k+1)} \qquad ...(5.56)$$

or $$T_1 = \frac{4k}{(k+1)}\sqrt{\frac{T_0}{3m}} \qquad ...(5.57)$$

and $$T_2 = \frac{2}{k+1}(T_0 - T_c) = \frac{4}{k+1}(3T_c - T_c)$$

or $$T_2 = \frac{4T_c}{k+1} = \frac{4}{k+1}\sqrt{\frac{T_0}{3m}} \qquad ...(5.58)$$

Problem 5.11. *An open belt drive connects two pulleys 90 cm and 60 cm diameters mounted on two parallel shafts 3 m apart. The maximum belt tension is 2000N. The co-efficient of friction is 0.3. The driving pulley of diameter 90 cm runs at 300 rpm. Find*

(a) the power transmitted,

(b) torque acting on each shaft

(c) initial tension in the drive.

Sol. Given : Diameter of driving pulley, $d_1 = 90$ cm

$\therefore \quad r_1 = 45 \text{ cm} = 0.45 \text{ m.}$

Diameter of driven pulley, $d_2 = 60$ cm.

$\therefore \quad r_2 = 30 \text{ cm} = 0.3 \text{ m}$

Distance between centres, $l = 3$ m.

Maximum tension, $T_1 = 2000$ N

$\mu = 0.3$

Speed of driver $= N_1 = 300$ rpm

$$\therefore \quad \text{Belt velocity, } v = \frac{\pi d_1 N_1}{60} = \frac{\pi \times 0.9 \times 300}{60} \text{ m/s}$$
$$= 14.137 \text{ m/s}$$

$$\alpha = \sin^{-1}\left(\frac{r_1 - r_2}{l}\right) = \sin^{-1}\left(\frac{0.45 - 0.3}{3}\right) = \sin^{-1} 0.05 = 2.866°$$

$\therefore$ Angle of lap on smaller pulley is given by

$$\theta_2 = (180 - 2\alpha)° = (180 - 2 \times 2.866)° = 174.268°$$

$$\therefore \quad \text{Tension ratio } \frac{T_1}{T_2} = e^{\mu\theta}$$

or
$$T_1 = T_2 \left(e^{0.3 \times 174.268 \times \frac{\pi}{180}} \right) = T_2 \times e^{0.912465}$$

or
$$T_1 = 2.49\, T_2$$

$$\therefore \quad T_2 = \frac{T_1}{2.49} = \frac{2000}{2.49} = \mathbf{802.21\ N}$$

(*a*) **Power transmitted :**

$$P = (T_1 - T_2)\, v$$
$$= (2000 - 802.21) \times 14.137 \text{ watts}$$
$$= 16933.16 \text{ watts}$$
$$= \mathbf{16.93317\ kW}$$

(*b*) **Torque on each shaft :**

Torque on driving shaft $= (T_1 - T_2) \times r_1$

$= (2000 - 802.21) \times 0.45$

$=$ **539.01 Nm. Ans.**

Torque of driven pulley $= (T_1 - T_2)\, r_2$

$= (2000 - 802.21) \times 0.3$ Nm

$=$ **359.337 Nm. Ans.**

(*c*) **Initial tension :**

$$T_0 = \frac{T_1 + T_2}{2} = \frac{2000 + 802.21}{2} = \mathbf{1401.105\ N} \quad \textbf{Ans.}$$

Problem 5.12. *The angle of lap at the smaller pulley running at 1500 m/min is 170°. Belt width is 18 cm and thickness is 10 mm. Density of belt material is 1 gm/cm³. Find the power that can be transmitted at 1500 rpm speed of smaller pulley. Also find the maximum power transmitted if the permissible stress in the belt is not to exceed 240 N/cm². Take μ = 0.3.*

Sol. Given :

Angle of contact at smaller pulley, $\theta = \theta_2 = 170°$.

Belt speed, $v = 1500 \text{ m/min} = \frac{1500}{60} = 25 \text{ m/s}$

Width of belt, $b = 18 \text{ cm} = 0.18 \text{ m}$

Thickness of belt, $t = 10 \text{ mm} = 0.01 \text{ m}$

Density of belt, $\rho = 1 \text{ gm/cm}^3 = \frac{1}{1000} \text{ kg/cm}^3$

Maximum stress in belt $= \sigma = 240 \text{ N/cm}^2$

$\mu = 0.3.$

Maximum tension, T_t = stress × Area of belt section $= \sigma bt$

$$= 240 \times 18 \times \frac{10}{10} = 4320 \text{ N}$$

m = mass of belt per metre length

$= \rho \times$ volume of belt of 1 m length

$= \rho\,(bt \times L)$

$= \rho bt \times 100$ $\qquad L = 1 \text{ m} = 100 \text{ cm}$

$$= \frac{1}{1000} \times 18 \times 1 \times 100 = 1.8 \text{ kg/m}$$

∴ Centrifugal tension

$$T_c = mv^2 = 1.8 \times (25)^2 \text{ N} = 1125 \text{ N}$$

∵ $T_t = T_1 + T_c$, ∴ $T_1 = T_t - T_c = 4320 - 1125 = 3195 \text{ N}$

$$\frac{T_1}{T_2} = e^{\mu\theta} = e^{0.25 \times (170 \times \pi/180)} = e^{0.741765} = 2.0996$$

∴ $$T_2 = \frac{T_1}{2.0996} = \frac{3195}{2.0996} = 1521.72 \text{ N}$$

Effective tension $= T_{eff} = T_1 - T_2 = 3195 - 1521.72 = 1673.28 \text{ N}$

∴ Power transmitted $= P = (T_1 - T_2)v = 1673.28 \times 25$ watts

$= 41832 \text{ W} = \mathbf{41.832\ kW}$ **Ans.**

For maximum power, $T_t = 3T_c$

or $$T_c = mv^2 = \frac{T_t}{3}$$

∴ $$v^2 = \frac{T_t}{3m} = \frac{4320}{3 \times 1.8} = 800$$

∴ $$v = \sqrt{800} = 28.284 \text{ m/s} = v_1 \text{ (say) for } P_{max}.$$

Also, $$T_1 = T_t - T_c = T_t - \frac{T_t}{3} = \frac{2}{3}T_t = \frac{2}{3} \times 4320 \text{ N} = 2880 \text{ N}$$

and $$T_2 = \frac{T_1}{2.0996} = 1371.7 \text{ N}$$

$\therefore$ $$T_{\text{eff}} = T_1 - T_2 = 2880 - 1371.7 = 1508.3 \text{ N}$$

$\therefore$ Maximum power,

$$P_{max} = T_{\text{eff}} \times v_1 = 1508.3 \times 28.284 \text{ W}$$
$$= 42660.76 \text{ W}$$
$$= \mathbf{42.66076\ kW} \quad \textbf{Ans.}$$

Problem 5.13. *A leather belt is required to transmit 9 kW from a pulley 120 cm in diameter running at 200 rpm. In angle embraced is 165° and the co-efficient of friction between leather belt and pulley is 0.3. If the safe working stress for the leather belt is 140 N/cm², the density of leather is 1 gm/cm³ and the thickness of the belt is 10 mm, determine the width of the belt taking the centrifugal tension into account.* [AMIE, Summer 1976]

Sol. Given :

Power, $P = 9$ kW ;

Diameter of Pulley, $d = 120$ cm $= 1.2$ m

Speed of pulley, $N = 200$ rpm,

Angle embraced, $\theta = 165°$,

Co-efficient of friction, $\mu = 0.3$

Safe working stress, $\sigma = 140$ N/cm²

Density, $\rho = 1 \text{ gm/cm}^3 = \frac{1}{1000} \text{ kg/cm}^3$

Belt thickness, $t = 10$ mm $= 1$ cm

To find, belt width $= b = ?$

$$\frac{T_1}{T_2} = e^{\mu\theta} = e^{0.3 \times \frac{165 \times \pi}{180}} = e^{0.86394} = 2.3725$$

Maximum tension $= T_t$ = safe stress × (Area of belt section)

or $$T_t = \sigma \times b \times t = 140 \times b \times 1 = 140b$$
$$T_c = mv^2$$

where $$v = \frac{\pi dN}{60} = \frac{\pi \times 1.20 \times 200}{60} = 4\pi \text{ m/s}$$

m = mass of 1 m length of belt

= Density × Area of belt section × 1 m

$$= \frac{1}{1000} \times bt \times 100 \text{ cm}$$

$$= \frac{1}{1000} \times b \times 1 \times 100 = 0.1b \text{ kg/m}$$

$\therefore$ Centrifugal tension, $T_c = mv^2 = (0.1b) \times (4\pi)^2 \text{ N} = 15.7914b$ N

But $$T_t = T_1 + T_c$$

$\therefore$ $$T_1 = T_t - T_c = 140b - 15.7914b = 124.2059b$$

and using $$T_1/T_2 = e^{\mu\theta} = 2.3725$$

$$T_2 = \frac{T_1}{2.3725} = 52.3523b$$

$\therefore$ Effective tension $= T_{eff} = T_1 - T_2 = (124.2059 - 52.3523)\ b$

$= 71.8536\ b$

Since power $P = (T_1 - T_2)\ v$ | $P = 9$ kW $= 9000$ W

$\therefore$ $9 \times 1000 = (71.8536\ b) \times 4\ \pi$

$\therefore$ $b = \frac{9000}{71.8536 \times 4\pi} = \mathbf{9.967\ cm}$ **Ans.**

Problem 5.14. *An open belt running over two pulleys 24 cm and 60 cm diameters connects two parallel shafts 3 m apart and transmits 3.75 kW from the smaller pulley that rotates at 300 rpm. Co-efficient of friction between the belt and the pulleys is 0.3 and the safe working tension is 100 N/cm width. Determine :*

(i) minimum width of belt,

(ii) initial belt tension, and

(iii) length of the belt required. [AMIE, Winter 1978]

Sol. Given :

Diameter, $d_1 = 60$ cm ; $d_2 = 24$ cm

distance between shafts, $l = 3$ m,

Power transmitted, $P = 3.75$ kW $= 3750$ watts

Speed of smaller pulley, $N_2 = 300$ rpm

Co-efficient of friction, $\mu = 0.3$,

Safe working tension, $f = 100$ N/cm width of belt

Let b = width of belt.

Velocity, $v = \frac{\pi N_2 d_2}{60} = \frac{\pi \times 300 \times (24/100)}{60} = 3.77$ m/s

For open drive $\sin \alpha = \frac{r_1 - r_2}{l} = \frac{60/2 - 24/2}{3 \times 100} = \frac{36}{600} = 0.06$

$\therefore$ $\alpha = 3.44°$

$\therefore$ Angle of lap at smaller pulley, $\theta_2 = (180 - 2\alpha)°$

or $\theta_2 = \theta = 180 - 2 \times 3.44 = 173.13° = 173.13 \times \frac{\pi}{180} = 3.0215$ radians

$\therefore$ $\frac{T_1}{T_2} = e^{\mu\theta} = e^{0.3 \times 3.0215} = e^{0.90645} = 2.476$

$\therefore$ $T_1 = 2.476\ T_2$ ...(*i*)

Power transmitted $= P = (T_1 - T_2)\ v$

$\therefore$ $3750 = (2.476\ T_2 - T_2) \times 3.77$

$= 1.476\ T_2 \times 3.77$

$\therefore$ $T_2 = \frac{3750}{1.476 \times 3.77} = 672.91$ N

and $T_1 = 2.476\ T_2 = 1666.13$ N from (*i*)

(*i*) **Minimum width of belt**

Neglecting the effect of centrifugal tension

$$T_c = 0$$

$\therefore$ $$T_t = T_1 + T_c$$

or $$T_1 = T_t = 1666.13 \text{ N}$$

But $$T_t = T_{max} = f \times b = 100\, b$$

$$100\, b = 1666.13$$

$\therefore$ $$b = \text{width of belt} = \frac{1666.13}{100} = \mathbf{16.7\ cm} \quad \textbf{Ans.}$$

(*ii*) **Initial belt tension**

Let T_0 = initial tension,

$\because$ $$T_0 = \frac{T_1 + T_2}{2}$$

$\therefore$ $$T_0 = \frac{1666.13 + 672.91}{2} = \frac{2339.04}{2} = \mathbf{1169.52\ N} \quad \textbf{Ans.}$$

(*iii*) **Length of the belt required**

(*a*) Approximate length for open drive

$$L_o = \pi\,(r_1 + r_2) + \frac{(r_1 \sim r_2)^2}{l} + 2l$$

$$= \pi\left(\frac{0.60}{2} + \frac{0.24}{2}\right) + \frac{(0.3 \sim 0.12)^2}{3.0} + 2 \times 3 \text{ m}$$

$$= \mathbf{7.3303\ m} \quad \textbf{Ans.}$$

(*b*) Exact length of belt for open drive

$$= L_o)_{\text{exact}} = r_1\theta_1 + r_2\,\theta_2 + 2l \cos\alpha$$

where $$\theta_1 = 180 + 2\alpha = 180° + 2 \times 3.44° = 186.88°$$

$$\theta_2 = 180 - 2\alpha = 180 - 2 \times 3.44 = 173.13°$$

$\therefore$ $$L_o)_{\text{exact}} = \frac{0.60}{2}\left(186.88 \times \frac{\pi}{180}\right) + \frac{0.24}{2}\,\frac{(173.13) \times \pi}{180} + 2 \times 3 \cos 3.44°$$

$$= \mathbf{7.3302\ m} \quad \textbf{Ans.}$$

Problem 5.15. *A pulley, used to transmit power by means of ropes, has a diameter of 3.6 m and has 15 grooves of 45°. The angle of contact is 170° and the co-efficient of friction between the ropes and the groove side is 0.28. The maximum possible tension in the ropes is 960 N and the mass of the rope is 1.5 kg per m length. What is the speed of the pulley in rpm and the power transmitted if the condition of maximum power prevails ?* [AMIE, Summer 1983]

Sol. Given :

Diameter of pulley, $d = 3.6$ m ;

Angle of contact, $\theta = 170°$,

Groove angle, $2\beta = 45°$,

$\therefore$ $\beta = 22.5°$

Number of grooves or number of ropes = 15

$\mu = 0.28$,

Maximum tension, $T_t = 960$ N

Mass of rope, $m = 1.5$ kg/m length

To find , (*i*) speed (rpm), and (*ii*) maximum power

For maximum power :

Centrifugal tension, $T_c = \frac{T_t}{3} = \frac{960}{3} = 320$ N.

But, $T_c = mv^2$

$\therefore$ $v = \sqrt{\frac{T_c}{m}} = \sqrt{\frac{360}{1.5}} = \mathbf{14.61\ m/s}$

Also, $v = \frac{\pi dN}{60}$

$\therefore$ $N = \frac{60v}{\pi d} = \frac{60 \times 14.61}{\pi \times 3.6} =$ **77.51 rpm. Ans.**

Tension ratio, $\frac{T_1}{T_2} = e^{\mu\theta/\sin\beta} = e^{(0.28) \times (170 \times \pi/180)/\sin 22.5°}$

or $T_1/T_2 = e^{2.1709} = 8.766$ ...(1)

$\because$ $T_t = T_1 + T_c$ and $T_c = T_t/3$ for maximum power

$\therefore$ $T_1 = T_t - T_t/3 = \frac{2}{3}T_t = \frac{2}{3} \times 960 =$ **640 N**

$\therefore$ $T_2 = \frac{T_1}{8.766} = \frac{640}{8.766} = 73.01$ N

$\therefore$ $T_1 - T_2 = 640 - 73.01 = 566.99$ N

Hence, maximum power transmitted per rope

$P_{max} = (T_1 - T_2)\, v = 566.99 \times 14.61 = 8283.72$ watts

$= 8.28372$ kW

$\therefore$ Total maximum power transmitted by 15 ropes

$= 8.28372 \times 15 =$ **124.26 kW Ans.**

Problem 5.16. *A capstan and rope are used in a railway goods-yard for moving trucks through short distance. The capstan runs at 60 rpm. The rope is wound around the capstan of 30 cm diameter through 3.5 turns and a force of 200 N is applied on the free end of the rope. Find the pull on the trucks and power supplied by the capstan. Take $\mu = 0.3$.*

Sol. See Fig. P. 5.16.

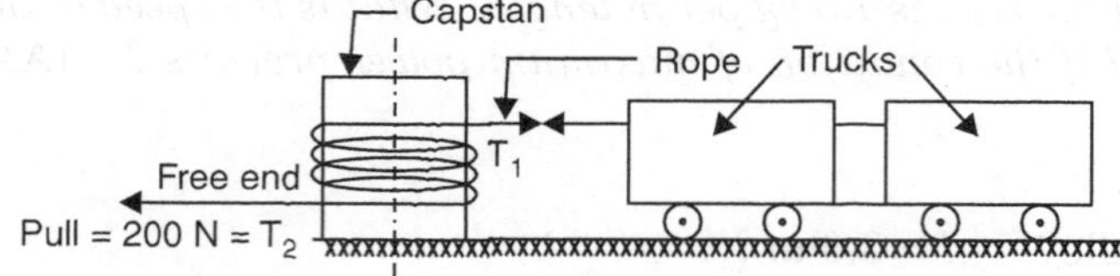

Fig. P. 5.16

Angle of lap, $\theta = 3.5$ truns

$= 3.5 \times 2\pi$ radians

$= 7\pi$ radians

Pull on the free end $= T_2 = 200$ N.
Speed of capstan $= 60$ rpm
Coefficient of friction $= 0.3$
diameter of capstan $= d = 30$ cm $= 0.3$ m.

Let T_1 be pull in rope on the side of trucks.

Tension ratio $T_1/T_2 = e^{\mu\theta} = e^{7\pi \times 0.3} = 733.146$

$\therefore$ $T_1 = 733.146\ T_2 = 733.146 \times 200 =$ **146629.2 N Ans.**

Velocity of rope $= \dfrac{\pi d N}{60} = \dfrac{\pi \times 0.3 \times 60}{60} = 0.3\ \pi$ m/s

$\therefore$ Power supplied, $P = (T_1 - T_2)\ v$

$= (146629.2 - 200) \times 0.3\pi$ watts

$= 138006.3$ watts

$=$ **138.0063 kW Ans.**

Problem 5.17. *A shaft which rotates at a constant speed of 160 rpm is connected by belting to a parallel shaft 72 cm apart which has to run at 60, 80 and 100 rpm. The smallest pulley on the driver shaft is 4 cm in radius. Determine the remaining radii of the two stepped pulleys for : (i) a crossed belt, and (ii) an open belt.* [AMIE, Winter 1976]

Sol. Given :

Speed of driving shaft $= N_1 = 160$ rpm

$\therefore$ Speed of pulleys 1, 3 and 5

$= 160$ rpm.

i.e., $N_1 = N_3 = N_5 = 160$ rpm

Speed of pulley 2 $= N_2 = 60$ rpm,

Speed of pulley 4 $= N_4 = 80$ rpm and that of pulley 6 $= N_6 = 100$ rpm.

Centre distance between two shafts,

$l = 72$ cm

Radius of pulley $1 = r_1 = 4$ cm.

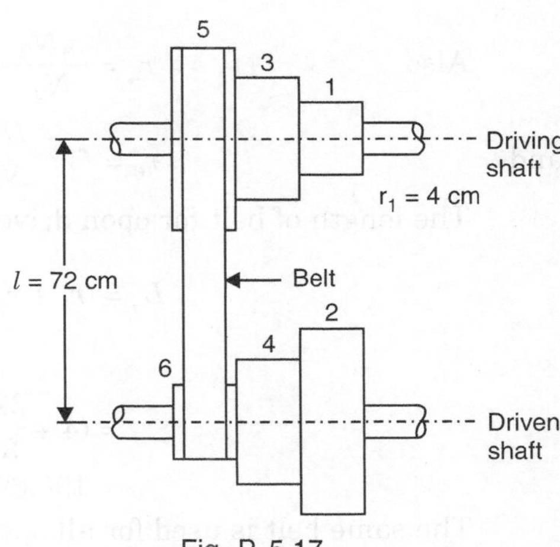

Fig. P. 5.17

$\therefore$ Using $\dfrac{N_2}{N_1} = \dfrac{r_1}{r_2}$, we get

$$r_2 = \frac{r_1 \times N_1}{N_2} = 4 \times \frac{160}{60} = \frac{32}{3} \text{ cm} = 10.67 \text{ cm}$$

For pulleys 3 and 4, we have

$$\frac{N_4}{N_3} = \frac{r_3}{r_4}, \quad \therefore \quad r_4 = r_3 \times \frac{N_3}{N_4} = r_3 \times \frac{160}{80} = 2r_3 \quad \text{...(1)}$$

(*i*) **For Crossed Belt Drive**

Since the same belt is used between all pulleys, therefore

$$r_1 + r_2 = r_3 + r_4 = r_5 + r_6 = \text{ constant}$$

But $r_1 + r_2 = 4 + \dfrac{32}{3} = \dfrac{44}{3}$ cm

$$\therefore \quad r_3 + r_4 = \frac{44}{3} \quad \text{...(2)}$$

Using eqn. (1), $r_3 + 2r_3 = \dfrac{44}{3}$

$$\therefore \quad r_3 = \frac{44}{3 \times 3} = \frac{44}{9} \text{ cm.} = 4.89 \text{ cm}$$

and $$r_4 = 2r_3 = \frac{88}{9} \text{ cm} = 9.78 \text{ cm.}$$

For pulleys 5 and 6, $r_6 = r_5 \times \dfrac{N_5}{N_6} = r_5 \times \dfrac{160}{100} = 1.6\, r_5$ ...(3)

Also, $r_5 + r_6 = \dfrac{44}{3}$

From (3), $r_5 + 1.6\, r_5 = \dfrac{44}{3}$

$$\therefore \quad r_5 = \frac{44}{3 \times 2.6} = \mathbf{5.64\ cm} \quad \textbf{Ans.}$$

$$\therefore \quad r_6 = 1.6\, r_5 = 1.6 \times 5.64 = \mathbf{9.02\ cm} \quad \textbf{Ans.}$$

(ii) **For Open Belt Drive**

$$r_1 = 4 \text{ cm}, \quad \text{and } r_2 = \frac{r_1 N_1}{N_2} = 4 \times \frac{160}{60} = \frac{32}{3} \text{ cm} = 10.67 \text{ cm.}$$

Also $$r_4 = \frac{r_3 N_3}{N_4} = r_3 \times \frac{160}{80} = 2r_3$$ as before [see eqn. (1)]

and $$r_6 = r_5 \times \frac{N_5}{N_6} = r_5 \times \frac{160}{100} = 1.6\, r_5$$ as before [see eqn. (2)]

The length of belt for open drive is given by

$$L_o = (r_1 + r_2)\,\pi + \frac{(r_1 - r_2)^2}{l} + 2l$$

$$= (4 + \frac{32}{3})\pi + \frac{\left(4 - \frac{32}{3}\right)^2}{72} + 2 \times 72 \text{ cm}$$

$$= 190.69 \text{ cm}$$

The same belt is used for all open drives between pulleys 1 and 2 ; pulleys 3 and 4 and pulleys 5 and 6. Thus, for second drive for pulleys 3 and 4, we get

$$L_o = (r_3 + r_4)\pi + \frac{(r_3 - r_4)^2}{l} + 2l$$ where $r_4 = 2r_3$

or $$190.69 = (r_3 + 2r_3)\pi + \frac{1}{l}(r_3 - 2r_3)^2 + 2l$$

$$190.69 = 3r_3\pi + \frac{r_3^2}{72} + 2 \times 72$$

$$\therefore \quad r_3^2 + 3 \times 72\, r_3\, \pi + 2 \times 72 \times 72 - 190.69 \times 72 = 0$$

or $$r_3^2 + 678.58 r_3 - 8545.68 = 0$$

$$\therefore \quad r_3 = \frac{1}{2}[-678.58 \pm \sqrt{(678.58)^2 + 4 \times 8545.68}] \text{ cm}$$

$$= \frac{1}{2}[-678.58 \pm 703.32] \text{ cm}$$

$$= 12.37 \text{ and } -690.95 \text{ cm}$$

Adopting +ve value of r_3, we get

$$r_3 = \mathbf{12.37 \text{ cm}} \quad \textbf{Ans.}$$

And, therefore $r_4 = 2r_3 = \mathbf{24.74 \text{ cm}}$ **Ans.**

For the third open drive connecting pulleys 5 and 6, we get

$$L_o = (r_5 + r_6)\,\pi + \frac{(r_5 - r_6)^2}{l} + 2l \qquad \text{where } r_6 = 1.6\, r_5$$

i.e.,
$$190.69 = (r_5 + 1.6\, r_5)\pi + \frac{1}{l}(r_5 - 1.6\, r_5)^2 + 2l$$

i.e.,
$$190.69 = 2.6\pi\, r_5 + \frac{(0.6 r_5)^2}{72} + 2 \times 72$$

or
$$190.69 \times 72 = 2.6\pi \times 72\, r_5 + 0.36\, r_5^2 + 2 \times 72 \times 72$$

or
$$r_5^2 + 1633.63\, r_5 - 9338 = 0$$

$$\therefore \quad r_5 = \frac{1}{2}\left[-1633.63 \pm \sqrt{(1633.63)^2 + 4 \times 9338}\right] \text{ cm}$$

$$= \frac{1}{2}[-1633.63 \pm 1645.02] \text{ cm}$$

$$= 5.695,\ -819.663 \text{ cm}$$

Adopting +ve value, we get

$$r_5 = 5.695 \text{ cm or } 5.7 \text{ cm}$$

and $r_6 = 1.6\, r_5 = \mathbf{9.12 \text{ cm}}$ **Ans.**

EXERCISE V

A. Theoretical Questions

1. Describe the types of belt and rope drives.
2. Explain open and crossed belt drives and clearly compare the salient features of the two.
3. What do you mean by simple and compound belt drives ? Explain with neat sketches ?
4. Derive expressions for velocity ratio for simple and compound belt drives without considering slip between the belt and the pulleys.
5. Derive expressions for velocity ratio for simple and compound belt drives when there is slip between belt and the pulleys.
6. What is effect of thickness of the belt on the velocity ratio ? Derive expression for VR for simple belt drive with and without slip.
7. Derive expressions for finding (*a*) exact length of belt and (*b*) approximate length of an open belt drive.
8. Derive expression for finding (*a*) exact length of belt and (*b*) approximate length of belt for a crossed belt drive.
9. Derive an expression for finding ratio of belt tensions for flat pulleys.
10. Derive an expression for finding ratio of belt tensions for a grooved pulley.
11. Derive an expression for finding centrifugal tension in belt.

12. Plot a curve showing variation of power (P) transmitted by a belt drive as a function of belt velocity v.

13. Derive expression for belt speed for maximum power transmission.

14. Why does the driven pulley rotate in direction opposite to that of driving pulley in case of a crossed belt drive ?

15. What is initial tension in belt drives ?

16. Explain creep in belt drives.

17. Prove that maximum power $P_{max} = \frac{4(k-1)}{(k+1)}\left(\frac{T_0}{3m}\right)^{3/2}$ by considering effect of initial tension (T_0) and centrifugal tension (T_c). k is belt tension ratio.

B. Numerical Problems

5.1. Speed of a shaft is 300 rpm and it carries a pulley of 60 cm diameter. What is speed of the machine shaft driven by the belt drive through a pulley of 45 cm diameter ? Consider (*a*) the belt of negligible thickness, and (*b*) belt of 12 mm thickness . [**Ans.** (*a*) 400 rpm, (*b*) 397.40 rpm]

5.2. Speed of shaft carrying a 60 cm diameter pulley rotates at 300 rpm. Another shaft with a pulley of 45 cm diameter is connected by a belt. Find (*a*) speed of belt reaching driven pulley if there is no slip, (*b*) speed of belt leaving driven pulley if there is no slip, (*c*) speed of belt reaching driven pulley if there is 2% slip between belt and driving pulley, (*d*) speed of second pulley and speed of belt leaving driven pulley if there is slip of 3% between belt and the driven pulley.

[**Ans.** (*a*) 18.85 m/s ; (*b*) 18.85 m/s ; (*c*) 18.47 m/s ; (*d*) 380 rpm ; 17.91 m/s]

5.3. Two pulleys of diameters 0.6 m and 0.48 m are mounted on two parallel shafts 3 m apart. The two pulleys are connected by an open belt drive. Find (*a*) exact length of belt, and (*b*) approximate length of belt. [**Ans.** (*a*) 7.6976 m ; (*b*) 7.6975 m]

5.4. Two shafts 3 m apart with pulleys 0.6 m and 0.48 m diameter are connected by crossed belt drive. Find (*a*) exact length of belt, and (*b*) approximate length of belt required.

[**Ans.** (*a*) 7.7939 m, (*b*) 7.7937 m]

5.5. A casting weighing 9000 N hangs freely from a rope which makes 2.5 turns around a drum of 30 cm diameter revolving at 20 rpm. The other end of the rope is pulled by a man. The co-efficient of friction is 0.25. Determine : (*i*) the force required by the man, and (*ii*) the power to raise the casting. [*AMIE, Winter 1986*]

[**Ans.** (*i*) 156.413 N ; (*ii*) 2.778 kW]

5.6. A shaft which rotates at a constant speed of 160 rpm is connected by belting to a parallel shaft 180 cm apart which has to run at 60, 80 and 100 rpm. The smallest pulley on the driver shaft is 15 cm in radius. Determine the remaining radii of the two stepped pulleys for : (*a*) a crossed belt, and (*b*) an open belt. Neglect belt thickness. [*AMIE, Summer 1982*]

[**Ans.** (*a*) r_2 = 40 cm, r_3 = 18.333 cm, r_4 = 36.667 cm, r_5 = 21.15 cm, r_6 = 33.85 cm ; (*b*) r_2 = 40 cm, r_3 = 18.5 cm, r_4 = 37 cm , r_5 = 21.465 cm and r_6 = 34.344 cm]

5.7. A belt transmits 28 kW power from a pulley 90 cm diameter which runs at 300 rpm. The angle of embrace of belt and pulley is 160° and co-efficient of friction between belt and pulley is 0.25. The belt is 6 mm thick and has a density of 1 g/cm^3. Determine the minimum width for maximum belt stress of 2 MPa. [*AMI Mech E (London), Jabalpur University 1982*]

[**Ans.** 205 mm]

5.8. A belt is required to transmit 33.33 kW power from a pulley 1.4 m diameter running at 300 rpm. The angle of contact between the belt and pulley is 160° and the co-efficient of friction is 0.25. The safe working stress for the belt is 3.15 MPa, the thickness of belt is 6.3 mm and density of belt material is 1.1 g/cm^3. What width of belt is required ? [*Punjab University*]

[**Ans.** 104 mm]

5.9. What maximum power can be transmitted per square cm of cross-section, if the tension in belt is not to exceed 2.5 MPa and the ratio of the tension in the tight side to the tension in the slack side is 1.8 ? Assume density of belt material as 1.1 g/cm^3. [**Ans.** 3.6 kW]

5.10. A pulley 30 cm in diameter running at 200 rpm is connected by belt to another pulley at a distance of 3 m. The second pulley has to run at 120 rpm. The belt is 5 mm thick and 15 cm wide and has density of 1 g/cm^3. Allowing slip of 3% between belt and pulley determine (*a*) the size of the second pulley, (*b*) the stress in the belt due to centrifugal tension, and (*c*) power transmitted. Take $\mu = 0.25$ [**Ans.** (*a*) 472 mm, (*b*) 10.4 kN/m^2, (*c*) 42.67 kW]

5.11. A belt transmitting power from a motor to a machine weighs 24 N/m length and the maximum permissible tension for it is 5500 N. The angle of contact of belt with pulley is 200° and co-efficient of friction is 0.28. If the belt runs under maximum power conditions, determine the maximum power transmitted and the initial tension in the belt so that drive can fulfil the above conditions. [*Bombay University*]

[**Ans.** 352 kW, 2.5236 kN]

5.12. A rope drive transmits 186.7 kW power at 225 rpm by ropes, each 25 mm diameter and density 6.8 g/cm^3. The maximum rope tension is 1500 N and it is designed for maximum power conditions. The angle of lap is 160° and co-efficient of friction between ropes and the pulley is 0.2. The groove angle is 45°. Determine the diameter of the pulley and the number of ropes.

[*Punjab University*]

[**Ans.** 2.268 m ; 6]

5.13. A *V*-belt having a lap angle of 180° has a cross-sectional area of 250 mm^2 runs in a groove of included angle 45°. The density of belt is 1.5 g/cm^3, the maximum stress is limited to 4 MPa, the co-efficient of friction is 0.15.

Find the maximum power that can be transmitted, if the pulley has a mean diameter of 0.3 m and runs at 1000 rpm. [*London University*]

[**Ans.** 17.973 kW]

5.14. A leather belt of thickness 6 mm and width 80 mm connects two pulleys of 0.5 m and 0.9 m diameter, their centres being 2.5 m apart. If the speed of the larger pulley is 300 rpm, the co-efficient of friction between belt and pulleys is 0.3 and maximum belt stress is 1.4 MPa, determine the maximum power transmitted and the initial belt tension. Prove any formula used and neglect centrifugal tension [*AMI Mech E (London)*]

[**Ans.** 10 kW ; 473.5 N]

5.15. The ratio of tension of the tight side of a belt to the tension on slack side of the transmitting power is 2.5. The belt is 20 cm wide and 10 mm thick. The maximum stress in the belt is not to exceed 5.65 MPa. The belt connects two pulleys, the diameter being 0.6 m for the driver and 0.45 m for the follower pulley. The shaft on which the driven pulley is mounted rotates at 120 rpm. Find : (*a*) Power transmitted, (*b*) speed of the driven pulley. [**Ans.** (*a*) 46 kW, (*b*) 160 rpm]

6

Lifting Machines

6.1. INTRODUCTION

A lifting machine is a device for lifting heavy load by applying small effort applied in a convenient direction. A simple wedge and a crowbar are examples of some lifting machines. Similarly, pulley used for lifting water from the well, the pedal and axle of a bicycle and screw jack are also examples of very commonly used lifting machines.

6.2. DEFINITIONS

Some common terms used in the study of lifting machines are as follows :

Load. This is the weight W of the block to be lifted. This is also treated as the resistance to be overcome by the machine.

Effort. This is the small force P applied at certain point in the lifting machine to over come the load and any other resistances against movement.

Mechanical Advantage. This is the ratio of the load lifted to the effort applied to the machine.

Thus, mechanical advantage $= M.A. = \dfrac{W}{A}$ ...(6.1)

Velocity Ratio. This is the ratio of the distance moved by the effort (y) to that moved by the load (x) during the same interval of time.

Thus, velocity ratio $= V = \dfrac{y}{x}$ ...(6.2)

The work input. It is workdone by the effort and is given by $W_{in} = P.y$ ...(6.3)

The work output. It is the workdone by the load, and is given by

$$W_{out} = W.x \qquad ...(6.4)$$

6.3. AN IDEAL MACHINE

An ideal machine is frictionless machine. Thus, in an ideal machine there is no loss of energy in overcoming frictional resistance and work input will be fully utilised to lift the load. Thus,

For ideal machine $\quad W_{in} = W_{out}$ ...(6.5)

$$\therefore \qquad \frac{W}{P} = \frac{y}{x} = V \qquad ...(6.6)$$

Thus for an ideal machine: **Mechanical Advantage = Velocity Ratio** ...(6.7)

Or, the greater the mechanical advantage, the greater is the velocity ratio and hence what we gain in mechanical advantage, we lose in speed.

6.4. EFFICIENCY OF A LIFTING MACHINE

In an actual machine frictional resistance is not equal to zero. Hence, a part of the work-in by the effort is lost in overcoming the frictional resistance, and therefore the useful workout will always be less than the work supplied by the effort.

Thus, for an actual lifting machine:

$$W_{out} < W_{in} \qquad ...(6.8)$$

or, $$W.x < P.y$$

or $$\frac{W}{P} < \frac{y}{x}$$

or, Mechanical Advantage < Velocity Ratio ...(6.9)

The ratio of the useful workoutput of the machine to the work input by effort is known as the **efficiency** of the lifting machine, generally expressed as a percentage.

Thus, $$\text{Efficiency, } \eta = \frac{\text{Work}_{out}}{\text{Work}_{in}} = \frac{W \cdot x}{P \cdot y} \qquad ...(6.10)$$

or, $$\eta = \frac{W}{P\left(\frac{y}{x}\right)} = \frac{W}{PV} \qquad ...(6.11)$$

or $$\eta = \left(\frac{W}{P}\right) \times \frac{1}{V} = \frac{\text{Mechanical advantage}}{\text{Velocity ratio}} \qquad ...(6.12)$$

This expression of efficiency can be written in various other useful forms. For this let us define two different terms:

Let $$(VP) = \text{ideal load} = W_i \qquad ...(6.13)$$

and let $$\frac{W}{V} = \text{ideal effort} = P_i \qquad ...(6.14)$$

Ideal load is the load that the machine will lift with the given effort if there is no friction.

Ideal effort is the effort required to lift the given load if there is no friction.

Then $$\eta = \frac{W}{PV} = \frac{W}{W_i} = \frac{\text{Actual load}}{\text{Ideal load}} \qquad ...(6.15)$$

Also, $$\eta = \frac{W}{VP} = \frac{W/V}{P} = \frac{P_i}{P} = \frac{\text{Ideal effort}}{\text{Actual effort}} \qquad ...(6.16)$$

6.5. IDEAL MACHINE

The maximum efficiency of a machine will be equal to 1.0, or 100%. This is when there are no friction losses, and the work output is equal to the work input. We have seen that this machine is called an **ideal machine.**

Efficiency of a machines, as defined above, is the ratio of mechanical advantage to the velocity ratio. *Since efficiency of a machine is less than 100%, therefore velocity ratio is always greater than its mechanical advantage. Also, velocity ratio always depends on the geometrical features of the machine and hence it is the same whether machine is ideal one (without friction) or actual one (with friction). This shows that mechanical advantage for a simple machine with friction is always less than that for an ideal machine.*

6.6. FRICTION LOSSES

Friction present in a machine can be described and determined in two ways.

If P is the **actual effort** for a given load W, the ideal effort is equal to $W/V = P_i$

$\therefore$ The effort lost in friction $= P_{\text{friction}} = P_{\text{actual}} - P_{\text{ideal}}$ ...(6.17)

or, $$P_{\text{friction}} = \left(P - \frac{W}{V}\right) \quad ...(6.18)$$

Similarly, for a given effort P, the actual load lifted by a machine will always be less than that which can be lifted by an ideal or frictionless machine.

Let W_{actual} = actual load lifted by the given effort P

W_{ideal} = load lifted by the same effort P when the machine is ideal = Ideal load

Therefore, frictional loss in load =

$$W_{\text{friction}} = W_{\text{ideal}} - W_{\text{actual}} = PV - W \quad ...(6.19)$$

6.7. LAW OF THE MACHINE OR PERFORMANCE

There is a straight line relation between the effort P and the load being lifted W by a simple machine. It is given as

$$P = mW + c \quad ...(6.20)$$

Comparing the relation (6.20) with the equation of the straight line $y = mx + c$, we note that if we take the effort (P) along y-axis and the load (W) along the x-axis then,

m = slope of the line P *vs* W,

and c = intercept along P-axis.

m = slope $= \dfrac{\Delta P}{\Delta W}$, and

c = effort when load lifted is zero

$= P_0$

A plot of eqn. (6.20) is shown in Fig. 6.7.1.

Since intercept $OA = c = P_i$ = effort required to overcome frictional resistance

$\therefore$ The law of friction yields

$$P = mW + P_i \quad ...(6.21)$$

As clear from the Fig. 6.7.1, we find that the effort P_i for overcoming the frictional resistance is zero for an ideal machine. Hence, for an ideal machine the law of machine will become

$$P = mW \quad ...(6.22)$$

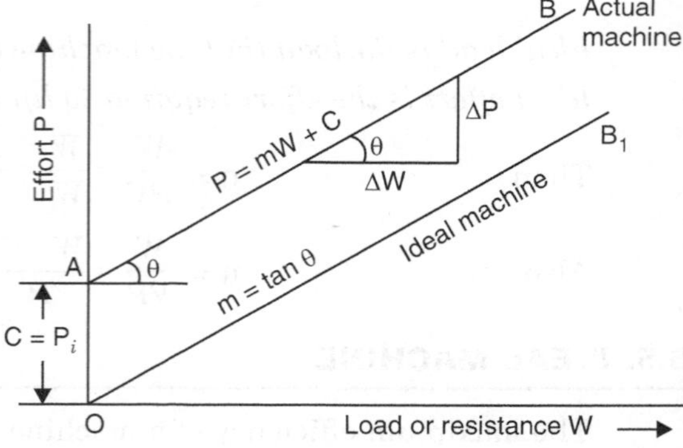

Fig. 6.7.1. Law of machine.

In Fig. 6.7.1, the line in AB is for actual machine and line OB_1 is for an ideal machine. The distance between them is equal to the effort lost in overcoming frictional resistance given by eqn. 6.2.

The slope of the line AB is given by

$$\tan \theta = m = \frac{\Delta P}{\Delta W} \quad ...(6.23)$$

Let us draw Fig. 6.7.1 again as shown in Fig. 6.7.2.

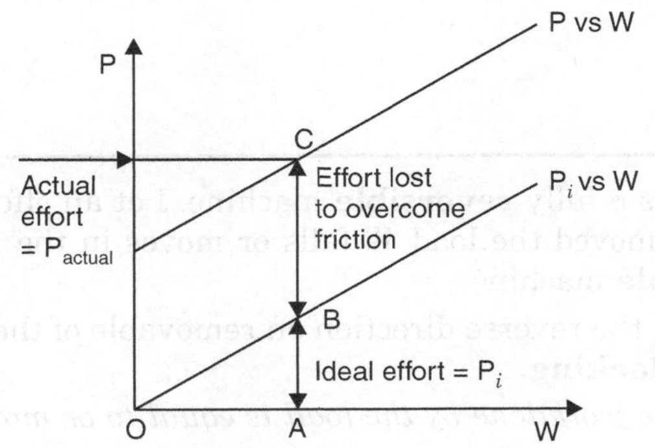

Fig. 6.7.2. Effort *vs* load.

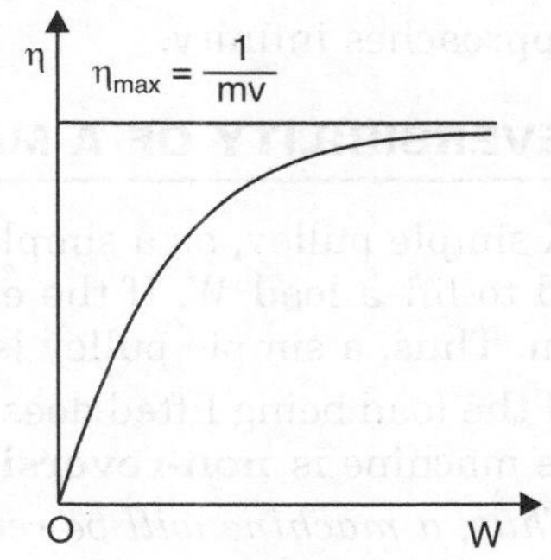

Fig. 6.7.3. Efficiency *vs* load.

In Fig. 6.7.2 ; let OA = load W being lifted and

$$AC = \text{actual effort}$$

Then, $$AB = \text{ideal effort}$$

and $$BC = \text{effort lost in overcoming friction.}$$

Since $$AC = BC + AB$$

$\therefore$ $$BC = AC - AB$$

or, $$P_{\text{friction}} = P_{\text{actual}} - P_i \quad \text{...(6.24)}$$

or, $$P_{\text{friction}} = \left(P - \frac{W}{V}\right) \quad \text{...(6.25)}$$

Let effort P_1 be required for load W_1 and effort P_2 be required for the load W_2.

Then, using the law of eqn. (6.20), we write

$$P_1 = mW_1 + c \quad \text{...}(a)$$

and $$P_2 = mW_2 + c \quad \text{...}(b)$$

$\therefore$ we get $$m = \frac{P_2 - P_1}{W_2 - W_1} = \frac{\Delta P}{\Delta W} \quad \text{...}(c)$$

6.7.1. Efficiency of Machine

Efficiency of a machine is given by

$$\eta = \frac{W}{PV} \quad \text{...}(d)$$

Substituting the value of effort $P = mW + c$,

we get $$\eta = \frac{W}{V\,[mW + c]}$$

or, $$\eta = \frac{1}{V\left[m + \dfrac{c}{W}\right]} \quad \text{...(6.26)}$$

In eqn. (6.26), m, c and V are constants.

As W increases the denominator of eqn. (6.26) decreases and therefore the efficiency η increases, as shown in Fig. 6.7.3. The maximum possible efficiency will approach

$$\eta_{max} = \frac{1}{mV} \quad \text{...(6.27)}$$

as, W approaches infinity.

6.8. REVERSIBILITY OF A MACHINE

A simple pulley, as a simple machine, is a fully **reversible** machine. Let an effort P be required to lift a load W. If the effort P is removed the load W falls or moves in the reverse direction. Thus, a simple pulley is a **reversible** machine.

If the load being lifted does not move in the reverse direction on removable of the effort, then the machine is **non-reversible** or **self-locking.**

Thus, a machine will be reversible if the workdone by the load is equal to or more than the frictional work.

Let W be the resistance or load moving through a distance d,

and P be the applied effort which moves through a distance D.

Then Work input $= PD$,

and Work output $= Wd$.

$\therefore$ Work lost in friction = (work input – work output)

$= (PD - Wd)$.

For a reversible machine, when effort is zero

the output should be > frictional losses

or, $$Wd \geq (PD - Wd)$$

or, $$2Wd \geq PD$$

$\therefore$ $$\frac{Wd}{PD} \geq \frac{1}{2}$$

or, $$\eta \geq \frac{1}{2}$$

For a reversible machine.

That is, a machine will be reversible if its efficiency is equal to or more than 50%.

Problem 6.1. *A load of 12.8 kN moves through a distance of 0.8 m by an effort of 0.8 kN which moves through 32 m.*

Determine the velocity ratio, the mechanical advantage and efficiency of the machine. Also determine the ideal effort, effort lost in friction, ideal load and the frictional resistance of the machine.

Sol. Given: $W = 12.8$ kN; $d = 0.8$ m, $P = 0.8$ kN and $D = 32$ m.

$\therefore$ **Mechanical advantage,**

$$MA = \frac{W}{P} = \frac{12.8 \text{ kN}}{0.8 \text{ kN}} = 16. \quad \textbf{Ans.}$$

Velocity Ratio, $$V = \frac{D}{d} = \frac{32}{0.8} = 40. \quad \textbf{Ans.}$$

Efficiency, $$\eta = \frac{\text{Mechanical advantage}}{\text{Velocity ratio}}$$

$$= \frac{16}{40} = 0.4. \quad \textbf{Ans.}$$

$\therefore \quad \eta = \mathbf{40\%}.$

Ideal effort; $\quad P_i = \dfrac{W}{V} = \dfrac{12.8}{40} = 0.32$ kN

Effort lost in friction $= P_{friction} = P - P_i = 0.8 - 0.32 = 0.48$ kN. **Ans.**

Ideal load, $\quad W_i = PV = 0.8 \times 40 = 32$ kN. **Ans.**

$\therefore$ Frictional resistance,

$$W_{friction} = W_i - W = 32 - 12.8 \text{ kN} = \mathbf{19.2\ kN}. \quad \textbf{Ans.}$$

Problem 6.2. *In a simple machine with velocity ratio of 40, an effort of 200 N lifts a load of 2400 N and an effort of 240 N lifts a load of 3200 N. Find, the law of the machine and calculate the load that will be lifted by an effort of 300 N. Also find*

(*i*) *the effort lost in friction,*

(*ii*) *the mechanical advantage,*

(*iii*) *the efficiency,*

(*iv*) *the frictional resistance, and*

(*v*) *the maximum efficiency of the machine.*

Sol. Let the law of friction be $P = mW + c$.

Given: When $\quad P = 200$ N; $W = 2400$ N

and when $\quad P = 240$ N; $W = 3200$ N.

$\therefore$ we get $\quad 200 = 2400\,m + c$ ...(1)

and $\quad 240 = 3200\,m + c$ ...(2)

$\therefore$ we get $\quad m = \dfrac{240 - 200}{3200 - 2400} = \dfrac{40}{800} = 0.05$

$\therefore$ From (1), $\quad 200 = 2400 \times 0.05 + c$

$\therefore \quad c = 200 - 2400 \times 0.05$

or, $\quad c = 200 - 120 = 80$

$\therefore$ The law of machine is

$$P = 0.05W + 80. \quad \textbf{Ans.} \qquad ...(3)$$

$\therefore$ For $P = 300$ N, we get

$$300 = 0.05W + 80$$

$\therefore \quad W = \dfrac{300 - 80}{0.05} = \dfrac{220}{0.05} = \mathbf{4400\ N.}$ **Ans.**

(*i*) Ideal effort $\quad = P_i = \dfrac{W}{V} = \dfrac{4400}{40} = \mathbf{110\ N.}$ **Ans.**

Effort lost in friction $\quad = P_{friction} = P - P_i = (300 - 110)$ N

or, $\quad P_{friction} = 190$ N. **Ans.**

(*ii*) Mechanical advantage, $MA = \dfrac{W}{P} = \dfrac{4400}{300} = \mathbf{14.67}$ **Ans.**

(*iii*) Efficiency, $\quad \eta = \dfrac{W}{PV} = \dfrac{4400}{300 \times 40} = 0.367 = \mathbf{36.7\%.}$ **Ans.**

(*iv*) Ideal load, $\quad W_i = PV = 300 \times 40 = 12{,}000$ N

$\therefore$ Frictional resistance, $W_{friction} = W_i - W$

$= 12000 - 4400$ N

or, $\quad W_{friction} = \mathbf{7600\ N.}$ **Ans.**

(*v*) The maximum efficiency is given by

$$\eta_{max} = \frac{1}{mV} = \frac{1}{0.05 \times 40} = \frac{1}{200} = 0.5$$

or, $\eta_{max} = 50\%$. **Ans.**

Problem 6.3. *In a lifting machine an effort of 150 N raised a load of 9000 N. What is the mechanical advantage. If efficiency of the machine is 48% what is the velocity ratio of the machine ?*

If with the same machine an effort of 215 lifted a load of 13500 N, find efficiency of the machine at this load. Also find the maximum mechanical advantage, the maximum efficiency and the law of the machine.

Sol. Given: Effort, $P = 150$ N, Load, $W = 9000$ N.

$\therefore$ Mechanical advantage, $MA = \frac{W}{P} = \frac{9000}{150} = 60$. **Ans.**

For efficiency, $\eta = 48\%$, we find effort by using the relation,

$$\eta = \frac{\text{Mechanical Advantage}}{\text{Velocity Ratio}}$$

i.e., Velocity Ratio, $V = \frac{MA}{\eta} = \frac{60}{0.48} = \mathbf{125}$. **Ans.**

When an effort of 215 N raised a load of 13500 N,

Efficiency, $\eta = \frac{W}{PV} = \frac{13500}{215 \times 125} = \mathbf{0.5023}$ or $\mathbf{50.23\%}$. **Ans.**

Let the law of the machine be

$$P = mW + c$$

Given: When, $P = 150$ N, $W = 9000$ N

and when $P = 215$ N, $W = 13500$ N

$\therefore$ we get $150 = 9000\,m + c$ ...(1)

and $215 = 13500\,m + c$ ...(2)

Solving, $m = \frac{215 - 150}{13500 - 9000} = \frac{65}{4500} = \frac{13}{900}$

$\therefore$ $c = 150 - 9000\,m$ from (1)

or $c = 150 - 9000 \times \frac{13}{900} = \mathbf{20}$

$\therefore$ Law of the machine becomes

$$P = \frac{13}{900} W + 20. \quad \textbf{Ans.} \qquad ...(3)$$

Maximum mechanical advantage $= \frac{1}{m}$

$$= \frac{1}{13/900} = \frac{900}{13} = \mathbf{69.23}. \quad \textbf{Ans.}$$

Maximum efficiency, $\eta_{max} = \dfrac{1}{mV} = \dfrac{1}{\left(\dfrac{13}{900}\right) \times 125}$

$= 0.55385$ or **55.385%. Ans.**

Problem 6.4. *Efforts required for lifting various loads by using a lifting machine are given below:*

Load lifted inN :	*100*	*200*	*300*	*400*	*500*	*600*	*700*
Effort required inN :	*22*	*35*	*48*	*61*	*74*	*87*	*100*

Plot the lift vs load diagram and find the law of the machine. Calculate efficiency and plot a graph between efficiency and the load. Find maximum efficiency if the velocity ratio of the machine is 32. What is the maximum mechanical advantage of the machine ?

Sol. Fig. P.6.4(a) shows the graph between the load W and the effort P.

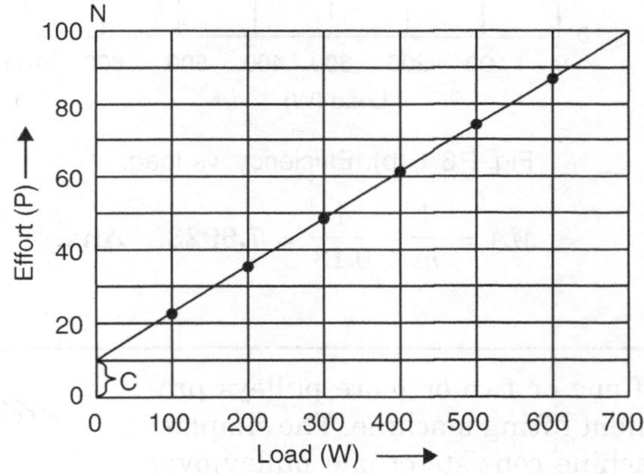

Fig. P. 6.4 (a). Effort *vs* load.

From the graph we find that $c = 9$ N and the slope can be determined from the values of effort 48 and 35 N and the load equal to 300 and 200 N, respectively.

$$\therefore \qquad m = \frac{48-35}{300-200} = \frac{13}{100} = \mathbf{0.13.}$$

∴ The law of the machine becomes

$$P = 0.13\,W + 9 \qquad \text{...(1)}$$

We find $\eta = \dfrac{MA}{V} = \dfrac{W}{P \times V} = \dfrac{W}{32P}$ (Given $V = 32$)...(2)

∴ We find that:

W in N :	100	200	300	400	500	600	700
P in N :	22	35	48	61	74	87	100
and, η in % :	14.2	17.86	19.53	20.49	21.11	21.55	22.86

Using these values a curve between efficiency and the load is plotted as shown in Fig. P. 6.4 (b).

The maximum efficiency, $\eta_{max} = \dfrac{1}{mV} = \dfrac{1}{0.13 \times 32} =$ **24.04%. Ans.**

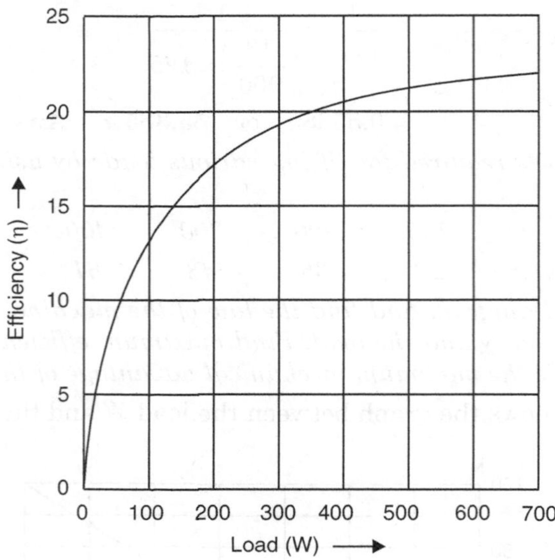

Fig. P.6.4 (b). Efficiency *vs* load.

Maximum $$MA = \frac{1}{m} = \frac{1}{0.13} = \mathbf{7.6923.} \quad \textbf{Ans.}$$

6.9. PULLEYS

An arrangement of one or two or more pulleys provides a simple and convenient lifting machine. The simplest form of this type of a machine consists of one pulley over which a rope or a chain runs, as shown in Fig. 6.9.1. In this case, velocity ratio is equal to one because both the load W and the effort P move the same distance. If the pulley surface is rough the effort will be less than the load, otherwise both will be equal. Depending on the arrangement, the systems of pulleys are classified as below:

(*a*) First order pulley system,

(*b*) Second order pulley system, and

(*c*) Third order pulley system.

Here all pulleys will be considered to be smooth.

In an ideal pulley system MA is equal to the velocity ratio.

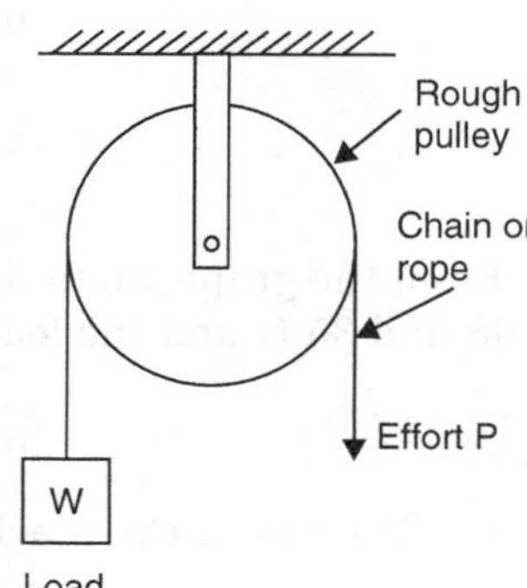

Fig. 6.9.1 The simplest lifting machine a pulley system

(*a*) First Order Pulley System

The first order pulley system is shown in Fig. 6.9.2. Pulley 1 is fixed to the support through the pulley hanger. A rope passes over this pulley and also pulley 2. On one side of the pulley effort P is applied while the other end of the rope after passing over the pulley 2 is attached to the fixed support. One end of another rope is attached to the pulley and another end is attached to the fixed support. Third rope has one end attached to pulley 3 and after passing over the pulley 4 the other end of the third rope is also attached to the fixed support. Thus, there may be an arrangement of n pulleys. As shown in Fig. 6.9.2, the pulley 1 is fixed and other pulleys are movable. The load W is suspended to the last pulley, while the effort P is applied to the free end of the rope passing over the pulley 1.

For an ideal (or frictionless) pulley system, the tension in each end of rope passing over pulley 4 is $W/2$. The load, at the centre of pulley 3 is $W/2$ and the tension in each end of rope passing over pulley 3 is $W/4$. Similarly, tension in each end of rope passing over pulley 2 is $W/8$ and that in the rope passing over pulley 1 is also $W/8$.

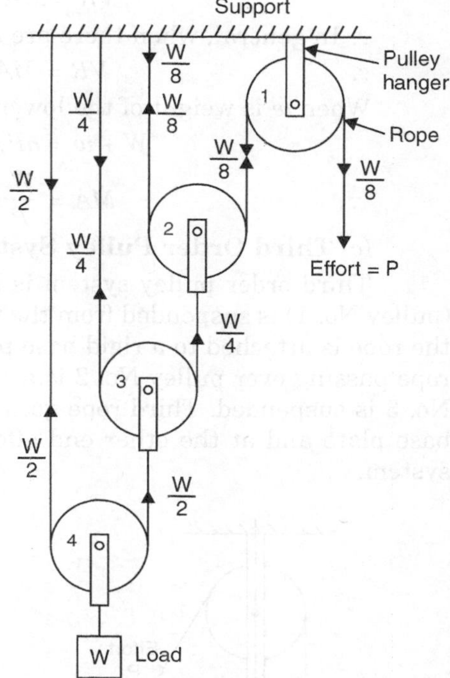

Fig. 6.9.2. First order pulley system.

Hence, effort P is equal to $W/8$.

$\therefore \quad P = \frac{W}{8}$, and hence

$$MA = \frac{W}{P} = \frac{W}{\left(\frac{W}{8}\right)} = 8 = 2^3$$

When there are 3 movable pulleys

$$MA = 2^3.$$

Since, the pulley system is smooth,

The velocity ratio, $\quad V = MA$

$\therefore \quad V = 2^3.$

If there are n movable pulleys,

Velocity ratio, $\quad V = 2^n$ ...(6.28)

Also, $\quad \because \quad \eta = 100\%, \quad \therefore \quad MA = 2^n$ ...(6.29)

When weight of each pulley is w, the $MA = \dfrac{2^n W}{W + w(2^n - 1)}$ ...[6.29 (a)]

(b) Second Order Pulley System

Fig. 6.9.3, shows a second order pulley system. In this pulley block, it is important to note that pulleys may be arranged either side by side or one below the other, as shown in Fig. 6.9.3. In Fig. 6.9.3, the top pulley block is fixed to the top support and the bottom pulley block is movable and can move vertically along with the load, attached to it. Load is suspended to the bottom pulley block with the help of a hook and the effort is applied to the other end of the rope which passes over all pulleys of the system. Let the load W be raised by the effort P.

When friction between the rope and pulleys is neglected, the tension in each portion of the rope will be equal to P. Let us pass a section 1-1 and consider the equilibrium of the lower pulley block.

By $\Sigma F_v = 0$, we find that, the load $W = 6P$

$$\therefore \quad P = \frac{W}{6},$$

Hence $$MA = \frac{W}{P} = \frac{W}{\left(\frac{W}{6}\right)} = 6 \quad ...(i)$$

For an ideal pulley system, the velocity ratio = mechanical advantage,

We note that there are 3 pulleys in each of the lower and the upper pulley blocks, and

$$VR = MA = 2 \times 3.$$

∴ In general, when there are n pulleys in the upper as well as the lower pulley blocks,

$$\therefore \qquad VR = MA = 2n \qquad ...(6.30)$$

When w is weight of the lower block

$$W + w = nP,$$

$$\therefore \qquad MA = \frac{W}{P} = n - \frac{w}{P} \qquad ...[6.30\ (a)]$$

(c) **Third Order Pulley System**

Third order pulley system is shown in Fig. 6.9.4. In this system, the top most pulley (pulley No. 1) is suspended from the top fixed support. Over pulley 1 a rope passes. One end of the rope is attached to a rigid base plate while its other end carries the pulley No. 2. Another rope passing over pulley No. 2 is attached to the rigid base plate and to the other end pulley No. 3 is suspended. Third rope goes over pulley No. 3 whose one end is attached to the rigid base plate and at the other end effort P is applied. Many pulleys may be arranged in this system.

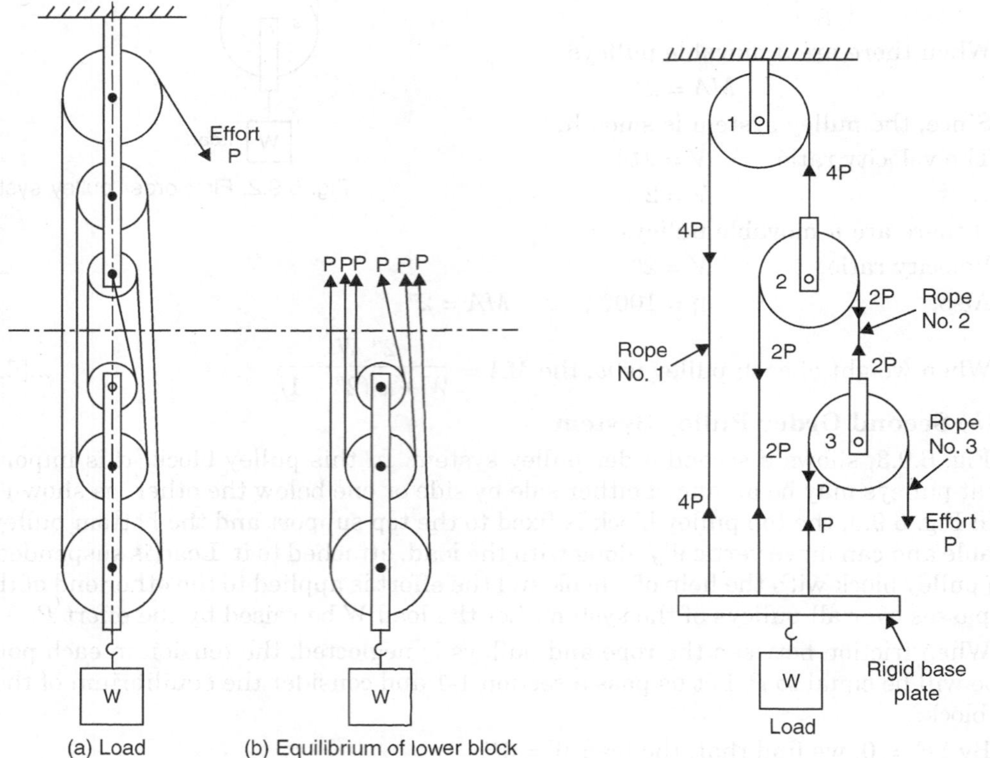

Fig. 6.9.3. Second order pulley system.

Fig. 6.9.4. Third order pulley system.

By considering the equilibrium of the rigid base plate we find that $W = 4P + 2P + P = 7P$

$$\therefore \quad MA = \frac{W}{P} = 7$$

Since this system of pulleys is ideal

hence, $VR = MA$

$\therefore$ Velocity ratio $V = 7 = 2^3 - 1$

If there are n pulleys each carrying a rope whose lower end is attached to the block of the system, then

$$\text{Velocity ratio,} \quad V = 2^n - 1 \qquad \text{...(6.31)}$$

When weight of each pulley is w, then

$$MA = (2^n - 1) + \frac{w}{P}(2^n - n - 1) \qquad \text{...[6.31 (a)]}$$

Problem 6.5. *Five movable pulleys are arranged as in the first order system. If weight of each pulley is 8 N, calculate the effort which is required for lifting a weight of 15 kN. Also find effort if the movable pulleys are of negligible weight.*

Sol. When weight of each movable pulley is w,

$$\text{then} \quad MA = \frac{2^n W}{W + w(2^n - 1)}$$

where, W = the load to be lifted, and

n = number of movable pulleys.

$$\therefore \quad MA = \frac{2^5 \times 15000}{15000 + 8(2^5 - 1)} = \frac{32 \times 15000}{15000 + 8 \times 33} = \frac{32 \times 15000}{15264}$$

or $= 31.447$

$$\text{But} \quad MA = \frac{W}{P}$$

$$\therefore \quad \frac{W}{P} = 31.447$$

$$\therefore \quad P = \frac{W}{31.447} = \mathbf{477\ N.} \quad \textbf{Ans.}$$

When weight of movable pulley is negligible,

$$MA = \frac{W}{P} = 2^n$$

Here, $\because n = 5$, $\therefore$ $MA = \frac{W}{P} = 2^5 = 32$

$$\therefore \text{ Effort,} \quad P = \frac{W}{32} = \frac{15000}{32} = \mathbf{468.75\ N.} \quad \textbf{Ans.}$$

Problem 6.6. *A lathe machine weighing 18000 N is lifted with the help of a 5-pulley second order system whose lower block weighs 200 N. Find the necessary effort required.*

Sol. Total number of pulleys = n = 5

W = 18000 N ; weight of lower block = 200 N

Total weight = $W_1 = 18000 + 200 = 18200$ N,

$MA = n = 5$

$$\therefore \quad \frac{W_1}{P} = 5$$

$\therefore \qquad P = \dfrac{W_1}{5} = \dfrac{18200}{5}$ = **3640 N. Ans.**

Problem 6.7. *A man weighing 1000 N sits on the plateform of the 5-pulley second order pulley system used to lift a machine weighing 10,000 N. The weight of the lower block is 200 N and the man supports him self by exerting a downward pull at the free end of the rope.*

(*a*) *Neglecting friction, find the minimum pull the man exerts.*

(*b*) *If friction increases pull at the rate of 5% of the load, find the pull required.*

(*c*) *Efficiency of the system.*

Sol. Total number of pulleys = n = 5,

(*a*) Total weight $\quad W$ = weight of machine + weight of man + weight of lower block

= 10000 + 1000 + 200 N

= 11200 N

Let P be the pull exerted by the man.

Due to this pull the effective load on the lower block will reduce to $W_{effective} = (11200 - P)$

Velocity ratio $= MA = 5$.

Since there is no friction,

$$VR = MA \quad \text{or} \quad \frac{MA}{VR} = 1$$

$$\therefore \qquad \frac{(11200 - P)}{P} = 5$$

$$\therefore \qquad P(5 + 1) = 11200$$

$$\therefore \qquad P = \frac{11200}{6} = \textbf{1866.67 N.} \quad \textbf{Ans.}$$

(*b*) When friction is present additional effort is ΔP equal to 5% of the total load.

$$\therefore \qquad \Delta P = \frac{5}{100} \times 11200 = 560 \text{ N.}$$

$\therefore \qquad$ Total pull = 1866.67 + 560 = **2426.67 N. Ans.**

$$(c) \qquad \text{Efficiency} = \frac{\text{Pull without friction}}{\text{Pull with friction}} \times 100$$

$$= \frac{1866.67}{2426.67} \times 100 = \textbf{76.92\%.} \quad \textbf{Ans.}$$

Problem 6.8. *A 6-pulley third order system, including the fixed pulley, is used to lift a load weighing W by exerting a pull of 200 N.*

(*a*) *Find the weight which can be lifted if weight of pulleys is neglected.*

(*b*) *If each pulley weighs 20 N, find the weight which can be lifted.*

(*c*) *What weight can be supported without any effort when each pulley weighs 20 N.*

Sol. (*a*) When pulleys are weightless,

For third order system of pulleys,

$$\frac{W}{P} = 2^n - 1 \qquad \text{...(1)}$$

Here, $\qquad P = 200$ N; and $n = 6$.

$$\therefore \qquad W = P \times (2^6 - 1) = 200 \times [63] \text{ N}$$

= **12600 N. Ans.**

(*b*) When weight of each pulley is w, we have

$$\frac{W}{P} = (2^n - 1) + \frac{w}{P}[2^n - n - 1] \qquad ...(2)$$

$$\therefore \quad W = P[(2^6 - 1) + 20 \times (2^6 - 6 - 1)]$$

$$= 200\,[63] + [20 \times 57] = 12600 + 1140$$

$$= \mathbf{13740\ N.} \quad \textbf{Ans.}$$

From eqn. (2), $W = (0)\,[2^6 - 1] + 20 \times (2^6 - 6 - 1) =$ **1140 N. Ans.**

Problem 6.9. *Find the pull required to lift a load of 16000 N with the system of pulleys shown in Fig. P. 6.9 whose efficiency is 80%. Also find pull when efficiency is 100%.*

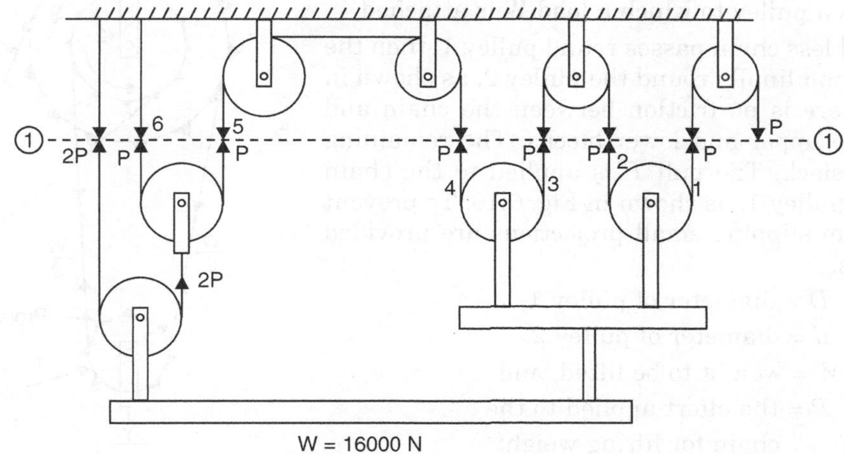

Fig. P. 6.9. The given system of pulleys.

Sol. If the given system is cut by a section 1-1, then we find that the tension in each of ropes 1 to 6 cut by the section 1-1 is equal to P, and the tension in 7 is $2P$, where P is the required pull. Finally, we find that

$$W = 6P + 2P = 8P$$

$$\therefore \quad MA = \frac{W}{P} = 8.$$

When the system is smooth, $VR = MA = 8$

But here the given system has an efficiency equal to 0.80.

$$\therefore \quad \eta = \frac{MA}{VR}$$

or, $$0.80 = \frac{MA}{VR}$$

$$\therefore \quad MA = 0.80\ VR = 0.8 \times (8)$$

$$\therefore \quad \frac{W}{P} = 0.80 \times 8$$

$$\therefore \quad P = \frac{W}{0.80 \times 8} = \frac{16000}{0.8 \times 8} = \frac{16000}{6.4}$$

or, Pull = **2500 N. Ans.**

when $\eta = 100\%$, we have $\dfrac{MA}{VR} = 100\% = 1.0$

or, $$\frac{16000}{8P} = 1$$

$$P = \frac{16000}{8} = 2000 \text{ N.}$$

6.10. WESTON DIFFERENTIAL PULLEY BLOCK (FIG. 6.10)

It consists of two blocks *A* and *B*. The upper block *A* has two integral pulleys-pulley 2 having diameter slightly smaller than that of the pulley 1. Both pulleys rotate with the same angular velocity. The lower block *B* also carries a pulley to which a load *W* is attached.

An end less chain passes round pulley 1, then the lower pulley and finally round the pulley 2, as shown in Fig. 6.10. There is no friction between the chain and pulleys at the upper and lower blocks. The remaining chain hangs slack. The pull *P* is applied to the chain passing over pulley 1, as shown in Fig. 6.10. To prevent the chain from slipping small projections are provided to the pulleys.

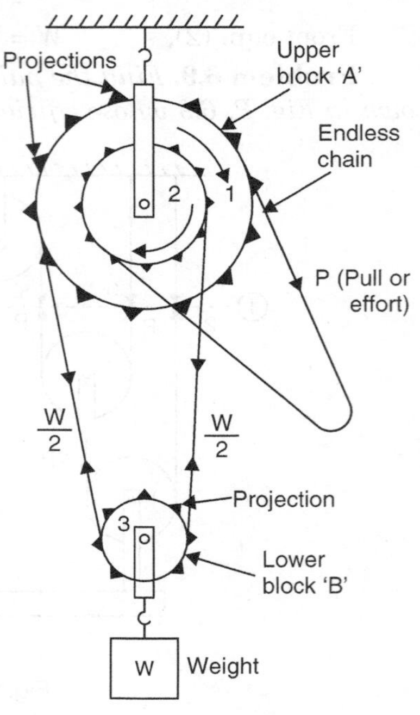

Fig. 6.10. Weston differential pulley block.

Let D = diameter of pulley 1,

d = diameter of pulley 2,

W = weight to be lifted, and

P = the effort applied to the chain for lifting weight

For one complete revolution of the pulley 1 the effort moves through the distance πD. The chain moves over the pulley 2 through πd = length of chain released by pulley 2.

Hence, the net shortening of the chain = $\pi D - \pi d = \pi(D - d)$.

This shortening of the chain is equally divided into two portions of chain passing round the lower pulley 3 and supporting the weight.

$\therefore$ Velocity ratio, $$VR = \frac{\text{Distance moved by } P}{\text{Distance moved by } W}$$

or $$VR = \frac{\pi D}{\frac{\pi}{2}(D - d)} = \frac{2D}{D - d} \quad \ldots(6.32)$$

Sometimes recesses are provided to the pulleys, then velocity ratio can be expressed in terms of recesses.

Let n_1 = number of recesses on pulley 1,

n_2 = number of recesses on pulley 2,

then, velocity ratio $$= VR = \frac{2n_1}{n_1 - n_2} \quad \ldots(6.33)$$

And, mechanical advantage, $$MA = \frac{W}{P} \quad \ldots(6.34)$$

And, finally, the efficiency, $\eta = \dfrac{W/P}{V} = \dfrac{W}{VP}$ ...(6.35)

Problem 6.10. *A Weston differential pulley blocks has two logged pulleys in the upper block having diameters of 400 mm and 300 mm. If efficiency of the system is 60%, find the pull required to lift a weight of 3000 N.*

Sol. Velocity ratio $= VR = \dfrac{2D}{D-d} = \dfrac{2 \times 400}{400 - 300}$

or $VR = \dfrac{800}{100} = 8$

$\therefore$ Efficiency, $\eta = \dfrac{\text{M.A.}}{VR} = \dfrac{W}{VP}$

$\therefore$ We write $0.60 = \dfrac{3000}{8 \times P}$

$\therefore$ The pull, $P = \dfrac{3000}{0.6 \times 8}$ = **625 N. Ans.**

6.11. SIMPLE WHEEL AND AXLE (FIG. 6.11)

Fig. 6.11 shows a simple wheel and axle lifting machine. The wheel and axle are keyed to the same spindle which runs in ball or roller bearings to reduce the friction. One rope is fixed to the wheel and is wound a few turns to the wheel. On the other end of this rope the pull P is applied. One end of the second rope is fixed to the axle and wound in the opposite direction to the axle and to the other end of this rope the load W is hung. When the wheel rope unwinds the rope on the axle winds on the axle and the load is raised.

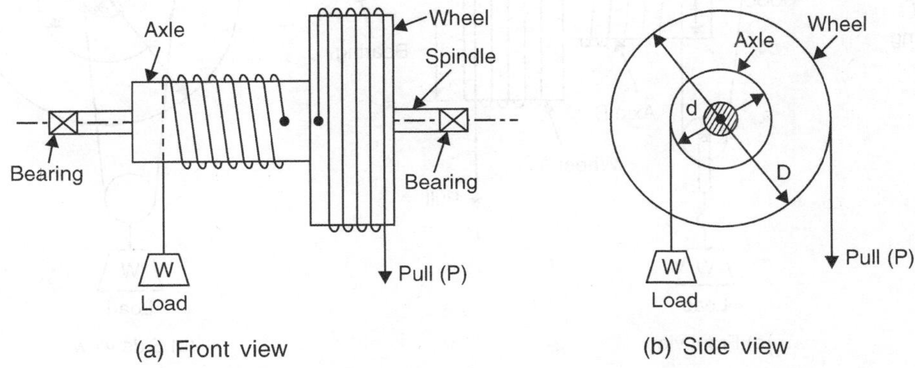

Fig. 6.11. Simple wheel and axle.

Let W = load to be lifted,
P = pull or effort
D = diameter of the wheel
d = diameter of the axle.

In one revolution, the effort P moved a distance $= \pi D$
and the load is lifted through a distance $= \pi d$

$\therefore$ The velocity ratio, $VR = \dfrac{\pi D}{\pi d} = \dfrac{D}{d}$ ...(6.36)

When the diameters of the two ropes are not negligible,

then velocity ratio, $$VR = \frac{\pi(D+d_1)}{\pi(d+d_2)} = \frac{D+d_1}{d+d_2} \quad ...(6.37)$$

where d_1 = diameter of rope wound on the wheel

and d_2 = diameter of rope wound on the axle.

Also $$MA = \frac{W}{P}, \text{ and}$$

$$\eta = \frac{MA}{VR}.$$

6.12. WHEEL AND DIFFERENTIAL AXLE (Fig. 6.12)

Fig. 6.12 shows wheel and differential axle as a lifting machine. There is a wheel A of diameter D and a rope is wound a few turns on it. One end of the rope is fixed on the wheel and on the other end a pull P is applied. There is a stepped (or differential) axle of diameters d_1 and d_2. One end of the second rope is tied to axle C of diameter d_2 and is wound a few turns in the same direction as the first rope is wound on the wheel. The remaining portion of second rope is wound on the axle B of diameter d_1 in the opposite direction and the other end of the second rope is fixed on the axle B. The load W to be lifted is hung on the a pulley mounted on the second rope.

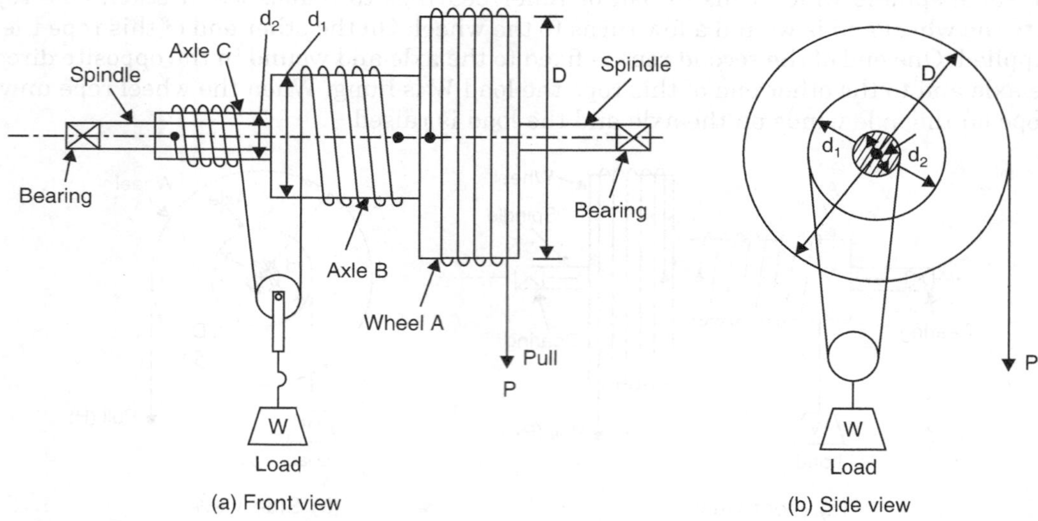

Fig. 6.12. Wheel and differential axle.

Let D = wheel diameter

d_1 = diameter of axle B

d_2 = diameter of axle C.

In one revolution of wheel effort P lowers through πD

rope wound on axle $B = \pi d_1$

rope unwound from axle $C = \pi d_2$

$\therefore$ The length of rope wound on the axle $= \pi d_1 - \pi d_2$

and therefore, the load W is lifted up through the distance $= \frac{\pi(d_1 - d_2)}{2}$

$$\therefore \quad \text{Velocity ratio,} \quad VR = \frac{\text{Distance moved by effort}}{\text{Distance moved by load}} = \frac{\pi D}{\frac{\pi(d_1 - d_2)}{2}}$$

or
$$VR = \frac{2D}{d_1 - d_2} \quad ...(6.38)$$

If d_{r_1} and d_{r_2} are diameters of rope wound on wheel and the rope wound on axles B and C, then

$$\text{Velocity ratio,} \quad VR = \frac{\pi(D + d_{r_1})}{\frac{\pi}{2}[(d_1 + d_{r_2}) - (d_2 + d_{r_2})]}$$

or
$$VR = \frac{2(D + d_{r_1})}{(d_1 - d_2)} \quad ...(6.39)$$

Problem 6.11. *A wheel and differential axle has the following particulars:*
Diameter of wheel = 400 mm;
Diameters of the differential axles = 200 mm and 150 mm.
Pull required to lift a load of 2 kN is 500 N, while a pull of 1000 N is required to lift a load of 5 kN.
Find: (a) The law of the machine.
(b) Effort lost in friction for the load of 5 kN.
(c) The maximum efficiency.
(d) The maximum load.
(e) Frictional resistance for 5 kN load, and
(f) Efficiency for 5 kN load.

Sol. Let the law of the machine be
$$P = mW + c \quad ...(1)$$
The two sets of values of P and W are :
(*i*) $P_1 = 500$ N, and $W_1 = 2\text{kN} = 2000$ N
(*ii*) $P_2 = 1000$ N, and $W_2 = 5$ kN $= 5000$ N
Substituting in eqn. (1), we get
$$500 = m \times 2000 + c \quad ...(2)$$
and
$$800 = m \times 5000 + c \quad ...(3)$$
$$\therefore \quad (800 - 500) = m(5000 - 2000)$$
$$\therefore \quad m = \frac{300}{3000} = 0.10$$
$\therefore$ From (2) $\quad c = 500 - 2000 \times 0.1 = 300$
(*a*) $\therefore$ The law of machine is
$$P = 0.1 \times W + 300. \quad \textbf{Ans.} \quad ...(4)$$
(*b*) Velocity ratio
$$= VR = \frac{2D}{d_1 - d_2} = \frac{2 \times 400}{(200 - 150)} = \frac{800}{50} = 16.$$
$\therefore$ Effort lost in friction for 5 kN load
$$= P_{\text{friction}} = P - \frac{W}{VR} = 1000 - \frac{5000}{16}$$
$$= \mathbf{687.5\ N.}\ \ \textbf{Ans.}$$

(*c*) Maximum efficiency,

$$\eta_{max} = \frac{1}{m(VR)} = \frac{1}{0.1 \times 16} = 0.625$$

$$= \mathbf{62.5\%.} \quad \textbf{Ans.}$$

(*d*) Maximum mechanical advantage

$$= \frac{1}{m} = \frac{1}{0.1} = \mathbf{10.} \quad \textbf{Ans.}$$

(*e*) Frictional resistance at 5 kN load,

$$W_{friction} = W_i - W$$
$$= PV - W$$
$$= 1000 \times 16 - 5000 = \mathbf{9000\ N.} \quad \textbf{Ans.}$$

(*f*) Efficiency for 5 kN load

$$= \frac{W}{P \times V} = \frac{5000}{1000 \times 16}$$

$$= 0.31.25 \quad \text{or} \quad \mathbf{31.25\%.} \quad \textbf{Ans.}$$

6.13. WINCH CRABS

The winch crabs are the lifting machines in which spur gears are used for increasing the velocity ratio. If there is one set of spur gears the winch crab is called the **single purchase winch crab** and when there are two sets of spur gears the winch crab is called the double purchase winch crab.

6.14. SINGLE PURCHASE WINCH CRAB (Fig. 6.14)

It consists two axles A and B. On axle A a handle of radius R and a spur gear G_1 of teeth T_1 are mounteds (Fig. 6.14). On the axle B are mounted a large spur gear or wheel G_2 of teeth T_2 and a load drum of radius r. Gear wheel G_2 meshes with the pinion G_1. A rope which is attached to the load drum is wound round the drum. A load W is attached the second end of the rope. Effort P is applied at the handle.

When handle is turned through one turn the effort P moves a distance $= 2\pi R$

The gear G_1 also turn 1 revolution.

The angle turned by the gear $G_2 = 1 \times \dfrac{T_1}{T_2}$ revolutions

The distance through which the load W moves is equal to the length of the rope wound on the load drum

$$= \left(1 \times \frac{T_1}{T_2}\right) \times 2\pi r.$$

$$\therefore \quad \text{Velocity ratio,} \quad VR = \frac{\text{Distance moved by the effort } P}{\text{Distance moved by the load } W}$$

$$= \frac{2\pi R}{2\pi r \times \left(1 \times \dfrac{T_1}{T_2}\right)}$$

or $$VR = \frac{T_2}{T_1} \times \frac{R}{r} \quad ...(6.40)$$

Efficiency, $$\eta = \frac{MA}{VR} = \frac{W/P}{\frac{T_2}{T_1} \times \frac{R}{r}} = \left(\frac{T_1 . r}{T_2 . R}\right) \times \frac{W}{P} \quad ...(6.41)$$

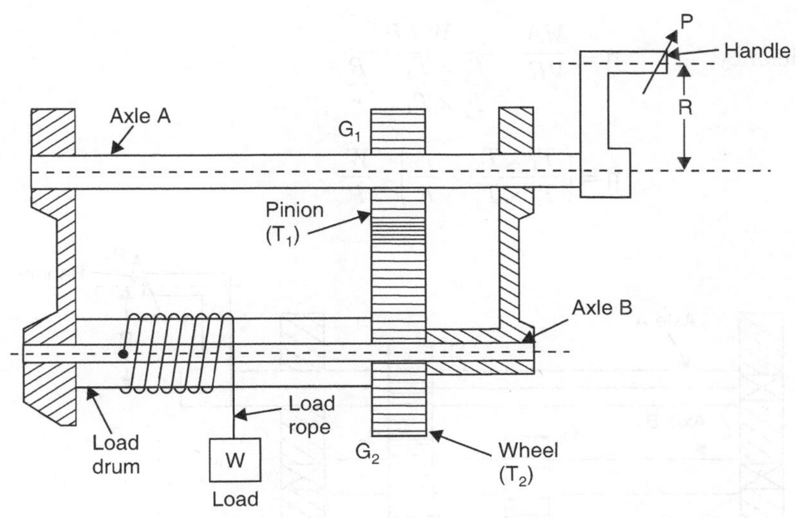

Fig. 6.14. Single purchase winch crab.

6.15. DOUBLE PURCHASE WINCH CRAB (Fig. 6.15)

There are three axles *A*, *B* and *C*. An effort handle of length *R* and a spur gear (pinion) G_1 of teeth T_1 are mounted on axle *A*. Axle *B* carries a spur gear G_2 having T_2 teeth and another spur gear G_3 having teeth T_3. Gear G_2 meshes with gear G_1 and gear G_3 meshes with gear G_4 of teeth T_4 mounted on axle *C*. A load drum of radius *r* is also mounted on the axle *C*. A rope is wound a few turns on the load drum. One end of the rope is fixed to the load drum and a load *W* is suspended at its other end.

When the effort handle is turned 1 turn, the distance moved by the effort $P = 2\pi R$.

For one turn of the gear G_1 the gear G_2 turns through $\left(1 \times \frac{T_1}{T_2}\right)$ turns.

Gear G_3 also turns through an angle $= \left(1 \times \frac{T_1}{T_2}\right)$ turns.

And the angle turned by gear G_4 $= \left(1 \times \frac{T_1}{T_2}\right) \times \frac{T_3}{T_4}$ turns.

∴ Distance moved by the load $= \left(1 \times \frac{T_1}{T_2} \times \frac{T_3}{T_4}\right) \times 2\pi r$

∴ $$\text{Velocity ratio} = \frac{\text{Distance moved by the effort } P}{\text{Distance moved by the load } W}$$

$$= \frac{2\pi R}{\left(1 \times \frac{T_1}{T_2} \times \frac{T_3}{T_4}\right) \times 2\pi r}$$

or $$VR = \left(\frac{T_2 \times T_4}{T_1 \times T_3}\right) \times \frac{R}{r} \quad ...(6.42)$$

And, efficiency, $$\eta = \frac{MA}{VR} = \frac{W/P}{\frac{T_2 \times T_4}{T_1 \times T_3} \times \frac{R}{r}}$$

or $$\eta = \left(\frac{T_1 \times T_3}{T_2 \times T_4} \times \frac{r}{R}\right) \times \frac{W}{P} \quad ...(6.43)$$

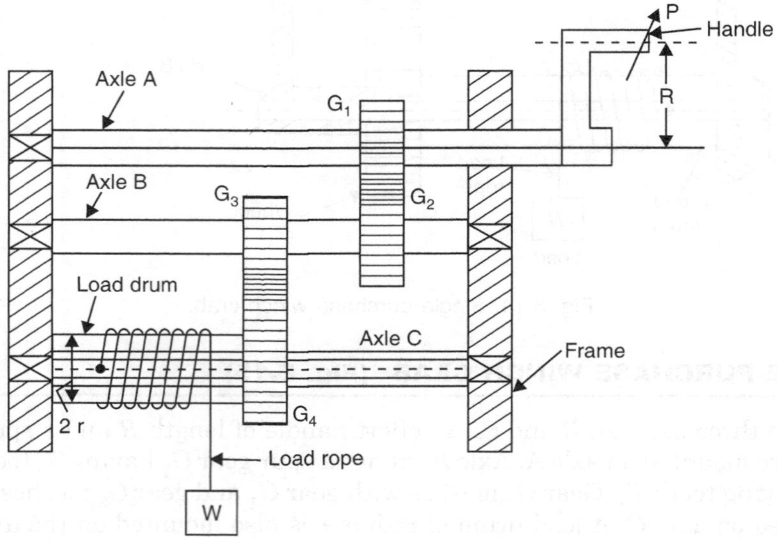

Fig. 6.15. Double purchase winch crab.

Problem 6.12. *The following are particulars of a single purchase winch crab:*

Length of lever = 1 m, Diameter of the load drum = 100 mm, Number of teeth on spur wheel = 112 ; Number of teeth on pinion = 16.

Find velocity ratio of the machine. A load of 4 kN requires a pull of 100 N and for a load of 10 kN a pull of 160 N is required. Find the law of machine and efficiency for both loads. Also find the maximum efficiency, the maximum mechanical advantage and effort lost in friction.

Sol. Given: Length of arm = R = 1 m = 100 cm.

Load drum diameter, $d = \frac{200}{2} = 100$ mm = 10 cm.

$T_1 = 16; T_2 = 112.$

$$\therefore \quad VR = \frac{2R}{d} \times \frac{T_2}{T_1} = \frac{2 \times 100}{10} \times \frac{112}{16} = \textbf{140.} \quad \textbf{Ans.}$$

Let law of machine be

$$P = mW + c \qquad ...(1)$$

Given: $P_1 = 100$ N; $W_1 = 4$ kN = 4000 N

and $P_2 = 160$ N; $W_2 = 10$ kN = 10,000 N.

$\therefore$ We get

$$100 = 4000\,m + c \qquad ...(2)$$

$$160 = 10000\,m + c \qquad ...(3)$$

Solving (2) and (3), we get

$$m = \frac{160 - 100}{10000 - 4000} = \frac{60}{6000} = \frac{1}{100} = 0.01$$

And from (2), $c = 100 - 4000\,m = 100 - 4000 \times 0.01 = 60$

$\therefore$ Law of machine is

$$P = 0.01\,W + 60. \quad \textbf{Ans.} \qquad ...(4)$$

For $W_1 = 4$ kN; $MA = \frac{W}{P} = \frac{4000}{100} = 40$

$$\therefore \quad \eta = \frac{MA}{VR} = \frac{40}{140} = 0.2857 \quad \text{or} \quad \textbf{28.57\%. Ans.}$$

For $W_2 = 10$ kN = 10,000 ; $MA = \frac{10000}{160} = 62.5$

$$\therefore \quad \eta = \frac{MA}{VR} = \frac{62.5}{140} = 0.4464 \quad \text{or} \quad \textbf{44.64\%. Ans.}$$

Maximum efficiency

$$\eta_{max} = \frac{1}{mV} = \frac{1}{0.01 \times 140} = 0.7143 \quad \text{or} \quad \textbf{71.43\%. Ans.}$$

Maximum mechanical advantage

$$= \frac{1}{m} = \frac{1}{0.01} = \textbf{100. Ans.}$$

Effort lost in friction $= P_{friction} = \left(P - \frac{W}{V}\right) \qquad ...(5)$

For $W_1 = 4000$ N; $P_1 = 100$ N

$$\therefore \quad P_{friction} = \left(100 - \frac{4000}{140}\right) = \textbf{71.43 N. Ans.}$$

For $W_2 = 10{,}000$ N, $P_2 = 160$ N,

$$\therefore \quad P_{friction} = \left(160 - \frac{10000}{140}\right) = \textbf{88.57 N. Ans.}$$

Problem 6.13. *In a double purchase crab, pinions have 16 and 24 teeth, while the gear wheels have 44 and 36 teeth. The effort handle is 600 mm long and the load drum diameter 90 mm. What load will be lifted by an effort of 250 N applied at the handle if efficiency of the machine is 44% ? What effort is lost in friction ?*

Sol. $T_1 = 16, \quad T_2 = 44, \quad T_3 = 24$ and $T_4 = 36$.

[Check $T_1 + T_2 = 16 + 44 = 60$ and $T_3 + T_4 = 24 + 36 = 60$]

Length of handle, $l = 600$ mm,

Radius of load drum, $r = \frac{90}{2} = 45$ mm.

$$\therefore \qquad VR = \frac{l}{r} \times \frac{T_2 \times T_4}{T_1 \times T_3} = \frac{600}{45} \times \frac{44 \times 36}{16 \times 24} = 55$$

Mechanical advantage, $\quad MA = \dfrac{W}{P} = \dfrac{W}{250}$

Efficiency, $\quad \eta = \dfrac{MA}{VR} = \dfrac{W/250}{55} = 0.44$

$\therefore \quad W = 0.44 \times 250 \times 55 = \mathbf{6050\ kN.}$ **Ans.**

Effort lost in friction, $\quad P_{\text{friction}} = \left(P - \dfrac{W}{V}\right) = \left(250 - \dfrac{6050}{55}\right) = \mathbf{140\ N.}$ **Ans.**

6.16. WORM AND WORM WHEEL (Fig. 6.16)

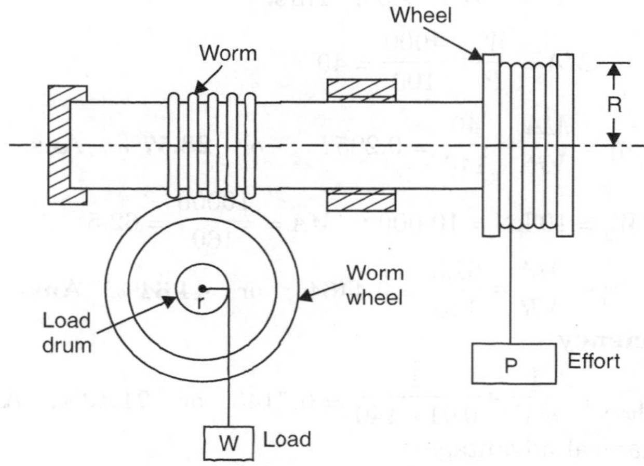

Fig. 6.16 Worm and worm wheel.

It consists of square threaded screw known as worm and a mating tooth wheel known as the worm wheel. On the worm axle an effort wheel of radius R is mounted. For one turn of the effort wheel and the worm the worm wheel advances only one tooth. There is a load drum keyed to the axle of the worm wheel, (Fig. 6.16). Let T be the number of teeth on the worm wheel and let the radius of the load drum be r. Let W be the load lifted by the effort. For one turn of the effort wheel, the distance moved by the effort

$$= y = 2\pi R. \qquad \text{...(1)}$$

The angle turned by the worm wheel

$$= \frac{1}{T} \text{ revolutions} = \frac{2\pi}{T} \text{ radians} = \theta$$

$\therefore$ The distance moved by the load $W = x = r\theta$

or $$x = \frac{2\pi r}{T} \qquad \text{...(2)}$$

$\therefore$ Velocity ratio, $$VR = \frac{y}{x} = \frac{2\pi R}{2\pi r / T}$$

or $$VR = \frac{RT}{r}$$

$$MA = \frac{W}{P}$$

$$\eta = \frac{MA}{VR} = \frac{W}{P} \cdot \frac{r}{RT} \qquad \text{...(6.44)}$$

If the worm is double threaded (*i.e.,* for one turn of the worm the worm wheel advances through two teeth or $\frac{2}{T}$ revolutions

or
$$\theta = \left(\frac{2\pi}{T}\right) \times 2 = \frac{4\pi}{T} \qquad \text{...(3)}$$

And, the load will be raised by $x = \theta r = \frac{\pi r}{T}$ ...(4)

$$\therefore \qquad VR = \frac{y}{x} = \frac{2\pi R}{\frac{4\pi r}{T}}$$

$$\therefore \qquad VR = \frac{RT}{2r} \qquad \text{...(6.45)}$$

$$MA = \frac{W}{P} \qquad \text{...(5)}$$

and
$$\eta = \frac{MA}{VR} = \frac{W}{P} \times \frac{2r}{RT} \qquad \text{...(6.46)}$$

as usual.

Problem 6.14. *In a worm and worm wheel, the effective diameter of the effort wheel is 280 mm. The worm has single start threads and the worm wheel has 100 teeth. The effective diameter of the load drum is 80 mm. What is the efficiency if a load if 15 kN is lifted by an effort 100 N ?*

Sol. Effort wheel radius = $R = \frac{280}{2} = 140$ mm,

Number of teeth on the worm wheel, $T = 100$

Load drum radius, $r = \frac{80}{2} = 40$ mm;

Load, $W = 15$ kN $= 15000$ N and effort $= P = 100$ N

$$\therefore \qquad VR = \frac{RT}{r} = \frac{140 \times 100}{40} = 350$$

$$MA = \frac{W}{P} = \frac{15000}{100} = 150$$

$$\therefore \qquad \eta = \frac{MA}{VR} = \frac{150}{350} = 0.42857 \quad \text{or} \quad \mathbf{42.857\%.} \quad \textbf{Ans.}$$

6.17. SCREW JACK (Fig. 6.17)

This is a device commonly used to lift heavy loads. It works on the principle of inclined plane with effort applied in horizontal direction.

[**Note.** Screw Jack has also been discussed in details in chapter on Friction in this book.]

Fig. 6.17 (*a*) shows a screw jack. It has a spindle with square threads and a frame or body in which internal threads are cut to provide a nut to the spindle. The load is carried by the screw head or a cup onto the top of the spindle as shown in the Fig. 6.17 (*a*). The spindle is turned by the pull P applied to a lever of length l.

In case of single start threads the load moves axially, up or down, through a distance known as **pitch (*p*)** of the screw threads cut on spindle when the lever turns through one turn.

For n-start threads for one turn of the spindle load moves axially through a distance called the **lead** which is equal to $n \times p$.

When one turn of the screw thread is developed we get a plane inclined at an angle α (called the helix angle) as shown in Fig. 6.17 (b).

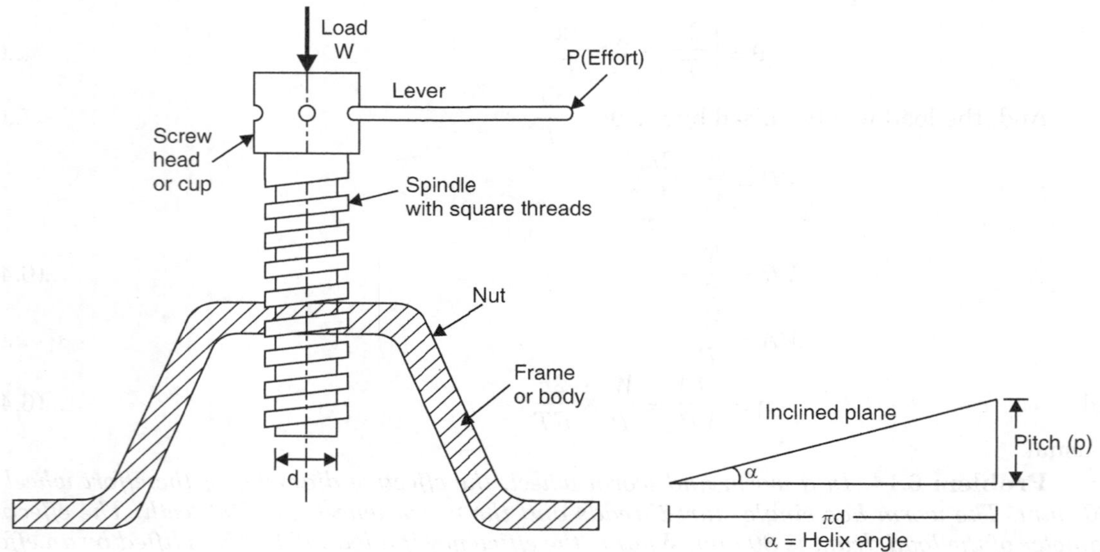

Fig. 6.17. (a) Screw jack

Fig. 6.17. (b) Inclined plane

Helix angle α is found from the relation:

$$\tan \alpha = \frac{\text{Pitch}}{\text{Circumference}} = \frac{p}{\pi d} \qquad ...(6.47)$$

for a single start threads.

However, the helix angle α for n-start threads is given by

$$\tan \alpha = \frac{\text{Lead}}{\text{Circumference}} = \frac{np}{\pi d} \qquad ...(6.48)$$

where d = mean diameter of threads on the spindle.

Load W lying on the cup of the screw jack may be treated as the load lying on the inclined plane (Fig. 6.17 (c))

Let P be the effort for raising the load.

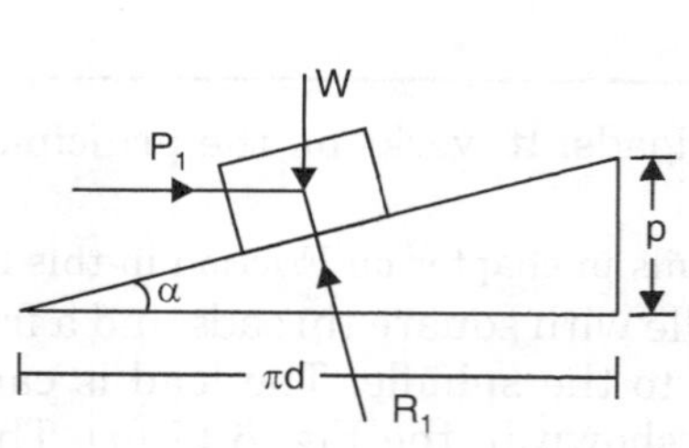

Fig. 6.17 (c). Forces acting on the load

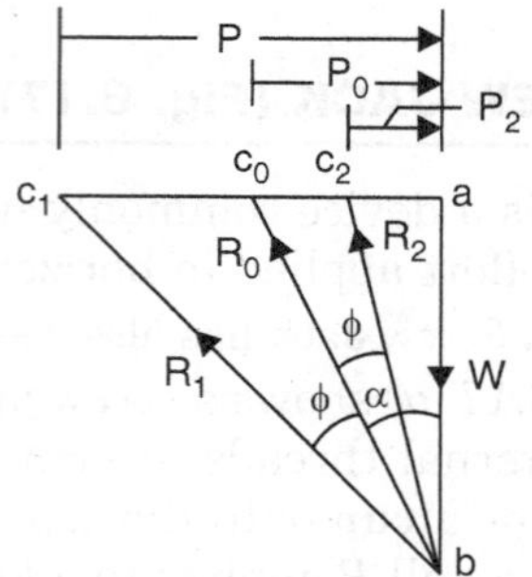

Fig. 6.17 (d). Triangles of forces.

Fig. 6.17 (c) shows the forces acting on the load. Fig. 6.17 (d) shows the $\Delta\, abc_1$ of forces W, P_1 and the reaction R_1 for raising the load W.

Inclination of R_1 to the load W is $(\alpha + \phi)$

where α = helix angle and

$$\phi = \text{friction angle} = \tan^{-1} \mu. \quad ...(6.49)$$

We find that effort required at the mean surface of the spindle is:

$$P_1 = W \tan (\alpha + \phi) \quad ...(6.50)$$

And the moment of P_1 about the centre line of the spindle.

We get $$P_1 \times \frac{d}{2} = P_1' \times l$$

$$\therefore \quad P_1' = \frac{P_1 d}{2l} = \frac{W \tan (\alpha + \phi) \times d}{2l} \quad ...(6.51)$$

where P_1' = effort required at the end of lever.

For lowering the load, from Δabc_2, we get the effort at the spindle,

$$P_2 = W \tan (\alpha - \phi) \quad ...(6.52)$$

And, effort at the lever will be

$$P_2' = \frac{P_2 d}{2l} = [W \tan (\alpha - \phi)] \times \frac{d}{2l} \quad ...(6.53)$$

We also find that the velocity ratio,

$$VR = \frac{\text{distance moved by effort}}{\text{distance moved by the load}}$$

or $$VR = \frac{2\pi l}{p} \quad ...(6.54)$$

Mechanical advantage, $$MA = \frac{W}{P_1} \quad ...(6.55)$$

Efficiency of the screw jack as a lifting machine,

$$\eta = \frac{MA}{VR} = \frac{W/P}{\left(\frac{2\pi l}{p}\right)}$$

or $$\eta = \frac{Wp}{2\pi l P} \quad ...(6.56)$$

Alternatively, the efficiency of the screw jack can be found by using the concept of ideal effort. When there is no friction, effort,

$$P_0 = W \tan \alpha \quad | \text{ (from } \Delta abc_0) \quad ...(6.57)$$

and $$\eta = \frac{P_0}{P_1} = \frac{\tan \alpha}{\tan (\alpha + \phi)} \quad ...(6.58)$$

[For details see chapter on Friction]

Problem 6.15. *A screw jack has square threads 60 mm diameter and 12 mm pitch. A load of 8 kN being raised by the screw jack rotates with the spindle. The coefficient of friction at the spindle is 0.05.*

(*a*) *Find the horizontal effort required at the end of a 450 mm lever.*

(*b*) *State whether the screw jack is self-locking. If not, find the torque which must be applied to keep the load from descending.*

Sol. Here, $\phi = \tan^{-1} \mu = \tan^{-1} 0.05 = 2.8624°.$

Helix angle $$\alpha = \tan^{-1}\left(\frac{p}{\pi d}\right) = \tan^{-1}\left(\frac{12}{\pi \times 60}\right) = \tan^{-1} 0.06366$$

$\therefore \quad \alpha = 3.6426°$

$\therefore \quad \alpha + \phi = 3.6426 + 2.8624 = \mathbf{6.5050°}$ and $\alpha - \phi = 0.7802°$

(*a*) Tangential force, to be applied on the surface of screw threads

$$= P = W \tan(\alpha + \phi)$$

or
$$P = 8000 \tan 6.5050 = \mathbf{912.19\ k}$$

Force to be applied at the lever

$$= P' = \frac{P \times d}{2l} = \frac{912.19 \times 60}{2 \times 450} \text{ N} = \mathbf{60.813\ N.} \quad \textbf{Ans.}\ (a)$$

(*b*)
$$VR = \frac{2\pi l}{p} = \frac{2\pi \times 450}{12} = \mathbf{235.62}$$

$$MA = \frac{W}{P'} = \frac{8000}{60.813} = \mathbf{131.55}$$

$$\therefore \quad \eta = \frac{MA}{VR} = \frac{131.55}{235.62} = 0.5583 \text{ or } \mathbf{55.83\%}$$

Since η is more than 50%, **the screw jack is not self-locking. Ans.**

∴ Torque required to keep the load from descending

$$= \frac{d}{2} \times W \tan(\alpha - \phi)$$

$$= \frac{60}{2} \times 8000 \tan(0.7802°)$$

$$= \mathbf{3268.3\ Nmm. = 3.2683\ Nm.\ Ans}$$

6.18. DIFFERENTIAL SCREW JACK (Fig. 6.18)

A differential screw jack is shown in Fig. 6.18. It is an improvement over the simple screw jack. It consists of two threaded spindles both having the right-handed threads. The spindle *B* is hollow and has both internal and external threads. External threads on *B* fit into the internal threads of the body *A* of the screw jack. The external threads on *C* fit into the threads cut on the inner surface of the *B*. Thus, the spindle *B* acts as a screw for the nut *A* and also as a nut for the element *C*. With the help of a lever inserted in the holes made on the top of the block *D*, which is attached to the spindle *C*, block *D* can be rotated. When *D* is rotated, *B* rotates with it. Rotation of *C* is prevented by suitable arrangement.

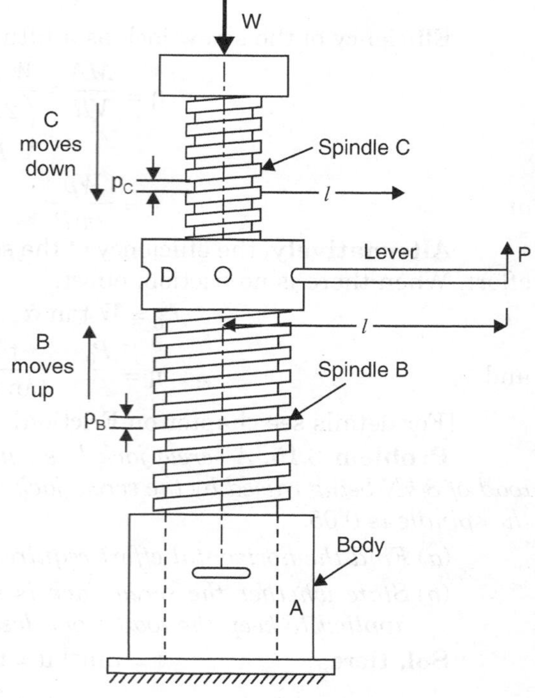

Fig. 6.18. Differential screw jack.

Let *D* and *d* be the mean diameter of spindles *B* and *C*, respectively. Also, let p_B and p_C be the pitch of threads on spindles *B* and C, respectively.

If lever is rotated through one turn, the spindle *B* moves up through the pitch p_B. In the mean time the spindle *C* moves down relative to *A*, through a distance equal to the pitch p_C.

$\therefore$ The net distance through which the load is raised = $(p_B - p_C)$.

If l is the length of the effort lever.

$$\therefore \quad \text{V.R.} = \frac{\text{Distance moved by the effort}}{\text{Distance moved by the load}}$$

or
$$VR = \frac{2\pi l}{p_B - p_C} \quad \text{...(6.59)}$$

In the simple screw jack the velocity ratio is given by

$$VR = \frac{2\pi l}{p} \quad \text{...(6.60)}$$

where, $p = p_B$ and $p_C = 0$ because there is one spindle.

Comparing Eqns. (6.59) and (6.60), we note that the velocity ratio of the differential screw jack is very large compared to that of the simple screw jack.

Problem 6.16. *The following are particulars of a differential screw jack:*
Pitches of the two spindles are 6 mm and 12 mm.
The length of effort arm = 600 mm.
An effort of 200 N raises a load of 20 kN.
Find velocity ratio and the efficiency of the screw jack.
If the above screw jack raises a load of 40 kN with an effort of 350 N find law of the machine. Also find maximum efficiency and maximum MA.

Sol. Let the two spindles have pitches p_B and p_C equal to 12 mm and 6 mm, respectively.

P = 200 N;

W = 20 kN;

Length of lever $= l = 600$ mm.

$$VR = \frac{2\pi l}{p_B - p_C} = \frac{2\pi \times 600}{12 - 6} = 200\pi = \mathbf{628.32.} \quad \textbf{Ans.}$$

$$MA = \frac{W}{P} = \frac{20 \text{ kN}}{200 \text{ N}} = \frac{20{,}000}{200} = 100$$

$$\therefore \quad \eta = \frac{MA}{VR} = \frac{100}{628.32} = 0.1592 \text{ or } \mathbf{15.92\%.} \quad \textbf{Ans.}$$

Let law of the machine be

$$P = mW + c \quad \text{...(1)}$$

Given for $W_1 = 20 \text{ kN} = 20{,}000 \text{ N}; \; P_1 = 200 \text{ N}$

for $W_2 = 40 \text{ kN} = 40{,}000 \text{ N}; \; P_2 = 350 \text{ N}$

$\therefore$ From (1), we get

$$200 = 20{,}000\, m + c \quad \text{...(2)}$$

and
$$350 = 40{,}000\, m + c \quad \text{...(3)}$$

Solving, we get
$$m = \frac{350 - 200}{40{,}000 - 20{,}000} = \frac{150}{20{,}000}$$

or
$$m = \frac{3}{400}$$

and from (1),
$$c = 200 - 20{,}000 \times \frac{3}{400}$$
$$= 200 - 150 = 50 \text{ N}.$$

∴ The law of machine is

$$P = \frac{3}{400} W + c. = \frac{\mathbf{3}}{\mathbf{400}} \mathbf{W + 50.} \quad \textbf{Ans.} \qquad ...(4)$$

Maximum efficiency, η_{max}

$$= \frac{1}{m \times VR} = \frac{1}{\frac{3}{400} \times 628.32}$$

$$= 0.2122 \text{ or } \mathbf{21.22\%.} \quad \textbf{Ans.}$$

Maximum $$MA = \frac{1}{m} = \frac{1}{3/400} = \mathbf{133.33.} \quad \textbf{Ans.}$$

Problem 6.17. *The spindle B which moves in the body A of the differential screw jack has a pitch of 10 mm and mean diameter 60 mm. The other spindle C which turns into the spindle B has threads of pitch 5 mm and mean diameter 35 mm. If coefficient of friction for threads is 0.12 find the torque required to be applied to the spindle B for raising a load of 8 kN.*

Sol. For spindle B which fits into the body A we have p_B = 10 mm and d_B = 60 mm

For spindle C:

p_C = 5 mm and d_C = 35 mm.

μ = 0.12, W = 8 kN = 8000 kN.

For spindle B:

$$\tan \alpha = \frac{p_B}{\pi d_B} = \frac{10}{\pi \times 60} = \frac{1}{6\pi} = 0.053052$$

∴ $$\alpha_B = 3.0368°$$

Similarly, for spindle C:

$$\tan \alpha_C = \frac{p_C}{\pi d_C} = \frac{5}{\pi \times 35} = \frac{1}{7\pi} = 0.045473$$

∴ $$\alpha_C = 2.6036°$$

and $$\phi = \tan^{-1} \mu = \tan^{-1} 0.12 = 6.8428°$$

∴ $$\phi + \alpha_B = 6.8428° + 3.0368° = 9.8796°$$

and $$\phi + \alpha_C = 6.8428° + 2.6036° = 9.4464°$$

Let P_1 = effort required at the mean surface of spindle B for raising load

∴ $P_1 = W \tan (\phi + \alpha_B) = 8000 \times \tan 9.8796°$

$= \mathbf{1393.29 \ N}$

Let T_1 = torque to be applied to the mean surface of spindle B for raising load

∴ $$T_1 = P_1 \times \frac{d_B}{2} = 1393.29 \times \frac{60}{2} = 41798.7 \text{ Nmm}$$

or $$T_1 = \mathbf{41.8 \ Nm.}$$

Similarly, for spindle C:

Effort, $$P_2 = W \times \tan (\phi + \alpha_C)$$

$$= 8000 \tan 9.4464°$$

$$= 1331.049 \text{ N}$$

∴ Torque to be applied to the spindle B,

$$T_2 = P_2 \times \frac{d_B}{2} = 1331.049 \times \frac{60}{3} = 39931.47 \text{ Nmm}$$

or $$T_2 = \mathbf{39.93 \ Nm}$$

$\therefore$ Total torque, $\quad T = T_1 + T_2 = 41.8 + 39.93$ Nm

$= \mathbf{81.73\ Nm.}$ **Ans.**

EXERCISE VI

A. Theoretical Questions

1. What is a lifting machine ?
2. Define : Load ; Effort ; Mechanical Advantage ; Velocity Ratio ; Work input ; Work output and Efficiency of a lifting machine.
3. What is an ideal machine ? Explain ideal effort and ideal load.
4. Establish expressions for efficiency of a lifting machine, ideal effort and ideal load.
5. What do you mean by friction losses ?
6. What is law of a lifting machine ? Find expression for effort lost in overcoming friction in a lifting machine.
7. What is a reversible and a non-reversible machine ? What is the maximum efficiency of a lifting machine ?
8. What is a pulley ? List the types of the pulley systems.
9. Describe one form of a first order pulley system. Derive expression for the velocity ratio for this lifting machine.
10. Describe one form of a second order pulley system. Derive expression for velocity ratio for this.
11. Describe one form of a third order pulley system. Derive expression for velocity ratio for this lifting machine.
12. Describe Weston differential pulley block. Establish an expression for velocity ratio for this.
13. Describe simple wheel and axle as a lifting machine. Establish an expression for the velocity ratio and efficiency for this.
14. Describe the working of a wheel and differential axle as a lifting machine. Derive expressions for velocity ratio and efficiency of the wheel and differential axle.
15. What do you understand by winch crabs ? What are single and double purchase crabs ?
16. Describe single purchase winch crab and derive expressions for its velocity ratio and efficiency.
17. Describe double purchase winch crab and derive expressions for velocity ratio and efficiency for this lifting machine.
18. Describe worm and worm wheel as a lifting machine. Also derive expressions for velocity ratio and efficiency for this.
19. Describe screw jack. Derive expressions for finding efforts for raising and for lowering a load by a screw jack.
20. Describe a differential screw jack. Derive expressions for velocity ratio and efficiency of a differential screw jack.

B. Numerical Problems

6.1. In a lifting machine an effort of 120 N raises a load of 6 kN and the efficiency of the machine is 60. If for the same machine an effort of 200 N raises a load of 12 kN, what are the maximum efficiency and the maximum mechanical advantage ? [**Ans.** $MA_{max} = 7.5$; $\eta_{max} = 90\%$]

6.2. In a lifting machine, an effort of 320 N raised a load of 105 kN, what is the mechanical advantage ? If efficiency is 70%, what is the velocity ratio ? If for the same machine an effort of 620 N raised a load of 20 kN, what is efficiency at this load ? Also find the maximum efficiency and the maximum mechanical advantage for the lifting machine. Also find loss of effort and loss of *MA* for a load of 6 kN. [**Ans.** $VR = 44.64$, $\eta = 72.26\%$; $MA_{max} = 33.33$, $\eta_{max} = 74.67\%$]

6.3. A machine lifts a load of 4.8 kN with an effort of 300 N. The velocity ratio of the machine is 32. Determine the machine friction in terms effort and the load lost due to friction. What is the efficiency of the machine ? [**Ans.** F_P = 150 N; F_W = 4800 N; η = 50%]

6.4. The following readings were noted in an experiment with a lifting machine whose velocity ratio is 40:

Load W(N)	:	60	120	180	240	300	360
Effort P(N)	:	7.2	10.7	14.2	17.7	21.2	24.7

Find law of the machine. Also find efficiency of the machine when the loads of 200 N and 320 N are lifted. Also find maximum efficiency and the maximum mechanical advantage.

[**Ans.** $P = 0.5833\ W + 3.7$, η = 32.54%; 35.77%; $M_{max}\ \eta$ = 42.86%, $M.A_{max}$ = 17.144]

6.5. The law of a machine, with velocity ratio of 80, is $P = 0.02\ W + 32$, where P is effort and W the load in Newtons. Find the maximum mechanical advantage and maximum efficiency. Is this machine reversible ? [**Ans.** MA_{max} = 50; η_{max} = 62.5%; The machine is reversible]

6.6. The initial frictional resistance of a machine is 30 N. The frictional resistance increases at the rate of 15 N per kN of load lifted by the machine. The velocity ratio is 24. Find the effort required to lift a load of 12 kN and the efficiency. Also find the maximum efficiency and the maximum mechanical advantage of the machine. [**Ans.** P = 710 N; 70.42%; M_{max} = 17.64; η_{max} = 73.49%]

6.7. In a certain machine of velocity ratio 80 an effort of 150 N lifts a load of 3000 N and an effort of 180 N lifts a load of 4000 N through the same distance. What effort will lift a load of 5400 N ? Find the maximum efficiency and the maximum mechanical advantage.
Also find the initial frictional effort and the rate at which the frictional effort increases.

[**Ans.** P = 222 N; η_{max} = 41.67%; MA_{max} = 33.33; Initial frictional resistance = 60 N; 17.5 N per kN of load lifted]

6.8. Four movable pulleys are arranged in the first system of pulleys. If the efficiency of the system is 80% find the load that can be lifted by an effort of 200 N. If weight of each pulley is 8 N calculate the effort for lifting a load of 6 kN. [**Ans.** 2016 N; 382.5 N]

6.9. In a third order system of pulleys, the weight of the lower movable block of pulleys is 90 N and there are three pulleys in each block. Find the greatest load that can be lifted with an effort of 300 N. [**Ans.** W = 1700 N]

6.10. In a third order system of pulleys there are 5 pulleys in all. Find the velocity ratio. Also find the total load lifted by an effort of 3000 N, if the efficiency of the machine is 75%. [**Ans.** 69.75 kN]

6.11. In a simple wheel and axle, the diameter of the wheel is 160 mm and that of the axle is 32 mm. If efficiency of the machine is 72% find the effort for lifting a load of 720 N. [**Ans.** 200 N]

6.12. In a wheel and differential axle of velocity ratio of 32, a load of 2400 N can be lifted by an effort of 120 N and a load of 3200 N lifted by an effort of 150 N. Find law of the machine and then determine (*a*) effort required for lifting a load of 4200 N. (*b*) effort wasted in friction for 4200 N load (*c*) maximum efficiency of the machine and (*d*) maximum mechanical advantage.

$$\left[\textbf{Ans. } P = \frac{3}{80} W + 30;\ (a)\ 187.5\ N;\ (b)\ 56.25\ N;\ (c)\ 83.33\%\ (d)\ 26.67\right]$$

6.13. In a Weston's differential pulley block there are 12 and 11 recesses on the two pulleys. Find the efficiency of the machine if an effort of 600 N lifts a load of 7200 N. [**Ans.** 50%]

6.14. In a differential pulley block, the diameters of the concentric pulleys are 180 mm and 120 mm. An effort of 300 N lifts a load of 1200 N. Calculate (*a*) efficiency of the machine, (*b*) effort lost in friction, and (*c*) friction load. [**Ans.** (*a*) 66.67%; (*b*) 100 N; (*c*) 600 N]

6.15. In a single purchase crab the length of handle is 300 mm and gear ratio is 8. Find the velocity ratio and efficiency of the machine if an effort of 250 N lifts a load of 6 kN. Diameter of load drum is 120 mm. [**Ans.** VR = 40; η = 60%]

6.16. A single purchase crab has a 300 mm long handle and teeth on gears are 20 and 180. Wheel drum diameter is 250 mm. Find velocity ratio. If a load of 4000 N is lifted by and effort of 160 N. An effort of 190 N lifts a load of 5000 N. Determine (*a*) the law of machine, (*b*) maximum MA (*c*) maximum efficiency, and (*d*) effort lost in friction for 5000 N load.

[**Ans.** VR = 43.2, η = 57.87%, (*a*) $P = 0.03\ W + 50$; (*b*) 33.33; (*c*) 77.16%, (*d*) 74.26 N]

6.17. In a double purchase winch crab following specifications are given:

Length of lever = 1 m; Number of teeth on effort pinion = 10, teeth of load gear = 160, Number of teeth on the intermediate axle = 100 and 16, respectively. Diameter of wheel drum = 300 mm. Find the velocity ratio of the machine and the load which can be lifted by an effort of 60 N if the efficiency at this load is 60%. [**Ans.** 666.67; 24 kN]

6.18. Find mechanical advantage and efficiency of a screw jack if an effort applied at the end a 200 mm long lever to lift a load of 3.2 kN. Pitch of threads on spindle is 6 mm and mean diameter of threads is 60 mm. Coefficient of friction is 0.04. [**Ans.** MA = 92.69; η = 44.26%]

6.19. In a differential screw jack the pitches on the two spindles are 12 mm and 9 mm. Find the load which can be raised by an effort of 120 N if the efficiency of the machine is 80% length of lever is 600 mm. [**Ans.** W = 120.64 kN]

6.20. A worm and worm wheel has effort wheel of 300 mm diameter, load drum diameter of 120 mm, worm wheel has 100 teeth. The following observation were noted during an experiment on this machine:

Load (W) :	200	400	600	800	1000
Effort (N) :	25	40	55	70	85

Find:

(*a*) the law of machine,

(*b*) efficiency of machine at a load of 900 N,

(*c*) maximum mechanical advantage,

(*d*) maximum efficiency

(*e*) loss of effort due to friction, at 900 N load,

(*f*) loss of load due to effort a 900 N load, and

(*g*) initial effort of friction.

$$\left[\textbf{Ans.}\ (a)\ P = \frac{3W}{40} + 10;\ (b)\ 4.65\%;\ (c)\ 13.33;\ (d)\ 5.33\%;\ (e)\ 73.9\text{ N};\ (f)\ 18475\text{ N};\ (g)\ 10\text{ N}\right]$$

7

Kinematics

7.1. INTRODUCTION

In **statics** we have considered rigid bodies that are at rest. In **dynamics** we shall consider bodies that are in motion. While statics is a very old science, dynamics, on the contrary, is a comparatively new one.

In statics there are only two kinds of units to be dealt with: that of **length** for measuring the dimensions or distances between bodies and that of **force** for measuring the actions and reactions between them. Precision instruments for the measurement of length and force are simple and were well developed in early times. In dynamics, we need, in addition to the units of length and force, a unit of measuring **time.**

Time measuring devices such as pendulum clock and balance wheel watch are comparatively new.

Dynamics is commonly divided into two branches called **kinematics** and **kinetics.** In kinematics we are concerned only with the space–time relationships of a given motion of a body and not at all with the forces that cause the motion. Thus, for example, if we see that a wheel rolls along a straight path with uniform speed, the determination of the shape of the path described by a point on its rim and of the position along its path that the chosen point will occupy at any given instant are problems of kinematics. Another example of kinematics may be the shape of path of a body thrown in the space. This path is called projectile. When we do not consider the weight of the rolling wheel, or the body thrown in space, we deal with kinematics.

In kinetics we are concerned with finding the kind of motion of a body or a system of bodies will have under the action of given forces. Thus, the study of motion of a body under the action of the given forces requires the consideration of the mass of the body itself. Hence, in kinetics we study the effect of mass of the bodies and also of the external forces acting on the body.

The whole science of dynamics is based on the natural laws governing the motion of a **particle** under the action of a given force. A **particle** is defined as a material point without dimensions but containing a definite quantity of matter. Strictly speaking, there can be no such thing as a particle. However, when the size of a body is extremely small compared to its range of motion it may be considered as a particle.

When a particle moves it describes a curve that is called the **path.** The path of a particle may be either a space curve called a **tortuous path,** or a plane curve called a **plane path.** In the simplest case the path of a particle will be a straight line, and the particle is said to have **rectilinear motion.**

PART I—RECTILINEAR MOTION OF PARTICLES

7.2. POSITION, VELOCITY, AND ACCELERATION

Consider a particle P moving along a straight line, Fig. 7.2.1. The position of P at any instant of time t may be specified by its distance s measured from some convenient reference point O on the line. At time $t + \Delta t$ the particle has moved to P' and its co-ordinate becomes $s + \Delta s$.

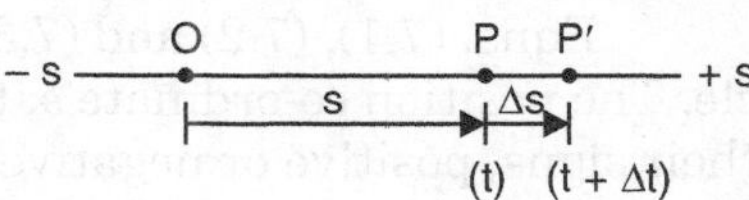

Fig. 7.2.1. Rectilinear motion of a particle.

The change in **position co-ordinate** (as called position vector) during the time interval Δt is called the **displacement** Δs of the particle. The displacement would be negative if the particle moves in the negative s–direction.

The **average velocity** of the particle during the time interval Δt is the displacement divided by the time interval or

$$v_{av} = \frac{\Delta s}{\Delta t}$$

As Δt becomes smaller and smaller and approaches zero in the limit, the average velocity approaches the **instantaneous velocity** of the particle, which is $v = \lim_{\Delta t \to 0} \frac{\Delta s}{\Delta t}$, or

$$\boxed{v = \frac{ds}{dt} = \dot{s}} \quad ...(7.1)$$

Thus, the velocity is defined as the time rate of change of position co-ordinate (or position vector) s. The velocity is positive or negative depending on whether the corresponding displacement of the particle is positive or negative. **Magnitude of v is known as speed of the particle.**

The **average acceleration** of the particle P during the time interval Δt is the change in its velocity divided by the time interval

Fig. 7.2.2

or $$a_{av} = \frac{\Delta v}{\Delta t}. \text{ (See Fig. 7.2.2)}$$

As Δt becomes smaller and smaller and approaches zero in the limit, the average acceleration approaches the **instantaneous acceleration** of the particle, which is $a = \lim_{\Delta t \to 0} \frac{\Delta v}{\Delta t}$ or

$$\boxed{a = \frac{dv}{dt} = \dot{v}} \quad \text{or} \quad \boxed{a = \frac{d^2 s}{dt^2} = \ddot{s}} \quad ...(7.2)$$

Again, the acceleration is positive or negative depending on whether the velocity is increasing or decreasing. If the particle is slowing down, the particle is said to be **retarding** or **decelerating.**

From eqn. (7.1), the time dt is given as

$$dt = \frac{ds}{v}$$

$\therefore$ By substituting the value of dt in eqn. (7.2), we get

$$adt = dv$$

or $$a \times \frac{ds}{v} = dv$$

$\therefore$ $$\boxed{vdv = ads} \quad \text{or} \quad \boxed{\dot{s}d\dot{s} = \ddot{s}\, ds} \qquad ...(7.3)$$

Eqns. (7.1), (7.2) and (7.3) are differential equations for the rectilinear motion of a particle. The position co-ordinate s, the velocity v, and the acceleration a are algebraic quantities, so their signs, positive or negative, must be carefully observed. Note that the positive directions of v and a are the same as the positive direction of the displacement, s.

7.3. NATURE OF PROBLEMS IN RECTILINEAR MOTION

If the position co-ordinate s is a known function of time $s = f(t)$, then successive differentiation of the function with respect to time t gives the velocity v and acceleration a.

In many problems, however, the functional relationship between position co-ordinate s and time t is unknown, and we determine it by successive integration from the acceleration. Acceleration is determined by the forces that act on the moving bodies and is determined from the equations of dynamics (or kinetics), to be discussed in subsequent chapters. Thus, depending on the nature of forces, the acceleration may be specified as a function of time, velocity, position co-ordinate, or a combined function of these quantities. That is, $a = f(t)$, $a = f(v)$, $a = f(s)$ or $a = f(v, s, t)$.

The procedure for integrating the differential equations of different cases is indicated as follows:

Case I. Given Displacement in terms of time, i.e., s = f(t), to find velocity-time (v-t) and acceleration-time (a-t) relations.

This is the simplest case and is easily solved by successive differentiation of the s-t relation. Hence, we get

$$v = \frac{ds}{dt} \quad \text{and} \quad a = \frac{dv}{dt} = \frac{d^2 s}{dt}$$

Case II. Given v = f(t), to find a-t and s-t relations.

Using the relation $a = \frac{dv}{dt}$, we get

$$a = \frac{df(t)}{dt}$$

While from relation $v = \frac{ds}{dt}$,

we get $$ds = vdt$$

or $$ds = f(t)\, dt$$

$\therefore$ Integrating $\int ds = s = \int f(t)dt$, we get the s-t relation.

Case III. Given acceleration in terms of time, i.e., a = f(t), to find s-t and v-t relations.

From $a = f(t) = \frac{dv}{dt}$, we get

$$dv = f(t)\, dt.$$

Integrating $\int dv = \int f(t)\, dt$, we get v-t relation.

$$v = \int f(t)\, dt = \frac{ds}{dt}$$

and finally $\int ds = \int\{\int f(t)\, dt\}\, dt$, we get

$$s = \int (\int f(t)\, dt)\, dt, \text{ which is the required } s\text{-}t \text{ relation.}$$

Case IV. Given a = f (v), to find the *a-t*, *v-t* and *s-t* relation.

Starting with $a = f(v) = \dfrac{dv}{dt}$, we get

$$\frac{dv}{f(v)} = dt.$$

Thus, integrating, we get the *v-t* relation *i.e.*, $\int dt = t = \int \dfrac{dv}{f(v)}$

And finally using the relation $v = \dfrac{ds}{dt}$,

We get *s-t* relation, as discussed in Case II, above.

Case V. Given v = f(s), to find *s-t*, *v-t* and *a-t* relations.

From the relation $v = \dfrac{ds}{dt} = f(s)$, we get $dt = \dfrac{ds}{f(s)}$. Integrating this we get *s-t* relation.

Thus, after getting *s-t* relation, by using the relation $a = \dfrac{d^2s}{dt^2}$, we get the *a-t* relation.

Case VI. Given a = f(s), to find *s-t*, *v-t* and *a-t* relations.

Using the relation $a ds = v dv$, we get

$$f(s) \,.\, ds = v\, dv.$$

Integrating this, we get $v = f(s)$, relation.

Again using the relation $v = \dfrac{ds}{dt} = f(s)$.

And, then we get $dt = \dfrac{ds}{f(s)}$. Integrating this, we get $s = f(t)$, *i.e.*, *s-t* relation.

Differentiating the $s = f(t)$, we get

$$v = \frac{ds}{dt} \quad \text{and} \quad a = \frac{dv}{dt} = \frac{d^2s}{dt^2}, \text{ as discussed in Case V, above.}$$

7.4. UNIFORM RECTILINEAR MOTION

This type of motion is frequently encountered in practice. In this motion, the acceleration *a* is zero for all values of time *t*. Therefore, the velocity *v* is constant. Hence from eqn. (7.1), we get

$$\frac{ds}{dt} = v = \text{constant}$$

The displacement or position co-ordinate *s* is obtained by integrating this equation. Let $s = s_0$ at $t = 0$. Thus,

$$\int_{s_0}^{s} ds = v\int_0^t dt$$

or $\quad s - s_0 = vt$

or $\quad s = s_0 + vt \qquad \text{...(7.4)}$

s_0 is called initial displacement at $t = 0$.

7.5. UNIFORMLY ACCELERATED RECTILINEAR MOTION

In this case the acceleration a is constant. Thus, from eqn. (7.2), we get

$$\frac{dv}{dt} = a = \text{constant}$$

Thus, we get $dv = a \, . \, dt$

Hence, by integration, we get

$$\int_{v_0}^{v} dv = \int_0^t dt$$

or $$v - v_0 = at$$

or $$v = v_0 + at \qquad ...(7.5)$$

where the velocity v_0 is the velocity at $t = 0$. v_0 is also called initial velocity.

Substituting eqn. (7.5) into eqn. (7.1), we get

$$\frac{ds}{dt} = v_0 + at$$

or $$ds = (v_0 + at)\, dt$$

Integrating this, we get

$$\int_{s_0}^{s} ds = \int_0^t (v_0 + at)\, dt$$

or $$s - s_0 = v_0 t + \frac{1}{2} at$$

or $$s = s_0 + v_0 t + \frac{1}{2} at^2 \qquad ...(7.6)$$

Also, using eqn. (7.3), we have

$$a = v . \frac{dv}{ds} = \text{constant}$$

$\therefore$ we get $$ads = vdv$$

Hence, integrating, we get

$$\int_{s_0}^{s} ads = \int_{v_0}^{v} vdv$$

or $$a(s - s_0) = \frac{v^2}{2} - \frac{v_0^2}{2}$$

or $$v^2 = v_0^2 + 2a\,(s - s_0) \qquad ...(7.7)$$

Freely Falling Body

An important case of uniformly accelerated motion is the motion of a **freely falling body**. The acceleration of a freely falling body is equal to 9.81 m/s^2 and is usually denoted by g. Thus eqns. (7.5), (7.6) and (7.7), respectively, become:

$$v = v_0 + gt \qquad ...(7.8)$$

$$y = y_0 + v_0 t + \frac{1}{2} gt^2 \qquad ...(7.9)$$

$$v^2 = v_0^2 + 2g\,(y - y_0) \qquad ...(7.10)$$

When a body is thrown up, it undergoes a constant retardation equal to g. Thus, eqns. (7.8), (7.9) and (7.10) for a body projected vertically up become :

$$v = v_0 - gt \quad ...[7.8(a)]$$

$$y = y_0 + v_0 t - \frac{1}{2}gt^2 \quad ...[7.9(a)]$$

$$v^2 = v_0^2 - 2g(y - y_0) \quad ...[7.10(a)]$$

Problem 7.1. *A particle moves along a straight line so that its displacement in metres, from a fixed point is given by,*

$$s = 2t^3 + 4t^2 - 6t + 8.$$

Find:

(i) Velocity at start. *(ii) Velocity after 4 seconds.*

(iii) Acceleration at start. *(iv) Acceleration after 4 seconds.*

Sol. We know that

Velocity, $$v = \frac{ds}{dt} \quad ...(1)$$

Thus, $$v = \frac{d}{dt}(2t^3 + 4t^2 - 6t + 8)$$

$$v = 6t^2 + 8t - 6 \quad ...(2)$$

(*i*) Velocity of the body at start, t

$$v\Big|_{t=0} = 6\,(0)^2 + 8\,(0) - 6 = -6 \text{ m/s.}$$

$\therefore$ $\boxed{v = -6 \text{ m/s.}}$ **Ans.**

(*ii*) Velocity of the body after 4*s*.

$$t = 4$$

$$v\Big|_{t=4} = 6 \times (4)^2 + 8 \times (4) - 6 = 96 + 32 - 6$$

$\therefore$ $\boxed{v = 122 \text{ m/s.}}$ **Ans.**

Acceleration. We know that

Acceleration, $$a = \frac{dv}{dt} = \frac{d^2s}{dt^2} \quad ...(3)$$

Thus, $$a = \frac{d^2s}{dt^2} = (12t + 8) \quad ...(4)$$

(*iii*) Acceleration at start, $a\big|_{t=0} = 12\,(0) + 8$

$\therefore$ $\boxed{a = 8 \text{ m/s}^2.}$ **Ans.**

(*iv*) Acceleration at 4*s*, $t = 4$

$$a\big|_{t=4} = 12 \times (4) + 8$$

$\therefore$ $\boxed{a = 56 \text{ m/s}^2.}$ **Ans.**

Problem 7.2. *The velocity of a moving particle is given by the relation*

$$V = 15 - t + \left(\frac{1}{24}\right)t^3,$$

where the velocity V is in m/s, and time t is in seconds. Determine the initial velocity and velocity after 3 seconds. Also determine the distance travelled in this time, average velocity and average acceleration.

Sol. Velocity, $V = 15 - t + \left(\frac{1}{24}\right)t^3$ (given)

(*i*) **Initial velocity,** we get $v|_{t=0} = 15 - (0) + \left(\frac{1}{24}\right)(0)^3$

$\therefore$ $\boxed{V|_{t=0} = 15 \text{ m/s.}}$ **Ans.**

(*ii*) Velocity after 3 seconds, $t = 3S$

$$V|_{t=3} = 15 - (3) + \left(\frac{1}{24}\right)(3)^3 = 15 - 3 + \frac{1}{24} \times 27$$

$\therefore$ $\boxed{V = 13.125 \text{ m/s.}}$ **Ans.**

We know that

Velocity, $V = \frac{ds}{dt}$

$\therefore$ $ds = V\,dt$

or $s = \int V\,dt$

(*iii*) Distance travelled in 3 seconds, $s = \int_0^3 \left\{15 - t + \left(\frac{1}{24}\right)t^3\right\} dt$

$$= \left[15t - \frac{t^2}{2} + \frac{1}{24} \times \frac{t^4}{4}\right]_0^3$$

$$s = \left[15t - \frac{t^2}{2} + \frac{t^4}{96}\right]_0^3$$

$\therefore$ $\boxed{s|_{t=3} = 15(3) - \frac{(3)^2}{2} + \frac{(3)^4}{96} = 41.34 \text{ m.}}$ **Ans.**

(*iv*) Average velocity

$$V_{av} = \frac{\text{Displacement in given time interval}}{\text{Time interval}}$$

$\Rightarrow$ $V_{av} = \frac{41.34}{3} = 13.78 \text{ m/s}$

$\boxed{V_{av} = 13.78 \text{ m/s.}}$ **Ans.**

(v) Average acceleration.

$$\text{Average acceleration} = \frac{(\text{Final velocity} - \text{Initial velocity})}{(\text{Time interval})}$$

$$V_{\text{initial}} = V|_{t=0} = 15$$

$$V_{\text{final}} = V|_{t=3} = 15 - 3 + \frac{(3)^3}{24}$$

Thus,

Average acceleration

$$a_{av} = \frac{\left(15 - 3 + \frac{3^3}{24}\right) - (15)}{3} = \frac{13.125 - 15}{3} = \frac{-1.875}{3} \text{ m/s}^2$$

or, $\boxed{a_{av} = -0.625 \text{ m/s}^2.}$ **Ans.**

Problem 7.3. *An automobile accelerates uniformly from rest on a straight level road. A second automobile starting from the same point 3 seconds later with zero initial velocity accelerates at 4 m / s² overtakes the first automobile 400 m from the starting point. What is the acceleration of first automobile ?*

Sol. Given:

u_1 = initial velocity of first automobile
= 0

u_2 = initial velocity of second automobile
= 0

$a_1 = ?$, $a_2 = 4$ m/s², $s_1 = s_2 = 400$ m

According to the question,

Distance travelled by first vehicle in t_1 seconds
= distance travelled by second vehicle in $(t_1 - 3)$ seconds.

For first automobile

$$s_1 = u_1 \times t_1 + \frac{1}{2}a_1 t_1^2 = 0 + \frac{1}{2}a_1 t_1^2$$

$$s_1 = \frac{1}{2}a_1 t_1^2 \qquad \text{...(1)}$$

For second automobile

$$s_2 = u_2 \times (t_1 - 3) + \frac{1}{2}a_2 \times (t_1 - 3)^2$$

$$s_2 = 0 + \frac{1}{2}a_2 \times (t_1 - 3)^2 \qquad \text{...(2)}$$

$$\Rightarrow \qquad 400 = \frac{1}{2} \times 4 \times (t_1 - 3)^2$$

$$t_1 = 17.14 \text{ seconds.}$$

Again

$$400 = \frac{1}{2} \times a_1 \times t_1^2$$

$$400 = \frac{1}{2} \times a_1 \times (17.14)^2$$

∴ $a_1 = 2.72 \text{ m/s}^2$. **Ans.**

Problem 7.4. *A vehicle starts from rest at a point O and travels along a straight line with acceleration 2 m/s². Another vehicle starts from rest at the same point 4 second later and, travels along the same path with an acceleration of 3 m/s². How far from O will B overtake A ?*

Sol. Given: Initial velocity of first vehicle = $u_1 = 0$

Initial velocity of second vehicle = $u_2 = 0$

t_1 = time taken by first vehicle.

$t_2 = (t_1 - 4)$ = time taken by second vehicle.

$a_1 = 2 \text{ m/s}^2, \quad a_2 = 3 \text{ m/s}^2$

According to question

Distance travelled by first vehicle in t_1 seconds = distance travelled by second vehicle in $(t_1 - 4)$ seconds.

$$\therefore \quad u_1 \times t_1 + \frac{1}{2} a_1 t_1^2 = u_2 \times t_2 + \frac{1}{2} a_2 t_2^2, \text{ where } u_1 = u_2 = 0$$

$$0 + \frac{1}{2} \times 2 \times t_1^2 = 0 + \frac{1}{2} \times 3 \times (t_1 - 4)^2$$

$$\therefore \quad t_1 \simeq 22 \text{ seconds.}$$

Thus,

$$\text{Total distance travelled} = \frac{1}{2} \times 2 \times (22)^2$$

$s = 484$ m. **Ans.**

Problem 7.5. *A particle moves from rest with an acceleration of 6 m/s² until its velocity is 12 m/s to the right. It is then subjected to an acceleration of 12 m/s to the left until its total distance travelled is 36 m. Determine the total elapsed time.*

Sol. Let t_1 be the time in which velocity of body becomes 12 m/s.

By using $v_1 = u_1 + a_1 t_1$

$$12 = 0 + 6 \times t_1$$

∴ $\mathbf{t_1 = 2}$ **seconds.**

If s_1 = distance travelled in t_1 sec. then,

$$s_1 = u_1 t_1 + \frac{1}{2} a_1 t_1^2 = 0 + \frac{1}{2} \times 6 \times (2)^2$$

$$\therefore \quad s_1 = 12 \text{ m.}$$

It can be seen that due to the retardation (– 12 m/s²) the body stops and then it returns back.

Let t_2 be the time elapsed when the body stops.

∴ By using $v_2 = u_2 + a_2 t_2$, where $a_2 = -12 \text{ m/s}^2$,

we get $0 = 12 + (-12) \times t_2$

∴ $\mathbf{t_2 = 1}$ **second**

∴ Total time elapsed = $t_1 + t_2$ = **3 seconds. Ans.**

And, $$s_2 = u_2 t_2 + \frac{1}{2} a_2 t_2^2 = 12 \times 1 + \frac{1}{2} \times (-12) \times 1^2 = 12 - 6$$

$$\therefore \quad s_2 = 6 \text{ m.}$$

Therefore, distance travelled by the body before it comes to rest

$$= s = s_1 + s_2 = (12 + 6) = \textbf{18 m.} \quad \textbf{Ans.}$$

Since the total distance travelled by the body is given to 36 m,

$\therefore$ The distance travelled to the left = 36 – 18 = **18 m.** **Ans.**

Let time required to travel a distance of 18 m on the left = t_3.

$\therefore$ Using $S = u_2 t_3 + \frac{1}{2} a_3 t_3^2$, where $u_2 = 0$, and $a_3 = 12$ m/s^2

We get $$18 = 0 + \frac{1}{2} \times 12 \times t_3^2$$

or, $$t_3^2 = \frac{18}{6} \quad \text{or} \quad t_3 = \sqrt{3} \text{ seconds} = 1.732 \ s.$$

$\therefore$ Total time elapsed to cover the total distance of 36 m

$$= t_1 + t_2 + t_3 = 2 + 1 + 1.732 \text{ seconds} = \textbf{4.732 seconds.} \quad \textbf{Ans.}$$

Problem 7.6. *A man in a balloon rising with a constant velocity 4 m/s propels a ball upwards with a velocity of 1.2 m/s relative to the balloon. After what time interval will the ball return to the balloon ?*

Sol. With respect to a stationary observer the ball appears to be thrown with a velocity $u_1 = 4 + 1.2 = 5.2$ m/s.

The balloon moves upward with uniform velocity

$$v = 4 \text{ m/s}.$$

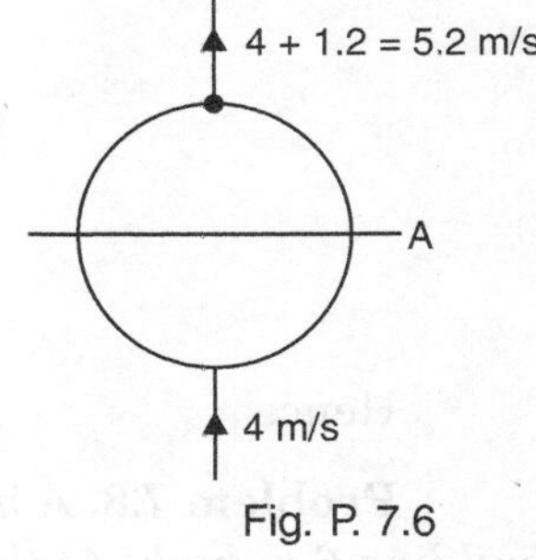

Fig. P. 7.6

It can be observed that when the ball return to the balloon, the displacement of ball will be same as that of balloon.

Let t_1 = time taken by ball to return back.

Displacement of balloon $= 4 \times t_1$ ...(1)

Displacement of ball $= 5.2 \times t_1 - \frac{1}{2} \times 9.8 \, t_1^2$ ...(2)

Equating (1) and (2)

$$4t_1 = 5.2 \, t_1 - \frac{1}{2} \times 9.8 \times t_1^2$$

$$\therefore \quad \boxed{t_1 = 0.245 \text{ s.}} \quad \textbf{Ans.}$$

Problem 7.7. *A stone is dropped with zero velocity into a well. The sound of the splash is heard after 4 seconds. How far below the ground surface is the surface of the water ? Assume the velocity of sound as 331 m/s.*

Sol. Let the water surface be at a distance h below from the ground surface.

t_1 = time taken by stone to reach water surface.

t_2 = time in which sound of splash reaches the ground surface.

t = total time elapsed between dropping of stone and hearing of splash = 4s. (given)

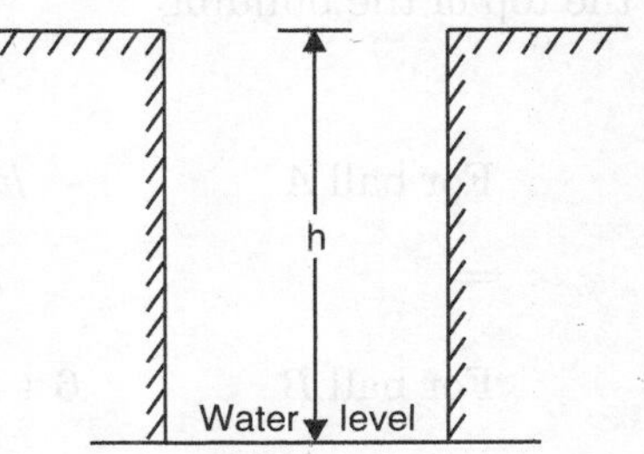

Fig. P. 7.7

By using equation of motion: $s = ut + \frac{1}{2} gt^2$ for stone, where $u = 0$, $s = h$ and $g = 9.8$ m/s^2

we have $$h = 0 \times t_1 + \frac{1}{2} \times 9.8 \times t_1^2$$

$$h = 4.9\, t_1^2$$

$$\Rightarrow \quad t_1 = \sqrt{\frac{h}{4.9}} \qquad \text{...(1)}$$

Speed of sound (splash) $= \dfrac{h}{t_2} = 331$ m/s

$$\therefore \quad t_2 = \frac{h}{331} \qquad \text{...(2)}$$

$$t_1 + t_2 = 4 \qquad \text{(given)}$$

$\therefore$ From (1) and (2)

$$\sqrt{\frac{h}{4.9}} + \frac{h}{331} = 4$$

$$\sqrt{\frac{h}{4.9}} = 4 - \frac{h}{331} \qquad \text{...(3)}$$

Squaring (3), we get $\dfrac{h}{4.9} = 16 - 2 \times 4 \times \dfrac{h}{331} + \dfrac{h^2}{(331)^2}$

$$\therefore \quad h^2 - h \times (331)^2 \left(\frac{1}{4.9} + \frac{8}{331}\right) + 16 \times (331)^2 = 0$$

or

$$h^2 - 25007\, h + 1752976 = 0$$

$$\therefore \quad h = \frac{1}{2}\left[25007 \pm \sqrt{(25007)^2 - 4 \times 1752976}\right]$$

$$= \frac{1}{2}[25007 \pm 24866.4] = 24936.7 \quad \text{or} \quad 70.3$$

Hence $\boxed{h \simeq 70 \text{ m.}}$ **Ans.**

Problem 7.8. *A man A throws a ball vertically up with a speed 10 m/s from the top of a building 6 m high. At the same instant man B throws a ball vertically up with a speed 15 m/s. Determine the time at which the two balls will be at the same height from the ground. What is the height ?*

Sol. Initial velocity of ball $A = u_A = 10$ m/s (↑)

Initial velocity of ball $B = u_B = 15$ m/s (↑)

Acceleration due to gravity $= g = -\,9.8$ m/s (↓)

Let us suppose the two balls pass at a height h from the top of the building

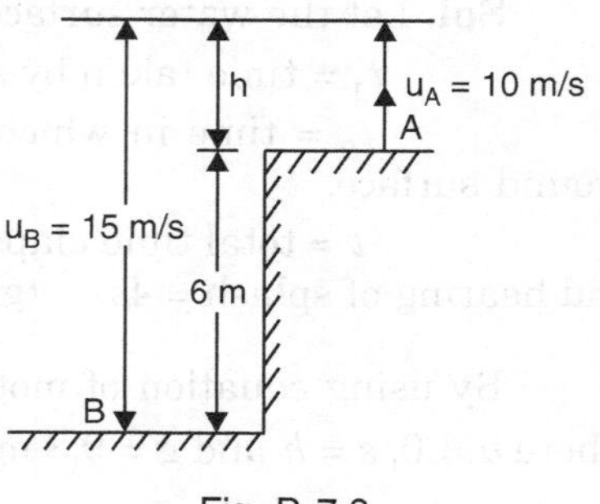

Fig. P. 7.8

$$t_1 = \text{time elapsed.}$$

For ball A $$h = 10 \times t_1 - \frac{1}{2} \times 9.8\, t_1^2$$

$$\Rightarrow \quad h = 10t_1 - 4.9t_1^2 \qquad \text{...(1)}$$

For ball B $$6 + h = 15\, t_1 - \frac{1}{2} \times 9.8 \times t_1^2$$

$$6 + h = 15\, t_1 - 4.9\, t_1^2 \qquad \text{...(2)}$$

Subtracting (1) from (2)

$$6 = 5t_1$$

$$t_1 = \frac{6}{5}$$

$$\boxed{t_1 = 1.2 \text{ s.}} \quad \textbf{Ans.}$$

$$h = 10t_1 - 4.9t_1^2$$

$$= 10 \times 1.2 - 4.9 \times 1.2^2$$

$$h = \mathbf{4.944}$$

Thus height above the ground at which collision takes place

$$= 6 + h = 6 + 4.944 = \mathbf{10.944 \text{ m.}} \quad \textbf{Ans.}$$

Problem 7.9. *A ball is thrown vertically upwards with a velocity 9 m/s from the edge of a cliff 15 m above the sea level. What is the highest point above sea level reached ? How long does it take to hit the water ? With what velocity does it hit the water ?*

Sol. Let h be the maximum height attained by the ball.

u = initial velocity of ball = 9 m/s (upwards)

g = – 9.8 m/s^2 (downwards)

(*i*) Maximum height:

s = height attained by the ball above the cliff.

At maximum elevation, velocity of ball = 0.

By equation of motion

$$v^2 = u^2 - 2gs.$$

$$0^2 = 9^2 - 2 \times 9.8 \times s.$$

$$\boxed{s = 4.128 \text{ m.}}$$

Maximum height attained above sea level

$$h = s + 15$$

$$= 4.128 + 15$$

$$\boxed{h = 19.128 \text{ m.}} \quad \textbf{Ans.}$$

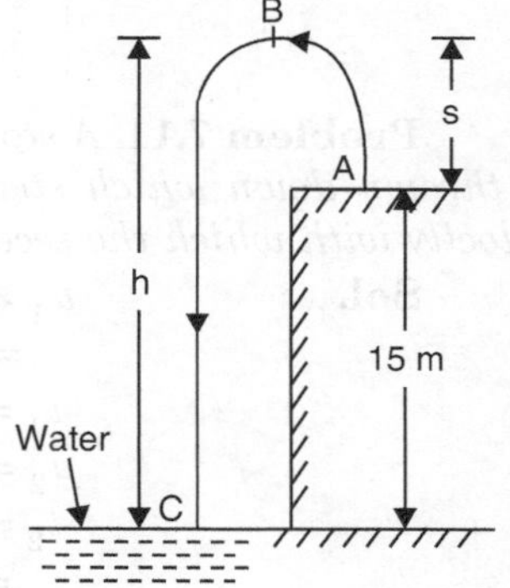

Fig. P. 7.9

(*ii*) Time taken to hit the water.

t = time taken by ball to hit the water.

It should be noted that when ball strikes the water, its displacement is – 15 m (negative sign shows that displacement is downward).

By the equation $\quad s = ut - \frac{1}{2}gt^2$

$$\Rightarrow \quad -15 = 9 \times t - \frac{1}{2} \times 9.81\, t^2$$

$$\Rightarrow \quad 4.9t^2 - 9t - 15 = 0$$

$$\boxed{t = 2.89 \text{ s.}} \quad \textbf{Ans.}$$

(*iii*) Velocity with which ball hits the ground,

By the equation $\quad v = u - g \times t$

$$v = 9 - 9.81 \times 2.89$$

$$\boxed{v = -19.35 \text{ m/s.}} \quad \textbf{Ans.}$$

Note. Negative sign shows velocity of the ball is downward.

Problem 7.10. *A body was thrown vertically down from a tower and traversed a distance of 75 metres during 7th second of its fall. Find the initial velocity of the body.*

Sol. Let u = initial velocity.

Taking downward velocity, acceleration etc. as positive.

$$g = +9.81 \text{ m/s}^2$$

Distance travelled in 7th second = Distance travelled in 7 seconds – Distance travelled in 6 seconds.

$$= s_7 - s_6$$

By equation of motion

$$s_7 = u \times 7 + \frac{1}{2} \times 9.81 \times (7)^2$$

$$s_7 = 7u + 240.3 \quad \text{...(1)}$$

$$s_6 = u \times 6 + \frac{1}{2} \times 9.81 \times 6^2$$

$$s_6 = 6u + 176.58 \quad \text{...(2)}$$

$$s_7 - s_6 = 75 \quad \text{(given)}$$

$$\Rightarrow \quad (7u + 240.3) - (6u + 176.58) = 75.$$

$$\boxed{u = 11.28 \text{ m/s.}} \quad \textbf{Ans.}$$

Problem 7.11. *A stone is dropped from a height of 150 m and after 1.5 sec. another stone is thrown down which strikes the first stone when it has just reached the ground. Find the velocity with which the second stone was thrown.*

Sol. u_1 = initial velocity of first stone
= 0 (given)
t_1 = time taken by first stone to reach the ground
u_2 = initial velocity of second stone. (unknown)
t_2 = time taken by second stone to reach the ground
= $t_1 - 1.5$ (given)

Both the stones cover the same distance (150 m), but the time taken by second stone is 1.5 sec less than that of first. Taking downward velocity, acceleration etc. as +ve.

For first stone

$$h = 150 = u_1 \times t_1 + \frac{1}{2} \times 9.81 \times t_1^2$$

$$150 = 0 + 4.9t_1^2$$

$$\therefore \quad t_1 = \sqrt{\frac{150}{4.9}}$$

$$\boxed{t_1 = 5.53 \text{ sec.}}$$

Thus $t_2 = t_1 - 1.5$
$= 4.03$ s

For second stone

$$h = 150 = u_2 \times t_2 + \frac{1}{2} \times 9.81 \times t_2^2$$

$$\Rightarrow \quad 150 = 4.03\, u_2 + 4.9 \times (4.03)^2$$

$$\boxed{u_2 = 17.47 \text{ m/s.}} \quad \textbf{Ans.}$$

Problem 7.12. *A ball is projected vertically upward with a velocity of 15 m/s. Two seconds later a second ball is projected vertically upward with a velocity 25 m/s. At what height will they meet above the earth surface ?*

Sol. Initial velocity of first stone $u_1 = 15$ m/s

Initial velocity of second stone $u_2 = 25$ m/s

Let after t seconds of projection of the first stone, both the stones are at the same height.

Time during which second stone travelled = $(t - 2)$ seconds.

Displacement of first stone in t second = Displacement of second stone in $(t - 2)$ seconds.

$$\Rightarrow \quad h = u_1 \times t - \frac{1}{2} \times 9.80\, t^2$$

Also

$$h = u_2 \times (t - 2) - \frac{1}{2}\, 9.8\, (t - 2)^2$$

$$15t - 4.9t^2 = 25(t - 2) - 4.9\, (t - 2)^2$$

$$4.9\, [(t - 2)^2 - t^2] = 25\, (t - 2) - 15t$$

$$\Rightarrow \quad 4.9\, [t^2 + 4 - 4t - t^2] = 25t - 50 - 15t$$

$$19.6 - 19.6\, t = 10t - 50$$

$$29.6t = 69.6$$

$$\boxed{t = 2.35 \text{ s.}}$$

$\therefore$ Height at which they meet

$$h = 15 \times 2.35 - \frac{1}{2} \times 9.81 \times (2.35)^2$$

$$\boxed{h = 8.16 \text{ m.}} \quad \textbf{Ans.}$$

Problem 7.13. *A stone is thrown vertically up into the air from the ground. At the same time second stone is thrown up from a 50 m high tower. The initial velocity of first stone is 40 m/s while that of the second stone is 24 m/s. When and where will the two stone be at the same height from the ground.*

Sol. Since both stones initially move upward which will be the positive sense of s, v and a for both stones.

Also, both stones are subjected with a retardation

$$a = -g = -9.81 \text{ m/s}^2.$$

Let the initial velocity of the first stone be $u_1 = 40$ m/s, and for the second stone $u_2 = 24$ m/s.

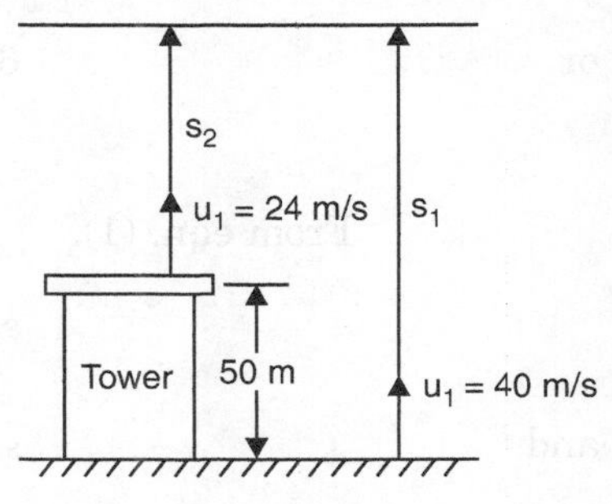

Fig. P. 7.13

Using $s = ut + \frac{1}{2}at^2$, we get

For first stone: $s_1 = 40t - \frac{1}{2} \times 9.81 \times t^2$ ...(1)

For second stone : $s_2 = 24t - \frac{1}{2} \times 9.81 \times t^2$ ...(2)

Also, $s_1 - s_2 = 50$ m = height of tower

$\therefore$ By eqn. (1) – (2), we get

$$s_1 - s_2 = 50 = (40 - 24)t$$

or $t = \frac{50}{16}$ seconds $= \frac{25}{8}$ **= 3.125 seconds. Ans.**

Therefore, from (1), $s_1 = 40 \times \left(\frac{25}{8}\right) - \frac{1}{2} \times 9.81 \times \left(\frac{25}{8}\right)^2 = 77.1$ m

and from (2), $s_2 = 24 \times \left(\frac{25}{8}\right) - \frac{1}{2} \times 9.81 \times \left(\frac{25}{8}\right)^2 = 27.1$ m

Hence the two stones pass each other at 77.1 m above the ground and after 3.125 seconds from their time of projection into air.

Problem 7.14. *A stone is thrown vertically up into the air from the ground. At the same instant second stone is thrown up from a 60 m high tower. The initial velocity of first stone is 40 m/s and that of the second stone is 30 m/s. When and where will the two stones be at the same height from the ground.*

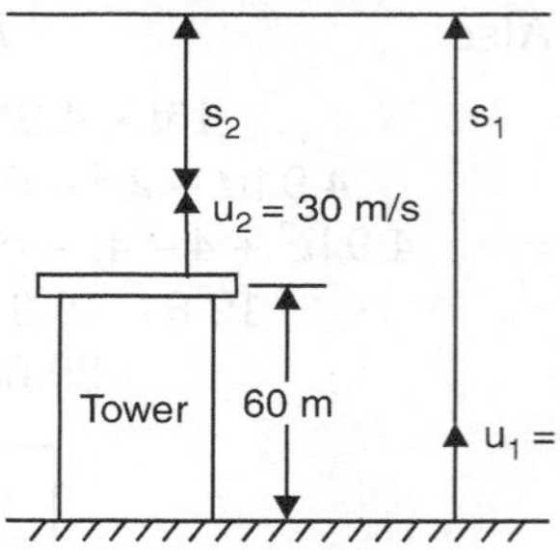

Fig. P. 7.14

Sol. Since both stones initially move upward which will be the positive sense of s, v and a.

Also, since both stones move upward they are subjected to a retardation at rate

$$a = -9.81 \text{ m/s}^2$$

Let initial velocities of stones be u_1 and u_2. Then, we get

For stone 1,

$$s_1 = u_1 t - \frac{1}{2} \times 9.81\, t^2$$

or $s_1 = 40\, t - \frac{1}{2} \times 9.81\, t^2$ ...(1)

and, $s_2 = 30t - \frac{1}{2} \times 9.81 \times t^2$ ...(2)

Given $s_1 - s_2$ = height of the tower = 60 m

$\therefore$ Subtracting eqn. (2) from (1),

$$s_1 - s_2 = (40t - \frac{1}{2} \times 9.81\, t^2) - (30t - \frac{1}{2} \times 9.81\, t^2)$$

or $60 = (40 - 30)\, t$

$\therefore$ $t = \frac{60}{10} = 6$ seconds.

$\therefore$ From eqn. (1),

$$s_1 = 40 \times (6) - \frac{1}{2} \times 9.81 \times (6)^2 = 63.42 \text{ m}$$

and $s_2 = 30 \times (6) - \frac{1}{2} \times 9.81 \times (6)^2 = +\ 3.42$ m

Thus, since S_1 and S_2 are both positive, when the two stones are at the same height, they are moving upward.

Problem 7.15. *A particle moves following the relation $a = 6t$ where a is in m/s^2 and t in seconds. At t = 1, given that s = 4 m, v = 4 m/s, find s-t, and v-t relations. Determine s, v and a at 2 seconds and 3 seconds.*

Sol. Given : $a = 6t$ And, at $t = 1$ second, $s = 4$ m, $v = 4$ m/s.

$$\therefore \quad \frac{dv}{dt} = 6t \quad \text{or } dv = (6t)\, dt$$

Integrating $$\int_{v_0}^{v} dv = \int_1^t 6t\, dt$$

or $$v - v_0 = \left(\frac{6t^2}{2}\right)_1^t$$

i.e., $$v - v_0 = 3t^2 - 3$$

or $$v - 4 = 3t^2 - 3$$

$$\therefore \quad v = 3t^2 - 3 + 4$$

or $$v = 3t^3 + 1 \qquad ...(1)$$

From eqn. (1), we get $\frac{ds}{dt} = 3t^2 + 1$, we get

$$ds = (3t^2 + 1)\, dt$$

Integrating, we get $$\int_{s_0}^{s} ds = \int_1^t (3t^2 + 1)\, dt$$

or $$s - s_0 = (t^3 + t)_1^t$$

or $$s = s_0 + (t^3 + t) - (1 + 1)$$

or $$s = s_0 + t^3 + t - 2 \qquad \text{where } s_0 = 4 \text{ m}$$

$$\therefore \quad s = 4 + t^3 + t - 2$$

or $$s = t^3 + t + 2 \qquad ...(2)$$

At $t = 2$ seconds, we get

$s = (2)^3 + (2) + 2 = 12$ m. **Ans.**

$v = 3\,(2)^2 + 1 = 13$ m/s. **Ans.**

$a = 6 \times (2) = 12$ m/s^2. **Ans.**

At $t = 3$ seconds, we get

$s = (3)^3 + (3) + 2 = 32$ m. **Ans.**

$v = 3\,(3)^2 + 1 = 28$ m/s. **Ans.**

$a = 6(3) = 18$ m/s^2. **Ans.**

Problem 7.16. *The resistance in a shock absorber is proportional to the velocity of the piston of the shock absorber which moves in viscous oil. The retardation of the piston is given by $a = -kv$. If at t = 0, s = 0 and $v = v_0$, derive expressions for velocity and displacement in terms of the time. Also express velocity in terms of displacement. Plot these relations. Also plot s-t relation when $s = s_0$ at t = 0.*

Sol. (*i*) **Velocity in terms of time**

Given : $a = -kv$, we get $\frac{dv}{dt} = -kv$

$$\therefore \quad \frac{dv}{v} = -k\,dt$$

Integrating we get $\int_{v_0}^{v} \frac{dv}{v} = \int_0^t - kdt$

or $\log v - \log v_0 = - kt$

or $\log v/v_0 = - kt$

Hence $v = v_0 e^{-kt}$ ...(1) **Ans.**

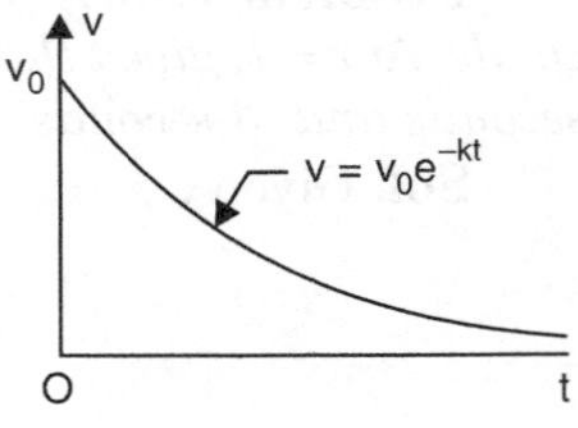

Fig. P. 7.16 (i) v-t relation.

Fig. P. 7.16 (*i*) shows the *v*-*t* relation

(*ii*) **Displacement in terms of time**

Using the relation $v = \frac{ds}{dt}$ and eqn. (1), we get

$$\frac{ds}{dt} = v_0 e^{-kt}$$

$\therefore$ $ds = v_0 e^{-kt}$

Integrating we have $\int_{s_0}^{s} ds = \int_0^t v_0 e^{-kt}$

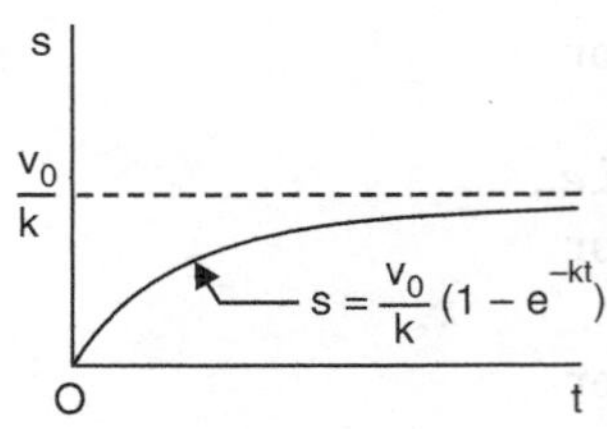

Fig. P. 7.16 (ii) s-t relation when $s = 0$ at $t = 0$.

or $s - s_0 = -\frac{v_0}{k}\left[e^{-kt}\right]_0^t = -\frac{v_0}{k}\left[e^{-kt} - 1\right].$

$\therefore$ $s = s_0 - \frac{v_0}{k}\left[e^{-kt} - 1\right]$ ...(2) where $s_0 = 0$ at $t = 0$

$\therefore$ we get $s = -\frac{v_0}{k}\left(e^{-kt} - 1\right) = \frac{v_0}{k}\left(1 - e^{-kt}\right)$ ...(3) **Ans.**

Fig. P.7.16 (*ii*) shows the *s*-*t* plot

when $s_0 = 0$ at $t = 0$, the *s*-*t* relation is given by eqn. (2) which is plotted as shown in Fig. P.7.16 (*iii*)

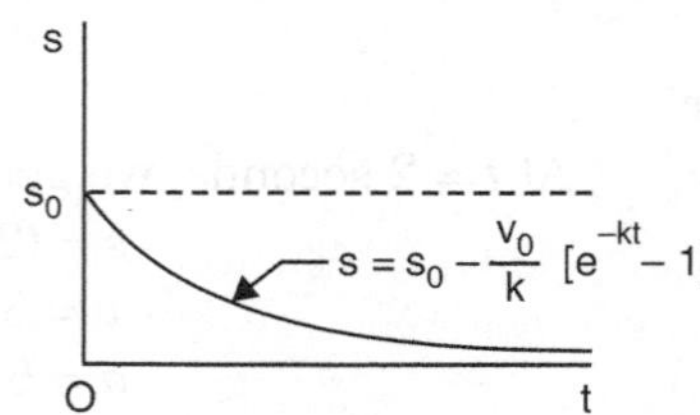

Fig. P. 7.16 (iii) s-t relation when $s = s_0$ at $t = 0$.

(*iii*) **Velocity in terms of displacement**

Writing $a = -kv$ in eqn. $a = v.\frac{dv}{ds}$, we get

$$- kv = v\frac{dv}{ds}$$

$\therefore$ $dv = - kds$

Integrating, we get

$$\int_{v_0}^{v} dv = - k\int_0^s ds$$

$\therefore$ $v = v_0 - ks$ ...(4)

Fig. P.7.16 (*iv*) shows the plot of eqn. (4).

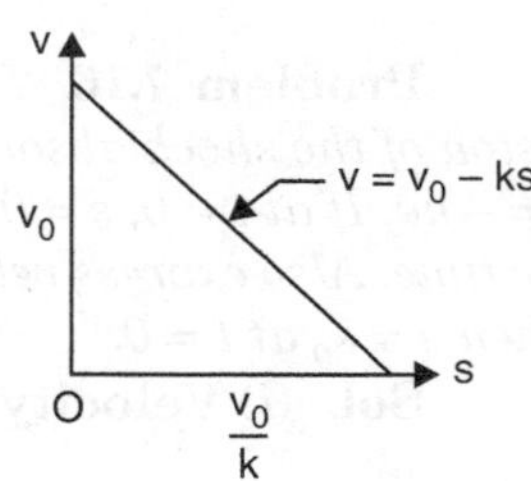

Fig. P.7.16 (iv) v-s relation.

7.6. NORMAL AND TANGENTIAL COMPONENTS OF ACCELERATION

In many situations the most useful components of acceleration are directed along the tangent and along the normal to the path. These components denote the rate of **change of magnitude** of the velocity of the particle and the rate of **change of direction** of its velocity, respectively. These components are particularly useful in relating the velocity and acceleration of a particle to its path.

Consider a particle moving along path shown in Fig. 7.6 (*a*). When particle is at point *A*, let its velocity be v and the radius of curvature of the particle be r. After a short interval of time δt, let the particle be at point *B* and let the velocity of the particle be $v + \delta v$. If the time interval δt is very small the radius of curvature of path at both points is almost the same.

The angle between the two radii is $\delta\theta$.

Let the velocity at point *A* be given by vector $\overline{oa}$.

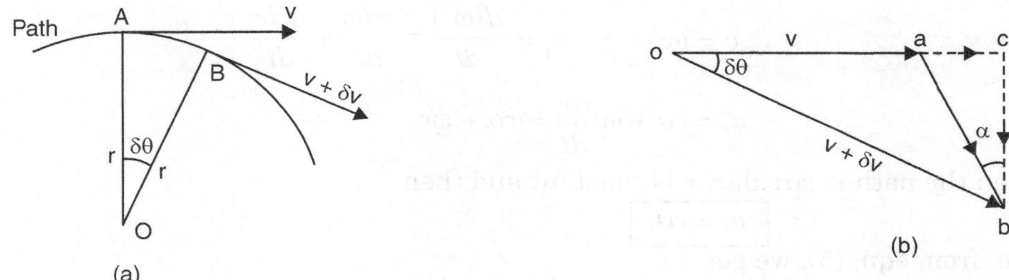

Fig. 7.6 Normal and tangential Acceleration components.

The velocity at *B* is $v + \delta v$, and it is given by $\overline{ob}$.

Then change in velocity of particle between points *A* and *B* is given by $\overline{ab}$. Resolving $\overline{ab}$ in two components, one along the radial line *AO i.e.*, along $\overline{cb}$ and the other along the tangent to the path at *A* (*i.e.*, along the velocity v at *A*) given by $\overline{ac}$. Let angular velocity of particle at *A* be ω and at *B* $(\omega + d\omega)$ rad/s.

Thus, $$\overline{ab} = \overline{ac} + \overline{cb} \quad ...(1)$$

where $$\overline{ac} = \overline{oc} - \overline{oa} = \overline{ob} \cos \delta\theta - \overline{oa}$$

or $$\overline{ac} = (v + \delta v) \cos \delta\theta - v \quad ...(2)$$

As $\delta\theta <<$, $\therefore$ $\cos \delta\theta \simeq 1$

$\therefore$ Eqn. (2) yields

$$\overline{ac} = (v + \delta v) . 1 - v = \delta v \quad ...(3)$$

and $$\overline{cb} = \overline{ob} \sin \delta\theta$$

or $$\overline{cb} = (v + \delta v) \sin \delta\theta \quad ...(4)$$

As $\delta\theta <<$, $\therefore$ $\sin \delta\theta \simeq \delta\theta$ in radians,

Hence eqn. (4) yields

$$\overline{cb} = (v + \delta v) . \delta\theta = v . \delta\theta + \delta v . \delta\theta$$

Neglecting the product $\delta v \,.\, \delta\theta$ of two small quantities,

we get $$\overline{cb} = v\, \delta\theta \quad ...(5)$$

Dividing eqns. (3) and (5) by the time interval, δt we get

$$\frac{\overline{ac}}{\delta t} = \frac{\delta v}{\delta t} = a_t$$

= Tangential component of acceleration

= rate of change of velocity with respect to time, parallel to v.

Thus $$a_t = \frac{\delta v}{\delta t}$$

or $$\boxed{a_t = \lim_{\delta t \to 0} \frac{\delta v}{\delta t} = \frac{dv}{dt}} \quad ...(7.11)$$

Now, $\because$ $$v = \omega r, \quad \therefore \quad a_t = \frac{d(\omega r)}{dt} = \frac{r d\omega}{dt} + \frac{\omega dr}{dt}$$

$\therefore$ $$a_t = r\alpha + \omega \frac{dr}{dt} = r\alpha + \omega r \quad ...(7)$$

When the path is circular, r is constant and then,

$$\boxed{a_t = r\alpha} \quad ...(7.12)$$

Also, from eqn. (5), we get

$$\frac{\overline{cb}}{\delta t} = \frac{v\delta\theta}{\delta t} = a_n \text{ Normal component of acceleration}$$

(Directed along AO or directed towards the centre of curvature)

or $$a_n = \lim_{\delta t \to 0} \frac{v\delta\theta}{\delta t} = v \,.\, \frac{d\theta}{dt}$$

or $$\boxed{a_n = v\omega} \quad ...(7.13)$$

But $v = \omega r$, therefore other forms of a_n become :

$$\boxed{a_n = v(v/r) = v^2/r} \quad ...(7.14)$$

Also, $$\boxed{a_n = (\omega r) \,.\, \omega = \omega^2 r} \quad ...(7.15)$$

Substituting values of $\overline{ac}$ and $\overline{cb}$ into eqn. (1), we get

$$\overline{ab} = \overline{ac} + \overline{cb} = \delta v + v\delta\theta$$

Dividing by δt, we get

$$\frac{\overline{ab}}{\delta t} = \frac{\overline{ac}}{\delta t} + \frac{\overline{cb}}{\delta t} = \text{Total acceleration}$$

$\therefore$ $$\overline{a} = \overline{a_t} + \overline{a_n} = \frac{dv}{dt} + v\omega \quad ...(7.16)$$

The angle α between the total acceleration (a) and the normal acceleration (a_n) as shown in Fig. 7.6 (b) is obtained as follows.

$$\tan\alpha = \frac{\overline{ac}}{\overline{cb}} = \frac{a_t}{a_n}$$

When r is constant, $\quad a_t = \dfrac{d}{dt}(r\omega) = r\alpha$

$$\therefore \qquad \tan \alpha = \frac{r\alpha}{r\omega^2} = \alpha / \omega^2$$

or $$\alpha = \tan^{-1} (\alpha/\omega^2) \qquad ...(7.17)$$

The vector representing the total acceleration is called the **Acceleration Image**, see Fig. 7.6 (*c*).

Important Cases :

(*a*) When the particle has rectilinear motion, r is equal to infinity. Therefore, $\quad a_n = \dfrac{v^2}{r} = 0$

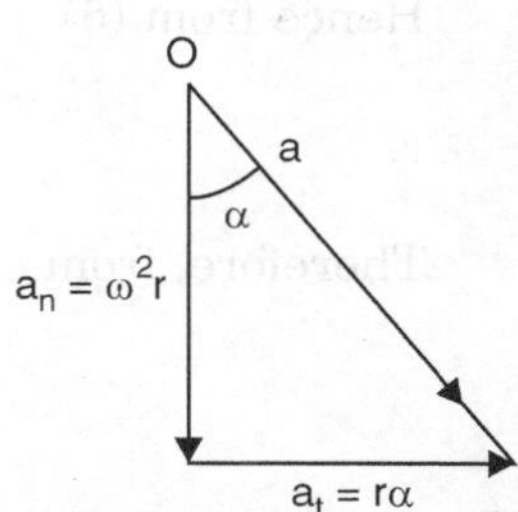

Fig. 7.6 (c) The acceleration image.

(*b*) When angular velocity ω is uniform, α = 0 and therefore, $a_t = 0$

(*c*) When the path of the particle is circular, r is constant, and then $\quad a_t = \dfrac{d}{dt}(v) = \dfrac{d}{dt}(r\omega) = r\alpha.$

(*d*) If the path is curvilinear, r is not constant and then

$$a_t = \frac{d(v)}{dt} = \frac{d}{dt}(r\omega) = \frac{rd\omega}{dt} + \frac{\omega dr}{dt} = \alpha r + \omega \dot{r}.$$

This is a case of a general motion.

(*e*) The tangential acceleration is parallel to the velocity of the particle.

7.7. PROJECTILE MOTION OR TRAJECTORY

An important application of two-dimensional kinematic theory is the problem of projectile motion. For a first treatment of the subject, the aerodynamic drag and the effects of curvature and rotation are neglected, and assume that the altitude range is small enough so that the acceleration due to gravity can be considered constant. With these assumptions, the conventional rectangular co-ordinates are useful for the analysis of the path of the projectile known as **trajectory.**

Thus, for the axes shown in Fig. 7.7 the acceleration components are :

$$a_x = 0 \quad \text{and} \quad a_y = -g \qquad ...(1)$$

Let the particle be projected at point O with initial velocity u inclined at α to x-axis. x- and y-components of initial velocity at point O are :

$$u_x = u \cdot \cos \alpha = \text{constant} \qquad ...(2)$$

and $$u_y = u \sin \alpha \qquad ...(3)$$

After time t let the projectile be at point P, where its velocity is v, inclined at θ to the x-axis.

Then, $$v_x = v \cos \theta = u \cos \alpha \qquad ...(4)$$

and $$v_y = v \sin \theta = u \sin \alpha - gt \qquad ...(5)$$

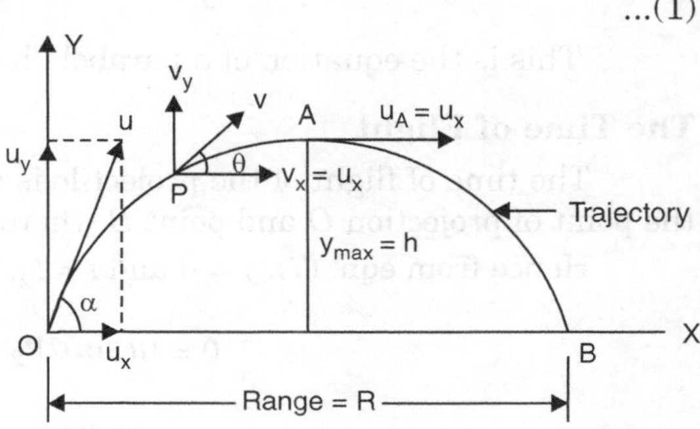

Fig. 7.7 Projectile motion.

The position of point p is given by $P(x, y)$, where

$$x = (u \cos \alpha) \cdot t \quad \text{...(6)}$$

and
$$y = u_y \cdot t - \frac{1}{2} gt^2 = (u \sin \alpha)t - \frac{1}{2} gt^2 \quad \text{...(7)}$$

The Maximum Height : $h = y_{max}$

At point A, the particle is at the maximum height above x-axis. At point A, $v_A = 0$ and let time = t_1

Hence from (5), $v_y = 0 = u. \sin \alpha - gt_1$

$$\therefore \quad t_1 = \frac{u \sin \alpha}{g} \quad \text{...(7.18)}$$

Therefore, from eqn. (7)

$$y\Big]_{t=t_1} = u_y t_1 - \frac{1}{2} g t_1^2 = y_{max} \text{ or } h.$$

$$\therefore \quad y_{max} = (u . \sin \alpha) \times \frac{u \sin \alpha}{g} - \frac{1}{2} \times g \times \left(\frac{u \sin \alpha}{g} \right)^2$$

$$= \frac{u^2 \sin^2 \alpha}{g} - \frac{u^2 \sin^2 \alpha}{2g}$$

or
$$y_{max} = \frac{u^2 \sin^2 \alpha}{2g} \quad \text{...(7.19)}$$

Equation of Path of Projectile

From eqn. (6), we get $t = \dfrac{x}{u \cos \alpha}$ ...(8)

Hence from eqn. (7), we get

$$y = (u \sin \alpha) \times \frac{x}{u \cos \alpha} - \frac{1}{2} g \times \left(\frac{x}{u \cos \alpha} \right)^2$$

or
$$y = x \tan \alpha - \frac{g}{2u^2 \cos^2 \alpha} x^2 \quad \text{...(7.20)}$$

This is the equation of a parabola having a vertical axis.

The Time of Flight

The time of flight of the projectile is defined as the time taken by the projectile between the point of projection O and point B where it hits the ground. Thus, at point B, $y = 0$.

Hence from eqn. (7), $y = 0$ and $t = t_2$, we get

$$0 = (u \sin \alpha) t_2 - \frac{1}{2} g \times t_2^2$$

$$\therefore \quad t_2 = \frac{2u \sin \alpha}{g} \quad \text{...(7.21)}$$

It is important to note that when OB is horizontal $t_2 = 2t_1$ ...(7.22)

Range of Projectile

Substituting the value of time t_2 from eqn. (7.22) into eqn. (6), we get

The range, $R = (u \cos \alpha) \times t_2$

$$= (u \cos \alpha) \times \frac{2u \sin \alpha}{g}$$

$$= \frac{2u^2 \sin \alpha \cos \alpha}{g}$$

or $$R = \frac{u^2 \sin 2\alpha}{g} \quad ...(7.23)$$

The Maximum Range

From eqn. (7.23), it is clear that the range will be maximum when $\sin 2\alpha = 1 = \sin 90°$ or when $\alpha = 45°$.

Thus, the maximum range $= R_{max} = \dfrac{u^2}{g}$ ...(7.24)

when the angle of projection of the projectile is 45° to the x-axis.

Problem 7.17. *A projectile is shot with an initial velocity of 8 m/s. If the inclination from the horizontal is 30° find, its time of flight, range and maximum height attained ?*

Sol. Given: $u = 8$ m/s

$\alpha = 30°$.

Let the projectile be shot with initial velocity of 8 m/s.

Time of flight,

$$t = \frac{2u \sin \alpha}{g}$$

$$= \frac{2(8)(\sin 30°)}{9.81} = 0.8155 \text{ sec.} \quad \textbf{Ans.}$$

Range, $$R = \frac{u^2 \sin 2\alpha}{g} = \frac{(8)^2 \,.\, \sin 60°}{9.81}$$

$$= 5.65 \text{ m.} \quad \textbf{Ans.}$$

Maximum height, $$H = \frac{u^2 \sin^2 \alpha}{2g}$$

$$= \frac{(8)^2 \,.(\sin 30°)^2}{2 \times 9.81} = 0.82 \text{ m.} \quad \textbf{Ans.}$$

Problem 7.18. *The pilot of an aeroplane flying horizontally at a height of 1000 m with a constant speed of 270 km/hr wishes to hit a target on the ground. At what distance from the target should he release the bomb in order to hit the target ? At what angle would the target appear to him from that distance ?*

Sol. Initial velocity of the bomb is the same as that of the aeroplane that is 270 km/hr.

Initial velocity of the plane is 270 km/hr in the horizontal direction is

$$= 270 \times \frac{1000}{3600} = 75 \text{ m/s.}$$

Initial velocity of the bomb in the vertical direction, is zero.

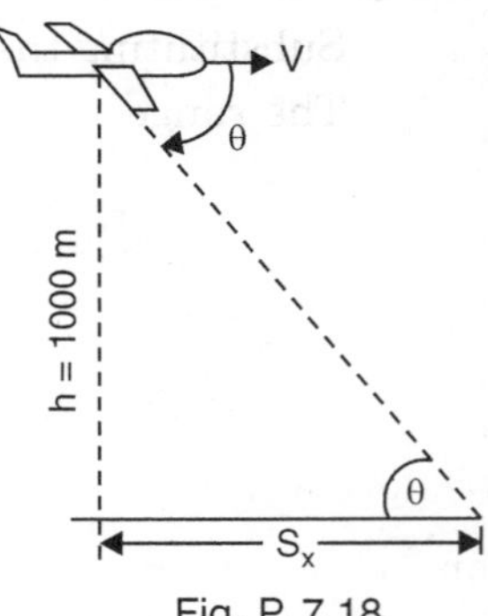

Fig. P. 7.18

Motion in vertical direction

Initial velocity, u_y = zero

Distance $= h = 1000$ m

Acceleration, $a_y = g = 9.81 \text{ m/s}^2$

Using $$h = s_y = u_y \,.\, t + \frac{1}{2} gt^2$$

We get $$1000 = 0 + \frac{1}{2} \times 9.81 \times t^2$$

$$t = \sqrt{\frac{2000}{9.81}} = \textbf{14.28 seconds.}$$

It is the time required to travel down a height 1000 m.

Motion in the horizontal direction.

Initial velocity = 75 m/s

Acceleration, $a_x = 0$

Using, $s_x = ut + \frac{1}{2} a_x \,.\, t^2$, the horizontal distance travelled by the bomb in the time $t = 14.28$ sec is,

$$s_x = 75 \times 14.28 + \frac{1}{2} \times (0) \times t^2$$

$$\therefore \quad s_x = 1071 \text{ m}$$

So, the bomb should be released when the aeroplane is at a horizontal distance of 1071 m from the target and angle is θ is given by

$$\tan \theta = \frac{1000}{1071} \approx 0.9337 \qquad \text{(See Fig. P.7.18)}$$

Hence θ = **43.037° from horizontal. Ans.**

Problem 7.19. *Two guns are pointed at each other, one upward at an angle of 30° and other at the same angle of depression the muzzle being 30 m apart. The guns are shot with velocities of 350 m/s upward and 280 m/s downward respectively. Find when and where the shots will meet.* [Oxford University]

Sol. Given : $\alpha = 30°$, $v_C = 350$ m/s, $v_D = 280$ m/s

Distance between the muzzle = 30 m. = CD

Time when the shots meet after they leave the guns :

Let the two shots meet at E as shown in Fig. P. 7.19.

t = time in seconds, when the two shots meet after they leave the guns.

x = horizontal distance between E and C = CF

y = vertical distance between E and C = FE.

∴ Horizontal distance between C and D (*i.e.*, CG)

$$= 30 \cos 30° = \frac{30\sqrt{3}}{2} = 15\sqrt{3} \text{ m.}$$

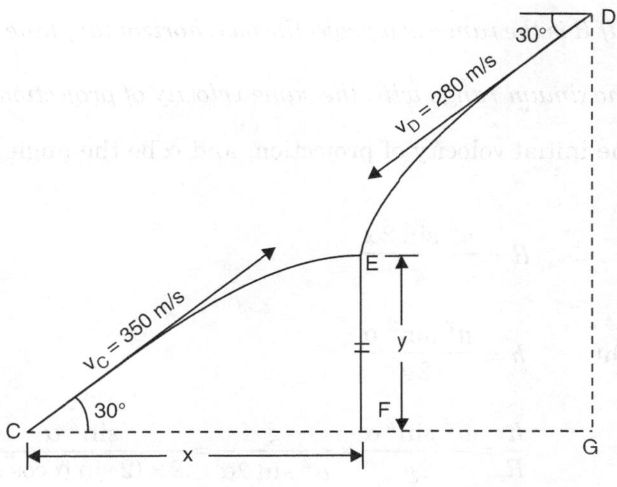

Fig. P.7.19

Distance covered by the shot C in t seconds

$$x = \text{horizontal component of } v_C \times t$$
$$= 350 \cos 30° \times t$$
$$= 350 \frac{\sqrt{3}}{2} \times t = 175\sqrt{3}\, t \qquad \text{...(1)}$$

Similarly, distance covered by the shot D in t seconds.

$$15\sqrt{3} - x = 280 \cos 30° \times t = 280 \frac{\sqrt{3}}{2} t = 140\sqrt{3}\, t \qquad \text{...(2)}$$

Adding (1) and (2), we have

$$15\sqrt{3} = 175\sqrt{3}\, t + 140\sqrt{3}\, t$$
$$15 = 175t + 140t$$
$$15 = 315\, t$$
$$t = \frac{15}{315}$$
$$t = 0.0476 \text{ sec.} \quad \textbf{Ans.}$$

Point where the two shots meet : Substituting this value of t in eqn. (1), we get

$$x = 175\sqrt{3}\,(0.0476)$$
$$= 14.42 \text{ m.} \quad \textbf{Ans.}$$

We know that vertical component of v_C

$$= 350 \sin 30°$$
$$= 175 \text{ m/sec}$$

Therefore vertical distance between C and E

$$y = ut - \frac{1}{2} gt^2 = 175 \times 0.0476 - \frac{1}{2} (9.81)(0.0476)^2$$
$$y = \mathbf{8.32 \text{ m.}} \quad \textbf{Ans.}$$

Problem 7.20. *If R is the range of a projectile on a horizontal plane and h is its maximum height show that the maximum range with the same velocity of projection is* $R_{max} = 2h + \dfrac{R^2}{8h}$.

Sol. Let u be the initial velocity of projection, and α be the angle of projection with the horizontal.

Then, Range, $$R = \frac{u^2 \sin 2\alpha}{g} \quad ...(1)$$

and, maximum height, $$h = \frac{u^2 \sin^2 \alpha}{2g} \quad ...(2)$$

$$\therefore \quad \frac{h}{R} = \frac{u^2 \sin^2 \alpha}{2g} \times \frac{g}{u^2 \sin 2\alpha} = \frac{\sin^2 \alpha}{2 \times (2 \sin \alpha \cos \alpha)} = \frac{\tan \alpha}{4} \quad ...(3)$$

$$\therefore \quad \tan \alpha = \frac{4h}{R} \quad ...(4)$$

Hence $$\sin \alpha = \frac{4h}{\sqrt{R^2 + (4h)^2}} = \frac{4h}{\sqrt{R^2 + 16h^2}}$$

and $$\cos \alpha = \frac{R}{\sqrt{R^2 + 16h^2}}$$

$\therefore$ Hence $$\sin 2\alpha = 2 \sin \alpha \,.\, \cos \alpha$$

or $$\sin 2\alpha = 2 \times \frac{4h}{\sqrt{R^2 + 16h^2}} \times \frac{R}{\sqrt{R^2 + 16\,h^2}} = \frac{8hR}{R^2 + 16h^2} \quad ...(5)$$

$\therefore$ Form (1) $$u^2 = \frac{gR}{\sin 2\alpha} = \frac{gR}{(8hR)/(R^2 + 16h^2)} = \frac{g(R^2 + 16h^2)}{8h} \quad ...(6)$$

Maximum Range, $$R_{\max} = \frac{u^2}{g}$$

$\therefore$ From (6), $$R_{\max} = \frac{1}{g} \times \frac{g(R^2 + 16h^2)}{8h}$$

$$= \frac{R^2 + 16h^2}{8h}$$

or $$R_{\max} = 2h + \frac{R^2}{8h}$$ **Proved**

Problem 7.21. *A boy throws a ball so that it may just clear a wall 3.6 m high. The boy is at a distance of 4.8 m from the wall. The ball was found to hit the ground at a distance of 3.6 m on the other side of wall. Find the least velocity with which the ball is thrown ?*

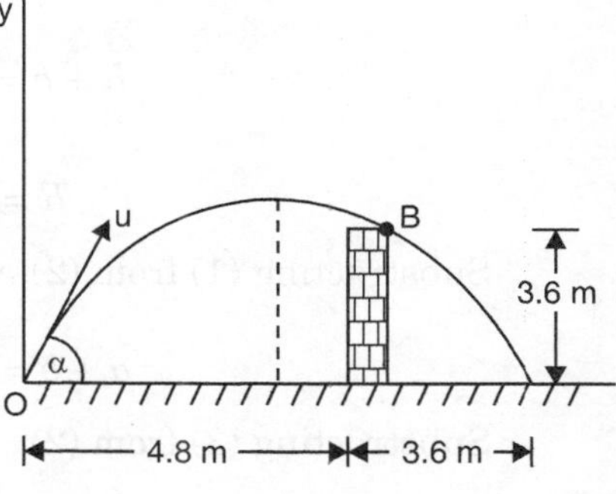

Fig. P. 7.21

Sol. Let velocity of projection be u, and angle of projection = α

Range = 4.8 + 3.6 = 8.4 m.

The top of the wall must lie in the projectile path.

The equation of path with the point of projection as origin is,

$$y = x \tan \alpha - \frac{g x^2}{2u^2 \cos^2 \alpha} \qquad ...(1)$$

Substituting value of point (x, y) of point B in (1), we get

$$3.6 = 4.8 \tan \alpha - \frac{g \times (4.8)^2}{2u^2 \cos^2 \alpha} \qquad ...(2)$$

The range of projectile,

$$R = \frac{u^2 \sin 2\alpha}{g} = 8.4 \qquad ...(3)$$

$$\therefore \quad \frac{u^2}{g} = \frac{8.4}{\sin 2\alpha} \qquad ...(4)$$

Substituting u^2/g in (2), we have

$$3.6 = 4.8 \tan \alpha - \frac{(4.8)^2}{2 \cos^2 \alpha} \cdot \frac{\sin 2\alpha}{8.4}$$

$$3.6 = 4.8 \tan \alpha - \frac{(4.8)^2}{2 \times 8.4} \cdot \frac{2 \sin \alpha \cos \alpha}{\cos^2 \alpha}$$

$$3.6 = 4.8 \tan \alpha - 2.74 \tan \alpha$$

$$\therefore \quad \tan \alpha = \frac{3.6}{2.06} = 1.7476$$

$$\therefore \quad \alpha = 60.2°$$

Substituting in (4), we have

$$u^2 = g \frac{8.4}{\sin 2\alpha} = \frac{9.81 \times 8.4}{\sin 120.4°} = 95.54$$

$$\therefore \quad u = 9.774 \text{ m/s.} \quad \textbf{Ans.}$$

Problem 7.22. *A projectile is aimed at a target which lies in the horizontal plane through the point of projection. It falls a metres short of target and goes b metres too far off when angles of projections are α and β, respectively.*

If a = b = 2m, α = 30°, β = 45° show that proper angle of projection θ = 34.45° if velocity of projection is the same in all cases.

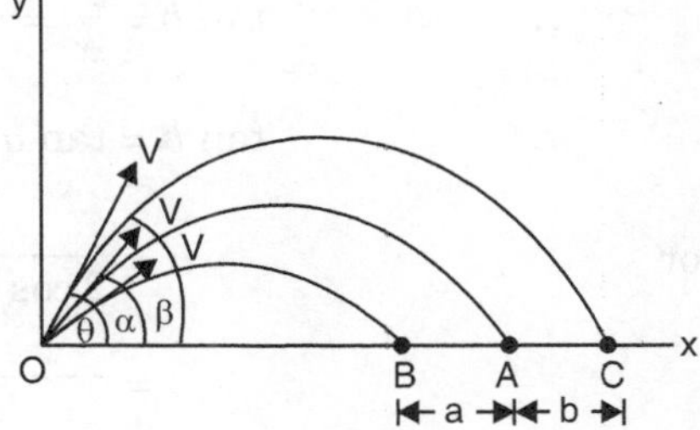

Fig. P. 7.22

Sol. Let u be the velocity of projection, θ be the angle of projection, R be the range.

$$R - a = \frac{u^2 \sin 2\alpha}{g} = OB \quad \text{(See Fig. P. 7.22)} \quad ...(1)$$

$$R + b = \frac{u^2 \sin 2\beta}{g} = OC \quad ...(2)$$

$$R = \frac{u^2 \sin 2\theta}{g} = OA \quad ...(3)$$

Substracting (1) from (2), we have

$$a + b = \frac{u^2 (\sin 2\beta - \sin \sin 2\alpha)}{g} \quad ...(4)$$

Substracting (3) from (2)

$$b = \frac{u^2 (\sin 2\beta - \sin 2\theta)}{g} \quad ...(5)$$

From (4),
$$\frac{u^2}{g} = \frac{a+b}{(\sin 2\beta - \sin 2\alpha)} \quad ...(6)$$

Also, from (5),
$$\frac{u^2}{g} = \frac{b}{(\sin 2\beta - \sin 2\theta)}$$

or
$$\frac{a+b}{(\sin 2\beta - \sin 2\alpha)} = \frac{b}{(\sin 2\beta - \sin 2\theta)}$$

On solving we get

$$\sin 2\theta = \frac{a \sin 2\beta + b \sin 2\alpha}{(a+b)} \quad ...(7)$$

On substituting $a = b = 2m$; $\alpha = 30°$, $\beta = 45°$, we get

$$\theta = 34.45°. \quad \textbf{Ans.}$$

Problem 7.23. *A particle is projected to an angle of elevation 'a' and after t seconds it appears to have angle of elevation b as seen from point of projection. Prove that initial velocity is* $\frac{gt \cos b}{2 \sin (a - b)}$.

Sol. Let coordinates of the particle after t-seconds be (x, y) referred to the point of projection as origin. Let u be the velocity of projection. See Fig. P. 7.23.

$$\therefore \quad y = (u \sin a)t - \frac{1}{2} gt^2$$

and
$$x = (u \cos a)\, t$$

$$\therefore \quad \tan b = \frac{y}{x} = \frac{(u \sin a)t - \frac{1}{2} gt^2}{(u \cos a)t.}$$

$$\therefore \quad \tan b = \tan a - \frac{gt}{2u \cos a}$$

or
$$u = \frac{gt}{2 \cos a (\tan a - \tan b)}$$

$$= \frac{gt}{2 \cos a \left(\dfrac{\sin a}{\cos a} - \dfrac{\sin b}{\cos b} \right)}$$

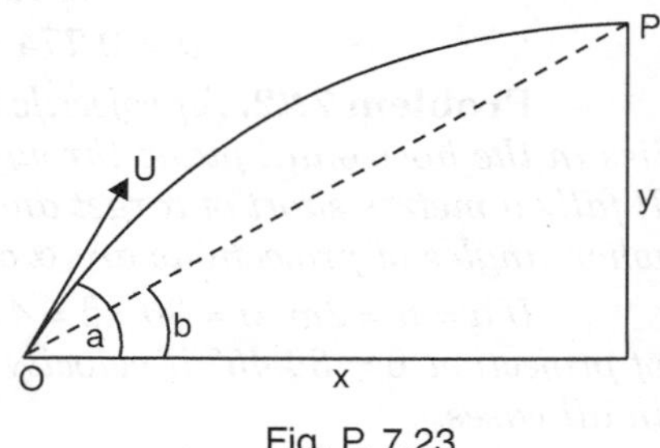

Fig. P. 7.23

$$= \frac{gt \cos a \cos b}{2 \cos a (\sin a \cos b - \cos a \sin b)}$$

or $$u = \frac{gt \cos a \cos b}{2 \cos a . \sin (a - b)}$$

or $$u = \frac{gt \cos b}{2 \sin (a - b)}.\quad \textbf{Proved}$$

Problem 7.24. *A ball of mass 1 kg starts from the position as shown in Fig. P. 7.24 and slides down a frictionless tube under its own weight. After leaving the tube where will the ball hit the wall ? Also calculate the time interval between the instant of leaving the tube and hitting the wall ?*

Sol. Mass of the ball = 1 kg

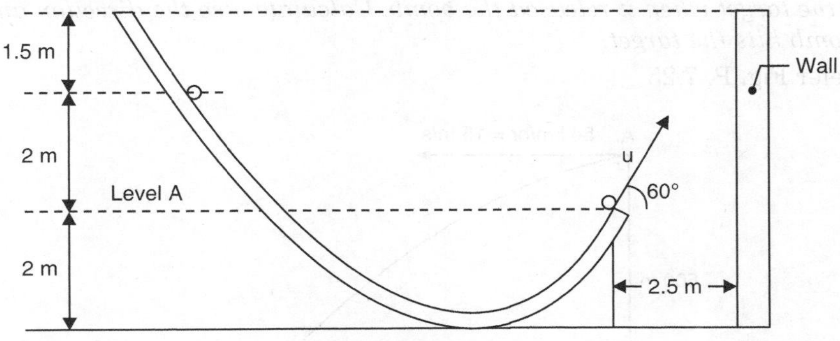

Fig. P. 7.24.

Potential energy of the ball in rest = mgh

$= 1 \times g \times 2 = 2g$ joule from Level A ...(1)

Let u be velocity of the ball at outlet of the tube. See Fig. P. 7.24.

Kinetic energy of ball at outlet of the tube

$$= \frac{1}{2} mu^2 = \frac{1}{2} (1) (u^2) = \frac{u^2}{2} \text{ joule} \quad ...(2)$$

Equating Eqns. (1) and (2), we get

$$\frac{u^2}{2} = 2g$$

$$u = \sqrt{2 \times 9.81 \times 2}$$

$$u = 6.264 \text{ m/sec}$$

Horizontal component of the velocity at outlet of tube

$= u \cos \alpha$

$= 6.264 \cos 60°$

$= 3.132$ m/s

After leaving the tube the ball will describe a parabolic path. But the horizontal component of the velocity during the motion will remain constant.

Time taken to hit the wall = t

$$t = \frac{2.5}{3.132}$$

$$t = 0.798$$

Vertical displacement in this interval

$$= (u \sin \alpha)t - \frac{1}{2}gt^2$$

$$= 6.264 \sin 60° \times 0.798 - \frac{1}{2} (9.81) (0.798)^2$$

$$= 4.329 - 3.124$$

$$= 1.205 \text{ m}$$

Hence, the ball will hit the wall at a point 1.205 m above the outlet of the tube, or above level *A*.

Problem 7.25. *An aircraft moving horizontally at 54 km/hr at a height of 500 m towards a target on the ground and releases a bomb which hits it. Estimate the horizontal distance of the aircraft from the target when it released the bomb. Calculate also the direction and the velocity with which bomb hits the target.*

Sol. Refer Fig. P. 7.25

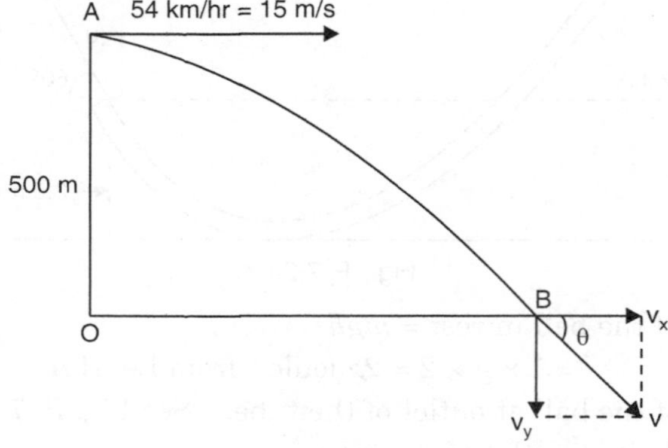

Fig. P. 7.25

Let *A* be the point where the aircraft releases the bomb and point *B* be the target.

Considering the vertical motion of the bomb:

$$h = ut + \frac{1}{2}gt^2$$

$$500 = 0 + \frac{1}{2} 9.81 t^2$$

$$t = 10.1 \text{ seconds}$$

Considering horizontal motion of the bomb:

$$OB = \text{horizontal velocity} \times \text{time}$$

Horizontal velocity, $u_x = 54 \times \frac{5}{18} = 15$ m/s

Horizontal distance, $x = 15 \times 10.1 = 151.5$ m

Hence, the horizontal distance of the aircraft from the target when it released the bomb is 151.5 m.

Now again considering vertical motion of the bomb:

$$v = u + gt = 0 + 9.81 \times 10.1 \text{ m/s}$$

$$= 99.1 \text{ m/s}$$

Resultant velocity, $v = \sqrt{v_x^2 + v_y^2} = \sqrt{(15)^2 + (99.1)^2} = 100.23$ m/s. **Ans.**

Hence the velocity with which the bomb hits the target is 100.23 m/s

Direction of velocity is given by:

$$\tan\theta = \frac{v_y}{v_x} = \frac{99.1}{15} = 6.607$$

$$\therefore \quad \theta = 81.39° \text{ from horizontal. } \textbf{Ans.}$$

Problem 7.26. *Find the maximum speed with which a scooter rider should leave a 15° ramp at B in order to clear the ditch. See Fig. P.7.26*

Sol. Let

Horizontal distance $x = 3$ m.

Vertical distance $y = 1.6$ m.

Here, $a_x = 0$ and $a_y = g = 9.81$ m/s^2

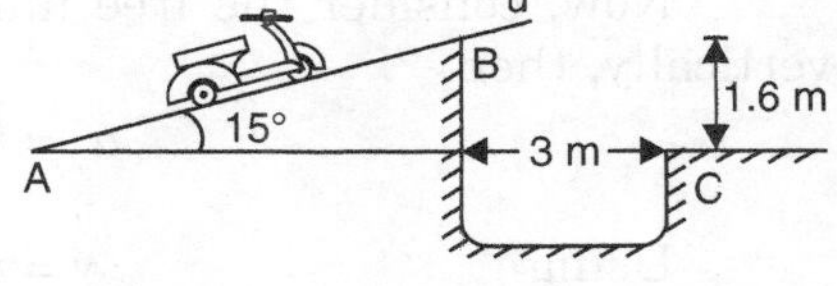

Fig. P. 7.26

$\therefore$ Using $x = u_x.t + \frac{1}{2}a_x.t^2$, we get

$$3 = (u\,.\cos 15°) \times t + 0$$

or $$3 = (u\cos 15°) \times t \quad ...(1)$$

Also, $$y = u_y t - \frac{1}{2}gt^2$$

$$-1.6 = u\sin 15° \times t - \frac{1}{2}gt^2 \quad ...(2)$$

From (1), $$t = \frac{3}{u\cos 15°},$$

$$\therefore \quad -1.6 = (\tan 15° \times 3) - \left[\frac{1}{2}g \times \frac{9}{(u\cos 15°)^2}\right]$$

or $$2.4038 = \frac{44.145}{u^2\cos^2 15°}$$

or $$u = 4.436 \text{ m/s.} = 15.9696 \text{ km/hr. } \textbf{Ans.}$$

Problem 7.27. *An arrow is shot from the top of a 110 m high building with a velocity of 100 m/s and at an elevation of 25° with the horizontal. Neglecting the air resistance find:*

(*i*) *The maximum height of the arrow from the ground.*

(*ii*) *How far must be the target on the ground so that the arrow strikes it.*

Fig. P. 7.27

Sol. Given: $u = 100$ m/s, $\alpha = 25°$,

h_1 is the height of the building = 110 m

Maximum height = h, above the building.

We know maximum height of the projectile from the top of building is

$$h = \frac{u^2\sin^2\alpha}{2g}$$

$$= \frac{(100)^2.(\sin 25°)^2}{2(9.81)}$$

$$h = 91.033 \text{ m.}$$

Therefore the maximum height of the arrow from the ground is

$$h' = h + h_1$$
$$= 91.033 + 110$$
$$= 201.033 \text{ m. } \textbf{Ans.}$$

Horizontal distance:

We know time of ascent of the arrow from the top of the building to maximum height attained is

$$t_a = \frac{u \sin \alpha}{g} = \frac{100 (\sin 25°)}{9.81} = 4.31 \text{ sec.}$$

Now, consider the free fall of the arrow from the maximum height of the projectile vertically, then

$$u_y = 0 \quad \text{and} \quad y = 201.033 \text{ m}$$

Using $$y = u_y \, . \, t + \frac{1}{2} g.t^2$$

and let $t = t_d$. Therefore, we get

$$201.033 = 0 + \frac{1}{2}(9.81)\,(t_d)^2$$
$$t_d = 6.4 \text{ sec.}$$

Therefore total time taken by the arrow to strike the ground is

$$t = t_a + t_d = 4.31 + 6.4$$
$$= 10.71 \text{ sec.}$$

Now, the horizontal component of velocity is

$$u_x = u \cos \alpha = 100 (\cos 25°) = 90.63 \text{ m/s}$$

∴ Horizontal distance covered is

$$R = u_x.t$$
$$= 90.63 \times 10.71$$
$$= 970.65 \text{ m. } \textbf{Ans.}$$

Problem 7.28. *A ball is thrown from the base of a hill which is in a shape of right circular cone. The ball just touches the top of the hill and falls back to the ground on the other side of the hill. If the angle of ball from the ground be 30° and the vertical height of the hill be 40 m, then find the semicone angle of the hill and the initial velocity with which the ball is being projected ?*

Sol. Given: $\alpha = 30°, h = 40 \text{ m}$

OAB is a right circular cone. The ball is projected from point O and falls at point B.

Now, maximum height

$$h = \frac{u^2 \sin^2 \alpha}{2g}$$

∴ $$u^2 = \frac{2gh}{\sin^2 \alpha}$$
$$= \frac{2(9.81)\,(40)}{(\sin 30°)^2}$$

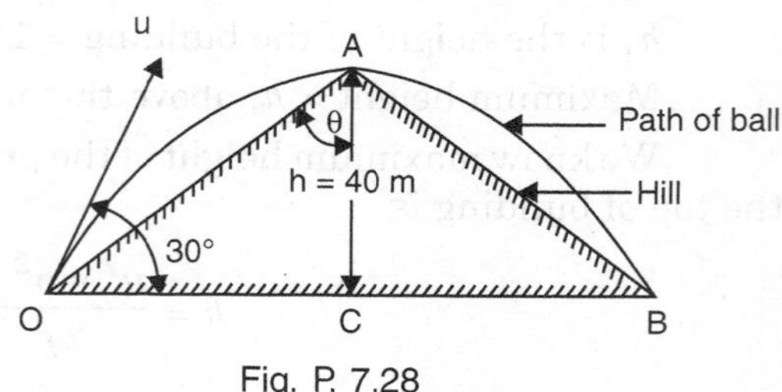

Fig. P. 7.28

$$u^2 = 3139.2 \text{ m/s}$$
$$u = 56.03 \text{ m/s.} \quad \textbf{Ans.}$$

$\therefore$ Range $\quad R = OB = \dfrac{u^2 \sin 2\alpha}{g}$

$$= \frac{(56.03)^2 \times \sin 2 \times 30°}{9.81} = 277.14 \text{ m}$$

Now, since the hill is in the shape of a right circular cone, therefore, from ΔOAC, the semicone angle θ is given by :

$$\tan \theta = \frac{R/2}{h} = \frac{(277.14)/2}{40} = 3.464$$

$\therefore$ $\quad \theta = \mathbf{73.9°}$ **Ans.**

Problem 7.29. *A bus starts from rest on a curve of 250 m radius and accelerates at a constant rate of 1.25 m/s². Determine the distance and time that the bus will travel before the magnitude of its total acceleration is 1.75 m/s².*

Sol. Radius of curve $= R = 250$ m.

Magnitude of total acceleration $= a_t$

$\therefore$ $\quad a_t = \dfrac{dv}{dt} = 1.25 \text{ m/s}^2$

Initially when the bus starts the total acceleration of the bus is the same as the tangential acceleration, (as at $t = 0$, $u = 0$).

Let the velocity of the bus be v when the total acceleration is 1.75 m/s².

We know that $\quad a_{total} = \sqrt{a_t^2 + a_n^2}$

$\therefore$ $\quad 1.75 = \sqrt{(1.25)^2 + a_n^2}$

$$a_n = \sqrt{1.75^2 - 1.25^2} = \sqrt{1.5} \text{ m/s}^2$$

But $a_n = v^2/R$, $\therefore$ $\quad v = \sqrt{R \cdot a_n}$

$$= \sqrt{250 \times \sqrt{1.5}} = \sqrt{306.19}$$

$$= 17.5 \text{ m/s} \quad \textbf{Ans.}$$

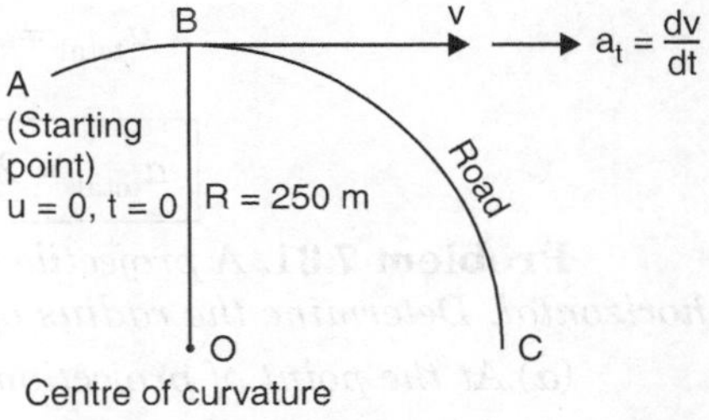

$$\boxed{v = 17.5 \text{ m/s.}}$$

Fig. P. 7.29

Let the time required to attain this velocity be t.

$\therefore$ $\quad v = u + \left(\dfrac{dv}{dt}\right) t.$

$$17.5 = 0 + 1.25 \times t.$$

$$\boxed{t = 14 \text{ s.}} \quad \textbf{Ans.}$$

Distance travelled $\quad s = \dfrac{1}{2} \times 1.25 \times (14)^2 \text{ m}$

$$\boxed{s = 122.5 \text{ m.}} \quad \textbf{Ans.}$$

Problem 7.30. *A motorist is travelling on a curved path of radius 700 m at a speed of 90 km/hr. The brakes are suddenly applied causing the speed to reduce at a constant rate of 2 m/s². Determine the magnitude of total acceleration of the automobile.*

(*a*) *immediately after the brakes are applied.*

(*b*) *7 seconds later.*

Sol. Initial velocity of automobile, u = 90 km/hr = 25 m/s.

Radius of curvature of path = R = 700 m

Magnitude of tangential acceleration,

$$a_t = \frac{dv}{dt} = -2 \text{ m/s}^2 \qquad \text{(retardation)}$$

(*a*) Total acceleration when brakes were applied.

$$a_{\text{total}} = \sqrt{\left(\frac{dv}{dt}\right)^2 + \left(\frac{u^2}{R}\right)^2} = \sqrt{(-2)^2 + \left(\frac{25^2}{700}\right)^2}$$

$$\boxed{a_{\text{total}} = 2.2 \text{ m/s}^2.} \quad \textbf{Ans.}$$

(*b*) Total acceleration after 7 seconds.

Let velocity of automobile after 7 seconds be v

$$v = u + \left(\frac{dv}{dt}\right) \times t$$

$$v = 25 - 2 \times 7$$

$$v = 11 \text{ m/s.}$$

$$a_{\text{total}} = \sqrt{\left(\frac{dv}{dt}\right)^2 + \left(\frac{v^2}{R}\right)^2} = \sqrt{(-2)^2 + \left(\frac{11^2}{700}\right)^2}$$

$$\boxed{a_{\text{total}} = 2.007 \text{ m/s}^2.} \quad \textbf{Ans.}$$

Problem 7.31. *A projectile was fired with a velocity of 75 m/s at an angle of 60° to the horizontal. Determine the radius of curvature of its path :*

(*a*) *At the point of projection.*

(*b*) *At the maximum height.*

(*c*) *At a position when it has travelled a distance of 35 m from point of projection.*

Sol. Initial velocity, u = 75 m/s.

Angle of projection, $\alpha = 60°$

We know that in case of projectile motion, only gravitational acceleration acts (in vertically downward direction). Thus the horizontal component of velocity is always constant and its magnitude is equal to $u \cos \alpha$.

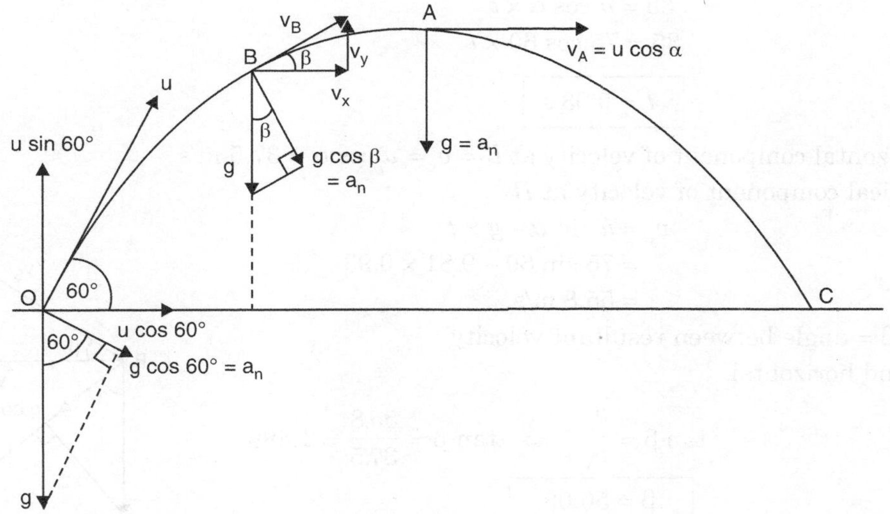

Fig. P. 7.31 (a)

In order to find the radius of curvature at any point on the projectile path we must know the component of gravitational acceleration in a direction normal to that of resultant velocity at the given point.

(*a*) **Radius of curvature at point of projection**

The resultant velocity at O is u

The component of g in the direction normal to $u = g \cos 60° = a_n$

Thus $$g \cos 60° = \frac{u^2}{R_0}$$

where R_0 = radius of curvature at O.

$$9.81 \cos 60° = \frac{(75)^2}{R_0}$$

$$\boxed{R_o = 1146.8 \text{ m.}} \quad \textbf{Ans.}$$

(*b*) **Radius of curvature at maximum height**

At maximum height the resultant velocity is $u \cos \alpha$.

Normal component of acceleration at point $A = a_n = g$

$$\therefore \quad g = \frac{(u \cos \alpha)^2}{R_A} \Rightarrow 9.81 = \frac{(75 \cos 60°)^2}{R_A}$$

$$\boxed{R_A = 143.35 \text{ m.}} \quad \textbf{Ans.}$$

(*c*) **Radius of curvature at point B**

Let t be the time taken by projectile to cover a horizontal distance of 35 m.

$$35 = u \cos \alpha \times t$$
$$35 = 75 \cos 60 \times t$$
$$\boxed{t = 0.93 \text{ s.}}$$

Horizontal component of velocity at $B = v_x = u \cos \alpha = 37.5$ m/s

Vertical component of velocity at B

$$v_y = u \sin \alpha - g \times t$$
$$= 75 \sin 60 - 9.81 \times 0.93.$$
$$= 55.8 \text{ m/s.}$$

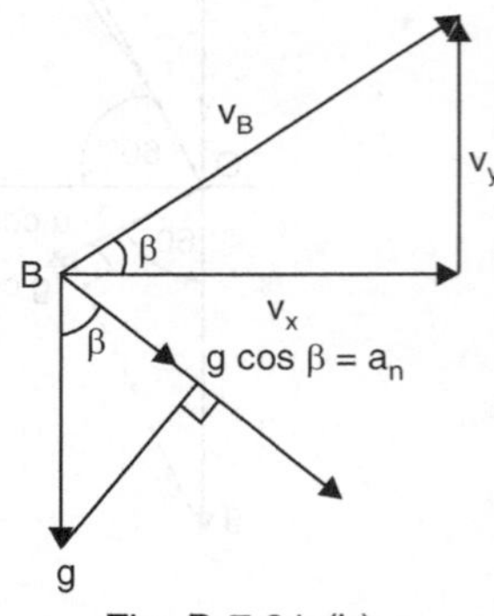

Fig. P. 7.31 (b)

Let β = angle between resultant velocity v_B and horizontal.

$$\tan \beta = \frac{v_y}{v_x} \quad \Rightarrow \quad \tan \beta = \frac{55.8}{37.5} = 1.488$$

$\Rightarrow$ $$\boxed{\beta = 56.09°}$$

Resultant velocity

$$v_B = \sqrt{v_x^2 + v_y^2} = \sqrt{37.5^2 + 55.8^2}$$
$$= 67.23 \text{ m/s}$$

Normal component of acceleration g at point $B = g \cos \beta = a_n$

$\therefore$ $$g \cos \beta = \frac{v_B^2}{R_B} \quad \Rightarrow \quad 9.81 \cos 56.09 = \frac{(67.23)^2}{R_B}$$

$\Rightarrow$ $$\boxed{R_B = 825.86 \text{ m.}}$$ **Ans.**

EXERCISES VII

A. Theoretical Questions

1. Explain the difference between (*a*) statics, (*b*) kinematics and (*c*) kinetics.
2. Define the terms : (*a*) position co-ordinate, (*b*) average velocity, (*c*) instantaneous velocity, (*d*) average acceleration, and (*e*) instantaneous acceleration.
3. Define the terms: plane path, tortuous path, rectilinear motion and curvilinear motion.
4. Describe the nature of problems generally encountered in rectilinear motion.
5. Develop equations for velocity, acceleration and displacement for rectilinear motion with uniform acceleration. Also express velocity in terms of displacement.
6. Derive expressions for normal and tangential acceleration component for a particle describing plane curvilinear path.
7. What assumptions are made while studying the motion of a projectile.
8. Derive the equation of a trajectory of a particle projected in space.
9. What do you mean by (*a*) the time of flight, (*b*) the range of the projectile.

B. Numerical Problems

7.1. What is the smallest radius which should be used for a highway curve if the normal component of the acceleration of a vehicle travelling at 90 km/hr is not to exceed 0.8 m/s^2.

[**Ans.** 781.25 m]

7.2. A car driver is moving along a curved path of radius 750 m with a speed of 120 km/hr. He suddenly applies brakes so that the speed of the car decreases at a constant rate from 120 km/hr to 50 km/hr over a distance of 150 m. Determine.

(*a*) The magnitude of total acceleration when brakes were applied.

(*b*) Total acceleration when car travels a distance of 100 m from the point of application of brakes.

[**Ans.** (*a*) 3 m/s^2, (*b*) 3.076 m/s^2]

7.3. A ball was projected in air at angle of 70° with horizontal, the velocity of projection being 15 m/s. Find the radius of curvature of the path

(*i*) At the point of projection.

(*ii*) At maximum height.

(*iii*) After 1 second from the projection. [**Ans.** (*i*) 67.06 m, (*ii*) 2.683 m, (*iii*) 4.286 m]

7.4. The initial velocity of a projectile, fired from a cliff 100 m high is, 150 m/s. The angle of projection being 30° with horizontal, find:

(*a*) The horizontal distance between the point of launch and the point where the projectile strikes the ground.

(*b*) The maximum height achieved by the projectile. [**Ans.** (*a*) 2864.72 m, (*b*) 286.69 m]

7.5. Find the firing angle 'α' of a projectile fired with initial velocity of 200 m/s at a point located 500 m above the firing point and at a horizontal distance 3500 m. Neglect the air resistance.

[**Ans.** 53.05°; 45.07°]

7.6. Given that the velocity of a projectile be *u* m/s and the direction of motion is at an angle θ inclined to the horizontal. Prove that it moves perpendicular to its direction after $\frac{u}{g}$ cosec θ sec.

Show that the velocity of projection '*u*' is given by

$$u^2 = \frac{g}{4}\left[\frac{4a^2}{h} + gh\right]$$

when the particle is projected from ground level and its height is *h* when it is at horizontal distances *a* and 2*a* from its point of projectile.

7.7. A boy kicks a football on a hill at an angle 60° upwards from horizontal. The ball hit with an initial velocity of 100 m/s hits a point 100 m below him. Find,

(*a*) Maximum height to which the ball will rise above the position of boy.

(*b*) Velocity with which ball hits the point.

(*c*) Time required to hit the target. [**Ans.** (*a*) 382.263 m, (*b*) 89.65 m/s, (*c*) 16.41 sec.]

7.8. An incline has a rise of 6 in 12. Find the range covered by a man who is standing on the incline and can throw a stone with an initial velocity of 100 m/s at an angle 30° with the vertical shown in Fig. E. 7.8. [**Ans.** *BC* = 1972.41]

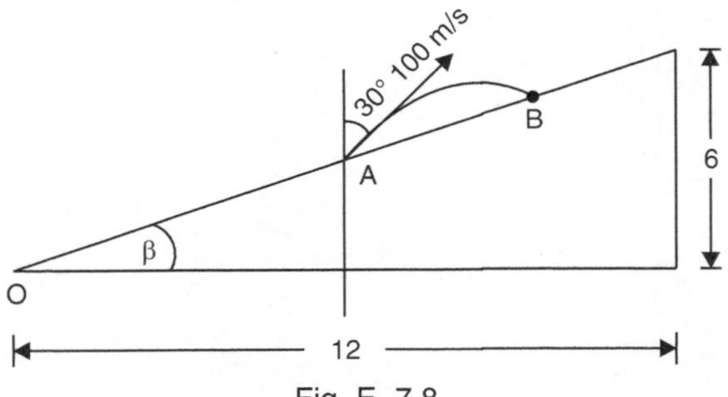

Fig. E. 7.8

7.9. In the Fig. E. 7.9 shown, a metallic brick weighing 20 N starts sliding from rest down a slope ($\mu_{friction}$ = 0.2). Find the distance of the point *G* where the brick hits the floor. [**Ans.** x = 7.32]

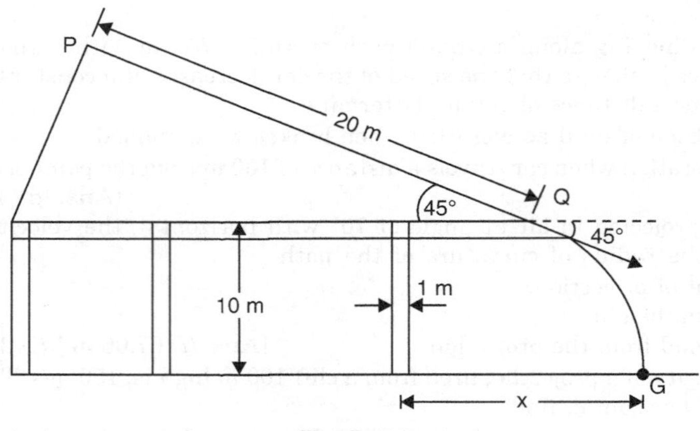

Fig. E. 7.9

7.10. A bomb shell thrown, splits in two parts at a the peak point of its path, which is 800 m from the point where the shell is thrown and 20 m high as shown in Fig. E 7.10. Two seconds after the splitting one part fall on ground at a point vertically below the point where it has splitted. Determine

(*a*) Velocity of shell just before splitting.

(*b*) Velocity of first part immediately after splitting.

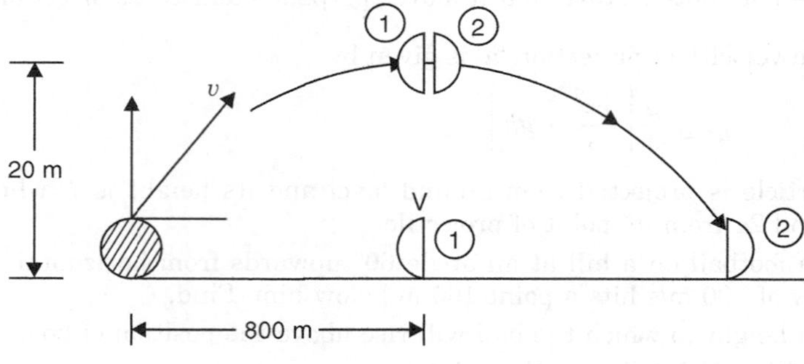

Fig. E. 7.10

(*c*) Velocity of second part immediately after splitting.

(*d*) The distance between throwing point and the point where second part of shell fell on ground.

[**Ans.** (*a*) 400 m/s, (*b*) 0 m/s (*c*) 800 m/s, (*d*) 2424 m/s]

8

Shear Force and Bending Moment Diagrams

8.1. BEAMS

A structural member designed to support transverse external loads applied at various points along the length of the member is known as a **beam.** In most cases external loads are perpendicular to the axis of the beam and will cause only *shear force* and *bending moment* in the beam. When the loads are inclined to the axis they also produce axial forces in the beam.

Beams are generally straight, prismatic (*i.e.,* of the same cross-sectional area throughout the beam length) and long. Designing a beam consists of two parts. Firstly, determination of shear forces and bending moments produced by the loads, and secondly, the selection of the cross-section best suited to resist the shearing forces and bending moments determined in the first part. This chapter deals with the determination of shear force and the bending moment at various sections of the beam and with the plot of the variation of shear force and bending moment, known as *shear force* and *bending moment diagrams.* For the second part, design and selection of beams, the reader is suggested to refer to some good test book on *Strength of Materials* or *Mechanics of Materials.*

8.2. TYPES OF SUPPORTS

There are three types of supports commonly used in planar structures. These are identified by the kind of resistance they offer to the external forces. These supports are :

(*a*) A roller or a link support,

(*b*) A pin or a hinge support,

(*c*) A fixed support.

(*a*) **A Roller or a Link Support.** It is capable of resisting a force in only *one specific line of action.* A link shown in Fig. 8.2.1 (*a*) can resist a force only in the direction of line *AB*. The roller in Fig. 8.2.1 (*b*) can resist only a vertical force, whereas the rollers in Fig. 8.2.1(*c*) can resist only a force that acts perpendicular to the plane *CD*. This type of support will be usually represented by rollers as shown in Fig. 8.2.1 (*b*) and (*c*), and it will be understood that *a roller support is capable of offering one reaction component that is perpendicular to the roller surface*. Only one equation of equilibrium, is required to find one unknown reaction component.

(*b*) **A Pin or a Hinge Support.** *A pin or a hinge support* is similar to the hinge used in fixing panels of a door or of a window. A hinge support offers two reaction components, *viz.,* a vertical reaction component and a horizontal reaction component, as shown in Fig. 8.2.2 (*a*). To find two unknown reaction components, two equations of equilibrium will be used.

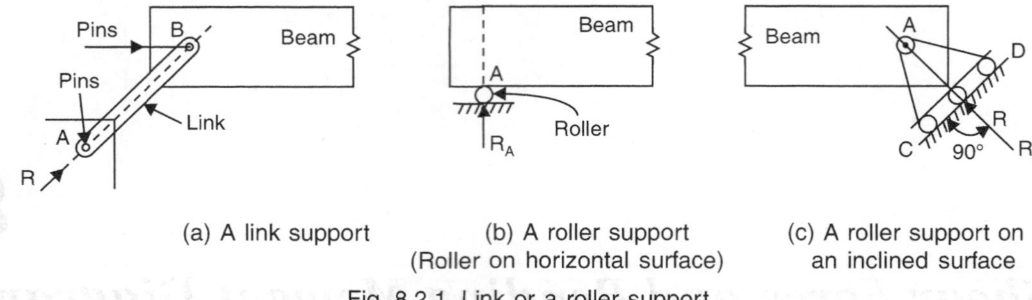

(a) A link support (b) A roller support (Roller on horizontal surface) (c) A roller support on an inclined surface

Fig. 8.2.1. Link or a roller support.

(*c*) **A Fixed Support.** A *fixed support is similar to a nail or a peg inserted in a wall.* It is capable of offering a horizontal reaction component, a vertical reaction component and also a moment or a couple. This support is also known as a *built-in* end. A fixed support is represented as shown in Fig. 8.2.2 (*b*).

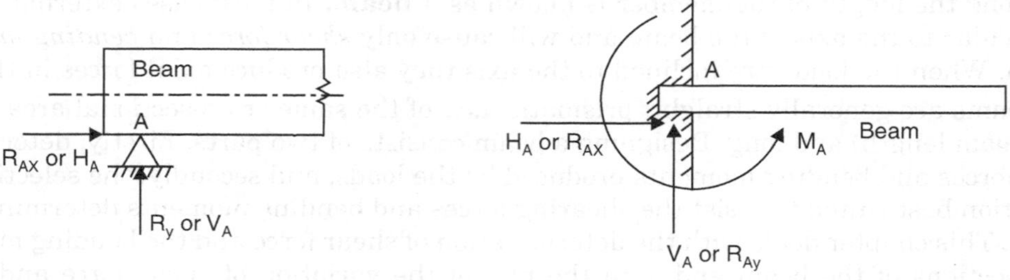

Fig. 8.2.2 (a) A Hinge support (The two unknown reaction components are H_A and V_A or R_{Ax} and R_{Ay})

Fig. 8.2.2 (b) A fixed support (The three unknown reaction components at the fixed support A are : H_A, V_A and M_A or R_{Ax}, R_{Ay} and M_A)

Note. When the external loads applied to the beam are vertical then the hinge support at *A* offers only one reaction component V_A or R_{Ay} whereas the fixed support offers only two reaction components V_A and M_A. In each case the horizontal reaction component H_A will not be set up.

8.3. THE TYPES OF BEAMS

Beams are classified into several groups,

(1) depending on the kind of supports

(2) depending on the position of supports

(3) depending on the number of supports

(4) depending on the number of unknown reactions components.

8.3.1. Types of beams depending on kind of supports such as :

(*a*) **Simply supported beams** having roller support at one end and a hinge or the pin support at the other end. See Fig. 8.3.1 (*a*). Distance between supports is called span of the beam.

(*b*) **Cantilever.** A cantilever has one end fixed and the other free, as shown in Fig. 8.3.1(*b*).

(*c*) **Fixed Beams.** A fixed beam or a built-up beam has both ends fixed as shown in Fig. 8.3.1 (*c*).

(*d*) **Propped cantilever.** A propped cantilever is fixed at one end and has a roller or hinge at the other end as shown in Fig. 8.3.1 (*d*).

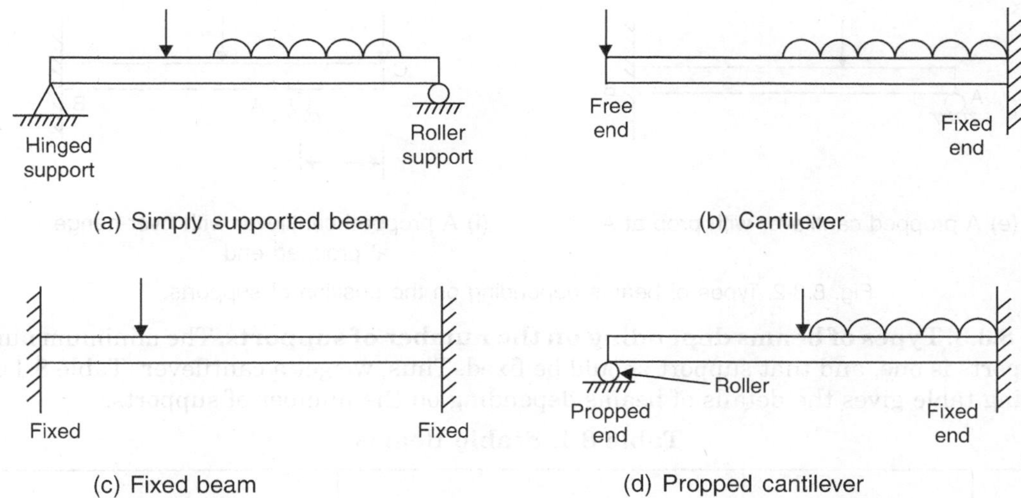

Fig. 8.3.1 Types of beams depending on the kind of supports.

8.3.2. Types of beams depending on the position of support :

(*a*) **Simply supported beams.** Here supports are at the two ends of the beam. See Fig. 8.3.2 (*a*).

(*b*) **Over-hanging beams.** Here a small portion of the beam over-hangs either one or both supports of a simply supported beam. In Fig. 8.3.2 (*b*), the beam over-hangs on the left of the left support, while in Fig. 8.3.2 (*c*), the over-hang is on the right end. In Fig. 8.3.2 (*d*), the beam over-hangs are at both ends.

(*c*) **Propped cantilevers.** The two propped cantilevers are shown in Figs. 8.4 (*e*) and (*f*).

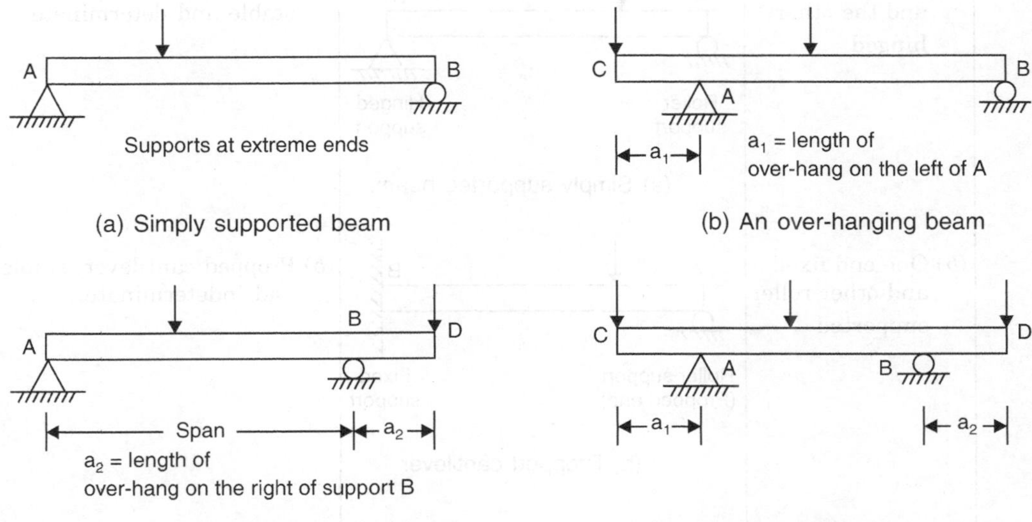

(e) A propped cantilever with prop at A

(f) A propped cantilever with over-hange at propped end

Fig. 8.3.2. Types of beams depending on the position of supports.

8.3.3. Types of beams depending on the number of supports. The minimum number of supports is one, and that support should be fixed. Thus, we get a cantilever (Table 8.1). The following table gives the details of beams depending on the number of supports.

Table 8.1. Stable Beams

No. of Supports	*Type of Supports*	*Figure of beam*	*Description of beam*
One	Fixed	A B Free end cantilever Fixed end	Cantilever, stable and determinate.
Two	(*a*) One roller end and the other hinged	A B Roller support Hinged support (a) Simply supported beam.	(*a*) Simply supported beam stable and determinate.
	(*b*) One end fixed and other roller supported	A B Roller support (Propped end) Fixed support (b) Propped cantilever	(*b*) Propped cantilever, stable and indeterminate.

No. of Supports	*Type of Supports*	*Figure of beam*	*Description of beam*
	(*c*) One end fixed and other hinged	A B Hinge support Fixed support (c) Propped cantilever	(*c*) Propped cantilever stable and indeterminate.
	(*d*) Both ends fixed	A B Fixed support Fixed support (d) Fixed beam	(*d*) Fixed beam, stable and indeterminate.
	(*e*) One end on roller support and the other on hinge supported	C A B Hinge Roller CA = over hang (*i*) A B D Hinge Roller BD = over hang (*ii*) C D A B Hinge Roller CA and BD = over hangs (*iii*) (e) Over hanging beams	(*e*) Over hanging beams, stable and determinate
More than two	Various combinations of roller, hinge and fixed supports	Hinge Roller Roller (*i*) Fixed Roller Hinge (*ii*)	Continuous beams, stable and indeterminate.

No. of Supports	*Type of Supports*	*Figure of beam*	*Description of beam*
		Hinge Roller Roller Hinge (*iii*) Fixed Roller Roller Hinge etc., (*iv*) etc., Continuous beams	

A structure supported on two roller supports is the simplest example of an unstable simply supported beam subjected to general loading system. If there is an inclined force P acting on the beam, the beam will not offer a horizontal reaction at any support and the beam starts moving in the direction of horizontal component $P \cos \theta$ of the applied force. See Fig. 8.3.3

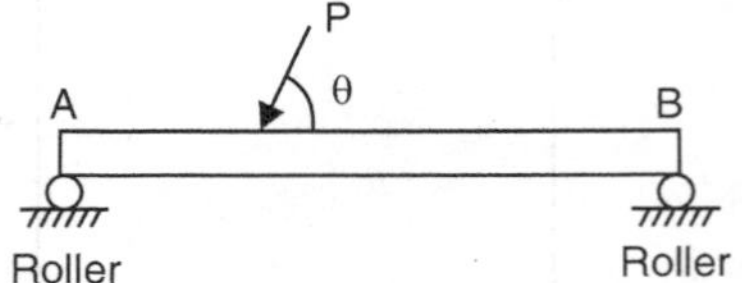

Fig. 8.3.3. Unstable simply support beam.

The horizontal force equal to $P \cos \theta$ remains unbalanced. The beam starts moving towards left under the action of the force $P \cos \theta$. This force is called the *tractive effort*. All bicycles, handcarts, scooters, motorcycles, cars, buses, trucks, overhead cranes etc. are supported on rollers and in which the unbalanced applied force is the tractive effort. They move in the direction of the unbalanced force $P \cos \theta$ acting parallel to the surface on which the rollers move.

8.3.4. Types of beams depending upon number of unknown reaction components. The reader is advised to see Art. 8.5.

8.4. TYPES OF LOADS

Various loads that act on beams are :

1. Concentrated or point loads
2. Uniformly distributed loads (*udl*)
3. Uniformly varying loads (*uvl*)
4. Pure couple
5. Special loads.

A *concentrated load or a point load* is one that acts over so small a distance that can be assumed to act at a point, as in Fig. 8.4 (*a*). A *uniformly distributed load* acts over a considerable length of the beam. It may be uniformly distributed over the entire length Fig. 8.4 (*b*), or over a part of the length as in Fig. 8.4 (*c*). Distributed loads may be uniformly varying or non-uniform. In a uniformly varying or triangular load, the intensity of loading increases or decreases at a constant rate, as shown in Fig. 8.4 (*d*). The trapezoidal loading [Fig. 8.4 (*e*)] is a combination of a uniformly distributed load (*udl*) and a uniformly varying load (*uvl*).

A pure couple M_0 shown in Fig. 8.4 (*f*) may also act on a beam. Some times a special loads as shown in Figs. 8.4 (*g* and *h*) may act on the beam.

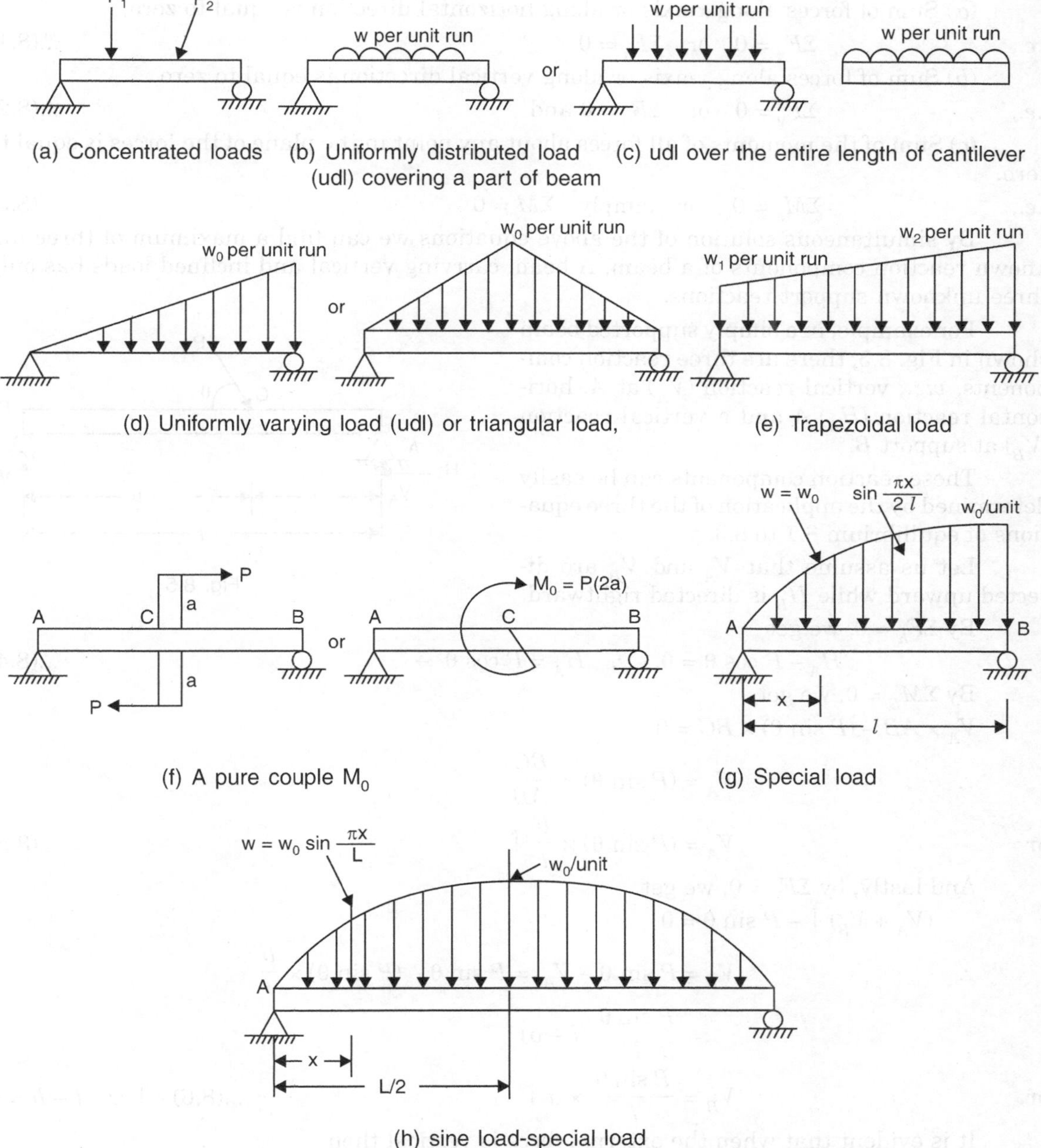

Fig. 8.4. Various types of loads.

8.5. EQUATIONS OF EQUILIBRIUM USED FOR FINDING SUPPORT REACTIONS

External forces acting at different points of a beam and the support reactions constitute a coplanar non-concurrent force system. The following equations of equilibrium are available for such a system of forces :

(*a*) Sum of forces along *x*-axis or along horizontal direction is equal to zero,

i.e., $\Sigma F_x = 0$ or $\Sigma F_h = 0$ ...(8.1)

(*b*) Sum of forces along *y*-axis or along vertical direction is equal to zero,

i.e., $\Sigma F_y = 0$ or $\Sigma F_v = 0$ and ...(8.2)

(*c*) Sum of the moments of all forces about any point in the plane of the forces is equal to zero,

i.e., $\Sigma M_z = 0$, or simply $\Sigma M = 0$ ...(8.3)

By simultaneous solution of the above equations we can find a maximum of three unknown reaction components of a beam. A beam carrying vertical and inclined loads has only three unknown support reactions.

For example, in a simply supported beam shown in Fig. 8.5, there are three reaction components, *viz.*, vertical reaction (V_A) at *A*, horizontal reaction (H_A) *A* and a vertical reaction (V_B) at support *B*.

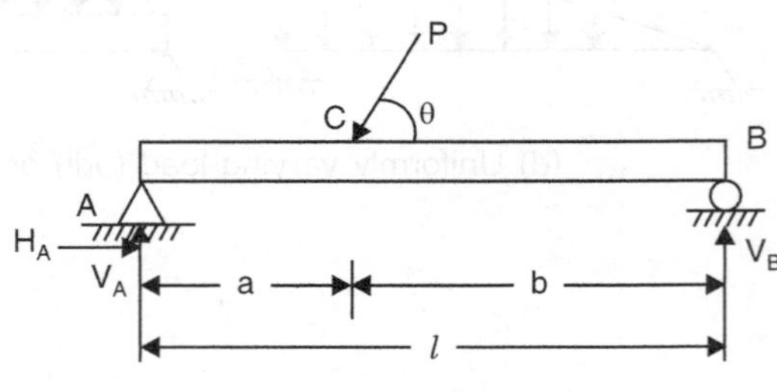

Fig. 8.5

These reaction components can be easily determined by the application of the three equations of equilibrium 8.1 to 8.3.

Let us assume that V_A and V_B are directed upward while H_B is directed rightward.

By $\Sigma F_h = 0$, we get

$$H_A - P\cos\theta = 0, \quad \therefore \quad H_A = P\cos\theta \rightarrow \qquad ...(8.4)$$

By $\Sigma M_B = 0$, we get

$V_A \times AB - (P\sin\theta) \times BC = 0$

$$\therefore \qquad V_A = (P\sin\theta) \times \frac{BC}{AB}$$

or

$$V_A = (P\sin\theta) \times \frac{b}{l} \uparrow \qquad ...(8.5)$$

And lastly, by $\Sigma F_v = 0$, we get

$(V_A + V_B) \uparrow - P\sin\theta = 0$

$$\therefore \qquad V_B = P\sin\theta - V_A = P\sin\theta - (P\sin\theta) \times \frac{b}{l}$$

$$= \frac{P\sin\theta}{l}(l - b)$$

or

$$V_B = \frac{P\sin\theta}{l} \times a \uparrow \qquad ...(8.6) \quad | \because \; l - b = a$$

It is evident that when the external force is vertical then

$$H_A = 0 \quad \text{and} \quad V_A = \frac{Pb}{l} \quad \text{and} \quad V_B = \frac{Pa}{l}.$$

Then eqns. 8.2 and 8.3 will be sufficient to find V_A and V_B.

A beam in which unknown reaction components (*R*), are equal to the number of equations of equilibrium available/necessary is called a ***determinate beam***. A simply supported beam; a cantilever and over-hanging beams (with a roller at one support and a hinge at the other support) are determinate beams and as such are in the course of Engineering Mechanics.

Beams in which unknown reaction components (R) are more than the equations of equilibrium (E) used, are called ***indeterminate beams.*** Thus, propped cantilevers fixed beams and the continuous beams are indeterminate beams and are not to be studied in a course on Engineering Mechanics.

A curious student may refer to some standard text on Strength of Materials or Mechanics of Materials for the analysis of indeterminate beams.

8.6. SHEAR FORCE AND BENDING MOMENT

Fig. 8.6.1 (a) shows a simple beam AB that carries a point load P at point C and is held in equilibrium by the reactions V_A and V_B. Mass of the beam is neglected and only the effect of P is considered. Assume that a cutting plane 1-1 at a distance x from V_A divides the beam into two segments. The free-body diagram of the left segment in Fig. 8.6.1 (b) shows that the externally applied load is V_A. For equilibrium of this segment of the beam the fibres in the section 1-1 must supply the resisting forces necessary to satisfy the conditions of equilibrium. Hence, the condition $\Sigma F_h = 0$ is automatically satisfied because there is no horizontal force.

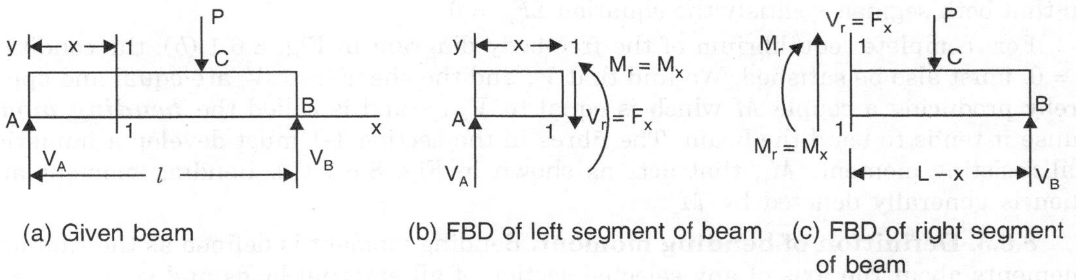

(a) Given beam (b) FBD of left segment of beam (c) FBD of right segment of beam

Fig. 8.6. 1. Equilibrium of segments to the left and right of a section 1-1 of the beam.

To satisfy $\Sigma F_v = 0$, the vertical unbalance caused by the V_A requires the fibres in section 1-1 to create a resisting force. This is shown as V_r and is called resisting shear force. Here, for the loading shown, V_r is numerically equal to V_A , but if additional loads are applied between V_A and the section 1-1, the net vertical unbalance would be found by the summation of their vertical components.

8.6.1. Definition of shear force. *We define the net vertical unbalance as the shearing force at any section 1-1 at a distance x from* V_A *as the shearing force in the beam.* It is denoted by F_x and may be determined by the summation of the vertical components of external loads on either side of the section. From this definition, it is simpler to take a positive shear force as one which is directed along the + y axis and vice-versa.

Thus, for any section from the origin we can write shear force as :

$$F_x = \Sigma F_v \quad ...(8.7)$$

Procedure for writing equation for shear force. It is important to note that :

(1) For simply supported and overhanging beam we take the left end of the beam as the origin for the distance x. Thus, + x distance is towards right of origin.

(2) For cantilever, the free end may be taken as origin.

(3) After selecting the origin the equation of shear force at any section x from the origin can be written.

(4) A positive vertical force is directed along the + y direction, and vice-versa.

8.6.2. Sign convention for shear forces. The above sign convention shows that a positive shear force tends to move the beam segment on the left of section *upward* with respect to the right segment, and vice-versa see Fig. 8.6.2.

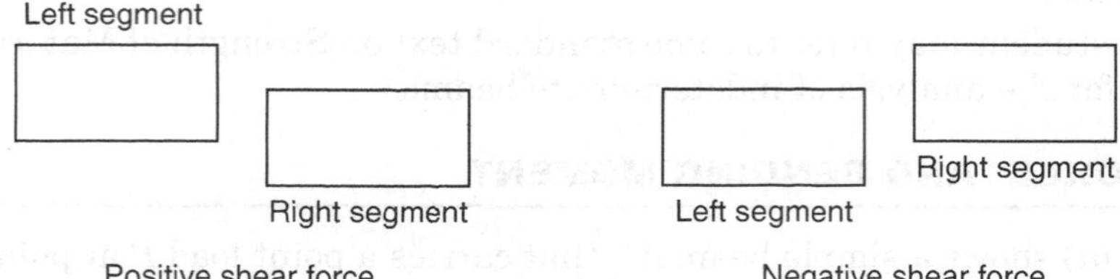

Fig. 8.6.2. Relative moments of the beam segments at any sections corresponding to +ve and –ve shear.

Since the entire beam AB is in equilibrium it satisfies the equations of equilibrium $\Sigma F_v = 0$ and $\Sigma M = 0$. (For the vertical loads). Any portion of the beam cut from this beam must also be in equilibrium, and should satisfy these equations of the equilibrium. We have seen that both segments satisfy the equation $\Sigma F_v = 0$.

For complete equilibrium of the free-body diagram in Fig. 8.6.1 (*b*), the condition of $\Sigma M = 0$, must also be satisfied. We find that V_A and the shear force V_r are equal and opposite thereby producing a couple M which is equal to $V_A \,.\, x$ and is called the ***bending moment*** because it tends to bend the beam. The fibres in the section 1-1 must develop a numerically equal resisting moment, M_r, that acts as shown in Fig. 8.6.1 (*b*). Bending moment at the section is generally denoted by M_x.

8.6.3. Definition of bending moment. Bending moment is defined as the summation of moments about the axis of any selected section of all external loads and support reaction acting either to the left or to the right side of the section, and is expressed as :

$$M_x = (\Sigma M)_L = (\Sigma M)_R \qquad ...(8.8)$$

The subscript L indicates that the bending moment is computed in terms of the loads and support reaction to the left of the selected section, and subscript R refers to the loads and the support reaction to the right of the selected section. Both shear force and bending moment are internal resistances of the beam material against cutting of the beam.

8.6.4. Sign convention of bending moment. It is generally assumed that the bending moment is positive if it produces bending of beam concave upward, as shown in Fig. 8.6.3 (*a*). However, we prefer to use an equivalent convention which states that

"Upward acting external forces cause positive bending moments with respect to any section ; downward forces cause negative bending moments". See Fig. 8.6.3.

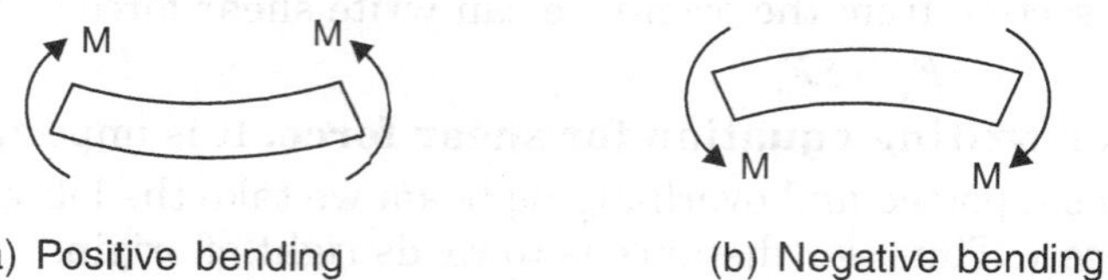

Fig. 8.6.3. Curvature of beam due to positive and negative bending.

Hence, for a left segment of a beam the clockwise moments about any section are positive, as indicated by moment of V_A. With respect to the right segment of a beam (Fig. 8.6.1 (*a*) (*c*)) this convention means that the moment of upward reaction V_B is positive in a counter clockwise direction. Thus, upward acting forces always cause positive bending moments regardless of whether they act to the left or to the right of the selected section.

The definition of shearing force (F_x) and bending moment (M_x) may be summarized mathematically as under :

For shear force : $F_x = (\Sigma F_v)_L$ ...(8.9)

For bending moment : $M_x = (\Sigma M)_L = (\Sigma M)_R$ ...(8.10)

in which positive effects are produced by upward forces and negative effects by downward forces.

It is important to note that an upward force is positive because it is directed along + y direction and a negative force is one which is directed along the – y direction.

Thus, a bending moment is positive if both force and the arm of the moment in the product are either positive or negative. See Fig. 8.6.4 (a) and (b).

Similarly, if either the force or the arm of the moment is negative and the other positive, their product, and hence the bending moment, is negative. Negative F_x and M_x are conventionally shown in Figs. 8.6.4 (c) and (d).

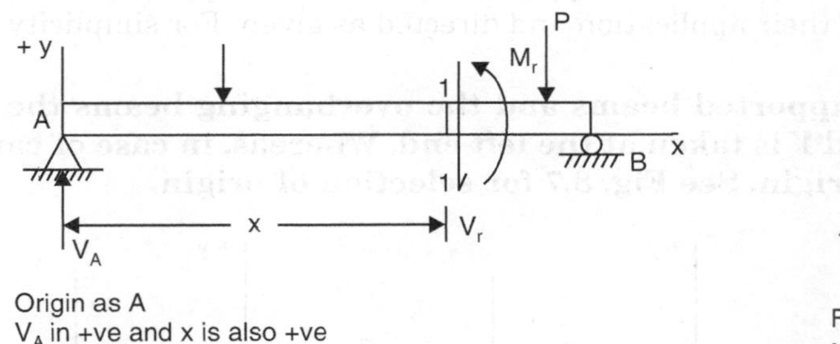

(a) +ve F_x and + M_x with A as origin

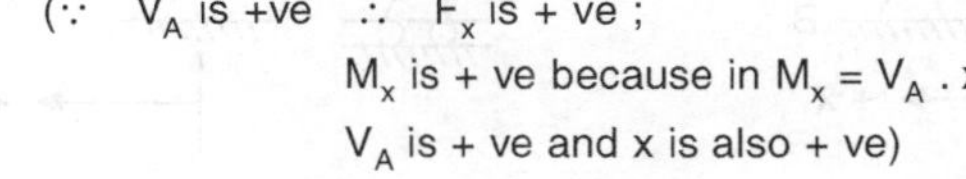
(∵ V_A is +ve ∴ F_x is + ve ;
M_x is + ve because in $M_x = V_A \cdot x$
V_A is + ve and x is also + ve)

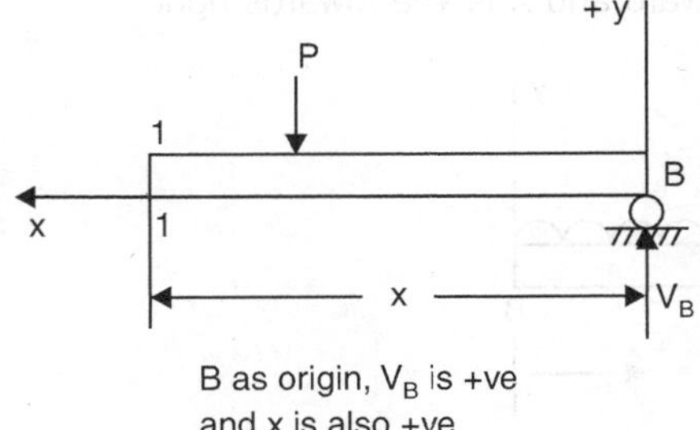

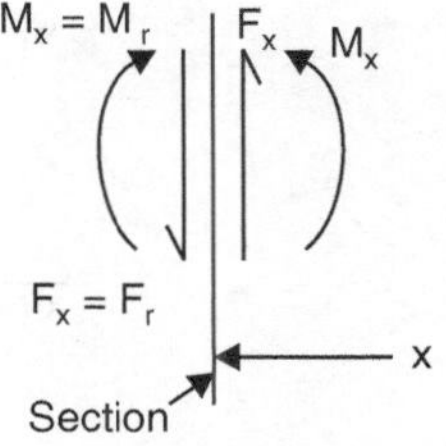

F_x is + ve because it is directed along + y axis and with B as origin the distance x is +ve and hence M_x is +ve

(b) + ve F_x and + M_x with right end of the beam as origin

(c) – ve F_x and –ve M_x origin at left end of beam (d) –ve F_x and –ve M_x origin at right end of beam

Fig. 8.6.4. Positive and Negative F_x and M_x.

8.7. SHEAR FORCE AND BENDING MOMENT DIAGRAMS

A shear force diagram is the plot of variation of shear force from one end to the other end of the given beam. All forces, externally applied forces and the support reactions are force vectors acting at the point of their application and directed as given. For simplicity of calculations we follow as under.

In case of simply supported beams and the overhanging beams the origin of the co-ordinate axes *X* and *Y* is taken at the left end. Whereas, in case of cantilevers, the free end is taken as origin. See Fig. 8.7 for selection of origin.

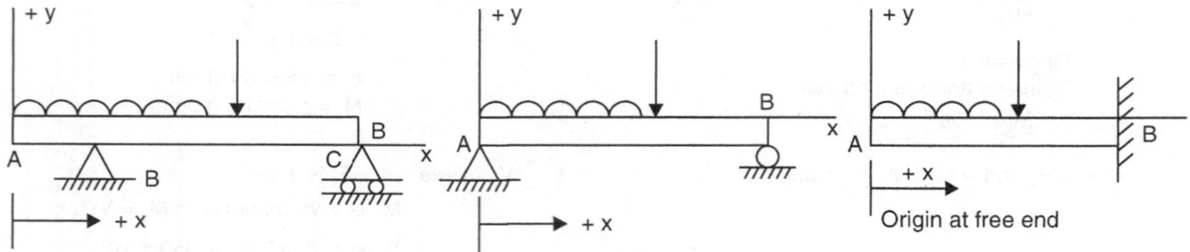

(a) Origin at the extreme left end. Y is +ve upward and X is +ve towards right

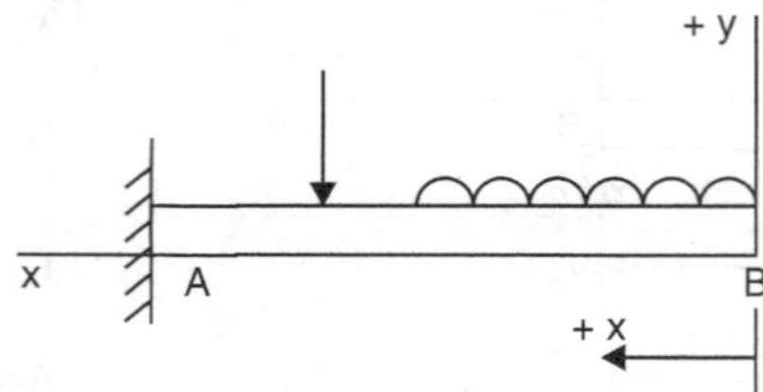

(b) Free end of cantilever is taken origin
Y is +ve upwards X is +ve left wards

Fig. 8.7. Selection of origin and convention for +ve X and +ve Y distances.

Thus, we can say that a shear force diagram is a force polygon drawn starting from a base line (where $F_x = 0$) and the last vector will close the force polygon for coplanar non-concurrent parallel forces acting perpendicular to the beam axis.

Similarly, variation of the bending moment M_x over the length of the given beam is called *Bending Moment Diagram.*

The procedure of drawing the shear force and bending moment diagrams is explained in the following examples.

Some standard cases are described below.

8.8. SOME STANDARD CASES

Case I. A simply supported beam carrying an eccentric load. [Fig. 8.8.1]

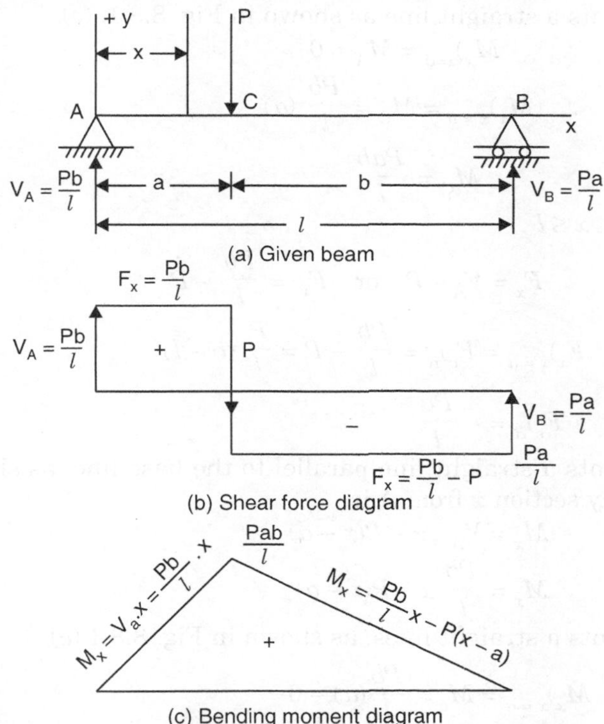

Fig. 8.8.1 Simply supported beam with a point load.

Since the given load P is vertical, therefore there is no horizontal reaction component. By $\Sigma M_B = 0$, we get

$$\curvearrowleft V_A \times l - P \times b = 0$$

$$V_A = \frac{Pb}{l}\uparrow \quad ...(8.11)$$

By $\Sigma F_v = 0$, we get

$$(V_A + V_B)\uparrow - P = 0$$

$$\therefore \quad V_B = P - V_A = P - \frac{Pb}{l} = \frac{P(l-b)}{l}$$

or

$$V_B = \frac{Pa}{l}\uparrow \quad ...(8.12)$$

Portion AC. $0 \le x \le a$:Taking A as origin, at any section x form A, we get shear force, $F_x = V_A$

or

$$F_x = \frac{Pb}{l} \quad ...(1)$$

Eqn. (1) represents a constant shear force, $\therefore$ shear force diagram is represented by a line parallel to the base line as shown in Fig. 8.8.1 (*b*).

From eqn. (1), we get $F_{x)x=0} = F_A = \dfrac{Pb}{l}$ and $F_x)_{x=a} = F_c)_L = \dfrac{Pb}{l}$

Bending moment at any section is

$$M_x = V_A \cdot x = \frac{Pb}{l} \cdot x \qquad \text{...(2)}$$

Eqn. (2) represents a straight line as shown in Fig. 8.8.1. (*c*).

At $A, x = 0,$ $\therefore$ $M_x)_{x=0} = M_A = 0$

At $C, x = a,$ $\therefore$ $M_x)_{x=a} = M_C = \dfrac{Pb}{l}(a)$

$$\therefore \qquad M_C = \frac{Pab}{l} \qquad \text{...(8.13)}$$

Portion BC. $a \le x \le l$

$$F_x = V_A - P \quad \text{or} \quad F_x = \frac{Pb}{l} - P \qquad \text{...(3)}$$

$\therefore$ At $x = a,$ $F_x)_{x=a} = F_c)_R = \dfrac{Pb}{l} - P = \dfrac{P}{l}(b - l)$

or $$F_c)_R = -\frac{Pa}{l}$$

Eqn. (2) represents a straight line parallel to the base line, as shown in Fig. 8.8.1 (*b*). Bending moment at any section *x* from *A* is

$$M_x = V_A \cdot x - P(x - a)$$

or $$M_x = \frac{Pb}{l}x - P(x - a) \qquad \text{...(4)}$$

Eqn. (4) represents a straight lines, as shown in Fig. 8.8.1 (*c*).

At $C, x = a,$ $\therefore$ $M_x)_{x=a} = M_c = \dfrac{Pb}{l}(a) - 0$

$$M_c = \frac{Pab}{l}$$

This is the same as eqn. (8.15) **(Check)**

At $B, x = l,$ $\therefore$ Bending moment,

$$M_x)_{x=l} = M_B = \frac{Pb}{l}(l) - P(l - a) \quad \text{or} \quad M_B = Pb - P(b) = 0 \quad \textbf{(Check)}$$

We determined V_A by using $\Sigma M_B = 0$, and again at the end we get $M_x)_{x=l} = M_B = 0$.

Hence, the correctness of the treatment of the entire procedure is checked. **The reader is advised to perform the "Check" before shifting to next question.**

Note. When $a = b = l/2$, we find from $V_A = \dfrac{P}{l}(l/2) = \dfrac{P}{2}$ from eqn. (8.11).

Also, $$V_B = P - V_A = P - \frac{P}{2} \quad \text{or} \quad V_B = \frac{P}{2}$$

Also, $$M_C = V_A \times (l/2) = \frac{P}{2}\frac{l}{2} \quad \text{or} \quad M_C = \frac{Pl}{4}$$

or $$M_C = \frac{pl}{4} \qquad \text{...(8.14)}$$

The shear force and bending moment diagram are shown in Fig. 8.8.2 for this case.

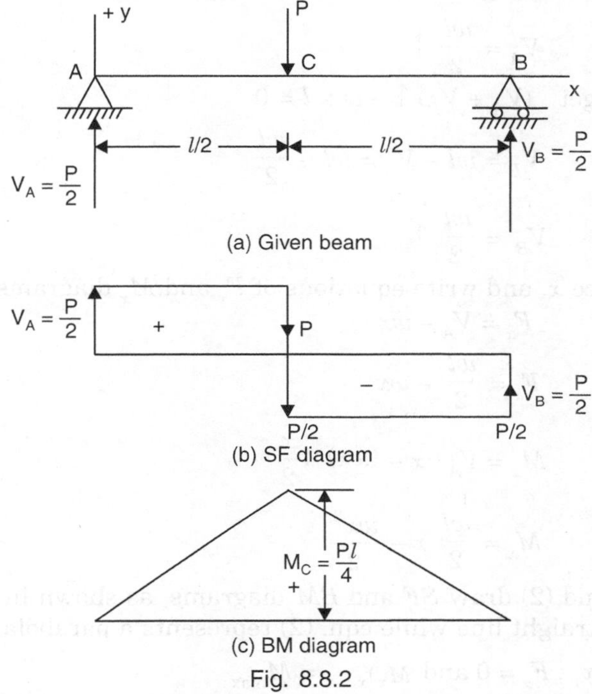

Fig. 8.8.2

Case II. A simply supported beam carrying *udl* on its entire span. [Fig. 8.8.3]

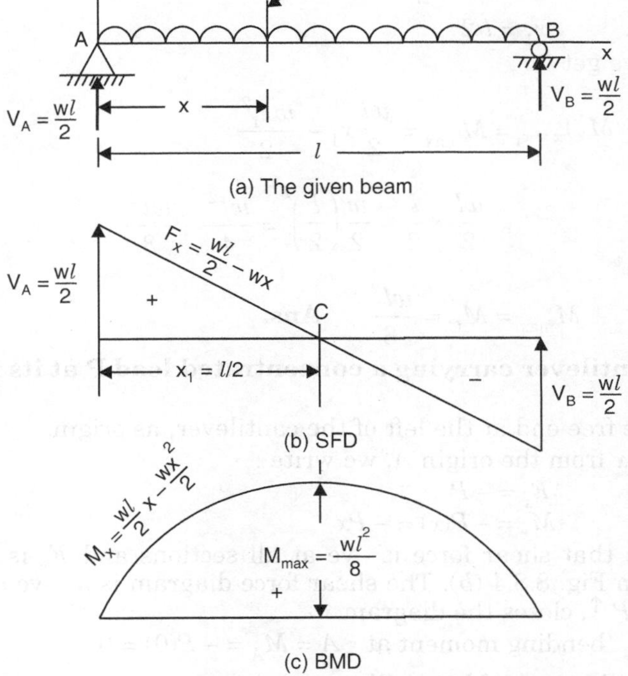

Fig. 8.8.3. A simply supported beam carrying a udl w per m length.

By $\Sigma M_B = 0$, we get $\circlearrowleft V_A \times l - (w \times l)(l/2) = 0$

$$\therefore \quad V_A = \frac{wl}{2} \uparrow \qquad ...(8.15)$$

By $\Sigma F_V = 0$, we get $(V_A + V_B)\uparrow - w \times l = 0$

$$\therefore \quad V_B = wl - V_A = wl - \frac{wl}{2}$$

$$\therefore \quad V_B = \frac{wl}{2} \uparrow \qquad ...(8.16)$$

Take any distance x, and write equations of F_x and M_x diagrams

$$\therefore \quad F_x = V_A - wx$$

or
$$F_x = \frac{wl}{2} - wx \qquad ...(1)$$

And,
$$M_x = V_A \cdot x - (wx) \times \frac{x}{2}$$

or,
$$M_x = \frac{wl}{2}x - \frac{wx^2}{2} \qquad ...(2)$$

Using eqns (1) and (2) draw *SF* and *BM* diagrams, as shown in Figs. 8.8.3 (*b*) and (*c*). Eqn. (1) represents a straight line while eqn. (2) represents a parabola.

At point (1), $x = x_1$; $F_x = 0$ and $M_x)_{x=x_1} = M_{max}$

$\therefore$ We get from eqn. (1),

$$F_x)_{x=x_1} = 0 = \frac{wl}{2} - wx_1$$

$$\therefore \quad x_1 = l/2$$

and from eqn. (2), we get

$$\therefore \quad M_x)_{x=x_1} = M_{max} = \frac{wl}{2}x_1 - \frac{wx_1^2}{2}$$

$$= \frac{wl}{2} \times \frac{l}{2} - \frac{w}{2}\left(\frac{l}{2}\right)^2 = \frac{wl^2}{4} - \frac{wl^2}{8}$$

$$\therefore \quad M_{max} = M_C = \frac{wl^2}{8}. \quad \textbf{Ans.} \qquad ...(8.17)$$

Case III. A cantilever carrying a concentrated load P at its free end. [Figs. 8.8.4 and 8.8.5].

Let us take the free end at the left of the cantilever, as origin.

At any section x from the origin A, we write :

$$F_x = -P \qquad ...(1)$$

and
$$M_x = -P(x) = -Px \qquad ...(2)$$

Eqn. (1) shows that shear force is –ve at all sections and F_x is constant. Eqn. (2) is plotted as in shown in Fig. 8.8.4 (*b*). The shear force diagram is a –ve diagram, and the last vector, equal to $V_B = P\uparrow$, closes the diagram.

At A, $x = 0$ $\therefore$ bending moment at $A = M_A = -P(0) = 0$

and at B, $x = l$, $\therefore$ $M_x)_{x=l} = M_B = -Pl$

It is important to note that the shear force and bending moment diagrams are both negative. The closing vector at point B, drawn to close the shear force diagram, is equal to P and is in the direction (↑) of the support reaction (V_B).

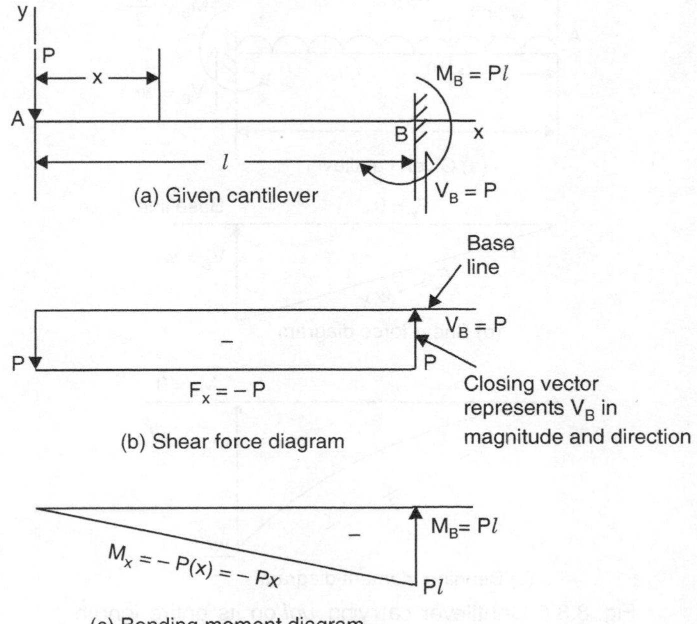

(a) Given cantilever

(b) Shear force diagram

(c) Bending moment diagram

Fig. 8.8.4. Left end of cantilever is free.

Eqns. (1) and (2) and the above calculations are also true when the right end of the cantilever is free, as shown in Fig. 8.8.5. Here x is +ve towards left.

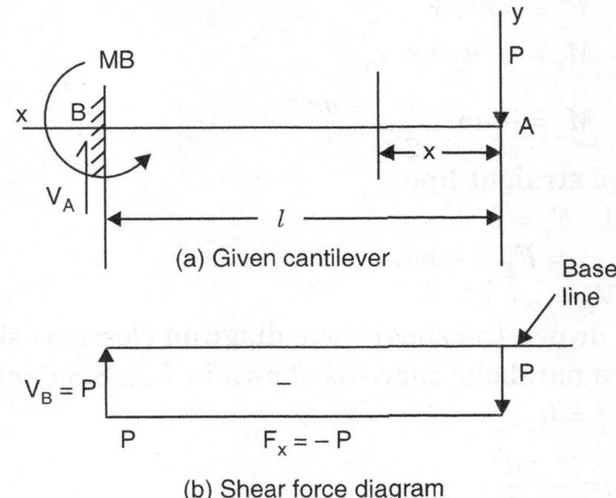

(a) Given cantilever

(b) Shear force diagram

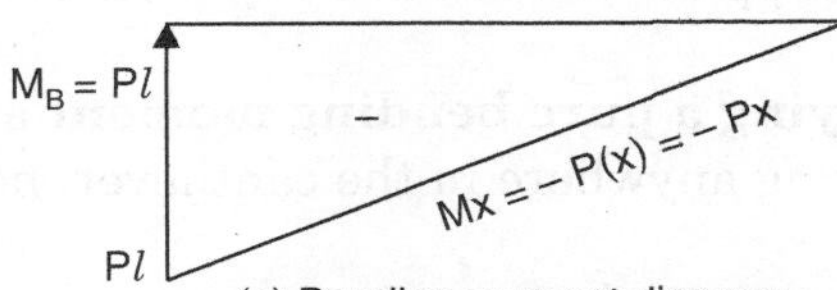

(c) Bending moment diagram

Fig. 8.8.5. The right end of cantilever is free.

Case IV. A cantilever carrying a uniformly distributed load over its entire length. See Fig. 8.8.6.

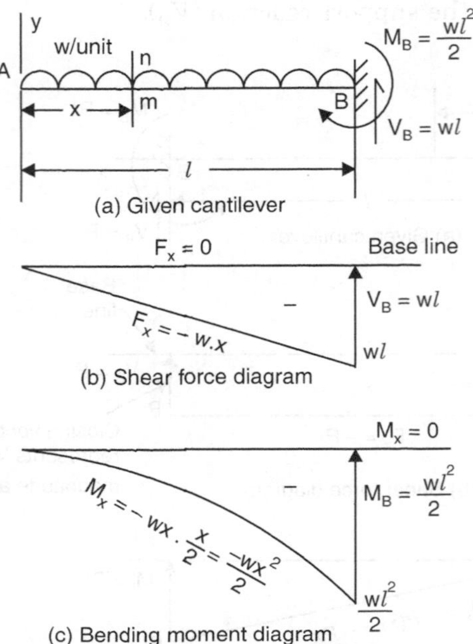

Fig. 8.8.6 Cantilever carrying *udl* on its entire length.

Taking A as origin, at any section m-n at a distance x we can write expressions for F_x and M_x.

Thus, we get

$$F_x = -w\,.\,x \qquad \text{...(1)}$$

$$M_x = (-wx) \times \bar{x}_m$$

or

$$M_x = -wx\,.\,\frac{x}{2} = \frac{-wx^2}{2} \qquad \text{...(2)}$$

Eqn. (1) represents a straight line.

At A, $x = 0$, and $F_x = 0$

At B, $x = l$, $F_x)_{x=l} = F_B = -wl$.

The last vector is $V_B = wl \uparrow$.

When this vector is drawn the shear force diagram closes as shown in Fig. 8.8.6 (*b*).

Eqn. (2) represents a parabolic curve as shown in Fig. 8.8.6 (*c*)

$$M_A = M_x)_{x=0} = 0$$

$$M_B = M_x)_{x=l} = -\frac{wl^2}{2} \qquad \text{...(8.18)}$$

$$F_B = wl \uparrow \quad \text{and } M_B = wl^2/2$$

are equal to the reactions at the support B. Reaction $V_B = wl$ is upward and M_B is in the clockwise direction.

Case V. A cantilever carrying a pure bending moment at its free end. Fig. 8.8.7.

There is no vertical load acting anywhere in the cantilever, hence at any section

$$F_x = 0 \qquad \text{...(1)}$$

and

$$M_x = -M_0 \qquad \text{...(2)}$$

Eqn. (1) shows that shear force is zero over the entire length of the cantilever. The bending moment is constant and equal to $-M_0$ at all sections of the cantilever, as shown in Fig. 8.8.7 (*c*). We find that the reaction at *B* is $V_B = 0$, and moment reaction at support $B = M_B = M_0$, as shown in Fig. 8.8.7 (*a*).

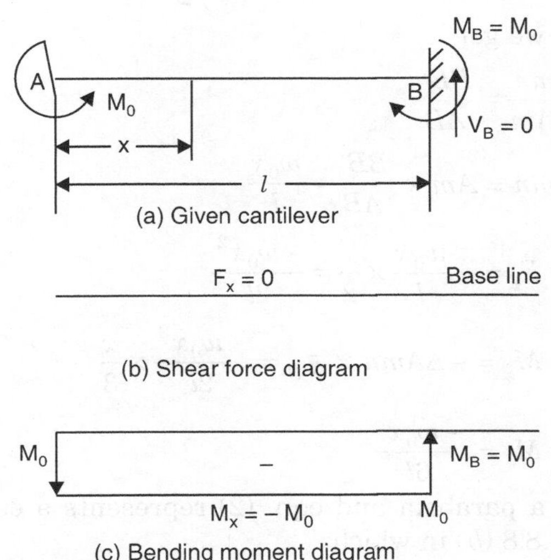

Fig. 8.8.7. A Cantilever carrying a pure moment M_0 at its free end.

Case VI. A cantilever carrying a uniformly varying load zero at the free end and w_0 per unit length at the fixed end. Fig. 8.8.8.

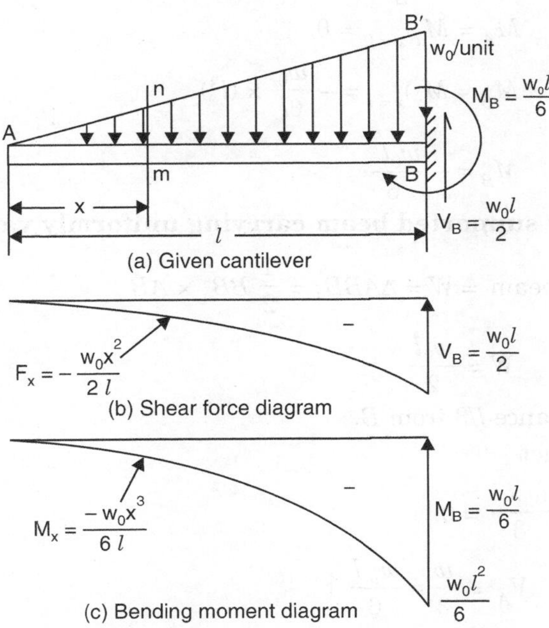

Fig. 8.8.8. A Cantilever carrying a *uvl* on its entire length zero at the free end and w_o per unit length at the fixed end.

Taking the free end (end A) as the origin, at any section $m - n$ at a distance x from A, we write

$$F_x = -\text{ Area of } \Delta Amn = -\frac{1}{2}(mn).(Am)$$

$\because$ $\Delta Amn \equiv \Delta ABB'$, we get

$$\frac{mn}{Am} = \frac{BB'}{AB}$$

$$\therefore \quad mn = Am \times \frac{BB'}{AB} = \frac{w_0 x}{l}$$

$$\therefore \quad F_x = -\frac{w_0 x}{l} \times \frac{x}{2} = \frac{-w_0 x^2}{2l} \qquad \text{...(1)}$$

And, $$M_x = -\Delta Amn \times \bar{x}_B = \frac{-w_0 x^2}{2l} \times \frac{x}{3}$$

$$\therefore \quad M_x = \frac{-w_0 x^3}{6l} \qquad \text{...(2)}$$

Eqn. (1) represents a parabola and eqn. (2) represents a cubic parabola. Shear force diagram is shown in Fig. 8.8.8 (*b*) in which

$$F_A = F_x)_{x=0} = 0 \quad \text{and} \quad F_B = F_x)_{x=l} = -\frac{w.l^2}{2l}$$

or $$F_B = -\frac{w_0 l}{2} \qquad \text{...(8.19)}$$

Similarly, $$M_A = M_x)_{x=0} = 0.$$

And, $$M_B = M_x)_{x=l} = -\frac{w_0}{6l} \times (l^3)$$

or $$M_B = -\frac{w_0 l^2}{6} \qquad \text{...(8.20)}$$

Case VII. Simply supported beam carrying uniformly varying load. (Fig. 8.8.9).

Total load on the beam $= W = \Delta ABB_1 = \frac{1}{2} BB_1 \times AB$

or $$W = \frac{w_0 l}{2}$$

It's CG is at a distance $l/3$ from B.

By $\Sigma M_B = 0$, we get

$$\circlearrowleft V_A \times l - \frac{w \times l}{3} = 0$$

$$\therefore \quad V_A = \frac{w}{3} = \frac{w_o l}{6} \uparrow \qquad \text{...(8.21)}$$

And, by $\Sigma F_v = 0$, we get

$$(V_A + V_B) \uparrow - W = 0$$

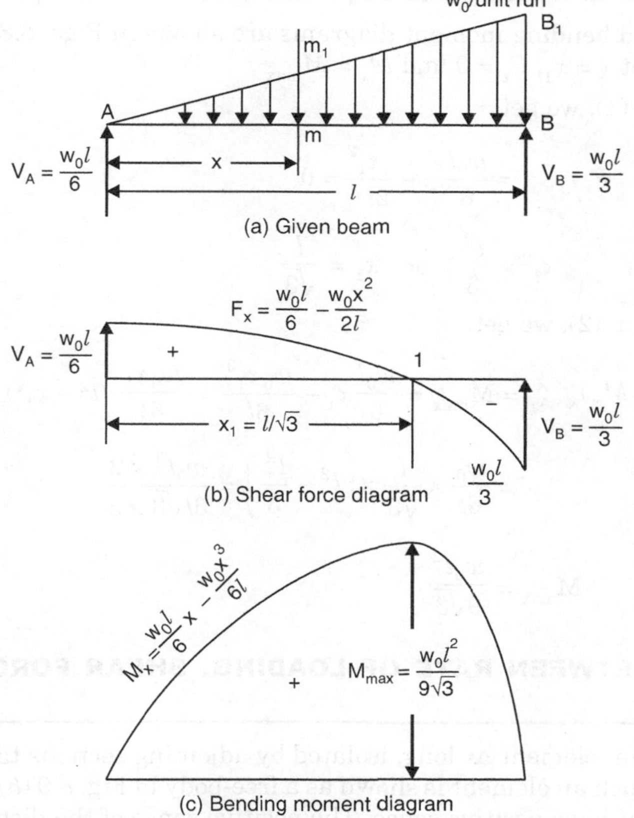

Fig. 8.8.9. A simply supported beam carrying a *uvl* on its entire span.

$$\therefore \qquad V_B = W - V_A = W - \frac{W}{3}$$

or $$V_B = \frac{2W}{3} = \frac{w_0\, l}{3} \uparrow \qquad \text{...(8.22)}$$

We note that, here $V_B = 2V_A$

Consider a section mm, at a distance x from A (the origin).

$$\because \qquad \Delta Amm_1 \equiv \Delta ABB_1$$

$$\therefore \qquad \frac{mm_1}{Am} = \frac{BB_1}{AB} \quad \therefore \quad mm_1 = \frac{BB_1}{AB} \times Am = \frac{w_0}{l}\, x$$

At this section, $F_x = V_A$ – external load between A and m

or $$F_x = \frac{w_0\, l}{6} - \frac{mm_1 \times Am}{2} = \frac{w_0\, l}{6} - \frac{w_0\, x}{l} \times \frac{x}{2}$$

or $$F_x = \frac{w_0\, l}{6} - \frac{w_0\, x^2}{2l} \qquad \text{...(1)}$$

And, $$M_x = V_A \,.\, x - (\Delta Amm_1) \times \frac{Am}{3} = V_A \,.\, x - \frac{w_0 x^2}{2l} \,.\, \frac{x}{3}$$

or, $$M_x = \frac{w_0 l}{6}\, x - \frac{w_0\, x^3}{6l} \qquad \text{...(2)}$$

Eqn. (1) represents a parabola and eqn. (2) represents a cubic parabola.

Shear force and bending moment diagrams are shown in Figs. 8.8.9 (*b*) and (*c*), respectively. At point (1), let $x = x_1$, $F_x = 0$ and $M_x = M_{max.}$

$\therefore$ From eqn. (1), we get

$$F_x)_{x = x_1} = \frac{w_0 l}{6} - \frac{w_0 {x_1}^2}{2l} = 0$$

$$\therefore \quad {x_1}^2 = \frac{l^2}{3} \quad \text{or} \quad x_1 = \frac{l}{\sqrt{3}} \qquad ...(3)$$

Hence from eqn. (2), we get

$$M_x)_{x = x_1} = \text{M}_{\text{max}} = \frac{w_0 l}{6} x_1 - \frac{w_0 {x_1}^3}{6l} = \frac{w_0 x_1}{6l} (l^2 - {x_1}^2)$$

$$= \frac{w_0}{6l} \times \frac{l}{\sqrt{3}} \times \left(l^2 - \frac{l^2}{3} \right) = \frac{w_0 l^3 \times 2}{6l\sqrt{3} \times 3}$$

or
$$\text{M}_{\text{max}} = \frac{w_0 l^2}{9\sqrt{3}} \qquad ...(8.23)$$

8.9. RELATION BETWEEN RATE OF LOADING, SHEAR FORCE AND BENDING MOMENT

Consider a beam element Δx long, isolated by adjoining sections taken perpendicular to its axis Fig. 8.9 (*a*). Such an element is shown as a free-body in Fig. 8.9 (*b*). All the forces shown acting on this element have positive sense. The positive sense of the distributed external load q is taken to coincide with the direction of the + y axis. As the shear and moment may each change from one section to the next, note that on the right side of the element, these quantities are, respectively, designated by $F_x + \Delta F_x$ and $M_x + \Delta M_x$.

From the condition of vertical forces, we get by $\Sigma F_y = 0$,

$$\uparrow F_x + q\Delta x - (F_x + \Delta F_x) = 0$$

or
$$\frac{\Delta F_x}{\Delta x} = q \qquad ...(8.24)$$

And, by summation of moments around A, equal to zero,

i.e., by $\Sigma M_A = 0$, we get

$$M_x + F_x \,.\, \Delta x + (q \times \Delta x) \frac{\Delta x}{2} - (M_x + \Delta M_x) = 0$$

or
$$M_x + F_x \,.\, \Delta F_x + \frac{q \,.\, \Delta x^2}{2} - M_x - \Delta M_x = 0$$

$$\therefore \quad \Delta M_x - F_x \,.\, \Delta F_x - \frac{q\Delta x^2}{2} = 0$$

Since Δx <<, $\therefore$ Neglecting Δx^2, we get

$$\Delta M_x - F_x \Delta x = 0$$

$$\therefore \qquad \frac{\Delta M_x}{\Delta x} = F_x \qquad \text{...(8.25)}$$

Eqns. (8.24) and (8.25) in the limit as $\Delta x \to 0$ yield the following differential equations :

$$\frac{dF_x}{dx} = q = \text{slope of shear force diagram} \qquad \text{...(8.26)}$$

And,

$$\frac{dM_x}{dx} = F_x = \text{slope of bending moment diagram} \qquad \text{...(8.27)}$$

By substituting eqn. (8.27) into eqn. (8.26), another useful relation is obtained :

$$\frac{d}{dx}\left(\frac{dM_x}{dx}\right) = \frac{d^2 M_x}{dx^2} = q \qquad \text{...(8.28)}$$

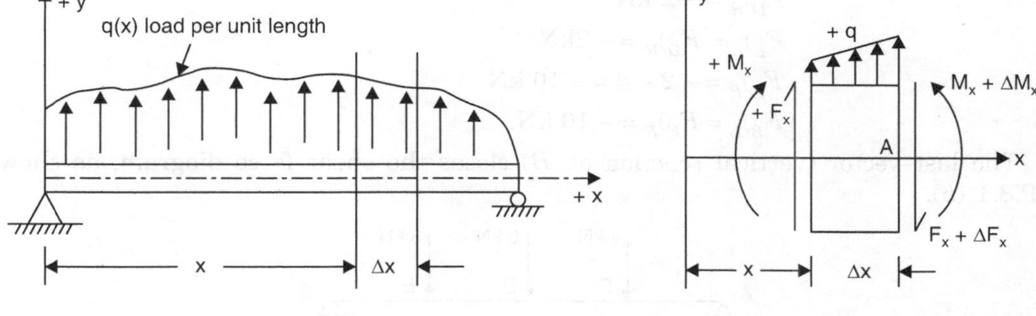

(a) The given beam (b) Free-body diagram of the element

Fig. 8.9. Beam and the element for differential relations between q, F_x and M_x.

The above differential equations can be used for determining reactions of determinate beams from the boundary conditions. These relations can also be used for the construction of the shear force and bending moment diagrams. The following signs for slope are useful in drawing shear force and bending moment diagrams.

+ slope = ↗

– slope = ↘

zero slope = →

Problem 8.1. *Draw shear force and bending moment diagrams for the simply supported beam shown in Fig. P. 8.1.*

Sol. By summing moments about *B*, (*i.e.*, by $\Sigma M_B = 0$), we get

$$\curvearrowleft V_A \times 8 - 4 \times 6 - 6 \times 4 - 8 \times 2 = 0$$

$$\therefore \quad V_A = \frac{1}{8}[24 + 24 + 16] = 8 \text{ kN}\uparrow$$

By $\Sigma F_v = 0$, we get↑

$$(V_A + V_B)\uparrow - (4 + 6 + 8) = 0$$

$$\therefore \quad (8 + V_B) = 18$$

$\therefore \qquad V_B = 18 - 8 = 10 \text{ kN} \uparrow$

S.F. Diagram :

Shear force at $\qquad A = F_A = V_A = 8 \text{ kN}$

Shear force on the left of point C

$$= F_C)_L = V_A = 8 \text{ kN}.$$

Shear on the right of point C

$$= F_C)_R = V_A - 4$$

or $\qquad F_C)_R = 8 - 4 = 4 \text{ kN}$

Similarly, $\qquad F_D)_L = F_C)_R = 4 \text{ kN}$

$$F_D)_R = F_D)_L - 6 = 4 - 6$$

or $\qquad F_D)_R = -2 \text{ kN}$

$$F_E)_L = F_D)_R = -2\text{kN}$$

or $\qquad F_E)_R = -2 - 8 = -10 \text{ kN}$

$$F_B)_L = F_E)_R = -10 \text{ kN}.$$

The last vector (vertical reaction at B) closes the shear force diagram, as shown in Fig. P.8.1 (*b*).

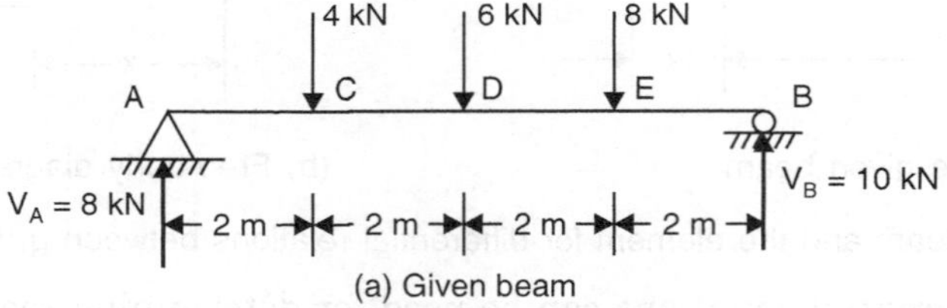

(a) Given beam

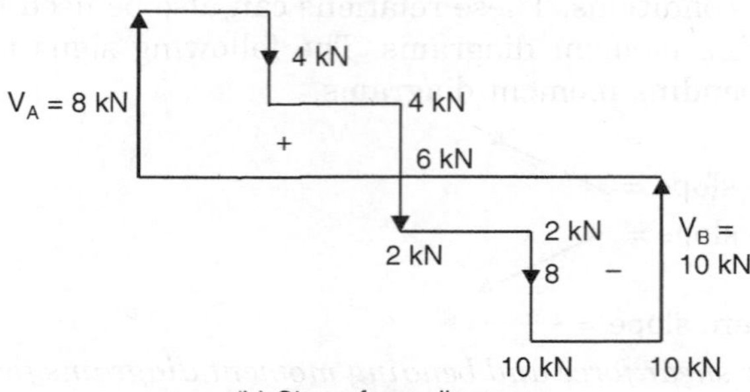

(b) Shear force diagram

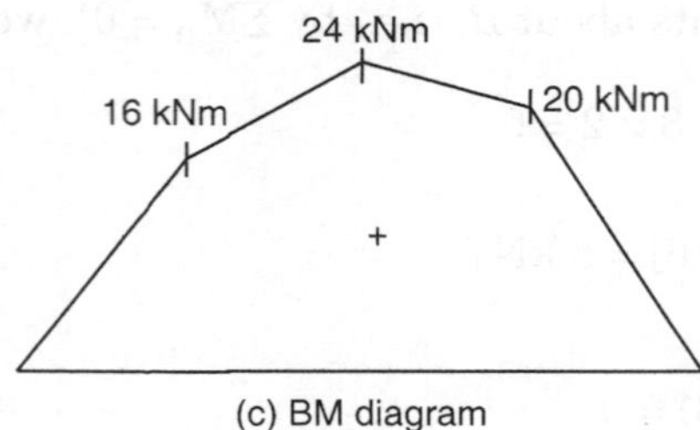

(c) BM diagram

Fig. P. 8.1

B.M. Diagram :

$$M_A = 0\,;$$

$$M_C = V_A \times 2 = 8 \times 2 = 16 \text{ kNm.}$$

$$M_D = V_A \times 4 - 4 \times 2$$

$$= 8 \times 4 - 4 \times 2 = 24 \text{ kNm.}$$

$$M_E = V_A \times 6 - 4 \times 4 - 6 \times 2$$

$$= 8 \times 6 - 16 - 12 = 20 \text{ kNm.}$$

$$M_B = V_A \times 8 - 4 \times 6 - 4 \times 4 - 8 \times 2 = 0 \quad \textbf{(Check)}$$

B.M. diagram is shown in Fig. P. 8.1 (*c*).

Problem 8.2. *Draw shear force and bending moment diagrams for the simply supported beam shown in Fig. P. 8.2 (a). Also find maximum bending moment.*

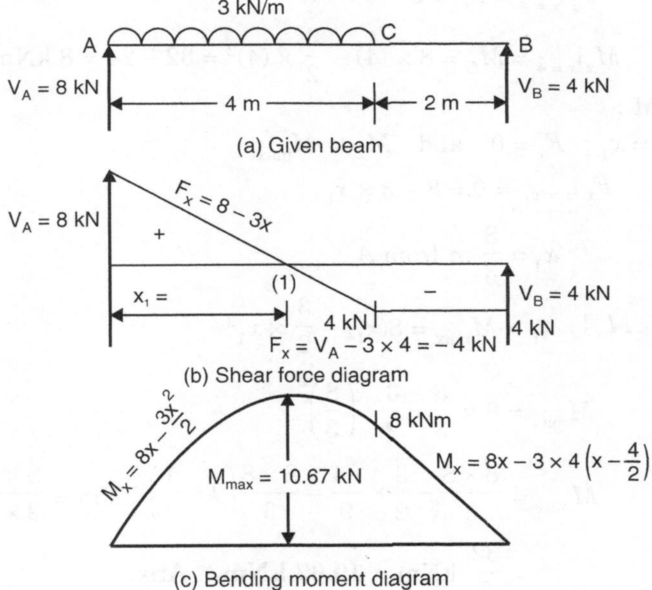

Fig. P. 8.2

Sol. By $\Sigma M_B = 0$, we get

$$\curvearrowleft V_A \times AB - 3 \times (AC)\left(\frac{AC}{2} + BC\right) = 0 \quad \text{or} \quad 6V_A - 3 \times 4 \times \left(\frac{4}{2} + 2\right) = 0$$

$$\therefore \qquad V_A = \frac{1}{6}\,(12 \times 4) = 8 \text{ kN} \uparrow$$

By $\Sigma F_V = 0$, we get

$$(V_A + V_B) \uparrow - 3 \times 4 = 0$$

$$\therefore \qquad V_B = 12 - V_A = 12 - 8 = 4 \text{ kN} \uparrow$$

For portion AC : $0 \le x \le 4m$

At any section x from A :

Shear force, $F_x = V_A - 3x$

or, $$F_x = 8 - 3x \quad \text{...(1)}$$

And, bending moment $M_x = V_A \cdot x - (3x)\left(\frac{x}{2}\right)$

or, $$M_x = 8x - \frac{3x^2}{2} \quad \text{...(2)}$$

Eqn. (1) represents a straight line while eqn. (2) represents a parabola.

At A, $x = 0$,

$\therefore$ $F_x]_{x=0} = 8 \text{ kN} = V_A$ and $M_x]_{x=0} = 0 = M_A$

At C, $x = 4$ m,

$\therefore$ From (1), $F_x]_{x=4} = F_C = 8 - 3 \times (4) = -4$ kN

And, from (2), $M_x]_{x=4} = M_C = 8 \times (4) - \frac{3}{2} \times (4)^2 = 32 - 24 = 8$ kNm

Maximum BM :

At point (1), $x = x_1$; $F_x = 0$ and $M_x = M_{max}$

$\therefore$ From (1), $F_x]_{x=x_1} = 0 = 8 - 3 \times x_1$

$\therefore$ $$x_1 = \frac{8}{3} \, m \text{ from } A$$

And, from (2), $M_x]_{x=x_1} = M_{max} = 8(x_1) - \frac{3}{2} \times x_1^2$

or $$M_{max} = 8 \times \frac{8}{3} - \frac{3}{2} \times \left(\frac{8}{3}\right)^2$$

or $$M_{max} = \frac{8 \times 8}{3} - \frac{3}{2} \times \frac{64}{9} = \frac{8 \times 8}{3}\left(1 - \frac{1}{2}\right) \text{ kNm} = \frac{8 \times 8}{2 \times 3} \text{ kNm}$$

$$= \frac{32}{3} \text{ kNm} = \mathbf{10.67 \text{ kNm}.} \quad \textbf{Ans.}$$

In portion CB. $4 \le x \le 2$ m from A

$$F_x = V_A - 3 \times 4 = -4 \text{ kN.} \quad \text{(Constant)} \quad \text{...(3)}$$

And, $$M_x = V_A \times x - 3 \times 4\left(x - \frac{AC}{2}\right)$$

or $$M_x = 8x - 3 \times 4\left(x - \frac{4}{2}\right) \quad \text{...(4)}$$

It is clear from eqn. (3) that in the portion CB the shear force F_x is constant and equal to 4 kN.

From eqn. (4),

$$M_x]_{x=4} = 8 \times (4) - 3 \times 4 \times \left(4 - \frac{4}{2}\right) \text{ kNm}$$

$$= 32 - 24 = 8 \text{ kNm} = M_C \quad \text{(Check-I)}$$

Above is the check on correctness of eqn. (4). Since both eqns.(2) and (4) give M_C = 8 kNm, therefore, eqn.(4) is correct.

Hence, from eqn. (4),

$$M_x]_{x=6} = M_B = 8 \times 6 - 3 \times 4 \times \left(6 - \frac{4}{2}\right) \text{ kNm}$$

$$= 48 - 12 \times 4 = 0 \quad \text{(Check-II)}$$

This was the first step in the solution of this problem, in which the support reaction V_A was calculated by using the condition $\Sigma M_B = 0$.

This last check-II confirms the correctness of the entire solution.

Shear force and bending moments diagrams are shown in Figs. P.8.2 (*b*) and (*c*).

Problem 8.3. *Draw shear force and bending moment diagrams for the simply supported beam shown in Fig. P.8.3 (a).*

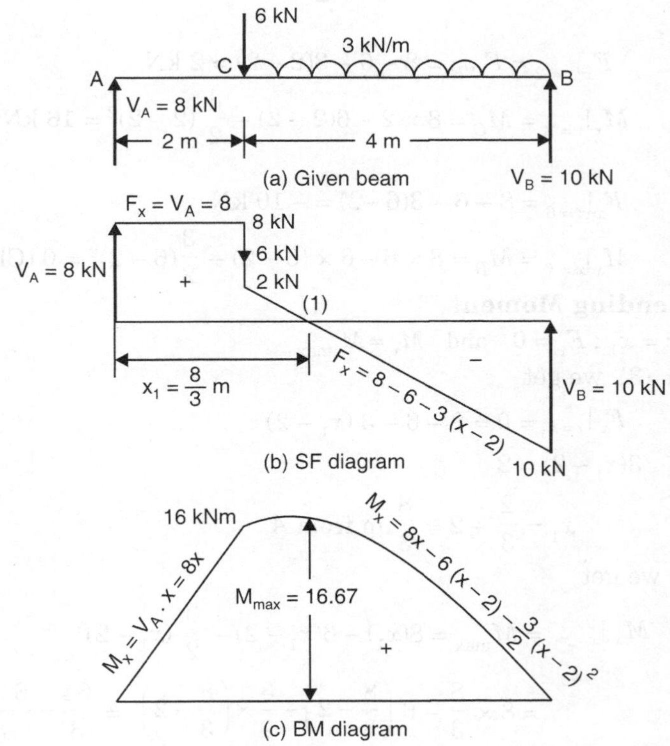

Fig. P. 8.3

Sol. Take *A* as origin.

By $\Sigma M_B = 0$, we get

$$\circlearrowleft V_A \times 6 - 6 \times 4 - \frac{3}{2} \times 4^2 = 0$$

$\therefore$
$$V_A = \frac{1}{6}[24 + 24] = \mathbf{8\ kN} \uparrow$$

By $\Sigma F_V = 0$, we get $(V_A + V_B) \uparrow - (6 + 3 \times 4) = 0$

$\therefore$
$$V_B = 18 - V_A = 18 - 8 = 10 \text{ kN} \uparrow$$

Portion AC. $0 \le x \le 2$ m

At any section x from A, we have

$$F_x = V_A = 8 \quad ...(1)$$

and
$$M_x = V_A \cdot x = 8x \quad ...(2)$$

At C, $x = 2$ m,

$$\therefore \quad F_x]_{x=2} = F_C)_L = 8 \text{ kN}$$

and
$$M_x]_{x=2} = 8 \times 2 = 16 \text{ kNm}$$

Portion CB : $2 \le x \le 6$ m

At any section x from A, we get

$$F_x = 8 - 6 - 3\,(x - 2) \quad ...(3)$$

and
$$M_x = 8x - 6(x-2) - \frac{3}{2}(x-2)^2 \quad ...(4)$$

At C, $x = 2$ m ;

$\therefore$ From (3), $F_x]_{x=2} = F_C)_R = 8 - 6 - 3(2-2) = 2$ kN

$$M_x]_{x=2} = M_C = 8 \times 2 - 6(2-2) - \frac{3}{2}\,(2-2)^2 = 16 \text{ kNm (Check)}$$

At B, $x = 6$ m,

$$\therefore \quad F_x]_{x=6} = 8 - 6 - 3(6-2) = -10 \text{ kN}$$

$$\therefore \quad M_x]_{x=6} = M_B = 8 \times 6 - 6 \times (6-2) - \frac{3}{2}(6-2)^2 = 0 \text{ (Check)}$$

Maximum Bending Moment

At point (1), $x = x_1$; $F_x = 0$ and $M_x = M_{max}$

$\therefore$ From eqn. (3), we get

$$F_x]_{x=x_1} = 0 = 8 - 6 - 3\,(x_1 - 2)$$

$$\therefore \quad 3(x_1 - 2) = 2$$

$$\therefore \quad x_1 = \frac{2}{3} + 2 = \frac{8}{3} \text{ m from } A$$

From eqn.(4), we get

$$M_x]_{x=x_1} = M_{max} = 8(x_1) - 6(x_1 - 2) - \frac{3}{2}\,(x_1 - 2)^2$$

$$= 8 \times \frac{8}{3} - 6\left(\frac{8}{3} - 2\right) - \frac{3}{2} \times \left(\frac{8}{3} - 2\right)^2 = \frac{64}{3} - \frac{6 \times 2}{3} - \frac{3}{2}\left(\frac{2}{3}\right)^2$$

$$= \frac{1}{3}\,(64 - 12 - 2) = \frac{50}{3} = 16.67 \text{ kNm.}$$

Shear force and bending moment diagrams are shown in Figs. P. 8.3 (*b*) and (*c*), respectively

Problem 8.4. *Draw SF and BM diagrams for the beam shown in Fig. P.8.4 (a).*

Sol. By $\Sigma M_B = 0$, we get

$$V_A \times 5 - (3 \times 2)\left(\frac{2}{2} + 1\right) = 0$$

$$\therefore \quad V_A = \frac{1}{5} \times 6 \times 2 = 2.4 \text{ kN} \uparrow$$

By $\Sigma F_V = 0$, we get

$$(V_A + V_B) \uparrow - 3 \times 2 = 0$$

$$\therefore \quad V_B = 6 - V_A = 6 - 2.4 \text{ kN} \quad \text{or } V_B = 3.6 \text{ kN} \uparrow$$

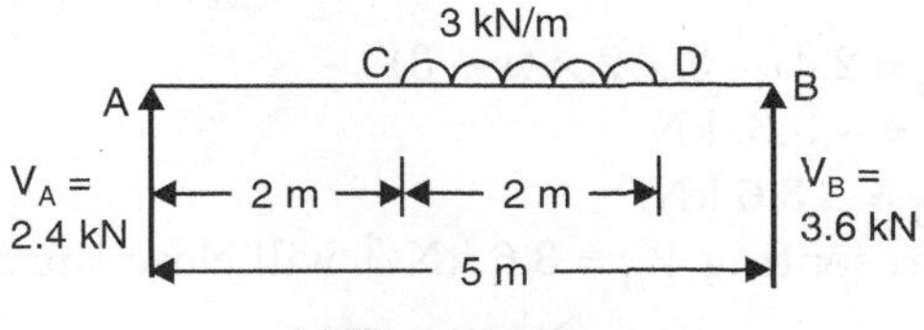

(a) The given beam

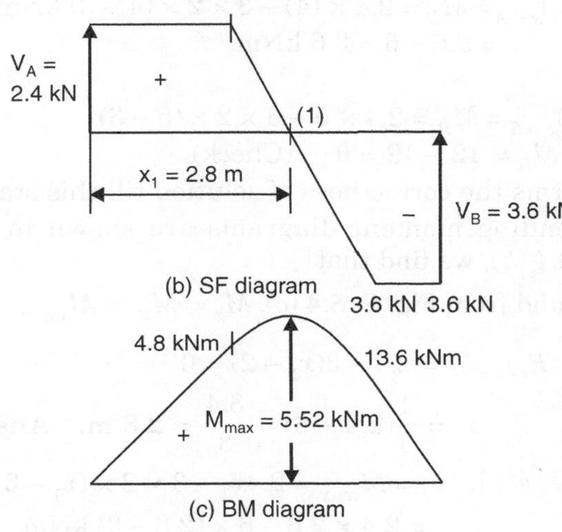

(b) SF diagram

(c) BM diagram

Fig. P.8.4

For Portion AC : With A as origin, $0 \le x \le 2$ m

$$F_x = 2.4 \text{ kN} \quad \text{...(1)}$$

$$M_x = 2.4\, x \quad \text{...(2)}$$

Eqn. (1) represents a straight line parallel to the base and 2.4 kN above it.

$$\therefore \quad F_A = F_C = 2.4 \text{ kN}$$

Eqn. (2) represents a straight line. At $A, x = 0$

$$M_x]_{x=0} = M_A = 0$$

At C , $x = 2$ m,

$$\therefore \quad M_x]_{x=2} = M_C = 2.4 \times 2 = 4.8 \text{ kNm}$$

For Portion CD : $2 \le x \le 4m$, A being origin for any section,

$$F_x = 2.4 - 3(x - 2) \quad \text{...(3)}$$

and
$$M_x = 2.4x - \frac{3}{2}(x - 2)^2 \quad \text{...(4)}$$

From eqn. (3), At $x = 2$, $F_x]_{x=2} = F_C = 2.4$ kN (Check of eqn. (3))

At $x = 4$ m ; $F_x]_{x=4} = F_D = 2.4 - 3 \times (4 - 2) = -3.6$ kN.

From eqn. (4) ; at $x = 2$,

$$M_x]_{x=2} = M_C = 2.4 \times 2 - 0 = 4.8 \text{ kN} \quad \text{(Check of eqn. (4))}$$

At D, $x = 4$ m,

$$\therefore \quad M_x]_{x=4} = M_D = 2.4 \times 4 - \frac{3}{2}(4 - 2)^2$$

$$= 9.6 - 6 = 3.6 \text{ kNm.}$$

For portion DB : $4 \le x \le 5$ m, with origin at A.

At any section, $F_x = 2.4 - 3 \times 2 = -3.6$ kN ...(5)

and, $M_x = 2.4 \times x - 3 \times (2) \times \left(x - 2 - \frac{2}{2}\right)$

or $M_x = 2.4x - 3 \times 2 \times (x - 3)$ ...(6)

From eqn. (5), $F_D = -3.6$ kN

also, $F_B = -3.6$ kN.

$\therefore$ The last vector representing $V_B = 3.6$ kN $\uparrow$ will close the *SF* diagram.

From eqn. (6),

At $x = 4$ m, $M_x)_{x=A} = M_D = 2.4 \times (4) - 3 \times 2 \times (4 - 3)$ kNm
$= 9.6 - 6 = 3.6$ kNm. (Check of eqn. (6))

And, for B, $x = 5$ m,

and $M_x)_{x=5} = M_B = 2.4 \times 5 - 3 \times 2 \times (5 - 3)$

or $M_B = 12 - 12 = 0$ (Check).

The last check confirms the correctness of solution till this stage .

Shear force and bending moment diagrams are shown in Figs. P.8.4 (*b*) and (*c*), respectively. From Fig. P.8.4 (*b*), we find that at point (1), $x = x_1$; $F_x = 0$, and from Fig. P. 8.4 (*c*), $M_x = M_{x_1} = M_{max}$.

$\therefore$ From eqn. (3), $F_x]_{x=x_1} = 2.4 - 3(x_1 - 2) = 0$

$\therefore$ $x_1 = \frac{1}{3}(2.4 + 6) = \frac{8.4}{3} = \mathbf{2.8\ m.}$ **Ans.**

$\therefore$ From eqn. (4), $M_x]_{x=x_1} = M_{max} = 2.4x_1 - 3 \times 2 \times (x_1 - 3)$
$= 2.4 \times 2.8 - 6 \times (2.8 - 3)$ kNm
$= 6.72 - 1.2$ kN-m $=$ **5.52** kNm. **Ans.**

Problem 8.5. *Draw shear force and bending moment diagrams for the beam carrying a pure moment M_0 at point C, shown in Fig. P. 8.5 (a).*

Sol. By $\Sigma M_B = 0$, we get

$V_A \times l - M_0 = 0$

$\therefore$ $V_A = \frac{M_0}{l} \uparrow$

By $\Sigma F_V = 0$, we get

$(V_A + V_B) \uparrow + 0 = 0$

$\therefore$ $V_B = -V_A = -\frac{M_0}{l}$

or $V_B = \frac{M_0}{l} \downarrow$

The shear force diagram is shown in Fig. P.8.5(*b*).

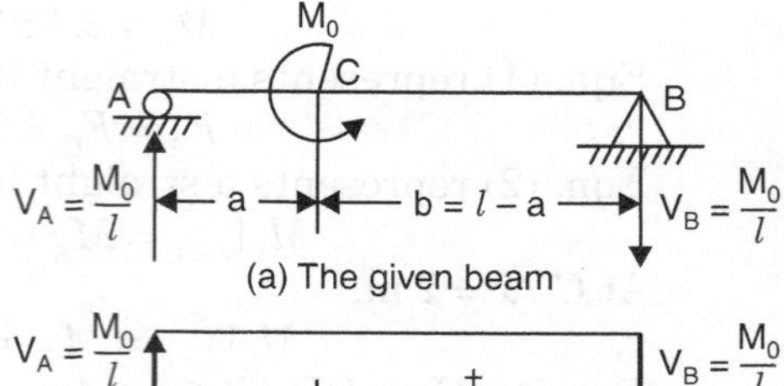

(a) The given beam

(b) SF diagram

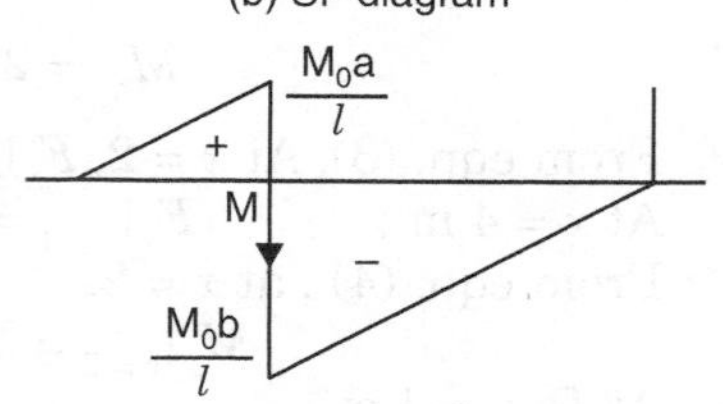

(c) BM diagram

Fig. P. 8.5

Bending moment on the left of C is

$$M_C)_L = V_A \times a = \frac{M_0}{l} a \quad ...(1)$$

And, bending moment on the right of C is

$M_C)_R = V_A \times a - M_0$

$$= \frac{M_0}{l} \cdot a - M_0 = \frac{M_0}{l} (a - l)$$

$$= \frac{M_0}{l} (-b)$$

or $$M_C)_R = -\frac{M_0 b}{l} \quad ...(2)$$

Also, $$M_C)_R = V_B \times b = -\frac{M_0}{l} \times b$$

Bending moment diagram is shown in Fig. P.8.5 (*c*).

Problem 8.6. *Draw shear force and bending moment diagrams for the beam shown in Fig. P.8.6 (a).*

Sol. By $\Sigma M_B = 0$, we get

$$\circlearrowleft V_A \times 4 - 2 \times 3 - 4 = 0$$

$$\therefore \quad V_A = \frac{1}{4} \times (6 + 4) = 2.5 \text{ kN} \uparrow$$

By $\Sigma F_V = 0$, we get

$$(V_A + V_B) \uparrow - 2 = 0$$

$$\therefore \quad V_B = 2 - V_A = 2 - 2.5 = -0.5 \text{ kN}$$

or $$V_B = 0.5 \text{ kN} \downarrow$$

Take A as origin.

$\therefore$ For portion AC : $0 \le x \le 1$ m

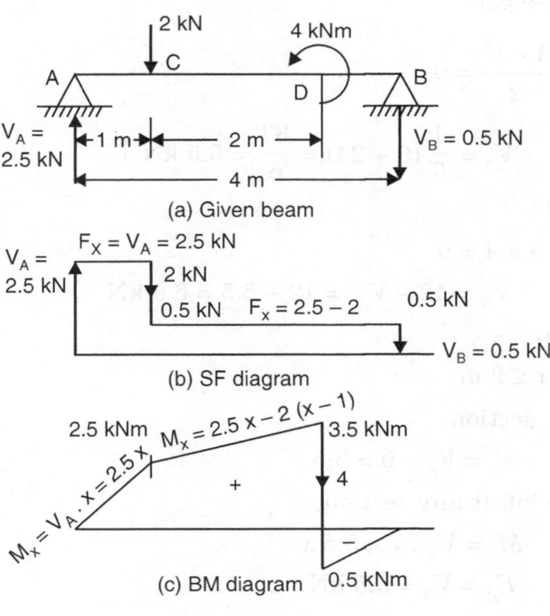

(a) Given beam

(b) SF diagram

(c) BM diagram

Fig. P. 8.6

$$F_x = V_A = 2.5 \text{ kN} \quad ...(1)$$

and $$M_x = V_A \cdot x = 2.5\, x \quad ...(2)$$

$$\therefore \quad F_A = 2.5 \text{ kN}$$

$F_C)_L = 2.5$ kN

$M_A = 2.5\ x)_{x=0} = 0$ and $M_C = 2.5 \times x)_{x=1} = 2.5$ kNm.

Portion CD : $1 \leq x \leq 3$ m

$$F_x = 2.5 - 2 \qquad ...(3)$$

$\therefore$ $F_x)_{x=1\text{ m}} = F_C)_R = 0.5$ kN.

and $$M_x = 2.5\ x - 2\ (x - 1) \qquad ...(4)$$

$\therefore$ $M_x]_{x=1\text{ m}} = M_C = 2.5 \times 1 - (0) = 2.5$ kNm (Check of eqn. (4))

At $x = 3$, we get $M_x]_{x=3} = M_D)_L = 2.5 \times 3 - 2(3 - 1) = 3.5$ kN

Portion DB : $3 \leq x \leq 4$ m.

At any section, $$F_x = 2.5 - 2 \qquad ...(5)$$

$\therefore$ $F_C)_R = 0.5$ kN ; and $F_B = 0.5$ kN.

And, $$M_x = 2.5\ x - 2(x - 1) - 4 \qquad ...(6)$$

$\therefore$ $M_x]_{x=3\text{ m}} = M_D)_R = 2.5 \times (3) - 2\ (3 - 1) - 4 = -\ 0.5.$ kN

and $M_x]_{x=4} = M_B = 2.5 \times (4) - 2 \times (4 - 1) - 4 = 0$ (Check of eqn.(6))

SF and *BMDs* are given in Fig. P. 8.6 (*c*).

Problem 8.7. *Draw SF and BM diagrams for the beam shown in Fig. P. 8.7 (a). Find the position and magnitude of maximum bending moment. Also, locate the position of point of contraflexion, if any.*

Sol. By $\Sigma M_B = 0$, we get

$$\circlearrowleft V_A \times 6 - 9 - \frac{3 \times 4^2}{2} = 0$$

$\therefore$ $$V_A = \frac{1}{6}\ (9 + 24) = \frac{33}{6} = 5.5 \text{ kN} \uparrow$$

By $\Sigma F_V = 0$, we get

$$(V_A + V_B) \uparrow - 3 \times 4 = 0$$

$\therefore$ $$V_B = 12 - V_A = 12 - 5.5 = 6.5 \text{ kN}$$

Take A as origin.

Portion AC : $0 \leq x \leq 2$ m

Shear force at any section,

$$F_x = V_A - 0 = 5.5 \qquad ...(1)$$

And bending moment at any section,

$$M_x = V_A\ .\ x = 5.5\ x \qquad ...(2)$$

From (1), $F_A = V_A = 5.5$ kN

or $F_C = 5.5$ kN

From (2), $M_x]_{x=0} = M_A = 0$

and $M_x\ ]_{x=2} = M_C)_L = 5.5 \times 2 = 11$ kNm

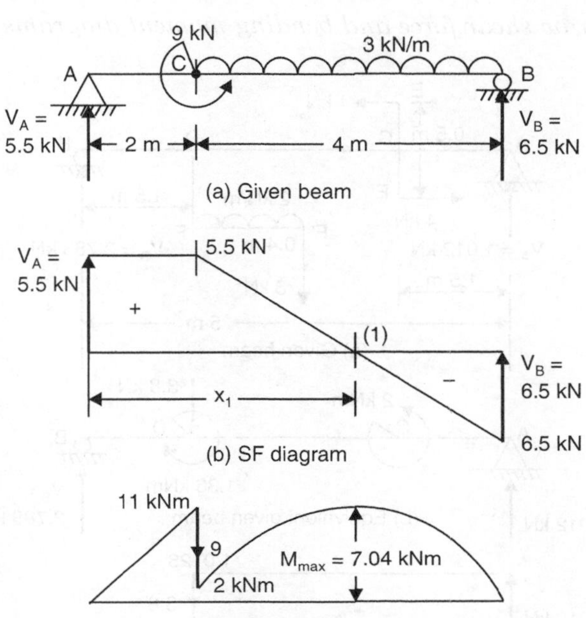

Fig. P.8.7

Portion CB : $2 \le x \le 6$ m

$$F_x = 5.5 - 3(x - 2) \qquad ...(3)$$

and
$$M_x = 5.5x - 9 - \frac{3}{2}(x - 2)^2 \qquad ...(4)$$

From (3), $F_x]_{x=2} = F_C = 5.5 - 0 = 5.5$ kN (Check of eqn.(3))

From (4), $M_x]_{x=2} = M_C)_R = 5.5\,(2) - 9 - 0 = 2$ kNm.

$$F_x]_{x=6\text{ m}} = F_B = 5.5 - 3(6 - 2) = 5.5 - 12 = -6.5 \text{ kN}$$

$$M_x]_{x=6} = M_B = 5.5 \times (6) - 9 - \frac{3(6-2)^2}{2} \text{ kNm} = 33 - 9 - 24$$

$\therefore$ $M_B = 0$ (Check of eqn. (4))

Shear force diagram is shown in Fig. P.8.7 (*b*).

At point (1). $x = x_1$ and shear force $F_x = 0$ and $M_x = M_{max}$.

$\therefore$ From Eqn. (3) $F_x]_{x=x_1} = F_{x_1} = 0$

$\therefore$ From eqn. (3), we get

$$F_x]_{x=x_1} = 5.5 - 3(x_1 - 2) = 0$$

$\therefore$
$$x_1 = \frac{1}{3}(5.5 + 6) = \frac{11.5}{3} = 3.83 \text{ m from } A$$

At point (1), $M_x]_{x=x_1} = M_{max}$

$\therefore$ From eqn. (4), we get

$$M_x]_{x=x_1} = M_{max} = 5.5x_1 - 9 - \frac{3}{2}(x_1 - 2)^2$$

$$= 5.5 \times (3.83) - 9 - \frac{3}{2}(3.82 - 2)^2$$

$$= \mathbf{7.04\ kNm.} \quad \textbf{Ans.}$$

The BMD is shown in Fig. P. 8.7 (*c*). There is no point of contraflexion in the beam. **Ans.**

Problem 8.8. *Draw shear force and bending moment diagrams for the beam shown in Fig. P. 8.8 (a).*

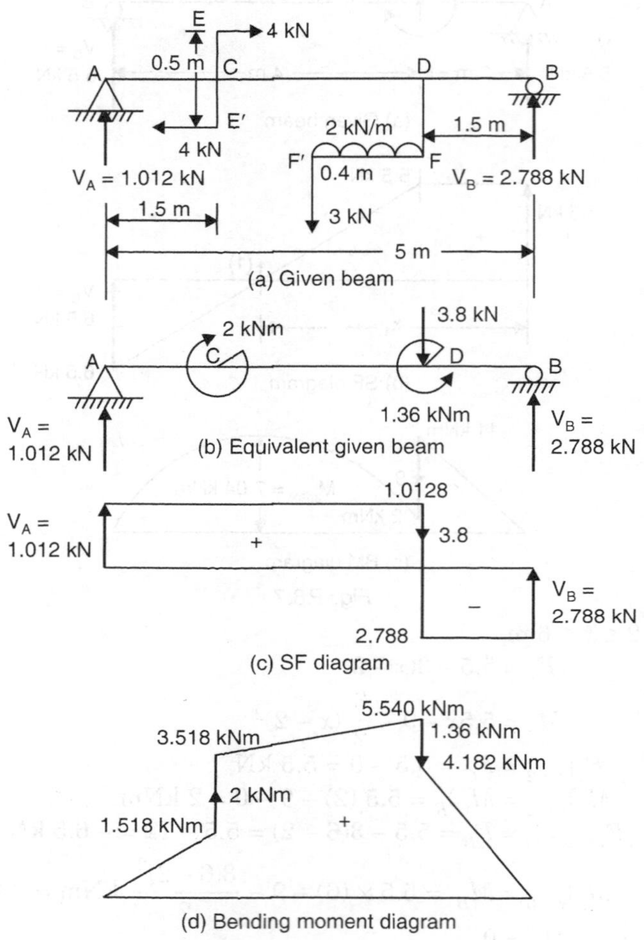

(a) Given beam

(b) Equivalent given beam

(c) SF diagram

(d) Bending moment diagram

Fig. P. 8.8

Sol. Parallel horizontal forces at *E* and *E′* of 4 kN can be replaced by a couple of magnitude 4 × 0.5 = 2 kNm (Clockwise). Similarly, forces on the portion *FF′* can be replaced by a vertical force equal to 3 + 2 × 0.4 = 3.8 kN ↓, and a pure couple equal to $\left(3 \times 0.4 + \frac{2 \times 0.4^2}{2}\right) = (1.2 + 0.16)$ = 1.36 kNm counter-clockwise, as shown in Fig. P.8.8 (*b*).

By $\Sigma M_B = 0$, we get

$$V_A \times 5 + 2 - 3.8 \times 1.5 - 1.36 = 0$$

$$\therefore \quad 5V_A + 2 - 5.70 - 1.36 = 0$$

$$\therefore \quad V_A = \frac{7.06 - 2}{5} = \frac{5.06}{5} = 1.012 \text{ kN} \uparrow$$

By $\Sigma F_V = 0$, we get $(V_A + V_B) \uparrow - 3.8 = 0$

$\therefore \quad V_B = 3.8 - V_A = 3.8 - 1.012 \text{ kN} \quad \text{or} \quad V_B = 2.788 \text{ kN} \uparrow$

$F_A = V_A = 1.012 \text{ kN} = F_D)_L$

$F_D)_R = 1.012 - 3.8 = -2.788 \text{ kN}$

$F_B)_L = -2.788 \text{ kN}$

The last vector equal to V_B closes the shear force diagram.

$M_A = 0, \quad M_C)_L = 1.012 \times 1.5 = 1.5180 \text{ kNm.}$

$M_C)_R = 1.518 + 2 = 3.518 \text{ kNm.}$

$M_D)_L = 1.012 \times 3.5 + 2 = 5.542 \text{ kNm.}$

$M_D)_R = 5.542 - 1.36 = 4.182 \text{ kNm.}$

$M_B = 1.012 \times 5 + 2 - 1.36 - 3.8 \times 1.5 = 0.$ (Check)

Shear force and bending moment diagrams are shown in Figs. P.8.8 (*c*) and (*d*).

Problem 8.9. *Draw shear force and bending moment diagrams for the simply supported beam shown in Fig. P.8.9 (a).*

Sol. Total load on the beam = W = Area of $\Delta CBB_1 = \frac{1}{2} \times 4 \times 6 = 12$ kN

Let V_A and V_B be the vertical reactions at A and B. By using the condition $\Sigma M_B = 0$, we get

$$\circlearrowleft V_A \times 8 - W \times \frac{BC}{3} = 0$$

$$\therefore \quad V_A \times 8 = \frac{12 \times 6}{3}$$

$$\therefore \quad V_A = 3 \text{ kN} \uparrow$$

And, by $\Sigma F_V = 0$, we get

$$(V_A + V_B)\uparrow - W = 0$$

$\therefore \quad V_B = W - V_A = 12 - 3 \text{ kN} \quad \text{or} \quad V_B = 9 \text{ kN} \uparrow$

$F_A = V_A = 3 \text{ kN}$

$F_C = 3 \text{ kN}$

$F_B = F_C - W = 3 - 12 = -9 \text{ kN}$

$M_A = 0,$

$M_C = V_A \times 2 = 3 \times 2 = 6 \text{ kNm}$

$$M_B = V_A \times 8 - W \times \frac{6}{3} = 3 \times 8 - 12 \times \frac{6}{3} = 0$$

At any section in the portion CB : [Fig. P. 8.9 (*b*)],

$$F_x = V_A - \Delta Cmm_1 = 3 - \frac{1}{2} \times mm_1 \times Cm \qquad \because \Delta Cmm_1 \equiv \Delta CBB_1$$

$$= 3 - \frac{1}{2} \times \left(\frac{4}{6}x\right) \times x \qquad \because \frac{mm_1}{Cm} = \frac{BB_1}{CB}$$

$$\therefore \quad mm_1 = \frac{BB_1}{CB} \times C_m = \frac{4}{6}x$$

$$\therefore \quad F_x = 3 - \frac{4x^2}{2 \times 6} \qquad \text{...(1)}$$

And, bending moment

$$M_x = V_A \times Am - \Delta Cmm_1 \times \frac{Cm}{3}$$

or $$M_x = 3(x + 2) - \frac{4x^2}{2 \times 6} \times \frac{x}{3}$$

or $$M_x = 3(x + 2) - \frac{4x^3}{6 \times 6} \quad ...(2)$$

Eqn. (1) represents a parabola and eqn. (2) represents a cubic parabola. Shear force and bending moment diagrams are shown in Figs. P.8.9 (*b*) and (*c*).

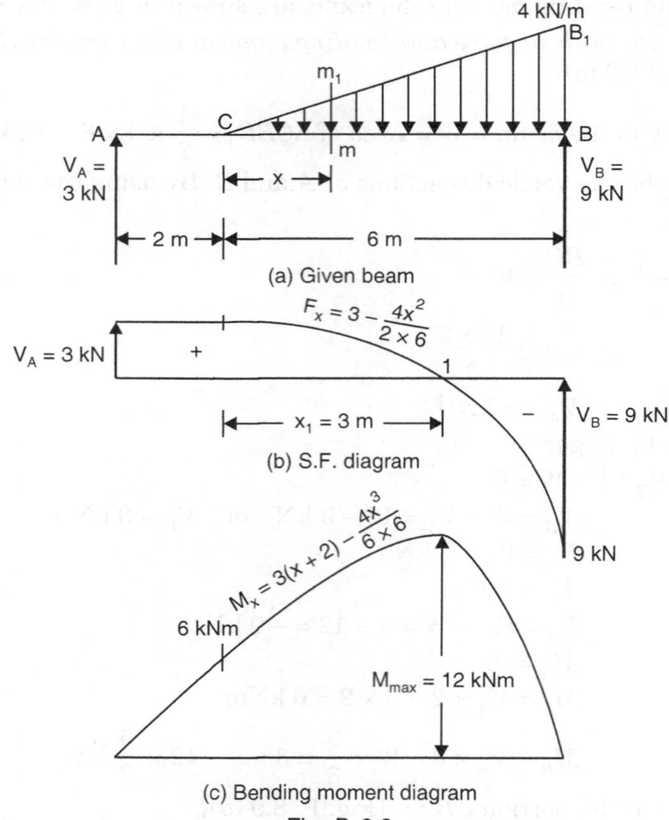

Fig. P. 8.9

At point (1) let $x = x_1$, shear force = F_{x_1} = 0 and bending moment = $M_{x_1} = M_{max.}$

Hence, from eqn. (1), we get

$$F_x]_{x = x_1} = 0 = 3 - \frac{4x_1^2}{2 \times 6}$$

$\therefore$ $$x_1^2 = \frac{3 \times 2 \times 6}{4} = 9 \text{ m}^2$$

$\therefore$ $$x_1 = 3 \text{ m from } C$$

Hence, from eqn. (2), we get

$$M_x]_{x=x_1} = M_{max} = 3(x_1 + 2) - \frac{4x_1^3}{6 \times 6}$$

or
$$M_{max} = 3 \times (3 + 2) - \frac{4 \times (3)^3}{6 \times 6} \text{ kNm} = 15 - 3 = \textbf{12 kNm.} \quad \textbf{Ans.}$$

Check :

From eqn. (2), for point B, we get

$$M_x]_{x=6} = M_B = 3(6 + 2) - \frac{4 \times (6)^3}{6 \times 6} = 24 - 24 = 0$$

For finding V_A, we used the condition $\Sigma M_B = 0$. Hence, the entire solution, as above, is correct.

Problem 8.10. *Draw shear force and bending moment diagrams for the simply supported beam shown in Fig. P. 8.10 (a).*

Sol. For the loading shown in Fig. P.8.10 (a), **it is advisable to take B as origin.**

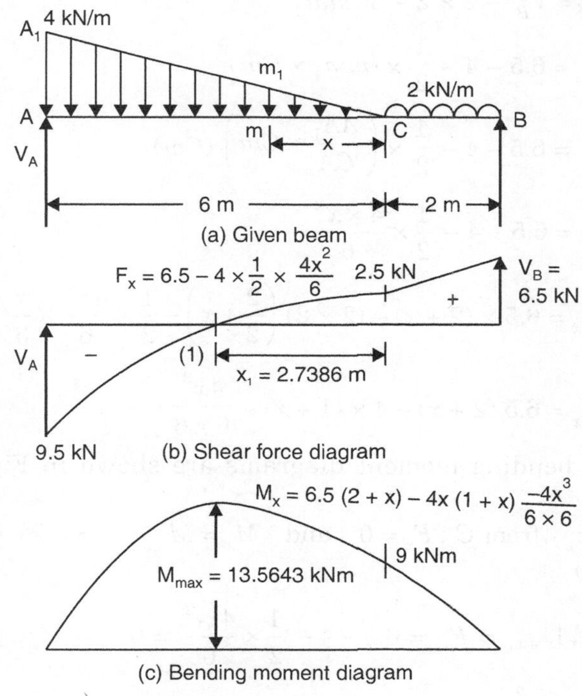

Fig. P. 8.10

By $\Sigma M_A = 0$, we get

$$V_B \times 8 - (2 \times 2)\left(\frac{2}{2} + 6\right) - (\Delta CAA_1) \times \frac{AC}{3} = 0$$

or
$$8V_B - 4 \times 7 - \left(\frac{1}{2} \times 4 \times 6\right) \times \frac{6}{3} = 0$$

or $$8V_B - 28 - 24 = 0$$

$$\therefore \quad V_B = \frac{52}{8} = 6.5 \text{ kN} \uparrow$$

By $\Sigma F_V = 0$, we get

$$(V_A + V_B) \uparrow - \left(2 \times 2 + \frac{4 \times 6}{2}\right) = 0$$

$$\therefore \quad (V_A + 6.5) - 16 = 0$$

$$\therefore \quad V_A = 16 - 6.5 = 9.5 \text{ kN} \uparrow$$

$$F_B = V_B = 6.5 \text{ kN}$$

$$F_C = 6.5 - 2 \times 2 = 2.5 \text{ kN}$$

$$M_B = 0$$

$$M_C = V_B \times 2 - \frac{2 \times 2^2}{2} = 6.5 \times 2 - 4 = 9 \text{ kNm.}$$

For the portion CA :

Taking C as origin, for any section in CA, we get

$$F_x = V_B - 2 \times 2 - \Delta Cmm_1$$

$$= 6.5 - 4 - \frac{1}{2} \times (mm_1 \times Cm)$$

$$= 6.5 - 4 - \frac{1}{2} \times \left(\frac{AA_1}{CA} \times Cm\right)(Cm)$$

or $$F_x = 6.5 - 4 - \frac{1}{2} \times \frac{4 \times x^2}{6} \quad \text{...(1)}$$

And, $$M_x = 6.5 \times (2 + x) - (2 \times 2)\left(\frac{2}{2} + x\right) - \frac{1}{2} \times \frac{4x^2}{6} \times \frac{x}{3}$$

or, $$M_x = 6.5\,(2 + x) - 4 \times (1 + x) - \frac{4x^3}{6 \times 6} \quad \text{...(2)}$$

Shear force and bending moment diagrams are shown in Figs. P.8.10 (*b*) and (*c*), respectively.

At point (1), $x = x_1$ from C ; $F_x = 0$ and $M_x = M_{max.}$

$\therefore$ From eqn. (1)

$$F_x]_{x = x_1} = F_{x_1} = 6.5 - 4 - \frac{1}{2} \times \frac{4x_1^2}{6} = 0$$

$$\therefore \quad 2.5 - \frac{4x_1^2}{2 \times 6} = 0$$

$$\therefore \quad x_1^2 = 2.5 \times 3 = 7.5$$

$\therefore$ $x_1 = 2.7386$ m from C or 4.7386 m from B. **Ans.**

∴ From eqn. (2), we get

$$M_x]_{x=x_1} = M_{max} = 6.5(x_1 + 2) - (2 \times 2)(1 + x_1) - \frac{4x_1^3}{6 \times 6}$$

or
$$M_{max} = 6.5 \times 4.7386 - 4 \times 3.7386 - \frac{4}{6 \times 6} \times (2.7386)^3 \text{ kNm}$$

$$= \mathbf{13.5643\ kNm.}\quad \textbf{Ans.}$$

Problem 8.11. *A simply supported beam of span 8 m carries a point load of 3 kN at a distance of 2 m from left support and a uniformly varying load of rate zero at 2 m and 4 kN/m at the right support. Draw shear force and bending moment diagrams for the beam and also find the magnitude of maximum bending moment.*

Sol. Fig. P. 8.11 (*a*) shows the given beam and the loading system.

By $\Sigma M_B = 0$, we get

$$V_A \times 8 - 3 \times 6 - \frac{4 \times 6}{2} \times \frac{6}{3} = 0$$

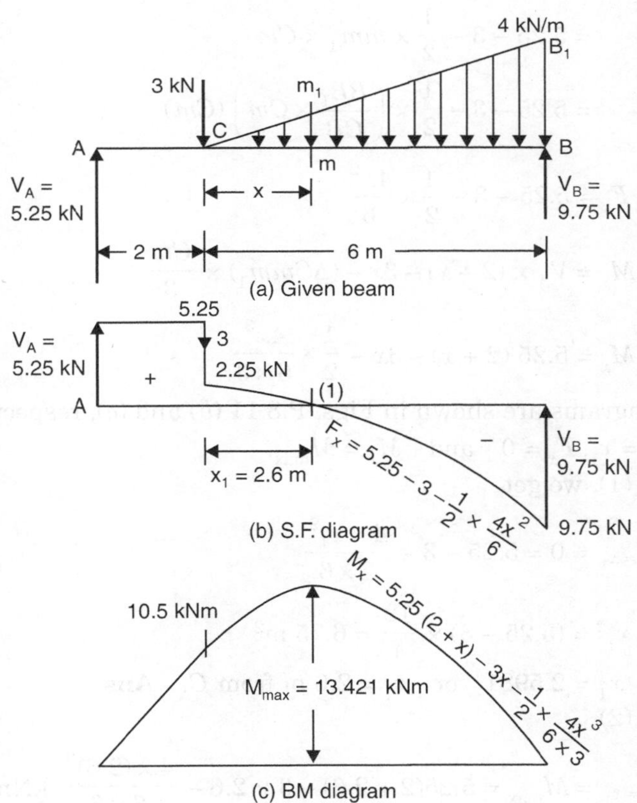

Fig. P. 8.11

∴ $8V_A - 18 - 24 = 0$

∴ $V_A = \frac{42}{8} = 5.25 \text{ kN} \uparrow$

By $\Sigma F_V = 0$, we get

$$(V_A + V_B) \uparrow - \left(3 + \frac{4 \times 6}{2}\right) = 0$$

$\therefore$ $$(5.25 + V_B) - 15 = 0$$

$\therefore$ $$V_B = 9.75 \text{ kN} \uparrow$$

Shear force on the left of $C = F_C)_L = V_A = 5.25$ kN

Shear force on the right of $C = F_C)_R = 5.25 - 3 = 2.25$ kN

$$F_B = F_C)_R - \Delta CBB_1 = 2.25 - \frac{1}{2} \times 4 \times 6 = -9.75 \text{ kN}$$

$$M_A = 0, \quad M_B = 0$$

and $$M_C = V_A \times 2 = 5.25 \times 2 = 10.5 \text{ kNm}$$

For portion CB :

At any section x form C, we write :

$$F_x = V_A - 3 - \Delta Cmm_1$$

$$= 5.25 - 3 - \frac{1}{2} \times mm_1 \times Cm$$

$$= 5.25 - 3 - \frac{1}{2} \times \left(\frac{BB_1}{CB} \times Cm\right)(Cm)$$

or $$F_x = 5.25 - 3 - \frac{1}{2} \times \frac{4x^2}{6} \qquad \text{...(1)}$$

And, $$M_x = V_A \times (2 + x) - 3x - (\Delta Cmm_1) \times \frac{Cm}{3}$$

$\therefore$ $$M_x = 5.25\,(2 + x) - 3x - \frac{1}{2} \times \frac{4x^3}{6 \times 3} \qquad \text{...(2)}$$

SF and *BM* diagrams are shown in Figs. P.8.11 (*b*) and (*c*), respectively.

At point (1), $x = x_1$, $F_x = 0$ and $M_x = M_{max}$.

$\therefore$ From eqn. (1), we get

$$F_x]_{x = x_1} = 0 = 5.25 - 3 - \frac{4 \times x_1^2}{2 \times 6}$$

$\therefore$ $$x_1^2 = (5.25 - 3) \times \frac{12}{4} = 6.75 \text{ m}^2$$

$\therefore$ $x_1 = 2.5981$ or $x_1 = 2.6$ m from C. **Ans.**

$\therefore$ From eqn. (2),

$$M_x]_{x = x_1} = M_{max} = 5.25(2 + 2.6) - 3 \times 2.6 - \frac{4 \times (2.6)^3}{6 \times 6} \text{ kNm}$$

or $$M_{max} = 5.25 \times 4.6 - 3 \times 2.6 - \frac{(2.6)^3}{6} \text{ kNm}$$

or $M_{max} =$ **13.421 kNm. Ans.**

Problem 8.12. *Draw shear force and bending moment diagrams for the beam shown in Fig. P. 8.12 (a).*

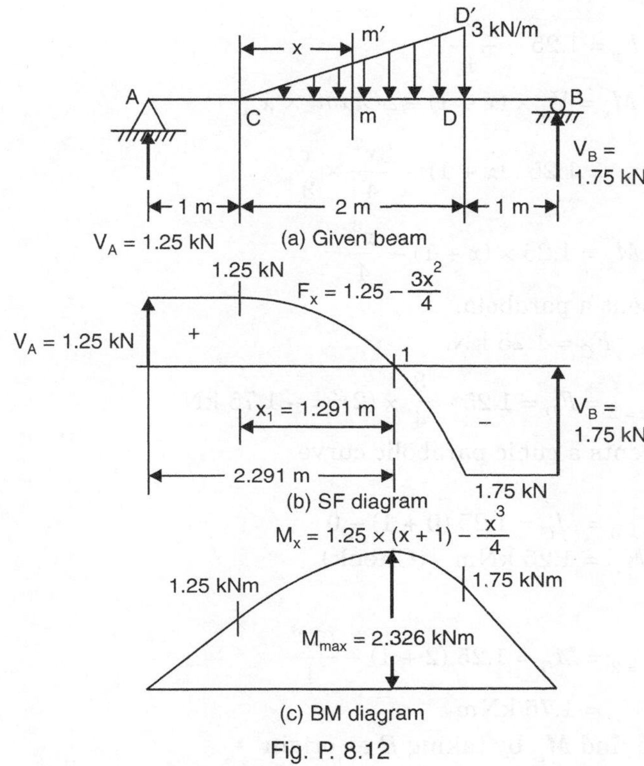

Fig. P. 8.12

Sol. Total external load = $W = \Delta CDD'$

$$= \frac{1}{2} \times 3 \times 2 = 3 \text{ kN}.$$

Its CG is at a distance of $\frac{1}{3} \times (2)$ m from D.

$\therefore$ By $\Sigma M_B = 0$, we get

$$V_A \times 4 - W \times \left(\frac{2}{3} + 1\right) = 0$$

$$\therefore \quad V_A = \frac{W}{4}\left(\frac{5}{3}\right) = \frac{5W}{12} = \frac{5 \times 3}{4 \times 3} = 1.25 \text{ kN} \uparrow$$

By $\Sigma F_V = 0$, we get

$$(V_A + V_B) \uparrow - W = 0$$

$$\therefore \quad V_B = W - V_A = 3 - 1.25 = 1.75 \text{ kN} \uparrow$$

$$F_A = V_A = 1.25 \text{ kN}$$

$$F_C = 1.25 \text{ kN}$$

$$M_A = 0,$$

$$M_C = V_A \times 1 = 1 \times 1.25 \text{ kNm}$$

In portion CD. $0 \le x \le 3$ m, with C as origin, we get

$$F_x = V_A - \Delta Cmm' = V_A - \frac{1}{2} \times mm' \times Cm$$

$$= 1.25 - \frac{1}{2} \times \left(\frac{DD'}{CD} \times Cm \right) \times Cm = 1.25 - \frac{1}{2} \times \frac{3 \times x}{2} \times x$$

or $$F_x = 1.25 - \frac{3x^2}{4} \quad ...(1)$$

And, $$M_x = V_A \times (x + 1) - \Delta Cmm' \times \bar{x}_m$$

$$= 1.25 \,.\, (x + 1) - \frac{3x^2}{4} \times \frac{x}{3}$$

$\therefore$ $$M_x = 1.25 \times (x + 1) - \frac{x^3}{4} \quad ...(2)$$

Eqn. (1) represent a parabola.

At C, $x = 0$, $\therefore$ $F_C = 1.25$ kN.

At $x = 2m$, $F_x]_{x=2} = F_D = 1.25 - \frac{3}{4} \times (2)^2 = -1.75$ kN

Eqn. (2) represents a cubic parabolic curve.

At C, $x = 0$,

$\therefore$ $M_x)_{x=0} = M_C = 1.25\ (0 + 1) - 0$

$\therefore$ $M_C = 1.25$ kNm **(Check)**

At D, $x = 2$m,

$\therefore$ $$M_x)_{x=2} = M_D = 1.25\ (2 + 1) - \frac{(2)^3}{4}$$

$\therefore$ $M_D = 1.75$ kNm

Check. We can find M_D by taking B as origin

$\therefore$ We get $M_D = V_B \times 1 = 1.75 \times 1 = 1.75$ kNm.

Shear force diagram is shown in Fig. P. 8.12 (*b*) and bending moment diagram is shown in Fig. P. 8.12.

From Fig. P.8.12 (*b*), we find that $F_x = 0$ at point (1).

Also, at point (1), $x = x_1$ from C, $F_{x_1} = 0$ and $M_{x_1} = M_{max}$ (Fig. P. 8.12(*c*)).

$\therefore$ From eqn. (1), we can write

$$F_x]_{x=x_1} = 1.25 - \frac{3}{4} \times x_1^2 = 0$$

$\therefore$ $$x_1^2 = \frac{1.25 \times 4}{3} = \frac{5}{3}$$

$\therefore$ $x_1 = \sqrt{5/3} = 1.291$ m from point C. **Ans.**

$\therefore$ From eqn. (2), we get

$$M_x]_{x=x_1} = M_{max} = 1.25\ (1 + x_1) - \frac{x_1^3}{4}$$

$$= 1.25 \times (1 + 1.291) - \frac{(1.291)^3}{4}$$

$\therefore$ M_{max} = **2.326 kNm. Ans.**

Problem 8.13. *Draw shear force and bending moment diagrams of the beam shown in Fig. P.8.13 (a).*

Sol. This problem can best be solved by using the differential relations between the rate of loading q, shear force and bending moment.

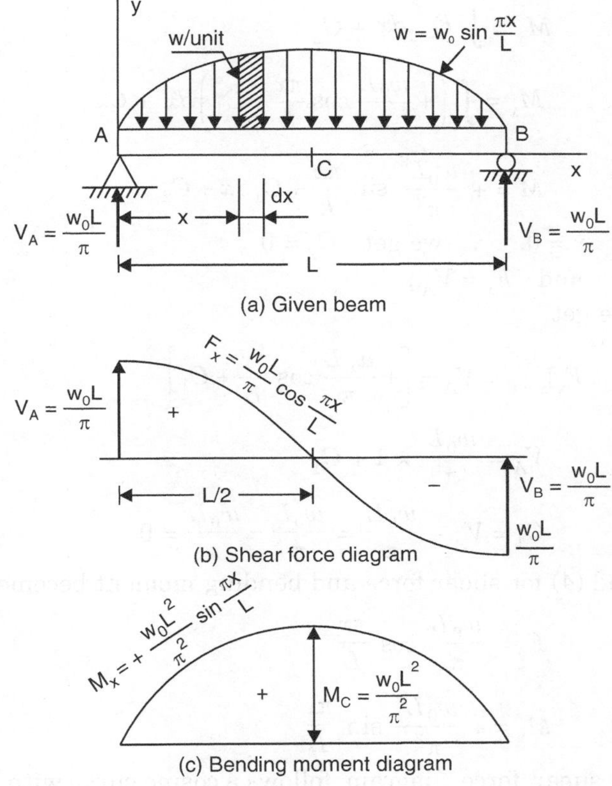

Fig. P. 8.13

Let us first find the total load W on the beam. Let us consider the shaded element at a distance x from the origin at end A and of width dx. If w denotes load per unit length, then the load corresponding the shaded element is $w.\ dx$, and the total load on the beam is

$$W = \int_0^L w\,.\,dx = \int_0^L w_0 \sin\frac{\pi x}{L}\,.\,dx$$

$$W = \frac{2w_0L}{\pi} \qquad \text{...(1)}$$

Since the load is symmetrical about mid-span, therefore reactions

$$V_A = V_B = \frac{W}{2} = \frac{w_0L}{\pi} \qquad \text{...(2)}$$

The shear force at the section is, therefore, given by

$$\therefore \quad \frac{dF_x}{dx} = -w \qquad\qquad \therefore \quad \int dF_x = \int -w\,dx + C_1$$

$$\therefore \quad F_x = \int dF_x = \int -w\,dx + C_1$$

Here w is –ve because it is acting downward

or $$F_x = \int - w_0 \sin \frac{\pi x}{L} dx + C_1$$

$$F_x = \frac{w_0 L}{\pi} \cos \frac{\pi x}{L} + C_1 \quad ...(3)$$

Similarly, $$M_x = \int F_x \,.\, dx + C_2$$

$\therefore$ $$M_x = \int \left(+ \frac{w_0 L}{\pi} \cos \frac{\pi x}{L} + C_1 \right) dx + C_2$$

$\therefore$ $$M = + \frac{w_0 L^2}{\pi^2} \sin \frac{\pi x}{L} + C_1 \,.\, x + C_2 \quad ...(4)$$

$\because$ $BM = 0$ at $x = 0$, $\therefore$ we get $C_2 = 0$

Also, at A, $x = 0$ and $F_x = V_A$,

$\therefore$ From (3), we get

$$F_x]_{x=0} = V_A = \left[+ \frac{w_0 L}{\pi} \cos \frac{\pi x}{L} + C_1 \right]$$

$\therefore$ $$V_A = \frac{w_0 L}{\pi} \times 1 + C_1$$

$\therefore$ $$C_1 = V_A - \frac{w_0 L}{\pi} = \frac{w_0 L}{\pi} - \frac{w_0 L}{\pi} = 0$$

$\therefore$ Eqns. (3) and (4) for shear force and bending moment become :

$$F_x = \frac{w_0 L}{\pi} \cos \frac{\pi x}{L} \quad ...(5)$$

And, $$M_x = + \frac{w_0 L^2}{\pi^2} \sin \frac{\pi x}{L} \quad ...(6)$$

It is clear that shear force diagram follows a cosine curve with ordinate at A equal to $+ \frac{w_0 L}{\pi}$ and that at B equal to $+ \frac{w_0 L}{\pi} \cos \pi = - \frac{w_0 L}{\pi}$.

Also, shear force at $x = \frac{L}{2}$ is zero. Fig. P.8.13 (*b*) shows shear force diagram.

Bending moment diagram is a sine curve.

At $x = 0$, $M_x]_{x=0} = 0$; also, $M_x]_{x=L} = M_B = 0$.

At mid span, $M_x]_{x=l/2} = M_C = \frac{w_0 L^2}{\pi^2} \times \sin \frac{\pi}{2} = + \frac{w_0 L^2}{\pi^2}$

$\therefore$ Maximum bending moment $= M_{max} = + \frac{w_0 L^2}{\pi^2} \quad ...(7)$

Shear force and bending moment diagrams are shown in Fig. P.8.13.

Problem 8.14. *Draw SF and BM diagrams for the beam shown in Fig. P.8.14. (a).*

Sol. By $\Sigma M_B = 0$, we get

$$V_A \times 8 - 8 \times 7 - \left(\frac{6 \times 3}{2} \right) \left(\frac{3}{3} + 4 \right) + 13 = 0$$

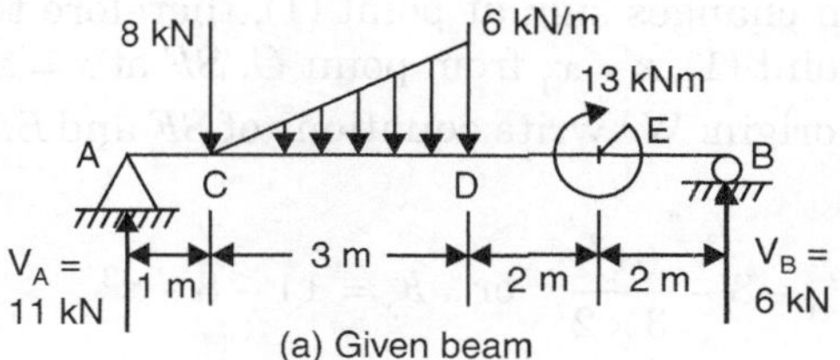

(a) Given beam

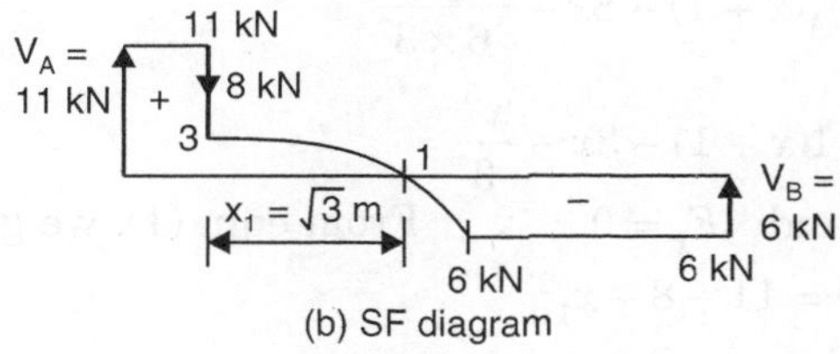

(b) SF diagram

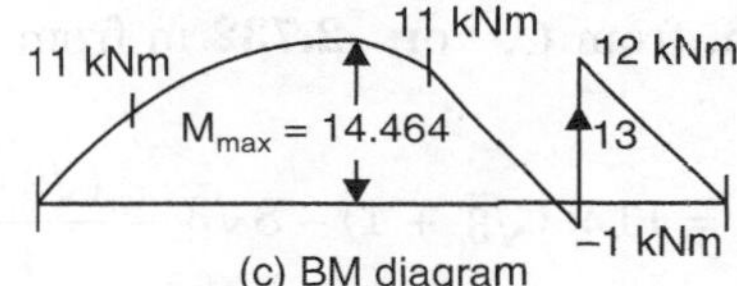

(c) BM diagram

Fig. P. 8.14

$$\therefore \qquad V_A = \frac{1}{8}[56 + 45 - 13] = \frac{88}{8} \text{ kN} \uparrow = 11 \text{ kN} \uparrow$$

By $\Sigma F_v = 0$, we get $(V_A + V_B) \uparrow - \left(8 + \frac{6 \times 3}{2}\right) = 0$

$$\therefore \qquad V_B = 17 - V_A = 17 - 11 = 6 \text{ kN} \uparrow$$

$$F_A = V_A = 11 \text{ kN}$$

$$F_C)_L = 11 \text{ kN}$$

$$F_C)_R = 11 - 8 = 3\text{kN}$$

$$F_D = 3 - \frac{6 \times 3}{2} = -6 \text{ kN}$$

$$F_E = -6 \text{ kN}$$

$$F_B = -6 \text{ kN}$$

The last vector V_B = + 6 kN will close the *SF* diagram.

$M_A = 0$, $M_B = 0$ because A and B are simply supported ends.

$$M_C = V_A \times 1 = 11 \text{ kNm}$$

$$M_E = V_B \times 2 = 6 \times 2 = 12 \text{ kNm}$$

$$M_D = V_A \times 4 - 8 \times 3 - \frac{6 \times 3}{2} \times \frac{3}{3} = 44 - 24 - 9 = 11 \text{ kNm}$$

$$M_E)_L = 11 \times 6 - 8 \times 5 - \frac{6 \times 3}{2} \times \left(\frac{3}{3} + 2\right)$$

$$= 66 - 40 - 27 = 66 - 67 = -1 \text{ kNm}$$

$$M_E)_R = M_E)_L + 13 = -1 + 13 = 12 \text{ kNm}$$

Since the *SF* diagram changes sign at point (1), therefore the bending moment will be maximum at point (1). At point (1), $x = x_1$ from point *C*, *SF* at $x = x_1$ is zero.

Thus, by taking *C* as origin. We write equations of *SF* and *BM* for position *CD*. Thus, we get

$$F_x = V_A - 8 - \frac{6x^2}{3 \times 2} \quad \text{or} \quad F_x = 11 - 8 - x^2 \qquad ...(1)$$

And,
$$M_x = V_A(x + 1) - 8x - \frac{6 \times x^3}{6 \times 3}$$

or
$$M_x = 11(x + 1) - 8x - \frac{x^3}{3} \qquad ...(2)$$

At point (1), $x = x_1$ and $F_x = 0$, $\therefore$ From eqn. (1), we get

$$F_x]_{x = x_1} = 0 = 11 - 8 - x_1^2$$

$$\therefore \quad x_1^2 = 3$$

$\therefore$ $x_1 = \sqrt{3}$ **m.** from *C*, or **2.732** m from *A*. **Ans.**

$\therefore$ From eqn. (2), we get

$$M_x]_{x = x_1} = M_{max} = 11 \times (\sqrt{3} + 1) - 8\sqrt{3} - \frac{(\sqrt{3})^3}{3}$$

$$= \sqrt{3}(11 - 8 - 1) + 11 = \sqrt{3} \times 2 + 11 \text{ kNm} = \mathbf{14.464 \text{ kNm.}} \quad \textbf{Ans.}$$

Figs. P.8.14 (*b*) and (*c*) represent the shear force and bending moment diagrams, respectively.

Problem 8.15. *Draw shear force and bending moment diagrams for the simply supported beam shown in Fig. P.8.15 (a).*

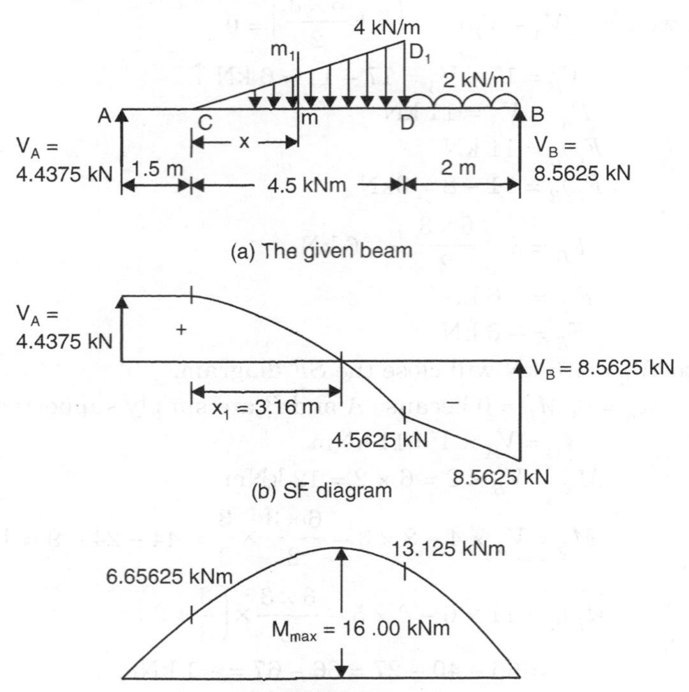

(a) The given beam

(b) SF diagram

(c) BM diagram

Fig. P. 8.15

Sol. By $\Sigma M_B = 0$, we get

$$\curvearrowleft V_A \times 8 - \frac{4 \times 4.5}{2} \times \left(\frac{4.5}{3} + 2\right) - (2 \times 2)\left(\frac{2}{2}\right) = 0$$

or $$8V_A - 9 \times 3.5 - 4 = 0$$

$$\therefore \quad V_A = \frac{1}{8} \times 35.5 = 4.4375 \text{ kN} \uparrow$$

By $\Sigma F_V = 0$, we have $(V_A + V_B) \uparrow - \dfrac{4 \times 4.5}{2} - 2 \times 2 = 0$

$$\therefore \quad V_B = 13 - V_A = 8.5625 \text{ kN} \uparrow$$

$$F_A = V_A = 4.4375 \text{ kN}$$

$$F_C = 4.4375 \text{ kN}$$

$$F_D = 4.4375 - \Delta CDD_1 = 4.4375 - \frac{4 \times 4.5}{2} = -4.5625 \text{ kN}$$

$$F_B = -4.5625 - 2 \times 2 \text{ kN} = -8.5625 \text{ kN}$$

$$M_A = 0, \quad M_B = 0,$$

$$M_C = V_A \times 1.5 = 4.4375 \times 1.5 \text{ kNm} = 6.65625 \text{ kNm}$$

$$M_D = V_A \times 6 - \frac{4 \times 4.5}{2} \times \frac{4.5}{3}$$

$$= 4.4375 \times 6 - \frac{4 \times 4.5}{2} \times \frac{4.5}{3} = 13.125 \text{ kNm}$$

Check. M_D can also determine as follows :

$$M_D = V_D \times 2 - \frac{2 \times 2^2}{2} = 13.125 \text{ kNm}$$

From shear force diagram, Fig. P.8.15(*b*), it is clear that point (1), where *F* shear force is zero, lies in portion *CD*. Hence, we consider the portion *CD* of the beam.

At any section *x* from *C*, we write :

$$F_x = V_A - \Delta Cmm_1 = 4.4375 - \frac{1}{2}\left(\frac{DD_1}{CD} \times Cm\right) . Cm$$

or $$F_x = 4.4375 - \frac{1}{2} \times \frac{4 \times x^2}{4.5} \qquad \text{...(1)}$$

And, $$M_x = V_A \times (x + 1.5) - \Delta Cmm_1 \times \frac{Cm}{3}$$

$$= 4.4375\,(x + 1.5) - \frac{4x^2}{2 \times 4.5} \times \frac{x}{3}$$

$$\therefore \quad M_x = 4.4375\,(x + 1.5) - \frac{4x^3}{6 \times 4.5} \qquad \text{...(2)}$$

At point (1), $\quad x = x_1$, $F_{x_1} = 0$ and $M_{x_1} = M_{max}$

$\therefore$ From eqn. (1), $F_x]_{x = x_1} = F_{x_1} = 4.4375 - \dfrac{4{x_1}^2}{2 \times 4.5} = 0$

$$\therefore \qquad x_1^2 = \frac{4.4375 \times (2 \times 4.5)}{4} = 9.984375 \text{ m}^2$$

$$\therefore \qquad x_1 = 3.1598 \text{ m} \quad \text{or} \quad 3.16 \text{ m from } C. \quad \textbf{Ans.}$$

And, from eqn. (2), we get

$$M_x]_{x=x_1} = M_{max} = 4.4375\,(x_1 + 1.5) - \frac{4x_1^3}{6 \times 4.5} \text{ kNm}$$

or $$M_{max} = \textbf{16.00 kNm. Ans.}$$

Shear force and bending moment diagrams are shown in Figs. P.8.15 (*b*) and (*c*), respectively.

Problem 8.16. *Draw SF and BM diagrams for the beam shown in Fig. P.8.16.*

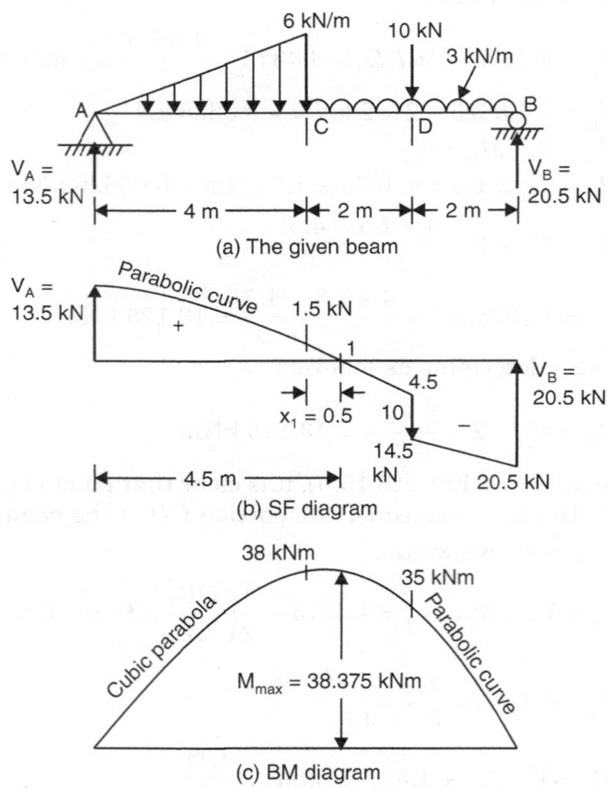

Fig. P. 8.16

Sol. By $\Sigma M_B = 0$, we get

$$\circlearrowleft V_A \times 8 - \frac{6 \times 4}{2} \times \left(\frac{4}{3} + 4\right) - 10 \times 2 - \frac{3 \times 4^2}{2} = 0$$

$$\therefore \qquad V_A = \frac{1}{8}\left[\frac{12 \times 16}{3} + 20 + 24\right] \text{kN} = \frac{108}{8} = 13.5 \text{ kN} \uparrow$$

By $\Sigma F_V = 0$, we get $(V_A + V_B) \uparrow - \dfrac{6 \times 4}{2} - 10 - 3 \times 4 = 0$

$$\therefore \qquad V_B = 34 - V_A = 34 - 13.5 \text{ kN} \quad \text{or} \quad V_B = 20.5 \text{ kN} \uparrow$$

$$F_A = V_A = +\ 13.5 \text{ kN}$$

$$F_C = 13.5 - \frac{1}{2} \times 6 \times 4 = 1.5 \text{ kN}$$

$$F_D)_L = 1.5 - 3 \times 2 = -\ 4.5 \text{ kN}$$

$$F_D)_R = -\ 4.5 - 10 = -\ 14.5 \text{ kN}$$

$$F_B = F_D)_R - 3 \times 2 = -\ 20.5 \text{ kN} \quad \textbf{(Check)}$$

The ordinates have been marked in *SF* Diagram.

The last vector $V_B = 20.5$ kN ↑ closes the shear force diagram as shown in Fig. P.8.16 (*b*).

$M_A = 0$, $M_B = 0$, by observation.

$$M_C = V_A \times 4 - \frac{6 \times 4^2}{6} = 13.5 \times 4 - 16 = 54 - 16 = 38 \text{ kNm.}$$

$$M_D = 13.5 \times 6 - \frac{6 \times 4}{2} \times \left(\frac{4}{3} + 2\right) - \frac{3 \times 2^2}{2}$$

$$= 81 - \frac{12 \times 10}{3} - 6 = 81 - 46 = 35 \text{ kNm.}$$

The *SF* diagram between *A* and *C* is parabolic, that between points *C* and *B* is straight line.

The *BM* diagram is cubic parabola between *A* and *C*. At point (1) shear force diagram changes sign.

∴ At point (1) bending moment is maximum.

In order to find position of point (1) equate the shear force at any section in portion *CD* equal to zero.

Taking origin at *C*, we write

$$F_x = V_A - \frac{6 \times 4}{2} - 3x$$

or $$F_x = 13.5 - 12 - 3x \qquad \text{...(1)}$$

and $$M_x = V_A\,(x + 4) - \frac{6 \times 4}{2} \times \left(\frac{4}{3} + x\right) - \frac{3x^2}{2} \qquad \text{...(2)}$$

At point (1), $x = x_1$, $F_x]_{x = x_1} = 0$ and $M_x]_{x = x_1} = M_{max}$

∴ From (1) we get, $13.5 - 12 - 3x_1 = 0$

∴ $$x_1 = \frac{1}{3} \times (13.5 - 12) = \frac{1.5}{3} = \mathbf{0.5\ m.} \quad \textbf{Ans.}$$

∴ From (2),

$$M_x]_{x = x_1} \quad M_{max} = V_A(x_1 + 4) - \frac{6 \times 4}{2} \times \left(\frac{4}{3} + x_1\right) - \frac{3{x_1}^2}{2}$$

or $$M_{max} = 13.5\,(0.5 + 4) - \frac{6 \times 4}{2} \times \left(\frac{4}{3} + 0.5\right) - \frac{3}{2} \times (0.5)^2 = \mathbf{38.375\ kNm.} \quad \textbf{Ans.}$$

SF and *BM* diagrams are shown in Fig. P.8.16.

Problem 8.17. *Draw shear force and bending diagrams for the cantilever shown in Fig. P.8.17. (a).*

Sol. Take A (the free end) as the origin. See Fig. P. 8.17.

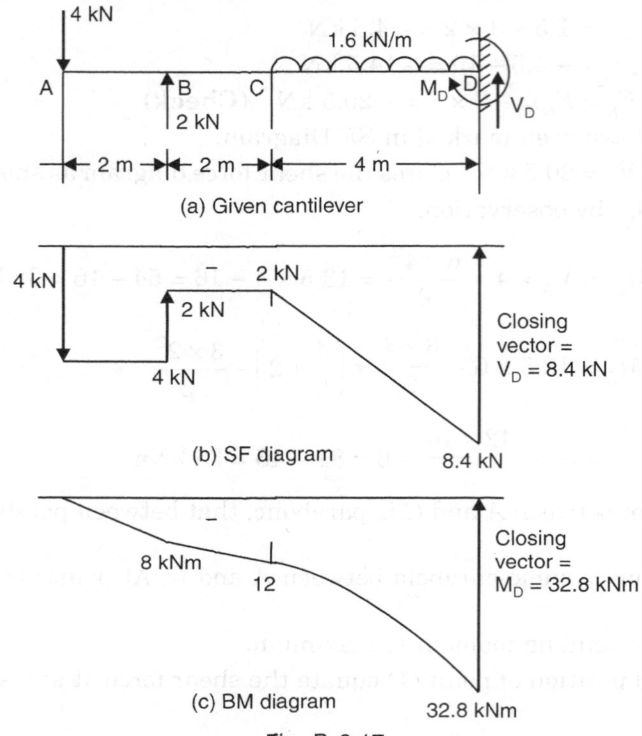

(a) Given cantilever

(b) SF diagram

(c) BM diagram

Fig. P. 8.17

$\therefore$ We get

$$F_A = -4 \text{ kN}$$
$$F_B)_L = -4 \text{ kN}$$
$$F_B)_R = -4 + 2 = -2 \text{ kN}$$
$$F_C = F_B)_R = -2 \text{ kN}$$
$$F_D = F_C - 1.6 \times 4 = -2 - 6.4 \text{ kN} = -8.4 \text{ kN}$$

$\therefore$
$$V_D = F_D)_R = 8.4 \text{ kN} \uparrow$$
$$M_A = 0, M_B = -4 \times 2 = -8 \text{ kNm}$$
$$M_C = -4 \times 4 + 2 \times 2 = -12 \text{ kNm}$$
$$M_D = -4 \times 8 + 2 \times 6 - \frac{1.6 \times 4^2}{2} \text{ kNm}$$
$$= -32 + 12 - 12.8 \text{ kNm} = -32.8 \text{ kNm}$$

$\therefore$ Support moment = M_D = 32.8 kNm. See Fig. P.8.17.

Problem 8.18. *Draw shear force and bending moment diagrams for the cantilever shown in Fig. P.8.18 (a).*

Sol.
$$F_A = -4 \text{ kN}$$
$$F_B)_L = -4 - 3 \times 1 = -7 \text{ kN}$$
$$F_B)_R = -7 + 10 = 3 \text{ kN}$$

$$F_C)_L = 3 \text{ kN}$$
$$F_C)_R = 3 - 6 = -3 \text{ kN}$$
$$F_D = -3 \text{ kN.}$$

(a) Given cantilever

(b) SF diagram

(c) BM diagram

Fig. P. 8.18

$\therefore$ Reaction at $D = V_D = 3 \text{ kN} \uparrow$ = closing vector of SF diagram at D.

$$M_A = 0$$

$$M_B = -4 \times 1 - \frac{3 \times 1^2}{2} = -5.5 \text{ kNm}$$

$$M_C = -4 \times 2 - (3 \times 1)\left(\frac{1}{2} + 1\right) + 10 \times 1 = -2.5 \text{ kNm}$$

$$M_D = -4 \times 3.8 - (3 \times 1)\left(\frac{1}{2} + 2.8\right) + 10 \times 2.8 - 6 \times 1.8 \text{ kNm} = -7.9 \text{ kNm}$$

$\therefore$ Moment reaction of support $D = M_D = 7.9$ kNm clockwise.

Shear force and bending moment diagrams are shown in Figs. P.8.18 (*b*) and (*c*), respectively.

Problem 8.19. *Draw shear force and bending moment diagrams for the cantilever shown in Fig. P.8.19.*

Sol. With origin at A :

$$F_A = -4 \text{ kN}$$
$$F_B = -4 \text{ kN}$$
$$F_C)_R = -4 \text{ kN}$$
$$F_C)_L = -4 + 2 = -2 \text{ kN}$$
$$F_D = -2 \text{ kN} \quad \therefore \quad V_D = 2\text{kN} \uparrow$$
$$M_A = 0$$

$M_B)_R = -4 \times 1 = -4$ kNm

$M_B)_L = M_B)_R + 8$ kNm $= -4 + 8 = +4$ kNm

$M_C = -4 \times 2 + 8 = 0$

$M_D = -4 \times 4 + 8 + 2 \times 2$ kNm $= -16 + 8 + 4$ kNm $= -4$ kNm

From the shear force diagram [Fig. P.8.19 (*b*)] and the bending moment [Fig. P.8.19 (*c*)], it is clear that support reaction V_D at support $D = 2$ kN ↑, and, moment reaction, $M_D = -4$ kNm, as shown in Fig. P.8.19 (*a*). From the *B.M.* diagram it is clear that the point of contra-flexure occurs at *C*, where *B.M.* = 0 and the *B.M.* diagram changes the sign.

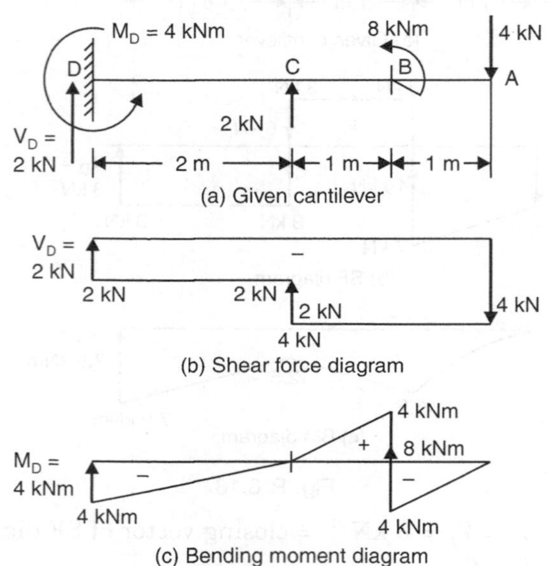

(a) Given cantilever

(b) Shear force diagram

(c) Bending moment diagram

Fig. P. 8.19

Problem 8.20. *Draw shear force and bending moment diagrams for the cantilever shown in Fig. P.8.20 (a).*

Sol. For problems on cantilever, when the left end is fixed, it is advisable to take origin at the free end and not the fixed end. With this procedure the fixed-end reaction components (vertical reaction, and the moment) need not be calculated.

Here with *A* as origin, $F_A = -3$ kN ;

$F_C = -3 - 2 \times 4 = -11$ kN

$F_D)_R = -11$ kN

$F_D)_L = -11 - 4 = -15$ kN, also $F_B)_R = -15$ kN

And, since when the last vector is added it closes the shear force diagram.

Therefore, the last vector = $V_B = +15$ kN ↑ closes the *SF* diagram, as shown in Fig. P.8.20 (*b*).

$$M_A = 0 \; ; \quad M_C)_R = -3 \times 4 - \frac{2 \times 4^2}{2} = -28 \text{ kN}$$

Bending moment diagram in portion *AC* is parabolic.

$M_C)_L = M_C)_R + 6$ kNm $= -28 + 6 = -22$ kNm

$$M_D = -3 \times 6 - (2 \times 4)\left(\frac{4}{2} + 2\right) + 6 = -18 - 32 + 6 \text{ kNm} = -44 \text{ kNm}$$

$$M_B = -3 \times 8 - (2 \times 4)\left(\frac{4}{2} + 4\right) + 6 - 4 \times 2 \text{ kNm}$$

$$= -24 - 48 + 6 - 8 \text{ kN} = -74 \text{ kNm } (CW)$$

$\therefore$ Moment reaction at B = 74 kN (*CCW*).

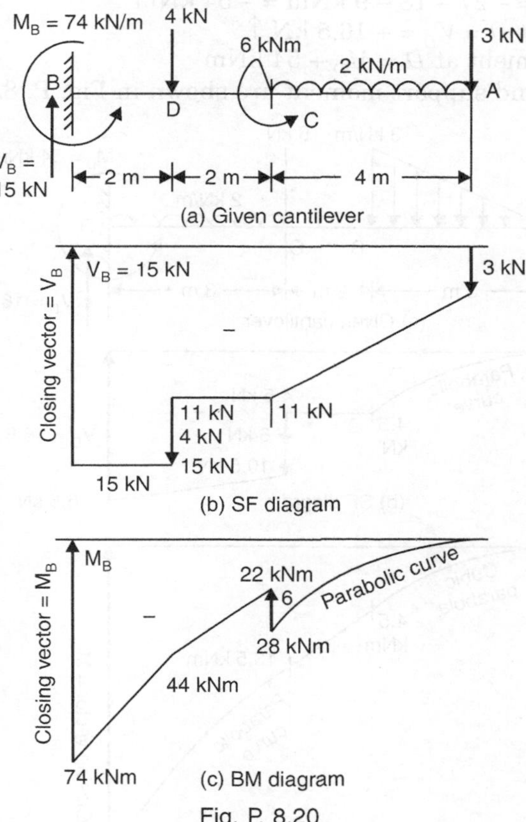

Fig. P. 8.20

The bending moment diagram is shown in Fig. P.8.20 (*c*). The fixed-end reaction components are V_B and M_B.

Problem 8.21. *Draw shear force and bending moment diagrams for the cantilever shown in Fig.P.8.21 (a).*

Sol. With the free end as origin, we get

$$F_A = 0$$

$$F_B = -\frac{3 \times 3}{2} = -4.5 \text{ kN}$$

$$F_C)_L = -4.5 \text{ kN}$$

$$F_C)_R = -4.5 - 6 = -10.5 \text{ kN}$$

$$F_D)_L = -10.5 - 2 \times 3 = -16.5 \text{ kN. Hence } V_D = 16.5 \text{ kN} \uparrow$$

Also, $M_A = 0,$

$$M_B = -\frac{3 \times 3}{2} \times \frac{3}{3} = -4.5 \text{ kNm}$$

$$M_C = -\frac{3\times 3}{2}\times\left(\frac{3}{3}+2\right)\text{ kNm} = -13.5\text{ kNm}$$

$$M_D = -\frac{3\times 3}{2}\times\left(\frac{3}{3}+5\right) - 6\times 3 - \frac{2\times 3^2}{2}\text{ kNm}$$

$$= -27 - 18 - 9\text{ kNm} = -54\text{ kNm}$$

Support reaction at $D = V_D = +16.5$ kN ↑

And supported moment at $D = M_D + 54$ kNm

Support reaction and support moment are shown in Fig. P. 8.21 (*a*).

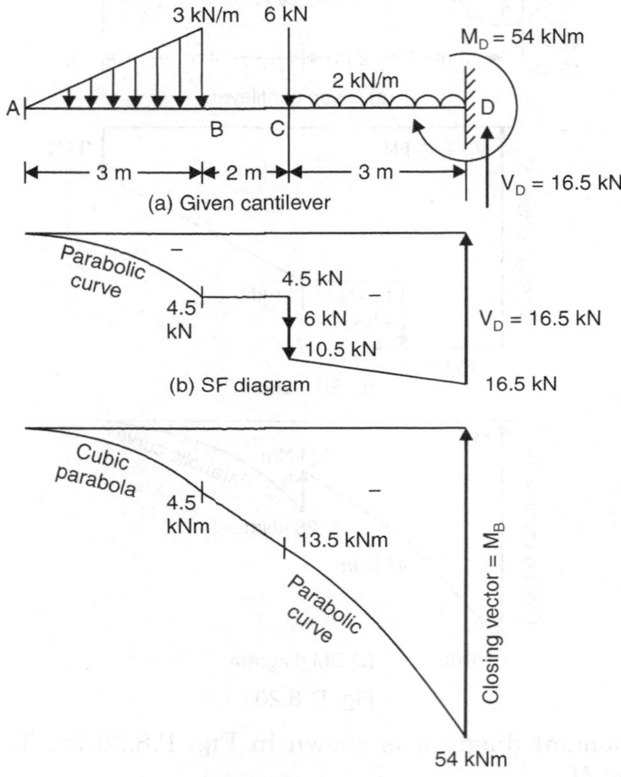

Fig. P. 8.21

Problem 8.22. *Draw shear force and bending moment diagrams for the cantilever shown in Fig. P.8.22 (a). Also find maximum bending moment and point of contraflexure.*

Sol. Starting from the free end A, we get $F_A = -4$ kN

$$F_B)_L = -4 - 3\times 2 = -10\text{ kN}$$

$$F_B)_R = -10 + 18 = 8\text{ kN}$$

$$F_C = +8 - 3\times 2 = 2\text{ kN}$$

$$F_D = 2 - 3\times 2 = -4\text{ kN}$$

$$\therefore \quad V_D = 4\text{ kN}\uparrow$$

$$M_A = 0$$

$$M_B = -4\times 2 - \frac{3\times 2^2}{2} = -14\text{ kNm}$$

$$M_C)_L = -4 \times 4 - \frac{3 \times 4^2}{2} + 18 \times 2 \text{ kNm} = -4 \text{ kNm}$$

$$M_C)_R = -4 + 5 = +1 \text{ kNm}$$

$$M_D = -4 \times 6 - \frac{3 \times 6^2}{2} + 18 \times 4 + 5 \text{ kNm} = -1 \text{ kNm}$$

∴ Support moment (or moment reaction) at D = 1 kN **Ans.**

Shear force and bending moment diagrams are shown in Figs. P.8.22 (*b*) and (*c*), respectively.

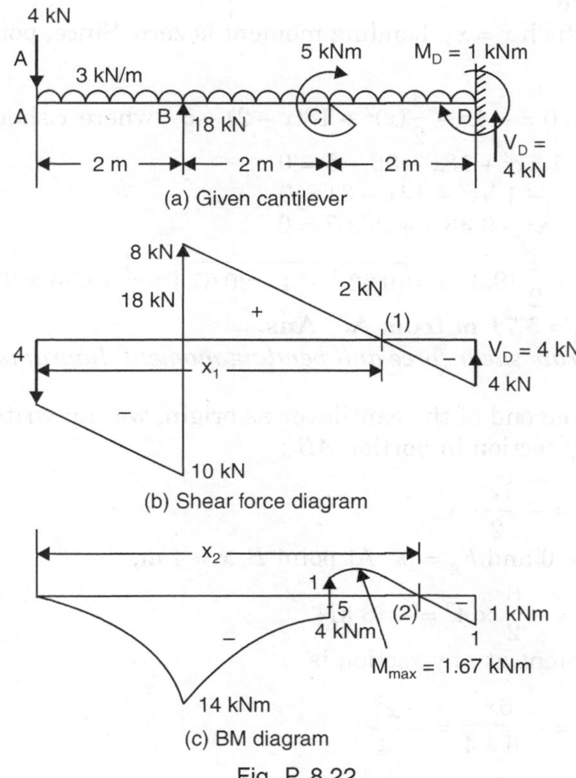

Fig. P. 8.22

From the shear force diagram it is clear that shear force is zero at point (1) and at point C. The bending moment at $B = M_B = -14$ kNm.

∴ Minimum bending moment = 14 kNm at point B. At point 1 bending moment will be maximum positive. At point 1, let $x = x_1$, $F_{x_1} = 0$ and $M_{x_1} = M_{max}$. See Fig. P.8.22 (*b*).

Point 1 lies in portion CD.

$$\therefore \qquad F_x = -4 - 3x + 18 \qquad \text{...(1)}$$

and

$$M_x = -4x - \frac{3x^2}{2} + 18(x - 2) + 5 \qquad \text{...(2)}$$

At point 1, from eqn. (1), we get

$$F_x]_{x = x_1} = 0 = -4 - 3x_1 + 18$$

$\therefore \qquad x_1 = \frac{18-4}{3} = \frac{14}{3} = 4.67$ m from A. **Ans.**

And, from eqn. (2), we get

$$M_x]_{x=x_1} = M_{max} = -4x_1 - \frac{3x_1^2}{2} + 18(x_1 - 2) + 5$$

$$= -4 \times \left(\frac{14}{3}\right) - \frac{3}{2}\left(\frac{14}{3}\right)^2 + 18\left(\frac{14}{3} - 2\right) + 5 \text{ kNm} = \mathbf{1.67 \text{ kNm.}} \quad \textbf{Ans.}$$

Point of Contraflexure

At point (2), for which $x = x_2$, bending moment is zero. Since, point 2 lies in portion CD, $\therefore$ from eqn. (2), we get

$$M_x = 0 = -4x - \frac{3}{2}(x)^2 + 18(x-2) + 5 \quad \text{where } x \text{ should be} > 4 \text{ m.}$$

$$\therefore \qquad -4x - 1.5x^2 + 18x - 36 + 5 = 0$$

$$\therefore \qquad -1.5x^2 + 14x - 31 = 0$$

or

$$x^2 - 9.33\,x + 20.67 = 0$$

$$\therefore \qquad x = \frac{1}{2}\,[9.33 \pm \sqrt{(9.33)^2 - 4 \times 20.67}\,] = \frac{1}{2}\,[9.33 \pm 2.09] = 5.71 \text{ m} \; ; \; 3.62 \text{ m}$$

$\therefore \qquad x_2 =$ **5.71 m from A. Ans.**

Problem 8.23. *Draw shear force and bending moment diagrams for the beam shown in Fig. P.8.23 (a).*

Sol. Taking the free end of the cantilever as origin, we can write :

Shear force at any section in portion AB :

$$F_x = -\frac{6x^2}{2} \qquad \text{...(1)}$$

$\therefore$ At point A, $x = 0$ and $F_A = 0$ At point B, $x = 4$ m,

$$\therefore \qquad F_B = -\frac{6}{2} \times 4^2 = -48 \text{ kN.}$$

And, bending moment at any section is

$$M_x = -\frac{6x^3}{6 \times 4} = -\frac{x^3}{4} \qquad \text{...(2)}$$

At A, $x = 0$,

$$M_x]_{x=0} = M_A = 0$$

And, at B, $M_x]_{x=4} = M_B$

$$\therefore \qquad M_B = -\frac{6 \times (4)^3}{6 \times 4} = -16 \text{ kNm}$$

$$F_C]_L = -48 \text{ kN}$$

$$F_C]_R = -48 - 4 = -52 \text{ kN}$$

$$M_C = -\frac{1}{2} \times (6 \times 4) \times \left(2 + \frac{4}{3}\right) = -\frac{6 \times 4 \times 10}{2 \times 3} = -40 \text{ kNm}$$

$$F_D = -52 \text{ kN}$$

$$M_D = -\frac{1}{2} \times (6 \times 4) \times \left(4 + \frac{4}{3}\right) - 4 \times 2$$

$$= -\frac{6 \times 4 \times 16}{6} - 8 = -72 \text{ kNm.}$$

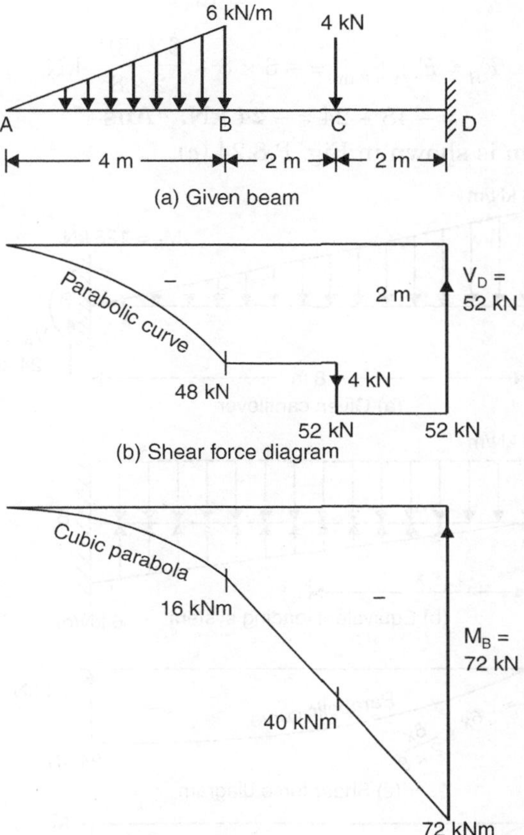

(a) Given beam

(b) Shear force diagram

(c) Bending moment diagram

Fig. P. 8.23

Problem 8.24. *Draw shear force and bending moment diagrams for the cantilever shown in Fig. P.8.24 (a).*

Sol. The given uniformly varying load of 6 kN/m can be considered as a 6 kN/m *udl* minus a *uvl* zero at *A* and 6 kN/m at *B* as shown in Fig. P. 8.24 (*b*). At any section *x* from *A*, the shear force is

$$F_x = -6x + \left(\frac{6x}{8}\right)\frac{x}{2}$$

or $$F_x = -6x + \frac{6x^2}{2\times 8} \quad ...(1)$$

and, bending moment $$M_x = -\frac{6x^2}{2} + \frac{6x^2}{2\times 8}\times\frac{x}{3}$$

or $$M_x = -\frac{6x^2}{2} + \frac{6x^3}{6\times 8} \quad ...(2)$$

Eqn. (1) represents a parabola and eqn. (2) represent a cubic parabola.

At point B :

$$F_B = F_x]_{x = 8\text{ m}} = -6 \times 8 + \frac{6 \times (8)^2}{2 \times 8}\text{ kN}$$

$$= -48 + 24 = -\mathbf{24\ kN.}\quad \mathbf{Ans}$$

Shear force diagram is shown in Fig. P.8.24 (c).

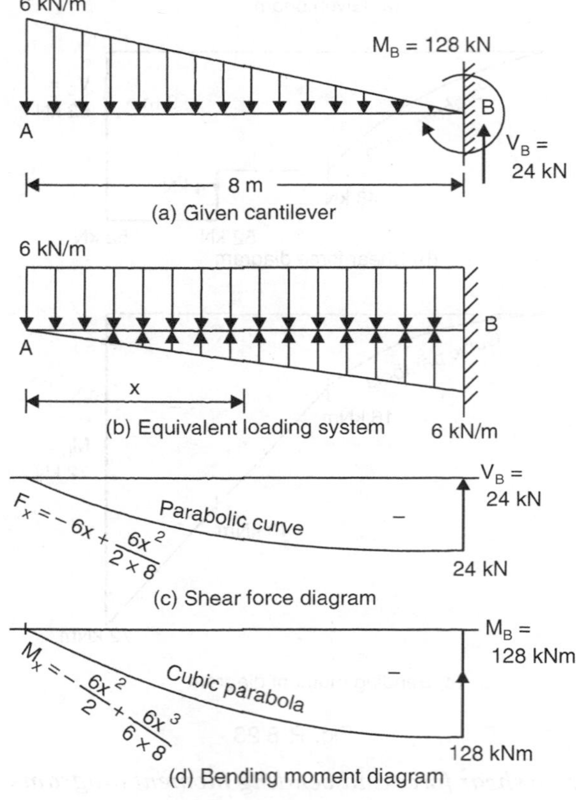

Fig. P.8.24

Bending moment at B is

$$M_B = M_x]_{x = 8\text{m}} = -\frac{6 \times (8)^2}{2} + \frac{6 \times (8)^3}{6 \times 8}\text{ kNm}$$

$$= -192 + 64 = -\mathbf{128\ kNm.}\quad \mathbf{Ans.}$$

Problem 8.25. *Draw shear force and bending moment diagrams for the beam shown in Fig. P.8.25 (a). Also locate the position of points of contraflexure.*

Sol. By $\Sigma M_A = 0$, we get

$$V_B \times 6 - (800 \times 6) \times \left(\frac{6}{2} + 2\right) = 0$$

$\therefore$ $$6V_B = 800 \times 6 \times 5$$

$\therefore$ $$V_B = 4000\text{ N} \uparrow$$

By $\Sigma F_V = 0$, we get $(V_A + V_B) \uparrow - 800 \times 6 = 0$

$\therefore$ $$V_A = 4800 - 4000 = 800\text{ N} \uparrow$$

$$F_A = V_A = 800\text{ N}$$

$$F_C = 800 \text{ N}$$
$$F_B)_L = 800 - 800 \times 4 = -2400 \text{ N}$$
$$F_B)_R = F_B)_L + V_B = -2400 + 4000 = +1600 \text{ N}$$
$$F_D = F_B)_R - 800 \times 2 = 1600 - 1600 \text{ N} = 0$$
$$M_A = 0, M_C = V_A \times 2 = 1600 \text{ Nm}$$
$$M_B)_L = V_A \times 6 - \frac{800 \times 4^2}{2} = 800 \times 6 - 400 \times 16$$
$$= 4800 - 6400 = -1600 \text{ Nm}$$

Also, $$M_B)_R = -\frac{800 \times 2^2}{2} = -1600 \text{ Nm}.$$

800 N/m
A C B D
V_A = 800 N V_A V_B = 4000 N
2 m 4 m 2 m

(a) Given beam

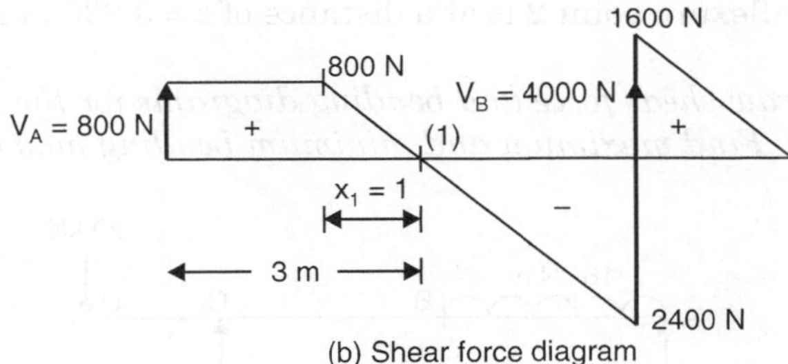

(b) Shear force diagram

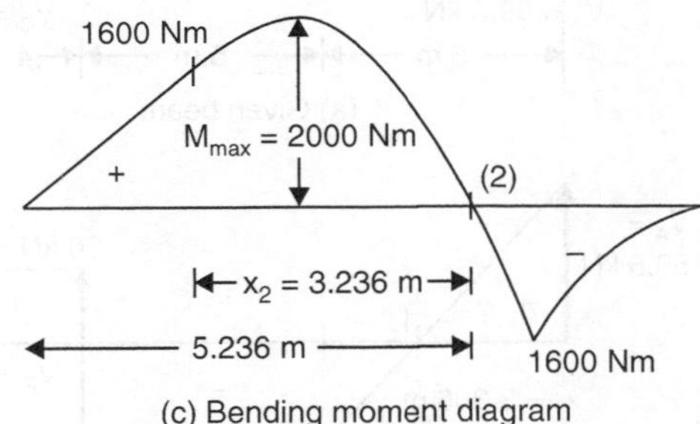

(c) Bending moment diagram

Fig. P. 8.25

From shear force diagram, Fig. P.8.25 (*b*), we find that at point (1) shear force is equal to zero.

$\therefore$ We get at point (1), $x = x_1$ from C, $F_x = 0$, and $M_x = M_{max}$. Since point (1) lies in portion CB, therefore we write equations for shear force and bending moment.

$\therefore$ In portion CB,

$$F_x = V_A - 800\,x = 800 - 800\,x \quad \text{...(1)}$$

and $$M_x = 800 \times (2 + x) - \frac{800x^2}{2} \quad \text{...(2)}$$

At point (1), $F_x]_{x = x_1} = 0 = 800 - 800\,x_1$ from eqn. (1)

$\therefore \quad x_1 = 1$ m

And, from eqn. (2), we get

$$M_x]_{x=x_1} = 800\,(x_1 + 2) - \frac{800}{2}\,x_1^2 = 800\,(1 + 2) - \frac{800}{2} \times 1^2 \text{ Nm}$$

$$= 2400 - 400 \text{ Nm}$$

$\therefore \quad M_{max} = \mathbf{2000\ Nm}$

Point of Contraflexure :

Point 2 in the bending moment diagram, where bending moment is zero, is called the point of contraflexure or contraflexion.

$\because \; M_x = 0$ at point 2, $\therefore$ Equating M_x in eqn. (2) equal to zero. We get

$$M_x = 800\,(2 + x) - \frac{800}{2}x^2 = 0$$

$$\therefore \quad (2 + x) - \frac{x^2}{2} = 0$$

or $\quad 4 + 2x - x^2 = 0 \quad \text{or} \quad x^2 - 2x - 4 = 0 \qquad \ldots(3)$

$$\therefore \quad x = \frac{1}{2}\,[2 \pm \sqrt{2^2 + 4 \times 4}\,] = \frac{1}{2}\,[2 \pm \sqrt{20}\,] = 3.236 \text{ m} ; -1.236 \text{ m}$$

$\therefore$ Point of contraflexure point 2 is at a distance of $x = 3.236$ m from point C or 5.236 m from support A. **Ans.**

Problem 8.26. *Draw shear force and bending diagrams for the simply supported beam shown in Fig. P. 8.26 (a). Find maximum and minimum bending moments and locate point of contraflexion.*

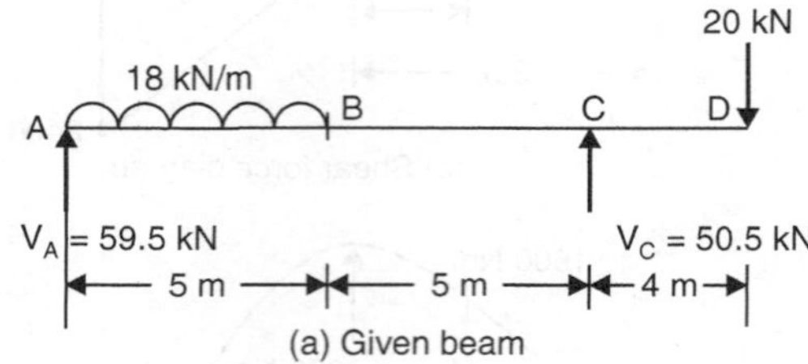

(a) Given beam

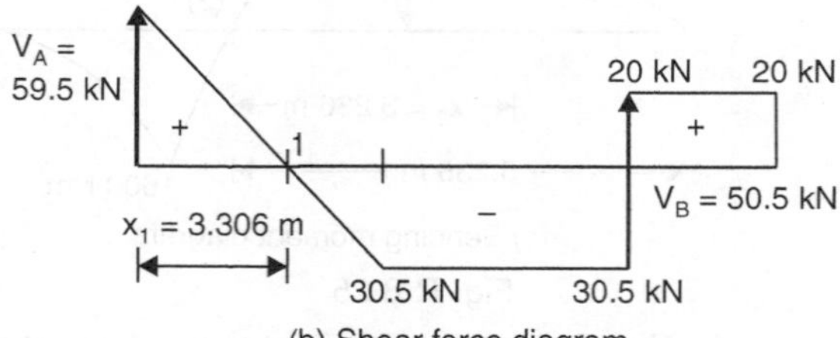

(b) Shear force diagram

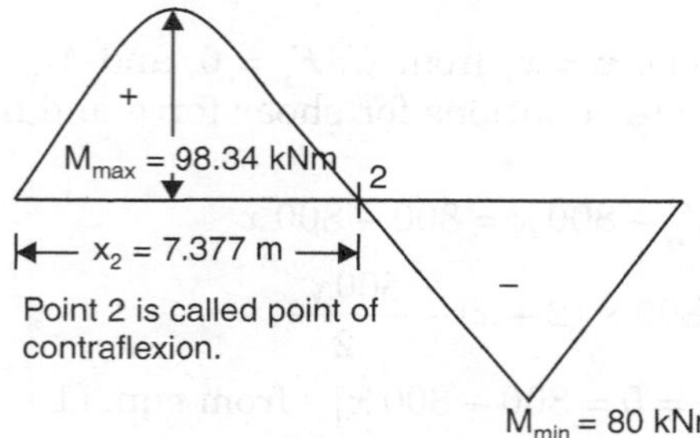

(c) Bending moment diagram

Fig. P. 8.26

Sol. To find vertical reaction (V_C) at support C, use the condition $\Sigma M_A = 0$, $\therefore$ We get

$$V_C \times 10 - 20 \times 14 - \frac{18 \times 5^2}{2} = 0$$

$\therefore$ $V_C = 50.5 \text{ kN} \uparrow$

By $\Sigma F_V = 0$, we get $(V_A + V_C)\uparrow - 18 \times 5 - 20 = 0$

$\therefore$ $V_A = 110 - V_C = 110 - 50.5 \text{ kN}$ or $V_A = 59.5 \text{ kN} \uparrow$

With A as origin, in portion AB, $0 \le x \le 5$ m,

Shear force at any section,

$$F_x = V_A - 18x \quad \text{or} \quad F_x = 59.5 - 18x \qquad \text{...(1)}$$

and bending moment at the section,

$$M_x = 59.5x - \frac{18x^2}{2} \qquad \text{...(2)}$$

Eqn. (1) represents a straight line and eqn. (2) represents a parabola.

For point B, $x = 5$ m,

$\therefore$ From eqn. (1), $F_B = 59.5 - 18 \times 5 = -30.5$ kN

and from eqn. (2),

$$M_B = 59.5 \times 5 - \frac{18 \times 5^2}{2} = 72.5 \text{ kNm.}$$

In portion BC, $5 \le x \le 10$ m.

Since there is no external load between B and C, therefore shear force remains constant in portion BC. Hence shear force on the left of point $C = F_C)_L = -30.5$ kN and, bending moment is given by

$$M_x = 59.5\,x - (18 \times 5)\left(x - \frac{5}{2}\right) \qquad \text{...(3)}$$

$$\therefore \quad M_C)_L = 59.5 \times 10 - 18 \times 5 \times \left(10 - \frac{5}{2}\right) = -80 \text{ kNm.}$$

In portion CD : $10 \le x \le 14$ m.

$\therefore$ $F_C)_R = 59.5 - 18 \times 5 + V_C$

or $F_C)_R = 59.5 - 90 + 50.5 = +20 \text{ kN} \uparrow$

$F_D = 20$ kN

And, $$M_x = 59.5 \times x - 18 \times 5 \times \left(x - \frac{5}{2}\right) + V_C \times (x - 10) \qquad \text{...(4)}$$

$\therefore$ $M_C = 59.5 \times 10 - 18 \times 5 \times (10 - 2.5) + 50.5 \times (10 - 10)$

$\therefore$ $M_C = -80$ kNm

Also, $M_C = -20 \times 4 = -80$ kNm **(Check)**

Shear force and bending moment diagrams are shown in Figs. P.8.26 (*b*) and (*c*).

M_{max} and M_{min} :

From the shear force diagram, it is clear that shear force changes sign at points (1) and C. At these points bending moment will have the maximum numerical values.

Position of point (1) :

Point (1) lies in portion AB. At point (1), $\therefore$ We have :

$$x = x_1 \,;\ F_x = 0 \quad \text{and} \quad M_x = M_{max}.$$

The values of F_x and M_x can be determined with the help of eqns. (1) and (2).

From eqn. (1),

$$F_x]_{x=x_1} = 0 = 59.5 - 18x_1$$

$$\therefore \quad x_1 = \frac{59.5}{18} = 3.306 \text{ m from } A.$$

$\therefore$ From eqn. (2), we get

$$M_x]_{x=x_1} = M_{max} = 59.5\, x_1 - \frac{18x_1^2}{2}$$

$$= 59.5 \times 3.306 - \frac{18}{2} \times (3.306)^2 \text{ kN-m} = 98.34 \text{ kNm.} \quad \textbf{Ans.}$$

From the *BM* diagram, it is clear that the maximum negative value of bending moment = 80 kNm occurs at point *C*.

$$\therefore \quad M_{min} = 80 \text{ kNm.} \quad \textbf{Ans.}$$

Point of Contraflexion or Point of Contraflexure. At this point the bending moment diagram changes sign and bending moment is zero. Thus, point (2) is the point of contraflexion and it lies in the portion *BC* of the beam.

Therefore, from eqn. (3),

$$M_x]_{x=x_2} = 0 = 59.5\, x_2 - 18 \times 5\,(x_2 - 2.5)$$

$$\therefore \quad (59.5 - 90)x_2 + 90 \times 2.5 = 0$$

$$\therefore \quad x_2 = \frac{225}{30.5} = \mathbf{7.377 \text{ m}} \text{ from } A. \quad \textbf{Ans.}$$

Problem 8.27. *Draw shear force and bending moment diagrams for the beam shown in Fig. P. 8.27 (a) . Also locate point of contraflexure.*

Sol. By $\Sigma M_B = 0$

$$\circlearrowleft V_A \times 4 - (15 \times 2)\left(\frac{2}{2} + 2\right) + 10 \times 1 = 0$$

$$\therefore \quad V_A = \frac{1}{4}[90 - 10] = 20 \text{ kN} \uparrow$$

By $\Sigma F_V = 0$, we get $(V_A + V_B) \uparrow - 15 \times 2 - 10 = 0$

$$\therefore \quad V_B = 40 - V_A = 40 - 20 \text{ kN} \quad \text{or} \quad V_B = 20 \text{ kN} \uparrow$$

Shear Force Diagram :

$$F_A = V_A = 20 \text{ kN}$$

$$F_C = 20 - 15 \times 2 = -10 \text{ kN}$$

$$F_B)_L = -10 \text{ kN}$$

$$F_B)_R = -10 + V_B \quad \text{or} \quad F_B)_R = -10 + 20 = +10 \text{ kN}$$

$$F_D)_L = 10 \text{ kN}$$

The last vector (10 kN ↓) , when added, the shear force diagram is closed.

Bending Moment Diagram :

$$M_A = 0$$

$$M_C = V_A \times 2 - \frac{15 \times 2^2}{2} = 20 \times 2 - 30 = 10 \text{ kN}$$

$$M_B)_L = 20 \times 4 - (15 \times 2)\left(\frac{2}{2}+2\right) \text{ kN} = 80 - 90 = -10 \text{ kNm.}$$

Also, $\quad M_B)_R = -10 \times 1 = -10$ kN (check)

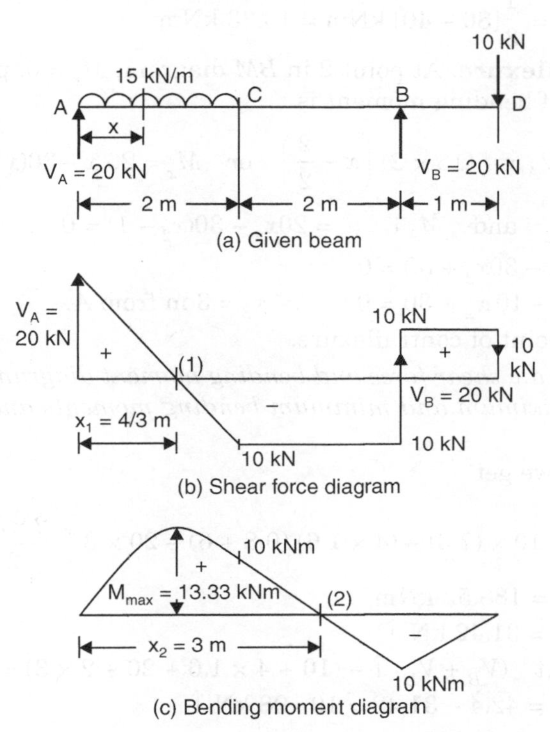

Fig. P. 8.27

From the shear force diagram it is clear that the shear force at point 1 is zero. This point lies in portion AC. Bending moment at this point is maximum. To locate the position of point 1 and value of M_{max}, we write the equations of shear force F_x and bending moment M_x for the portion AC.

At any section x from the origin A, for the portion AC, we have

$$F_x = V_A - 15x \quad \text{or} \quad F_x = 20 - 15x \qquad ...(1)$$

And, $$M_x = V_A \cdot x - \frac{15x^2}{2} \quad \text{or} \quad M_x = 20x - \frac{15x^2}{2} \qquad ...(2)$$

At point 1, $x = x_1$, $F_x = 0$ and $M_x]_{x=x_1} = M_{max}$

∴ From eqn. (1), we get

$$0 = 20 - 15\,x_1$$

∴ $$x_1 = \frac{20}{15} \text{ m} = \frac{4}{3} \text{ m} = 1.333 \text{ m from } A$$

And, from eqn. (2), we get

$$M_x]_{x=x_1} = M_{max} = V_A \cdot x - \frac{15{x_1}^2}{2}$$

$$= 20 \times \frac{20}{15} - \frac{15}{2} \times \left(\frac{20}{15}\right)^2 \text{ kNm} = \frac{80}{3} - \frac{15}{2} \times \frac{16}{9} \text{ kNm}$$

$$= \frac{1}{3}[80 - 40] \text{ kNm} = 13.33 \text{ kNm}$$

Point of Contraflexure. At point 2 in *BM* diagram, $M_x = 0$, point 2 lies in portion *CB* for which the equation of bending moment is :

$$M_x = V_A\, x - (15 \times 2)\left(x - \frac{2}{2}\right) \quad \text{or} \quad M_x = 20\,x - 30(x - 1) \qquad ...(3)$$

At point 2, $x = x_2$ and $M_x]_{x = x_2} = 20x_2 - 30(x_2 - 1) = 0$

$\therefore$ We get $20x_2 - 30x_2 + 30 = 0$

$-10\,x_2 + 30 = 0\,;\quad \therefore\quad x_2 = 3$ m from *A*.

Point 2 is called point of contraflexure.

Problem 8.28. *Draw shear force and bending moment diagrams for the beam shown in Fig. P. 8.28 (a). Find maximum and minimum bending moments and the position of point of contraflexure, if any.*

Sol. By $\Sigma M_D = 0$, we get

$$\circlearrowleft V_B \times 6 - 10 \times (7.6) - (4 \times 1.6)(0.8 + 6) - 20 \times 3 - \frac{2 \times 3^2}{2} = 0$$

$\therefore$ $V_B \times 6 = 188.52$ kNm

$\therefore$ $V_B = 31.42$ kN $\uparrow$

By $\Sigma F_V = 0$, we get $(V_B + V_D)\uparrow - (10 + 4 \times 1.6 + 20 + 2 \times 3) = 0$

$\therefore$ $V_D = 42.4 - 31.42 = 10\,.\,98$ kN $\uparrow$

$F_A = -10$ kN

$F_B)_L = -10 - 4 \times 1.6 = -16.4$ kN

$F_B)_R = -16.4 + V_B = -16.4 + 31.42$ kN $= 15.02$ kN

$F_C)_L = 15.02$ kN

$F_C)_R = 15.02 - 20 = -4.98$ kN

$F_D = -4.98 - 2 \times 3 = -10.98$ kN

The last vector, $V_D = 10.98$ kN $\uparrow$ closes the shear force diagram (Fig. P.8.28 (*b*)).

$$M_A = 0\,;\, M_B = -10 \times 1.6 - \frac{4 \times 1.6^2}{2} = -16 - 5.12 = -21.12 \text{ kNm}$$

$$M_C = -10 \times (1.6 + 3) - 4 \times 1.6 \times (0.8 + 3) + 31.42 \times 3 \text{ kNm} = 23.94 \text{ kNm}$$

$$M_D = -10 \times (1.6 + 6) - (4 \times 1.6)(0.8 + 6) - 20 \times 3 - \frac{2 \times 3^2}{2} + V_B \times 6$$

$$= -76 - 43.52 - 60 - 9 + 31.42 \times 6 = 0 \textbf{ (Check)}$$

From the bending moment diagram it is clear that $M_{max} = 23.94$ kNm at *C*,

$M_{min} = -21.12$ kNm at *B*. **Ans.**

Point of Contraflexure. Point of contraflexure occurs at point (1) as shown in Fig. P.8.28(*c*). Point 1 lies in portion *BC*. Its position can be found by equating bending moment M_x equal to zero.

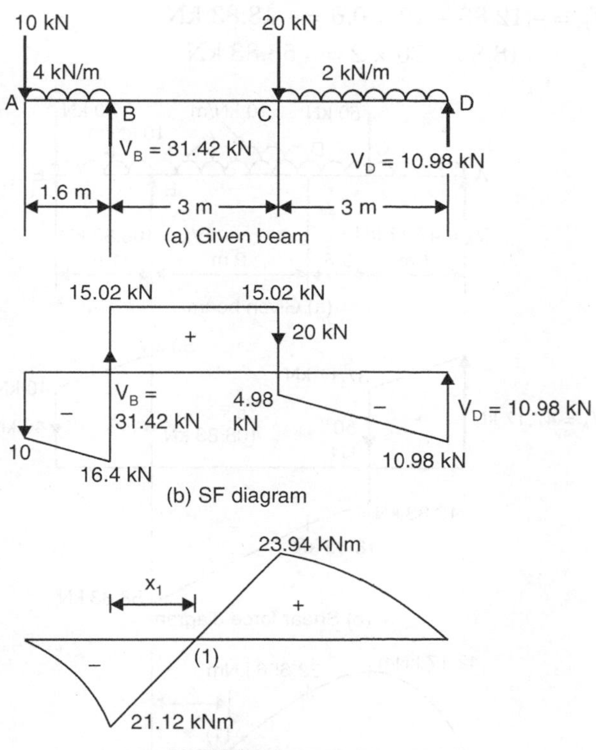

Fig. P. 8.28

Therefore, we get $M_x = -10 \times (1.6 + x_1) - 4 \times (1.6) \times (0.8 + x_1) + 31.42 \times x_1 = 0$

or $\qquad -16 - 10\,x_1 - 6.4 \times 0.8 - 6.4\,x_1 + 31.42\,x_1 = 0$

$\therefore \quad x_1\,(-10 - 6.4 + 31.42) = 16 + 6.4 \times 0.8$

$$x_1 = \frac{21.12}{15.02} = 1.406 \text{ m from } B. \quad \textbf{Ans.}$$

Problem 8.29. *Draw shear force and bending moment diagrams for the overhanging beam shown in Fig. P.8.29 (a). Determine maximum and minimum bending moments and locate prints of contraflexure.*

Sol. By $\Sigma M_B = 0$, we get

$$\circlearrowleft V_A \times 3.6 - \frac{10 \times (3.6)^2}{2} - 50 \times 2.6 - \frac{10 \times 2^2}{2} + 40 \times 1 + \frac{10 \times 1^2}{2} = 0$$

$$\therefore \quad V_A \times 3.6 - 64.8 - 130 - 20 + 40 + 5 = 0$$

$$\therefore \quad V_A = \frac{169.8}{3.6} = 47.17 \text{ kN} \uparrow$$

and by $\Sigma F_V = 0$, we get

$$(V_A + V_B) \uparrow - (10 \times 4.6 + 50 + 40 + 10 \times 2) = 0$$

$$\therefore \quad V_B = 156 - V_A = 156 - 47.17 \text{ kN} = \mathbf{108.83 \text{ kN}} \uparrow$$

$$F_A = V_A = 47.17 \text{ kN}$$

$$F_C)_L = 47.17 - 10 \times 1 = 37.17 \text{ kN}$$

$$F_C)_R = 37.17 - 50 = -12.83 \text{ kN}$$

$$F_D = -12.83 - 10 \times 0.6 = -18.83 \text{ kN}$$
$$F_B)_L = -18.83 - 20 \times 2 = -58.83 \text{ kN}$$

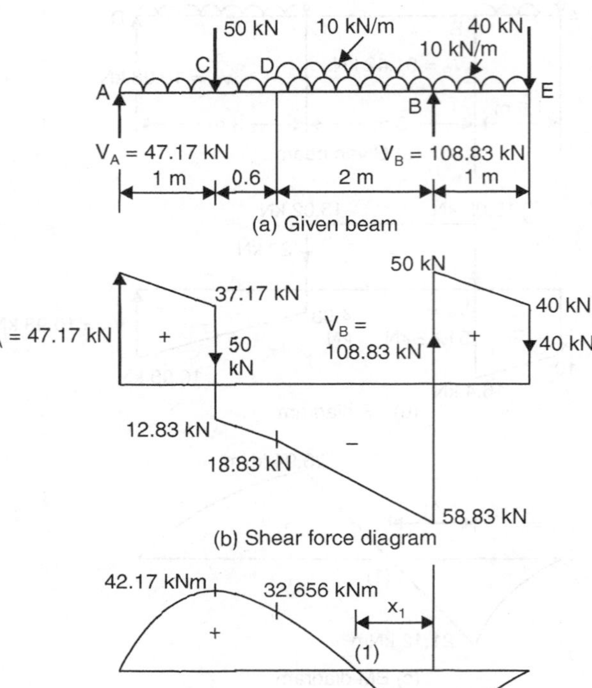

Fig. P. 8.29

$$F_B)_R = -58.83 + V_B = -58.83 + 108.83 = +50 \text{ kN}$$
$$F_E)_L = 50 - 10 \times 1 = 40 \text{ kN}$$
$$F_E)_R = 40 - 40 = 0 \quad \textbf{(Check)}$$

Thus, the shear force diagram closes.

$$M_A = 0\ ;\ M_B)_R = -40 \times 1 - \frac{10 \times 1^2}{2} = -45 \text{ kNm.}$$

$$M_C = 47.17 \times 1 - \frac{10 \times 1^2}{2} = 42.17 \text{ kNm}$$

$$M_D = 47.17 \times 1.6 - \frac{10}{2} \times (1.6)^2 - 50 \times 0.6 = 32.656 \text{ kNm}$$

$$M_B)_L = 47.17 \times 3.6 - \frac{10}{2} \times (3.6)^2 - 50 \times 2.6 - \frac{10 \times 2^2}{2} \text{ kNm}$$

$$= -44.988 \text{ or } \mathbf{-45\ kNm.} \text{ as before } \textbf{(Check)}$$

From bending moment diagram we note that

$$M_{max} = 42.17 \text{ kNm at } C$$
$$M_{min} = -50 \text{ kNm at } B.$$

Point of Contraflexure :

Point of contraflexure lies between B and D.

We write with origin at B,

$$M_x = -40 \times (x + 1) - \frac{10}{2}(x_1 + 1)^2 - \frac{10}{2}x^2 + V_B.x \qquad ...(1)$$

$\therefore$ For point (1), we get

$$0 = -40(x_1 + 1) - 5(x_1 + 1)^2 - 5(x_1)^2 + 108.83x_1$$

or $$0 = -40x_1 - 40 - 5(x_1^2 + 2x_1 + 1) - 5x_1^2 + 108.83x_1$$

or $$0 = +x_1(-40 - 10 + 108.83) - (40 + 5) - 5x_1^2$$

$\therefore$ $$5x_1^2 - 58.83x_1 + 45 = 0$$

$$x_1^2 - 11.766x_1 + 9 = 0$$

$$x_1 = \frac{1}{2}[11.766 \pm \sqrt{(11.766)^2 - 4 \times 9}\,] \text{ m}$$

$$= \frac{1}{2}[11.766 \pm 10.121] \text{ m} = 10.9435, + 0.8224 \text{ m}$$

$\therefore$ Point of contraflexure (point 1) lies at a distance of 0.8224 m from B. **Ans.**

Problem 8.30. *Draw shear force and bending moment diagrams for the beam shown in Fig. P.8.30 (a).*

Sol. By $\Sigma M_E = 0$, we get

$$V_B \times 5 - 60 \times 6 - (50 \times 1) \times (0.5 + 5) - (40 \times 3)(1.5 + 2) + 120 = 0$$

$\therefore$ $$V_B = \frac{1}{5}[360 + 275 + 420 - 120] = \frac{935}{5} = 187 \text{ kN} \uparrow$$

By $\Sigma F_V = 0$, we get $(V_B + V_E)\uparrow - (60 + 50 + 40 \times 3) = 0$

$\therefore$ $$V_E = 230 - 187 = 43 \text{ kN} \uparrow$$

$$F_A = -60 \text{ kN}$$

$$F_B)_L = -60 - 50 \times 1 = -110 \text{ kN}$$

$$F_B)_R = -110 + V_B = -110 + 187 = 77 \text{ kN}$$

$$F_C = 77 - 40 \times 3 = -43 \text{ kN}$$

$$F_D = -43 \text{ kN}$$

$$F_E)_L = -43 \text{ kN}$$

The last vector (= V_E = 43 kN) closes the shear force diagram.

$$M_A = 0,$$

$$M_B = -60 \times 1 - \frac{50 \times 1^2}{2} = -85 \text{ kNm}$$

$$M_C = -60 \times 4 - (50 \times 1)\left(\frac{1}{2} + 3\right) + 187 \times 3 - \frac{40 \times 3^2}{2} = -34 \text{ kNm}$$

$$M_D)_L = -60 \times 5 - (50 \times 1)\left(\frac{1}{2} + 4\right) + 187 \times 4 - (40 \times 3)\left(\frac{3}{2} + 1\right) \text{ kN} = -77 \text{ kN}$$

$$M_D)_R = M_D)_L + 120 = -77 + 120 = 43 \text{ kN}$$

Shear force is zero at point 1 which lies in portion *BC*.

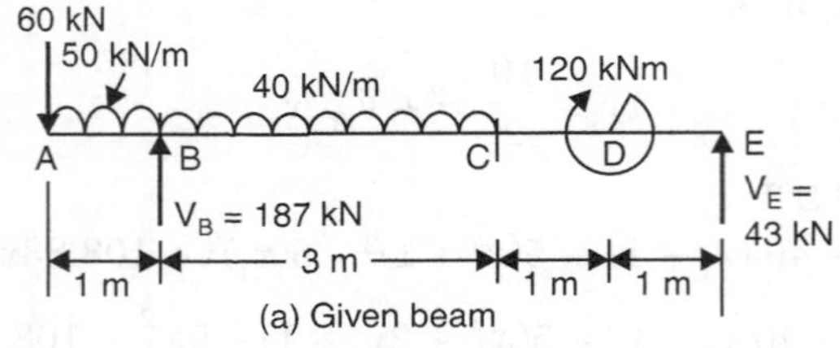

(a) Given beam

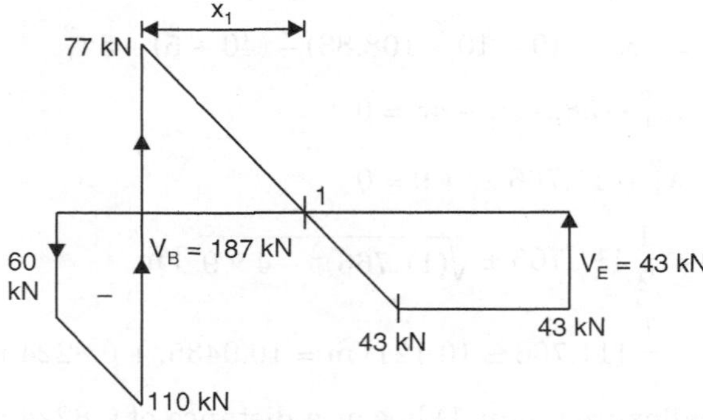

(b) Shear force diagram

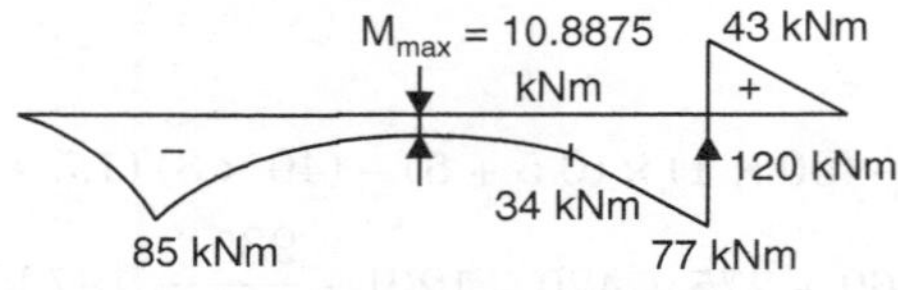

(c) Bending moment diagram

Fig. P.8.30

Therefore for the portion *BC* :

$$F_x = -60 - 50 \times 1 + 187 - 40x = 77 - 40x \quad \text{...(1)}$$

And, $$M_x = -60 \times (1 + x) - (50 \times 1) \times (0.5 + x) + 187x - \frac{40x^2}{2} \quad \text{...(2)}$$

From eqn. (1) for point 1, $x = x_1$ we get $F_x = 0$ and $M_x = 0$,

$\therefore$ We get $F_x]_{x = x_1} = 0 = 77 - 40\,x_1$

$$\therefore \quad x_1 = \frac{77}{40} = 1.925 \text{ m from } B$$

$\therefore$ From eqn. (2), for $x = x_1$, we get

$$M_x]_{x = x_1} = M_{max} = -60\,(1 + 1.925) - (50 \times 1)\,(0.5 + 1.925) + 187 \times 1.925 - \frac{40}{2} \times 1.925^2 \text{ kNm} = \mathbf{-10.8875 \text{ kNm}}.$$

Problem 8.31. *Draw shear force and bending diagrams for the beam shown in Fig. P.8.31 (a).*

Sol. By $\Sigma M_B = 0$; $\circlearrowleft V_A \times 5 - 3000 + 750 \times 2 = 0$

$$\therefore \quad V_A = \frac{1}{5}[3000 - 1500] = 300 \text{ N} \uparrow$$

By $\Sigma F_V = 0$, we get $(V_A + V_B) \uparrow - 750 = 0$

$\therefore \qquad V_B = 750 - V_A = 750 - 300 = 450 \text{ N} \uparrow$

Shear Force Diagram :

$$F_A = V_A = 300 \text{ N}$$
$$F_C = 300 \text{ N}$$
$$F_B)_L = 300 \text{ N}$$
$$F_B)_R = 300 + V_B = 300 + 450 = 750 \text{ N}$$
$$F_D = 750 \text{ N}$$

The last vector (750 N) will close the shear force diagram shown in Fig.P. 8.31 (*b*).

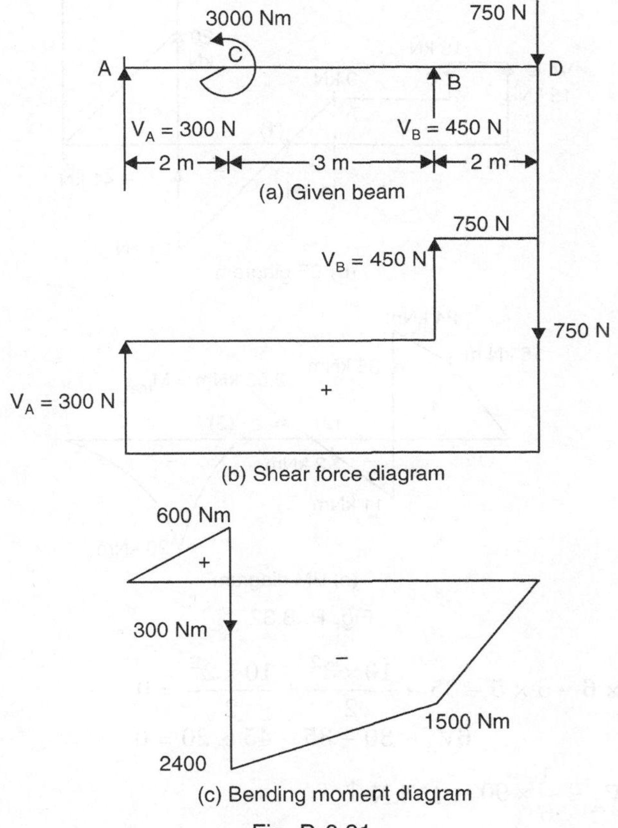

Fig. P. 8.31

Bending Moment Diagram :

$$M_A = 0$$
$$M_C)_L = V_A \times 2 = 300 \times 2 = 600 \text{ Nm}$$
$$M_C)_R = M_C)_L - 3000 = -2400 \text{ Nm}$$
$$M_B)_L = 300 \times 5 - 3000 = -1500 \text{ Nm}$$

Also $\qquad M_B)_R = -750 \times 2 = -1500 \text{ Nm}$ **(Check)**

The bending moment diagram is shown in Fig. P.8.31 (*c*).

Problem 8.32. *Draw shear force and bending moment diagrams for the beam shown in Fig. P.8.32 (a). Find maximum and minimum values of bending moment. Also locate points of contraflexion.*

Sol. By $\Sigma M_B = 0$, we get

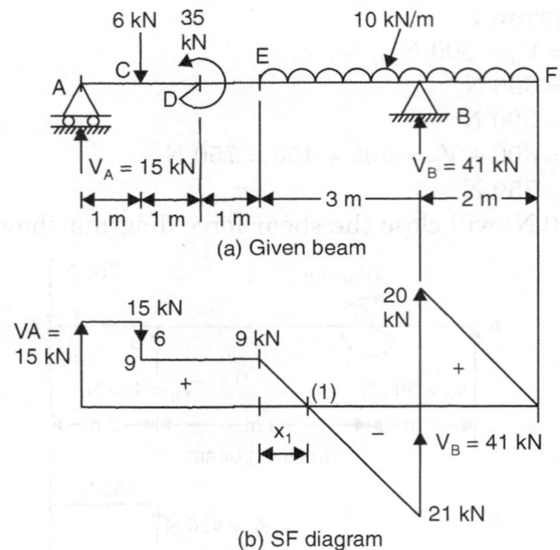

(a) Given beam

(b) SF diagram

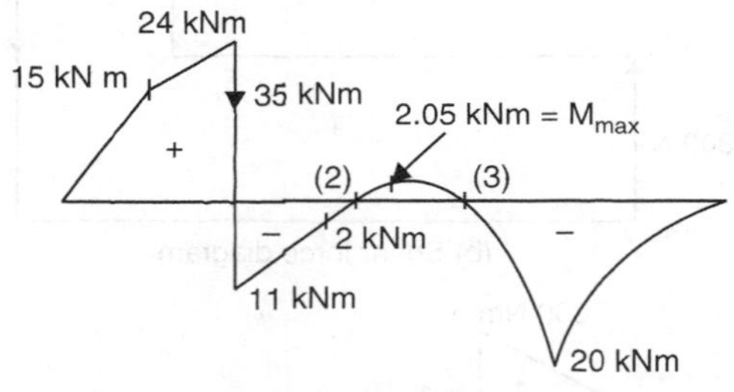

(c) BM diagram

Fig. P. 8.32

$$\circlearrowleft V_A \times 6 - 6 \times 5 - 35 - \frac{10 \times 3^2}{2} + \frac{10 \times 2^2}{2} = 0$$

$$\therefore \quad 6V_A - 30 - 35 - 45 + 20 = 0$$

$$\therefore \quad V_A = \frac{1}{6} \times 90 = 15 \text{ kN} \uparrow$$

By $\Sigma F_V = 0$, we get $(V_A + V_B) \uparrow - (6 + 10 \times 5) = 0$

$$\therefore \quad V_B = 56 - V_A = 41 \text{ kN} \uparrow.$$

$$F_A = V_A = 15 \text{ kN}$$

$$F_C)_L = 15 \text{ kN}$$

$$F_C)_R = 15 - 6 = 9 \text{ kN}$$

$$F_D = F_C)_R = 9 \text{ kN}$$

$$F_E = 9 \text{ kN}$$

$$F_B)_L = F_E - 10 \times 3 = 9 - 30 = -21 \text{ kN}$$

$$F_B)_R = -21 + V_E = -21 + 41 = 20 \text{ kN}$$

$$F_F = F_B)_R - 10 \times 2 = 20 - 20 = 0 \quad \textbf{(Check)}$$

$$M_A = 0,$$

$$M_B = -\frac{10 \times 2^2}{2} = -20 \text{ kNm}$$

$$M_C = V_A \times 1 = 15 \times 1 = 15 \text{ kNm}$$

$$M_D)_L = 15 \times 2 - 6 \times 1 = 24 \text{ kNm}$$

$$M_D)_R = M_D)_L - 35 = 24 - 35 = -11 \text{ kN}$$

$$M_E = 15 \times 3 - 6 \times 2 - 35 = -2 \text{ kNm}$$

$$M_B = 15 \times 6 - 6 \times 5 - 35 - \frac{10 \times 3^2}{2} = 90 - 30 - 35 - 45$$

$= -20$ kNm as before **(Check)**

At point 1, shear force, $F_x = 0$, $M_x = M_{max}$. To locate point 1, we write the equations for F_x and M_x for the portion EB.

$\therefore$ For the portion EB,

$$F_x = V_A - 6 - 10x \quad \text{or} \quad F_x = 15 - 6 - 10x \qquad ...(1)$$

and

$$M_x = V_A \times (x + 3) - 6 \times (x + 2) - 35 - \frac{10}{2}(x)^2 \qquad ...(2)$$

At point (1) we have $x = x_1$, $F_x = F_{x_1} = 0$ and $M_x]_{x = x_1} = M_{max}$.

$\therefore$ We get, from eqn. (1),

$$F_x]_{x = x_1} = 15 - 6 - 10x_1 = 0$$

$\therefore$
$$x_1 = \frac{1}{10} \times 9 = 0.9 \text{ m from } E$$

$\therefore$ From (2) when, $x = x_1 = 0.9$ m, we get

$$M_x]_{x = x_1} = M_{max} = 15 \times (x_1 + 3) - 6 \times (x_1 + 2) - 35 - \frac{10x_1^2}{2}$$

or

$$M_{max} = 15 \times (0.9 + 3) - 6 \times (0.9 + 2) - 35 - \frac{10}{2} \times (0.9)^2$$

= **2.05 kNm** in portion EB. **Ans.**

$M_{min} = M_B = -20$ kNm. **Ans.**

Points 2 and 3 are points of contraflexure, which lie in portion EB.

From eqn. (2), we get

$$M_x = 15 \times (x + 3) - 6(x + 2) - 35 - \frac{10}{2}(x)^2 = 0$$

or

$$15x + 45 - 6x - 12 - 35 + 5x^2 = 0$$

$\therefore$
$$5x^2 + 9x - 2 = 0$$

$\therefore$
$$x^2 + 1.8x - 0.4 = 0$$

$\therefore$
$$x = \frac{1}{2}[-1.8 \pm \sqrt{(1.8)^2 + 4 \times 0.4}] = \frac{1}{2}[-1.8 \pm \sqrt{4.84}] \text{ m} = \frac{1}{2}(-1.8 \pm 2.2) \text{ m}$$

$= 0.2$ m and -2 m

$\therefore$ $x = 0.2$ m from E for point of contraflexure 2.

Point (2) is at a distance of $(x_1 - 0.2) = 0.9 - 0.2 = 0.7$ m on the left of point 1 (the point of M_{max}). The parabolic curve passing through points 2, 1 and 3 is symmetrical about point 1. Hence point 2 is 0.7 m on the right of point 1. Therefore, point 2 lies at a distance of $0.9 + 0.7 = 1.6$ m from E. **Ans.**

Problem 8.33. *Draw shear force and bending moment diagrams for the beam shown in Fig. P. 8.33 (a).*

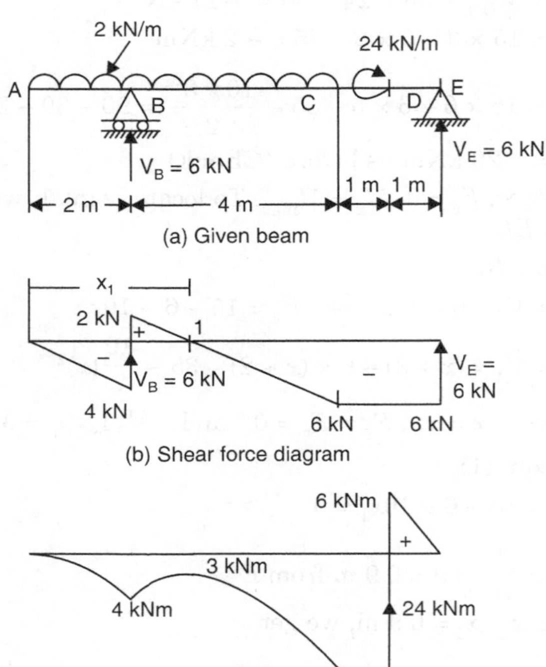

Fig. P. 8.33

Sol. By $\Sigma M_E = 0$, we get $\quad V_B \times 6 - (2 \times 6)\left(\frac{6}{2} + 2\right) + 24 = 0$

$$\therefore \quad V_B = \frac{1}{6}(60 - 24) = 6 \text{ kN} \uparrow$$

By $\Sigma F_V = 0$, we get $\quad (V_B + V_E) \uparrow - 2 \times 6 = 0$

$$\therefore \quad V_E = 12 - V_B = 12 - 6 = 6 \text{ kN} \uparrow$$

$$F_A = 0\,;$$

$$F_B)_L = -2 \times 2 = -4 \text{ kN}$$

$$F_B)_R = F_B)_L + V_B = -4 + 6 = 2 \text{ kN}$$

$$F_C = F_B)_R - 2 \times 4 = 2 - 8 = -6 \text{ kN}$$

$$F_D = -6 \text{ kN}$$

$$F_E = -6 \text{ kN}$$

The last vector ($V_E = +6$ kN) closes the shear force diagram [Fig.P.8.33 (*b*)].

$$M_A = 0\,;$$

$$M_B = -\frac{2 \times 2^2}{2} = -4 \text{ kNm}$$

$$M_C = -\frac{2\times 6^2}{2} + V_A \times 4 \text{ kNm} = -36 + 6\times 4 = -12 \text{ kN}$$

$$M_D)_L = -(2\times 6)\left(\frac{6}{2}+1\right) + V_A \times 5 = -12\times 4 + 6\times 5 \text{ kNm}$$

$$= -48 + 30 \text{ kNm} = -18 \text{ kNm}.$$

$$M_D)_R = -18 + 24 = 6 \text{ kNm}$$

$$M_E = 0, \text{ (by inspection)}$$

Also, $M_E = -2\times(6)\times\left(\frac{6}{2}+2\right) + 6\times 6 + 24 = -60 + 36 + 24 = 0$ **(Check)**

For portion BC, $(2 \le x \le 6 \text{ m})$

and $F_x = -2x + V_B$

or $F_x = -2x + 6$ ...(1)

And, $M_x = -\frac{2x^2}{2} + V_B(x-2)$

or $M_x = -\frac{2x^2}{2} + 6(x-2)$ ...(2)

At point (1) in Fig. P.8.33 (*b*), $x = x_1$, $F_x = 0$, $\therefore$ $M_x = M_{max}$

$\therefore$ From eqn. (1), $F_x]_{x=x_1} = 0 = -2x_1 + 6$

$\therefore$ $x_1 = \frac{6}{2} = 3$ m.

$\therefore$ From eqn. (2), we get

$$M_x]_{x=x_1} = M_{x_1} = -\frac{2{x_1}^2}{2} + 6(x_1 - 2) \text{ kNm} = -\frac{2\times 3^2}{2} + 6\times(3-2) \text{ kNm}$$

$\therefore$ $M_{x_1} = -9 + 6 = -3$ kN.

From the bending moment diagram, Fig. P.8.33 (*c*), we notice that

$$M_{min} = M_D)_L = -18 \text{ kNm, and } M_{max} = M_D)_R = +6 \text{ kNm.} \quad \textbf{Ans.}$$

Problem 8.34. *Draw shear force and bending moment diagrams for the overhanging beam shown in Fig. P. 8.34 (a). Find maximum and minimum bending moments. Also locate points of contraflexure.*

Sol. By $\Sigma M_D = 0$, we get $\curvearrowleft V_B \times 8 - (2\times 2)\times\left(\frac{2}{2}+8\right) - 10\times 4 + 4\times 2 = 0$

$\therefore$ $8V_B - 36 - 40 + 8 = 0$

$\therefore$ $V_B = \frac{68}{8} = 8.5 \text{ kN} \uparrow$

By $\Sigma F_V = 0$, we get $(V_B + V_D)\uparrow - (2\times 2 + 10 + 4) = 0$

$\therefore$ $V_D = 18 - V_B = 18 - 8.5 \text{ kN} = 9.5 \text{ kN} \uparrow$

Shear Force Diagram :

$$F_A = 0,$$

$$F_B)_L = -2\times 2 = -4 \text{ kN}$$

$$F_B)_R = -4 + V_B = -4 + 8.5 = 4.5 \text{ kN}$$

$$F_C)_L = 4.5 \text{ kN}$$

$$F_C)_R = 4.5 - 10 = -5.5 \text{ kN}$$

$$F_D)_L = -5.5 \text{ kN}$$

$F_D)_R = -5.5 + V_D = -5.5 + 9.5 = 4$ kN

$F_E)_L = 4$ kN.

The last vector of 4kN ↓ will close the shear force diagram shown in Fig. P. 8.34 (*b*).

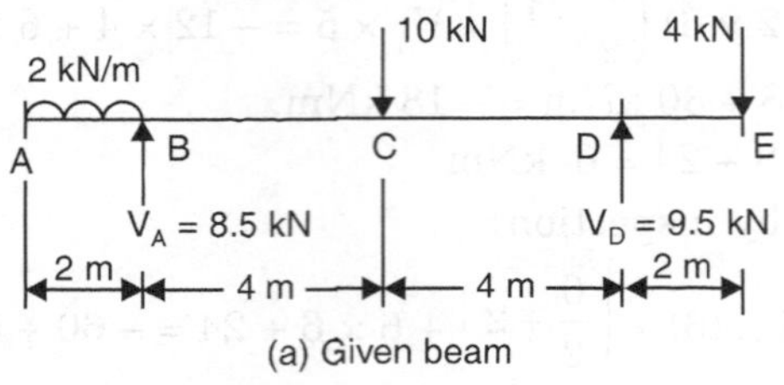

(a) Given beam

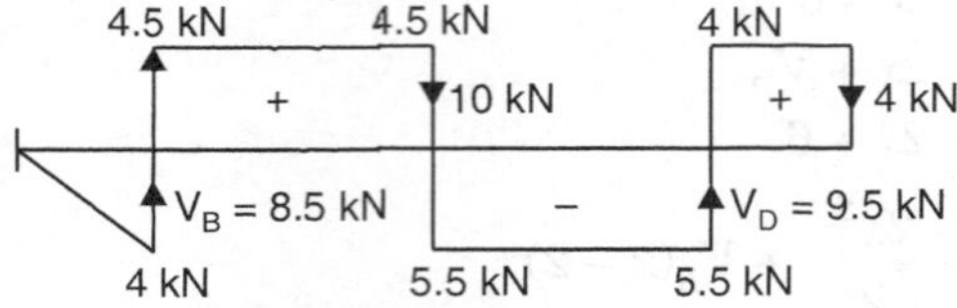

(b) Shear force diagram

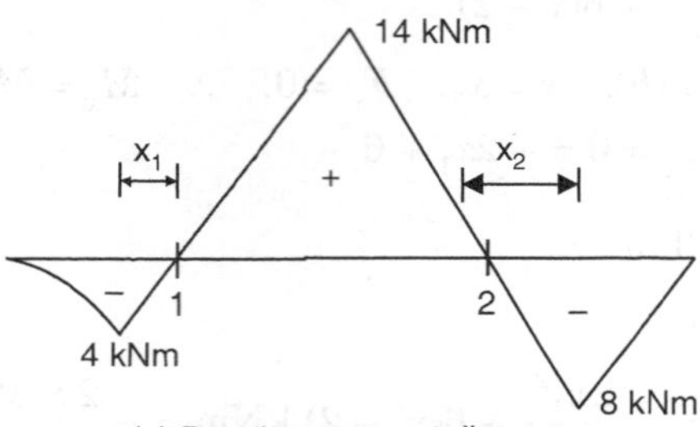

(c) Bending moment diagram

Fig. P. 8.34

Bending Moment Diagram :

$$M_A = 0 ;$$

$$M_B = -2 \times 2 \times \left(\frac{2}{2}\right) = -4 \text{ kNm} ;$$

$$M_D)_R = -4 \times 2 = -8 \text{ kNm}$$

$$M_C = -2 \times 2 \times \left(\frac{2}{2} + 4\right) + 8.5 \times 4 = -20 + 34 = +14 \text{ kNm}.$$

$$M_D)_L = -(2 \times 2)\left(\frac{2}{2} + 8\right) + 8.5 \times 8 - 10 \times 4$$

$$= -36 + 68 - 40 = -8 \text{ kNm } \textbf{(Check)}$$

Bending moment diagram is shown in Fig. P.8.34 (*c*).

The maximum bending moment occurs at *C*. Thus, $M_{max} = +14$ kNm. **Ans.**

And the minimum bending moment occurs at *D*. Thus,

$$M_{min} = -8 \text{ kNm}. \quad \textbf{Ans.}$$

Points of Contraflexure. From the bending moment diagram shown in Fig. P.8.34 (*c*), we find that points of contraflexure are points 1 and 2.

For point 1, we equate the equation for M_x for the portion *BC* equal to zero. Thus, we get

$$M_{x_1} = -2 \times 2 \times \left(\frac{2}{2} + x_1\right) + V_B x_1 = 0$$

$\therefore \quad x_1 (V_B - 4) - 4 \times 1 = 0$

$$\therefore \quad x_1 = \frac{4}{V_B - 4} = \frac{4}{8.5 - 4} = \frac{4}{4.5} \text{ m} = \mathbf{0.889 \text{ m}} \text{ from } B. \quad \textbf{Ans.}$$

For point 2, we write the equation for M_x for the portion CD.

$$\therefore \quad M_{x_2} = -4(x_2 + 2) + V_D \times x_2 = 0$$

$\therefore \quad x_2 (V_D - 4) - 8 = 0$

$$\therefore \quad x_2 = \frac{8}{V_D - 4} = \frac{8}{9.5 - 4} = \frac{8}{5.5} \text{ m} = \mathbf{1.455 \text{ m}} \text{ from } D. \quad \textbf{Ans.}$$

Problem 8.35. *Draw shear force and bending moment diagrams for the beam shown in Fig. P.8.35 (a).*

Sol. By $\Sigma M_B = 0$, we get $\quad V_A \times 6 - \frac{4 \times (7.6)^2}{2} - 8 \times 3 + \frac{4 \times 1^2}{2} + 2 \times 1 = 0$

$$\therefore \quad 6V_A - 115.52 - 24 + 2 + 2 = 0$$

$$\therefore \quad V_A = \frac{135.52}{6} = 22.60 \text{ kN} \uparrow$$

By $\Sigma F_V = 0$, we get $\quad (V_A + V_B) \uparrow - 4 \times 8.6 - 8 - 2 = 0$

$$\therefore \quad V_B = 44.4 - V_A = 44.4 - 22.6 = 21.80 \text{ kN} \uparrow$$

$$F_A)_L = -4 \times 1.6 = -6.4 \text{ kN}$$

$$F_A)_R = -6.4 + V_A = -6.4 + 22.6 \text{ kN} = +16.2 \text{ kN}$$

$$F_D)_L = 16.2 - 4 \times 3 = 4.2 \text{ kN}$$

$$F_D)_R = 4.2 - 8 = -3.8 \text{ kN}$$

$$F_B)_L = -3.8 - 4 \times 3 = -15.8 \text{ kN}$$

$$F_B)_R = -15.8 + V_B = -15.8 + 21.8 = +6.0 \text{ kN}$$

$$F_E)_L = +6.0 - 4 \times 1 = 2 \text{ kN}$$

The last vector of 2 kN closes the SF diagram.

$$M_C = 0,$$

$$M_E = 0;$$

$$M_A = -\frac{4 \times (1.6)^2}{2} = -5.12 \text{ kNm}$$

$$M_D = -\frac{4}{2} \times (4.6)^2 + 22.6 \times 3 = +25.48 \text{ kNm}$$

$$M_B)_R = -2 \times 1 - \frac{4 \times 1^2}{2} = -4 \text{ kNm}$$

Shear force diagram is shown in Fig.P. 8.35 (*b*) while bending moment diagram is shown in Fig. P.8.35 (*c*).

$$M_{max} = 25.48 \text{ kNm at } D. \quad \textbf{Ans.}$$

$$M_{min} = M_A = -5.12 \text{ kNm.} \quad \textbf{Ans.}$$

At point 1, $\quad M_x = 0.$

For the portion AD,

$$M_x = -\frac{4x^2}{2} + V_A (x - 1.6) \quad \text{...(1)}$$

or $$M_x = -2x^2 + 22.6x - 1.6 \times 22.6$$

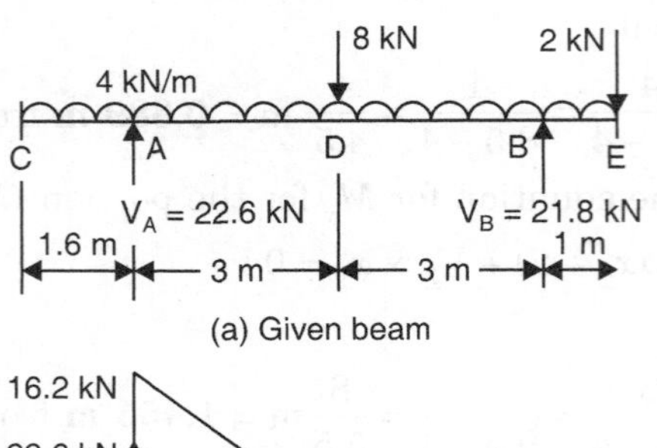

(a) Given beam

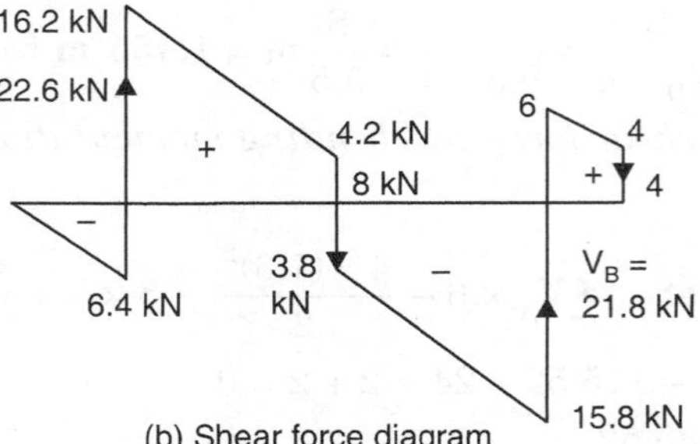

(b) Shear force diagram

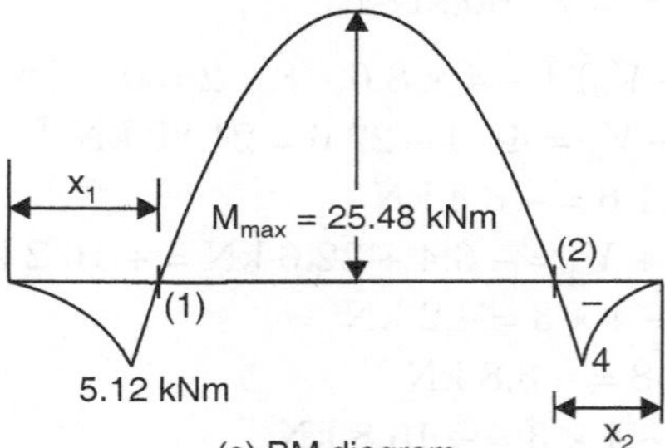

(c) BM diagram

Fig. P. 8.35

or we get $$-2x^2 + 22.6x - 36.16 = 0 \quad \text{...(2)}$$

At point (1), $x = x_1$, $M_x = 0$

$\therefore$ From eqn. (2) we get

$$-2x_1^2 + 22.6x_1 - 36.16 = 0$$

$$\therefore \quad x_1^2 - 11.3x_1 + 18.08 = 0$$

$$\therefore \quad x_1 = \frac{1}{2}\left[+11.3 \pm \sqrt{(11.3)^2 - (18.08) \times 4 \times 1}\right] = \frac{1}{2}[11.3 \pm 7.44] = 9.37, 1.93$$

$\therefore$ x_1 = 1.93 m from *C*. **Ans.**

Similarly, for point 2, with origin *E*, we get

$$M_{x_2} = -2x_2 - \frac{4x_2^2}{2} + 21.8(x_2 - 1) \quad \text{...(3)}$$

$M_{x_2} = 0$ at point (2), $\therefore$ we get

$$-2x_2 - \frac{4}{2}x_2^2 + 21.8(x_2 - 1) = 0$$

or $$-2x_2^2 + 19.8x_2 - 21.8 = 0$$

$$\therefore \quad x_2^2 - 9.9x_2 + 10.9 = 0 \quad \text{...(4)}$$

$$\therefore \quad x_2 = \frac{1}{2}\left[9.9 \pm \sqrt{(9.9)^2 - 4 \times 10.9}\right] = \frac{1}{2}[9.9 \pm 7.376] = 8.638 ; 1.262 \text{ m}$$

$\therefore$ x_2 = **1.262 m from E. Ans.**

Problem 8.36. *Draw shear force and bending moment diagrams for the beam shown in Fig. P.8.36 (a). Find maximum bending moment and the position of point of contraflexure.*

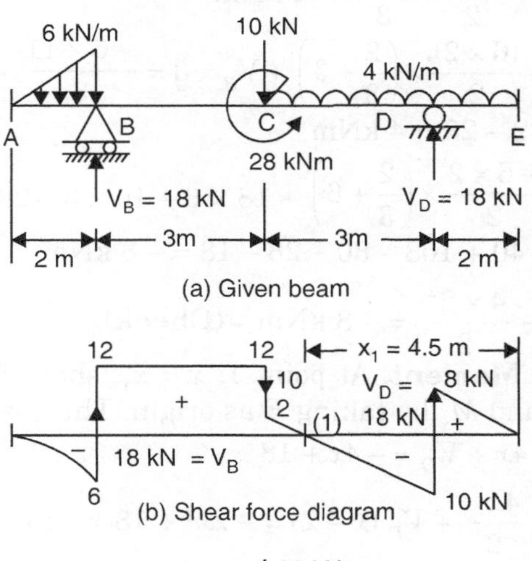

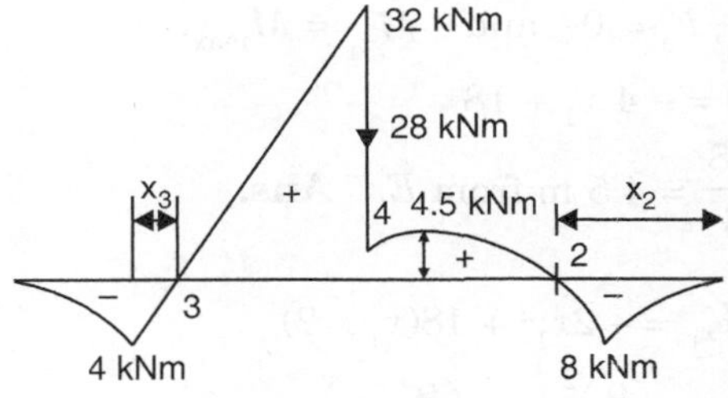

(c) BM diagram

Fig. P. 8.36

Sol. By $\Sigma M_D = 0$, we get

$$\circlearrowleft V_B \times 6 - \left(\frac{6\times 2}{2}\right)\left(\frac{2}{3}+6\right) - 10\times 3 - 28 - \frac{4\times 3^2}{2} + \frac{4\times 2^2}{2} = 0$$

$$6V_B - 40 - 30 - 28 - 18 + 8 = 0$$

$$V_B = \frac{108}{6} = 18 \text{ kN} \uparrow$$

By $\Sigma F_V = 0$, we get

$$(V_B + V_D)\uparrow - \frac{6\times 2}{2} - 10 - 4\times 5 = 0$$

$$\therefore \quad V_D = (6 + 10 + 20) - 18 \text{ kN} = 18 \text{ kN} \uparrow$$

$$F_A = 0\ ;$$

$$F_B)_L = -\frac{6\times 2}{2} = -6 \text{ kN}$$

$$F_B)_R = -6 + V_B = -6 + 18 = 12 \text{ kN}$$

$$F_C)_L = 12 \text{ kN}$$

$$F_C)_R = 12 - 10 = 2 \text{ kN}$$

$$F_D)_L = 2 - 4\times 3 = -10 \text{ kN}$$

$$F_D)_R = -10 + V_D = -10 + 18 = 8 \text{ kN}$$

$F_E = 8 - 4\times 2 = 0$ **(Check)**

$$M_A = 0\,;$$

$$M_B = -\frac{(6\times 2)}{2}\times\frac{2}{3} = -4 \text{ kNm}$$

$$M_C)_L = -\frac{(6\times 2)}{2}\times\left(\frac{2}{3}+3\right) + V_B\times 3 = \frac{-6\times 11}{3} + 18\times 3 = -22+54 = 32 \text{ kNm}$$

$$M_C)_R = 32 - 28 = 4 \text{ kNm}$$

$$M_D)_L = \frac{-6\times 2}{2}\times\left(\frac{2}{3}+6\right) + 18\times 6 - 10\times 3 - 28 - \frac{4\times 3^2}{2} \text{ kNm}$$

$$= -40 + 108 - 30 - 28 - 18 = -8 \text{ kNm}$$

Also, $\qquad M_D)_R = -\dfrac{4\times 2^2}{2} = -8 \text{ kNm}$ **(Check)**

Maximum Bending Moment. At point 1, $x = x_1$, shear force is zero and $M_x = M_{max}$. Thus write equations of F_x and M_x by taking E as origin. Thus, we get

$$F_x = -4x + V_D = -4x + 18 \qquad \text{...(1)}$$

and $$M_x = -\frac{4x^2}{2} + V_D\,(x-2) = -2x^2 + 18(x-2) \qquad \text{...(2)}$$

At point (1), $\quad x = x_1,\ F_x = 0$ and $M_{x_1} = M_{max}$.

$\therefore$ We get $\quad F_{x_1} = 0 = -4\,x_1 + 18$

$$x_1 = \frac{18}{4} = 4.5 \text{ m from } E. \quad \textbf{Ans.}$$

$\therefore$ From (2), we get

$$M_x]_{x=x_1} = M_{x_1} = -2x_1^{\,2} + 18(x_1-2)$$

or $$M_{x_1} = -2\times\left(\frac{9}{2}\right)^2 + 18\left(\frac{9}{2}-2\right) \text{ kNm}$$

$$= -40.5 + 18\times\frac{5}{2} \text{ kNm} = -40.5 + 45 \text{ kNm} = 4.5 \text{ kNm.} \quad \textbf{Ans.}$$

Hence $\quad M_{max} = 32$ kNm. **Ans.**

Point of Contraflexure. At point of contraflexure (point 2), bending moment is zero. Therefore from eqn. (2), we get for point (2).

$$M_x]_{x=x_2} = 0 = -2x_2^{\,2} + 18(x_2-2)$$

$$\therefore \quad -2x_2^{\,2} + 18x_2 - 36 = 0 \quad \text{or} \quad x_2^{\,2} - 9x_2 + 18 = 0$$

$$\therefore \quad x_2 = \frac{1}{2}[9 \pm \sqrt{(9)^2 - 4\times 18}\,] = \frac{1}{2}\,[9\pm 3] = 6 \text{ and } 3$$

$$\therefore \quad x_2 = 3 \text{ m from } E$$

For Point (3), we have $M_x = -\dfrac{6\times 2}{2}\left(x_3+\dfrac{2}{3}\right) + V_B\,x_3 = 0$

$$\therefore \quad -6\left(x_3+\frac{2}{3}\right) + 18x_3 = 0$$

$$\therefore \quad x_3\,(18-6) = \frac{6\times 2}{3}$$

$$\therefore \quad x_3 = \frac{12}{12\times 3} = \frac{1}{3} = 0.33 \text{ m from } B.$$

Problem 8.37. *Find support reactions V_A, V_B and V_C for the beam shown in Fig.P 8.37 (a). Also plot its shear force and bending moment diagrams. Bending moment at the hinge is zero.*

Sol. Hinge at E does not take any moment. The given beam can be split into two beams AE and EC (Fig.P. 8.37 (*b*)). For beam AE, support reactions $V_A = V_E = 20$ kN. For the overhanging beam EC, by using

$$\Sigma M_C = 0, \text{ we get } \quad V_B \times 6 - 20 \times 8 - \frac{4 \times 6^2}{2} = 0$$

$$\therefore \qquad V_B = \frac{1}{6}\,[160 + 72] \text{ kN} = \frac{116}{3} \text{ kN} \uparrow$$

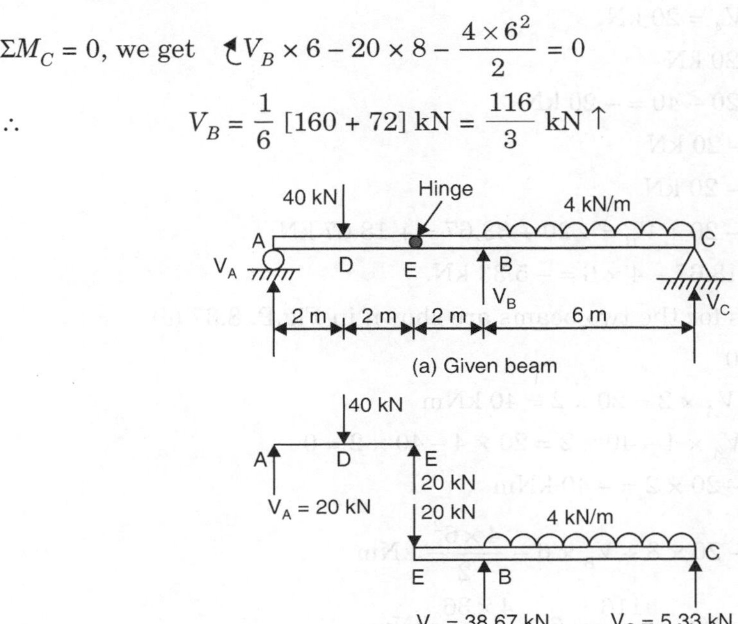

(a) Given beam

(b) Given beam split into two beams

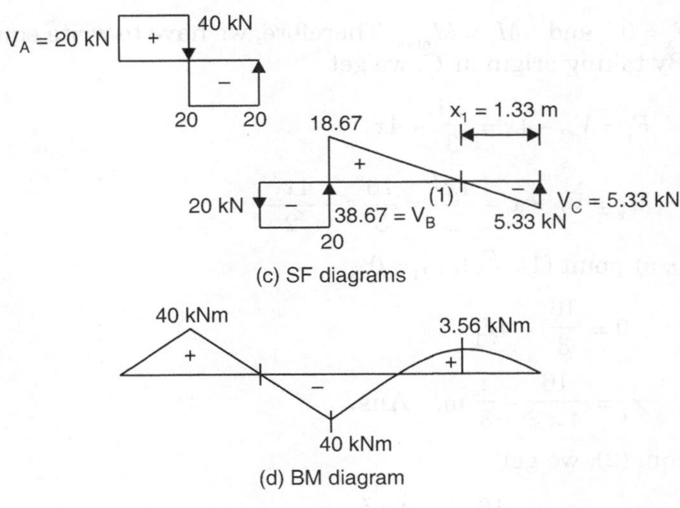

(c) SF diagrams

(d) BM diagram

Fig. P. 8.37

or $\qquad V_B = 38.67 \text{ kN}\uparrow$

By $\Sigma F_V = 0$, we get $(V_B + V_C)\uparrow - (20 + 4 \times 6) = 0$

$$V_C = 44 - \frac{116}{3} = \frac{16}{3} \text{ kN} \uparrow = 5.33 \text{ kN} \uparrow$$

$$F_A = V_A = 20 \text{ kN}$$

$$F_D)_L = 20 \text{ kN}$$

$$F_D)_R = 20 - 40 = -20 \text{ kN}$$

$$F_E = -20 \text{ kN}$$

$$F_B)_L = -20 \text{ kN}$$

$$F_B)_R = -20 + V_B = -20 + 38.67 = +18.67 \text{ kN}$$

$$F_C)_L = 18.67 - 4 \times 6 = -5.33 \text{ kN.}$$

Shear force diagrams for the two beams are shown in Fig.P. 8.37 (*c*)

$$M_A = 0$$

$$M_D = V_A \times 2 = 20 \times 2 = 40 \text{ kNm}$$

$$M_E = V_A \times 4 - 40 \times 2 = 20 \times 4 - 40 \times 2 = 0$$

$$M_B = -20 \times 2 = -40 \text{ kNm}$$

$$M_C = -20 \times 8 + V_B \times 6 - \frac{4 \times 6^2}{2} \text{ kNm}$$

$$= -160 + \frac{116}{3} \times 6 - \frac{4 \times 36}{2} \text{ kNm}$$

$$= -160 + 232 - 72 \text{ kNm} = 0 \quad \textbf{(Check)}$$

At point (1) $F_x = 0$ and $M_x = M_{max}$. Therefore, we have to write equations of F_x and M_x in the portion BC. By taking origin at C, we get

$$F_x = V_C - 4x = \frac{16}{3} - 4x \qquad ...(1)$$

and

$$M_x = V_C \times x - \frac{4x^2}{2} = \frac{16}{3}x - \frac{4x^2}{2} \qquad ...(2)$$

From eqn. (1), at point (1), $F_x]_{x = x_1} = 0$,

$\therefore$ We get $\qquad 0 = \frac{16}{3} - 4x_1$

$\therefore \qquad x_1 = \frac{16}{4 \times 3} = \frac{4}{3}$ m. **Ans.**

And, from eqn. (2), we get

$$M_x]_{x = x_1} = M_{max} = \frac{16}{3}x_1 - \frac{4x_1^2}{2}$$

or $$M_{max} = \frac{16}{3} \times \frac{4}{3} - \frac{4}{2} \times \left(\frac{4}{3}\right)^2 = \frac{64}{9} - \frac{32}{9} \text{ kNm}$$

$$= \frac{64-32}{9} = \frac{32}{9} \text{ kNm} = \mathbf{+\ 3.56\ kNm.} \quad \textbf{Ans.}$$

Problem 8.38. *Draw shear force and bending moment diagrams for the beam shown in Fig. P. 8.38 (a).*

Sol. Total load W = Area of the trapezium $ABB''A'$.

$$= \text{Area of the } \square ABB'A' + \text{Area of } \Delta A'B'B'' = W_1 + W_2$$

The CG of the rectangle is at a distance $l/2$ from B and CG of the triangle is at a distance $l/3$ from B.

$$W_1 = BB' \times AB = w_1 l$$

$$W_2 = \frac{B'B'' \times AB}{2} = \frac{(w_2 - w_1)l}{2}$$

$\therefore$ By $\Sigma M_B = 0$, we get

$$V_A \times l - W_1 \times \frac{l}{2} - W_2 \times \frac{l}{3} = 0$$

$\therefore$ $$V_A = \frac{W_1}{2} + \frac{W_2}{3} \quad \text{or} \quad V_A = \frac{w_1 l}{2} + \frac{(w_2 - w_1)l}{6}$$

And, by $\Sigma F_V = 0$, we get $(V_A + V_B)\uparrow - (W_1 + W_2) = 0$

$\therefore$ $$V_B = W_1 + W_2 - V_A$$

$$= W_1 + W_2 - \left(\frac{W_1}{2} + \frac{W_2}{3}\right) = \frac{W_1}{2} + \frac{2W_2}{3}$$

or $$V_B = \frac{w_1 l}{2} + \frac{2}{3} \times \frac{(w_2 - w_1)l}{2}$$

or $$V_B = \frac{w_1 l}{2} + \frac{(w_2 - w_1)l}{3}$$

At any section x from A, we get

$$F_x = V_A - \square Amm'A' - \Delta A'm'm''$$

$\therefore$ $$F_x = \left(\frac{w_1 l}{2} + \frac{(w_2 - w_1)l}{6}\right) - w_1 x - \frac{(w_2 - w_1)x^2}{2l} \quad \text{...(1)}$$

and, $$M_x = V_A - x - \square Amm'A' \times \bar{x}_m - \Delta A'm'm''.\bar{x}_m$$

or $$M_x = \left(\frac{w_1 l}{2} + \frac{(w_2 - w_1)l}{6}\right)x - \frac{w_1 x^2}{2} - \frac{(w_2 - w_1)x^3}{6l} \quad \text{...(2)}$$

Shear force and bending moment diagrams are shown in Figs. P.8.38 (*b*) and (*c*), respectively.

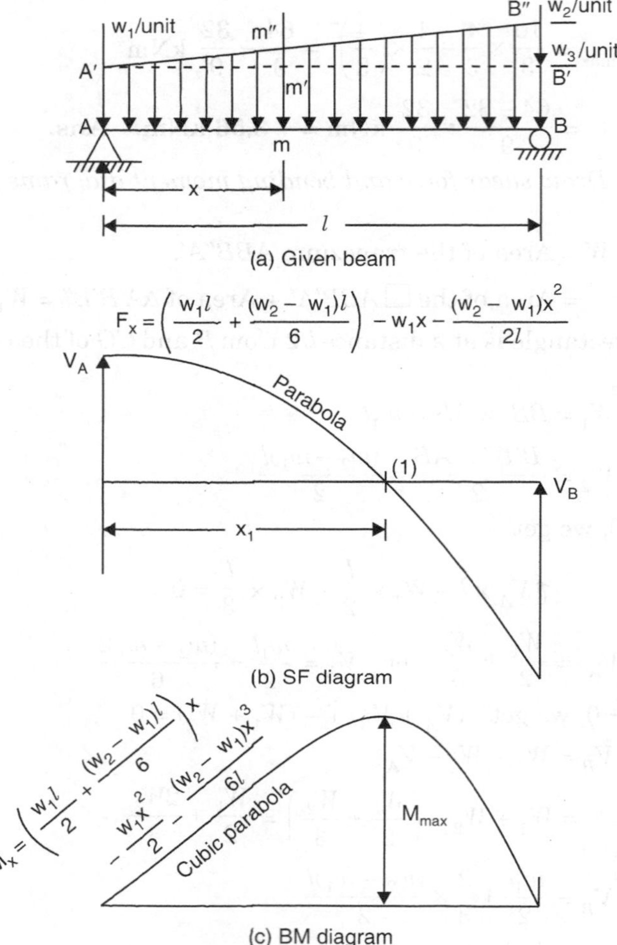

(a) Given beam

(b) SF diagram

(c) BM diagram

Fig. P. 8.38

At section (1), $x = x_1, F_x = 0$ and $M_x = M_{\max}$.

For numericals refer to the problem (8.39), below.

Problem 8.39. *Draw shear force and bending moment diagrams for the beam shown in Fig. P. 8.39 (a).*

Sol. The given load on the beam

$$= W = \text{Area } ABB''A' \quad \text{or} \quad W = \square ABB'A' + \Delta A'B'B''$$

$$= W_1 + W_2 = BB' \times AB + \frac{(B'B'')AB}{2}$$

or $$W = 2 \times 6 + \frac{(5-2) \times 6}{2} \text{ kN} = 12 + \frac{3 \times 6}{2} \text{ kN} = 12 + 9 \text{ kN} \quad \text{or} \quad W = 21 \text{ kN}$$

By $\Sigma M_B = 0$, we get

$$\circlearrowleft V_A \times AB - \frac{W_1 \times AB}{2} - \frac{W_2 \times AB}{3} = 0$$

$$\therefore \quad V_A = \frac{W_1}{2} + \frac{W_2}{3} = \frac{12}{2} + \frac{9}{3} \text{ kN} \quad \text{or} \quad V_A = 9 \text{ kN} \uparrow$$

By $\Sigma F_V = 0$, we get $(V_A + V_B) \uparrow - W = 0$

$\therefore \quad V_B = W - V_A = 21 - 9 = 12 \text{ kN} \uparrow$

At any section x from the origin A, we get

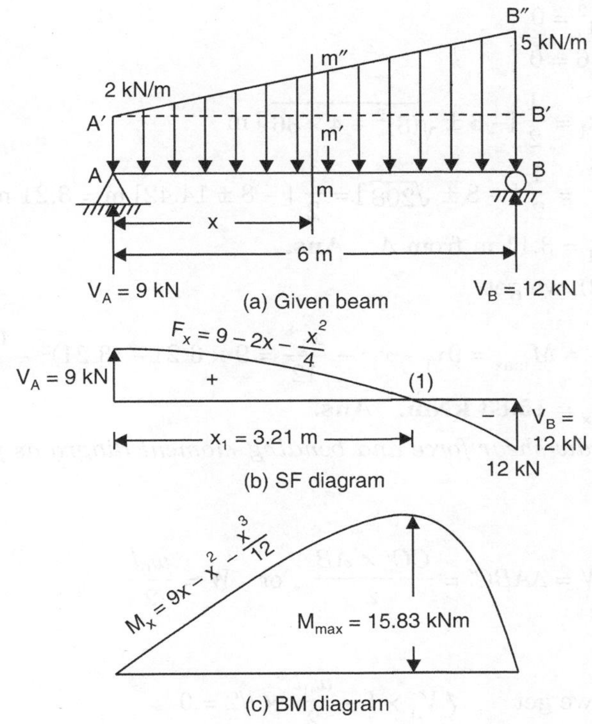

(a) Given beam

(b) SF diagram

(c) BM diagram

Fig. P. 8.39

$$F_x = V_A - \square Amm'A' - \Delta A'm'm'' = 9 - AA' \times Am - \frac{1}{2} \times m'm'' \times A'm'$$

$$= 9 - 2 \times x - \frac{1}{2}\left(\frac{B'B''}{A'B'} \times Am'\right) A'm' = 9 - 2x - \frac{1}{2}\frac{(5-2)}{6} \times x^2$$

or $$F_x = 9 - 2x - \frac{x^2}{4} \qquad \text{...(1)}$$

And, $$M_x = V_A \,.\, x - (\square Amm'A') \times \frac{Am}{2} - \Delta A'm'm'' \times \frac{A'm'}{3}$$

or $$M_x = 9x - \frac{2x^2}{2} - \frac{x^2}{4} \times \frac{x}{3}$$

or $$M_x = 9x - x^2 - \frac{x^3}{12} \qquad \text{...(2)}$$

Eqn. (1) is plotted as shear force diagram as shown in Fig. P. 8.39 (*b*). Similarly, bending moment diagram is the plot of eqn. (2).

From *SF* diagram, we find that at point (1), $F_x = 0$, $x = x_1$ and $\therefore \quad M_{x_1} = M_{max.}$

$\therefore$ From eqn. (1)

$$F_x]_{x=x_1} = 9 - 2x_1 - \frac{x_1^{\,2}}{4} = 0$$

$\therefore$ $36 - 8x_1 - x_1^{\,2} = 0$

$\therefore$ $x_1^{\,2} + 8x_1 - 36 = 0$ ...(3)

$$x_1 = \frac{1}{2}\,[-8 \pm \sqrt{(8)^2 + 4 \times 36}\,]\text{ m}$$

$$= \frac{1}{2}\,[-8 \pm \sqrt{208}\,] = \frac{1}{2}\,[-8 \pm 14.42]\text{ m} = 3.21\text{ m} ; -11.21\text{ m}$$

$\therefore$ $x_1 = 3.11$ m from A. **Ans.**

$\therefore$ From eqn. (2), we get

$$M_x]_{x=x_1} = M_{\max} = 9x_1 - x_1^{\,2} - \frac{x_1^{\,3}}{12} = 9 \times 3.21 - (3.21)^2 - \frac{(3.21)^3}{12}\text{ kNm}$$

$\therefore$ $M_{\max}$ = **15.83 kNm. Ans.**

Problem 40. *Draw shear force and bending moment diagrams for the beam shown in Fig. P.8.40 (a).*

Sol. Total load,

$$W = \Delta ABC' = \frac{CC' \times AB}{2} \quad\text{or}\quad W = \frac{w_0 l}{2}$$

Its CG is at C.

$\therefore$ By $\Sigma M_B = 0$, we get $\quad \circlearrowleft V_A \times l - \frac{w_0 l}{2} \times l/2 = 0$

$$\therefore \quad V_A = \frac{w_0 l}{4} = \frac{W}{2}$$

and by $\Sigma F_V = 0$, we get $(V_A + V_B) \uparrow - W = 0$

$$\therefore \quad V_B = W - V_A = W - \frac{W}{2} \quad\text{or}\quad V_B = \frac{W}{2} = \frac{w_0 l}{4}$$

Take origin at A. Therefore SF at any section is

$$F_x = V_A - \text{external load}$$

$$= V_A - \Delta Amm'$$

$$= \frac{w_0 l}{4} - \frac{1}{2} \times (mm' \times Am)$$

$$= \frac{w_0 l}{4} - \frac{1}{2} \times \left(\frac{CC'}{AC} \times Am\right)(Am) = \frac{w_0 l}{4} - \frac{1}{2} \times \left(\frac{w_0}{l/2} \times x\right)x$$

or $$F_x = \frac{w_0 l}{4} - \frac{w_0 x^2}{l} \qquad \text{...(1)}$$

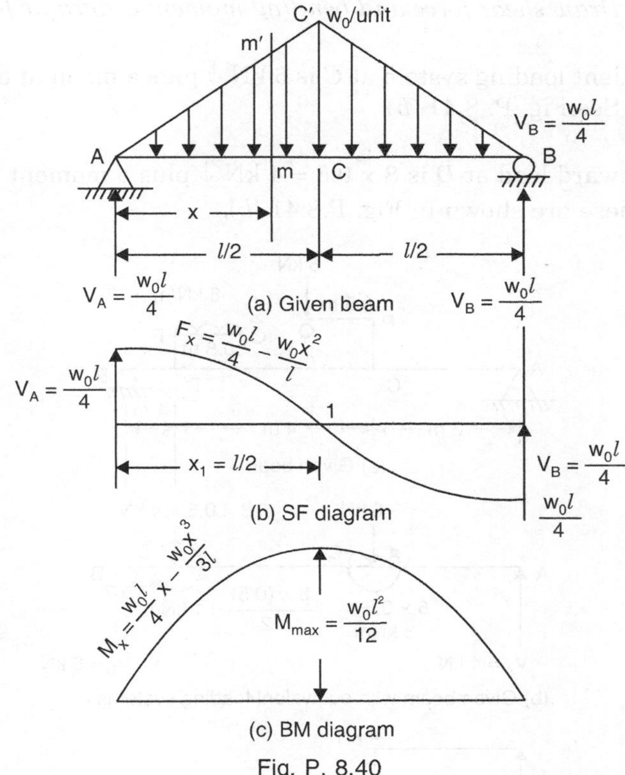

(a) Given beam

(b) SF diagram

(c) BM diagram

Fig. P. 8.40

And, $$M_x = V_A \,.\, x - \Delta Amm' \times \bar{x}_m = \frac{w_0 l}{4} x - \frac{w_0 x^2}{l} \cdot \frac{x}{3}.$$

$$\therefore \quad M_x = \frac{w_0 l}{4} x - \frac{w_0 x^3}{3l} \qquad \text{...(2)}$$

Eqn. (1) is parabolic and eqn. (2) represents a cubic parabola.

From (1), we find the point where shear is zero. At the section $x = x_1$, shear force is zero and M_x will be M_{max}.

$\therefore$ From eqn. (1), we get

$$F_x]_{x = x_1} = \frac{w_0 l}{4} - \frac{w_0 x_1^{\,2}}{l} = 0$$

$$\therefore \quad x_1^{\,2} = \frac{l^2}{4}, \quad \therefore \quad x = \frac{l}{2}.$$

$\therefore$ From eqn. (2), we get

$$M_x]_{x = x_1} = M_{max} = \frac{w_0 l}{4} x_1 - \frac{w_0 x_1^{\,3}}{3l}$$

or $$M_{max} = \frac{w_0 x_1}{12l}[3l^2 - 4x_1^{\,2}] = \frac{w_0}{12l} \times \left(\frac{l}{2}\right) [3l^2 - 4 \times (l/2)^2] = \frac{w_0 l}{24l} \times l^2[3 - l].$$

$$\therefore \quad M_{max} = \frac{w_0 l^2}{12}. \quad \textbf{Ans.}$$

Shear force and bending moment diagrams are shown in Figs. P. 8.40 (*b*) and (*c*).

Problem 8.41. *Draw shear force and bending moment diagrams for the beam shown in Fig. P.8.41 (a).*

Sol. The equivalent loading system at C is 6 kN ↓ plus a moment equal $6 \times 0.5 = 3$ kNm clockwise acting at C. See Fig. P. 8.41 (*b*).

Similarly, downward load at D is $8 \times 0.5 = 4$ kN ↓ plus a moment $= \dfrac{8 \times (0.5)^2}{2} = 1$ kNm (counterclockwise). These are shown in Fig. P.8.41 (*b*).

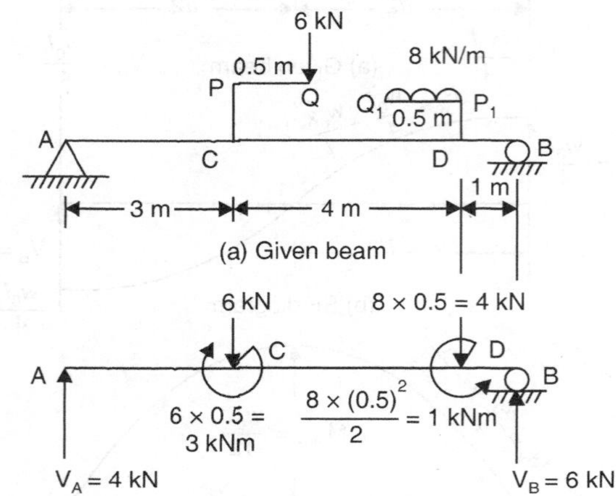

(a) Given beam

(b) Given beam with equivalent loading systems

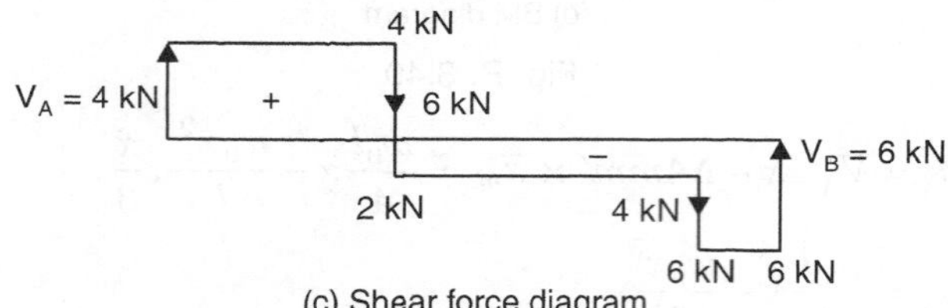

(c) Shear force diagram

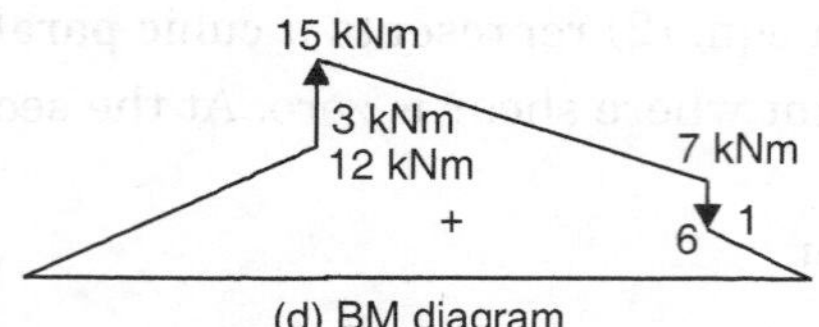

(d) BM diagram

Fig. P.8.41

By $\Sigma M_B = 0$, we get $\circlearrowleft V_A \times 8 - 6 \times 5 + 3 - 4 \times 1 - 1 = 0$

$$\therefore \quad V_A = \frac{1}{8}(30 - 3 + 4 + 1) = \frac{32}{8} = 4 \text{ kN} \uparrow$$

By $\Sigma F_V = 0$, we get $(V_A + V_B) \uparrow - (6 + 4) = 0$

$$\therefore \quad V_B = 10 - V_A = 10 - 4 = 6 \text{ kN} \uparrow$$

$$F_A = V_A = 4 \text{ kN}$$

$$F_C)_L = 4 \text{ kN};$$

$$F_C)_R = 4 - 6 = -2 \text{ kN}$$

$F_D)_L = -2$ kN

$F_D)_R = -2 - 4 = -6$ kN

$F_B = -6$ kN. The last vector ($V_B = 6$ kN) closes the shear force diagram.

$M_A = 0,$

$M_B = 0,$

$M_C)_L = V_A \times 3 = 4 \times 3 = 12$ kNm

$M_C)_R = 12 + 3 = 15$ kNm

$M_D)_L = 4 \times 7 + 3 - 6 \times 4 = 7$ kNm

$M_D)_R = M_D)_L - 1 = 6$ kNm

$M_B = 4 \times 8 - 6 \times 5 + 3 - 4 \times 1 - 1$

$= 32 - 30 + 3 - 4 - 1 = 35 - 35 = 0$ **(check).**

Problem 8.42. *Draw shear force and bending moment diagrams for the beam shown in Fig. P.8.42 (a).*

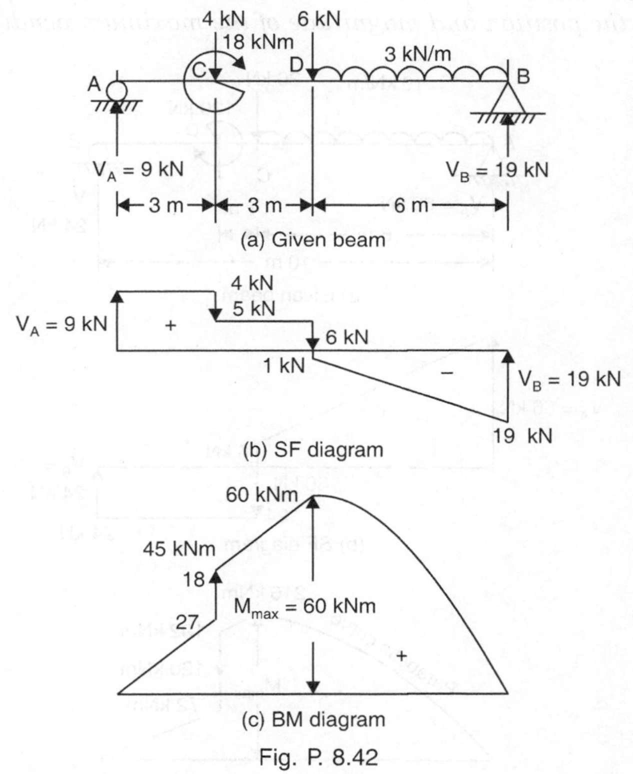

Fig. P. 8.42

Sol. By $\Sigma M_B = 0$, we get $\circlearrowleft V_A \times 12 - 4 \times 9 + 18 - 6 \times 6 - \dfrac{3 \times 6^2}{2} = 0$

$$\therefore \quad V_A = \frac{1}{12}[36 - 18 + 36 + 54] = 9 \text{ kN} \uparrow$$

By $\Sigma F_V = 0$ we get $(V_A + V_B) \uparrow - (4 + 6 + 3 \times 6) = 0$

$$\therefore \quad V_B = 28 - V_B = 28 - 9 = 19 \text{ kN} \uparrow$$

$F_A = V_A = 9$ kN

$F_C)_L = 9$ kN ;

$F_C)_R = F_C)_L - 4 = 5$ kN

$F_D)_L = 5$ kN ;

$F_D)_R = 5 - 6 = -1$ kN

$F_B = -1 - 3 \times 6 = -19$ kN.

The last vector $V_B = 19$ kN ↑ will close the shear force diagram as shown in Fig. P.8.42 (*b*).

$M_A = 0,$

$M_C)_L = V_A \times 3 = 9 \times 3 = 27$ kNm

$M_C)_R = M_C)_L + 18 = 27 + 18 = 45$ kNm

$M_D = 9 \times 6 - 4 \times 3 + 18 = 54 - 12 + 18 = 60$ kNm

$$M_B = 9 \times 12 - 4 \times 9 + 18 - 6 \times 6 - \frac{3 \times 6^2}{2} \text{ kNm}$$

$= 108 - 36 + 18 - 36 - 54$ kN $= 126 - 126 = 0.$ **(Check)**

Problem 8.43. *Draw shear force and bending moment diagrams for the beam shown in Fig. P. 8.43 (a). Find the position and magnitude of the maximum bending moment.*

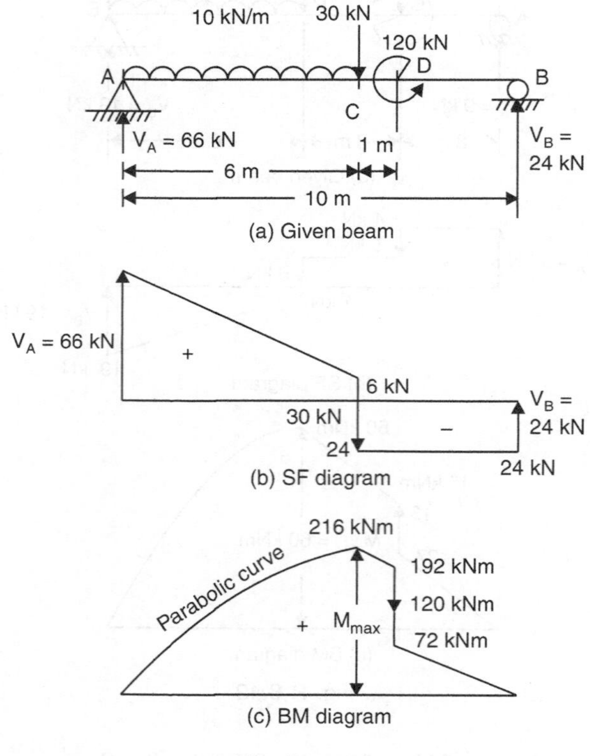

(a) Given beam

(b) SF diagram

(c) BM diagram

Fig. P. 8.43

Sol. By $\Sigma M_B = 0$, we get $\circlearrowleft V_A \times 10 - 10 \times 6 \times \left(\frac{6}{2} + 4\right) - 30 \times 4 - 120 = 0$

$$\therefore \quad V_A = \frac{1}{10}[420 + 120 + 120] \text{ kN} = 66 \text{ kN} \uparrow$$

By $\Sigma F_V = 0$, we get $(V_A + V_B) \uparrow - 10 \times 6 - 30 = 0$

$\therefore \quad V_B = 90 - V_A = 90 - 66$ kN $= 24$ kN ↑

or $F_A = V_A = +\,66 \text{ kN}$

$$F_C)_L = 66 - 10 \times 6 = 6 \text{ kN}$$

$$F_C)_R = F_C)_L - 30 = 6 - 30 = -\,24 \text{ kN}$$

$$F_D = -\,24 \text{ kN}$$

$$F_B = -\,24 \text{ kN}$$

The last vector $V_B = 24 \text{ kN} \uparrow$ will close the shear force diagram shown in Fig. P. 8.43(*b*).

$$M_A = M_B = 0$$

$$M_C = V_A \times 6 - \frac{10 \times 6^2}{2} \text{ kNm} = 66 \times 6 - 180 \text{ kNm} = 396 - 180$$

$$\therefore \quad M_C = 216 \text{ kNm}$$

$$M_D)_L = 66 \times 7 - (10 \times 6) \times \left(\frac{6}{2} + 1\right) - 30 \times 1 \text{ kNm}$$

$$= 462 - 240 - 30 = 192 \text{ kNm}$$

$$M_D)_R = M_D)_L - 120 = 192 - 120 = 72 \text{ kNm}$$

From Fig.P. 8.43 (*c*), maximum bending occurs at C and $M_{max} = M_C$ = **216 kNm. Ans.**

Problem 8.44. *Draw shear force and bending moment diagrams for the beam shown in Fig. P. 8.44 (a). Find maximum bending moment and the position of point of contraflexure.*

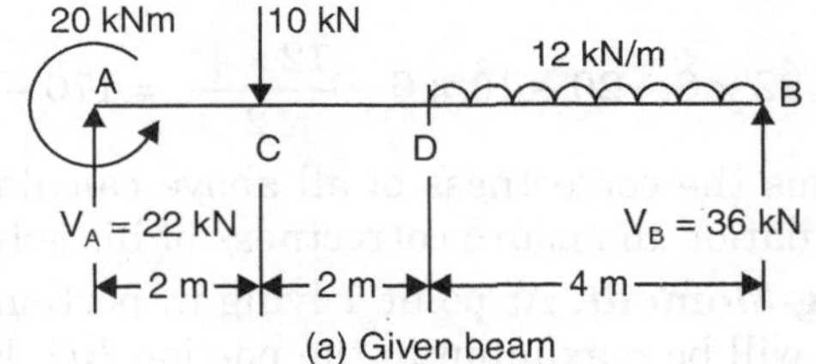

(a) Given beam

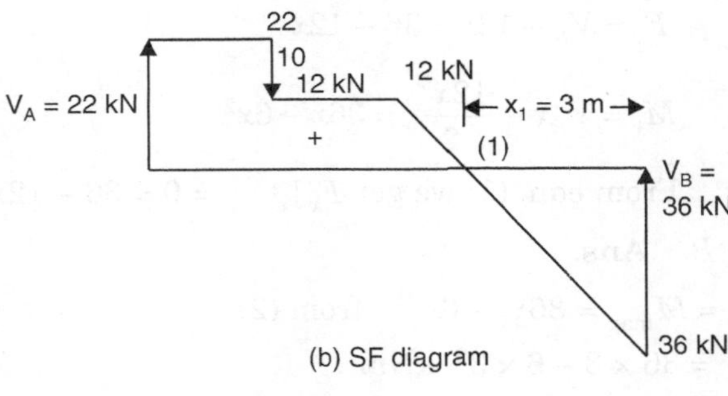

(b) SF diagram

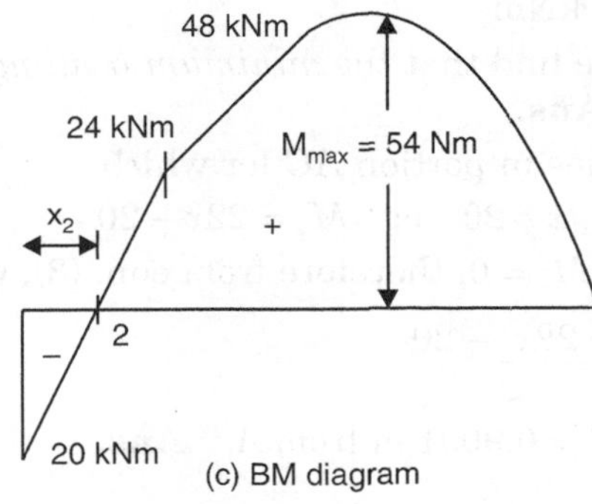

(c) BM diagram

Fig. P. 8.44

Sol. By $\Sigma M_B = 0$, we get $\circlearrowleft V_A \times 8 - 20 - 10 \times 6 - \frac{12 \times 4^2}{2} = 0$

$$\therefore \quad V_A = \frac{1}{8}[20 + 60 + 96] = \frac{176}{8} = 22 \text{ kN} \uparrow$$

By $\Sigma F_V = 0$, we get $(V_A + V_B) \uparrow - (10 + 12 \times 4) = 0$

$$\therefore \quad V_B = 58 - V_A = 58 - 22 = 36 \text{ kN} \uparrow$$

Shear Force Diagram

$$F_A = V_A = 22 \text{ kN}$$
$$F_C)_L = 22 \text{ kN}$$
$$F_C)_R = 22 - 10 = 12 \text{ kN}$$
$$F_D = 12 \text{ kN}$$
$$F_B)_L = 12 - 12 \times 4 = -36 \text{ kN}$$

When the last vector representing $V_B = 36$ kN ↑ is added the shear force diagram closes.

Bending Moment Diagram

$$M_A = -20 \text{ kNm}$$
$$M_C = V_A \times 2 - 20 = 22 \times 2 - 20 \text{ kNm} = 24 \text{ kNm}$$
$$M_D = 22 \times 4 - 20 - 10 \times 2 = 88 - 20 - 20 = 48 \text{ kNm}$$
$$M_B = 22 \times 8 - 20 - 10 \times 6 - \frac{12 \times 4^2}{2} = 176 - 20 - 60 - 96 = 0 \quad \textbf{(Check)}$$

[This "check" confirms the correctness of all above calculations. Student is advised to apply "check" in the examination to ensure correctness of his solution].

Maximum Bending Moment. At point 1 lying in portion *DB* shear force is zero and therefore bending moment will be maximum in the portion *BD*. For the portion *BB*,

Shear force $\quad F_x = V_B - 12x = 36 - 12x \qquad$ ...(1)

and, bending moment $\quad M_x = V_B x - \frac{12x^2}{2} = 36x - 6x^2 \qquad$ ...(2)

when $x = x_1$, $F_x = 0$, $\therefore$ From eqn. (1) we get $F_x]_{x = x_1} = 0 = 36 - 12x$,

$\therefore \; x_1 = 3$ m from *B*. **Ans.**

And, $\quad M_x]_{x = x_1} = M_{max} = 36x_1 - 6x_1^2 \quad$ from (2)

$$= 36 \times 3 - 6 \times 3^2 \text{ kNm}$$

or $\quad M_{max} = 54$ kNm

From *BM* diagram, we find that the *minimum bending moment* (M_{min}) occurs at point *A*. $\therefore$ $M_{min} = -20$ kNm. **Ans.**

Point of contraflexure lies in portion *AC* for which

$$M_x = V_A \,.\, x - 20 \quad \text{or} \quad M_x = 22x - 20 \qquad ...(3)$$

At point (2), $x = x_2$ and $M_x = 0$, therefore from eqn. (3), we get

$$M_x]_{x = x_2} = 0 = 22x_2 - 20$$

$$\therefore \quad x_2 = \frac{20}{22} = 0.9091 \text{ m from } A. \quad \textbf{Ans.}$$

Problem 8.45. *Draw shear force and bending moment diagrams for the simply supported beam shown in Fig. P. 8.45 (a).*

Sol. By $\Sigma M_B = 0$, we get $\circlearrowleft V_A \times 8 - 6 + 6 - 8 = 0$

$$\therefore \quad V_A = \frac{8}{8} = 1 \text{ kN} \uparrow$$

By $\Sigma F_V = 0$, we get $(V_A + V_B) \uparrow = 0$

$$\therefore \quad V_B = -V_A = -1 \text{ kN} \quad \text{or} \quad V_B = 1 \text{ kN} \downarrow$$

$$F_A = V_A = 1 \text{ kN}$$

$$F_C = 1 \text{ kN};$$

$$F_D = F_E = 1 \text{ kN}$$

$$F_B = 1 \text{ kN},$$

By the addition of last vector the shear force diagram closes, as shown in Fig. P. 8.45 (*b*).

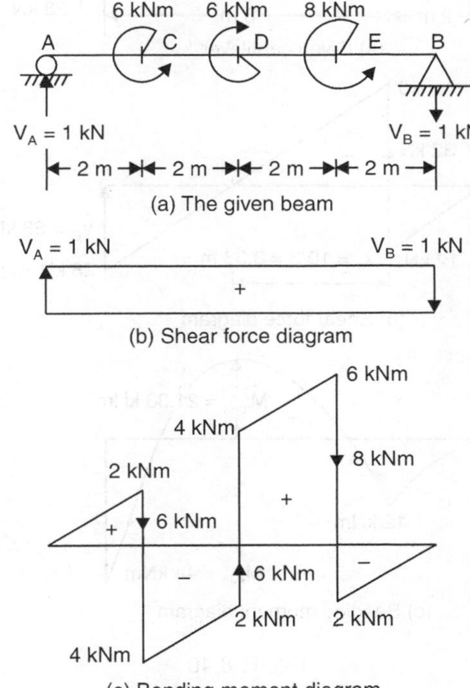

Fig. P. 8.45

$$M_A = 0,$$

$$M_B = 0$$

$$M_C)_L = V_A \times 2 = 1 \times 2 \text{ kNm} = 2 \text{ kNm}$$

$$M_C)_R = 2 - 6 = -4 \text{ kNm}$$

$$M_D)_L = V_A \times 4 - 6 = 1 \times 4 - 6 \quad \text{or} \quad M_D)_L = -2 \text{ kNm}.$$

$$M_D)_R = -2 + 6 = +4 \text{ kNm}$$

$$M_E)_L = 1 \times 6 - 6 + 6 = 6 \text{ kNm}$$

$$M_E)_R = M_E)_L - 8 = 6 - 8 = -2 \text{ kNm}$$

$$M_B = 1 \times 8 - 6 + 6 - 8 = 0 \quad \textbf{(Check).}$$

Bending moment diagram is shown in Fig. P.8.45 (*c*).

Problem 8.46. *Draw shear force and bending moment diagrams for the cantilever shown in Fig. P.8.46 (c).*

Sol. $F_A = 0$

$F_B)_L = -6 \times 2 = -12$ kN

$F_B)_R = -12 + 32 = +20$ kN

$F_C)_L = -6 \times 10 + 32 = -28$ kN

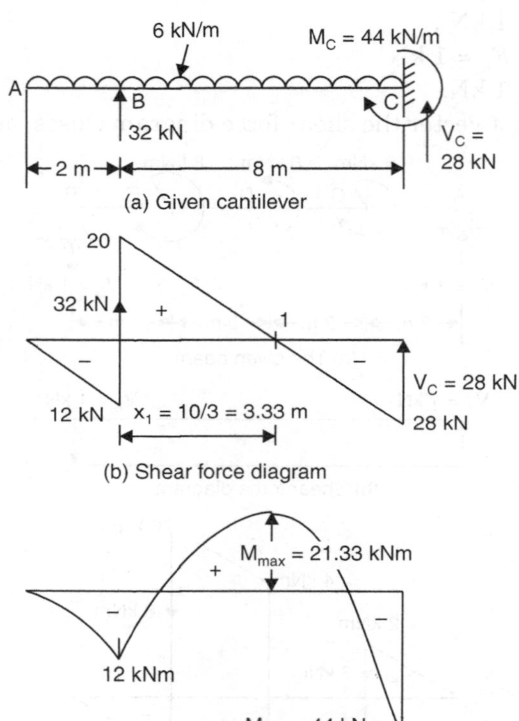

(a) Given cantilever

(b) Shear force diagram

(c) Bending moment diagram

Fig. P. 8.46

$$M_A = 0\,;$$

$$M_B = -\frac{6 \times 2^2}{2} = -12 \text{ kNm},$$

$$M_C = \frac{-6 \times 10^2}{2} + 32 \times 8 = -300 + 256 \text{ kNm} = -44 \text{ kNm}$$

and support moment at $C = 44$ kNm

At point (1), $x = x_1$ from B,

$$F_x = 0 \quad \text{and} \quad M_x = M_{max}.$$

Point (1) lies in portion *BC*.

To find x_1 and M_{max}, we first write down the equations of F_x and M_x for the portions *BC*.

Now $$F_x = -6(x + 2) + 32 \quad ...(1)$$

$$M_x = -\frac{6}{2}(2 + x)^2 + 32x \quad ...(2)$$

From eqn. (1) for point (1),

$$F_x]_{x = x_1} = 0 = -6(x_1 + 2) + 32,$$

$$\therefore \quad x_1 = \frac{1}{6}(32 - 12) = 3.3 \text{ m}$$

$\therefore$ From eqn. (2), $M_x]_{x=x_1} = M_{max} = -\frac{6}{2}(2 + x_1)^2 + 32x_1$

or $$M_{max} = -\frac{6}{2}\left(2 + \frac{10}{3}\right)^2 + 32 \times \frac{10}{3}$$

$$= -3 \times \left(\frac{16}{3}\right)^2 + 32 \times \frac{10}{3} = 21.33 \text{ kNm}$$

Shear force and bending moment diagrams are shown in Figs. P. 8.46 (*b*) and (*c*) respectively. We find that M_{min} occurs at point *C*.

$$\therefore \quad M_{min} = -44 \text{ kNm.}$$

Problem 8.47. *Draw shear force and bending moment diagrams for the beam shown in Fig. P. 8.47 (a).*

Sol. By $\Sigma M_B = 0$, we get

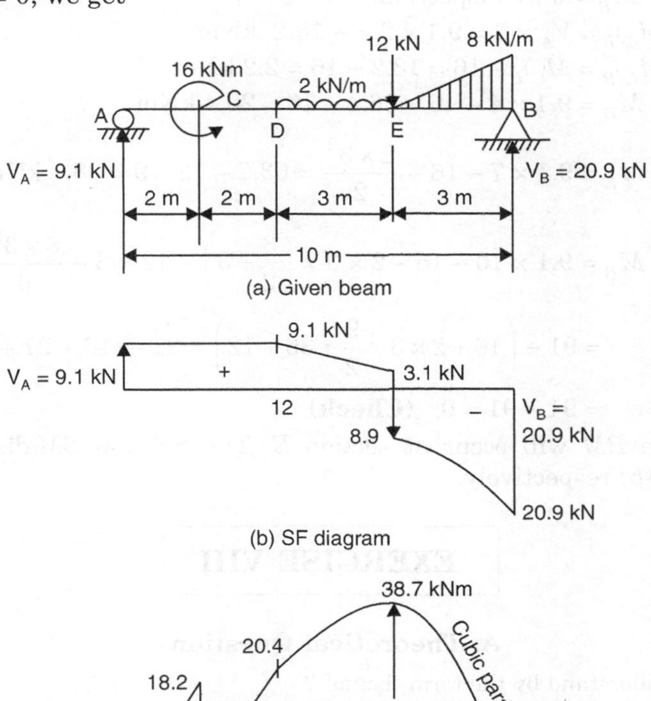

(a) Given beam

(b) SF diagram

(c) BM diagram

Fig. P.8.47

$$\curvearrowleft V_A \times 10 - 16 - (2 \times 3)\left(\frac{3}{2} + 3\right) - 12 \times 3 - \frac{8 \times 3}{2} \times \frac{3}{3} = 0$$

or $$10V_A - 16 - 27 - 36 - 12 = 0$$

$\therefore$ $$V_A = \frac{1}{10} \times (91) = 9.1 \text{ kN} \uparrow$$

By $\Sigma F_V = 0$, we get $(V_A + V_B) \uparrow - \left(2 \times 3 + 12 + \frac{8 \times 3}{2}\right) = 0$

$\therefore$ $$V_B = 30 - V_A = 30 - 9.1 \text{ kN} \quad \text{or} \quad V_B = 20.9 \text{ kN} \uparrow$$

$$F_A = V_A = 9.1 \text{ kN}$$

$$F_C = 9.1 \text{ kN}$$

$$F_D = 9.1 \text{ kN}$$

$$F_E)_L = 9.1 - 2 \times 3 = 3.1 \text{ kN}$$

$$F_E)_R = F_E)_L - 12 = 3.1 - 12 = -8.9 \text{ kN}$$

$$F_B = 8.9 - \frac{1}{2} \times 8 \times 3 = -20.9 \text{ kN}$$

The last vector $V_B = 20.9$ kN $\uparrow$ will close the *SF* diagram.

$$M_A = 0\,;$$

$M_B = 0$ by inspection.

$$M_C)_L = V_A \times 2 = 9.1 \times 2 = +18.2 \text{ kNm}$$

$$M_C)_R = M_C)_L - 16 = 18.2 - 16 = 2.2 \text{ kNm}$$

$$M_D = 9.1 \times 4 - 16 = 36.4 - 16 = 20.4 \text{ kNm}$$

$$M_E = 9.1 \times 7 - 16 - \frac{2 \times 3^2}{2} = 63.7 - 16 - 9 = 38.7 \text{ kNm}$$

$$M_B = 9.1 \times 10 - 16 - 2 \times 3 \times \left(\frac{3}{2} + 3\right) - 12 \times 3 - \frac{8 \times 3^2}{6} \text{ kN}$$

$$= 91 - \left(16 + 2 \times 3 \times \frac{9}{2} + 36 + 12\right) = 91 - (16 + 27 + 36 + 12)$$

$$= 91 - 91 = 0 \quad \textbf{(Check)}$$

The maximum *BM* will occur at section *E*. The *SF* and *BM* diagrams are shown in Figs. P.8.47 (*a*) and (*b*) respectively.

EXERCISE VIII

A. Theoretical Questions

1. What do you understand by the term "beam" ?
2. How many types of supports are known to you ? Explain with the help of neat sketches and suitable examples.
3. How many equations of equilibrium are used in finding support reactions ?
4. Explain the term "types of beams".
5. What do you mean by determinate beams ? Explain.

6. What is a simply supported beam ?
7. What do you understand by propped "cantilever" ?
8. What do you understand by "indeterminate beams" ? Explain with the help of suitable examples.
9. What is a "Fixed beam"?
10. What do you understand by "continuous beams" ?
11. What "types of loads" are known to you ? Explain.
12. What do you understand by the terms "Shear force" and "bending moment" in beams ?
13. Write the procedure for writing of equation of shear force in a beam.
14. What are the sign conventions for positive shear force and positive bending moment ?
15. What are the sign conventions for negative shear force and negative bending moment ?
16. Establish relationships between rate of loading, shear force and bending moment at any section in a beam.
17. Where does the maximum bending moment occur in a beam ?
18. What do you mean by the term "point of contraflexure" or the "point of contraflexion" in a beam ?
19. Draw shear force and bending moment diagrams for a simply supported beam carrying a load P inclined at an angle θ to the horizontal and applied at a distance a from left support and b from the right support. Take span equal to l.
20. Draw shear force and bending moment diagrams for a simply supported beam of span l carrying a uniformly varying load, rate of loading being zero at the left support and w_0 per unit at the right support.

B. Unsolved Problems

Exercises 8.1 to 8.42.

Draw shear force and bending moment diagrams for the beams shown in Fig. E. 8.1 to Fig. E. 8.42 Find M_{max} and the position of point of contraflexure, if any.

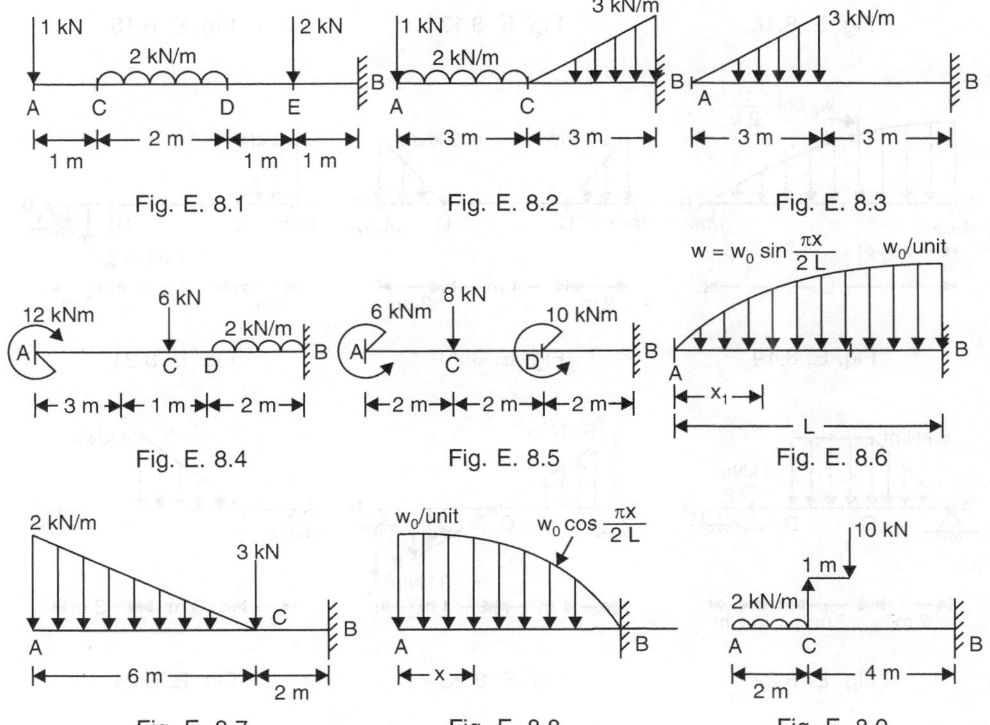

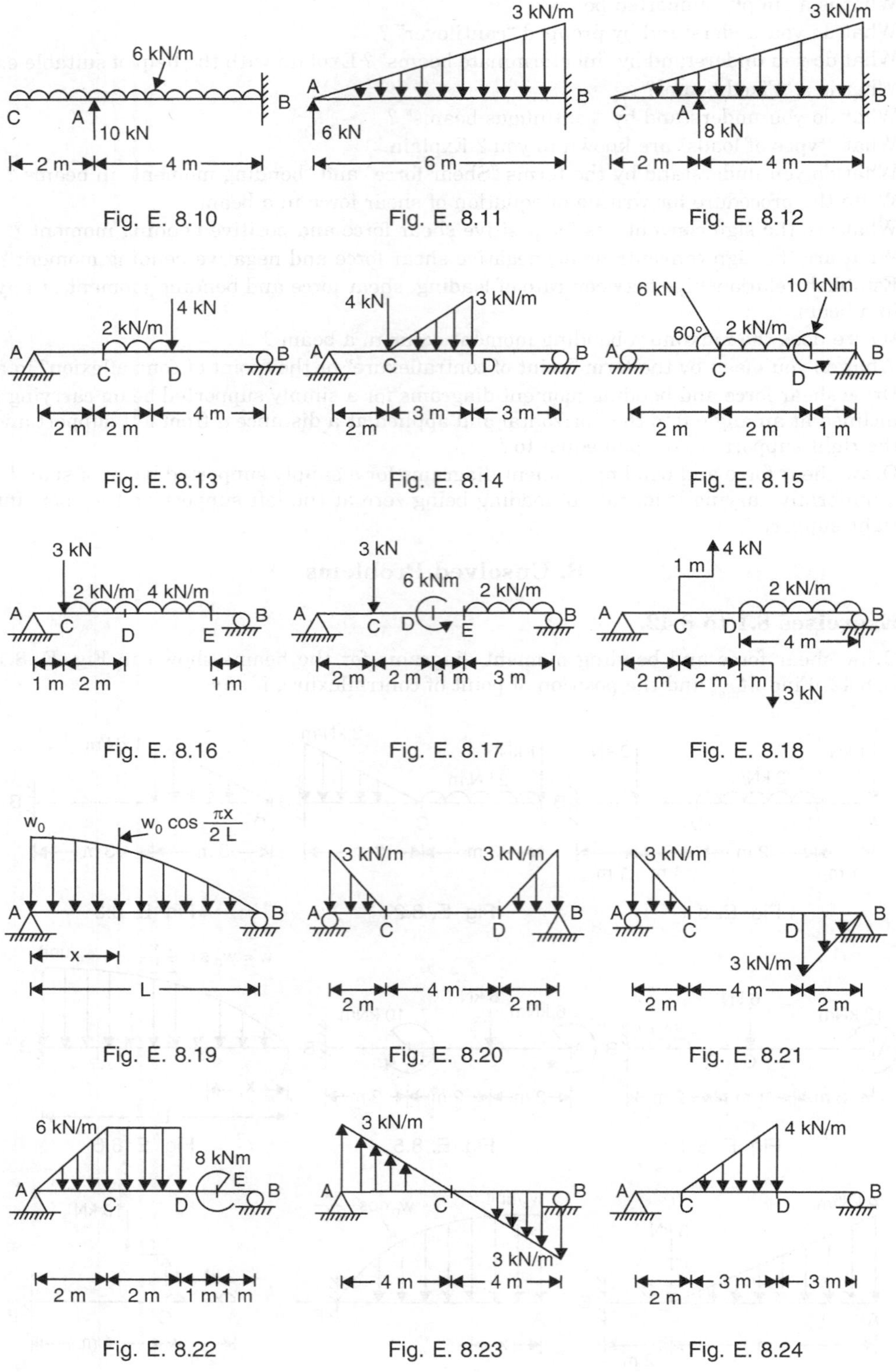

Fig. E. 8.10 Fig. E. 8.11 Fig. E. 8.12

Fig. E. 8.13 Fig. E. 8.14 Fig. E. 8.15

Fig. E. 8.16 Fig. E. 8.17 Fig. E. 8.18

Fig. E. 8.19 Fig. E. 8.20 Fig. E. 8.21

Fig. E. 8.22 Fig. E. 8.23 Fig. E. 8.24

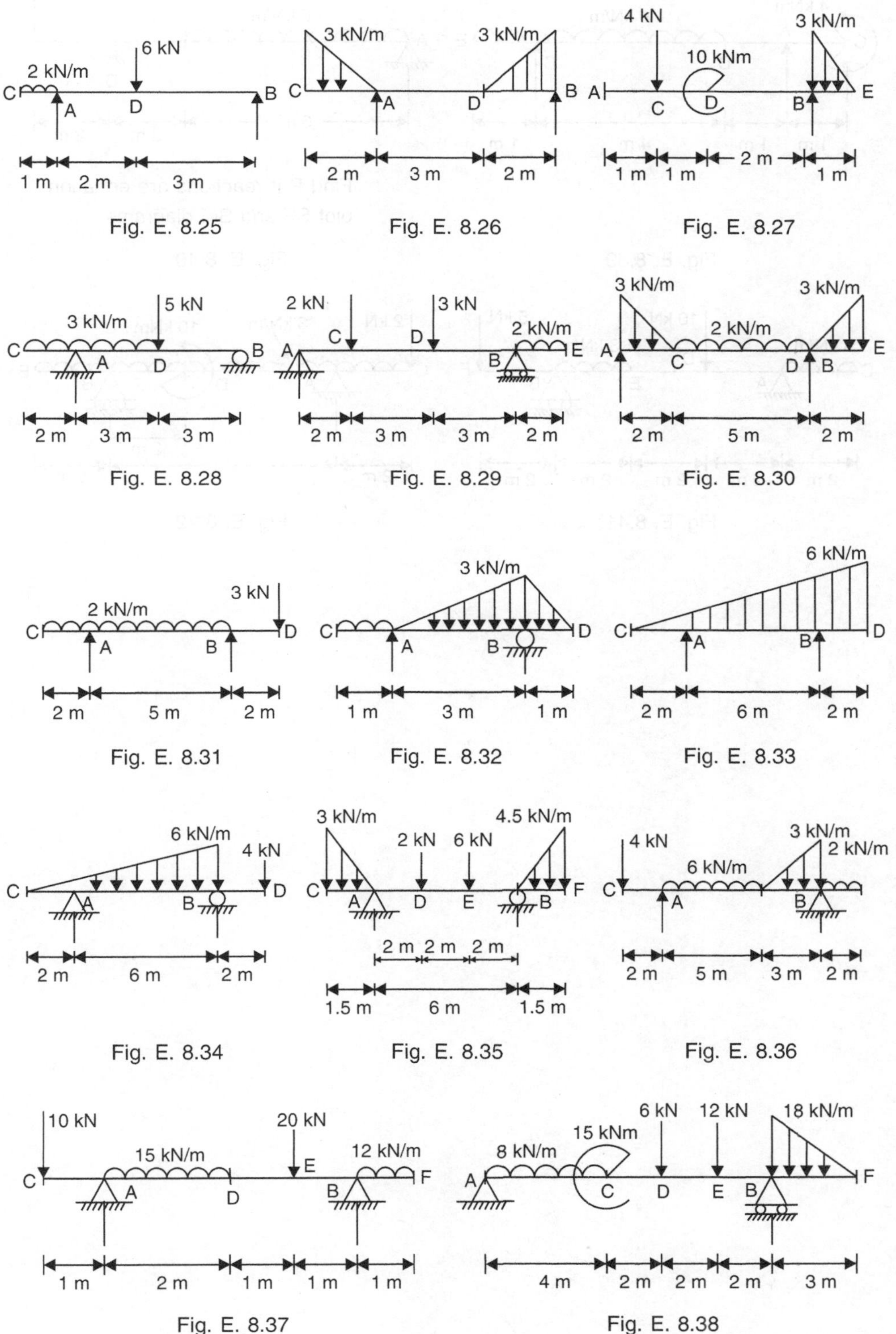
2 kN/m
6 kN
C A D B
1 m 2 m 3 m
Fig. E. 8.25
3 kN/m
3 kN/m
C A D B
2 m 3 m 2 m
Fig. E. 8.26
4 kN
10 kNm
3 kN/m
A C D B E
1 m 1 m 2 m 1 m
Fig. E. 8.27
3 kN/m
5 kN
C A D B
2 m 3 m 3 m
Fig. E. 8.28
2 kN
3 kN
2 kN/m
A C D B E
2 m 3 m 3 m 2 m
Fig. E. 8.29
3 kN/m
2 kN/m
3 kN/m
A C D B E
2 m 5 m 2 m
Fig. E. 8.30
2 kN/m
3 kN
C A B D
2 m 5 m 2 m
Fig. E. 8.31
3 kN/m
C A B D
1 m 3 m 1 m
Fig. E. 8.32
6 kN/m
C A B D
2 m 6 m 2 m
Fig. E. 8.33
6 kN/m
4 kN
C A B D
2 m 6 m 2 m
Fig. E. 8.34
3 kN/m
2 kN
6 kN
4.5 kN/m
C A D E B F
2 m 2 m 2 m
1.5 m 6 m 1.5 m
Fig. E. 8.35
4 kN
6 kN/m
3 kN/m
2 kN/m
C A B
2 m 5 m 3 m 2 m
Fig. E. 8.36
10 kN
15 kN/m
20 kN
12 kN/m
C A D E B F
1 m 2 m 1 m 1 m 1 m
Fig. E. 8.37
8 kN/m
15 kNm
6 kN
12 kN
18 kN/m
A C D E B F
4 m 2 m 2 m 2 m 3 m
Fig. E. 8.38

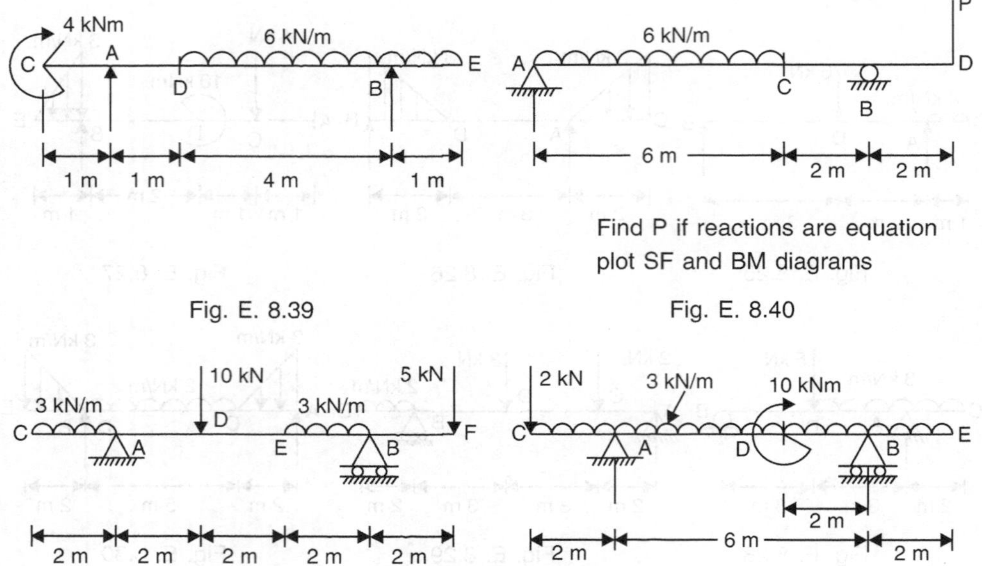

Fig. E. 8.39

Fig. E. 8.40

Fig. E. 8.41

Fig. E. 8.42

9

Kinetics of a Particle-Rectilinear Motion

9.1. INTRODUCTION

As discussed earlier, Kinematics is that branch of Engineering Mechanics in which we study the motion of a particle or rigid bodies of negligible mass. In other words, we study motion of bodies without considering the forces causing the motion. In this chapter we shall study the forces causing motion of a particle and of a rigid body.

When the resultant of the force system or an unbalanced force acts on a particle, it moves in the direction of this applied force. Newton's second law of motion states that the particle will accelerate or retard under the action of an unbalanced force. A large number of engineering problems can be solved by the application of Newton's second law of motion or the equation of motion. The problems of kinetics can be solved by using the following three methods :

1. Force and acceleration method (or D'Alembert's Principle)
2. Work and energy method, and
3. Impulse and momentum method.

In this book, the study of kinetics is sub-divided as under.

(*a*) Kinetics of particles :

(*i*) Chapter 9 deals with the kinetics of a particle in rectilinear motion.

(*ii*) Kinetics of a particle in curvilinear motion is discussed in Chapter 10.

(*iii*) Special methods of work and energy and impulse and momentum for particle are dealt with in Chapters 13 and 14, respectively.

(*b*) Collision or impact of elastic bodies is discussed in Chapter 15, and

(*c*) Kinetics of rigid bodies is given in Chapter 12.

9.2. EQUATIONS OF MOTION FOR RECTILINEAR MOTION

Consider a particle A of mass m having an acceleration $\overline{a}$ when acted upon by forces F_1 and F_2. Let ΣF be the resultant of these forces, Fig. 9.2.

Applying second law of motion, we can write

$$\Sigma \overline{F} \propto \overline{a} \quad \text{...(9.1)}$$

or

$$\Sigma \overline{F} = C \times \overline{a} \quad \text{...(9.2)}$$

where C is a proportionality constant.

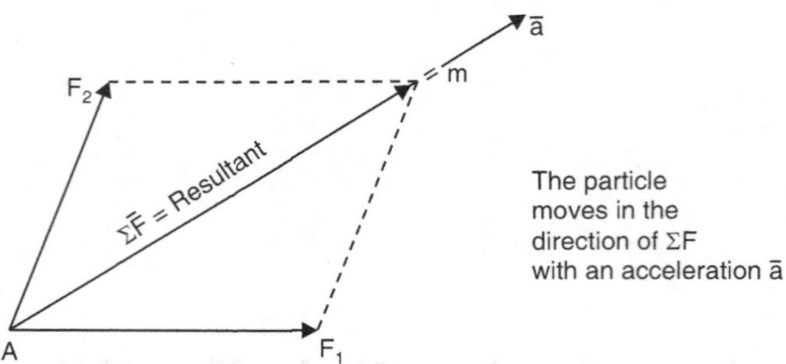

Fig. 9.2. Particle *A* subjected to the resultant force ΣF.

C is a property, known as **inertia** of the particle, **which is defined as the resistance to the rate of change of velocity.**

Definition. Inertia may be defined as that property of a particle (or a body) whereby the particle (or the body) will not move by itself unless it is acted upon an external force to do so. Inertia is measured by **mass** which is defined as the quantity of matter. Let us write $C = Km$, where m is mass and K is a constant which takes into account the system of units for measuring length, force, etc. In SI system, the unit of force is Newton (N).

Thus, 1 newton = force that produces an acceleration of 1 m/s^2 in a body of mass 1 kg.

$\therefore$ $$\overline{F} = Km\overline{a} \quad \text{...(9.3)}$$

Gives $$1 \text{ N} = K \times (1 \text{ kg} \times 1 \text{ m/s}^2) \quad \text{or} \quad 1 \text{ N} = K \times (1 \text{ N})$$

$\therefore$ $$K = 1.$$

$\therefore$ Eqn. (9.3) can be written as

$$\overline{F} = m\overline{a} \quad \text{...(9.4 (a))}$$

And, where there are a number of forces eqn. 9.4(*a*) becomes

$$\Sigma\overline{F} = m\overline{a} \quad \text{...(9.4 (b))}$$

Eqns. (9.4) are known as **equations of motion.**

Since $\Sigma\overline{F}$ and $\overline{a}$ are vectors,

Eqn. (9.4 (*a*)) can be written as

$$\overline{i}\,\Sigma F_x + \overline{j}\,\Sigma F_y + \overline{k}\,\Sigma F_z = m\overline{i}\,a_x + m\overline{j}a_y + m\overline{k}\,a_z$$

Or, simply, we have

$$\Sigma F_x = ma_x = m\ddot{x} \quad \text{...(9.5)}$$

$$\Sigma F_y = ma_y = m\ddot{y} \quad \text{...(9.6)}$$

$$\Sigma F_z = ma_z = m\ddot{z} \quad \text{...(9.7)}$$

For plane motion in *x-y* plane, $a_z = \ddot{z} = 0$.

9.3. D'ALEMBERT'S PRINCIPLE OR EQUATIONS OF DYNAMIC EQUILIBRIUM

A French mathematician, Jean Le Rouad D'Alembert (1717-1783), first published a corollary of Newton's second law of motion known as D'Alembert's principle which may be stated as :

The resultant of external forces acting on a particle (or a body) is equal to the product of the mass (m) of the particle (or the body) and the resultant acceleration (a) of the particle (or the body).

Let a particle of mass m be subjected to the forces F_1, F_2 and F_3 shown in Fig. 9.3 (*a*). The resultant of these forces $\Sigma\bar{F}$ is shown in Fig. 9.3 (*b*). These forces or their resultant, cause an acceleration a. Then, from Newton's second law, $\Sigma\bar{F}$ is equal to the vector $m\bar{a}$ Fig. 9.3 (*c*). The vector $m\bar{a}$ is known as **effective force.**

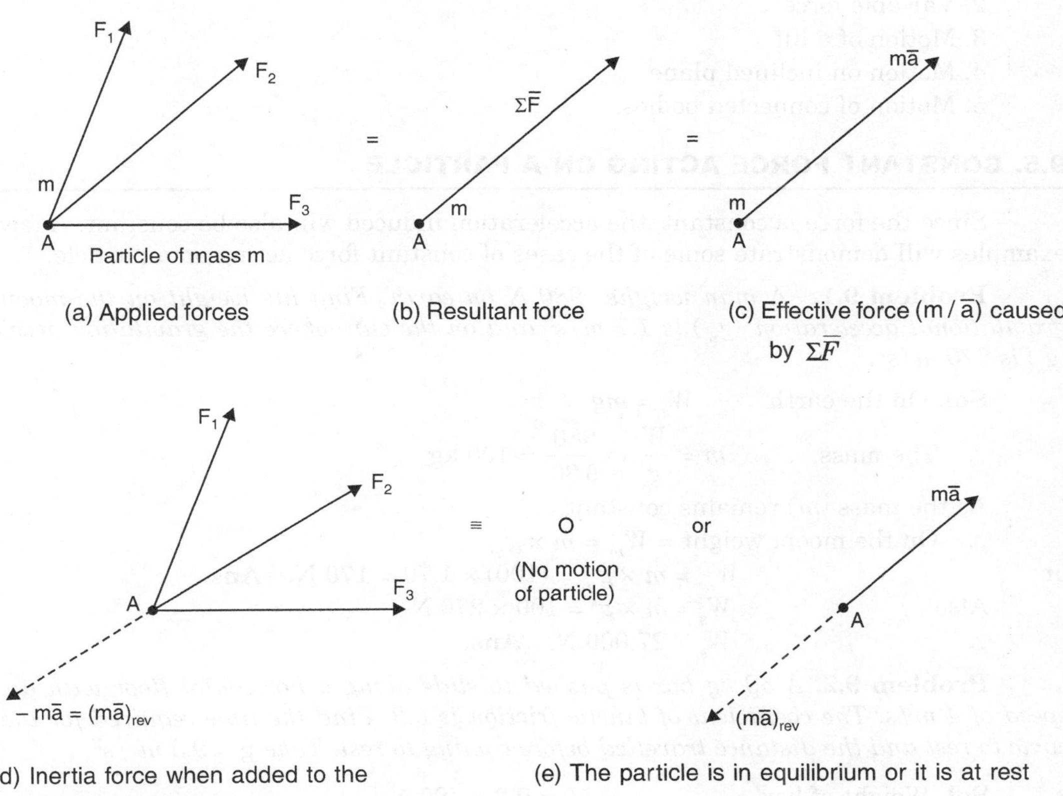

Fig. 9.3. D'Alembert's principle.

The second law of motion can, thus, be expressed as :

$$\Sigma\bar{F} = m\bar{a} \qquad \text{...(9.8)}$$

Now apply a force $-m\bar{a} = (m\bar{a})_{rev.}$ to the particle as shown in Fig. 9.3 (*d*) or (*e*).

Thus from eqn. (9.8), we get

$$\Sigma\bar{F} + (-m\bar{a}) = (m\bar{a}) - m\bar{a}$$

or
$$\Sigma\bar{F} + (m\bar{a})_{rev.} = 0 \qquad \text{...(9.9)}$$

or
$$\Sigma\bar{F} - m\bar{a} = 0 \qquad \text{...(9.10)}$$

The force $(m\bar{a})_{rev.}$ or $(-m\bar{a})$ is called as **inertia force.**

Thus, from eqns. (9.9) and (9.10) it is clear that if an inertia force is added to a particle subjected to a system of forces (Fig. 9.3 (*a*)), it is brought to equilibrium. And, the particle is said to be in **dynamic equilibrium.**

9.4. TYPES OF PROBLEMS TO BE SOLVED

In this chapter we shall deal with the rectilinear motion in a plane. The following cases will be considered :

1. Constant force acting on a body
2. Variable force
3. Motion of a lift
4. Motion on inclined plane
5. Motion of connected bodies.

9.5. CONSTANT FORCE ACTING ON A PARTICLE

Since the force is constant, the acceleration induced will also be constant. A few solved examples will demonstrate some of the cases of constant force acting on a particle.

Problem 9.1. *A man weighs 980 N on earth. Find his weight on the moon where gravitational acceleration* (g_m) *is 1.7 m/s² and on the sun where the gravitation acceleration* (g_s) *is 270 m/s².*

Sol. On the earth $\because \quad W_e = mg_e$

$\therefore$ The mass, $$m = \frac{W_e}{g_e} = \frac{980}{9.80} = 100 \text{ kg}$$

As the mass (m) remains constant.

$\therefore$ On the moon, weight $= W_m = m \times g_m$

or $$W_m = m \times g_m = (100) \times 1.70 = \mathbf{170\ N.} \quad \textbf{Ans.}$$

Also $$W_s = m \times g_s = 100 \times 270 \text{ N}$$

$\therefore$ $$W_s = 27{,}000 \text{ N.} \quad \textbf{Ans.}$$

Problem 9.2. *A 50 kg box is pushed to slide along a horizontal floor with an initial speed of 4 m/s. The coefficient of kinetic friction is 0.3. Find the time required for the box to come to rest and the distance travelled before coming to rest. Take* g = *9.8 m /s².*

Sol. Weight of box $= 50 \times 9.8 = 490$ N,

Normal reaction of the floor $= R = 490$ N,

$\therefore$ Frictional force, $F_k = \mu_k R = 0.3 \times 490 = 147$ N.

Let the retardation of the box be a. Then by the equation of motion, we get

$$F = ma$$

or, $$147 = 50a$$

$\therefore$ $$a = \frac{147}{50} = 2.94 \text{ m/s}^2$$

Using the relation $v = u + at$, we have

$$0 = 4 + (-2.94) \times t$$

$\therefore$ $$t = \frac{4}{2.94} = \mathbf{1.361\ sec.} \quad \textbf{Ans.}$$

The distance travelled by the box before stopping is given by

$$v^2 = u^2 + 2as,$$

$$0 = (4)^2 + 2 \times (-2.94) \times s$$

$\therefore$ $$s = \frac{4^2}{2 \times 2.94} = \mathbf{2.72\ m.} \quad \textbf{Ans.}$$

Problem 9.3. *A locomotive of mass 150 tonnes hauls a train of mass 600 tonnes. The coefficient of adhesion between the train and the track is 0.5. If 80% of the weight of the locomotive is carried by the driving wheels and the friction resistance to motion is 10 Newton per kilotonne of weight, find the maximum permissible acceleration without slipping.*

Sol. Weight of locomotive

$$W = 150 \times 9.8 \text{ kN} = 1470 \text{ kN}$$

80% of this weight is carried by the driving wheels of the locomotive. Therefore the maximum frictional force,

$$= F_{max} = 0.5 \times (0.8 \times 1470) \text{ kN}$$

or $$F_{max} = 588 \text{ kN.}$$

The driving force, P must not exceed F_{max}.

$\therefore$ $$P \le 588 \text{ kN}$$

or $$P \le 588{,}000 \text{ N.}$$

Total weight of the train and the locomotive

$$= (150 + 600) \times 9.8 \text{ kN} = 750 \text{ tonnes} \times 9.8$$
$$= 7350 \text{ kN}$$

$\therefore$ Frictional resistance $= R_F = 10$ N/tonne

or $$R_F = 10 \times 7350 = 73{,}500 \text{ N}$$

$\therefore$ Acceleration force $= P - R_F$

$$= 588{,}000 - 73{,}500 = 514{,}500 \text{ N}$$

$\therefore$ By the equation of motion, we have

$$P - R_F = ma$$

$\therefore$ $$514{,}500 = (750 \times 1000) \times a$$

$\therefore$ $$a = \frac{514500}{750 \times 1000} = \mathbf{0.686\ m/s^2}. \quad \textbf{Ans.}$$

Problem 9.4. *A man of 80 kg mass, jumps into a swimming pool from a height of 7 m. After going down in water by 1.5 m he started rising. Find the average resistance of water for the downward motion of the man.*

Sol. Weight of man, $W = 80 \times 9.8 = 784$ N. Velocity of man when reaches surface of water is found by using the relation,

$$v^2 = u^2 + 2gy$$

$\therefore$ $$v^2 = 0 + 2 \times 9.8 \times 7 = 1372$$

$\therefore$ $$v = \mathbf{11.713\ m/s.}$$

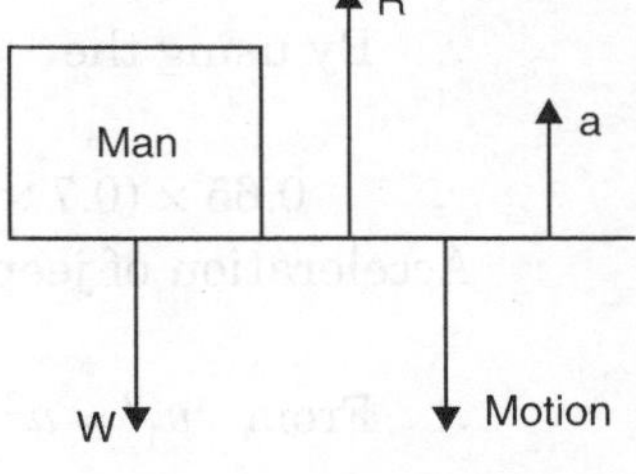

Fig. P. 9.4

Let R be the resistance of water and a be the retardation of man. When man goes down by 1.5 m, his downward velocity becomes zero and due to the buoyancy of water he starts rising out of water. The retardation a can be obtained by using :

$$v^2 = u^2 - 2ay$$

$\therefore$ $$0 = (11.713)^2 - 2a \times 1.5$$

$\therefore$ $$a = \frac{(11.713)^2}{2 \times 1.5} = 45.731 \text{ m/s}^2$$

Net upward force [Fig. P.9.4] on man $= (R - W)$

By using the relation

$$\Sigma F = ma, \text{ we get}$$

$$(R - W) = ma$$

$$\therefore \quad R - 784 = 80 \times 45.731$$

$$\therefore \quad R = \mathbf{4442.48\ N.} \quad \textbf{Ans.}$$

Problem 9.5. *What maximum speed a jeep can attain over a distance of 100 m starting from rest, if the coefficient of cohesion between the road and the tyres is 0.65 ? Consider the following conditions :*

(*a*) *Four-wheel drive.*

(*b*) *Front-wheel drive when 70% of total weight being transferred to front wheels.*

(*c*) *Rear-wheel drive when 30% of total weight being transferred to the rear wheels.*

Sol. Let the total mass of the jeep be m and

$\therefore$ Total weight, $W = mg$

and total force of adhesion $= F = \mu W = 0.65\ mg$

where μ = coefficient of adhesion = 0.65.

(*a*) **Four-wheel drive.** For the 4-wheel drive, the effective adhesion takes place at all the four wheels.

$\therefore$ By using the relation $\Sigma F = ma$, we get

$$(0.65) \times mg = ma$$

$\therefore$ Acceleration of jeep $= a = 0.65\ g = 0.65 \times 9.8 = 6.37$ m/s^2

$\therefore$ By using the relation

$$v^2 = u^2 + 2as, \text{ we get}$$

$$v^2 = 0 + 2 \times 6.37 \times 100 \qquad | \ u = 0,$$

$$\therefore \quad v = \sqrt{1274} = 35.693 \text{ m/s}$$

$$\therefore \quad \text{Speed} = 35.693 \times \left(\frac{60 \times 60}{1000}\right) \text{kmph} = \mathbf{128.5\ kmph.} \quad \textbf{Ans.}$$

(*b*) **Front-wheel drive.** Total weight on the front wheels $= W_1 = 0.7$ W

$$\therefore \quad F = \mu \times W_1 = \text{Force of cohesion}$$

$\therefore$ By using the $\Sigma F = mg$, we get

$$\mu W_1 = ma_1$$

$$\therefore \quad 0.65 \times (0.7 \times mg) = ma_1$$

Acceleration of jeep $= a_1 = 0.65 \times 0.7 \times g$

$= 4.459$ m/s^2

$\therefore$ From $v_1{}^2 = u^2 + 2a_1 s$, we get

$$v_1{}^2 = 0 + 2 \times (4.459) \times 100 = 891.8$$

$\therefore$ Velocity of jeep $= v_1 = \sqrt{891.8} = 29.863$ m/s

$\therefore$ Speed of jeep $= 29.863 \times \left(\dfrac{3600}{1000}\right) = \mathbf{107.51\ kmph.}$ **Ans.**

(*c*) **Rear-wheel drive.** Total weight on the rear wheels $= W_2 = 0.3$ W

$\therefore$ Total force of cohesion $= F_2 = \mu \times W_2$

or $F_2 = 0.65 \times (0.3 \times mg)$

∴ By using the formula $\Sigma F = ma$, we get

$$0.65 \times (0.3 \times mg) = ma_2,$$

∴ $$a_2 = \text{acceleration of jeep} = 0.65 \times 0.3 \times g$$
$$= 0.195\, g = 1.911 \text{ m/s}^2$$

∴ From the relation $v_2^2 = u^2 + 2a_2 s$, we get

$$v_2^2 = 0 + 2 \times (1.911) \times 100 = 382.2$$

∴ $$v_2 = \sqrt{382.2} = 19.55 \text{ m/s}$$

∴ Speed of jeep = 19.55 × 3.6 = **70.38 kmph. Ans.**

9.6. VARIABLE FORCE ACTING ON A PARTICLE

The force may vary as a function of time, distance or velocity of the particle. Following examples will demonstrate some of the cases of variable applied force.

Problem 9.6. *A body of a mass 120 kg is acted upon by a force of 60 N which increases uniformly till it becomes 100 N in 2 minutes. If the body is initially at rest, determine :*

(*a*) *the velocity of the body after 1.5 minutes and 2 minutes from the start,*

(*b*) *the acceleration of the body after 1.5 minutes and 2 minutes from the start, and*

(*c*) *the distance travelled after 1.5 minutes and 2 minutes from the start.*

Sol. Let the applied force be P.

Then, $P = P_1 + \dfrac{P_2 - P_1}{\Delta t}.t$, where t is in seconds.

∴ $$P = 60 + \frac{100 - 60}{2 \times 60}.t, N$$

or $$P = \left(60 + \frac{t}{3}\right) N \quad ...(1)$$

But, $$P = ma = 120a \quad ...(2)$$

∴ Equating the RHS of eqns. (1) and (2), we get

$$120a = 60 + \frac{t}{3} \quad ...(3)$$

∴ $$120\frac{dv}{dt} = 60 + \frac{t}{3} \quad \text{or} \quad 120\, dv = \left(60 + \frac{t}{3}\right) dt$$

Integrating, we get

$$120 \int_0^v dv = \int_0^t \left(60 + \frac{t}{3}\right) dt$$

or $$v = \frac{t}{2} + \frac{t^2}{6 \times 120} \text{ m/s} \quad ...(4)$$

or $$120\frac{ds}{dt} = 60t + \frac{t^2}{6}$$

∴ By integrating, we get

$$120 \int_0^s ds = \int_0^t \left(60t + \frac{t^2}{6}\right) dt$$

or $$120s = \frac{60\,t^2}{2} + \frac{t^3}{18} \quad \text{or} \quad s = \frac{t^2}{4} + \frac{t^3}{120 \times 18}, m \qquad ...(5)$$

(*a*) **Velocity of the body**

From eqn. (4), we get $v_{90} = \left(\frac{t}{2} + \frac{t^2}{6 \times 120}\right)_{t=90}$ in t = 1.5 minutes = 90 sec.

$$= \frac{90}{2} + \frac{90^2}{6 \times 120} = \textbf{56.25 m/s.} \quad \textbf{Ans.}$$

And, $v_{120} = \left(\frac{t}{2} + \frac{t^2}{6 \times 120}\right)_{t=120}$ in t = 2 minutes = 120 sec.

$$= \frac{120}{2} + \frac{(120)^2}{6 \times 120} = 80 \text{ m/s.} \quad \textbf{Ans.}$$

(*b*) **Acceleration of the body**

From eqn. (2), we get

$$120a = 60 + \frac{t}{3}$$

For t = 90 sec ; $a = \frac{60}{120} + \frac{90}{3 \times 120} = \textbf{0.75 m/s}^{\textbf{2}}\textbf{.}$ **Ans.**

For t = 120 sec ; $a = \frac{60}{120} + \frac{120}{3 \times 120} = \textbf{0.833 m/s}^{\textbf{2}}\textbf{.}$ **Ans.**

(*c*) **Distance travelled by the body**

From eqn. (5), we get

For t = 90 sec ; $s = \frac{90^2}{4} + \frac{90^3}{120 \times 18}$ m = **2362.5 m. Ans.**

And for t = 120 sec.

$$s = \frac{120^2}{4} + \frac{(120)^3}{120 \times 18} \text{ m} = \textbf{4400 m.} \quad \textbf{Ans.}$$

Problem 9.7. *A body of mass 100 kg moves along a straight line with an initial velocity of 4 m/s under the action of a force,* $P = \left(50 + \frac{x}{40}\right)$, *Newtons where x is the distance in metres from the rest. Find the velocity and acceleration of the body when it has travelled a distance of 80 m from start. Also find time for travelling 80 m from start.*

Sol. Here the applied force varies as the distance travelled.

Given that $$P = \left(50 + \frac{x}{40}\right) \qquad ...(1)$$

∵ $a = v.\frac{dv}{dx}$, and $P = ma$, we get

$$P = ma = m\left(v.\frac{dv}{dx}\right) \qquad ...(2)$$

Equating the RHS of eqns. (1) and (2), we get

$$m\left(v.\frac{dv}{dx}\right)=\left(50+\frac{x}{40}\right) \quad \text{or} \quad 100.v.\frac{dv}{dx}=\left(50+\frac{x}{40}\right) \qquad ...(3)$$

∴ By integrating eqn. (3), we get

$$\int_4^v 100\, v\, dv = \int_0^x \left(50+\frac{x}{40}\right)dx$$

or $$\left(\frac{100\, v^2}{2}-\frac{100\times 4^2}{2}\right)=50x+\frac{x^2}{80} \qquad ...(4)$$

or $$50(v^2-16)=50x+\frac{x^2}{80}$$

∴ We get $$v^2=\frac{1}{50}\left[50x+\frac{x^2}{80}\right]+16$$

or $$v^2=x+\frac{x^2}{4000}+16 \qquad ...(5)$$

At $x = 80$ m,

$$v^2=80+\frac{(80)^2}{4000}+16=80+1.6+16$$

or $$v^2=97.6$$

∴ $$v = 9.8793 \text{ m/s} \quad \text{or} \quad 9.88 \text{ m/s.}$$ **Ans.**

Also, from $v^2 = u^2 + 2ax$, we get

Acceleration, $$a=\frac{v^2-u^2}{2x}=\frac{97.6-(4)^2}{2\times 80}$$

a = **0.51 m/s²**. **Ans.**

Using $v = u + at$, we get

$$t=\frac{v-u}{a}=\frac{9.88-4}{0.51}=\mathbf{11.53}$$ **seconds. Ans.**

Problem 9.8. *A car moving at a speed 144 kmph is subjected to air resistance given by $F = 0.4v^2$ Newtons. Determine the distance moved by the car before its speed becomes 40% of the initial speed. Mass of the car is 400 kg.*

Sol. Initial speed = 144 kmph

∴ Initial velocity $= u = \frac{144}{3.6} = 40$ m/s

Final speed, v = 40% of 40 = 16 m/s

Given, $$F = ma = m.\frac{dv}{dt}$$

∴ $$m.\frac{dv}{dt}=-0.4\, v^2 \quad \text{(the car is retarding)}$$

∴ $$dv=-\frac{0.4}{m}v^2 dt \qquad | \; m = 400 \text{ kg}$$

Integrating, we get $$\int_{40}^{v}\frac{dv}{v^2}=\int_0^t -\frac{0.4}{m}.dt$$

or $$\left[-\frac{1}{v}\right]_{40}^{v} = \left[-\frac{0.4t}{400}\right]_{0}^{t}$$

$$\therefore \quad -\frac{1}{v} + \frac{1}{40} = -\frac{0.4}{400}t$$

$$\therefore \quad -\frac{1}{v} = -\frac{t}{1000} - \frac{1}{40}$$

$$\therefore \quad \frac{1}{v} = \frac{t}{1000} + \frac{1}{40} \qquad ...(1)$$

or $$v = \left[\frac{t}{1000} + \frac{1}{40}\right]^{-1} \qquad ...(2)$$

When $v = 16$ m/s, we get

$$\frac{1}{16} = \frac{t}{1000} + \frac{1}{40}$$

$$\frac{t}{1000} = \frac{1}{16} - \frac{1}{40} = \frac{3}{80}$$

$$\therefore \quad t = \frac{3}{80} \times 1000 = \frac{300}{8} \text{ seconds} = 37.5 \text{ seconds.}$$

Eqn. (2) can be written

$$\frac{dx}{dt} = \left[\frac{t}{1000} + \frac{1}{40}\right]^{-1}$$

$$\therefore \quad dx = \frac{dt}{\left[\frac{t}{1000} + \frac{1}{40}\right]}$$

Integrating, we get

$$\therefore \quad \int dx = \int \frac{dt}{0.001t + 0.025}$$

$$\int dx = \int \frac{dt}{(0.001t + 0.025)}$$

or $$x = \frac{1}{0.001}[\log(0.001\,t + 0.025) + C_1 \text{ from (2)}$$

When $t = 0$, $x = 0$

$$\therefore \quad C_1 = -\log 0.025 = \log\frac{1}{(0.025)}$$

$\therefore$ Eqn. (2) becomes

$$x = \frac{1}{0.001}\log\frac{0.001t + 0.025}{0.025}$$

$$\therefore \quad x = 1000\log(0.04t + 1) \qquad ...(3)$$

$\therefore$ At $t = 37.5$ sec.

$$x = 1000\log(0.04 \times 37.5 + 1)$$

$$x = 916.29 \text{ m.} \quad \textbf{Ans.}$$

Problem 9.9. *A particle of mass m, starting from rest, moves on a horizontal plane. It is subjected to a force given by (P – ct) and the maximum velocity is attained after t_1 seconds. Show that the displacement in t_1 seconds is $\frac{ct_1^3}{3m}$.*

Sol. Using $\Sigma F = ma$, we get

$$(P - ct) = ma \quad ...(1)$$

When, $t = t_1$, velocity is maximum. Also, when velocity is maximum, $a = 0$

$\therefore$ Eqn. (1) gives,

$$(P - ct)_{t=1} = 0 \quad \text{or} \quad P = ct_1 \quad ...(2)$$

Again eqn. (1) can be written as

$$(P - ct) = m\frac{dv}{dt}$$

$$\therefore \quad mdv = (P - ct)\,dt$$

or $$m\int_0^v dv = \int_0^t (P - ct)\,dt$$

or $$mv = Pt - \frac{ct^2}{2} \quad ...(3)$$

or $$m\frac{ds}{dt} = Pt - \frac{ct^2}{2}$$

Integrating, $$\int_0^s mds = \int_0^t \left(Pt - \frac{ct^2}{2}\right)dt$$

$$\therefore \quad ms = \left(\frac{Pt^2}{2} - \frac{ct^3}{6}\right) \quad ...(4)$$

But $P = ct_1$, in eqn. (1),

$\therefore$ We get $$ms = \left(\frac{ct_1 . t^2}{2} - \frac{ct^3}{6}\right) \quad ...(5)$$

At $t = t_1$, $v = v_{max}$ and therefore, distance travelled when the maximum velocity is attained is given by

$$ms = \frac{ct_1\,(t_1)^2}{2} - \frac{ct_1^{\,3}}{6} = \frac{ct_1^{\,3}}{3}$$

$\therefore$ From (4), Displacement = $\frac{ct_1^{\,3}}{3m}$. **Proved.**

9.7. MOTION OF A LIFT

Consider a lift of total weight W and moving with a uniform acceleration. Let T be the tension in the cable or the reaction of lift. The mass of lift is $m = \frac{W}{g}$. Two cases arise, *viz* ; (*a*) lift moving up, and (*b*) lift moving down. These cases are shown in Fig. 9.7.

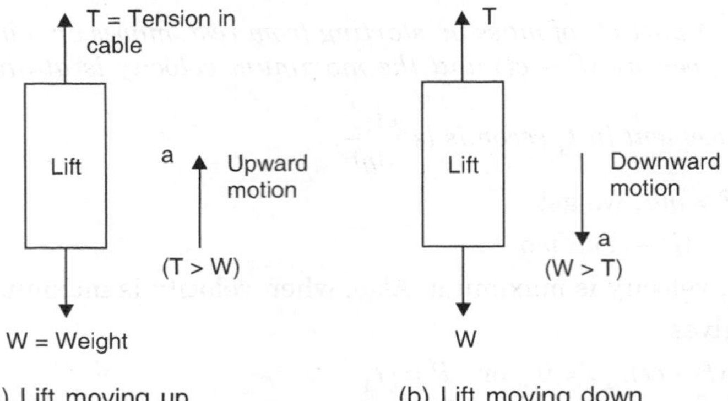

Fig. 9.7. Motion of a lift.

Case (a)

Upward motion : Fig. 9.7(*a*)

Net upward force $= T - W$

$$\text{Mass of lift} = \frac{W}{g} \; ; \quad \text{Acceleration} = a$$

Using $\Sigma F = \frac{W}{g} a$, we get

$$T - W = \frac{W}{g} a \qquad \text{...(9.11)}$$

$$\therefore \qquad T = W\left(1 + \frac{a}{g}\right) \qquad \text{...(9.12)}$$

Case (b)

Downward motion : Fig. 9.7(*b*)

Net downward force $= W - T$,

Using $\Sigma F = \frac{W}{g} a$, we get $\quad W - T = \frac{W}{g} a \qquad \text{...(9.13)}$

or
$$T = W\left(1 - \frac{a}{g}\right) \qquad \text{...(9.14)}$$

Problem 9.10. *An elevator of mass 500 kg starts moving upwards at a constant acceleration and attains a velocity of 3 m/s after travelling a distance of 3 m.*

(*a*) *Determine the tension in the cable.*

(*b*) *If after attaining the velocity of 3 m/s the elevator stops in 2.5 seconds, find the pressure exerted by the lift to the man weighing 600 N.*

Sol. Given $m = 500$ kg,

$\therefore \qquad W = mg = 500 \times 9.8 \text{ N} = 4900 \text{ N}$

(*a*) **Motion of elevator,** See Fig. P. 9.10 (*a*)

$$u = 0 \; ; \quad v = 3 \text{ m/s} \; ; \quad y = 3 \text{ m}$$

$\therefore$ Using $v^2 = u^2 + 2ay$, we get

$$(3)^2 = 0 + 2a \times 3$$

$$\therefore \qquad a = \frac{(3)^2}{2 \times 3} = 1.5 \text{ m/s}^2$$

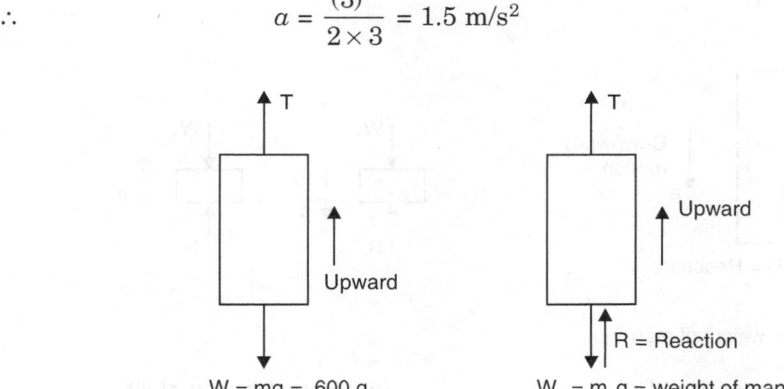

Case (a) Motion of lift Case (b) Motion of man.

Fig. P. 9.10

Net upward force on lift = $(T - W)$

$$\therefore \qquad T - W = ma \quad \text{or} \quad T - 600 \text{ g} = 600 \times 1.5$$

$$\therefore \qquad T = 600 \times 1.5 + 600 \times 9.8 = \textbf{6780 N.} \quad \textbf{Ans.}$$

(*b*) **Motion of man.** Let R be the reaction between lift and the man. See Fig. P. 9.10(*b*)

$\therefore$ Net upward force = $(R - W_1)$

$$u = 3.0 \text{ m/s} \; ; \quad v = 0 \quad \text{and} \quad t = 2.5 \text{ sec.}$$

$\therefore$ Using $v = u + at$, we get

$$0 = 3 + a \times 2.5$$

$$\therefore \qquad a = -\frac{3}{2.5} = -1.2 \text{ m/s}^2$$

$\therefore$ From $\Sigma F = ma$, we get

$$R - W_1 = \frac{W_1}{g} a$$

$$\therefore \qquad R = W_1 \left(1 + \frac{a}{g}\right)$$

$$R - 600 = \frac{600}{9.80} \times (-1.2)$$

$$\therefore \qquad R = 600\left(1 - \frac{1.2}{9.8}\right), \text{ N} = \textbf{526.53 N.} \quad \textbf{Ans.}$$

Problem 9.11. *A lift has a downward acceleration of 1.2 m/s².*

(*a*) *Find pressure between the floor of lift and the man weighing 600 N and standing in the lift.*

(*b*) *Find the pressure if the lift had an upward acceleration of 1.2 m/s².*

Sol. See Fig. P.9.11.

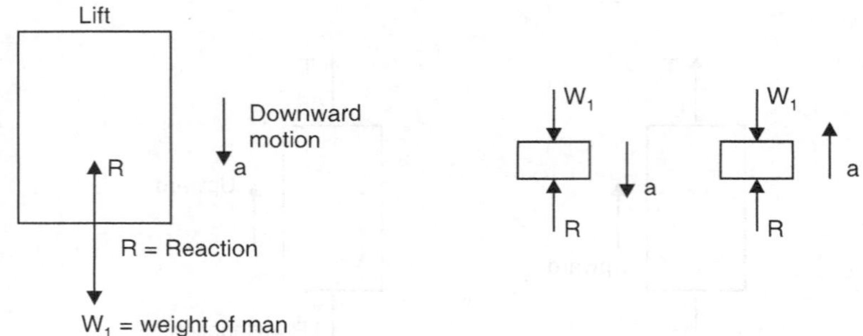

(a) Downward motion of lift (b) Upward motion of lift

Fig. P. 9.11

Motion of man

(*a*) **For downward motion:** See Fig. P.9.11(*a*)

By using $\Sigma F = \frac{W}{g} a$, we have

$$W_1 - R = \frac{W_1}{g} a.$$

$$\therefore \quad R = W_1 \left(1 - \frac{a}{g}\right) = 600 \left(1 - \frac{1.2}{9.8}\right)$$

$$= \mathbf{526.53\ N.} \quad \textbf{Ans.}$$

(*b*) **For upward motion:** See Fig. P. 9.11 (*b*).

Net upward force on man $= R - W_1$

$\therefore$ Using $\Sigma F = \frac{W}{g} a$, we get

$$R - W_1 = \frac{W_1}{g} a$$

$$\therefore \quad R = W_1 \left(1 + \frac{a}{g}\right) = 600 \left(1 + \frac{1.2}{9.8}\right) \text{ N}$$

$$= \mathbf{673.47\ N.} \quad \textbf{Ans.}$$

Problem 9.12. *A spring operated platform (or weighing balance) of negligible mass is placed on the floor of a lift. A box placed on the platform, the scale reads 600 N when the lift is at rest.*

(*a*) *Find the acceleration of the lift when the scale shows a reading of 750 N, and*

(*b*) *What load will be indicated by the scale when the lift has a downward acceleration of 1.25 m/s² ?*

Sol. See Fig. P. 9.12

Mass of box $= m = \frac{W}{g} = \frac{600}{9.8}$ kg

(*a*) **Upward motion :**

Net force $= R - W = 750 - 600 = 150$ N

$\therefore$ By using, $\Sigma F = \dfrac{W}{g} a$, we get

$$150 = \frac{600}{9.8} \times a$$

$$\therefore \quad a = \frac{150 \times 9.8}{600} = \mathbf{2.45\ m/s^2}. \quad \textbf{Ans.}$$

Box
600 N
Upward motion
R = 750 N
= Reaction of platform
(a) Upward motion

Box
600 N
$a = 1.25\ m/s^2$
R
(b) Downward motion

Fig. P. 9.12.

(*b*) **For downward motion :** The net downward force = $W - R = 600 - R$

$\therefore$ By using the relation

$$\Sigma F = \frac{W}{g} a, \text{ we get } \quad (a = 1.25\ m/s^2)$$

$$600 - R = \frac{600}{9.8} \times \propto$$

$$\therefore \quad R = W\left(1 - \frac{a}{g}\right) \quad ...(1)$$

When $a = 1.25\ m/s^2$

R = **523.47 N. Ans.**

$\therefore$ The reading of the scale = **523.47 N. Ans.**

Note. From eqn. (1), it is clear that when $a = 9.8\ m/s^2$ R will be zero, or the scale will not show any reading and the box will appear to be **weightless.**

9.8. MOTION ALONG A ROUGH INCLINED PLANE

Consider a body of mass m moving up a rough inclined plane under the action of a pull, P parallel to the plane. Let the plane be inclined at an angle α to the horizontal. See Fig. 9.8.

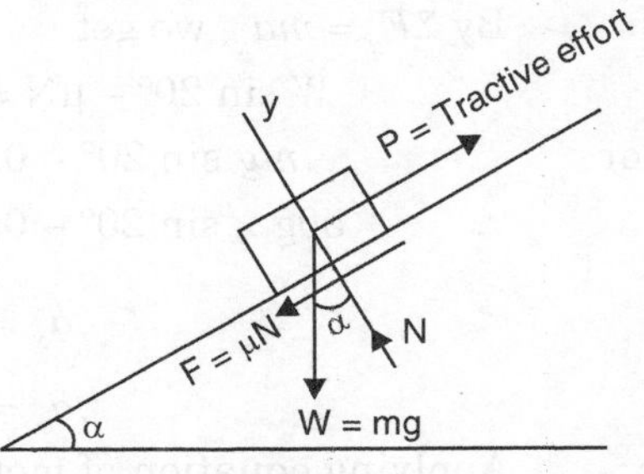

Fig. 9.8

Following forces are acting on the body :

1. The weight $W = mg$, vertically downward.
2. The normal reaction N of the plane.
3. The frictional force, $F = \mu N$, down the plane.
4. The tractive effort P, up along the plane.

Let us consider motion of the body :

Since, there is no motion normal to the plane (or along y-axis), $\therefore$ $a_y = 0$ and only static equilibrium condition will be used, *i.e.*, by $\Sigma F_y = 0$, we get

$$N - W \cos \alpha = 0 \quad \text{or} \quad N = W \cos \alpha = mg \cos \alpha \qquad ...(9.15)$$

Since the body is capable to move along the force P, by applying the law of dynamic equilibrium, we get

$$\Sigma F = ma_x$$

or $$P - (mg \sin \alpha + \mu N) = ma_x$$

or $$P - (mg \sin \alpha + \mu\, mg \cos \alpha) = ma_x \qquad ...(9.16)$$

or $$P - (R_G + R_F) = ma_x \qquad ...(9.17)$$

where R_G = resistance for gravity, and

R_F = frictional resistance = μN

Three cases arise :

(*a*) When $P = (R_G + R_F)$, then $a_x = 0$, and

(*i*) If the body is at rest, it remain at rest, and

(*ii*) If the body is moving it continues to move with zero acceleration or at uniform velocity.

(*b*) When $P < (R_G + R_F)$, the body will not move and will remain at rest.

(*c*) When $P > (R_G + R_F)$, then the body will move along P with acceleration a_x.

Note 1. If a_x is constant, the force P will also be constant, and vice-versa, and

2. If a_x is variable, the applied force P will also be variable and vice-versa.

3. $R_G = 0$ for horizontal plane ; $R_G = + W \sin \alpha$ for upward motion and $R_G = - W \sin \alpha$ for downward motion on the inclined plane.

Problem 9.13. *A 50 kg body is initially at rest lying on a 20° rough inclined plane, μ = 0.25. Find the distance the body travels before attaining a speed of 15 m/s. Also find the time ellapsed.*

Sol. See Fig. P. 9.13.

There is no external force P applied to the body.

Since, there is no motion along y-axis

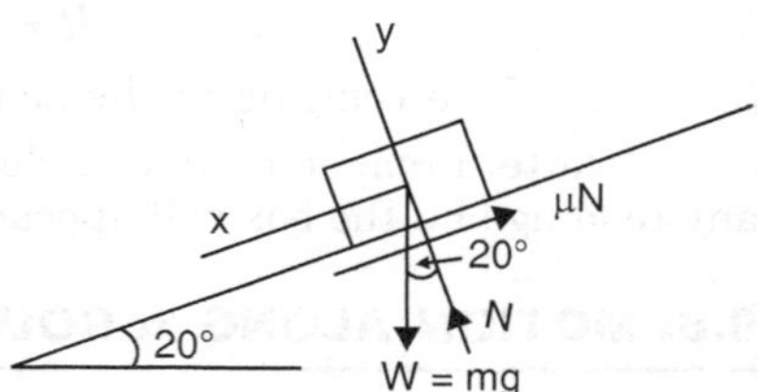

Fig. P. 9.13

∴ By $\Sigma F_y = 0$, we have

$$N - W \cos 20° = 0$$

or $$N = W \cos 20° = mg \cos 20°$$

or $$N = 50\, g \cos 20° = 460.45 \text{ N}$$

By $\Sigma F_x = ma_x$, we get

$$W \sin 20° - \mu N = ma_x$$

or $$mg \sin 20° - 0.25 \times 460.45 = 50a_x$$

∴ $$50g \times \sin 20° - 0.25 \times 460.45 = 50a_x$$

∴ $$a_x = g \sin 20° - \frac{0.25 \times 460.45}{50}$$

$$a_x = \mathbf{1.04955\ m/s^2.} \quad \textbf{Ans.}$$

Applying equation of motion $v^2 = u^2 + 2a_x.s$, we get

$$(15)^2 = (0)^2 + 2 \times 1.04955 \times s,$$

∴ $$s = \mathbf{107.19\ m.} \quad \textbf{Ans.}$$

And, by using equation

$$v = u + at, \text{ we get}$$

$$15 = 0 + 1.04955 \times t$$

$$\therefore \quad t = \textbf{14.29 seconds.} \quad \textbf{Ans.}$$

Problem 9.14. *A locomotive weighing 1 MN is moving up a 1% railway track and pulling a 2.4 MN train. The speed of train is 54 kmph and its acceleration is 0.2 m/s². Frictional resistance to the train amounts to 8 N per kN.*

(a) Find the power exerted by the engine.

(b) What power is to be exerted when the train moves on the level track ?

Sol. Inclination of track, $\alpha = 1\%$

i.e.,

$$\sin \alpha \simeq \tan \alpha = \frac{1}{100} = 0.01$$

$$\therefore \quad \alpha = \sin^{-1} 0.01 = \mathbf{0.573°}$$

Fig. P. 9.14

Fig. P. 9.14 shows the train moving up along the 1% track.

By $\Sigma F_y = 0$, we get

$$N - W \cos \alpha = 0$$

where, W = weight of engine + weight of train = (1 + 2.4) = 3.4 MN

$$\therefore \quad N = W \cos \alpha = 3.4 \cos \alpha$$

$$= 3.4 \cos 0.573° = 3.4 \text{ MN}$$

$\therefore$ Friction resistance $(R_F) = 8 \times W$

$$= 8 \times \left(\frac{3.4 \times 10^6}{10^3} \right) = 8 \times 3.4 \times 10^3 \text{ N} = \mathbf{27{,}200 \text{ N}}$$

Gravitational resistance,

$$R_G = W \sin \alpha = (3.4 \times 10^6) \times \frac{1}{100} \text{ N} = 34{,}000 \text{ N}.$$

Force required for accelerating the train,

$$R_a = m.a_x \qquad | \; m = \text{mass of train and locomotive}$$

$$= \frac{3.4 \times 10^6}{9.8} \times 0.2 \qquad \text{N} = \frac{3.4 \times 10^6}{9.8} \text{ kg.}$$

$$= \mathbf{69387.76 \text{ N}} \qquad a_x = 0.2 \text{ m/s}^2$$

$\therefore$ Total force to be exerted by the engine

$$= P = R_F + R_G + R_a = (27{,}200 + 34{,}000 + 69{,}387.76) \text{ N}$$

$$= 130{,}587.76 \text{ N}$$

Speed of train = 54 kmph

$$= \frac{54 \times 1000}{60 \times 60} \text{ m/s}$$

or $\quad v = \mathbf{15 \text{ m/s}}$

(*a*) Power exerted by the locomotive = Pv

or $\quad$ Power = 130587.76 × 15 watts = **1958.8164 kW. Ans.**

(*b*) When the track is level, the gravitational resistance (R_G) will be zero.

Then, the tractive effort required

$$= P = R_F + R_a = 27{,}200 + 69{,}387.76 \text{ N}$$
$$= 96{,}587.76 = 96.58776 \text{ kN}$$

$\therefore$ Power $= Pv = 96.58776 \times 15$ kW $=$ **1448.8164 kW. Ans.**

Problem 9.15. *A train weighing 4 MN moves down a slope of 1 in 160 at 45 km/hr with engine developing 80 kW.*

(*a*) *What power will be required to pull the train up the same track at the same speed ?*

(*b*) *Also find engine power for pulling the train on the level track at the same speed.*

Sol. Slope of the track is given by :

$$\sin \alpha = \tan \alpha = \frac{1}{160}$$

Gravity resistance parallel to track $= R_G = W \sin \alpha$

or $$R_G = 4 \times 10^6 \times \frac{1}{160} = 25000 \text{ N} = 25 \text{ kN}$$

Since the train is moving at constant speed, therefore the accelerating force, $R_a = 0$.

Speed = 45 km/hr = 12.5 m/s

When the engine develops 80 kW power, then the tractive efforce P is given by

Power $= P \times$ speed of train

or 80 kW $= P \times 12.5$

$\therefore$ $$P = \frac{80}{12.5} = 6.4 \text{ kN} = \text{pull of engine}$$

But $$P = R_F + R_a + R_G = R_F + 0 + (-W \sin \alpha)$$

$\therefore$ $$R_F = 6.4 + 25 = 31.4 \text{ kN}$$

Note. For downward motion R_G is – ve. For level track $R_G = 0$ and for upward motion $R_G = W \sin \alpha$

(*a*) When the train is moving up the track, at a speed of 45 km/hr, the total tractive effort $= R = R_F + R_G + R_a$

or $$R = 31.4 + 25 + 0 = 56.4 \text{ kN.}$$

$\therefore$ Engine power $= R \times v = 56.4 \times 12.5$ kW

$=$ **705 kW. Ans.**

(*b*) For the level track, $R_G = 0$, and total tractive effort,

$$P = R_F + R_a + R_G$$
$$= 31.4 + 0 + 0 = 31.4 \text{ kN}$$

$\therefore$ Engine power $= Pv = 31.4 \times 12.5 =$ **392.5 kW. Ans.**

Problem 9.16. *A train weighing 4 MN moves on a level track at a constant speed of 45 km/hr with engine developing 90 kW power.*

(*a*) *What will be the speed of the train after 5 minutes when it started moving down a 1 in 160 track with the engine still developing 90 kW power ?*

(*b*) *What will be the speed of the train after 3 minutes when it started moving up a 1 in 160 track with engine still developing 90 kW power ?*

Sol. Weight of train = 4 MN $= 4 \times 10^6$ N = 4000 kN.

Constant speed of train on level track $= v =$ 45 km/hr = **12.5 m/s**

$\because$ Power = Pv

$\therefore$ 90 kW = $P \times 12.5$ m/s

$\therefore$ $P = \dfrac{90}{12.5} = \mathbf{7.2\ kN}$

$P = R_G + R_F + R_a = 0 + R_F + 0$;

$\therefore$ $R_F = P = 7.2$ kN.

(*a*) **For downward motion of the train**

$u = 12.5$ m ; $t = 5$ min = 300 sec.

Gravitational resistance,

$$R_G = -W \sin \alpha = 4000 \times \frac{1}{160} = -25 \text{ kN.}$$

$\because$ Power of engine = 90 kW = $P \times$ velocity

Initial velocity of train when it starts moving down the track = u = 12.5 m/sec.

And $P = 7.2$ kN

But $P = R_G + R_a + R_F$

$\therefore$ $7.2 = -W \sin \alpha + R_a + (7.2)$

$\therefore$ $R_a = W \sin \alpha = \dfrac{4000}{160} = 25$ kN.

But $R_a = \Sigma F_x = \dfrac{W}{g} \times a$

$\therefore$ $25 = \dfrac{4000}{9.8} \times a$

$\therefore$ $a = \dfrac{25 \times 9.8}{4000} = \mathbf{0.06125\ m/sec^2}$ = acceleration

$\therefore$ Speed after 5 min. or 300 sec. is given by

$v = u + at = 12.5 + 0.06125 \times 300 = 12.5 + 18.375$

$= 30.875$ m/s = **66.15 km/hr. Ans.**

(*b*) **Motion of train up the track**

Here, $u = 12.5$ m/s ;

And $P = 7.2$ kN at the start of upward motion of train.

But $P = R_G + R_F + R_a$

$\therefore$ $7.2 = +W \sin \alpha + 7.2 + R_a$

$\therefore$ $R_a = -W \sin \alpha = -25$ kN

But $R_a = \dfrac{W}{g} . a$

$\therefore$ $-25 = \dfrac{4000}{9.8} \times a$

$\therefore$ $a = -\dfrac{25 \times 9.8}{4000} = -0.06125$ m/s^2 (Retardation)

$\therefore$ After 3 minutes the speed of train will be

$v = u + at = 12.5 + (-0.06125) \times 180$

$= 12.5 - 11.025 = 1.475$ m/sec.

= **5.31 km/hr. Ans.**

9.9. MOTION OF CONNECTED BODIES

When connected bodies move, their motions are inter-related. Some constraints in the motion of connected bodies are set-up. The following examples will illustrate the kinetics of the connected bodies.

Problem 9.17. *Two masses m_1 and m_2 ($m_1 > m_2$) are suspended at the ends of a light string passing over a smooth pulley. Find acceleration of the two masses and the tension in the string. Find acceleration of mass and string tension if $m_1 = 120$ kg and $m_2 = 60$ kg.*

Sol. Fig. P.9.17 shows the given system of connected masses and their free-body-diagrams. Since the pulley is smooth, the tension in both ends of the string will be the same, and let it be equal to T.

The heavier mass, m_1, will move down and the lighter mass m_2 will move up.

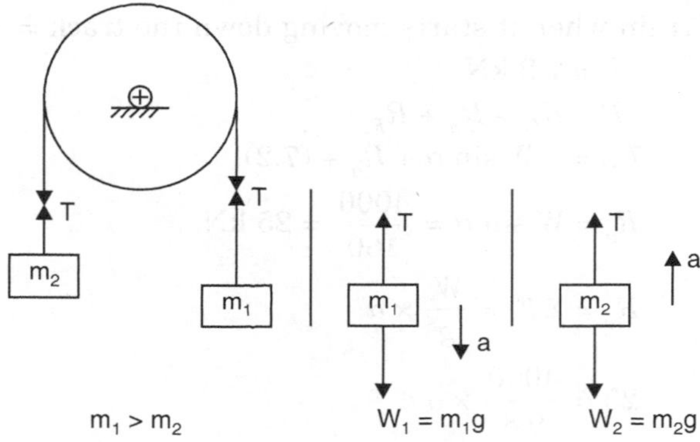

(a) Given system (b) and (c) FBDS of m_1 and m_2

Fig. P. 9.17

Equations of motion of the two masses are as under :

For mass m_1

$$m_1g - T = m_1.a \quad ...(1)$$

For mass m_2

$$T - m_2g = m_2a \quad ...(2)$$

Adding (1) and (2), we get

$$m_1g - m_2g = a(m_1 + m_2)$$

$$\therefore \quad \boxed{a = \left(\frac{m_1 - m_2}{m_1 + m_2}\right)g} \quad ...(3)$$

And, from eqn. (2), tension

$$T = m_2a + m_2g = \frac{m_2\,(m_1 - m_2)\,g}{m_1 + m_2} + m_2g$$

$$= \frac{g\,[m_2\,m_1 - m_2^2 + m_1m_2 + m_2^2]}{(m_1 + m_2)}$$

or

$$\boxed{T = \frac{2m_1m_2}{m_1 + m_2}.g} \quad ...(4)$$

When $m_1 = 120$ kg and $m_2 = 60$ kg,
from eqn. (3)

$$a = \frac{m_1 - m_2}{m_1 + m_2}.g = \frac{120 - 60}{120 + 60} g = \frac{60}{180}.g$$

or $$a = \frac{1}{3} g = 3.27 \text{ m/s}^2. \quad \textbf{Ans.}$$

And, tension $$T = \frac{2m_1 m_2}{m_1 + m_2} \times g = \frac{2 \times 120 \times 60}{120 + 60} \times g = 80\, g = \mathbf{784\ N.} \quad \textbf{Ans.}$$

Problem 9.18. *A stepped pulley of radii $r_1 = 2r_2$ supports two masses m_1 and m_2 ($m_2 > m_1$). Find acceleration of the two masses. Neglect friction and inertia of the pulleys. Find the numerical value of acceleration when $m_1 = 8$ kg and $m_2 = 12$ kg.*

Sol. Fig. P. 9.18.

Let acceleration of mass $m_1 = a_1$ and acceleration of mass $m_2 = a_2$.

For relation between a_1 and a_2, see Fig. 9.12 (*b*)

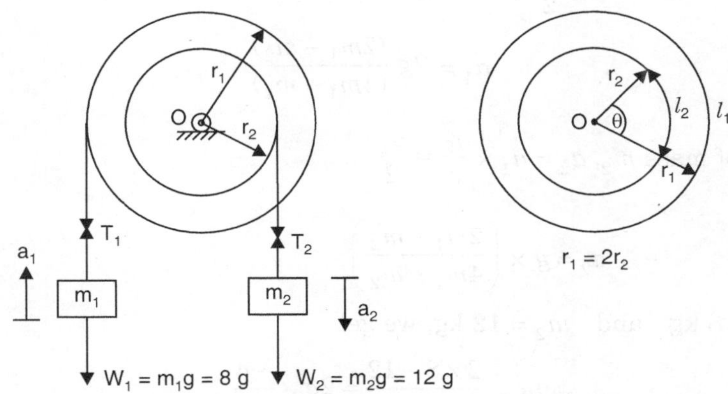

(a) Given system of masses (b) Relation between a_1 and a_2

Fig. P. 9.18

We find that the angle

$$\theta = \frac{l_1}{r_1} = \frac{l_2}{r_2}$$

$\therefore$ $$l_1 r_2 = l_2 r_1 \qquad \text{...(1)}$$

Differentiating (1), we get

$$r_2 . \frac{d^2(l_1)}{dt^2} = r_1 . \frac{d^2(l_2)}{dt^2}$$

or $$r_2 . a_1 = r_1 . a_2 \qquad \text{...(2)}$$

$\therefore$ $$a_2 = a_1 \frac{r_2}{r_1} \qquad \text{...(3)}$$

Consider the dynamic equilibrium of the two masses. By taking moments about the centre O, and by using the condition.

$$\Sigma M_0 = 0, \quad \Sigma I_G . \alpha. \quad I_G = 0 = \text{Given. Hence,}$$

we get $$r_1 \times (W_1 - m_1.a_1) - r_2(W_2 + m_2a_2) = 0$$

or $$r_1.[m_1g - m_1.a_1] - r_2\,[m_2g + m_2a_2] = 0 \quad ...(4)$$

Since, $r_1 = 2r_2$ (given) and $a_2 = a_1 . \dfrac{r_2}{r_1} = a_1 . \dfrac{r_2}{2r_2} = \dfrac{a_1}{2}$

$\therefore$ We get

$$2r_2\,[m_1g - m_1 \times a_1] - r_2\left[m_2g + m_2 \times \frac{a_1}{2}\right] = 0$$

$\therefore$ $$(2m_1g - 2m_1.a_1) - \left(m_2g + \frac{m_2 . a_1}{2}\right) = 0$$

$\therefore$ $$a_1(2m_1 + m_2/2) = g(2m_1 - m_2)$$

or $$\frac{a_1\,(4m_1 + m_2)}{2} = g(2m_1 - m_2)$$

or $$a_1 = 2g\,\frac{(2m_1 - m_2)}{(4m_1 + m_2)} \quad ...(5)$$

and acceleration of mass m_2, $a_2 = a_1 \times \dfrac{r_2}{r_1} = \dfrac{a_1}{2}$

or $$a_2 = g \times \left(\frac{2m_1 - m_2}{4m_1 + m_2}\right) \quad ...(6)$$

When $m_1 = 8$ kg and $m_2 = 12$ kg, we get

$$a_1 = 2g \times \frac{2 \times 8 - 12}{4 \times 8 + 12} = 2g \times \frac{+4}{44}$$

$$= +\frac{2g}{11}. \quad \textbf{Ans.}$$

i.e., the mass m_1 moves upwards and

$$a_2 = \frac{a_1}{2} = -\frac{g}{11}. \quad \text{downwards.} \quad \textbf{Ans.}$$

Problem 9.19. *Find tension in the string and acceleration of blocks A and B of mass m_A and m_B, respectively. Assume pulleys to be massless and smooth. See Fig. P. 9.19. Take $m_A = 24$ kg and $m_B = 8$ kg.*

Sol. Acceleration of A and B. If mass *A moves* down by a distance y_A than the mass *B* will move up by $2y_A$.

Thus, we have $$y_B = 2y_A \quad ...(1)$$

Differentiating we get

$$\dot{y}_B = 2\dot{y}_A \quad \text{or} \quad v_B = 2v_A \quad ...(2)$$

And, $$\ddot{y}_B = 2\ddot{y}_A \quad \text{or} \quad a_B = 2a_A \quad ...(3)$$

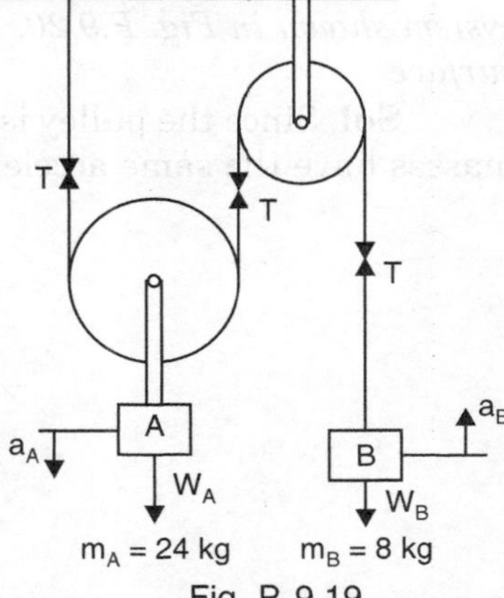

Fig. P. 9.19

Equation of motion of mass A is

$$W_A - 2T = m_A.a_A$$

or $$m_A.g - 2T = m_A.a_A \quad ...(4)$$

Equation of motion of mass B is

$$T - m_B.g = m_B.a_B \quad ...(5)$$

By eqn. (4) + 2 × eqn. (5), we get

$$m_A.g - 2m_B.g = m_A.a_A + 2m_B.a_B$$

$$(m_A - 2m_B)g = m_A.a_A + m_B \times (2a_A)$$

$$\therefore \quad (m_A - 2m_B)g = a_A(m_A + 2m_B)$$

$$\therefore \quad a_A = \left(\frac{m_A - 2m_B}{m_A + 2m_B}\right).g \quad ...(6)$$

And, $$a_B = 2a_A = \left(\frac{m_A - 2m_B}{m_A + 2m_B}\right) \times 2g \quad ...(7)$$

And from eqn. (5), the tension

$$T = m_B.a_B + m_B.g$$

$$= m_B \times 2g \times \left(\frac{m_A - 2m_B}{m_A + 2m_B}\right) + m_B.g$$

$$= \frac{m_B.g}{m_A + 2m_B} \times [2m_A - 4m_B + m_A + 2m_B]$$

or $$T = \frac{m_B.g}{m_A + 2m_B} \times (3m_A - 2m_B) \quad ...(8)$$

Substituting $m_A = 24$ kg and $m_B = 8$ kg, we get

$$a_A = \frac{24 - 2\times 8}{24 + 2\times 8} g = \frac{8}{40} g$$

$$\therefore \quad a_A = \frac{g}{5} = 1.96 \text{ m/s}^2. \quad \textbf{Ans.}$$

And $$a_B = 2a_A = \frac{2g}{5} = 3.92 \text{ m/s}^2. \quad \textbf{Ans.}$$

And, from eqn. (8), tension in the string

$$= T = \frac{8 \times g}{24 + 2\times 8} \times (3 \times 24 - 2 \times 8)$$

or $$T = \frac{8g}{40} \times 56 = 11.2 \text{ g}$$

or $$T = \textbf{109.76 N.} \quad \textbf{Ans.}$$

Problem 9.20. *Find acceleration of the two masses and tension in the string for the system shown in Fig. P.9.20. Pulley is smooth and μ_k = 0.2 between mass A and the horizontal surface.*

Sol. Since the pulley is smooth the rope has the same tension T throughout and both the masses have the same acceleration a.

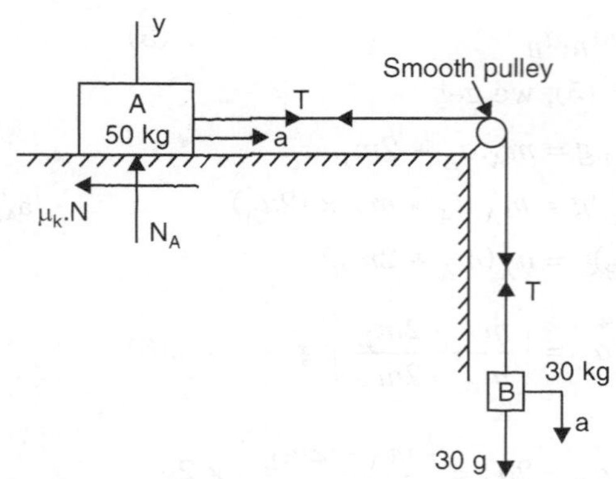

Fig. P. 9.20

Fig. P.9.20 shows various forces acting on the two masses, and also the acceleration a.

For mass A :

By $\Sigma F_y = 0$, we get

$$N_A - m_A.g = 0$$

$\therefore$ $$N_A = m_A.g = 50\ g \qquad ...(1)$$

By $\Sigma F_x = m.a_x$, we get

$$T - \mu_k.N_A = m_A.a \quad \text{or} \quad T - 0.2 \times (50\ g) = 50a \qquad ...(2)$$

For mass B :

By $\Sigma F_y = ma_y$, we get

$$m_B.g - T = m_B.a \quad \text{or} \quad 30g - T = 30a \qquad ...(3)$$

Adding (2) and (3), we get

$$30g - 0.2\ (50g) = 80a$$

$\therefore$ $$a = \frac{20g}{80} = \frac{g}{4} = \mathbf{2.45\ m/s^2}. \quad \textbf{Ans.}$$

From (2), $$T = 50a + 0.2 \times (50g) = \mathbf{220.5\ N}. \quad \textbf{Ans.}$$

Problem 9.21. *Find maximum and minimum tensions in the rope and the acceleration of the two masses shown in Fig. P.9.21 (a). Coefficient of friction between mass A and the horizontal surface is 0.2 and that between the pulley and the rope is 0.18.*

Sol. Since the pulley is not smooth, tensions in the two ends will be T_A and T_B, ($T_B > T_A$). Let the acceleration of two masses be equal to a.

Angle of lap at pulley = $\theta = \pi/2$

$\therefore$ $$T_B/T_A = e^{\mu\theta} = e^{(0.18) \times \pi/2} = e^{0.28274}$$

$\therefore \quad T_B/T_A = 1.33,$

$\therefore \quad T_B = 1.33\ T_A$ ...(1)

For mass A :

By $\Sigma F_y = 0$, we get $N_A - W_A = 0$

$\therefore \quad N_A = W_A = 50\,g$

By $\Sigma F_x = m_A.a_A$, we get

$T_A - \mu_A.N_A = m_A.a$

or $\quad T_A - 0.2 \times 50\,g = 50a$ ...(2)

For mass B :

By $\Sigma F_v = m_B.a_B$, we get

$m_B.g - T_B = m_B.a$

or $\quad 30g - T_B = 30a$ ...(3)

Eqn. (3) be rewritten as

$30g - 1.33T_A = 30a$ ...(3a)

By (2) × 1.33 + (3a), we get

$(T_A - 10g) \times 1.33 + (30\,g - 1.33\,T_A) = 50 \times 1.33\,a + 30\,a$

$\therefore \quad -13.3g + 30g = 96.5a$

$\therefore \quad a = \dfrac{16.7g}{96.5} = \mathbf{1.698\ m/s^2}$. **Ans.**

$\therefore$ From (2) ; $\quad T_A = 50 \times (1.696) + 10g$

$= \mathbf{182.98\ N}$. **Ans.**

And from (1), $\quad T_B = 1.33 \times T_A = \mathbf{243.37\,N}$. **Ans.**

(a) Given system

(b) Mass A

(c) Pulley

(d) Mass B

Fig. P. 9.21

Problem 9.22. *Find acceleration of the two masses and tension in the string for the system shown in Fig. P.9.22. Assume pulley to be smooth and coefficient of friction of the rough inclined plane to be 0.3. m_A = 20 kg and m_B = 60 kg. (Take g = 9.80 m/s²)*

Sol. Let T be the tension in the string and a be the acceleration. Forces acting on the masses A and B are shown in Fig. P.9.22.

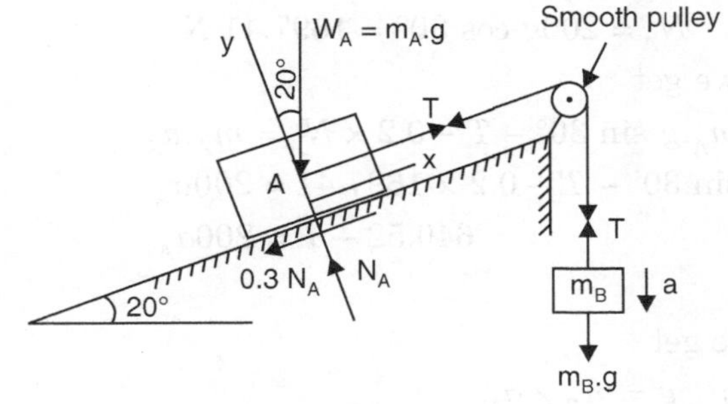

Fig. P. 9.22

For mass A :

By $\Sigma F_y = m_A \cdot a_y$, we get

$\therefore \quad N_A - W_A \cdot \cos 20° = 0$ $\quad | \because \quad a_y = 0$

$\therefore \qquad N_A = W_A \,.\, \cos 20° = (20g) \cos 20°$

or $\qquad N_A = 184.18$ N

By $\Sigma F_x = ma_x$, we get

$$T - W_A \,.\, \sin 20° - 0.3N_A = m_A.a \qquad | \because \; a_x = a$$

or $\qquad T - 20g \sin 20 - 0.3 \times 184.18 = 20a$

$$T - 122.29 = 20a \qquad ...(1)$$

For mass B :

By $\Sigma F_y = m_B.a$, we get

$$m_B.g - T = m_B.a \quad \text{or} \quad 60g - T = 60a \qquad ...(2)$$

Adding eqns. (1) and (2), we get

$$60g - 122.29 = 80a$$

$\therefore \qquad 80a = 465.71$

$\therefore \qquad a =$ **5.821 m/s². Ans.**

And from eqn. (1), we get

$T = 20a + 122.29 =$ **238.72 N. Ans.**

Problem 9.23. *For the system shown in Fig. P. 9.23 find the tension in the string and acceleration of the masses m_A and m_B. Take $m_A = 200$ kg ; $m_B = 120$ kg, $\mu = 0.2$. Assume pulley to be smooth.*

Sol. Let the tension in string be T and let the mass m_A move down the plane. When mass A moves down by x_A, the mass B moves up by $x_B = \dfrac{x_A}{2}$.

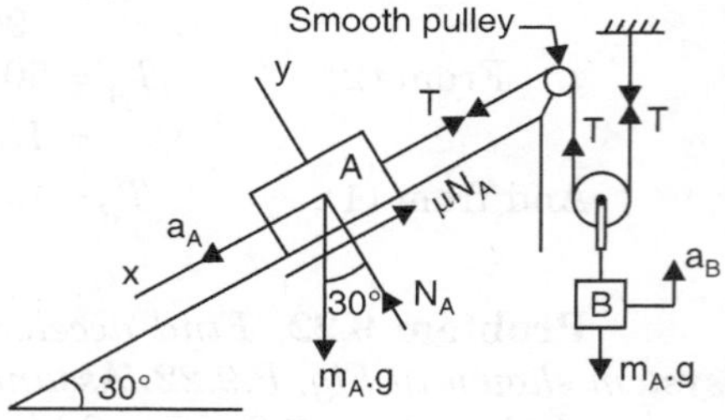

Smooth pulley
y
T
T
T
A
μN_A
a_A
x
a_B
$30°$
N_A
B
$m_A \cdot g$
$30°$
$m_A \cdot g$

Fig. P. 9.23

$\because \qquad x_A = 2x_B$

$\therefore \qquad a_A = 2a_B \qquad ...(1)$

For mass A :

By $\Sigma F_y = 0$, we get

$$N_A - m_A\, g \cos 30° = 0$$

$\therefore \qquad N_A = m_A g \cos 30°$

or $\qquad N_A = 200g \cos 30° = 1697.41$ N,

By $\Sigma F_x = m_A.a_x$, we get

$$m_A.g \sin 30° - T - 0.2 \times N_A = m_A.a_A$$

or $\qquad 200 \text{ g} \sin 30° - T - 0.2 \times 1697.41 = 200a_A$

or $\qquad 640.52 - T = 200a_A \qquad ...(2)$

For mass B :

By $\Sigma F_y = m_B.a_B$, we get

$$2T - m_B.g = m_B \times a_B$$

or $\qquad 2T - 120 \times g = 120 \times \left(\dfrac{a_A}{2}\right) \qquad ... | \because \; a_A = 2a_B$

Dividing by 2, we get

$$T - 60g = 30a_A \qquad ...(3)$$

Adding (2) and (3), we get

$$640.52 - 60g = 230a_A$$

$\therefore$ $$52.52 = 230a_A$$

$\therefore$ $$\mathbf{a_A = 0.2283\ m/s^2.}\quad \textbf{Ans.}$$

And $$a_B = \frac{a_A}{2} = \mathbf{0.1142\ m/s^2.}\quad \textbf{Ans.}$$

Finally, tension, $T = 30a_A + 60g =$ **594.849 N. Ans.**

Problem 9.24. *Two masses m_1 and m_2 ($m_1 > m_2$) placed on two inclined planes of inclinations α_1 and α_2 are connected by a string passing over a smooth pulley. Coefficient of friction between the planes and the masses is μ. Assume : $m_1 = 10$ kg, $m_2 = 20$ kg, $\alpha_1 = 30°$, $\alpha_2 = 60°$ and $\mu = 0.2$. Find acceleration of both masses and the string tension.*

Sol. Let us assume that the mass m_2 lying on the steeper plane move down. Various forces acting on the two masses are shown in Fig. P. 9.24. The acceleration of both masses is the same.

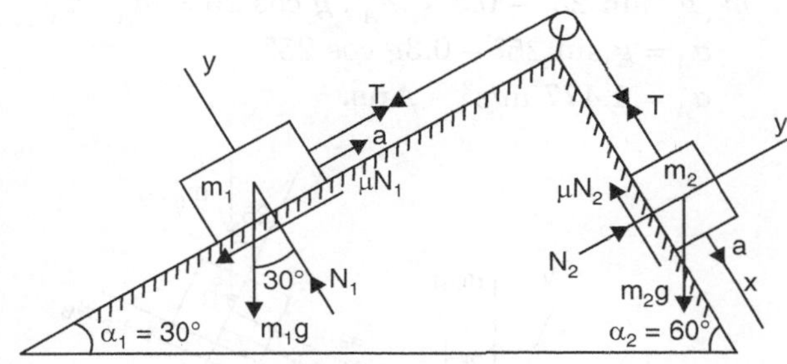

Fig. P. 9.24.

For mass $\mathbf{m_1}$:

Since there is no motion of mass m_1 normal to the plane ;

$\therefore$ By $\Sigma F_y = 0$, we get

$$N_1 - m_1 g \sin 30° = 0$$

$\therefore$ $$N_1 = m_1 g \cos 30°$$
$$= 10 \times 9.80 \cos 30°$$
$$= 84.87 \text{ N}$$

By $\Sigma F_x = m_1.a$, we get

$$T - m_1 g \sin 30° - \mu N_1 = m_1.a$$

$\therefore$ $$T - 10g \times \sin 30° - 0.2 \times 84.87 = 10a$$

$\therefore$ $$T - 65.97 = 10a \qquad \text{...(1)}$$

For mass $\mathbf{m_2}$:

By $\Sigma F_y = 0$, we get

$$N_2 - m_2.g \cos 60° = 0$$

$\therefore$ $$N_2 = 20 \times g \cos 60° = 98$$

And by $\Sigma F_x = ma$, we get

$$m_2 g \sin 60° - T - N_2 \mu = m_2 a$$

or $$20g \sin 60° - T - 98 \times 0.2 = 20a$$

or $$150.14 - T = 20a \qquad \text{...(2)}$$

Adding (1) and (2), we get

$$30a = 150.14 - 65.97$$

$\therefore$ $$30a = 84.17$$

$\therefore$ $$\mathbf{a = 2.81\ m/s^2.}\quad \textbf{Ans.}$$

And from (1), $$T = 10a + 65.97 = \mathbf{94.07\ N.}\quad \textbf{Ans.}$$

Problem 9.25. *Two blocks A and B are placed 10 m apart on a rough 25° plane.* $\mu_A = 0.3$ *and* $\mu_B = 0.15$. *If the two blocks start sliding down at the same time, find time and the distance moved by them before collision. Also find velocity of each block at the time of collision.*

Sol. The two blocks are acted upon by various forces shown in Fig. P. 9.25.

For block A :

By $\Sigma F_y = 0$, we get

$$N_A - m_A . g \cos 25° = 0$$

$$N_A = m_A . g \cos 25°$$

By $\Sigma F_x = ma_x$, we get

$$m_A g \sin 25° - \mu_A N_A = m_A . a_A \qquad | \ a_x = a_A$$

$\therefore$ $$m_A g . \sin 25° - 0.3 \times m_A . g \cos 25 = m_A . a_A$$

$\therefore$ $$a_A = g \sin 25° - 0.3 g \cos 25°$$

$$a_A = 1.477 \text{ m/s}^2. \quad \textbf{Ans.}$$

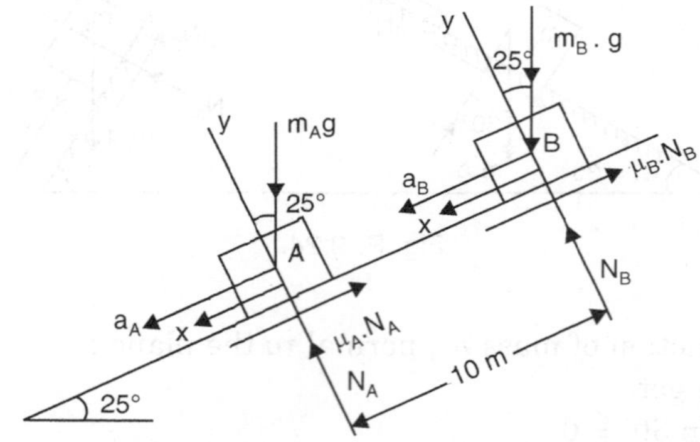

Fig. P. 9.25.

For mass $\mathbf{m_B}$:

By $\Sigma F_y = 0$, we get

$$N_B - m_B g \cos 25° = 0$$

$\therefore$ $$N_B = m_B . g \cos 25°$$

By $\Sigma F_x = ma_x$, we get

$$m_B . g \sin 25° - \mu_B \times N_B = m_B . a_B \qquad | \because \ a_B = a_x$$

$\therefore$ $$m_B . g \sin 25° - 0.15 \times m_B . g \cos 25° = m_B . a_B$$

$\therefore$ $$\mathbf{a_B = 2.81\ m/s^2.}\quad \textbf{Ans.}$$

Let distance travelled by m_A be s_A before the two masses meet each other. Then $s_B = (s_A + 10)$ m. Let t = time of travel.

By using the relation $s = ut + \frac{1}{2} at^2$, we get

For mass m_A : We have

$$s_A = u_A.t + \frac{1}{2}a_A.t^2 \qquad | u_A = 0$$

or $$s_A = 0 + \frac{1}{2} \times 1.477t^2 \qquad ...(1)$$

For mass m_B, : We have

$$s_B = u_B.t + \frac{1}{2}a_B.t^2$$

or $$s_A + 10 = 0 + \frac{1}{2} \times 2.81t^2 \qquad ...(2)$$

By substituting eqn. (1) into (2), we get

$$\left(\frac{1}{2}\times 1.477t^2 + 10\right) = \frac{1}{2} \times 2.81t^2$$

$$\therefore \quad t^2\left(\frac{2.81 - 1.477}{2}\right) = 10$$

$$\therefore \quad t^2 = 15.004$$

$$\therefore \quad \mathbf{t = 3.8735\ seconds.}$$

$\therefore$ From eqn. (1),

$$s_A = 11.08 \text{ m.} \quad \textbf{Ans.}$$

And $$s_B = s_A + 10 = \mathbf{21.08\ m.} \quad \textbf{Ans.}$$

Velocity at collision :

By using the relation $v = u + at$, we get

$$v_A = 0 + a_A.t = 0 + 1.477 \times 3.8735 = \mathbf{5.721\ m.} \quad \textbf{Ans.}$$

And, $$v_B = 0 + a_B \times t = 0 + 2.81 \times 3.8735$$

$$= \mathbf{10.885\ m/s.} \quad \textbf{Ans.}$$

EXERCISE IX

A. Theoretical Questions

1. Explain Newton's laws of motion for linear motion.
2. State and define D' Alembert's principle for linear motion.
3. What do you understand by inertia force ?
4. State the types of problems which may encounter in the study of rectilinear motion.
5. Explain the procedure for the solution of a problem in which a particle moving in a rectilinear path is acted upon by a constant force.
6. How will you solve a problem of a variable force acting on a particle translating a straight path ?
7. Explain the procedure of solving a particle moving on an inclined plane and acted upon by a variable force.
8. Analyse the motion of a lift when (*a*) it is moving up, and (*b*) it is moving down.
9. Analyse the motion of two masses m_A and m_B ($m_A > m_B$) hanging at the two ends of a weightless string lying on a smooth pulley.

B. Numerical Problems

9.1. A force of 200 N acts on a body of mass 5 kg for 8 seconds. If the initial velocity is zero, find :

(*a*) acceleration produced in the direction of the force.

(*b*) distance travelled by the body in 8 seconds. [**Ans.** (*a*) 40 m/s^2 ; (*b*) s = 1280 m]

9.2. A force of 200 N acts on a body of mass 5 kg for 8 seconds. If the body is initially moving with a velocity of 40 m/s, find :

(*a*) acceleration of the body.

(*b*) distance travelled by the body after 5 seconds.

(*c*) the velocity of the body after 5 seconds. [**Ans.** (*a*) 40 m/s^2 ; (*b*) 700 m ; (*c*) 240 m/s]

9.3. A body of mass 10 kg falls in the pond of mud from a height of 4.9 m. Find the depth through which the body will penetrate into the pond of mud if the mud offers a constant resistance of 2450 N. Also find the velocity of the body when it falls on mud. Take g = 9.8 m/s^2.

[**Ans.** 20.42 cm ; 9.8 m/s]

9.4. A man weighing 637 N dives into a swimming pool from a tower of height 19.6 m. He was found to go down in water by 2 m and then started rising. Find the average resistance of water. Neglect the resistance of air. [*AMIE, Summer 1982*]

[**Ans.** 6879.73 N]

9.5. A bullet of mass 81 gm moving with a velocity of 300 m/s is fired into a log of wood and it penetrates to a depth of 10 cm. If the bullet moving with the same velocity, were fired into a similar piece of wood 5 cm thick, with what velocity would it emerge ? Find also the force of resistance assuming it to be uniform. [*AMIE, Winter 1981*]

[**Ans.** 36450 N ; 212.132 m/s]

9.6. A particle of mass 0.5 kg is falling freely under the action of gravity. A force of 20 N inclined at 30° upwards to the right is applied to it. (*a*) Find the horizontal and vertical acceleration components after the application of the force, (*b*) Find the total acceleration of the particle after the force is applied. [**Ans.** (*a*) a_x = 34.64 m/s^2→, a_y = 10.19 m/s^2↓

(*b*) a = 36.11 m/s^2 ↘ θ_x ; $\theta_x = -16.26°$]

9.7. An elevator cage of a high tower weighing 10 kN when empty, is lifted by means of a steel rope. Four persons each weighing an average of 750 N entered the cage and the cage attained a velocity of 20 m/s when it covered a distance of 50 m. Find :

(*a*) for upward motion of the cage : (*i*) reaction of cage to each man, (*ii*) tension in rope

(*b*) also solve the problem for downward motion of the cage.

[**Ans.** (*a*) (*i*) 18300.7 N, (*ii*) 1055.81 N ; (*b*) (*i*) 7699.3 N, (*ii*) 444.2 N]

9.8. A motorist travelling at a speed of 90 kmph suddenly applies brakes a halts after skidding through 20 m. Find :

(*a*) the time required for the motor to stop,

(*b*) the average frictional resistance offered by the road.

[**Ans.** (*a*) t = 1.5815 ; (*b*) 1.593 N / unit total weight of motor]

9.9. A 2000 N crate rests on a 600 N cart. The coefficient of friction between crate and the cart is 0.28 and that between the cart and the road is 0.32. What pull should be applied so that the crate does not slip. Also find the corresponding acceleration of the cart.

[**Ans.** 1560 N ; 0.28 g = 2.747 m/s^2]

9.10. Two weights 1000 N and 600 N are connected by a string and they move on a rough horizontal plane under the action of a horizontal force of 800 N applied to the 1000 N weight. Find (*a*) the tension in the string and (*b*) acceleration of the two weights if the coefficient of friction between weights and the plane is 0.25. [**Ans.** 300 N ; $\frac{g}{4}$ = 2.4525 m/s^2]

9.11. A body weighing 2000 N rests on a rough 20° plane. A light rope connecting this weight runs over a smooth pulley carries another weight of 1000 N. If coefficient of friction between the 2000 N weight and the plane is 0.18, find :

(*a*) tension in rope,

(*b*) acceleration of the two weights, and

(*c*) the distance moved by the two weights in 5 seconds if the system starts from rest.

[**Ans.** (*a*) 1133.4 N ; (*b*) 0.0555 g = 0.5445 m/s^2 ; (*c*) 6.81 m]

9.12. Two weights of 800 N and 240 N connected with the help of a light string are placed on smooth pulley. The initial velocity of heavier weight is 3 m/s. Find the weight *W* attached on the lighter side of the string required to bring the weights to rest in an interval of 5 seconds.

[**Ans.** 664 N]

9.13. The three weights have mass m_A = 10 kg, m_B = 20 kg and m_C = 20 kg. The coefficients of friction between the blocks *A* and *C* and the horizontal table are 0.20 and 0.24, respectively. Find : (*a*) tension in the string, and (*b*) the acceleration of the three blocks. See Fig. E.9.13. Assume pulleys to be smooth. [**Ans.** a_A = 4.878 m/s^2 ; a_C = 1.066 m/s^2 ; a_B = 2.972 m/s^2]

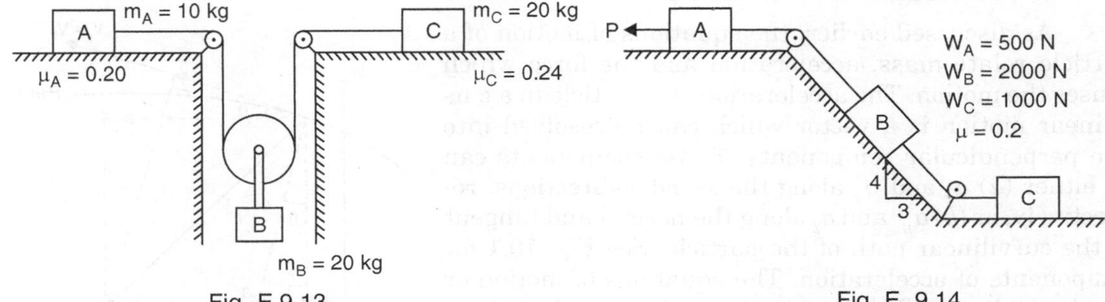

Fig. E.9.13

Fig. E. 9.14

9.14. For the system of blocks shown in Fig. E. 9.14, find the pull *P* that will give system a velocity of 4 m/s after moving a distance of 8 m starting from rest. μ = 0.2 for all surfaces. Also, find tension in the cord. [**Ans.** *P* = 2496.8 N ; *T* = 301.94 N]

9.15. Determine the tension in the strings and acceleration of blocks *A* and *B* weighing 150 N and 50 N connected by a string and a frictionless and weightless pulley as shown in Fig. E. 9.15.

[*AMIE, Nov. 1970*]

[**Ans.** *T* = 64.3 N ; a_A = 1.402 m/s^2 , a_B = 2.80 m/s^2]

9.16. Neglecting friction and the inertia of the two pulleys in Fig. E. 9.16 determine the acceleration *a* of the weight *Q* assuming that *P* = *Q* [**Ans.** *a* = 0.4 g]

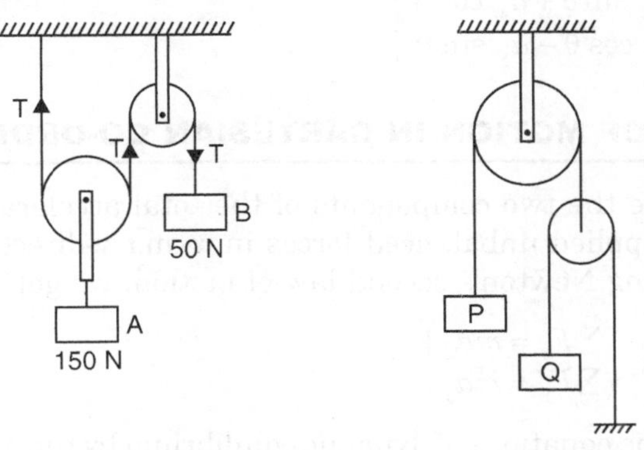

Fig. E. 9.15

Fig. E 9.16

9.17. A particle of weight *W* starts from rest at the origin and moves rectilinearly under the action of a force $P = P_0 e^{-kt}$, where P_0 is the initial magnitude of the force and *k* is a constant. Establish expressions for displacement and velocity of the particle.

$$\left[\textbf{Ans. } x = \frac{P_0 \cdot g}{Wk^2}\left(kt + e^{-kt} - 1\right) \quad \dot{x} = v = \frac{P_0 \cdot g}{Wk}(1 - e^{-kt})\right]$$

10

Kinetics of a Particle-Curvilinear Motion

10.1. INTRODUCTION

As discussed earlier, the equations of motion of a particle relate mass, acceleration and the force which causes the motion. The acceleration of a particle in a curvilinear motion is a vector which can be resolved into two perpendicular components. These components can be either (*a*) a_x and a_y along the *x*-and *y*-directions, respectively, or (*b*) a_n and a_t along the normal and tangent to the curvilinear path of the particle. See Fig. 10.1 for components of acceleration. The equations of motion or the dynamic equilibrium equations following the principle of D' Alembert can be written in terms of either of the above two components of acceleration. From Fig. 10.1, if a_n and a_t are known, we get

$$\left.\begin{aligned} a_x &= a_n \sin\theta + a_t \cos\theta \\ a_y &= a_n \cos\theta - a_t \sin\theta \end{aligned}\right\} \quad ...(10.1)$$

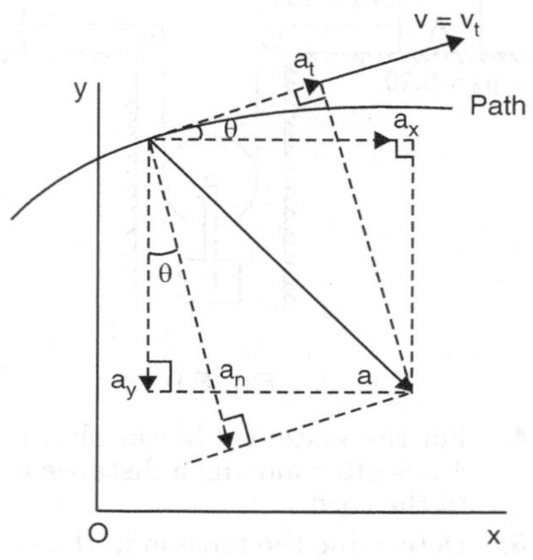

Fig. 10.1. Relations between the tangential and normal components of acceleration and the rectangular acceleration components.

Also if a_x and a_y are known, then,

$$\left.\begin{aligned} a_n &= a_x \sin\theta + a_y \cos\theta \\ a_t &= a_x \cos\theta - a_y \sin\theta \end{aligned}\right\} \quad ...(10.2)$$

10.2. EQUATIONS OF MOTION IN CARTESIAN CO-ORDINATES

Let a_x and a_y be the two components of the total acceleration of a particle of mass *m*. Also, let the sum of applied unbalanced forces in *x*-and *y*-directions be ΣF_x and ΣF_y, respectively. Then by applying Newton's second law of motion, we get

and $$\left.\begin{aligned} \Sigma F_x &= ma_x \\ \Sigma F_y &= ma_y \end{aligned}\right\} \quad ...(10.3)$$

Alternatively, the equations of dynamic equilibrium by the application of the D' Alembert's principle, we get

and $$\left.\begin{aligned} \Sigma F_x - ma_x &= 0 \\ \Sigma F_y - ma_y &= 0 \end{aligned}\right\} \quad ...(10.4)$$

10.3. EQUATIONS OF MOTION IN TANGENTIAL AND NORMAL COMPONENTS OF ACCELERATION

Equations of motion in the normal and tangential direction are as under :

$$\Sigma F_n = ma_n \quad ...(10.5)$$

and $$\Sigma F_t = ma_t \quad ...(10.6)$$

Similarly, equations of dynamic equilibrium are :

$$\Sigma F_n - ma_n = 0 \quad ...(10.7)$$

and $$\Sigma F_t - ma_t = 0 \quad ...(10.8)$$

where $$\left.\begin{aligned} a_n &= \rho\dot{\theta}^2 = \rho\omega^2 \\ a_n &= v^2/\rho = v\omega \end{aligned}\right\} \quad ...(10.9)$$
or

$$v = \rho\omega = \rho\dot{\theta} \quad ...(10.10)$$

and $$a_t = \frac{dv}{dt} \quad ...(10.11)$$

where ρ = radius of curvature of the path,

v = velocity of the particle along the tangent of the path.

10.4. EQUATIONS OF MOTION IN POLAR CO-ORDINATES

Equations of motion in the polar co-ordinates are :

$$\Sigma F_r = ma_r \quad ...(10.12)$$

$$\Sigma F_\theta = ma_\theta \quad ...(10.13)$$

where a_r = acceleration component along the radial direction, (also known as centripetal accelerations)

and a_θ = tangential or circumferential acceleration (This is along the velocity of the particle)

In the above equations

$$a_r = \ddot{r} - r\dot{\theta}^2 = \frac{d^2r}{dt^2} - r\left(\frac{d\theta}{dt}\right)^2 \quad ...(10.14)$$

and $$a_\theta = r\ddot{\theta} + 2\dot{r}\dot{\theta} = r\alpha + 2v\omega \quad ...(10.15)$$

where, α = angular acceleration, and

$2v\omega = 2\dot{r}\dot{\theta}$ = Corioli's component acceleration.

r = radius vector of motion (also known as position vector)

If particle moves in a circular path r is constant. Then $a_r = -r\dot{\theta}^2$ (it is directed towards the centre of curvature).

When the particle moves in a straight line then

$$a_\theta = \frac{dv}{dt} \quad \text{and} \quad a_r = 0.$$

The following examples will help students to understand the kinetics of a particle in the curvilinear motion.

Problem 10.1. *A motor cycle of weight w travels with a constant speed of 72 km/hr over a vertical parabolic curve $x^2 = ky$, as shown in Fig. P. 10.1. Find the pressure between the motor cycle and the road at the point C, and also at point A or B.*

Sol. Let equation of the parabola be $x^2 = ky$ ...(1)

Where k is a constant. At mid-point C,

$$y = h \text{ and } x = l/2,$$

$\therefore$ from eqn. (1), we get

$$(l/2)^2 = k \,.\, h,$$

$$\therefore \quad k = \frac{l^2}{4h},$$

$\therefore$ Eqn. (1) gives $\quad x^2 = \dfrac{l^2}{4h}\, y$ ...(1a)

y
R
$x^2 = ky$
C
v = 20 m/s
v
W
h = 2 m
y
x
A
B
x
l/2 = 60 m
l/2

Fig. P. 10.1

Curvature of the curve is given by

$$\frac{1}{\rho} = \frac{d^2y/dx^2}{[1+(dy/dx)^2]^{3/2}} \quad ...(2)$$

$$\because \quad x^2 = ky \; ; \quad \therefore \; y = \frac{x^2}{k} \quad ...(3)$$

$$\therefore \quad \frac{dy}{dx} = \frac{2x}{k} \quad \text{and} \quad \frac{d^2y}{dx^2} = \frac{2}{k}$$

At point C, $\quad \dfrac{dy}{dx} = 0 \quad \text{and} \quad \dfrac{d^2y}{dx^2} = \dfrac{2}{k}$

$$\therefore \quad \frac{1}{\rho_C} = \frac{2/k}{[1+0]^{3/2}} = \frac{2}{k}$$

$$v = 72 \text{ km/hr} = 20 \text{ m/s}$$

$\because$ The motor-cycle is moving with constant speed,

$$\therefore \quad \frac{dv}{dt} = a_t = 0, \quad \text{and} \quad a_n = v^2/\rho.$$

$$\therefore \quad \frac{1}{\rho} = \frac{2/k}{(1+0)^2} = \frac{2}{k} = \frac{2}{l^2/4h} = \frac{8h}{l^2} = \frac{2h}{(l/2)^2}$$

$$\therefore \quad \frac{1}{\rho_c} = \frac{2\times 2}{(60)^2} = \frac{2\times 2}{60\times 60} = \frac{1}{900}$$

By the second law of motion, we get

$$R - W = ma_n \quad | \; R = \text{Reaction} \quad \text{and} \quad W = \text{weight} \quad \text{and} \quad m = W/g$$

$$\therefore \quad R - W = \frac{W}{g} \times a_n$$

where $\quad a_n = \dfrac{v^2}{\rho} = \dfrac{(20)^2}{900}\, m/s^2$

$$W - R = \frac{W \times (20)^2 \times 1}{9.8 \times 900} = 0.0454\, W$$

$\therefore$ Pressure $= R = W - 0.0454\, W =$ **0.9546 W. Ans.**

For point A :

We know that $x^2 = \dfrac{l^2 y}{4h}$

$\therefore \quad y = \dfrac{4hx^2}{l^2}, \quad \therefore \quad \dfrac{dy}{dx} = \dfrac{8h}{l^2}x$ and $\dfrac{d^2y}{dx^2} = \dfrac{8h}{l^2}$

For point A : $x = 0$, $\therefore \quad \dfrac{dy}{dx} = 0$ and $\dfrac{d^2y}{dx^2} = \dfrac{8h}{l^2} = \dfrac{8 \times 2}{(120)^2} = \dfrac{1}{900}$

From $\dfrac{1}{\rho} = \dfrac{d^2y/dx^2}{[1+(dy/dx)^2]^{3/2}}$

for point A, we get $\dfrac{1}{\rho_A} = \dfrac{1}{900}$

$$\therefore \quad a_{n_A} = \frac{v^2}{\rho_A} = \frac{(20)^2}{900} = \frac{4}{9} \text{ m/s}^2$$

$$\therefore \quad W - R = \frac{W}{g} \times a_n$$

$$\therefore \quad W - R = \frac{W}{9.8} \times \frac{4}{9} = 0.0454\,W$$

$\therefore \quad R = W(1 - 0.0454) =$ **0.9546 W. Ans.**

Problem 10.2. *A ball of mass 1.2 kg moves over a sinusoidal curve $y = y_0 \sin \dfrac{\pi x}{l}$, as shown in Fig. P. 10.2. Find the reaction of the path at points A and C. Assume $y_0 = 1.2$ m and $l = 6$ m.*

Sol. Equation of path of the ball is $y = y_0 \sin \dfrac{\pi x}{l}$ ...(1)

At C, $y_0 = 1.2$ m and $x = l/2 = 3$ m.

$\therefore$ we get $y = 1.2 \sin \dfrac{\pi x}{l}$ ...(2)

$$\therefore \quad \frac{dy}{dx} = 1.2\left(\frac{\pi}{l}\right)\cos\frac{\pi x}{l} = \frac{1.2\pi}{l}\cos\frac{\pi x}{l}$$

and
$$\frac{d^2y}{dx^2} = -\left(\frac{1.2\pi}{l}\right)\left(\frac{\pi}{l}\right)\sin\frac{\pi x}{l} = -\frac{1.2\pi^2}{l^2}\sin\frac{\pi x}{l}$$

Fig. P. 10.2

Curvature,
$$\frac{1}{\rho} = \frac{d^2y/dx^2}{[1+(dy/dx)^2]^{3/2}} = \frac{-(1.2\pi^2/l^2)\sin \pi x/l}{\left[1+\left(\dfrac{\pi \times 1.2}{l}\right)^2 \cos^2\left(\dfrac{\pi x}{l}\right)\right]^{3/2}}$$

At C, $x = l/2 = 3$ m ; and $y = 1.2$ m.

$$\therefore \quad \frac{1}{\rho_C} = \frac{-1.2\pi^2/(3)^2 \sin \pi \times (3/6)}{\left[1+\left(\dfrac{1.2\pi}{l}\right)^2 \times \cos^2\left(\dfrac{\pi \times 3}{6}\right)\right]^{3/2}} = -\frac{(1.2\pi^2/l^2)\times 1}{1} = -\frac{1.2 \times \pi^2}{36} = \frac{\pi^2}{30}$$

At point C, $v = \sqrt{2gy_0}$

$$\therefore \quad a_n = \frac{v^2}{\rho_C} = (2gy_0) \times \frac{\pi^2}{30} = \frac{\pi^2 gy_0}{15} \quad \text{where } y_0 = 1.2 \text{ m}$$

$$\therefore \quad a_n = \frac{\pi^2 \times 9.8 \times 1.2}{15} = 7.738 \text{ m/s}^2$$

$\therefore$ By $\Sigma F_n = ma_n$, we get $R - W = \dfrac{W}{g} \times a_n$,

$$\therefore \quad R = W\left(1 + \frac{\pi^2 y_0}{15}\right). \quad \textbf{Ans.}$$

or

$$R = W\left(1 + \frac{\pi^2 \times 1.2}{15}\right) \quad \text{or} \quad R = 1.79\ W. \quad \textbf{Ans.}$$

Substituting, $m = 1.2$ kg, we get

$$R = 1.79 \times (1.2\ g) = \mathbf{21.05\ N.} \quad \textbf{Ans.}$$

Problem 10.3. *A particle of mass 2 kg moves along the path whose equation is $r = 4\theta$ m. If the angle $\theta = t^2$ radians determine velocity and acceleration components in x-and y-direction when $\theta = 60°$. Also find the inertia force components in x-and y-directions for $\theta = 60°$.*

Sol. The path of the particle is shown in Fig. P. 10.3 (a) with unit vectors $\bar{e}_r$ and e_θ for polar co-ordinates.

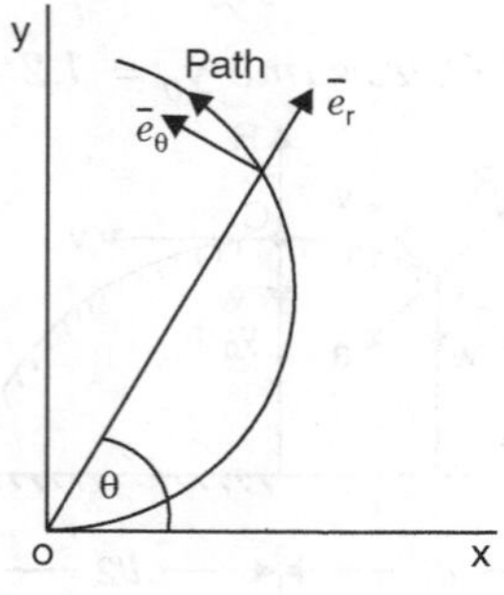

(a) Path

(b) Velocity components

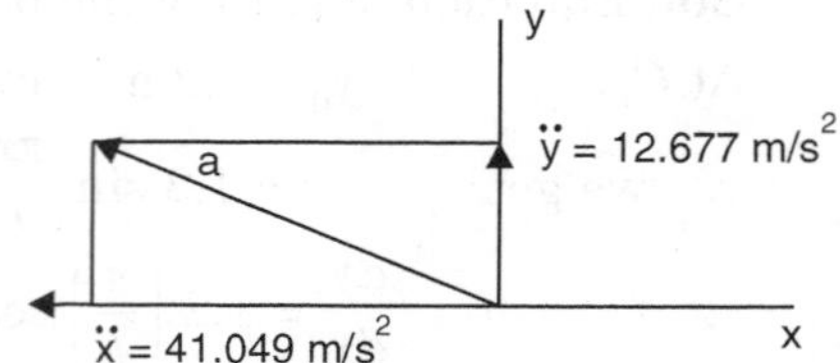

(c) Acceleration components

Fig. P. 10.3

Given : $\theta = t^2$, $\therefore$ $r = 4\theta = 4t^2$

$$x = r \cos \theta = 4\theta \cos \theta = 4t^2 \cos t^2$$

$$y = r \sin \theta = 4\theta \sin \theta = 4t^2 \sin t^2$$

$$\therefore \quad \dot{x} = \frac{dx}{dt} = 4\ .\ (2t) \cos t^2 + 4t^2 (-\sin t^2)\ (2t)$$

$$\dot{x} = 8t \cos t^2 - 8t^3 \sin t^2 \qquad \text{...(1)}$$

$$\dot{y} = \frac{dy}{dt} = \frac{d}{dt}(4t^2 \sin t^2) = 4(2t) \sin t^2 + 4t^2 (2t) \cos t^2$$

or $$\dot{y} = 8t \sin t^2 + 8t^3 \cos t^2 \qquad ...(2)$$

Also $$\ddot{x} = \frac{d}{dt}[8t \cos t^2 - 8t^3 \sin t^2]$$

$$= [8 \times \cos t^2 + 8t\,(2t)\,(-\sin t^2)] - [8\,(3t^2) \sin t^2 + 8t^3 (2t) \cos t^2]$$

$$= (8 - 16\,t^4) \cos t^2 - (16t^2 + 24t^2) \sin t^2$$

or $$\ddot{x} = (8 - 16t^4) \cos t^2 - 40t^2 \sin t^2 \qquad ...(3)$$

and $$\ddot{y} = \frac{d}{dt}(8t \sin t^2 + 8t^3 \cos t^2)$$

$$= [8 \sin t^2 + 8t \times (2t) \cos t^2] + [8 \times (3t^2 \cos t^2 + 8t^3 \times (2t) \times (-\sin t^2)]$$

$$= (8 - 16\,t^4) \sin t^2 + (16t^2 + 24\,t^2) \cos t^2$$

or $$\ddot{y} = (8 - 16\,t^4) \sin t^2 + 40t^2 \cos t^2 \qquad ...(4)$$

When $\theta = 60° = \pi/3$ radians $= t^2$, $\therefore \quad t = \sqrt{\frac{\pi}{3}}$ sec

$\therefore$ From (1) $$(\dot{x})_{\frac{\pi}{3}} = 8\sqrt{\frac{\pi}{3}}\,.\cos 60° - 8 \times \left(\frac{\pi}{3}\right) \times \sqrt{\frac{\pi}{3}} \sin 60° \text{ m/s}$$

or $$(\dot{x})_{\frac{\pi}{3}} = 4.0933 - 7.4244 = -\mathbf{3.3311\ m/s} \quad \textbf{Ans.}$$

From (2), $$(\dot{y})_{\frac{\pi}{3}} = 8\sqrt{\frac{\pi}{3}} \sin 60° + 8 \times \frac{\pi}{3} \times \sqrt{\frac{\pi}{3}} \cos 60°$$

or $$(\dot{y})_{\frac{\pi}{3}} = 7.0898 + 4.2865 = \mathbf{11.3763\ m/s} \quad \textbf{Ans.}$$

From (3), $$(\ddot{x})_{\frac{\pi}{3}} = \left[8 - 16 \times \left(\frac{\pi}{3}\right)^2\right] \cos 60° - 40 \times \frac{\pi}{3} \times \sin 60° \text{ m/s}^2$$

or $$(\ddot{x})_{\frac{\pi}{3}} = -4.7730 - 36.276 = -\mathbf{41.049\ m/s^2}$$

and from (4), $$(\ddot{y})_{\frac{\pi}{3}} = \left[8 - 16 \times \left(\frac{\pi}{3}\right)^2\right] \sin 60° + 40 \times \frac{\pi}{3} \times \cos 60° \text{ m/s}^2$$

$$= -8.2670 + 20.944 \text{ m/s}^2 = \mathbf{12.677\ m/s^2}. \quad \textbf{Ans.}$$

Inertia force $$F_x = m\,.\,\ddot{x} = 2 \times (-41.049) \text{ N} = -\mathbf{82.098\ N}. \quad \textbf{Ans.}$$

$$F_y = m\ddot{y} = 2 \times 12.677 = \mathbf{25.354\ N}. \quad \textbf{Ans.}$$

Problem 10.4. *A small block of mass 2 kg lies on a smooth plane inclined at 15°. The plane is free to turn about the y-axis as shown in Fig. P. 10.4 (a). Length of the chord l is 500 mm. What is the tension in the chord when the angular velocity of the plane is 30 revolutions per minute ?*

Sol. From the Fig. P. 10.4 (*b*), we find that

$$r = l \sin 75° = 500 \sin 75° = 482.96 \simeq 0.483 \text{ m.}$$

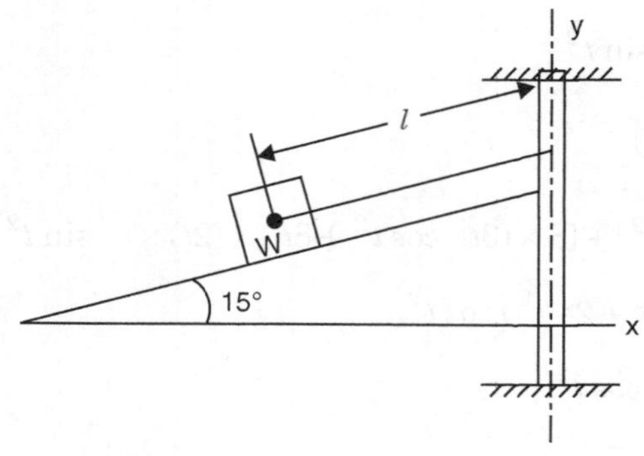

(a)

(b) Free body diagram of mass

Fig. P. 10.4

The only acceleration present is $a_n = \omega^2 \cdot r$

where $$\omega = \frac{2\pi N}{60} = \frac{2\pi \times 30}{60} = \pi \quad \text{rad/s}$$

$$\therefore \quad a_n = \omega^2 r = (\pi)^2 \times 0.483 = 4.77 \text{ m/s}^2$$

By $\Sigma F_n = m a_n$, we get

$$T \cos 15° - N \sin 15° = 2 \times 4.77 \qquad ...(1)$$

And, $\because \; a_y = 0, \;\therefore$ By $\Sigma F_y = m a_y$, we get

$$N \cos 15° - W + T \sin 15° = 0 \qquad ...(2)$$

From eqn. (2), $N = \dfrac{W}{\cos 15°} - T \tan 15°$

$$\therefore \quad N = \frac{2 \times 9.8}{\cos 15°} - T \tan 15°$$

or $$N = 20.29 - T \tan 15° \qquad ...(3)$$

$\therefore$ From eqn. (1), we get

$$T \cos 15° - (20.29 - T \tan 15°) \sin 15° = 2 \times 4.77$$

or $$T(\cos 15° + \tan 15° \sin 15°) = 2 \times 4.77 + 20.29 \sin 15°$$

or $$T(1.0353) = 14.79$$

$\therefore$ T = **14.79 N. Ans.**

Problem 10.5. *A small block of mass 2 kg lies on a smooth plane inclined at 15°. The plane is free to turn about the y-axis, as shown in Fig. P. 10.5 (a). Length of the cord l is 500 mm. What is the angular velocity about y-axis necessary to cause the mass to just float above the plane ? Then, what is the tension in the cord under these conditions ?*

Sol. When the block floats just above the inclined plane, the normal reaction is equal to zero. (*i.e.*, $N = 0$, as in Fig. P. 10.5 (*b*)).

$\therefore$ The radius of rotation, $r = l \cos 15°$

or $$r = 500 \cos 15° = 483 \text{ mm} \quad \text{or} \quad 0.483 \text{ m}.$$

$\therefore$ $$a_n = \omega^2 r$$

$\therefore$ By $\Sigma F_n = ma_n$, we get

$$T \cos 15° = ma_n \quad \text{or} \quad T \cos 15° = 2 \times r\omega^2$$

or $$T \cos 15° = 2 \times 0.483\ \omega^2 \qquad ...(1)$$

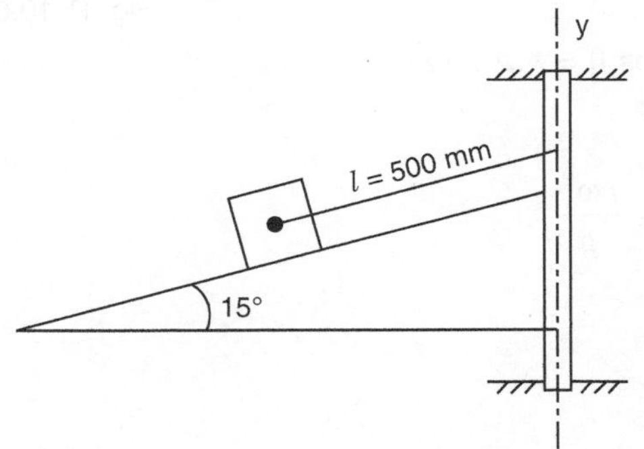

(a) Given system

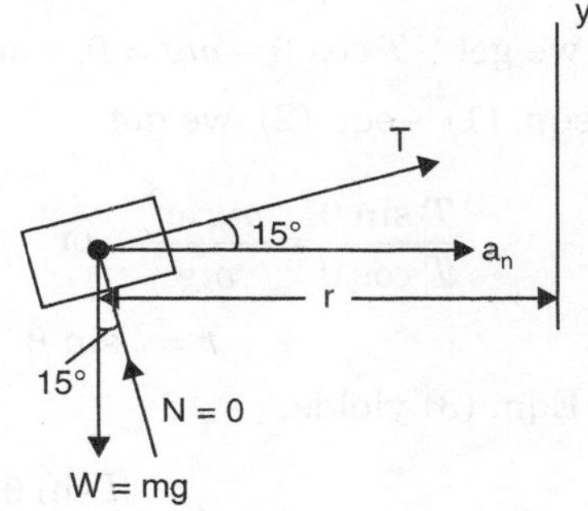

(b) Free body diagram of mass when it leaves the plane

Fig. P. 10.5

And by $\Sigma F_y = 0$, we get

$$T \sin 15° - W = 0$$

$\therefore$ $$T \sin 15° = W = 2 \times 9.8 \qquad ...(2)$$

$\therefore$ By eqn. (2) ÷ (1), we get

$$\frac{\sin 15°}{\cos 15°} = \frac{2 \times 9.8}{2 \times 0.483\ \omega^2}$$

$\therefore$ $$\omega^2 = \frac{2 \times 9.8}{2 \times 0.483} \times \frac{1}{\tan 15°} = 7.572$$

$\therefore$ $$\boldsymbol{\omega = 8.70 \text{ rad/s} \quad \text{or} \quad 83.08 \text{ rev./min.}} \quad \textbf{Ans.}$$

And from eqn. (2), the tension in the cord = $T = \dfrac{2 \times 9.8}{\sin 15°}$ **= 75.73 N. Ans.**

Problem 10.6. *A particle of mass m, tied up by a cord of length l to a verticle axle which rotates at a uniform angular velocity ω. The mass describes a horizontal circular path, Fig. P. 10.6. This arrangement is known as spherical or conical pendulum. Find expression of frequency of the pendulum.*

Sol. Let ω be the angular velocity when r is the radius of rotation. Let the cord make an angle θ with the y-axis (the axis of rotation). Let T be the tension in the cord. The acceleration $a_n = r\omega^2$

∴ By $\Sigma F_n = ma_n$ we get

$$T \sin \theta = ma_n$$

or $$T \sin \theta = mr\omega^2 \quad ...(1)$$

By $\Sigma F_y = ma_y$, we get

$$T \cos \theta - mg = ma_y$$

and $$a_y = 0,$$

∴ we get $T \cos \theta - mg = 0.$ or $T \cos \theta = mg$ ...(2)

Fig. P. 10.6

By eqn. (1) ÷ eqn (2), we get

$$\frac{T \sin \theta}{T \cos \theta} = \frac{m\,r\omega^2}{mg} \quad \text{or} \quad \tan \theta = \frac{r\omega^2}{g} \quad ...(3)$$

But $$r = l \sin \theta \quad ...(4)$$

∴ Eqn. (3) yields,

$$\tan \theta = \frac{(l \sin \theta)\omega^2}{g}$$

∴ $$\omega^2 = \frac{g \tan \theta}{l \sin \theta} = \frac{g}{l \cos \theta}$$

∴ $$\omega = \sqrt{\frac{g}{l \cos \theta}}. \quad \textbf{Ans.} \quad ...(5)$$

Since ω is constant for the given angle θ, therefore we get frequency

$$f = \frac{\omega \text{ rad/s}}{2\pi \text{ rad/rev}} = \frac{1}{2\pi}\sqrt{\frac{g}{l \cos \theta}}$$

or, $$f = \frac{1}{2\pi}\sqrt{\frac{g}{l \cos \theta}} \text{ Hz.} \quad \textbf{Ans.} \quad ...(6)$$

And, lastly, the tension in cord $T = \dfrac{mg}{\cos \theta}$. **Ans.** ...(7)

Problem 10.7. *The bob of mass m of a simple pendulum is suspended on a cord of length l, as shown in Fig. P.10.7 (a). Derive expressions for the frequency and period of this pendulum.*

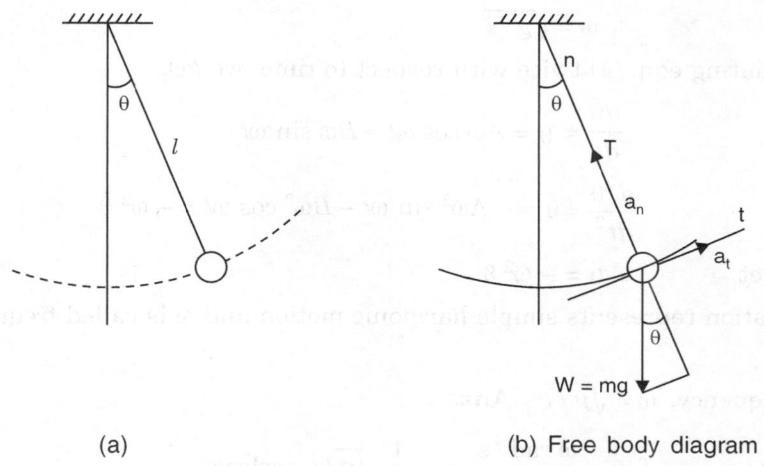

Fig. P. 10.7

Sol. The path of the mass is an arc of circle of radius l. For the position shown in Fig. P. 10.7 (*a*), the forces acting on the mass are : The weight $W = mg$ and tension T in the cord. The normal and tangential directions are shown in Fig. P. 10.7(*b*).

Since pendulum is oscillating, at the lowest position its velocity is maximum (and hence, its acceleration is zero), while at the highest position the velocity is zero.

The only acceleration present in the tangential direction $= a_t = l \times \dfrac{d^2\theta}{dt^2}$

$\therefore$ by $\Sigma M_t = ma_t$, we get

$$-W \sin \theta = ma_t$$

or,
$$-mg \sin \theta = m\left(l \frac{d^2\theta}{dt^2}\right) \quad ...(1)$$

Now,
$$\sin \theta = \theta - \frac{\theta^3}{3!} + \frac{\theta^5}{5!} - \frac{\theta^7}{7!} + ...$$

Since θ is small, therefore terms of higher powers are negligible. Hence, we get $\sin \theta \simeq \theta$ in radians.

$\therefore$ Equation (1) becomes,

$$-mg\theta = m\left(l \frac{d^2\theta}{dt^2}\right) \quad \text{or} \quad \ddot{\theta} + \frac{g\theta}{l} = 0 \quad ...(2)$$

Let solution of eqn. (2) be,

$$\theta = A \sin\left(\sqrt{\frac{g}{l}}\right) t + B \cos\left(\sqrt{\frac{g}{l}}\right) t \quad ...(3)$$

Eqn. (3) can be written in form given below :

$$\theta = A \sin \omega t + B \cos \omega t \quad ...(4)$$

where $$\omega = \sqrt{g/l}. \quad ...(5)$$

Differentiating eqn. (4) twice with respect to time, we get

$$\frac{d\theta}{dt} = \dot{\theta} = A\omega \cos \omega t - B\omega \sin \omega t$$

and, $$\frac{d^2\theta}{dt^2} = \ddot{\theta} = -A\omega^2 \sin \omega t - B\omega^2 \cos \omega t = -\omega^2 \theta$$

∴ We get $$\ddot{\theta} = -\omega^2 \theta \quad ...(6)$$

This equation represents simple harmonic motion and ω is called frequency in radians per second.

Thus, frequency, $\omega = \sqrt{g/l}$ **Ans.** ... (7)

or, frequency, $$f = \frac{\omega \text{ rad/s}}{2\pi \text{ rad/cycle}} = \frac{1}{2\pi}\sqrt{g/l} \text{ cycles/s}$$

or frequency, $$f = \frac{1}{2\pi}\sqrt{g/l} \text{ Hz} \quad ...(8)$$

The time period $= \tau = \dfrac{1}{f} = 2\pi\sqrt{l/g}$ second ...(9)

τ is expressed in seconds per cycle.

Problem 10.8. *A particle of mass m rests on the top of a smooth circular disc of radius r, as shown in Fig. P. 10.8(a). Assuming that the particle starts to move from rest, at what position will it leave the disc ?*

Sol. Let the particle move to the position B after time t. The free body diagram of the particle is shown in Fig. P. 10.8(*b*). Here $a_n = r\omega^2$ and $a_t = r\alpha$.

If N = normal reaction of the disc to the particle, we get

$$\Sigma F_r = \Sigma F_n = -N + mg \cos \theta$$

$$\Sigma F_t = mg \sin \theta,$$

∴ By $\Sigma F_n = ma_n$, we get

$$-N + mg \cos \theta = m\, r\omega^2 \quad ...(1)$$

By $\Sigma F_t = ma_t$, we get

$$mg \sin \theta = mr\alpha \quad ...(2)$$

A third equation is developed as follows.

We know that $$\omega = \frac{d\theta}{dt} \quad ...(i)$$

and $$\alpha = \frac{d\omega}{dt} \quad ...(ii)$$

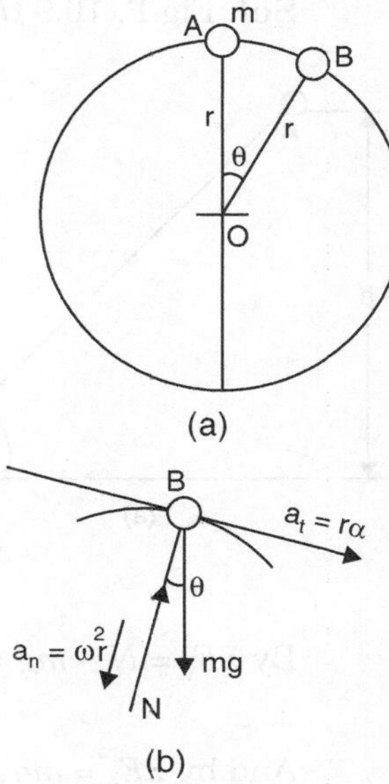

∴ By eliminating t from (i) and (ii), we get

∵ $$dt = \frac{d\theta}{\omega}, \text{ from } (i),$$

∴ Eqn. (ii) gives $$\alpha = \frac{d\omega}{\left(\frac{d\theta}{\omega}\right)} \quad \text{or} \quad \alpha\, d\theta = \omega \,.\, d\omega \qquad ...(3)$$

Fig. P. 10.8 — (a): A, m, B, r, r, θ, O. (b): B, $a_t = r\alpha$, θ, $a_n = \omega^2 r$, mg, N.

From eqn. (2), $$\alpha = \left(\frac{g}{r}\right)\sin\theta \qquad ...(4)$$

Substituting eqn. (4) into eqn. (3), we get

$$\left(\frac{g}{r}\right)\sin\theta \,.\, d\theta = \omega .\, d\omega \qquad ...(5)$$

Integrating eqn. (5), we get

$$\int_0^{\theta}\left(\frac{g}{r}\right)\sin\theta\, d\theta = \int_0^{\omega}\omega \,.\, d\omega,$$

∴ $$-\frac{g}{r}\cos\theta\Big|_0^{\theta} = \frac{\omega^2}{2}\Big|_0^{\omega}$$

i.e., $$-\left(\frac{g}{r}\cos\theta - \frac{g}{r}\times 1\right) = \frac{1}{2}\,\omega^2 \quad \text{or} \quad \frac{g}{r}(1-\cos\theta) = \frac{\omega^2}{2}\Bigg| \qquad ...(6)$$

When the particle leaves the disc, normal reaction N is zero. Therefore, from eqn. (1), we get

$$-0 + mg\cos\theta = m\,r\omega^2 \quad \text{or,} \quad \frac{g}{r}\cos\theta = \omega^2 \qquad ...(7)$$

By substituting the value of ω^2 from eqn. (7) into eqn. (6), we get

$$\frac{g}{r}(1-\cos\theta) = \frac{1}{2}\,\frac{g}{r}\cos\theta$$

∴ $$1-\cos\theta = \frac{1}{2}\cos\theta \quad \text{or,} \quad \frac{3}{2}\cos\theta = 1$$

∴ $$\cos\theta = \frac{2}{3} \qquad ...(8)$$

or, $\theta = 0.841$ radian, or **48.2°. Ans.**

Problem 10.9. *A particle of mass m slides down a frictionless incline and entres a loop-the-loop of diameter d as shown in Fig. P. 10.9 (a). What should be the height h at the start in order the particle may make a complete turn in the loop ?*

Sol. Fig.P. 10.9 (*b*) shows the free body diagram of the particle at any time from start.

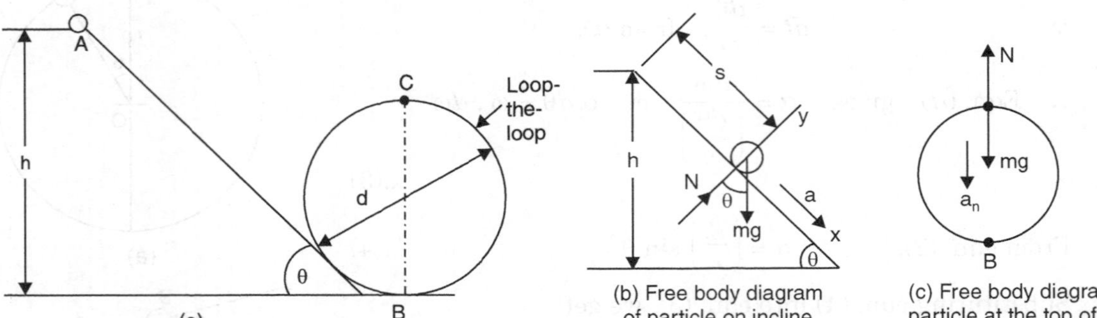

Fig. P. 10.9

By $\Sigma F_y = N - mg \cos \theta = 0$ ...(1)

$\therefore \qquad N = mg \cos \theta$

And by $\Sigma F_x = ma$, we get

$$mg \sin \theta = ma \qquad ...(2)$$

$$\therefore \qquad a = g \sin \theta \qquad ...(3)$$

At start, $u = 0$, $\therefore$ at B, $v_B^2 = u^2 + 2as$

$$\therefore \qquad v_B^2 = 0 + (2g \sin \theta) \times (AB) \qquad | \; s = AB = \frac{h}{\sin \theta}$$

$$= 2g \sin \theta \times \frac{h}{\sin \theta}$$

$$\therefore \qquad v_B^2 = 2gh \qquad ...(4)$$

At the top of the loop (point C) ; the particle has acceleration $a_n = v_C^2/(d/2) = \dfrac{2v_C^2}{d}$

By using the relation $v_C^2 = v_B^2 - 2g(d)$, $\qquad | \; s = d,$

we have $\qquad v_C^2 = 2gh - 2gd = 2g(h - d)$ ...(5)

$$\therefore \qquad a_n = \frac{2v_C^2}{d} = \frac{2}{d} \times 2g(h - d)$$

The normal reaction of the loop at point C is zero $\therefore \; N_C = 0$.

$\therefore$ By $\Sigma F_n = ma_n$, we get

$$mg = m \times \frac{4g}{d}(h - d)$$

$\therefore$ we get $\qquad h = \dfrac{5d}{4}$. **Ans.**

Problem 10.10. *Small objects are released from rest at A and slide down the smooth circular surface of radius R to a belt conveyor B. Find expression for the normal contact force N between an object and the guide and also find the correct angular velocity ω of the conveyor pulley of radius r to prevent any sliding of objects on the conveyor.*

Sol. The free body diagram of an object is shown in Fig. P. 10.10 (*b*) with co-ordinates *n* and *t*.

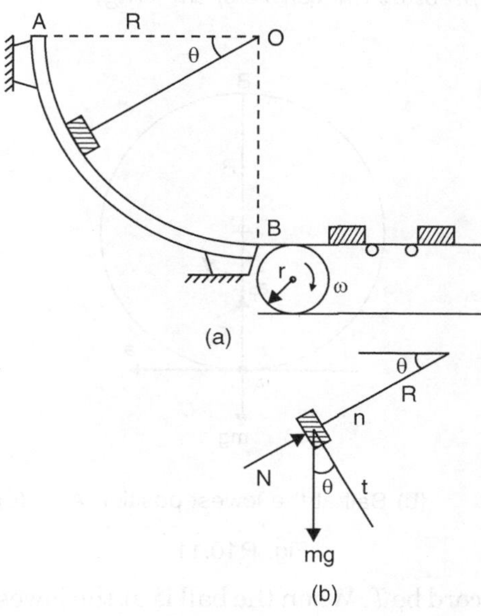

Fig. P. 10.10

By $\Sigma F_t = ma_t$, we get

$$mg \cos \theta = ma_t$$

$\therefore$ $$a_t = g \cos \theta \qquad \text{...(1)}$$

$\because$ $$v dv = a_t \, ds \; ; \quad ds = d(R\theta) = R d\theta$$

$\therefore$ We get $$\int v \, dv = \int g \cos \theta \, d(R\theta)$$

$\therefore$ $$\frac{v^2}{2} = g R \sin \theta$$

or $$v^2 = 2gR \sin \theta \qquad \text{...(2)}$$

By $\Sigma F_n = ma_n$, we get $[a_n = v^2/R]$.

$$N - mg \sin \theta = \frac{mv^2}{R}$$

$\therefore$ $$N = mg \sin \theta + \frac{m \times 2gR \sin \theta}{R}$$

or $$N = 3 \, mg \sin \theta. \quad \textbf{Ans.} \qquad \text{...(3)}$$

The conveyor pulley must turn at the velocity

$$v = r\omega$$

for $\theta = \pi/2$ (at point B),

$$\omega = \frac{v}{r} = \left[\frac{\sqrt{2gR \sin \theta}}{r} \right]_{\theta = \pi/2} = \frac{\sqrt{2gR}}{r} \text{ rad/s.} \quad \textbf{Ans.}$$

Problem 10.11. *A ball of mass m rotates at uniform velocity in vertical circular ring of radius r, as shown in Fig. P.10.11 (a). Determine the maximum and minimum tension in the cord connecting the ball to a pivot at the centre of the ring.*

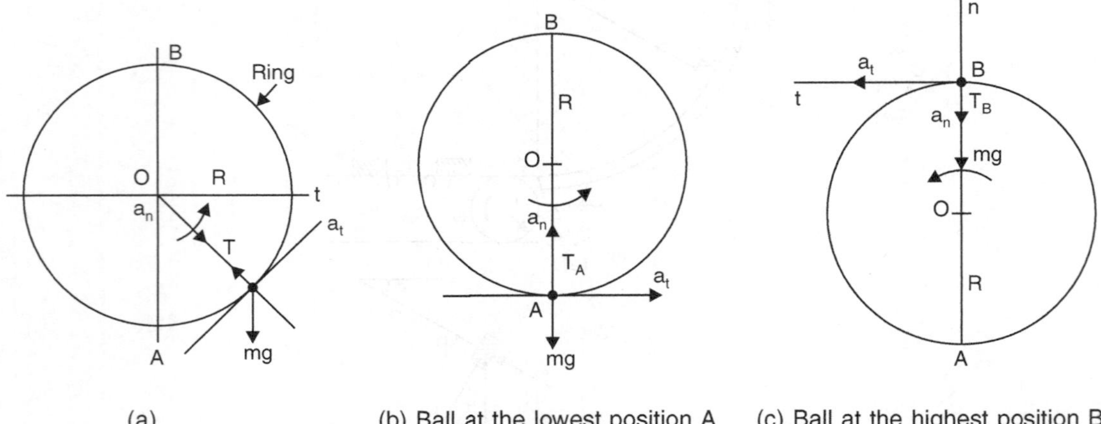

(a) (b) Ball at the lowest position A (c) Ball at the highest position B

Fig. P.10.11

Sol. Let tension in the cord be T. When the ball is at the lowest position A [Fig. P.10.11(a)], the ball is subjected to tension T_A, tangential acceleration $a_t = \dfrac{dv}{dt} = 0$ ($\because$ v is constant) ; normal acceleration $a_n = \dfrac{v^2}{R}$, directed towards the centre.

By $\Sigma F_n = ma_n$, we get,

$$T_A - mg = m \,.\, v^2/R$$

$$\therefore \quad T_A = mg + mv^2/R = m(g + v^2/R) \quad ...(1)$$

When, the ball is at the highest position B, the forces acting on it are shown in Fig. 10.11 (c).

By $\Sigma F_n = ma_n$, we get

$$T_B + mg = ma_n \quad \text{or} \quad T_B + mg = mv^2/R$$

$$\therefore \quad T_B = mv^2/R - mg = m(v^2/R - g) \quad ...(2)$$

From eqns. (1) and (2), it is clear that

$T_{max} = T_A = m(v^2/R + g)$, when the ball at A. **Ans.**

And, $T_{min} = T_B = m(v^2/R - g)$, when the ball is at the highest position. **Ans.**

Problem 10.12. *The motion of a particle of mass 2 kg on a horizontal plane is described by the relations, $r = 3(1 + \cos 2\pi t)$ and $\theta = 2\pi t$, where r is in m and t in seconds. Find transverse and tangential force components when (a) t = 0, and (b) t = 1.4 seconds.*

Sol. $m = 2$ kg

$$F_r = ma_r = m(\ddot{r} + r\dot{\theta}^2) \quad ...(1)$$

and

$$F_\theta = ma_\theta = m(r\ddot{\theta} + 2\dot{r}\dot{\theta}) \quad ...(2)$$

Given : $r = 3 + 3\cos 2\pi t$ and $\theta = 2\pi t$

$\therefore$ $\dot{r} = \dfrac{dr}{dt} = -6\pi \sin 2\pi t$ and $\dot{\theta} = 2\pi$

and $\ddot{r} = -12\pi^2 \cos 2\pi t$ and $\ddot{\theta} = 0$

$\therefore$ (*a*) At **t = 0,**

$r = 3 + 3 \times 1 = 6\ m$ $\theta = 0$

$\dot{r} = 0$ $\dot{\theta} = 2\pi$

$\ddot{r} = -12\pi^2$ $\ddot{\theta} = 0$

$\therefore$ $F_r = 2 \times [-12\pi^2 + 6 \times (2\pi)^2]$, from (1),

$= \mathbf{+ 236.871\ N.}$ **Ans.**

And $F_\theta = 2[6 \times (0) + 2 \times (0) \times (2\pi)] = 0.$ **Ans.**

(*b*) At **t = 1.4 s.**

$r = 3 + 3\cos 2\pi \times (1.4) = 0.573$ m, $\theta = 2\pi \times (1.4) = 2.8\ \pi$

$\dot{r} = -6\pi \sin 2\pi \times 1.4 = -11.08$ m/s ; $\dot{\theta} = 2\pi$

$\ddot{r} = -12\pi^2 \cos \times 2\pi \times 1.4 = 95.82$ m/s^2 ; $\ddot{\theta} = 0$

$\therefore$ $F_r = 2[95.82 - 0.573 \times (2\pi)^2]$ N

or $F_r = 146.4$ N. **Ans.**

And $F_\theta = 2[0.573 \times (0) + 2 \times (-11.08) \times (2\pi)] = \mathbf{-278.47\ N.}$ **Ans.**

Problem 10.13. *A block B of mass m slides freely on a frictionless arm OA which rotates in a horizontal plane at a uniform angular velocity ω_0 rad/s counterclock as shown in Fig. P. 10.13 (a). Initial distance of the block from O is r_0. Find (a) velocity of sliding of block, (b) the magnitude of the resultant horizontal force F exerted on the block by the arm.*

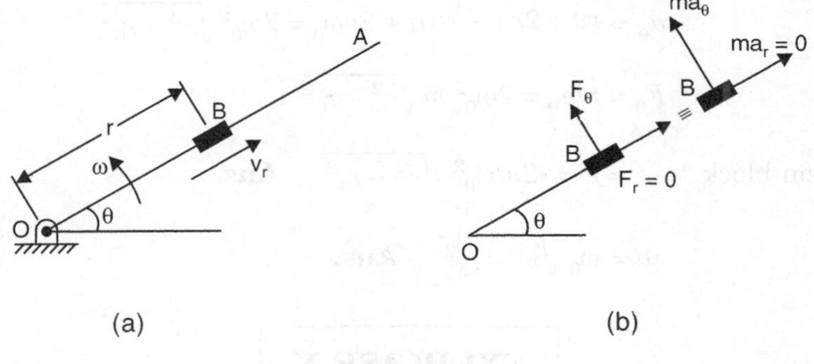

Fig. P. 10.13

Sol. The path of the block *B* is known as Archimedian spiral.

Equations of motion of the mass can be written as :

$$F_r = ma_r = m(\ddot{r} - r\dot{\theta}^2) \qquad \text{...(1)}$$

$$F_\theta = ma_\theta = m(r\ddot{\theta} + 2\dot{r}\,\dot{\theta}) \quad ...(2)$$

Since the block is free to slide on the arm OA,

$\therefore \quad F_r = 0.$

It is therefore subjected to F_θ only, as shown in Fig. P. 10.13 (*b*).

$\therefore$ From eqn. (1), we get

$$F_r = ma_r = m(\ddot{r} - r\dot{\theta}^2) = 0$$

$$\therefore \quad \ddot{r} = r\dot{\theta}^2$$

$$\therefore \quad \frac{dv}{dt} = r\omega_0^{\,2} \qquad | \because \dot{\theta} = \dot{\theta}_0 = \omega = \text{constant}$$

$$\text{or} \quad \frac{dv}{dr}\cdot\frac{dr}{dt} = r\omega_0^{\,2}$$

$$\text{or} \quad v.\left(\frac{dv}{dr}\right) = r\omega_0^{\,2}$$

$$\text{or} \quad vdv = r\,\omega_0^{\,2}\,dr \quad ...(3)$$

Integrating, $\displaystyle\int_0^v v\,dv = \int_{r_0}^r r\omega_0^{\,2}\,dr \qquad | \because \text{ at } t = 0, v_0 = 0, r = r_0$

$$\therefore \quad \frac{1}{2}[v^2 - 0^2] = \frac{\omega_0^{\,2}}{2}(r^2 - r_0^{\,2})$$

$$\therefore \quad v^2 = \omega_0^{\,2}(r^2 - r_0^{\,2}) \quad ...(4)$$

$$\text{or} \quad v = \omega_0\sqrt{r^2 - r_0^{\,2}} \quad ...(5)$$

Since $\dot{\theta} = \omega_0 = \text{constant}$, $\therefore \ \ddot{\theta} = 0$

Therefore $\quad a_\theta = r\ddot{\theta} + 2\dot{r}\dot{\theta} = r(0) + 2v\omega_0 = 2\omega_0^{\,2}\sqrt{r^2 - r_0^{\,2}}$

$$\therefore \quad F_\theta = ma_\theta = 2\omega_0^{\,2}\,m\sqrt{r^2 - r_0^{\,2}}$$

$\therefore$ Force on block $= F = 2m\omega_0^{\,2}\sqrt{r^2 - r_0^{\,2}}$. **Ans.**

and its velocity, $\quad v = \omega_0\sqrt{r^2 - r_0^{\,2}}$. **Ans.**

EXERCISE X

A. Theoretical Questions

1. Establish relationships between the rectangular acceleration components and the curvilinear acceleration components, *viz.*, tangential and normal acceleration components.
2. Write down the equations of motion in rectangular co-ordinates.

3. Write down the equations of motion in curvilinear motion.
4. Derive an expression for finding the radius of curvature of a path of a particle motion.
5. Describe the analysis of a conical pendulum.
6. Derive an expression for finding the acceleration of a thin disc of radius r, mass m rolling down a rough inclined plane.
7. Find acceleration of a sphere of radius r and mass m rolling down a rough inclined plane.
8. A vertical rope of negligible mass wound round a reel of mass m and radius r unwinds itself, the upper end of the rope being fixed. Find expressions for the tension in rope and acceleration of reel. Assume k to be the radius of gyration of the reel.

B. Numerical Problems

10.1. The drum shown in Fig. E. 10.1 has a radius of gyration of 36 cm and weighs 3 kN. It is supported by means of bearings. A weight of 2 kN is attached to one end of a rope while the other end is wrapped around the drum. Determine the acceleration of the weight, the angular acceleration of the drum, and the tension in the rope. Neglect friction in bearings.

[**Ans.** a = 4.467 m/s^2 ; α = 11.112 rad/s^2 ; T = 1048.5 N]

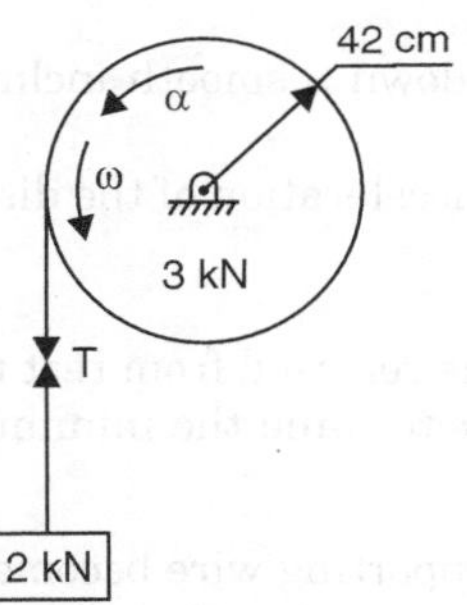

Fig. E. 10.1

10.2. Solve question No. 10.1 above by taking the coefficient of friction at the bearings as 0.18. Also assume that the drum rests in the bearings through hub of 6 cm.

[**Ans.** a = 4.5444 m/s^2 ; α = 10.82 m/s^2 ; T = 1073.52 N]

10.3. A heavy block weighing 12 kN is raised by a rope wound around a drum of a crane weighing 6 kN and having a radius of gyration of 75 cm and diameter 1.2 m. The drum is rotated by 36 kW electric motor at a uniform speed of 15 rpm. Neglecting friction at the bearings of drum and also by assuming the rope to be of negligible mass find (*a*) acceleration of the block, (*b*) tension in rope, (*c*) angular acceleration of the drum, and (*d*) the time for raising the load through 10 m.

[**Ans.** (*a*) 12.03 ; (*b*) 26715.58 N ; (*c*) 20.04 rad/s^2 ; (*d*) 1.29 sec.]

10.4. Two weights W_1 and W_2 of 12 kN and 7.5 kN are attached at the ends of an inextensible cable. The cable passes over a pulley of 120 cm diameter, weight of pulley is 900 N and its radius of gyration is 48 cm. Determine the torque which must be applied to the pulley to raise the 12 kN weight with an acceleration of 2 m/s^2. Neglect friction at the bearings of the pulley. Also find tensions in the rope. [**Ans.** Torque = 5155.785 Nm ; T_1 = 14446.48 N, T_2 = 5970.95 N]

10.5. A block of weight W rests on a horizontal turn table at a distance of 0.6 m from the centre of the table. Find the maximum uniform velocity of the turn table at which the block does not slip. Take μ = 0.3. [**Ans.** v = 1.33 m/s ; N = 21.17 rpm]

10.6. A sphere of mass 120 gm is attached to an inextensible string of length 1 m whose upper end is fixed to the horizontal plane. The sphere describes a horizontal circular path of radius 54 cm. Calculate :

(*a*) the tension in the string

(*b*) frequency and the time period of the sphere.

[**Ans.** (*a*) 1.4 N, (*b*) 3.414 rad/s ; τ = 1.84 s ; n = 0.5434 Hz]

10.7. A bob of a 2.1 m pendulum describes an arc of a circle in a vertical plane. If the tension in the cord is 3 times its weight of the bob when the pendulum is inclined at an angle of 25° from its vertical equilibrium position ; find velocity and acceleration of the bob in this position.

[**Ans.** a_t = 4.146 m/s^2 ; a_n = 20.54 m/s^2 ; a = 20.954 m/s^2 ; v = 6.568 m/s]

10.8. A pulley of mass M has a mass-moment of inertia I_G about its axis and radius r. A mass m is suspended by a light cord wound round the pulley, the end being attached to the periphery of the pulley. Find the tension in the cord and the acceleration of the mass.

[**Ans.** Tension = $\left(\dfrac{Mm}{M+2m}\right)g$; $a = \left(\dfrac{m}{m+M/2}\right)g$]

10.9. A uniform disc of radius R rolls down a smooth inclined plane inclined at an angle α with the horizontal. Prove that the linear acceleration of the disc is given by $a = \dfrac{\sin\alpha}{I+(k/R)^2}\cdot g$, where k is radius of gyration of the disc.

10.10. A sphere of mass m and radius r is released from rest to roll on a rough inclined plane of inclination of 18° with the horizontal. Determine the minimum value of coefficient of friction.

[**Ans.** μ = 0.0928]

10.11. A small ball of mass m and its supporting wire becomes a simple pendulum when the horizontal wire breaks. Find the ratio k of the tension T in the supporting wire immediately after the cord breaks to that in the wire before the cord breaks. See Fig. E. 10.11. [**Ans.** $k = \cos^2\theta$]

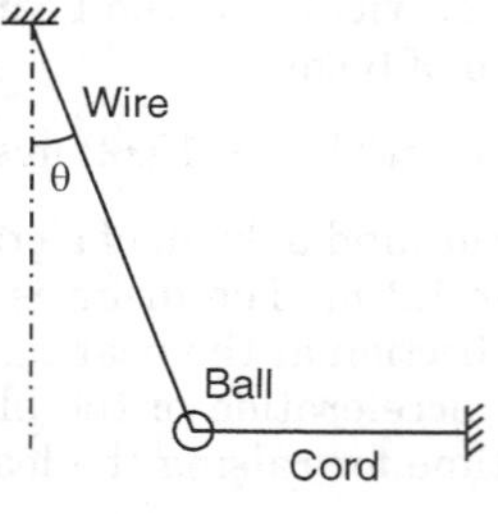

Fig. E.10.11

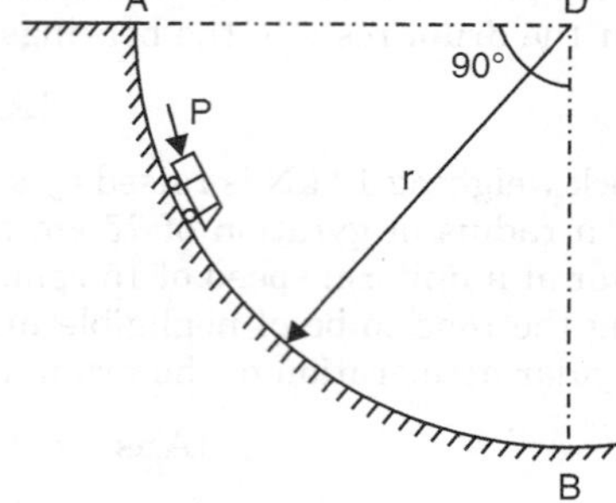

Fig. E. 10.12

10.12. A small toy car of mass m travels down a circular path of radius r under a constant thrust P from its motor. If the car starts from rest at A, find its velocity and the reaction of the path as it reaches B. See Fig. E.10.12.

[**Ans.** $v_B = \sqrt{2gr + \dfrac{\pi r}{m}}$; $R = (3mg + \pi P)$]

10.13. A small toy car reaches the top A of the circular path with a horizontal velocity v_0 and moves down the path. Find the angle α to the position where the toy car leaves the path. Find magnitude of α for $v_0 = 0$. Neglect friction. See Fig. E. 10.13. [**Ans.** $\alpha = \cos^{-1}\left(\frac{2}{3} + \frac{v_0^2}{3gR}\right)$; $\alpha = 48.2°$]

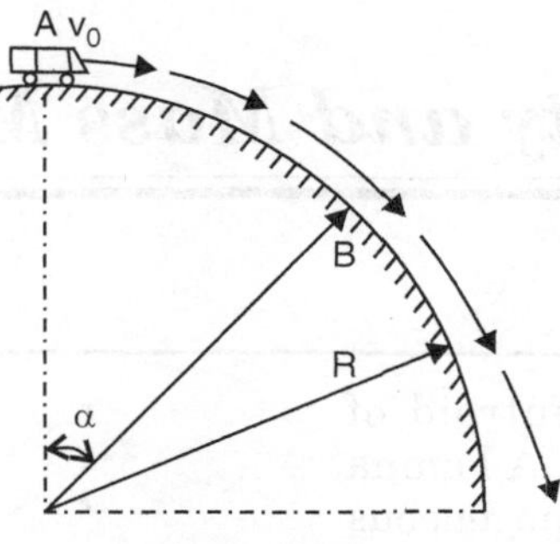

Fig. E. 10.13

10.14. If the pendulum in Fig. E. 10.14 is released from rest in its vertical position as shown, find the angle ϕ defining the position in its downward fall at which the axial force in the rod changes from compression to tension. [**Ans.** $\phi = \cos^{-1}(2/3)$]

10.15. Find the value of the angle ϕ defining the position of the point B (Fig. E.10.15) where the particle will jump clear the cylindrical surface after the string OA is cut. Neglect friction.

[**Ans.** $\cos\phi = \frac{2}{3}\cos\alpha$]

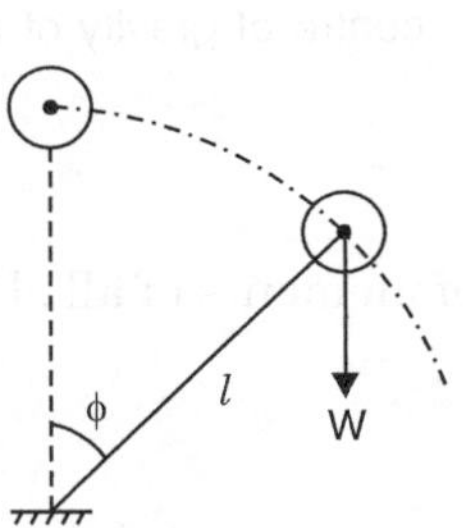

Fig. E. 10.14

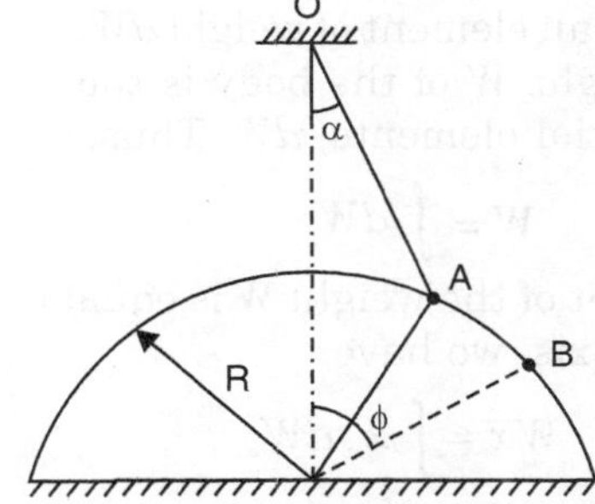

Fig. E. 10.15

11

Centre of Gravity and Mass Moment of Inertia

11.1. CENTRE OF GRAVITY

Earlier in Chapter 6, centroid of lamina was considered, at length. A lamina is a special case of very thin homogeneous plate. We shall now discuss the method of determination of the centre of gravity of solid bodies.

The principle of moments, used for the determination of centroid of the areas in Chapter 6, is readily extended to find the position of the centre of gravity or centroids of volumes or solid bodies.

Consider the solid body of weight W and of uniform thickness, as shown in Fig. 11.1. Consider an element of weight dW. Then the total weight W of the body is the sum of all differential elements, dW. Thus,

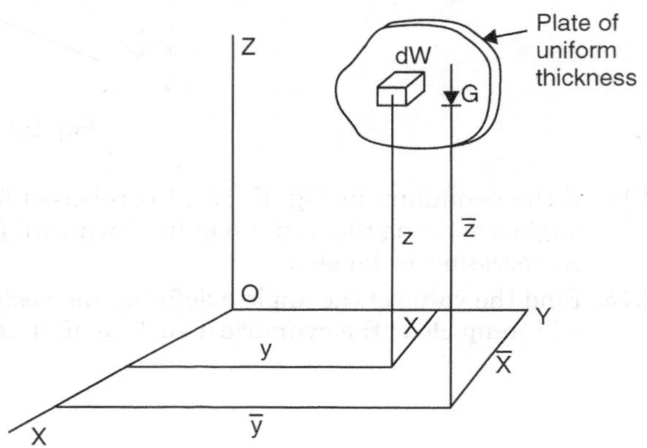

Fig. 11.1. Determination of location of centre of gravity of solids

$$W = \int dW \qquad ...(11.1)$$

Also, moment of the weight W is equal to the sum of moments of all elements. Therefore, with respect to Y-axis, we have

$$W\bar{x} = \int x \,.\, dW \qquad ...(11.2)$$

And, with respect to X-axis, we have

$$W\bar{y} = \int y \,.\, dW \qquad ...(11.3)$$

When we rotate the body about its reference frame (see Fig. 11.1) through 90° so that Z-axis becomes horizontal, a moment summation about Z-axis gives

$$W\bar{z} = \int z \,.\, dW \qquad ...(11.4)$$

If the body in Fig. 11.1 is composed of the same material throughout, its weight W is given by

$$W = \gamma V \qquad ...(11.5)$$

where γ = weight density in N/m^3 and V = volume in m^3.

Then, weight of the element dW will be given by

$$dW = \gamma \,.\, dV \qquad ...(11.6)$$

Therefore eqn. (11.2) becomes

$$[W\,\bar{x} = \int x\,dW] \quad \text{or} \quad (\gamma V)\bar{x} = \gamma \int x\,dV$$

And, cancelling γ, we get

$$V\,\bar{x} = \int x\,dV \qquad \text{....(11.7)}$$

And similarly, equations (11.3) and (11.4) yield :

$$V\,\bar{y} = \int y\,dV \qquad \text{...(11.8)}$$

and

$$V\,\bar{z} = \int z\,dV \qquad \text{...(11.9)}$$

Note that equations (11.7), (11.8) and (11.9) are applicable to bodies of one material, and are not applicable to materials of variable density.

When we deal with composite bodies or volumes we use above equations written in the form of summation, given below :

$$\left.\begin{aligned} W\,\bar{x} &= \sum_{i=1}^{n} x_i\,.\,w_i \\ W\,\bar{y} &= \sum_{i=1}^{n} y_i\,.\,w_i \\ W\,\bar{z} &= \sum_{i=1}^{n} z_i\,.\,w_i \end{aligned}\right\} \qquad \text{...(11.10)}$$

and

$$\left.\begin{aligned} V\,\bar{x} &= \sum_{i=1}^{n} x_i\,.\,v_i \\ V\,\bar{y} &= \sum_{i=1}^{n} y_i\,.\,v_i \\ V\,\bar{z} &= \sum_{i=1}^{n} z_i\,.\,v_i \end{aligned}\right\} \qquad \text{...(11.11)}$$

Note : 1. If a body has a plane of symmetry, the centre of gravity lies in that plane.

2. If a body has two planes of symmetry, the centre of gravity lies on the line of intersection of these planes.

3. If a body has three planes of symmetry, the centre of gravity will lie at the intersection of the three planes.

11.2. CENTRE OF GRAVITY OF A RIGHT SOLID CIRCULAR CONE

Consider a right circular solid cone of base radius r and altitude h as shown in Fig. 11.2. The axis of the cone is taken as X-axis. The cone is symmetrical with respect to the X and Z-axis, hence $\bar{y} = 0$ and $\bar{z} = 0$. To find $\bar{x}$, consider an element at a distance x from its vertex O. Let the base radius of the element be y and its thickness be dx. The volume dV of the element $= \pi y^2\ dx$.

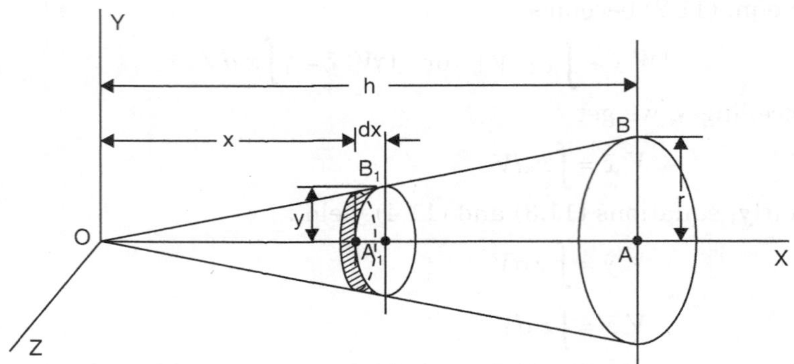

Fig. 11.2. A right circular solid cone

Total volume of the cone

$$V = \int dV = \int \pi y^2 \, dx \qquad \text{...(11.12)}$$

Let us express y with the help of similar triangles ΔOA_1B_1 and ΔOAB. Thus, we get

$$\frac{A_1B_1}{OA_1} = \frac{AB}{OA} \quad \text{or} \quad \frac{y}{x} = \frac{r}{h}$$

$$\therefore \qquad y = \frac{r}{h}\,x\,; \quad \therefore \quad dV = \pi\left(\frac{r}{h}x\right)^2 . \, dx$$

Hence, volume of the cone,

$$V = \int_0^h \pi\left(\frac{r}{h}x\right)^2 . \, dx \quad \text{or} \quad V = \frac{\pi r^2}{h^2} \times \frac{h^3}{3} = \frac{\pi r^2 h}{3}$$

And, using the relation

$$\bar{x}V = \int x \, dV \text{, we get}$$

$$\bar{x} \times \left(\frac{\pi r^2 h}{3}\right) = \int_0^h x\left\{\pi\left(\frac{r}{h}x\right)^2\right\} dx$$

$$= \frac{\pi r^2}{h^2}\int_0^h x^3 . \, dx$$

$$\therefore \qquad \bar{x} \times \frac{\pi r^2 h}{3} = \frac{\pi r^2}{h^2} \times \frac{h^4}{4}$$

$$\therefore \qquad \bar{x} = \frac{3h}{4} \qquad \text{...(11.13)}$$

from the vertex.

Therefore, CG of the cone is at a distance

$$\bar{x} = h - \frac{3h}{4} = \frac{h}{4} \qquad \text{...(11.14)}$$

from the base.

11.3. CENTRE OF GRAVITY OF A THIN HOLLOW CIRCULAR RIGHT CONE

Consider a cone of base radius r and height h, as shown in Fig. 11.3 and surface density ρ_s (N/unit surface area).

Let α be the semi-cone angle. Consider an element at a distance y from the base. The thickness of the element is dy. The surface area of the element

$$= ds = 2\pi(A_1B_1) \times B_1B_2$$

where $B_1B_2 = A_1A_2 \sec\alpha = (dy)(\sec\alpha)$

and since $\Delta OA_1B_1 \equiv \Delta OAB$, we get

$$\frac{A_1B_1}{OA_1} = \frac{AB}{OA}$$

or $$A_1B_1 = \frac{AB}{OA} \times OA_1$$

or $$x = \frac{r}{h}(h - y)$$

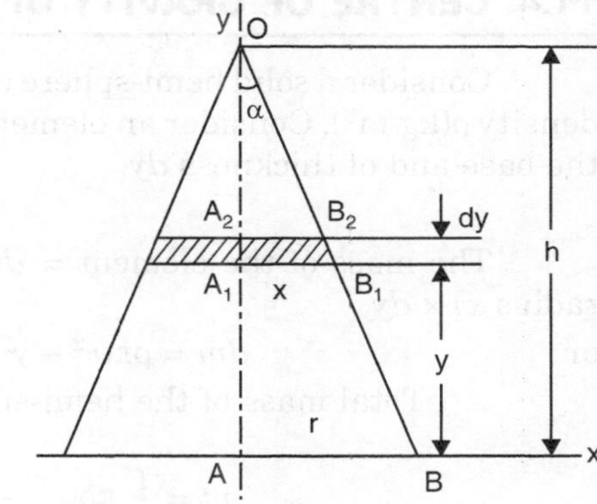

Fig. 11.3. Thin hollow right circular cone

$\therefore$ $$ds = 2\pi\left[\frac{r}{h}(h - y)\right] \times \sec\alpha \,.\, dy$$

and weight of the element

$$= dW = ds \times \rho_s = 2\pi\rho_s \frac{r}{h}(h - y) \,.\, \sec\alpha \,.\, dy$$

$\therefore$ $$W = \int dW = 2\pi\rho_s \,.\, \frac{r}{h} \sec\alpha \int_0^h (h - y)\, dy$$

$$= 2\pi\rho_s \,.\, \frac{r}{h} \sec\alpha \left[hy - \frac{y^2}{2}\right]_0^h = 2\pi\rho_s \,.\, \frac{r}{h} \,.\, \sec\alpha \times \frac{h^2}{2}$$

$\therefore$ $$W = \pi\rho_s \,.\, rh \sec\alpha \qquad \text{...(11.15)}$$

And $$\int y \,.\, dW = \int_0^h y\left[2\pi \frac{r}{h}\rho_s \,.\, \sec\alpha(h - y)\, dy\right]$$

$$= \frac{2\pi r}{h}\rho_s \,.\, \sec\alpha \int_0^h (hy - y^2)\, dy$$

$$= \frac{2\pi r}{h}\rho_s \,.\, \sec\alpha \left[\frac{hy^2}{2} - \frac{y^3}{3}\right]_0^h = \frac{2\pi r}{h}\rho_s \,.\, \sec\alpha \left[\frac{3h^3 - 2h^3}{6}\right]$$

$$\int y\, dW = \frac{\pi r h^2 \rho_s \,.\, \sec\alpha}{3} \qquad \text{...(11.16)}$$

By using $\bar{x} = \dfrac{\int y\, dW}{\int dW}$, we get

$$\bar{x} = \frac{(\pi r h^2 \rho_s \,.\, \sec\alpha)/3}{(\pi r h \rho_s \,.\, \sec\alpha)} = \frac{h}{3}$$

$\therefore$ $$\bar{x} = \frac{h}{3} \qquad \text{...(11.17)}$$

from the base

or, $$\bar{x} = \frac{2h}{3} \qquad \text{...(11.18)}$$

from the vertex.

11.4. CENTRE OF GRAVITY OF SOLID HEMI-SPHERE

Consider a solid hemi-sphere of radius r and of mass density ρ(kg/m^3). Consider an element at a distance y from the base and of thickness dy.

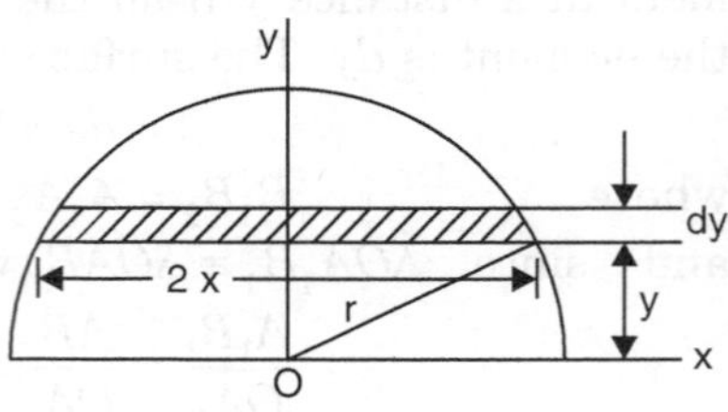

Fig. 11.4. Solid hemi-sphere

The mass of the element = dm = ρ(area of disc of radius x) × dy

or $$dm = \rho\pi(r^2 - y^2) \,.\, dy \qquad ...(11.19)$$

∴ Total mass of the hemi-sphere

$$m = \int_0^r \pi\rho(r^2 - y^2)\,.\,dy = \pi\rho\left(r^2 y - \frac{y^3}{3}\right)_0^r$$

∴ $$m = \frac{2\pi\rho r^3}{3} = \frac{2}{3}\pi r^3 \rho \qquad ...(11.20)$$

And $$\int y\,dm = \int_0^r \rho\pi(r^2 - y^2) y\,dy \quad \text{from eqn. (11.19)}$$

or $$\int y\,dm = \rho\pi \int_0^r (r^2 y - y^3)\,dy$$

$$= \rho\pi\left(r^2 \frac{y^2}{2} - \frac{y^4}{4}\right)_0^r = \frac{\rho\pi r^4}{4} \times (2 - 1)$$

or $$\int y\,dm = \frac{\rho\pi r^4}{4} \qquad ...(11.21)$$

∴ $$\bar{y} = \frac{\int y\,dm}{\int dm} = \frac{\frac{\rho\pi r^4}{4}}{\frac{2}{3} \times \rho\pi r^3}$$

∴ $$\bar{y} = \frac{3}{8} r \qquad ...(11.22)$$

from the base.

11.5. CENTRE OF GRAVITY OF A THIN HOLLOW HEMI-SPHERE

Let base radius be r, and the surface density of the thin hollow hemi-sphere be ρ_s (kg/m^2).

Consider an element at a distance y from the base. The diameter of the element will be = AB = $2x$.

By mensuration, the surface area ds of the element = $2\pi r\,.\,dx$

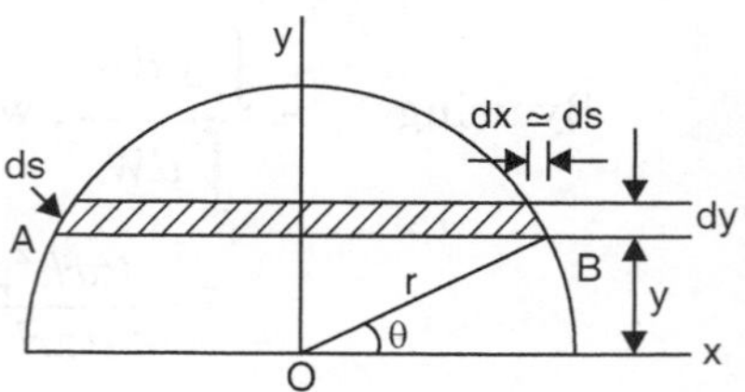

Fig. 11.5. Thin hollow hemi-sphere

∴ $$\int dm = \int \rho_s \,.\, ds = \int_0^1 \rho_s \, 2\pi r \,.\, dx$$

∴ $$m = \int dm = 2\pi\rho_s \,.\, r^2 = 2\pi r^2 \rho_s \qquad ...(11.23)$$

And $$\int x \, . \, dm = \int_0^r (2\pi r \rho_s) \times x \, dx = 2\pi r \left(\frac{x^2}{2}\right)_0^r \times \rho_s$$

$$\therefore \quad \int x \, . \, dm = \pi \rho_s \times r^3 \quad \text{...(11.24)}$$

$$\therefore \quad \bar{x} = \frac{\int x \, dm}{\int dm} = \frac{\pi \rho_s r^3}{2\pi \rho_s r^2}$$

$$\therefore \quad \bar{x} = \frac{r}{2} \quad \text{...(11.25)}$$

from the base.

11.6. CENTRE OF GRAVITY OF THE SOLID RIGHT CIRCULAR CYLINDER

Consider a solid right circular cylinder of base radius r and length l. Let the mass density be ρ (kg/m^3). Consider an element of thickness dy, as shown in Fig. 11.6.

Volume of the element = $dv = \pi r^2 \, dy$

$$\therefore \quad v = \int_0^l \pi r^2 \, dy = \pi r^2 l$$

and $$y dv = \int_0^l \pi r^2 \, . \, y \, dy = \frac{\pi r^2 l^2}{2}$$

$$\therefore \quad \bar{y} = \frac{\int y \, dv}{\int dv} = \frac{\pi r^2 l^2/2}{\pi \, r^2 l} = \frac{l}{2}$$

$$\therefore \quad \bar{y} = \frac{l}{2} \quad \text{...(11.26)}$$

from each of the two flat faces.

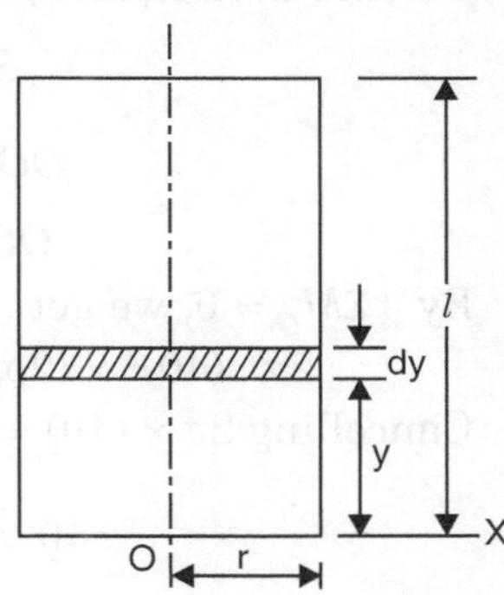

Fig. 11.6. Right circular solid cylinder

Problem 11.1. *A solid hemi-sphere is joined with the right circular cylinder as shown in Fig. P. 11.1. The density of material of hemi-sphere is twice that of the cylinder. Find the greatest height of the cylinder so that composite body may rest with any point of the hemi-sphere in contact with a horizontal plane when (a) the two parts are solid, and (b) the two parts are hollow and thin.*

Sol. (*a*) **Both cylinder and hemi-sphere are solid :**

The composite body will rest in any position if its centre of gravity coincides with O, as shown in Fig P.11.1. $\rho_1 = 2\rho_2$

Mass of hemi-sphere = $M_1 = \rho_1 \times$ volume

or $$M_1 = (2\rho_2) \times \frac{2}{3} \times \pi \times (40)^3$$

$$OO_1 = -\frac{3}{8} r = -\frac{3}{8} \times 40 = -15 \text{ mm}$$

Mass of cylinder $= M_2 = \rho_2 \times \text{volume} = \rho_2 \times \pi r^2 h$

$= \rho_2 \times \pi \times (40)^2 \times h$

$$OO_2 = h/2$$

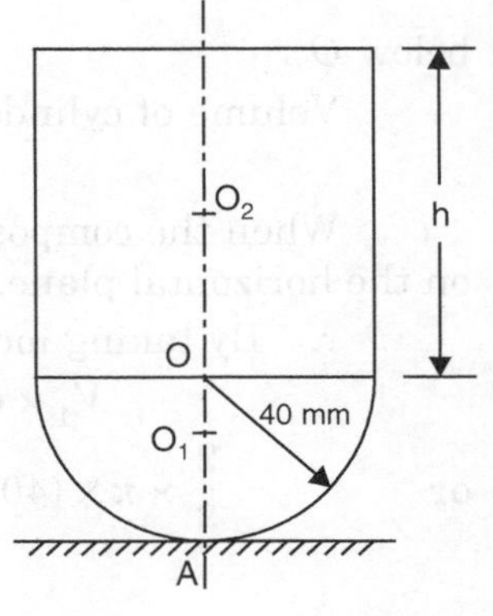

Fig. P. 11.1.

By moments about O, we get $\quad M_2 \times O_2 O - M_1 \times OO_1 = 0$

$$\therefore \quad [(\rho_2)\pi \times (40)^2 h] \times h/2 - \left[(2\rho_2) \times \frac{2}{3} \times \pi \times (40)^3\right] \times 15 = 0$$

Cancelling $\rho_2 \times \pi \times (40)^2$, we get

$$\frac{h^2}{2} = 2 \times \frac{2}{3} \times 40 \times 15$$

$$\therefore \quad h^2 = \frac{2 \times 2}{3} \times 40 \times 15 \times 2 = 40^2$$

$$\therefore \quad h = \mathbf{40\ mm.} \quad \textbf{Ans.}$$

(*b*) **When cylinder and hemi-sphere are hollow :**

Surface area of cylinder $\quad = 2\pi rh = S_2 = 2\pi \times 40h$

Surface area hemi-sphere $\quad = S_2 = \int_0^r 2\pi r \,.\, dx$

or $\quad S_2 = 2\pi r^2 = 2\pi \times (40)^2$

$$OO_1 = \frac{r}{2} = \frac{-40}{2} = -20 \text{ mm}$$

$$OO_2 = h/2$$

By $\Sigma M_O = 0$, we get $\quad (S_1 \times \rho_1) \times OO_1 + S_2\rho_2 \times OO_2 = 0$

or $\quad 2\pi \,.\, (40)^2 \times (2\rho_2) \times (-20) + (2\pi \times 40 \times h) \times \rho_2 \times (h/2) = 0$

Cancelling $2\pi \times (40) \times \rho_2$, we get

$$40 \times 2 \times (-20) + \frac{h^2}{2} = 0$$

$$\therefore \quad h^2 = 40 \times 40 \times 2$$

$$\therefore \quad h = 40\sqrt{2} = \mathbf{56.5685\ mm.} \quad \textbf{Ans.}$$

Problem 11.2. *Solve problem 11.1 when cylinder and hemi-sphere are of the same material.*

Sol. (*a*) **Both parts are solid :**

Volume of hemi-sphere $= V_1 = \frac{2}{3} \times \pi \times (40)^3$

$$OO_1 = -\frac{3}{8} \times 40 = -15 \text{ mm}$$

below O

Volume of cylinder $\quad = V_2 = \pi \times 40^2 \times h$

$$OO_2 = h/2$$

When the composite body rest at any point of hemi-sphere on the horizontal plane, its CG should be at O.

$\therefore$ By taking moments about O, we get

$$V_1 \times OO_1 + V_2 \times OO_2 = 0$$

or $$\frac{2}{3} \times \pi \times (40)^3 \times (-15)\rho + \pi \times (40)^2 \times h \times h/2 \times \rho = 0$$

$$\therefore \quad -\frac{2}{3} \times 40 \times 15 + \frac{h^2}{2} = 0$$

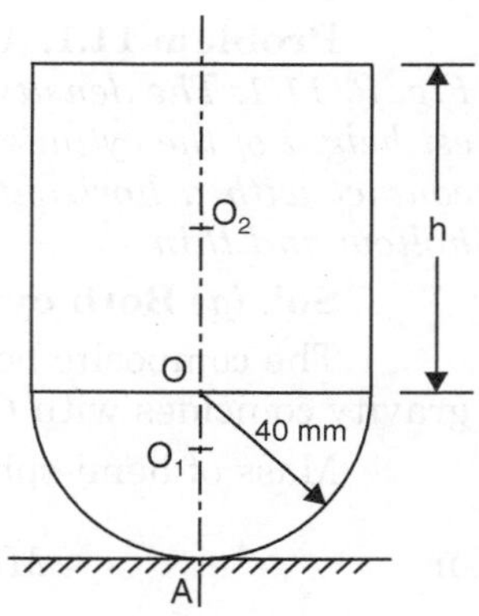

Fig. P. 11.2

$\therefore \quad h^2 = \frac{2}{3} \times 40 \times 15 \times 2 = 40 \times 20$

$\therefore \quad h = 20\sqrt{2}$ = **28.2843 mm. Ans.**

(*b*) **Both parts hollow :** Surface area of hemi-sphere = $S_1 = 2\pi r^2 = 2\pi \times (40)^2$

$$OO_1 = \frac{r}{2} = -\frac{40}{2} = -20 \text{ mm (below } O)$$

Surface area of cylinder = $S_1 = 2\pi \times rh = 2(40)\pi h$

$$OO_2 = h/2$$

By $\Sigma M_0 = 0$, we get

$$2\pi \times (40)^2 \times (-20) \times \rho_s + 2(40)\ \pi h \times h/2 = 0$$

$$h^2 = 2 \times 40 \times 20 = 40^2$$

$\therefore \quad h$ = **40 mm. Ans.**

Problem 11.3. *Find the position of centre of gravity of the composite body shown in Fig. P. 11.3. Density of the material of hemi-sphere is twice that of cone.*

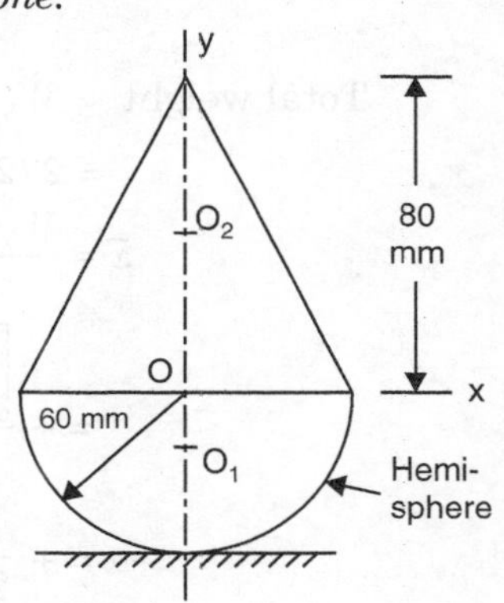

Fig. P. 11.3

Sol. Let density of cone be ρ.

Mass of hemi-sphere = $M_1 = \rho_1 V_1$

$$= (2\rho) \times \frac{2}{3} \times \pi r^3 = \frac{2}{3}(2\rho) \times \pi \times (60)^3$$

$$= 32\ \pi \times 9000\ \rho = 288000\ \pi\rho$$

Mass of cone $= M_2 = \frac{\pi \times r^2 h}{3} \rho$

$$= \frac{\pi \times (60)^2 \times 80}{3} \rho = 96000\ \pi\rho$$

$$OO_1 = -\frac{3}{8} r = -\frac{3}{8} \times 60 = -\frac{45}{2} \text{ mm below, } O$$

$$OO_2 = \frac{h}{4} = \frac{80}{4} = 20 \text{ mm above } O.$$

$$\therefore \quad \bar{y} = \frac{M_1 y_1 + M_2 y_2}{M_1 + M_2} = \frac{(32\pi \times 9000\rho) \times \left(-\frac{45}{2}\right) + 96000\pi\rho \times 20}{32\pi \times 9000\rho + 96000\pi\rho}$$

$$= -13.26 \text{ mm}$$

$\therefore$ *CG* of the composite body is 13.26 mm below *O*. **Ans.**

Problem 11.4. *A wooden block shown in Fig. P. 11.4. is 120 × 120 × 90 mm. A hole 40 mm in diameter and 60 mm deep is drilled in it. The hole is filled by a metal pin of density 20 times the density of wood. Locate centre of gravity of the body.*

Sol. Let ρ be density of wood

Density metal pin = 20ρ

Weight of block = W_b $= \rho \times 120 \times 120 \times 90 = W_b = 1296 \times 1000\rho$ N,

Weight of wooden pin $= W_w = -\rho \times (\pi \times 40^2 \times 60) \times \frac{1}{4}$

or $\quad W_w = -301.6\rho \times 1000 \times \frac{1}{4}$ N

Weight of metal $= W_m = +(20\rho) \times \pi \times \frac{40^2 \times 60}{4} = \frac{6031.86}{4} \rho$N

$x_b = 60$ mm ; $x_w = 40$ mm ; $x_m = 40$ mm. from $y - z$ plane.

$y_b = -45$ mm ; $y_w = -30$ mm and $y_m = -30$ mm from $x - z$ plane

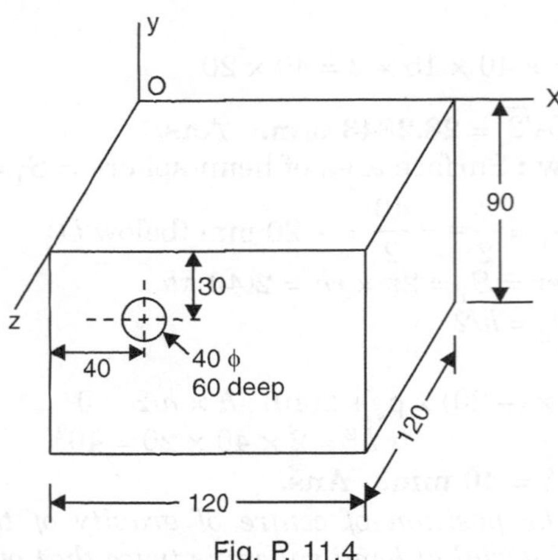

Fig. P. 11.4

$$\text{Total weight} = W_b + W_w + W_m = 1296 \times 1000\rho + \left(-\frac{301.6\,\rho}{4} \times 1000\right) + \frac{6031.8\,\rho}{4} \times 1000$$

$$= 2728.55\,\rho \times 1000 \text{ N}$$

$$\therefore \quad \bar{x} = \frac{W_b \cdot x_b + W_w \cdot x_w + W_m \cdot x_m}{W_b + W_w + W_m}$$

$$= + \frac{\left[1.296 \times 60 + \frac{1}{4}(-301.6) \times 40 + 6031.8 \times 40) \times \frac{1}{4}\right] \times 1000\,\rho}{7026.2\rho \times 1000}$$

$$= + \frac{135062}{2728.55}$$

$= +$ **49.5 mm** to the right of y–z plane. **Ans.**

$$\text{And} \quad \bar{y} = \frac{W_b \times y_b + W_w \cdot y_w + W_m \cdot y_m}{W_b + W_w + W_m}$$

$$= \frac{[1296 \times (-45) + (-301.86)\frac{1}{4} \times (-30) + \frac{6031.8}{4} \times (-30)] \times \rho \times 1000}{2728.55 \times \rho \times 1000}$$

$$= -\frac{101296.5}{2728.55} = -37.12 \text{ mm.}$$

$\therefore \quad \bar{y} = 37.12$ mm below x–z plane. **Ans.**

$$z_b = 60 \text{ mm} ; \quad z_w = +60 + \frac{60}{2} = 90 \text{ mm} ; z_m = 60 + \frac{60}{2} = 90 \text{ mm}$$

from x–y plane

$$\therefore \quad \bar{z} = \frac{W_b \cdot z_b + W_w \cdot z_w + W_m \cdot z_m}{W_b + W_w + W_m}$$

$$= \frac{\left[1296 \times 60 - \frac{1}{4}(301.86) \times 90 + (6031.8)\frac{1}{4} \times 90\right] \times 1000\,\rho}{7026.2 \times 1000\,\rho}$$

$$= \frac{206689.5}{2728.55}$$

$= $ **75.75 mm** to the front of x–y plane **Ans.**

11.7. MASS MOMENT OF INERTIA

Moment of inertia of the mass about an axis is defined as the second of moment of the mass about that axis. Thus, the moment of inertia of an elemental mass dm about the centroidal. x_0-axis is defined as :

$$d\bar{I}_{xx} = y^2\, dm \quad ...(11.27)$$

Similarly, $$d\bar{I}_{yy} = x^2\, dm \quad ...(11.28)$$

For the entire mass $m = \int dm$, eqns. (11.27) and (11.28), respectively yield

$$\bar{I}_{xx} = \int y^2\, dm \quad ...(11.29)$$

$$\bar{I}_{yy} = \int x^2\, dm \quad ...(11.30)$$

See Fig. 11.8 which gives the definition of $\bar{I}_{xx}$ and $\bar{I}_{yy}$.

11.8. PERPENDICULAR-AXIS THEOREM

Let Z_0 be the centroidal axis perpendicular to the plate to the X_0 and Y_0 axes. [Fig. 11.8].

Adding equations (11.27) and (11.28), we get

$$d\bar{I}_{xx} + d\bar{I}_{yy} = (y^2 + x^2)\, dm$$

or $$d\bar{I}_{zz} = d\bar{I}_{xx} + d\bar{I}_{yy} = \rho^2 \,.\, dm \quad ...(11.31)$$

where ρ = the shortest distance of element dm from the Z_0-axis

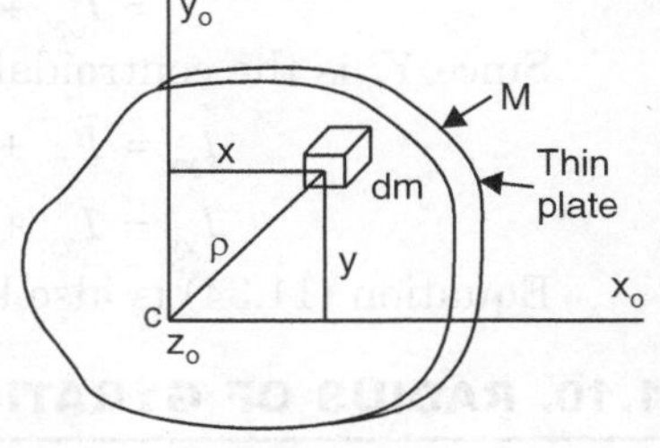

Fig. 11.8. Definitions of centroidal moments of inertia $\bar{I}_{xx}$ and $\bar{I}_{yy}$

Integrating eqn. (11.31), we get

$$\bar{I}_{zz} = \int d\bar{I}_{zz} = \bar{I}_{xx} + \bar{I}_{yy} = \int \rho^2\, dm \quad ...(11.32)$$

Eqn. (11.32) is known as the perpendicular axis theorem for the mass.

It states that the sum of the centroidal mass moment of inertia $\bar{I}_{xx}$ and $\bar{I}_{yy}$ is equal to the centroidal mass moment of inertia $\bar{I}_{zz}$ about the centroidal Z_0-axis which is perpendicular to the X_0-and Y_0-axes.

11.9. PARALLEL AXES THEOREM

Consider a thin plate of mass M. Let C be the centre of gravity of the plate. Let OX be parallel to OX_0 and at a distance of h from it.

By definition [11.1], we can write,

$$dI_{xx} = (y + h)^2\, dm$$

Integrating, we get

$$I_{xx} = \int dI_x = \int (y + h)^2\, dm$$

$$= \int y^2\, dm + \int 2hy\, dm + \int h^2\, dm$$

$$= \bar{I}_{xx} + 2h(\bar{x}m) + mh^2$$

Since X_0 is the centroidal axis, in $\bar{y} = 0$

$\therefore$ we get $\quad I_{xx} = \bar{I}_{xx} + 2h(0 \times m) + mh^2$

or $$I_{xx} = \bar{I}_{xx} + mh^2 \quad ...(11.33)$$

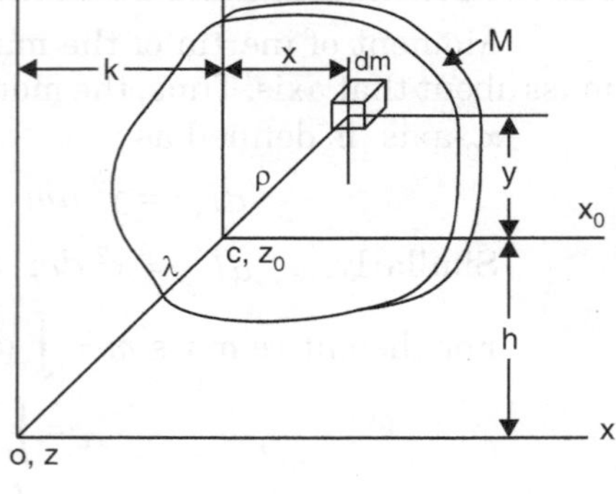

Fig. 11.9. Parallel axes theorem

Equation (11.33) is called parallel axes theorem. It states that the mass moment of inertia of a mass about X-axis is equal to sum of mass-moment of inertia about the centroidal axis X_0 plus the product of mass m and square of the distance between the two parallel axes X and x_0.

Similarly, by definition as in eqn. (11.28), we can write $dI_{yy} = (x + k)^2\, dm$

Integrating, we get

$$I_{yy} = \int dI_{yy} = \int x^2\, dm + \int 2kx\, dm + \int k^2 m$$

$$= \bar{I}_{yy} + 2k\,(\bar{x}m) + km^2$$

Since Y_0 is the centroidal axis, $\therefore \quad \bar{x} = 0$. Hence, we get :

$$I_{yy} = \bar{I}_{yy} + 2k(0 \times m) + mk^2$$

or $$I_{yy} = \bar{I}_{yy} + mk^2 \quad ...(11.34)$$

Equation (11.34) is also known as the "parallel axes theorem" for Y_0 and Y-axes.

11.10. RADIUS OF GYRATION

Mass moment of inertia is also expressed as the product of the mass (m) and square of a distance k, as given below :

$$I = mk^2 \quad ...(11.35)$$

or $$k = \sqrt{I/m} \quad ...(11.36)$$

Here k is called the *radius of gyration*. It may be defined as the distance from the axis of inertia to the point at which the entire mass of the body may be assumed to be concentrated.

Thus, from equations (11.35) and (11.36), we get

$$\bar{I}_{xx} = \bar{k}_{xx}^{\ 2}\, m \quad ...(11.37)$$

$\therefore$ $$\bar{k}_{xx} = \sqrt{\frac{\bar{I}_{xx}}{m}} \quad ...(11.38)$$

Also, $$\bar{I}_{yy} = \bar{k}_{yy}^{\ 2}\, m \quad ...(11.39)$$

$\therefore$ $$\bar{k}_{yy} = \sqrt{\frac{\bar{I}_{yy}}{m}} \quad ...(11.40)$$

Also, $$\bar{I}_{zz} = \rho_0^{\ 2}\, m \quad ...(11.41)$$

where $$\rho_0 = \sqrt{\bar{I}_{zz}/m} \quad ...(11.42)$$

In the above equations $\bar{k}_{xx}$ or k_{x0} is the radius of gyration of the mass m with respect to X_0-axis. Similarly, k_{y_0} and ρ_0 may be defined.

11.11. UNITS OF MASS-MOMENT OF INERTIA

Mass moment of inertia has units of kg-m². It is generally used to analyse problems on rotation of rigid bodies.

11.12. COMBINED PARALLEL AND PERPENDICULAR AXES THEOREM

From Fig. 11.9. we see that z_0 and z and two parallel axes at a distance λ. We note that

$$I_{xx} = \bar{I}_{xx} + mh^2$$

$$I_{yy} = \bar{I}_{yy} + mk^2$$

Adding, we get

$$I_{xx} + I_{yy} = (\bar{I}_{xx} + mh^2) + (\bar{I}_{yy} + mk^2)$$

or

$$I_{xx} + I_{yy} = (\bar{I}_{xx} + \bar{I}_{yy}) + m(h^2 + k^2)$$

By perpendicular axes theorem, we get

$$I_{zz} = \bar{I}_{zz} + m\lambda^2 \qquad ...(11.43)$$

Equation (11.43) is known as the *parallel axes theorem* with respect to z_0 and z axes. Using the concept of radius of gyration, we see that

$$I_{z_0} = mk_0^2$$

$$h^2 + k^2 = \lambda^2 \quad \text{and let} \quad I_{zz} = mk^2$$

∴ Equation (11.43) becomes

$$mk^2 = mk_0^2 + m\lambda^2 \quad \text{or} \quad k^2 = k_0^2 + \lambda^2 \qquad ...(11.44)$$

11.13. NOTATIONS FOR MASS MOMENT OF INERTIA

(1) For plane areas we used the single suffix system for representing I_x , I_y, I_z, etc. (2) For mass moment of inertia, hence forth we shall use double suffix system in each of I_{xx} , I_{yy}, I_{zz} etc.

In all equations in sections 11.7, 11.8, 11.9, 11.10, 11.11 and 11.12 this convention has been strictly followed. It is further to be carefully noted that $I_{x_0x_0} = \bar{I}_{xx}$; $I_{y_0y_0} = \bar{I}_{yy}$, $I_{z_0z_0} = \bar{I}_{zz}$ for solids.

11.14. MASS MOMENT OF INERTIA OF A THIN PLATE

Let ρ be the mass per unit volume. Fig. 11.14 shows a thin plate of uniform thickness t. The X-and Y-axes are in the surface of the plate while the Z-axis is perpendicular to the plate.

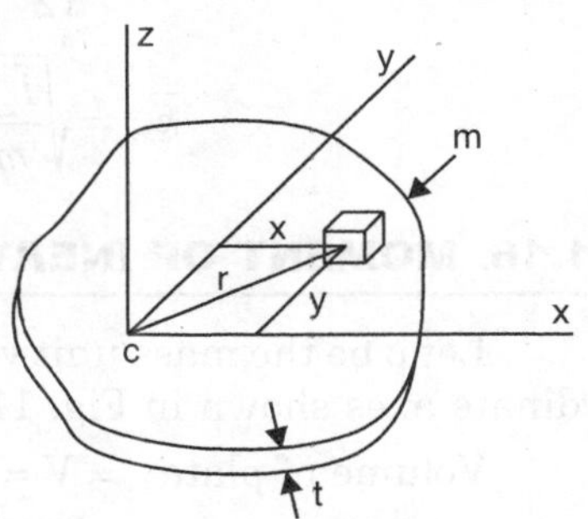

Fig. 11.14. Thin plate of uniform thickness

Then for the moment of inertia of the plate with respect to the X-axis in the plane of the plate, we have

$$I_{xx} = \int y^2\, dm = \rho t. \int y^2\, dA \quad [dm = \rho t\, dA]$$

or

$$I_{xx} = \rho t I_x \qquad ...(11.45)$$

Similarly,

$$I_{yy} = \int x^2\, dm = \rho t. \int x^2\, dA = \rho t.\, I_y \qquad ...(11.46)$$

And, $I_{zz} = \int r^2\, dm = \rho t. \int (x^2 + y^2)\, dA$

or $I_{zz} = \rho t I_z$ ...(11.47)

Above relations are very important and useful.

11.15. MASS MOMENT OF INERTIA OF THIN RECTANGULAR PLATE

Consider a thin rectangular plate of length b, height d and thickness t. X_0, Y_0 and Z_0 are centroidal axes, passing through the centre of gravity CG of the plate. Let ρ be the mass density per unit volume of the plate. If m is mass of the plate in kg, then $m = \rho V = \rho t\, bd$.

We know that for the rectangular are

$$\bar{I}_x = \frac{bd^3}{12}, \quad \bar{I}_y = \frac{db^3}{12} \quad \text{and} \quad \bar{I}_z = \bar{I}_x + \bar{I}_y.$$

By using the relation

$\bar{I}_{xx} = \rho t.\, \bar{I}_x$, we have

$$\bar{I}_{xx} = \rho t \times \frac{bd^3}{12} = \frac{m}{tbd} \times t \times \frac{bd^3}{12} \qquad \left| \; \rho = \frac{m}{V} = \frac{m}{tbd} \right.$$

or, $$\bar{I}_{xx} = \frac{md^2}{12} \qquad \text{...(11.48)}$$

Radius of gyration, $$\bar{k}_{xx} = \sqrt{\frac{d^2}{12}} = \frac{d}{2\sqrt{3}} \qquad \text{...(11.49)}$$

Fig. 11.15. Thin rectangular plate

Similarly,

$$\bar{I}_{yy} = \rho t. \bar{I}_y = \frac{m}{tbd} \times t \times \frac{db^3}{12} \quad \text{or} \quad \bar{I}_{yy} = \frac{mb^2}{12} \qquad \text{...(11.50)}$$

$$k_{yy} = \sqrt{\frac{b^2}{12}} = \frac{b}{2\sqrt{3}} \qquad \text{...(11.51)}$$

And, $$I_{zz} = \rho t.\, (I + I_y) = \frac{m}{tbd}.\, t\left(\frac{bd^3}{12} + \frac{db^3}{12}\right)$$

or, $$\bar{I}_{zz} = \frac{m}{12}(d^2 + b^2) \qquad \text{...(11.52)}$$

$\therefore$ $$\bar{k}_{zz} = \sqrt{\frac{\bar{I}_{zz}}{m}} = \sqrt{\frac{d^2 + b^2}{12}}. \qquad \text{...(11.53)}$$

11.16. MOMENT OF INERTIA OF A THIN CIRCULAR PLATE

Let ρ be the mass/unit volume of the material of plate of radius r and thickness t. The co-ordinate axes shown in Fig. 11.16 are passing through the centre of gravity CG of the plate.

Volume of plate $= V = \pi r^2 t$

$\therefore$ Mass, $m = \rho V = \rho \pi r^2 t$

or $$\rho = \frac{m}{\pi r^2 t}.$$

Using the relation : $\bar{I}_{xx} = \rho t \, \bar{I}_x$, we get

$$\bar{I}_{xx} = \frac{m}{\pi r^2 t} \times t \times \frac{\pi r^2}{4} = \frac{mr^2}{4} \qquad ...(11.54)$$

Similarly, $$\bar{I}_{yy} = \rho t \, \bar{I}_y = \frac{m}{\pi r^2 t} \times t \times \frac{\pi r^4}{4}$$

or $$\bar{I}_{yy} = \frac{mr^2}{4} \qquad ...(11.55)$$

Fig. 11.16. Thin circular plate

And, $$\bar{I}_{zz} = \rho t \, \bar{I}_z = \frac{m}{\pi r^2 t} \times t \times \frac{\pi r^4}{2}$$

$\therefore$ $$\bar{I}_{zz} = \frac{mr^2}{2} \qquad ...(11.56)$$

Radius of gyration, $$\bar{k}_{xx} = \sqrt{\frac{\bar{I}_{xx}}{m}} = \frac{r}{2} \qquad ...(11.57)$$

$$\bar{k}_{yy} = \frac{r}{2} \qquad ...(11.58)$$

and $$\bar{k}_{zz} = \frac{r}{\sqrt{2}} \qquad ...(11.59)$$

11.17. MASS MOMENT OF INERTIA OF SOLID SPHERE

Let r be the radius of the sphere. Consider an element at a distance y from the centroidal axis X_0. [Fig. 11.17].

The length PQ = diameter of the element

$$= 2x = 2\sqrt{r^2 - y^2}.$$

$\therefore$ Radius $= x = \sqrt{r^2 - y^2}$

Volume of element $= \delta V = \pi x^2 \, dy = \pi(r^2 - y^2) \, dy$

$\therefore$ Total volume,

$$V = \int_{-r}^{+r} \pi(r^2 - y^2) \, dy = 2\pi \int_0^r (r^2 - y^2) \, dy = \frac{4\pi r^3}{3}$$

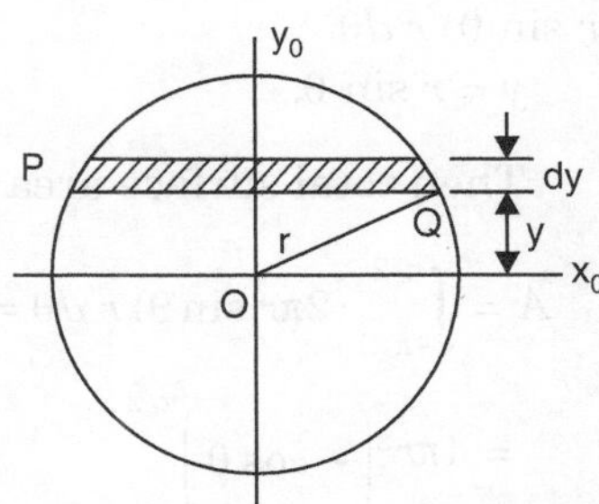

Fig. 11.17. Solid sphere

If M be the mass of the sphere then mass per unit volume $= \rho = \frac{M}{V} = \frac{M}{4\pi r^3/3} = \frac{3M}{4\pi r^3}.$

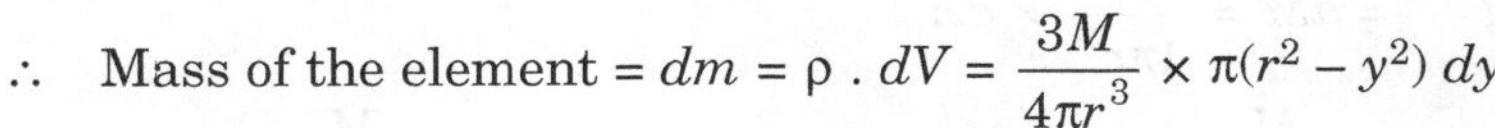

$\therefore$ Mass of the element $= dm = \rho \, . \, dV = \frac{3M}{4\pi r^3} \times \pi(r^2 - y^2) \, dy$

$\therefore$ Mass moment of inertia of element $= d\bar{I}_{yy} = \frac{X^2}{2} \, dm$

or $$\bar{I}_{yy} = \int_{-r}^{r} \frac{3M}{4\pi r^3} \times \pi(r^2 - y^2) \, dy \times \frac{x^2}{2} = \int_{-r}^{r} \frac{3M}{4r^3} (r^2 - y^2) \times \left(\frac{r^2 - y^2}{2}\right) dy$$

$$= \frac{3}{8}\frac{M}{r^3} \times 2\int_0^r (r^2 - y^2)^2\, dy = \frac{3}{4}\frac{M}{r^3}\int_0^r (r^4 - 2r^2y^2 + y^4)\, dy$$

$$= \frac{3}{4}\frac{M}{r^3}\left[r^5 - \frac{2r^5}{3} + \frac{r^5}{5}\right]$$

$$\therefore \quad \frac{3}{4}\frac{M}{r^3} \times r^5\left[\frac{15 - 10 + 3}{15}\right] \quad \text{or} \quad \bar{I}_{yy} = \frac{2}{5}Mr^2 \qquad ...(11.60)$$

By symmetry, $\bar{I}_{xx} = \bar{I}_{yy} = \frac{2}{5}Mr^2$ ...(11.61)

And, $\bar{I}_{zz} = \bar{I}_{xx} + \bar{I}_{yy} = \frac{4}{5}Mr^2$ ...(11.62)

Also, $\bar{I}_{yy} = \bar{k}_{yy}^{\,2} \times M$

$$\bar{k}_{yy} = \sqrt{\frac{2}{5}r^2} = r\sqrt{\frac{2}{5}} \qquad ...(11.63)$$

Also, $\bar{k}_{xx} = r\sqrt{\frac{2}{5}}$ ...(11.64)

and $\bar{k}_{zz} = \frac{2r}{\sqrt{5}}$ ...(11.65)

11.18. MASS MOMENT OF INERTIA OF A THIN HOLLOW SPHERE

Consider a thin hollow sphere of outer radius r, as shown in Fig. 11.18. Consider an element PQ of length $ds = r\, d\theta$. Surface area of the ring $= dA = (2\pi r \sin\theta)\, r\, d\theta$.

$y = r \sin \theta$.

Then total surface area $= A = \int dA$

or $$A = \int_{-\pi/2}^{\pi/2} (2\pi r \sin\theta)\, r\, d\theta = 2\pi r^2 \times 2\int_0^{\pi/2} \sin\theta\, d\theta$$

$$= 4\pi r^2 \left(-\cos\theta\right)_0^{\pi/2}$$

$$A = 4\pi r^2$$

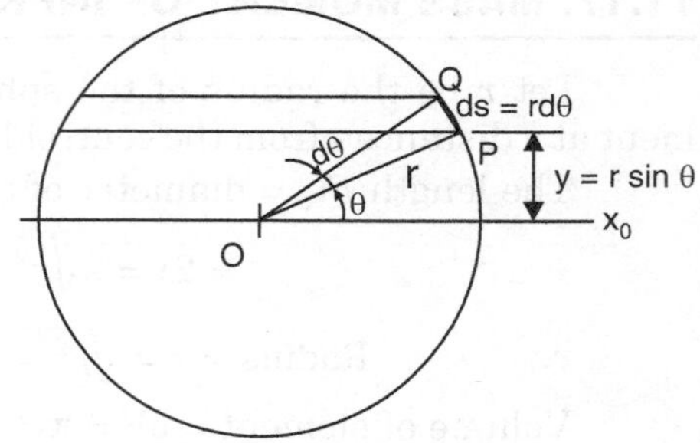

Fig. 11.18. Moment of inertia of a thin sphere

Let M be the mass of the sphere, the mass per unit area $= \frac{M}{4\pi r^2}$

$\therefore$ Mass of the ring $= dM = \frac{M}{4\pi r^2} \times dA$

or $$dM = \frac{M}{4\pi r^2}\, dA = \frac{M}{4\pi r^2} \times (2\pi r \sin\theta) \,.\, r\, d\theta = \frac{M}{2}\sin\theta \,.\, d\theta$$

And, mass moment of inertia of ring about the X_0-axis

$$= d\bar{I}_{xx} = y^2(dM) = \left(\frac{M}{2}\sin\theta\right) d\theta \times (r\sin\theta)^2$$

$\therefore \quad \bar{I}_{xx} = \int d\bar{I}_{xx} = 2\int_0^{\pi/2} \frac{Mr^2}{2} \sin^3\theta \, d\theta.$

or $\quad \bar{I}_{xx} = \frac{2}{3} Mr^2$ ...(11.66)

By symmetry, $\bar{I}_{yy} = \bar{I}_{xx} = \frac{2}{3} Mr^2$ ...(11.67)

And, $\quad \bar{I}_{zz} = \bar{I}_{xx} + \bar{I}_{yy} = \frac{4}{3} Mr^2$ ...(11.68)

Since $I = Mk^2$, where k = radius gyration,

$\therefore \quad \bar{I}_{xx} = M\bar{k}_{xx}^{\,2},$

hence $\quad \bar{k}_{xx} = \sqrt{\frac{\bar{I}_{xx}}{M}} = \sqrt{\frac{2Mr^2}{3}\Big/M} = r\sqrt{\frac{2}{3}}$ ...(11.69)

Also, $\quad \bar{k}_{yy} = r\sqrt{\frac{2}{3}}$ ...(11.70)

And, $\quad \bar{k}_{zz} = \frac{2r}{\sqrt{3}}.$ ...(11.71)

11.19. MOMENT OF INERTIA OF A RIGHT SOLID CONE

Let the base of cone be $b = 2r$, and its height be h. Consider an element at a distance y from the vertex O and of thickness dy. See Fig. 11.19 (a)

Volume of the element $= dV = \pi(A_1C_1)^2 \,.\, dy$

Since $\quad \Delta OC_1A_1 \equiv \Delta OCA,$

$\therefore \quad \frac{C_1A_1}{OC_1} = \frac{CA}{OC}$

$\therefore \quad C_1A_1 = \frac{CA}{OC} \times OC_1 = \frac{r}{h} y$ ...(11.72)

$\therefore \quad dV = \pi \times \left(\frac{ry}{h}\right)^2 dy = \frac{\pi r^2}{h^2} y^2 \, dy$

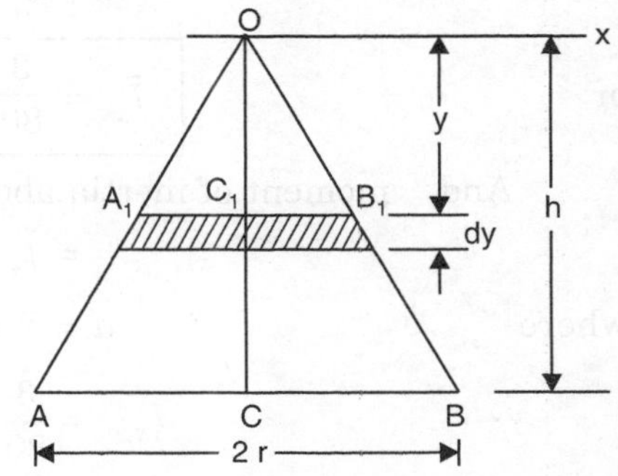

Fig. 11.19. (a) M.I. of a solid right cone

Hence, total volume $= V = \int dV = \int_0^h \frac{\pi r^2}{h^2} y^2 \, dy$

or $\quad V = \frac{\pi r^2 h}{3}$

Hence, mass per unit volume $= \frac{M}{V} = \frac{3M}{\pi r^2 h}$

Hence mass of the element $= dM = \frac{3M}{\pi r^2 h} \times dV = \frac{3M}{\pi r^2 h} \times \frac{\pi r^2 y^2}{h^2} \, dy$

or $\quad dM = \frac{3M}{h^3} y^2 \, dy$ ...(11.73)

Hence, moment of inertia of the element

$$= dI_{xx} = y^2 \times dM = (y^2)\left(\frac{3M}{h^3}y^2\right)dy$$

$$\therefore \qquad I_{xx} = \int dI_{xx} = \int_0^h \frac{3M}{h^3}y^4\,dy = \frac{3M}{h^3}\times\frac{h^5}{5}$$

or $$\boxed{I_{xx} = \frac{3}{5}Mh^2} \qquad ...(11.74)$$

Applying the parallel-axis theorem, we have

$$I_{xx} = \bar{I}_{xx} + M(d)^2 \qquad ...(11.75)$$

where d = distance between OX and the centroidal axis GX_0 through the centre of gravity. From Fig. 11.19 (*b*).

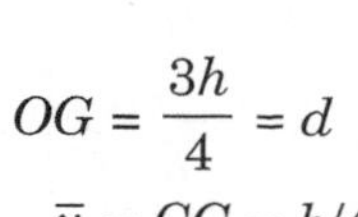

$$OG = \frac{3h}{4} = d$$

and $$\bar{y} = CG = h/4.$$

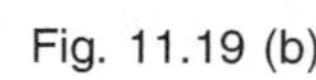

Fig. 11.19 (b)

$\therefore$ By parallel-axis theorem,

$$\bar{I}_{xx} = I_{xx} - M(d)^2$$

or $$\bar{I}_{xx} = \frac{3}{5}Mh^2 - M\times\frac{9h^2}{16} = Mh^2\,\frac{48-45}{80}$$

or $$\boxed{\bar{I}_{xx} = \frac{3}{80}Mh^2} \qquad ...(11.76)$$

And, moment of inertia about the base of the cone

$$= I_{x_1x_1} = \bar{I}_{xx} + M(d)^2$$

where $$d = \bar{y} = h/4$$

$$\therefore \qquad I_{x_1x_1} = \frac{3}{80}Mh^2 + M(h/4)^2 = \frac{Mh^2}{80}(3+5)$$

or $$\boxed{I_{x_1x_1} = \frac{Mh^2}{10}} \qquad ...(11.77)$$

Also, $$\bar{I}_{xx} = \bar{k}_{xx}^{\,2}\times M$$

$$\therefore \qquad \bar{k}_{xx} = \sqrt{\frac{\bar{I}_{xx}}{M}} = \sqrt{\frac{3}{80}\frac{Mh^2}{M}}$$

or $$\boxed{\bar{k}_{xx} = h\sqrt{\frac{3}{80}}} \qquad ...(11.78)$$

Similarly, with respect to the vertex

$$k_{xx} = \sqrt{\frac{I_{xx}}{M}} = h\sqrt{\frac{3}{5}} \quad ...(11.79)$$

Also, with respect to the base

$$k_{x_1x_1} = \sqrt{\frac{I_{x_1x_1}}{M}} = \frac{h}{\sqrt{10}} \quad ...(11.80)$$

Moment of inertia about the axes of symmetry (or Y-axis)

Moment of inertia of the element about the Y-axis $= dI_{yy} = dM \times \frac{(C_1A_1)^2}{2}$

Hence from eqns. (11.72) and (11.73) we have

$$dI_{yy} = \left(\frac{3M}{h^3} y^2\, dy\right) \times \frac{1}{2}\left(\frac{ry}{h}\right)^2 = \frac{3}{2}\frac{Mr^2}{h^5} \cdot y^4\, dy$$

Hence $$I_{yy} = \int dI_{yy} = \frac{3Mr^2}{2h^5}\int_0^h y^4\, dy = \frac{3Mr^2}{10h^5} \times (h^5)$$

$$\therefore \quad I_{yy} = \frac{3Mr^2}{10} = \bar{I}_{yy} \quad ...(11.81)$$

Since, $$I_{yy} = Mk_{yy}^2$$

$$\therefore \quad k_{yy} = \sqrt{\frac{I_{yy}}{M}} = r\sqrt{\frac{3}{10}} = \bar{k}_{yy} \quad ...(11.82)$$

11.20. M.I. ABOUT AXIS THROUGH VERTEX AND PERPENDICULAR TO THE AXIS OF SYMMETRY

Consider an element A_1B_1 as shown in Fig. 11.20.

$$\because \quad \Delta OC_1A_1 \equiv \Delta OCA,$$

$$\frac{C_1A_1}{OC_1} = \frac{CA}{OC}$$

$$\therefore \quad C_1A_1 = \frac{CA}{OC} \times OC_1 = \frac{r}{h}x$$

$\therefore$ Volume of the element

$$= dV = \pi C_1A_1^2\, dx$$

or $$dV = \frac{\pi r^2 x^2}{h^2} \times dx \quad ...(11.83)$$

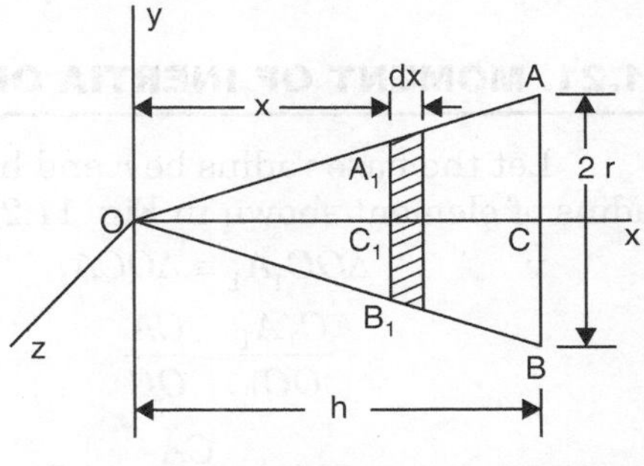

Fig. 11.20. M.I. about an axis perpendicular to the axis of symmetry and passing through vertex

$\therefore$ Total volume $V = \frac{1}{3}\pi r^2 h$

If M is total mass of cone, then mass per unit volume $= \frac{M}{V} = \frac{3M}{\pi r^2 h}$

Hence, mass of element

$$= dM = \frac{3M}{\pi r^2 h} \times dV$$

or $$dM = \frac{3M}{\pi r^2 h} \cdot \frac{\pi r^2 x^2}{h^2}\ dx = \frac{3M}{h^3}\ x^2\ dx \quad ...(11.84)$$

Hence, moment of inertia of the element about Z-axis through the vertex

$$= dI_{zz} = dI_{xx} + dI_{yy}$$

$$= (dM)\left(\frac{C_1A_1}{2}\right)^2 + (dM)\ .\ x^2$$

$$= \left(\frac{3M}{h^3}x^2\ dx\right)\left(\frac{1}{2}\frac{r}{h}x\right)^2 + \left(\frac{3M}{h^3}x^2\ dx\right)x^2$$

$$= \frac{3M}{h^3}\left(\frac{r^2}{4h^2}x^4 + x^4\right).\ dx$$

$$\therefore \quad I_{zz} = \int dI_{zz} = \frac{3M}{h^5}\int_0^h \left[\frac{r^2}{4} + h^2\right] x^4\ .\ dx$$

$$= \frac{3M}{h^5}\left(\frac{r^2}{4} + h^2\right)\left(\frac{h^5}{5}\right)$$

$$\therefore \quad \boxed{I_{zz} = \frac{3M}{5}\left(\frac{r^2}{4} + h^2\right)} \quad ...(11.85)$$

Also since, $$I_{zz} = k_{zz}^2\ M$$

$$\therefore \quad k_{zz} = \sqrt{\frac{3}{5}\left(\frac{r^2}{4} + h^2\right)} \quad ...(11.86)$$

11.21. MOMENT OF INERTIA OF A THIN HOLLOW CONE

Let the base radius be r and height be h. The radius of element shown in Fig. 11.21 is C_1A_1.

$\because \quad \Delta OC_1A_1 \equiv \Delta OCA,$

$$\therefore \quad \frac{C_1A_1}{OC_1} = \frac{CA}{OC}$$

$$\therefore \quad C_1A_1 = \frac{CA}{OC} \times OC_1 = \frac{r}{h}\ x = r_1 \text{ (say)} \quad ...(11.87)$$

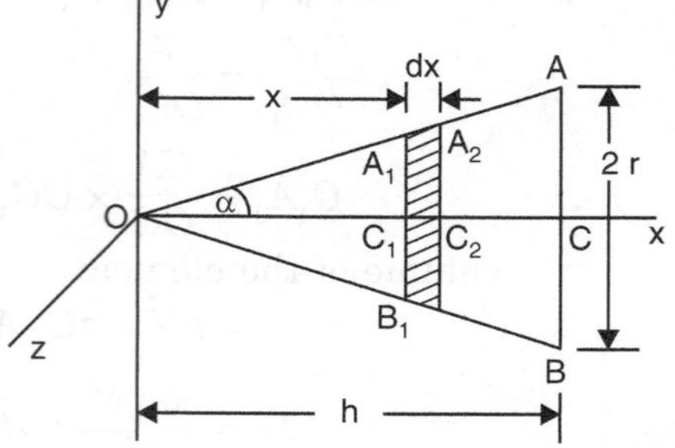

Fig. 11.21. Moment of inertia of a thin hollow cone about the axis of symmetry

$\therefore$ Surface area of the element,

$$= 2\pi r_1 \times (A_1A_2)$$

or $$dA = \left(\frac{2\pi r x}{h}\right) \times \frac{C_1C_2}{\cos\alpha}$$

or $$dA = \frac{2\pi r}{h\cos\alpha}\ x\ dx = \frac{2\pi r}{h} x \sec\alpha\ dx \quad ...(11.88)$$

Hence, total surface area

$$A = \int dA = \int_0^h \frac{2\pi r}{h} x \sec \alpha \, dx$$

or $$A = \left(\frac{2\pi r}{h} \sec \alpha\right)\frac{h^2}{2}$$

or $$\boxed{A = \pi r h \sec \alpha} \quad ...(11.89)$$

Let M be the mass of the cone, then mass of cone per unit surface area

$$= \frac{M}{A} = \frac{M}{\pi r h \sec \alpha} \quad ...(11.90)$$

∴ Mass of the element

$$= dM = \frac{M}{\pi r h \sec \alpha} \times dA = \frac{M}{\pi r h \sec \alpha} \times \left(\frac{2\pi r}{h} x \sec \alpha \, . \, dx\right)$$

or $$dM = \frac{2M}{h^2} x \, dx \quad ...(11.91)$$

∴ M.I. of element about $OX = dI_{xx} = (dM)$ (radius)^2

or $$dI_{xx} = \left(\frac{2M}{h^2} x \, dx\right)\left(\frac{r}{h} x\right)^2$$

∴ $$I_{xx} = \int dI_{xx} = \int_0^h \frac{2Mr^2}{h^4} x^3 \, dx = \frac{2Mr^2}{h^4} \times \frac{h^4}{4}$$

or $$\boxed{I_{xx} = \frac{1}{2} M r^2} \quad ...(11.92)$$

Problem 11.5. *Find the moment of inertia of a thin long rod (or slander rod) of length L and mass m with respect to an axis through one end to the rod shown in Fig. P.11.5. Also find M.I. through CG.*

Sol. Let A be the area of the cross-section of the rod. Let γ be density of the rod. Consider an element of length dx.

Then, the mass of the element = $dm = \gamma A \, dx$.

∴ Total mass of rod = $m \int_0^L \gamma A dx = \gamma A L$ ...(1)

∴ Moment of inertia of element about Y-axis

$$= dI_{yy} = r^2 \, dm = (x \sin \theta)^2 \, dm$$
$$= x^2 \sin^2 \theta \times \gamma A \, . \, dx$$

∴ $$I_{yy} = \int dI_{yy} = \int_0^L \gamma A \, x^2 \sin^2 \theta \, dx$$

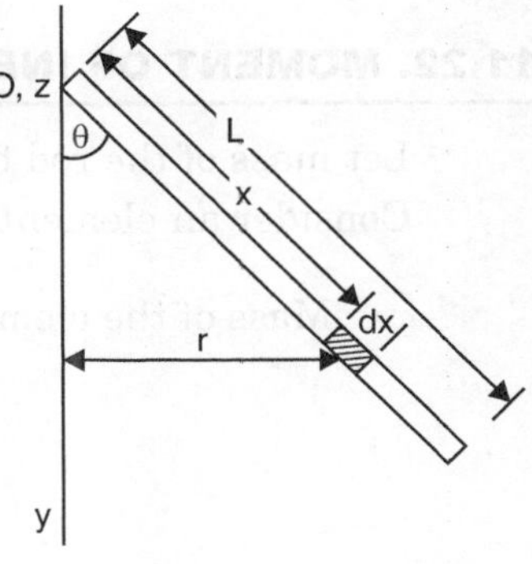

Fig. P. 11.5

$$\boxed{I_{yy} = \frac{\gamma A L^3}{3} \sin^2 \theta} \quad ...(2)$$

but from (1), $\gamma = \frac{m}{AL}$

$$\therefore \quad I_{yy} = \frac{1}{3}\left(\frac{m}{AL}\right) \times AL^3 \sin^2 \theta = \frac{m}{3}(L^2 \sin^2 \theta)$$

or $$\boxed{I_{yy} = \frac{1}{3} mL^2 \sin^2 \theta} \quad ...(3)$$

Let $L \sin \theta = L_e$, then,

$$\boxed{I_{yy} = \frac{1}{3} mL_e^2} \quad ...(4)$$

Using parallel-axes theorem, we have

$$I_{yy} = \bar{I}_{yy} + m\bar{x}^2 \quad \text{where} \quad \bar{x} = \frac{L}{2} \sin \theta$$

$$\therefore \quad \bar{I}_{yy} = I_{yy} - m\bar{x}^2 = \frac{1}{3}mL^2 \sin^2 \theta - m\left(\frac{L}{2} \sin \theta\right)^2$$

$$\bar{I}_{yy} = \frac{mL^2}{3} \sin^2 \theta - \frac{mL^2}{4} \sin^2 \theta = \frac{mL^2 \sin^2 \theta}{12}$$

or $$\boxed{\bar{I}_{yy} = \frac{mL^2 \sin^2 \theta}{12}} \quad ...(5)$$

or $$\boxed{\bar{I}_{yy} = \frac{mL_e^2}{12}} \quad ...(6)$$

11.22. MOMENT OF INERTIA OF A THIN LONG ROD

Let mass of the rod be M, $\therefore$ Mass per unit length = M/L

Consider an element of length dx.

$\therefore$ Mass of the element = $dM = \frac{M}{L} . dx$

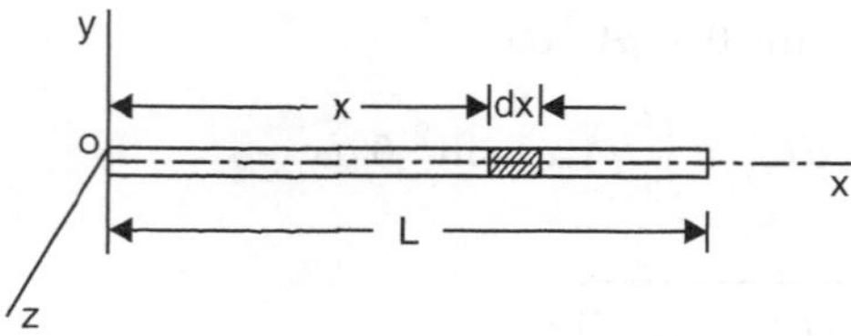

Fig. 11.22. Moment of inertia of a thin long rod

Mass moment of inertia about an axis perpendicular to the rod = $dI_{yy} = dI_{zz} = (dM)\, x^2$

or $$I_{yy} = I_{zz} = \int x^2\, dM = \int_0^L x^2 \frac{M}{L} . dx = \frac{ML^3}{3L}$$

or $$\boxed{I_{yy} = I_{zz} = \frac{1}{3}ML^2} \qquad ...(11.93)$$

By parallel-axis theorem, we get

$$I_{yy} = \bar{I}_{yy} + M(\bar{x})^2 \qquad \text{where } (\bar{x}) = L/2$$

$$\therefore \quad \bar{I}_{yy} = \frac{ML^2}{3} - M\left(\frac{L}{2}\right)^2$$

or $$\boxed{\bar{I}_{yy} = \frac{ML^2}{12}} \qquad ...(11.94)$$

$$\because \quad I_{yy} = Mk^2{}_{yy},$$

$$\therefore \quad \boxed{k_{yy} = \sqrt{\frac{ML^2}{3M}} = L/\sqrt{3}} \qquad ...(11.95)$$

Also $$\boxed{\bar{k}_{yy} = \frac{L}{\sqrt{12}}.} \qquad ...(11.96)$$

11.23. MASS MOMENT OF INERTIA OF HOLLOW CYLINDER

Consider a hollow cylinder of inner and outer radii r and R, respectively, and length L. Let m be the mass of the cylinder.

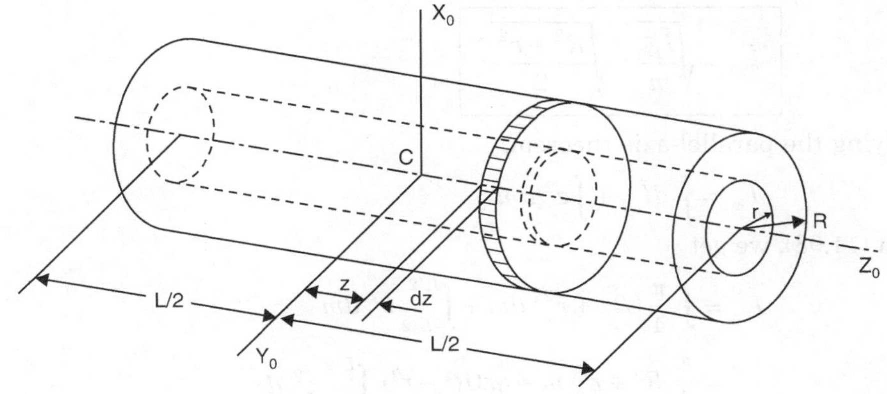

Fig. 11.23. M.I. of a hollow cylinder

Total volume of cylinder = $V = \pi(R^2 - r^2) . L$

$\therefore$ Mass per unit volume of cylinder = $\dfrac{M}{V} = \dfrac{M}{\pi(R^2 - r^2)L}$.

Hence for the element of length dz [Fig. 11.23], the mass

$$= dm = \frac{M}{\pi(R^2 - r^2)L} . dz \quad ...(11.97)$$

We know that for the circular ring

$$\bar{I}_x = \bar{I}_y = \frac{\pi}{4}(R^4 - r^4)$$

We know that $\bar{I}_{xx} = \gamma t . \bar{I}_x$

$$\therefore \quad d\bar{I}_{zz} = \gamma dz(d\bar{I}_x) = [(\gamma dz)\bar{I}_x] \times \frac{dm}{\gamma dV}$$

where $dV = \pi(R^2 - r^2)\, dz$ and $\bar{I}_z = \dfrac{\pi}{2}(R^4 - r^4) = \bar{I}_x + \bar{I}_y$

$\therefore$ we get $d\bar{I}_{zz} = \left[\gamma dz \times \dfrac{\pi}{2} \times (R^4 - r^4)\right] \times \dfrac{dm}{\gamma \times \pi(R^2 - r^2)\, dz}$

or $$d\bar{I}_{zz} = \frac{(R^2 + r^2)}{2}\, dm \quad ...(11.98)$$

Integrating, $\bar{I}_{zz} = \displaystyle\int \frac{R^2 + r^2}{2}\, dm = \left(\frac{R^2 + r^2}{2}\right) . m$

or, $$\boxed{\bar{I}_{zz} = \frac{R^2 + r^2}{2}\, m} \quad ...(11.99)$$

Since $\bar{I}_{zz} = m\bar{k}_{zz}^{\ 2}$

$$\therefore \quad \boxed{\bar{k}_{zz} = \sqrt{\frac{\bar{I}_{zz}}{m}} = \sqrt{\frac{R^2 + r^2}{2}}} \quad ...(11.100)$$

Applying the parallel-axis theorem.

$$\bar{I}_{xx} = \int d\bar{I}_{xx} + \int z^2\, dm$$

From (11.98), we get

$$I_{xx} = \int \frac{\pi}{4}(R^2 + r^2)\, dm + \int_{-L/2}^{L/2} z^2\, dm$$

$$= \frac{\pi}{4}(R^2 + r^2)\, m + \gamma\pi(R^2 - r^2) \int_{-L/2}^{L/2} z^2\, dz$$

or $$\bar{I}_{xx} = \frac{\pi}{4}(R^2 + r^2)\, m + \gamma\pi(R^2 - r^2)\left[\frac{m}{\gamma\pi(R^2 - r^2)L}\right] \times \frac{(L/2)^3}{3}$$

$$\bar{I}_{xx} = \frac{\pi}{4}(R^2 + r^2)m + \frac{mL^2}{12} \quad ...(11.101)$$

$$\because \bar{I}_{xx} = \bar{I}_{yy}, \quad \therefore \quad \bar{I}_{xx} = \bar{I}_{yy} = \frac{m}{12}[L^2 + 3(R^2 + r^2)] \quad ...(11.102)$$

For a solid cylinder, $r = 0$. Therefore, we get from eqns. (11.99) and (11.102)

$$\bar{I}_{zz} = \frac{mR^2}{2} \quad ...(11.103)$$

and,

$$\bar{I}_{xx} = \bar{I}_{yy} = \frac{m}{12}[L^2 + 3R^2] \quad ...(11.104)$$

For a slander rod, R is negligibly small hence, we get

$$\bar{I}_{zz} = 0 \quad \text{and} \quad \bar{I}_{xx} = \bar{I}_{yy} = \frac{mL^2}{12} \quad ...(11.105)$$

11.24. MOMENT OF INERTIA OF A PARABOLOID OF REVOLUTION

A paraboloid of revolution is generated by rotating a parabola $y^2 = kx$ about its axis (*i.e.*, about X-axis), as shown in Fig. 11.24. Let $y = h$ at $x = a$, $\therefore$ from $y^2 = kx$, we get

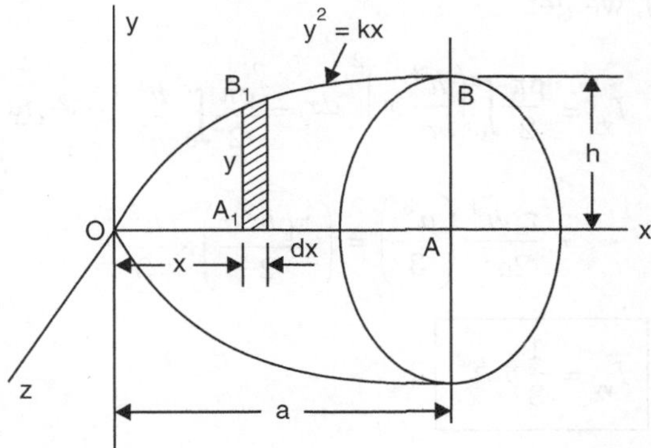

Fig. 11.24. Moment of inertia of a paraboloid of revolution

$$h^2 = ka \quad \therefore \quad k = \frac{h^2}{a}$$

Hence, equation of parabola becomes $y^2 = \dfrac{h^2}{a}x$ ...(11.106)

The elementary strip of height $y = A_1 B_1$ and thickness dx has volume dV given by,

$$dV = \pi y^2\, dx = \pi\left(\frac{h^2}{a}x\right)^2 dx \qquad ...(11.107)$$

If ρ is the density, then mass of the element = $dM = \rho dV$

∴ Total mass of the solid = $\int dm$

or $$m = \int \rho\pi y^2\, dx = \int_0^a \rho\pi\left(\frac{h^2}{a}x\right) dx \text{ from (11.106),}$$

or $$m = \frac{\rho\pi h^2}{a}\int_0^a x\, dx = \frac{\rho\pi h^2 a}{2} \qquad ...(11.108)$$

Volume of the solid $= V = \int \pi y^2\, dx = \dfrac{\pi h^2 a}{2}$ ...(11.109)

Hence, mass of the element = $dm = \rho\pi . y^2\, dx$

∴ M.I. of the element about X-axis = $dI_{xx} = dm \times \left(\dfrac{y^2}{2}\right)$ $\quad \because \; I = \dfrac{Mr^2}{2}$

$$= \frac{1}{2}(\rho\pi y^2\, dx)\, y^2$$

or $$dI_{xx} = \frac{\rho\pi y^4}{2} dx$$

Using eqn. (11.106), we get

$$I_{yy} = \frac{\rho\pi}{2}\int_0^a\left(\frac{h^2}{a}x\right)^2 dx = \frac{\rho\pi}{2}\int_0^a \frac{h^4}{a^2}\times x^2\, dx$$

$$= \frac{\pi\rho h^4}{2a^2}\left(\frac{a^3}{3}\right) = \left(\frac{\pi\rho h^2 a}{2}\right)\times\frac{h^2}{3}$$

$$\boxed{I_{yy} = \frac{1}{3}mh^2} \qquad ...(11.110)$$

EXERCISE XI

Numerical Problems

11.1. A thin semi-circular plate of radius r has mass m and thickness t. Prove that the mass moment of inertia about the normal to the plane of the plate and passing through its centre of gravity is given by $\dfrac{mr^2}{\pi}$.

11.2. For the thin semi-circular ring of mass m shown in Fig. E.11.2. Find I_{zz}. [**Ans.** 0.4984 mr^2]

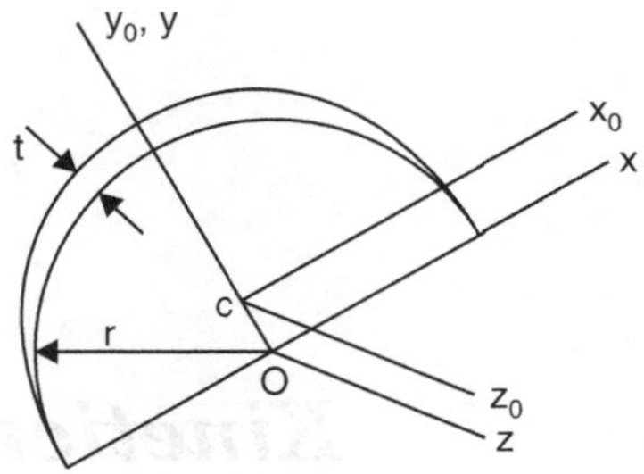

Fig. E. 11.2

11.3. Prove that the mass moment of inertia of the homogeneous parallelopiped of mass m shown in Fig. E. 11.3. with respect to the centroidal axes parallel to the edges are given by

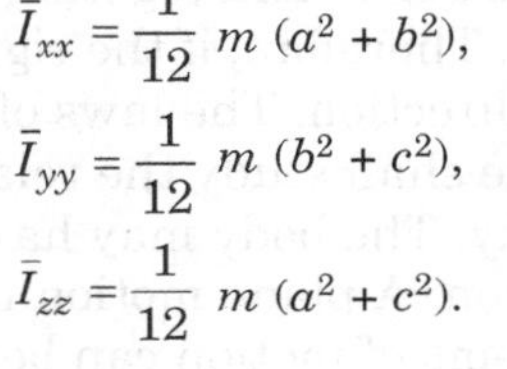

$$\bar{I}_{xx} = \frac{1}{12}\, m\, (a^2 + b^2),$$

$$\bar{I}_{yy} = \frac{1}{12}\, m\, (b^2 + c^2),$$

and $$\bar{I}_{zz} = \frac{1}{12}\, m\, (a^2 + c^2).$$

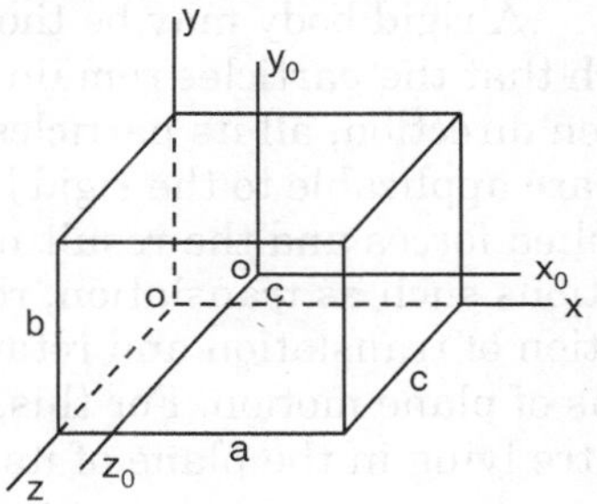

Fig. E. 11.3

11.4. Prove that mass moment of inertia of a right circular cone of base radius r, altitude h and mass m with respect to any diametral axes in the b as is equal to $I = \frac{m}{20}\,(3r^2 + 2h^2)$.

11.5. For the semi-circular cylindrical plate shown in Fig. E. 11.2 find the mass moment of inertia with respect to (*a*) X_0-axis, (*b*) Y_0-axis, (*c*) Z_0-axis, (*d*) X-axis, (*e*) Y-axis and (*f*) Z-axis.

[**Ans.** (*a*) $I_{x_0x_0} = 0.07\, mr^2$; (*b*) $I_{y_0y_0} = 0.32\, mr^2$; (*c*) $I_{z_0z_0} = \frac{mr^2}{\pi}$;

(*d*) $I_{xx} = 0.95\, mr^2$; (*e*) $I_{yy} = 0.32\, mr^2$ and $I_{zz} = 0.4984\, mr^2$]

11.6. A right cylinder of length l has an elliptical cross-section with semi-axes a and b. Find its moment of inertia with respect to the longitudinal centroidal axis. $\left[\textbf{Ans. } \bar{I}_{zz} = \frac{m}{4}(a^2 + b^2)\right]$

11.7. Calculate the moment of inertia of a homogeneous rectangular lamina of dimensions a and b with respect to a diagonal of the rectangle. The mass of lamina is M.

$$\left[\textbf{Ans. } I_d = \frac{1}{6} \cdot M\, \frac{a^2b^2}{a^2 + b^2}\right]$$

11.8. Determine the moment of inertia of a homogeneous rectangular parallelopiped of dimensions a, b, c and mass M with respect to a diagonal. $\left[\textbf{Ans. } I_d = \frac{1}{6} \cdot M\, \frac{a^2b^2 + b^2c^2 + c^2a^2}{a^2 + b^2 + c^2}\right]$

11.9. Calculate moment of inertia of a right circular cone of mass M, altitude h and base radius r with respect to a generator. $\left[\textbf{Ans. } I_g = \frac{3}{20}\, Mr^2\, \frac{r^2 + 6h^2}{r^2 + h^2}\right]$

12

Kinetics of Rigid Bodies

12.1. INTRODUCTION

A rigid body may be thought of a collection of particles held together by internal forces such that the particles remain the same distance apart. Therefore, if the rigid body moves in a given direction, all its particles will move in the same direction. The laws of motion of a particle are applicable to the rigid bodies. In this chapter, we shall study the relations between the applied forces and the resulting motion of the rigid body. The body may have various types of motions such as translation, rotation or the plane motion. A plane motion is a combination of motion of translation and rotation. The study of problems of motion can be idealized as problems of plane motion. For this, the body can be thought of a thin lamina or slab with its mass centre lying in the plane of its motion.

Various types of problems involving rigid body motion are represented in Fig. 12.1.

Here a rigid body is subjected to various forces whose resultant is equal to $\Sigma P = R$, as shown in Fig. 12.1 (*a*).

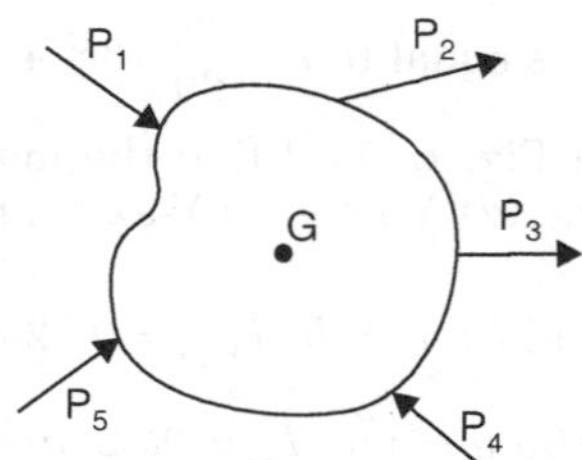

(a) Given rigid body subjected to a given force system

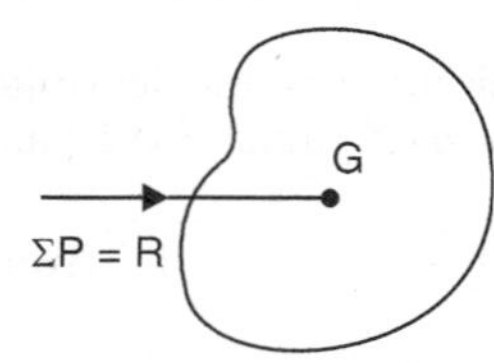

(b) Case 1. The resultant R = ΣP passes through the mass centre G (Pure translation)

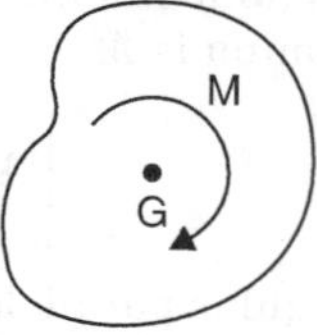

(c) Case 2. The pure resultant couple M acting on the body (Pure rotation)

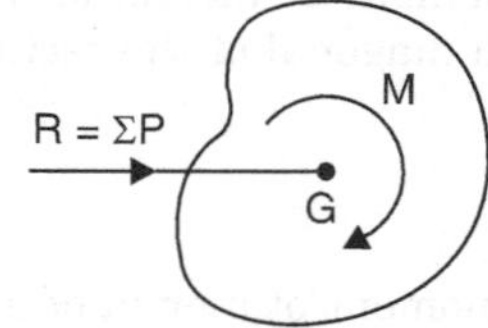

(d) Case 3. The resultant force R = ΣP acting at some distance from G accompanied by pure couple M (Plane motion)

Fig. 12.1. Various cases of force systems

Case 1. When the resultant force $R = \Sigma P$ passes through the mass centre G, it causes pure translation as shown in Fig. 12.1 (*b*).

Case 2. When the given rigid body is subjected to pure couple (M) it has pure rotational motion as shown in Fig. 12.1 (*c*).

Case 3. When the resultant force $R = \Sigma P$ does not pass through the mass centre of the body it produces translation accompanied by a rotation. This is known as a plane motion, as shown in Fig. 12.1 (*d*).

(A) TRANSLATION

12.2. MOTION OF TRANSLATION

If a rigid body has translation motion, every line of the body remains parallel to its original position at all time of motion. The translation motion may be rectilinear or curvilinear.

If a rigid body of mass moves with acceleration a_x and a_y then the following equations of motion will be applicable :

$$\boxed{\Sigma P_x = ma_x \quad \text{and} \quad \Sigma P_y = ma_y} \qquad ...(12.1)$$

These equations are obtained from Newton's Laws of motion, as discussed earlier. These equations determine the magnitude and direction of the resultant force in terms of mass and acceleration but they do not locate the point of application of the resultant force. To locate the resultant force. We use the D' Alembert's principle given below :

"The resultant of all external forces acting on the body and is equal to the vector summation of the effective forces acting on all particles $\left(\sum_{i=1}^{n} m_i a_i\right)$.

Let resultant of external forces $= \sum_{i=1}^{n} F_i = \overline{R}$

and the sum of effective force be $\sum_{i=1}^{n} m_i \, . \, a_i = m\overline{a}$

Then, D' Alembert's principle states that

$$\boxed{\overline{R} = m\overline{a}} \qquad ...(12.2)$$

or,
$$\Sigma F_x = ma_x \qquad ...(12.2\ (a))$$
$$\Sigma F_y = ma_y \qquad ...(12.2\ (b))$$
and
$$\Sigma M_G)_x = 0 \qquad ...(12.2\ (c))$$
and
$$\Sigma M_G)_y = 0 \qquad ...(12.2\ (d))$$

Both $\overline{R}$ and $m\overline{a}$ pass through the mass centre m_G.

D' Alembert's principle for rectilinear and curvilinear motion are graphically explained in Fig. 12.2.

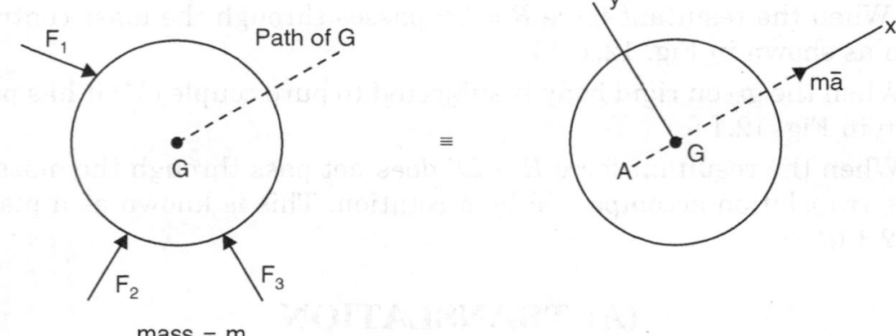

(i) Free body diagram of rigid body (ii) Kinetics diagram of the rigid body

(a) Rectilinear motion ($\alpha = 0$, $\omega = 0$ *i.e.*, no rotation of body)

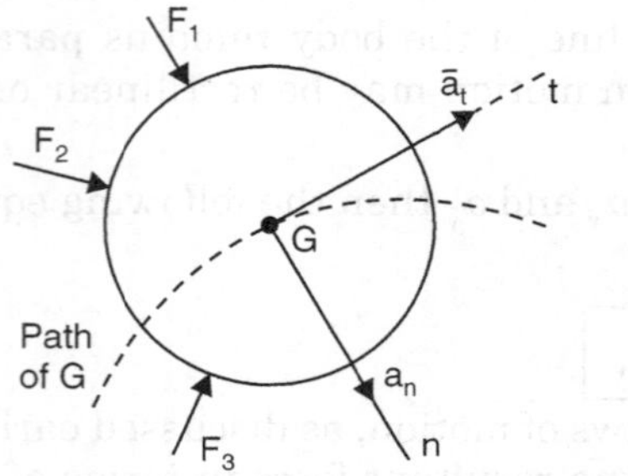

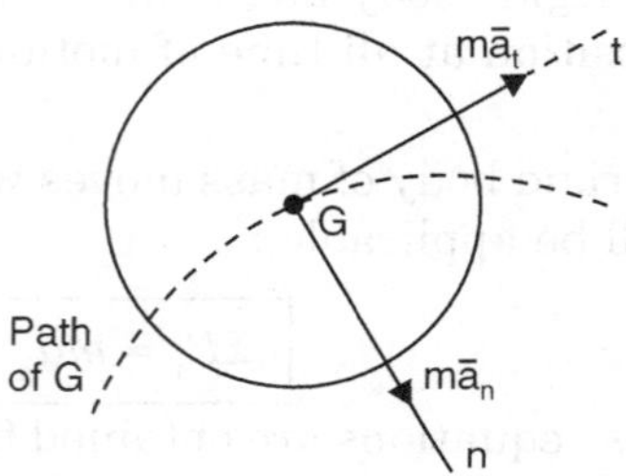

(i) Free body diagram of rigid body (ii) Kinetic diagram of the rigid body

(b) Curvilinear motion ($\alpha = 0$, $\omega = 0$, *i.e.*, no rotation)

Fig. 12.2.1 Kinetics of a rigid body (a) Rectilinear motion, (b) Curvilinear motion.

The above equations are valid for a rigid body having curvilinear translation motion. The kinematics equations for curvilinear translation motion are given as under :

(*a*) **Rectangular co-ordinates.** See Fig. 12.2.2 (*a*).

$$\bar{v} = \frac{d\bar{s}}{dt}\ ; \qquad \therefore \qquad v_x = \frac{dx}{dt} = \dot{x} \quad \text{and} \quad a_x = \frac{d^2x}{dt^2} = \ddot{x},$$

$$v_y = \frac{dy}{dt} = \dot{y} \quad \text{and} \quad a_y = \frac{d^2y}{dt^2} = \ddot{y}$$

$$\bar{V} = \bar{V}_x + \bar{V}_y = \bar{i}\,v_x + \bar{j}v_y$$

and

$$\bar{a} = \bar{a}_x + \bar{a}_y = \bar{i}\,a_x + \bar{j}a_y$$

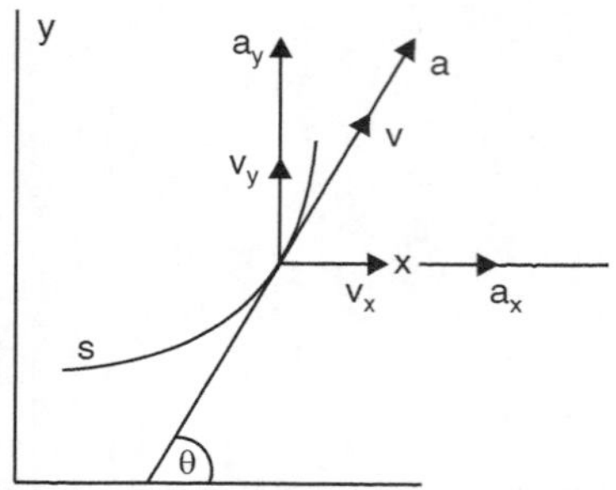

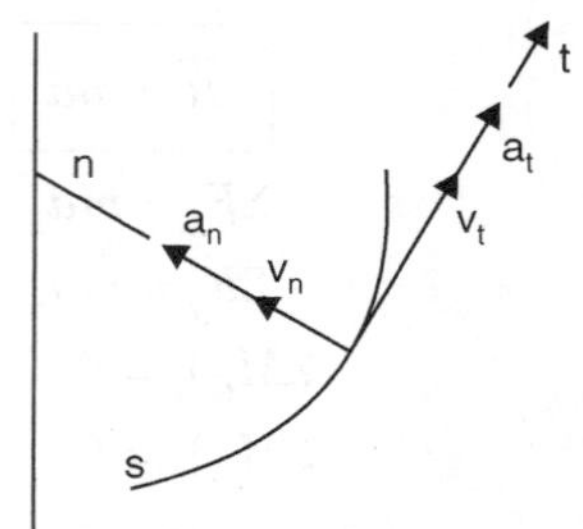

(i) Rectangular components (ii) Normal and tangential components

Fig. 12.2.2. Components of velocity and acceleration in

(a) Rectangular co-ordinates, (b) Intrinsic (n – t) co-ordinates.

(*b*) **Normal-tangential co-ordinates.** Fig. 12.2.2(*b*).

$$\bar{v}_t = \frac{ds}{dt} = \bar{\dot{s}}_t = \bar{e}_t \cdot v_t \;; \quad \bar{v}_n = \bar{e}_n \cdot v_n$$

$$\bar{a} = \frac{d\bar{v}}{dt} = \frac{d}{dt}(v\bar{e}_t) = \frac{dv}{dt} \cdot \bar{e}_t + v \cdot \frac{d\bar{e}_t}{dt}$$

where $$\frac{dv}{dt} = \frac{d^2 s}{dt^2}$$

Now $$\frac{d\bar{e}_t}{dt} = \frac{v}{\rho} \times \bar{e}_n$$

∴ We get $$\bar{a} = \frac{dv}{dt} \cdot \bar{e}_t + \frac{v^2}{\rho} \bar{e}_n = \bar{a}_t + \bar{a}_n \quad \text{...(12.3)}$$

∴ We can write $$\boxed{a_t = \frac{dv}{dt} \quad \text{and} \quad a_n = \frac{v^2}{\rho}} \quad \text{...(12.4)}$$

where ρ = radius of curvature of the paths s, Hence, the equations of motion become :

and $$\boxed{\begin{aligned} \Sigma F_n &= ma_n \\ \Sigma F_t &= ma_t \\ \Sigma M_G)_n = \Sigma M_G)_t &= 0 \end{aligned}} \quad \text{...(12.5)}$$

The terms ma_n and ma_t in eqn. (12.5) represent inertia force components in the normal and the tangential directions.

Similarly, the terms ma_x and ma_y in eqn. (12.2) represent inertia force components in x- and y-directions.

The following examples will explain the procedure for analysis of kinetics of rigid bodies.

Problem 12.1. *The frame of a machine part is accelerating rightwards at 0.6 g. The frame carries a uniform bar ABC weighing 160 N/m. The uniform rod DE, pinned at D and E to the frame, weighs, 200 N. Find the components of the forces at the pins D, C and E. Refer Fig. P.12.1 (a)*

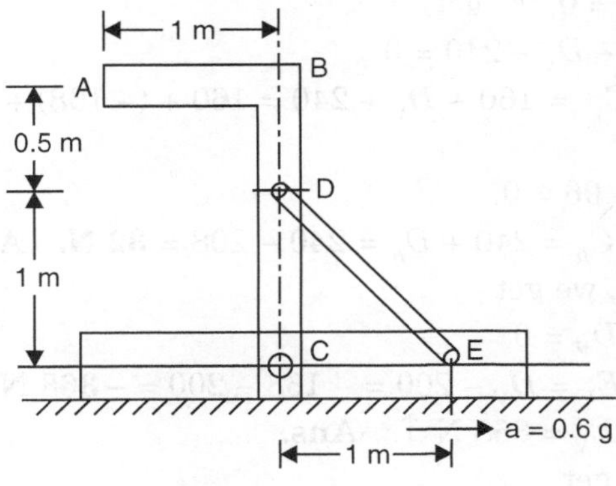

(a) Given frame

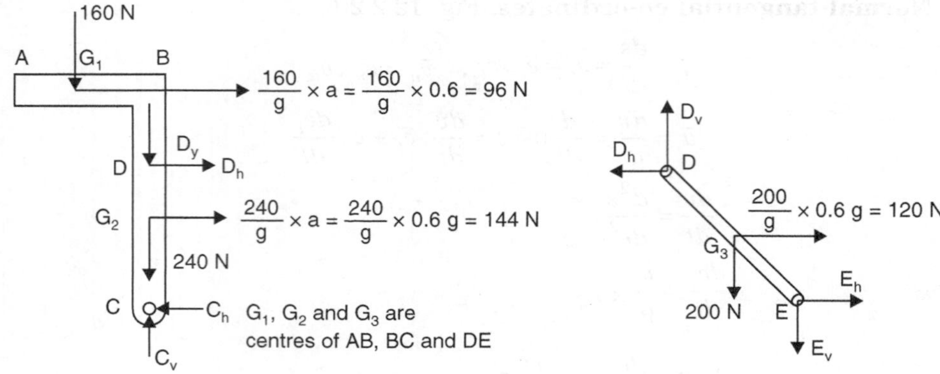

(b) Free body diagrams

Fig. P. 12.1

Sol. The free-body diagrams of the three portions *AB, BC* and *DE* are shown in Fig. P.12.1 (*b*). Inertia reactions are :

For AB : $\frac{W}{g}a = \frac{160}{g} \times 0.6\,g = 96$ N, acting at G_1

For BC : $\frac{W}{g}a = \frac{240}{g} \times 0.6\,g = 144$ N, acting at G_2

and **for DE :** $\frac{W}{g}a = \frac{200}{g} \times 0.6\,g = 120$ N, acting at G_3

Consider the free-body-diagram for ABC :

By $\Sigma M_C = 0$, we get

$$D_h \times 1 - 160 \times 0.5 + 96 \times 1.5 + 144 \times 1 = 0$$

$\therefore$ $D_h = 80 - 114 - 144 = \mathbf{-208}$ **N,** $\therefore$ $\mathbf{D_h = 208}$ **N ←. Ans.**

By $\Sigma M_E = 0$ for the bar *DE*, we get

$$D_V \times 1 - D_h \times 1 - 200 \times 0.5 + 120 \times 0.5 = 0$$

$\therefore$ $D_V - D_h = 40$ N

$\therefore$ $D_V = 40 + D_h = 40 - 208 = -168$ N, $\therefore$ $\mathbf{D_v = 168}$ **N ↓. Ans.**

For ABC : *By* $\Sigma F_V = 0$, we get

$$C_V - 160 - D_V - 240 = 0$$

$\therefore$ $C_V = 160 + D_V + 240 = 160 + (-168) + 240 = \mathbf{232}$ **N. Ans.**

and $\Sigma F_h = 0$, we get

$$C_h - D_h - 144 - 96 = 0,$$

$\therefore$ $C_h = 240 + D_h = 240 - 208 = \mathbf{32}$ **N. Ans.**

For DE : By $\Sigma F_V = 0$, we get

$$E_V + 200 - D_V = 0$$

$\therefore$ $E_V = D_V - 200 = -168 - 200 = \mathbf{-368}$ **N**

i.e., $E_V = 368$ N ↑. **Ans.**

And by $\Sigma F_h = 0$, we get

$$E_h + 120 - D_h = 0$$

$\therefore \qquad E_h = D_h - 120 = 50 - 208 = -328$ N

or $\qquad E_h = 328$ N ←. **Ans.**

Problem 12.2. (*a*) *A steel almirah of 80 kg mass is standing on a horizontal floor as shown in Fig. P. 12.2 (a). Find the minimum force P required to tip it at corner A (a) when the floor is smooth, and (b) when the coefficient of kinetic friction between the almirah and the floor is 0.2. Also find acceleration if no tipping takes place.*

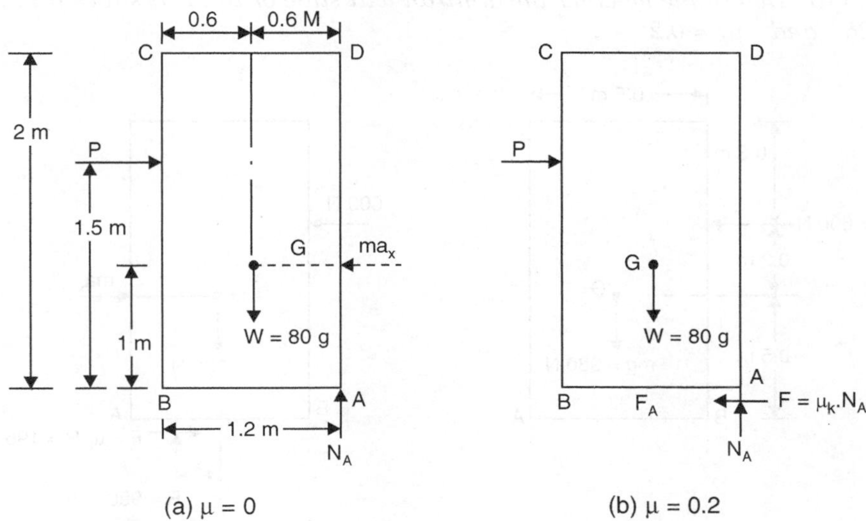

Fig. P. 12.2

Sol. (*a*) When the almirah tips or overturns about *A*, the normal reaction will pass through *A*.

By $\Sigma F_V = 0$, we get

$$N_A - 80 \text{ g} = 0$$

$$\therefore \qquad N_A = 80 \text{ g} = 80 \times 9.81 = \mathbf{784.8 \text{ N}} \uparrow$$

By $\Sigma M_G = 0$, we get

$$P \times (1.5 - 1) - NA \times 0.6 = 0$$

$$\therefore \qquad P \times 0.5 = 784.8 \times 0.6$$

$$\therefore \qquad P = \frac{784.8 \times 0.6}{0.5} = \mathbf{941.76 \text{ N.}} \quad \textbf{Ans.}$$

Thus, tipping will take place.

(*b*) If the body is in dynamic equilibrium and the acceleration is equal to a_x then, the reversed inertia force $ma_x = 80\ a_x$ acts through the mass centre *G*. And, the almirah starts moving before tipping takes place. Thus, by $\Sigma F_x = ma_x$, we get

If $\mu_k = 0.2$, $\quad F_k = \mu_k N_A = 0.2 \times 80 \text{ g} = 156.96$ N,

$\therefore$ By $\Sigma M_G = 0$, we get

$$P \times (1.5 - 1) - N_A \times 0.6 + 156.96 \times 1 = 0$$

$\therefore \quad P \times 0.5 = 313.92$

$\therefore \quad$ **P = 627.84 N Ans.**

$\therefore$ By $\Sigma F_x = ma_x$, we get

$$(P - \mu_k N_A) = 80\ a_x \quad \text{or} \quad (627.84 - 156.96) = 80\ a_x$$

$\therefore \quad a_x =$ **5.886 m/s². Ans.**

Problem 12.3. *A steel almirah of weight 980 N is subjected to a force of 600 N as shown in Fig. P. 12.3(a). Determine whether the almirah will slide or tip. If it slides find its acceleration.* $\mu_s = 0.25$ *and* $\mu_k = 0.2$.

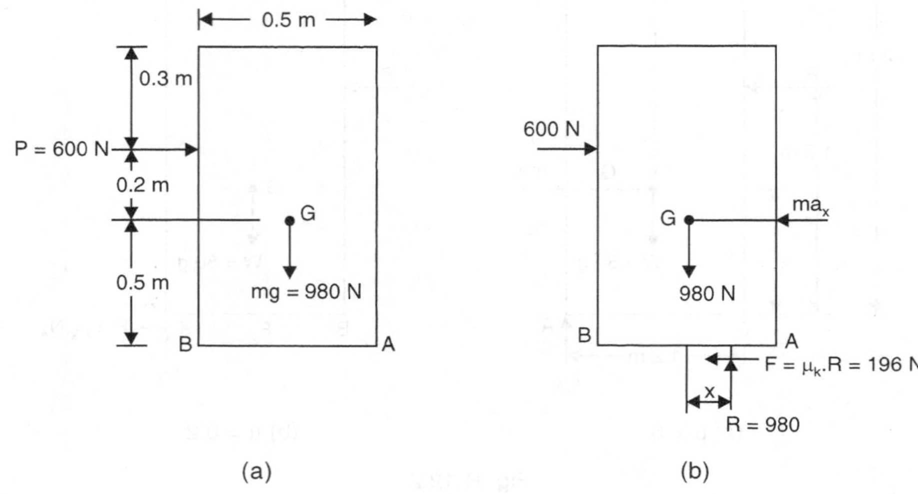

Fig. P. 12.3

Sol. Case I.

By $\Sigma F_y = ma_y$, we get $(a_y = 0)$

$$980 - R = 0, \quad \therefore \quad R = 980 \text{ N}$$

Maximum force of friction.

$$F_s = \mu_s \,.\, R = 0.25 \times 980 = 245 \text{ N}$$

The force of kinetic friction $= F_k = \mu_k \times R = 0.2 = 196$ N

Let us check whether the almirah tips about A.

$\therefore$ We get by $\quad \Sigma M_A = 0,$

$\therefore \quad P_x (0.5 + 0.2) - 980 \times 0.25 = 0$

$$\therefore \quad P = \frac{980 \times 0.25}{0.7} = \mathbf{350\ N}$$

As $\quad F_s = 245$ N, and $P = 350$ N,

$\therefore$ Almirah slides before tipping about P.

$\therefore$ As the almirah starts sliding the force of friction induced $= F_k = 196$ N.

$\therefore$ By $\Sigma F_x = ma_x$, we get

$$P - F_x = ma_x$$

or, $$600 - 196 = \frac{980}{9.80} a_x = 100\ a_x$$

$$\therefore \quad a_x = \frac{404}{100} = \mathbf{4.04\ m/s^2.} \quad \textbf{Ans.}$$

Case II. Here we find whether the almirah, will slide or tip about A when the motion has started.

Let us assume that the vertical reaction, $R = 980$ N acts at a distance x from centre line, as shown in Fig. 12.3 (*b*).

By $\Sigma M_G = 0$, we get

$$600 \times 0.2 + (\mu_k \,.\, R) \times 0.5 - 980\,x = 0$$

$$\therefore \quad 100 + 196 \times 0.5 - 980\,x = 0$$

$$\therefore \quad x = \frac{100 + 98}{980} = 0.222 \text{ m}$$

As $x < b/2$ or $x < 0.25$ m, $\therefore$ The almirah will continue sliding.

The force P required for tipping about A during motion can be found by using $\Sigma M_G = 0$.

$$\therefore \quad P \times 0.2 + 196 \times 0.5 - 980 \times 0.25 = 0$$

$$\therefore \quad P = \frac{1}{0.2}(980 \times 0.25 - 196 \times 0.5) = 0 = \frac{147}{0.2} = \mathbf{735\ N}$$

As this force P is greater than 600 N therefore, the almirah will slide before tipping

$\therefore$ By $\Sigma F_x = ma_x$, we get

$$a_x = \frac{P - F_k}{m} = \frac{735 - 196}{80} = \frac{539}{80} = \mathbf{6.7375\ m/s^2} \quad \textbf{Ans.}$$

Problem 12.4. *Fig. P.12.4(a) shows a small cart weighing 1470 N. A rod AB of weight 147 N is hinged at B and rests against a smooth vertical edge at A. Determine the value of the force P to give the system an acceleration 0.80 m/s² towards right and the reaction at A. At what acceleration to the left will the rod AB exert no pressure at A ? Also find the acceleration at which pressure at A is doubled.*

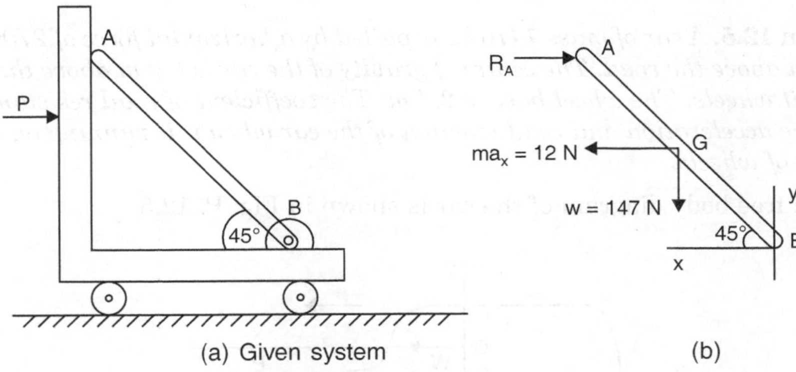

Sol. Total mass of cart $= \dfrac{1470 + 147}{9.8} = 165$ kg

$$P = ma_x = 165 \times 0.80$$

$$= \mathbf{115.5\ N.} \quad \textbf{Ans.}$$

Effective force on the rod AB

$$= ma_x = \frac{147}{9.8} \times 0.8 = 12 \text{ N}$$

The reversed inertia force acts to the left at the CG of rod AB as shown in Fig. P. 12.4 (*b*).

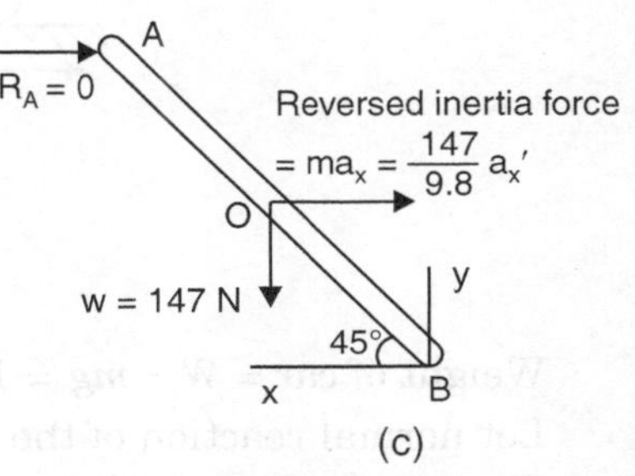

Fig. P. 12.4

By $\Sigma M_B = 0$, we get

$$R_A \times AB \sin 45° - 147 \times \frac{AB}{2} \cos 45° - 12 \times \frac{AB}{2} \sin 45° = 0$$

$$\therefore \quad R_A = \frac{147}{2} \cot 45° + \frac{12}{2} = 79.5 \text{ N}$$

Let at an acceleration a'_x to the **left** (or retardation) the pressure at A becomes zero. For this case the free body diagram of rod AB is shown in Fig. P.12.4 (*c*).

By $\Sigma M_B = 0$, we get

$$\left(\frac{147}{9.8} a'_x\right) \times \frac{AB}{2} \sin 45° - 147 \times \frac{AB}{2} \cos 45° = 0$$

$$\therefore \quad a'_x = 9.8 \text{ m/s}^2 \text{ towards left.} \quad \textbf{Ans.}$$

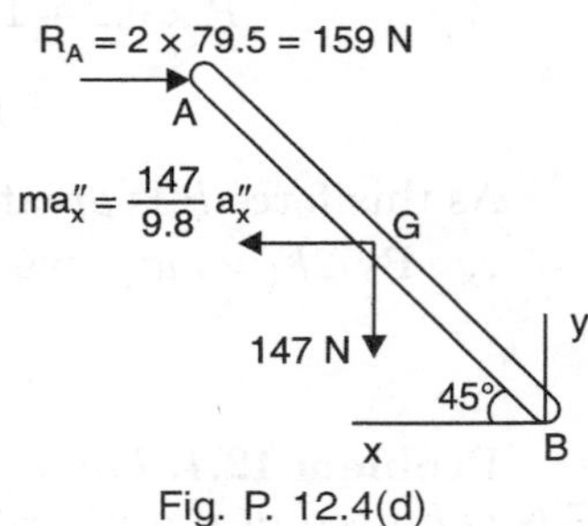

Fig. P. 12.4(d)

Let at an acceleration a''_x the pressure at A is doubled. The free-body diagram for this condition is shown in Fig. P. 12.4 (*d*).

Here $R_A = 2 \times 79.5$ N.

By $\Sigma M_B = 0$, we get

$$(2 \times 79.5) AB \sin 45° - 147 \times \frac{AB}{2} \cos 45° - \frac{147}{9.8} a''_x \cdot \frac{AB}{2} \sin 45° = 0$$

$$\therefore \quad \frac{147}{9.8} \frac{a''_x}{2} = (2 \times 79.5) - \frac{147 \times \cot 45°}{2} = 85.5 \text{ N}$$

$$\therefore \quad a''_x = \textbf{11.4 m/s}^2. \quad \textbf{Ans.}$$

Problem 12.5. *A car of mass 1470 kg is pulled by a horizontal force of 2700 N applied at a point 600 mm above the road. The centre of gravity of the car is 0.9 m above the road and 1 m behind the front wheels. The wheel base is 2.4 m. The coefficient of road resistance is 0.1 at all wheels. Find the acceleration and road reaction of the car when it is running on the level track. Neglect inertia of wheels.*

Sol. The free body diagram of the car is shown in Fig. P. 12.5.

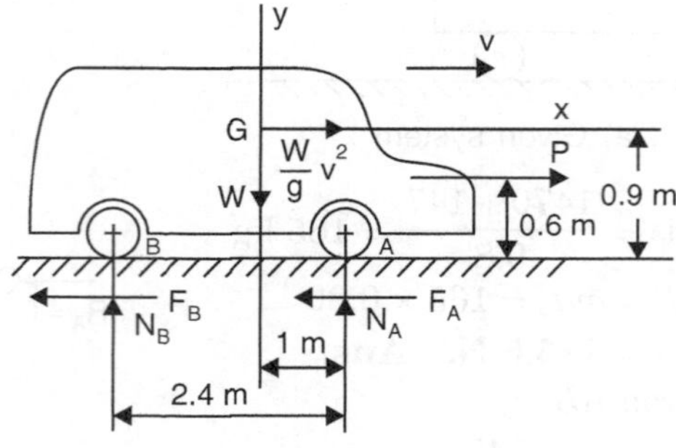

Fig. P. 12.5

Weight of car = $W = mg = 1470 \times 9.8 = 14{,}406$ N

Let normal reaction of the front wheels = N_A and that of the rear wheels be N_B.

Therefore, force of friction at A and B will be $0.1\, N_A = F_A$ and $F_B = 0.1\, N_B$.

Total force of friction $= F = F_A + F_B = 0.1 \times (N_A + N_B) = \mathbf{0.1\ W}$.

$\therefore \quad F = 0.1 \times 14406 = 1440.6$ N.

The effective force $= ma_x = 1470\, a_x$

By $\Sigma F_x = ma_x$, we get

$P - F = ma_x$

$$\therefore \quad a_x = \frac{P - F}{m} = \frac{2700 - 1440.6}{1470} = \frac{1259.4}{1470} = 0.8567 \text{ m/s}^2. \quad \textbf{Ans.}$$

By taking moments of all forces about the centre of gravity G, we get

$$N_B \times (2.4 - 1) - N_A \times 1 + F \times 0.9 - 2700 \times (0.9 - 0.6) = 0$$

$$\therefore \quad N_B \times 1.4 - N_A + 1440.6 \times 0.9 - 2700 \times 0.3 = 0$$

$$\therefore \quad 1.4N_B - N_A = -\,486.54 \qquad \text{...(1)}$$

By $\Sigma F_V = 0$, we have

$$N_A + N_B = W$$

or, $$N_A + N_B = 14406 \qquad \text{...(2)}$$

Adding (1) and (2) we have

$$2.4\, N_B = 14406 - 486.54$$

$$\therefore \quad N_B = \frac{13919.46}{2.4} = \mathbf{5799.8\ N.} \quad \textbf{Ans.}$$

and from (2), $N_A = 14406 - 5799.8 = \mathbf{8606.2\ N.}$ **Ans.**

Problem 12.6. *Two weights 100 N and 30 N are connected by a rope and move towards right on a rough horizontal plane under a pull of 60 N applied to the 100 N weight. The coefficient of friction between the weights and the surface is 0.2.*

Find the acceleration of weights and the tension in the rope.

Sol. The system of weights is shown in Fig. P. 12.6 (*a*) and the free-body diagrams of the two weights are shown in Fig. P. 12.6 (*b*).

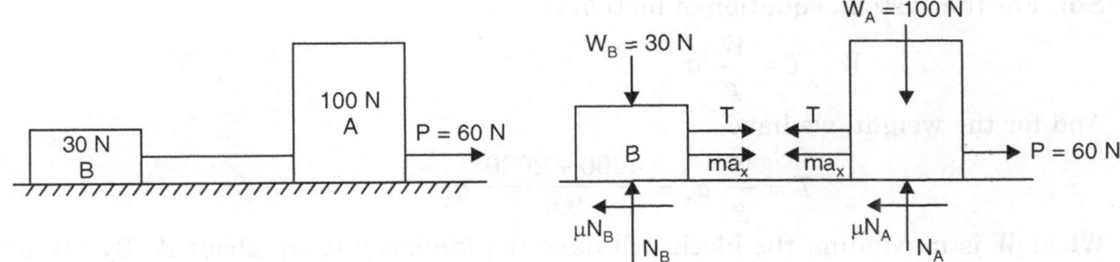

(a) Given system of two weights (b) Free-body diagrams

Fig. P. 12.6

Normal reactions at A and B are 100 N and 30 N, respectively.

Total frictional force $= F = \mu N_A + \mu N_B = 0.2 \times (100 + 30) = 26$ N

For the entire system shown in Fig. P.12.6 (*a*), by $\Sigma F_x = (m_A + m_B)\, a_x$, we get

$$\therefore \quad 60 - 26 = ma_x \quad \text{where} \quad m = \left(\frac{100 + 30}{9.8}\right)\text{N} = \frac{130 \text{ N}}{9.8}$$

$$\therefore \quad a_x = \frac{60 - 26}{(130/9.8)} = \mathbf{0.262\ m/s^2}. \quad \textbf{Ans.}$$

For weight A :

By $\Sigma F_x = m_A a_x$, we get

$$60 - T - \mu N_A = m_A \cdot a_x$$

or,
$$60 - T - 0.2 \times 100 = \frac{100}{9.8} \times 0.262$$

$$\therefore \quad T = 60 - 0.2 \times 100 - \frac{100}{9.8} \times 0.262$$

$$\therefore \quad T = \mathbf{37.33\ N.} \quad \textbf{Ans.}$$

Problem 12.7. *A block P of weight 4000 N rests on an open car Q weighing 2000 N, which may roll on the horizontal surface without friction. The car and the block are accelerated by a weight W as shown in Fig. P. 12.7 (a). The coefficient of friction between the block and the car is 0.2. Find the maximum value of W which accelerates the car. Also find the tension in the cord, the acceleration of the block and normal reaction between the block and the car.*

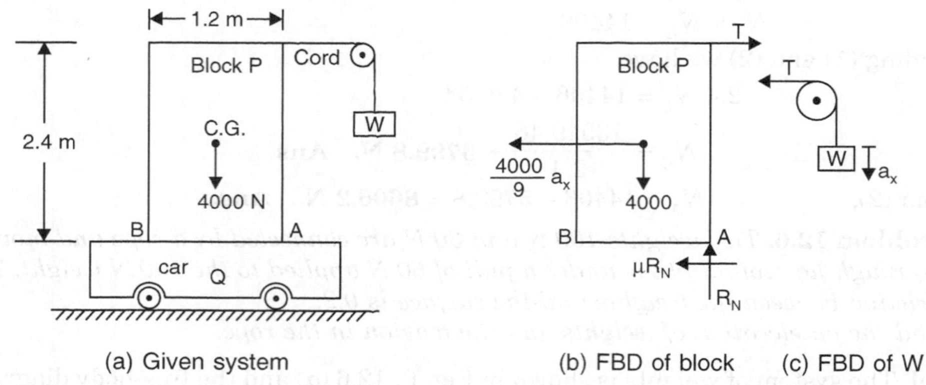

(a) Given system (b) FBD of block (c) FBD of W

Fig. P. 12.7

Sol. For the system, equation of motion is

$$W - T = \frac{W}{g} a_x \qquad \text{...(1)}$$

And for the weight, we have

$$T = \frac{W}{g} a_x = \frac{4000 + 2000}{9.8} a_x \qquad \text{...(2)}$$

When W is maximum, the block will have the tendency to tip about A. By taking moments of forces acting on the block about A, we get

$$T \times 2.4 - 4000 \times 0.6 - \frac{4000}{9.8} \times a_x \times 1.2 = 0 \qquad \text{...(3)}$$

From (2) and (3), we have

$$\left(\frac{6000}{9.8} \times a_x\right) \times 2.4 - 4000 \times 0.6 - \frac{4000}{9.8} a_x \times 1.2 = 0$$

$$\therefore \quad \frac{a_x}{9.8}(6000 \times 2.4 - 1.2 \times 4000) = 4000 \times 0.6$$

$$\therefore \quad a_x = \frac{(4000 \times 0.6) \times 9.8}{9600} = \mathbf{2.45\ m/s^2.} \quad \textbf{Ans.}$$

From (2), $T = \frac{6000}{9.8} \times 2.45 =$ **1500 N. Ans.**

$\therefore$ From eqn. (1),

$$W - 1500 = \frac{W}{9.8} a_x, \quad \text{we get}$$

$$W\left(1 - \frac{a_x}{9.8}\right) = 1500$$

or $$W\left(1 - \frac{2.45}{9.8}\right) = 1500$$

$$\therefore \quad W = \frac{1500}{0.75} = \textbf{2000 N. Ans.}$$

By taking moments about C.G. of the block [Fig. P. 12.7 (*b*)], we get

$$R_N \times 0.6 - 0.2\, R_N \times 1.2 - T \times 1.2 = 0$$

$$\therefore \quad R_N\,(0.6 - 0.2 \times 1.2) = 1.2 \times T$$

$$\therefore \quad R_N\,(0.6 - 0.24) = 1.2 \times 1500$$

$$\therefore \quad R_N = \frac{1.2 \times 1500}{0.36} = \textbf{5000 N.}$$

Force of friction = $\mu R_N = 0.2 \times 5000 =$ **1000 N.**

$\therefore$ Resultant reaction between block and the car

$$= R = \sqrt{R_N{}^2 + (\mu R_N)^2} = \sqrt{(5000)^2 + (5000 \times 0.2)^2} = \textbf{5200 N.} \quad \textbf{Ans.}$$

$$\tan \theta_x = \frac{\mu R_N}{R_n} = \textbf{0.2}$$

$$\therefore \quad \theta_x = \tan^{-1} 0.2 = \textbf{11.31°} \text{ from horizontal.} \quad \textbf{Ans.}$$

Problem 12.8. *A uniform almirah shown in Fig. P. 12.8 (a) weighs 1800 N. It is pulled up the inclined plane by a weight W of 3000 N. Find the maximum and minimum values of the distance h so that the almirah does not tip as it slides up the plane. Take* μ *= 0.25.*

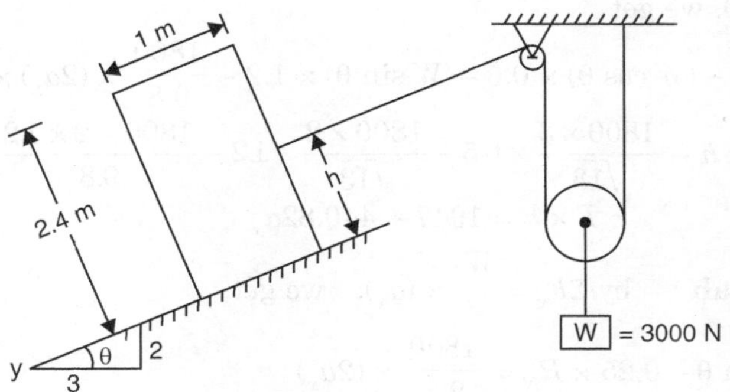

(a) Given system

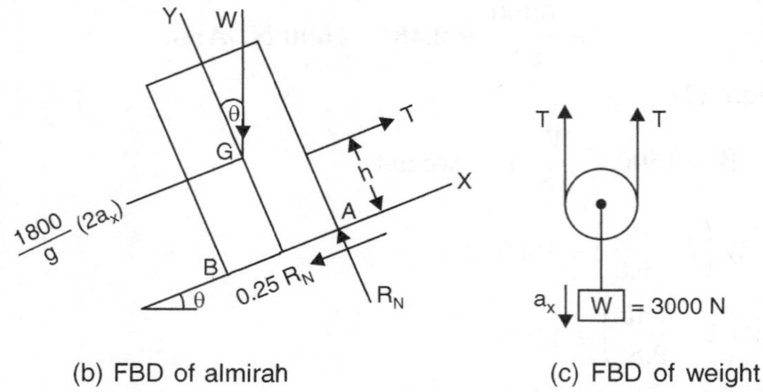

(b) FBD of almirah (c) FBD of weight

Fig. P.12.8

Sol. Given : Inclination of the plane is given by

$$\tan\theta = \frac{2}{3}\ ; \quad \therefore \quad \sin\theta = \frac{2}{\sqrt{13}} \quad \text{and} \quad \cos = \frac{3}{\sqrt{13}}.$$

Let T be the tension in the string and a_x be acceleration of the almirah. From the free-body diagram of the almirah, Fig. P.12.8(*b*), we get

by $\Sigma F_y = 0 : R_n - W\cos\theta = 0$

$$\therefore \qquad R_n = 1800 \times \frac{3}{\sqrt{13}} = 1497.7 \text{ N}$$

$$\therefore \qquad \mu R_N = 0.25 \times 1497.7 = 374.425 \text{ N}.$$

Equation of motion for the weight is [Fig. P.12.8(*c*)] :

$$3000 - 2T = \frac{3000}{9.8} \times a_x \qquad ...(1)$$

Acceleration of almirah = $2a_x$

$\therefore$ For the almirah the inertia force $= \frac{1800}{9.8} \times (2a_x)$

For the maximum value of the distance h the almirah will be on the point of tipping over the corner A.

$\therefore$ By $\Sigma M_A = 0$, we get

$$T \times h - (W\cos\theta) \times 0.5 - (W\sin\theta) \times 1.2 - \frac{1800}{9.8} \times (2a_x) \times 1.2 = 0$$

or,
$$T \times h - \frac{1800 \times 3}{\sqrt{13}} \times 0.5 - \frac{1800 \times 2}{\sqrt{13}} \times 1.2 - \frac{1800 \times 2 \times 1.2}{9.8} \times a_x = 0$$

or,
$$T \times h = 1947 + 440.82a_x \qquad ...(2)$$

For the almirah : by $\Sigma F_x = \frac{W}{g} \times (a_x)$, we get

$$T - 1800\sin\theta - 0.25 \times R_N = \frac{1800}{9.8} \times (2a_x)$$

or,
$$T = 1372.9 + 367.35\, a_x \qquad ...(3)$$

Multiplying (3) × 2 + (1), we get

$$(3000 - 2T) + 2T = \frac{3000 \times a_x}{9.8} + 2[1372.9 + 367.35\, a_x]$$

$\therefore \quad 3000 = a_x(1040.82) + 2 \times 1372.9$

$\therefore \quad a_x = \dfrac{3000 - 2 \times 1372.9}{1040.82} = \mathbf{0.2442\ m/s^2}$

$\therefore$ From (3), $\quad T = 1372.9 + 367.35 \times 0.2442 = 1462.6$ N

and from (2), $\quad h = \dfrac{1}{1462.6}[1947 + 440.82 \times 0.2442]$ m

$\therefore$ Maximum $\quad h = \mathbf{1.4048\ m.}$ **Ans.**

For the minimum value of *h* the almirah would tip about *B*. Therefore, by $\Sigma M_B = 0$, we get

$$T \times h + (1800 \times \cos\theta) \times 0.5 - (1800 \times \sin\theta) \times 1.2 - \frac{1800}{9.8} \times (2a_x) \times 1.2 = 0$$

$$\therefore \quad T \times h = 1800\left[1.2 \sin\theta - 0.5\cos\theta + \frac{2.4a_x}{9.8}\right]$$

or, $\quad T \times h = 449.31 + 440.82\, a_x \quad$ ...(4)

where the values $\quad T = 1462.6$ N and $a_x = 0.2442$ m/s^2, as above

$\therefore$ From eqn. (4), $\quad h = \dfrac{1}{1462.6}[449.31 + 440.82 \times 0.2442]$ m

$\therefore$ Minimum $\quad h = \mathbf{0.381\ m.}$ **Ans.**

12.3. CURVILINEAR TRANSLATION

We have seen earlier that when a particle moves along a curved path it is subjected to normal (or radial) and tangential accelerations. The radial acceleration is also called the centripetal acceleration and is directed towards the centre of rotation. And, the centrifugal force is always directed away from the centre of rotation.

The tangential acceleration is perpendicular to the radius or it is tangential to the path of the particle. The corresponding force is called the tangential force. We have seen that the centripetal or radial or normal acceleration

$$\boxed{a_c \text{ or } a_r = \frac{v^2}{r} = r\omega^2} \quad ...(12.6)$$

and centripetal force $\quad F_c = \dfrac{mv^2}{r} = m\omega^2 r \quad$ ...(12.7)

directed towards the centre of rotation.

Also, centrifugal force, $F_c = \dfrac{mv^2}{r} = m\omega^2 r \quad$ ...(12.8)

directed away from the centre of rotation.

In the above equations

m = mass, in kg,

v = linear velocity, m/s

r = radius, m

and ω = angular velocity, radians/s.

12.4. DIFFERENTIAL EQUATIONS OF CURVILINEAR MOTIONS

Consider the motion of a particle traversing a curved path, as shown in Fig. 12.4. The particle at P has mass m. Let ΣF be the resultant force acting on P and ΣF_x and ΣF_y are its components along X-and Y-directions. Then, the following differential equations of motion will be used in solving problems of this nature :

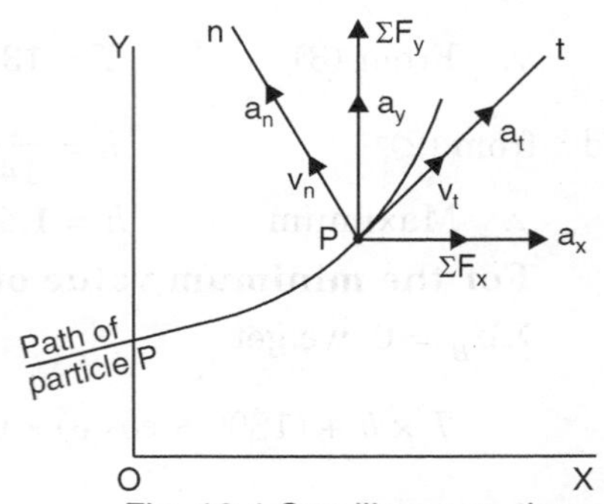

Fig. 12.4.Curvilinear motion–equations of motion

$$\left.\begin{aligned}\Sigma F_x &= ma_x \\ \text{and} \quad \Sigma F_y &= ma_y\end{aligned}\right\} \qquad \text{...(12.9)}$$

Sometimes the tangential and normal acceleration components a_t and a_n, respectively of the body in curvilinear motion are known. In such a case, the following equations of motion will be preferred :

$$\left.\begin{aligned}\Sigma F_t &= ma_t = m\frac{dv}{dt} \\ \text{and} \quad \Sigma F_n &= ma_n = m\frac{v^2}{r}\end{aligned}\right\} \qquad \text{...(12.10)}$$

Problem 12.9. *A particle of mass m attached to a string of length l swings in a vertical plane with uniform speed v. Find tension T in the string.*

Sol. Fig. P.12.9 shows the mass swinging in the vertical plane. Since the speed v is uniform, therefore,

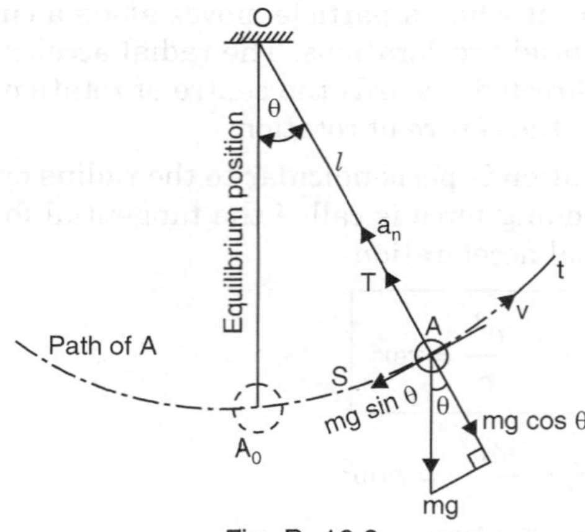

Fig. P. 12.9

$$a_t = \frac{dv}{dt} = 0,$$

and
$$a_n = \frac{v^2}{l}$$

where l is the radius of curvature of the path of particle A.

a_n is directed towards the centre of curvature O.

By $\Sigma F_n = ma_n$, we get

$$T = m\frac{v^2}{l}. \quad \textbf{Ans.}$$

We note that the tension in the string is proportional to the square of the velocity v of the particle.

Problem 12.10. *The simple pendulum of mass m suspended with string of length l swings in a vertical plane. Find the period τ for small amplitudes of swing. Also find the frequency of oscillations.*

Sol. Fig. P. 12.10 shows the simple pendulum. Let the displacement of the bob at position A be s from the equilibrium position A_0. The forces acting on the bob are weight $W = mg$ of the bob and the tension T of the string. Resolving forces along the tangent t to the path and using the condition $\Sigma F_t = ma_t$, we have the equation of motion :

$$-mg \sin\theta = m\frac{dv}{dt} \quad ...(1)$$

where $v = \dfrac{ds}{dt}$ and $s = A_0 A$

For the small amplitude of oscillations, $\sin\theta \simeq \theta \simeq \dfrac{s}{l}$ ...(2)

Therefore, eqn. (1) becomes

$$m\frac{d^2 s}{dt^2} + mg \times \frac{s}{l} = 0$$

$$\left| \frac{dv}{dt} = \frac{d}{dt}\left(\frac{ds}{dt}\right) = \frac{d^2 s}{dt^2} \right.$$

Fig. P. 12.10

$$\therefore \quad \frac{d^2 s}{dt^2} + \frac{gs}{l} = 0 \quad ...(3)$$

Eqn. (3) is the differential equation of simple harmonic motion, for which, we get

$$\omega = \sqrt{g/l} \text{ rad/s.} \quad \textbf{Ans.} \quad ...(4)$$

$$\tau = 2\pi\sqrt{\frac{l}{g}}. \quad \textbf{Ans.} \quad ...(5)$$

And, frequency $f = \dfrac{1}{\tau} = \dfrac{1}{2\pi}\sqrt{\dfrac{g}{l}}$. **Ans.** ...(6)

Problem 12.11. *A particle of mass m moving with uniform velocity v traverses along a sinusoidal speed breaker shown in Fig. P. 12.11 (a). Find expression for the pressure R exerted by the road when the particle is at the top of its motion.*

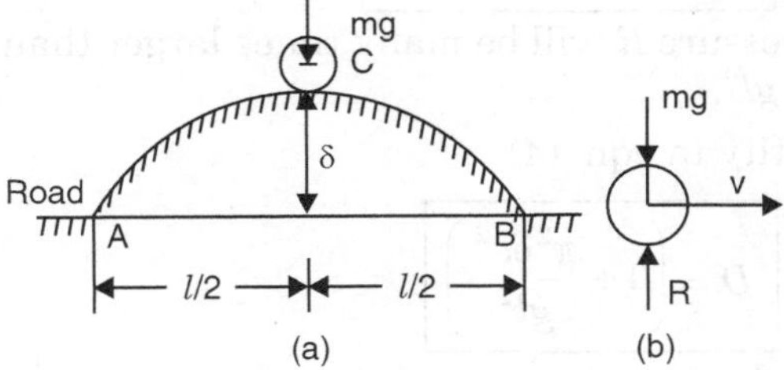

Fig. P. 12.11. Speed breaker

Sol. Let the equation of curve be $y = \delta \sin \dfrac{\pi x}{l}$ ...(1)

The normal acceleration of the particle at the highest point $C = a_n = \dfrac{v^2}{\rho}$, directed downwards the centre of curvature of the speed breaker.

The free-body diagram of the particle is shown in Fig. P.12.11 (*b*).

The equation of motion of the particle is $W - R = ma_n$

or, $$mg - R = \frac{mv^2}{\rho} \qquad ...(2)$$

where, ρ = radius of curvature of the path, which can be found by using the relation :

$$\frac{1}{\rho} = \frac{d^2y/dx^2}{\left[1+\left(\frac{dy}{dx}\right)^2\right]^{3/2}} \qquad ...(3)$$

Differentiating eqn. (1), we get

$$\frac{dy}{dx} = \frac{\delta\pi}{l}\cos\frac{\pi x}{l}$$

and $$\frac{d^2y}{dx^2} = -\frac{\delta\pi^2}{l^2}\sin\frac{\pi x}{l}$$

Substituting $x = l/2$ in the above expressions, we get

$$\left.\frac{dy}{dx}\right]_{x=l/2} = \frac{\delta\pi}{l}\cos\frac{\pi}{2} = 0$$

and $$\left.\frac{d^2y}{dx^2}\right]_{x=l/2} = -\frac{\delta\pi^2}{l^2}\sin\frac{\pi}{2} = -\frac{\delta\pi^2}{l^2}$$

and, therefore eqn. (3) becomes

$$\left.\frac{1}{\rho}\right]_{x=l/2} = \frac{-\delta\pi^2/l^2}{[1+0]^{3/2}} = -\frac{\pi^2\delta}{l^2}$$

$\therefore$ Eqn. (2) yields,

$$mg - R = mv^2\left(\frac{-\pi^2\delta}{l^2}\right).$$

$\therefore$ $$\boxed{R = mg\left(1+\frac{\pi^2\delta v^2}{gl^2}\right)} \qquad ...(4)$$

Thus, we find the pressure R will be many times larger than the force of gravity (mg) if δv^2 is large compared with gl^2.

The bracketed quantity in eqn. (4)

$$\boxed{D = \left(1+\frac{\pi^2\delta v^2}{gl^2}\right)} \qquad ...(5)$$

is called the *impact* or the *dynamic factor*.

The similar expression gives R when the particle traverses a cosine pit in the road surface given by

$$y = \delta\cos\frac{2\pi x}{l} \qquad ...(6)$$

as shown in Fig. P. 12.11 (*c*).

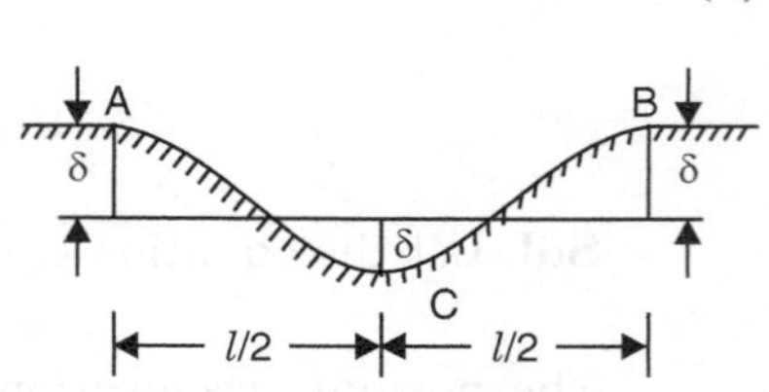

Fig. P.12.11 (c) Road pit

And, expression for R is

$$R = mg\left(1+\frac{4\pi^2\,\delta v^2}{gl^2}\right) \quad ...(7)$$

12.5. CONICAL PENDULUM

Let a body of mass m be suspended by light inextensible cord or a light arm of length l from fixed smooth point O, as shown in Fig. 12.5. This arrangement is known as a "Conical Pendulum". OC is the axis of rotation and ω is the angular speed of rotation. The mass follows a horizontal circular path of radius $CA = CB = r$. Let $h = CO$ be the vertical distance of the mass from the point of suspension.

The centripetal acceleration of the mass is $a_n = \frac{v^2}{r} = \omega^2 r$. The centripetal force $P_c = ma_n = mv^2/r = m\omega^2 r$ is directed to the centre C of the circle.

Fig. 12.5. Conical pendulum

Three forces are acting on the mass, *viz* :

(1) weight of the mass, $W = mg$.

(2) tension T in the cord, and

(3) the centrifugal force $P_c = m\omega^2 r = \frac{mv^2}{r}$.

By taking moments about O, we get

$$P_c \times h - Wr = 0$$

$$\therefore \quad (m\omega^2 r)h = mgr$$

$$\therefore \quad \omega^2 = \frac{g}{h} \quad \text{or} \quad \omega = \sqrt{g/h} \quad ...(12.11)$$

or,

$$h = g/\omega^2 \quad ....(12.12)$$

The periodic time of one complete oscillation $= \tau = \frac{2\pi}{\omega}$

But since, $\omega^2 = \frac{g}{h}$, therefore periodic time $\tau = 2\pi\sqrt{\frac{h}{g}}$ ...(12.13)

The total tension $T = \sqrt{W^2 + P_c^2} = \sqrt{W^2 + \left(\frac{W}{g}\times\frac{v^2}{r}\right)^2}$

or,

$$T = W\sqrt{1+\frac{\omega^4}{g^2 r^2}} \quad ...(12.14)$$

But from eqn. (12.11),

$$\omega^2 \times r = \frac{gr}{h} = \frac{v^2}{r}$$

$$\therefore \quad v^4/r^2 = g^2 r^2/h^2$$

∴ Eqn. (12.14.) becomes,

$$T = W\sqrt{1+(r/h)^2}\,. \qquad ...(12.15)$$

or,
$$T = \frac{W}{h}\sqrt{h^2+r^2} = \frac{Wl}{h} = \frac{mg.l}{h} \qquad ...(12.16)$$

Problem 12.12. *A mass of 1.5 kg, suspended by a mass-less rod of 180 mm, rotates at a speed of 300 rpm. Determine :*

(a) The periodic time,

(b) the height of the mass above the lowest level,

(c) The tension in the rod, and

(d) the speed when the mass rises by 40 mm above the mean level.

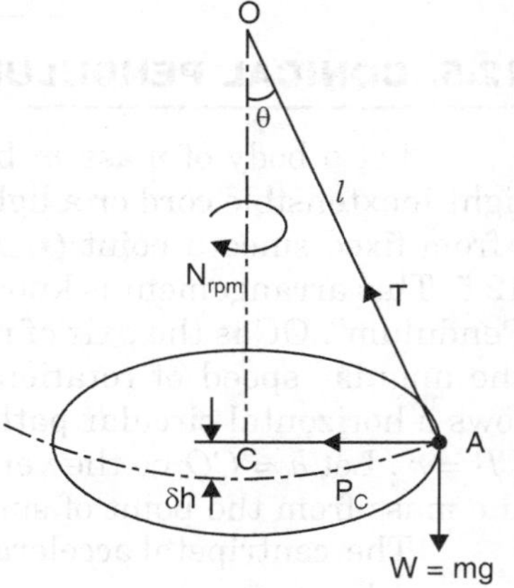

Fig. P. 12.12

Sol. Fig. P.12.12. shows the conical pendulum,

$$W = mg = 1.5 \times 9.8 \text{ N};$$

$$\omega = \frac{2\pi N}{60} = \frac{2\pi \times 300}{60} = 10\,\pi \text{ rad/s.}$$

The height $h = OC = \dfrac{g}{\omega^2} = \dfrac{9.81}{(10\,\pi)^2}$ m = 0.00994 m = **9.94 mm**

(*a*) Period, $\tau = \dfrac{2\pi}{\omega} = \dfrac{2\pi}{10\pi}$ = **0.20** seconds. **Ans.** (*a*)

(*b*) The rise of mass $= \delta_h = l - h$ = (180 – 9.94) mm = **170.06 mm. Ans.**

(*c*) Tension, $T = W\sqrt{1+\left(\dfrac{r}{h}\right)^2} = \dfrac{Wl}{h} = (1.5 \times 9.81) \times \dfrac{180}{9.94}$ = **266.47 N. Ans.**

(*d*) Using $h = g/\omega^2$, we get the rise $= \delta_h = (l - h)$

or,
$$\left(\frac{40}{1000}\right) = \frac{180}{1000} - \frac{g}{\omega^2}$$

$$\therefore \quad \frac{g}{\omega^2} = \frac{180-40}{1000} = \frac{140}{1000} = 0.140$$

$$\therefore \quad \omega^2 = \frac{g}{0.140} = \frac{9.81}{0.140} = 70.07$$

$$\therefore \quad \omega = \sqrt{70.07} = 8.37 \text{ rad/s}$$

or,
$$\frac{2\pi N}{60} = 8.37$$

$$\therefore \quad N = \frac{8.37 \times 60}{2\pi} = \mathbf{79.93 \text{ rpm.}} \quad \textbf{Ans.}$$

Problem 12.13. *A particle is projected from point O with initial velocity v_0 at an angle of elevation α. Prove that the velocity of the particle when it has raised by y is given by*

$$v = \sqrt{v_0^2 - 2gy}.$$

Sol. Fig P. 12.13 shows the path of the particle which is projected from point *O* with initial velocity v_0. Let the particle be at *A* or *B* after it has raised *y* above the point of projection.

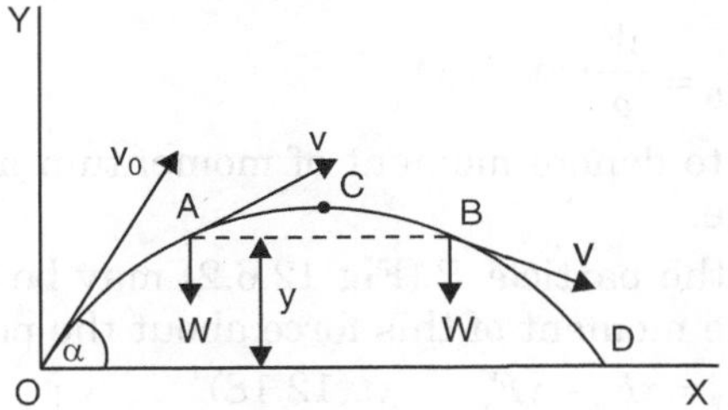

Fig. P.12.13

Sum of horizontal forces acting on particle $= \Sigma F_x = 0$

Sum of vertical forces acting on the particle $= \Sigma F_y = -W$

Loss of kinetic energy $= \dfrac{W}{2g} v^2 - \dfrac{W}{2g} {v_0}^2$

Change in potential energy $= (\Sigma F_y) \times \delta y = -W(y - 0) = -W \cdot y$

Neglecting air resistance, we have $\Delta KE = \Delta PE$

$$\therefore \quad \frac{W}{2g} v^2 - \frac{W}{2g} {v_0}^2 = -W \cdot y$$

$$\therefore \quad v^2 = {v_0}^2 - 2yg$$

$$\therefore \quad v = \sqrt{{v_0}^2 - 2gy} \qquad \text{Hence proved.} \qquad ...(1)$$

At point D, we find that $y = 0$,

$\therefore$ From eqn. (1), we get $v_D = \sqrt{{v_0}^2 - 0} = v_0$

Hence velocity at D = velocity at $O = v_0$

12.6. MOMENT OF MOMENTUM IN CURVILINEAR MOTION

A particle P of weight $W = mg$ moves along a curved path AB in the xy plane, as shown in Fig 12.6.1 and has the velocity v. The momentum of the particle is $\left(\dfrac{W}{g} v\right)$ and is given by the vector $\overline{PC}$, which coincides with the direction of velocity v.

The components of velocity v are $v_x = \dot{x}$ and $v_y = \dot{y}$.

Hence the components of the momentum are $\dfrac{W}{g}\,\dot{x}$ and $\dfrac{W}{g}\dot{y}$ as shown in Fig 12.6.1.

Fig. 12.6.1. Moment of momentum about point O.

The *moment of momentum* of the particle about the origin is defined as the product of the momentum and the perpendicular distance OQ.

Since, moment of a resultant about a point is equal to the sum of moments of its component about the same point. Hence we get

$$H_0 = \frac{W}{g}(\dot{y}x - \dot{x}y) \qquad ...(12.17)$$

where the symbol H is used to denote moment of momentum and the counter clockwise moment is considered as positive.

The force F acting on the particle P (Fig 12.6.2) may be resolved into the rectangular components F_x and F_y and the moment of this force about the point O is

$$M_0 = xF_y - yF_x \qquad ...(12.18)$$

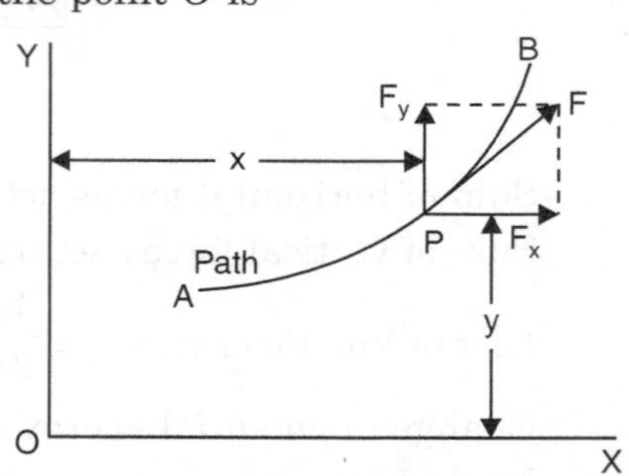

Fig. 12.6.2 Moment of a force about O

Equations of motion of the particle are therefore :

$$F_x = \frac{W}{g}\, a_x \qquad ...(12.19)$$

and

$$F_y = \frac{W}{g}\, a_y \qquad ...(12.20)$$

Multiplying eqn. (12.19) by y and eqn. (12.20) by x, we get

$$y\left(\frac{W}{g}\ddot{x}\right) = yF_x \qquad ...(12.21)$$

$$x\left(\frac{W}{g}\ddot{y}\right) = xF_y \qquad ...(12.22)$$

By (12.22) – (12.21), we get

$$\frac{W}{g}[\ddot{y}x - \ddot{x}y] = x\dot{F}_y - yF_x \qquad ...(12.23)$$

or,

$$\frac{d}{dt}\left[\frac{W}{g}(\dot{y}x - \dot{x}y)\right] = xF_y - yF_x \qquad ...(12.24)$$

From eqn. (12.17), we find that

$$\frac{d}{dt}\left[\frac{W}{g}(\dot{y}x - \dot{x}y)\right] = \frac{d}{dt}H_0$$

and from eqn. (12.18), $xF_y - yF_x = M_0$

Then, eqn. (12.24) becomes

$$\frac{d}{dt}H_0 = M_0 \qquad ...(12.25)$$

where H_0 = moment of momentum with respect to the point O, and

M_0 = moment of the resultant force with respect to the point O.

Eqn. (12.25) states that the **"rate of change of moment of momentum of a particle with respect to any point in the plane of motion is equal to the moment of the resultant force acting with respect to the same point".**

12.7. WORK-ENERGY EQUATION IN CURVILINEAR MOTION

The differential equation along x-axis is :

$$\frac{W}{g}\ddot{x} = F_x$$

This can be written as

$$\frac{W}{g}\frac{d\dot{x}}{dt} = F_x$$

Multiplying both sides by dx, we get

$$\frac{W}{g} . (dx) \frac{d\dot{x}}{dt} = F_x . dx$$

or, $$\frac{W}{g} . \frac{dx}{dt} . d\dot{x} = F_x . dx$$

or, $$d\left(\frac{W}{g} . \frac{\dot{x}^2}{2}\right) = F_x . dx \quad ...(12.26)$$

In eqn. (12.26), the force F_x is component of force in x-direction and is a function of displacement x as represented in Fig. 12.7.1.

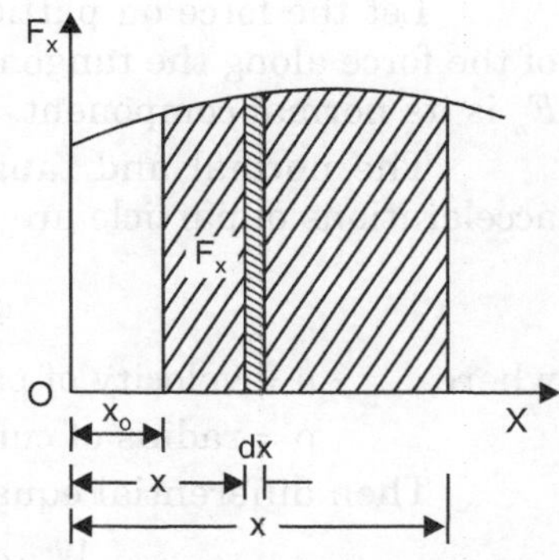

Fig. 12.7.1. Work of force F_x as function of displacement

The area of the shadded strip = $F_x . dx$

Integrating (12.26), we get

$$\frac{W}{g} . \frac{\dot{x}^2}{2} + C = \int_0^x F_x . dx \quad ...(12.27)$$

where C is the constant of integration which can be found by using the initial condition :

$$\text{at } x = x_0, \ \dot{x} = \dot{x}_0$$

$\therefore$ From (12.27), we get

$$\frac{W}{g}\frac{\dot{x}_0{}^2}{2} + C = \int_0^{x_0} F_x . dx$$

$$\therefore \quad C = -\frac{W}{g}\frac{\dot{x}_0{}^2}{2} + \int_0^{x_0} F_x . dx$$

Hence, eqn. (12.27) becomes

$$\frac{W}{g}\frac{\dot{x}^2}{2} - \frac{W}{g}\frac{\dot{x}_0{}^2}{2} = \int_{x_0}^{x} F_x . dx \quad ...(12.28)$$

Eqn. (12.28) can be applied for the curvilinear motion between points 1 and 2 on the path as shown in Fig. 12.7.2.

Let a point P on the path subjected to a force F_x move to P'. Then arc $PP' = ds$. Its components of displacement in X- and Y-axes are dx and dy respectively. Let velocity of particle at state 1 be v_1 and v_2 at state 2.

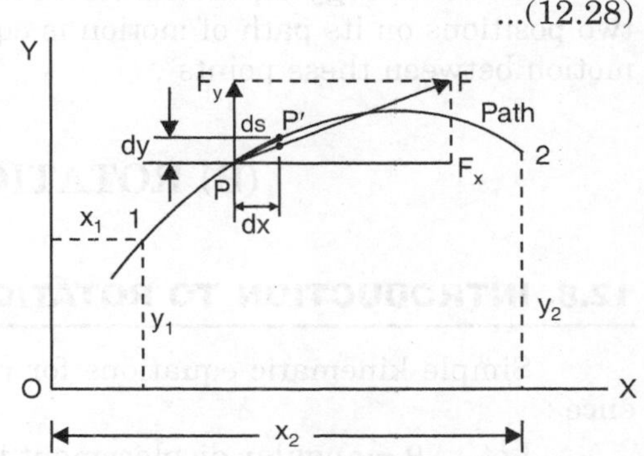

Fig. 12.7.2. Work-energy equation for curvilinear motion rectangular co-ordinates

The total workdone for a displacement

$$Fds = F_x . dx + F_y . dy$$

Then eqn. (12.26) becomes

$$d\left(\frac{W}{g}\frac{v^2}{2}\right) = F_x . dx + F_y . dy \quad ...(12.29)$$

And eqn. (12.28) gives

$$\frac{W}{g}v_2^2 - \frac{W}{g}v_1^2 = \int_{x_1}^{x_2} F_x \,.\, dx + \int_{y_1}^{y_2} F_y \,.\, dy \qquad ...(12.30)$$

Eqn. (12.30) is used when the path is not prescribed

When the path between points 1 and 2 is prescribed, we use different form of work-energy equation. We proceed as follows.

Let displacement of particle at point 1 on the path be S_1 and that for point 2 be S_2, both measured from the origin O [Fig 12.7.3]

Let the force on particle P be F. Component of the force along the tangent to the curve is F_t and F_n is its normal component.

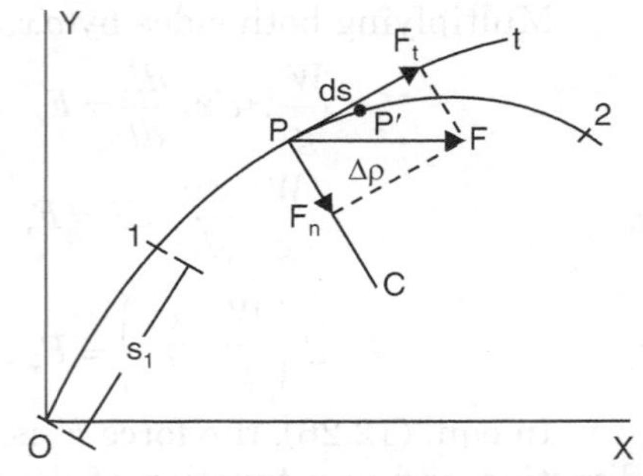

Fig. 12.7.3. Work-energy equation for curvilinear motion in a general form

The normal and tangential components of accelerations of particle are :

$$a_n = v^2/\rho \quad \text{and} \quad a_t = \frac{dv}{dt}$$

where v = velocity of particle and

ρ = radius of curvature of the path.

Then differential equations of motion become :

$$\left.\begin{aligned} \frac{W}{g}\cdot\frac{dv}{dt} &= F_t \\ \text{and} \quad \frac{W}{g}\frac{v^2}{\rho} &= F_n \end{aligned}\right\} \qquad ...(12.31)$$

Then eqn. (12.28) yields $\frac{W}{g}\left(\frac{v_1^2}{2} - \frac{v_0^2}{2}\right) = \int_{s_1}^{s_2} F \,.\, ds$ ...(12.32)

Eqns. (12.28), (12.30) and (12.32) are known as work-energy equation.

Work-energy equation states that "the change in kinetic energy of a particle between two positions on its path of motion is equal to the work of all forces acting upon it during its motion between these points".

(B) ROTATION OF RIGID BODIES

12.8. INTRODUCTION TO ROTATION OF RIGID BODIES

Simple kinematic equations for rotation of a particle are given below for ready reference :

Let θ = angular displacement in radians,

$\dot{\theta}_0 = \omega_0$ = initial angular velocity in radian/second.

$\dot{\theta} = \omega$ = angular velocity after t sec in rad/sec.

$\alpha = \ddot{\theta}$ = angular acceleration in radians/sec^2.

N_0, N = initial and final angular speeds in revolutions/min.

Then, $$\omega_0 = \frac{2\pi N_0}{60},\ \omega = \frac{2\pi N}{60}. \quad ...(12.33)$$

(*a*) When ω_0 is uniform, then displacement $\theta_0 = \omega_0 t$ ...(12.34)

(*b*) When ω_0 is not constant,

angular velocity after t seconds is

$$\omega = \omega_0 + \alpha t \quad ...(12.35)$$

and angular displacement

$$\theta = \text{average angular velocity } \times t$$

or $$\theta = \frac{\omega_0 + \omega}{2} \times t \quad ...(12.36)$$

By using eqn. (12.35), eqn. (12.36) becomes

$$\theta = \frac{\omega_0 + (\omega_0 + \alpha t)}{2} \times t$$

or, $$\theta = \omega_0 t + \frac{1}{2}\alpha t^2 \quad ...(12.37)$$

From eqn. (12.35), we have angular acceleration

$$\alpha = \frac{\omega - \omega_0}{t},$$

or, time $$t = \frac{\omega - \omega_0}{\alpha}$$

∴ Eqn. (12.36) gives

$$\theta = \frac{\omega_0 + \omega}{2} \times t = \frac{\omega_0 + \omega}{2} \times \frac{\omega - \omega_0}{\alpha}$$

or, $$\theta = \frac{\omega_0^2 - \omega^2}{2\alpha}$$

or $$\omega^2 = \omega_0^2 + 2\alpha\theta \quad ...(12.38)$$

when a particle is moving along a circular path of radius r with variable angular velocity ω then its tangential velocity $v_t = 0$, $v = \omega r$ ...(12.39)

and its radial or normal velocity $v_n = 0$;

Also, tangential acceleration $$= a_t = \frac{d}{dt}(v) = \frac{d}{dt}(r\omega) = r\frac{d\omega}{dt} = r\alpha = r\ddot{\theta} \quad ...(12.40)$$

and normal acceleration, $$a_n = \frac{v^2}{r} = \omega^2 r = \dot{\theta}^2 r \quad ...(12.41)$$

Problem 12.14. *Angular displacement θ of a body is given as a function of time by the equation*

$$\theta = a + bt + ct^2.$$

Find the general expressions for the angular velocity and angular acceleration of the body when the initial angular velocity is 2π rad/sec and 2 sec. later it is 6π rad/s and also let initial angular displacement be θ_0.

Sol. Given : $$\theta = a + bt + ct^2 \quad ...(1)$$

Differentiating, we have

$$\frac{d\theta}{dt} = \omega = b + 2ct \quad ...(2)$$

and $$\frac{d^2\theta}{dt^2} = \alpha = 2c \qquad \text{...(3)}$$

At $t = 0$, $\omega = \omega_0 = 2\pi$ rad/s.

$\therefore$ From eqn. (2), we get

$$2\pi = b + 2c \times (0)$$

and $\therefore$ $$b = 2\pi \text{ rad/s}$$

and at $t = 2$ sec ; $$\omega = 6\pi \text{ rad/s,}$$

$\therefore$ From (2) $$6\pi = [b + 2c \times t]_{t=2}$$

$\therefore$ $$6\pi = 2\pi + 2c \times (2)$$

$\therefore$ $$c = \frac{6\pi - 2\pi}{4} = \pi \text{ rad/s}^2$$

$\therefore$ Eqn. (1) yields $$\theta = a + 2\pi t + \pi t^2$$

At $t = 0$; $\theta = \theta_0$, (given)

$\therefore$ We get $$\theta_0 = a + 0 + 0$$

$\therefore$ $$a = \theta_0$$

Hence eqn. (1) yields

$$\boldsymbol{\theta} = \theta_0 + 2\pi t + \pi t^2 \qquad \text{...(4)}$$

$$\omega = \frac{d\theta}{dt} = b + 2ct = 2\pi + 2\pi t$$

$\therefore$ $$\boldsymbol{\omega} = 2\pi + 2\pi t \qquad \text{...(5)}$$

and $$\alpha = \frac{d\omega}{dt} = \ddot{\theta} = 2c$$

or, $$\boldsymbol{\alpha} = 2\pi \qquad \text{...(6)}$$

Problem 12.15. *A circular disc attached to a long rod, as shown in Fig. P.12.15, has torsional oscillations in such a manner that its angular displacement of rotation measured from the mean equilibrium position is given by the equation :*

$$\theta = \theta_0 \cos \omega t.$$

Find (a) general expressions for the angular velocity and acceleration of the disc and also (b) the general expressions for the radial and tangential accelerations of a point on the periphery of radius 10 cm of the disc. (c) Also find the maximum values of these components if the disc makes 4 oscillations/sec and its maximum angular velocity is 4π rad/sec.

Sol. Given : $$\theta = \theta_0 \cos \omega t \qquad \text{...(1)}$$

$\therefore$ $$\dot{\theta} = -\theta_0 \omega \sin \omega t \qquad \text{...(2)}$$

and, $$\ddot{\theta} = -\dot{\theta}_0 \omega^2 \cos \omega t \qquad \text{...(3)}$$

At the radius r, $$a_t = r\ddot{\theta} = -\theta_0\, \omega^2\, r \cos \omega t \qquad \text{...(4)}$$

and $$a_n = \dot{\theta}^2\, r = (-\theta_0\, \omega \sin \omega t)^2 r = \theta_0^{\,2}\, \omega^2\, r \sin^2 \omega t \qquad \text{...(5)}$$

The given conditions are :

(*i*) That the disc makes 4 oscillations per second.

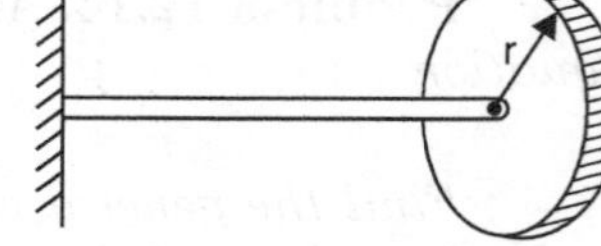

Fig. P. 12.15

$\therefore$ Time period, $$\tau = \frac{2\pi}{\omega} = \frac{1}{4} \text{ sec.}$$

$\therefore$ $$\omega = \frac{2\pi}{\tau} = \frac{2\pi \times 4}{1} = 8\pi \text{ rad/sec.}$$

(*ii*) That, the maximum angular velocity is equal to 4π rad/s.

From eqn. (2), we get

$$\dot{\theta}_{max} = \theta_0 \cdot \omega = 4\pi \text{ rad/sec.}$$

$$\therefore \qquad \theta_0 = \frac{4\pi}{\omega} = \frac{4\pi}{8\pi} = \frac{1}{2} \text{ radians.}$$

(*a*) Substituting values of θ_0 and ω into eqns. (2) and (3), we get

$$\dot{\theta} = -\left(\frac{1}{2}\right)(8\pi) \sin 8\pi t$$

or, $$\dot{\theta} = -4\pi \sin 8\pi t. \quad \textbf{Ans.} \qquad ...(6)$$

and $$\ddot{\theta} = -\left(\frac{1}{2}\right)(8\pi)^2 \cos 8\pi t$$

or $$\ddot{\theta} = -32\,\pi^2 \cos 8\pi t. \quad \textbf{Ans.} \qquad ...(7)$$

(*b*) From eqns. (4) and (5), we get

$$a_t = -\left(\frac{1}{2}\right)(8\pi)^2 \times (10) \cos 8\pi t$$

or $$a_t = -320\,\pi^2 \cos 8\pi t \text{ cm/s}^2. \quad \textbf{Ans.} \qquad ...(8)$$

and $$a_n = \left(\frac{1}{2}\right)^2 (8\pi)^2 \times (10) \sin^2 (8\pi)t$$

or $$a_n = 160\,\pi^2 \sin^2 8\pi t \text{ cm/s}^2. \quad \textbf{Ans.} \qquad ...(9)$$

(*c*) $$a_t)_{max} = \theta_0 \omega^2 r = -320\,\pi^2 \text{ cm/s}^2. \quad \textbf{Ans.}$$

and $$a_n)_{max} = \theta_0^2 \omega^2 r = 160\,\pi^2 \text{ cm/s}^2. \quad \textbf{Ans.}$$

12.9. EQUATION OF MOTION FOR RIGID BODY ROTATION

By the use of D' Alembert's principle, we shall derive the equation of motion for the rigid body rotating about a fixed axis. We think a rigid body as a system of rigidly connected particles and apply to each particle its inertia force. These inertia forces and all external forces acting on the body form a system of forces in equilibrium. Internal forces (actions and reactions between various particles) always occur in pairs of equal and opposite collinear forces and are themselves balanced.

Hence *the algebraic sum of moments of external forces and inertia forces about the axis of rotation is zero.*

Consider an element of mass *dm* at a radius *r* from the axis of rotation [Fig. 12.9] and find the moment of the tangential force about the axis of rotation. The tangential force = $a_t.dm$ = $(r\ddot{\theta})\, dm$ and the inertia force = $-(r\ddot{\theta})\, dm$. Its moment = $(-r\ddot{\theta}\, dm)\, r$. Thus, by summing the moments equal to zero we get

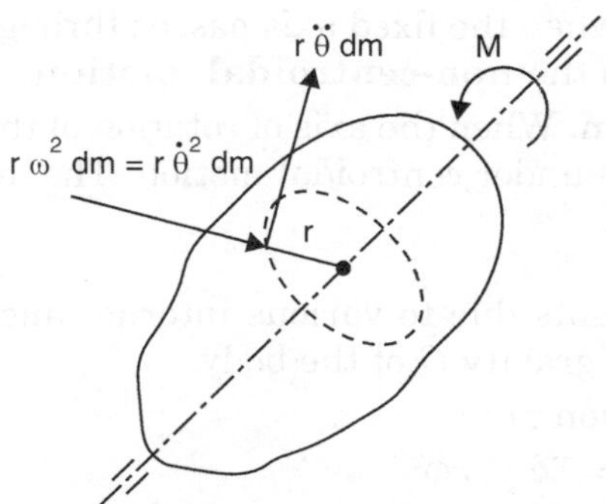

Fig. 12.9. Equation of motion for rigid body rotation.

$$\Sigma M - \int_m r^2 \ddot{\theta} \, dm = 0 \qquad ...(12.42)$$

Since, the angular acceleration $\ddot{\theta}$ is the same for all particles, Eqn. (12.42) yields

$$\Sigma M - \ddot{\theta} \int r^2 \, dm = 0$$

But $\int r^2 \, dm$ = mass moment of inertia I with respect to the axis of rotation.

∴ We get $\Sigma M - I\ddot{\theta} = 0$

or, $$\boxed{\Sigma M = I\ddot{\theta}} \qquad ...(12.43)$$

12.10. MOMENT OF MOMENTUM OR ANGULAR MOMENTUM

Consider a rigid body of mass m rotating about a fixed axis normal to the plane of paper and passing through O, Fig 12.10. Let the angular velocity be ω. Consider an element of mass dm at a distance r from O. The linear velocity of the element is $r\omega = v$.

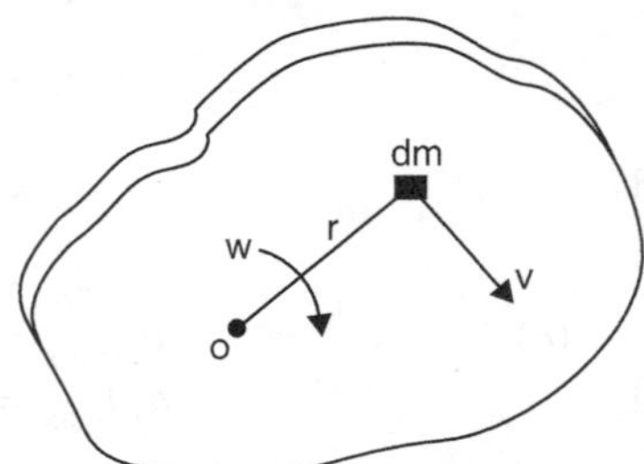

Fig. 12.10. Moment of momentum or angular momentum

Momentum of the element about O

$$= (dm)v = (dm)r\omega$$

Moment of momentum of the element,

$$dH_0 = r \times [dm \, . \, r\omega] = dm \, r^2\omega$$

Moment of momentum of the whole body about O

$$H_0 = \int dH_0 = \int dm \times (r^2 \, \omega) = \omega \int r^2 \, dm$$

But $\int r^2 \, dm = I_0$

Hence moment of momentum about O or the angular momentum = $\boxed{H_0 = I_0\omega}$ ...(12.44)

where $I_0 = mk_0^{\,2}$ ...(12.45)

and k_0 is called radius of gyration with respect to O.

12.11. CENTROIDAL AND NON-CENTROIDAL ROTATION

As discussed earlier, a rigid body has pure rotation when its all particles move in a circular path about a fixed axis normal to the plane of motion. When the fixed axis passes through the centre of gravity or the mass centre of the body, the motion is known as **centroidal rotation.** On the other hand, when the fixed axis passes through a point other than the mass-centre, the motion is known as the **non-centroidal motion.**

(*a*) **Centroidal Rotation.** When the axis of rotation of the rigid body passes through its centre of gravity G, the body is under centroidal motion. The eqn. (12.43) will become

$$\Sigma M_G = I_G \, . \, \alpha = I_G \, . \, \ddot{\theta} \qquad ...(12.46)$$

where ΣM_G = sum of all moments due to various internal and external forces acting on the body taken about the centre of gravity G of the body.

Other equations of motion are :

$$\Sigma F_n = \Sigma \delta_m \, . \, r\omega^2 \qquad ...(12.47)$$

and $$\Sigma F_t = \Sigma \delta_m \, . \, r\alpha \qquad ...(12.48)$$

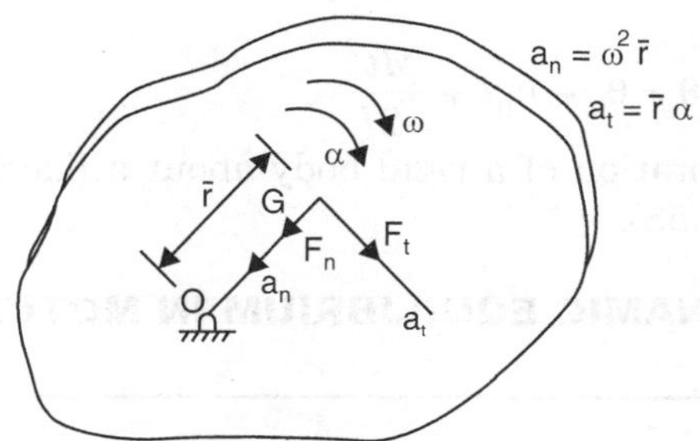

Fig. 12.11. Non-centroidal rotation of a rigid body

(*b*) **Non-Centroidal Rotation.** Let a rigid body have a motion of rotation about an axis through point O, which is at a distance $\bar{r}$ from the centre of gravity G of the body, Fig. 12.11. Let ΣF_t and ΣF_n be the sum of external forces along the tangential and normal directions, respectively. Let I_G be the mass moment of inertia with respect to the centre of gravity. Then the equations of motion are as given below :

$$\Sigma F_t = ma_t = m\bar{r}\alpha \qquad ...(12.49)$$

$$\Sigma F_n = ma_n = m\omega^2\bar{r} \qquad ...(12.50)$$

and $$\Sigma M_G = I_G \times \alpha \qquad ...(12.51)$$

The moment of force about point O is

$$M_0 = (m\bar{r}\alpha)\bar{r} + \bar{I}_G \times \alpha = \alpha(m\bar{r}^2 + \bar{I}_G)$$

or $$M_0 = I_0\,\alpha \qquad ...(12.52)$$

In eqn. (12.52) I_0 can be obtained by using the parallel axis of inertia :

$$I_0 = (I_G + m\bar{r}^2) \qquad ...(12.53)$$

12.12. ROTATION UNDER THE ACTION OF A CONSTANT MOMENT

Here the moment M of external force with respect to the axis of rotation is given and it is required to derive an expression for the angle of rotation θ as a function of time. In the particular case where the moment of all external forces acting on the body with respect to the axis of rotation is constant the necessary integration of equation

$$I\ddot{\theta} = M, \qquad ...(12.54)$$

can be made. Thus, We get,

$$I\dot{\theta} = Mt + C_1 \qquad ...(12.55)$$

Constant of integration C_1 can be found by using the initial conditions of motion. If, for example we have $\dot{\theta} = \dot{\theta}_0$ when $t = 0$, then we get $C_1 = I\dot{\theta}_0$ and eqn. (12.55) gives

$$\dot{\theta} = \frac{M}{I}t + \dot{\theta}_0 \qquad ...(12.56)$$

Integrating further, we get

$$\theta = \frac{M}{I}\frac{t^2}{2} + \dot{\theta}_0 t + C_2 \qquad ...(12.57)$$

To find C_2, we use the initial condition : $\theta = \theta_0$ at $t = 0$

$\therefore$ $$C_2 = \theta_0$$

$\therefore$ Eqn. (12.57) becomes

$$\theta = \frac{Mt^2}{2I} + \dot{\theta}_0 t + \theta_0$$

or $$\theta = \theta_0 + \dot{\theta}_0 t + \frac{Mt^2}{2I} \quad ...(12.58)$$

Various problems of rotation of a rigid body about a fixed axis can be solved by using eqns. (12.56), (12.57) and (12.58).

12.13. EQUATION OF DYNAMIC EQUILIBRIUM IN MOTION OF ROTATION ABOUT A FIXED AXIS

Writing eqn. (12.54) ; $M = I\ddot{\theta}$, governing equation of rotation of a rigid body about a fixed axis, in the form :

$$M - I\ddot{\theta} = 0 \quad ...(12.59)$$

We conclude that if we apply a couple . $I\ddot{\theta}$ to the rotating body in addition to the moment of the active or given forces, we get the equation of ***dynamic equilibrium***. In eqn. (12.59) the couple $-I\ddot{\theta}$ which balances or neutralises the moment M of the active forces is called the *inertia couple*.

As an example, consider the circular disc shown in Fig. 12.13 which rotates about the geometric axis through O by a falling weight $W = mg$ attached to a string wound on a shaft of radius r. Let $\ddot{\theta}$ be the angular acceleration of the disc. The falling weight W has a vertical acceleration $a_y = r\ddot{\theta}$ if the string is inextensible. The inertia force on the weight is $mr\ddot{\theta} = -\frac{W}{g} r\ddot{\theta}$, and the inertia couple on the disc is $-I\ddot{\theta}$, as shown in Fig. 12.13.

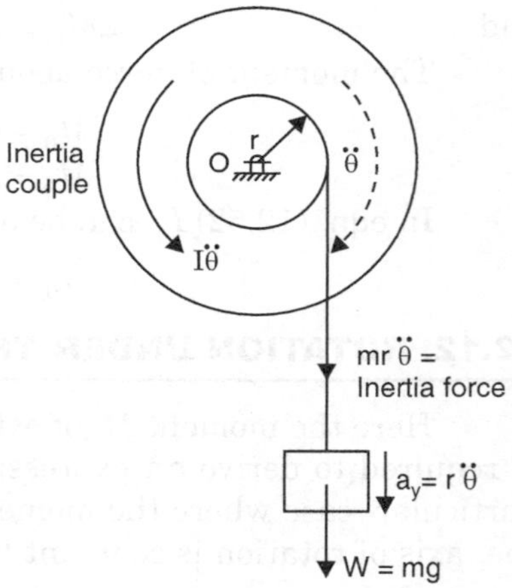

Inertia couple

O

r

$\ddot{\theta}$

$I\ddot{\theta}$

$mr\ddot{\theta}$ = Inertia force

$a_y = r\ddot{\theta}$

W = mg

Fig. 12.13. Equation of dynamic equilibrium for rotation about a fixed axis

By equating to zero the algebraic sum of moments, with respect to the point O, of all forces (including the inertia force and the inertia couple), we get [by $\Sigma M_0 = 0$]

$$Wr - \frac{W}{g} r^2\ddot{\theta} - I\ddot{\theta} = 0$$

or $$\left(I + \frac{W}{g} r^2\right)\ddot{\theta} = Wr \quad ...(12.60)$$

where I = moment of inertia of the disc with respect to the axis of rotation.

We note that eqn. (12.60) has the same form as eqn. (12.43) if we write the quantity

$$I_e = \left(I + \frac{W}{g} r^2\right) \quad ...(12.61)$$

as an *equivalent moment of inertia* of the system of Fig 12.13.

That is, the system behaves as a body of fictitious or imaginary moment of inertia I_e rotating about a fixed axis under the action of a constant external moment Wr, and eqns. (12.56) and (12.58) may be used directly to find the angular velocity and angular displacement as functions of the time t.

Problem 12.16. *A cord is wrapped around a thin homogeneous disc of mass m = 10 kg and radius 30 cm. If the cord is pulled upward with a force T = 200 N, find (a) acceleration of the centre of the disc (b) angular acceleration of the disc, and (c) the acceleration of the cord. See Fig. P 12.16.*

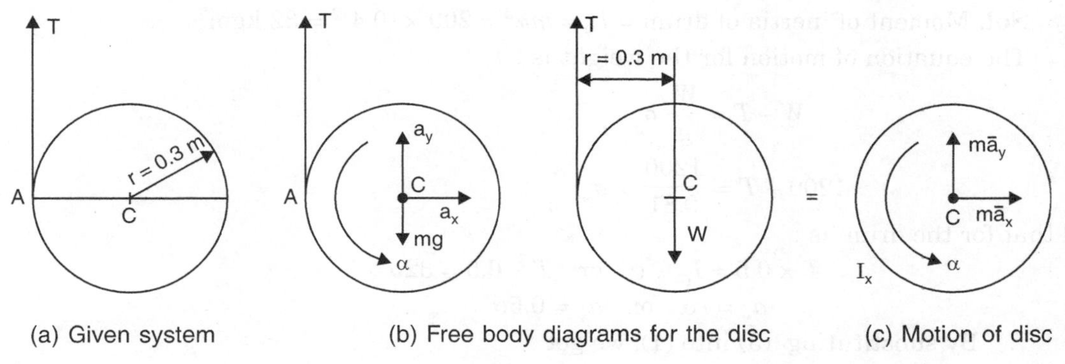

(a) Given system (b) Free body diagrams for the disc (c) Motion of disc

Fig. P. 12.16

Sol. (*a*) **Acceleration of disc.** Let the acceleration of *C.G.* of the disc be as shown in Fig. P.12.16 (*b*) and $\bar{a}_x$ and $\bar{a}_y$ of centre of gravity be $\bar{a}_x$ and $\bar{a}_y$, as shown.

For the disc the equations of motion are by $\Sigma F_x = m\bar{a}_x$, ∴ $0 = m\bar{a}_x$, ∴ $\bar{a}_x = 0$

by $\Sigma F_y = m\ \bar{a}_y$, we get $T\text{-}mg = m\bar{a}_y$ [∴ $T = m\ (\bar{a}_y + g)$]

$$\therefore \qquad \boxed{\bar{a}_y = \frac{T - mg}{m}}$$

Since $T = 200$ N ; $m = 10$ kg and $W = mg = 10 \times 9.81 = 98.1$ N

$$\therefore \qquad \bar{a}_y = \frac{200 - 10 \times 9.81}{10} = +\frac{101.9}{10} \text{ m/s}^2, \quad \therefore \quad \bar{a}_y = \mathbf{10.10\ m/s^2} \uparrow. \quad \textbf{Ans.}$$

(*b*) **Angular acceleration of disc**

By $\Sigma M_G = I.\alpha$, we get

$$-T \times r = \left(\frac{1}{2}mr^2\right) \times \alpha$$

$$\therefore \qquad \alpha = \frac{-2Tr}{mr^2} = \frac{-2T}{mr} = \frac{-2 \times 200}{10 \times (0.3)}$$

$$= -133.33 \text{ rad/sec}^2$$

$$\therefore \qquad \alpha = -133.33 \text{ rad/sec}^2.$$

(*c*) **Acceleration of cord** See Fig. P.12.16(*d*)

Acceleration of cord, $a_{cord} = a_A)_t \uparrow + (a_{AG})_t \uparrow$

$= 10.19 \text{ m/s}^2 + r\alpha \uparrow$

$= 10.19 \uparrow + 0.3 \times 133.33 \uparrow \text{m/s}^2$

$= 50.189 \text{ m/s}^2$. **Ans.**

Fig. P. 12.16 (d). Acceleration of cord

Problem 12.17. *A drum of mass 200 kg and radius of gyration 0.4 m is shown in the Fig. P.12.17 (a).*

A weight of 1.2 kN is attached to one end of the rope whose other end is wrapped around the drum. Determine :

(a) the acceleration of the weight,

(b) tension in rope, and

(c) angular acceleration of the drum.

Neglect friction in the bearings of the drum.

Sol. Moment of inertia of drum = $I_G = mk^2 = 200 \times (0.4)^2 = 32$ kgm^2

The equation of motion for the weight is :

$$W - T = \frac{W}{g}\, a_y$$

or

$$1200 - T = \frac{1200}{9.81} \times a_y \quad \text{...(1)}$$

and that for the drum is :

$$T \times 0.5 = I_G \,.\, \alpha \quad \text{or} \quad T \times 0.5 = 32\alpha \quad \text{...(2)}$$

but

$$a_y = r\alpha \quad \text{or} \quad a_y = 0.5\alpha \quad \text{...(3)}$$

$\therefore$ By substituting (3) into (1), we get

$$1200 - T = \frac{1200}{9.81} \times 0.5\alpha$$

$$\therefore \quad 1200 - T = 61.16\alpha \quad \text{...(4)}$$

Multiplying (2) by 2 and adding to (4), we get

$$(1200 - T) + (0.5 \times T) \times 2 = 61.16\,\alpha + (32\alpha) \times 2$$

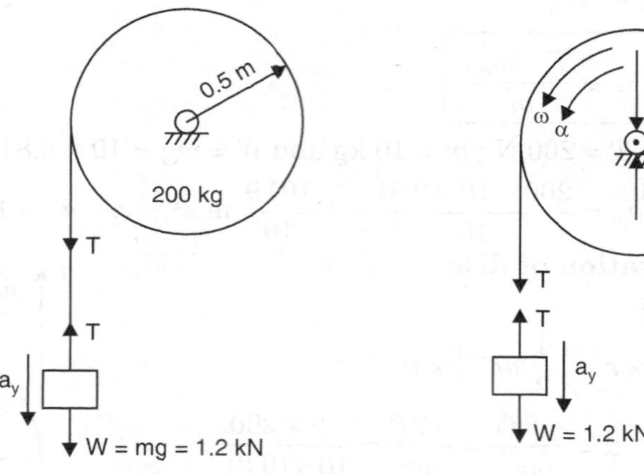

(a) Given system (b) Free body diagrams

Fig. P. 12.17

$$\therefore \quad 1200 = 125.16\,\alpha$$

$$\therefore \quad \alpha = \mathbf{9.59\ rad/s^2}. \quad \textbf{Ans. (c)}$$

From (3), acceleration of weight

$$a_y = 0.5\,\alpha = 0.5 \times 9.59 = \mathbf{4.795\ m/s^2}. \quad \textbf{Ans. (a)}$$

And from (4), tension in rope

$$T = 1200 - 61.16\,\alpha = 1200 - 61.16 \times 9.59 \text{ N}$$

$$= \mathbf{613.48\ N} \quad \textbf{Ans.}$$

Problem 12.18. *A drum of mass 200 kg and radius of gyration 0.4 m is shown in Fig. P.12.18 (a). A weight of 1.2 kN is attached to one end of the rope whose other end is wrapped around the drum. The hub diameter is 0.06 m and the coefficient of friction in bearing is 0.12. Determine :*

(*a*) *the acceleration of the weight,*

(*b*) *tension in the rope and*

(*c*) *the angular acceleration of drum.*

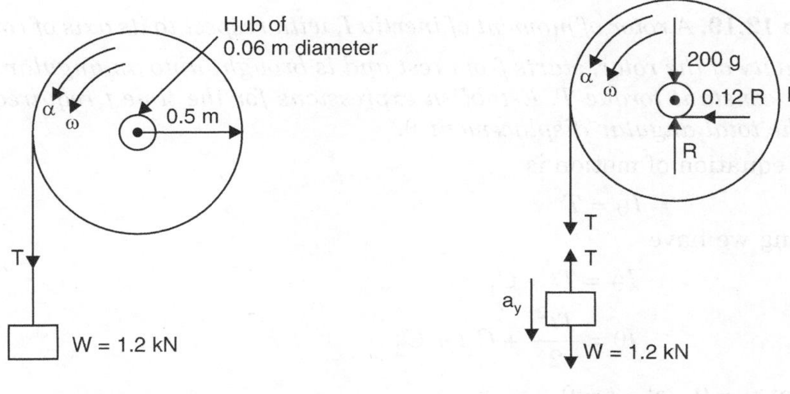

Fig. P. 12.18

Sol. Let R be the normal reaction of the bearing to the drum. Then by $\Sigma F_y = 0$ for the drum, we have

$$R - (T + 200 \times 9.81) = 0 \quad \text{...(1)}$$

And force of friction, $F = 0.12\,R = (T + 200 \times 9.81) \times 0.12$

or, $$F = 0.12\,T + 235.44 \quad \text{...(2)}$$

Equation of motion of weight :

By $\Sigma F_y = ma_y$, we get

$$1200 - T = \frac{1200}{9.81} \times a_y \quad \text{...(3)}$$

For the drum, by using $\Sigma M_G = I_G\,\alpha$, we get

$$T \times 0.5 - F \times 0.03 = I_G \times \alpha$$

or, $$0.5T - (0.12\,T + 235.44) \times 0.03 = \{200 \times (0.4)^2\}\alpha$$

or $$0.4964\,T - 7.0632 = 32\alpha \quad \text{...(4)}$$

But, $$a_y = \alpha \times r$$

$\therefore$ $$a_y = \alpha \times 0.5 \quad \text{...(5)}$$

By substituting (5) into (3), we get

$$1200 - T = \frac{1200}{9.81} \times 0.5\alpha$$

$\therefore$ $$1200 - T = 61.16\alpha \quad \text{...(6)}$$

By (4) + (6) × 0.4964, we get

$$(0.4964\,T - 7.0632) + 0.4964\,(1200 - T) = 32\alpha + (61.16\alpha) \times 0.4964,$$

i.e.,
$$-\,7.0632 + 0.4964 \times 1200 = (32 + 61.16 \times 0.4964)\alpha$$

or
$$588.62 = 62.36\alpha$$

$\therefore$ $\alpha = \mathbf{9.44\ rad/s^2}$. **Ans. (c)**

From (5), $a_y = 0.5 \times 9.44 = 4.72\ \text{m/s}^2$. **Ans. (a)**

And, from (6), $T = 1200 - 61.16\,\alpha$

$= 1200 - 61.16 \times 9.44 = \mathbf{622.65\ N.}$ **Ans.**

Problem 12.19. *A rotor of moment of inertia I, with respect to its axis of rotation through the centre of gravity of the rotor, starts from rest and is brought upto an angular speed $\dot{\theta}$ by the application of a constant torque T. Establish expressions for the time t required to reach this speed and for the total angular displacement θ.*

Sol. The equation of motion is

$$I\ddot{\theta} = T \qquad ...(1)$$

Integrating we have

$$I\dot{\theta} = Tt + C_1 \qquad ...(2)$$

and
$$I\theta = \frac{Tt^2}{2} + C_1 t + C_2 \qquad ...(3)$$

Given that $\dot{\theta} = 0$ at $t = 0$,

$\therefore$ From eqn. (2) we get

$$I\dot{\theta}_0 = T(0) + C_1 = 0$$

$\therefore$ $C_1 = 0.$

$\therefore$ Eqn. (2) becomes $I\dot{\theta} = T \,.\, t \quad \therefore \quad t = \dfrac{I\dot{\theta}}{T}$. **Ans.** ...(4)

Now since $\theta = \theta_0 = 0$ at $t = 0$,

$\therefore$ From eqn. (3), we get

$$I\theta_0 = \frac{T}{2}(0)^2 + 0 + C_2 = 0, \quad \therefore \quad C_2 = 0.$$

Hence eqn. (3) becomes

$$I\theta = \frac{Tt^2}{2} \quad \therefore \quad \theta = \frac{Tt^2}{2I} \qquad ...(5)$$

Using eqn. (4), eqn. (5) gives

$$\theta = \frac{T}{2I} \times \left(\frac{I\dot{\theta}}{T}\right)^2 = \frac{I\dot{\theta}^2}{2T}. \quad \textbf{Ans.} \qquad ...(6)$$

Note. If it is given that a rotor with initial angular speed $\dot{\theta}_0$ is brought to rest with the application of a constant torque T, then above eqns. (4) and (6) reduce to

$$t = \frac{I}{T}\dot{\theta}_0 \qquad ...(i)$$

and
$$\theta = \frac{I\dot{\theta}_0^{\,2}}{2T} \qquad ...(ii)$$

respectively.

Problem 12.20. *A cage of mass 2000 kg is raised by a rope wound over a drum of mass 600 kg and radius of gyration of 0.6 m and diameter 1.5 m. The drum is rotated by a 25 kW electric motor at a speed of 15 rpm. Assume the rope to be tight when the drum begins to rotate and neglecting the friction and the self weight of the rope, find :*

(a) tension in rope,

(b) acceleration of rope, and

(c) the time required for an ascend of 20 m of the cage.

Sol. Fig. P.12.20 shows the system.

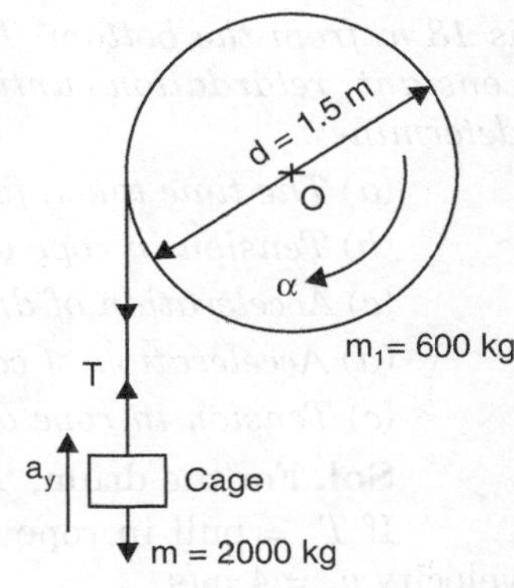

Fig. P.12.20

Equation of the motion of the cage is written by using $\Sigma F_x = ma_y$.

$\therefore$ We get

$$T - 2000\,g = 2000\,a_y \quad \text{...(1)}$$

For the drum, we have

$$\Sigma M_G = I_G \, . \, \alpha \quad \text{...(2)}$$

where $I_G = mk^2 = 600 \times (0.6)^2 \text{ kg-m}^2 = 216 \text{ kg-m}^2$

Also, $a_y = \alpha \times r = \dfrac{1.5\alpha}{2}$ m/s^2

$$\text{Power} = \frac{2\pi N}{60} \times M_t$$

$$\therefore \quad M_t = \frac{60 \times \text{Power}}{2\pi N} = \frac{60 \times (25 \times 1000)}{2\pi \times 15} = \frac{50{,}000}{\pi} \text{ Nm}$$

Also $\Sigma M_G = M_t - T \times \dfrac{1.5}{2} = \dfrac{50{,}000}{\pi} - 0.75\,T$

$$\therefore \quad \text{Eqn. (2) becomes, } \frac{50{,}000}{\pi} - 0.75\,T = 216\,\alpha \quad \text{...(3)}$$

And eqn. (1) gives

$$T - 2000\,g = 2000 \times (0.75\,\alpha) \quad \text{...(4)}$$

Multiplying (4) by 0.75 and adding (3), we get

$$(T - 2000\,g) \times 0.75 + \left(\frac{50{,}000}{\pi} - 0.75\,T\right) = (2000 \times 0.75\,\alpha) \times 0.75 + 216\,\alpha$$

or,

$$-1500\text{ g} + \frac{50{,}000}{\pi} = \alpha(1125 + 216)$$

$$\therefore \quad \alpha = \frac{\dfrac{50000}{\pi} - 1500 \times 9.81}{1341} = \frac{1200.494}{1341} = 0.895 \text{ rad/s}^2. \quad \textbf{Ans.}$$

$\therefore$ Acceleration of rope $= a_y = \alpha r = 0.895 \times 0.75$ m/s^2 = **0.67125 m/s²**. **Ans. (b)**

And from (4), tension in rope

$$= T = 2000 \times 0.75\,\alpha + 2000\,g$$
$$= 2000(0.75 \times 0.895 + 9.81) \text{ N}$$
$$= 20{,}962.5 \text{ N} = \textbf{20.9625 kN}. \quad \textbf{Ans. (a).}$$

Using the relation

$$S = ut + \frac{1}{2}at^2, \text{ we get}$$

$$20 = 0 + \frac{1}{2} \times 0.67125 \times t^2$$

$$\therefore \qquad t = \sqrt{\frac{20 \times 2}{0.67125}} = \textbf{7.72 seconds.} \quad \textbf{Ans. (c).}$$

Problem 12.21. *A cage of mass 2000 kg is attached to a wire wound round drum of 1 m diameter. The drum has a breaking attachment with the combined mass of 600 kg and radius of gyration of 0.6 m. The cage starts from rest and is allowed to descend until it attains a velocity of 4 m/s. The brake is then used and the speed is maintained constant until the cage is 18 m from the bottom. The brake is now tightened further so that the cage moves under a constant retardation until it comes to rest at the bottom. If the total displacement is 60 m, determine :*

(*a*) *The time taken for the descent. (Take g = 9.8 m/s*^{2}*)*

(*b*) *Tension in rope during free descent of cage.*

(*c*) *Acceleration of drum during free descent of cage.*

(*d*) *Acceleration of cage during constant retardation period.*

(*e*) *Tension in rope during retardation period.*

Sol. For the drum, $I_G = mk^2 = 600 \times (0.6)^2 = 216$ kg-m^2.

If T_1 = pull in rope during the first stage of motion of pulley when the cage attains a velocity v_1 = 4 m/s,

Let displacement = y_1, and t_1 = time for the first stage.

Let y_2, t_2 be the distance travelled and time taken during the second stage of descent at the constant speed of 4 m/s.

Let y_3, t_3 be the distance travelled and time taken during the third stage of motion under retardation. Here y_3 = 18 m.

And $$y_1 + y_2 + y_3 = 60 \text{ m} \qquad \text{...(1)}$$

For the first stage : The equation of motion of the cage is :

$$mg - T_1 = ma_1 \quad \text{where } a_1 = \alpha_1 \cdot r$$

or $$2000 \times 9.8 - T_1 = 2000 \times a_1 \qquad \text{...(2)}$$

For the drum, $$\Sigma M = \bar{I}_G \times \alpha,$$

or $$T_1 \times r = \bar{I}_G \times \frac{a_1}{r}$$

$$\therefore \qquad T_1 \times r^2 = 216 \times a_1 \quad \text{where } r = \frac{1.0}{2} = 0.5 \text{ m}$$

or $$T_1 \times (0.5)^2 = 216\, a_1$$

or, $$T_1 = \frac{216}{(0.5)^2} \times a_1 = 864\, a_1 \qquad \text{...(3)}$$

Adding (2) and (3), we get

$$2000 \times 9.8 = 2000 a_1 + 864 a_1$$

$$\therefore \qquad a_1 = \frac{2000 \times 9.8}{2864} = \textbf{6.844 m/s}^2$$

$$\therefore \qquad \alpha_1 = \frac{a_1}{r} = \frac{6.844}{0.5} = \textbf{13.688 rad/s}^2. \quad \textbf{Ans. (c)}$$

Tension in rope, $$T_1 = 864\, a_1 = 864 \times 6.844 \text{ N} = \textbf{5913.216 N.} \quad \textbf{Ans. (b).}$$

Since $v_1 = a_1 t_1$ during first stage of motion

$$\therefore \quad t_1 = \frac{v_1}{a_1} = \frac{4}{6.844} = \mathbf{0.5845\ seconds}$$

And, $$y_1 = \frac{v_1}{2} \times t = \frac{4}{2} \times 0.5845 \text{ m} = \mathbf{1.1690\ m.}$$

Using eqn. (1), we have $y_1 + y_2 + y_3 = 60$

$$\therefore \quad 1.1690 + y_2 + 18 = 60$$

$$\therefore \quad y_2 = 60 - 19.169 = 40.831 \text{ m.}$$

$$t_2 = \frac{y_2}{v_2} = \frac{40.831}{4} = \mathbf{10.208\ seconds}$$

$v = u + at$

$\therefore \quad 0 = 4 + at$

In third stage of motion of cage,

$y_3 = 18$ m, and using equation v $\quad y_3 = \frac{4+0}{2} \times t_3$

we get $$t_3 = \frac{y_3 \times 2}{(4+0)} = \frac{y_3 \times 2}{4} = \frac{18 \times 2}{4} = 9 \text{ seconds}$$

$\therefore$ Total time of descent $= t = t_1 + t_2 + t_3$

$= 0.5845 + 10.208 + 9$ seconds $=$ **19.7925 seconds. Ans. (a)**

Retardation a_3 during the thirds stage is given by $v = u + at$,

$$\therefore \quad 0 = 4 + a_3 \times t_3$$

$$\therefore \quad a_3 = \frac{-4}{t_3} = \frac{-4}{9} \text{ m/s}^2$$

or $$a_3 = \frac{4}{9} \text{ m/s}^2 \text{ (retardation).}$$ **Ans. (d)**

Equation of motion of cage during third stage is :

$$T_3 - 2000g = 2000a_3$$

$$\therefore \quad T_3 = 2000 \times (a_3 + g)$$

$$= 2000 \times \left(-\frac{4}{9} + 9.80\right) \text{ N} = \mathbf{18711.1\ N.}$$ **Ans. (e)**

Problem 12.22. *Two weights W_1 and W_2 of 9 kN and 5 kN are attached at the ends of a rope passing over a pulley of 1.2 m diameter and weighing 7.2 kN. The radius of gyration of the pulley is 0.5 m. Find torque which must be applied to the pulley to raise the weight W_1 with an acceleration of 1.2 m/s². Neglect friction in the bearing and slip of rope over the pulley. Also find tensions in the two ends of rope.*

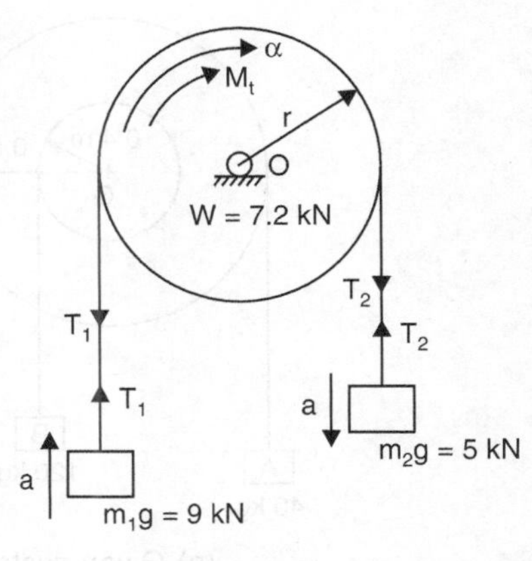

Fig. P. 12.22

Sol. Fig. P.12.22 shows the system of the two weights and the pulley. Let a be the acceleration of W_1, and W_2. Let T_1 and T_2 be the tensions in the two ends of the rope.

Equations of motion :

For W_1 : $\quad T_1 - W_1 = m_1 a \quad$ ...(1)

For W_2 : $\quad W_2 - T_2 = m_2 a \quad$ ...(2)

For pulley $\quad \Sigma M_G = I_G \alpha$

or, $$M_t + T_2 \times r - T_1 r = \left(\frac{W}{g}k^2\right)\alpha \quad \text{or,} \quad M_t - (T_1 - T_2)\, r = \frac{W}{g} \times k^2 \times \alpha$$

where $$\alpha = \frac{a}{r}$$

$\therefore$ We have $$M_t - (T_1 - T_2)r = \frac{W}{g}k^2 \times \frac{a}{r}$$

or, $$M_t - (T_1 - T_2) \times 0.6 = \frac{7.2}{9.8} \times (0.5)^2 \times \frac{a}{0.6}$$

or, $$M_t - (T_1 - T_2) \times 0.6 = 0.306\, a \qquad ...(3)$$

Adding (1) and (2), we have

$$(T_1 - T_2) - (W_1 - W_2) = (m_1 + m_2)\, a$$

$\therefore$ $$T_1 - T_2 = (m_1 + m_2)\, a + (W_1 - W_2)$$

$$= \frac{9+5}{9.8}\, a + (9 - 5)$$

or, $$T_1 - T_2 = 1.43\, a + 4 \qquad ...(4)$$

$\therefore$ Eqn. (3) gives

$$M_t - (1.43\, a + 4) \times 0.6 = 0.306\, a$$

$\therefore$ $$M_t = \{0.306 + 1.43 \times 0.6\} \times a + 0.6 \times 4$$

or, $$M_t = 1.164\, a + 2.4 \qquad ...(5)$$

For $a = 1.2$ m/s^2 (given)

$$M_t = 1.164 \times 1.2 + 2.4 = \mathbf{3.7968\ kNm}$$

$$= \mathbf{3796.8\ Nm.}\ \ \textbf{Ans.}$$

From (1), $$T_1 = m_1 a + W_1 = \frac{9}{9.8} \times 1.2 + 9 = \mathbf{10.10\ kN.}\ \ \textbf{Ans.}$$

From (2), $$T_2 = W_2 - m_2 a = 5 - \frac{5}{9.8} \times 1.2 = \mathbf{4.388\ kN.}\ \ \textbf{Ans.}$$

Problem 12.23. *Two masses m_A and m_B are connected to the ends of an inextensible rope wrapped around a compound pulley as shown in Fig. P.12.23 (a). The mass moment of inertia of the pulley is 20 kg-m^2. Determine the acceleration of each mass and tensions in the rope.*

Sol. Fig. P. 12.23(*b*) shows the free body diagram of m_A, m_B and the pulley. Equations of motion are :

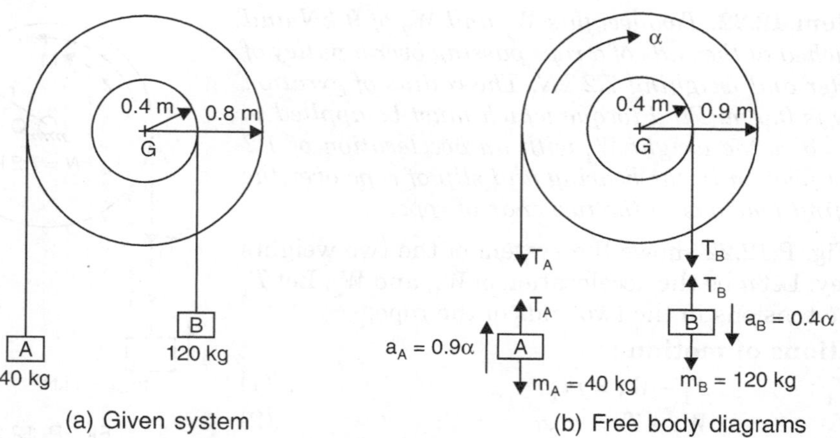

(a) Given system (b) Free body diagrams

Fig. P. 12.23

For mass A : $T_A - 40g = 40\, a_A = 40 \times (0.9\alpha)$

or, $T_A = 36\alpha + 40g$ ...(1)

For mass B : $120g - T_B = 120\, a_B = 120 \times (0.4 \times \alpha)$

or, $T_B = 120g - 48\alpha$ ...(2)

For pulley : By $\Sigma M_G = I_G \cdot \alpha$

$T_B \times 0.4 - T_A \times 0.9 = 20\alpha$ ...(3)

Substituting values of T_A and T_B into eqn. (3), we get

$$(120g - 48\,\alpha) \times 0.4 - (36\,\alpha + 40g) \times 0.9 = 20\alpha$$

or, $\alpha(20 + 48 \times 0.4 + 36 \times 0.9) = g(120 \times 0.4 - 40 \times 0.9)$

or, $71.6\,\alpha = 12g$ for $g = 9.80$ m/s^2

$\therefore$ $\alpha = \mathbf{1.642\ rad/s^2}$. **Ans.**

$\therefore$ $a_A = 0.9\alpha = \mathbf{1.4778\ m/s^2}$. **Ans.**

and $a_B = 0.4\alpha = \mathbf{0.6568\ m/s^2}$. **Ans.**

From eqn. (1), $T_A = 36\alpha + 40g = 36 \times 1.642 + 40 \times 9.8$ N

or, $T_A = \mathbf{451.112\ N}$. **Ans.**

From eqn. (2) $T_B = 120g - 48\alpha = 120 \times 9.8 - 48 \times 1.642$ N $= \mathbf{1097.184\ N}$. **Ans.**

Problem 12.24. *A turbine rotor of mass 2000 kg and radius of gyration of 0.75 m starts from rest and attains a speed of 6000 rpm in 6 minutes under the action of a steady torque of 2500 Nm. Find the constant frictional torque at the bearings of the rotor. Also find the time to come to rest when the steam supply is cut off.*

Sol. Initial angular speed $= \omega_0 = 0$

Final angular speed $= \omega = \dfrac{2\pi \times 6000}{60} = 200\,\pi$ rad/s

Using $\omega = \omega_0 + \alpha t$, we get

$$\alpha = \frac{\omega - \omega_0}{t} = \frac{200\pi - 0}{6 \times 60} = \frac{5\pi}{9} \text{ rad/s}^2$$

Let M_f be the constant frictional torque. Then net torque $= (2500 - M_f)$ Nm

For the rotor $I_G = mk^2 = 2000 \times (0.75)^2$ kg-m^2 = 1125 kg. m^2

By $\Sigma M_G = I_G \cdot \alpha$, we get

$$2500 - M_f = 112.5 \times \left(\frac{5\pi}{9}\right)$$

$\therefore$ $M_f = 2500 - 1125 \times \dfrac{5\pi}{9} = 2500 - 1963.5 = \mathbf{536.50\ Nm}$. **Ans.**

When steam supply is cut off the rotor will retard under the action of the frictional torque of 536.50 Nm. Let α_1 be the angular retardation.

Then by $\Sigma M_a = I_G \cdot \alpha$, we get

$$M_f = I_G \times \alpha_1$$

$\therefore$ $\alpha_1 = \dfrac{M_f}{I_G} = \dfrac{536.5}{1125} = \mathbf{0.477\ rad/s^2}$

Time taken to come to rest

$$= \frac{\omega}{\alpha} = \frac{200\pi}{0.477} = \mathbf{1317.23\ seconds} \text{ or } \mathbf{21.954\ minutes.}$$ **Ans.**

12.14. WORK DONE BY A COUPLE AND KINETIC ENERGY OF ROTATION

(*a*) **Work of a Couple.** A force P acts on a body through a point A at a distance r from the $C.G.$ (G) of the body, as shown in Fig. 12.14. Let point A moves to A' after time δt. Then displacement δs = arc $AA' = r.\ \delta\theta$. Then work done = $\delta U = P \times \delta s$. or $\delta U = P \times \delta s$.

$\therefore$ Total workdone for an angular displacement θ is

$$U = \int dU = \int Pr\, d\theta$$

or, $$U = Pr\theta$$

when r is perpendicular to P_x, then $P\,.\,r$ = couple = C

And, $\therefore$ workdone by a couple will be

$$U = C\,.\,\theta \quad ...(12.62)$$

$$\text{Power} = \frac{dU}{dt} = C\,.\frac{d\theta}{dt} = C \times \omega \quad ...(12.63)$$

where ω = angular speed of rotation.

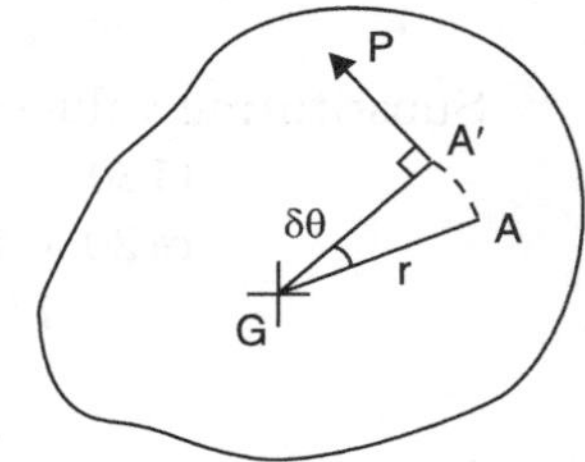

Fig. 12.14. Work done by a couple.

(*b*) **Kinetic Energy of Rotation.** A body of mass m rotating with an angular velocity ω has the capacity of doing work by virtue of its kinetic energy of rotation. The kinetic energy of rotation is measured by the amount of workdone by the body before coming to rest. Let a couple C be applied to bring body to rest, and the angular displacement be θ. Then by using eqn. (12.62), we find the workdone = $C\theta$ ...(12.64)

Let the initial velocity be ω. Therefore, by using the relation

$$\omega^2 = \omega_0^{\,2} - 2\alpha\,.\,\theta, \text{ we get}$$

$$\alpha = \frac{\omega^2}{2\theta} \quad ...(12.65)$$

The equation of motion of the body is $\Sigma M_G = I_G\,.\,\alpha$

$$\therefore \quad C = I_G\,.\,\alpha = I_G\,.\,\frac{\omega^2}{2\theta} \quad ...(12.66)$$

$$\therefore \quad \text{Workdone} = U = C\theta = \left(I_G\,.\,\frac{\omega^2}{2\theta}\right) \times \theta$$

or, $$U = \frac{1}{2} I_G\,.\,\omega^2 \quad ...(12.67)$$

$$\therefore \quad \text{K.E.} = U = \frac{1}{2} I_G\,.\,\omega^2. \quad ...(12.68)$$

12.15. ANGULAR IMPULSE AND ANGULAR MOMENTUM

(*a*) **Angular Momentum.** In case of rectilinear motion we found that linear momentum = mv where m is the mass and v is the linear velocity of the body.

In rotational motion, the corresponding expression for the angular momentum will be

$$\text{Angular momentum} = I_G\,.\,\omega \quad ...(12.69)$$

where I_G = mass moment of inertia about C.G. of the body, and

ω = angular velocity.

Since $$C = I_G\,.\,\alpha = I_G \times \frac{d\omega}{dt}$$

or,
$$C = \frac{d}{dt}(I_G \cdot \omega) \quad \text{...(12.70)}$$

$\therefore$ Torque = C = rate of change of angular momentum.

(*b*) **Angular Impulse.** In rectilinear motion we discussed that impulse is a very large force acting for a small time interval, and the change of momentum ($mv - mu$) is equal to the impulse.

Similarly, for the rotational motion, we can write :

Angular impulse = change of angular momentum = $I_G(\omega_1 - \omega_2)$ ...(12.71)

12.16. FREELY ROLLING BODY

When a body rolls without slipping, the acceleration of the centre of gravity is equal to $r\alpha = \bar{a}$, where α is the angular acceleration and r is the radius of the body, as shown in Fig. 12.16 (*a*).

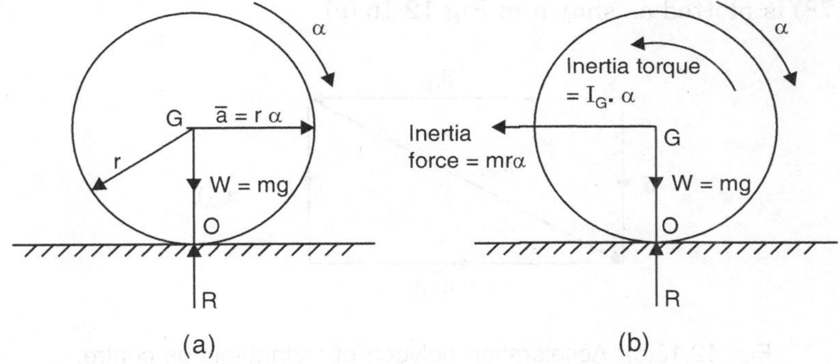

Fig. 12.16. Freely rolling body

The equations of motion of the body are :

$$\Sigma F_y = ma_y \quad \text{where} \quad a_y = 0, \quad \text{...(12.72)}$$

$$\therefore \quad W - R = 0$$

where R = reaction of surface

and $W = mg$ = weight of body

$$\Sigma F_x = ma_x = mr\,\alpha \quad \text{...(12.73)}$$

$$\Sigma M_G)_{\text{ext. force}} = \Sigma M_G)_{\text{eff. forces}}$$

$$\therefore \quad \Sigma M_G = I_G \cdot \alpha \quad \text{...(12.74)}$$

If P is externally applied force in x-direction and F is the force of friction between the body and the surface of contact at O,

Then eqn. (12.73) becomes

$$P - F = mr\alpha \quad \text{or} \quad F = P - mr\alpha \quad \text{...(12.75)}$$

And, eqn. (12.74) gives

$$F \times r = I_G \cdot \alpha$$

$$\therefore \quad \boxed{F = \frac{I_G \cdot \alpha}{r}} \qquad ...(12.76)$$

Fig. 12.16. (*b*) shows the dynamic equilibrium of the rolling body in which the inertia force = $mr\alpha$ and the inertia couple = $I_G \cdot \alpha$.

The point O in Fig. 12.16 (*a*) is known as *instantaneous centre* or *instant centre.* The total acceleration of point O is given by

$$\bar{a}_O = \bar{a}_G + \bar{a}_{OG} \qquad ...(12.77)$$

where a_G = given acceleration of the mass centre G.

and, a_{OG} = acceleration of O relative to G

$= a_{OG})_n + a_{OG})_t$

$= r\omega^2 + r\alpha.$

$$\therefore \quad \bar{a}_O = \bar{a}_G \rightarrow (r\alpha^2 + r\omega^2) \qquad ...(12.78)$$

Eqn. (12.78) is plotted as shown in Fig 12.16 (*c*).

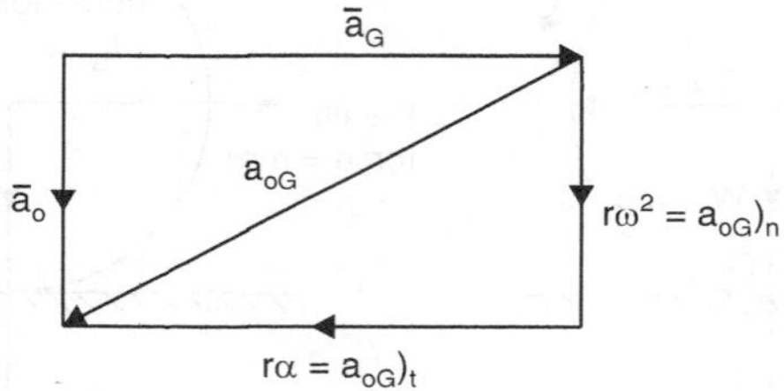

Fig. 12.16(c). Acceleration polygon of instantaneous centre.

Thus, the equation of motion with respect to point G will be :

$$\Sigma M_O = I_O \cdot \alpha \qquad ...(12.79)$$

where I_O = mass moment of inertia with respect to point O.

$= m{k_O}^2$

$$I_O = m({k_G}^2 + d^2) \qquad ...(12.80)$$

where k_G = radius of gyration of the body and d = distance between G and $O = r$ (here)

The term ${k_G}^2 + d^2$ may be written as k_O and it is radius of gyration with respect to O.

$$\therefore \quad {k_O}^2 = {k_G}^2 + d^2 \qquad ...(12.81)$$

Eqns. (12.80) and (12.81) can be derived with the help of parallel-axis theorem as discussed in Chapter 11.

12.17. PLANE MOTION OF A RIGID BODY

A plane motion of a rigid body is the combination of its rectilinear motion and its rotation about an axis perpendicular to the plane of motion. Or in other words, a general plane motion can be regarded equivalent to a combination of translation of the centre of gravity and rotation about an axis passing through the centre of mass. Therefore, the equations of motion will be :

$$F_x = ma_x \qquad ...(12.82)$$

$$F_y = ma_y \qquad ...(12.83)$$

for translation of the centre of mass,

and $$M_G = I_G \,.\, \alpha \qquad ...(12.84)$$

for rotation about an axis passing through centre of mass of the body.

Problem 12.25. *A cricket ball hits the bat at point A. The mass of the bat is m and G is its centre of gravity, as shown in Fig. P. 12.25 (a). Find the position of point A when the horizontal force exerted on the hands of the batsman at O is zero.*

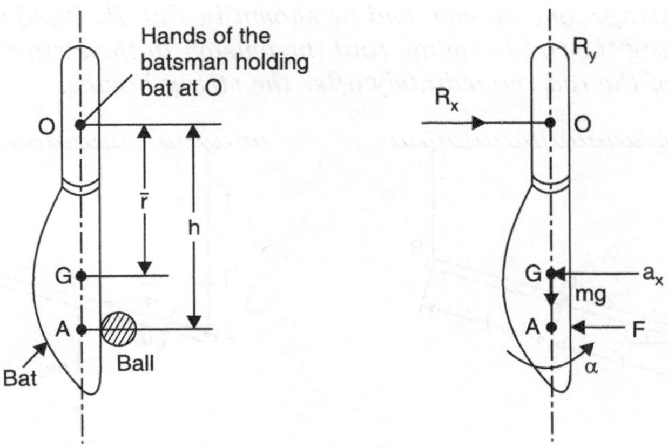

(a) Ball hitting the bat at A (b) Free-body diagram of bat

Fig. P. 12.25

Sol. Let the horizontal force exerted by the ball on the bat be F which acts at A. Let a_x be horizontal acceleration of the bat, and let R_x and R_y be reactions in x- and y-direction acting at O.

The equations of motion for the bat are :

$\Sigma F_x = ma_x$ gives

$$R_x - F = ma_x \qquad ...(1)$$

$$\Sigma M_0)_{ext.} = I_0 \,.\, \alpha)_{inertia,} \quad \text{gives}$$

$$-Fh = (I_G + m\bar{r}^2)\,.\,\alpha$$

or, $$-Fh = (mk_G^2 + m\bar{r}^2)\,\alpha \qquad ...(2)$$

or $$-F = \frac{m(k_G^{\,2} + \bar{r}^2)\alpha}{h} \qquad ...(3)$$

$\therefore$ Eqn. (1) gives $$R_x = ma_x - \frac{m(k_G^{\,2} + \bar{r}^2)\alpha}{h}$$

But, $$a_x = \alpha \,.\, \bar{r}$$

$\therefore$ We get $$R_x = m\alpha\bar{r} - \frac{m(k_G^{\,2} + \bar{r}^2)\alpha}{h} \qquad ...(4)$$

when, $R_x = 0$, we get

$$m\alpha\bar{r} = \frac{m(k_G{}^2 + \bar{r}^2)\alpha}{h}$$

or,
$$\boxed{h = \frac{k_G{}^2 + \bar{r}^2}{\bar{r}}} \quad ...(5)$$

Note. When the ball hits, the bat at point A such that $R_x = 0$, then point A is called the *point of percussion*.

Problem 12.26. *A thin uniform rod of length 2L and mass m is held by means of two inextensible vertical strings, one at each end as shown in Fig. P. 12.26 (a). If the string at the right end breaks, leaving the rod to swing, find the tension in the string at the left end and the angular acceleration of the rod immediately after the string breaks.*

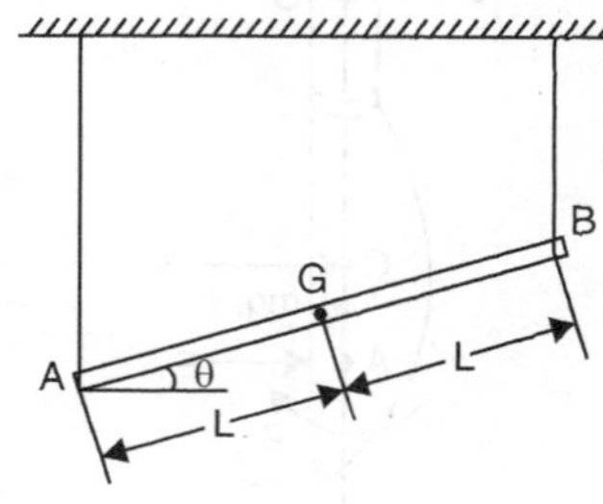

(a) Given system

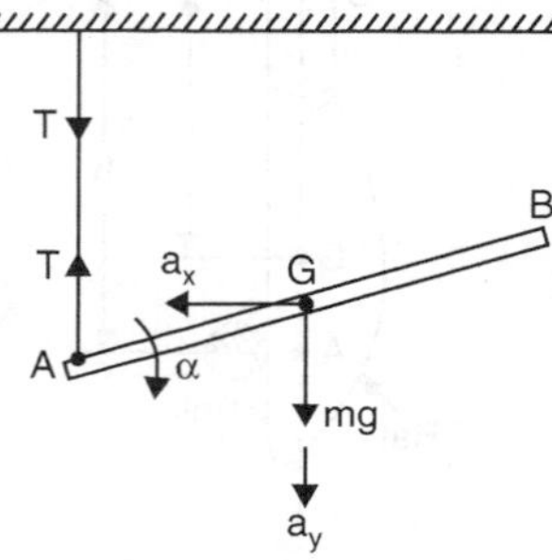

(b) Forces and acceleration components of the rod AB immediately after the string at B breaks.

Fig. P. 12.26

Sol. Let the tension in string at the left end A be T. The rod swings clockwise about A with angular acceleration α. a_x and a_y are accelerations of rod along x-and y-axes.

Equations of motion of the rod are :

By $\Sigma F_y = ma_y$, we get

$$mg - T = ma_y \quad ...(1)$$

By $\Sigma F_x = ma_x$, we get

$$0 = ma_x \quad ...(2)$$

and by $\Sigma M_G)_{ext.} = I_G \cdot \alpha)_{inertia}$, we get

$$T(L \cos \theta) = I_G \cdot \alpha \quad ...(3)$$

where
$$I_G = \frac{m(2L)^2}{12} = \frac{mL^2}{3} \quad ...(4)$$

We also know that $a_y = \alpha(L \cos \theta)$ ...(5)

Hence, eqn. (1) becomes

$$mg - T = m\alpha\,(L \cos \theta) \quad ...(6)$$

And eqn. (3) becomes

$$T \,.\, L \cos \theta = \frac{mL^2}{3}\alpha$$

Dividing by $L \cos \theta$, we get

$$T = \frac{mL\alpha}{3 \cos \theta} \qquad \text{...(7)}$$

Adding (6) and (7), we get

$$mg = m\alpha(L \cos \theta) + \frac{mL\alpha}{3 \cos \theta}$$

$$\therefore \qquad mg = \frac{\alpha m}{3 \cos \theta} [\, 3L \cos^2 \theta + L]$$

$$\therefore \qquad \alpha = \frac{3g \cos \theta}{L(3 \cos^2 \theta + 1)}. \quad \textbf{Ans.} \qquad \text{...(8)}$$

And, from (7), we get

$$T = \frac{mL}{3 \cos \theta} \times \frac{3g \cos \theta}{L(1 + 3 \cos^2 \theta)}$$

$$\therefore \qquad T = \frac{mg}{(1 + 3 \cos^2 \theta)}. \quad \textbf{Ans.} \qquad \text{...(9)}$$

Problem 12.27. *A uniform bar of length L and weight W rests on smooth horizontal and vertical surfaces as shown in Fig. P. 12.27(a). Derive expressions for angular velocity, angular acceleration of the bar and reactions at its ends. Also find the angle θ at which the bar no longer touches the vertical surface.*

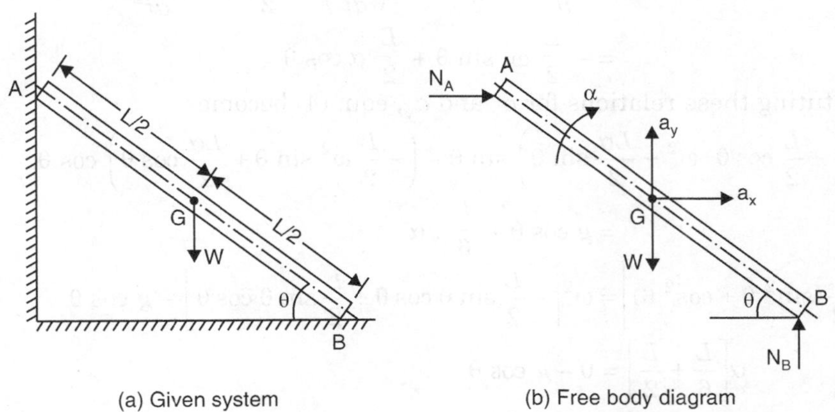

Fig. P. 12.27

Sol. Since the contact surfaces are smooth the frictional forces F_A and F_B are zero and only normal reactions N_A and N_B exist.

The equations of motion of the bar are :

By $\Sigma F_x = ma_x$, we get $N_A = ma_x$ ...(1)

By $\Sigma F_y = ma_y$, we get $N_B - W = ma_y$ ...(2)

and by $\Sigma M_G = I_G \cdot \alpha$, we get

$$N_A \cdot \frac{L}{2} \sin \theta - N_B \cdot \frac{L}{2} \cos \theta = I_G \cdot \alpha \qquad \text{...(3)}$$

where $$I_G = \frac{mL^2}{12} \quad \text{and} \quad W = mg$$

Substituting for N_A, N_B, W and I_G, eqn. (3) becomes,

$$(ma_x)\frac{L}{2}\sin\theta - (mg + ma_y)\frac{L}{2}\cos\theta = \frac{mL^2}{12}\alpha$$

or, $$a_x \sin\theta - a_y \cos\theta = g\cos\theta + \frac{L}{6}\alpha \qquad \text{...(4)}$$

From Fig. P. 12.27 (*b*), we have

$$x = \frac{L}{2}\cos\theta \quad \text{and} \quad y = \frac{L}{2}\sin\theta,$$

Differentiating we get

$$v_x = \frac{dx}{dt} = \frac{dx}{d\theta}.\frac{d\theta}{dt} = -\frac{L}{2}\sin\theta.\left(\frac{d\theta}{dt}\right) = -\frac{L}{2}\omega\sin\theta$$

and $$v_y = \frac{dy}{dt} = \frac{dy}{d\theta}.\frac{d\theta}{dt} = +\frac{L}{2}\cos\theta.\left(\frac{d\theta}{dt}\right) = \frac{L}{2}\omega\cos\theta$$

Also, $$a_x = \frac{dv_x}{dt} = -\frac{L}{2}\cos\theta.\left(\frac{d\theta}{dt}\right)^2 - \frac{L}{2}\sin\theta.\frac{d^2\theta}{dt^2}$$

$$= -\frac{L}{2}\omega^2\cos\theta - \frac{L}{2}\alpha\sin\theta$$

and $$a_y = \frac{dv_y}{dt} = -\frac{L}{2}\sin\theta.\left(\frac{d\theta}{dt}\right)^2 + \frac{L}{2}\cos\theta.\frac{d^2\theta}{dt^2}$$

$$= -\frac{L}{2}\omega^2\sin\theta + \frac{L}{2}\alpha\cos\theta$$

Substituting these relations for a_x and a_y, eqn. (4) becomes

$$\left(-\frac{L}{2}\cos\theta.\omega^2 - \frac{L\alpha}{2}\sin\theta\right)\sin\theta - \left(-\frac{L}{2}\omega^2\sin\theta + \frac{L\alpha}{2}\cos\theta\right)\cos\theta$$

$$= g\cos\theta + \frac{L}{6}.\alpha$$

or $$\alpha\left[\frac{L}{6} + \frac{L}{2}(\sin^2\theta + \cos^2\theta)\right] = \omega^2\left[-\frac{L}{2}\sin\theta\cos\theta + \frac{L}{2}\sin\theta\cos\theta\right] - g\cos\theta$$

or $$\alpha\left[\frac{L}{6} + \frac{L}{2}\right] = 0 - g\cos\theta$$

$\therefore$ $$\alpha = \frac{-3g}{2L}\cos\theta. \quad \textbf{Ans.} \qquad \text{...(5)}$$

And, the angular speed can be obtained by integrating eqn. (5), $\left[\because \alpha = \omega\frac{d\omega}{d\theta}\right]$

Since $$\alpha = \frac{\omega.d\omega}{d\theta}$$

$\therefore$ Eqn. (5) becomes

$$\omega\, d\omega = -\frac{3g}{2L}\cos\theta.d\theta$$

Integrating, we get

$$\int \omega\, d\omega = \int -\frac{3g}{2L}\cos\theta.d\theta + C$$

or, $$\frac{\omega^2}{2} = -\frac{3g}{2L} \sin\theta + C \qquad \text{...(6)}$$

where C is a constant of integration.

Let $\omega = 0$ at $\theta = 90° = \pi/2$, then

we get $$0 = -\frac{3g}{L} \sin \pi/2 + C \quad \therefore \quad C = \frac{3g}{2L}$$

∴ Eqn. (6) becomes

$$\omega^2 = \frac{3g}{L}(1 - \sin\theta)$$

or, $$\omega = \sqrt{\frac{3g}{L}(1-\sin\theta)}. \quad \textbf{Ans.} \qquad \text{...(7)}$$

Hence eqn. (1) gives

$$N_A = ma_x = \frac{-mL}{2}(\omega^2 \cos\theta + \alpha \sin\theta) \qquad \text{...(8)}$$

From (2), we get

$$N_B = ma_y + mg = \frac{mL}{2}(-\omega^2 \sin\theta + \alpha\cos\theta) + mg$$

or, $$N_B = m\left[g + \frac{L}{2}(\alpha\cos\theta - \omega^2 \sin\theta)\right] \qquad \text{...(9)}$$

When there is no contact of the bar at A,

$N_A = 0$, we get $a_x = 0$

Hence, from eqn. (8), we get

$$\omega^2 \cos\theta + \alpha \sin\theta = 0 \qquad \text{...(10)}$$

Substituting values of ω^2 and α into eqn. (10), we get

$$\frac{3g}{L}(1 - \sin\theta)\cos\theta + \left(-\frac{3g}{2L}\cos\theta\right)\sin\theta = 0$$

or, $$2\cos\theta(1 - \sin\theta) - \cos\theta \sin\theta = 0$$

or, $$2(1 - \sin\theta) - \sin\theta = 0$$

∴ $$2 - 3\sin\theta = 0$$

or $$\sin\theta = \frac{2}{3}$$

∴ $$\theta = 41.81°. \quad \textbf{Ans.}$$

Problem 12.28. *At what distance p should the horizontal force F be applied to the homogeneous bar, homogeneous disc and homogeneous sphere Fig. P. 12.28 so that the horizontal component of the reaction at the point of suspension is zero.*

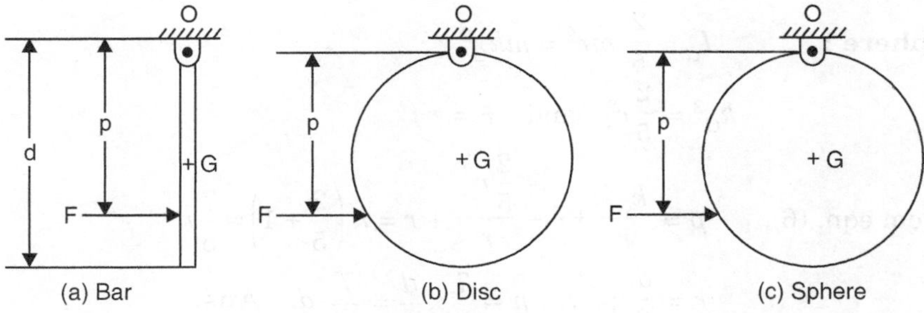

Fig. P. 12.28

Sol. For bar :

By $\Sigma F_x)_{ext.} = ma_x,$ we have

$$R_x - F = ma_x \quad ...(1)$$

By $\Sigma M_0 = I_0\,\alpha$, we have

$$-pF = I_0\,.\,\alpha = (mk_G^2 + m\bar{r}^2)\,.\,\alpha \quad ...(2)$$

where $\bar{r}$ = distance $OG = d/2$

and $a_x = \alpha\,.\,\bar{r} = \dfrac{d}{2}\,\alpha$; and $k_G^2 = \dfrac{I_G}{m} = \dfrac{md^2}{12m} = \dfrac{d^2}{12}$

$\therefore$ Eqn. (2) gives $-pF = m\left(\dfrac{d^2}{12} + \left(\dfrac{d}{2}\right)^2\right).\,\alpha$

or,

$$-pF = \frac{md^2}{3}\,\alpha \quad ...(3)$$

$\therefore$ Eqn. (1) gives $R_x = ma_x + F$

or

$$R_x = m\left(\frac{d}{2}\alpha\right) - \frac{md^2}{3p}\,.\,\alpha \quad ...(4)$$

when $R_x = 0$, we get $\dfrac{md\alpha}{2} - \dfrac{md^2\alpha}{3p} = 0$

$$\therefore \quad p = \frac{2}{3}d. \quad \textbf{Ans.} \quad ...(5)$$

From above we can write

$$\boxed{p = \frac{k_G^2 + \bar{r}^2}{\bar{r}} = \frac{k_G^2}{\bar{r}} + \bar{r}} \quad ...(6)$$

For disc : $I_G = \dfrac{mr^2}{4}$ where $r = d/2$

$$I_G = \frac{m}{4}(d/2)^2 = \frac{md^2}{16} = mk_G^2\,;$$

$$\therefore \quad k_G^2 = \frac{d^2}{16} \quad \text{and} \quad \bar{r} = \frac{d}{2}$$

$\therefore$ Using eqn. (6), $p = \dfrac{k_G^2}{\bar{r}} + \bar{r} = \dfrac{d^2}{16 \times (d/2)} + \dfrac{d}{2} = d\left(\dfrac{1}{8} + \dfrac{1}{2}\right)$

$$\therefore \quad p = \frac{5}{8}\,d. \quad \textbf{Ans.}$$

For sphere : $I_G = \dfrac{2}{5}\,mr^2 = mk_G^2$

$$\therefore \quad k_G^2 = \frac{2}{5}r^2 \quad \text{and} \quad \bar{r} = r$$

$\therefore$ From eqn. (6), $p = \dfrac{k_G^2}{\bar{r}} + \bar{r} = \dfrac{\frac{2}{5}r^2}{r} + r = r\left(\dfrac{2}{5} + 1\right) = \dfrac{7}{5}r$

But $r = \dfrac{d}{2}$; $\therefore$ $p = \dfrac{7}{5} \times \dfrac{d}{2} = \dfrac{7}{10}\,d.$ **Ans.**

Problem 12.29. *A cord passing over a light frictionless pulley carries a weight W suspended at its one end and is wrapped around a thin cylinder of weight W_c as shown in Fig. P. 12.29 (a). If the cylinder rolls without slip on a horizontal surface find the acceleration of the suspended weight and the cylinder. Also find tension in the cord.*

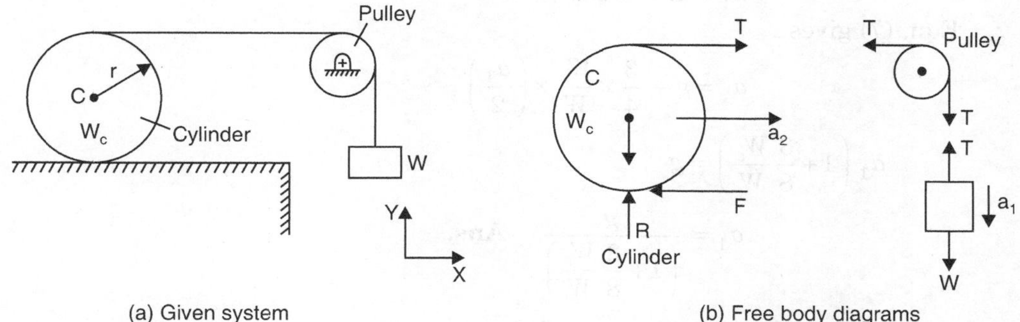

(a) Given system (b) Free body diagrams

Fig. P. 12.29

Sol. Equations of motion are :

For the suspended weight $W - T = \dfrac{W}{g}\, a_1$ ...(1)

For the cylinder :

By $\Sigma F_x = ma_x$, we have

$$T - F = \frac{W_c}{g}\, a_2 \qquad ...(2)$$

By $\Sigma F_y = ma_y$, we have

$$R - W_c = 0 \qquad ...(3)$$

and by $\Sigma M_c = I_c \,.\, \alpha$, we have

$$(T + F)\, r = \frac{1}{2}\left(\frac{W_c}{g}\, r^2\right) . \,\alpha \qquad ...(4)$$

In equation (2), $a_2 = r\alpha$,

$\therefore$ Eqn. (4) becomes

$$(T + F)\, r = \frac{1}{2}\left(\frac{W_c}{g}\, r^2\right) \times \frac{a_2}{r}$$

or,

$$T + F = \frac{W_c}{2g} a_2 \qquad ...(5)$$

Adding (2) and (5), we get

$$2T = \left(\frac{W_c}{g} + \frac{W_c}{2g}\right) a_2$$

or

$$T = \frac{3}{4}\frac{W_c}{g}\, a_2 \qquad ...(6)$$

Therefore, eqn. (1) gives

$$W - \frac{3}{4}\frac{W_c}{g}\, a_2 = \frac{W}{g}\, a_1$$

or, $$a_1 = g - \frac{3}{4}\frac{W_c}{W}\, a_2 \quad ...(7)$$

From the geometry of the system, we have

$$a_1 = 2r\alpha = 2\alpha_2 \quad ...(8)$$

$\therefore$ Eqn. (7) gives

$$a_1 = g - \frac{3}{4} \times \frac{W_c}{W} \times \left(\frac{a_1}{2}\right)$$

or $$a_1\left(1 + \frac{3}{8}\frac{W_c}{W}\right) = g$$

or $$a_1 = \frac{g}{\left(1 + \frac{3}{8}\frac{W_c}{W}\right)}. \quad \textbf{Ans.} \quad ...(9)$$

$$a_2 = \frac{a_1}{2} = \frac{g}{2\left(1 + \frac{3}{8}\frac{W_c}{W}\right)}. \quad \textbf{Ans.} \quad ...(10)$$

And the tension in cord

From eqn. (6), $$T = \frac{3}{4} \times \frac{W_c}{g} \times \left[\frac{g}{2\left(1 + \frac{3}{8}\frac{W_c}{W}\right)}\right]$$

or $$T = \frac{3}{8}\frac{W_c}{\left(1 + \frac{3}{8}\frac{W_c}{W}\right)}. \quad \textbf{Ans.} \quad ...(11)$$

Problem 12.30. *A thin uniform bar of length L pivoted about a horizontal axis through its lower end is allowed to fall from the vertical position. Find angular velocity and angular acceleration of the bar.*

Sol. After sometime the bar occupies a position shown in Fig. P. 12.30. Let R_x and R_y be the reaction components at the pivot O.

By $\Sigma M_0 = I_0 \cdot \alpha$, we get $W \times \frac{L}{2} \sin\theta = \left(\frac{W}{3g} L^2\right) \alpha$

$$\therefore \quad \alpha = \frac{3g}{2L}\sin\theta \quad ...(1)$$

or, $$\omega \cdot \frac{d\omega}{d\theta} = \frac{3g}{2L}\sin\theta$$

or, $$\omega \cdot d\omega = \frac{3g}{2L}\sin\theta \cdot d\theta$$

Integrating, we get

$$\frac{\omega^2}{2} = \frac{3}{2}\frac{g}{L}(-\cos\theta) + C \quad ...(2)$$

where C is the constant of integration.

Given that at $\theta = 0$, $\omega = 0$,

$\therefore$ Eqn. (2) gives

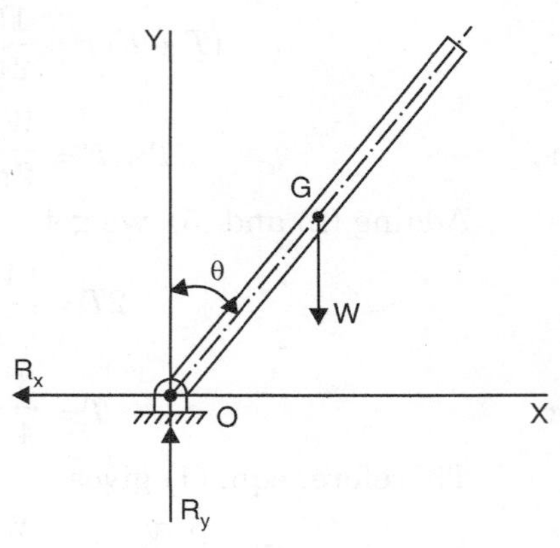

Fig. P. 12.30

$$0 = \frac{3}{2}\frac{g}{L}(-\cos 0) + C \quad \therefore \quad C = \frac{3}{2}\frac{g}{L}$$

$\therefore$ Eqn. (2) gives

$$\frac{\omega^2}{2} = -\frac{3}{2}\frac{g}{L}\cos\theta + \frac{3}{2}\frac{g}{L} = \frac{3}{2}\frac{g}{L}(1-\cos\theta)$$

$\therefore$ $$\omega = \sqrt{\frac{3g}{L}(1-\cos\theta)}. \quad \textbf{Ans.} \qquad ...(3)$$

When $\theta = 90°$, the bar becomes horizontal

$\therefore$ We get $$\omega_{90°} = \sqrt{\frac{3g}{L}} \qquad ...(4)$$

and from eqn. (1) $$\alpha_{90°} = \frac{3g}{2L} \qquad ...(5)$$

when $\theta = 180°$, the bar become vertical, and we get

$$\omega_{180°} = \sqrt{\frac{6g}{L}} \qquad ...(6)$$

and $$\alpha_{180°} = 0.$$

12.18. MOTION OF A VEHICLE ROLLING DOWN THE ROAD

Let M be the total mass of the vehicle (including axles, wheels and the mass of the passengers) acting at the centre of gravity G of the vehicle. Let m be the mass of axles and wheels, r the radius of wheels and k the radius of gyration.

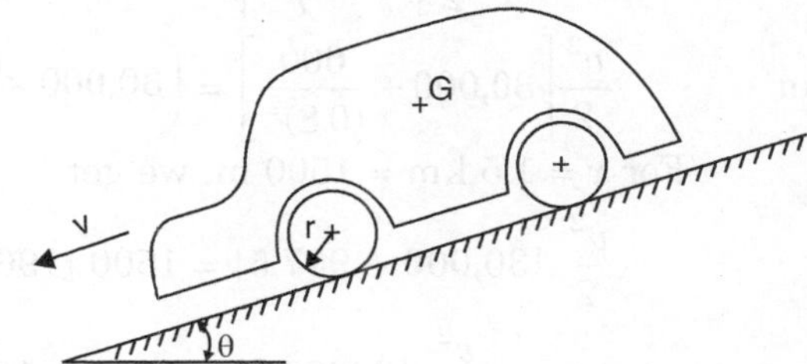

Fig. 12.18

Let v the velocity attained by the vehicle after rolling down a distance x. Then by using the principle of conservation of energy, we get

$$\boxed{\frac{Mv^2}{2} + \frac{I\omega^2}{2} = (Mg)x\sin\theta} \qquad ...(12.85)$$

when there is no loss of energy or work due to friction.

Since $v = \omega r$ and $I = mk^2$

$\therefore$ Eqn. (12.85) gives,

$$\frac{Mv^2}{2} + \left(\frac{mk^2}{2}\right)\times\left(\frac{v}{r}\right)^2 = Mgx\sin\theta$$

$\therefore$ $$\frac{v^2}{2}\left[M + \frac{mk^2}{r^2}\right] = Mg\,x\sin\theta \qquad ...(12.86)$$

Also, using $v^2 = u^2 + 2ax$, and since $u = 0$,

we have $$v^2 = 2ax \qquad ...(12.87)$$

$\therefore$ $$a = \frac{v^2}{2x} = \frac{Mg\sin\theta}{\left(M + \frac{mk^2}{r^2}\right)} \qquad ...(12.88)$$

Considering friction, the workdone due to friction = $R_F \cdot x$

And, eqn. (12.85) becomes, $$\left(\frac{Mv^2}{2} + \frac{I\omega^2}{2}\right) = (Mg)\,x \,.\, \sin\theta - R_F \,.\, x \quad ...(12.89)$$

Problem 12.31. *A wagon of mass 30 tonnes moves down a gradient of 1 in 150. Find its velocity when it has moved 1.5 km, the friction at axles being 60 N/tonne. The mass of the axles and wheels is 2.4 tonnes, having a radius of 0.8 m and radius of gyration of 0.5 m.*

Sol. Given : M = 30 tonnes = 30000 kg,

m = 2.4 tonnes = 2400 kg ;

r = 0.8 m ; k = 0.5 m

$$I_G = mk^2 = 2400 \times (0.5)^2 = 600 \text{ kg-m}^2$$

$$\omega = \frac{v}{r} = \frac{v}{0.8} = 1.25\, v \text{ rad/s ;}$$

Friction resistance R_f = 60 N/tonne = 60 × 30 = 1800 N.

Energy equation for the wagon is

$$\frac{Mv^2}{2} + \frac{1}{2} I_G \,.\, \omega^2 = Mg\,x \sin\theta - R_F \,.\, x$$

or, $$\frac{v^2}{2}\left[M + \frac{I_G}{r^2}\right] = Mg\,x \sin\theta - R_F \,.\, x$$

or $$\frac{v^2}{2}\left[30{,}000 + \frac{600}{(0.8)^2}\right] = \left(30{,}000 \times 9.81 \times \frac{1}{150} - 1800\right) x$$

For x = 1.5 km = 1500 m, we get

$$\frac{v^2}{2}\,[30{,}000 + 937.5] = 1500\,[1962 - 1800]$$

or, $$\frac{v^2}{2}\,[30937.5] = 1500 \times 162$$

$\therefore$ $v^2 = 15.71$

$\therefore$ **v = 3.964 m/s = 14.2704 km/hr. Ans.**

Problem 12.32. *A mini truck 4 tonne mass has its centre of gravity 1.2 m above the road. The distance between the axles (or wheel base) is 2.4 m and the centre of gravity is 0.8 m in front of the rear axle. The truck has a speed of 45 km/h on a level road. If the coefficient of adhesion is 0.4, find the shortest distance in which the mini truck will stop (a) by applying front brakes, (b) by applying rear brakes and (c) by applying front and rear brakes. Neglect rotational inertia.*

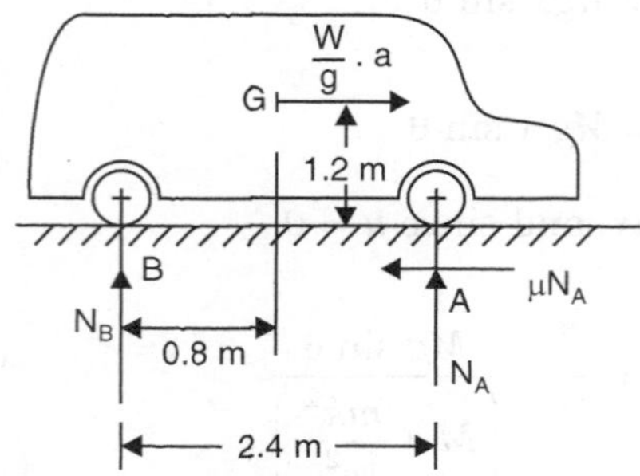

Fig. P. 12.32. Brakes applied on front wheels only

Sol. (*a*) **Brake applied on front wheels :**

Velocity of mini truck = $\dfrac{45 \times 1000}{3600}$ = 12.5 m/s

If brakes are applied on front wheels, the retarding force = μN_A = 0.4 N_A.

By $\Sigma F_y = 0$, we get $N_A + N_B = 4000\, g$ ...(1)

By taking moments about G, we get

$$N_B \times 0.8 + (0.4\, N_A) \times 1.2 - N_A \times (2.4 - 0.8) = 0,$$

or, $$0.8\, N_B = 1.12\, N_A$$

or $$N_B = 1.4\, N_A \quad ...(2)$$

From (1) and (2), we get

$$N_A + 1.4\, N_A = 4000 \text{ g}$$

$$\therefore \quad N_A = \frac{4000 \text{ g}}{2.4} = 16,350 \text{ N}$$

and $$N_B = 4000 \text{ g} - 16350 = 22,890 \text{ N}$$

And the retarding force = 0.4 N_A = 6540 N

If the mini truck is brought to rest in x_1 metres, the K.E. of the mini truck at 12.5 m/s is absorbed by the retarding force in x_1 m.

∴ Equating the K.E. and the work of frictional force, we get

$$\frac{1}{2} \times 4000 \times (12.5)^2 = 6540\, x_1$$

$$\therefore \quad x_1 = \frac{4000 \times (12.5)^2}{2 \times 6540} = \mathbf{47.783\ m.} \quad \textbf{Ans.}$$

(*b*) **Brakes applied to rear wheels :**

Force of friction = 0.4 $N_B = R_F$

By $\Sigma F_V = 0$, we get

$$N_A + N_B = 4000 \text{ g} \quad ...(3) \quad \text{[it is the same as eqn. (1)]}$$

By taking moments about G, we get

$$N_B \times 0.8 + R_F \times 1.2 - N_A \times (2.4 - 0.8) = 0$$

or, $$0.8N_B + 0.4N_B \times 1.2 = 1.6\, N_A$$

or, $$1.28\, N_B = 1.6\, N_A$$

$$N_A = 0.8\, N_B \quad ...(4)$$

∴ From (3), we get $0.8\, N_B + N_B = 4000$ g

$$\therefore \quad N_B = \frac{4000 \text{ g}}{1.8} = 21,800 \text{ N} \quad \text{and} \quad N_A = 17440 \text{ N},$$

$$\therefore \quad R_F = 0.4 \times N_B = 8720 \text{ N}$$

By equating the K.E. and the workdone by R_F, we get

$$\frac{1}{2} \times \frac{4000}{g} \times (12.5)^2 = 8720\, x_2$$

$$\therefore \quad x_2 = \frac{4000}{2} \times \frac{(12.5)^2}{8720} = \mathbf{35.84\ m.} \quad \textbf{Ans.}$$

(*c*) **Brakes applied both front and rear wheels :**

Force of friction $= R_F = 0.\ 4(N_A + N_B) = 0.4 \times (4000\ g) = 15696$ N

By equating K.E. and work of friction, we get

$$\frac{1}{2} \times [4000 \times (12.5)^2] = 15{,}696 \times x_3$$

$\therefore \qquad x_3 =$ **19.91 m. Ans.**

12.19. MOTION OF A VEHICLE GOING ROUND A CURVE

We have seen that when a particle moves in a circular path of radius of curvature r with a uniform velocity v, its centripetal or normal acceleration $a_n = v^2/r$ and the corresponding force $F = ma_n = mv^2/r$.

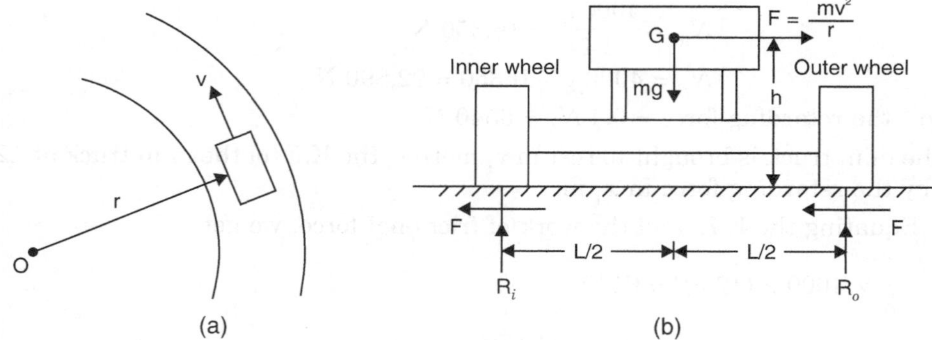

Fig. 12.19. Vehicle moving along a curved road

When a vehicle goes along a curved road the central force acting through its *C.G.* is $\frac{mv^2}{r}$ and there is a force of friction between the wheels and the road surface. The central force is supplied by the frictional force.

The vehicle experiences a centrifugal force mv^2/r acting outwards through its centre of gravity.

Let m be the mass of the vehicle. Let the centre of gravity G be at a distance h above the road.

$$\text{The overturning moment} = \frac{mv^2}{r} \times h \qquad ...(1)$$

By $\Sigma F_v = 0$, we have

$$R_i + R_o = mg \qquad ...(2)$$

By taking moments about G, as there is no side-skid motion, we have

$$(R_i - R_o) \times \frac{L}{2} = Fh = \frac{mv^2}{r} \times h$$

$$\therefore \qquad R_i - R_o = \frac{mv^2}{r} \times \frac{2h}{L} \qquad ...(3)$$

Adding (2) and (3), we have

$$2R_i = mg + \frac{mv^2}{r} \times \frac{2h}{L}$$

$$\therefore \qquad R_i = \frac{1}{2}\left[mg + \frac{mv^2}{r} \times \frac{2h}{L}\right] \qquad ...(4)$$

and $$R_o = \frac{1}{2}\left[mg - \frac{mv^2}{r} \times \frac{2h}{L}\right] \qquad ...(5)$$

We have seen that the necessary central force $\frac{mv^2}{r}$ is supplied by the frictional force F. There is a limit to the maximum force of friction that can be called into play. Therefore,

the limiting force of friction = $F_{max} = \mu_s \times (R_i + R_o)$

or $$F_{max} = \mu_s \times (R_i + R_g) = \mu_s \times mg \qquad ...[12.89(a)]$$

In order that the above relations are true,

$$F < \mu_s \,.\, mg$$

or, $$\frac{mv^2}{r} < \mu_s\, mg$$

$$\therefore \qquad v^2 < \mu_s \,.\, gr$$

and $$v < \sqrt{\mu_s \,.\, gr} \qquad ...(6)$$

or skidding speed = $\sqrt{\mu_s \,.\, gr}$ ...(12.90)

If the vehicle attains this velocity it will skid.

Therefore, in order to increase the limiting speed the coefficient of friction μ_s should be increased. Hence, it is evident that there is more skidding of the vehicle with worn-out tyres. Also, the road surface is made rough for higher speeds.

The overturning moment = $\frac{mv^2}{r} \times h$ ...(12.91)

and the restoring moment = $mg \times \frac{L}{2}$ ...(12.92)

The vehicle will overturn about the outer wheel if R_i is negative, *i.e.*, if the overturning moment exceeds the restoring moment. To ensure that the vehicle does not overturn, it is necessary that

$$\frac{mv^2}{r} \times h < mg \times \frac{L}{2}$$

or, $$v^2 < \frac{glr}{2h}$$

or $$v < \sqrt{\frac{glr}{2h}} \qquad ...(7)$$

or, overturning speed $= \sqrt{\frac{glr}{2h}}$ ...(12.93)

Equating eqns. (12.90) and (12.93), we get

$$\sqrt{\mu_s \,.\, gr} = \sqrt{\frac{glr}{2h}}$$

$$\therefore \qquad \mu_s = \frac{l}{2h} = \text{Limiting value of } \mu_s \text{ for a vehicle} \qquad ...(12.94)$$

and $$h = \frac{l}{2\mu_s} = \text{Limiting value of } h \text{ for a vehicle} \qquad ...(12.95)$$

Problem 12.33. *A car of mass 1600 kg is travelling along an unbanked curve of 80 m radius. The wheel base is 1.6 m and the C.G., is 0.64 m above the ground and mid-way between the two wheels. The coefficient of friction between the wheels and the road is 0.4. Find :*

(a) reactions at wheels at a speed of 63 km/hr ;

(b) the maximum speed at which the car will overturn, and

(c) the skidding speed.

Sol. Referring to Fig. P.12.33, we have

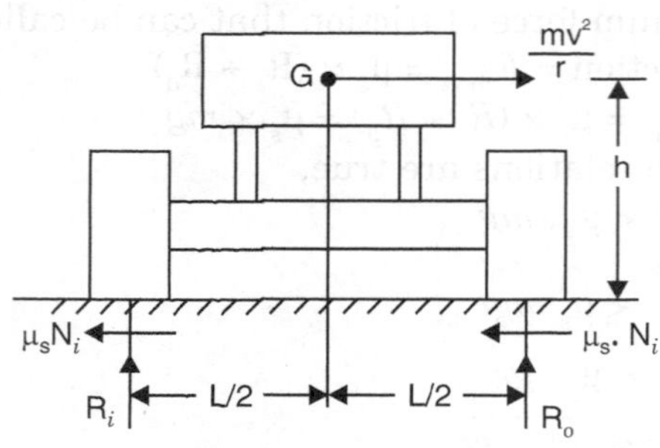

Fig. P. 12.33

$$m = 1600 \text{ kg},$$
$$L = 1.6 \text{ m} ; h = 0.64 \text{ m}$$
$$r = 80 \text{ m}.$$

At 63 km/hr, $v = \dfrac{63}{3.6} = 17.5$ m/s

(*a*) $F = \dfrac{mv^2}{r} = \dfrac{1600 \times (17.5)^2}{80} = 6125$ N

By $\Sigma F_v = 0$, we have

$$R_i + R_o = mg$$

$\therefore$ $R_i + R_o = 1600 \times 9.81 = 15696$ ...(1)

Overturning moment $= \dfrac{mv^2}{r} \times h = 6125 \times 0.64 = 3920$ Nm

Restoring moment $= (R_i - R_o) \times \dfrac{L}{2}$

Equating the overturning moment and the restoring moment, we have

$$(R_i - R_o) \times 0.8 = 3920$$

$\therefore$ $R_i - R_o = \dfrac{3920}{0.8} = 4900$ N ...(2)

Adding (1) and (2), we get

$$2R_i = 15696 + 4900$$

$\therefore$ $R_i = 10298$ N ;

and $R_o = 15696 - 10298 = 5398$ N.

(*b*) Maximum speed for overturning

$$v = \sqrt{\frac{glr}{2h}} = \sqrt{\frac{9.81 \times 1.6 \times 80}{2 \times 0.64}} = \mathbf{31.321 \text{ m/s.}} \quad \textbf{Ans.}$$

(*c*) The skidding speed $= \sqrt{\mu_s \cdot gr} = \sqrt{0.4 \times 9.81 \times 80} = \mathbf{17.72 \text{ m/s.}}$ **Ans.**

12.20. BANKING A CURVE ; SUPER-ELEVATION

Banking a curve is a term used for road ways and super-elevation is used for railway tracks.

(*a*) **Banking a curve.** As we have seen in Art. 12.18 that, when a vehicle is rounding a curved track, an outward centrifugal force $\frac{mv^2}{r}$ is acting on the vehicle which is resisted by the friction between the tyres and the road. The force of friction causes wear and tear in tyres. In order to reduce friction and to increase the permissible speed of the vehicle the outer side of the road is raised by some amount compared to the inner side of the road. This is known as **banking a curve**, as shown in Fig. 12.20.1.

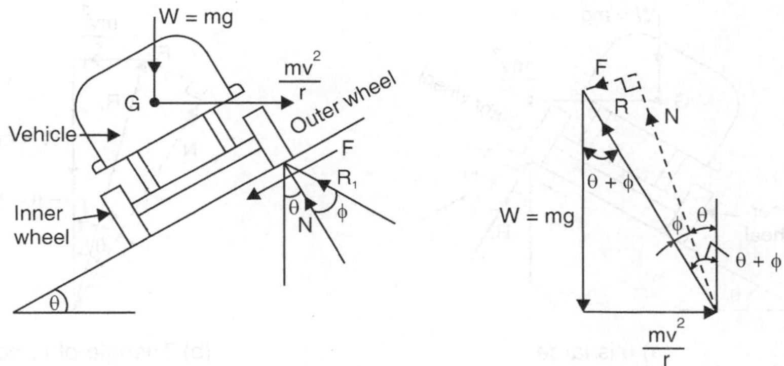

(a) Track as an inclined plane : θ is small (b) Force polygon

Fig. 12.20.1. Banking a curve motion impending upward.

The road in cross-section becomes an inclined plane inclined at an angle θ with the horizontal and it is known as *angle of banking*.

The forces acting on the vehicle are :

(*a*) the weight $W = mg$,

(*b*) the centrifugal force $= \frac{mv^2}{r}$ and

(*c*) the resultant reaction R_1.

Fig. 12.20.1 (*b*) shows the triangle of these forces.

(*i*) **Motion impending upward.** From Fig. 12.20.1 (*b*), we get

$$\tan(\theta + \phi) = \frac{mv^2/r}{mg} = \frac{v^2}{gr} \quad ...(12.96)$$

Hence, the maximum permissible speed to avoid skidding is,

$$v = \sqrt{gr \tan(\theta + \phi)} \quad ...(12.97)$$

Here when angle θ is small and the speed v of the vehicle is large, the vehicle impends to move up along the inclined plane. The force of friction is down the plane and helps in avoiding skidding.

(*ii*) **Motion impending downward.** When the angle θ is large and the speed v of the vehicle is not large, the vehicle has motion impending down the inclined plane. The force of friction F acts upwards. Fig. 12.20.2 shows the vehicle and the triangle of forces.

From the Fig. 12.20.2 (*b*), we have

$$\tan(\theta - \phi) = \frac{mv^2/r}{mg} = \frac{v^2}{rg} \quad ...(12.98)$$

Eqn. (12.105) gives the minimum speed which can be permitted to avoid the skidding inwards.

The minimum speed of the curve is

$$v_{\min} = \sqrt{gr\tan(\theta - \phi)} \quad ...(12.99)$$

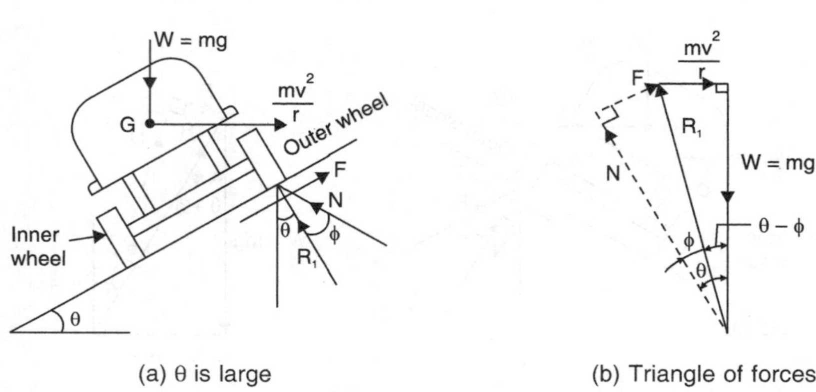

(a) θ is large (b) Triangle of forces

Fig. 12.20.2. Motion impending down the plane.

If v and θ are adjusted such that the vehicle does not slip up or down the plane of the road, θ is known as *ideal angle of banking* (θ_i) and the corresponding speed v is known as the *rated speed*.

At rated speed, thus we find that, the frictional force F is zero, *i.e.*, $\phi = 0$.

Then from eqns. (12.96) and (12.97), we have

$$\tan\theta_i = \frac{v_r^2}{gr} \quad ...(12.100)$$

$$\therefore \quad v_r = \sqrt{gr\tan\theta_i} \quad ...(12.101)$$

where v_r = rated speed

and θ_i = ideal angle of banking.

(*b*) **Super-elevation.** In case of railway track the outer rail in the curved track is raised compared to the inner rail. This amount of difference in the levels of the two rails is known as the *super-elevation*. Let R_o and R_i be the reactions of railway track at the inner outer and inner wheels. The vertical component of $(R_1 + R_2)$ support the weight of the wagon while the horizontal component of $(R_i + R_o)$ is equal to the centrifugal force $\frac{mv^2}{r}$ acting through the *C.G.* of the wagon. Fig. 12.20.3 shows super-elevation of the track and the triangle of forces. As there is no impending motion perpendicular to the rail track, therefore the force of friction is zero, hence $\phi = 0$.

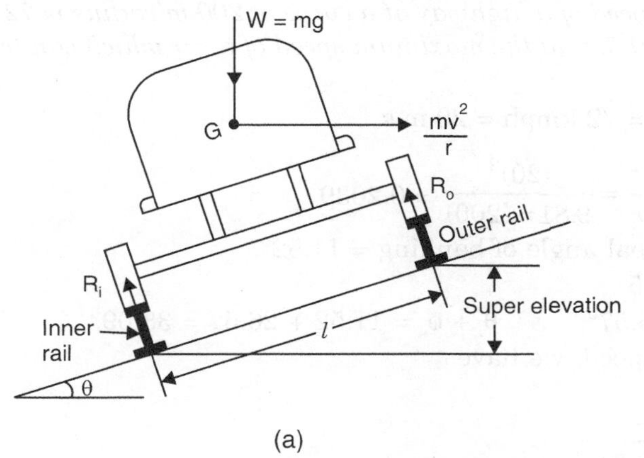

(a)

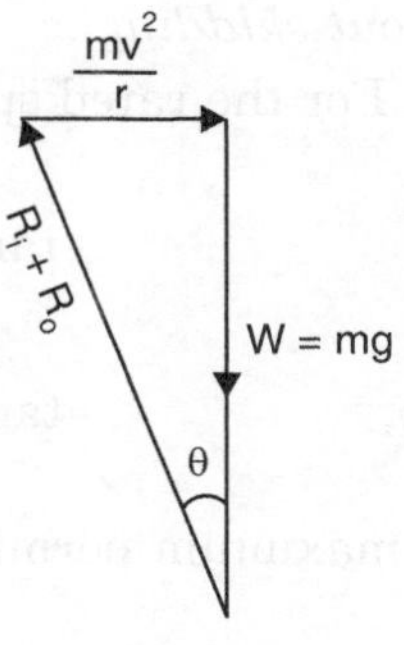

(b) Triangle of forces

Fig. 12.20.3. Super-elevation

From the triangle of forces, by $\Sigma F_V = 0$, we get

$$(R_i + R_o) \cos \theta = mg \qquad ...(i)$$

and by $\Sigma F_h = 0$, we get

$$(R_i + R_o) \sin \theta = \frac{mv^2}{r} \qquad ...(ii)$$

By (ii) ÷ (i), we get

$$\tan \theta = \frac{mv^2/r}{mg} = \frac{v^2}{rg} \qquad ...(12.102)$$

or, directly from Fig. 12.20.3 (b), we get

$$\tan \theta = \frac{mv^2/r}{mg} = \frac{v^2}{rg}$$

$\therefore$ $\sin \theta \simeq \tan \theta$

Angle θ is very small. And if l is distance between two wheels,

$$\text{super-elevation} = l \sin \theta \simeq l \tan \theta = \frac{lv^2}{gr} \qquad ...(12.103)$$

Problem 12.34. *A racing car of mass m travels at a speed of 216 km/hr on a track banked at 25° to the horizontal. The coefficient of static friction μ_s = 0.4. Find the minimum radius of curvature of the track if the car does not skid outwards.*

Sol. Given : $\theta = 25°$, $\tan \phi_s = \mu_s = 0.4$

$\therefore$ $\phi_s = \tan^{-1} 0.4 = 21.80°$

$\therefore$ $\theta + \phi_s = 25 + 21.80 = 46.8°$

$$v = 216 \text{ kmph} = 216 \times \frac{1000}{3600} = 60 \text{ m/s}$$

For the maximum permissible speed, we have $\tan (\theta + \phi_s) = \dfrac{v^2}{gr}$

$$\therefore \quad r = \frac{v^2}{g \tan (\theta + \phi_s)} = \frac{(60)^2}{9.81 \tan (46.8°)} = \mathbf{344.61\ m.} \quad \textbf{Ans.}$$

Problem 12.35. *The rated speed of a highway at a curve of 200 m radius is 72 km/hr. If the coefficient of static friction μ_s = 0.5, find the maximum speed of a car which can traverse the curve without skidding.*

Sol. For the rated speed, v_r = 72 kmph = 20 m/s

and $$\tan \theta_i = \frac{v_r^2}{gr} = \frac{(20)^2}{9.81 \times (200)} = 0.2039$$

$\therefore$ θ_i = ideal angle of banking = 11.52°.

Also, $\tan \phi_s = 0.5$

$\therefore$ $\phi_s = 26.57°$. $\therefore$ $\theta_i + \phi_s = 11.52 + 26.57 = 38.09°$

For maximum permissible speed, we have

$$\tan (\theta_i + \phi_s) = \frac{v^2}{gr}$$

or, $$\tan 38.09° = \frac{v^2}{9.81 \times 200}$$

$\therefore$ $v^2 = 1537.85$

$\therefore$ $v = \sqrt{1537.85}$ = 39.215 m/s = **141.174 kmph. Ans.**

Problem 12.36. *A road track is banked at an angle θ such that tan θ = 0.14. If μ = 0.6 and the distance the wheels of a vehicle be 2m, find the height of the centre of gravity so that the vehicle may overturn and simultaneously skid and the speed at which this may happen. Take the track radius to be 200 m. Also find the minimum speed at the road curve.*

Sol. Let l be the distance between wheels. The height of *C.G.*, is given by :

$$h = \frac{l}{2\mu} = \frac{2}{2 \times 0.6} = \textbf{1.67 m.} \quad \textbf{Ans.}$$

$\theta = \tan^{-1} 0.14 = 7.97°$,

and $\phi = \tan^{-1} \mu = \tan^{-1} 0.6 = 30.96°$

The speed at which the vehicle may skid and overturn simultaneously is given by

$$v^2 = gr \tan (\theta + \phi)$$
$$= 9.81 \times 200 \tan (7.97 + 30.96)$$
$$= 9.81 \times 200 \tan 38.93° = 1584.83$$

$\therefore$ v = **39.81 m/s = 143.32 km/hr. Ans.**

$$v_{min}^2 = gr \tan (\theta \sim \phi).$$
$$= gr \tan (30.96 - 7.97)$$
$$= 9.81 \times 200 \times \tan 22.99° = 832.415$$

$\therefore$ v_{min} = 28.852 m/s = **103.87 km/hr. Ans.**

Problem 12.37. *A locomotive of 150 tonnes mass travels round a curve of 800 m radius at a speed of 72 km/hr. If the gauge is 1.676 m, find the amount by which the outer rail be raised above the level of the inner rail so that there is no thrust between the flanges of the outer wheels and the outer rails.*

Sol. Speed, v = 72 km/hr = 20 m/s.

$$\tan \theta = \frac{v^2}{gr} = \frac{(20)^2}{9.81 \times 800} = 0.05097$$

$\therefore$ Super-elevation $= l \tan \theta = 1.676 \times 0.05097$

= **0.0854 m = 8.54 cm. Ans.**

12.21. COMPOUND PENDULUM

A rigid body of mass m swings about a fixed horizontal axis through a point O. G is centre of gravity. If the angle of oscillation θ is small, it is required to find the periodic time of oscillations. Let the line OG make an angle θ with the vertical as shown in Fig. 12.21.

Let the weight $W = mg$ act through the $C.G.$, G and let the reaction at O be R.

By taking moments about O, we get torque producing rotation given as

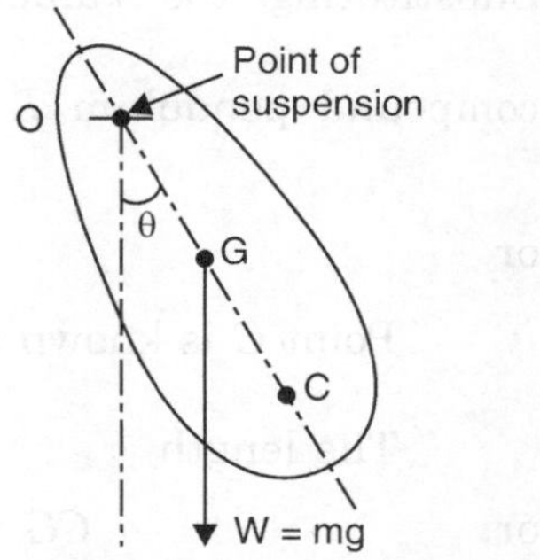

$$C = mg \times OG \sin \theta$$

Point of suspension
O
θ
G
C
W = mg

Fig. 12.21. Compound pendulum

if θ is small, $\sin \theta = \theta$ in radian,

and $$C = mg \times OG \times \theta \quad \text{...(1)}$$

By using relation $\Sigma M_o)_{ext.} = I_o \alpha$,

we get $$C = I_o \alpha = \bar{I}_o \frac{d^2\theta}{dt^2} \quad \text{...(2)}$$

$\therefore$ $$- mg \times OG \,.\, \theta = I_o \, d^2\theta/dt^2 \quad \text{...(3)}$$

[**Note.** Since +ve sign is for CCW couple, the – ve sign is used in eqn. (3)]

where $\bar{I}_o = mk_o^2$

$\therefore$ We get, $- mg \times OG \times \theta = mk_o^2 \times \dfrac{d^2\theta}{dt^2}$

or, $$\frac{d^2\theta}{dt^2} + \frac{g\,.\,OG}{k_o^2}\,.\,\theta = 0 \quad \text{...(4)}$$

or $$\frac{d^2\theta}{dt^2} + \omega^2 \theta = 0 \quad \text{...(5)}$$

Eqns. (4) and (5) are the standard equations for simple harmonic motion, for which

$$\omega = \sqrt{\frac{g\,.\,OG}{k_o^2}} \quad \text{...(12.104)}$$

and the time period, $$\tau = \frac{2\pi}{\omega} = 2\pi\sqrt{\frac{k_o^2}{g\,.\,OG}} \quad \text{...(12.105)}$$

For a simple pendulum of length l, the time period, $\tau = 2\pi\sqrt{l/g}$

Hence, the rigid body of m is equivalent to a simple pendulum of length

$$L = \frac{k_o^2}{OG} \quad \text{...(12.106)}$$

Thus, in problems on compound pendulum we have to find the moment of inertia about the point of suspension O. Thus, by the use of parallel-axis theorem, we get

$$I_o = I_G + m\,.\,OG^2$$

or, $$mk_o^2 = mk_G^2 + m\,.\,OG^2$$

$\therefore$ $$k_o^2 = k_G^2 + OG^2 \quad \text{...(12.107)}$$

where k_G is the radius of gyration of the body about the centre of gravity G of the body. Substituting the value of k_o^2 in eqn. (12.107), we get the equivalent length OC of compound pendulum $L = \dfrac{OG^2 + k_G^2}{OG} = OG + \dfrac{k_G^2}{OG}$

or $$OC = L = OG + \frac{k_G^2}{OG} \quad ...(12.108)$$

Point C is known as centre of oscillation.

The length $$CG = \frac{k_G^2}{OG}$$

or, $$CG \times OG = k_G^2 \quad ...(12.109)$$

12.22. CENTRE OF PERCUSSION

In Fig. 12.22, G is the mass centre and O is the point of support. Let the body be subjected to an impulsive force P acting for a small interval of time. The impulse of the blow will result into the change of momentum in the body which is initially at rest. The impulsive force P at C is equivalent to a force P at G plus a couple $P \times CG$. The force P at G will cause translation and the couple $P \times CG$ will cause rotation about G. Let after a small time t the velocity of point G be v and the angular velocity of rotation about G be ω.

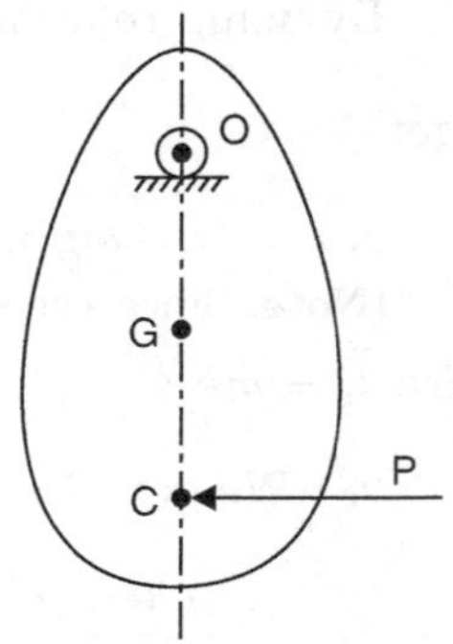

Fig. 12.22. (a) Centre of percussion

The combined motion of translation at G and rotation about G is equivalent to *pure rotation* about some point which is known as the *instantaneous centre of rotation*. The instantaneous centre is a point at rest. Thus the instantaneous centre also receives a blow of impulse $P \times t$.

Join CG and produce it to some point O, then the linear velocity of $O = v_o = (\omega \,.\, OG - v)$.

In order to have velocity v_o of point to be zero, we have

$$OG = v/\omega \quad ...(1)$$

and O will be the instantaneous centre of rotation. To evaluate v and ω,

we have at point G :

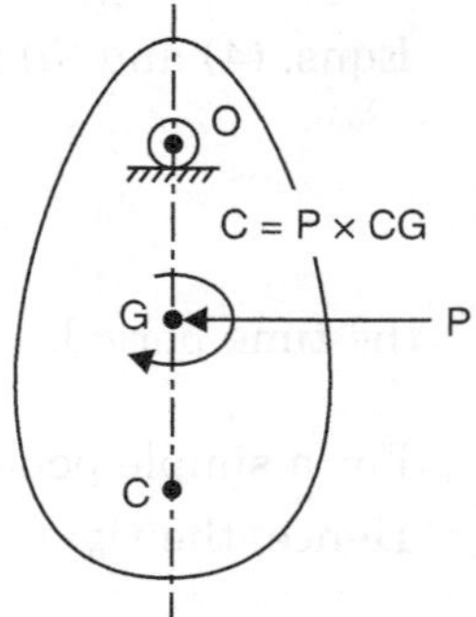

Fig. 12.22. (b) Equivalent system

$$ma_x = m \times \frac{v}{t}$$

But $$ma_x = P$$

$\therefore$ $$P \times t = mv \quad ...(2)$$

The couple $P \times GC$ produces angular acceleration α given by

$$\text{the couple} = C = P \times GC = I_G \times \alpha = I_G \times \frac{\omega}{t}$$

or $$P \times t = I_G \times \frac{\omega}{CG} = mk_G^2 \times \frac{\omega}{CG} \quad ...(3)$$

Equating (2) and (3)

$$mv = mk_G^2 \times \frac{\omega}{CG}$$

or,
$$\frac{v}{\omega} = \frac{k_G^2}{CG} \quad \text{...(4)}$$

But $\frac{v}{\omega} = OG$ from eqn. (1),

$\therefore$ Eqn. (4) gives
$$OG = \frac{k_G^2}{CG} \quad \text{...(12.109(a))}$$

or
$$OG \,.\, CG = k_G^2 \quad \text{...(12.110)}$$

The instantaneous centre O is at rest, and hence there will be no active or reactive force at O, even if there is a hinged support. Point C is called the *centre of percussion* for the blow for the centre of oscillation O.

We have
$$OC = OG + CG = OG + \frac{k_G^2}{OG} \quad \text{...(12.111)}$$

Eqn. (12.111) is similar to the eqn. (12.108) for the compound pendulum.

12.23. FLYWHEELS

A flywheel is a heavy wheel fitted on the shaft of an engine or that of a forging press or a punching machine and stores a large amount of kinetic energy of rotation. The flywheel absorbs kinetic energy when it is in excess of the mean kinetic energy of the engine and gives away the same to the engine when the engine develops less energy. Thus, flywheels acts like a reservoir of kinetic energy, or a bank of kinetic energy from which energy can be drawn when necessary and into which energy can be stored.

When kinetic energy is stored into flywheels its speed increases, while its speed decreases when kinetic energy is taken from it.

Let N_1 be the maximum speed in rpm and N_2 be the minimum speed during a cycle. The corresponding maximum and minimum kinetic energies are E_1 and E_2, respectively.

Now,
$$E_1 = \frac{1}{2} I\omega_1^2, \quad \text{where} \quad \omega_1 = \frac{2\pi N_1}{60}$$

and
$$E_2 = \frac{1}{2} I\omega_2^2$$

$\therefore$ Change in
$$\text{K.E.} = \Delta E = E_1 - E_2 = \frac{1}{2} I(\omega_1^2 - \omega_2^2) \quad \text{...(12.112)}$$

or
$$\Delta E = \frac{1}{2} I(\omega_1 + \omega_2)(\omega_1 - \omega_2)$$

or
$$\Delta E = I\omega \times (\omega_1 - \omega_2). \quad \text{...(12.113)}$$

because
$$\omega_1 + \omega_2 = 2\omega.$$

Also, from eqn. (12.113),

$$\Delta E = I\omega^2 \times \left(\frac{\omega_1 - \omega_2}{\omega}\right) = I\omega^2 \times q \quad \text{...(12.114)}$$

or
$$\Delta E = I\omega^2 \, \frac{k_s}{100} \quad \text{...(12.115)}$$

where $q = \frac{\omega_1 - \omega_2}{\omega}$ = coefficient of fluctuation of speed and $$k_s = \frac{\omega_1 - \omega_2}{\omega} \times 100 = \text{percentage coefficient of fluctuation of speed.}$$

Let $$E = \frac{1}{2} I\omega^2 \quad ...(12.116)$$

be the mean K.E. of flywheel.

Then from eqn. (12.115), we get

$$\frac{\Delta E}{I\omega^2} = \frac{2ks}{100}$$

or, $$\Delta E = \frac{2ks \,.\, E}{100} \quad ...(12.117)$$

Let W = workdone per cycle

Then, $\frac{\Delta E}{W} = k_e$ = coefficient of fluctuation of energy ...(12.118)

Problem 12.38. *A flywheel of mass 4 tonnes is mounted on a 120 mm shaft. Find kinetic energy of rotation at a speed of 150 rev/min, if its radius of gyration is 0.8 m.*

If the coefficient of friction between the shaft and the bearings is 0.018, find the amount of work absorbed by the friction in one revolution and also find the number of revolutions made by the flywheel before coming to rest and the time it will take.

Sol. $$\omega = \frac{2\pi N}{60} = \frac{2\pi \times 150}{60} = 5\,\pi \text{ rad/s}$$

$$I = mk^2 = (4000) \times (0.8)^2 = 2560 \text{ kgm}^2$$

$$E = \frac{I\omega^2}{2} = \frac{1}{2} \times 2560 \times (5\pi)^2 = \mathbf{315827.34\ Nm.}\quad \textbf{Ans.}$$

The frictional torque at the bearing = $T_f = \mu W \times r$

or $$T_f = \left[0.018 \times (4000 \times 9.81) \times \frac{0.120}{2}\right] = 42.38 \text{ Nm}$$

Energy absorbed per revolution = $T_f \times 2\pi$ = 266.16 N-m.

∴ Number of revolutions made to absorb 315827.34 Nm of energy

$$= \frac{315827.34}{266.16} = \mathbf{1186.61\ revolutions.}\quad \textbf{Ans.}$$

$$\text{Time taken} = \frac{1186.61}{\text{average speed}} = \frac{1186.61}{\left(\frac{150}{2}\right)} = \mathbf{15.82\ min.}\quad \textbf{Ans.}$$

Problem 12.39. *A flywheel of 4000 kg and with a radius of gyration 0.6 m gains a speed of 216 rpm in 3 minutes. Calculate the average torque exerted on it.*

Sol. $$\omega = \frac{2\pi \times 216}{60} = 7.2\,\pi \text{ rad/s}$$

$$t = 3 \text{ min.} = 180 \text{ seconds}\ ;\ \omega_0 = 0$$

Using $$\omega = \omega_0 + \alpha t$$

we get $$\alpha = \frac{7.2\pi - 0}{180} = 0.1257 \text{ rad/s}$$

∴ $$\text{Torque} = I\alpha = [4000 \times (0.6)^2] \times 0.1257 \text{ Nm} = \mathbf{181.01\ Nm.}\quad \textbf{Ans.}$$

12.24. CIRCUMFERENTIAL OR HOOP STRESS DUE TO ROTATION

Consider a thin ring of radius r and radial thickness t rotating at an angular velocity ω. Let γ be the specific density of ring material in N/unit volume. Let b be the width of the ring.

The centrifugal force $= ma_n = \dfrac{mv^2}{r}$ acts in radially outward direction as shown in Fig. 12.24 (*a*). The ring is cut in two halves by a cutting plane 1-1. The upper half is considered as shown in Fig. 12.24 (*b*). Let σ be the tensile stress induced normal to the section 1-1. Consider two elements shown in figure, each subtending an angle $\delta\theta$ at the centre O. The centrifugal force on each element is $\delta F = \delta m . \dfrac{v^2}{r}$.

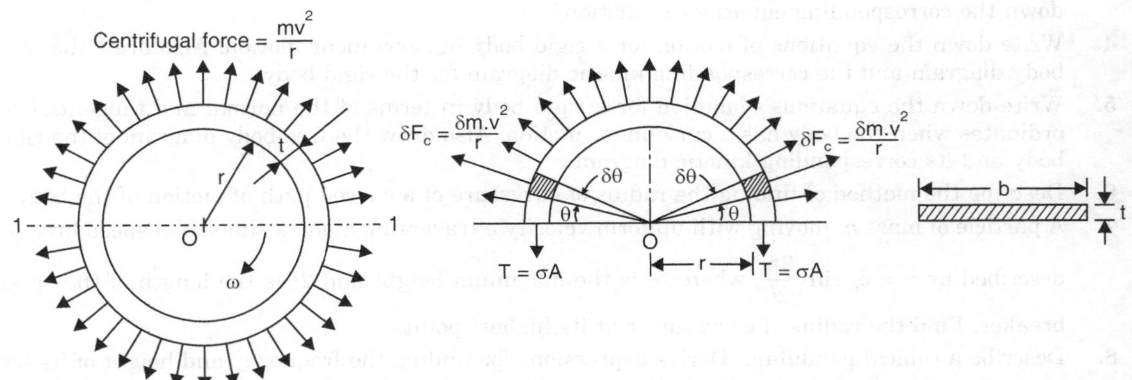

(a) Centrifugal force in a thin ring due to rotation (b) hoop stress (σ) due to rotation (c) Cross-section of ring

Fig. 12.24. Circumferential or hoop stress in a thin ring due to rotation

where $$\delta m = \frac{\delta w}{g} = \frac{(\gamma A . r . \delta\theta)}{g}$$

Hence $$\delta F = \frac{\gamma A r . v^2 \delta\theta}{rg} = \frac{\gamma A v^2}{g}\ \delta\theta \qquad ...(1)$$

By $\Sigma F_x = 0$, we get $\int dF_x = 0$, by symmetry.

and $\Sigma F_y = 0$ gives $2\int_0^{\pi/2} \delta F \sin\theta - 2\sigma . A = 0$

or $$2\int_0^{\pi/2} \left(\frac{\gamma A v^2}{g}\ d\theta\right) \sin\theta = 2\sigma A$$

Cancelling $2A$, we get

$$\frac{\gamma v^2}{g}\bigg(-\cos\theta\bigg)_0^{\pi/2} = \sigma$$

$\therefore$ Hoop stress $$\sigma = \frac{\gamma v^2}{g} \qquad ...(12.119)$$

And, tension $= T = \sigma A = \dfrac{\gamma v^2}{g} A = \dfrac{\gamma v^2 bt}{g}$ ...(12.120)

Here A = area of cross-section of the ring = bt. ...(12.121)

EXERCISES XII

A. Theoretical Questions

1. State equations of motions for translation motion of a rigid body.
2. Describe, with suitable sketches, the various cases of force systems.
3. Draw free body diagram and the kinetic diagram of the rigid body in translation. Also write down the corresponding equations of motion.
4. Write down the equations of motion for a rigid body in curvilinear motion. Also draw the free-body diagram and the corresponding kinetic diagram for the rigid body.
5. Write down the equations of motion for a rigid body in terms of the normal and tangential co-ordinates when the body has a curvilinear motion. Also draw the free-body diagram of the rigid body and its corresponding kinetic diagram.
6. Describe the method of finding the radius of curvature of a curved path of motion of the body.
7. A particle of mass m moving with uniform velocity v traversing along a sinusoidal speed breaker described by $y = \delta_0 \sin \dfrac{2\pi}{2l}$, where δ_0 is the maximum height and $2l$ is the length of the speed breaker. Find the radius of curvature r at its highest point.
8. Describe a conical pendulum. Derive expressions for finding the frequency and height of its bob of mass m. Also find the tension in the suspension cord.
9. Define moment of momentum of particle describing a curvilinear motion.
10. Derive the expression $\dfrac{d}{dt} H_o = M_o$ for curvilinear motion, where H_o is the moment of momentum with respect to point O, and M_o is the moment of the resultant of forces about point O.
11. Establish the work-energy equation for curvilinear motion.
12. State the work-energy equation.
13. Derive, and explain, the equation $\sum M = I\ddot{\theta}$ for rigid body rotation.
14. Derive the expression for moment of momentum for the rigid body rotation.
15. What do you understand by centroidal and non-centroidal rotation ?
16. Derive expressions for angular displacement (θ), and angular velocity for rotation of a rigid body under the action of a constant acceleration.
17. Derive equation of dynamic equilibrium in motion of rotation of a rigid body about a fixed axis.
18. Derive an expression for (*a*) work done and (*b*) kinetic energy due to a couple acting on a rigid body.
19. What do you understand by angular momentum and angular impulse ?
20. What is plane motion ?
21. Derive expressions for skidding speed and overturning speed of a vehicle moving along a curved rod.
22. What do you understand by the terms "limiting value of coefficient of adhesion" in the light of the motion of a vehicle traversing a road curve ? Establish the suitable expression for this.
23. What do you mean by the term "banking a curve" for motion of a vehicle impending upward a curved road.

24. What is "rated speed" of a high way ?

25. Explain super-elevation. Derive suitable expression for super-elevation.

26. What is a compound pendulum ? Derive expression for equivalent length of a compound pendulum.

27. Define centre of precession. Derive expression for finding its position.

28. What is the function of a flywheel ?

29. Define terms : "coefficient of fluctuation of speed", and "coefficient of fluctuation of energy" of flywheel.

30. Derive expression for finding hoop stress developed in the rim of a rotating ring.

B. Numerical Problems

12.1. A simple wheel and axle h as a mass m and a radius of gyration k [Fig. E.12.1] . The system is accelerated by a descending weight W attached to a cord wound round the axle and radius r. Prove that acceleration of the descending mass is $\alpha = \dfrac{W}{W + mg\ k^2/r^2} \times g$ and the tension of the cord is $T = \dfrac{mg\ Wk^2}{Wr^2 + mgk^2}$.

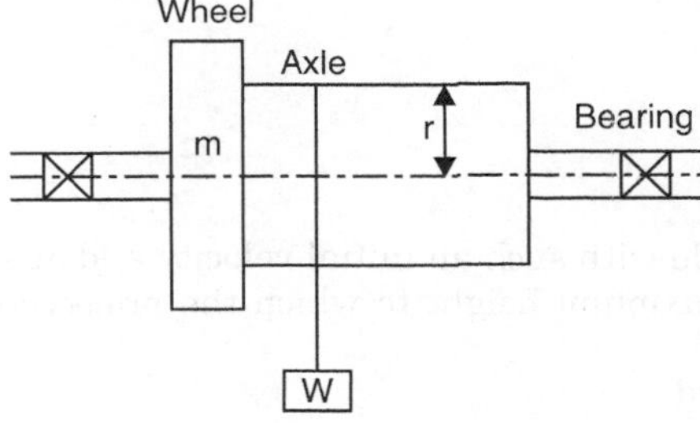

Fig. E. 12.1

12.2. An inextensible string passes over a pulley of mass m, radius and radius of gyration k and carries suspended weights W_1 and W_2 at its ends. If there is no slip of string over pulley, prove that the acceleration of the string is

$$a = \frac{W_1 - W_2}{W_1 + W_2 + mgk^2/r^2} \times g$$

and the difference in the tensions in the two ends of the string is

$$T_1 - T_2 = \left[\frac{(W_1 - W_2)\ mgk^2/r^2}{W_1 + W_2 + mgk^2/r^2}\right].$$

12.3. A circular disc (2) is driven through frictional contact with disc (1) as shown in Fig. E. 12.3. External torque T acts on disc 1. Prove that the resultant angular acceleration of disc 2 is

$$\alpha = \frac{Tr_1 . r_2}{I_1 r_2^{\ 2} + I_2 . r_1^{\ 2}}.$$

Fig. E. 12.3

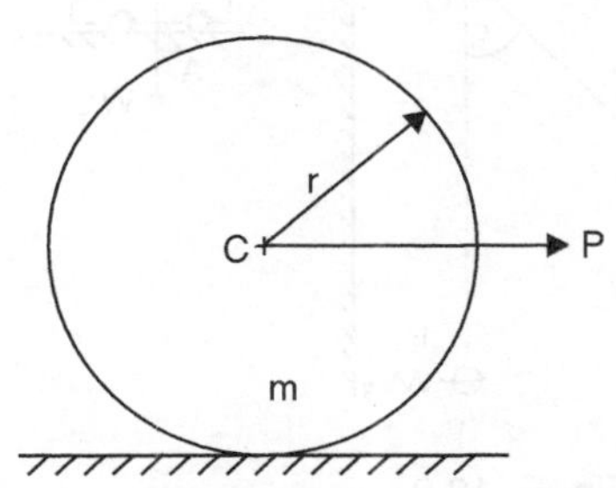

Fig. E.12.4

12.4. A horizontal force P applied to the centre of the mass of a sphere of radius r and mass m causes it to roll without slip [Fig. E. 12.4]

(*a*) Find the acceleration of the sphere, and

(*b*) If sphere is replaced by a disc of the same mass and radius find the acceleration of the disc.

Take μ as coefficient of friction in both cases. $\left[\textbf{Ans. } (a)\ a = \frac{5}{7}P\ ;\ (b)\ a = \frac{2P}{3}\right]$

12.5. Small oscillations of a simple pendulum of weight W and length l are defined by the equation $S = S_o \cos \omega t$, where S_o is the amplitude of swing and ω is a constant equal to $\sqrt{g/l}$. Find the maximum value of the tensile force T induced in the string. [**Ans.** $T_{max} = W(1 + S_o^2/l^2)$]

12.6. A small block of weight W rests on a horizontal turntable at a distance r from the axis of rotation. If μ is the coefficient of friction between block and the surface of the turntable, find the maximum uniform speed v_{max} that the block can have owing to rotation of the turntable without slip. See Fig. E. 12.6.

[**Ans.** $v_{max} = \sqrt{ugr}$]

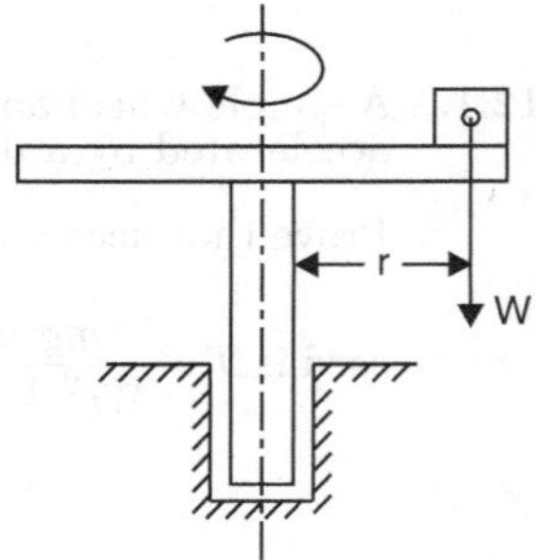

Fig. E. 12.6

12.7. A cannon fires its projectile with such an initial velocity and at such an angle of elevation α that the range is r and the maximum height to which the projectile rises is h. Find the maximum range that can be obtained. $\left[\textbf{Ans. } R_{max} = 2k + \frac{r^2}{8h}\right]$

12.8. Prove that maximum bending moment in the side rod BC of length l of an engine shown in Fig. E. 12.8 is given by

$$M_{max} = \left(q + \frac{qv^2}{gr}\right)\frac{l^2}{8}.$$

where q is the weight of the rod BC per unit length.

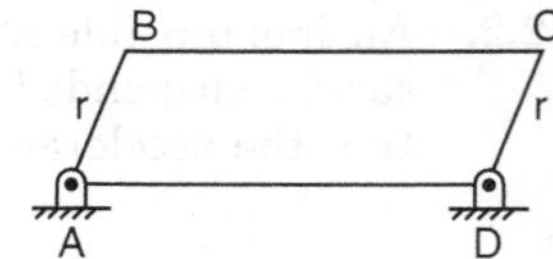

Fig. E. 12.8

12.9. A simple pendulum of weight W and length l, as shown in Fig. E. 12.9, is released from rest at A, swings, downward and strikes a spring of stiffness k at B. Neglecting the mass of the spring, prove that if α = 60° the maximum compression of the spring = $\delta = \sqrt{\frac{wl}{k}}$.

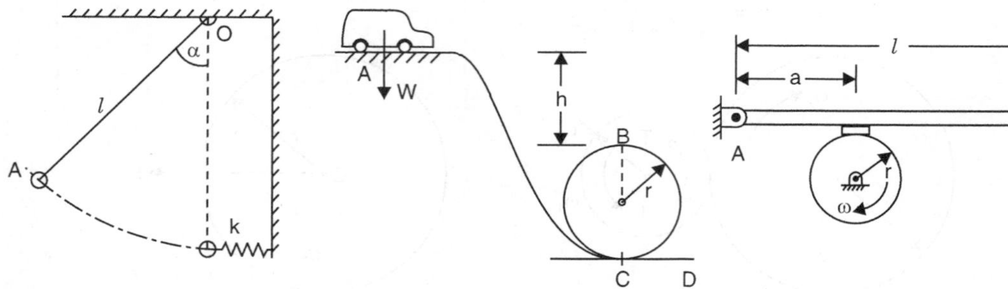

Fig. E. 12.9 Fig. E. 12.10 Fig. E. 12.11

12.10. A small car of weight A starts from rest at A rolls without friction along the loop-the-loop $ACBD$ as shown in Fig. E.12.10. What is the least height h above the top of the loop at which car should start without falling off the track at point B, and what is velocity of the car in straight horizontal portion CD of the track ? [**Ans.** $h_{min} = r/2$; $v = \sqrt{5gr}$]

12.11. A solid circular drum of radius r and weight W rotates in its bearings at ω rad/second. The drum is braked by the application of the force P as shown in Fig. E.12.11. If m is the coefficient of friction, prove that the number of revolutions made by the drum before coming to rest can be obtained from

$$n = \frac{W\omega^2 \, ra}{8\pi \, \mu gl}.$$

12.12. Find the period of small oscillations of a circular cylindrical bar OA of length L and radius r suspended from O, Fig. E.12.12. [**Ans.** $\tau = 2\pi\sqrt{L/g}$, and $L = 2l/3 + r^2/2l$]

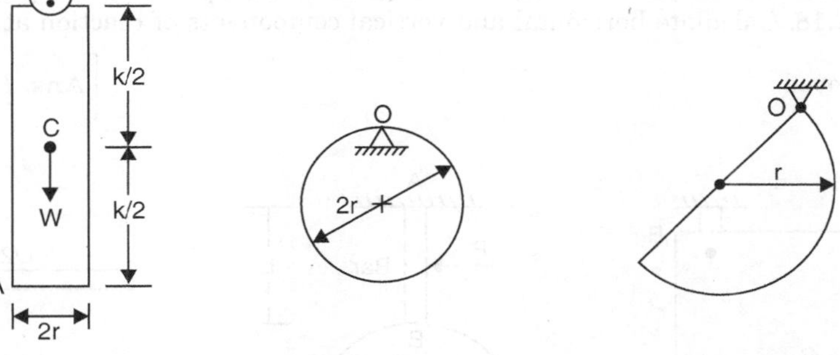

Fig. E. 12.12 Fig. E. 12.13 Fig. E. 12.14

12.13. A thin circular hoop (a hollow ring) of diameter $d = 2r$ hangs over a knife edge at O as shown in Fig. E.12.13. Prove that period τ for small oscillations is given by $\tau = 2\pi\sqrt{\dfrac{2r}{g}}$.

12.14. A homogeneous semi-circular disc of radius r is suspended at O, as shown in Fig. E.12.14, and swings with small amplitude. Calculate the period. [**Ans.** $\tau = 2\pi\sqrt{1.38\, r/g}$]

12.15. Prove that period τ for small oscillations of the compound pendulum shown in Fig. E. 12.15 is $\tau = 2\pi\sqrt{17l/18g}$. Treat OD and AB as identical bars of uniform cross-section.

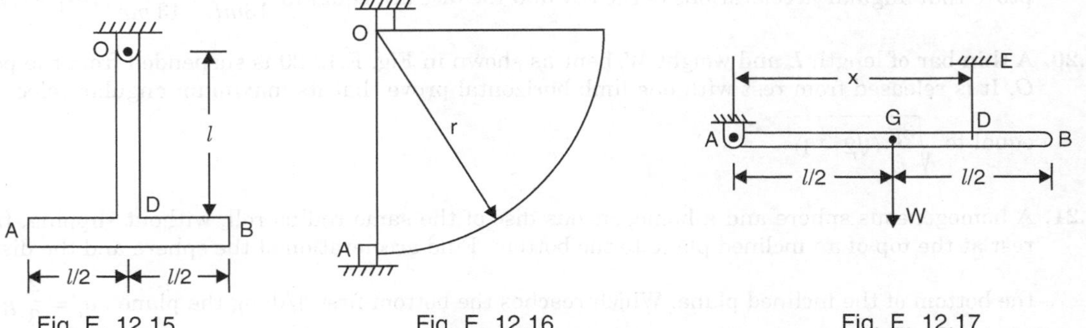

Fig. E. 12.15 Fig. E. 12.16 Fig. E. 12.17

12.16. A homogeneous thin plate in the form of a circular quadrant of radius r is supported in the vertical plane by a hinge at O and a stop at A shown in Fig. E.12.16. Find the horizontal and vertical components of reaction at the hing O : (a) before the stop A is removed, and (b) just after it is removed, (c) angular acceleration 3π of the plate.

[**Ans.** (a) $R_{A_x} = \dfrac{4W}{3\pi}$; $R_{A_y} = W$; (b) $R_{A_x} = \dfrac{1}{\sqrt{2}}\dfrac{W}{g}c\,\ddot{\theta}$;

$R_{A_y} = W - \dfrac{1}{\sqrt{2}}\dfrac{W}{g}c\,\ddot{\theta}$, where $c = \sqrt{2}\,(4r/3\pi)$; (c) $\ddot{\theta} = \dfrac{32g}{3\pi r}$]

12.17. A thin prismatic bar AB is supported in a horizontal position as shown in Fig. E.12.17. At what distance x from the hinge A should the vertical cord DE be attached to the bar in order that when it is cut there will be no immediate change in the reaction at A ? $\left[\textbf{Ans. } x = \dfrac{2l}{3}\right]$

12.18. A homogeneous square plate of weight W hangs in vertical plane by two pins A and B as shown in Fig. E.12.18. Calculate horizontal and vertical components of reaction at B just after the pin at A is removed. $\left[\textbf{Ans. } H_B = \dfrac{3W}{8} ; V_B = \dfrac{5W}{8}\right]$

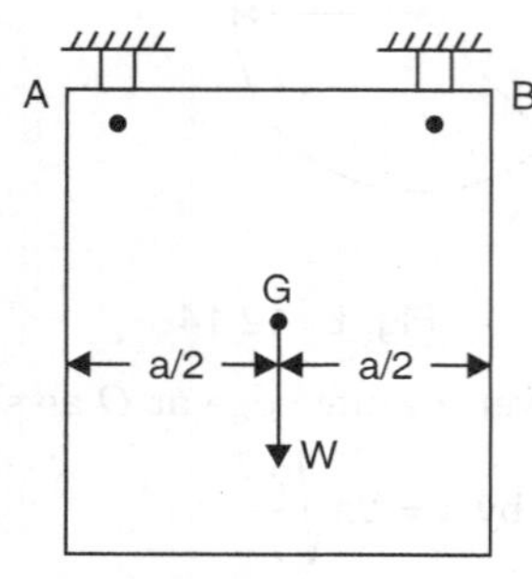

Fig. E. 12.18

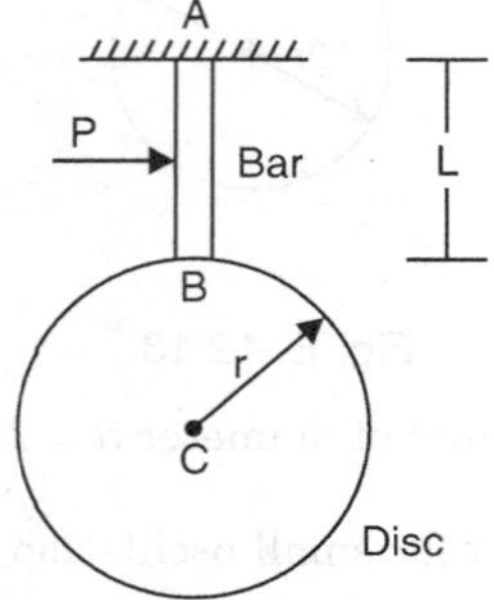

Fig. E. 12.19

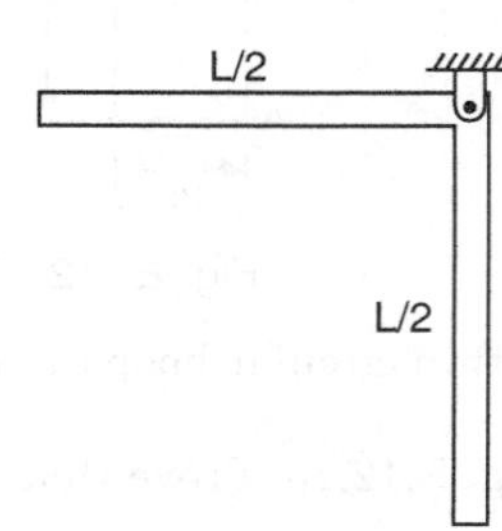

Fig. E. 12.20

12.19. A bar of mass M and length L is pinned to a disc of mass m and radius r and their assembly is suspended at point A as shown in Fig. E.12.19. If a force P is applied at the centre of the bar prove that angular accelerations of the bar and the disc are equal to $\dfrac{24P}{13mL}$; $\dfrac{12P}{13\,mr}$, respectively.

12.20. A thin bar of length L and weight W, bent as shown in Fig. E.12.20 is suspended from the point O. It is released from rest with one limb horizontal prove that its maximum angular velocity is equal to $\sqrt{\dfrac{3g}{L}(\sqrt{2}-1)}$.

12.21. A homogeneous sphere and a homogeneous disc of the same radius roll, without slipping, from rest at the top of an inclined plane to the bottom. Find acceleration of the sphere and the disc at the bottom of the inclined plane. Which reaches the bottom first. [Along the plane : $a_c = \dfrac{2}{3}\,g$ sin θ, $a_s = \dfrac{5}{7}\,g \sin\theta$; sphere reaches the bottom first].

12.22. In Fig. 12.22, a block is moving to the right with an acceleration of 1.5 m/s^2. A homogeneous bar 1 m long and of mass 1 kg is suspended as shown. The pin A at the point of suspension is frictionless. Find the angle θ. [**Ans.** θ = 8.7°]

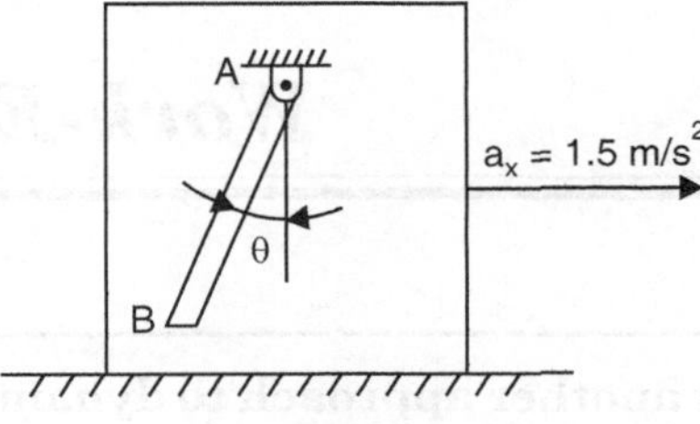

Fig. E. 12.22

12.23. A small block moves with an acceleration a_x as shown in Fig. E.12.23. The bob at the end of the string maintains a constant angle θ. Prove that $a_x = g \tan \theta$.

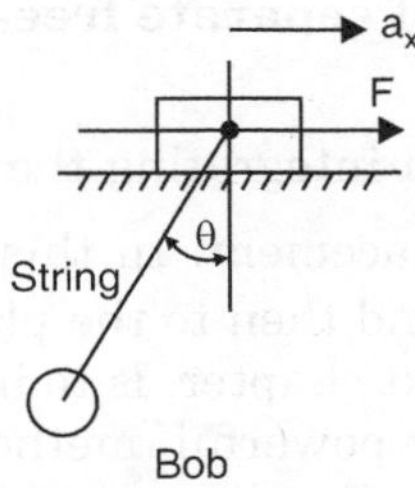

Fig. E. 12.23

13

Work-Energy Method

13.1. INTRODUCTION

In this chapter we present another approach to dynamics which directly relates forces, displacements and velocities, and there is no need to calculate accelerations. This approach, known as work-energy method, is faster and easier to apply than the force-inertia method discussed in the preceding chapters. Work-energy method involves only scalar quantities. Also, the work-energy method can be applied directly to the whole system of connected bodies thereby eliminating forces which are internal to the system and hence the drawing of separate free-body diagrams of each part of the system is not required.

The work-energy method is based on integrating the fundamental equations $\overline{R} = m\overline{a}$ and $\Sigma\overline{M} = \overline{H}$ directly in terms of the displacement. In this chapter, we shall first apply the work-energy equations to the translation and then to the plane motion of rigid body. Another approach, which we shall discuss in the next chapter, is to integrate the above equations with respect to time, thereby leading to the very powerful method, called the impulse-momentum method. Many a times, a combination of work-energy and impulse-momentum methods will solve problems that are otherwise very difficult.

13.2. WORK-ENERGY EQUATION FOR TRANSLATION

The general problem in dynamics is to determine the relation between the force system and the motion of the body or a system of bodies. For a particle or a translating rigid body, we have to solve the equation $R = ma$ for the acceleration and then integrate one of the equations $a = \dfrac{dv}{dt}, v = \dfrac{ds}{dt}$ or $a \,.\, ds = v.\, dv$, where R = resultant of the force system.

Hence, thus using the relation

$$R = ma \quad \text{for the motion of translation,}$$

$$\therefore \qquad R = ma = m \,.\, \frac{dv}{dt} = m \,.\, \frac{dv}{ds} \times \frac{ds}{dt} = m \,.\, \frac{dv}{ds} \times v$$

and, by transposing, we get

$$R \,.\, ds = mv.\, dv \qquad \text{...}(a)$$

Integrating eqn. (a), we get

$$\int_{s_1}^{s_2} R\, ds = \int_{v_1}^{v_2} mv\, dv = \frac{1}{2} m({v_2}^2 - {v_1}^2) \qquad \text{...(13.1)}$$

The left-hand side is known as the resultant work (U_{1-2}) done upon the particle and the term $\dfrac{1}{2}\, mv^2$ is called the kinetic energy (abbreviated as KE). Thus, we get

$$U_{1-2} = \frac{1}{2} m(v_2^2 - v_1^2) = \Delta KE \qquad ...(13.2)$$

This is known as work-energy equation for a particle. It states that the resultant workdone upon a particle between positions 1 and 2 is equal to the change in kinetic energy.

13.3. WORKDONE BY A CONSTANT FORCE

A force does work when its point of application moves in any direction, but not perpendicular to the direction of the force. Work is the product of the force and the displacement in the direction of the force. **Work is a scalar quantity, therefore it has sign but no direction.** Work of a force is equal to the product of the displacement and the component of force in the direction of the displacement [Fig. 13.3.1 (*a*)].

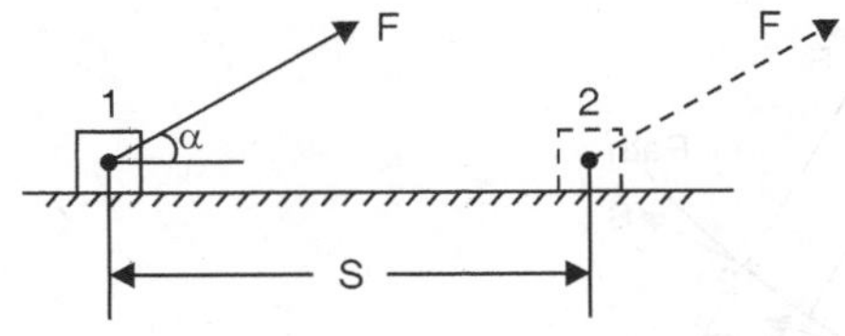

α = angle between F and S
(F projected along displacement)

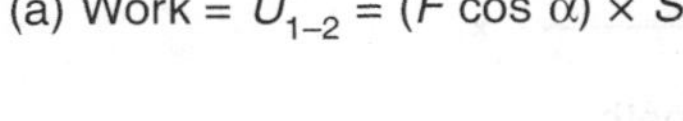

(a) Work = $U_{1-2} = (F \cos \alpha) \times S$

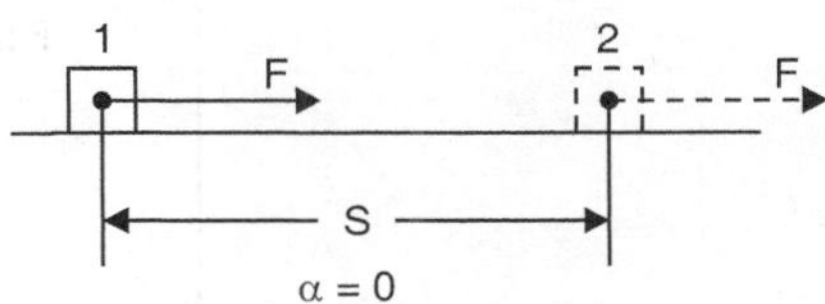

(b) Work = $U_{1-2} = F \cos (0) \times S = FS$

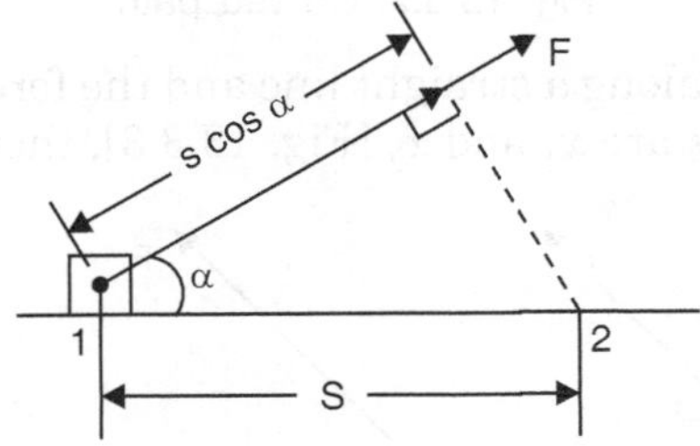

(*S* projected along this force *F*)

(c) Work = $F \times (S \cos \alpha)$

Fig. 13.3.1

If the angle between the force F and the resultant displacement s is α, then workdone by the force $= U_{1-2} = (F \cos \alpha) \, . \, S$ ...(13.3)

When the force F is collinear with displacement S [Fig. 13.3.1 (*b*)] $\alpha = 0$. Then the work done by the force when the point of application move from position 1 to position 2 is

$$U_{1-2} = F \cos (0) \times S = FS \qquad ...(13.4)$$

In Fig. 13.3.1 (*c*), the component of displacement projected along the force is $(S \cos \alpha)$ and the workdone is given by :

$$U_{1-2} = F \times (S \cos \alpha) \qquad ...(13.5)$$

Eqns. (13.3) and (13.5) are the same.

Thus, the work done is the dot product of force vector $\overline{F}$ and the displacement vector $(\overline{S})$,

$\therefore \qquad U_{1-2} = \overline{F} \,.\, \overline{S}$ ...(13.6)

Hence, when $\alpha = 90°$,

$$U_{1-2} = (F \cos 90°) \times S U_{1-2} = 0$$

When $\alpha = 180°$, $U_{1-2} = (F \cos 180°) \times S = -\overline{F} \,.\, \overline{S}$

i.e., the displacement is in the direction opposite to the direction of the force F. The example of negative work is the workdone by the force of friction.

When the path is curved, as shown in Fig. 13.3.2. Then workdone by the force F when moves its point of application from S_1 to S_2 is given by :

$$U_{1-2} = \int_{S_1}^{S_2} (F \cos \alpha)\, ds \qquad ...(13.7)$$

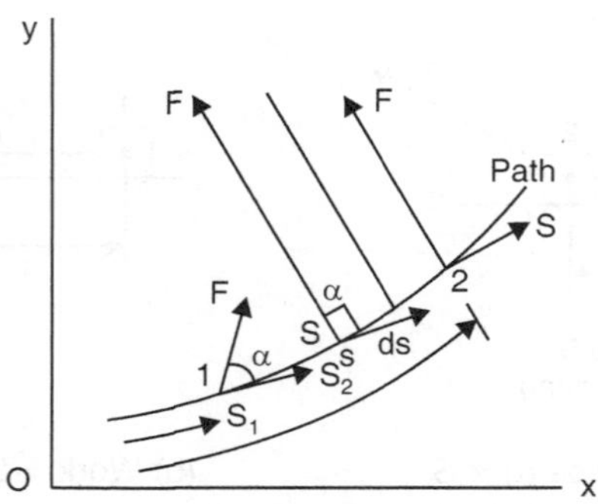

Fig. 13.3.2. Curved path

When the particle moves along a straight line and the force is inclined at different angles θ_1 and θ_2 and the displacements are x_1 and x_2 [Fig. 13.3.3], then the workdone is given by

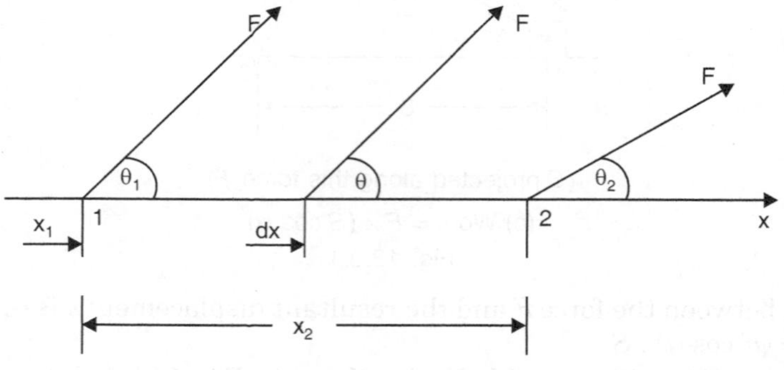

Fig. 13.3.3. Straight path

$$U_{1-2} = \int_{x_1}^{x_2} F_x \,.\, dx \qquad ...(13.8)$$

13.4. WORKDONE BY A VARIABLE FORCE

When the force F is variable and the displacement is along x-axis, the workdone by the force can be obtained by integrating the elemental workdone $dU = Fdx$.

The total workdone $= U_{1-2} = \int_{x_1}^{x_2} F\,dx$ ...(13.9)

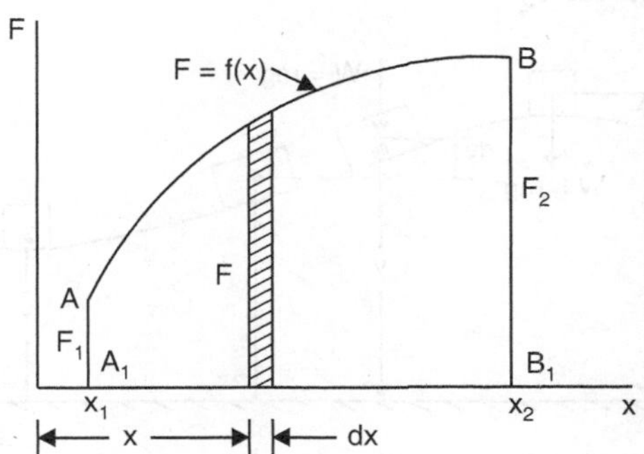

Fig. 13.4.1

The force F in eqn. (13.9) should be a known function of the displacement x.

The workdone U_{1-2} in Fig. 13.4.1 is equal to the area $AA_1\,BB_1$ under the curve AB.

One example of the variable force is workdone by the expansion of a perfect gas following the law $pv^n = c$ or $pv = c$. [See Fig. 13.4.2]

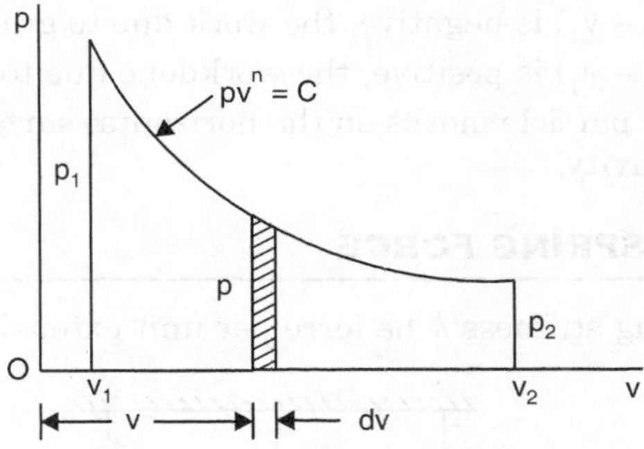

Fig. 13.4.2

Consider an element of pressure = p and of width dv

$\therefore$ Workdone for the elemental expansion of the gas = $dW = p.dv$

$$\therefore \quad \text{Workdone} = \int dW = \int_{v_1}^{v_2} p\,dv\ ; \quad \text{writing } p = \frac{C}{v^n}, \text{ we get}$$

$$W = \int_{v_1}^{v_2} \left(\frac{C}{v^n}\right).dv = \frac{C}{1-n}\Big[v^{1-n}\Big]_{v_1}^{v_2}$$

$$= -\frac{p_1 v_1{}^n \times v_1{}^{1-n} - p_2 v_2{}^n\, v_2{}^{1-n}}{1-n} = -\frac{p_1 v_1 - p_2 v_2}{1-n}$$

or, $$\text{workdone, } W = \frac{p_2 v_2 - p_1 v_1}{n-1}.$$

13.5. WORKDONE BY THE FORCE OF GRAVITY

Consider a particle of weight $W = mg$ moving under the force of gravity as shown in Fig. 13.5. The workdone is given by

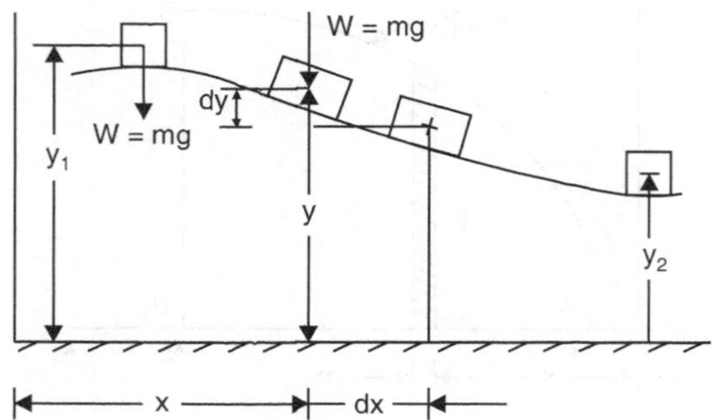

Fig. 13.5. Workdone by force of gravity.

$$U_{1-2} = \int_{y_1}^{y_2} P_y \,.\, dy = \int_{y_1}^{y_2} W\, dy = -\, mg \int_{y_1}^{y_2} dy$$

[As dy is +ve upward and –ve downward].

or $$U_{1-2} = -\, mg\, (y_2 - y_1) \qquad ...(13.10)$$

(*a*) When $(y_2 - y_1)$ is negative, the work due to gravity will be positive.

(*b*) When $(y_2 - y_1)$ is positive, the workdone due to gravity is negative.

(*c*) When the particle moves on the horizontal surface $(y_2 - y_1) = 0$, and hence no work is done by force of gravity.

13.6. WORK OF SPRING FORCE

Let the spring stiffness k be force per unit extension (or contraction).

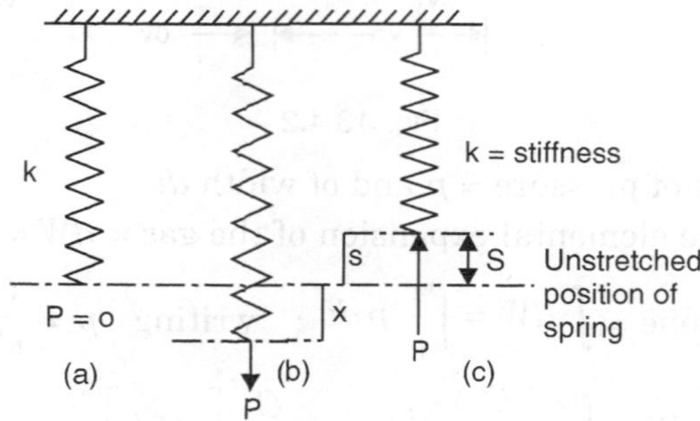

Fig. 13.6.1. Work of spring force (a) No external force (F = 0) ;

(b) spring stretched a distance s, (c) spring compressed a distance s.

Fig. 13.6.1 shows spring in three different cases. Force P is zero in Fig. 13.6.1 (*a*) and the spring is in unstretched position. At Fig. 13.6.1 (*b*), spring is stretched by a distance s by the force P. And, in Fig. 13.6.1 (*c*) the spring is compressed a distance s.

$\therefore$ Stiffness, $k = \dfrac{\text{Spring Force}}{\text{Deformation}} = \dfrac{P}{s}$ ...(13.11)

k is also known as spring modulus or spring constant.

$\therefore$ Spring force, $P = ks$ ...(13.12)

The force-deformation diagram is linear, as shown in Fig. 13.6.2

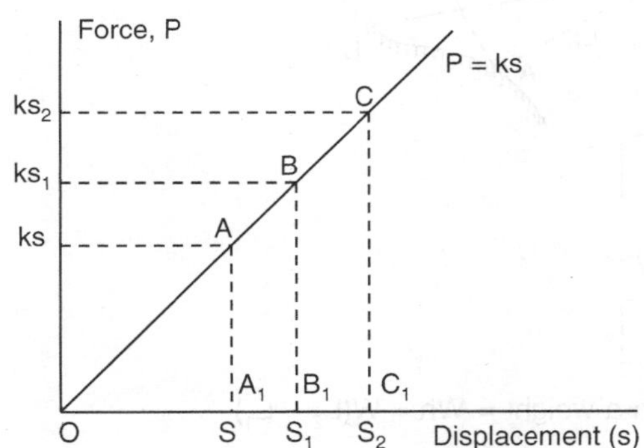

Fig. 13.6.2. Force-displacement diagram for a spring

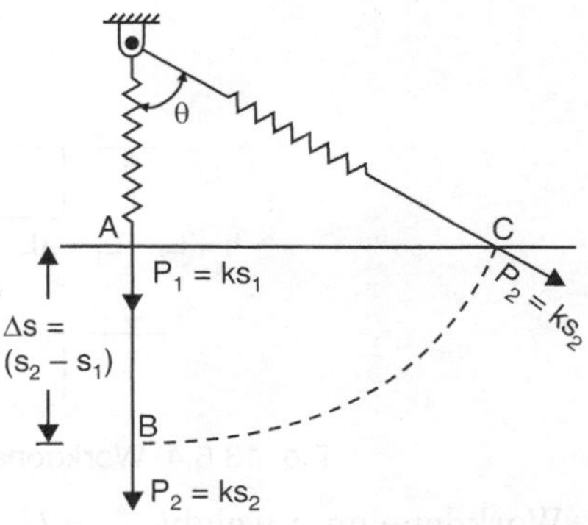

Fig. 13.6.3. Work = $P_{av} \times (\Delta s)$

$= \frac{1}{2}(P_1 + P_2)(\Delta s)$

When the spring deforms through a small distance ds, the workdone in deforming the spring is equal to

$$dU = P.\, ds = (ks)\, .\, ds \qquad ...(a)$$

$\therefore$ Total workdone in deforming spring through S units is equal to

$$U = \int dU = \int_0^s (ks)\, .\, ds = \frac{1}{2}\, ks^2 \qquad ...(13.13)$$

And, when spring is deformed from S_1 to S_2, the workdone is equal to

$$U_{1-2} = \int dU = \int_{s_1}^{s_2} (ks)\, ds = \frac{1}{2} k\, ({s_2}^2 - {s_1}^2) \qquad ...(13.14)$$

The work as expressed in eqn. (13.13) is given by the area of the ΔOAA_1, or area below the line OA in the Fig. 13.6.2.

The workdone U_{1-2} as expressed in eqn. (13.14) is given by area of the trapezium CC_1BB_1, which is equivalent to the average force multiplied by the change in deformation of the spring,

or $$U_{1-2} = \int_{s_1}^{s_2} ks\, ds = \frac{ks_1 + ks_2}{2} \times (s_2 - s_1) \qquad ...(13.15)$$

Since the force exerted by a spring depends only on the spring stiffness and the magnitude of the spring deformation, the workdone on a spring is due only to the change in length of the spring and is not dependent on any rotation of the spring. For example, consider positions A and B of the spring (Fig. 13.6.3), when it rotates through angle θ. The spring has deformations S_1 and S_2 at positions A and B. As one end of the spring moves from position A to C, imagine that the spring first moves from position A to position B and then it rotates through an angle θ to the position C. During this rotation, the spring force is normal to the circular path B to C and therefore *does no work during rotation.*

Note. A similar situation is shown in Fig. 13.6.4. The weight W at one end of an inextensible cord is raised as the other end moves along a curved path from position A to position B. Here, the change in elevation of weight is equal to the difference ΔL between length L_1 and L_2

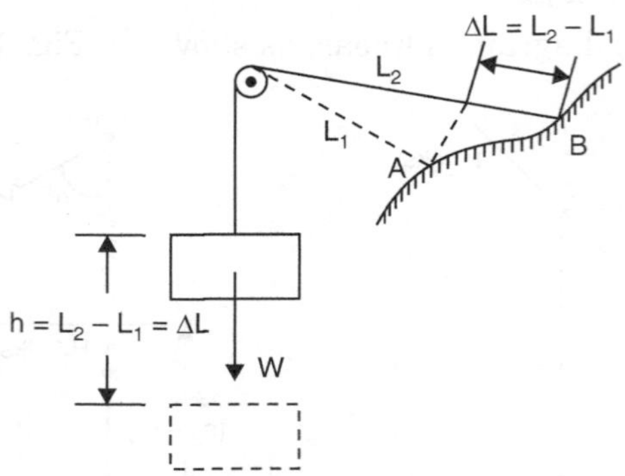

Fig. 13.6.4. Workdone on a weight = Wh = W(L2 – L1)

$\therefore$ Workdone on a weight $= U_{1-2} = Wh = W(L_2 - L_1)$ ...(13.16)

13.7. WORKDONE BY A FLUID UNDER CONSTANT PRESSURE

Fig. 13.7 shows a liquid under a constant pressure p. The liquid is filled in a cylinder of cross-sectional area A, and a piston moves through a distance x. Then the workdone by the fluid = force × displacement

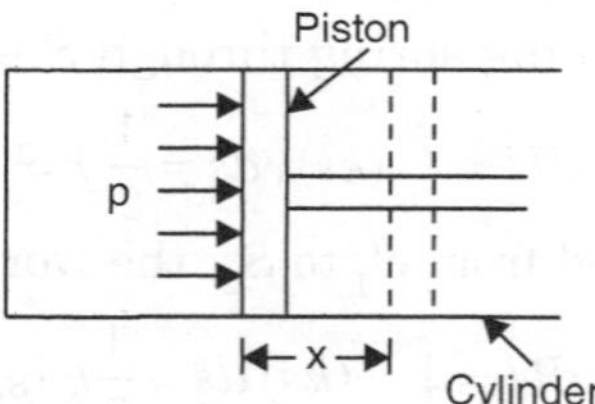

Fig. 13.7. Workdone by a fluid under pressure

or U = (pressure × area) × displacement = $(pA) \times x$

or $U = p \times (v)$ = pressure × displacement volume ...(13.17)

13.8. WORKDONE BY A GAS UNDER VARIABLE PRESSURE

Consider a p-v diagram for expansion of a gas as shown in Fig. 13.8. When the pressure on the piston is a variable quantity, the workdone on the piston during a small displacement dx, is equal to

$$dU = pA \times (dx) = p(A\,dx)$$

or $dU = p\,dv$ ...(*a*)

where A is area of piston.

Therefore, the workdone on the gas is

$$U_{1-2} = \int dU = \int_{v_1}^{v_2} p\,dv \qquad \text{...(13.18)}$$

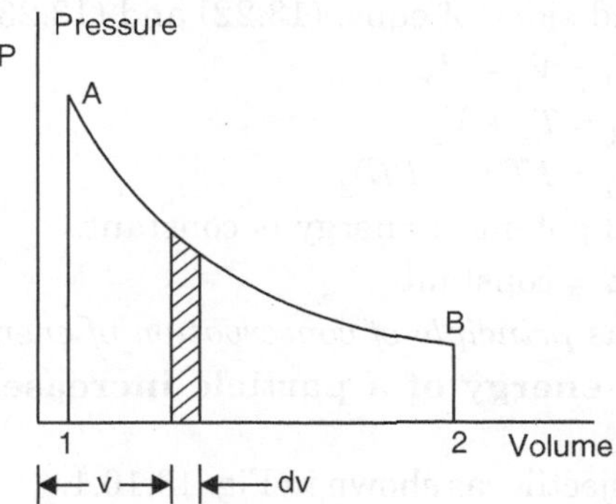

Fig. 13.8. Workdone on the gas= ∫pdv.

The area and the curve between A to B is equal to the workdone U_{1-2} given by eqn. (13.18).

13.9. ENERGY OF A PARTICLE

Energy is defined as the capacity of doing work. Thus energy has the units of work. Energy is in various forms, like mechanical energy, thermal energy, chemical energy, electric energy etc. Here we are interested in only mechanical energy of a particle.

Mechanical energy consists of two types of energies kinetic energy and potential energy.

Kinetic energy will be denoted by T. The energy possessed by a particle of mass m, which is moving with velocity v, is given by

Kinetic energy, $$T = \frac{1}{2} mv^2 \quad ...(13.19)$$

Potential energy is the energy which a particle possesses by virtue of its position. The potential energy will be denoted by V.

(*a*) Thus, if a particle is held at a height h above the ground, it possesses a potential energy.

$$V = (mg)h \quad ...(13.20)$$

(*b*) A compressed spring possesses a potential energy as the force exerted by the spring can do work as the spring extends a distance s. Thus,

$$V = \frac{1}{2} ks^2 \quad ...(13.21)$$

(*c*) Steam stored in the boiler possesses potential energy due to its pressure and can do work, etc.

13.10. PRINCIPLE OF CONSERVATION OF ENERGY

The principle of work-energy states that

Workdone = change in kinetic energy

i.e., $$U_{1-2} = T_2 - T_1 \quad ...(13.22)$$

If a particle moves down under gravity, then

Workdone = change in potential energy

or $$U_{1-2} = V_1 - V_2 \quad ...(13.23)$$

Equating the right hand sides of eqns. (13.22) and (13.23),

we get $$T_2 - T_1 = V_1 - V_2$$

or $$T_1 + V_1 = T_2 + V_2 \qquad ...(13.24)$$

or, $$KE)_1 + PE)_1 = KE)_2 + PE)_2 \qquad ...(13.25)$$

or sum of kinetic energy and potential energy is constant.

That is, $$KE + PE = \text{constant} \qquad ...(13.26)$$

Eqn. (13.26) is known as *principle of conservation of energy.*

Thus, as the kinetic energy of a particle increases, its potential energy decreases and vice-versa.

Consider motion of a projectile, as shown in Fig. 13.10.1.

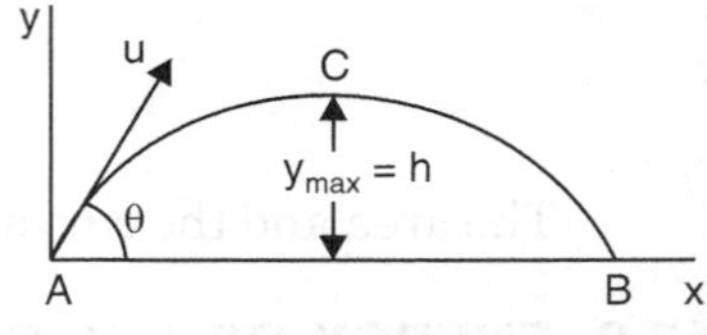

Fig. 13.10.1

When the particle of weight $W = mg$ is projected at point A, with velocity u, it has two components of velocity, *viz.*,

$$u_x = u \cos \theta \qquad ...(a)$$

and $$u_y = u \sin \theta \qquad ...(b)$$

u_x is constant and u_y decreases at the rate of g m/s^2. $u_y = 0$ at point C, where the height

$$h = y_{max}.$$

$\therefore$ At C, $\because$ $v_y = 0$, $\therefore$ $KE)_v = 0$ and $PE = W \,.\, h = mg\, y_{max}$ ...(c)

$\therefore$ Total energy at $A = KE)_A + PE)_A$

$$\therefore \text{ Total energy at } A = E_A = \frac{1}{2} u_y^2\, m + 0 = KE)_{max.} \qquad ...(d)$$

And, at point C, total vertical energy = $E_C = KE)_C + PE)_C$

or $$E_C = 0 + mg \times y_{max} = PE)_{max} \qquad ...(e)$$

$\therefore$ At any other instance,

Total energy, $$E = KE + PE = \text{constant} \qquad ...(f)$$

Thus, from eqns. (*d*), (*e*) and (*f*), we conclude that

Total energy $$= KE + PE = KE)_{max.} = PE)_{max} \qquad ...(13.27)$$

or, $$E = T + V = T_{max} = V_{max} \qquad ...(13.28)$$

Since, $$E = KE + PE = \text{constant} \qquad ...(g)$$

$\therefore$ By differentiating eqn. (*g*), we find

$$\frac{d}{dt}(KE) + \frac{d}{dt}(PE) = 0 \qquad ...(13.29)$$

or $$\frac{d}{dt}(T + V) = 0 \qquad ...(13.30)$$

Eqn. (13.30) is known as *Energy Method* and is used for solving problems on Mechanical Vibrations.

Eqn. (13.27) is called *Rayleigh's Method* and is very useful in the analysis of problems on Mechanical Vibrations.

As another example of conservation of energy is the motion of a simple pendulum, as shown in Fig. 13.10.2.

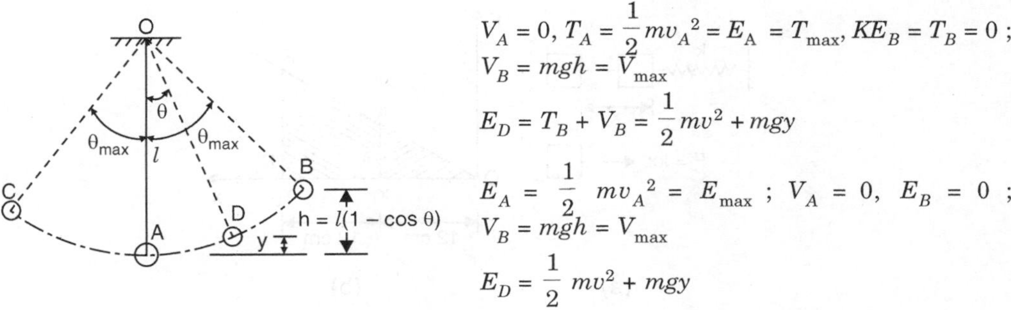

$$V_A = 0,\ T_A = \frac{1}{2} m{v_A}^2 = E_A = T_{max},\ KE_B = T_B = 0\ ;$$
$$V_B = mgh = V_{max}$$
$$E_D = T_B + V_B = \frac{1}{2} mv^2 + mgy$$
$$E_A = \frac{1}{2} m{v_A}^2 = E_{max}\ ;\ V_A = 0,\ E_B = 0\ ;$$
$$V_B = mgh = V_{max}$$
$$E_D = \frac{1}{2} mv^2 + mgy$$

Fig. 13.10.2. Motion of a simple pendulum

At point A, $v = v_{max}$,

$\therefore$ $$KE)_A = T_A = T_{max} = \frac{1}{2} m{v_{max}}^2$$

and P.E. = 0

At point B : $v_B = 0$; $\therefore$ $T_B = 0$

The potential energy at $B = V_B = mgh = V_{max}$

B and C are the extreme positions.

At a point D, between A and B, $E = \frac{1}{2} mv^2 + mg\,y$

We find that total energy, $$E_A = E_B = E_C = E_D \quad ...(h)$$

Also, $$E_D = T_{max} = V_{max} \quad ...(i)$$

Thus, we conclude that the sum of kinetic energy and potential energy at any instance is constant and it is equal to maximum kinetic energy and also it is equal to maximum potential energy.

Problem 13.1. *Determine the potential energy of a body acted upon by a force which is inversely proportional to the square of the radius r from the source of the force.*

Sol. When the body is at a radial distance r from the source, the force acting on it is equal to $F = C/r^2$, where C is a constant.

When the distance decreases by δr, the potential energy (δV) becomes

$$\delta V = F\delta r = -(C/r^2) \times \delta r.$$

Hence the δV total potential energy

$$= V = \int dV = \int_0^r \frac{C}{r^2}\, dr = -\frac{C}{r}. \quad \textbf{Ans.}$$

Problem 13.2. *A spring of stiffness 800 N/m is stretched by 12 cm from the unstretched position. Find the workdone by the spring force. Also find the workdone by another 12 cm stretching.*

Sol. Let a body be connected to spring of stiffness k N/m. When spring stretches x, the spring force is equal to $-kx$. See Fig. P. 13.2 (a).

$\therefore$ Workdone by the spring

$$= U_{0-1} = -\frac{k}{2}({x_1}^2 - {x_0}^2)$$

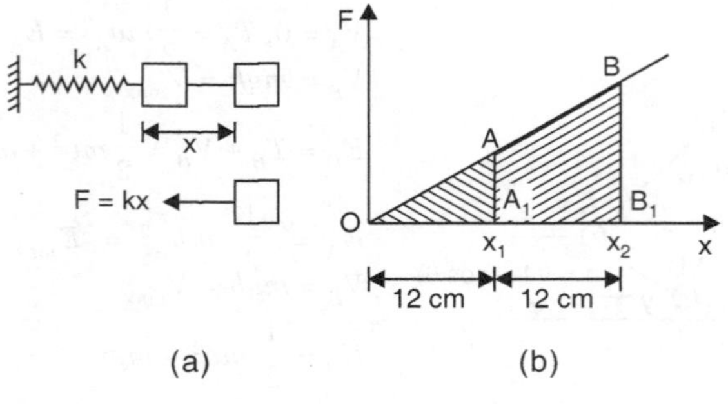

Fig. P. 13.2

$$= -\frac{k}{2}(x_1^2 - 0) = -\frac{k}{2}(12^2 - 0) \qquad | \; x_1 = 12 \text{ cm} = 0.12 \text{ m}$$

$$= \frac{-800}{2} \times (0.12^2 - 0) = \mathbf{-5.76\ Nm.} \quad \textbf{Ans.}$$

U_{0-12} is equal to the area of ΔOAA_1 in Fig. P. 13.2 (*b*). Workdone by the spring for another stretch of 12 cm of the spring is equal to the area of the trapezoid $AA_1 BB_1$.

$$\therefore \qquad U_{12-24} = -\frac{k}{2}(x_2^2 - x_1^2) = -\frac{800}{2}\left[\left(\frac{24}{100}\right)^2 - \left(\frac{12}{100}\right)^2\right] = \mathbf{-17.28\ Nm.} \quad \textbf{Ans.}$$

Problem 13.3. *Find the workdone when a block of mass 20 kg slides down an inclined plane shown in Fig. P. 13.3 through a distance of 5 m. The coefficient of kinetic friction* μ_k *is 0.25.*

Also find velocity of the block after it has travelled a distance of 5 m.

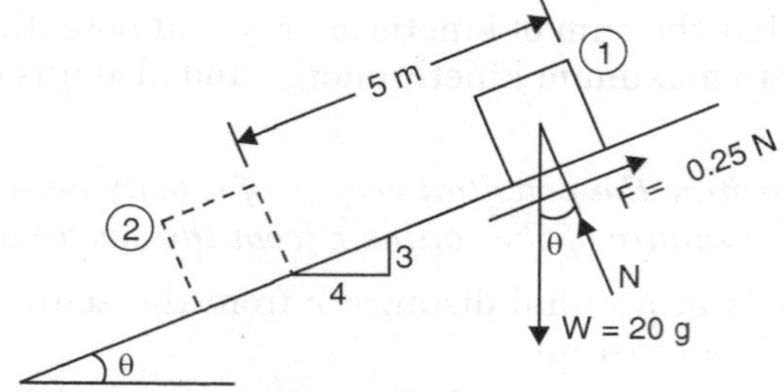

Fig. P. 13.3

Sol. At position (1) the block is at rest. Let the velocity of the block after it has travelled to position (2) be v.

The forces involved in the motion of the block are :

(*i*) The weight, $W = 20$ g ;

(*ii*) The normal reaction, $N = W \cos \theta = 20 \text{ g} \times \frac{4}{5} = 16$ g, and

(*iii*) The frictional force, $F = \mu_k N = 0.25 \times 16 = 4$ g.

$$\text{Workdone} = u_{1-2} = (W \sin \theta) \times x - F \times x$$

or

$$U_{1-2} = \left(20 \text{ g} \times \frac{3}{5}\right) \times 5 - 4 \text{ g} \times 5 = 40 \text{ g Nm} = \mathbf{392\ Nm.} \quad \textbf{Ans.}$$

Problem 13.4. *A body of weight W slides along the inclined plane AB, starting from an elevation to above point B. Find velocity of the body at point B, when (a) the plane is smooth, and (b) when the coefficient of friction between body and the plane is μ.*

Sol. (*a*) When the plane is smooth, the weight W of the body at A can be resolved into two components, along and normal to the plane, as shown in Fig. P. 13.4 (*a*).

Let α be the inclination of the plane.

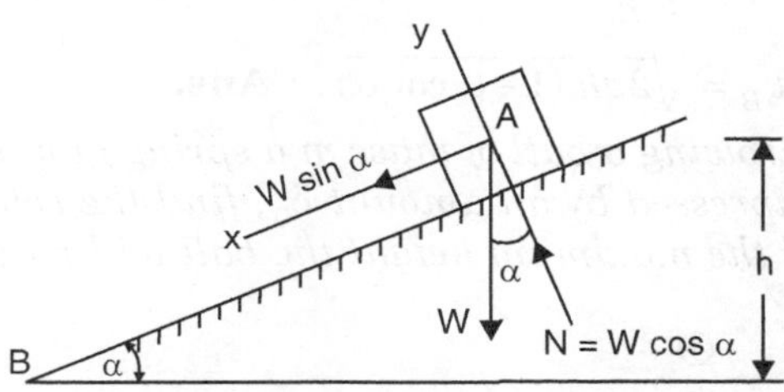

Fig. P. 13.4 (a) Smooth plane

The x-component of $W = W_x = W \sin \alpha$ does work.
Therefore the workdone

$$U = W_x \times AB = (W \sin \alpha) \frac{h}{\sin \alpha} \qquad ...(1)$$

The component W_y of the weight W which is perpendicular to the plane is always balanced by the reaction $N = W \cos \alpha$, and therefore, it does not do work.

Applying the work-energy equation

$$U_{1-2} = \Delta E$$

where ΔE = change of kinetic energy which is equal to the loss of potential energy Wh, and also equal to the gain in $KE = \frac{1}{2}\frac{W}{g} v^2$ ...(2)

∴ From (1) and (2), we get

$$\therefore \quad (W \sin \alpha) \times \frac{h}{\sin \alpha} = Wh = \left(\frac{1}{2}\frac{W}{g}\right) v^2$$

$$\therefore \quad v^2 = 2gh, \; \therefore \; v = \sqrt{2gh}. \quad \textbf{Ans.} \qquad ...(3)$$

(*b*) When the body slides on the rough plane the frictional force $F = \mu N = \mu W \cos \alpha$

This force also does work

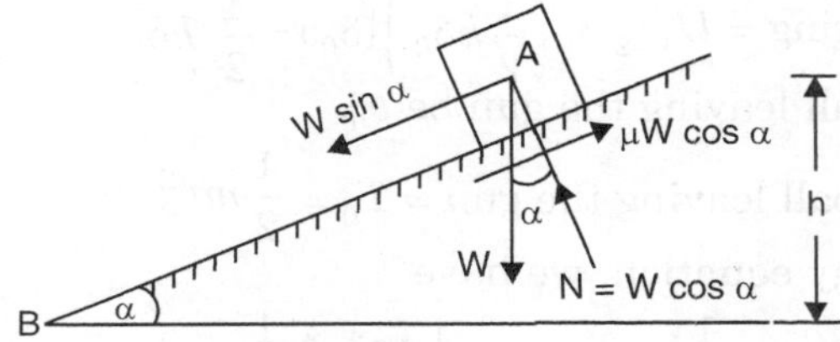

Fig. P. 13.4 (b) Rough inclined plane

$$\therefore \quad U_{1-2} = (W \sin \alpha - \mu W \cos \alpha) \times AB$$

or

$$U_{1-2} = W (\sin \alpha - \mu \cos \alpha) \times \frac{h}{\sin \alpha} = W (1 - \mu \cot \alpha)\, h \qquad ...(4)$$

The change in $K.E. = \Delta K.E. = \frac{1}{2}\frac{W}{g} v_B^2 - \frac{1}{2}\frac{W}{g} v_A^2$

where $v_A = 0$

$$\therefore \qquad T_B - T_A = \frac{1}{2}\frac{W}{g}(v_B^2 - 0) \qquad ...(5)$$

Using the work-energy equation, we get

$$W(1 - \mu \cot \alpha)\, h = \frac{1}{2}\frac{W}{g} v_B^2$$

$$\therefore \qquad v_B = \sqrt{2gh(1 - \mu \cot \alpha)}. \quad \textbf{Ans.} \qquad ...(6)$$

Problem 13.5. *For throwing a ball of mass m a spring gun is used. If the vertical spring of constant k is initially compressed by an amount δ_0, find the velocity v_0 with which the ball will leave the gun. Also find the maximum height the ball will reach.*

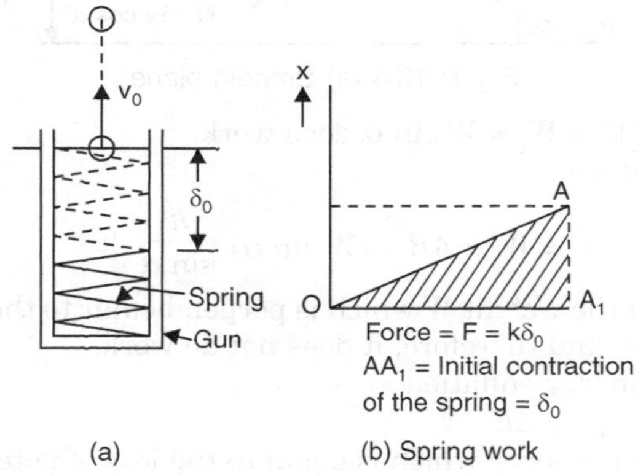

Fig. P. 13.5

Sol. The spring force F exerted on the ball = $F = kx$, where x is the compression. The corresponding force, displacement diagram $(F - x)$ of the spring is shown in Fig. P. 13.5 (*b*).

The workdone in compressing the spring is equal to the area of the ΔOAA_1. For the initial compression of an amount $AA_1 = \delta_0$

$$\text{area of the } \Delta OAA_1 = \frac{1}{2} \times OA_1 \times AA_1 = \frac{1}{2} \times OA_1 \times \delta_0$$

$$\therefore \quad \text{Workdone by spring} = U_{1-2} = \left(\frac{1}{2}k\delta_0\right)(\delta_0) = \frac{1}{2}k\delta_0^2$$

Let the velocity of ball leaving the gun be v_0.

Then the KE of the ball leaving the gun $= T_0 = \frac{1}{2}mv_0^2$

Applying work-energy equation, we have

$$U_{1-2} = \frac{1}{2}mv_0^2 \quad \text{or} \quad \frac{1}{2}k\delta_0^2 = \frac{1}{2}mv_0^2$$

$$\boxed{v_0 = \delta_0\sqrt{\frac{k}{m}}} \quad \textbf{Ans.} \qquad ...(1)$$

Let h be the maximum height reached by the ball.

Applying the law of conservation of energy, *i.e.,* Maximum P.E. = Maximum K.E.,

we get $$mgh = \frac{1}{2} m v_0^2$$

$$\therefore \quad \boxed{h = \frac{v^2}{2g}}. \quad \textbf{Ans.} \qquad ...(2)$$

From (1), $$h = \frac{1}{2g} \times \delta_0^2 \times \frac{k}{m}$$

$$= \frac{\delta_0^2 k}{2W}$$

or, $$\boxed{h = \frac{k\,\delta_0^2}{2W}}. \quad \textbf{Ans.} \qquad ...(3)$$

Problem 13.6. *A ball of weight W is dropped on to a spring of stiffness k from a height h. Find the maximum deflection of the spring.*

Assume $m = 2kg$; $k = 1000\ N/m$ *and* $h = 15\ cm$.

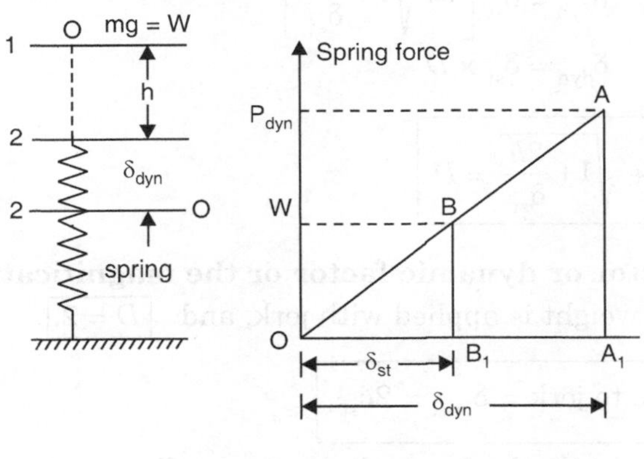

Fig. P. 13.6

Sol. Let the maximum compression of the spring be δ_{dyn}.

The potential energy lost by ball = $\Delta V = W(h + \delta_{dyn.})$ ...(1)

The spring force during the maximum compression = (Average Force) × δ_{dyn}

or spring work $$U_{1-2} = \frac{1}{2}(0 + k\,.\,\delta_{dyn}) \times \delta_{dyn}$$

or $$U_{1-2} = \frac{k\,\delta_{dyn}^2}{2} \qquad ...(2)$$

When the same weight W is applied statically (*i.e.,* simply placed on the spring), the spring compresses an amount δ_{st}, known static deflection. Then, $\delta_{st} = \frac{W}{k}$ ...(3)

$$\therefore \text{ Eqn. (2) gives } U_{1-2} = \frac{1}{2} \times \left(\frac{W}{\delta_{st}}\right) \delta_{dyn}^2 \qquad ...(4)$$

Applying work-energy equation, we get

$$W(h + \delta_{dyn}) = \frac{W \cdot \delta_{dyn}^2}{2\delta_{st}}$$

or $$\delta_{dyn}^2 - 2\delta_{st} \cdot \delta_{dyn} - 2h \cdot \delta_{st} = 0 \qquad ...(5)$$

Solution of eqn. (5) is

$$\delta_{dyn} = \frac{1}{2}\left[2\delta_{st} \pm \sqrt{\delta_{st}^2 + 4 \times 2h \cdot \delta_{st}}\right]$$

$$= \frac{1}{2}\left[2\delta_{st} \pm 2\delta_{st}\sqrt{1 + \frac{2h}{\delta_{st}}}\right]$$

$$\delta_{dyn.} = \delta_{st}\left[1 + \sqrt{1 + \frac{2h}{\delta_{st}}}\right] = \text{Dynamic or instantaneous deflection} \qquad ...(6)$$

This is the case of impact loading and the maximum deflection δ_{dyn} is always greater than the static deflection, therefore eqn. (6) becomes

$$\delta_{dyn} = \delta_{st}\left[1 + \sqrt{1 + \frac{2h}{\delta_{st}}}\right] \qquad ...(7)$$

or $$\delta_{dyn} = \delta_{st} \times D \qquad ...(8)$$

The quantity $$\boxed{1 + \sqrt{1 + \frac{2h}{\delta_{st}}} = D} \qquad ...(9)$$

is called the **impact factor or dynamic factor or the magnification factor.**

When $h = 0$, the weight is applied with jerk, and $\boxed{D = 2.}$

$$\therefore \quad \boxed{\delta_{dyn.} \text{ due to jerk} = \delta_{max} = 2\delta_{st}.}$$

Since the spring is elastic, the force – displacement diagram is as shown in Fig. P.13.6 (*b*).

Let $P_{dyn.}$ be the maximum dynamic force which when applied to the spring gradually (or statically) produces the same deflection δ_{dyn} in the spring.

Let $P_{dyn.}$ be represented by the AA_1 and W by BB_1.

$P_{dyn.}$ produces deflection δ_{dyn} given by OA_1 and W produces deflection δ_{st} given by OB_1.

$$\because \quad \Delta OAA_1 \equiv \Delta OBB_1$$

$$\therefore \quad \frac{AA_1}{OA_1} = \frac{BB_1}{OB_1} \quad \text{or} \quad \frac{P_{dyn.}}{\delta_{dyn.}} = \frac{W}{\delta_{st}}$$

$$\therefore \quad P_{dyn.} = \text{maximum load} = W.\frac{\delta_{dyn.}}{\delta_{st}} = WD \qquad ...(10)$$

when, $m = 2$ kg, $k = 1000$ N/m, and $h = 15$ cm,

$$\delta_{st} = \frac{W}{k} = \frac{2 \times 9.81}{1000} = 0.01962 \text{ m}$$

$$\therefore \quad D = 1 + \sqrt{1 + \frac{2h}{\delta_{st}}} = 1 + \sqrt{1 + \frac{2 \times 0.15}{0.01962}} = 17.29$$

$\therefore$
$$\delta_{max} = \delta_{dyn.} = \delta_{st} \times D$$
$$= 0.01962 \times 17.29 \text{ m}$$
$$= \mathbf{0.3392 \text{ m.}} \quad \textbf{Ans.}$$

Problem 13.7. *A body of mass m travelling with a velocity of v strikes a spring of stiffness k and stops. Find the maximum compression of the spring.*

Sol. Let the compression of spring be δ, when the body stops after striking the spring. See Fig. P. 13.7

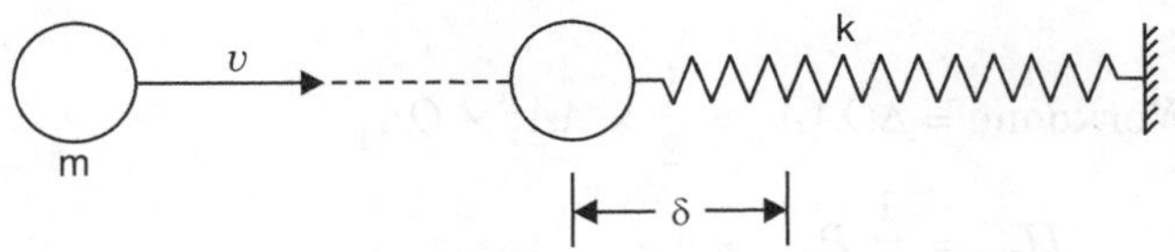

Fig. P. 13.7

Loss of KE of body $= \Delta T = \dfrac{1}{2} mv^2$ ...(1)

The energy gained by spring = spring work or spring work $= U_{1-2} = \dfrac{1}{2} k\delta^2$ ...(2)

Equating U_{1-2} and ΔT, we get

$$\frac{1}{2}k\delta^2 = \frac{1}{2} mv^2$$

$$\therefore \quad \boxed{\delta = v\sqrt{\frac{m}{k}}.} \quad \textbf{Ans} \qquad ...(3)$$

Problem 13.8. *A body of weight W falls freely through height h on the collar attached to a vertical bar of length l and cross-sectional area A. Find the maximum elongation and the impact load applied on the bar. E is the modulus of elasticity of the material of bar. Assume that the elastic limit of material is not exceeded.*

Sol. Let the static elongation of the bar when the load W is applied statically be δ_{st}. Also, let the maximum elongation of the bar be $\delta_{max} = \delta_{dyn}$ produced by the load of W when it falls on the collar from a height h. See Fig. P. 13.8.

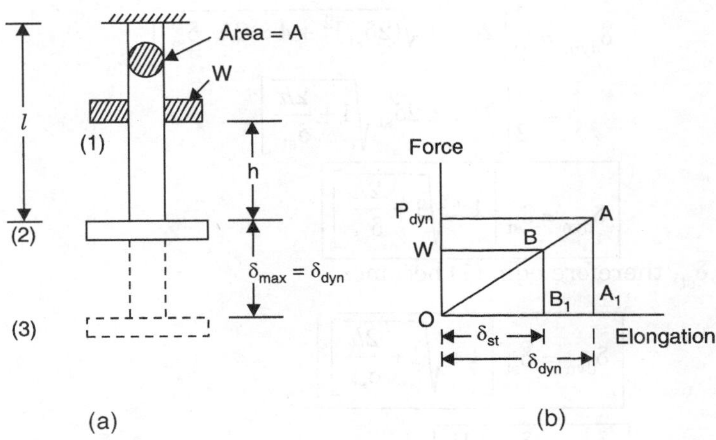

Fig. P. 13.8

$$\delta_{st} = \frac{Wl}{AE} \quad \text{...(1)}$$

when the weight W falls on the collar from the height h, the collar moves to the dotted position (3) δ_{dyn} below the position (2).

Total potential energy lost = $\Delta V = W(h + \delta_{dyn.})$ ...(2)

When a load $P_{dyn.}$ is applied gradually it produces the same elongation δ_{dyn}.

The workdone on the collar is equal to the strain energy stored in the bar. Since the elastic conditions prevail therefore the workdone on bar is recovered when the load P_{dyn} is removed.

$$\therefore \quad \text{Workdone} = \Delta OAA_1 = \frac{1}{2} \times AA_1 \times OA_1$$

or $$U_{1-3} = \frac{1}{2} P_{dyn.} \times \delta_{dyn.} \quad \text{....(3)}$$

Since OA is a straight line, therefore $\Delta OAA_1 \equiv \Delta OBB_1$

or $$\frac{AA_1}{OA_1} = \frac{BB_1}{OB_1} \quad \text{or} \quad \frac{P_{dyn}}{\delta_{dyn}} = \frac{W}{\delta_{st}}$$

$$\therefore \quad P_{dyn} = W \times \frac{\delta_{dyn}}{\delta_{st}} \quad \text{...(4)}$$

$\therefore$ Eqn (3) becomes

$$U_{2-3} = \frac{1}{2}\left(\frac{W \,.\, \delta_{dyn}}{\delta_{st}}\right) \times \delta_{dyn} = \frac{1}{2} \text{W} \frac{\delta_{dyn^2}}{\delta_{st}} \quad \text{...(5)}$$

Using the work-energy equation, we get

$$U_{2-3} = \Delta V$$

or $$\frac{1}{2} W \,.\, \frac{\delta_{dyn^2}}{\delta_{st}} = W(h + \delta_{dyn})$$

Rearranging, we get

$$\delta_{dyn}{}^2 - 2\delta_{st} \,.\, \delta_{dyn.} - 2h\delta_{st} = 0 \quad \text{...(6)}$$

Solution of eqn. (6) is

$$\delta_{dyn.} = \frac{1}{2}\left[2\delta_{st} \pm \sqrt{(2\delta_{st})^2 + 4 \times 2h \,.\, \delta_{st}}\right]$$

$$= \frac{1}{2}\left[2\delta_{st} \pm 2\delta_{st}\sqrt{1 + \frac{2h}{\delta_{st}}}\right]$$

or $$\boxed{\delta_{dyn} = \delta_{st}\left[1 \pm \sqrt{1 + \frac{2h}{\delta_{st}}}\right]} \quad \text{...(7)}$$

Since $\delta_{dyn} > \delta_{st}$, therefore eqn. (7) becomes

$$\boxed{\delta_{dyn} = \delta_{st}\left[1 + \sqrt{1 + \frac{2h}{\delta_{st}}}\right]} \quad \text{...(8)}$$

or, $$\boxed{\delta_{dyn} = \delta_{st} \times D} \quad \text{...(9)}$$

where D = dynamic or impact factor

$$\boxed{D = 1 + \sqrt{1 + \frac{2h}{\delta_{st}}}} \quad \text{...(10)}$$

$$\therefore \quad \delta_{dyn} \text{ or } \delta_{max} = \delta_{st}\left[1 + \sqrt{1 + \frac{2h}{\delta_{st}}}\right] \quad \textbf{Ans.}$$

And, from eqn. (4), we get

$$\boxed{P_{dyn} = W \times \frac{\delta_{dyn}}{\delta_{st}} = W \times D}$$

or, dynamic load $= P_{dyn} = W \times \left[1 + \sqrt{1 + \frac{2h}{\delta_{st}}}\right]$ **Ans.**

Problem 13.9. *Starting from rest, a collar of mass 2 kg slides along a vertical bar through 20 cm. The spring connected to the collar has unstretched length (l_o) equal to 25 cm. The other end of the spring is connected to a fixed point. [See Fig. P. 13.9]. Spring constant k is equal to 800 N/m. Find velocity of the collar when it is at the lowest position.*

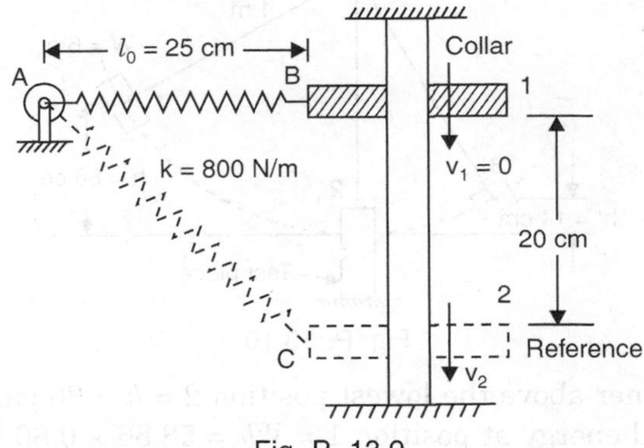

Fig. P. 13.9

Sol. Here $m = 2$ kg and $k = 800$ N/m.

Take position (2) as the reference position.

Then at position (1) :

Total energy = E_1 = (P.E. of mass + P.E. of spring) + K.E.

or $E_1 = [W(20) + 0] + 0 = 20\text{ W} = mg \times 20 = 2 \times 9.8 \times 0.20 = 3.92$ Nm

At position (2) :

Length of spring $= AC = \sqrt{AB^2 + BC^2}$

or, $l_2 = \sqrt{25^2 + 20^2} = 32.02$ cm = **0.3202 m**

Let v_2 = velocity of collar at position (2).

Total P.E. = P.E. of mass + P.E. of spring

or $= 0 + \frac{1}{2}k\delta^2$ | δ = extension of spring

$$= 0 + \frac{1}{2} \times 800 \times (0.3202 - 0.25)^2 = 1.9712 \text{ Nm}$$

K.E. = K.E. of spring + K.E. of weight

$$= 0 + \frac{1}{2} m v_2^2 = \frac{1}{2} \times 2\, v_2^2 = v_2^2$$

$\therefore$ Total energy at (2) = $E_2 = v_2^2 + 1.9712$

Equating E_1 and E_2, we get

$$1.9712 + v_2^2 = 3.92$$

$$\therefore \quad v_2^2 = 3.92 - 1.9712$$

$$\therefore \quad v_2 = \sqrt{1.9488} = \mathbf{1.396 \text{ m/s.}} \quad \textbf{Ans.}$$

Problem 13.10. *In an impact testing machine the hammer of mass 6 kg swings in a circular arc. The arm of the hammer is 1 m long and is released from rest at a point 86 cm higher than the lowest point of the circle. Find the kinetic energy and velocity of the hammer at the lowest point. If the hammer breaks a metal piece (called test piece) placed at the lowest point and rises though 14 cm find the energy lost in breaking the test piece (See Fig. P. 13.10).*

Sol. Weight of hammer, W = 6 g = 58.86 N

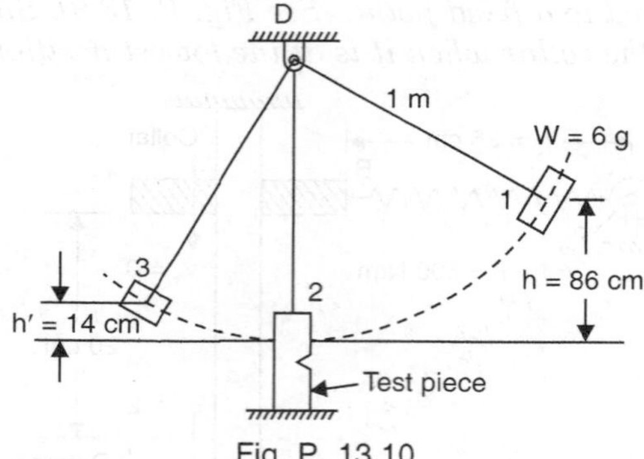

Fig. P. 13.10

Height of the hammer above the lowest position 2 = h = 86 cm = 0.86 m.

Maximum potential energy at position 1 = Wh = 58.86 × 0.86 Nm = 50.62 Nm

K.E. at position 1 = 0

$\therefore$ Total energy at position 1 = E_1 = 58.86 × 0.86 = 50.62 Nm

Let v_2 be the velocity of hammer at position 2.

$$\therefore \quad \text{K.E., } T_2 = \frac{m v_2^2}{2} = \frac{6 \times v_2^2}{2} = T_{max}\,; \; PE = 0, \quad \therefore \quad E_2 = \frac{6 \times v_2^2}{2}$$

$\because$ There is no loss of energy between positions 1 and 2, we can equate E_1 and E_2

$$\therefore \quad \text{We get} \quad \frac{6 v_2^2}{2} = 50.62, \quad \therefore \quad v_2 = \sqrt{\frac{50.62 \times 2}{6}} = \mathbf{4.108 \text{ m/s.}} \quad \textbf{Ans.}$$

Kinetic energy at the lowest point = T_1 = **50.62 Nm. Ans.**

After breaking the test piece hammer stops at position 3 after rising through 14 cm.

$\therefore$ Total energy at position 3 = $E_3 = T_3 + V_3$

or $$E_3 = 0 + Wh'$$

or $$E_3 = 58.86 \times 0.14 \text{ Nm} = 8.24 \text{ Nm.}$$

$\therefore$ Energy lost in breaking test piece = $E_1 - E_3$

$$\therefore \quad \Delta E = (50.62 - 8.24) \text{ Nm} = \mathbf{42.38 \text{ Nm.}} \quad \textbf{Ans.}$$

Problem 13.11. *(a) A small block at rest released from point 1 and slides down the rough inclined plane as shown in Fig. P. 13.11 (a) . What distance along the horizontal plane will it travel before coming to rest ? μ = 0.3 for both inclined and horizontal planes.*

(b) Solve part (a) when the inclined plane is smooth and for the horizontal plane μ = 0.3.

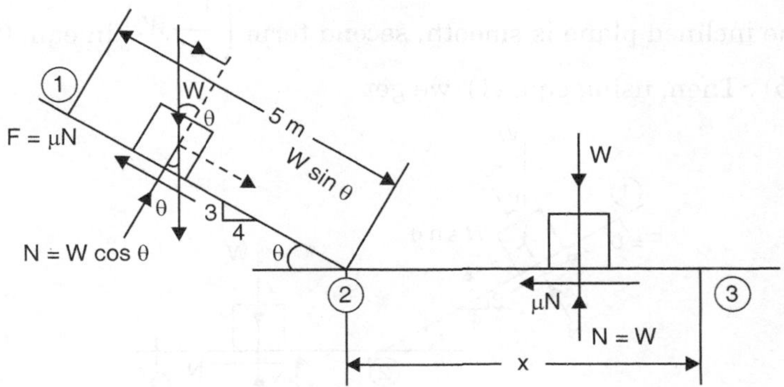

Fig. P. 13.11 (a) Motion on rough planes

Sol. *(a)* Let weight of the block = $W = mg$ [See Fig. P. 13.11.]

For the inclined plane tan θ = 3/4,

∴ sin θ = 3/5 and cos θ = 4/5.

At position 1, $v_0 = 0$;

At position 3, $v = 0$

∴ Change of kinetic energy $\Delta T = \text{KE})_3 - \text{KE})_1 = 0$

When the block is sliding down the rough inclined plane 1-2, the various forces acting on the block are shown in Fig. P.13.11 *(a)*

By $\Sigma F_n = 0$, we get $N - W\cos\theta = 0$

$$N = W\cos\theta = \frac{4}{5}W$$

∵ $F = \mu N$, we have, $F = 0.3$ N

or
$$F = 0.3\text{ N} = 0.3 \times \frac{4}{5}W = \frac{1.2}{5}W$$

The net force acting on the block along the plane = $W\sin\theta - F = W \times \frac{3}{5} - \frac{1.2}{5}W.$

The reaction of the horizontal plane = $N = W$

The only force on the block when it slides along the horizontal plane = $\mu N = 0.3 \times W$

∴ Workdone on the block from position 1 to position 3

$$= U_{1-3} = U_{1-2} + U_{2-3} \quad \text{...(1)}$$

$$= (W\sin\theta - \mu \times N) \times 5 + (\mu \times N)x$$

or
$$U_{1-3} = \left(\frac{3}{5}W - \frac{1.2}{5}W\right) \times 5 - 0.3 \times Wx \quad \text{...(2)}$$

Applying work-energy equation we have

$$U_{1-3} = \text{KE})_3 - \text{KE})_1 \quad \text{...(3)}$$

or $\left(\frac{3}{5}W - \frac{1.2}{5}W\right) \times 5 - 0.3\ Wx = 0$

$\therefore$ $1.8\ W = 0.3\ Wx$

$\therefore$ **x = 6 m. Ans.**

(*b*) When the inclined plane is smooth, second term $\left(\frac{1.2}{5}W\right)$ in eqn. (2) becomes zero. See Fig.P. 13.11 (*b*) . Then, using eqn. (1), we get

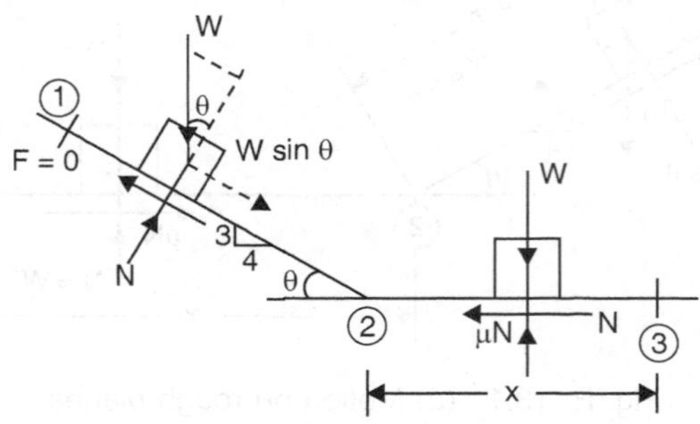

Fig. P. 13.11 (b)

$$U_{1-3} = U_{1-2} + U_{2-3} = (W \sin \theta) \times 5 - 0.3\ Wx \quad ...(4)$$

$\therefore$ By using eqn. (3), we get

$$\left(\frac{3}{5}W\right) \times 5 - 0.3\ Wx = 0$$

$\therefore$ $0.3\ x = 3 \quad \therefore$ **x = 10 m. Ans.**

Problem 13.12. *Determine the force P required that will give the system of bodies in Fig. P. 13.12, a velocity of 4 m/s after moving 6 m from rest. Take μ = 0.25 and assume pulleys to be smooth.*

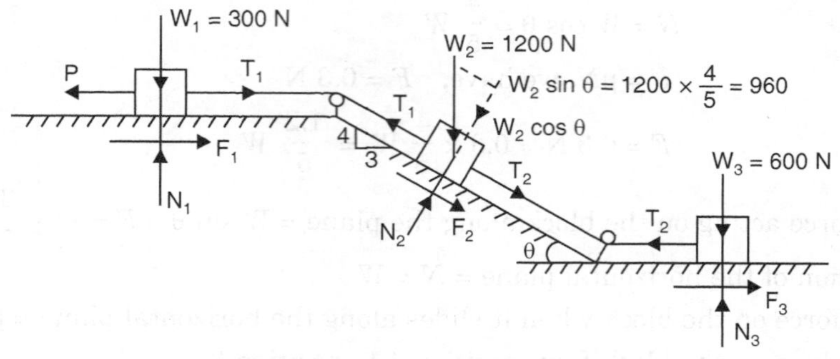

Fig. P. 13.12

Sol. Given : $\tan \theta = 4/3, \quad \therefore \quad \sin \theta = 4/5$ and $\cos \theta = 3/5$.

Normal reactions are :

$$N_1 = W_1 = 300\ \text{N}\ ;$$

$$N_2 = W_2 \cos \theta = 1200 \times \frac{3}{5}\ ;$$

$$N_3 = W_3 = 600\ \text{N}.$$

Frictional forces are :

$$F_1 = 0.25\ N_1 = 75\ N\ ;$$

$$N_2 = 0.25 \times \frac{1200 \times 3}{5} = 180\ N\ ;$$

$$F_3 = 0.25\ N_3 = 600 \times 0.25 = 150\ N.$$

Workdone, $U_{1\text{-}2} = [P - (F_1) - (F_2 + W_2 \sin \theta) - (F_3)] \times 6$

$$= [P - 75 - 180 - 960 - 150] \times 6$$

$$= 6P - 1365 \times 6$$

$\therefore$ $U_{1-2} = 6P - 8190$, Nm

Note. The tensions T_1 and T_2 do not do any work.

Velocity of all mass is 4 m/s after a displacement of 6 m. Initial velocity is zero.

$\therefore$ Change in K.E. $= T_2 - T_1 = \left[\frac{1}{2}\frac{W_1}{g} + \frac{1}{2}\frac{W_2}{g} + \frac{1}{2}\frac{W_3}{g}\right] v^2 - 0$

$\therefore$ $= \dfrac{(4)^2}{2 \times 9.81}\ [300 + 1200 + 600] =$ **1712.54 Nm.**

Using, $U_{1-2} = T_2 - T_1$, we get

$$6P - 8190 = 1712.54$$

$\therefore$ $P =$ **1650.42 N. Ans.**

Problem 13.13. *A 98 N block moves the two inclines which have the coefficient of friction is 0.2. (a) If the block starts from rest at A and slides down the 30° incline, how far will it move along the other incline ? (b) What will be the velocity of the block when it returns to the position B. See Fig. P. 13.13 (a).*

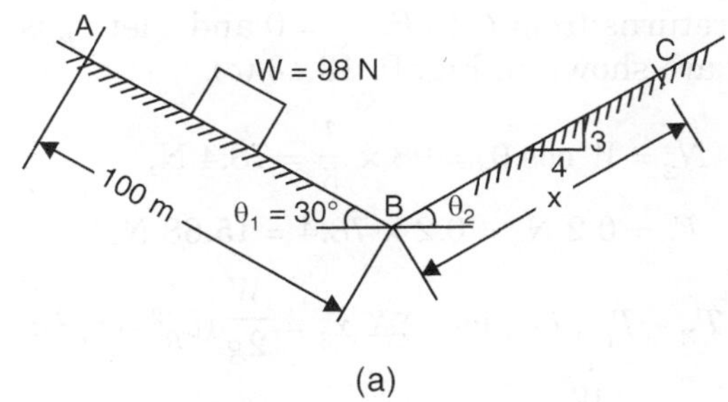

(a)

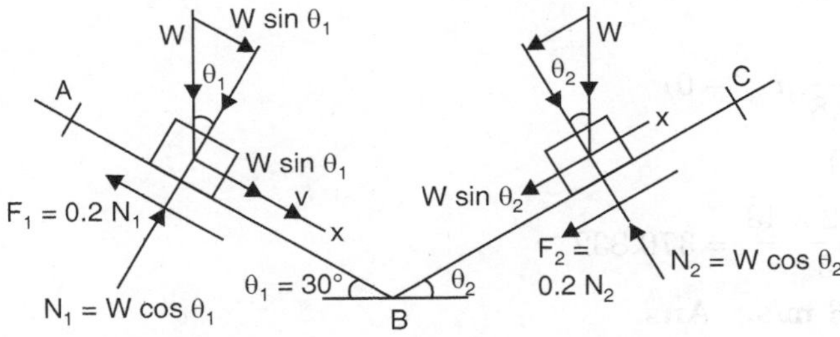

(b) Forces acting on the block when moving from A to B and from B to C.

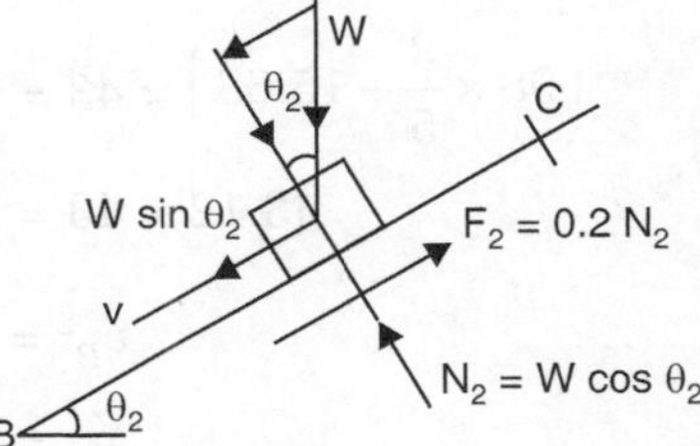

(c) Forces acting on the block when moving from C to B.

Fig. P. 13.13

Sol. (*a*) The free-body diagrams of the block moving down the plane *AB* and also when the block moves up the inclined plane *B* to *C* are shown in Fig. P.13.13 (*b*).

Normal reactions are :

$$N_1 = W \cos \theta_1 = 98 \cos 30°,$$

$$N_2 = W \cos \theta_2 = 98 \times \frac{4}{5} = 78.4 \text{ N}$$

Frictional forces : $F_1 = N_1 \times \mu = (98 \cos 30°) \times 0.2 = 19.6 \cos 30°$

And, $F_2 = \mu N_2 = 0.2 \times 78.4 = 15.68$ N

As the block starts from rest from *A* and stops at *C*, therefore $v_A = 0$ and $v_C = 0$.

Applying work-energy equation, [*i.e.*, by $U_{A-C} = \Delta T$], we get

$$\therefore \quad U_{AB} + U_{BC} = T_C - T_A$$

or $$\Sigma X_1 x_1 + \Sigma X_2 x_2 = \frac{W}{2g}(v_C^2 - v_A^2)$$

or $$(W \sin \theta_1 - F_1) \times 100 - (W \sin \theta_2 + F_2) \times x = 0$$

or $$(98 \sin 30° - 19.6 \cos 30°) \times 100 - \left(98 \times \frac{3}{5} + 15.68\right) x = 0$$

$$\therefore \quad \left(98 \times \frac{3}{5} + 15.68\right) x = (98 \sin 30° - 19.6 \cos 30°) \times 100$$

$$\therefore \quad 74.48\, x = 32.025 \times 100$$

$$\therefore \quad x = 42.999 \text{ m} \quad \text{or} \quad \mathbf{43\ m.} \quad \mathbf{Ans.}$$

(*b*) When the block returns from *C* to *B*, $v_C = 0$ and let v_B be its velocity at position *B*. Forces acting on the block are shown in Fig. P. 13.13 (*c*).

$$N_2 = W \cos \theta_2 = 98 \times \frac{4}{5} = 78.4 \text{ N},$$

$$F_2 = 0.2 \text{ N}_2 = 0.2 \times 78.4 = 15.68 \text{ N},$$

By $U_{B-C} = T_B - T_C$, *i.e.*, by $\Sigma X_3 x_3 = \frac{W}{2g}(v_B^2 - v_C^2)$, where $x_3 = x = 43$ m

$$\therefore \quad (W \sin \theta_2 - F_2) \times x = \frac{W}{2g}(v_B^2 - v_C^2)$$

or $$\left(98 \times \frac{3}{5} - 15.68\right) \times 43 = \frac{98}{2 \times 9.8}(v_B^2 - 0)$$

$$\therefore \quad 43.12 \times 43 = 5(v_B^2)$$

$$\therefore \quad v_B^2 = \frac{43.12 \times 43}{5} = 370.832$$

or $$v_B = \mathbf{19.26\ m/s.} \quad \mathbf{Ans.}$$

Problem 13.14. *A 30 kg block slides down an incline having slope of 4 vertical and 3 horizontal, as shown in Fig. P. 13.14 (a). It starts from rest, and after moving 3 m strikes a spring whose stiffness (or modulus) is 200 N/cm. determine (a) the maximum compression of the spring and (b) the maximum velocity of block.* $\mu = 0.20$.

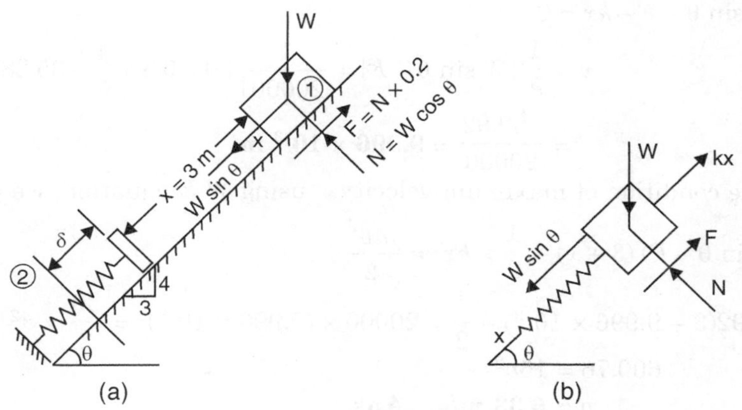

Fig. P. 13.14

Sol. Here, $\tan\theta = \frac{4}{3}$;

$\therefore$ $\sin\theta = \frac{4}{5}$ and $\cos\theta = \frac{3}{5}$

$\therefore$ Forces acting on the block are :

$$N = W\cos\theta = (30 \times 9.8) \times \frac{3}{5} = 176.4 \text{ N}$$

$$F = 0.2 \text{ N} = 0.2 \times 176.4 = 35.28 \text{ N}$$

The net force acting along the x-axis during free sliding $= \Sigma X = W\sin\theta - F = (30 \times 9.8) \times \frac{4}{5} - 35.28 = 199.92$ N. When the block compresses the springs by a distance x, the forces on the block in the x-direction are shown in Fig. P. 13.14 (*b*) : Therefore,

$$\Sigma F_x = W\sin\theta - F - kx \quad ...(1)$$

(*a*) **Maximum compression of spring :**

If the maximum compression of the spring when the block comes to rest is δ, then $v_2 = 0$.

$\therefore$ From (1), we get

$$(W\sin\theta - F)(3 + \delta) - \frac{1}{2}(0 + k\delta)\delta = \frac{m}{2}(v_2^2 - v_1^2)$$

or $$\left(30 \times 9.8 \times \frac{4}{5} - 35.28\right)(3 + \delta) - \frac{1}{2} \times (200 \times 100) \times \delta^2 = 0$$

$\therefore$ $$199.92\,(3 + \delta) - 10000\,\delta^2 = 0$$

or $$\delta^2 - 599.76 \times 10^{-4} - 199.92 \times 10^{-4}\,\delta = 0$$

$\therefore$ $$\delta = \frac{1}{2}\left[199.92 \times 10^{-4} \pm \sqrt{(199.92 \times 10^{-4})^2 + 4 \times 599.76 \times 10^{-4}}\right]$$

$$= \frac{1}{2}[199.92 \times 10^{-4} \pm 0.49]$$

$$= 0.255, -0.215$$

$\therefore$ Maximum compression $\delta =$ **0.2551 m. Ans.**

(*b*) **Maximum velocity of block :**

When the velocity of the block is maximum, the net force acting on it is zero. Therefore, by equating the net force given in eqn. (1) equal to zero, we get

$$W \sin \theta - F - kx = 0$$

$$\therefore \quad x = \frac{1}{k}[W \sin \theta - F] = \frac{1}{20000}\left[30 \times 9.8 \times \frac{4}{5} - 35.28\right]$$

$$= \frac{199.92}{20000} = \mathbf{9.996 \times 10^{-3} \ m.}$$

$\therefore$ For the condition of maximum velocity, using *W-E* equation, we get

$$(W \sin \theta - F)(3 + x) - \frac{1}{2} \times kx^2 = \frac{mv^2}{2}$$

or $$199.92(3 + 9.996 \times 10^{-3}) - \frac{1}{2} \times 20000 \times (9.996 \times 10^{-3})^2 = \frac{30}{2} \times v^2$$

$$\therefore \quad 600.76 = 15v^2$$

$$\therefore \quad v = \mathbf{6.33 \ m/s.} \quad \textbf{Ans.}$$

Problem 13.15. *A 500 N weight rotates in a vertical circle of 1.2 m radius. Velocity at the top most position is 4 m/s (a). Find the tension in the chord tied to the weight when it is at 120° to the top position, and (b) the least velocity of the weight at the bottom which will keep it in the circular path at the top.*

Sol. (*a*) The circular path and forces acting on the weight at different positions are shown in Fig. P.13.15.

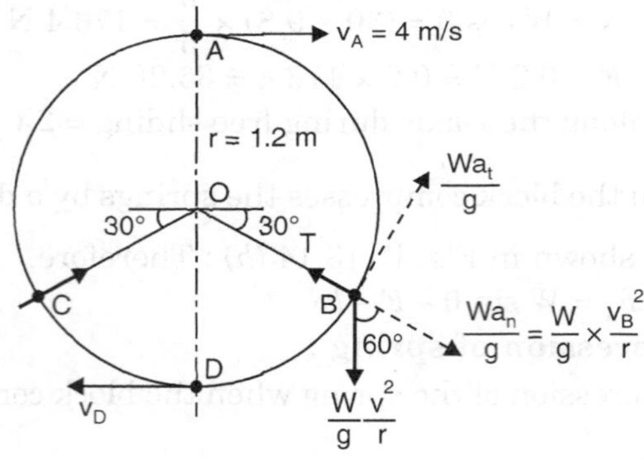

Fig. P. 13.15

The workdone between positions *A* and *B* = $U_{A-B} = W(r + r \sin 30°) = 500 \times 1.2 (1 + 0.5)$ $= 500 \times 1.2 \times 1.5 = 900$ Nm

The change in K.E. $= T_B - T_A$

$$= \frac{500}{2 \times 9.8} \times (v_B^2 - v_A^2) = \frac{500}{2 \times 9.8} \times (v_B^2 - 4^2)$$

By W.E. equation $\left[i.e., \text{ by } Wh = \frac{W}{2g}(v_B^2 - v_A^2) \right]$, we get

$$500 \times (1.2 + 1.2 \times \sin 30°] = \frac{500}{2 \times 9.8} \times (v_B^2 - 4^2)$$

$$\therefore \quad v_B^2 - 16 = 2 \times 9.8 \times (1.2)(1.5)$$

$$\therefore \quad v_B^2 = 35.28 + 16 = 51.28$$

$$\therefore \quad v_B = \mathbf{7.16 \ m/s}$$

By $\Sigma F_r = 0$, we get, $T - W \cos 60° = \dfrac{W}{g} \times a_n$ $\quad\Big|\quad a_n = \dfrac{v_B^2}{r}$

or, $$T - 500 \cos 60° = \frac{500}{9.8} \times \frac{v_B^2}{r}$$

$$\therefore \quad T = 500 \cos 60° + \frac{500}{9.8} \times \frac{(7.16)^2}{1.2} = \mathbf{2429.66\ N.}\quad \textbf{Ans.}$$

(*b*) When the weight has just the circular path at *A*, the tension in chord at *A* is zero. And, at *A* the centrifugal force is equal to the weight (*W*)

or $$\frac{W}{g} \times \frac{v_A^2}{r} = W$$

$$\therefore \quad v_A^2 = gr \quad ...(1)$$

By the formula $v_D^2 = v_A^2 + 2gS$,

where $S = 2r$, we get

$$v_D^2 = v_A^2 + 2g(2r) = (gr) + 4gr$$

$$\therefore \quad v_D^2 = 5gr$$

$$\therefore \quad v_D = \sqrt{5gr}. \quad \textbf{Ans.}$$

or the least velocity at $D = v_D = \sqrt{5 \times 9.8 \times 1.2}$ m/s

or $$v_D = \sqrt{58.8} \text{ m/s} = \mathbf{7.668\ m/s.}\quad \textbf{Ans.}$$

Problem 13.16. *A truck, which carries a 100 kg crate, starts from rest and attains a speed of 72 km/hr in a distance of 100 m on a level road with constant acceleration. Calculate the workdone by the force of friction acting on the crate during this interval if* μ_s *and* μ_k *between the crate and the truck bed are (a) 0.30 and 0.25 and (b) 0.20 and 0.18, respectively.*

Sol. If the crate does not slip on the truck bed, its acceleration will be equal to that of the truck, which is given by

$$v^2 = u^2 + 2as,$$

where $u = 0$; $v = 72$ km/hr $= 72/3.6$ m/s $= 20$ m/s

and $s = 100$ m,

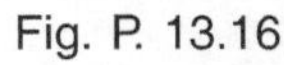

Fig. P. 13.16

$$\therefore \quad a = \frac{v^2}{2s} = \frac{(72/3.6)^2}{2 \times 100} = 2 \text{ m/s}^2$$

(*a*) This acceleration requires a force of friction on the crate $= F = ma$

or $$F = 100 \times 2 = 200 \text{ N.}$$

The maximum force of friction $= F = \mu_s \cdot N$

where normal reaction, $N = W = 100 \times 9.81 = 981$

$\therefore$ The maximum friction force, $F_{max} = \mu_s \cdot N$

or $$F_{max} = 0.30 \times 981 = 294.3 \text{ N}$$

Since F(= 200 N) is less than F_{max}(= 294.3 N), therefore the crate does not slip and the workdone by the **actual static friction force** of 200 N is equal to $U_{1-2} = 200 \times 100 = 20{,}000$ J or **20 kJ. Ans.**

(*b*) For $\mu_s = 0.2$, the maximum possible friction force is $F_{max} = 0.2 \times (100 \times 9.81) = 196.2$ N, which is slightly less than the value of 200 N required for no slipping. Here, the crate slips and force of friction is governed by the kinetic coefficient of friction ($\mu_k = 0.18$), and therefore

$$F = \mu_k \,.\, N = 0.18 \times (100 \times 9.81) = 176.58 \text{ N.}$$

Hence, the acceleration of the crate

$$a' = \frac{F}{m} = \frac{176.58}{100} = 1.7658 \text{ m/s}^2$$

Therefore, the distance travelled by the crate

$$= s = 100 \times \frac{a'}{a} = 100 \times \left(\frac{1.7658}{2}\right) = \textbf{88.29 m}$$

Therefore, workdone by force of friction = $U_{1-2} = F \times S$

or $$U_{1-2} = 176.58 \times 88.29 = \textbf{15590.25 J} \text{ or } \textbf{15.59 kJ. Ans.}$$

Problem 13.17. *The 60 kg block at A is mounted on rollers so that it moves on the fixed smooth horizontal platform under a constant force of 360 N applied to the cable. The block starts from rest at A, with the spring connected to it initially extented by an amount x_1 = 0.25 m. Find velocity of the block when it reaches B. Take spring stiffness k = 100 N/m. (See Fig. P.13.17.)*

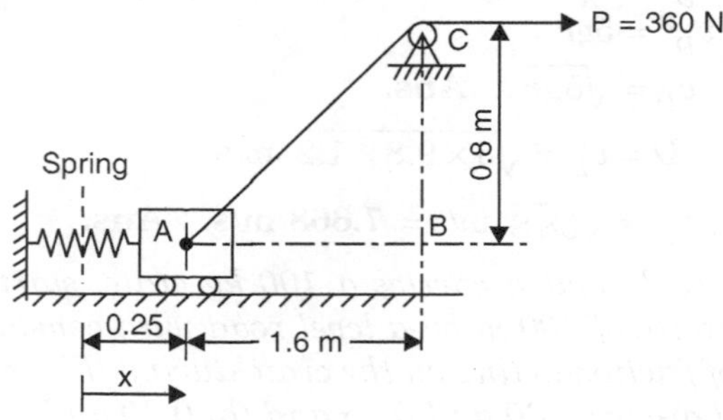

Fig. P. 13.17

Sol. At A the extension of spring is x_1 = 0.25 m and x_2 when the block is at B, the spring extension, $x_2 = x_1 + AB$ = 0.25 + 1.6 = 1.85 m. The workdone by the spring force acting on the block is negative and is equal to

$$U_{\text{spring}} = -\int_{x_1}^{x_2} kx\,dx = -\int_{0.25}^{1.85} 100\,x\,dx = -\frac{100}{2}\left(x^2\right)_{0.25}^{1.85} = \textbf{– 168 N}$$

when the block is at A, the length $l_1 = AC = \sqrt{1.6^2 + 0.8^2}$ or $l_1 = 0.8\sqrt{5}$ m

when the block is at B, the length of the cable $l_2 = BC$ = 0.8 m

Hence the workdone by the constant force P is = $P(l_1 - l_2) = 360 \times [0.8\sqrt{5} - 0.8]$ = **356 J**

Applying the work-energy equation, we get

$$[U_{1-2} = \Delta T] - 168 + 356 = \frac{1}{2} \times 60\,(v^2 - 0)$$

$$\therefore \qquad v^2 = \frac{1}{30} \times (356 - 168) = \frac{188}{30}$$

$$\therefore \qquad v = \textbf{2.503 m/s. Ans.}$$

Problem 13.18. *A 120 g tablet is released from rest at A when the spring is initially compressed 80 mm. The tablet travels around the loop ABCDE. Find the value of smallest spring stiffness (k) for which the tablet will travel around and in contact with the loop. See Fig. P. 13.18 (a).*

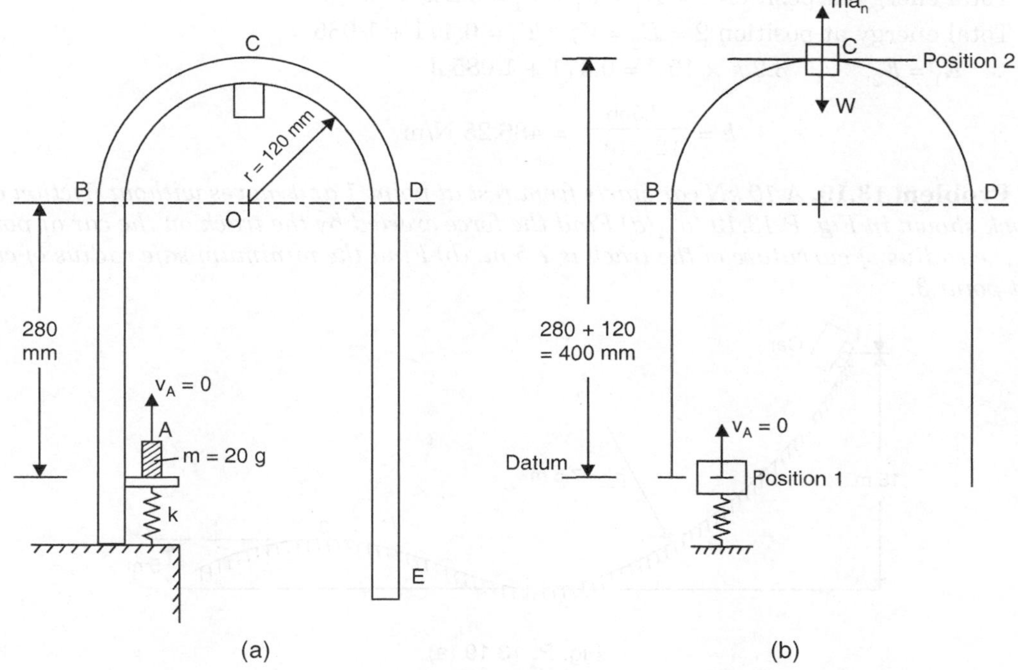

Fig. P. 13.18

Sol. Velocity at C :

As the tablet passes through the highest point (C) of the loop its velocity should be such that the centrifugal force acting on it is equal to its weight.

$\therefore$ By $\Sigma F_n = ma_n$, we have, $F_n = W = ma_n$ | a_n = normal acceleration = v^2/r

$\therefore$ $$W = \frac{W}{g} \times a_n \; ; \; \therefore \quad a_n = g$$

or $$a_n = \frac{v_c^2}{r} = g \quad \text{or} \quad v_c = \sqrt{gr}$$

or $$v_c = \sqrt{gr} = \sqrt{(9.81) \times \left(\frac{120}{1000}\right)} = \mathbf{1.085\ m/s.}$$

Position 1. When the tablet is at A, $v_A = 0$ (the tablet is released from rest) ;

$\therefore$ Kinetic energy $T_A = T_1 = 0$

Potential energy, $V_1 = V_k + V_g$ with datum at A, $V_g = 0$;

and $$V_k = \frac{1}{2}k(x_1)^2 = \frac{1}{2}k\left(\frac{80}{1000}\right)^2 = \frac{1}{2}k \times (0.08)^2 = 3.2\,k \times 10^{-3}\text{ J.}$$

Position 2. Potential energy of tablet at C

$$V_C = Wy = \left(\frac{120}{1000}\right) \times g \times \frac{400}{1000}$$

or $$V_C = \frac{120}{1000} \times 9.81 \times 0.4 = \mathbf{0.471\ J.}$$

Total energy at position 1 = $E_1 = V_1 + T_1 = 3.2\,k \times 10^{-3} + 0$

Total energy at position 2 = $E_2 = V_2 + T_2 = 0.471 + 1.085$ J

$\because \quad E_1 = E_2, \quad \therefore \quad 3.2\,k \times 10^{-3} = 0.471 + 1.085$ J

$$\therefore \qquad k = \frac{1.556}{3.2 \times 10^{-3}} = 486.25 \text{ N/m}$$

Problem 13.19. *A 10 kN car starts from rest at point 1 and moves without friction down the track shown in Fig. P.13.19 (a). (a) Find the force exerted by the track on the car at point 2, where the radius of curvature of the track is 7.5 m. (b) Find the minimum safe radius of curvature at point 3.*

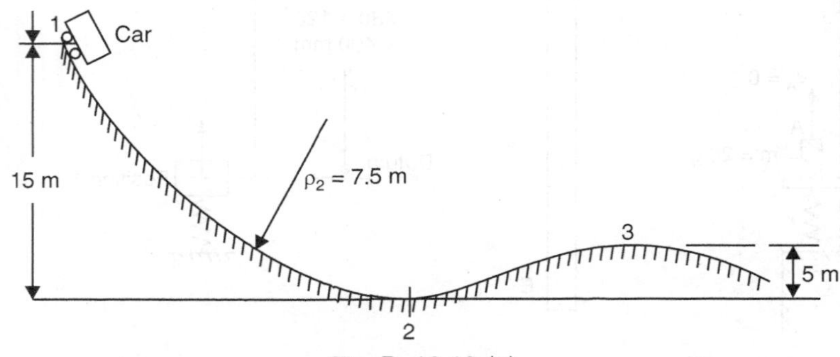

Fig. P. 13.19 (a)

Sol. (*a*) **Track reaction at point 2**

At point 1, $v_1 = 0, \quad \therefore \quad T_1 = 0$

The only force which does work is the weight W of the car. With point 2 as datum, the workdone by the weight = $U_{1-2} = +\,Wh = 15\,W$.

Using the W-E equation $U_{1-2} = T_2 - T_1$, we have

$$15\,W = \frac{m}{2}\,v_2^2 - 0$$

$$\therefore \qquad 15\,W = \frac{W}{2g}\,v_2^2$$

$$\therefore \qquad v_2^2 = 30\,g = 30 \times 9.81 = 294.3$$

$$\therefore \qquad v_2 = \mathbf{17.16 \text{ m/s}}$$

$\therefore$ Normal acceleration of car at point 2

$$= a_n = \frac{v_2^2}{\rho_2} = \frac{294.3}{7.5} = 39.24 \text{ m/s}^2$$

$\therefore$ Using Newton's second law of motion, we have [By $\Sigma F_n = ma_n$], see Fig. P.13.19 (*b*)

$$N_2 - W = ma_n$$

$$\therefore \qquad N_2 = ma_n + W$$

$$= \frac{W}{g} \times \frac{v_2^2}{\rho_2} + W$$

$$= W\left(\frac{v^2}{g\rho_2} + 1\right)$$

W

a_{n_2}

N_2

Fig. P. 13.19 (b)

or $$N_2 = 10{,}000 \times \left(\frac{294.3}{9.81 \times 7.5} + 1\right) \times N$$

$$= 50{,}000 = \mathbf{50 \text{ kN}.} \quad \textbf{Ans.}$$

(*b*) **Minimum radius of curvature at point 3 :**

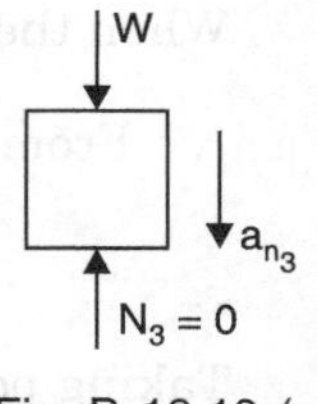

Fig. P. 13.19 (c)

Using the work-energy equation between points 1 and 3, we have

$$U_{1-3} = T_3 + T_1$$

or, $$W(15-5) = \frac{W}{2g} \times v_3^2 + 0$$

$\therefore$ $$v_3^2 = 2g \times 10 = 9.81 \times 20$$

or $$v_3^2 = 196.2$$

or $$v_3 = 14.007 \text{ m/s}$$

Applying the second law of motion at point 3, we find the minimum safe radius of curvature ρ_3. For the minimum curvature the normal reaction $N_3 = 0$.

$\therefore$ By $\Sigma F_n = ma_n$, we get $W - N_3 = ma_n$ $\qquad \left| \; a_n = \dfrac{v_3^2}{\rho_3} \right.$

or $$W - 0 = \frac{W}{g} \times \frac{v_3^2}{\rho_3}$$

$\therefore$ $$\rho_3 = \frac{v_3^2}{g} = \frac{9.81 \times 20}{g} = 20 \text{ m}$$

$\therefore$ The least safe radius of curvature = 20 m. **Ans.**

Problem 13.20. *A bar OA, hinged at O, carries a mass m at end A. The bar is released from the vertical position. Find the angle θ at which the force in the bar becomes zero. See Fig. P. 13.20.*

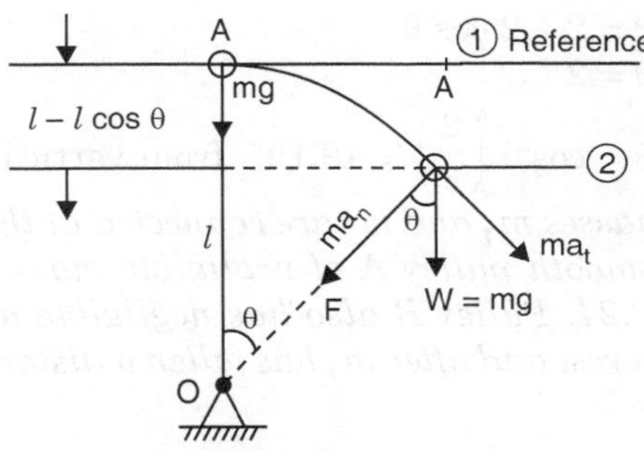

Fig. P. 13.20

Sol. When the bar is in vertical position it is subjected to compressive force equal to the weight $W = mg$ of the mass at A.

When the bar is at position (2), let a_t and a_n be the tangential and normal acceleration components of A. At this position the mass is subjected to

(*a*) vertical force equal to the weight $W = mg$,

(*b*) tangential force $\qquad F_t = ma_t = m\dfrac{dv}{dt}$,

(*c*) normal or radial force $F_n = ma_n = \dfrac{mv^2}{r} = \dfrac{mv^2}{l}$,

(*d*) the axial force F in the bar.

By $\Sigma F_n = ma_n$, we get

$$W \cos \theta - F = ma_n$$

or $$mg \cos \theta - F = \frac{mv^2}{l}$$

$\therefore$ $$F = mg \cos \theta - \frac{mv^2}{l} \qquad \text{...(1)}$$

When there is no force in the bar $F = 0$

$\therefore$ From eqn. (1), we get $0 = mg \cos \theta - \dfrac{mv^2}{l}$

$$\therefore \quad \cos \theta = \frac{v^2}{gl} \qquad ...(2)$$

Taking position (1) as the reference,

$$PE)_1 = V_1 = 0, \quad KE)_1 = T_1 = 0,$$

$\therefore$ Total energy, $E_1 = V_1 + T_1 = 0$

For position (2), $PE)_2 = V_2 = -mg(l - l \cos \theta) = -mgl(1 - \cos \theta)$

$$KE)_2 = T_2 = \frac{mv^2}{2}$$

$$\therefore \quad E_2 = T_2 + V_2 = \frac{mv^2}{2} - mgl\,(1 - \cos \theta)$$

$$\because \quad E_2 = E_1, \quad \therefore \text{ we get}$$

$$\frac{mv^2}{2} - mgl(1 - \cos \theta) = 0$$

$$\therefore \quad v^2 = 2gl(1 - \cos \theta) \qquad ...(3)$$

Substituting the value of v^2 from eqn. (3) into eqn. (2), we get

$$\cos \theta = \frac{2gl(1 - \cos \theta)}{gl}$$

$$\therefore \quad \cos \theta = 2 - 2 \cos \theta$$

$$\therefore \quad 3 \cos \theta = 2$$

$\therefore$ $\theta = \cos^{-1}\left(\dfrac{2}{3}\right) =$ **48.19°** from vertical position. **Ans.**

Problem 13.21. *Two masses m_1 and m_2 are connected at the ends of an inextensible cord which is lying on a vertical smooth pulley A of negligible mass. Masses m_1 and m_2 are connected as shown in Fig. P. 13.21. Pulley B also has negligible mass. Find the velocity of the mass m_2 which is released from rest and after m_1 has fallen a distance of 1.5 m. Take $m_1 = 1.6\, m_2$. Take $m_2 = 5$ kg.*

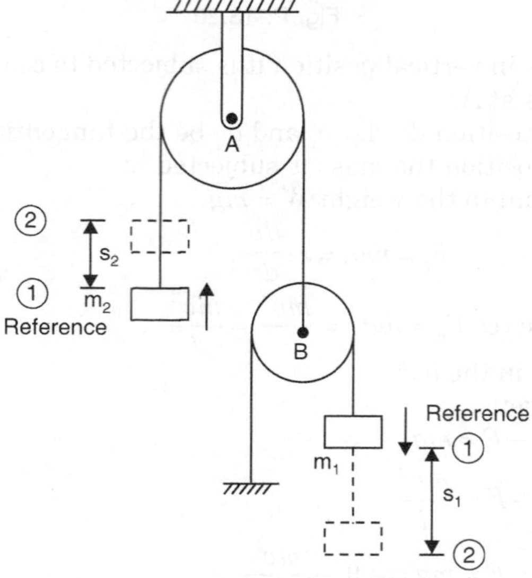

Fig. P. 13.21

Sol. When m_1 has fallen a distance S_1, the mass m_2 rises through a distance $S_2 = -\dfrac{S_1}{2}$

$\therefore \quad S_1 = -2S_2$ ...(1)

Differentiating with respect to time, we get

$v_1 = -2v_2$ ...(2)

and $a_1 = -2a_2$ ...(3)

When at position (1), since the system is at rest,

$$KE)_1 = T_1 = 0 \ ; \ PE)_1 = V_1 = 0,$$

$\therefore$ Total energy E_1 for both masses = 0.

At position (2). $PE)_2 = V_2 = PE$ of mass $m_1 + PE$ of mas m_2

or
$$V_2 = m_1gS_1 + m_2gS_2$$
$$= m_1g \times S_1 + m_2g(S_2) \qquad | \ S_1 = -2S_2$$
$$= g(S_1m_1 + S_2m_2)$$
$$= g\left[1.6m_2 \times (-S_1) + m_2 \times \left(+\frac{S_1}{2}\right)\right] = m_2g\left(-1.6 \times S_1 + \frac{S_1}{2}\right)$$
$$= 5 \times g(-1.6 \times 1.5 + 0.5) = 1.5 \times 5g \times 1.1 \text{ J}$$

$\therefore \quad V_2 = -8.25\,g$ J

$$KE)_2 = T_2 = \frac{1}{2}m_1v_1^2 + \frac{1}{2}m_2v_2^2$$
$$= \frac{1}{2}m_1v_1^2 + \frac{1}{2}m_2\left(-\frac{v_1}{2}\right)^2 = \frac{1}{2}v_1^2\left[m_1 + \frac{1}{4}m_2\right]$$
$$= \frac{1}{2} \times v_1^2\left[1.6\,m_2 + \frac{1}{4} \times m_2\right] = \frac{1}{2}v_1^2 \times m_2\,(1.6 + 0.25)$$
$$= \frac{1}{2}m_2v_1^2 \times 1.85 = \frac{1}{2} \times 5v_1^2 \times 1.85 = 4.625\,v_1^2 \text{ J}$$

$\therefore \quad E_2 = V_2 + T_2 = -8.25\,g + 4.625\,v_1^2$

Since, there is no friction, therefore the system is conservative.

Hence $\quad E_1 = E_2,$

$\therefore$ We get $\quad 0 = -8.25\,g + 4.625\,v_1^2$

$\therefore \quad v_1^2 = \dfrac{8.25 \times g}{4.625} = 17.5$

$\therefore \quad v_1 =$ **4.180 m/s. Ans.**

$\therefore \quad v_2 =$ **– 2.09 m/s. Ans.**

Problem 13.22. *A uniform flexible chain AB of length l is held on a smooth table with a portion h over hanging as shown in Fig.P. 13.22 (a). If released from rest find the velocity at which the chain will leave the table. w is the weight of per unit length of the chain.*

Sol. Let v be the velocity with which the chain leaves the table.

With reference to position (1), we get

$\because$ The chain is at rest, $KE)_1 = T_1 = 0$

$$PE)_1 = V_1 = -(Wh)\,\bar{y}_1 = -(mgh)\frac{h}{2}$$

At position (2), $\quad KE)_2 = T_2 = \dfrac{1}{2}(ml) \times v^2$

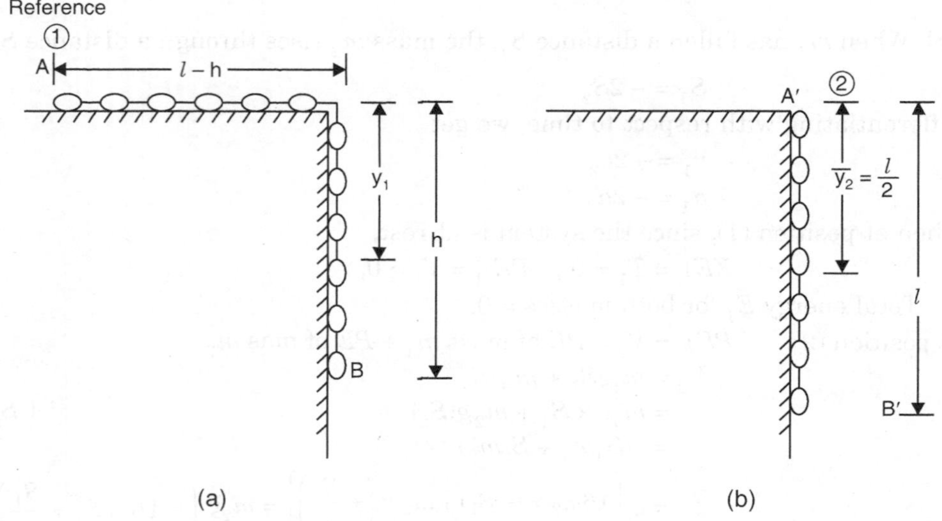

Fig. P. 13.22

$$PE)_2 = V_2 = -(Wl)(\bar{y}_2) = -(mgl)(l/2)$$

Since the system is conservative $E_1 = E_2$

$$\therefore \quad \text{We get} \quad 0 + \left(\frac{-mgh^2}{2}\right) = \frac{mlv^2}{2} + \left(\frac{-mgl^2}{2}\right)$$

$$\therefore \quad lv^2 = gl^2 - gh^2$$

$$\therefore \quad v = \sqrt{\frac{g}{l}(l^2 - h^2)}. \quad \textbf{Ans.}$$

Problem 13.23. *A uniform flexible chain of length* $\frac{\pi R}{2}$ *is held on a smooth cylindrical surface as shown in Fig. P.13.23 (a). When released from rest find the velocity at which the chain will leave the cylindrical surface. Let w be the weight of a unit length of the chain.*

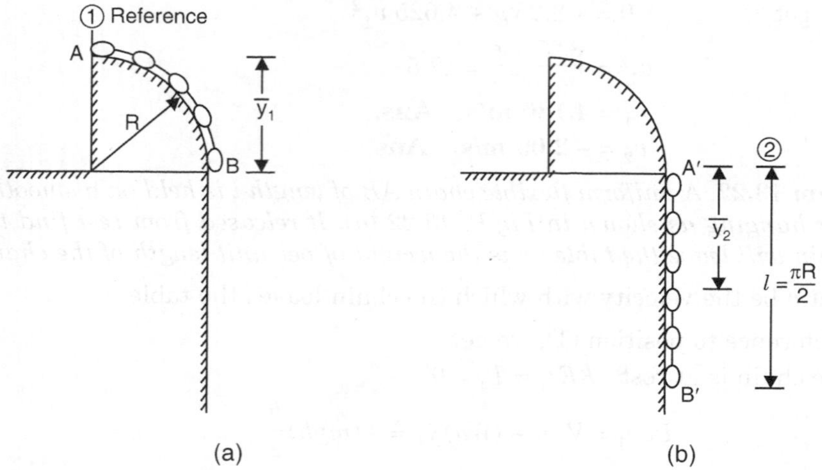

Fig. P. 13.23

Sol. Since chain bends in the form of a quarter circular arc, its C.G. is at a distance $\bar{y}_1 = \frac{2R}{\pi}$ above point B. At the reference position (1),

$$KE)_1 = T_1 = 0\ ;$$

$$PE)_1 = w\left(\frac{\pi R}{2}\right)\bar{y}_1 = w\left(\frac{\pi R}{2}\right)\left(\frac{2R}{\pi}\right)$$

$\therefore$ Total energy $$E_1 = T_1 + V_1 = 0 + w\left(\frac{\pi R}{2}\right)\left(\frac{2R}{\pi}\right)$$

At position (2), let the velocity of the chain be v. Then,

$$KE)_2 = T_2 = \frac{w \times \left(\frac{\pi R}{2}\right)}{2g} \times v^2 = \frac{w\pi R v^2}{4g}$$

$$PE)_2 = V_2 = \left(w \times \frac{\pi R}{2}\right)\bar{y}_2 = \frac{w\pi R}{2} \times \left(\frac{-\pi R}{4}\right) = -\frac{w\pi^2 R^2}{8}$$

$\therefore$ $$E_2 = T_2 + V_2 = \frac{w\pi R v^2}{4g} - \frac{w\pi^2 R^2}{8}$$

$\because$ $$E_2 = E_1,$$

$\therefore$ $$\frac{w\pi R v^2}{4g} - \frac{w\pi^2 R^2}{8} = w\left(\frac{\pi R}{2}\right)\left(\frac{2R}{\pi}\right)$$

$\therefore$ $$v^2 = \frac{4g}{\pi R}\left(\frac{\pi^2 R^2}{8} + R^2\right)$$

$\therefore$ $$v = \sqrt{gR\left(\frac{\pi}{2} + \frac{4}{\pi}\right)}.$$ **Ans.**

Problem 13.24. *A uniform chain of length l rests on the surface of a smooth sphere of radius R as shown in Fig. P.13.24 (a). When released from rest, find the velocity when the chain leaves the surface of the sphere.*

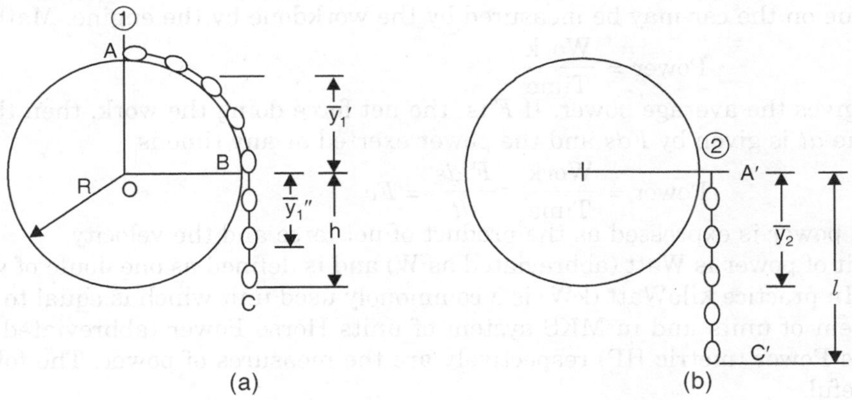

Fig. P. 13.24

Sol. Let w be the weight of a unit length of chain.

With reference to position (1),

$$KE)_1 = T_1 = 0 \text{ (chain being at rest),}$$

$$PE)_1 = V_1 = (w \times AB)\ \bar{y}_1' + wBC \times \bar{y}_1'' = \left(w \times \frac{\pi R}{2}\right) \times \frac{2R}{\pi} + (wh)\left(-\frac{h}{2}\right)$$

$$= w\left(R^2 - \frac{h^2}{2}\right)$$

$$\therefore \quad E_1 = T_1 + V_1 = 0 + w\left(R^2 - \frac{h^2}{2}\right)$$

At position (2), we have

$$KE)_2 = \frac{(w \times l)}{2g} v^2$$

$$PE)_2 = V_2 = (wl)\,(\bar{y}_2) = wl\left(\frac{-l}{2}\right) = -\frac{wl^2}{2}$$

$$\therefore \quad E_2 = T_2 + V_2 = \frac{wl}{2g} v^2 - \frac{wl^2}{2}$$

$$\because \quad E_2 = E_1$$

$$\therefore \quad \frac{wlv^2}{2g} - \frac{wl^2}{2} = 0 + w\left(R^2 - \frac{h^2}{2}\right)$$

$$\therefore \quad \frac{lv^2}{2g} = \frac{l^2}{2} + R^2 - \frac{h^2}{2}$$

$$\therefore \quad v^2 = \frac{2g}{l}\left(\frac{l^2}{2} + R^2 - \frac{h^2}{2}\right)$$

$$\therefore \quad v = \sqrt{\frac{g}{l}(l^2 + 2R^2 - h^2)}. \quad \textbf{Ans.}$$

13.11. POWER AND EFFICIENCY

Power is the time rate of doing work. For example, if a car is being pulled by an engine, the work done on the car may be measured by the workdone by the engine. Mathematically.

$$\text{Power} = \frac{\text{Work}}{\text{Time}} \quad \text{...(13.31)}$$

This gives the average power. If F is the net force doing the work, then the workdone during a time dt is given by Fds and the power exerted at any time is

$$\text{Power} = \frac{\text{Work}}{\text{Time}} = \frac{F.ds}{dt} = Fv \quad \text{...(13.32)}$$

Thus, power is expressed as the product of net force and the velocity.

SI unit of power is Watt (abbreviated as W) and is defined as one Joule of work done in one second. In practice kiloWatt (kW) is a commonoly used unit which is equal to 1000 Watts. In FPS system of units and in MKS system of units Horse Power (abbreviated as HP) and metric Horse Power (metric HP) respectively are the measures of power. The following relations are useful.

1 HP = 550 ft-lb/sec. = 33000 ft-lb/min

1 metric HP = 75 kg_f-m/sec = 4500 kg_f-m/min.

1 metric HP = 735.75 Watts

1 HP or 1 British HP = 745.8 Watts.

1 kW = 1.34 HP

Sometimes work is also expressed in units such as : Watt-hour.

The large quantities of work are, therefore, horsepower-hour (hp-hr) and kilowatt-hour (kW-hr). These units indicate the amount of workdone in one hour at a constant rate of 1 hp or 1 kW, respectively. Power is generally abbreviated as P.

Efficiency. All machines which are used to do some useful work have internal friction. Hence, some power is lost due to friction and due to some other causes. Hence, the power delivered by these machines and almost by all other devices is never equal to the power put into it. *Efficiency is the ratio of power output to the power input.* Efficiency is generally expressed in percentage. Since the input power and output power are measured simultaneously at the same time therefore efficiency of a device may also be expressed as the ratio of energy output to the energy input or of the work output to the work input. Efficiency is represented by the Greek letter η (Eta). Hence, we write

$$\text{Efficiency,} \quad \eta = \frac{\text{Work output}}{\text{Work input}} \qquad ...(13.33)$$

$$\text{or} \quad \eta = \frac{\text{Energy output}}{\text{Energy input}} \qquad ...(13.34)$$

The *overall efficiency* of a number of machines placed in series is equal to the product of their individual efficiencies.

Thus, if there are n machines in a system of machines then the overall efficiency of the system is given as :

$$\text{Overall efficiency, } \eta_{\text{overall}} = \eta_1 \cdot \eta_2 \cdot \ldots\ldots \eta_n \qquad ...(13.35)$$

Many other types of efficiencies used in the study of mechanical systems, (like engines, power plants, etc.) are thermal efficiency, volumetric efficiency, indicated thermal efficiency, etc. These are beyond the scope of this book.

Problem 13.25. *A 1.5 MN train is accelerated at a constant rate up a 3% grade track. The train resistance is constant at 8 N per kN of train. The velocity increases from 36 kmph to 72 kmph in a distance of 800 m. Calculate the maximum power developed by the locomotive.*

[**Note.** Inclination of the railway track is expressed as the *rise in metres measured in a horizontal distance of 100 m.* Thus, a 3% grade means a rise of 3 m in a horizontal distance of 100 m.

For grades upto 10% the length of the track is practically equal to its horizontal projection so that cosine of angle is almost equal to unity. Hence the normal track reaction (or pressure) **remains** constant and may be taken as equal to the weight of the train. Therefore the normal pressure remains constant for upto 10% grade. Further, for small angles the sine may be taken equal to the tangent, so that for 3% grade, the sine may be equal to $\frac{3}{100}$].

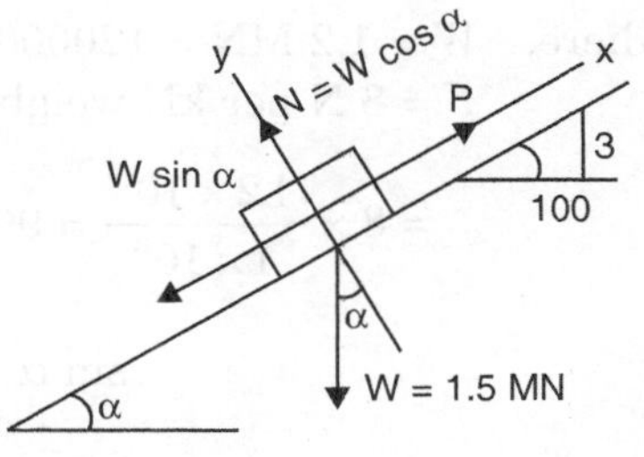

Fig. P. 13.25

Sol. The free-body diagram of the train is shown in Fig. P. 13.25. P is the pull of the locomotive.

$$\sin \alpha = \frac{3}{100} \quad \text{and} \quad \cos \alpha \simeq 1$$

$$R = \text{normal reaction} = W \cos \alpha \simeq W = 1.5 \times 10^6 \text{ N},$$

$$W \sin \alpha = 1.5 \times \frac{3}{100} = 1.5 \times 10^6 \times \frac{3}{100} = 45000 \text{ N}$$

Initial speed v_0 = 36 kmph

$$= \frac{36 \times 1000 \text{ m/hr}}{60 \times 60 \text{ s/h}} = 10 \text{ m/s}$$

Final speed v = 72 kmph = **20 m/s.**

Resistance = 8 N per kN of weight = $\dfrac{8 \times 1.5 \times 10^6}{1 \times 10^3}$ = 12,000 N

Thus, by applying the work-energy equation $\left[\Sigma F_x \times (s) = \dfrac{W}{2g}(v^2 - v_0{}^2)\right]$, we get

$$[P - 45000 - 12000] \times 800 = \frac{1.5 \times 10^6}{9.81} \times (20^2 - 10^2)$$

or $[P - 45000 - 12000] \times 800 = 45871559.63$

$\therefore$ **P = 114339.45 N**

Hence the maximum power when its speed is 20 m/s is equal to

Power = Pv = 114339.45 × 20 = 2286788.991 watts = **2286.8 kW. Ans.**

Problem 13.26. *Find the power of a diesel locomotive drawing a train whose weight including that of the engine is 1.2 MN up an inclined track 1 in 240 at a steady speed of 54 km/hr. The of resistance to the train may be taken as 8 N/kN weight of the train. Take efficiency equal to 80%.*

If the diesel supply to the locomotive is cut off, find the distance during which the speed falls to 36 kmph.

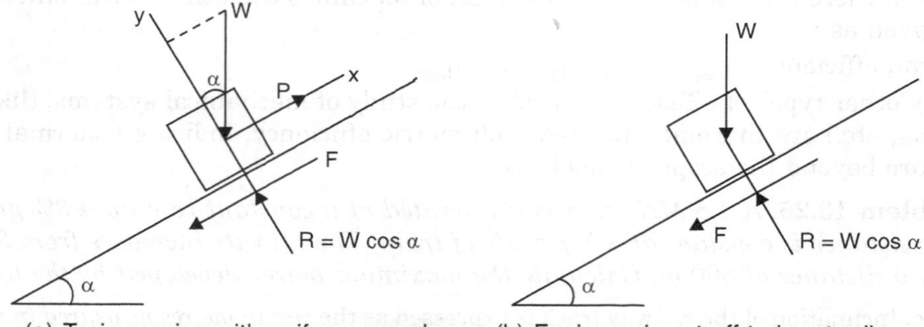

(a) Train moving with uniform speed (b) Fuel supply cut off train retarding.

Fig. P. 13.26

Sol. (*a*) When the train is moving with uniform speed of 54 kmph or 15 m/s, the free body diagram of the train is given in Fig. P. 13.26 (*a*).

By $\Sigma F_x = 0$, we get

$$P - F - W \sin \alpha = 0$$

where, W = 1.2 MN = 1200000 N,

F = 8 N per kN weight of train

$$= 8 \times \frac{1.2 \times 10^6}{1 \times 10^3} = 9600 \text{ N},$$

$$\sin \alpha \simeq \frac{1}{240}$$

$\therefore$ P = Pull of engine = $F + W \sin \alpha = 9600 + 1200000 \times \dfrac{1}{240}$

= 9600 + 5000 = **14600 N**

Speed of train, v = 54 kmph = 15 m/s,

$\therefore$ Power = Pv = 14600 × 15 = 219000 W = **219 kW**

when efficiency η = 80%, Power = $\frac{219}{\eta} = \frac{219}{0.8}$ = **273.75 kW. Ans.**

(*b*) When fuel (diesel) supply to the engine is the cut off, the pull P becomes zero, as shown in Fig. P. 13.26 (*b*).

By applying work-energy equation, we get

$$(F + W \sin \alpha) \times S = \frac{W}{2g}(v^2 - v_0^2)$$

$$(9600 + 5000) \times S = \frac{1200000}{2 \times 9.81} \times (10^2 - 15^2)$$

Here v_0 = 54 kmph = 15 m/s and v = 36 kmph = 10 m/s

$\therefore$ 14600 S = – 7645256

$\therefore$ S = **523.65 m. Ans.**

Problem 13.27. *At a certain instant during acceleration of car along a level road the engine pull is 1800 N. What power is developed by the engine when the speed of the car is 72 kmph ? When the efficiency is 75%, find the work input of engine.*

Sol. Power = pull × velocity, = Pv

where Pull = P = 1800 N,

velocity, v = 72 kmph = $\frac{72}{3.6}$ m/s = 20 m/s

$\therefore$ Power = 1800 × 20 = **36000 W** or **36 kW. Ans.**

Input power = $\frac{\text{Output power}}{\text{Efficiency}} = \frac{36000}{0.75}$ = **48000 W = 48 kW. Ans.**

Problem 13.28. *The force P acting on the Prony brake dynamometer shown in Fig. P.13.28 is 152 N when the brake drum rotates at 600 rpm. Find the power developed by the engine to which the dynamometer is mounted.*

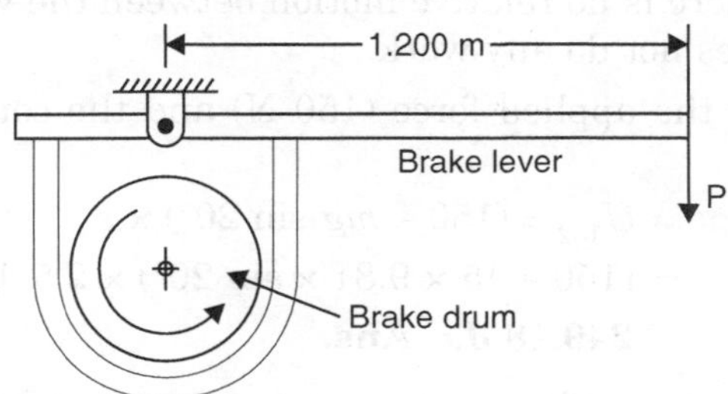

Fig. P. 13.28

Sol. The turning moment $T = P \times l$ = 152 × 1.2 = 182.4 Nm

Angular speed of drum, $\omega = \frac{2\pi N}{60} = \frac{2\pi \times 600}{60} = 20\,\pi$ rad/s

$\therefore$ Power = $T \,.\, w$ = 182.4 × 20 π Watts = **11460.53 Watts. Ans.**

Problem 13.29. *A block of mass m starts moving with initial velocity on a horizontal plane. Find the coefficient of friction μ if the block stops after covering a distance s.*

Sol. Normal reaction, $N = mg$,

Force of friction $= F = \mu N = \mu mg$

Workdone by the force of friction = $U_{1-2} = -Fx$

or $$U_{1-2} = -(\mu mg)s.$$

Change of kinetic energy $= T_2 - T_1 = \frac{1}{2} mv^2 - \frac{1}{2} mv_0^2 = 0 - \frac{1}{2} mv_0^2$

By applying work-energy equation, we have

$$U_{1-2} = T_2 - T_1$$

or $$-(\mu mg)s = 0 - \frac{1}{2} mv_0^2$$

$\therefore$ $$\boxed{\mu = \frac{v_0^2}{2gs}}. \quad \textbf{Ans.}$$

Problem 13.30. *Find the workdone in rolling a 15 kg wheel a distance 2.5 m up a plane inclined 20° with the horizontal as shown in Fig. P. 13.30 (a). Take the coefficient of friction as 0.2.*

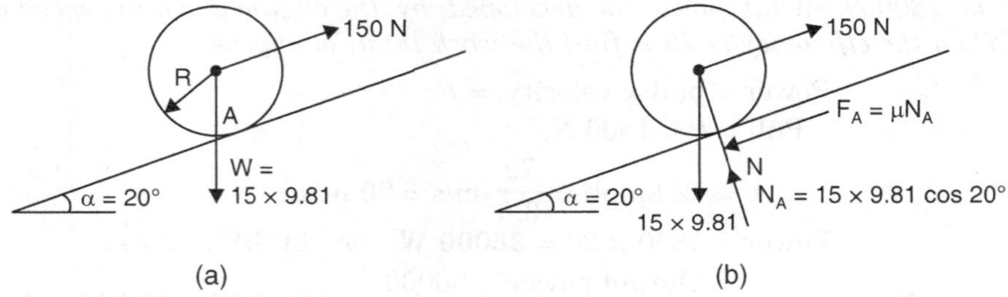

Fig. P. 13.30

Sol. The forces acting on the wheel are shown in the free-body-diagram shown in Fig. P. 13.30 (*b*).

As the wheel rolls, there is no relative motion between the wheel and the plane. Therefore, the force of friction does not do any work.

The work is done by the applied force (150 N) and the component mg sin 20° of the weight of the block.

Thus, workdone $= U_{1-2} = (150 - mg \sin 20°) \times s$

$$= (150 - 15 \times 9.81 \times \sin 20°) \times 2.5 \text{ J}$$

$$= \textbf{249.18 J.} \quad \textbf{Ans.}$$

Problem 13.31. *A long rod of length 3 m and of mass of 5 kg increases its speed about a vertical axis through one of its ends from 30 rpm to 60 rpm in 12 seconds. Find the constant moment M required to do this.*

Sol. Initial angular speed $\omega_1 = \frac{2\pi N_1}{60} = \frac{2\pi \times 30}{60} = \pi$ rad/s

final angular speed $\omega_2 = \frac{2\pi N_2}{60} = \frac{2\pi \times 60}{60} = 2\pi$ rad/s

Moment of inertia I_0 of the rod about an axis through one end $= I_0 = \frac{1}{3} ml^2$

$$= \frac{1}{3} \times 5 \times (3)^2 = 15 \text{ kg.m}^2$$

$\because$ Workdone = change in kinetic energy

$$\therefore \quad M\theta = \frac{1}{2}I(\omega_2^2 - \omega_1^2) \quad ...(1)$$

where θ is found by using the relation $\theta = \theta_1 t + \frac{1}{2}\alpha t^2$

We find angular acceleration $\alpha = \frac{\omega_2 - \omega_1}{t}$

$$\alpha = \frac{2\pi - \pi}{12} = \frac{\pi}{12}$$

and
$$\theta = \frac{\omega_2^2 - \omega_1^2}{2\alpha} = \frac{4\pi^2 - \pi^2}{2 \times \pi/12} = \frac{3\pi^2 \times 6}{\pi} = 18\,\pi$$

By using the work-energy equation $\Sigma M\,\theta = T_2 - T_1$,

we get
$$M\theta = \frac{1}{2}\,I_0(\omega_2^2 - \omega_1^2)$$

or
$$M \times 18\pi = \frac{1}{2} \times 15 \times (4\pi^2 - \pi^2)$$

$$\therefore \quad M = \frac{15}{2 \times 18\pi} \times 3\,\pi^2 = 1.25\,\pi \text{ Nm} = \mathbf{3.927\ Nm.} \quad \textbf{Ans.}$$

Problem 13.32. *A sphere of mass M and rolling with initial velocity 12 m/s starts moving up a 30° inclined plane. How far will the sphere roll before it stops ?*

Sol. The free-body diagram of the sphere is shown in Fig. P.13.32.

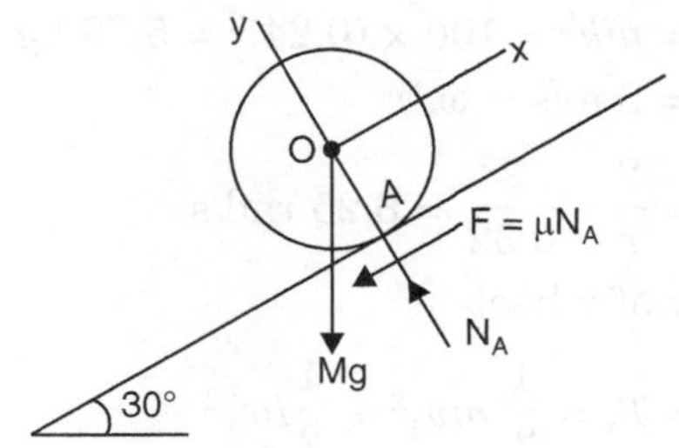

Fig. P. 13.32

Mass-moment of inertia, $I_0 = \frac{2}{5}\,MR^2$.

Initial kinetic energy

$$T_1 = \frac{1}{2}\,Mv_1^2 + \frac{1}{2}\,I\omega_1^2 = \frac{1}{2}M(\omega_1 R)^2 + \frac{1}{2}I\omega_1^2$$

$$= \frac{1}{2}M(\omega_1^2 \times R^2) + \frac{1}{2} \times \frac{2}{5}\,MR^2\,\omega_1^2$$

$$= \frac{M\omega_1^2}{10}R^2\,(5 + 2) = \frac{7}{10}\,M(\omega_1 R)^2 \qquad |\ v = \omega_1 R = 12 \text{ m/s}$$

$$\therefore \quad T_1 = \frac{7}{10}\,M \times (12)^2 = 100.8\,M$$

$$T_2 = 0.$$

The only force that does work is Mg sin 30° the workdone by Mg sin 30° is negative.

By using the work-energy equation, we get

$$U_{1-2} = T_2 - T_1$$

or $$-(Mg \sin 30°) \times x = 0 - 100.8\, M$$

$\therefore$ $$x = \frac{100.8}{g \sin 30°} = \mathbf{20.55\ m.}\quad \textbf{Ans.}$$

Problem 13.33. *A wheel of mass 100 kg, has radius of gyration 0.24 m with respect to its mass centre G. When the initial velocity of the mass centre G is 2 m/s downwards a 30° inclined plane the spring is stretched 0.2 m. If the spring modulus or stiffness is 900 N/m, what will be the total stretch of the spring when the wheel stops ? The radius of the wheel is 0.32 m.*

Sol. Fig. P.13.33 shows the free-body diagram of the wheel. When the spring stretches by x, the spring force is kx.

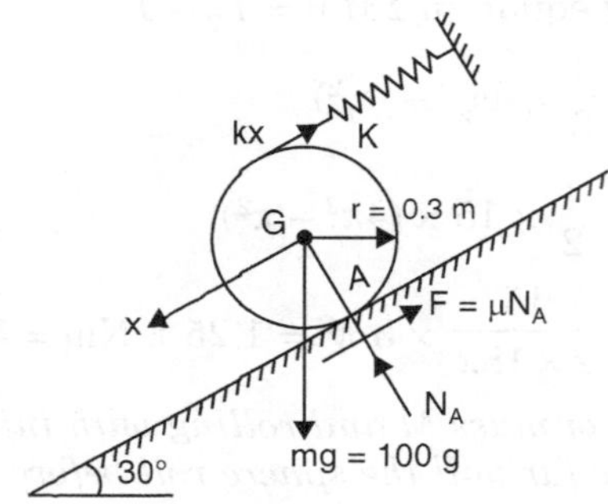

Fig. P. 13.33

$$I = mk^2 = 100 \times (0.24)^2 = 5.76\ \text{kg.m}^2$$

$$v_1 = 2\ \text{m/s} = \omega_1 r$$

$\therefore$ $$\omega_1 = \frac{2}{r} = \frac{2}{0.32} = 6.25\ \text{rad/s}$$

$\therefore$ Initial kinetic energy of wheel

$$= T_1 = \frac{1}{2} m v_1^2 + \frac{1}{2} I \omega_1^2$$

or $$T_1 = \frac{1}{2} \times 100 \times (2)^2 + \frac{1}{2} \times 5.76 \times (6.25)^2 = 200 + 112.5 = 312.5\ \text{Nm}$$

Final kinetic energy = $T_2 = 0$ because the wheel stops.

As the wheel moves down the plane, the workdone by $mg \sin \theta$ component of the weight of wheel will be positive.

Also, since the spring force acts in the opposite direction, the spring work will be negative.

Initial spring stretch $= x_1 = 0.2$ m.

Hence, total workdone $$= U_{1-2} = +\, mg \sin \theta \left(\frac{x - x_1}{2} \right) - \int_{0.2}^{x} kx\, dx$$

or, $$U_{1-2} = (100 \times 9.81) \sin 30° \times \left[\frac{x - 0.2}{2} \right] - \int_{0.2}^{x} 900\, x\, dx$$

$$= 245.25\,x - 49.05 - \left[\frac{900}{2} \times (x^2) - \frac{900}{2}(0.2)^2\right]$$

$$= 245.25\,x - 49.05 - 450\,x^2 + 18$$

$\therefore$ $$U_{1-2} = -\,450\,x^2 + 245.25\,x - 31.05$$

Using the work-energy relation, we get or by $U_{1-2} = T_2 - T_1$

or $$(-\,450\,x^2 + 245.25\,x - 31.05) = -\,312.5 - 0$$

$\therefore$ $$-\,450x^2 + 245.25\,x + 281.45 = 0$$

or $$x^2 - 0.545x - 0.6254 = 0$$

$\therefore$ $$x = \frac{1}{2}\,[0.545 \pm \sqrt{(0.545)^2 + 4 \times 0.6254}\,]$$

$$= \frac{1}{2}\,[0.545 \pm 1.5740] = 1.0595 \text{ m} \quad \text{and} \quad -\,0.5145 \text{ m}$$

$\therefore$ Maximum stretch of spring = x = **1.0595 m. Ans.**

Problem 13.34. *A long rod of mass m and length l is pinned at one end to a horizontal plane. The rod is initially vertical and is allowed to fall. At what linear velocity will the rod strike the floor ?*

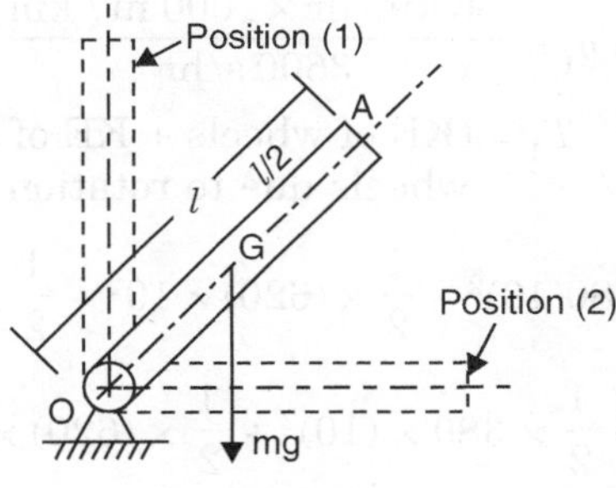

Fig. P. 13.34

Sol. When the rod occupies the position shown in Fig. P.13.34 only the gravity force, *mg*, acting through the mass centre *G* does the work. Workdone by gravity force,

$$U_{1-2} = (mg) \times \frac{l}{2} - mg\,(0) = mg(l/2)$$

Mass moment of inertia, $$I_0 = \left(\frac{1}{3} ml^2\right)$$

Since the rod is at rest initially, its $KE = T_1 = 0$

When the rod strikes the floor, $$KE = T_2 = \frac{1}{2} I_0\,\omega_2^{\,2}$$

or $$T_2 = \frac{1}{2}\left(\frac{1}{3} ml^2\right) \times \omega_2^{\,2}$$

Using the work-energy equation [*i.e.*, $U_{1-2} = T_2 - T_1$], we have

$$\frac{1}{2}\, mgl = \frac{1}{2}\left(\frac{1}{3} ml^2\right) \times \omega_2^{\,2}$$

$\therefore \quad \omega_2^2 = 3g/l$

$\therefore \quad \omega_2 = \sqrt{3g/l}$

$\therefore$ linear velocity of rod $v_2 = \omega_2 \,.\, l/2$

or $$v_2 = \left(\sqrt{\frac{3g}{l}}\right)\frac{l}{2} = \frac{1}{2}\sqrt{3gl}\,. \quad \textbf{Ans.}$$

Problem 13.35. *A car and four wheels have a mass of 1000 kg. Each wheel has a mass of 95 kg and is of 760 mm diameter. The car initially having a speed of 36 km/h stops on a level track in 2.4 km. Find the rolling resistance.*

Sol. When brakes are not applied, the frictional force between wheels and the road surface will not do any work. Work done by the rolling resistance $(F) = U = F \times S = F \times (2.4 \times 1000)$ Nm.

As the final velocity of car is equal to zero therefore the final kinetic energy, $T_2 = 0$.

Total mass of 4 wheels $= 4 \times 95 = 380$ kg

Mass of car $= 1000 - 380 = 620$ kg.

Fig. P. 13.35

Initial velocity $= v_1 = \dfrac{36 \text{ km/hr} \times 1000 \text{ m / km}}{3600 \text{ s/hr}} = 10$ m/s

$\therefore$ Initial kinetic energy T_1 = (KE of wheels + KE of car) due to translation + KE of wheels due to rotation

or $$T_1 = \frac{1}{2} \times 380(10)^2 + \frac{1}{2} \times (620) \times 10^2 + \frac{1}{2} \times \{(380) \times (0.760)^2\} \times \left(\frac{10}{0.760/2}\right)^2$$

$$= \frac{1}{2} \times 380 \times (10)^2 + \frac{1}{2} \times (620) \times (10)^2 + \frac{1}{2} \times 380 \times (20)^2$$

$$= 50000 + 76000 \text{ Nm}$$

or $$T_1 = 126000 \text{ Nm}$$

Applying work-energy equation, we get

$$U_{1-2} = T_2 - T_1$$

or $$F \times 2.4 \times 1000 = 0 - 126000$$

$\therefore \quad F = -52.5$

$\therefore$ Rolling resistance $=$ **52.5 N. Ans.**

Problem 13.36. *A 60 kg, 1.8 m long rod falls from rest from the horizontal position to the 45° position shown in Fig. P.13.36. The upper end of the rod is hinged. Find the support reactions at the position shown.*

Sol. Let n be the longitudinal axis and t be perpendicular to the length of the rod.

In Fig. P. 13.36, G is the mass centre, $OG = l/2 = 0.9$ m ;

and AP = distance of the centre of percussion P from $A = k_A^2/\bar{r}$,

where $$k_A^2 = I_A/m = \frac{\frac{1}{3}ml^2}{m} = \frac{l^2}{3}$$

and $$\bar{r} = \frac{l}{2}.$$

$$\therefore \quad AP = \frac{k_A^2}{\bar{r}} = \frac{l^2/3}{l/2} = \frac{2l}{3} = \frac{2}{3} \times 1.8 = 1.2 \text{ m.}$$

To find the angular acceleration α of the rod at position (2), we use the condition $\Sigma M_A = 0$.

$$(mg \times AG \cos 45°) - (m\bar{r}\ \alpha) \times AP = 0$$

$$\therefore \quad mg \times \frac{l}{2} \times \cos 45° - (m \times l/2 \times \alpha) \times 1.2 = 0$$

$$\therefore \quad g \cos 45° = 1.2\ \alpha$$

$$\therefore \quad \alpha = \frac{9.81}{1.2} \times \cos 45° = 5.781 \text{ rad/s}^2$$

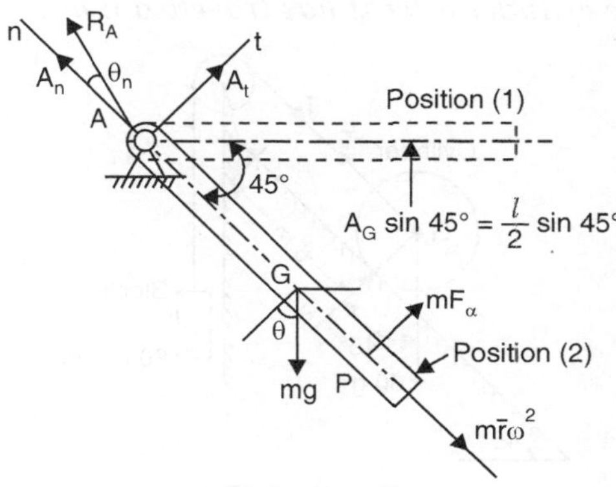

Fig. P. 13.36

To find the angular velocity ω_2 at position (2), we use the work-energy equation.

$\therefore$ by $U_{1-2} = T_2 - T_1$, we get

$$mg \times AG \sin 45° = \frac{1}{2} I_0\omega^2 - 0$$

or $$mg \times 0.9 \sin 45° = \frac{1}{2} \times \left[\frac{1}{3} \times m \times (1.8)^2\right] \times \omega^2$$

$$\therefore \quad 0.9\ g \sin 45° = \frac{1}{2} \times \frac{1}{3}\ (1.8)^2\ \omega^2$$

$\therefore$ ω^2 = **11.561.** Hence ω = **3.4 rad/sec.**

By summing forces along t-axis, we get $A_t - mg \cos 45° + m\bar{r}\alpha = 0$

$\therefore$ t–component of support reaction $= A_t = mg \cos 45° - m \times \frac{l}{2} \times 5.781$

$$= 60 \times 9.81 \cos 45° - 60 \times \frac{1.8}{2} \times 5.781$$

$$= 416.20 - 312.17$$

or A_t = **104.03 N. Ans.**

By summing forces along the n-axis, we get $A_n - mg \sin 45° - m\bar{r}w^2 = 0$

or n–component of support reaction $= A_n = mg \sin 45° + m \times \frac{l}{2} \times 11.561$

or A_n = **1040.50 N. Ans.**

$\therefore$ Resultant reaction at A

$$= R_A = \sqrt{A_n{}^2 + A_t{}^2} = \sqrt{(1040.50)^2 + (104.03)^2}$$

or R_A = 1045.7 N. **Ans.**

Inclination of R_A to the axis (or to the n-axis) is

$$= \theta_n = \tan^{-1}\left(\frac{A_t}{A_n}\right) = \tan^{-1}\left(\frac{104.03}{1040.50}\right) = \tan^{-1}(0.0999808)$$

$\therefore$ $\theta_n =$ **5.71°. Ans.**

Note. The force $m\bar{r}\omega^2$ is the reverse inertia force due to the acceleration a_n. Also $m\bar{r}\alpha$ is the reverse inertia force acting through the centre of percussion.

Problem 13.37. *A cylinder of mass 60 kg is pulled up a 45° plane by the tension in a cord which passes over a frictionless pulley and is attached to a block of mass 80 kg, as shown in Fig. P. 13.37. The cylinder has a diameter of 1500 mm, and moves from rest up a distance of 6 m. What is the speed of the cylinder after it has travelled 6 m ?*

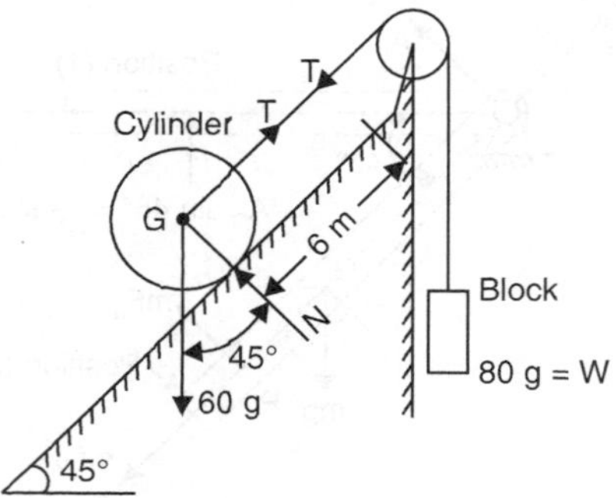

Fig. P. 13.37

Sol. Initial velocity of cylinder = 0, $\therefore$ $T_1 = 0$

KE at position (2)

$$T_2 = T_{2cyl.} + T_{2\text{ block}}$$
$$= \left[\frac{1}{2} m_c \,.\, v_2^2 + \frac{1}{2}\bar{I}_c \,.\, \omega^2\right] + \left[\frac{1}{2} m_b \times v_2^2\right]$$
$$= \left[\frac{1}{2} \times 60 \times v_2^2 + \frac{1}{2}\left(\frac{1}{2} \times 60 \times R^2\right) \times \frac{v_2^2}{R^2}\right] + \left[\frac{1}{2} 80 \times v_2^2\right]$$
$$= \frac{v_2^2}{2}(60 + 30 + 80)$$
$$= 85\, v_2^2$$

Initial potential energy of cylinder is zero and it will have final PE equal to (9.81 × 60) (6 sin 45°).

The initial PE of 80 kg block is zero, and final PE of block becomes (– 9.81 × 80) × (6). Thus, by applying the conservation of energy principle, we can write

$$V_1 + T_1 = V_2 + T_2$$

$\therefore$ $$0 + 0 = -(9.81 \times 80) \times (6) + 9.81 \times 60 \times 6 \sin 45° + 85\, v_2^2$$

$\therefore$ $$v_2^2 = \frac{9.81 \times 80 \times 6 - 9.81 \times 60 \times 6 \sin 45°}{85} = 26.02$$

$\therefore$ $v_2 =$ **5.101 m/s. Ans.**

Problem 13.38. *A solid cylinder of mass 150 kg and radius 200 mm rolls with slipping on the horizontal plane as shown in Fig. P. 13.38. The spring stiffness is 600 N/m and its unstretched length is 500 mm. Find the angular speed of the cylinder when its centre moves 600 mm to the left.*

Sol. At I position, the cylinder is at rest, $\therefore \quad KE)_1 = T_1 = 0.$

Final $KE = T_2 = \frac{1}{2} m\bar{v}_2^{\,2} + \frac{1}{2}\bar{I}\,\omega_2^{\,2}$ where $\bar{v}_2 = \omega_2 r_2 = 0.200\omega_2,$

$$\bar{I} = \frac{1}{2} mr^2 = \frac{1}{2} \times 150 \times (0.2)^2 = 3 \text{ kg m}^2,$$

$\therefore$ We get $\quad T_2 = \frac{1}{2} \times 150 \times (0.2\,\omega_2)^2 + \frac{1}{2} \times (3) \times \omega_2^{\,2} = 4.5\,\omega_2^{\,2}$

Since the point of application of force of friction moves with the cylinder the work of friction is zero.

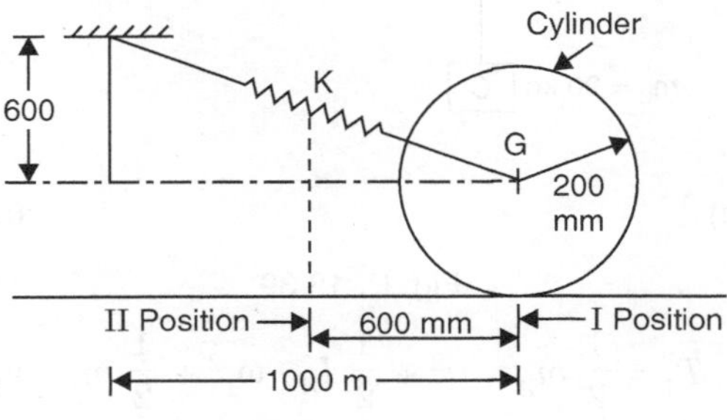

Fig. P. 13.38

Hence, the system is conservative. The law of conservation of energy is therefore, used. The system possesses kinetic and potential energies. The potential energy is due to elongation or contraction of spring.

Initial length of spring at I position

$$= S_1 = \sqrt{(0.6)^2 + (1.000)^2} = 1.1662 \text{ m.}$$

The final length of spring at II position

$$= S_2 = \sqrt{(0.6)^2 + (1 - 0.6)^2} = 0.7211 \text{ m}$$

Unstretched length, $S_0 = 0.5$ m.

$\therefore$ Extension of spring at I position

$$= \Delta x_1 = S_1 - S_0 = 1.1662 - 0.5 = 0.6662 \text{ m}$$

and $\quad \Delta x_2 = S_2 - S_0 = 0.7211 - 0.5 = 0.2211$ m.

$\therefore$ Potential energy, $V_1 = \frac{1}{2} k\Delta x_1^{\,2} = \frac{1}{2} \times 600 \times (0.6662)^2$

$$= 133.17 \text{ Nm}$$

And P.E. $\quad V_2 = \frac{1}{2} k\Delta x_2^{\,2} = \frac{1}{2} \times 600 \times (0.2211)^2 = 14.67$ N.m

Applying the law of conservation of energy

$$T_1 + V_1 = T_2 + V_2, \text{ we get}$$

$$0 + 133.17 = 4.5\,\omega_2^{\,2} + 14.67$$

$$\therefore \quad \omega_2^{\,2} = \frac{133.17 - 14.67}{4.5} = 26.33$$

$\therefore \quad \omega_2 =$ **5.132 radian/s. Ans.**

Problem 13.39. *In the system shown in Fig. P.13.39 (a) the block C moves down at 1.8 m/s. The cylinder B is homogeneous and moves in frictionless bearings. The spring is initially compressed 200 mm and has stiffness of 950 N/m. Find the velocity of the block C when it has dropped 1.5 m.*

Sol. The free-body diagram of the system is shown in Fig. P. 13.39 (*b*).

The initial kinetic energy

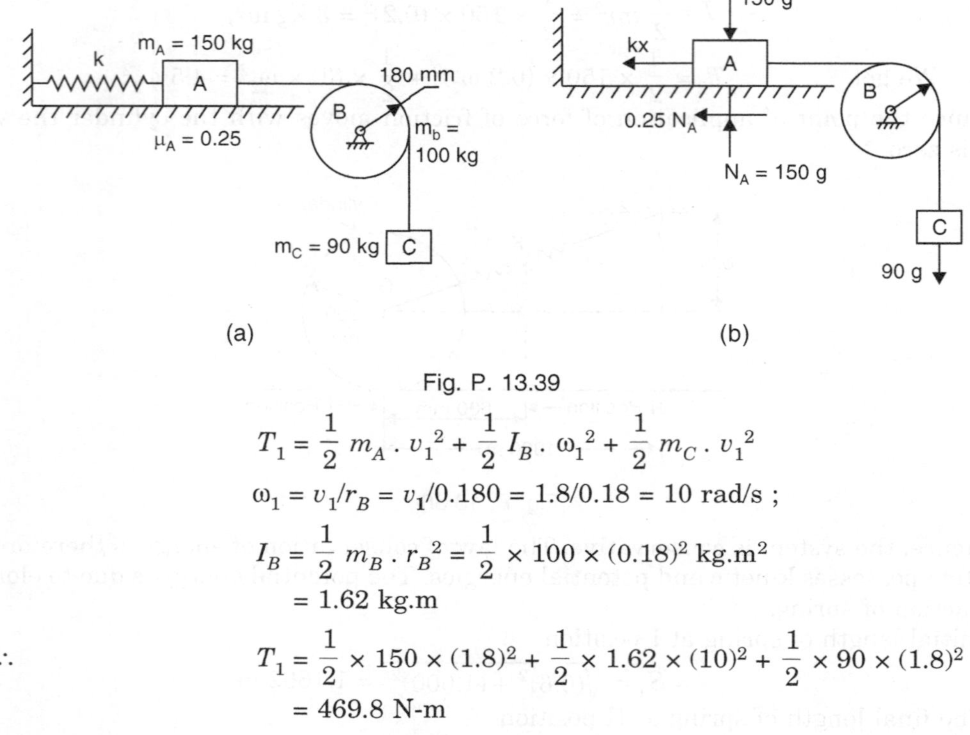

Fig. P. 13.39

$$T_1 = \frac{1}{2} m_A \cdot v_1^2 + \frac{1}{2} I_B \cdot \omega_1^2 + \frac{1}{2} m_C \cdot v_1^2$$

where $\omega_1 = v_1/r_B = v_1/0.180 = 1.8/0.18 = 10$ rad/s ;

$$I_B = \frac{1}{2} m_B \cdot r_B^2 = \frac{1}{2} \times 100 \times (0.18)^2 \text{ kg.m}^2$$
$$= 1.62 \text{ kg.m}$$

$$\therefore \quad T_1 = \frac{1}{2} \times 150 \times (1.8)^2 + \frac{1}{2} \times 1.62 \times (10)^2 + \frac{1}{2} \times 90 \times (1.8)^2$$
$$= 469.8 \text{ N-m}$$

Final kinetic energy, $T_2 = \frac{1}{2} m_A \cdot v_2^2 + \frac{1}{2} I_B \cdot \omega_2^2 + \frac{1}{2} m_C \times v_2^2$

i.e.,
$$T_2 = \frac{1}{2} \times 150 \times v_2^2 + \frac{1}{2} \times \frac{1.62 \times v_2^2}{(0.18)^2} + \frac{1}{2} \times 90 \times v_2^2$$
$$= 145 \times v_2^2.$$

Workdone. The mass m_C does positive work (= 90 g × 1.5) ;

B does not do any work because it moves in smooth bearings ;

the workdone by the force of friction $\mu_A N_A$ acting on block A is negative and is equal to $(\mu_A N_A) \times 1.5$.

Since the spring is initially compressed 200 mm or 0.2 m, the spring work $\left(= + \int_0^{0.2} ks\, ds\right)$ for the first 0.2 m is positive.

For the remaining stretching of the spring (= 1.5 – 0.2 = 1.3 m) the spring work is negative and equal to $-\int_0^{1.3} ks\, dx$. The total workdone

$$= U = (0.25 \times 150\, g) \times 1.5 + (90\, g) \times 1.5 + \int_0^{0.2} k \times s\, ds - \int_0^{1.3} ks\, dx$$

$$= (0.25 \times 150 \times 9.81 \times 1.5) + (90 \times 9.81 \times 1.5) + \left[950 \times \frac{S^2}{2}\right]_0^{0.2} - \left[\frac{950 \times S^2}{2}\right]_0^{1.3}$$

$$= 551.81 + 1324.35 + 19 - 802.75 \text{ Nm}$$

$$= 1092.41 \text{ Nm}$$

Applying the work-energy equation, we get

$$U_{1-2} = T_2 - T_1$$

or $$1092.41 = 145v_2^2 - 469.8$$

$$v_2^2 = (1092.41 + 469.8) \times \frac{1}{145} = 10.774$$

$\therefore$ $$v_2 = \mathbf{3.2824\ m/s.}\ \ \textbf{Ans.}$$

Problem 13.40. *A solid homogeneous cylinder of radius r and mass m is initially at rest position with G vertically above O. The cylinder freely turns through angle θ about the horizontal axis through O. Find the final angular velocity of the cylinder. See Fig. P. 13.40.*

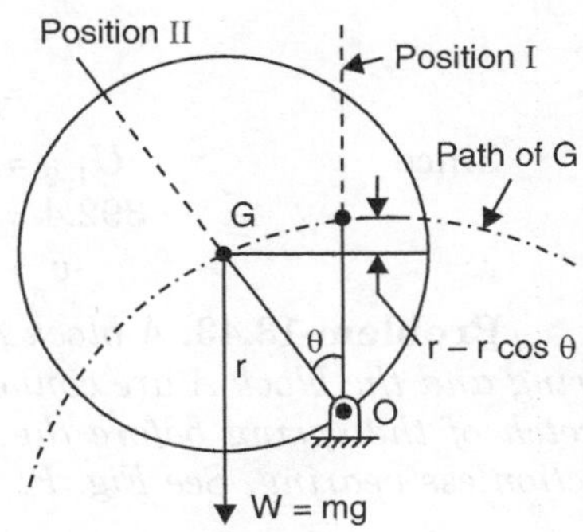

Fig. P. 13.40

Sol. *CG* of cylinder falls through $(r - r\cos\theta)$ as it turns from position I to position II.

$\therefore$ Work of gravity $= mgr\,(1 - \cos\theta)$

Kinetic energy at position I $= T_1 = 0$ because the cylinder is at rest.

Final $$\text{K.E.} = T_2 = \frac{1}{2} I_0\, \omega_2^2$$

where $$I_0 = \bar{I} + mr^2 = \frac{1}{2}mr^2 + mr^2 = \frac{3}{2}mr^2$$

$\therefore$ $$T_2 = \frac{1}{2} \times \left(\frac{3}{2}mr^2\right) \times \omega_2^2 = \frac{3mr^2}{4}\,\omega_2^2$$

Applying $U_{1-2} = T_2 - T_1$, we get

$$mgr(1 - \cos\theta) = \frac{3mr^2}{4}\,\omega_2^2$$

$\therefore$ $$\omega_2^2 = \frac{4g}{3r}(1 - \cos\theta)$$

$\therefore$ $$\omega_2 = \sqrt{\frac{4g}{3r}(1 - \cos\theta)}.\ \ \textbf{Ans.}$$

Problem 13.41. *A mass m falls through 2.5 m on the top of the spring of stiffness 4 kN/m and compresses it through 300 mm. Find the magnitude of mass.*

Sol. Work due to gravity $= mg \times (h + \delta_{max})$

$$= m \times 9.81 \times (2 + 0.3) = 22.563\ m \text{ Nm.}$$

Work of spring $$= -\frac{1}{2}k\delta^2 = -\frac{1}{2} \times 4000 \times (0.3)^2 = -180 \text{ Nm.}$$

Change of K.E. $= T_2 - T_1 = 0$, because $T_1 = 0$ and $T_2 = 0$, because $v_1 = 0$ and also $v_2 = 0$.

Applying $U = T_2 - T_1$, we get

$$22.563\ m - 180 = 0$$

$\therefore$ $$m = \frac{180}{22.563} = \mathbf{7.98\ kg.}\ \ \textbf{Ans.}$$

Problem 13.42. *A rope of negligible mass is wrapped around a 10 kg solid homogeneous cylinder of diameter 600 mm, as shown in Fig.P. 13.42. Find the speed of the centre C of the cylinder after it has dropped 4 m from rest.*

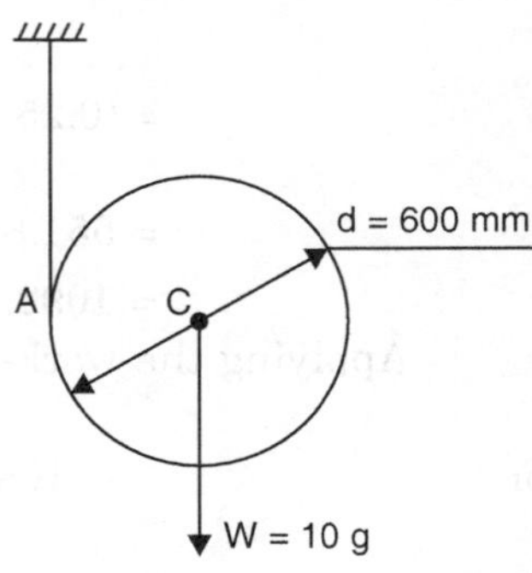

Fig. P. 13.42

Sol. The only force which does work is the weight of the cylinder equal to 10 g. And, work done, $U = W \times s = (10 \times 9.81) \times 4 = 392.4$ N.m

Initial K.E., $T_1 = 0,$

Final K.E., $T_2 = \frac{1}{2} mv^2 + \frac{1}{2} I\omega^2$

$$= \frac{1}{2} mv^2 + \frac{1}{2}\left(\frac{1}{2}mr^2\right) \times \left(\frac{v}{r}\right)^2$$

$$= \frac{3}{4} mv^2 = \frac{3}{4} \times (10) \times v^2 = 7.5v^2$$

Since $U_{1-2} = T_2 - T_1$

$\therefore$ $392.4 = 7.5\, v^2 - 0$

$\therefore$ $v =$ **7.233 m/s. Ans.**

Problem 13.43. *A block A of mass 100 kg rests on a spring of stiffness 4000 N/m. The spring and the block A are connected by an inextensible cord of loose length of 1.5 m. Find the stretch of the spring before the system comes to rest. The cylinder of mass 80 kg rotates in frictionless bearing. See Fig. P. 13.43.*

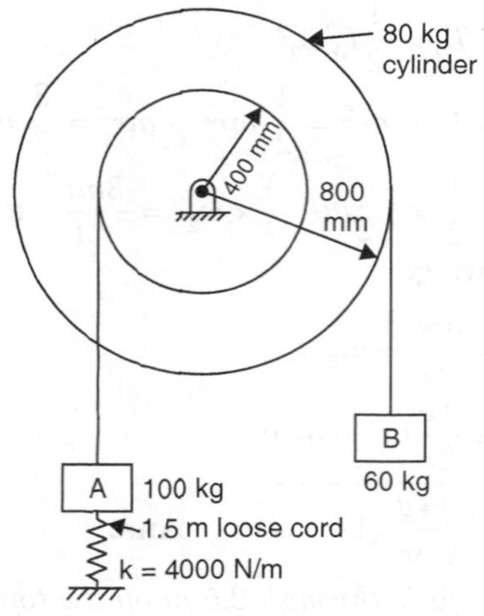

Fig. P. 13.43

Sol. When the block *A* rises through the height equal to the length of the loose rope (S_A), the block *B* drops through $S_B = \frac{S_A}{0.4} \times 0.8 = 2\, S_A$

$\therefore$ The mass *A* rises through $S_A = 1.5$ m and the block *B* drops through

$$S_B = 2 \times 1.5 = 3 \text{ m}.$$

Hence the initial kinetic energy T_1 is given as

$$T_1 = (m_B \times g)\, S_B - (m_A \times g)\, S_A$$

$$= (60 \times 9.81)\,(3) - (100 \times 9.81) \times (1.5) = 294.3 \text{ Nm}$$

Final K.E. = $T_2 = 0$

Workdone U_{1-2} = workdone on the system by gravity of A and B + workdone by the stretching of spring

$$= +(60 \times 9.81) \times \left(\frac{1.5}{0.4}x\right) - (100 \times 9.81)(x) - \frac{1}{2}kx^2$$

$$= 1226.25\,x - \frac{1}{2} \times 4000\,x^2 = 1226.25\,x - 2000\,x^2$$

$\because$ $U_{1-2} = T_2 - T_1,$

$\therefore$ $1226.25x - 2000x^2 = 0 - 294.3$

or, $2000x^2 - 1226.25\,x - 294.3 = 0$

or $x^2 - 0.613x - 0.147 = 0$

$$x = \frac{1}{2}\left[0.613 \pm \sqrt{(0.613)^2 - 4 \times (-0.147)}\right] = \frac{1}{2}\,[0.613 \pm 0.982]$$

$$= 0.7975,\ -0.1845$$

Selecting the positive value of x, the stretch of the spring = **0.7975 m. Ans.**

Problem 13.44. *Find the mass m_B of the block to cause the cylinder of mass-moment of inertia $I = 200$ kg.m² to attain an angular velocity of 3 radian/s after rotating counter clockwise 15 radians from rest. See Fig. P. 13.44.*

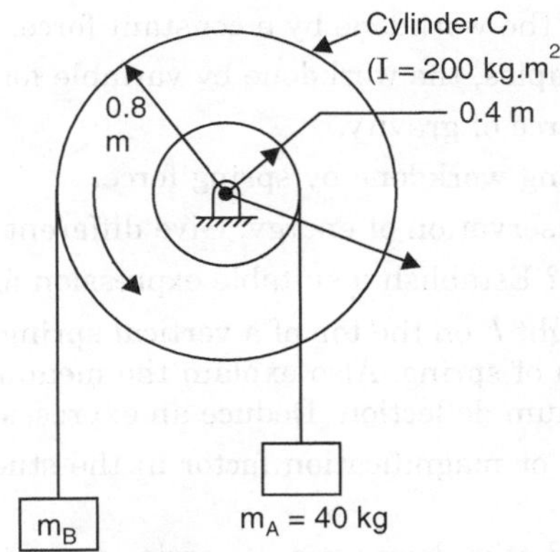

Fig. P. 13.44

Sol. Torque on the bearing centre

$$= (m_B \times g)\,0.8 - (m_A \times g) \times 0.4$$

$$= m_B \times 9.81 \times 0.8 - 40 \times 9.81 \times 0.4$$

$$= 7.848\,m_B - 156.96 \text{ N.m (Counter clockwise)}$$

$\therefore$ Workdone = U_{1-2} = torque × angle

$$= (7.848\,m_B - 156.96) \times 15$$

or $U_{1-2} = 117.72\,m_B - 2354.4$

Initial kinetic energy of the system, $T_1 = 0$.

Final K.E. = $T_2 = \frac{1}{2}\,m_A\,v_A^2 + \frac{1}{2}\,m_B \cdot v_B^2 + \left(\frac{1}{2}I\omega^2\right)_{\text{cylinder}}$

where $v_A = \omega \times r_A = 3 \times \left(\frac{0.4}{2}\right) = 0.6$ m/s

$$v_B = \omega \times r_B = 3 \times 0.8 = 2.4 \text{ m/s}$$
$$\omega = 3 \text{ rad/s}$$

$\therefore$ $$T_2 = \frac{1}{2} \times (40) \times (0.6)^2 + \frac{1}{2} \times m_B \times (2.4)^2 + \frac{1}{2} \times (200) \times 3$$
$$= 107.2 + 2.88\, m_B$$

$\because$ $$U_{1-2} = T_2 - T_1,$$

$\therefore$ we get

$$117.72\, m_B - 2354.4 = 107.2 + 2.88\, m_B$$

$\therefore$ $$m_B\,(117.72 - 2.88) = 107.2 + 2354.4$$

or $$m_B = \frac{2461.6}{120.6} = \mathbf{20.41\ kg.} \quad \textbf{Ans.}$$

EXERCISES XIII

A. Theoretical Questions

1. What do you understand by work-energy method ? How is it different from other methods ?
2. Establish the work-energy equation for translation.
3. Explain with neat sketches the workdone by a constant force.
4. Explain, with suitable examples, the workdone by variable force.
5. Explain the workdone by force of gravity.
6. Derive expressions for finding workdone by spring force.
7. What is the principle of conservation of energy. Give different expressions.
8. What is Rayleigh's method ? Establish a suitable expression for this method.
9. A weight *W* falls from a height *h* on the top of a vertical spring of stiffness *k*. Derive expressions for instantaneous deflection of spring. Also explain the meaning of an equivalent dynamic load P_{dyn} which gives the maximum deflection. Deduce an expression for finding P_{dyn}.
10. What is impact or dynamic or magnification factor in the study of impact loading ? Derive expression for this.
11. A light vertical bar is hinged at its lower end. It carries a mass *m* at its top end. When the rod is released from its vertical position, find the position of the bar at which there is no stress induced in the bar.
12. Define power and efficiency. What is overall efficiency ?
13. Establish work-energy equation for rotation.
14. Establish work-energy equation for plane motion.

B. Numerical Problems

13.1. A 15 kg block resting on 30° rough plane slides up the plane through 6 m under the action of a 400 N force acting on the block parallel to the plane. Take $\mu_k = 0.18$. Also find its velocity after 6 m. [**Ans.**1820.92 Nm, $v = 15.582$ m/s]

13.2. A 15 kg block resting on a 30° rough plane slides up the plane through 6 m under the action of a horizontal 400 N force. Take $\mu_k = 0.18$. Also find its velocity after 6 m.

[**Ans.** 1283.38 Nm, $v = 13.081$ m/s]

13.3. Two blocks A and B are resting at rest as shown in Fig. E. 13.3. Find the velocity of the system after the block B has travelled 6 m down the plane. Also find the workdone.

[**Ans.** $v = 4.281$ m/s ; $U_{1-2} = 934.17$ Nm]

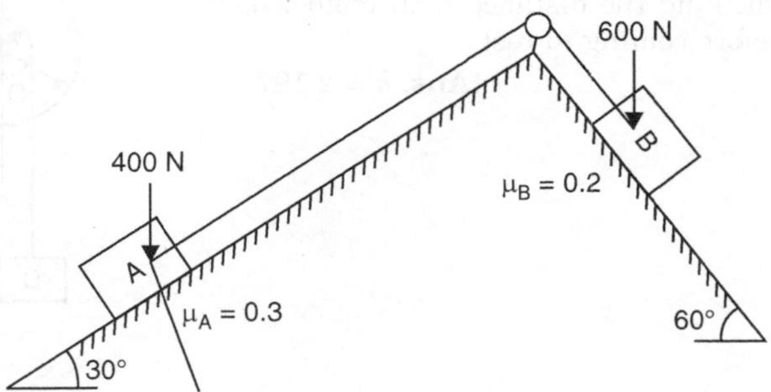

Fig. E. 13.3

13.4. A railway wagon of mass 12000 kg moving at 27 km/hr strikes 6 buffer springs and stops. If stiffness of each spring is 600 kN/m, find the maximum contraction of the springs before the wagon comes to rest. Also find force on each spring [**Ans.** 0.433 m ; 259.81 kN]

13.5. A block of weight 3 kN rests on a smooth horizontal plane, as shown in Fig. E. 13.5. The stiffness of spring is 1500 N/m and the free length is 0.8 m. If the block has zero velocity when its displacement s = 1 m, find its velocity when s = 2 m. [**Ans.** 0.98 m/s]

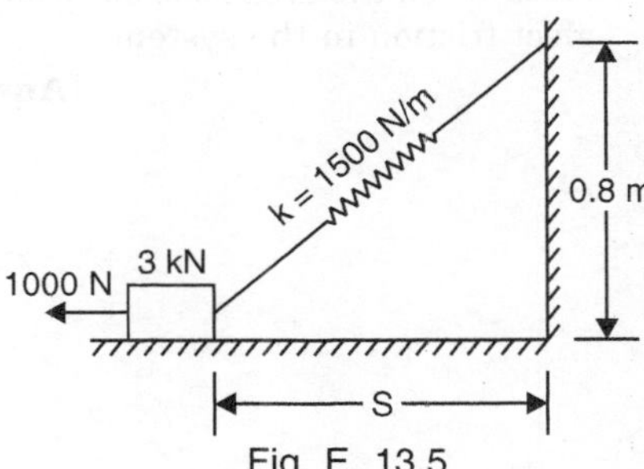

Fig. E. 13.5

13.6. Block weighs 1000 N while block B weighs 400 N. Find velocity of each block after B has moved 5 m up the inclined plane shown in Fig. E. 13.6. Also find the tension in the cord attached to block B. Assume μ for the plane as 0.18 and pulleys C *and* D to be smooth.

[**Ans.** $v_A = 2v_B = 12.198$ m/s ; $T = 485.23$ N]

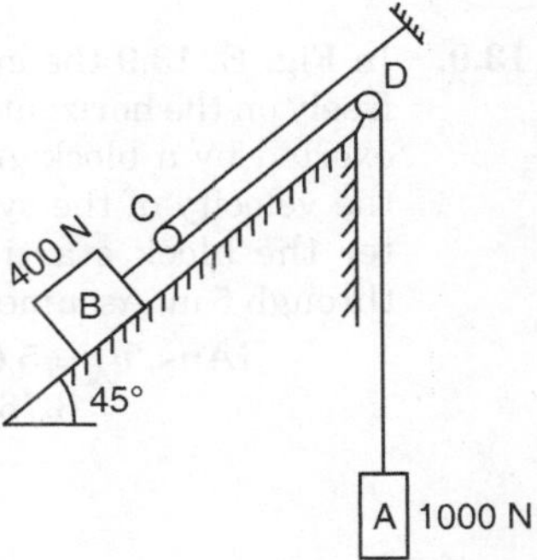

Fig. E. 13.6

13.7. The block D weighs 400 N and move down with a velocity of 10 m/s when the braking effort P acting at B is 1.5 kN. The coefficient of friction between the arm AB and the wheel C is 0.35. Weight of wheel is 1000 N and its radius of gyration is 0.7 m. Find the distance h through which the block D falls before coming to rest.

[**Ans.** h = 2.197 m]

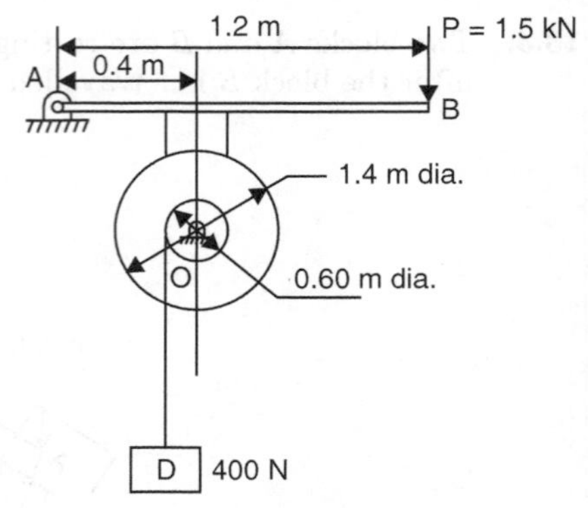

Fig. E. 13.7

13.8. The cylinder in Fig. E. 13.8 weighs 2 kN and is 1.6 m in diameter. The rod is 1.8 m long and weighs 750 N and is rigidly attached to the cylinder. The block A is 1 kN. When the rod is vertical (*i.e.*, θ = 0°) the system is at rest. Find the velocity of the block when the angle θ is 60° from the vertical. Neglect friction in the system.

[**Ans.** v = 2.6572 m/s]

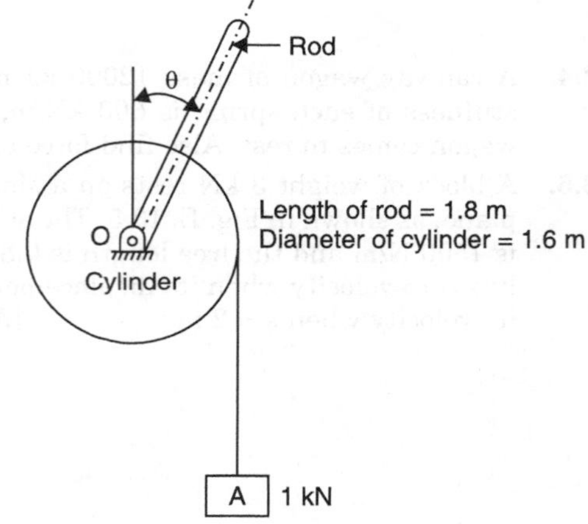

Fig. E. 13.8

13.9. In Fig. E. 13.9 the cylinder C is of 2 m rolls freely on the horizontal surface under the pull exerted by a block A of weight 800 N. Find the velocity of the cylinder and the block after the block starting rest has descended through 5 m. Assume the pulley to be smooth.

[**Ans.** v_A = 5.641 m/s ; For cylinder ω = 3.761 rad/s ; v_{cyl} = 1.88 m/s]

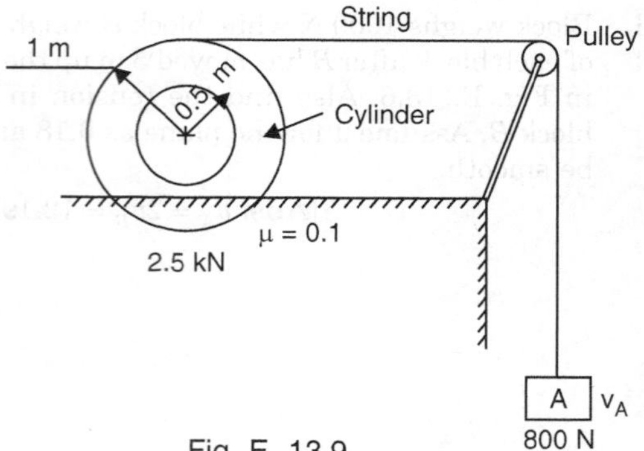

Fig. E. 13.9

13.10. A solid cylinder of radius r and mass m starts from rest and rolls down a rough inclined plane without slipping. Inclination of the plane is θ to the horizontal. Find the time taken to cover a distance s along the plane and also acceleration of the centre of the cylinder.

$$\left[\textbf{Ans. } t = \sqrt{\frac{3s}{g \sin \theta}} \; ; a = \frac{2}{3} g \sin \theta \right]$$

13.11. A railway wagon mounted on four wheels, weighing 200 kN rolls down a 1% grade track. Find the speed of the wagon after it has move 800 m. The track resistance is 8 N/kN weight of the wagon. Axles and the wheels weigh 30 kN. The wheels have a diameter of 1 m and a radius of gyration of 0.4 m. [**Ans.** v = 5.3519 m/s = 19.27 km/hr]

13.12. A body weighing 10 N slides along the curved ends and 5 m flat central portion. The curved portions are smooth and μ for the central flat portion is 0.2. The body is released from a left point *A* 2 m above the flat portion *BC*. Find the position of the body as it stops.

[**Ans.** The body stops at B]

13.13. A cylinder, a sphere and a hoop, each having the same mass and the same radius, and a flat block of the same mass, are released from rest on an inclined plane. Determine the velocity of each body after it has descended a height *h*.

[**Ans.** $v_{cyl} = 0.8165\sqrt{2gh}$; $v_{sph} = 0.8452\sqrt{2gh}$; $v_{hoop} = 0.7071\sqrt{2gh}$; $v_{flat} = \sqrt{2gh}$]

13.14. *A* gear a has a mass of 3 kg, a radius of 10 cm and a radius of gyration of 8 cm, meshes with gear *B* which has mass of 10 kg, a radius of 25 cm, and a radius of gyration of 20 cm . The system is at rest when a couple of magnitude 12 Nm is applied to the gear *A*. Determine (*a*) the number of revolutions made by gear *A* before its angular velocity becomes 600 rpm, (*b*) the tangential force exerted by gear *A* on gear *B*. [**Ans.** (*a*) 2.178 revolutions ; (*b*) *F* = 92.31 N]

14

Impulse-Momentum Method

14.1. INTRODUCTION

As discussed in chapter 13 *the work-energy method was based on integrating the kinetic equations of motion with respect to the displacement. The resulting displacement integrals were scalar quantities which were particularly useful in dealing with problems relating force, velocity and displacement.*

In this chapter, a third basic method for solving problems dealing with the motion of particles will be discussed. This method, known as *impulse-momentum method, directly relates force, velocity and time. It is based on integrating the kinetic equations of motion with respect to the time.*

The impulse momentum method is particularly useful in situations when forces act for a very small time intervals during which forces may vary, as in sudden blow, collision or impact. It is also useful in which the system gains or loses mass. Problems like the motion of satellites can, however, be solved by the combination of work-energy and impulse-momentum methods.

14.2. LINEAR IMPULSE-MOMENTUM METHOD

Let $\overline{F}$ be the force acting on the mass centre of any body, rigid or non-rigid or a particle. We have studied that the force is the product of mass of the body and the acceleration. Or, in other words, force is given as

$$\overline{F} = \frac{d}{dt}(mv)$$

$$\therefore \quad \overline{F}.dt = d(mv) \qquad \text{...(14.1)}$$

The product (mv) is called momentum (G).

From eqn. (14.1) yields

$$\overline{F}.dt = d(G)$$

Integrating $\displaystyle\int_{t_1}^{t_2} \overline{F}\,dt = \int_{V_1}^{V_2} d(mv)$

or $$\int_{t_1}^{t_2} \overline{F}\,dt = mv_2 - mv_1 \qquad \text{...(14.2)}$$

or $$mv_2 = mv_1 + \int_{t_1}^{t_2} \overline{F}\,dt \qquad \text{...(14.3)}$$

The integral $\int_{t_1}^{t_2} \overline{F}\, dt$ is a vector known as **linear-impulse** or simply **impulse** of a force $\overline{F}$ during the time interval $(t_2 - t_1)$ considered.

This is also represented by $\overline{I}_{imp}$ or $(\overline{I}_{imp})_{1-2}$.

The units of impulse are Force-second or Newton-seconds, (N-s) in SI units.

Since force $\overline{F}$ is a vector, we can write :

$$= \int_{t_1}^{t_2} \overline{F}\, dt = \overline{i} \int_{t_1}^{t_2} F_x .dt + \overline{j} \int_{t_1}^{t_2} F_y\, dt + \overline{k} \int_{t_1}^{t_2} F_z\, dt \qquad ...(14.4)$$

eqn. (14.3) can also be written in the form

$$mv_2 = mv_1 + \overline{I}_{imp})_{1-2} \qquad ...(14.5)$$

Eqns. (14.3) and (14.5) can be graphically represented as in Fig. 14.2.

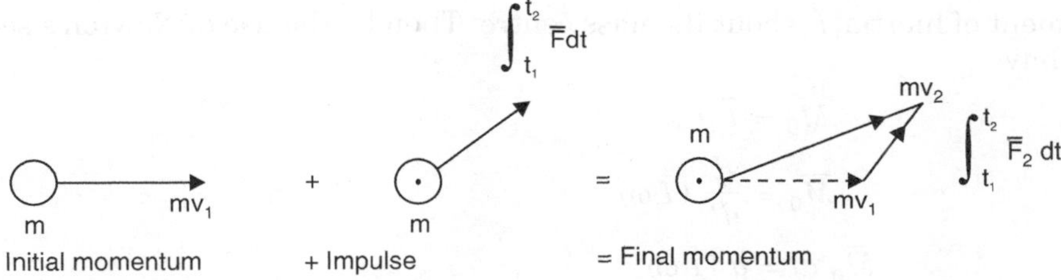

Fig. 14.2. Diagram of impulse-momentum equation

When several forces act on a particle, the impulse of each force must be considered. Then eqn. (14.5) becomes

$$mv_2 = mv_1 + \Sigma\, \overline{I}_{imp})_{1-2} = mv_1 + \Sigma \int_{t_1}^{t_2} \overline{F}\, dt \qquad ...(14.6)$$

When there are two or more particles in a system, each particle may be considered separately and eqn. (14.6) may be written as

$$\Sigma\, mv_2 = \Sigma\, mv_1 + \Sigma \int_1^2 \overline{F}\, dt \qquad ...(14.7\ (a))$$

or

$$\Sigma\, mv_2 = \Sigma\, mv_1 + \Sigma \overline{I}_{imp})_{1-2} \qquad ...(14.7\ (b))$$

If there is no external force acting on the system or if the sum of the external forces is zero, the second term in eqns. (14.5), (14.6) and (14.7) vanishes

And, we get $\quad \Sigma\, mv_2 = \Sigma\, mv_1 \qquad ...(14.8)$

Eqn. (14.8) is known as the **Principle of conservation of momentum. It states that initial momentum is equal to final momentum if there is no resistance offered to the motion. Or in other words, in a conservative system the initial momentum is equal to the final momentum.**

14.3. ANGULAR MOMENTUM

A comparison of linear and angular systems, (Fig. 14.3), will help to apply the relations of Art. 14.2 for linear motion to the cases of angular motion.

Table 14.1. Comparison of Linear and angular System

Quantity	*Linear motion*	*Angular motion*
Displacement	s, m	θ, radians
Velocity	v, m/s	ω, rad/s
Mass	m, kg	Mass moment of inertia, $\bar{I}$ kgm^2
Force	Force, F, N	Torque, $\overline{M_0}$ Nm
Impulse-momentum eqns.	$mv_2 = mv_1 + \int_{t_2}^{t_2} \overline{F} \, . \, dt$,	$\bar{I}\omega_2 = \bar{I}\omega_1 + \int_{t_1}^{t_2} \overline{M_0} \, dt$
Conservation of momentum equation	$mv_2 = mv_1$	$\bar{I}\omega_2 = \bar{I}\omega_1$

The study of the above table of comparison we can easily write the following relations for impulse-momentum for the angular motion. When an external moment $\overline{M}_0$ acts on a body of mass-moment of inertia $\bar{I}$ about its mass centre. Then by the use of Newton's second law of motion, we have

$$\overline{M}_0 = \bar{I}\,\alpha \qquad \text{...(14.9)}$$

or
$$\overline{M}_0 = \frac{d}{dt}(\bar{I}\,\omega)$$

or
$$\overline{M}_0\, dt = d\,(\bar{I}\,\omega)$$

Integrating, we get :
for rigid body

$$\int_{t_1}^{t_2} \overline{M}_0\, dt = \int_{\omega_1}^{\omega_2} d\,(\bar{I}\omega)$$

or
$$\int_{t_1}^{t_2} \overline{M}_0\, dt = (\bar{I}\,\omega_2 - \bar{I}\,\omega_1) \qquad \text{...(14.10)}$$

or
$$\bar{I}\,\omega_2 = \bar{I}\,\omega_1 + \int_{t_1}^{t_2} \overline{M}_0\, dt \qquad \text{...(14.11)}$$

Eqns. (14.10) and (14.11) are known as **moment of momentum equation.**

For a number of rigid bodies in a system eqns. (14.10) and (14.11) become

$$\Sigma \int_{t_1}^{t_2} \overline{M}_0\, dt = \Sigma\, \bar{I}\,\omega_2 - \Sigma\, \bar{I}\,\omega_1 \qquad \text{...(14.12)}$$

and
$$\Sigma \bar{I}\,\omega_2 = \Sigma\, \bar{I}\,\omega_1 + \Sigma \int_{t_1}^{t_2} \overline{M}_0\, dt \qquad \text{...(14.13)}$$

When, the body is not subjected to any external couple or moments $\overline{M}_0$, or when the sum of the external moments is zero or when there is no resistance to the motion or for conservative systems, eqn. (14.11) becomes

$$\bar{I}\omega_2 = \bar{I}\omega_1 \qquad \text{...(14.14)}$$

And eqn. (14.14) for a system of bodies becomes

$$\Sigma\bar{I}\omega_2 = \Sigma\bar{I}\omega_1 \qquad \text{...(14.15)}$$

14.4. ALTERNATIVE SYMBOLS USED

It is convenient to denote linear momentum mv by $\overline{G}$. Therefore linear momentum equation (14.6) can be written as

$$\overline{G}_2 = \overline{G}_1 + \int_{t_1}^{t_2} \overline{F}\, dt \qquad ...(14.16)$$

in general,

or, for a conservative system eqn. (14.16) becomes

$$\overline{G}_2 = \overline{G}_1 \qquad ...(14.17)$$

For a number of particles eqn. (14.16) becomes

$$\Sigma \overline{G}_2 = \Sigma \overline{G}_1 + \Sigma \int_{t_1}^{t_2} \overline{F}\, dt \qquad ...(14.18)$$

and eqn. (14.17) becomes

$$\Sigma \overline{G}_2 = \Sigma \overline{G}_1 \qquad ...(14.19)$$

For angular motion the angular momentum $(\bar{I}\,\omega)$ is represented by $\overline{H}_G$ with respect to mass centre G. Whereas, when the torque is applies about a point O, the angular momentum $= I_0\omega$ and it is generally represented by $\overline{H}_0$. Thus, eqns. (14.11) can be written as below :

$$\overline{H}_{G_2} = \overline{H}_{G_1} + \int_{t_1}^{t_2} M_G\, dt \qquad ...(14.20)$$

For conservative system eqn. (14.20) becomes

$$\overline{H}_{G_2} = \overline{H}_{G_1} \qquad ...(14.21)$$

when the torque is applied at a point O, eqn. (14.20) becomes

$$\overline{H}_0)_2 = \overline{H}_0)_1 + \int_{t_1}^{t_2} M.\, dt \qquad ...(14.22)$$

For a conservative system eqn. (14.22) becomes

$$\overline{H}_0)_2 = \overline{H}_0)_1 \qquad ...(14.23)$$

Problem 14.1. *A car weighing 2000 N is driven on a horizontal road at a speed of 108 kmph when brakes are applied, causing a constant braking force of 750 N between the tyres and the road. Determine the time required for the car to stop.*

Sol. We apply the principle of impulse momentum.

$$v_1 = 108 \text{ kmph} = \frac{108}{3.6} \text{ m/s} = 30 \text{ m/s}$$

$$\text{Initial momentum} = \overline{G}_1 = mv_1 = \left(\frac{2000}{9.81} \times 30\right) = 6116.21 \text{ kgm/s}$$

Final momentum $= \overline{G}_2 = mv_2 = 0$ along the line of motion.

$$\text{Impulse} = \Sigma F.dt = I_{mp1\text{-}2} = -F_{\text{resistance}} \times t$$
$$= -750\, t$$

[See Fig. P.14.1]

$$mv_2 = mv_1 + I_{mp1-2}, \text{ we get}$$
$$0 = 6116.21 + [-750\, t]$$

Fig. P. 14.1

$$t = \frac{6116.21}{750} = \mathbf{8.155\ s.}\quad \textbf{Ans.}$$

Problem 14.2. *A car weighing 2000 N is moving down a 10° road at a speed of 108 kmph when brakes are applied causing a constant braking force of 750 N. Determine the time required for the car to stop.*

Sol. Applying the impulse-momentum equation along the road (See Fig. P.14.2), we get

Fig. P. 14.2

$$mv_1 + I_{imp_{1-2}} = mv_2 \qquad | \; v_1 = 108 \text{ kmph} = 30 \text{ m/s}$$

or
$$\frac{2000}{9.81} \times 30 + [W \sin 10° - F] \times t = 0$$

or
$$\frac{2000}{9.81} \times 30 + (2000 \sin 10° - 750) \times t = 0$$

or
$$6116.21 + (-402.7) \times t = 0$$

$\therefore$
$$t = \frac{6116.21}{402.7} = \mathbf{15.188\ s.}\quad \textbf{Ans.}$$

Problem 14.3. *A steel ball of weight 0.1 N falls from a height 6 m and rebounds to a height of 4 m. Find the impulse and the average force between the ball and the floor if the contact between the ball and the floor lasts for $\frac{1}{12}$ th of a second.*

Sol. The ball strikes the floor at the velocity $v_1 = \sqrt{2gh_1} = \sqrt{2 \times 9.81 \times 6} = 10.85$ m/s. (downward). The velocity of the ball after impact $v_2 = \sqrt{2 \times g \times h_2} = \sqrt{2 \times 9.81 \times 4} = 8.59$ m/s. (upward)

$$\therefore \text{ Impulse } = m(v_2) - mv_1 = \frac{0.1}{9.81} \times \{10.85 - (-8.59)\}$$

$$= \frac{0.1 \times 19.44}{9.81} = \mathbf{0.1982\ N\text{-}s.}\quad \textbf{Ans.}$$

If F is the average force between the ball and the floor, then impulse = $F \times \Delta t$

$$\therefore \quad F \times \Delta t = 0.1982$$

$$\therefore \quad F \times \frac{1}{12} = 0.1982$$

$\therefore$ Average force, $F = 12 \times 0.1982 =$ **2.3784 N. Ans.**

Problem 14.4. *A coolie tosses a 20 kg suitcase with a horizontal velocity of 15 m/s on to an 80-kg luggage carrier. The carrier which is initially at rest, can slide freely. Find the velocity of the carrier after the suitcase has slided to a stop provided in the carrier.*

Sol. See Fig. P. 14.4.

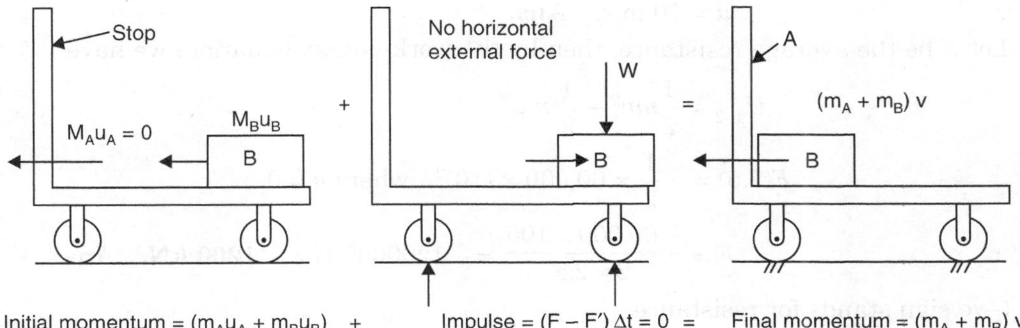

Fig. P. 14.4

Let m_A = mass of luggage carrier = 80 kg

m_B = mass of the suitcase = 20 kg

initial velocity of carrier = $u_A = 0$

initial velocity of suitcase = u_B = 15 m/s

Let the final velocity of suitcase and carrier = v = ?

There is no horizontal external force acting on the system,

therefore impulse $= \int_{t_1}^{t_2} F\, dt = 0.$

The system is conservative in horizontal direction.

$\therefore$ By $\Sigma\, mu = \Sigma\, mv$, we get

$$m_Au_A + m_Bu_B = (m_A + m_B)\, v$$

$$\therefore \quad 0 + \frac{20}{g} \times 15 = \frac{(80+20)\,v}{g}$$

$$\therefore \quad v = \frac{20 \times 15}{100} = \mathbf{3\ m/s.} \quad \textbf{Ans.}$$

Problem 14.5. *A bullet of 40 gm mass leaves the barrel of a gun with a velocity of 600 m/s. horizontal. Find the impulse produced by the bullet due to its discharge. Also find the average impulse force if the impulse lasts for* $\frac{1}{250}$ *seconds.*

Sol. Impulse = change of momentum = $mv_2 - mv_1$

$$= \frac{40}{1000} \times (600 - 0) \text{ kgm/s} = \textbf{24 kgm/s.} \quad \textbf{Ans.}$$

Also, impulse = F × t

$$\therefore \quad F = \frac{24}{t} = \frac{24}{\frac{1}{250}} = \mathbf{6000\ N.} \quad \textbf{Ans.}$$

Problem 14.6. *A shell of mass 1000 kg is fired horizontally with a velocity of 600 m/s from a gun of mass 60,000 kg. Find the velocity of recoil of the gun. If the gun stops in 2.5 m find the average resistance offered to the gun during recoil.*

Sol. Let the initial velocity of recoil of gun be u. By the principle of conservation of momentum $mu)_{gun} = mv)_{shell}$

$$\therefore \quad 60{,}000 \times u = 1000 \times 600$$

$$\therefore \quad u = 10 \text{ m/s.} \quad \textbf{Ans.}$$

Let F be the average resistance, then by the work-energy equation, we have

$$U_{1-2} = \frac{1}{2}mv^2 - \frac{1}{2}mu^2$$

or

$$F(2.5) = -\frac{1}{2} \times 60{,}000 \times (10)^2, \text{ where } v = 0$$

$$\therefore \quad F = \frac{-60{,}000 \times 100}{2 \times 2.5} = -1200000 \text{ N} = -\mathbf{1200\ kN.} \quad \textbf{Ans.}$$

(–ve sign stands for resistance)

Alternatively :

Using the relation $v^2 = u^2 + 2as$ for gun

we have, $0 = (10)^2 + 2 \times a \times (2.5)$

$\therefore$ Retardation of gun, $a = -\dfrac{100}{5} = -20 \text{ m/s}^2$

Retarding force, F = mass × retardation

$= 60{,}000 \times (-20)$

$= -1200{,}000$ N $= -$ **1200 kN. Ans.**

(–ve sign stands for resistance).

Problem 14.7. *A bullet of 30 g mass moving at a velocity u strikes a simple pendulum of mass 5 kg. The pendulum, and the bullet imbedded into it, rises through 300 mm from its lowest position. Find the velocity of the bullet.*

Sol. Let m_b = *mass of bullet*, m_p = mass of pendulum

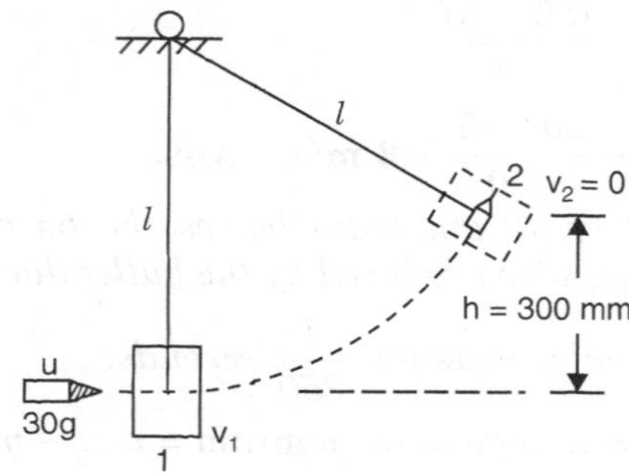

Fig. P. 14.7

The maximum petential energy at position 2 = $V = (m_b + m_p)\, g \times h$.

$$\therefore \quad V_2 = \left(\frac{30}{1000} + 5\right) \times 9.81 \times \frac{300}{1000} \text{ J.}$$

$$= 14.8 \text{ J}$$

Let v_1 = velocity of the bullet and the pendulum (moving together) at position 1

$$\therefore \quad \text{Maximum K.E. at position} = T_1 = \frac{1}{2}\left(\frac{30}{1000} + 5\right) v_1^2 = \frac{5.03}{2} v_1^2$$

By the law of conservation of energy
Maximum K.E. = Maximum P.E.

$$\therefore \quad \frac{5.03}{2} v_1^2 = 14.8$$

$$\therefore \quad v_1 = \mathbf{2.426 \text{ m/s}}$$

By applying the law of conservation of momentum, we get

$$m_b . u = (m_b + m_p)\, v_1$$

$$\therefore \quad \frac{30}{1000} \times u = \left(\frac{30}{1000} + 5\right) \times 2.426$$

$$\therefore \quad u = \frac{5.03 \times 2.426}{0.03} = \mathbf{406.76 \text{ m/s.}} \quad \textbf{Ans.}$$

Problem 14.8. *An inelastic pile of mass M is driven into the ground through a distance y by the blow of pile hammer of mass m which falls on the pile head through a height h. Prove that the average resistance offered by the ground is given by :*

$$R = \left[(m + M) + \frac{m^2}{(m + M)} \times \frac{h}{y}\right] \times g$$

and the kinetic energy lost during impact is

$$\Delta K.E. = \left(\frac{mM}{m + M}\right) \times gh$$

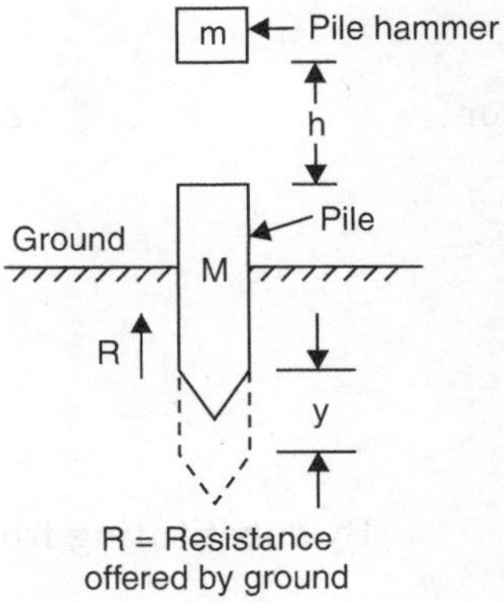

Sol. See Fig. P. 14.8. The velocity attained by pile hammer when it falls through h

$$= v = \sqrt{2gh} \qquad \text{...(1)}$$

When the pile hammer strikes the pile, both hammer and the pile move with a velocity, V. By conservation of momentum we get

$$mv = (m + M)\, V$$

$$V = \frac{mv}{(m + M)} \qquad \text{...(2)}$$

Fig. P. 14.8

Average Resistance :

Let R be the resistance of ground. When the pile moves through a distance y, the workdone $U_{1-2} = R \times y$ ...(3)

The K.E. attained by hammer and pile $= \frac{1}{2}(m + M) V^2$

The Potential Energy $= (m + M) gy.$

Total energy of pile and hammer $= E = \frac{1}{2}(m + M) V^2 + (m + M) gy$ ...(4)

By work-energy equation, we get

$$Ry = \frac{1}{2}(m + M) V^2 + (m + M) gy \quad \text{...(5)}$$

From eqn. (2), we get

$$R \times y = \frac{1}{2}(m + M) \times \left(\frac{mv}{m + M}\right)^2 + (m + M) \times gy$$

or

$$Ry = \frac{\frac{1}{2} m v^2}{(m + M)} + (m + M)gy$$

By substituting eqn. (1), we get

$$Ry = \frac{\frac{1}{2} \times m^2}{m + M} \times 2gh + (m + M)gy$$

$$= \left(\frac{m^2}{m + M}\right) \times gh + (m + M)gy$$

$$\therefore \quad R = \left(\frac{m^2}{m + M}\right) \times \frac{gh}{y} + (m + M)g$$

or

$$R = \left[(m + M) + \left(\frac{m^2}{m + M}\right) \times \frac{h}{y}\right] \times g \quad \textbf{Hence Proved} \quad \text{...(6)}$$

Loss of Kinetic Energy :

$$\text{Loss of K.E. during impact} = \frac{mv^2}{2} - \frac{1}{2}(m + M) V^2$$

or

$$\Delta\text{K.E.} = \frac{mv^2}{2} - \frac{1}{2}(m + M) \times \left(\frac{mv}{m + M}\right)^2$$

$$= \frac{mv^2}{2} - \left(\frac{m^2}{m + M}\right) \times \frac{v^2}{2} = \left[\frac{m(m + M) - m^2}{m + M}\right] \times \frac{v^2}{2}$$

$$= \left(\frac{m M}{m + M}\right) \times \frac{v^2}{2}$$

By substituting from eqn. (1), we get

$$\Delta\text{K.E.} = \left(\frac{m M}{m + M}\right) \times \frac{1}{2} \times 2gh$$

$$\therefore \quad \Delta\text{K.E.} = \left(\frac{mM}{m + M}\right) \times gh \quad \textbf{Hence Proved} \quad \text{...(7)}$$

Problem 14.9. *In an impact testing machine a hammer of mass 10.5 kg is attached to a 1.5 m long lever. The lever is held in a horizontal position and is released. The hammer swings in a vertical plane and strikes a specimen with V-notch held in the lowest position. The hammer strikes the test-piece and breaks it and the energy lost during the impact is 32.4 Joules. Determine :*

(*a*) *velocity of the hammer at impact.*

(*b*) *the height to which the hammer rises after breaking the test piece.*

(*c*) *the impulse, and*

(*d*) *the impulsive force exerted to the test piece if the impact lasts during* $\frac{1}{200}$ *seconds.*

Sol. (*a*) Mass of hammer = m = 10.5 kg.

At position 1, $v_1 = 0$

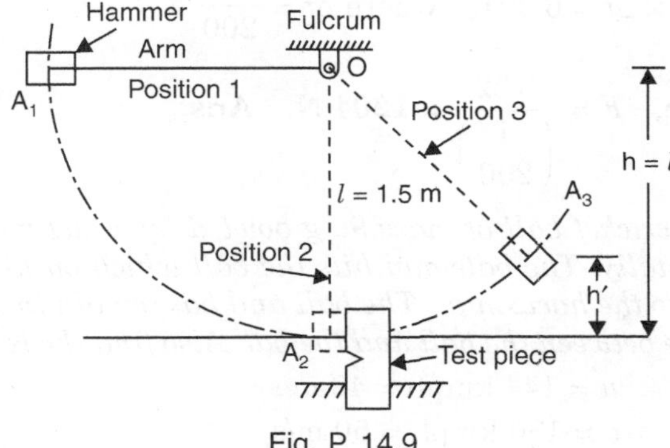

Fig. P. 14.9

Let velocity of hammer at position 2 be v_2

With reference to position 2, the maximum P.E. of hammer at position 1 = mgh, and maximum K.E. of hammer at position 2 = $\frac{mv_2^2}{2}$.

By conservation of energy, we get maximum K.E. = maximum P.E.

$$\therefore \quad \frac{mv_2^2}{2} = mgh$$

$$\therefore \quad v_2 = \sqrt{2gh} = \sqrt{2 \times 9.81 \times 1.5} = \mathbf{5.425\ m/s.} \quad \textbf{Ans.}$$

(*b*) K.E. of hammer after impact = $\frac{mv^2}{2} - \Delta\text{K.E.} = T_2$

or $$T_2 = mgh - \Delta\text{K.E.}$$

or $$T_2 = 10.5 \times 9.81 \times 1.5 - 32.4 = 122.108 \text{ J.}$$

Let the hammer rise through a height h_1 before it comes to rest at position 3 :

Then by principle of conservation of energy, we get

P.E. at position 3 = K.E. at position 2

$$\therefore \quad mgh' = T_2$$

$$\therefore \quad h' = \frac{T_2}{mg} = \frac{122.108}{10.5 \times 9.81} = \mathbf{1.185\ m.} \quad \textbf{Ans.}$$

(*c*) Let velocity of the hammer immediately after impact = v_2'

Then $$\frac{mv_2'^2}{2} = T_2$$

$\therefore$ $$\frac{1}{2} \times 10.5 \times v_2'^2 = 122.108$$

$\therefore$ $$v_2 = \sqrt{\frac{2 \times 122.108}{10.5}} = 4.823 \text{ m/s}$$

$\therefore$ impulse = $mv_2 - mv_2'$ = 10.5 (5.425 – 4.823) = 10.5 × 0.602

or impulse = **6.321 N-s** or kg-m/s. **Ans.**

(*d*) Let F be the impulsive force and Δt be the duration of impact.

Then $$F \times \Delta t = 6.321, \quad \text{where } \Delta t = \frac{1}{200} \text{ s.}$$

$\therefore$ Impulsive force, $$F = \frac{6.32}{\left(\frac{1}{200}\right)} = \mathbf{1264\ N.} \quad \textbf{Ans.}$$

Problem 14.10. *A cricket ball of mass 90 g bowled by a fast bowler reaches the bat at a velocity 144 kmph horizontally. The batsman hits the ball which on hitting attains a velocity of 180 kmph inclined at 45° to the horizontal. The ball and bat remain in contact for 0.0125 s. Find the average impulsive force between the ball and the bat. Also find the resultant impulse involved.*

Sol. Here, u = 144 kmph = 40 m/s

and v = 180 kmph = 50 m/s

The impulse-momentum equation $\int F\,dt = mv - mu$ can be resolved in the direction of x- and y-axis. Thus, we get

For x-direction : $$\int_{t_1}^{t_2} F_x \,.\, dt = mv_x - mu_x \qquad ...(1)$$

and for y-direction : $$\int_{t_1}^{t_2} F_y \,.\, dt = mv_y - mu_y \qquad ...(2)$$

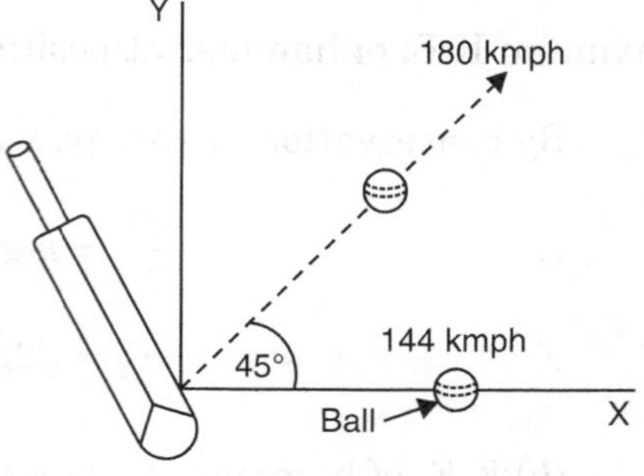

Fig. P. 14.10

Here $$m = 90 \text{ g} = \frac{90}{1000} = 0.09 \text{ kg.}$$

Using eqn. (1), we have

$$\int_{t_1}^{t_2} F_x \,.\, dt = 0.09\,\{(50 \cos 45°) - (-40)\}$$

or $$I_{imp)x} = \int_{t_1}^{t_2} F_x \,.\, dt = 6.782 \qquad ...(3)$$

And, from eqn. (2), we get

$$I_{imp)y} = \int_{t_1}^{t_2} F_y \,.\, dt = mv_y - mu_x$$

$$= 0.09 \times (50 \sin 45° - 0)$$

$\therefore \quad I_{imp)y} = 0.09 \times 50 \sin 45° = 3.182$ ...(4)

$\therefore$ Resultant Impulse

$$\int_{t_1}^{t_2} \bar{F}dt = \sqrt{I_{imp)_x^2} + I_{imp)_y^2}} = \sqrt{(6.782)^2 + (3.182)^2}$$

$\therefore$ Impulse = **7.4914 N-s**

And, angle $\theta_x = \tan^{-1}\left\{\frac{I_{imp)_y}}{I_{imp)_x}}\right\} = \tan^{-1}\left(\frac{3.182}{6.782}\right) = \tan^{-1}(0.46918)$

$\therefore \quad \theta_x = 25.14°$ from x-axis. **Ans.**

Average impulsive force,

$$F = \frac{\text{impulse}}{\text{time}} = \frac{7.4914}{0.0125}$$

or impulsive force, $F =$ **599.312 N. Ans.**

Problem 14.11. *Two blocks at rest connected to each other, are placed on the two inclined planes as shown in Fig. P. 14.11. Find the time required for the blocks to attain a velocity of 19.62 m/s and the tension in the chord. Take μ = 0.2 for both planes. The pulley may be assumed frictionless.*

Sol. Let the block A move down the plane. The free-body diagrams of the two blocks is shown in Fig. P. 14.11 (b).

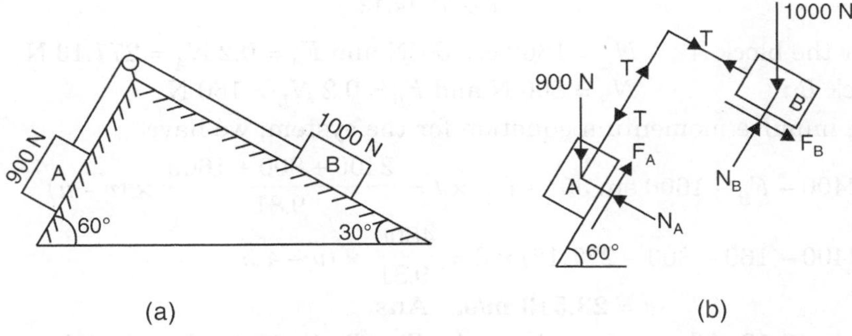

Fig. P. 14.11

For block A: $N_A = 900 \cos 60° = 450$ N ;

$F_A = \mu N_A = 0.20 \times 450 = 90$ N.

for block B: $N_B = 1000 \cos 30° = 500\sqrt{3}$ N.

and $F_B = \mu N_B = 0.2 \times 500\sqrt{3} = 100\sqrt{3}$ N.

Writing the impluse-momentum equation for the connected blocks, we have

$$(900 \sin 60° - F_A - 1000 \sin 30° - F_B) \times t = (m_A + m_B)(v - u)$$

or $$(450\sqrt{3} - 90 - 500 - 100\sqrt{3}) \times t = \frac{900 + 1000}{9.81} \times (19.62 - 0)$$

$$16.22\, t = \frac{1900}{9.81} \times (19.62 - 0)$$

$\therefore$ **t = 234.28 s. Ans.**

To find the tension in the chord write the impulse-momentum equation for any block. Thus, for block B, we have

$$(T - 1000 \sin 30° - F) \times t = \frac{1000}{9.81} \times (v - u)$$

or
$$(T - 500 - 100\sqrt{3}) \times 234.28 = \frac{1000}{9.81} \times (19.62 - 0)$$

$\therefore$
$$(T - 673.21) \times 234.28 = 2000$$

$\therefore$
$$T = \frac{2000}{234.28} + 673.21 \text{ N}$$

$\therefore$ Tension, $T =$ **681.75 N. Ans.**

Problem 14.12. *The system shown in Fig. P. 14.12 has a rightward velocity of 4.5 m/s. Find its velocity after 8 second. Take μ = 0.2 for all surfaces and assume the pulleys to be frictionless.*

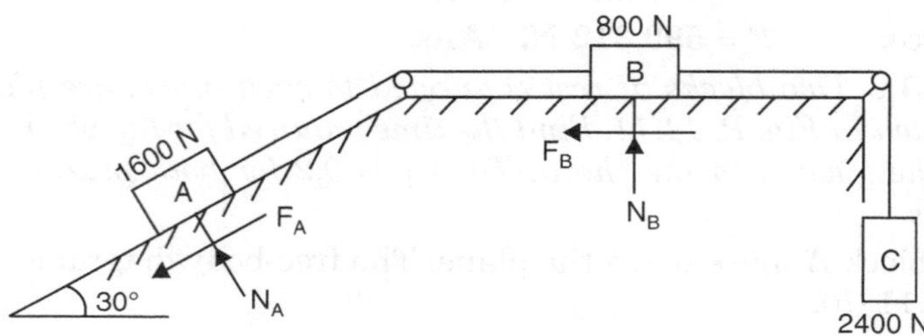

Fig. P. 14.12

Sol. For the block A : $N_A = 1600 \cos 30°\text{N}$ and $F_A = 0.2\ N_A = 277.13$ N

For block B : $N_B = 800$ N and $F_B = 0.2\ N_B = 160$ N

Writing impulse momentum equation for the system, we have

$$(2400 - F_B - 1600 \sin 30° - F_A) \times t = \frac{2400 + 800 + 1600}{9.81} \times (v - u)$$

$\therefore$
$$(2400 - 160 - 800 - 277.13) \times 8 = \frac{4800}{9.81} \times (v - 4.5)$$

$\therefore$ $v =$ **23.513 m/s. Ans.**

Problem 14.13. *The system shown in Fig. P. 14.13 (a) has a rightward velocity, of 6m/s just before the application of the force P. Find the value of P that will give a leftward velocity of 4 m/s in a time interval of 15 s. Take μ = 0.2 and assume the pulley to be frictionless. Also find when the system comes to rest.*

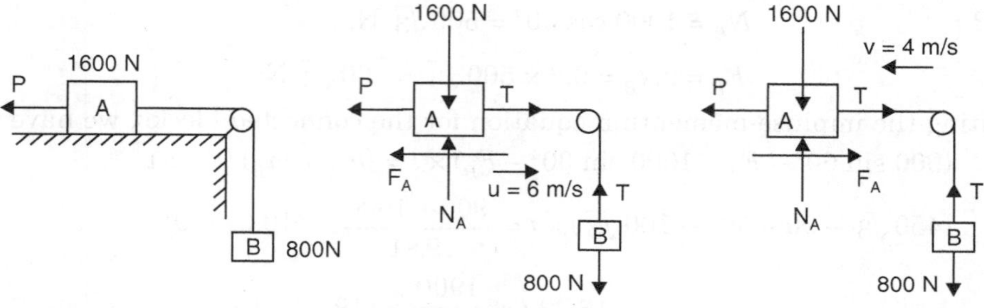

Fig. P. 14.13

Sol. When the system moves rightwards as shown in Fig. P. 14.13 (*b*), the frictional force F_B acts leftwards. The applied force P first brings the system to rest. After the system is brought to rest, the force P will cause the system to move towards left.

When the system comes to rest, by the application of impulse-momentum equation we get [from Fig. P.14.13 (*b*)] :

$$(800 - P - F_A)\, t_1 = \frac{1600 + 800}{9.81} \times (0 - u)$$

or $$(800 - 0.2 \times 1600 - P) \times t_1 = \frac{2400}{9.81} \times (0 - 6)$$

$$(480 - P)\, t_1 = -\frac{14400}{9.81}$$

or $$(P - 480)\, t_1 = \frac{14400}{9.81} \quad ...(1)$$

or $$t_1 = \frac{14400}{9.81} \times \frac{1}{P - 480} \quad ...(2)$$

When the system starts moving towards left and after 15 seconds the velocity becomes 4 m/s. See Fig. P. 14.13 (*c*).

∴ By applying the impulse-momentum equation, we get

$$(P - F_A - 800) \times t_2 = \frac{1600 + 800}{9.81} \times [4 - 0]$$

But $t_2 + t_1 = 15$ seconds,

∴ $$t_2 = (15 - t_1),$$

∴ We get $$(P - 0.2 \times 1600 - 800) \times (15 - t_1) = \frac{2400}{9.81} \times 4$$

∴ $$(P - 1120) \times (15 - t_1) = \frac{9600}{9.81} \quad ...(3)$$

Substituting the value of t, we get

$$(P - 1120) \times \left[15 - \frac{14400}{9.81} \times \left(\frac{1}{P - 480}\right)\right] = \frac{9600}{9.81}$$

or $$(P - 1120)\left[15 - \frac{1467.9}{P - 480}\right] = 978.6$$

or $$\frac{(P - 1120)}{P - 480}\,[15 \times (P - 480) - 1467.9] = 978.6$$

or $$15\,(P - 480)\,(P - 1120) - 1467.9 \times (P - 1120) = (P - 480) \times (978.6)$$

or $$15[P^2 - 1600P + 537600] - 1467.9P + 1644048 = 978.6P - 469728$$

or $$15P^2 - P\,[15 \times 1600 + 1467.9 + 978.6] + 15 \times 537600 + 1644048 + 469728 = 0$$

∴ $$15P^2 - 26446.5P + 10177776 = 0$$

∴ $$P^2 - 1763.1P + 678518.4 = 0 \quad ...(4)$$

∴ $$P = \frac{1}{2}\left[1763.1 \pm \sqrt{(1763.1)^2 - 4 \times 678518.4}\right]$$

or $$= \frac{1}{2}\,[1763.1 \pm 628.1]$$

or $P =$ **1195.5 N ; 567.5 N**

We have to select correct value of the force P.

From eqn. (2), $$t_1 = \frac{14400}{9.81} \times \frac{1}{P - 480}$$

$$\therefore \quad t_1 = \frac{14400}{9.81} \times \frac{1}{(1195.6 - 480)} \quad \text{or} \quad \frac{14400}{9.81} \times \frac{1}{567.5 - 480}$$

$$= 2.05 \text{ s} \quad \text{or} \quad 16.7 \text{ s}$$

Since $t_1 + t_2 = 15$ s, $\therefore$ The system comes to rest after **2.05 s**, and the corresponding value of the force

$$P = \mathbf{1195.6\ N.} \quad \textbf{Ans.}$$

Problem 14.14. *Find the velocity of the mass B in Fig. P.14.14 (a) after moving for 8 seconds (a) starting from rest, and (b) starting with a downward velocity of 3 m/s.*

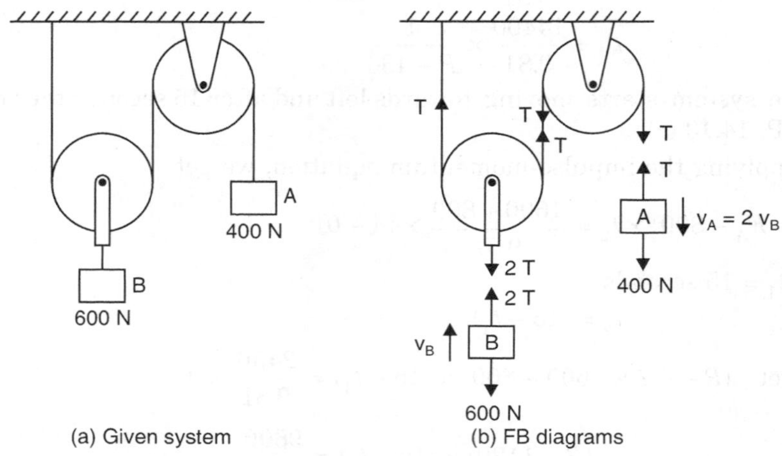

Fig. P. 14.14

Sol. In Fig. P.14.14 (*b*), it is clear that $v_A = 2v_B$. Applying impluse-momentum equation to both the blocks, we get $\left[\Sigma \bar{F}\,(t) = \frac{W}{g} \times (v - u)\right]$

For block A: $$(400 - T) \times 8 = \frac{400}{9.81} \times (v_A - u_A)$$

or $$(400 - T) \times 8 = \frac{400}{9.81}\,[2v_B - 2u_B] \quad ...(1)$$

And for block B: $$(2T - 600) \times 8 = \frac{600}{9.81}\,[v_B - u_B] \quad ...(2)$$

Multiplying eqn. (1) by 2 and adding to eqn. (2), we get

$$(800 - 600) \times 8 = \frac{(600 + 2 \times 400) \times 2}{9.81}\,(v_B - u_B)$$

or $$200 \times 8 = \frac{1400 \times 2}{9.81}\,(v_B - u_B)$$

or $$800 = \frac{1400}{9.81} \times (v_B - u_B) \qquad ...(3)$$

(*a*) When $u_B = 0$; we get $800 = \frac{1400}{9.81} \times (v_B - 0)$

$$\therefore \qquad v_B = \frac{800 \times 9.81}{1400} = \textbf{5.61 m/s. upwards.} \quad \textbf{Ans.}$$

(*b*) **When u_B = 3 m/s downwards** or $u_B = -3$ m/s.

∴ From eqn. (3), we get

$$800 = \frac{1400}{9.81} \times [v_B - (-3)]$$

$$\therefore \qquad v_B = \frac{800 \times 9.81}{1400} - 3 = +2.61 \text{ m/s.}$$

i.e., v_B = **2.61 m/s upwards. Ans.**

14.5. DYNAMIC ACTION OF A JET

A jet of water or other system of particles which impinges upon either stationary or moving object produces a force. This force can easily be determined by using the principle of impulse-momentum. Details of the application of impulse-momentum equation/principle can be understood by the study of the illustrative example given below.

Problem 14.15. *A jet of velocity 45 m/s, produced by a nozzle of 60 mm diameter, impinges tangentially upon a smooth vane and deflects through an angle of 15° to horizontal without any loss of velocity. Find the force exerted by the jet upon the vane. See Fig. P. 14.15 (a)*

(a) When vane is stationary,

(b) When the vane is moving with a velocity of 15 m/s rightwards.

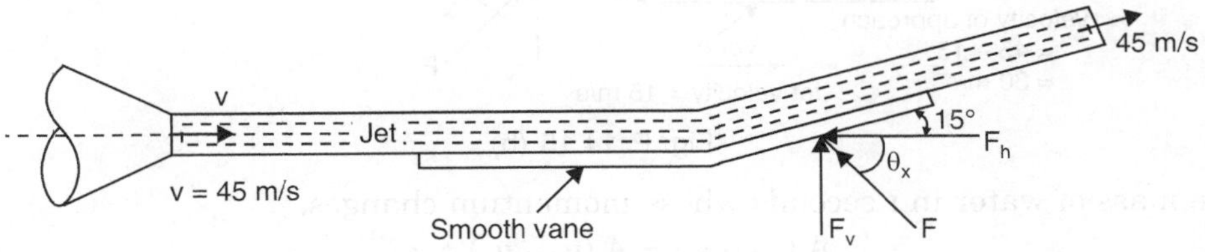

Fig. P. 14.15 (a)

Sol. Area of jet section $= \frac{\pi}{4} \times d^2 = \frac{\pi}{4} \times \left(\frac{60}{1000}\right)^2 = 9 \times \pi \times 10^{-4} \text{ m}^2$

mass of water discharged by the nozzle in t second

= (Area of jet section × velocity of jet) × mass density of water × t

= $(9 \times \pi \times 10^{-4} \text{ m}^2) \times 45 \text{ m/s} \times (1000 \text{ kg/m}^3) \times t$

$$m = 127.2345\, t \text{ kg}$$

weight of water discharged in t seconds

$$mg = 127.2345 \times 9.81\, t$$

(*a*) **Vane is stationary :**

Applying the impulse-momentum principle in *x*-direction, we get

$$-F_h \times t = m(v_x - u_x),$$

where F_h is horizontal force exerted on vane

or $$-F_h \times t = (127.2345\ t) \times (45 \cos 15° - 45)$$

$\therefore$ $$F_h = 195.09 \text{ N, towards left}$$

Similarly, by applying the impulse-momentum principle in the vertical direction, we get

$$F_v \times t = m(v_y - u_y)$$

where F_v is vertical force exerted on vane.

or $$F_v \times t = (127.2345 \times t) \times (45 \sin 15° - 0).$$

$\therefore$ $$F_v = 1481.88 \text{ N}$$

$\therefore$ Resultant force exerted by the jet on the vane = $F = \sqrt{F_h^2 + F_v^2}$

or $$F = \sqrt{(1481.88)^2 + (195.09)^2} = \mathbf{1494.67\ N.}\quad \textbf{Ans.}$$

Inclination of the force F with the horizontal = $\theta_x = \tan^{-1} \dfrac{F_v}{F_h}$

$$\theta_x = \tan^{-1}\left(\frac{1481.88}{195.09}\right) = \tan^{-1} 7.596 = \mathbf{82.5°.}\quad \textbf{Ans.}$$

(*b*) **Vane moving towards right with a velocity V :**

Velocity of approach of jet = jet velocity – vane velocity = $u_j - u_v$

or $$u = 45 - 15 = 30 \text{ m/s}.$$

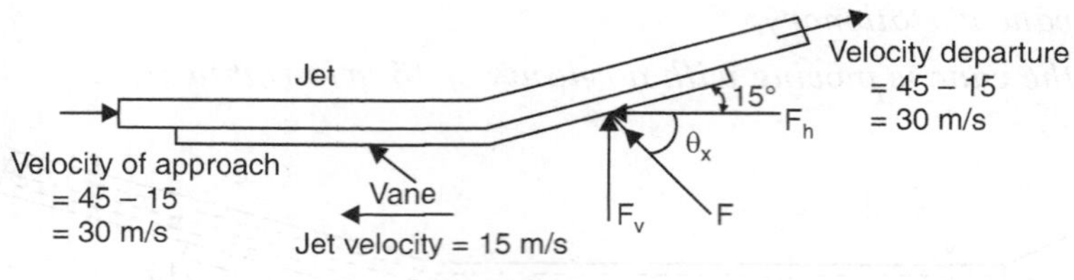

Fig. P. 14.15 (b)

The mass of water in *t* seconds whose momentum changes.

$$= A\,(u)\,\rho \times t = A\,(u_j - u_v)\,\rho \times t$$

$$= \frac{\pi}{4} \times \left(\frac{60}{1000}\right)^2 \times (30) \times 1000\ t \text{ kg} = 84.823t \text{ kg}.$$

$\therefore$ Writing impulse-momentum equation in horizontal direction [see Fig. P. 14.15 (*b*)], we get

velocity of approach = 45 – 15 = 30 m/s

and, velocity of departure = velocity of approach = 45 – 15 = 30 m/s

$\therefore$ $$-F_h \times t = (84.823 \times t)\,(30 \cos 15° - 30)$$

$\therefore$ $$F_h = \mathbf{86.71\ N}$$

And, by writing impulse momentum equation in vertical direction, we get

$$F_v \times t = (84.823 \times t) \times [30 \sin 15° - 0].$$

$\therefore$ $$F_v = \mathbf{658.61\ N}$$

$\therefore$ Resultant force, $\quad F = \sqrt{(86.71)^2 + (658.61)^2}$ = **664.3 N. Ans.**

$$\theta_x = \tan^{-1}\left(\frac{F_v}{F_h}\right) = \tan^{-1}\left(\frac{658.61}{86.71}\right) = \tan^{-1} 7.5725 = \mathbf{82.48°}. \quad \textbf{Ans.}$$

Note. The pressure P exerted by the jet is equal and opposite to the resultant force F.

Problem 14.16. *A man of mass m standing at one end of a stationary boat of length l and mass M. If the man walks toward the other end at a uniform velocity* v_{mb}*, determine : (a) velocity* v_b *of the boat as observed from the river banks, (b) the distance through which the boat gets shifted. Let m = 80 kg ; M = 500 kg,* v_{mb} *= 2m/s. towards left, and l = 8 m.*

Sol. Let the man be initially at A. (See Fig. P. 14.16)

Man walks towards left with the relative velocity v_{mb}.

Let $\quad v_m$ = absolute velocity of man, and

$\quad\quad v_b$ = absolute velocity of boat.

Initial velocity of boat and man is zero.

Therefore, the velocity of man relative to the boat

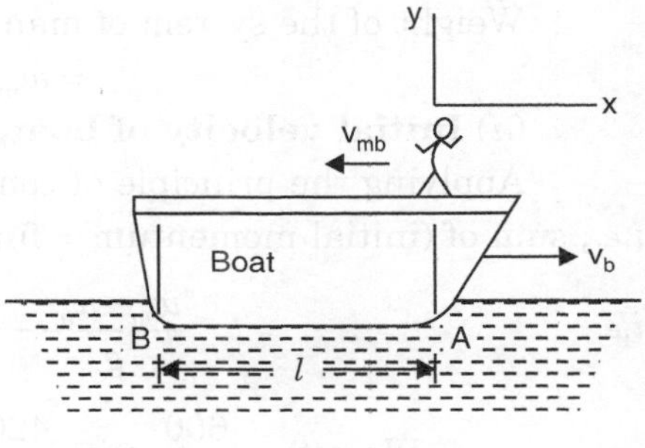

$$v_{mb} = v_m - v_b \quad \text{...(1)}$$

$\therefore$ Absolute velocity of man

$$= v_m = v_b + v_{mb} \quad \text{...(2)}$$

Initial momentum $\quad = (m + M) \times (0) = 0$

Final momentum $\quad = + mv_m + Mv_b$

Fig. P. 14.16

Since the momentum is conserved,

$\therefore$ we get $\quad 0 = + mv_m + M.v_b$ ...(3)

By substituting the values of v_m and v_{mb}, in eqn. (3), we get

$$0 = + m(v_{mb} + v_b) + M(v_b)$$

$$\therefore \quad v_b(M + m) = - m \,.\, v_{mb}$$

$$\therefore \quad v_b = -\frac{mv_{mb}}{M + m} \quad \text{...(4)}$$

– ve sign indicates that boat moves in opposite direction. **Ans.**

The distance travelled by man in time t seconds = l

$$\therefore \quad t = \frac{l}{v_{mb}} = \frac{l}{v_m - v_b} \quad \text{...(5)}$$

During this time, the boat gets shifted through a distance

$$l_b = v_b \times t = \left(\frac{l}{v_m - v_b}\right) \times v_b \quad \text{...(6)}$$

Given : $\quad m$ = 80 kg ; M = 500 kg and v_{mb} = 2 m/s

$\therefore$ Velocity of boat,

$$v_b = -\frac{mv_{mb}}{m + M} = -\frac{80 \times (2)}{80 + 500} \text{ m/s.}$$

$$= -0.276 \text{ m/s}$$

= 0.276 m/s towards right. **Ans.**

Time required for the man to cover a distance equal to the length (l) of the boat

$$= \frac{l}{v_{mb}} = \frac{8}{2} = 4 \text{ seconds}$$

$\therefore$ The boat gets shifted through by a distance $l_b = v_b \times t$

$= 0.276 \times 4 =$ **1.104 m** towards right. **Ans.**

Problem 14.17. *A man weighing 600 N, moving horizontally with a velocity of 4 m/s, jumps off the end of a pier into a 4200 N boat. Find the horizontal velocity of the boat (a) if it is initially at rest, and (b) if it is approaching pier with an initial velocity of 1.2 m/s, (c) if the boat is initially drifting the pier at a rate of 1.2 m/s.*

Sol. Weight of man, w_m = 600 N,

Velocity of man $= v_m = 4$ m/s

Weight of the system of man and boat when the man jumps into the boat

$= w_m + w_b = 600 + 4200 = 4800$ N.

(*a*) **Initial velocity of boat, $u_b = 0$**

Applying the principle of conservation of momentum,

i.e., sum of (initial momentum = final momentum),

i.e., $$\frac{w_m \cdot u_m}{g} + \frac{w_b \cdot u_b}{g} = \frac{w_m + w_b}{g} \times V_f \qquad ...(1)$$

$\therefore$ We get $$\frac{600}{g} \times 4 + \frac{4200}{g} \times (0) = \frac{4800}{g} \times V_f$$

$\therefore$ $$V_f = \frac{4 \times 600}{4800} = \textbf{0.5 m/s.} \quad \textbf{Ans.}$$

Boat moves away from pier.

(*b*) **Initial velocity of boat u_b = – 1.2 m/s.**

(– ve sign is used because boat is approaching pier.

Using equation (1), *i.e.*, the principle of conservation of momentum, we get

$$\frac{600}{g} \times 4 + \frac{4200}{g} (-1.2) = \frac{4800}{g} V_f$$

$\therefore$ $$V_f = \frac{1}{4800} [2400 - 5040]$$

$$= \frac{-2640}{4800} = \textbf{– 0.55 m/s.} \quad \textbf{Ans.}$$

i.e., the boat with man is moving towards the pier.

(*c*) **Initially boat is moving away the pier,**

$\therefore$ $u_b = 1.2$ m/s.

Using equation (1), above, we get

$$\frac{600}{g} \times 4 + \frac{4200}{g} \times 1.2 = \frac{4800}{g} \times V_f$$

$\therefore$ $$V_f = \frac{1}{4800} [600 \times 4 + 4200 \times 1.2] = \frac{7440}{4800} = \textbf{1.55 m/s.} \quad \textbf{Ans.}$$

The boat moves away from pier.

Problem 14.18. *Smugglers carrying three bags of smuggled goods each weighing 200 N are travelling in a 12000 N truck at an initial velocity of 72 km/hr. Each smuggler weighs 800 N. The smuggler picks up a bag and moves towards the rear end of the truck at a speed of 4 m/s and throws away the bag on the road. Find the speed of truck if the three smugglers throw bags :*

(a) in succession, and

(b) all together.

Sol. (*a*) **When bags are thrown in succession :**

Initial velocity of truck

$$= 72 \text{ km/hr} = \frac{72}{3.6} = 20 \text{ m/s}.$$

Let v_1 = velocity of truck after first bag is thrown

v_2 = velocity of truck after the second bag is thrown

v_3 = velocity of truck after the third bag is thrown

u_m = uniform velocity of each man with bag = 4 m/s.

Total weight of the truck and smugglers = 12000 + 4 × 800 = 15200 N.

Total weight of truck + 4 smugglers + 3 bags = W = 15200 + 3 × 200 = 15800 N.

Absolute velocity of the first bag $= (v_1 - 4)$

Initial momentum $= (15800 \times 20)/g$

Final momentum $= [(15200 + 3 \times 200)\, v_1 + 200\, (v_1 - 4)]/g$

∴ Initial momentum = final momentum

$$\therefore \quad \frac{1}{g}(15800 \times 20) = [(15200 + 3 \times 200)\, v_1 + 200\, (v_1 - 4)]/g$$

or $$15800 \times 20 = (15200 + 3 \times 200)\, v_1 - 200 \times 4$$

or $$15800 \times 20 = 15800\, v_1 - 200 \times 4$$

$$\therefore \quad v_1 = 20 + \frac{200 \times 4}{15800} = \mathbf{20.051 \text{ m/s.}} \quad \textbf{Ans.}$$

When the second bag is thrown away :

Initial momentum $= [(15200 + 200 \times 2)\, v_1]/g$

Final momentum $= [(15200 + 200) \times v_2 + 200\, (v_2 - 4)]/g$

$= [(15200 + 2 \times 200)\, v_2 - 200 \times 4]/g$

By conservation of momentum, we get

$$\frac{1}{g}[(15200 + 200 \times 2) \times v_1] = [(15200 + 2 \times 200)\, v_2 - 200 \times 4]/g$$

$$\therefore \quad v_2 = v_1 + \frac{200 \times 4}{15200 + 200 \times 2}$$

or $$v_2 = 20.051 + 0.051 = \mathbf{20.102 \text{ m/s.}} \quad \textbf{Ans.}$$

When the third bag is thrown away :

Initial momentum $= [(15200 + 200)\, v_2]/g$

Final momentum $= [15200 \times v_3 + 200 \times (v_3 - 4)]/g$

Equating the initial momentum and the final momentum, we get

$$\frac{1}{g}[(15200 + 200)]\, v_2 = [15200 \times v_3 + 200 \times (v_3 - 4)]/g$$

or $$(15200 + 200)\, v_2 = (15200 + 200)\, v_3 - 200 \times 4$$

$$\therefore \quad v_3 = v_2 + \frac{200 \times 4}{15200 + 200}$$

or $$v_3 = 20.102 + 0.052$$

$$\therefore \quad \mathbf{v_3 = 20.154\ m/s.} \quad \textbf{Ans.}$$

(*b*) **When all bags are thrown together :**

Let the velocity of the truck after the three bags have been thrown = v.

Then initial momentum = $[(15200 + 3 \times 200) \times 20]/g$

Final momentum = $[15200\,v + 3 \times 200\,(v - 4)]/g$

By the principle of conservation of momentum, we have

$$\frac{1}{g}[(15200 + 3 \times 200) \times 20] = [15200\,v + (3 \times 200)\,(v - 4)]/g$$

or, $$(15200 + 3 \times 200) \times 20 = (15200 + 3 \times 200)\,v - 3 \times 200 \times 4$$

$$\therefore \quad v = 20 + \frac{3 \times 200 \times 4}{15200 + 3 \times 200}\ \text{m/s}$$

$$= 20 + \frac{2400}{15800}\ \text{m/s}$$

i.e., $$\mathbf{v = 20.152\ m/s.} \quad \textbf{Ans.}$$

Problem 14.19. *A car weighing 60 kN and moving due east at 72 km/hr collides with a van of weight 90 kN and moving at 54 km / hr north wards. If the vehicle lock after collision find the velocity and direction of two vehicles.*

Sol. Fig. P.14.19 (*a*) shows the two vehicles before and after collision.

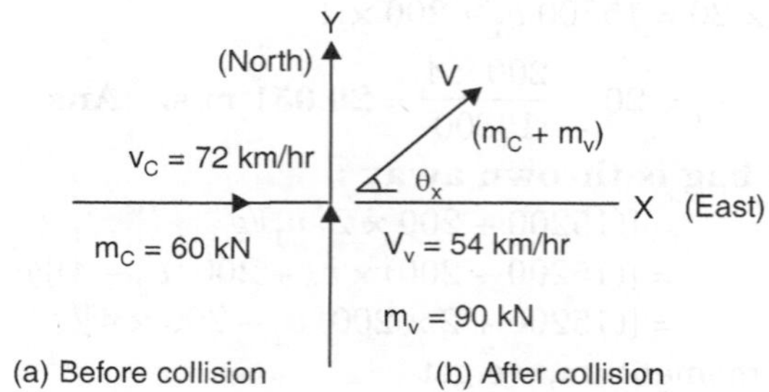

Fig. P. 14.19

Let m_c *and* v_c be the mass and initial velocity of car and, let m_v and v_v be the mass and initial velocity of van.

Let V be the final velocity of the locked vehicles after collision.

Given : $$v_c = 72\ \text{km/hr} = \frac{72}{3.6} = 20\ \text{m/s}, \quad m_c = 60\ \text{kN}$$

$$v_v = \frac{54}{3.6} = 15\ \text{m/s}\ ; \quad m_v = 90\ \text{kN}.$$

By the principle of conservation of momentum, we have

$$m_c \,.\, v_c + m_v \,.\, v_v = (m_c + m_v).\ V$$

$$\therefore \quad V = \frac{m_c . v_c + m_v . v_v}{m_c + m_v} = \frac{60 \times 20 + 90 \times 15}{60 + 90} = 17 \text{ m/s.}$$

$$= 17 \times 3.6 = \mathbf{61.2\ kmph.} \quad \textbf{Ans.}$$

$$\tan \theta_x = \frac{v_v}{v_c} = \frac{15}{20} = \frac{3}{4} = 0.75$$

$$\therefore \quad \theta_x = \tan^{-1} 0.75 = \mathbf{36.87°} \text{ as shown in Fig. P. 14.15 } (b). \quad \textbf{Ans.}$$

EXERCISES XIV

A. Theoretical Questions

1. Differentiate between work-energy method and the impulse-momentum method.
2. Illustrate the impulse-momentum method. Establish the impulse-momentum equation.
3. State the impulse-momentum equations for the linear and angular motions.
4. What do you mean by the terms : impulse and impulsive force.
5. Using the impulse-momentum method find the average resistance of the ground against the penetration of a pile when struck by a hammer.
6. Establish an expression for finding the loss of kinetic energy of a pile moving into the ground due to a blow from a hammer.
7. State the law of conservation of momentum.

B. Numerical Problems

14.1. A block of mass 25 kg is sliding due east on a smooth table with a velocity of 4 m/s . A 40 N force with a direction 30° NE acts on the body for 8*s*. Find the magnitude and direction of the final velocity of the block. [**Ans.** v = 15.2 m/s ; θ_x = 46.84° = 43.16° NE]

14.2. In the system shown in Fig. E. 14.2, the initial velocity of the block moving up the 30° incline is 8 m/s. The frictional resistance offered by the plane is 20 N/kN weight of the block. If the weight of block is 45 kN, find the time in which the block come to rest. The pulley may be assumed to be smooth. [**Ans.** t = 0.8153 second ; Tension, T = 18900 N]

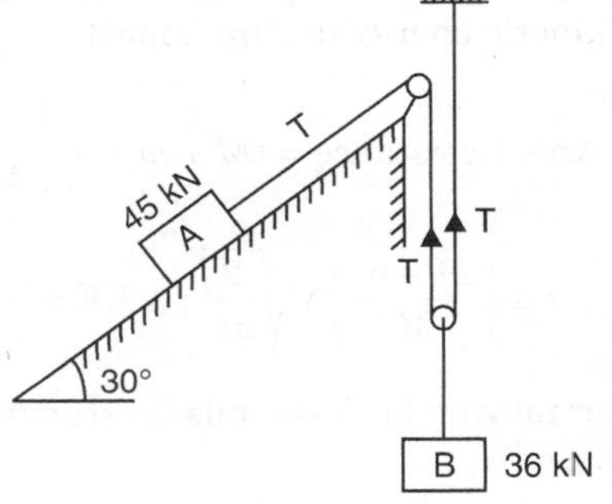

Fig. E. 14.2

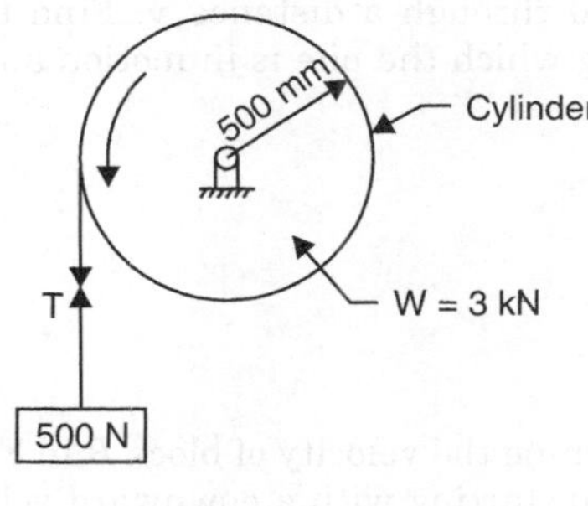

Fig. E. 14.4

14.3. A gun weighing 100 kN fires a 180 N shell with the muzzle velocity of 600 m/s. The shell shoots a 300 N projectile moving in the horizontal direction. The shell moves with one-half of muzzle velocity. Find the initial recoil velocity of the gun. If the gun recoils through 1 m find the average resistance for recoil. [**Ans.** 2.34 m/s, 27.91 kN]

14.4. A solid cylinder, as shown in Fig. E. 14.4 is supported on a frictionless journal bearings. A weight weighing 500 N is suspended from a rope wound round the cylinder. Find the angular velocity of the cylinder and the tension T in the rope, 6 seconds after the system is released from rest.

[**Ans.** 78.493 rad/s or 749.52 rpm ; 1000.26 N]

14.5. A flywheel weighing 7.5 kN has hub 15 cm in diameter supported in bearings. The radius of gyration of the flywheel is 60 cm. The initial velocity of flywheel is 600 rpm. If the co-efficient of friction of bearings is 0.10, find the time period for the flywheel to come to rest.

[**Ans.** 153.717 s]

14.6. A 4 kN wheel shown in Fig. E. 14.6 rolls freely on an inclined plane. Wheel has a radius of gyration of 0.60 times the radius of the wheel. Find velocity of the wheel and that of the weight and tension in the string, 6 seconds after the system is released from rest. Assume pulley to be frictionless. [**Ans.** v_1 = 6.235 m/s ; v_2 = $2v_1$; T = 1576.24 N]

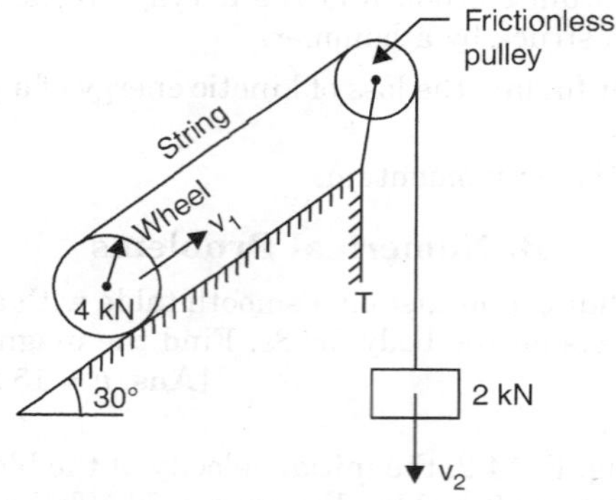

Fig. E. 14.6

14.7. A hammer of mass M falls from the height h on the top of a pile of mass m and drives it into the ground through a distance y. Find the average resistance of the ground. Also, find the time during which the pile is in motion and the loss in kinetic energy during impact.

[**Ans.** Resistance = $(M + m)g + \left(\frac{M^2}{M+m}\right)\frac{gh}{y}$;

$t = \left(\frac{M+m}{M}\right) y \sqrt{\frac{2}{gh}}$; $\Delta KE = \left(\frac{Mm}{M+m}\right) gh$]

14.8. Determine the velocity of block B in Fig. E. 14.8 after moving for 5 seconds (*a*) starting from rest, and (*b*) starting with a downward velocity of 2 m/s.

[**Ans.** (*a*) v_A = 8.92 m/s ; $v_B = \frac{v_A}{2}$ = 4.46 m/s ; (*b*) v_A = 4.92 m/s ; v_B = 2.92 m/s]

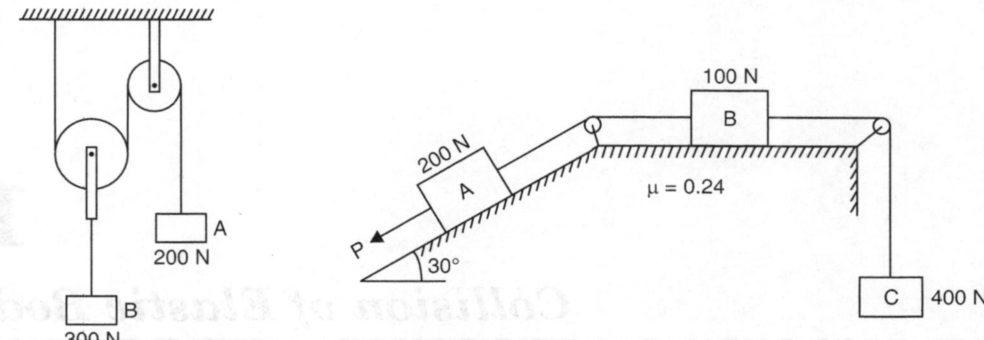

Fig. E. 14.8 Fig. E. 14.9

14.9. The system in Fig. E. 14.9 has a rightward velocity of 3 m/s. Determine the constant force P that will give it a leftward velocity of 6 m/s in a time interval of 30 seconds. [**Ans.** P = 382.24 N]

14.10. A 100 g base ball is thrown with a horizontal velocity of 30 m/s toward a batter. After being struck by the bat, the ball has a velocity of 42 m/s inclined at 36° to the horizontal. If the ball and the bat remain in contact for 0.018 second, find the average force exerted on the ball during the impact. [**Ans.** 380.98 N, θ_x = 21.1°]

15

Collision of Elastic Bodies

15.1. INTRODUCTION

When the two elastic bodies collide, they are first deformed, then they spring (or separate) apart because of the action of restoring elastic forces. Throughout this elastic impact, there occurs a perfect action and reaction of elastic forces. The velocity with which the two bodies separate after impact or collision depends on (*a*) the velocities with which the two bodies approach each other before impact, (*b*) the shape and size (*c*) the elastic properties of the colliding bodies, and (*d*) the direction (*i.e.,* the line) of impact.

The magnitude of the forces acting on the two bodies during the short time is so large that the nonimpulsive forces are negligible in comparison. Hence, the resultant force on the system is zero and **conservation of momentum** may be applied to write one relation between the unknown final velocities of the two bodies. Another relation between these velocities is obtained from the definition of the **coefficient of restitution.** The coefficient of restitution (denoted by e) is defined as the ratio of the relative velocity of the colliding bodies after impact of their relative velocity before impact. All velocities are measured along the **line of impact,** which is the **common normal** to the surfaces of the colliding bodies.

15.2. DEFINITIONS

(*a*) **Impact.** It is collision between bodies in very short interval of time during which they exert very large forces on each other.

(*b*) **Line of impact.** The common normal to the surfaces of the two colliding bodies is called **the line of impact.** Fig. 15.2.1 shows line of impact for various cases.

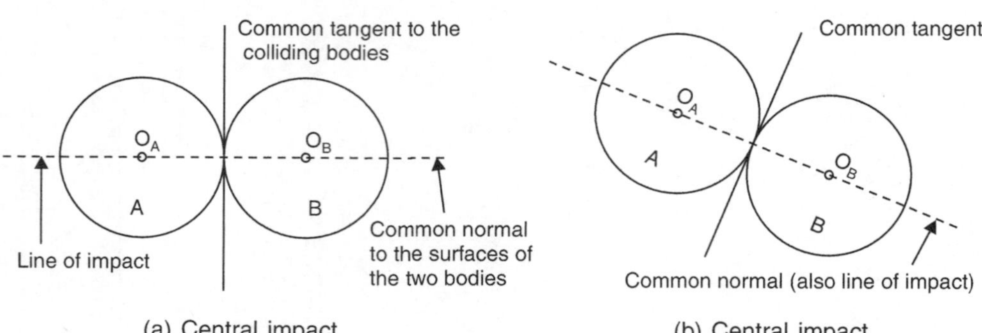

(a) Central impact

(b) Central impact

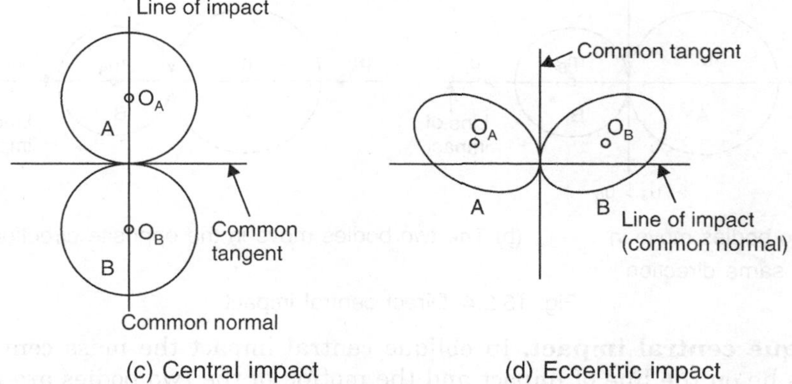

Fig. 15.2.1. Various cases showing line of impact. O_A and O_B are mass centres of bodies A and B

(*c*) **Central impact.** When the mass centres of the colliding bodies lie on the line of impact, it is termed as central impact. See Fig. 15.2.1 (*a*, *b*, *c*).

(*d*) **Eccentric impact.** In eccentric impact the mass centres O_A and O_B do not lie on the line of impact See Fig. 15.2.1 (*d*).

(*e*) **Direct impact.** If the motion of two colliding bodies is directed along the line of impact, it is known as direct impact. See Fig. 15.2.2.

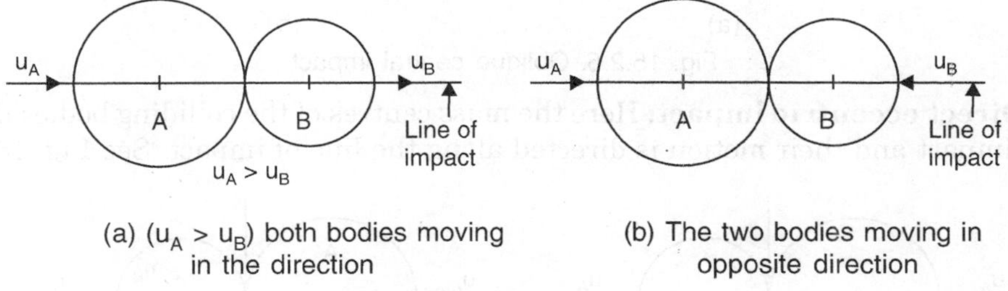

Fig. 15.2.2 Direct impact

(*f*) **Oblique impact.** In oblique impact the motion of one or both of the colliding bodies is not directed along the line of impact. See Fig. 15.2.3

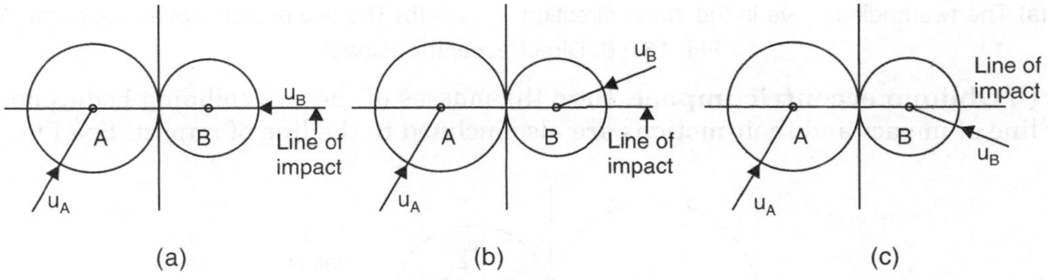

Fig. 15.2.3. Cases of oblique impact

(*g*) **Direct central impact.** In direct central impact the mass centres of the two colliding bodies lie on the line of impact and their velocities are also directed along the line of impact. See Fig. 15.2.4.

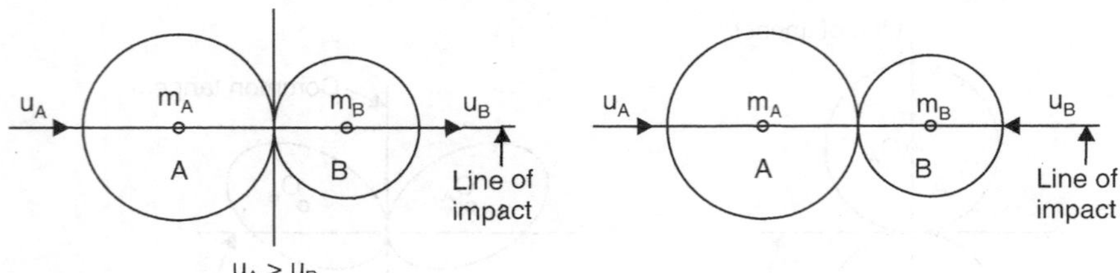

(a) The two bodies move in the same direction

(b) The two bodies move in the opposite direction

Fig. 15.2.4. Direct central impact

(*h*) **Oblique central impact.** In oblique central impact the mass centres of the two colliding bodies lie on the line of impact and the motion of the two bodies are inclined to the line of impact. See Fig. 15.2.5.

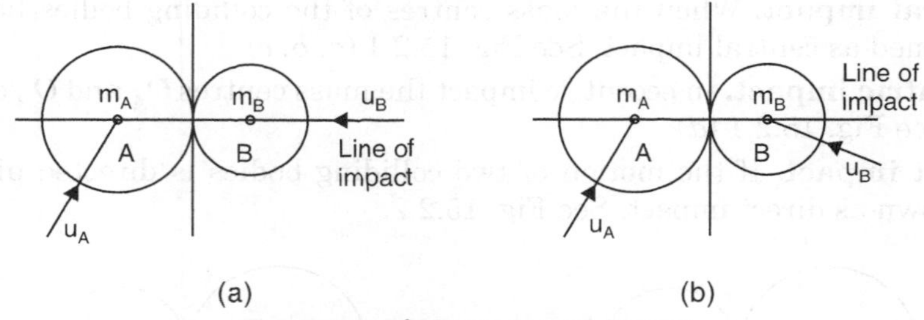

(a) (b)

Fig. 15.2.5. Oblique central impact

(*i*) **Direct eccentric impact.** Here the mass centres of the colliding bodies do not lie on the line of impact and their motion is directed along the line of impact. See Fig. 15.2.6.

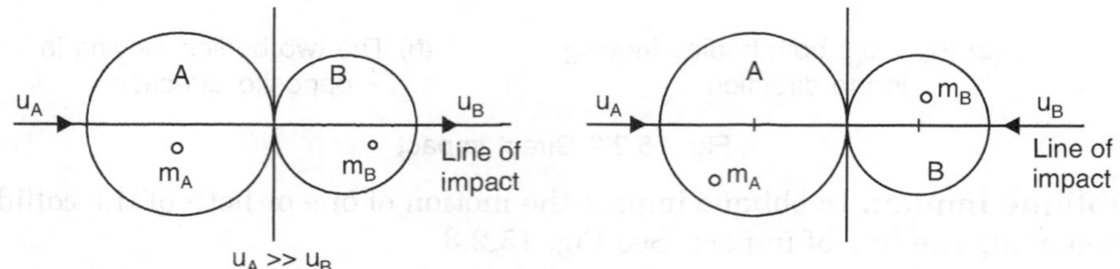

(a) The two bodies move in the same direction

(b) The two bodies move in opposite direction

Fig. 15.2.6. Direct eccentric impact

(*j*) **Oblique eccentric impact.** Here the masses of the two colliding bodies do not lie on the line of impact and their motions are also inclined to the line of impact. See Fig. 15.2.7.

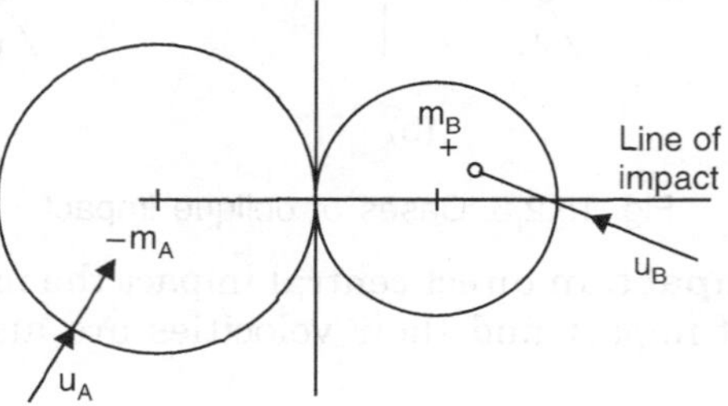

Fig. 15.2.7. Oblique eccentric impact

15.3. DIRECT CENTRAL IMPACT

Consider two bodies of mass m_A and m_B moving in the same straight line towards right with velocities u_A and u_B, respectively (Fig. 15.3). If u_A is larger than u_B, body *A* will collide with body *B*. The relative velocity before impact is $(u_A - u_B)$. After impact, assume that the two bodies continue moving in their original directions. In order for them to separate, body m_B must have larger final velocity than body m_A ; therefore, the relative velocity after impact is $(v_B - v_A)$. The two relative velocities before and after impact are related to each other by a ratio known as **coefficient of restitution** (*e*), given below :

$$e = \frac{\text{Relative velocity after impact}}{\text{Relative velocity before impact}} = \frac{v_B - v_A}{u_A - u_B} \quad ...(15.1)$$

or, $$e(u_A - u_B) = (v_B - v_A) \quad ...(15.2)$$

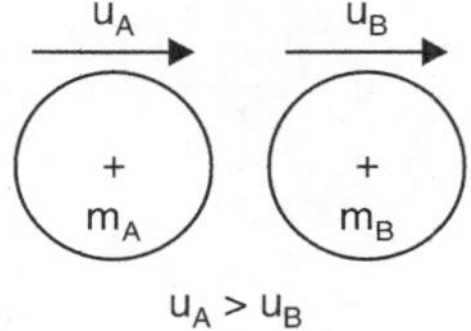

(a) Before impact

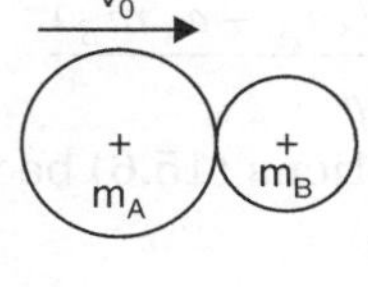

(b) Maximum deformation, the two bodies deform and move with the common velocity v_0

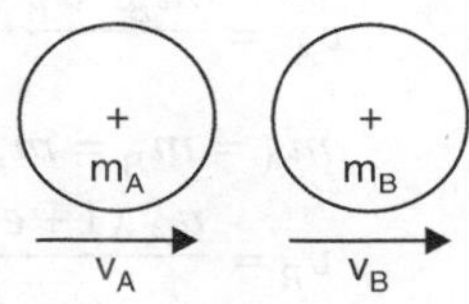

(c) Separation, the two bodies move in the same direction ($v_B > v_A$)

Fig. 15.3.1. The direct central impact—the two bodies move in the same direction

When the two bodies move in the opposite directions u_A and u_B, as shown in Fig. 15.3.2, their relative velocity before impact = $u_A - (-u_B) = u_A + u_B$. And, their relative velocity after impact = $v_B - (-v_A) = v_B + v_A$.

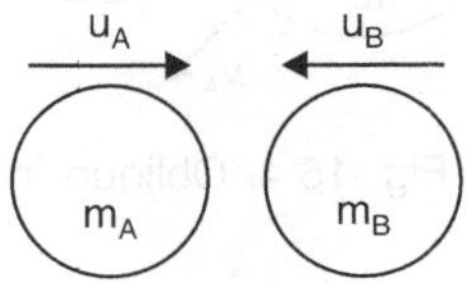

(a) Relative velocity before impact $u_A - (-u_B) = u_A + u_B$

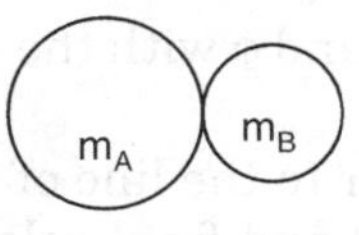

(b) Deformation

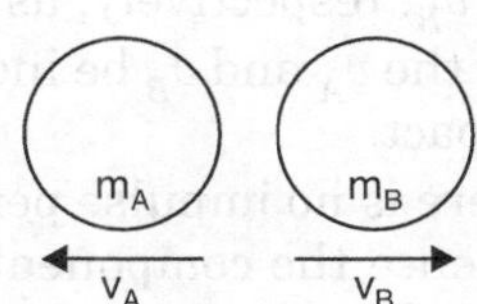

(c) After impact relative velocity $= v_B - (-v_A) = v_B + v_A$

Fig. 15.3.2. Direct central impact—the two bodies moving in opposite directions

Relative velocity before impact is also called the **velocity of approach,** and the relative velocity after impact is known as the **velocity of separation.** Thus, the velocity of approach,

$$= v_{\text{approach}} = u_A - (-u_B) = u_A + u_B \quad ...(15.3)$$

And, velocity of separation,

$$v_{\text{separation}} = v_B - (-v_A) = v_B + v_A \quad ...(15.4)$$

Applying the principle of conservation of momentum,

We have

$$m_A \,.\, u_A + m_B \,.\, u_B = m_A \,.\, v_A + m_B \,.\, v_B \qquad ...(15.5)$$

There are two unknown velocities v_A and v_B. These can be found by solving Equations (15.1) and (15.5), as given below.

Multiplying eqn. (15.2) by m_A, we get

$$m_A\,(u_A - u_B)\,.\,e = m_A \,.\, v_B - m_A \,.\, v_A \qquad ...(a)$$

Adding (a) and (15.5), we get

$$(m_A + m_B)\,v_B = m_A \,.\, u_A + m_B \,.\, u_B + m_A \,.e\,(u_A - u_B)$$

$$\therefore \qquad v_B = \frac{m_A \,.\, u_A + m_B \,.\, u_B + e \,.\, m_A\,(u_A - u_B)}{m_A + m_B}$$

or $$v_B = \frac{m_A \,.\, u_A\,(1+e) + u_B\,(m_B - e \,.\, m_A)}{m_A + m_B} \qquad ...(15.6(a))$$

Similarly, $$v_A = \frac{m_B \,.\, u_B\,(1+e) + u_A\,(m_A - e \,.\, m_B)}{m_A + m_B} \qquad ...(15.6(b))$$

If $m_A = m_B = m$, then equations (15.6) become

$$v_B = \frac{u_A\,(1+e) + u_B\,(1-e)}{2} \qquad ...(15.7(a))$$

and, $$v_A = \frac{u_A\,(1-e) + u_B\,(1+e)}{2} \qquad ...(15.7(b))$$

15.4. OBLIQUE IMPACT

Let A and B be the centres of two spheres of masses m_A and m_B as shown in Fig. 15.4.

Let the initial velocities of the two spheres be u_A and u_B, inclined at α and β, respectively to the line of impact.

Let the velocity of the two spheres after impact be v_A and v_B, respectively, as shown in Fig. 15.4.

Let the v_A and v_B be inclined at θ and ϕ with the line of impact.

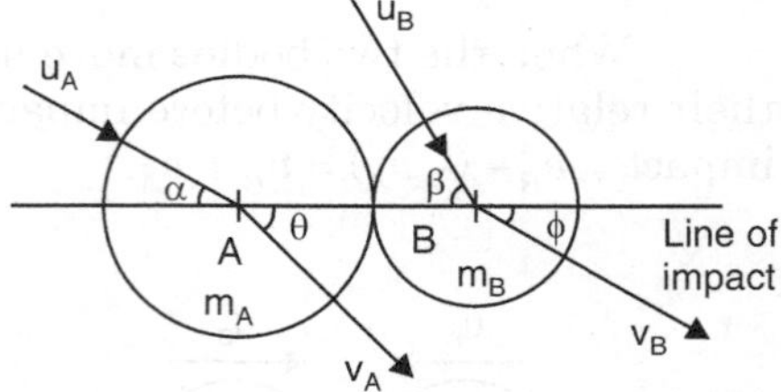

Fig. 15.4. Oblique impact

There is no impulse perpendicular to the line of impact. Hence the components of initial and final velocities of the two spheres perpendicular to the line of impact remain unchanged. Thus, we get

$$v_A \sin\theta = u_A \sin\alpha \qquad ...(15.8)$$

and $$v_B \sin\phi = u_B \sin\beta \qquad ...(15.9)$$

By the principle of conservation of momentum along the line of impact, we have

$$m_A.u_A \cos\alpha + m_B.u_B \cos\beta = m_A.v_A \cos\theta + m_B.v_B \cos\phi \qquad ...(15.10)$$

Also, by the definition of coefficient of restitution, we have

$$e = \frac{v_B \cos\phi - v_A \cos\theta}{u_A \cos\alpha - u_B \cos\beta} \qquad ...(15.11)$$

From these four equations, we can find the final velocities v_A, v_B and their angles θ and ϕ with the line of impact.

15.5. IMPACT AGAINST FIXED PLANE

A smooth sphere of mass m, moving with a velocity u strikes a smooth fixed horizontal plane making an angle α with the normal to the plane and rebounds with velocity v making an angle β with the normal as shown in Fig. 15.5.

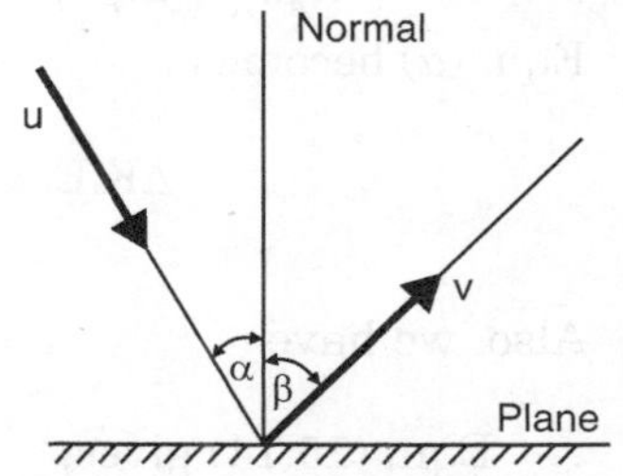

Fig. 15.5. Impact against a fixed surface

As the plane is smooth, therefore, change in velocity parallel to the plane is zero. We get,

$$u \sin \alpha = v \sin \beta \quad ...(a)$$

Also,
$$e = \frac{0 - (-v \cos \beta)}{u \cos \alpha - 0}$$

or
$$v \cos \beta = eu \cos \alpha \quad ....(b)$$

Squaring (a) and (b), and adding, we get

$$v^2 (\sin^2 \beta + \cos^2 \beta) = u^2 \sin^2 \alpha + e^2 u^2 \cos^2 \alpha.$$

or,
$$v^2 = u^2 (\sin^2 \alpha + e^2 \cos^2 \alpha) \quad ...(15.12)$$

By dividing (b) by (a), we get

$$\cot \beta = e \cot \alpha \quad ...(15.13)$$

15.6. LOSS OF KINETIC ENERGY

(a) **Direct impact**

Loss of K.E. = ΔK.E. = K.E. before impact – K.E. after impact

or,
$$\Delta\text{K.E.} = \frac{1}{2}(m_A.u_A^2 + m_B.u_B^2) - \frac{1}{2}(m_A.v_A^2 + m_B.v_B^2)$$

Multiplying and dividing by $(m_A + m_B)$, we get

$$\Delta\text{K.E.} = \left(\frac{m_A + m_B}{m_A + m_B}\right)\frac{1}{2}[(m_A.u_A^2 + m_B.u_B^2) - (m_A.v_A^2 + m_B.v_B^2)]$$

$$= \frac{1}{2(m_A + m_B)} [m_A^2.u_A^2 + m_A m_B.u_B^2 + m_A m_B.u_A^2 + m_B^2.u_B^2$$
$$- m_A^2 v_A^2 - m_A m_B v_B^2 - m_A m_B . v_A^2 - m_B^2 v_B^2]$$

$$= \frac{1}{2(m_A + m_B)} [m_A^2 u_A^2 + m_B u_B^2 + m_A . m_B (u_A^2 + u_B^2)$$
$$- (m_A^2 v_A^2 + m_B^2 v_B^2 + m_A m_B (v_A^2 + v_B^2)]$$

$$= \frac{1}{2(m_A + m_B)}[\{(m_A u_A + m_B u_B)^2 - 2m_A m_B u_A u_B$$
$$+ m_A m_B (u_A - u_B)^2 + 2m_A m_B u_A u_B\}$$
$$- \{(m_A v_A + m_B v_B)^2 - 2m_A m_B v_A v_B + m_A m_B (v_A - v_B)^2 + 2m_A m_B.v_A v_B\}]$$

or
$$\Delta\text{K.E.} = \frac{1}{2(m_A + m_B)} [((m_A u_A + m_B u_B)^2 + m_A m_B (u_A - u_B)^2\}$$
$$- \{(m_A v_A + m_B v_B)^2 + m_A m_B (v_A - v_B)^2\}] \quad ...(a)$$

But by conservation of momentum, we have

$$m_A u_A + m_B u_B = m_A v_A + m_B \cdot v_B \quad ...(b)$$

Eqn. (*a*) becomes :

$$\Delta\text{K.E.} = \frac{1}{2(m_A + m_B)} [m_A m_B\{(u_A - u_B)^2 - (v_A - v_B)^2\}] \quad ...(15.14)$$

Also, we have $$e = \frac{v_B - v_A}{u_A - u_B} \quad ...(c)$$

∴ Eqn. (15.14) gives,

$$\Delta\text{K.E.} = \frac{1}{2(m_A + m_B)} [m_A m_B\{(u_A - u_B)^2\}(1 - e^2)]$$

or $$\Delta\text{K.E.} = \frac{m_A m_B}{2(m_A + m_B)} [(u_A - u_B)^2 (1 - e^2)] \quad ...(15.15)$$

(*b*) **Oblique impact**

$$\Delta\text{K.E.} = \frac{1}{2}\{m_A u_A{}^2 + m_B . u_A{}^2\} - \frac{1}{2}\{m_A v_A{}^2 + m_B v_B{}^2\}$$

or $$\Delta\text{K.E.} = \frac{1}{2} m_A (u_A{}^2 \cos^2\alpha + u_A{}^2 \sin^2\alpha) + \frac{1}{2} m_B (u_B{}^2 \cos^2\beta + u_B{}^2 \sin^2\beta)$$

$$- \frac{1}{2} m_A (v_A{}^2 \sin^2\theta + v_A{}^2 \cos^2\theta) - \frac{1}{2} m_B (v_B{}^2 \sin^2\phi + v_B{}^2 \cos^2\phi) \quad ...(d)$$

Also, $$v_A \sin\theta = u_A \sin\alpha \quad ...(e)$$

or $$v_B \sin\phi = u_B \sin\beta \quad ...(f)$$

∴ Eqn. (*d*) gives

$$\Delta\text{K.E.} = \frac{1}{2} m_A u_A{}^2 \cos^2\alpha + \frac{1}{2} m_B . u_B{}^2 \cos^2\beta - \frac{1}{2} m_A . v_A{}^2 \cos^2\theta - \frac{1}{2} m_B v_B{}^2 \cos^2\phi \quad ...(g)$$

Proceeding as in case of direct impact, and writing

$$\left.\begin{aligned} u_A \cos\alpha &= u_A \\ u_B \cos\beta &= u_B \\ v_A \cos\theta &= u_A \\ \text{and, } v_B \cos\phi &= v_B \end{aligned}\right\} \quad ...(h)$$

Equation (*g*) gives

$$\Delta\text{K.E.} = \frac{m_A m_B}{2(m_A + m_B)} (u_A \cos\alpha - u_B \cos\beta)^2 (1 - e^2) \quad ...(15.16)$$

(*c*) **Impact against a fixed plane**

$$\Delta\text{K.E.} = \frac{1}{2} mu^2 - \frac{1}{2} mv^2 \quad ...(i)$$

We also know that $$e = \frac{0 + v\cos\beta}{u\cos\alpha - 0}$$

or, $$v\cos\beta = eu\cos\alpha \quad ...(j)$$

Also, $u \sin \alpha = v \sin \beta$ ...(k)

$$\therefore \quad \Delta K.E. = \frac{1}{2}mu^2 - \frac{1}{2}mu^2(\sin^2 \alpha + e^2 \cos^2 \alpha)$$

or,
$$\Delta K.E. = \frac{1}{2}mu^2(1 - e^2)\cos^2 \alpha \quad ...(15.17)$$

Problem 15.1. *Two bodies of weight 200 N and 120 N moving in opposite direction at initial velocities of 6 m/s and 8 m/s, respectively, collide with each other. Find their velocities after impact when the coefficient of restitution is 0.75.*

Sol. Fig. P. 15.1 shows the two bodies before and after impact. Let bodies by *A* and *B*.

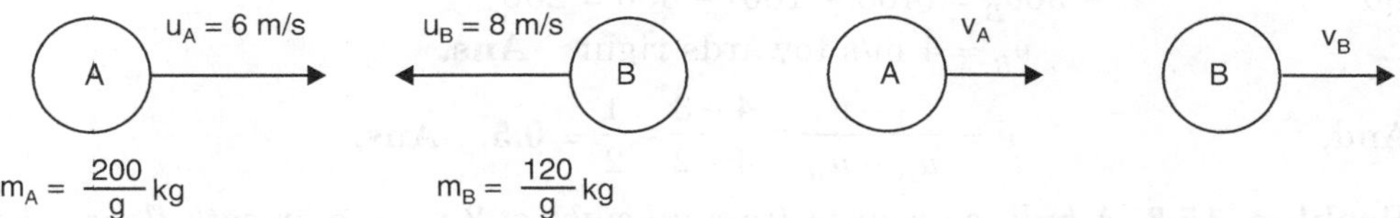

(a) Before impact (b) After impact

Fig. P. 15.1

Here, we note that :

$$m_A = \frac{200}{g} \text{ kg} \; ; \; u_A = 6 \text{ m/s towards right}$$

$$m_B = \frac{120}{g} \text{ kg} \; ; \; u_B = -8 \text{ m/s, (towards left)}$$

Let the final velocity of the mass *A* be v_A and that of mass *B* be v_B, both towards right. From the principle of conservation of momentum, we have

$$\frac{200}{g} \times 6 + \frac{120}{g} \times (-8) = \frac{200}{g} \times v_A + \frac{120}{g} \times v_B$$

or,
$$30 - 24 = 5v_A + 3v_B$$

or,
$$5v_A + 3v_B = 6$$

or,
$$v_A + 0.6v_B = 1.2 \quad ...(1)$$

From the definition of coefficient of restitution, we get

$$e = \frac{v_B - v_A}{u_A - u_B}, \quad \text{or} \quad 0.75 = \frac{v_B - v_A}{6 - (-8)}$$

$$\therefore \quad 14 \times 0.75 = v_B - v_A$$

or,
$$v_B - v_A = 10.50 \quad ...(2)$$

Adding (1) and (2), $1.6v_B = 1.2 + 10.5$

$$\therefore \quad v_B = \frac{11.7}{1.6} = 7.3125 \text{ m/s towards right.} \quad \textbf{Ans.}$$

And from (1),

$$v_A = 1.2 - 0.6v_B = 1.2 - 0.6 \times 7.3125$$

or
$$v_A = \mathbf{-3.1875 \text{ m/s}} \text{ towards left.} \quad \textbf{Ans.}$$

Problem 15.2. *A 100 N body moving at a speed of 4 m/s towards right collides with another body of weight 50 N moving at 2 m/s towards right. The 100 N body has a final* *velocity of 3 m/s towards right. Find the final velocity of the second body and the coefficient of restitution between the two bodies.*

Sol. Let $m_A = \dfrac{100}{g}$; $m_B = \dfrac{50}{g}$; u_A = 4 m/s (towards right) ; u_B = 2 m/s (rightwards) ; v_A = 3 m/s (rightward).

By conservation of momentum, we have

$$m_A.u_A + m_B.u_B = m_A.v_A + m_B.v_B$$

or,
$$\frac{100}{g} \times 4 + \frac{50}{g} \times 2 = \frac{100}{g} \times (3) + \frac{50}{g} \times v_B$$

So
$$+ 50v_B = (400 + 100) - 300 = 200$$

∴
$$v_B = 4 \text{ m/s towards right.} \quad \textbf{Ans.}$$

And,
$$e = \frac{v_B - v_A}{u_A - u_B} = \frac{4-3}{4-2} = \frac{1}{2} = \mathbf{0.5.} \quad \textbf{Ans.}$$

Problem 15.3. *A ball is dropped from a height of 2 m on a smooth floor. The height of ball after the first bounce is 1.62 m. Find the height after the second and third bounces and also find the coefficient of restitution between the ball and the floor. Take g = 9.8 m/s².*

Sol. Initial height, h_0 = 2 m,

Velocity of ball before impact

$$= u_A = \sqrt{2gh_0} = \sqrt{2g \times 2} \text{ m/s}$$
$$= 2\sqrt{g} \text{ m/s} \quad \text{(downwards)}$$

Since the floor is stationary, $u_B = 0$, and $v_B = 0$

Height after first impact = h_1 = 1.62 m.

∴ Final velocity of ball before first impact

$$= v_A = -\sqrt{2gh_1} = -\sqrt{2g \times 1.62} \text{ m (upwards)}$$

By definition,
$$e = \frac{v_B - v_A}{u_A - u_B}$$

or,
$$e = \frac{0 - \sqrt{2gh_1}}{-\sqrt{2gh_1} - 0}$$

∴
$$e = \sqrt{\frac{h_1}{h_0}} = \sqrt{\frac{1.62}{2}} = \sqrt{0.81} = \mathbf{0.9.} \quad \textbf{Ans.}$$

Therefore, height (h_2) after second bounce is given by

$$e = \frac{\sqrt{2gh_2}}{\sqrt{2gh_1}}$$

∴
$$h_2 = h_1(e)^2 = 1.62 \times (0.90)^2 = \mathbf{1.3122 \text{ m.}} \quad \textbf{Ans.}$$

And, height of third bounce is

$$h_3 = e^2 h_2$$
$$= (0.9)^2 \times 1.3122 \text{ m}$$
$$= \mathbf{1.062882} \text{ say } \mathbf{1.063 \text{ m.}} \quad \textbf{Ans.}$$

Problem 15.4. *A ball A of weight 16 N traversing a smooth tube falls through a height of 4 m and strike a ball B of weight 24 N hanging from a 2 m long rope. Find :*

The height to which the ball B will rise if

(a) The two balls are perfectly elastic, and

(b) If the coefficient of restitution is 0.8.

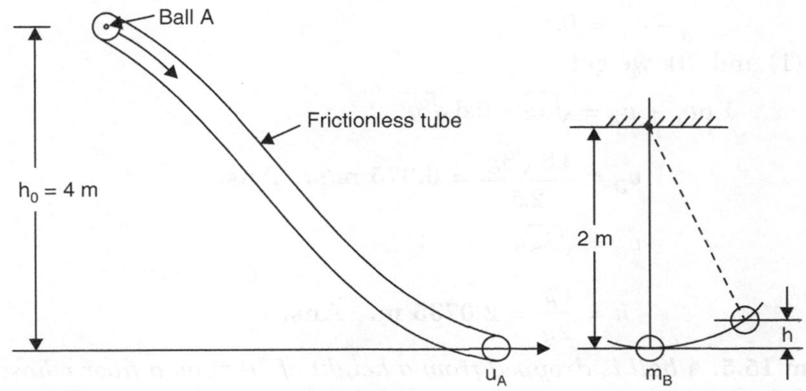

Fig. P. 15.4

Sol. Velocity of ball A before impact $= u_A = \sqrt{2gh_0} = \sqrt{2g \times 4}$ m/s

Velocity of ball B before impact $= u_B = 0$

By the law of conservation of momentum, we have

$$m_A.u_A + m_B.u_B = m_A.v_A + m_B.v_B$$

$$\therefore \quad \frac{16}{g} \times \sqrt{2g \times 4} + \frac{24}{g} \times (0) = \frac{16}{g} \times v_A + \frac{24}{g} \times v_B$$

$$\therefore \quad 16v_A + 24v_B = 16\sqrt{8g}$$

or, $$v_A + 1.5v_B = \sqrt{8g} \qquad ...(1)$$

(*a*) Given : $e = 1$ for perfectly elastic bodies,

$$\therefore \quad 1 = \frac{v_B - v_A}{u_A - u_B}$$

$$\therefore \quad v_B - v_A = 1 \times (u_A - u_B) = \sqrt{8g} - 0$$

$$\therefore \quad v_B - v_A = \sqrt{8g} \qquad ...(2)$$

$\therefore$ Adding (1) and (2), we get

$$2.5v_B = 2\sqrt{8g}$$

$$\therefore \quad v_B = \frac{2\sqrt{8g}}{2.5} = \mathbf{7.0835\ m/sec}$$

If the ball B rises through height h, then we have,

$$v_B = \sqrt{2gh} \ ; \quad \therefore \quad h = \frac{v_B^2}{2g} = \mathbf{2.56\ m}$$

(*b*) $e = 0.8$

$\therefore \quad e = \dfrac{v_B - v_A}{u_A - u_B}$

or, $\quad 0.8 = \dfrac{v_B - v_A}{\sqrt{8g} - 0}$

$\therefore \quad v_B - v_A = 0.8\sqrt{8g}$ ...(3)

Adding (1) and (3), we get

$$1.5v_B + v_B = \sqrt{8g} + 0.8\sqrt{8g}$$

$\therefore \quad v_B = \dfrac{1.8\sqrt{8g}}{2.5} = $ **6.375 m/s. Ans.**

Now, $\quad v_B = \sqrt{2gh}$

$\therefore \quad h = \dfrac{v_B^2}{2g} = $ **2.0735 m. Ans.**

Problem 15.5. *A ball is dropped from a height of 10 m on a floor whose coefficient of restitution is 0.9. Determine the maximum heights reached by the ball after first, fourth and tenth bounce.*

Sol. Initial velocity before the first bounce $= u = -\sqrt{2gh_0}$

Velocity of ball after first bounce $= v_1 = e\sqrt{2gh_0}$

$\therefore$ Maximum height reached after the first bounce $= h_1 = \dfrac{v_1^2}{2g}$

$\therefore \quad h_1 = \dfrac{v_1^2}{2g} = \dfrac{(e^2\sqrt{2gh_0})^2}{2g} = \dfrac{e^2 . 2gh_0}{2g} = e^2 h_0.$

i.e., $\quad h_1 = e^2 h_0$ ...(1)

$\therefore \quad h_1 = h_0 e^2 = 10 \times (0.9)^2 = 8.1$ m **Ans.**

From eqn. (1), we find that

$$e = \sqrt{\frac{h_1}{h_0}} \quad \therefore \quad h_1 = e^2 h_0.$$

$\therefore$ Maximum height after second bounce

$$= h_2 = e^2 h_1 = e^2[e^2 h_0] = e^4 h_0$$

$\therefore$ In general, maximum height reached by the ball after nth bounce

$$= h_n = e^{2n}.h_0 \qquad ...(2)$$

Hence, the maximum height after fourth bounce $= h_4 = e^{2 \times 4} h_0$

or $\quad h_4 = (0.9)^{2 \times 4} \times 10 = 0.43047 \times 10 = $ **4.3047 m. Ans.**

Similarly, the maximum height after 10th bounce

$= h_{10} = e^{2 \times 10} h_0 = (0.9)^{20} \times 10 = $ **1.2158 m. Ans.**

Problem 15.6. *A 20 N sphere is released from rest from the horizontal position as shown in Fig. P.15.6. It strikes a 60 N block after swinging through 90° about the point of* *suspension. If e = 0.8, and the coefficient of friction between the block and the horizontal surface is 0.2 find the distance moved by the block after impact.*

Sol. Initial velocity of sphere before impact = $u_S = \sqrt{2gh_0}$

$$= \sqrt{2g \times 2} = 2\sqrt{g} \text{ m/s}$$
$$= 6.261 \text{ m/s towards left}$$

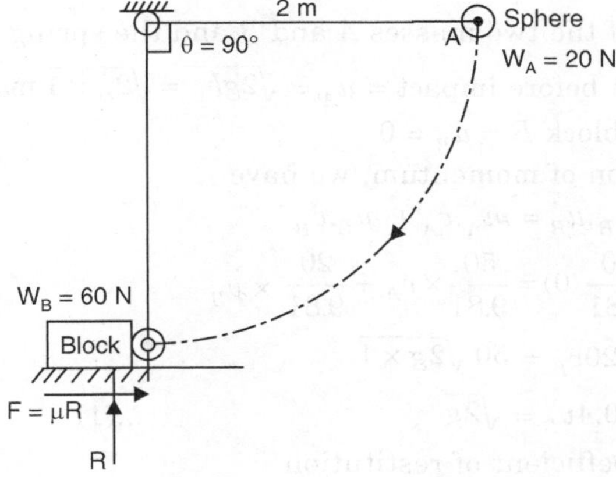

Fig. P. 15.6.

Initial velocity of block = $u_B = 0$

$$\because \quad e = \frac{v_B - v_S}{u_S - u_B}$$

$$\therefore \quad 0.8 = \frac{v_B - v_S}{6.261 - 0}$$

$$\therefore \quad v_B - v_S = 0.8 \times 6.261 = 5.0088 \text{ m/s} \quad ...(1)$$

From the conservation of momentum, we have

$$m_S.u_S + m_B.u_B = m_S \times v_S + m_B.v_B$$

or, $$\frac{20}{g} \times 6.261 + \frac{60}{g} \times (0) = \frac{20}{g} \times v_S + \frac{60}{g} \times v_B$$

$$\therefore \quad 20v_S + 60v_B = 20 \times 6.261 \quad \text{or,} \quad v_S + 3v_B = 6.261 \quad ...(2)$$

Adding (1) and (2), we get

$$4 \times v_B = 6.261 + 5.0088$$

$$\therefore \quad v_B = \frac{11.2698}{4} = \mathbf{2.8175 \text{ m/s}}$$

And from (1) ; $v_S = v_B - 5.0088 = 2.8175 - 5.0088 = -2.1913$

∴ Final velocity of sphere = v_S = **2.1913 m** towards right

Normal reaction of the horizontal surface = $R = W_B = 60$ N.

∴ Force of friction = $F = \mu R = 0.2 \times 60 = 12$ N

Applying work-energy equation, we have

$$F \times S = \frac{W_B}{2g} \times (v_B^2 - u_B^2)$$

$$\therefore \quad 12 \times S = \frac{60}{2 \times 9.81} [(2.8175)^2 - 0] \; ; \; \therefore \; S = \mathbf{2.023 \text{ m.}} \quad \textbf{Ans.}$$

Problem 15.7. *A block A of weight 50 N falls from a height of 1 m on to a horizontal block B of weight 20 N supported on a helical spring of stiffness of 1600 N/m. Neglecting the weight of spring and the air resistance to the motion of the system of spring and the two masses, find the maximum contraction of the spring. Block B is initially at rest. Take e = 0.8.*

Sol. The system of the two masses A and B and the spring is shown in Fig. P. 15.7.

Velocity of block A before impact $= u_A = \sqrt{2gh_0} = \sqrt{2g \times 1}$ m/s

Initial velocity of block $B = u_B = 0$.

By the conservation of momentum, we have

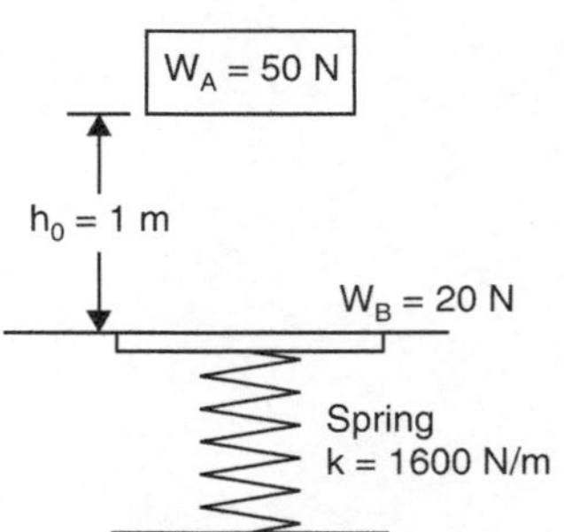

Fig. P. 15.7

$$m_A . u_A + m_B . u_B = m_A . v_A + m_B . v_B$$

i.e.,
$$\frac{50}{9.81} \times \sqrt{2g \times 1} + \frac{20}{9.81}(0) = \frac{50}{9.81} \times v_A + \frac{20}{9.81} \times v_B$$

$$\therefore \quad 50 v_A + 20 v_B = 50\sqrt{2g \times 1}$$

$$\therefore \quad v_A + 0.4 v_B = \sqrt{2g} \qquad \ldots(1)$$

By definition of coefficient of restitution

$$e = \frac{v_B - v_A}{u_A - u_B} \text{, we get}$$

$$0.8 = \frac{v_B - v_A}{\sqrt{2g \times 1}}$$

$$\therefore \quad v_B - v_A = 0.8\sqrt{2g} \qquad \ldots(2)$$

Adding (1) and (2), we get

$$1.4 v_B = (0.8 + 1)\sqrt{2g}$$

$$\therefore \quad v_B = \frac{1.8}{1.4}\sqrt{2g} = \mathbf{5.695 \text{ m/s.}} \quad \textbf{Ans.}$$

And from (1), $\quad v_A = \sqrt{2g} - 0.4 \times 5.695 = \mathbf{2.151 \text{ m/s.}}$ **Ans.**

Let the maximum contraction of spring be δ.

Then gain in energy stored in spring = loss in K.E. of block B

or,
$$\frac{1}{2} K\delta^2 = \frac{1}{2} \times \frac{W_B}{g} [v_B^{\,2} - u_B^{\,2}]$$

or,
$$\frac{1}{2} \times 1600 \times (\delta)^2 = \frac{1}{2} \times \frac{20}{9.81} \times [(5.695)^2 - 0]$$

$$\therefore \quad \delta^2 = \frac{20}{2 \times 9.81}[(5.695)^2] / \frac{1600}{2} = 0.0413265 \text{ m}^2$$

$$\therefore \quad \delta = \mathbf{0.2033 \text{ m.}} \quad \textbf{Ans.}$$

Problem 15.8. *A ball of mass (m_B) 4 kg moving with a velocity of 6 m/s strikes another body of mass (m_A) 2 kg and velocity 3 m/s, (a) both moving in the same direction and (b) the two balls moving in opposite directions. If e = 0.4, find the velocity of the two balls after impact. Also find the loss in kinetic energy during impact.*

Sol. Given : $m_A = 2$ kg ; $m_B = 4$ kg,

(*a*) $u_A = 3$ m/s ; $u_B = 6$ m/s.

(*b*) $u_A = 3$ m/s ; $u_B = -6$ m/s.

$e = 0.4.$

(*a*) By law of conservation of momentum, we get

$$m_A u_A + m_B u_B = m_A v_A + m_B v_B,$$

or $$2 \times 3 + 4 \times 6 = 2v_A + 4v_B.$$

$\therefore$ $$v_A + 2v_B = 15 \quad \text{...(1)}$$

Also, $$e = \frac{v_B - v_A}{u_A - u_B} \text{ gives, } 0.4 = \frac{v_B - v_A}{3 - 6}$$

$\therefore$ $$v_B - v_A = -1.2 \quad \text{...(2)}$$

Adding (1) and (2), we get

$$3v_B = 15 - 1.2$$

$\therefore$ $$v_B = \frac{13.8}{3} = 4.6 \text{ m/s.} \quad \textbf{Ans.}$$

$\therefore$ From (1), $$v_A = 15 - 2v_B = 15 - 2 \times 4.6 = 15 - 9.2$$

$\therefore$ $$v_A = 5.8 \text{ m/s.} \quad \textbf{Ans.}$$

$$\Delta\text{K.E.} = \frac{m_A . m_B}{2(m_A + m_B)}(u_A - u_B)^2 (1 - e^2)$$

$$= \frac{2 \times 4}{2(2 + 4)}(3 - 6)^2 (1 - 0.4^2)$$

$$= \frac{4}{6} \times 9 \times (1 - 0.16) = 6 \times 0.84 = \textbf{5.04 Nm.} \quad \textbf{Ans.}$$

(*b*) By the conservation of momentum :

$$m_A . u_A + m_B . u_B = m_A . v_A + m_B . v_B, \text{ we get}$$

$$2 \times 3 + 4 \times (-6) = 2 \times v_A + 4 \times v_B$$

$\therefore$ $$v_A + 2v_B = -9 \quad \text{...(3)}$$

Also $\because$ $$e = \frac{v_B - v_A}{u_A - u_B},$$

$\therefore$ We get $$0.4 = \frac{v_B - v_A}{3 - (-6)}$$

$\therefore$ $$v_B - v_A = 0.4 \times 9 = 3.6 \quad \text{...(4)}$$

Adding (3) and (4), we get

$$3v_B = -9 + 3.6 = -5.4$$

$\therefore$ $$v_B = -1.8 \text{ m/s.} \quad \textbf{Ans.}$$

and $$v_A = -9 - 2v_B = -9 - 2 \times (-1.8) = -9 + 3.6$$

$\therefore$ $$v_A = -5.4 \text{ m/s.} \quad \textbf{Ans.}$$

i.e., both balls move in opposite directions after impact.

And $$\text{K.E.} = \frac{m_A m_B}{2(m_A + m_B)}(u_A - u_B)^2 (1 - e^2)$$

$$= \frac{2 \times 4}{2 \times (2+4)} \times [3 - (-6)]^2 \{1 - (0.4)^2\} \text{ N-m}$$

$$= \frac{2 \times 4}{2 \times 6} \times 81 \times (1 - 0.16) \text{ Nm}$$

$$= \frac{2 \times 4}{2 \times 6} \times 81 \times 0.84 \text{ Nm}$$

$\therefore$ K.E. = **45.36 Nm. Ans.**

Problem 15.9. *A ball weighing 0.8 kg and moving with an initial velocity of 4 m/s impinges on a fixed horizontal plane at an angle of 30° to the plane. Find velocity and direction of motion of the ball after impact if the coefficient of restitution is 0.8. Also find the loss in kinetic energy and the impulse on the plane during the impact.*

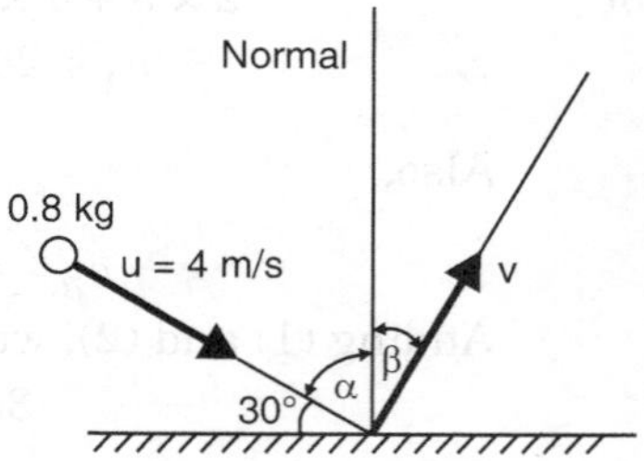

Fig. P. 15.9

Sol. See Fig. P. 15.9.

Here, $u = 4$ m/s ; $m = 0.8$ kg

$\alpha = 90° - 30° = 60°$, $e = 0.8$.

By conservation of momentum parallel to plane, we have

$$u \sin \alpha = v \sin \beta, \quad \therefore \quad 4 \sin 60° = v \sin \beta$$

or, $$\frac{4\sqrt{3}}{2} = v \sin \beta \quad \text{or} \quad v \sin \beta = 2\sqrt{3} \qquad \text{...(1)}$$

$$e = \frac{0 - (-v)\cos\beta}{u \cos\alpha - 0}\ ; \quad \therefore \quad 0.8 = \frac{0 + v\cos\beta}{4\cos 60° - 0}$$

$\therefore$ $$v \cos \beta = 0.8 \times 4 \times 0.5 = 1.6 \qquad \text{...(2)}$$

By (1) ÷ (2) we get $\tan \beta = \dfrac{2\sqrt{3}}{1.6} = 2.1651$

$\therefore$ $\beta = 65.21°$. **Ans.**

$\therefore$ $v = \dfrac{1.6}{\cos\beta} = \dfrac{1.6}{\cos 65.21°} =$ **3.82 m/s. Ans.**

Loss in kinetic energy,

$$\Delta\text{K.E.} = \frac{1}{2} m(u^2 - v^2) = \frac{1}{2} \times 0.8(4^2 - 3.82^2) \text{ Nm}$$

or ΔK.E. = **0.56304 Nm. Ans.**

Impulse = $m(1 + e)\, u \cos \alpha$

= 0.8(1 + 0.8)(4 cos 60°) kgm/s

= 0.8 × 1.8 × 2 kgm/s

$\therefore$ Impulse = 2.88 kgm/s. **Ans.**

Problem 15.10. *A ball falls on a horizontal. If the velocity of the ball after impact is u inclined at an angle θ to the plane, show that ball will stop rebounding after it has travelled a distance* $\dfrac{u^2 \sin 2\theta}{g(1-e)}$*, and the time elapsed is* $\dfrac{2u \sin\theta}{g(1-e)}$ *where, e is the coefficient of restitution.*

Sol. Fig. P. 15.10 shows the initial velocity u after the impact. It also shows the velocity after successive rebounds.

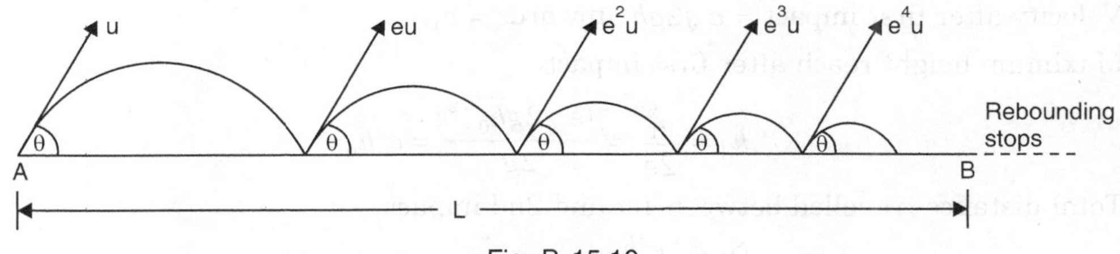

Fig. P. 15.10

Vertical component of initial velocity $= u_y = u \sin \theta$

Horizontal component of initial velocity $= u_x = u \cos \theta$

Due to impact only the vertical velocity (v_y) will be effected. The horizontal velocity v_x remains unchanged.

Range R, during 1st flight $= \dfrac{2u \sin \theta}{g}$.

The time T_1, for the first flight of ball $= \dfrac{2u \sin \theta}{g}$,

Time for 2nd flight $= T_2 = \dfrac{2\,(eu) \sin \theta}{g}$.

Time for 3rd flight $= T_3 = \dfrac{2\,(e^2u) \sin \theta}{g}$

$\therefore$ Total time of flight till the ball stops

$$T = T_1 + T_2 + T_3 + \ldots\ldots$$

$$= \frac{2u \sin \theta}{g} + \frac{2\,(eu) \sin \theta}{g} + \frac{2\,(e^2u) \sin \theta}{g} + \ldots\ldots$$

$$= \frac{2u \sin \theta}{g}\,[1 + e + e^2 + e^3 + \ldots\ldots]$$

$$\therefore \quad T = \frac{2u\,(\sin \theta)}{g\,(1-e)}. \quad \textbf{Proved.} \qquad \ldots(a)$$

Since the horizontal velocity $u_x = u \cos \theta$, remains unaffected, therefore the total distance travelled for the ball after it stops rebouncing.

$$L = T \times u \cos \theta$$

$$\therefore \quad L = \frac{2u\,(\sin \theta)}{g\,(1-e)} \times u \cos \theta$$

$$\therefore \quad L = \frac{u^2 \sin 2\theta}{g\,(1-e)}. \quad \textbf{Proved.} \qquad \ldots(b)$$

Problem 15.11. *A ball falls from a height h_0 on to a fixed horizontal plane. If the coefficient of restitution is e prove that the total distance travelled by the ball before it stops rebouncing is $\dfrac{1+e}{1-e}h_0$ and that the time elapsed is $\dfrac{1+e^2}{1-e^2}\sqrt{\dfrac{2h}{g}}$ seconds.*

Sol. Velocity before first impact = $u_0 = \sqrt{2gh_0}$, downwards.

Velocity after first impact = $e\sqrt{2gh}$ upwards = v_1.

Maximum height reach after first impact

$$h_1 = \frac{v_1^2}{2g} = \frac{(e\sqrt{2gh_0})^2}{2g} = e^2 h_0$$

Total distance travelled between 1st and 2nd impact

$$S_1 = 2e^2 h_0$$

Velocity before 2nd impact = $e\sqrt{2gh_0}$ downwards

Velocity after 2nd impact = $v_2 = e(e\sqrt{2gh_0}) = e^2 u$, upwards,

Maximum height reached after 2nd impact = $\dfrac{v_2^2}{2g} = e^4 u$

∴ Total distance travelled between 2nd and 3rd impacts = $S_2 = 2e^4 h_0$

Similarly, total distance travelled between 3rd and 4th impacts = $S_3 = 2e^6 h_0$

Distance travelled before the first impact = $S_0 = \dfrac{1}{2}\left[\dfrac{(\sqrt{2gh})^2}{2g}\right] = h$

∴ Total distance travelled before the ball stops rebouncing

$$S = S_0 + S_1 + S_2 + S_3 + \ldots\ldots \text{ at infinity}$$
$$= h + 2e^2 h + 2e^4 h + 2e^6 h + \ldots\ldots$$
$$= h + \frac{e^2}{1-e^2} h$$

∴ $$S = \left(\frac{1+e^2}{1-e^2}\right) h. \quad \textbf{Hence proved.}$$

Time to fall freely from $h_0 = t_0 = \sqrt{\dfrac{2h_0}{g}}$;

Time between 1st and 2nd impacts = $t_1 = 2e\dfrac{v_1}{g} = 2e \times \dfrac{\sqrt{2gh_0}}{g} = 2e\sqrt{\dfrac{2h_0}{g}} = 2et_0$

Time between 2nd and 3rd impacts = $t_2 = 2e^2\sqrt{\dfrac{2h_0}{g}} = et_1$

Similarly, time between 3rd and 4th impacts = $t_3 = et_2 = 2e^3\sqrt{\dfrac{2h_0}{g}}$

∴ Total time elapsed = $t_0 + t_1 + t_2 + t_3 + \ldots\ldots$ ad infinity

or $$T = \sqrt{\frac{2h_0}{g}} + 2e\sqrt{\frac{2h_0}{g}} + 2e^2\sqrt{\frac{2h_0}{g}} + 2e^3\sqrt{\frac{2h_0}{g}} + \ldots\ldots$$

$$= \sqrt{\frac{2h_0}{g}} + 2e\,(1 + e + e^2 + e^3 + ...)\sqrt{\frac{2h_0}{g}}$$

$$= \sqrt{\frac{2h_0}{g}} + 2e \times \frac{1}{1-e}\sqrt{\frac{2h_0}{g}}$$

$\therefore$
$$T = \sqrt{\frac{2h_0}{g}}\left[\frac{(1-e)+2e}{(1-e)}\right]$$

or
$$T = \left(\frac{1+e}{1-e}\right)\sqrt{\frac{2h_0}{g}}.\ \textbf{Hence proved.}$$

Problem 15.12. *A steel sphere is dropped from a height of 4.9 m on to a horizontal surface. If coefficient of restitution is 0.8 find the total distance travelled by the sphere before it stops rebouncing. Also find the time elapsed. Take g = 9.8 m/s².*

Sol. $\boxed{\text{Total distance travelled} = \frac{1+e^2}{1-e^2}h = S}$

or,
$$S = \frac{1+(0.8)^2}{1-(0.8)^2} \times 4.9 = 22.222 \text{ m.} \quad \textbf{Ans.}$$

$$\text{Total time} = T = \frac{1+e}{1-e}\sqrt{\frac{2h_0}{g}}$$

$$= \frac{1+0.8}{1-0.8} \times \sqrt{\frac{2 \times 4.9}{9.8}} \text{ seconds}$$

$$= 9 \text{ seconds.} \quad \textbf{Ans.}$$

Problem 15.13. *A ball falls vertically from a height of 4.9 m upon a 30° inclined. Find the resultant velocity of the ball after impact, if e = 0.9. See Fig. P. 15.13. Take g = 9.8 m/s².*

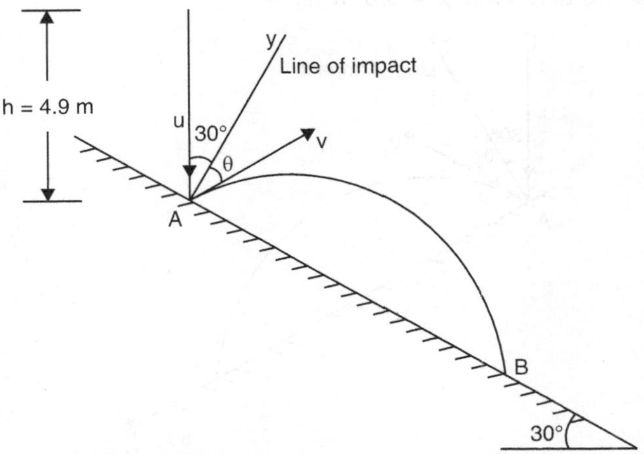

Fig. P. 15.13

Sol. This is a case of oblique central impact on a fixed plane. Initial velocity before impact = $u_1 = \sqrt{2gh_0} = \sqrt{2 \times 9.8 \times 4.9}$ = 9.8 m/s. The line of impact is perpendicular to the plane at *A*, *i.e.*, inclined at 30° to the vertical.

Let x-axis be parallel to plane and y-axis be perpendicular to the plane.

$\therefore \qquad u_{1x} = u_1 \sin 30° = 9.8 \times 0.5 = 4.9$ m/s

and $\qquad u_{1y} = u_1 \cos 30° = 9.8 \times \frac{\sqrt{3}}{2} = 4.9\sqrt{3}$ m/s

= initial velocity along line of impact

Let v_1 be the final velocity after impact.

We know that velocity in the direction parallel to the plane remains unaffected.

$\therefore \qquad v_{1x} = u_{1x}$ or $v_{1x} = 4.9$ m/s

For the fixed plane, the initial and final velocities are zero.

$$\because \qquad e = \frac{v_{2y} - v_{1y}}{u_{1y} - u_{2y}} \text{ along the line of impact}$$

$$\therefore \qquad 0.9 = \frac{0 - v_{1y}}{4.9\sqrt{3} - 0}$$

$$\therefore \qquad v_{1y} = -\,0.9 \times 4.9\sqrt{3} \text{ m/s upwards.}$$

$$\therefore \quad \text{Total velocity of the ball} = v_1 = \sqrt{v_{1x}^2 + v_{1y}^2}$$

or
$$v = \sqrt{(4.9)^2 + (-\,0.9 \times 4.9 \times \sqrt{3})^2}$$

$$= 4.9\sqrt{1 + 2.43} = 4.9\sqrt{3.43} = \mathbf{9.075 \text{ m/s.}} \quad \textbf{Ans.}$$

And,
$$\tan \theta_y = \frac{v_{1x}}{v_{1y}} = \frac{4.9}{0.9 \times 4.9 \times \sqrt{3}} = 0.6415$$

$$\theta_y = \mathbf{32.68°} \text{ from } y\text{-axis.} \quad \textbf{Ans.}$$

or, $\qquad (90° - 32.68°) =$ **59.32° from the plane. Ans.**

Problem 15.14. *A ball falls vertically after 5 second from rest from the point of release. It falls on a 30° incline. If e = 0.9 find the distance between the first two impacts of ball. Also find velocity after first impact. Take g = 9.8 m/s.*

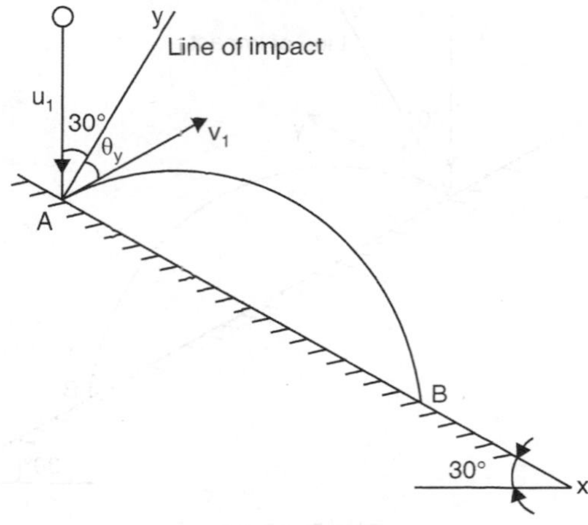

Fig. P. 15.14

Sol. Initial velocity of ball = $u_1 = 0 + gt = 9.8 \times 5$ m/s = 49.0 m/s vertically down

$$u_{1x} = u_1 \sin 30° = 49 \times 0.5 = 44.5 \text{ m/s, along the plane}$$

$$u_{1y} = u_1 \cos 30° = 49 \frac{\sqrt{3}}{2} = 44.5\sqrt{3} \text{ m/s, along the line of impact.}$$

Let v_1 be the velocity of ball after impact

$\therefore$ $v_{1x} = u_{1x} = 44.5$ m/s.

u_2 and v_2 for the plane are zero.

$\because$ $$e = \frac{v_{2y} - v_{1y}}{u_{1y} - u_{2y}}$$

$\therefore$ $$0.9 = \frac{0 - v_{1y}}{44.5\sqrt{3} - 0}$$

$\therefore$ $v_{1y} = -0.9 \times 44.5\sqrt{3}$ m/sec upwards.

Now, $g_y = g \cos 30° = -9.8 \cos 30° = 4.9\sqrt{3}$ m/sec^2

And $g_x = g \sin 30° = -4.9$ m/s^2

At point B, $x = AB$ and $y = 0$.

For point B, $$y = v_{1y}\, t + \frac{1}{2} g_y \times t^2$$

$\therefore$ $$0 = (0.9 \times 44.5\sqrt{3})t + \frac{1}{2}(-4.9\sqrt{3})t^2$$

$\therefore$ $$t = \frac{0.9 \times 44.5\sqrt{3}}{\frac{1}{2} \times 44.9\sqrt{3}} = \frac{0.9 \times 44.5 \times 2}{4.9} = 16.347 \text{ seconds.}$$

$\therefore$ $$AB = v_{1x} \times t + \frac{1}{2} g_x . t^2$$

$$= 44.5 \times 16.347 + \frac{1}{2} \times (-4.9) \times (16.347)^2 \text{ m}$$

$= \mathbf{72.74}$ **m. Ans.**

Also, $$v_1 = \sqrt{v_{1x}^2 + v_{1y}^2} = \sqrt{(44.5)^2 + (44.5 \times 0.9\sqrt{3})^2} \text{ m/s}$$

$= 44.5\sqrt{1 + 2.43} = 44.5 \times 1.852$ m/s $= \mathbf{82.414}$ **m/s. Ans.**

$$\tan \theta_y = \frac{v_{1x}}{v_{1y}} = \frac{44.5}{44.5 \times 0.9\sqrt{3}} = 0.6415$$

$\therefore$ $\theta_y = 32.68°$ from normal or $(90 - 32.68) = 57.32°$ from the plane. **Ans.**

Problem 15.15. *A sphere A of mass m_A and initial velocity u_A strikes another sphere B of mass m_B travelling in the same direction with velocity u_B. If after an elastic impact, spheres have equal and opposite velocities v_A and v_B determine the ratio of m_B/m_A.*

Sol. By the law of conservation of momentum, we have

$$m_A u_A + m_B . u_B = m_A v_A + m_B . v_B$$

But, it is given that v_A and v_B are opposite to each other and equal in magnitude,

∴ We get

$$m_A u_A + m_B . u_B = m_A . v + m_B (-v) \qquad ...(1)$$

Also, by $e = \dfrac{v_B - v_A}{u_A - u_B}$, we get

$$1 = \frac{v - (-v)}{u_A - u_B}, \quad \text{or} \quad u_A - u_B = 2v \qquad ...(2)$$

From (2), $$v = \frac{u_A - u_B}{2}$$

∴ Eqn. (1) becomes

$$m_A . u_A + m_B . u_B = -(m_A - m_B) \times v$$

∴ $$m_A(u_A + v) = m_B(v - u_B)$$

or, $$m_A \left(u_A + \frac{u_A - u_B}{2} \right) = m_B \left(\frac{u_A - u_B}{2} - v_B \right)$$

or $$\frac{m_A}{2}(3u_A - u_B) = m_B(u_A - 3u_B)$$

∴ $$\frac{m_B}{m_A} = \left(\frac{3u_A - u_B}{u_A - 3u_B} \right) \quad \textbf{Ans.}$$

EXERCISES XV

A. Theoretical Questions

1. Define the terms : impact, line of impact.
2. Illustrate with neat sketches : central impact, direct impact, oblique impact, eccentric impact, direct central impact, oblique central impact, direct eccentric impact and oblique eccentric impact.
3. Define the terms : coefficient of restitution, velocity of approach and velocity of separation.
4. Describe different stages of impact.
5. For direct central impact find the velocity of two masses after impact.
6. Derive an expression to find loss of kinetic energy during direct central impact.
7. Derive expressions for (*a*) coefficient of restitution and (*b*) velocity of the two bodies after oblique impact.
8. Derive expression for the loss of kinetic energy of the two masses after oblique impact.
9. Discuss the case of impact against the fixed plane. Also find the loss of kinetic energy during impact.

B. Numerical Problems

15.1. A golf ball is dropped from a height of 40 m on a steel plate. The coefficient of restitution is 0.9. Determine the height to which the ball rebounds on the first, second and fourth rebounces. Take $g = 9.8$ m/s^2. [**Ans.** $h_1 = 3.24$ m, $h_2 = 2.6244$ m, $h_3 = 2.126$ m, $h_5 = 1.3947$ m]

15.2. A 15 N sphere released from rest swings through 90° and strikes a 90 N block resting a horizontal plane. If $e = 0.8$ and $\mu = 0.2$ find the distance through which the block will move before coming to rest. The sphere is connected to a hinged support through a 1.6 m rod.

[**Ans.** 0.529 m]

15.3. 90 N and 180 N bodies are approaching each other with velocities of 10 m/s and 5 m/s, respectively. Find velocity of each body after impact. Also find the loss of kinetic energy. Take $e = 0.8$. [**Ans.** Velocity of 90 N body = v_1 = 4 m/s ; v_2 = 4 m/s, 90 N body moves in reversed direction ; ΔK.E. = 468.37 Nm]

15.4. Central impact occurs between a 50 N ball and a stationary block of 80 N. Find the velocity of the ball and the block after impact if the ball is moving at 10 m/s. Take $e = 0.6$. If $\mu = 0.2$ between the block and its plane, find the distance through which it moves before stopping. [**Ans.** Velocity of ball = v_1 = 0.154 m/s ; v_2 = 6.154 m/s S = 9.661 m]

15.5. A steel ball falls from the height h on a fixed floor and in its second rebounce it attains a height of $h/2$. Find the coefficient of restitution between the ball and the fixed floor. Also find the percentage of loss of kinetic energy after second impact. [**Ans.** e = 0.841 ; 50%]

15.6. A ball drops through a height h_0 on to a smooth horizontal floor and rebounds successively. If e is the coefficient of restitution find (*a*) R_n/R_{n-1}, (*b*) R_n/R_0 ; (*c*) t_n/t_1 and (*d*) the percentage loss of K.E. after fourth impact. R = Range and t_n = time of flight after nth impact ; t_1 = time for 1st impact. [**Ans.** (*a*) e ; (*b*) e^n, (*c*) e^n ; (*d*) $\frac{(1-e^8)}{1} \times 100\%$]

15.7. A ball of mass 1.2 kg moving rightwards with a velocity of 5 m/s strikes another ball of mass 3 kg moving with a velocity of 2.5 m/s leftwards. If the angles of inclination of the motion of the two ball are 30° upward and 30° down, respectively find their velocities after the impact, if $e = 0.8$. [**Ans.** For 1.2 kg ball v_1 = 2.526 m/s, θ = – 81.83° from line of impact ; For 3 kg ball v_2 = 1.588 m/s ; ϕ = – 82.66° from line of impact]

15.8. A ball of mass m moving with a velocity of 10 m/s strikes a vertical wall at 30° from the horizontal and rebounds. If $e = 0.6$ find the velocity and its direction after impact.
[**Ans.** 7.21 m ; 43.9° on the opposite side of normal to the plane]

15.9. A ball moving with a velocity of 4 m/s strikes another identical stationary ball such that after impact the direct of each ball makes an angle of 30° with original line of motion. Find the speed of the two balls after the impact. Find the loss of kinetic energy during collision.

15.10. The ram of a pile hammer driver has a mass of 600 kg and is released from rest 2 m above the top of the 1800 kg pile. If the ram rebounds to a height of 0.1 m after impact with the pile, determine (*a*) the velocity v_p of the pile just of impact, (*b*) the coefficient of restitution e, and (*c*) the percentage loss of energy due to impact.
[**Ans.** (*a*) 2.553 ; (*b*) e = 0.6315 ; (*c*) 45.10%]

15.11. Spherical particle 1 has a velocity u_1 = 9 m/s vertical and collides with spherical particle 2 of equal mass and equal diameter and initially at rest. If $e = 0.8$, find the resulting velocity of each particle after impact. Also find the percentage loss of energy due to impact. See Fig. E. 15.11. [**Ans.** v_1 = 4.57 m/s, θ_1 = 9.826 from common tangent ; v_2 = 7.015 m/s ; along the line of centres ; ΔK.E. = 14.74%]

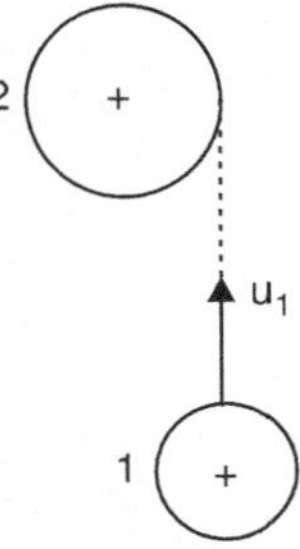

Fig. E.15.11

15.12. A man throws a ball from point A with a velocity of 20 m/s as shown in Fig. E. 15.12. It strikes the wall at point B and returns exactly to point A. Find the necessary angle α if the coefficient of restitution is 0.5. [**Ans.** 78.4° or 11.6°]

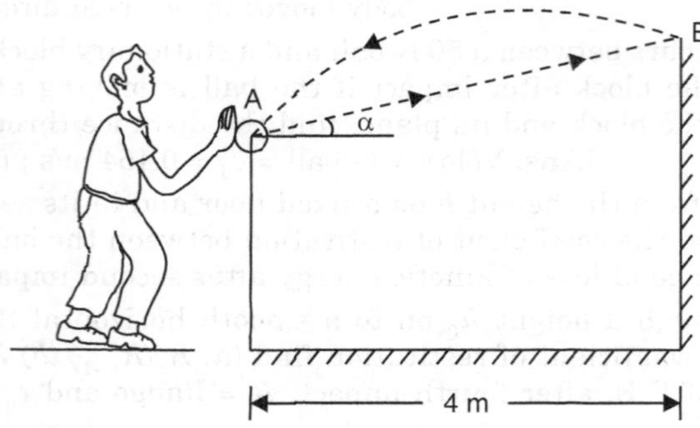

Fig. E. 15.12

15.13. Three balls A, B and C of mass 10, 20 and 40 respectively move along the same straight line and in the same directions with velocity 24 m/s, 6 m/s and 4.5 m/s. If A collides with B, and subsequently B collides with C, find the final velocity of the three balls. Assume perfectly elastic impacts $v_A = 15.75$ m/s ; $v_B = 5.25$ m/s in opposite direction and $v_C = 2.4$ m/s.

15.14. Two identical balls B and C are at rest when ball B is struck by a ball A of the same mass with a velocity 9 m/s. This causes a series of collisions between the three balls. Knowing that $e = 0.6$ find the velocity of each ball after all collisions have taken place. Final velocities $v_A = 1.504$ m/s ; $v_B = 2.966$ m/s and $v_C = 4.53$ m/s in the same directions.

15.15. A sphere impinges directly on another sphere of the size which is at rest. If e is coefficient of restitution, show that their velocities after impact are in the ratio $(1 - e) : (1 + e)$.

15.16. If m_1 and m_2 are masses of the two spheres ; u_1, u_2 and v_1, v_2 are the velocities of two sphere before and after impact, show that the kinetic energy lost due to the central impact is

$$\frac{m_1 m_2}{2(m_1 + m_2)} [(u_1 - u_2)^2 . (1 - e^2)]$$

16

Virtual Work

16.1. DEFINITION AND MEANING OF WORK

Consider a particle P moving along the path under the application of the force F, as shown in Fig. 16.1(a) (i). Particle moves through ds and inclination of force F with ds is equal to θ. Differential workdone dU is equal to the projection ($F \cos \theta$) of force along displacement multiplied by ds. Or, dU is equal to dot product of vector $\overline{F}$ and vector $\overline{ds}$. Thus, we have

From Fig. 16.1(a) (ii), $dU = (F \cos \theta)\, ds$ ...(a)

Also, from Fig. 16.1 (a) (iii), $dU = F\,(ds \cos \theta)$ ...(b)

Combining Eqns. (a) and (b), we can write

$$dU = (Fds) \cos \theta = \overline{F} \cdot \overline{ds} \qquad ...(16.1)$$

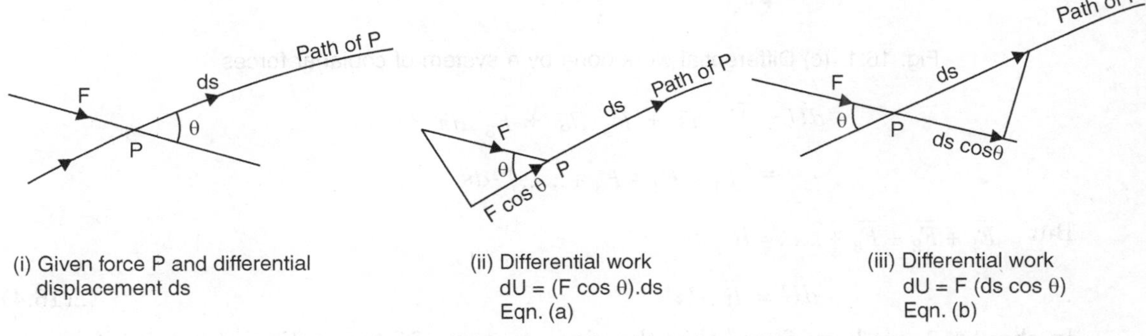

Fig. 16.1 (a) Definition and meaning of workdone by a force F.

In Eqn. (a), dU = Projection $F \cos \theta$ of force along displacement multiplied by displacement ds.

In eqn. (b), dU = force F multiplied by the projection ($ds \cos \theta$) of displacement of particle P.

As a second case consider a tangential force F_t which produces a displacement ds along the arc of radius r. As the length of arc $ds = r\, d\theta$, as shown in Fig. 16.1(b).

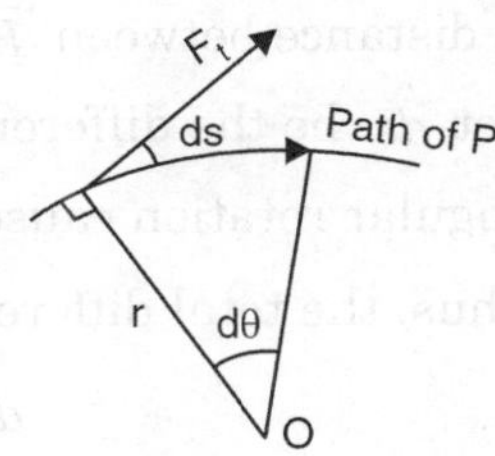

Fig. 16.1. (b) Definition and meaning of work by a couple or moment M.

Differential work (dU) done by the tangential force F_t is

$$dU = F_t \times ds = F_t\,(r\,d\theta)$$

But $F_t.r$ = moment M of F_t about the centre of curvature θ. Thus, we can write

$$dU = F_t(r\,.\,d\theta) = \overline{M}\,.\,d\overline{\theta} \qquad ...(16.2)$$

Equation (16.2) shows that the differential work dU is equal to the dot product of vector $\overline{M}$ and $d\overline{\theta}$.

Combining Eqn. (16.1) and (16.2), we have

$$dU = \overline{F}\,.d\overline{s} + \overline{M}\,.d\overline{\theta} \qquad ...(16.3)$$

Next let us consider a system of coplanar forces $\overline{F_1}$, $\overline{F_2}$, $\overline{F_3}$, etc. acting on a particle O, as shown in Fig. 16.1(c). Let $\overline{R}$ be the resultant of these forces and let $d\overline{s}$ be the differential displacement. Thus, the differential work dU done by these forces is

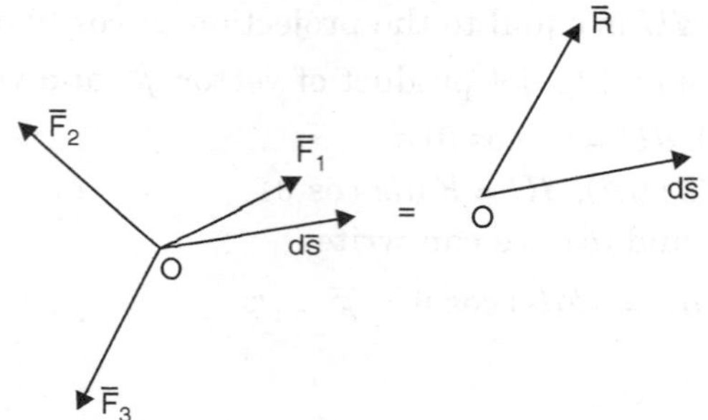

Fig. 16.1. (c) Differential work done by a system of coplanar forces.

$$dU = \overline{F_1}\,.d\overline{s} + \overline{F_2}\,.d\overline{s} + \overline{F_3}\,.d\overline{s} + ...$$

$$= (\overline{F_1} + \overline{F_2} + \overline{F_3} + ...)\cdot d\overline{s}$$

But $\overline{F_1} + \overline{F_2} + \overline{F_3} + ... = \overline{R}$

$$\therefore \qquad dU = \overline{R}\,.d\overline{s} \qquad ...(16.4)$$

In chapter 1, we have found that the given system of forces acting at a point A can be replaced by their resultant $\overline{R}$ plus a couple $\overline{M}$ applied to a point B, where $M = \overline{R}d$ and d is shortest distance between $\overline{R}$ and the point B, as shown in Fig. 16.1(d).

Let $d\overline{s}$ be the differential displacement at point B due to $\overline{R}$. And let $d\overline{\theta}$ be the differential angular rotation caused by the couple $M = \overline{R}\,.\,d$.

Thus, the total differential work done by $\overline{R}$ and the $\overline{M}$ is

$$dU = \overline{R}\,.d\overline{s} + \overline{M}\,.d\overline{\theta} \qquad ...(16.5)$$

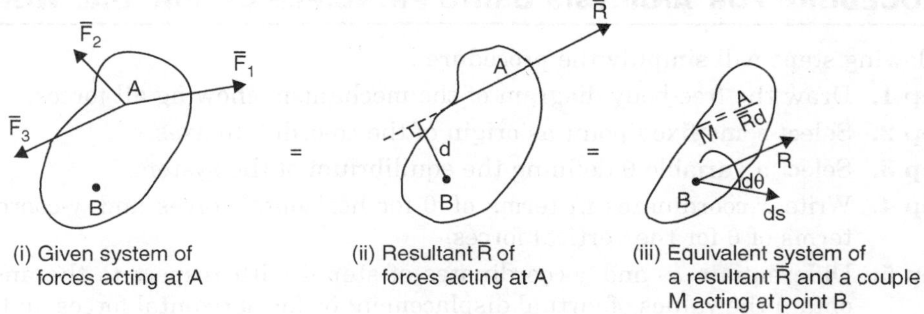

Fig. 16.1(*d*) Differential work done by an equivalent system of a force and a couple.

16.2. PRINCIPLE OF VIRTUAL WORK

Consider a system of concurrent coplanar forces acting on a particle O in equilibrium, as shown in Fig. 16.2.

Let the particle O, which is in equilibrium, be given a virtual (or an imaginary) displacement $ds = OO'$ along x-axis. The workdone by the system of given forces is called **virtual work** (dU).

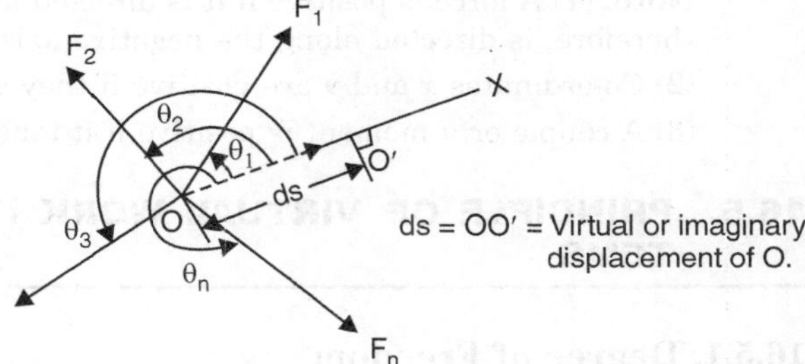

Fig. 16.2. Principle of Virtual Work.

$\therefore \quad dU = (F_1 \cos \theta_1)\, ds + (F_2 \cos \theta_2) \,.\, ds + \ldots\ldots + (F_n \,.\, \cos \theta_n) \,.\, ds$

or $\quad dU = (F_1 \cos \theta_1 + F_2 \cos \theta_2 + \ldots\ldots + F_n \cos \theta_n) \,.\, ds = \Sigma F_x \,.ds$

Since the particle O is in equilibrium $\therefore \; \Sigma F_x = 0$

Hence $\quad dU = \Sigma \mathbf{F_x} \,.\, \mathbf{ds} = \mathbf{0}$ ...(16.6)

Forces F_1, F_2, ..., F_n etc. are externally applied to the particle and these forces are called **active forces**.

When the system of forces and couples is applied on rigid bodies or connected bodies then the equation of virtual work becomes

$$dU = \sum_{i=1}^{n} \bar{F}_i \,.\, d\bar{s} + \sum_{i=1}^{n} \bar{M}_i \,.\, d\bar{\theta} = 0 \qquad ...(16.7)$$

where $d\theta$ is virtual rotation of the rigid body.

Principle of virtual work [Eqn. (16.6)] states that **if a particle is in equilibrium, the total work of the forces acting on the particle is zero for any virtual displacement of the particle.**

16.3. APPLICATIONS OF VIRTUAL WORK

The principle of virtual work may be applied to the following problems :

(1) Systems of connected bodies with internal hinges or to mechanisms having either (*a*) single degree of freedom, or (*b*) multiple degrees of freedom.

(2) Beams and compound beams.

16.4. PROCEDURE FOR ANALYSIS USING PRINCIPLE OF VIRTUAL WORK

Following steps will simplify the procedure :

Step 1. Draw the free-body diagram of the mechanism showing all forces.

Step 2. Select some fixed point as origin of the co-ordinate axes.

Step 3. Select a variable θ defining the equilibrium of the system.

Step 4. Write x-coordinates in terms of θ for horizontal forces and y-coordinates in terms of θ for the vertical forces.

Step 5. Differentiate x- and y-coordinates of step 4 with respect to the angle θ and obtain the values of virtual displacement δx for horizontal forces and δy for the vertical forces.

Step 6. Write down equation of virtual work $dU = 0$ for all applied forces.

Step 7. Now using equation $dU = 0$ find the required result for given value of θ or find θ for equilibrium of the given system.

Note. (1) A force is positive if it is directed along the position co-ordinate axis. A negative force, therefore, is directed along the negative axis.

(2) Co-ordinates x and y are positive if they are along positive axes.

(3) A couple or a moment is positive if it tends to increase angle θ.

16.5. PRINCIPLE OF VIRTUAL WORK FOR MULTI-DEGREE OF FREEDOM SYSTEMS

16.5.1. Degree of Freedom

The number of independent co-ordinate required to specify the position of a moving point is known as degrees of freedom of the system. Fig. 16.5.1 (*a*) shows systems of single degree of freedom. Only one angle θ is required to specify the position of any point.

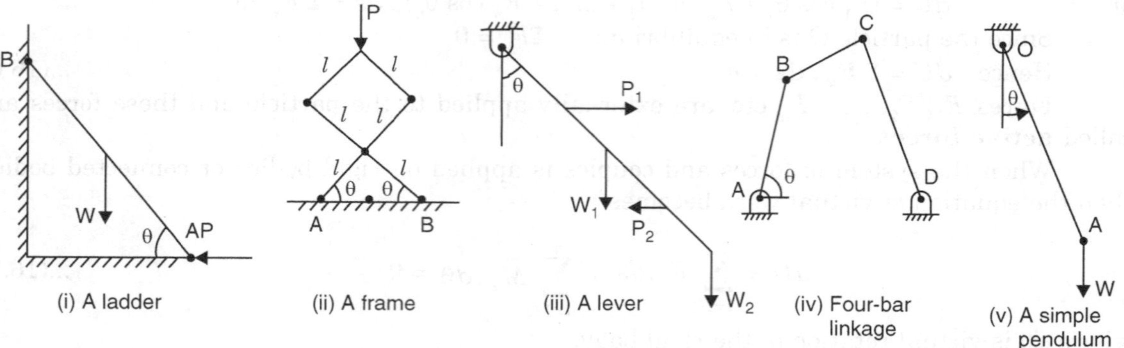

Fig. 16.5.1(*a*) Systems with one degree of freedom.

Fig. 16.5.1(b) shows systems with multi-degrees of freedom.

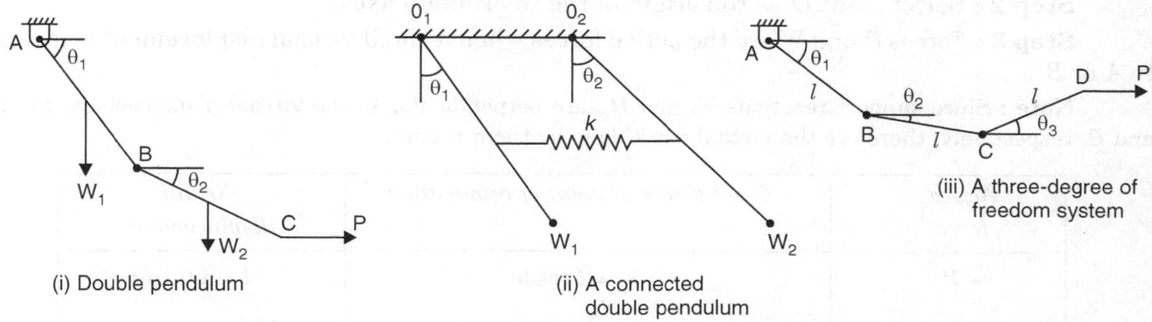

(i) Double pendulum (ii) A connected double pendulum (iii) A three-degree of freedom system

Fig. 16.5.1(b) Systems with multi-degree of freedom.

16.5.2. Principle of Virtual Work Applied to Systems with Multi-degree of Freedom.

The number of unknowns consisting of active forces or equilibrium positions, that can be determined by the virtual work method is equal to the degrees of freedom of the system since for each independent coordinate of position we can write a corresponding equation of virtual work. For multi-degree of freedom, we can express the total virtual work as

$$\delta U = \frac{\partial u}{\partial x} . \delta x + \frac{\partial u}{\partial y} . \delta y + \ldots\ldots \quad \ldots(16.6)$$

or

$$\delta U = \frac{\partial u}{\partial \theta} . \delta\theta_1 + \frac{\partial u}{\partial \theta_2} . \delta\theta_2 + \frac{\partial u}{\partial \theta_3} . \delta\theta_3 \quad \ldots(16.7)$$

Solved problems given below will help reader to understand the procedure.

Problem 16.1. *A uniform ladder AB of length l and weight W rests against a smooth vertical wall. Find the magnitude of the horizontal force P applied at the lower end A for the equilibrium of the ladder. Assume floor to be smooth. Use method of virtual work.*

Sol. Refer to Fig. P. 16.1.

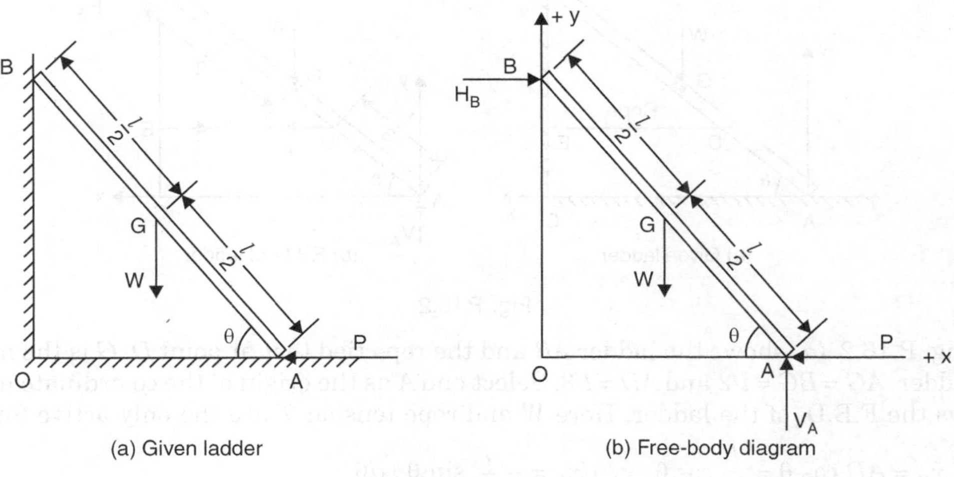

(a) Given ladder (b) Free-body diagram

Fig. P. 16.1

Step 1 : Draw free-body diagram (F.B.D.) of the ladder as shown in Fig. P.16.1(*b*).

Step 2 : Select point *O* as the origin of the co-ordinate axes.

Step 3 : Forces *P* and *W* are the active forces when a small virtual displacement is given to *A* or *B*.

Note : Since support reactions V_A and H_B are perpendicular to the virtual displacements at *A* and *B*, respectively, therefore the virtual work done by them is zero.

Active force	*Co-ordinate of point of application*	*Virtual displacement*
$-P$	$x_A = l\cos\theta$	$-l\sin\theta\,.\,d\theta$
$-W$	$y_G = \frac{1}{2}\,l\sin\theta$	$\frac{1}{2}\,l\cos\theta\,d\theta.$

By principle of virtual work, $dU = 0$

$$\therefore \quad -P(-l\sin\theta\,.\,d\theta) - W\left(\frac{1}{2}l\cos\theta\,.\,d\theta\right) = 0$$

$$\therefore \quad P\,l\sin\theta\,.\,d\theta = \frac{1}{2}\,W\,l\cos\theta\,.\,d\theta$$

or
$$P = \frac{1}{2}\,W\cot\theta \quad \textbf{Ans.}$$

Problem 16.2. *A uniform ladder of weight W and length l rests on a smooth floor and against a smooth wall. The slipping of the ladder is prevented by connecting a horizontal rope from a point l/3 from its lower end to the wall. Using principle of virtual work find tension induced in the rope.*

Sol. Refer to Fig. P.16.2.

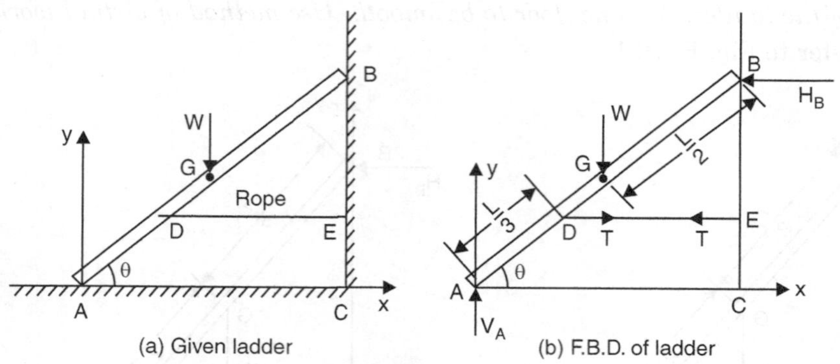

Fig. P.16.2

Fig. P.16.2. (*a*) shows the ladder *AB* and the rope tied to it at point *D*. *G* is the mid point of the ladder. $AG = BG = l/2$ and $AD = l/3$. Select end *A* as the origin of the co-ordinate axes. Fig. (*b*) shows the F.B.D. of the ladder. Here *W* and rope tension *T* are the only active forces. For point *D*, $x_D = AD\cos\theta = \frac{l}{3}\cos\theta$, $\therefore\ dx_D = -\frac{l}{3}\sin\theta\,.\,d\theta$

For point E, $x_E = AC - x_D = l\cos\theta - \frac{l}{3}\cos\theta = \frac{2l}{3}\cos\theta$, $\therefore\ dx_E = -\frac{2}{3}l\sin\theta\,.\,d\theta$

For point G, $y_G = AG\sin\theta = \frac{l}{2}\sin\theta$, $\therefore\ dy_G = \frac{l}{2}\cos\theta\,.\,d\theta$

Virtual work $dU = (+T)\,dx_D + (-T)\,d\,x_E + (-W)\,.\,dy_G$

$$= T\left(-\frac{l}{3}\sin\theta\,.\,d\theta\right) + (-T)\left(-\frac{2}{3}l\sin\theta\right) + (-W)\left(\frac{l}{2}\cos\theta\,d\theta\right)$$

or $$dU = \frac{Tl}{3}\sin\theta\,d\theta - \frac{Wl}{2}\cos\theta\,.\,d\theta$$

By the principle of virtual work $dU = 0$

$\therefore$ We get $\frac{Tl}{3}\sin\theta\,.\,d\theta - \frac{Wl}{2}\cos\theta\,.\,d\theta = 0$

$\therefore$ $$T = \frac{3}{2}W\cot\theta \quad \textbf{Ans.}$$

Problem 16.3. *The collapsible mechanism shown in Fig. P.16.3(a) is acted by the vertical force P. Find the magnitude of the horizontal force required at the roller support B for the equilibrium of the mechanism. End A is fixed.*

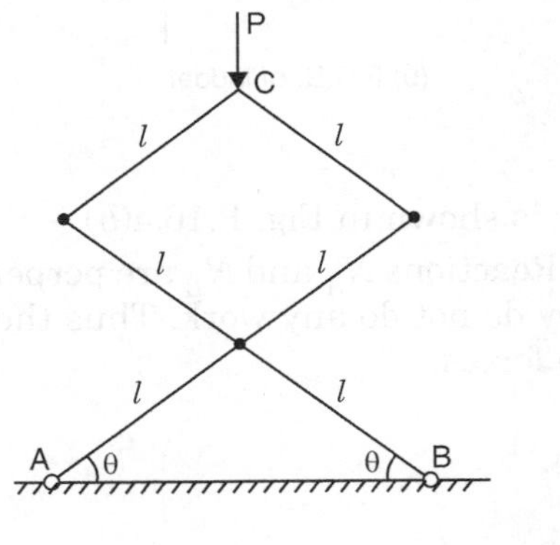

(a) Given mechanism

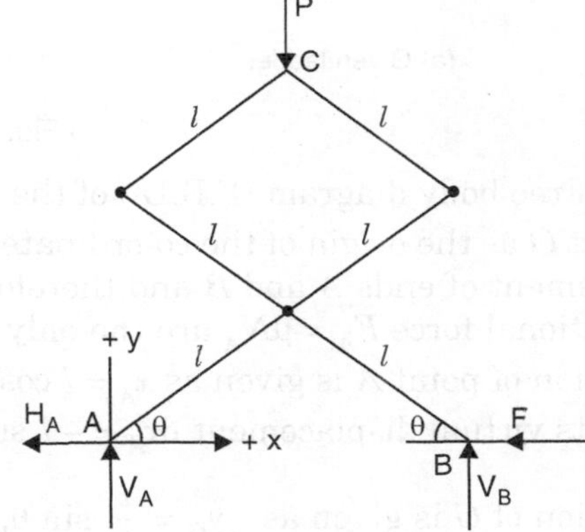

(b) F.B.D. of the mechanism

Fig. P.16.3.

Sol. Free-body diagram of the mechanism is shown in Fig. P.16.3(*b*).

Support A is fixed hence reaction components V_A and H_A do not do any work. Similarly V_B does not do any work. The applied force P and the force F are the only "active" forces.

Select A as the origin of co-ordinate axes.

For point B, $x_B = 2l\cos\theta$, $\therefore\ dx_B = -2l\sin\theta\,.\,d\theta$

For point C, $y_c = 3l\sin\theta$, $\therefore\ dy_c = 3l\cos\theta\,.\,d\theta$

Total virtual work done,

$$dU = (-F)\,dx_B + (-P)\,.\,dy_C$$

or $$dU = (-F)\times(-2l\sin\theta\,.\,d\theta) + (-P)\,(3l\cos\theta\,.\,d\theta)$$

By the principle of virtual work, $dU = 0$

$\therefore$ $(F \times 2l \sin\theta\, d\theta - P \times 3l \cos\theta \,.\, d\theta = 0$

$$\therefore \qquad F = \frac{3P}{2} \cot\theta \quad \textbf{Ans.}$$

Problem 16.4. *A uniform ladder of length l and weight W rests on the rough floor and leans against a smooth vertical wall, shown in Fig. P.16.4(a). Using principle of virtual work find the inclination θ of the ladder for its equilibrium.*

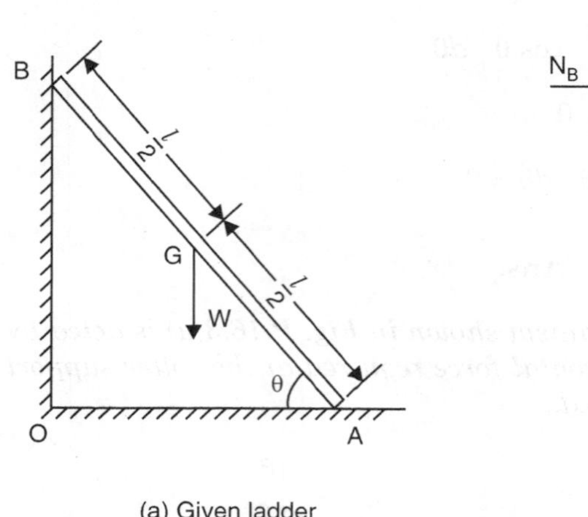

(a) Given ladder

(b) F.B.D. of ladder

Fig. P.16.4

Sol. Free body diagram (F.B.D.) of the ladder is shown in Fig. P.16.4(*b*).

Select O as the origin of the co-ordinate axes. Reactions N_A and N_B are perpendicular to the displacement of ends A and B and therefore they do not do any work. Thus the weight W and the frictional force $F_A = \mu N_A$ are the only active forces.

Position of point A is given as $x_A = l \cos\theta$,

and, $\therefore$ its virtual displacement $\delta x_A = -\, l \sin\theta\, \delta\theta$

Position of G is given as $y_G = \frac{l}{2} \sin\theta$, $\therefore$ $\delta y_G = \frac{l}{2} \,.\, \cos\theta.\ \delta\theta$

$\therefore$ Virtual work done $= \delta u = (-\,F_A)\ \delta x_A + (-\ \overline{W}\ )\ \delta y_G$

or $$\delta u = (-\,F_A)\,(-\,l \sin\theta \,.\, \delta\theta) + (-\,W)\,(l/2 \cos\theta \,.\, \delta\theta)$$

By the principle of virtual work, $\delta u = 0$

$$\therefore \quad +\,F_A\, l \sin\theta \,.\, \delta\theta - \frac{W \,.\, l}{2} \cos\theta \,.\, \delta\theta = 0$$

$$\therefore \qquad F_A \sin\theta = \frac{W}{2} \cos\theta \qquad \text{...(1)}$$

where $$F_A = \mu N_A \qquad \text{...(2)}$$

By using the condition $\Sigma F_V = 0$, we have

$$N_A - W = 0, \quad \therefore \quad N_A = W.$$

$\therefore$ From eqn. (2), $F_A = \mu W$

$\therefore$ Eqn. (1) gives $\mu.W \sin \theta = \frac{W}{2} \cos \theta.$

$$\therefore \qquad \tan \theta = \frac{1}{2\mu} \cos \theta.$$

$$\text{or,} \qquad \theta = \tan^{-1} \left(\frac{1}{2\mu}\right) \quad \textbf{Ans.}$$

Problem 16.5. *A uniform ladder of length i and weight W rests on a rough floor and leans against rough wall, as shown in Fig. P.16.5(a). Using principle of virtual work find the inclination of the ladder for its equilibrium. Co-efficient of friction at all surfaces is μ.*

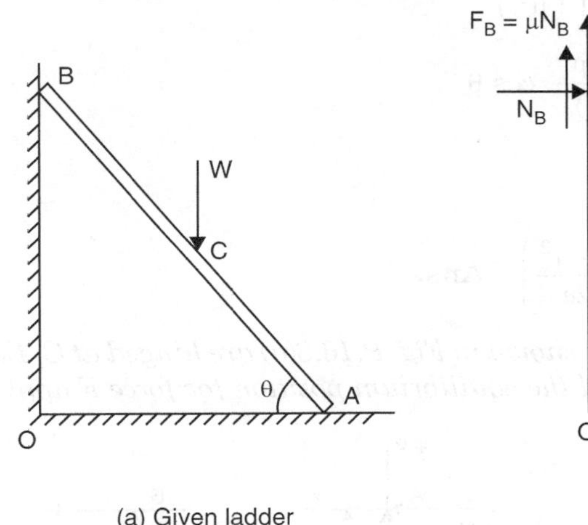

(a) Given ladder

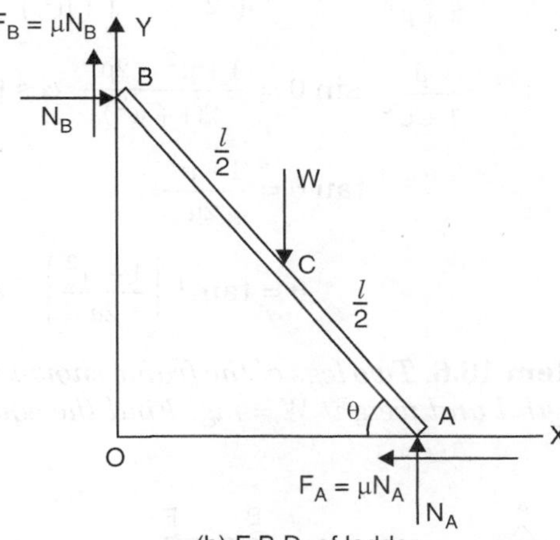

(b) F.B.D. of ladder

Fig. P.16.5.

Sol. Free body diagram of the ladder is shown in Fig. P.16.5(*b*).

Select point O as the origin of the co-ordinate axes. Normal component N_A and N_B of end reactions do not do any work because they are perpendicular to the movements of the ends A and B. Thus the only active forces are the weight W of the ladder, and the forces of friction F_A and F_B.

For point A, $\quad x_A = AB \cos \theta = l \cos \theta \quad \therefore \quad \delta x_A = - l \sin \theta \,.\, \delta\theta,$

For point B, $\quad y_B = l \sin \theta, \quad \therefore \quad \delta y_B = l \cos \theta \,.\, \delta\theta$

and for point C, $\quad y_C = l/2 \sin \theta, \quad \therefore \quad \delta y_C = \frac{l}{2} \cos \theta \,.\, \delta\theta.$

$\therefore$ Virtual work, $\quad \delta u = (- F_A) \,.\, \delta x_A + F_B \;\, \delta y_B + (- W) \;\, \delta y_C$

or $\quad \delta u = - \mu N_A \,.\, (- l \sin \theta \,.\, \delta\theta) + \mu N_B \,.\, (l \cos \theta \,.\, \delta\theta) + (- W) \, (l/2 \cos \theta \,.\, \delta\theta)$

$\therefore$ By using the principle or virtual work,

$$\delta u = 0, \text{ we get}$$

$$\therefore \quad + \mu N_A \sin \theta \; \delta\theta + \mu N_B \,.\, l \cos \theta \; \delta\theta - W(l/2 \cos \theta \; \delta\theta) = 0$$

$$\therefore \quad \mu N_A \sin \theta = (W/2 - \mu N_B) \cos \theta \qquad \text{...(1)}$$

By $\Sigma F_V = 0$, we have $\quad N_A + \mu N_B - W = 0$

$$\therefore \quad N_A + \mu N_B = W \qquad \text{...(2)}$$

By $\Sigma F_h = 0$, we get $\quad N_B - \mu N_A = 0$

$$\text{or,} \quad N_B = \mu N_A \qquad \text{...(3)}$$

By using Eqn. (3), and Eqn. (2), we have

$$N_A + \mu(\mu N_A) = W$$

$$\therefore \quad N_A = \frac{W}{1+\mu^2} \quad \text{...(4)}$$

$\therefore$ Eqn. (1) gives

$$\mu\left(\frac{W}{1+\mu^2}\right) \sin\theta = \left\{\frac{W}{2} - \mu(\mu N_A)\right\} \cos\theta$$

or,

$$\frac{\mu W}{1+\mu^2} \sin\theta = \left(\frac{W}{2} - \mu^2 \frac{W}{1+\mu^2}\right) \cos\theta$$

$$\therefore \quad \frac{\mu}{1+\mu^2} \sin\theta = \frac{1+\mu^2-2\mu^2}{2(1+\mu^2)} \cos\theta$$

$$\therefore \quad \tan\theta = \frac{1-\mu^2}{2\mu}$$

$$\therefore \quad \theta = \tan^{-1}\left(\frac{1-\mu^2}{2\mu}\right) \quad \textbf{Ans.}$$

Problem 16.6. *Two legs of the frame shown in Fig. P.16.6(a) are hinged at C. Each bar has the same length l and weight W = mg. Find the equilibrium position for force F applied at B.*

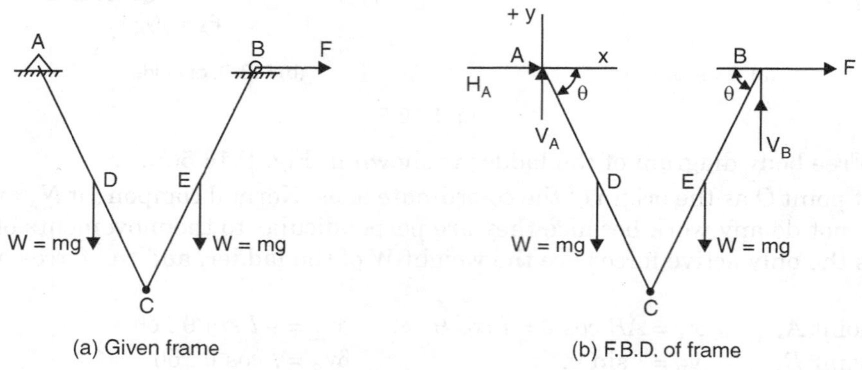

(a) Given frame

(b) F.B.D. of frame

Fig. P.16.6.

Sol. Free body diagram of the frame is shown in Fig. P.16.6(*b*)

Select A as the origin of the co-ordinate axes. Active forces are F, mg and mg.

No work is done by the forces H_A, V_A and V_B. Give small displacement δx_B to the roller B.

Active force	*Co-ordinate of the point of application*	*Virtual displacement*
+ F at B	$x_B = 2l\cos\theta$	$\delta x_B = -2l\sin\theta\,\delta\theta$
− mg at D	$y_D = -\frac{l}{2}\sin\theta$	$\delta y_D = -\frac{l}{2}\cos\theta\,.\,\delta\theta$
− mg at E	$y_E = -\frac{l}{2}\sin\theta$	$\delta y_E = -\frac{l}{2}\cos\theta\,.\,\delta\theta$

By principle of virtual work, $\delta u = 0$, we get

$$F(-2l \sin\theta\, \delta\theta) + (-mg)(-l/2 \cos\theta \,.\, \delta\theta) + (-mg)(-l/2 \cos\theta\, \delta\theta) = 0$$

$$\therefore \quad \frac{\sin\theta}{\cos\theta} = \tan\theta = \left(\frac{mg}{2F}\right)$$

$$\therefore \quad \theta = \tan^{-1}\left(\frac{mg}{2F}\right) \quad \textbf{Ans.}$$

Problem 16.7. *Find the horizontal component of reaction at the hinge C for the frame shown in Fig. P.16.7 (a).*

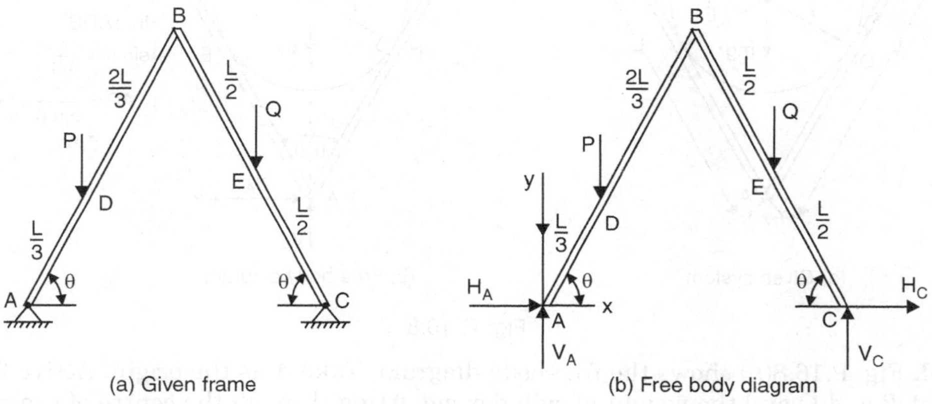

Fig. P.16.7.

Sol. Draw free body diagram of the frame as shown in Fig. P.16.7(*b*). *A* is taken as origin of the co-ordinate axes. Give a small displacement to *C*.

Active forces are P, Q and H_C. No work is done by forces V_A, V_C and H_A.

Active force	*Co-ordinate of point of application of force*	*Virtual displacement*
H_C	$x_C = 2L \cos\theta$	$\delta x_C = -2L \sin\theta \,.\, \delta\theta$
$-P$	$y_D = \frac{L}{3} \sin\theta$	$\delta y_D = \frac{L}{3} \cos\theta \,.\, \delta\theta$
$-Q$	$y_E = \frac{L}{2} \sin\theta$	$\delta y_E = \frac{L}{2} \cos\theta \,.\, \delta\theta$

By principle of virtual work, $du = 0$, we get

$$(H_C)(-2L \sin\theta) \,.\, \delta\theta + (-P)\left(\frac{L}{3}\cos\theta\right) . \,\delta\theta + (-Q)\left(\frac{L}{2}\cos\theta\right) . \,\delta\theta = 0$$

$$\therefore \quad H_C (2 \sin\theta) = -\left(\frac{P}{3} + \frac{Q}{2}\right)\cos\theta$$

$$\therefore \quad H_C = -\left(\frac{P}{6} + \frac{Q}{4}\right)\cot\theta. \quad \textbf{Ans.} \qquad ...(1)$$

The – ve sign in Eqn. (1) shows that the support reaction H_C is directed to the left. **Ans.**

Problem 16.8. *Find the force F to support a cylinder of mass m shown in Fig. P. 16.8(a). Neglect friction.*

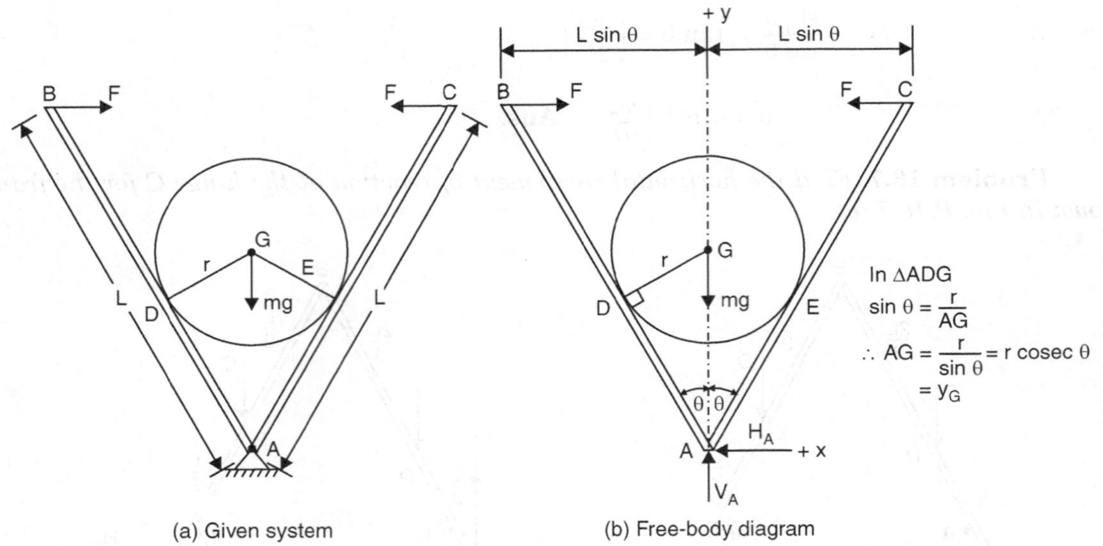

(a) Given system (b) Free-body diagram

Fig. P. 16.8

Sol. Fig. P.16.8(*b*) shows the free-body diagram. Take *A* as the origin. Active forces are *F* acting at *B* and *C* and the weight of cylinder *mg* acting through the centre of gravity *G*.

No work is done by H_A, V_A and reactions at *D* and *E*

Active force	*Co-ordinate of point of application of active force*	*Virtual displacement*	*Virtual work*
F	$x_B = -L \sin \theta$	$\delta x_B = -L \cos \theta \,.\, \delta\theta$	$F \,.\, (-L \cos \theta) \,.\, \delta\theta$
$-F$	$x_C = L \sin \theta$	$\delta x_E = L \cos \theta \,.\, \delta\theta$	$(-F)\,(L \cos \theta) \,.\, \delta\theta$
$-mg$	$y_G = r \text{ cosec } \theta$	$\delta y_G = -r \text{ cosec } \theta \,.\, \cot \theta \,.\, \delta\theta$	$(-mg)\,(-r \text{ cosec } \theta \,.\, \cot \theta)\,\delta\theta$

∴ Total virtual work

$$= \delta u = F(-L \cos \theta \,.\, \delta\theta) + (-F)\,(L \cos \theta \,.\, \delta\theta) + (-mg)\,(-r \text{ cosec } \theta \,.\, \cot \theta \,.\, \delta\theta)$$

$= 0$ by principle of virtual work

$$\therefore \quad 2Fl \cos \theta = mgr \times \text{cosec } \theta \,.\, \cot \theta$$

$$\therefore \quad F = \frac{mgr}{2L} \times \frac{1}{\cos \theta \,.\, \sin \theta} \; \frac{\cos \theta}{\sin \theta}$$

or

$$F = \frac{mgr}{2L \sin^2 \theta} \quad \textbf{Ans.}$$

Problem 16.9. *A uniform rod of weight W is acted upon by forces as shown in Fig. P.16.9(a). Using principle of virtual work, find the force Q required for the equilibrium of the rod. Also find magnitude of f the force Q when the rod is weightless.*

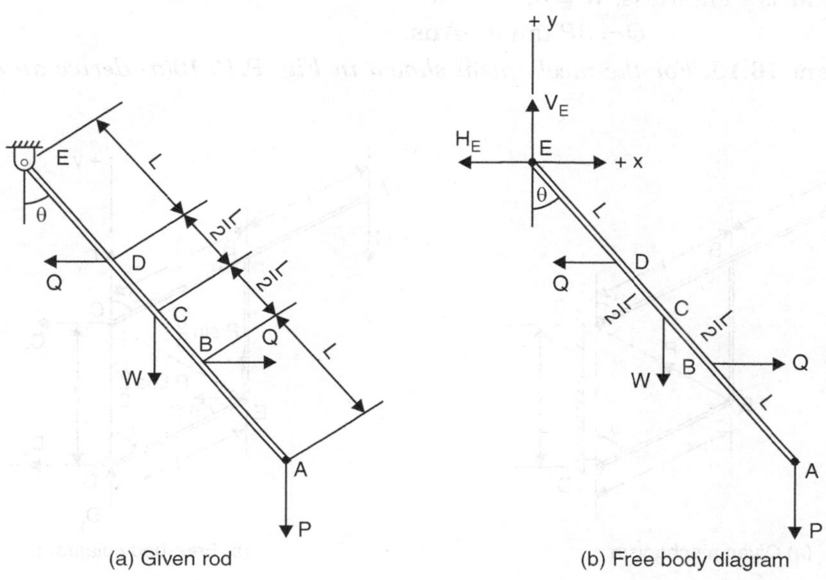

(a) Given rod (b) Free body diagram

Fig. P.16.9

Sol. Fig. P.16.9(*a*) shows the free body diagram of the rod. Take support *E* as the origin. Active forces are (*a*) *P* at point *A*, (*b*) + *Q* at point *B*, (*c*) – *Q* at point *D* and (*d*) – *W* at point *C*.

Reactions H_E and V_E do not do any work.

Active force	*Co-ordinate of point of application*	*Virutal displacement*	*Virtual work*
$-P$	$y_A = -3l\cos\theta$	$\delta y_A = 3\,l\sin\theta\,.\,\delta\theta$	$(-P)(3l\sin\theta)\,.\,\delta\theta$
$+Q$	$x_B = 2l\sin\theta$	$\delta x_B = 2\,l\cos\theta\,.\,\delta\theta$	$(Q)(2l\cos\theta)\,.\,\delta\theta$
$-Q$	$x_D = l\sin\theta$	$\delta x_D = l\cos\theta\,.\,\delta\theta$	$(-Q)(l\cos\theta)\,.\,\delta\theta$
$-W$	$y_C = -\dfrac{3}{2}\,l\cos\theta$	$\delta y_C = \dfrac{3}{2}\,l\sin\theta\,.\,\delta\theta$	$(-W)\left(\dfrac{3}{2}\,l\sin\theta\right).\,\delta\theta$

∴ Total virtual work,

$$\delta u = (-P)(3l\sin\theta)\,.\,\delta\theta + (Q)(2l\cos\theta)\,.\,\delta\theta + (-Q)(l\cos\theta)\,\delta\theta + (-W)\left(\frac{3}{2}\,l\sin\theta\right).\,\delta\theta$$

By principle of virtual work, $\delta u = 0$

$$\therefore\quad (-P)(3l\sin\theta) + Q(2l - l)\cos\theta + (-W)\left(\frac{3}{2}\,l\sin\theta\right) = 0$$

$$\therefore\qquad Q\cos\theta = \left(3P + \frac{3}{2}W\right)\sin\theta$$

$$\therefore \qquad Q = 3\left(P + \frac{W}{2}\right) \tan \theta \quad \textbf{Ans.}$$

When bar is weightless, $W = 0$,

$$\therefore \qquad Q = 3P \tan \theta \quad \textbf{Ans.}$$

Problem 16.10. *For the mechanism shown in Fig. P.16.10(a) derive an expression for finding P.*

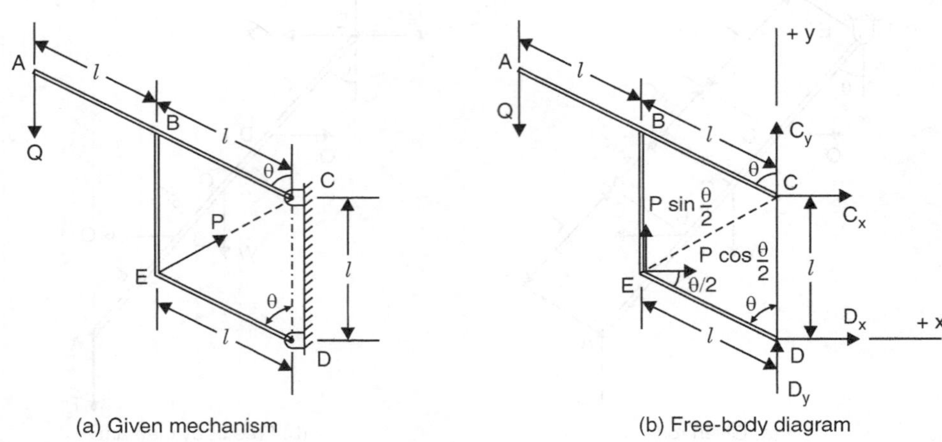

(a) Given mechanism (b) Free-body diagram

Fig. P.16.10

Sol. ΔCDE is an equilateral triangle.

$\therefore$ $\angle\theta = 60°$ and inclination of P with horizontal (or x axis) is $\theta/2 = 30°$.

Fig. P.16.10(*b*) shows the free-body diagram of the mechanism. Select D as the origin of the co-ordinate axes.

Give a small angular displacement $d\theta$ to the mechanism. Active forces are $P \cos \theta$, Q, $P \sin \theta/2$.

Forces (or reaction components) C_x, C_y, D_x and D_y do no work.

Active force	*Co-ordinate of point of application*	*Virtual displacement*	*Virtual work*
$-Q$	$y_A = l + 2l \cos \theta$	$dy_A = -2l \sin \theta \, . \, d\theta$	$2l\, Q \sin \theta \, . \, d\theta$
$+P \cos \theta/2$	$x_E = -l \sin \theta$	$dx_E = -l \cos \theta \, . \, d\theta$	$-P\, l \cos \theta \, . \cos \theta/2 \, . \, d\theta$
$+P \sin \theta/2$	$y_E = l \cos \theta$	$dy_E = -l \sin \theta \, . \, d\theta$	$-P\, l \sin \dfrac{\theta}{2} \, . \sin \theta = d\theta$

$\therefore$ Total virtual work, $du = 2lQ \sin \theta \, . \, d\theta - P\, l \cos \theta \, . \cos \theta/2 \, . \, d\theta - P\, l \sin \theta \, . \sin \theta/2 \, . \, d\theta]$

By principle of virtual work $du = 0$

$\therefore$ $(2l\, Q \sin \theta - P\, l \cos \theta \cos \theta/2 - Pl \sin \theta \sin \theta/2) \, . \, d\theta = 0$

$$\therefore \qquad 2Q \sin \theta = P[\cos \theta \cos \theta/2 + \sin \theta \sin \theta/2]$$

Using $\cos A \, . \cos B + \sin A \, . \sin B = \cos (A - B)$, we get

$$2Q \sin \theta = P \cos \theta/2$$

or $\qquad 2Q[2 \sin \theta/2 \, . \cos \theta/2] = P \cos \theta/2$

$$\therefore \qquad P = 4Q \sin \theta/2 \quad \textbf{Ans.}$$

When $Q = 60°$, $P = 4Q \sin \frac{60°}{2} = 2Q$ **Ans.**

Problem 16.11. *A long thin rod AB is attached to a collar A and rests on a small roller at C. Neglecting friction and weight of various bars derive an expression for finding the magnitude of the force Q required for the equilibrium of the rod. See Fig. P.16.11(a).*

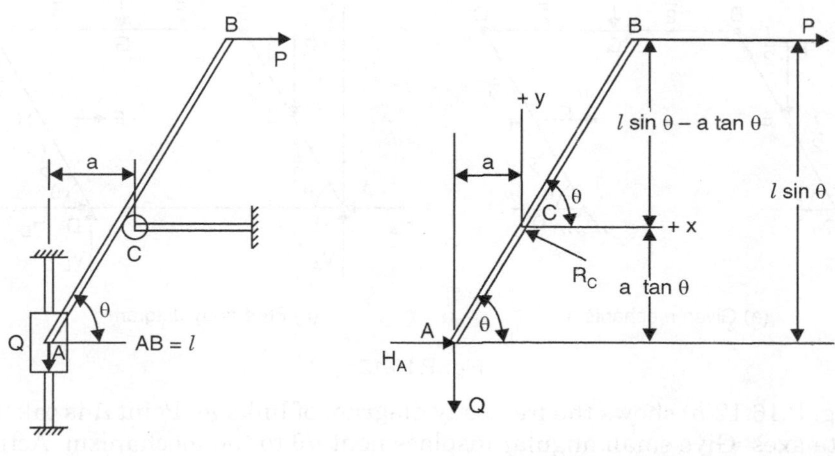

Fig. P.16.11

Sol. Fig. P.16.11(*b*) shows the free-body diagram of the rod *AB*. R_C is normal reaction of roller at *C*. H_A is horizontal reaction of collar *A* and *Q* is the applied force for maintaining equilibrium. H_A is perpendicular to the collar *A*. Forces R_C and H_A do no work.

Therefore, active forces are *P* and *Q*.

Select *C* as the origin of the co-ordinate axes.

Give a small angular displacement $d\theta$ to the rod *AB*.

Active force	*Co-ordinate of point of application*	*Virtual displacement*	*Virtual work*
$+P$	$x_B = l \cos \theta - a$	$dx_B = -l \sin \theta \,.\, d\theta$	$(+P)(-l \sin \theta \,.\, d\theta)$
$-Q$	$y_A = -a \tan \theta$	$dy_B = -a \sec^2 \theta \,.\, d\theta$	$(-Q)(-a \sec^2 \theta \,.\, d\theta)$

Total virtual work $= du = (P)(-l \sin \theta \,.\, d\theta) + (-Q)(-a \sec^2 \theta \,.\, d\theta)$

By principle of virtual work $du = 0$

$\therefore \quad -Pl \sin \theta \,.\, d\theta + Q \,.\, a \sec^2 \theta \,.\, d\theta = 0$

$$\therefore \qquad Q = \frac{Pl \sin \theta}{a \,.\, \sec^2 \theta} \quad \textbf{Ans.}$$

Problem 16.12. *The four-bar mechanism shown in Fig. P.16-12(a) forms a parallelogram. Forces P, F and Q are applied at the mid-points of links. Determine the force F for the equilibrium condition in terms of angle θ.*

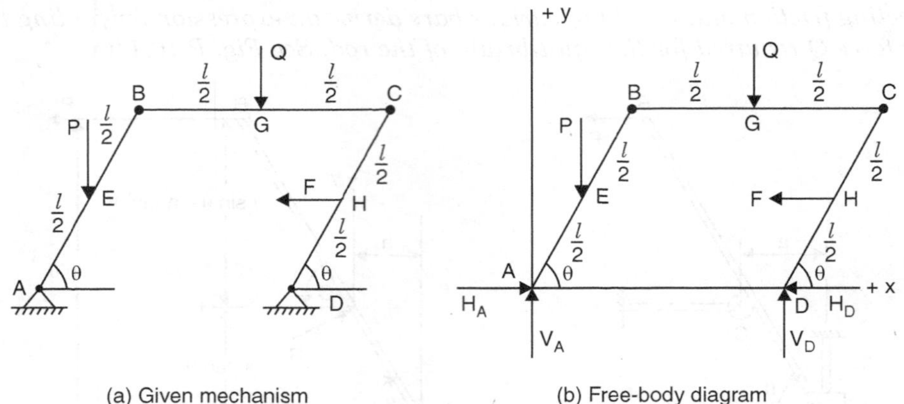

(a) Given mechanism (b) Free-body diagram

Fig. P.16.12

Sol. Fig. P.16.12(*b*) shows the free-body diagram of linkage. Point *A* is taken as origin of the co-ordinate axes. Give small angular displacement $d\theta$ to the mechanism. Active forces are *P*, *Q* and *F*.

Active force	*co-ordinate of point of application*	*Virtual displacement*	*Virtual work*
$-P$	$y_E = \frac{l}{2}\sin\theta$	$dy_E = \frac{l}{2}\cos\theta\,.\,d\theta$	$(-P)\left(\frac{l}{2}\cos\theta\,.\,d\theta\right)$
$-Q$	$Y_G = l\sin\theta$	$dy_G = l\cos\theta\,.\,d\theta$	$(-Q)(l\cos\theta\,.\,d\theta)$
$-F$	$x_H = l + \frac{l}{2}\cos\theta$	$dx_H = -\frac{l}{2}\sin\theta\,.\,d\theta$	$(-F)\left(-\frac{l}{2}\sin\theta\,.\,d\theta\right)$

Total virtual work

$$du = (-P)\left(\frac{l}{2}\cos\theta\,.\,d\theta\right) + (-Q)(l\cos\theta\,.\,d\theta) + (-F)\left(-\frac{l}{2}\sin\theta\,.\,d\theta\right)$$

Using principle of virtual work, $du = 0$

$$\therefore \quad (-P)\left(\frac{l}{2}\cos\theta\,.\,d\theta\right) + (-Q)(l\cos\theta\,.\,d\theta) + (-F)\left(-\frac{l}{2}\sin\theta\,.\,d\theta\right) = 0$$

$$\therefore \quad \frac{F}{2}\sin\theta = \frac{1}{2}P\cos\theta + Q\cos\theta$$

$$\therefore \quad F = (P + 2Q)\cot\theta \quad \textbf{Ans.}$$

Problem 16.13. *For the mechanism shown in Fig. P.16.13(a), find force P required to maintain equilibrium. All bars are of negligible mass.*

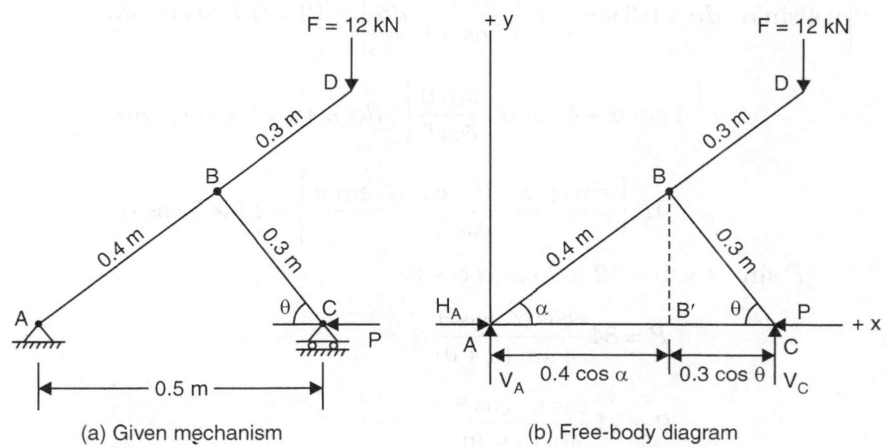

(a) Given mechanism (b) Free-body diagram

Fig. P.16.13

Sol. Free body diagram of the mechanism is shown in Fig. P.16.13(*b*). Take point *A* as the origin of the co-ordinate axes.

Give a small horizontal displacement.

Active forces are *P* and 12 kN

No work is done by reaction components V_A, H_A and V_C.

Active force	*Co-ordinate of point of application*	*Virtual displacement*	*Virtual work*
$-P$	$x_C = 0.4 \cos \alpha + 0.3 \cos \theta$	$dx_C = -0.4 \sin \alpha \, d\alpha - 0.3 \sin \theta \, . \, d\theta$	$(-P)(-0.4 \sin \alpha \, d\alpha - 0.3 \sin \theta \, . \, d\theta)$
-12	$y_D = 0.7 \sin \alpha$	$dy_D = 0.7 \cos \alpha \, . \, d\alpha$	$(-12)(0.7 \cos \alpha \, . \, d\alpha)$

∴ Total virtual work, $du = (-P)(-0.4 \sin \alpha \, . \, d\alpha - 0.3 \sin \theta \, . \, d\theta) + (-12)(0.7 \cos \alpha \, . \, d\alpha)$

By principle of virtual work $du = 0$.

∴ $(-P)(-0.4 \sin \alpha \, . \, d\alpha - 0.3 \sin \theta \, . \, d\theta) + (-12)(0.7 \cos \alpha \, d\alpha) = 0$

∴ $P(0.4 \sin \alpha \, . \, d\alpha + 0.3 \sin \theta \, . \, d\theta) = 12 \times 0.7 \cos \alpha \, . \, d\alpha$ ...(1)

Let us find relation between $d\alpha$ and $d\theta$.

From Fig. P.16.13(*b*), we note that $BB' = 0.4 \sin \alpha$ and also $BB' = 0.3 \sin \theta$

$$\therefore \quad 0.4 \sin \alpha = 0.3 \sin \theta$$

$$\therefore \quad 0.4 \cos \alpha \, . \, d\alpha = 0.3 \cos \theta \, . \, d\theta$$

$$\therefore \quad d\theta = \frac{0.4 \cos \alpha}{0.3 \cos \theta} \, . \, d\alpha$$

or

$$d\theta = \frac{4}{3} \frac{\cos \alpha}{\cos \theta} \, . \, d\alpha \qquad ...(2)$$

Substituting value of $d\theta$ in Eqn. (1), we get

$$P\left[0.4 \sin \alpha \,.\, d\alpha + 0.3 \sin \theta \,.\left(\frac{4}{3}\frac{\cos \alpha}{\cos \theta}\right).\, d\alpha\right] = 12 \times 0.7 \cos \alpha \,.\, d\alpha$$

$$\therefore \quad P\left[4 \sin \alpha + 4 \cos \alpha \,.\, \frac{\sin \theta}{\cos \theta}\right].\, d\alpha = 12 \times 7 \cos \alpha \,.\, d\alpha$$

$$4P\left[\frac{\sin \alpha \,.\, \cos \theta + \cos \alpha \,.\, \sin \theta}{\cos \theta}\right] = 12 \times 7 \cos \alpha$$

or $$4P \sin (\alpha + \theta) = 12 \times 7 \cos \alpha \cos \theta.$$

$$\therefore \quad P = 84\, \frac{\cos \alpha \,.\, \cos \theta}{4 \sin (\alpha + \theta)}$$

or $$P = 21 \frac{\cos \alpha \,.\, \cos \theta}{\sin (\alpha + \theta)} \qquad ...(3)$$

From the geometry of the given mechanism, we find that $\alpha + \theta = 90°$, and also

$$\cos \alpha = \frac{0.4}{0.5} = \frac{4}{5} \text{ and } \cos \theta = \frac{0.3}{0.5} = \frac{3}{5}$$

and $$\sin (\alpha + \theta) = \sin 90° = 1$$

$\therefore$ Eqn. (3) gives $$P = \frac{21 \times \frac{4}{5} \times \frac{3}{5}}{1} \text{ or } P = \frac{252}{25} = 10.08 \text{ kN} \quad \textbf{Ans.}$$

Problem 16.14. *Four identical bars each of weight W are joined together to form a square ABCD. The fifth bar BD also of weight W is connected as shown in Fig. P.16.14(a). Find force induced in the bar BD when the frame is suspended from A.*

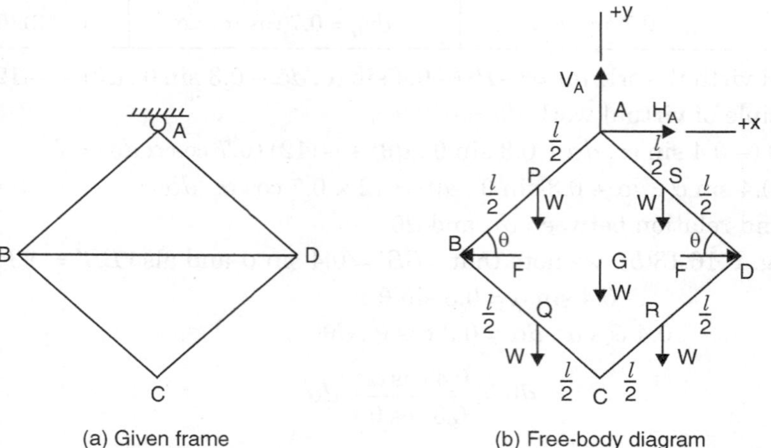

(a) Given frame (b) Free-body diagram

Fig. P.16.14

Sol. Free body diagram of the frame is shown in Fig. P.16.14(*b*).

Select *A* as the origin of the co-ordinate axes. Let force in member *BD* be *F*. Give a small displacement to the frame. Active forces are weights of bars and axial forces (*F*) is member *BP*.

Support reaction component V_A and H_A do no work.

Active force	*Co-ordinate of point of application*	*Virtual displacement*	*Virtual work*
$-W$	$y_P = -\frac{l}{2}\sin\theta$	$dy_P = -\frac{l}{2}\cos\theta \,.\, d\theta$	$(-W)\left(-\frac{l}{2}\cos\theta \,.\, d\theta\right)$
$-W$	$y_Q = -\frac{3l}{2}\sin\theta$	$dy_Q = -\frac{3l}{2}\cos\theta \,.\, d\theta$	$(-W)\left(-\frac{3l}{2} \,.\, \cos\theta \,.\, d\theta\right)$
$-W$	$y_R = -\frac{3l}{2}\sin\theta$	$dy_R = -\frac{3l}{2}\cos\theta \,.\, d\theta$	$(-W)\left(-\frac{3l}{2}\cos\theta \,.\, d\theta\right)$
$-W$	$y_S = -\frac{l}{2}\sin\theta$	$dy_S = -\frac{l}{2}\cos\theta \,.\, d\theta$	$(-W)\left(-\frac{l}{2}\cos\theta \,.\, d\theta\right)$
$-W$	$y_G = -l\sin\theta$	$dy_G = -l\cos\theta \,.\, d\theta$	$(-W)(-l\cos\theta \,.\, d\theta)$
$-F$	$x_B = -l\cos\theta$	$dx_B = l\sin\theta \,.\, d\theta$	$(-F)(l\sin\theta \,.\, d\theta)$
$+F$	$x_D = l\cos\theta$	$dx_D = -l\sin\theta \,.\, d\theta$	$(F)(-l\sin\theta \,.\, d\theta)$

$\therefore$ Total virtual work

$$du = (-W)\left(-\frac{l}{2}\cos\theta \,.\, d\theta\right) + (-W)\left(-\frac{3l}{2}\cos\theta \,.\, d\theta\right) + (-W)\left(-\frac{3l}{2}\cos\theta \,.\, d\theta\right)$$

$$+ (-W)\left(-\frac{l}{2}\cos\theta \,.\, d\theta) + (-W)(-l\cos\theta \,.\, d\theta) + (-F)(l\sin\theta \,.\, d\theta) + (F)(-l\sin\theta \,.\, d\theta)\right.$$

$\therefore$ $Wl(5\cos\theta) \,.\, d\theta - Fl(2)\sin\theta \,.\, d\theta$

By the principle of virtual work $du = 0$

$\therefore$ $5\,Wl\cos\theta \,.\, d\theta - 2\,Fl\sin\theta \,.\, d\theta = 0$

$$\therefore \qquad F = \frac{5W}{2}\cot\theta$$

But $\theta = 45°$,

$$\therefore \qquad F = \frac{5W}{2}\cot 45°$$

$$\therefore \qquad F = \frac{5}{2}W \quad \textbf{Ans.}$$

Problem 16.15. *A load W is applied to the linkage shown in Fig. P. 16.15(a). When no load is applied links AB and BC are horizontal and the spring force is zero. Find the value of angle θ for equilibrium. Take W = 500 N and k = 2500 N/m and l = 0.25 m.*

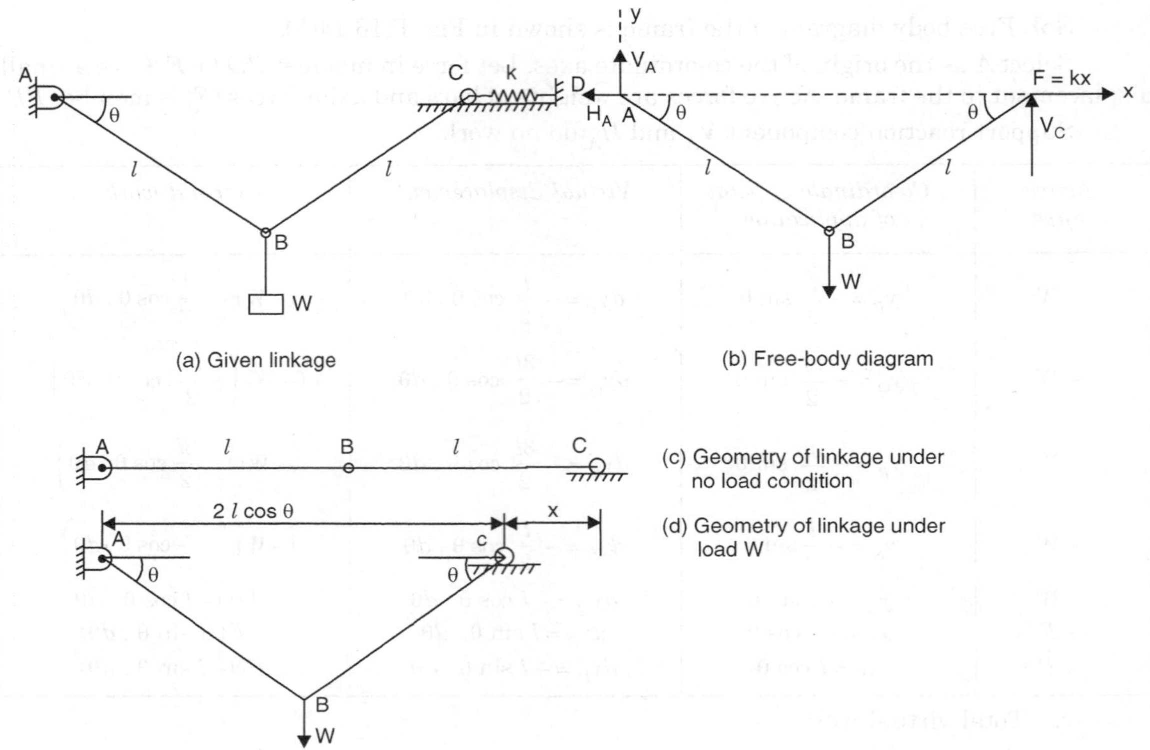

Fig. P.16.15

Sol. Fig. P.16.15(*c*) and (*d*), respectively, show the geometry of linkage under no load condition and with applied load.

Fig. P.16.15(*b*) shows the free-body diagram of the linkage at equilibrium.

Select *A* as origin of the co-ordinate axes.

Active forces are *W* and spring force $F = kx$

Reaction components V_A, H_A and V_C do no work.

Active force	*Co-ordinate of point of application*	*Virtual displacement*	*Virtual work*
$-W$	$y_B = -l \sin \theta$	$dy_B = -l \cos \theta \cdot d\theta$	$(-W)(-l \cos \theta \cdot d\theta)$
F	$x_C = 2l \cos \theta$	$dx_C = -2l \sin \theta \cdot d\theta$	$(F)(-2l \sin \theta \cdot d\theta)$

Total virtual work $du = (-W)(-l \cos \theta \cdot d\theta) + (F)(-2l \sin \theta \cdot d\theta)$

By principle of virtual work $du = 0$

$\therefore \quad (-W)(-l \cos \theta \cdot d\theta) + (F)(-2l \sin \theta \cdot d\theta) = 0$

$\therefore \quad W \cos \theta = 2F \sin \theta$ ...(1)

where $F = kx = k(2l - 2l \cos \theta)$ (see Figs. P.16.15(*c*) and (*d*)).

or $F = 2kl(1 - \cos \theta)$ ...(2)

$\therefore$ Eqn. (1) becomes

$$W \cos \theta = (2 \sin \theta)\,[2\,kl\,(1 - \cos \theta)]$$

or $W \cos \theta = 4kl \sin \theta\,(1 - \cos \theta)$ ...(3)

Substituting $W = 500$ N, $k = 2500$ N/m and $l = 0.25$ m,

we get $500 \cos \theta = 4 \times (2500)(0.25) \sin \theta (1 - \cos \theta)$

$0.2 \cos \theta = \sin \theta (1 - \cos \theta)$

or $0.2 = \tan \theta(1 - \cos \theta)$ ...(4)

Solving equation (4) for angle θ, we get

∴ $\theta =$ **40.25° Ans.**

[When $W = 1000$ N, $k = 2500$ N/m and $l = 0.25$ m, Eqn. (3) gives $0.4 = \tan \theta [1 - \cos \theta]$, which gives $\theta =$ **49.15°**]

Problem 16.16. *Derive an expression for θ for finding spring force for the equilibrium of the linkage shown in Fig. P.16.16(a). The unstretched length of the spring is b and the spring constant is k. Assume bars AB and AC to be weightless.*

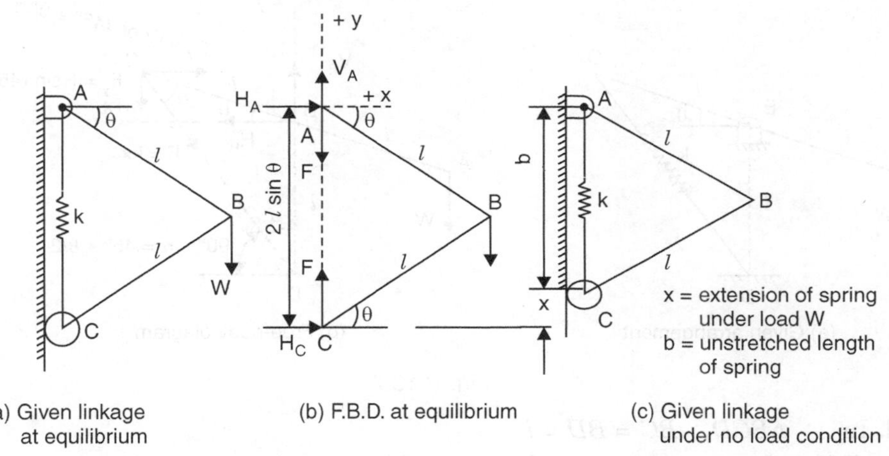

(a) Given linkage at equilibrium (b) F.B.D. at equilibrium (c) Given linkage under no load condition

Fig. P.16.16.

Sol. Fig. P.16.16(*b*) shows free-body diagram of the linkage at equilibrium. Fig. P.16.16(*c*) shows the mechanism under no load condition. Extension of spring when the load W is applied $= x = 2l \sin \theta - b$ ...(1)

∴ Spring force $F = kx$.

Active forces are W and F acting at point C.

No work is done by force and reaction components H_A, V_A and H_C.

Select A as origin of the co-ordinate axes.

Active force	*Co-ordinate of point of application*	*Virtual displacement*	*Virtual work*
$-W$	$y_B = -l \sin \theta$	$dy_B = -l \cos \theta \, . \, d\theta$	$(-W)(-l \cos \theta \, . \, d\theta)$
$+F$	$y_C = -2l \sin \theta$	$dy_C = -2l \cos \theta \, . \, d\theta$	$(F)(-2l \cos \theta \, . \, d\theta)$

∴ Total virtual work $du = (-W)(-l \cos \theta \, . \, d\theta) + (F)(-2l \cos \theta \, . \, d\theta)$

By the principle of virtual work $du = 0$,

∴ $(-W)(-l \cos \theta \, . \, d\theta) + (F)(-2l \cos \theta \, . \, d\theta) = 0$

∴ $W = 2F$ ...(2)

But $F = kx = k(2l \sin \theta - b)$

∴ Eqn. (2) gives $W = 2k(2l \sin \theta - b)$

or $$W + 2kb = 4kl \sin \theta \quad ...(3)$$

∴ $$\sin \theta = \frac{W + 2kb}{4kl}$$

∴ $$\theta = \sin^{-1}\left[\frac{W + 2kb}{4kl}\right]$$ **Ans.**

Problem 16.17. *For arrangement shown in Fig. P.16.17(a) the spring constant k = N/m and the spring is unstretched when lever ABC is horizontal and no load is applied at end A. Find the value of angle θ for the equilibrium when a load W is applied. Take W = 300 N, k = 4000 N/m and l = 0.5 m.*

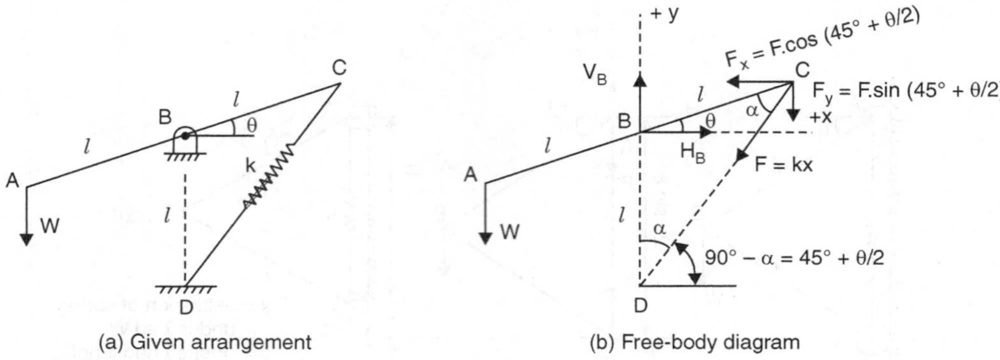

Fig. P.16.7

Sol. In ΔBCD, $BC = BD = l$

∴ ΔBCD is an isosceles triangle. ∴ $\angle D + \angle C + (90° + \theta) = 180°$

∴ $$\angle D = \angle C = \alpha = \frac{1}{2}(90° - \theta) = 45° - \theta/2$$

Hence components of the spring force F acting at C are :

$$F_y = F \sin (45° + \theta/2),$$

and $$F_x = F \cos (45° + \theta/2).$$

Select B as origin of the co-ordinate axes. Active forces are W and F or W, F_x and F_y. No work is done by reaction components V_B and H_B.

Active force	*Co-ordinate of point of application*	*Virtual displacement*	*Virtual work*
$-W$	$y_A = -l \sin \theta$	$dy_A = -l \cos \theta \,.\, d\theta$	$(-W)(-l \cos \theta \,.\, d\theta)$
$-F \cos (45° + \theta/2)$	$x_C = l \cos \theta$	$dx_C = -l \sin \theta \,.\, d\theta$	$[-F \cos(45° + \theta/2)] \,.\, (-l \sin \theta \, d\theta)$
$-F \sin (45° + \theta/2)$	$y_C = l \sin \theta$	$dy_C = l \cos \theta \,.\, d\theta$	$[-F \sin(45° + \theta/2)] \,.\, (l \cos \theta \,.\, d\theta)$

∴ Total virtual work

$$du = (-W)(-l \cos \theta \,.\, d\theta) + [-W \cos (45° + \theta/2)] \,.\, (-l \sin \theta \,.\, d\theta) + [-F \sin (45° + \theta/2)] \,.\, (l \cos \theta \,.\, d\theta)$$

By the principle of virtual work $du = 0$.

$\therefore$ $(-W)(-l\cos\theta \,.\, d\theta) + [-F\cos(45^\circ + \theta/2)] \,.\, (-l\sin\theta \,.\, d\theta)$
$+ [-F\sin(45^\circ + \theta/2)] (l\cos\theta \,.\, d\theta) = 0$

$\therefore$ $$W\cos\theta = F[-\cos(45^\circ + \theta/2) \,.\, \sin\theta + \sin(45^\circ + \theta/2) \,.\, \cos\theta]$$

or $$W\cos\theta = F[\sin(45^\circ + \theta/2 - \theta)]$$

or $$W\cos\theta = F\sin(45^\circ - \theta/2) \quad ...(1)$$

Spring Force:

Unstretched length of the spring = $l\sqrt{2}$ (when the lever *ABC* is horizontal).

From the ΔBCD, by using the sine rule, we get

$$\frac{BD}{\sin \angle BCD} = \frac{DC}{\sin \angle DBC}$$

or $$\frac{l}{\sin\alpha} = \frac{DC}{\sin(90^\circ + \theta)}$$

or $$DC = \frac{l\sin(90^\circ + \theta)}{\sin\alpha}$$

or $$DC = \frac{l\cos\theta}{\sin\alpha}$$

$\therefore$ Extension of spring = $x = \dfrac{l\cos\theta}{\sin\alpha} - l\sqrt{2}$

$\therefore$ $$\text{Spring force} = F = kx = kl\left(\frac{\cos\theta}{\sin\alpha} - \sqrt{2}\right)$$

or $$F = kl\left(\frac{\cos\theta}{\sin(45^\circ - \theta/2)} - \sqrt{2}\right) \quad ...(2)$$

By substituting value of F in Eqn. (1), we get

$$W\cos\theta = kl\left(\frac{\cos\theta}{\sin(45^\circ - \theta/2)} - \sqrt{2}\right) \,.\, \sin(45^\circ - \theta/2)$$

or $$W\cos\theta = kl\left[\cos\theta - \sqrt{2}\sin(45^\circ - \theta/2)\right]$$

$\therefore$ $$\frac{W}{kl} = \frac{1}{\cos\theta}[\cos\theta - (\cos\theta/2 - \sin\theta/2)]$$

$\therefore$ $$\frac{W}{kl} = 1 - \left[\frac{\cos\theta/2 - \sin\theta/2}{\cos\theta}\right] \quad ...(3)$$

Substituting $W = 300$ N; $k = 4000$ N/m, $l = 0.50$ m,

$\therefore$ Eqn. (3) gives $\dfrac{300}{4000 \times 0.50} = 1 - \dfrac{\cos\theta/2 - \sin\theta/2}{\cos\theta}$

or $$\frac{\cos\theta/2 - \sin\theta/2}{\cos\theta} = 1 - 0.15$$

$\therefore$ $$\frac{\cos\theta/2 - \sin\theta/2}{\cos\theta} = 0.85$$

Solving we get $\theta = 22.58^\circ$ **Ans.**

Problem 16.18. *Find the magnitude of the couple M required to maintain equilibrium of the link AB shown in Fig. P. 16.18(a).*

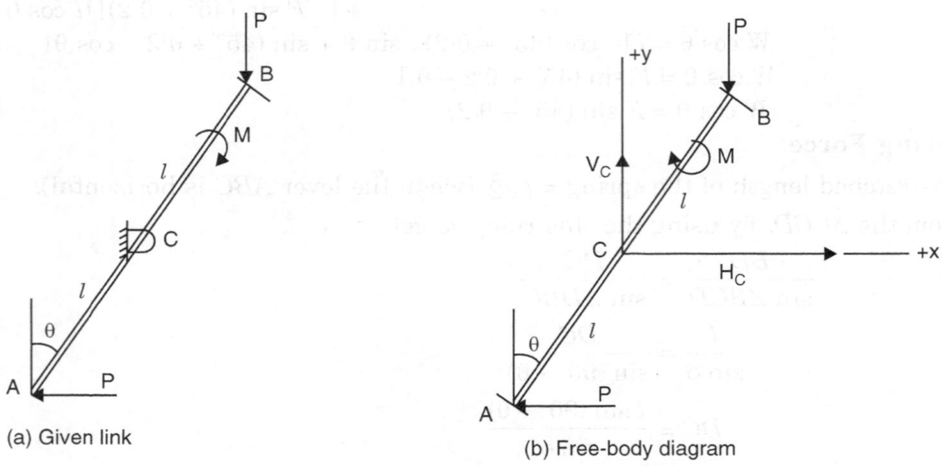

Fig. P.16.18

Sol. Free-body diagram of link is shown in Fig. P.16.18(*b*). Select point *C* as origin of the co-ordinate axes. Give virtual displacement $d\theta$ to the bar.

Active forces are *P* applied at points *A* and *B* and the couple *M*.

No work is done by reaction components at V_C and H_C.

Active force	*Co-ordinate of point of application*	*Virtual displacement*	*Virtual work*
$-P$	$x_A = -l \sin\theta$	$dx_A = -l\cos\theta \,.\, d\theta$	$(-P)(-l\cos\theta \,.\, d\theta)$
$-P$	$y_B = l\cos\theta$	$dy_B = -l\sin\theta \,.\, d\theta$	$(-P)(-l\sin\theta \,.\, d\theta)$
M	–	$d\theta$	$(M)\,(d\theta)$

Note : Work done by couple is positive as *M* tends to increase the angle θ.

Total virtual work, $du = (-P)(-l\cos\theta \,.\, d\theta) + (-P)(-l\sin\theta \,.\, d\theta) + M \,.\, d\theta$

But $dU = 0,$

$\therefore \quad M = -Pl(\cos\theta + \sin\theta)$ **Ans.**

Negative sign of couple *M* indicates that it will reduce the angle θ.

Problem 16.19. *A simply supported beam AB is 4.5 m long has an overhang BC 1.5 m. It carriers a load of 2.4 kN at point D 3m from A and another load of 1.2 kN at C. Using method of virtual work. Find reactions at supports A and B.*

Sol. Fig. P.16.19(*a*) shows the given beam.

To find reaction V_B, remove support *B* and give it a virtual displacement δ as shown in Fig. P.16.18(*b*).

The displacement $$CC' = \frac{\delta}{AB} \times AC = \delta \times \frac{6}{4.5}$$

or $$CC' = \frac{4}{3}\delta$$

And $DD' = \frac{\delta}{AB} \times AD = \frac{\delta}{4.5} \times 3$

or $DD' = \frac{2}{3}\ \delta$

$\therefore$ Total virtual work done $du = (+\ V_B)\ (+\ BB') + (-\ 1.2)\ (+\ CC') + (-\ 2.4) \times (DD')$

or $du = V_B \times \delta - 1.2 \times \frac{4\delta}{3} - 2.4 \times \frac{2}{3}\ \delta.$

By principle of virtual work $du = 0$.

$\therefore\ V_B \,.\, \delta - 1.2 \times \frac{4\delta}{3} - 2.4 \times \frac{2}{3} \,.\, \delta = 0$

$\therefore\ V_B = 1.6 + 1.6$ kN

or $V_B = \mathbf{3.2\ kN}$ **Ans.**

To find reaction V_A, remove support A and give it a virtual displacement $\delta = AA'$ as shown in Fig. (C).

$\therefore$ Displacement at D

$= DD' = \frac{\delta}{BA} \times BD$

or $DD' = \frac{\delta \times 1.5}{4.5} = \frac{\delta}{3}$

and displacement at $C = CC' = \frac{\delta}{BA} \times BC$

or $CC' = \frac{\delta \times 1.5}{4.5} = \frac{\delta}{3}$

Applying principle of virtual work, we get

$du = (+\ V_A)\ (AA') + (-\ 2.4)\ (DD') + (-\ 1.2)\ (-\ CC') = 0$

$\therefore\ V_A \times \delta + (-\ 2.4) \times \frac{\delta}{3} + (-\ 1.2)\left(-\frac{\delta}{3}\right) = 0$

$\therefore\ V_A = 0.4$ kN **Ans.**

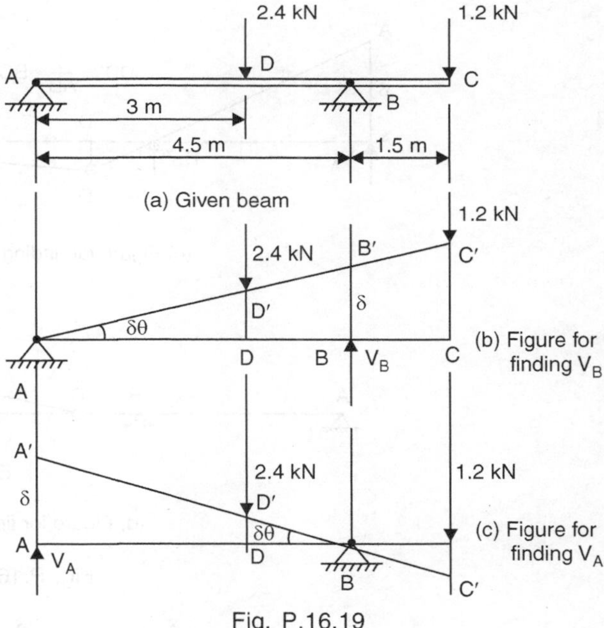

Fig. P.16.19

Problem 16.20. *A beam ABC, 9 m long, is hinged at A and freely supported at C with a prop at B, 3 m from A. It has another hinge at D, 4m from A and carries a load of 150 kN at E, 2 m from C. Using principle of virtual work find reaction at the support B. Also find reaction at supports A and C.*

Sol. The given beam is shown in Fig. P.16.20(*a*). Let V_B be the reaction at support B. Give the hinge D a slight vertical displacement $DD' = -\,\delta$, giving the beam the virtually displaced configuration shown in Fig. P.16.19(*b*). The vertical displacement at points B and E are :

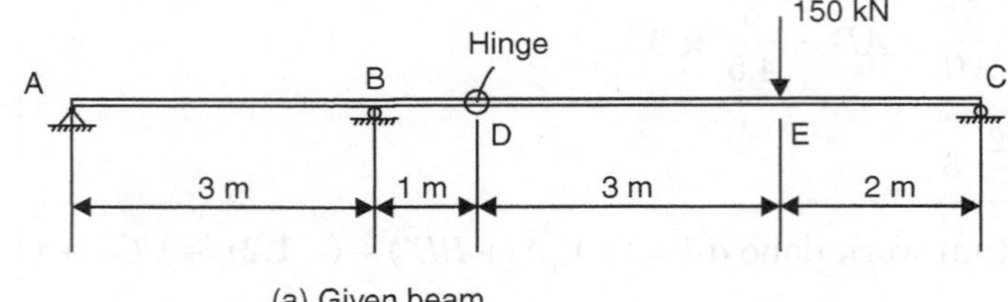

(a) Given beam

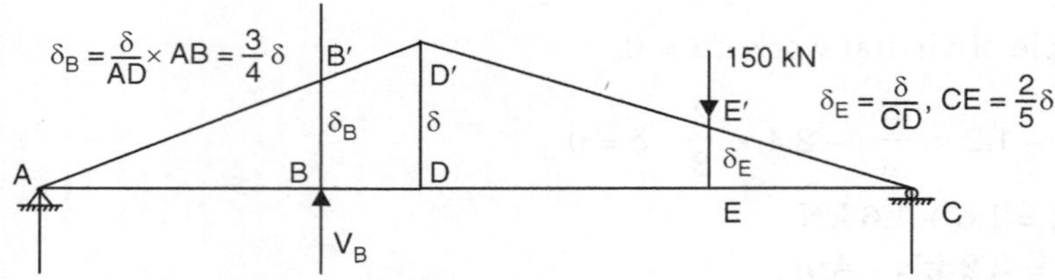

(b) Figure for finding V_B

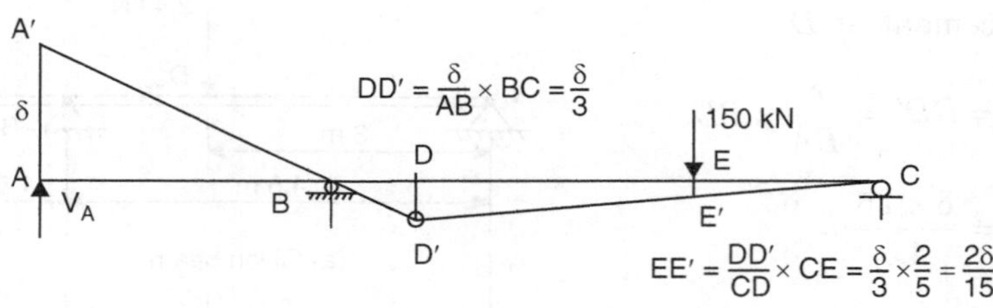

(c) Figure for finding V_A

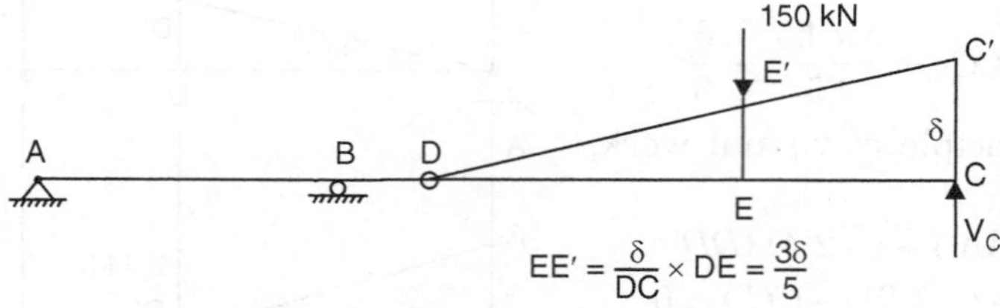

(d) Figure for finding V_C

Fig. P.16.20

At B, $$BB' = \frac{\delta}{AD} \times AB = \frac{3\delta}{4}$$

At E, $$EE' = \frac{\delta}{CD} \times CE = \frac{2\delta}{5}$$

The equation of virtual work is :

$$(+ V_B) \times BB' + (-150) \times EE' = 0$$

$$\therefore \quad V_B \times \frac{3\delta}{4} + (-150) \times \left(\frac{2\delta}{5}\right) = 0$$

$$\therefore \quad V_B = 150 \times \frac{2}{5} \times \frac{4}{3}$$

$\therefore \quad V_B = 80$ kN. **Ans.**

For finding reaction V_A virtually displace the support by $\delta = AA'$ shown in Fig. P.16.20(*c*). The displacements of points *D* and *E* are :

$$\text{At } D: \quad DD' = \frac{\delta}{AB} \times AD = \frac{\delta \times 1}{3}$$

$$\text{At } E: \quad EE' = DD' \times \frac{CE}{CD} = \frac{\delta \times 1}{3} \times \frac{2}{5} = \frac{2}{15}\,\delta.$$

Equation of virtual work is :

$$(+V_A)(AA') + (-150)(-EE') = 0$$

$$\therefore \quad V_A \times \delta + (-150) \times \left(\frac{-2\delta}{15}\right) = 0$$

$$\therefore \quad V_A = -150 \times \frac{2}{15} \text{ kN} \quad \text{or} \quad V_A = -20 \text{ kN}$$

$\therefore \quad$ **V_A = 20 kN downward Ans.**

For finding reaction V_C give a virtual displacement to $C = CC'$. The virtually displaced configuration of the beam is shown in Fig. P.16.20(*d*). The displacement of, point *E*

$$= EE' = \frac{\delta}{CD} \times CE \quad \text{or} \quad EE' = \frac{3}{5}\,\delta.$$

The equation of virtual work is :

$$(+V_C)(+CC') + (-150) \times \left(\frac{3}{5}\delta\right) = 0$$

$$\therefore \quad V_C \times \delta - 150 \times \frac{3\delta}{5} = 0$$

$\therefore \quad V_C$ = **90 kN Ans.**

Check: For the entire beam let us, find whether $\Sigma F_V = 0$ is satisfied.

$\therefore$ by $\Sigma F_V = 0$, we get

$$\therefore \quad V_A + V_B + V_C - 150 = 0$$

$$\therefore \quad (-20) + 80 + 90 - 150 = 0$$

$\therefore \quad \Sigma F_V = 0$ is satisfied.

Problem 16.21. *A beam AC, 11 m long is fixed at A and supported on rollers at C. It has a hinge at B at 6 m from A. It carries a load of 10 kN at D 4m from A and another load of 12 kN at E 2m from C. Using virtual work method find reaction at supports A and C.*

Sol. For finding V_C give virtual displacement $CC' = \delta$ as shown in Fig. P.16.21(*b*).

$$\therefore \quad EE' = \frac{CC'}{BC} \times BE = \frac{3}{5}\delta$$

$\therefore$ by virtual work principle, we get

$$du = (-12)(EE') + V_C\,(CC') = 0$$

$$(-12) \times \frac{3}{5}\delta + V_C \times \delta = 0$$

$\therefore \quad V_C$ = **7.2 kN Ans.**

For finding M_A give vertical

virtual displacement $\delta = BB'$ to the hinge B as shown in Fig. P.16.21(*c*) ;

Then $$DD' = \frac{BB'}{AB} \times AD = \frac{\delta \times 4}{6} = \frac{2}{3}\delta$$

And $$EE' = BB' \times \frac{CE}{BC} = \delta \times \frac{2}{5}$$

The virtual displacement δ at B gives a virtual angular rotation $d\theta_1$ at A.
Note that M_A is positive as it tends to increase angle $d\theta_1$.
The equation of virtual work is

$$(-10)\,(DD') + (-12)\,(EE') + M_A\,.\,d\theta_1 = 0, \quad \text{where} \quad d\theta_1 = \frac{BB'}{AB} = \frac{\delta}{6}$$

$$\therefore \quad (-10)\left(\frac{2\delta}{3}\right) + (-12) \times \left(\frac{2}{5}\delta\right) + (MA)\left(\frac{\delta}{6}\right) = 0$$

$$\therefore \quad \frac{1}{6}M_A = \frac{20}{3} + \frac{24}{5} = \frac{172}{15} \text{ kNm.}$$

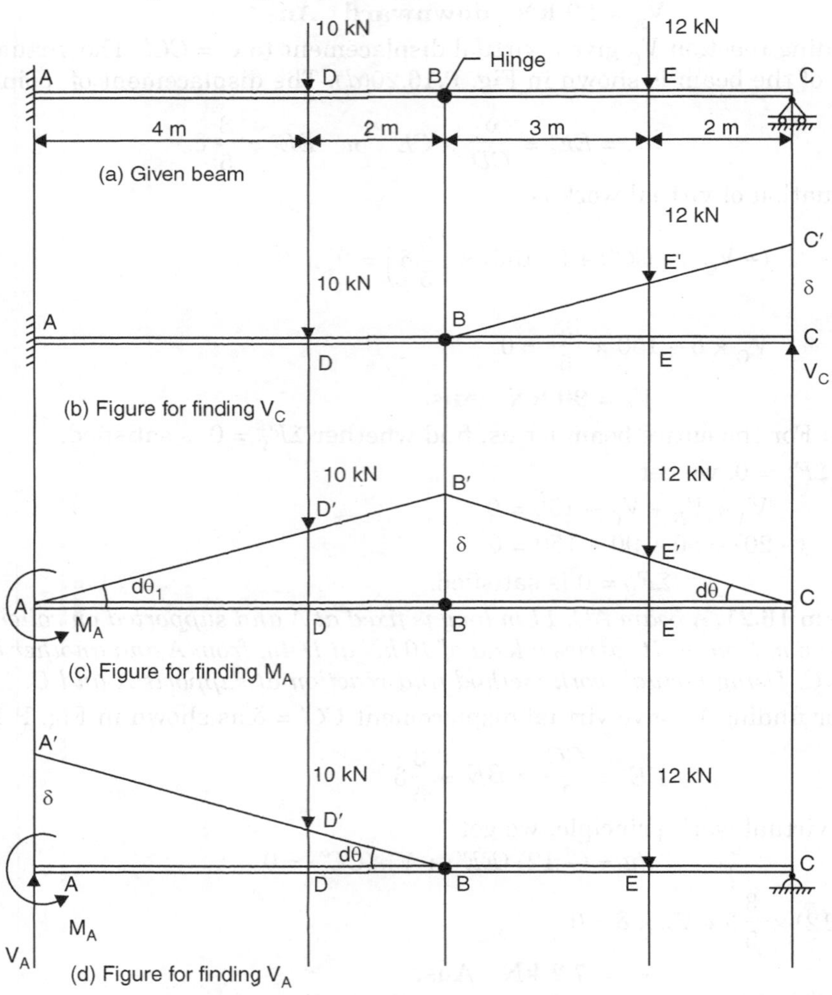

Fig. P.16.21

$$\therefore \qquad M_A = \frac{172}{15} \times 6 = \textbf{68.8 kNm, CCW} \quad \textbf{Ans.}$$

In order to find V_A give virtual displacement $AA' = \delta$ to the support A as shown in Fig. P. 16.21(*d*).

$$\therefore \qquad DD' = \frac{AA'}{AB} \times BD = \frac{\delta}{6} \times 2$$

And angle $\quad d\theta = \frac{AA'}{AB} = \frac{\delta}{6}$. Note that $d\theta$ is clockwise. Hence it –ve.

i.e., $\quad d\theta = -\frac{\delta}{6}$

The equation of virtual work is

$$V_A \times AA' + (M_A) \times (d\theta) + (-10) \times DD' = 0$$

or $$V_A \times \delta + M_A \left(-\frac{\delta}{6}\right) + (-10)\left(\frac{2\delta}{6}\right) = 0$$

or $$V_A \times \delta \left(\frac{172 \times 6}{15}\right)\left(\frac{-\delta}{6}\right) + (-10) \times \frac{2\delta}{6} = 0$$

or $$V_A - \frac{172}{15} - 10 \times \frac{1}{3} = 0$$

$$\therefore \qquad V_A = \frac{10}{3} + \frac{172}{15} \text{ kN} \quad \text{or} \quad V_A = 14.8 \text{ kN} \quad \textbf{Ans.}$$

[**Note :** For the entire beam $\Sigma F_V = -10 - 12 + V_A + V_C = -10 - 12 + 7.2 + 14.8 = 0$ (Check)].

Problem 16.22. *Using the principle of virtual work find reaction components of all supports of the beam shown in Fig. P.16.22(a).*

Sol. Free-body diagram of the beam is shown in Fig. P.16.22(*b*). For finding V_C, give vertical displacement to support *C* such that portion *BC* rotates by angle $\delta\theta$, as shown in Fig. P.16.22(*c*).

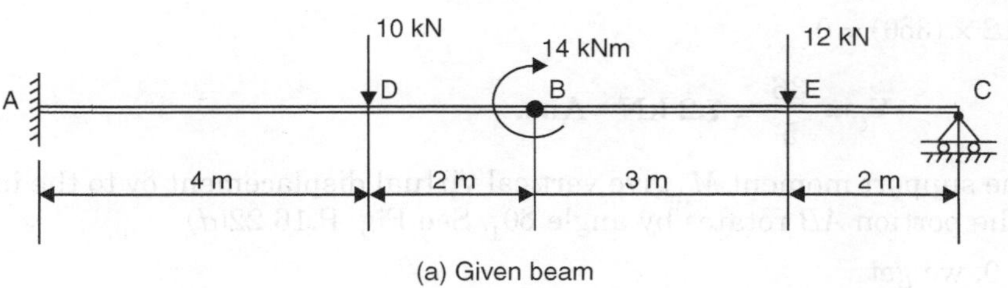

(a) Given beam

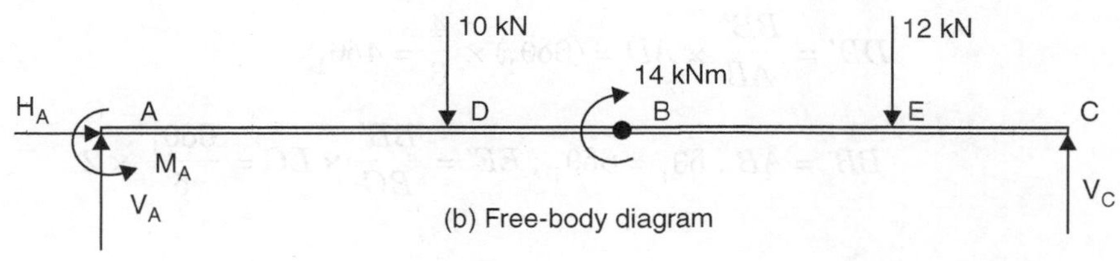

(b) Free-body diagram

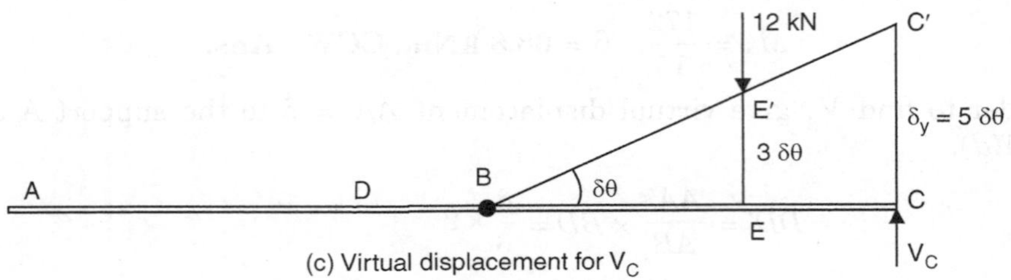

(c) Virtual displacement for V_C

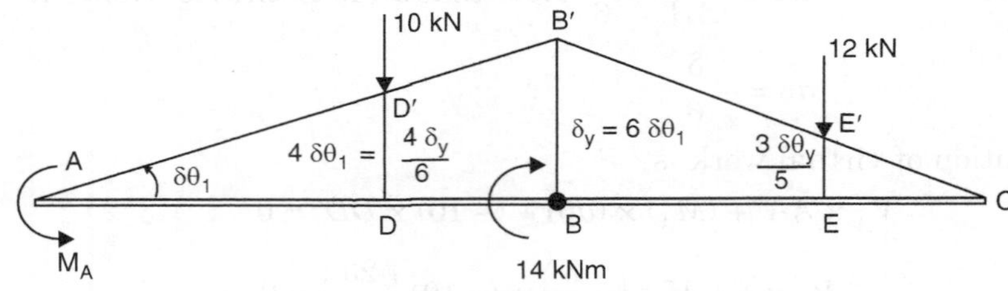

(d) Virtual displacement for M_A

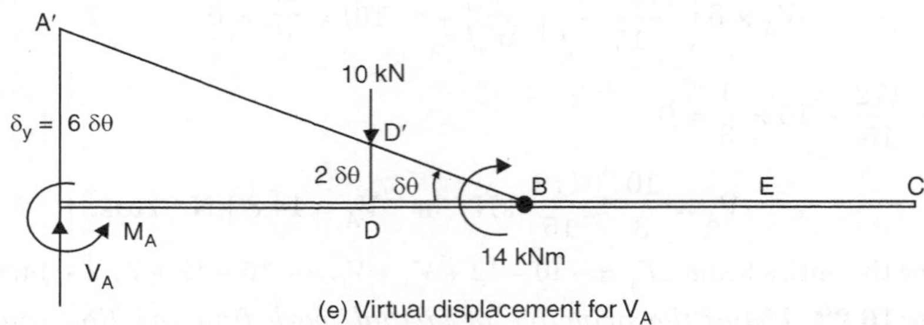

(e) Virtual displacement for V_A

Fig. P.16.22

∴ by using

$$\Sigma du = 0, \text{ we get}$$

$$V_C \times (CC') + (-12)(EE') = 0$$

or $\quad V_C \times (5 \times \delta\theta) - 12 \times (3\delta\theta) = 0$

or $$V_C = \frac{36}{5} = \mathbf{7.2\ kN} \quad \textbf{Ans.}$$

For finding the support moment M_A give vertical virtual displacement δy to the internal hinge B, such that the portion AB rotates by angle $\delta\theta_1$. See Fig. P.16.22(*d*).

∴ By $\Sigma du = 0$, we get

$$M_A \times (+\delta\theta_1) + (-10)(DD') + (-14)\,\delta\theta_1 + (-12)(EE') = 0$$

where $$DD' = \frac{BB'}{AB} \times AD = (6\delta\theta_1) \times \frac{4}{6} = 4\delta\theta_1$$

$$BB' = AB \,.\, \delta\theta_1 = 6\delta\theta_1\,;\; EE' = \frac{BB'}{BC} \times EC = \frac{6\delta\theta_1}{5} \times 2$$

$\therefore$ We get $M_A \times \delta\theta_1 + (-10)(4 \times \delta\theta_1) + (-14)(\delta\theta_1) + (-12)\left(\frac{6\delta\theta_1}{5} \times 2\right) = 0$

or $\quad M_A - 40 - 14 - \frac{12 \times 12}{5} = 0$

or $\quad M_A = \mathbf{82.8\ kNm}$ **Ans.**

For finding vertical reaction V_A, give a virtual vertical displacement δy to support A such that the portion AB rotates by $\delta\theta$. See Fig. P.16.23(*C*).

$\therefore \quad AA' = 6.\delta\theta\ ;\ DD' = 2 \times \delta\theta$

Since $\delta\theta$ is clockwise, it is negative.

Therefore, by using $\Sigma\ \delta u = 0$, we get

$$V_A \times AA' + (-10)(DD') + (-14)(-\delta\theta) + M_A(-\delta\theta) = 0$$

$$V_A(6\ \delta\theta) + (-10)(2\ \delta\theta) + 14\ \delta\theta - (82.8)\ \delta\theta = 0$$

$$\therefore \quad 6V_A - 20 + 14 - 82.8 = 0$$

$$\therefore \quad V_A = \frac{88.8}{6} = \mathbf{14.8\ kN} \quad \textbf{Ans.}$$

Check : By $\Sigma F_v = 0$, we get

$$V_A + V_C - 10 - 12 = 0$$

$14.8 + 7.2 - 22 = 0$ **satisfied.**

Since all applied forces are vertical, therefore, by $\Sigma F_h = 0$, we get

$H_A = 0$ **Ans.**

Problem 16.23. *Using principle of virtual work, find reaction components of the compound beam shown in Fig. 16.23(a).*

Sol. The *u.d.l.* of 5 kN/m acting on beam AB is replaced by a point load of 5 × 4 = 20 kN acting at the centre E of AB. Similarly, the u.d.l. of 5 kN/m acting the beam BD is replaced by a point load of 5 × 6 = 30 kN acting at the centre F of BD. The free-body diagram of the given beam is shown in Fig. P.16.23(*b*).

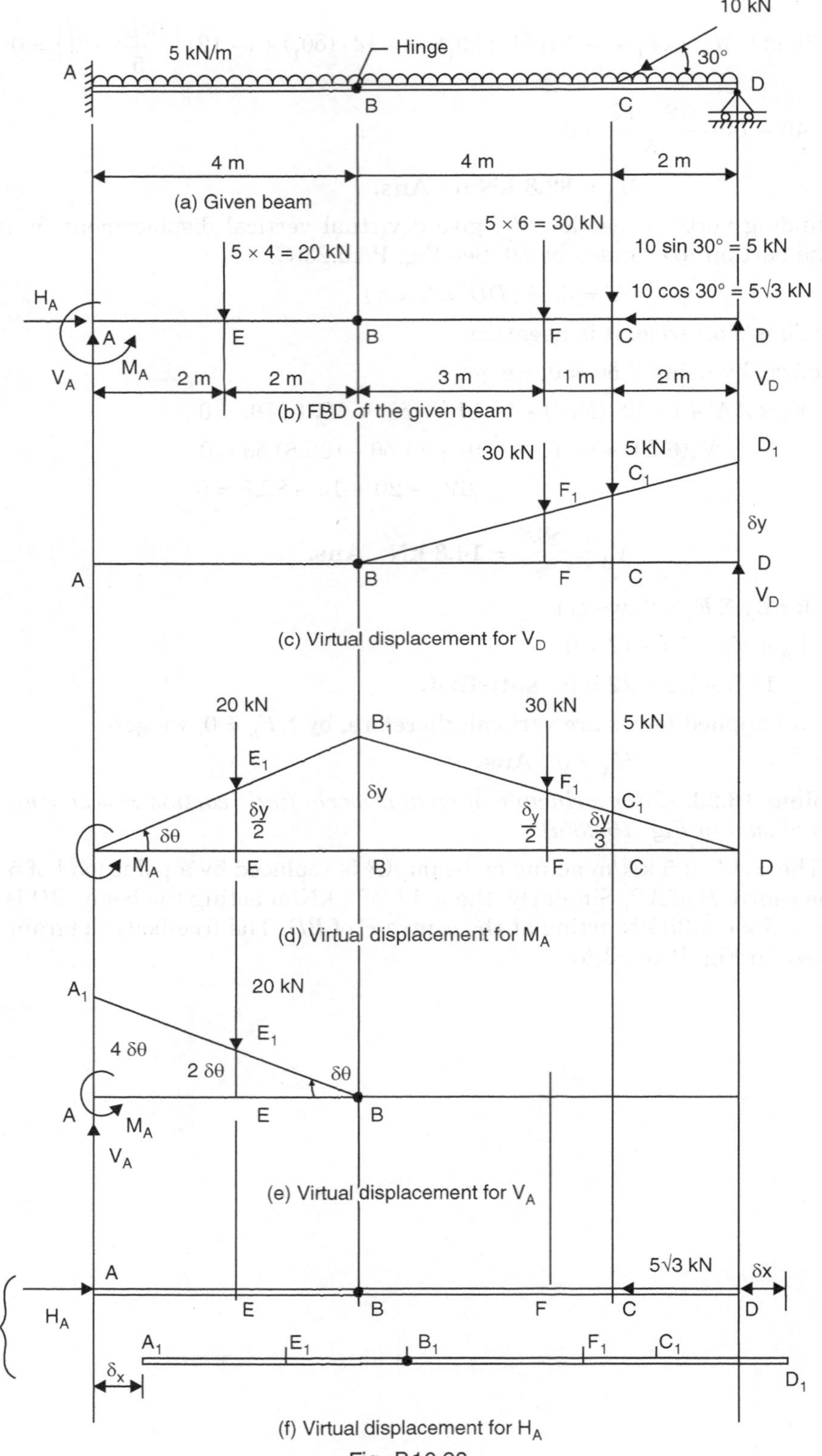
10 kN
5 kN/m
Hinge
30°
A
B
C
D
4 m
4 m
2 m
(a) Given beam
5 × 6 = 30 kN
5 × 4 = 20 kN
10 sin 30° = 5 kN
10 cos 30° = 5√3 kN
H_A
A
E
B
F
C
D
V_A
M_A
2 m
2 m
3 m
1 m
2 m
V_D
(b) FBD of the given beam
30 kN
5 kN
D_1
C_1
F_1
δy
A
B
F
C
D
V_D
(c) Virtual displacement for V_D
20 kN
30 kN
5 kN
B_1
E_1
F_1
δy
C_1
δy/2
δy/2
δy/3
δθ
M_A
E
B
F
C
D
(d) Virtual displacement for M_A
A_1
20 kN
E_1
4 δθ
2 δθ
δθ
A
M_A
E
B
V_A
(e) Virtual displacement for V_A
A
5√3 kN
δx
H_A
E
B
F
C
D
A_1
E_1
B_1
F_1
C_1
D_1
δx
(f) Virtual displacement for H_A
Fig. P.16.23

(*a*) **The Vertical Reaction V_D at D :**

Give a virtual displacement $DD_1 = \delta y$ to the support D, as shown in Fig. P. 16.23(*c*). Thus, $DD_1 = \delta y$;

$$FF_1 = \frac{3}{6}\,\delta y = \frac{\delta y}{2} \text{ and } CC' = \frac{4}{6}\,\delta y = \frac{2}{3}\,\delta y.$$

By using $\Sigma\delta u = 0$, we get

$$(+V_D)(+DD_1) + (-5)(+CC_1) + (-30)(FF_1) = 0$$

or
$$V_D \times \delta_y + (-5) \times \left(\frac{2\delta y}{3}\right) + (-30)\,.\left(\frac{1}{2}\,\delta y\right) = 0$$

∴
$$V_D = 5 \times \frac{2}{3} + 30 \times \frac{1}{2}$$

or
$$V_D = \frac{110}{6} = \mathbf{18.33\ kN} \quad \textbf{Ans.}$$

(*b*) **Moment Reaction Component M_A at A :**

Give a virtual vertical displacement $BB_1 = \delta y$ to hinge B as shown in Fig. P. 16.23(*d*).

Thus, we have $EE_1 = \dfrac{\delta y}{2} = 2.\delta\theta$; $FF_1 = \delta y \times \dfrac{3}{6} = \dfrac{\delta y}{2}$, and $CC_1 = \dfrac{\delta y \times 2}{6} = \dfrac{\delta y}{3}$

Thus, by using $\Sigma\delta u = 0$, we have

$$(+M_A)(\delta\theta) + (-20)\left(\frac{\delta y}{2}\right) + (-30)(FF_1) + (-5)(CC_1) = 0$$

where $\delta\theta = \dfrac{BB_1}{AB} = \dfrac{\delta y}{4}$. Since $\delta\theta$ is counter clockwise, therefore it is positive. Thus, we get

$$M_A \times \frac{\delta y}{4} + (-20) \times \left(\frac{\delta y}{2}\right) + (-30)\left(\frac{\delta y}{2}\right) + (-5)\left(\frac{\delta y}{3}\right) = 0$$

∴
$$M_A = 4\left[10 + 15 + \frac{5}{3}\right] = \frac{4 \times 80}{3} \text{ kNm} = \frac{320}{3} \text{ kNm}$$

or
$$M_A = \mathbf{106.67\ kNm} \quad \textbf{Ans.}$$

(*c*) **Vertical Reaction V_A :**

Give virtual angular displacement $\delta\theta$ to the beam AB at B as shown in Fig. P. 16.23(*e*). Thus, we get $AA_1 = AB\,.\,\delta\theta = 4\delta\theta$, and $EE_1 = 2\delta\theta$.

Thus, by using $\Sigma\delta u = 0$, we get

$$(+V_A) \times (AA_1) + (M_A)(-\delta\theta) + (-20)(EE_1) = 0$$

or
$$V_A \times (4\delta\theta) + \left(\frac{320}{3}\right)(-\delta\theta) + (-20)(2\delta\theta) = 0$$

∴
$$V_A = \frac{1}{4}\left[\frac{320}{3} + 40\right] = \frac{1}{12} \times 440 \text{ kN}$$

or
$$\mathbf{V_A = 36.67\ kN} \quad \textbf{Ans.}$$

(*d*) **Horizontal Reaction H_A at A :**

Give a virtual horizontal displacement δx to the support D such that all points move towards right by δx as shown in Fig. P. 16.23(*f*). Thus, by using $\Sigma\delta u = 0$, we get

$$(+ H_A)\,(\delta x) + (-\,5\sqrt{3})\,(\delta x) = 0$$

$\therefore$ $\quad$ $\mathbf{H_A = 5\sqrt{3}}$ **kN towards right** $\quad$ **Ans.**

Problem 16.24. *The double pendulum shown in Fig. P.16.24(a) consists of two equal links AB and BC pinned smoothly together and held in position by a horizontal force P. Find the equilibrium position as defined by angles θ_1 and θ_2.*

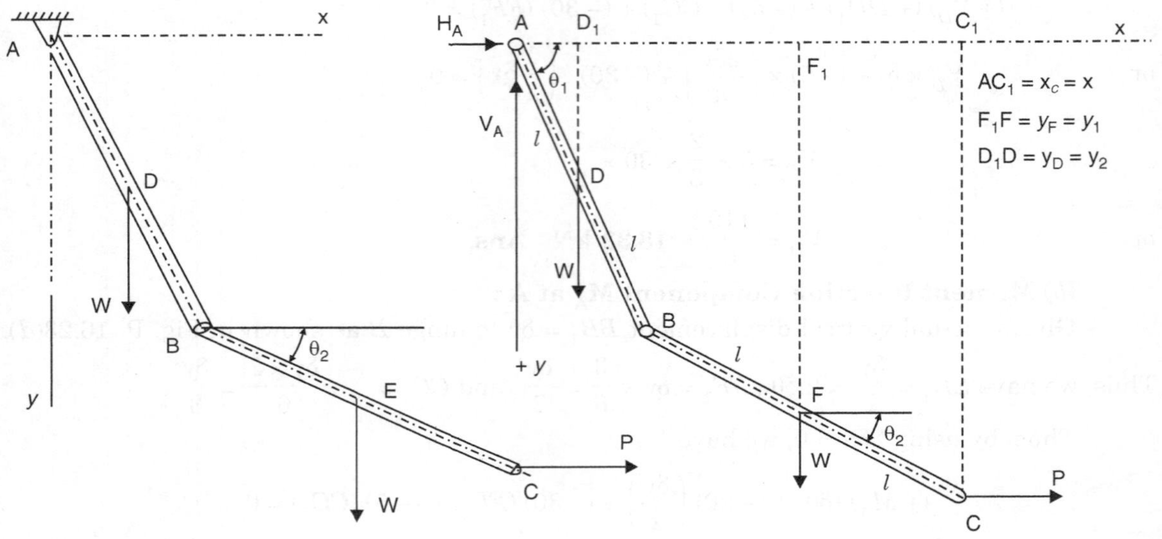

Fig. P.16.24

Sol. Fig. P.16.24(*b*) shows the free-body-diagram (F.B.D.) of the double pendulum. With respect to the cordinate axes shown, the co-ordinates of the points of application of the active forces are

$$y_D = y_1 = l \sin \theta_1$$
$$y_F = y_2 = 2\,l \sin \theta_1 + l \sin \theta_2$$
$$x_c = x = 2\,l \cos \theta_1 + 2l \cos \theta_2$$

Differentiating above equations we get :

$$\delta y_1 = l \cos \theta_1 \,.\, \delta\theta_1 \qquad \ldots(1)$$
$$\delta y_2 = 2l \cos \theta_1 \,.\, \delta\theta_1 + l \cos \theta_2 \,.\, \delta\theta_2 \qquad \ldots(2)$$
$$\delta x = -\,2l \sin \theta_1 \,.\, \delta\theta_1 - 2l \sin \theta_2 \,.\, \delta\theta_2 \qquad \ldots(3)$$

For multidegree of freedom system virtual work equation is written as

$$\delta u = \frac{\partial u}{\partial x} \,.\, \delta x + \frac{\partial u}{\partial y} \,.\, \delta y = 0 \qquad \ldots(4)$$

or
$$\delta u = \frac{\partial u}{\partial \theta_1} \,.\, \delta\theta_1 + \frac{\partial u}{\partial \theta_2} \,.\, \delta\theta_2 = 0 \qquad \ldots(5)$$

Keeping θ_1 constant ($\delta\theta_1 = 0$) and θ_2 to vary, we get

by $\quad (\delta u)_{\theta_2} = 0, \quad W\,.\,(\delta y_2)_{\theta_2} + P(\delta x)_{\theta_2} = 0 \qquad \ldots(6)$

where $(\delta y_2)_{\theta_2} = l \cos \theta_2 \,.\, \delta\theta_2$ [From Eqn. (2)]

and $(\delta x)_{\theta_2} = -2\, l \sin \theta_2 \,.\, \delta\theta_2$ [From Eqn. (3)]

$\therefore$ Eqn. (6) becomes $W \times l \cos \theta_2 \,.\, \delta\,\theta_2 - P \times 2l \sin \theta_2 \,.\, \delta\theta_2 = 0$

or $$\tan \theta_2 = \frac{W}{2P} \quad \textbf{Ans.}$$

Now, keeping θ_2 constant *i.e.*, ($\delta\theta_2 = 0$) and θ_1 to vary, Equations (2) and (3) give

$$(\delta y_2)_{\theta_1} = 2l \cos \theta_1 \,.\, \delta\theta_1$$

and $$(\delta x)_{\theta_1} = -2l \sin \theta \,.\, \delta\theta_1$$

And, using virtual work principle we can write :

$$(\delta U)_{\theta_1} = W \,.\, (\delta y_1)_{\theta_1} + W(\delta y_2)_{\theta_1} + P(\delta x)_{\theta_1} = 0$$

or $$(W \,.\, l \cos \theta_1) \,.\, \delta\theta_1 + W \,.\, \delta y_2 + P\delta x = 0$$

or $$Wl \cos \theta_1 \,.\, \delta\theta_1 + W \,.\, 2l \cos \theta_1\, \delta\theta_1 + P(-2l \sin \theta_1) \,.\, \delta\theta_1 = 0$$

$$2P \sin \theta_1 = 3\, W \cos \theta_1$$

$\therefore$ $$\tan \theta_1 = \frac{3W}{2P} \quad \textbf{Ans.}$$

Problem 16.25. *Using the principle of virtual work, determine the equilibrium position of the double pendulum with weightless links as shown in Fig. P.16.25.*

Sol. Since the independent co-ordinates θ_1 and θ_2 must be known to locate the position of both the links AB and BC, the system has two degrees of freedom. Select A as the origin of the co-ordinate axes, as shown in Fig. P. 16.25.

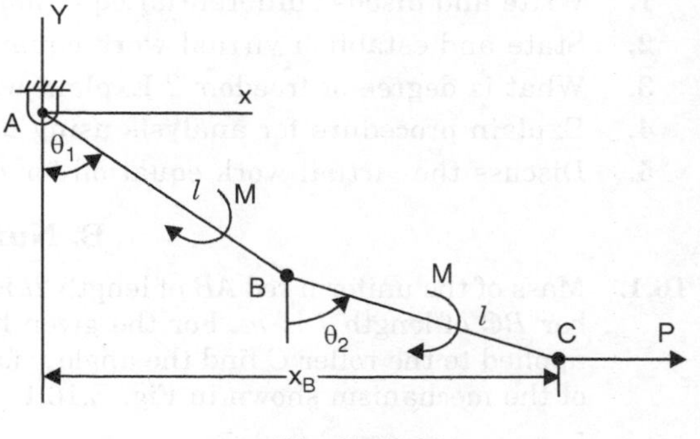

Fig. P.16.25

The position co-ordinate x_C may be expressed as

$$x_C = l \sin \theta_1 + l \sin \theta_2 \quad ...(1)$$

Differentiating Eqn. (1), we get

$$\delta x_C = l \,.\, \cos \theta_1 \,.\, \delta\theta_1 + l \cos \theta_2 \,.\, \delta\theta_2 \quad ...(2)$$

By keeping θ_1 constant, and since $\delta\theta_2 \neq 0$, from Eqn. (2) we get

$$(\delta x_C)_{\theta_2} = l \cos \theta_2 \,.\, \delta\theta_2 \quad ...(3)$$

And by keeping θ_2 constant and since $\delta\theta_1 \neq 0$, from Eqn. (2), we get

$$(\delta x_C)_{\theta_1} = l \cos \theta_1 \,.\, \delta\theta_1 \quad ...(4)$$

The virtual work equation is

$$\delta U = \frac{P \,.\, \partial x_C}{\partial \theta_1} \,.\, \delta\theta_1 + \frac{P \,.\, x_C}{\partial \theta_2} \,.\, \delta\theta_2 - M(\delta\theta_1) - M(-\delta\theta_1) + M(-\delta\theta_2) = 0 \quad ...(5)$$

or $$\delta U = P(\delta x_C)_{\theta_1} \,.\, \delta\theta_1 + P(\delta x_C)_{\theta_2} \,.\, \delta\theta_2 - 2M(\delta\theta_1) - M \,.\, (\delta\theta_2) = 0 \quad ...(6)$$

Since the couple M tends to decrease θ_1 and θ_2, therefore the last terms in equations (5) and (6) are negative.

When θ_2 is held constant (*i.e.*, $\delta\theta_2 = 0$), only θ_1 varies, therefore the virtual work equation can be obtained by using $(\delta U)_{\theta_1} = 0$ from Eqn. (6)

$\therefore \quad (\delta U)_{\theta_1} = P(\delta x_C)_{\theta_1} - 2M\,\delta\theta_1 = 0$

$\therefore \quad P(l \cos \theta_1) . \delta\theta_1 - 2M\,\delta\theta_1 = 0$

or $$\cos \theta_1 = \frac{2M}{Pl}$$

or $$\theta_1 = \cos^{-1}\left(\frac{2M}{Pl}\right) \quad \textbf{Ans.} \qquad ...(7)$$

When θ_1 is kept fixed $\delta\theta_1 = 0$, but $\delta\theta_2 \neq 0$, the virtual work equation can be written using Eqn. (6)

$$(\delta u)_{\theta_2} = P . (\delta x_C)_2 . \delta\theta_2 - M . \delta\theta_2 = 0$$

Using eqns. (3) we get $(P.l \cos \theta_2) . \delta\theta_2 - M\delta\theta_2 = 0$

or $$\cos \theta_2 = \frac{M}{PL}$$

$\therefore$ $$\theta_2 = \cos^{-1}\left(\frac{M}{PL}\right) \quad \textbf{Ans.} \qquad ...(8)$$

EXERCISE XVI

A. Theoretical Questions

1. Write and discuss differential equations of work due to a force and a couple.
2. State and establish virtual work equation.
3. What is degree of freedom ? Explain with illustrations.
4. Explain procedure for analysis using principle of virtual work.
5. Discuss the virtual work equation for multi-degree of freedom systems.

B. Numerical Problems

16.1. Mass of the uniform rod *AB* of length $2l$ is $2m$ while that of bar *BC* of length l is m. For the given horizontal force F applied to the roller C find the angle θ for the equilibrium of the mechanism shown in Fig. E.16.1.

$$\left[\textbf{Ans.} \quad \theta = 2\tan^{-1}\left(\frac{4F}{mg}\right)\right]$$

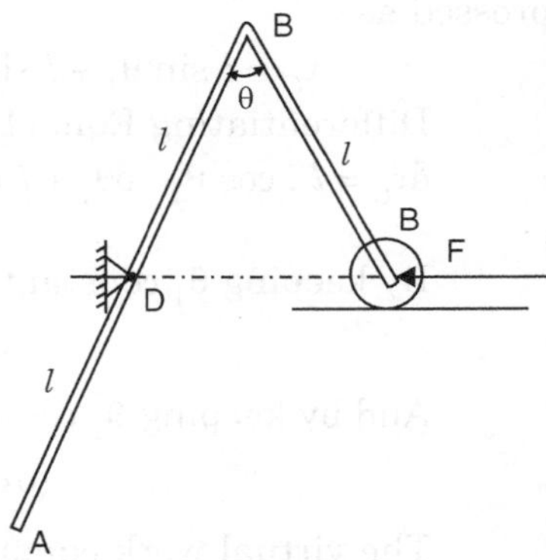

Fig. E.16.1

16.2. Establish a relation between the active forces P and Q for the equilibrium of the system of bars shown in Fig. E.16.2. Each bar is of length l. **[Ans.** $Q = P/4$**]**

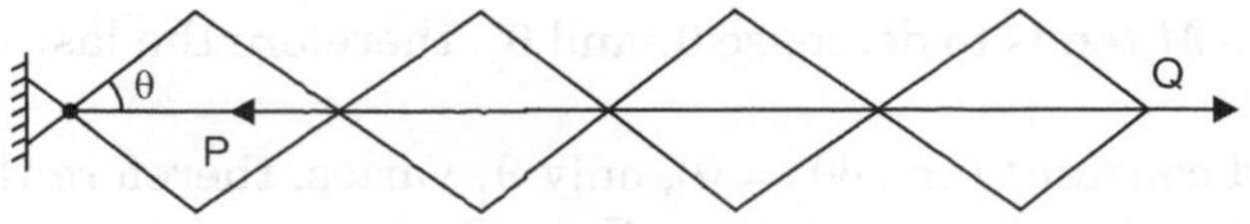

Fig. E.16.2

16.3. For the mechanism shown in Fig. E.16.3 find force P to maintain equilibrium.

[**Ans.** P = 2.94 kN]

16.4. Two rods of length $2L$ are hinged at B and supported as shown in Fig. E.16.4. Find magnitude of P to maintain equilibrium when θ = 30°.

[**Ans.** P = 4.928 kN]

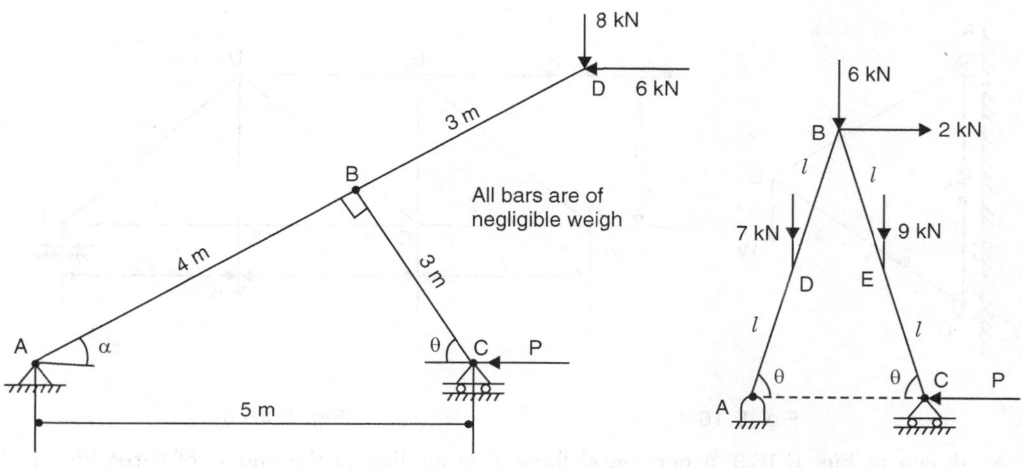

Fig. E.16.3 Fig. E.16.4

16.5. Four bars of the same length l and the same weight mg are hinged together to form a rhombus as shown in Fig. E.16.5. Find the relationship between P and Q for the equilibrium of the system.

[**Ans.** P/Q = tan θ]

16.6. A bar AB of negligible weight is placed on two inclined smooth planes as shown in Fig. E.16.6. Find angle θ for equilibrium when a load P is applied as shown. Also find θ when a = 1.5 b and α = 60°.

$$\left[\textbf{Ans.}\quad \theta = \tan^{-1}\left(\frac{a}{b}\tan\alpha\right); \theta = 68.95°\right]$$

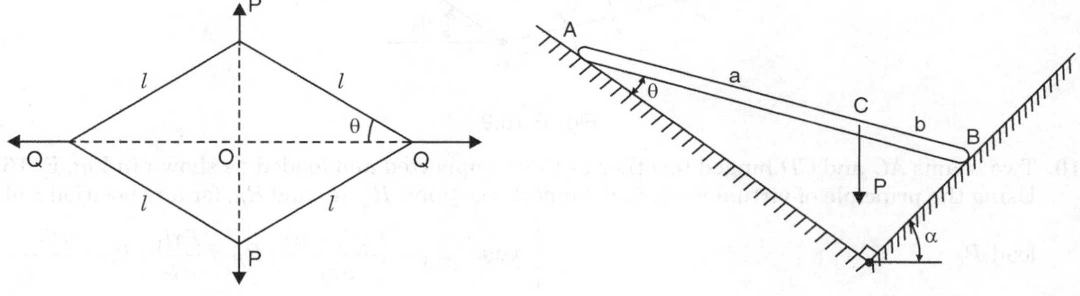

Fig. E.16.5 Fig. E.16.6

16.7. Two rods of length l and weight mg are held as shown in Fig. E.16.7. End A is hinged whereas end C carries a roller. A vertial load W is applied at the hinge B. Unstretched length of the spring connecting roller C and the hinge A is b. Determine the expression for the angle θ for the equilibrium of the system.

$$\left[\textbf{Ans.}\quad \theta = \tan^{-1}\left(\frac{W + 2kb + 2mg}{4kl}\right)\right]$$

16.8. Using the virtual work method find force in the member *BD* of the truss shown in Fig. E.16.8.

$$\left[\textbf{Ans. } F_{BD} = \frac{2}{3}F - \frac{Pl}{3h}\right]$$

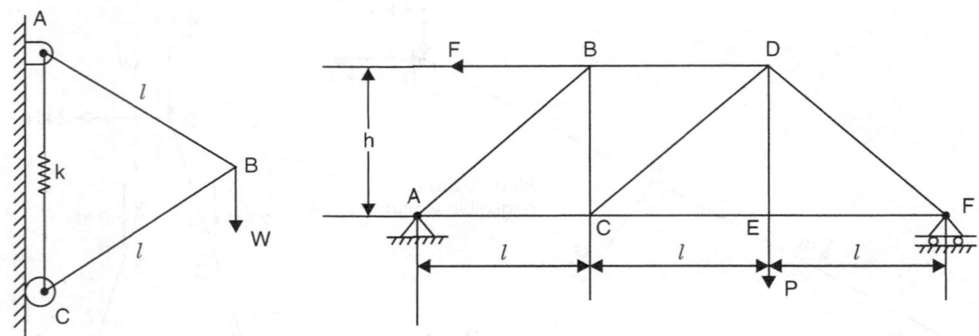

Fig. E.16.7 Fig. E.16.8

16.9. As shown in Fig. E.16.9, a horizontal force *F* is applied to the end *C* of three identical links smoothly pinned together. If each link is of weight *W*, find the equilibrium position as defined by the angles θ_1, θ_2 and θ_3.

$$\left[\textbf{Ans. } \theta_1 = \tan^{-1}\left(\frac{5W}{2F}\right); \theta_2 = \tan^{-1}\left(\frac{3W}{2F}\right); \theta_3 = -\tan^{-1}\left(\frac{W}{2F}\right)\right]$$

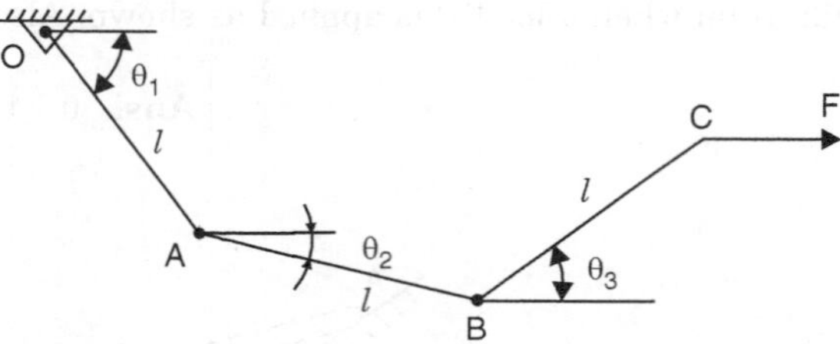

Fig. E.16.9

16.10. Two beams *AC* and *CD* hinged together at *C* are supported and loaded as shown in Fig. E. 16.10. Using the principle of virtual work find support reactions. R_A, R_B and R_C, for any position *x* of the load *P*.

$$\left[\textbf{Ans. } R_A = \frac{Px(l_1 - a)}{a\,l_2}; R_B = \frac{Pxl_1}{a\,l_2}; R_C = \frac{P(l_2 - x)}{l_2}\right]$$

Fig. E.16.10

16.11. A beam *ABCDEF*, as shown in Fig. E.16.11 is hinged at *A*, *C*, *D* and *F* and is freely supported at *B* and *E*. $AB = 3\text{m} = CD = EF$; $BC = DE = 2\text{m}$. The suspended span *CD* carries a point load of 100 kN at *G*, 1m from *C*. Using virtual work method. Find reactions at *A*, *B*, *E* and *F*.

$$\left[\textbf{Ans. } R_A = \frac{400}{9} \text{ kN} \downarrow ; R_B = \frac{1000}{9} \text{ kN} \uparrow ; R_E = \frac{500}{9} \text{ kN} \uparrow ; R_F = \frac{200}{9} \text{ kN} \downarrow \right]$$

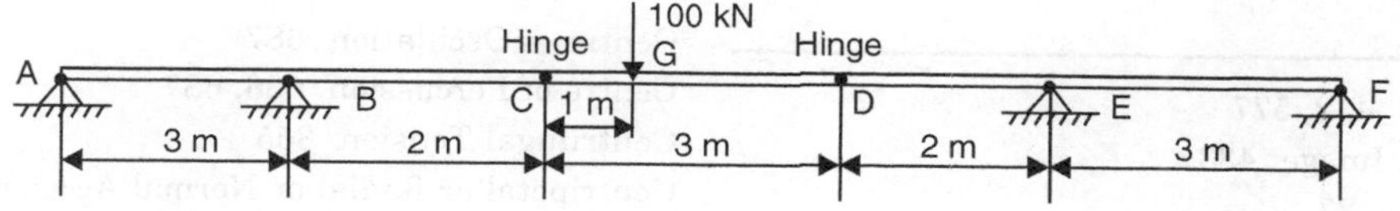

Fig. E. 16.11

16.12. A beam *ABC*, 8 m long, is freely supported at *B* and *C* 5 m apart, with an overhang *AB* of 1 m to left of *B* and 2 m to the right of *C*. It carries a point load of 12 kN at *A* and loads 18, 24 and 36 kN at 1 m intervals on *BC* and 48 kN at the end of right overhang. Using virtual work method find reactions at supports *B* and *C*.

[**Ans.** $R_B = 38.4$ kN ↑ ; $R_C = 70$ kN ↑]

16.13. A beam *ABCDE*, *m* long is hinged at *A* and freely supported at *B* and *D* . $AB = 3$ m, $BD = 6$ m ; the overhang $DE = 2$ m. There is a hinge at *C* mid-way between *B* and *D*. The loading consists of a point-load of 12 kN at the free end *E*, 24 kN at the middle of *BC*, 36 kN at the middle of *CD* and a uniformly distributed load of 6 kN/m on the span *AB*. Using principle of virtual work determine reactions at the supports. [$R_A = 13$ kN ↓, $R_B = 65$ kN ↑, $R_D = 38$ kN ↑]

16.14. Using principle of virtual work, find the horizontal force that the pin at *B* must exert in order to hold the mechanism shown in Fig. E.16.14 in equilibrium when $\theta = 45°$. Neglect the weight of members *AC* and *BC*.

[**Ans.** $H_B = 0.1373$ kN]

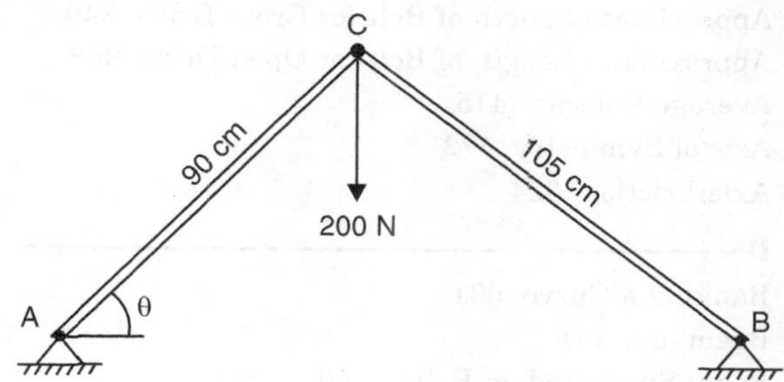

Fig. E.16.14

Index

D

E

F